Continued

Z	.00	.01	.02	.03	.04	.05	.06	.07	.08	.09
.0	.5000	.5040	.5080	.5120	.5160	.5199	.5239	.5279	.5319	.5359
.1	.5398	.5438	.5478	.5517	.5557	.5596	.5636	.5675	.5714	.5753
.2	.5793	.5832	.5871	.5910	.5948	.5987	.6026	.6064	.6103	.6141
.3	.6179	.6217	.6255	.6293	.6331	.6368	.6406	.6443	.6480	.6517
.4	.6554	.6591	.6628	.6664	.6700	.6736	.6772	.6808	.6844	.6879
.5	.6915	.6950	.6985	.7019	.7054	.7088	.7123	.7157	.7190	.7224
.6	.7257	.7291	.7324	.7357	.7389	.7422	.7454	.7486	.7517	.7549
.7	.7580	.7611	.7642	.7673	.7703	.7734	.7764	.7794	.7823	.7852
.8	.7881	.7910	.7939	.7967	.7995	.8023	.8051	.8078	.8106	.8133
.9	.8159	.8186	.8212	.8238	.8264	.8289	.8315	.8340	.8365	.8389
1.0	.8413	.8438	.8461	.8485	.8508	.8531	.8554	.8577	.8599	.8621
1.1	.8643	.8665	.8686	.8708	.8729	.8749	.8770	.8790	.8810	.8830
1.2	.8849	.8869	.8888	.8907	.8925	.8944	.8962	.8980	.8997	.9015
1.3	.9032	.9049	.9066	.9082	.9099	.9115	.9131	.9147	.9162	.9177
1.4	.9192	.9207	.9222	.9236	.9251	.9265	.9279	.9292	.9306	.9319
1.5	.9332	.9345	.9357	.9370	.9382	.9394	.9406	.9418	.9429	.9441
1.6	.9452	.9463	.9474	.9484	.9495	.9505	.9515	.9525	.9535	.9545
1.7	.9554	.9564	.9573	.9582	.9591	.9599	.9608	.9616	.9625	.9633
1.8	.9641	.9649	.9656	.9664	.9671	.9678	.9686	.9693	.9699	.9706
1.9	.9713	.9719	.9726	.9732	.9738	.9744	.9750	.9756	.9761	.9767
2.0	.9772	.9778	.9783	.9788	.9793	.9798	.9803	.9808	.9812	.9817
2.1	.9821	.9826	.9830	.9834	.9838	.9842	.9846	.9850	.9854	.9857
2.2	.9861	.9864	.9868	.9871	.9875	.9878	.9881	.9884	.9887	.9890
2.3	.9893	.9896	.9898	.9901	.9904	.9906	.9909	.9911	.9913	.9916
2.4	.9918	.9920	.9922	.9925	.9927	.9929	.9931	.9932	.9934	.9936
2.5	.9938	.9940	.9941	.9943	.9945	.9946	.9948	.9949	.9951	.9952
2.6	.9953	.9955	.9956	.9957	.9959	.9960	.9961	.9962	.9963	.9964
2.7	.9965	.9966	.9967	.9968	.9969	.9970	.9971	.9972	.9973	.9974
2.8	.9974	.9975	.9976	.9977	.9977	.9978	.9979	.9979	.9980	.9981
2.9	.9981	.9982	.9982	.9983	.9984	.9984	.9985	.9985	.9986	.9986
3.0	.9987	.9987	.9987	.9988	.9988	.9989	.9989	.9989	.9990	.9990
3.1	.9990	.9991	.9991	.9991	.9992	.9992	.9992	.9992	.9993	.9993
3.2	.9993	.9993	.9994	.9994	.9994	.9994	.9994	.9995	.9995	.9995
3.3	.9995	.9995	.9995	.9996	.9996	.9996	.9996	.9996	.9996	.9997
3.4	.9997	.9997	.9997	.9997	.9997	.9997	.9997	.9997	.9997	.9998

Commonly Used *Z*-Table Values

Confidence Intervals	**Confidence**	**80%**	**90%**	**95%**	**98%**	**99%**	**99.9%**	
	Z-value	±1.28	±1.65	±1.96	±2.33	±2.58	±3.29	
Hypothesis Testing	**Type I Error Risk = α**		**.20**	**.10**	**.05**	**.02**	**.01**	**.001**
	Two-tailed test *Z*-value		±1.28	±1.65	±1.96	±2.33	±2.58	±3.29
	One-tailed test *Z*-value*		.84	1.28	1.65	2.05	2.33	3.09

*Minus sign needs to be provided for lower-tailed tests.

INTRODUCTORY

BUSINESS STATISTICS

WITH MICROCOMPUTER APPLICATIONS

THE DUXBURY ADVANCED SERIES IN STATISTICS AND DECISION SCIENCES

Applied Nonparametric Statistics, Second Edition, Daniel
Applied Regression Analysis and Other Multivariable Methods, Second Edition, Kleinbaum, Kupper, and Muller
Classical and Modern Regression with Applications, Second Edition, Myers
Elementary Survey Sampling, Fourth Edition, Scheaffer, Mendenhall, and Ott
Introduction to Contemporary Statistical Methods, Second Edition, Koopmans
Introduction to Probability and Its Applications, Scheaffer
Introduction to Probability and Mathematical Statistics, Bain and Engelhardt
Linear Statistical Models, Bowerman and O'Connell
Probability Modeling and Computer Simulation, Matloff
Quantitative Forecasting Methods, Farnum and Stanton
Time Series Analysis, Cryer
Time Series Forecasting: Unified Concepts and Computer Implementation, Second Edition, Bowerman and O'Connell

THE DUXBURY SERIES IN STATISTICS AND DECISION SCIENCES

Applications, Basics, and Computing of Exploratory Data Analysis, Velleman and Hoaglin
A Course in Business Statistics, Second Edition, Mendenhall
Elementary Statistics, Fifth Edition, Johnson
Elementary Statistics for Business, Second Edition, Johnson and Siskin
Essential Business Statistics: A Minitab Framework, Bond and Scott
Fundamentals of Biostatistics, Third Edition, Rosner
Fundamentals of Statistics in the Biological, Medical, and Health Sciences, Runyon
Fundamental Statistics for the Behavioral Sciences, Second Edition, Howell
Introduction to Probability and Statistics, Seventh Edition, Mendenhall
An Introduction to Statistical Methods and Data Analysis, Third Edition, Ott
Introductory Business Statistics with Microcomputer Applications, Shiffler and Adams
Introductory Statistics for Management and Economics, Third Edition, Kenkel
Mathematical Statistics with Applications, Fourth Edition, Mendenhall, Wackerly, and Scheaffer
Minitab Handbook, Second Edition, Ryan, Joiner, and Ryan
Minitab Handbook for Business and Economics, Miller
Operations Research: Applications and Algorithms, Winston
Probability and Statistics for Engineers, Third Edition, Scheaffer and McClave
Probability and Statistics for Modern Engineering, Second Edition, Lapin
Statistical Experiments Using BASIC, Dowdy
Statistical Methods for Psychology, Second Edition, Howell
Statistical Thinking for Behavioral Scientists, Hildebrand
Statistical Thinking for Managers, Second Edition, Hildebrand and Ott
Statistics: A Tool for the Social Sciences, Fourth Edition, Ott, Larson, and Mendenhall
Statistics for Business and Economics, Bechtold and Johnson
Statistics for Management and Economics, Sixth Edition, Mendenhall, Reinmuth, and Beaver
Understanding Statistics, Fifth Edition, Ott and Mendenhall

INTRODUCTORY

BUSINESS STATISTICS

WITH MICROCOMPUTER APPLICATIONS

■ Ronald E. Shiffler
University of Louisville

■ Arthur J. Adams
University of Louisville

PWS-KENT PUBLISHING COMPANY
Boston

PWS-KENT
Publishing Company

20 Park Plaza
Boston, Massachusetts 02116

PWS-KENT Publishing Company is a division of Wadsworth, Inc.

Library of Congress Cataloging-in-Publication Data

Shiffler, Ronald E.
Introductory business statistics with microcomputer applications / Ronald E. Shiffler, Arthur J. Adams.
p. cm.
ISBN 0-534-92186-8
1. Commercial statistics. 2. Commercial statistics—Computer-assisted instruction. 3. Microcomputers. I. Adams, Arthur J. II. Title.
HF1017.S5 1989
519.5′02433—dc20 89-16375
CIP

Printed in the United States of America.
90 91 92 93 94 — 10 9 8 7 6 5 4 3 2 1

Sponsoring Editor: Michael Payne
Production Editor: Susan M. C. Caffey
Production: Technical Texts, Inc./Sylvia Dovner
Manufacturing Coordinator: Marcia A. Locke
Assistant Editor: Marcia Cole
Editorial Assistant: Susan Hankinson
Interior Designer: Susan M. C. Caffey
Cover Designer: Designworks
Interior Illustrator: Scientific Illustrators
Typesetter: Weimer Typesetting Company, Inc.
Cover Printer: New England Book Components, Inc.
Printer/Binder: R.R. Donnelley and Sons Company

Pages 1036 and 1039 contain charts reprinted with permission from Commodity Research Bureau.

Chart reprinted from:
CRB Futures Chart Service
a weekly publication of
COMMODITY RESEARCH BUREAU
a Knight-Ridder Business Information Service
100 Church St., Suite 1850
New York, N.Y. 10007

To my wife and confidante, Barbara —R.E.S.

For Sarah and Jan —A.A.

ABOUT THE AUTHORS

Ronald E. Shiffler is Associate Professor of Business Statistics at the University of Louisville. Previously, he was a full-time faculty member at Georgia State University where he taught quantitative courses in the BBA, MBA, and Ph.D. programs. He has won teaching awards at the University of Florida, Georgia State University, and the University of Louisville. In 1979 and again in 1984, Dr. Shiffler won the Presentation Award in the Section on Statistical Education at the American Statistical Association's annual meeting. His research has appeared in the *American Statistician, Teaching Statistics, International Journal of Mathematical Education in Science and Technology, International Journal of Operations and Production Management,* and *Journal of Marketing Research*. He is an active member of the Decision Sciences Institute and the American Statistical Association. Dr. Shiffler holds a Ph.D. in Statistics from the University of Florida, a masters degree in Mathematics from Bucknell University, and an undergraduate degree from the University of North Carolina at Greensboro.

Arthur J. Adams is Professor of Business Statistics at the University of Louisville. His undergraduate degree is in Chemistry (Knox College, in Illinois), and he spent two years as a production engineer in the Saran plants of Dow Chemical Company, Midland, Michigan. Dr. Adams taught at Illinois State University prior to obtaining a Ph.D. in Applied Statistics at the University of Iowa. He has published statistics-related articles in journals such as *Journal of Marketing Research, Journal of Advertising, Journal of Advertising Research, Journal of the Academy of Marketing Science,* and *Educational & Psychological Measurement*. Dr. Adams is a member of Decision Sciences Institute and is a regular participant at the DSI National Conference.

TO THE INSTRUCTOR

Special Features

There is no shortage of possible books for you to choose from for use in your business statistics course. We take it as given that most, if not all, such books cover a common base of topics and do so reasonably well. What makes this book different? We view the following as differentiating features of our offering.

Computer Software Our text can be ordered with a two-disk software package that operates on most IBM and compatible microcomputers. The package was written in C language by John K. Hedges, a graduate student with extensive programming experience, in consultation with the authors. The package contains a series of chapter-specific modules that we call processors. These processors can solve most of the problems in the book. Examples of output are provided in separate sections within most chapters.

If you wish to integrate the personal computer into your course, we think you will be excited about this feature. Students can be assigned problems to be done on the computer, or they can be directed to use the computer to verify answers worked by hand. The software is menu driven and user friendly. The package has file setup and retrieval capabilities so that students can create and save data files on their own data disks.

We anticipate that the software will be very appealing to the students in your course. Further, it is something that the students can keep throughout their college careers for potential use in other courses. If you prefer not to use the computer in the course, simply omit the optional (starred) computer sections.

"To Be Continued . . ." Sections At the conclusion of most chapters is a short section titled "To Be Continued . . . ." The specific purpose of these sections is twofold:

1. To alert the student to the fact that material just studied may be encountered again in future college courses (although perhaps with a slightly different name or symbol). We provide several connections to finance, management science, marketing, and accounting.
2. To alert the student to applications of chapter material in the media as well as in business practice.

More generally, the idea of "To Be Continued . . ." is to refute the notion that business statistics doesn't relate to other disciplines or that it is not useful. We think it is important to show where applications exist. To this end, we cite the popular press often; in fact nearly 100 citations were extensive enough to require permission from such sources as *USA Today, Business Week, Los Angeles Times,* and so forth. In addition, several statistics-related questions are reproduced verbatim from past CPA exams. Our intent is to provide ample opportunity for students to see applications of business statistics in the "real world."

Section on Opinion Polling As an extension of the idea of showing applications of business statistics, we offer a section (in Chapter 12) on opinion polling. We relate polling to making inferences about the binomial parameter and expand on the need for a random sample taken from a clearly defined target population. This section has worked well in class-testing, and there are always current polls in the news that can be used as fresh examples of well- or poorly-conducted or reported cases for class discussion.

Wide Problem Selection This book has over 1600 problems. Each chapter has a variety of exercises that range from drill work to more challenging problems. We provide exercises from a variety of scenarios. Most chapters have several problems that are taken from current literature, including various research journals, such as the *Journal of Marketing Research,* the *American Statistician,* and the *Journal of Advertising Research*. Many other problems make use of "real" data sets. We also provide 100 review problems that appear in sets of twenty after every fourth chapter. Since these problems are not tied to specific sections, they offer the student a chance to test himself or herself prior to an exam.

We use optional (starred) exercises to denote problems involving topics or finer points that we did not have space to discuss as a regular part of the chapter. Some examples of concepts treated in this fashion include the geometric mean, the hypergeometric distribution, the uniform distribution, a one-sided confidence interval, inferences using the coefficient of skewness, and various alternative computing formulas not usually seen.

Classroom Examples Sprinkled throughout the chapters are problems that are set aside and labelled "Classroom Examples." These multi-part problems can be used in several ways. One possibility is this: After the instructor has introduced a new topic (such as how to use the binomial formula and tables), she or he can assign the students a Classroom Example that bears directly on the new topic. The students can work in-class on the problem either individually or in small groups. The instructor is free to walk about the class and assist any student who needs help. After perhaps seven to ten minutes, the instructor can go over the solution with the whole class. We think the students benefit from getting immediate experience working a problem right after its presentation by the instructor. In addition, the instructor can assess the class's comprehension of the material.

Content and Flexibility

The first ten or twelve chapters cover what many instructors consider a proper amount of material for a first course in statistics. Instructors teaching a second term are likely to vary their coverage beyond this point to suit their own preferences and objectives.

Chapters 1 through 3 concentrate on problem structure, terms, organizing data, and descriptive statistics. We place some emphasis on the Empirical Rule in Chapter 3, using the ± 2 standard deviations statement as a preliminary device for setting up the notion of rejection regions later on in the course.

Chapter 4 discusses the ideas of correlation and the least-squares line in a descriptive context. This relatively brief treatment is intended to assure that those students who will take only one course in statistics in their career will have been exposed to these important concepts. Instructors whose schools require a two-term sequence may choose to omit this chapter or schedule it later on.

Chapter 5 introduces probability and probability distributions for discrete and continuous random variables. As other authors are beginning to do, we place a reduced emphasis on permutations, combinations, and Venn diagrams in the study of basic probability. In Chapters 6 and 7, we cover specific probability distributions: the binomial, Poisson, normal, and exponential. We also treat the Poisson approximation to the binomial and the normal approximation to the binomial. While we would expect all instructors to teach the binomial and normal distributions, personal taste and/or time constraints may influence the inclusion of the others.

Chapter 8 deals specifically with the sampling distribution of the mean, although it lays out the logic for developing other sampling distributions. Section 8.3 treats the case of sampling from finite populations and may be viewed as optional. After this chapter the student will be ready to consider statistical inference. (Some preliminary and informal examples of making an inference appear in Chapters 4, 6, and 7.) Chapter 9 is a short transitional chapter that sets forth the logic, framework, and vocabulary of inferential statistics. None of the problems posed in Chapter 9 need a calculator for solution.

Chapter 10 develops confidence intervals for μ and hypothesis tests about the value of μ for the large-sample case; Chapter 11 is for samples less than size 30. We include the signed ranks test in Chapter 11 as a parallel to the t-test, since we find this nonparametric alternative to the t-test receiving more attention in the literature. A section on Quality Control and its relation to hypothesis testing concludes Chapter 11. Chapter 12 demonstrates inferences for π, including the section on polling, and inferences for σ^2.

Chapter 13 covers the case of two-population sampling, including the concept of matched pairs. Chapter 14 introduces ANOVA and the F-distribution. Besides one-way, we discuss follow-up tests, the randomized block design, and two-factor ANOVA with interaction. Chapter 15 is the first of an extensive three-chapter regression sequence. Chapter 15 offers the simple regression and correlation models, while Chapter 16 generalizes to the multiple regression case. We provide an optional matrix algebra development of multiple regression and briefly discuss some commercial software for regression and compare their output to that of our own processors. Chapter 17 is concerned with model building for multiple regression.

Chapter 18 investigates forecasting for business and economic time series. A variety of forecasting methods are presented along with the concept of combining forecasts as a strategy for forecast improvement. Chapter 19 presents chi-square tests and several of the more useful nonparametric statistical methods.

Supplemental Materials

Some additional materials are available to instructors:

- *Solutions Manual:* Although the textbook includes answers (not solutions) to the odd-numbered problems, the *Solutions Manual* contains complete solutions to all exercises.
- *Test Bank:* Some of the items in the test bank are objective, while some are of the type where students are expected to show their work as well as give the answer.
- *Transparency Masters:* A set of transparency masters for classroom use primarily provides selected figures from the text.

All the materials above have been prepared by the authors.

Overall, we think that the intuitive orientation of the text and the emphasis on real applications, when combined with the supporting materials (including the software package), will enhance your classroom teaching and your students' appreciation of statistics.

Ronald E. Shiffler
Arthur J. Adams

ACKNOWLEDGMENTS

Creating a book is a team effort and we would like to acknowledge our team members here. At the head of our team is our coach and editor, Michael Payne, who provided support, guidance, and experience at all times. Production editor Susan Caffey and production coordinator Sylvia Dovner had the responsibility for turning our manuscript into a book. We feel fortunate to have worked with these professionals.

We would like to thank the following reviewers, listed in alphabetical order, for their constructive comments that helped to improve the manuscript: Harry C. Benham, *University of Oklahoma;* Michael E. Bohleber, *Georgia Southwestern College;* Bruce Bowerman, *Miami University;* Barbara J. Bulmahn, *Indiana University-Purdue University at Fort Wayne;* Myron K. Cox, *Wright State University;* Brett D. Crow, *Eastern Montana College;* John Daughtry, *East Carolina University;* George C. Dery, *University of Lowell;* Terry Dielman, *Texas Christian University;* David L. Eldredge, *Murray State University;* Deborah J. Gougeon, *University of Scranton;* John Hewitt, *Emory University;* Ronald Lucchesi, *Wittenberg University;* Barbara J. Mardis, *University of Northern Iowa;* Barbara Treadwell McKinney, *Western Michigan University;* Mehdi Mohaghegh, *Norwich University;* Joseph B. Murray, *Community College of Philadelphia;* Eduardo M. Ochoa, *California State University at Los Angeles;* David W. Pentico, *Duquesne University;* Dale Sauers, *York College of Pennsylvania;* Herman Senter, *Clemson University;* and Stanley R. Schultz, *Cleveland State University.*

We are especially grateful to our typists—Leesa Foster and Sandy Hartz—who processed and reprocessed our work without complaint. Also, we appreciate the typing support of Jan Pollard in the early going.

Mary Jane Burch, Joyce Hoffman, and Missy Mountz contributed their expertise in editing, class testing the manuscript, and working the exercises, respectively.

In addition, we would like to acknowledge the cooperation of those individuals and organizations who allowed us to reprint selected material.

Finally, to our families we wish to say: Thank you for your support!

TO THE STUDENT

As business statistics instructors, we have taught many students whose math experiences may be similar to yours. Most likely, your math background is not as strong as you would like it to be. In fact, it may have been several years since you have taken a quantitative course. But here you are, signed up for business statistics, knowing full well that you will *never* use statistics again in your life. Sound familiar?

Now, given that background, what would you do if you were the instructor preparing to teach a class of students with similar anxieties and attitudes? We think you would select a textbook that might calm the fears and change the attitudes. To that end, we decided to use our experience as instructors to develop a textbook specifically for you, the student.

Still speaking as instructors to students, we believe the following points represent some of the keys to success in mastering business statistics.

Work a Lot of Problems This point is no secret. Whether playing sports, making crafts, or teaching students, we sharpen our skills through practice. So it is with business statistics. You should work as many exercises in each chapter as you can to develop your problem-solving skills. Work all of them, if possible.

Use Your Math Skills Business statistics assumes certain math tools as prerequisites. The elements of basic algebra are mandatory; calculus is not. A background in matrix algebra will be helpful in Chapters 15 and 16, but it is not required. Arithmetic skills are needed, but could be replaced with the use of a handheld calculator, if your instructor permits them. There is an optional software package to accompany the book that your instructor may choose to use. The package is very user friendly; no programming skills are required. If your instructor has not elected to use the software but you would like your own copy, you can order it through your college bookstore from PWS-KENT Publishing Company.

Read the Book This point may sound obvious, but some students hope to pass the course by studying just the instructor's lecture notes. Reading the book is essential. Furthermore, reading is not a passive exercise. We challenge you to make it an active experience. Read sitting at a desk, not slouched on the couch. Have highlighters, pencils, and pens nearby. Write notes to yourself in the margin as you read. Highlight important points. Work the textbook over—don't treat it like fine china!

The following paragraphs tell how we have addressed these points to make the textbook easier to use.

Answers to Odd-Numbered Exercises We feel strongly about working lots of exercises and the need for feedback. Consequently, we have developed over 1600 exercises for the book. (You may run out of time before you run out of exercises!) And, the answers to the odd-numbered problems are provided at the back of the book.

Partial Solutions Manual and Math Review Sometimes an answer is not enough. In a separate supplement, *Partial Solutions Manual and Math Review,* we have written out full solutions to some (not all) of the odd-numbered exercises. In addition, we have identified ten math skills, such as solving an equation for an unknown and using summation notation, and prepared short review sections on each. Each skill is explained briefly and illustrated with several examples. Extra exercises (and answers) are provided for practice. If your math preparation is weak or if you think you would benefit by seeing the complete solution to an exercise, we recommend using the *Partial Solutions Manual*. It should also be available from your college bookstore.

Definitions, Formulas, and Tables To help you pick out the important points as you read the text, we have numbered key formulas, used boldface or italic type for salient words or phrases, and summarized material into boxes and tables.

We are proud of our book. It represents the culmination of many years' work. Imagine how proud you will feel after completing a similar long-term project such as graduating from college. By the way, you most likely *will* use statistics again in your life. To convince you, we have included sections titled "To Be Continued . . . " in most of the chapters. In these sections, we present examples of how and where statistics might reappear in other business courses in your college curriculum, in the print media, and in everyday use in the business world. In summary, we hope this textbook facilitates your current study of business statistics and serves as a reference book in your future career.

CONTENTS

*Optional section

INTRODUCTORY

BUSINESS STATISTICS

WITH MICROCOMPUTER APPLICATIONS

Chapter Maxim *Be wary of descriptions and inferences developed from nonrandom samples.*

CHAPTER 1

INTRODUCTION

*Optional

Objectives

After studying this chapter and working the exercises, you should be able to

1. Distinguish a variable from its values.
2. Classify data as qualitative or quantitative.
3. Differentiate the types (inferential and descriptive) of business statistics problems.
4. State the goal of descriptive statistics and the goal of inferential statistics.
5. Contrast a finite population with an infinite population.
6. Explain the benefits of taking a sample and conducting a census.
7. Differentiate random from nonrandom sampling methods.
8. Describe four types of random sampling.
9. Select a simple random sample from a finite population using a table of randomly arranged numbers.

*10. Execute the computer processor Simple Random Sample to obtain a random sample.

With your trusty calculator in hand, your anxiety level high, and your fear of mathematics firmly entrenched, you are ready to tackle business statistics. Welcome! We also welcome those without a calculator and/or those with no fear of mathematics. Welcome to your first business statistics lesson as well: The variability in the characteristics of students in this class makes it impossible to describe everyone by the same terms. Some students may have a calculator; some may not. A study of students' calculators would reveal a variety of brand names, costs, capabilities, power sources, and so forth. Students likewise will possess varying degrees of apprehension about mathematics.

Business statistics as a course of study exists because items of interest to us exhibit variability. **Variability** is the opposite of constancy. For instance, the closing value of the Dow Jones Industrial Average is reported each business day, and every day it is subject to change. Reporting this figure represents one of the two main branches of business statistics: descriptive statistics. Most of the figures you see on television and read in the print media are descriptive in nature. The early chapters of this book are devoted to descriptive statistics.

In addition to knowing today's closing Dow, for example, we also may wish to know its direction tomorrow. Where is the market headed? An answer to this question requires a judgment or prediction about an uncertain event. This is the second major branch of business statistics: inferential statistics. **Inference** implies a conclusion or

*Applies to optional section.

estimate, and the unavoidable consequence of risk. *Risk* refers to the possibility that an inference is wrong. From our study of inferential statistics, we will learn how to minimize risk in decision-making situations. The last eleven chapters of the book deal with inferential statistics.

Business statistics is concerned with description and inference. Before we can describe or infer, however, we need data. **Data** are numbers, like 1, 2, or 3, that represent values of a concept such as the number of dependents claimed on a tax return, or words like growth, income, or capital appreciation, representing types of mutual funds. Data are the genesis of business statistics.

In this chapter, we will focus on the nature of business statistics problems by discussing the different types of data and describing methods for data collection. Understanding data allows us to outline the structure of descriptive and inferential statistics and to state the goals of analysis for each.

1.1 DATA

Data drive the study of business statistics. In fact, business statistics can be characterized as the collection, organization, summary, analysis, and interpretation of data. If this is all you wish to know about business statistics, you may close the book and go home. But, if you would like to know *how* to collect, organize, summarize, analyze, and/or interpret data, please continue reading.

Still here? Good decision! Making a decision is perhaps the ultimate business activity, and most of the time it is based on judgment and information. Experience governs the former while business statistics can be used to provide the latter. To illustrate, in a study of 349 executives (vice-presidents and above) and their decision-making processes, it was found that 30.9 percent rely *exclusively* on numbers or statistics (Pinnacle Group, reported in *USA Today,* July 21, 1987). Information results from data. Before we can practice making decisions, though, we must learn to collect raw data and turn it into useful information.

Basic Terms

We begin with the basic building block in business statistics: an *entity*.

Definition

An **entity** is a person, place, date, time, or thing that supplies the attribute, count, or measurement of interest.

It is natural to think of an entity as being a person since we are often interested in a person's thoughts, abilities, actions, and so on. However, many inanimate or physical objects qualify as entities too. As examples, we may be interested in knowing the following:

- The number of companies locating their corporate headquarters per state
- The type of heat per house
- The number of miles driven by a sales representative per day
- The number of vehicles traveling a certain stretch of highway per hour

In these examples the entities are the states (places), the houses (things), the days (dates), and the hours (time), respectively. The use of "per" in the preceding examples makes the entity obvious. Initially this is helpful in determining the object(s) of interest. With exposure to more examples and exercises you may not need this convention to identify the entity. Instead, merely ask yourself what object is selected to yield the attribute, count, or measurement.

The entities in the first and last examples yield a number in the form of a count. The entities in the second example possess attributes like oil, gas, or electric. In the third example, the entities produce a number in the form of a measurement. The attribute or number represents the *value* for a concept. For example, the concept "type of heat" produces values like oil, gas, or electric when applied to an entity (house). The concept "number of miles driven by a sales representative" generates values like 37.4, 100, or 142.6 when applied to an entity (day). In business statistics we use the term *variable* to define a concept.

Definition

A **variable** is a quality or quantity having a changeable value among entities.

A variable always appears as a phrase or possibly a single word description, as Table 1.1 indicates. This table also lists possible values for each variable and the corresponding entity. A set of data is merely the collection of values of a variable for specific entities. If you note the sex of five people sitting in your classroom, then the set of five male/female values is a set of data. If you examine the telephone bills of a small business for the past three months and count the number of long distance phone calls for each month, then the resulting three numbers form a set of data. Figure 1.1 summarizes the relationship among the terms entity, variable, value of the variable, and a set of data.

COMMENTS

1. Synonyms for the word *entity* are element, individual unit, elementary unit, or experimental unit.
2. There may be dual entities when a variable involves two uses of "per." For example, the variable "number of hours worked per employee per day" suggests two entities: people and days.
3. When an entity involves a time factor and time is used to arrange the values of the

Table 1.1 Examples of Variables, Values, and Entities

Variable	Possible Values	Entity
Age	19, 20, 21, . . .	Person
Sex	Male, female	Person
Type of purchase per transaction	Cash, check, credit card	Transaction
Standard & Poor's municipal bond ratings	AAA, AA, A, BBB, etc.	Bond
Square feet of available office space	1900, 2500, etc.	Building
Advertising medium	Radio, TV, newspaper, magazine, etc.	Advertisement
Number of long distance phone calls made by a small business per month	0, 1, 2, . . .	Month
Flight number of airplane flight between two cities	12, 211, 359, . . .	Flight

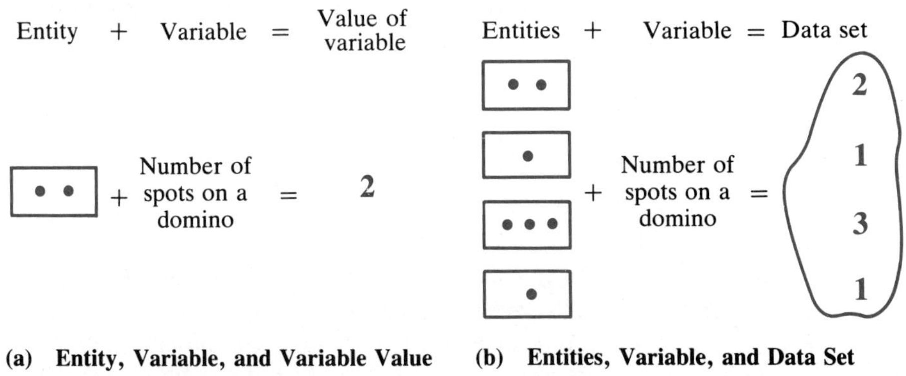

Figure 1.1 Relationships Among Terms

variable in sequence, we refer to the data as "time-related data" or as a "time series." Time-related data are treated separately in Chapter 18.

Data Classifications

Immediately we notice a difference in the possible values for the variable *sex* and for the variable *number of long distance phone calls per month*. In the former case, a possible value is a word; in the latter case, a possible value is a number. These two styles of values reflect the two types of data: quantitative and qualitative. Generally speaking, quantitative data are numbers; qualitative data are words. More specifically, when entities yield a label, category, attribute, or code, the resulting data are considered **qualitative.** When entities produce a count or measurement, the data are considered **quantitative.**

Codes are the exception to the word–number rule of classification. An example of a code is an identification number such as a Social Security number or a four-digit code that allows a person access to his or her bank account through an automated teller machine. The possible values of these variables are numbers, but the numbers are identifiers, not counts or measurements, and are therefore qualitative data.

Codes also are used to facilitate record-keeping or computer processing of the data. For example, the various departments in a grocery store—produce, meat, delicatessen, and so forth—may be assigned a number as a code. Consequently, sales by departments can be recorded by a number rather than a word; for example, meat = 1, produce = 2, and so on. This artificial numbering system does not turn the categorization of each grocery item into quantitative data. We treat the number codes as qualitative data.

A third form of codes is ranked data. Ranking implies the values of the variable are ordered in a rating scheme from "best" to "worst" (or vice versa). For example, movie critics may use a star system to assess new movies. A four-star movie is then perceived to be "better than" a three-star movie. The rankings or ratings, though numerical, are (coded) qualitative data, not quantitative data.

Table 1.2 shows the data classifications for the variables listed in Table 1.1. The second column of Table 1.2 lists the form of the data. Notice that the last variable—flight number—yields numbers in the form of coded data. Thus, the data are classified as qualitative. Any subsequent procedures that we develop for describing data must take into account the type of data we have.

Sample Versus Population

There are special terms to describe data sets in business statistics. If the set of data is exhaustive, then it is said to form a *population*. To be exhaustive, every entity must be present. When some, but not all, of the entities are selected, the resulting collection of values is a *sample* set of data.

Definitions

A **population** is a complete set of data.

A **sample** is a set of data representing a portion of the population.

To illustrate the concepts of a sample and a population, consider the variable *number of long distance phone calls made by a small business per month*. Possible

Table 1.2 Classification of Data for Variables Listed in Table 1.1

Variable	Form of the Data	Data Classification
Age	Count	Quantitative
Sex	Attribute	Qualitative
Type of purchase per transaction	Category	Qualitative
Standard & Poor's municipal bond ratings	Label	Qualitative
Square feet of available office space	Measurement	Quantitative
Advertising medium	Category	Qualitative
Number of long distance phone calls made by a small business per month	Count	Quantitative
Flight number of airplane flight between two cities	Code	Qualitative

values are the integers 0, 1, 2, and so on, while the entity is a month. At the end of each month, a value is recorded for the variable. For example, the business might have made two long distance phone calls in June 1988 (denoted by June–88 in Figure 1.2).

In terms of entities, the population is the collection of all months for which the firm has been in business. A sample might be the months June–88, Mar–91, Sept–89, Oct–90, and Dec–89. In terms of data, the population is the set of values of the variable corresponding to each month. A sample would be the values for June–88, Mar–91, Sept–89, Oct–90, and Dec–89.

Figure 1.2 depicts the relationship between a sample and the population in terms of entities in panel (a) and in terms of data in panel (b). The rectangle contains a set of entities or a set of data and represents the population. The sample is the set of entities or numbers inside the amoeba-like shape. Notice that the number 0, for example, appears in the population but *not* in the sample. Also note that the numbers differ from one another inside both sets of data. This is indicative of *variability* that we spoke of at the beginning of this chapter.

Populations and samples must be defined or identified carefully since each depends on a point of reference. To demonstrate, suppose we describe a variable as the gross revenues in a retail store per day. As each day passes the store records a figure for its gross revenues. If we define the population as the collection of daily revenues

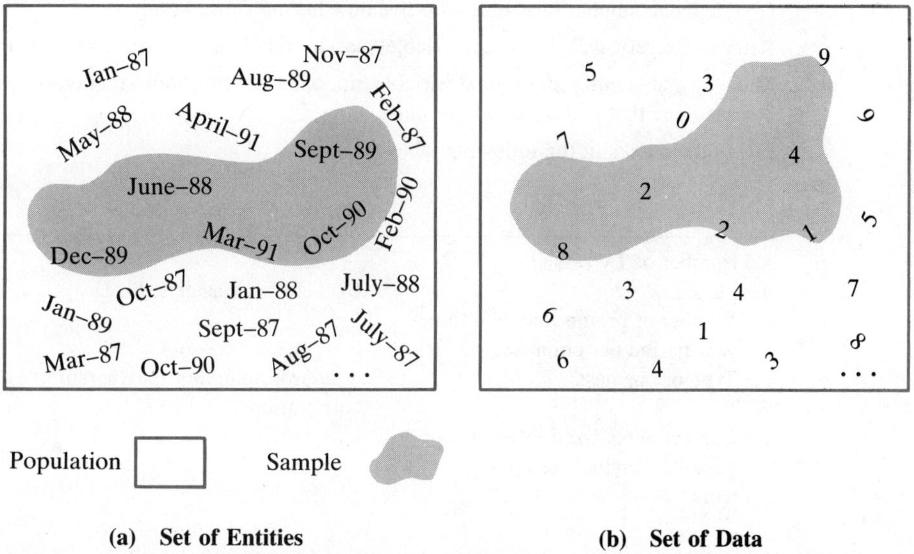

Figure 1.2 Sample Versus Population

for the month of March, the population consists of 31 numbers if we assume the store is open on Sundays. A sample would be the revenue figures for the first seven days of the month, for example. Alternatively, the collection of 31 daily revenue figures for March becomes a sample if the population is redefined to be the daily revenue figures for the current fiscal year.

Recording the value of a variable from an entity produces a single piece of data. Repeating this process with different entities generates a set of data. Depending on the perspective, the set of data is either a population or a sample. Although most business statistics problems involve data, analyzing the data depends on the nature of the problem. The next section introduces the two types of analysis: descriptive and inferential.

SECTION 1.1 EXERCISES

1.1 What are the two types of data?

1.2 Identify the entities and possible values of the following variables:

a. name of credit card used to make a purchase.
b. interest earned in your checking account per month.
c. number of first-class letters mailed per business day by a small firm.
d. type of seating (smoking/nonsmoking) requested per traveler on a TWA flight from Denver to San Francisco.

e. flight number per TWA flight from Denver to San Francisco.
f. telephone number of people who live on Charing Cross Road.

1.3 Refer to Exercise 1.2. For each variable, classify the data as qualitative or quantitative.

1.4 Make up and identify an original variable (not one from this book) that produces quantitative data and one that yields qualitative data.

1.5 Fill in the blanks in the following table:

Variable	Possible Values	Entity
a. Number of TV sets	_____	Household
b. _____	Bachelor's, Master's, Ph.D.	Person
c. Number of promotions in a three-year period per employee	0, 1, 2, . . .	_____
d. Type of business	Sole proprietorship, partnership, corporation	_____
e. Current annualized yield on passbook savings accounts per bank	_____	_____

1.6 Identify the entities and possible values of the following variables:
a. number of bedrooms per house.
b. percentage of departing flights delayed at an airport per month.
c. student classification (freshman, sophomore, etc.).
d. mode of transportation to get to work per employee of an insurance company.
e. subsidiary account number on a firm's ledger.
f. risk rating (1 = least amount of total risk; 9 = most amount of total risk) of mutual funds.

1.7 Refer to Exercise 1.6. For each variable, classify the data as quantitative or qualitative.

1.8 Refer to Exercise 1.2. Describe the entities in a possible sample if the entities in the population are as follows:
a. all transactions on a Friday.
b. the first 6 months of the year.
c. the last 8 business days of the month.
d. the 108 passengers on Flight #189.
e. the 12 different TWA flights from Denver to San Francisco.
f. all adults over 21 years of age.

1.9 Refer to Exercise 1.6. Describe the entities in a possible population if the entities in the sample are as follows:
a. all the odd-numbered houses on Galway Lane.
b. February and June of last year.
c. all the students in this class.
d. all employees whose last name begins with one of the letters E, J, O, or T.
e. the menswear and jewelry subsidiary accounts from the merchandise inventory account of a retail department store.
f. all growth-type mutual funds.

1.10 What property of data justifies the existence of business statistics as a course of study?

1.2 BUSINESS STATISTICS PROBLEMS

Dealing with data requires a purpose. Why should we collect and analyze data? The answer is to help solve problems through description or inference. If we view statistics as a decision-aiding tool that fits into the decision-making process of management rather than as a formula-plus-computational exercise, then we are studying *business* statistics.

Problem Structure

Wherever possible in this book, we shall emphasize the management *structure*—plan, execute, report, evaluate, and act—in our business statistics problems.

Planning involves statements of a general goal and achievable objectives. In his text *Management,* Robert Kreitner discusses the difference between a goal and an objective:

> *Just as a distant port serves as a target for a ship's crew, goals and objectives also give managers targets to steer toward. . . . A* goal *is a general statement of intention. . . . An* objective *. . . is a specific commitment to achieve a measurable result. . . . Goals are farsighted and general; objectives are nearsighted and specific.**

According to Kreitner, *planning* is the primary function of management. Planning is not restricted to personnel matters, financial activities, marketing strategies, or budgeting problems. Planning has a place in business statistics too.

Execution includes the specific methods and techniques of data analysis. *Reporting* the results of a data analysis in common terms is done through written and/or oral communication. Our primary goal in this book is to explain various data analysis techniques with secondary attention given to reporting the results.

Evaluation and action are more difficult to "teach" in a textbook like this. *Evaluating* a data analysis requires you to sift through the analysis with a critical eye to see if the results meet the stated objectives and support the goal. Occasionally we will provide some reasonability checks to help you evaluate the correctness of an analysis. *Action* refers to a decision based on the information from the data and other tangible and intangible inputs.

To illustrate this structure, consider the problem facing the marketing department of an automobile manufacturer: How does the staff develop an advertising campaign for a new luxury auto to appeal to potential customers? Before beginning to work on an ad campaign, the marketers must determine the traits of the "target" customer. Their goal is to develop a profile of the consumer interested in the luxury segment of the automobile market. Through the planning process, important demographic char-

*Kreitner, Robert. *Management,* Third Edition. Copyright 1986 by Houghton Mifflin Company. Used with permission.

acteristics—such as sex, age, annual income, and so forth—are identified that describe the type of person most likely to purchase a luxury auto. Specific objectives would be to collect and summarize information on the key variables. Selecting the entities (people) and analyzing the data is the execution stage of the process. Incorporating the results into a written summary or oral presentation is the reporting step. The evaluation of the report might reveal a previously unknown trend requiring further analysis. For example, the data could indicate that working women make up a large chunk of this market segment. Yet, the company historically may have directed most of its advertising toward males when introducing luxury autos. Refocusing future advertisements could be an action resulting from this data analysis.

As a second example of problem structure, consider a fast food company that has developed a new food item for its menu. The goal is to decide if the new item should be added to the menu in all the company's stores. To gauge the public's reaction to the product, the company *plans* to test market it in selected restaurants. The test markets generate key data that must be organized and analyzed (*execution*). A summary of the data analysis is *reported* and *interpreted*. A *decision* about the overall impact of the product is based on the results of the analysis.

Rather than being decision-oriented, some business problems are more exploratory in nature, with the primary objective to provide information. For example, an electric utility might collect comparable rate data from other utilities and publish an analysis of the data to spotlight its own, favorable (low) rates. Or, the government could report the inflation rate, unemployment rate, or the current yield on Treasury bills as public information.

Business statistics procedures used to analyze data are commonly labeled as descriptive or inferential. A problem may involve one or both procedures, depending on the objectives to be achieved. For instance, the automobile manufacturer relies on a descriptive procedure to characterize potential customers and may use the results to predict a market share (an inference). The utility's data analysis is primarily descriptive, while the fast food company's main objective is to infer the profitability of the new menu item.

In all situations data analyses are part of a problem-solving process that has a well-defined structure. The following box highlights the two types of business statistics procedures.

Types of Business Statistics Procedures

- Description
- Inference

Descriptive Statistics

Descriptive statistics is concerned with describing the observed set of data, be it a sample or a population. Almost all of the problems we encounter in this branch of business statistics involve describing sample data. Unless we specifically state otherwise, you may assume the data you encounter in examples or exercises are sample data.

Data description takes on many forms, including tables, graphs, numbers, and formulas. A descriptive data analysis may include

- Examining (informally) the data to check for reasonable and/or missing values
- Sorting or arranging the data in order
- Compiling the data into a table
- Graphing the data
- Computing summary values such as the average
- Deriving functional relationships between and among variables

These activities form the specific *objectives* to support the overall goal of description. The following definition formally states the purpose of descriptive statistics.

Definition

The **goal of descriptive statistics** is to describe a set of sample data.

The following examples illustrate business statistics problems with description as their goal.

EXAMPLE 1.1

A store tracks its daily sales figures. At the end of each quarter these figures must be synthesized into a report. The day-to-day results are too detailed and cumbersome to publish. Summary data, such as total sales, the percentage increase in total sales from the previous quarter, sales by category of goods, or the day of greatest sales, are needed.

□

EXAMPLE 1.2

Mutual funds are an attractive investment because they offer a better-than-average return for a minimum amount of risk. The "track records" of mutual funds are an important investment criterion for some investors. Lipper Analytical Services, Inc., records the performances of all mutual funds so that prospective investors can compare the historical returns in the short and long run. Compiling these figures is facilitated with graphs and numbers that summarize average gains (or losses), percentage gains (or losses), and the variability in each fund's share price relative to all mutual funds.

□

EXAMPLE 1.3

The human resource group in a large firm recognizes the need to attract and retain quality people. In so doing, the company maintains its differential advantage in the industry and reduces the cost of training and development. To aid in the selection process, a handwriting analysis of some current and former employees is performed, yielding scores from 1 to 100. These data are combined with each person's length of time with the company to develop a model or formula that relates the two variables. The analysis reveals that higher handwriting scores are associated with employees having longer tenures with the firm.

□

In each scenario the business problem involves describing data. The last example bridges both descriptive statistics and inferential statistics. The model or formula simultaneously describes the handwriting score and length of time with the company for current and former employees and can be used to forecast the tenure of potential employees based on their handwriting score. The problem has evolved into one of inference.

■ Inferential Statistics

Inferential statistics is concerned with reasoning from particular data to arrive at a general conclusion. The particular data are a sample; the general conclusion is a statement or judgment about the population from which the sample has been drawn.

Inferences or generalizations are of two types: estimates or decisions. We repeat that risk always accompanies an inference. Since we intend to make a statement about an entire population from which we have seen only a sample, we cannot expect our inferences always to be error-free. Defining and controlling risk is an integral part of our subsequent study of inferential statistics.

As inferential statistics is more sophisticated than descriptive statistics, we are not able, at this point, to identify specific objectives for an inferential data analysis. However, our overall goal is well-defined, as the following definition indicates.

Definition

The **goal of inferential statistics** is to make a statement about a population based on a sample.

The following are examples of inferential business statistics problems.

EXAMPLE 1.4

We have grown accustomed to watching election returns on television in the evening. Just as commonplace is the commentary that begins: "With 2% of the precincts reporting we project that candidate X will defeat candidate Y . . . ." And, to the dismay of those people who voted for "Y," the prediction is usually correct. The voters' preferences in the sample precincts are used to generalize about the population of voters and thus to make a decision about the election's outcome.

□

EXAMPLE 1.5

The Environmental Protection Agency selects several cars of each type at the beginning of each model year and test drives them to measure fuel economy. The resulting miles-per-gallon (mpg) data are used to forecast the city and highway mpg for each style of automobile. Likely you have noticed these mpg ratings posted in the windows of new cars.

□

EXAMPLE 1.6

In Example 1.3 we claimed that the model relating handwriting score to longevity with the company can be used to forecast the latter based on each applicant's handwriting score. Clearly this application of inferential statistics is fraught with risk. We cannot expect each applicant (who later becomes an employee) with a high handwriting score to stay with the company for a long time. Nor could we conclude that all applicants with low scores (if hired) would remain employed for only a short time. What we hope is that the model aids the human resource manager in his or her hiring decisions.

□

COMMENT Unless there is a math error in the analysis or a human error in transcribing the data, the description that results from a descriptive statistics problem is *not* subject to a risk of being wrong. A description is an objective statement. Even with error-free information, however, the inference resulting from an inferential statistics problem is subject to a risk of being wrong. Therefore, an inference is not a risk-free statement.

Examples 1.4–1.6 illustrate the need to generalize patterns or characteristics from the observed sample data to the unseen population data. This process of inductive reasoning is an attractive and efficient means of solving business statistics problems.

Problem solving is a structured process beginning with a plan, not a formula. Solutions proceed from the planning stage to the execution stage. Most of our work in this book is geared toward executing a data analysis. In the next section, we discuss collecting the data.

SECTION 1.2 EXERCISES

1.11 What are the five components of the problem-solving structure?

1.12 What is the difference between a goal and an objective?

1.13 Name the two branches of business statistics.

1.14 The play of the game in the popular television show "Wheel of Fortune" is analogous to one of the business statistics procedures. For instance, after several turns the following sample of letters is revealed:

TH__ __ __ __ N__ __ __ __ __H__N__ __ __TT__R

The contestant's job is to guess the phrase (the population) from this information. Which procedure does this resemble?

1.15 The process of arriving at the final grade that you earn in a course is analogous to one of the business statistics procedures. Through tests, quizzes, homework assignments, and so on, each student generates a set of data. At the completion of the final exam, the instructor's job is to summarize the data into a letter grade. Which procedure does this resemble?

1.16 Assuming the sample data are error free, identify the following as an inference or a description by deciding whether the statements are risk free.

a. The amount of precipitation recorded in the first 10 days of a month is 1.3 inches.

b. A retired couple enlisted the aid of an accountant to help them fill out the IRS form 1040–ES to estimate their tax liability for the coming year. On the basis of their tax return for last year, he estimates they will owe $5,126 in federal tax in the coming year.

c. A sample of several 30-year fixed rate mortgages was taken to estimate the current average rate of all fixed mortgages. A figure of 10.18 percent was reported.

d. Bond prices are expected to be lower in the next quarter due to a predicted rise in interest rates.

e. After an analysis of similar homes in the area to determine the fair market value of a specific house, a real estate agent believes the house will sell for $70,000.

1.17 Assuming the sample data are error free, identify the following as an inference or a description by deciding whether the statements are risk free.

a. On the basis of the amount of rain recorded in the first ten days of the month, a weatherperson predicts a below normal amount of precipitation for the month.

b. A survey of voters one week before an election shows the Republican candidate favored by 55 percent of those sampled.

c. A sample of 12 network programs reveals an average cost of a 30-second television commercial during prime time to be $140,000.

d. From historical data a manager believes her department will spend $26,000 for supplies in the next year.

1.3 COLLECTING DATA BY SAMPLING

Finite and Infinite Populations

As we have argued previously, the structure of a problem must be known before we can collect any data. Part of this framework is the form of the population: finite or infinite. Recall in an earlier definition that we described a population as a complete set of data or entities. If the population is **finite,** the first, second, third, and so on up to the last entity exist and are identifiable.

Table 1.3 Notation for the Numbers of Entities in a Set of Data

Set of Data	Symbol for Number of Entities
Population (finite)	N
Sample	n

For designating the number of entities in a finite population we will use the letter N. We let the lowercase letter n represent the number of entities in a sample. We sometimes refer to N as the *size* of the population and n as the *size* of the sample. Table 1.3 summarizes this notation.

By definition, an **infinite** population is one in which we cannot identify the "last" entity or one in which the last entity does not exist currently. The concept of infinity is troublesome to comprehend, so it may be helpful to associate an infinite population with the notion of a *process* that continually generates data. For instance, at the close of each business day a value occurs for the variable, closing price of a share of stock. The process of recording the closing share price is related to time, a never-ending process. Thus, the population of closing prices of a share of stock is infinite. The population becomes finite only if we restrict our attention to a historical period of time.

Examples in other areas of business of processes that continue without end are as follows:

Area	Entity	Process
Manufacturing	Shift	Record the number of components produced per shift
Banking	Day	Count the number of loans approved per day
Finance	Quarter	Report a stock's quarterly dividend

The infinite population in each case is a set of entities with a clearly identifiable first, second, and present entity, but without a "last" entity. Shifts continue around the clock; days are perpetual, and so forth. Since all entities (past, present, *and* future) are not currently available for inspection, the population cannot qualify as a finite population. If the process is halted and all entities exist, then we can regard the data as part of a finite population.

Taking a Sample Versus Conducting a Census

Sampling is the process by which we collect the data. If we collect data from some of the entities in the population, then we are "taking a sample." Collecting data from all of the entities in a finite population is called "conducting a census." It is impossible to conduct a census for an infinite population.

Our approach in this text will focus exclusively on taking a sample, or sampling. Although we recognize the advantages of conducting a census in certain situations, the opportunities for doing so are rare. Time and money are the two commonly cited reasons for not conducting a census. When time is short and/or money is tight, taking a sample is a viable alternative. In Table 1.4 we have summarized the benefits of both means of collecting data.

Sampling offers us a glimpse of the population but not an entire view. As a result, what we see in the sample will not reflect perfectly the population. As an analogy, consider a ranger scanning a distant mountainside with a pair of binoculars. He sees a portion of the mountain through the binoculars, but never the entire mountain in one glance.

Table 1.4 Major Benefits of Taking a Sample and Conducting a Census

Benefit	Comment
	Taking a Sample
1. Minimizes cost	Selecting some, but not all, of the entities usually costs less.
2. Yields current information	Sampling ordinarily ensures current data.
3. Decreases time involvement	Collecting data is time-consuming; a sample reduces the time involvement.
4. Reduces destructive testing	Studies to measure a car's ability to protect passengers in a crash at 35 miles per hour, for example, involve *destructive testing*. Taking a sample is clearly preferred in such situations.
	Conducting a Census
1. Eliminates uncertainty	An inferential analysis based on a sample always involves risk. Conducting a census removes this doubt. For example, information from the census conducted every 10 years in the United States is used to apportion federal funds and allocate Congressional seats to states.
2. Promotes quality	Conducting a census, especially on consumer goods that lend themselves to inspection or testing, projects an image of quality. For instance, a slip of paper stamped "Inspected by No. 3" in an article of clothing implies that someone examined the good and declared it suitable. As another example, most manufacturers of floppy disks certify one or both sides before shipping to guarantee all tracks and sectors are not damaged.

To illustrate, look closely at Figure 1.2 and determine the smallest number in the population and in the sample. The sample shows 1 as its minimum value, while 0 is the smallest value in the population. That these two minimum values are not identical is attributable to sampling error. **Sampling error** refers to the fact that no matter how properly we take the sample, it will not be a perfect reproduction, in miniature, of the population. Although "sampling error" is a standard term in business statistics, the word *error* may create an impression that something is wrong with the sampling and must be corrected. Not true. Whenever we sample, we expect slight differences between the sample and the population; that is, we *expect* some sampling error.

On the other hand, **nonsampling error** implies that something *is* wrong, though not as a result of sampling, and must be corrected if possible. Often the error lies in the data. Perhaps the recorded value of the variable is not the true value for the entity. For example, the recorded value of a sales transaction might be $616, when the actual value was $661 (transposition of digits), or an employee might make some erroneous entries on an expense account because of lost or misplaced receipts (spurious data). A nonsampling error also occurs when the value of the variable is nonexistent for an existing entity. This happens, for example, when a cashier overlooks an item and fails to ring it up (omitting data), or when only 10 percent of those included in a survey respond (perhaps the 90 percent not responding differ in a substantial way from the respondents). Nonsampling errors are hard to locate. But they must be identified and resolved, because the subsequent data analysis is only as accurate as the data on which it is based.

COMMENTS

1. Sampling error is also called experimental error.
2. In the publication "What Is a Survey?" by Robert Ferber et al., we find the following in-depth discussion of nonsampling error:

Nonsampling errors can be classified into two groups—random types or errors whose effects approximately cancel out if fairly large samples are used, and biases which tend to create errors in the same direction and thus cumulate over the entire sample. . . .

Biases can arise from any aspect of the survey operation. Some of the main contributing causes of bias are

1. Sampling operations. *There may be errors in sample selection, or part of the population may be omitted from the sampling frame, or weights to compensate for disproportionate sampling rates may be omitted.*

2. Noninterviews. *Information is generally obtained for only part of the sample. Frequently there are differences between the noninterview population and those interviewed.*

3. Adequacy of respondent. *Sometimes respondents cannot be interviewed and information is obtained about them from others, but the "proxy" respondent is not always as knowledgeable about the facts.*

4. Understanding the concepts. *Some respondents may not understand what is wanted.*

5. Lack of knowledge. *Respondents in some cases do not know the information requested or do not try to obtain the correct information.*
6. Concealment of the truth. *Out of fear or suspicion of the survey, respondents may conceal the truth. In some instances, this concealment may reflect a respondent's desire to answer in a way that is socially acceptable, such as indicating that s(he) is carrying out an energy conservation program when this is not actually so.*
7. Loaded questions. *The question may be worded to influence the respondents to answer in a specific (not necessarily correct) way.*
8. Processing errors. *These can include coding errors, data keying, computer programming errors, etc.*
9. Conceptual problems. *There may be differences between what is desired and what the survey actually covers. For example, the population or the time period may not be the one for which information is needed but had to be used to meet a deadline.*
10. Interviewer errors. *Interviewers may misread the question or twist the answers in their own words and thereby introduce bias.**

Although these comments are pertinent to data collected by surveys, many of the biases mentioned also are applicable to data collected by other means. Each sample may not contain all these sources of nonsampling error.

EXAMPLE 1.7

Perhaps one of the classic examples of undetected nonsampling error occurred in the mid-1980s when the Coca-Cola Company withdrew Old Coke from the market and introduced New Coke. Lest you think this was done at the whim of an executive, consider the following. Coca-Cola spent over two years and more than $4 million on marketing research and conducted over 200,000 taste tests involving several New Coke formulas. New Coke was preferred over Old Coke in 60 percent of these taste tests. With so much evidence supporting New Coke, why did Coca-Cola's marketing strategy fail?

There is not one simple answer, but some nonsampling errors definitely can be cited. The people participating in the taste tests may not have understood the purpose of the tests (to determine which New Coke formula would replace, not co-exist with, Old Coke). Coca-Cola shoulders some of the blame as well for not recognizing the conceptual problems between what it sought (a sweeter tasting soft drink to deflect the "Pepsi Challenge") and the actual results of the survey (60 percent preferred New Coke, but 40 percent still wanted Old Coke). Coca-Cola's research focused on one dimension only—taste—and not on other factors, such as (Old) Coke's successful history spanning almost 100 years or the public's emotional attachment to the product.

*From R. Ferber, P. Sheatsley, A. Turner, and J. Waksberg, "What Is a Survey?" (Washington, D.C.: American Statistical Association, 1980).

In retrospect, we clearly see these weaknesses. But, as the Coca-Cola case illustrates, money, time, and the size of the survey cannot eliminate nonsampling errors. A sound statistical analysis must involve considerable thought and investigation into latent sources of bias.*

□

■ Methods of Sampling

Taking a sample can be done in one of two ways: randomly or nonrandomly.

Nonrandom **Nonrandom sampling** is a method of selecting entities by personal choice rather than mathematical chance. Anytime we exercise control over whether to include or exclude an entity in the sample, our subconscious thoughts and feelings affect our decision. As a result, nonrandom sampling involves a criterion for selection that includes personal biases. There is no way to control or account for this partiality.

Nonrandom samples fall into one of three main categories: judgment, convenience, or quota. *Judgment* samples include entities that we think are "representative" of the population. In this selection process, the sample is intended to be a miniature population. Until recently, auditors of trust accounts in banks relied solely on judgment samples. "Trust Banking Circular 16," issued in 1979 by the Office of the Comptroller of the Currency, authorized the use of random sampling of trust assets.

A *convenience* sample contains entities that happened to be in the right place at the right time. Nearby entities are selected; those not in close physical or communication range are not considered. For example, if we wished to take a sample of college students from the population of all students at your college, then a convenience sample would result if only students in this class were chosen.

Quota samples are developed so that entities with certain known characteristics in the population appear in relatively similar proportions in the sample. For instance, suppose the ratio of girls to boys in the population of teenagers is roughly 6:4. If an interviewer, stationed inside the entrance to a shopping mall, is instructed to interview 120 teenage girls and 80 teenage boys, and if the choice of respondents is left to the discretion of the interviewer, then this nonrandom sampling plan qualifies as a quota sample.

Nonrandom samples are selected so as to be "representative" of the population, whenever possible. However, as personal choice may be the criterion for the selection of entities, nonrandom samples actually may be quite different from or unrepresentative of the population. While sampling error accounts for such differences, we cannot measure the *extent* of the sampling error for a nonrandom sample. This drawback of nonrandom sampling does not apply to random sampling.

Random **Random sampling** leaves everything to chance and nothing to preference—that is, if biases are removed, then every entity in the population has an equal chance of selection. This "equal chance" requirement distinguishes random sampling from nonrandom sampling.

*Adapted from Philip Kotler, *Marketing Management: Analysis, Planning, Implementation, and Control*, 6th Ed. (Englewood Cliffs, N.J.: Prentice-Hall, 1988).

There are four major types of samples associated with random sampling. The first and most common is called a *simple random sample* of size n. Not only must each entity have the same chance of selection, but every conceivable group of n entities also must have an equal chance of occurring in order to qualify as a simple random sample. The following definition summarizes this concept.

Definition

A **simple random sample of size** n from a finite population is a set of n entities chosen such that every collection of n entities has an equal chance of selection.

A second popular type of random sampling is a *stratified sample*. The entities in the population are segregated either physically or naturally into several homogeneous groups or strata. Treating the strata like separate populations, we select a simple random sample from each. To illustrate, from the population of students in this class, we could group each student according to his/her classification (freshman, sophomore, junior, senior, other), and then select a simple random sample within each classification. The resulting collection of five simple random samples is a stratified sample. A stratified sample ensures all strata are represented, often in proportion to the size of the stratum. A simple random sample cannot guarantee this property.

A third type of random sampling is a *cluster sample*. Again the entities in the population are classified into groups or clusters. Next a random sampling of clusters is taken. Then we conduct a census within the selected clusters. (Depending on the size of the cluster, a simple random sample is sometimes taken within the selected clusters instead.) For example, a computer store may receive a shipment of 100 boxes of 3½-inch disks with 10 disks per box. Suppose the store manager wished to spot check the quality of some of the disks. Neither a simple random sample nor a stratified sample (with each box representing a stratum) would be very efficient because the manager would have to open so many boxes. Rather, this situation is naturally conducive to a cluster sample. By selecting a couple of the boxes (now each box represents a cluster) and then examining all of the disks in each selected box, the manager quickly can verify a portion of the shipment. (In Chapter 6, we will learn that decisions to accept or refuse the shipment often can be made on the basis of such a sampling.)

A *systematic sample*, the fourth main type of random sampling, selects every kth entity in a finite population of size N. Usually k is the rounded-off value of N/n. To begin the sampling process, a randomly selected starting point between 1 and N is needed. As an example, suppose each student's name in this class is alphabetically listed (and numbered) on a roster from the registrar's office. Suppose there are 50 names (N). To generate a systematic sample of size $n = 8$, we would randomly pick a number between 1 and 50 and then consecutively add and/or subtract 6 ($50/8 \simeq 6$) to identify the sample entities. If our random starting point were 11, then these student numbers would constitute the sample entities: 5, 11, 17, 23, 29, 35, 41, and 47. In the next section, we will explain how to pick a random starting point.

The following box summarizes the types of samples produced from random and nonrandom sampling plans.

Types of Samples

Random	Nonrandom
Simple	Judgment
Stratified	Convenience
Cluster	Quota
Systematic	

COMMENTS

1. Random and nonrandom sampling also are referred to as probability and nonprobability sampling, respectively.
2. We will use the term *random sample* interchangeably with the term *simple random sample*.

While descriptive statistics problems can be solved with either nonrandomly selected or randomly selected data, only randomly selected data are appropriate for inferential statistics problems. As we shall see, a random sample permits description *and* allows us to identify and quantify the risk in an inferential analysis. We cannot assess the risk from a nonrandomly selected sample. Thus, we recommend random sampling wherever possible. All of our work in this book assumes that the data resulted from a simple random sample. Let us discuss how to obtain one.

Simple Random Sampling from a Finite Population

To ensure that each entity has the same chance of selection, we sometimes appeal to electronic aids such as a computer for help. The most popular device is called a *random number generator,* a software routine designed to randomly produce a number between 1 and N.

Random, loosely defined, means unpredictable—that is, we cannot predict, with any degree of accuracy other than chance, what specific number between 1 and N will occur. The attractiveness of a random number generator is that each time it is invoked, we are assured of obtaining a random or unpredictable number, regardless of whatever numbers already have been generated. People are not good random number generators. No matter how hard we try, we are not capable of reproducing the property of unpredictability over and over again. And, practically speaking, it is foolish to even try. Computers are more efficient at performing this chore than we are.

A string of numbers produced by a random number generator may be compiled into a table of random numbers. There are entire books devoted to such tables (see, for example, the *Table of 105,000 Random Decimal Digits,* cited in the References at the end of this chapter). The style of presenting the random numbers may vary, but all random number tables are developed from the proposition that each digit from 0 to 9 has an equal chance of occurring.

If so, then in a very long string of printed random numbers, each digit occurs with about the same frequency. However, when a portion of this string is abridged into a page or two and called a "random number table," the equal frequency requirement of each digit is not necessarily valid. Thus, if we use a page of excerpted random numbers, we cannot be certain of achieving a simple random sample.

The randomly arranged numbers in Tables 1.5 and 1.6 are variations of a random number table. In Table 1.5, for example, each two-digit number from 1 (listed as 01) to 100 (listed as 00) is randomly arranged within 10 rows and 10 columns, numbered consecutively from 1 to 10 (recorded as 0) in the margins of the table.

By randomly arranging each two-digit (and three-digit) number in a table, we guarantee each two-digit (and three-digit) number occurs once and only once. Hence, our tables ensure that a simple random sample from a finite population of size at most 1000 will result, provided we arbitrarily initialize the sampling process. (We will demonstrate this concept shortly.)

A finite population has exactly N existent and identifiable entities. If we were to number and list these entities on a sheet of paper or in a data file, we would call the list of entities a *frame*. It is easy to conceptualize the idea of a framed population, but quite another matter to actually produce one. For example, a telephone directory seems like a frame of people with telephones. Unfortunately, it excludes those with unlisted numbers and those who added telephone service after the directory was printed.

This example points out the difficulty of producing a current frame and the need for distinguishing the target population from the framed population. A **target population** is the set of entities for which the description or inference is intended. If we

Table 1.5 Randomly Arranged Two-Digit Numbers

Row	Column 1	2	3	4	5	6	7	8	9	0
1	10	54	93	57	92	65	53	40	83	25
2	22	48	26	16	56	19	18	61	50	81
3	24	52	41	01	30	68	67	97	90	87
4	42	32	86	13	47	62	80	12	38	59
5	37	29	70	36	91	64	15	58	49	31
6	77	51	21	04	55	39	46	27	00	20
7	99	07	23	75	11	34	06	84	45	74
8	96	09	71	79	08	88	72	76	82	98
9	89	63	03	95	02	33	14	73	78	17
0	85	28	94	35	66	05	69	43	60	44

Note: This table assumes sampling without replacement from a finite population with at most 100 entities. Each two-digit number from 00 to 99 appears exactly once.

Table 1.6 Randomly Arranged Three-Digit Numbers

Row	1	2	3	4	5	6	7	8	9	10	11	12	13	14	15	16	17	18	19	20	21	22	23	24	25
	Column																								
1	815	258	826	409	935	707	253	276	128	395	901	224	704	566	038	809	788	405	525	969	575	242	049	770	435
2	296	963	290	069	204	876	299	027	466	895	430	078	784	728	527	338	277	081	970	589	972	497	741	292	540
3	007	333	143	394	041	493	073	141	119	882	438	994	709	151	588	050	289	651	614	249	937	853	752	138	342
4	053	780	683	887	055	144	672	761	922	402	807	833	756	584	810	547	499	377	318	544	768	260	858	838	029
5	919	862	064	356	474	982	464	791	920	975	638	212	346	821	502	618	746	535	668	369	940	750	347	506	868
6	005	454	012	188	231	533	827	366	339	508	261	718	733	094	252	449	820	434	751	758	354	158	421	933	372
7	690	814	631	909	237	278	781	173	516	008	186	852	686	116	946	280	004	076	373	902	477	677	932	251	691
8	259	017	468	279	859	742	496	900	305	662	297	925	884	938	560	725	894	327	720	945	999	311	193	847	112
9	097	385	987	735	174	101	491	437	157	799	625	051	154	957	717	904	942	398	363	757	621	640	241	285	198
10	915	865	329	298	236	954	301	149	531	213	648	453	881	043	427	696	238	022	565	991	813	436	414	956	234
11	179	467	848	958	570	142	129	908	086	971	428	786	835	062	159	732	762	310	708	445	462	011	898	314	514
12	465	424	649	879	207	417	886	312	095	719	864	134	079	532	330	873	541	203	611	412	538	576	678	981	365
13	921	590	812	641	870	897	552	341	610	655	178	397	617	797	917	667	585	216	841	842	619	955	545	910	181
14	145	099	703	256	265	337	561	699	605	352	953	361	111	375	201	978	136	548	808	918	063	629	724	871	118
15	984	239	303	390	020	512	773	283	376	793	490	426	452	353	092	519	928	636	300	036	977	624	734	404	507
16	349	926	702	680	831	199	943	122	045	869	739	806	486	343	378	817	567	787	659	358	609	581	220	082	018
17	700	795	948	048	993	286	664	170	025	470	485	688	671	850	801	878	250	098	067	014	927	612	088	494	460
18	539	083	961	722	469	415	964	153	644	645	068	000	393	591	766	660	171	536	607	367	293	169	461	646	551
19	760	180	052	205	319	546	248	121	221	528	152	860	191	803	571	060	233	601	413	263	100	986	966	382	084
20	907	856	381	573	663	383	846	656	523	628	934	165	196	941	804	598	684	443	247	302	564	037	874	070	459
21	643	796	211	042	729	583	444	077	457	647	420	056	384	572	620	529	730	163	223	823	160	488	877	380	114
22	089	032	849	264	391	676	267	557	331	006	345	765	743	418	779	798	370	811	480	792	015	335	723	776	705
23	950	976	669	047	511	534	863	071	710	243	313	359	482	939	127	857	117	861	687	013	021	175	713	996	120
24	156	439	578	698	066	255	189	135	130	168	738	187	172	885	774	235	890	245	284	652	592	906	521	854	093
25	164	603	407	657	992	176	679	432	219	227	753	947	693	689	777	195	503	087	472	836	883	131	580	586	368
26	182	218	080	579	924	031	308	721	340	125	355	271	892	952	001	501	558	951	107	307	745	140	190	965	635
27	731	824	124	834	074	061	040	840	244	888	542	974	455	775	065	759	230	650	626	362	044	210	019	217	747
28	574	637	009	425	240	364	200	379	388	830	692	802	495	476	562	505	225	492	471	778	024	711	030	139	226
29	304	309	332	563	893	162	023	844	257	595	197	324	484	889	785	755	106	072	749	998	988	392	085	034	027
30	166	374	194	185	054	325	396	058	829	764	410	281	416	348	322	851	387	429	184	137	673	577	109	593	771
31	967	115	763	896	035	670	016	294	473	597	923	555	697	633	727	401	772	515	675	983	423	845	685	046	513
32	389	371	642	627	475	979	344	209	828	504	816	654	931	110	867	192	790	246	155	553	604	980	326	968	602
33	316	275	202	075	613	622	232	899	510	229	543	715	661	448	419	463	682	517	442	403	962	498	905	272	091
34	789	653	350	639	606	616	587	634	599	113	837	989	960	596	033	712	269	262	959	489	985	559	929	714	594
35	039	500	839	282	270	930	658	479	183	875	800	866	126	288	518	825	386	855	406	891	783	321	456	736	550
36	744	912	167	549	400	632	148	441	487	568	408	767	133	483	524	608	805	520	716	903	973	944	681	274	320
37	090	913	351	295	059	615	105	665	273	206	254	147	754	317	287	132	726	666	537	782	458	694	623	108	334
38	422	446	819	522	266	872	911	323	880	740	997	222	914	411	916	146	228	096	600	123	177	431	451	530	291
39	161	556	818	674	630	949	794	150	440	399	357	995	336	706	843	748	102	028	306	447	268	010	481	701	769
40	214	215	104	003	554	526	990	450	208	328	822	360	737	569	002	936	509	103	478	832	582	695	433	057	315

Note: This table assumes sampling without replacement from a finite population with at most 1000 entities. Each three-digit number from 000 to 999 appears exactly once.

were to sample from the telephone directory, then the framed population would be the collection of people with listed telephone numbers as of the day the directory went to press. The target population might be the set of people with current telephone service. Although most of those listed in the framed population are members of the target population, some people will be omitted. Discrepancies between the two populations are a form of nonsampling error. We highly recommend that any difference between the framed population and the target population be minimized *before* sampling.

A target population must be defined before it can be "hit" or sampled. Once we have established the frame, noted the value of N, and decided how many entities to sample, we are ready to collect the data. Taking a sample begins with identifying the entities to be selected from the population. Sampling is usually a sequential activity that is done by permanently removing each entity from the population as it is selected, and the process is called **sampling without replacement.** If each entity is removed temporarily—long enough to record the value of the variable and then returned to the population before the next entity is selected—then the process is called **sampling with replacement.** Our study will proceed on the basis of sampling *without* replacement unless otherwise noted.

Table 1.5 can be used to identify the entities to be sampled if $N \leq 100$. To demonstrate the use of this table, consider a finite population with 76 entities from which a simple random sample of size $n = 6$ is to be drawn.

The sampling process is initialized by making two declarations: One denotes the starting point; the other specifies the direction in which we plan to proceed within the table. To locate a starting point, we need to reference a row–column intersection. One way to do this is by arbitrarily opening the textbook to any page and noting the page number. Suppose it is page 158. We need *any two* of the digits 1, 5, and 8. Let us assume we predetermined to use the first and last digits, yielding 1 and 8, and that the first digit would correspond to a row and the last to a column. Locate the intersection of row 1 and column 8 in Table 1.5. There you should see the number 40. Entity 40 in the finite population is the first of the entities to be included in the sample. We can then identify the next five numbers in the eighth column (or in the first row) of the table, proceeding in some predescribed direction. For instance, if we had decided to work down the eighth column, we would read 40, 61, 97, 12, 58, and 27. These are the sample entities, except that we have no entity numbered 97 in the population. Therefore, skip 97 and obtain the next viable number in the column—76—to replace the 97. The simple random sample of size $n = 6$ includes the entities numbered 12, 27, 40, 58, 61, and 76. (A more involved way to obtain the six entities would be to start the process over once the first of the six entities is located, without ever employing a direction rule. In effect, six different starting points would then constitute the sample entities.)

A similar procedure is used to obtain three-digit entities from Table 1.6 whenever $N \leq 1000$. Now there are 40 rows and 25 columns to house 1000 different numbers. The initialization process involving the digits of an arbitrary page can be used with minor variations. For example, page 124 could be interpreted as row 12 and column 24; page 418 might be row 4 and column 18.

If you would prefer a "live" random number generator, we suggest that you use the computer processor, Simple Random Sample, accompanying the text. A brief description of how to use this processor is found in the next section.

COMMENTS

1. A framed population also is known as a sampled population.
2. Other ways of generating a random sample include *randomized response* and *random digit dialing*. For information on the randomized response technique, refer to Warner (1965) in the References at the end of this chapter. For information on random digit dialing, refer to Landon and Banks (1978) or Waksberg (1978).
3. Guidelines for determining the size of the sample depend on the specific problem and will be presented as needed.
4. To select a random starting point for a systematic sample use either Table 1.5 or Table 1.6, provided $N \leq 1000$.
5. If we run out of viable numbers in a row or column of the tables, we move down or over to an adjacent row or column, respectively, and continue reading. For example, if we reach the end of row 3 (at column 0) in Table 1.5, then we would move down to row 4 (still at column 0) and read across row 4 from right to left. When we arrive at a corner of the table and still need additional numbers, we recommend moving to the corner diagonally across from our present position. Then, continue to read in the same direction down or across the table. For instance, if $N = 30$, $n = 6$, starting position = row 9, column 9, and direction = across row 9, then the six entity numbers in the sample would be 17, 05, 28, 25, 10, and 22.

Please note that both Tables 1.5 and 1.6 and the computer processor merely provide the *entity numbers* to be included in the sample. Sampling is not complete until we obtain the value of the variable under study from each of the chosen entities. The values form the sample data.

CLASSROOM EXAMPLE 1.1

Using Random Numbers to Obtain a Sample

No-Load Fund Investor, a mutual fund advisory newsletter in Hastings-on-Hudson, N.Y., called the customer service departments at 44 major mutual fund companies via each fund's 800 number. Five calls were made to each fund during the week of June 9–13, 1987, to obtain data on the variable *number of seconds to reach a live voice*. Assume the list of average times in Table 1.7 represents the framed population.

a. Identify the entity in this population.

b. Suppose we wish to draw a simple random sample of size $n = 5$ without replacement. Table 1.5 is appropriate since the population contains fewer than one hundred entities. Let

   Starting point = Row 0, column 2
   Direction = Up the second column

   Circle the entity numbers that would be included in our sample.

c. Write down the set of sample data.

Table 1.7 Data for Classroom Exercise 1.1

	Fund Group	Seconds		Fund Group	Seconds		Fund Group	Seconds
1.	AARP	3	16.	Founders	11	30.	Quest	17
2.	AMA	30	17.	Gintel	9	31.	Reserve	6
3.	American Inv.	24	18.	GIT	65	32.	SAFECO	25
4.	Axe Houghton	18	19.	Gradison	39	33.	Scudder	327
5.	Babson	31	20.	Heine	63	34.	Selected	67
6.	Benham	34	21.	Ivy	83	35.	Stein Roe	9
7.	Boston Co.	3	22.	Janus	19	36.	Strong	16
8.	Bull & Bear	26	23.	Legg Mason	10	37.	T. Rowe Price	27
9.	Calvert	98	24.	Lexington	107	38.	Twentieth Century	33
10.	Columbia	13	25.	Loomis-Sayles	52	39.	United	62
11.	Dreyfus	40	26.	Neuberger &		40.	United Services	17
12.	Evergreen	44		Berman	46	41.	USAA	51
13.	Fidelity	14	27.	Newton	7	42.	Value Line	29
14.	Financial	75	28.	Nicholas	10	43.	Vanguard	35
15.	Flex	29	29.	North Star	7	44.	WPG	60

Source: "Newsletter adds wait to mutual fund phone survey," Copyright 1987, *USA Today*. Adapted with permission.

SECTION 1.3 EXERCISES

1.18 Each model year, automakers plan to build a certain number of vehicles of each type. The process of making cars proceeds throughout the model year. At any point *during* the model year, should we regard the vehicles as entities from a finite or infinite population? Why?

1.19 There is a popular political adage for presidential elections: As Maine goes, so goes the nation. For each presidential election, if we view the election results from Maine as a sample, what type of random or nonrandom sample is suggested by this saying? Why?

1.20 In shopping malls, marketing research companies often station interviewers with clipboards to intercept shoppers who happen along. The interviewee is asked to respond to a few questions or is perhaps shown a television commercial in a nearby kiosk and asked for feedback. The responses of those shoppers who participate in this interviewing process form what type of random or nonrandom sample?

1.21 For each type of sample listed, think of a possible benefit and a drawback:
a. stratified **b.** cluster **c.** judgment **d.** systematic

1.22 Make up an example of a finite population and an infinite population.

1.23 In selecting a simple random sample from Tables 1.5 or 1.6, identify the entity numbers to be included in the sample if the following information is given:
a. $N = 100$, $n = 12$, starting point = row 7, column 5, and direction = down column 5.
b. $N = 69$, $n = 5$, starting point = row 4, column 5, and direction = back across row 4 from right to left.
c. $N = 811$, $n = 10$, starting point = row 40, column 9, and direction = up column 9,

1.24 Refer to Classroom Example 1.1. Suppose the mutual funds are organized alphabetically into four groups as follows:

Group	Alphabet Groupings for Fund Names	Number of Funds in Group
1	A–D	11
2	E–G	8
3	H–P	10
4	Q–Z	15

a. By identifying the entity numbers select a stratified sample containing two randomly selected funds from each group. Use the following starting points and directions:

Group	Starting Point	Direction
1	Row 9, column 2	Across row 9
2	Row 7, column 6	Down column 6
3	Row 0, column 7	Across row 0 from right to left
4	Row 5, column 1	Down column 1

b. By identifying the group numbers select a cluster sample containing two randomly selected groups. Use a random starting point = row 1, column 3 and direction = down column 3.

c. Ignoring the grouping, select a systematic sample of size $n = 4$. Use a random starting point of row 2 and column 1 and identify the entity numbers for the sample.

1.25 Many foodstuffs are processed and packaged on a production line. At randomly selected times during a shift, a clerk will remove some of the packages from the line and test the contents to ensure that quality standards are being met. What type of random or nonrandom sample results from this activity?

1.26 In David Rosenbaum's article titled "Misplaced Decimals and Other Errors Raise Hospital Bills by Millions" (*New York Times* News Service, Louisville *Courier-Journal*, July 8, 1984), he reports on the concern of insurance company executives over incorrect hospital bills. Most insurance companies hire auditors to sample hospital bills and verify their accuracy. According to the auditors, over 90 percent of the large hospital bills contain overcharges. Rosenbaum writes that ". . . most errors involved computer keypunch mistakes, charges for services that were ordered but not performed, and items placed on the bill of the wrong patient." Classify each type of error as a sampling error or a nonsampling error.

1.27 A label found in the pocket of a new shirt reads: "I have personally examined every detail of this garment to make sure it meets our high quality standards. Inspector: 16." Does this label imply that this shirt's construction has been sampled, or has a census been conducted? What benefit is being highlighted?

***1.28** In marketing research, data sometimes are categorized as primary or secondary. *Primary data* are those numbers or words collected specifically for the current problem. Firms rarely

*Optional

destroy data so that primary data may be archived, included in company reports, or stored in a computer data base. *Secondary data* are data that already exist from a sampling for an unrelated problem or from a previous attack on the current problem. For example, a company offers a mail-in rebate on one of its products in exchange for the cash register receipt and some information about the purchase (for example, type of store, reason for buying, demographics on the household, and so on). This primary data is useful in identifying characteristics of consumers who take advantage of the rebate. The same demographic data also may be used as secondary data in a later study of advertising.

In the following situations, identify the sets of data as primary or secondary.

a. A marketer who uses census data to characterize the residents of a certain county.

b. A pollster who uses the data from her telephone poll to report on a candidate's popularity with voters.

c. A manager who incorporates earnings-per-share (EPS) figures from the company's annual reports in a study on the growth of the company's EPS.

***1.29** Refer to Exercise 1.28. Name a possible advantage and a disadvantage for primary data and for secondary data.

*1.4 PROCESSOR: SIMPLE RANDOM SAMPLE

Accompanying this text is a software package containing several microcomputer programs that we call processors. Each processor is explained in a starred section of the book as we go along. The processor for Chapter 1 is titled Simple Random Sample. It illustrates random sampling as discussed in Section 1.3.

To get started with this or any other processor, you need to have turned on your microcomputer and have read the disks into memory and/or placed them into the proper disk drives. Specific details on setting up the disks in order to arrive at the first menu of choices are found in Appendix N at the back of the book.

Calling Up the Simple Random Sample Processor

By following the startup instructions from Appendix N, you should obtain the first Overview screen (see Figure 1.3). At this point, you may obtain further overview information, or you may touch the F8 key to clear the screen. Then use the right arrow (→) key to move the highlight box to Processors on the row of menu selections. If you touch the Enter (or Return) key when Processors is highlighted, you will be provided a list of processor choices, as shown in Figure 1.4.

To obtain the desired processor, use the up and down arrow keys to locate the choice. In this case, the highlighted starting point is our choice: Simple Random Sample. By touching the Enter key again, you will get a listing of the two selections or subprograms within Simple Random Sample (see Figure 1.5).

Our purpose here is to illustrate the first subprogram, Sample Entity Numbers. (To run Sample Data Points, we would need to have created a data file previously from which to sample.) By touching the Enter key again, you will view the screen shown

*Optional

```
≡        Microcomputer Applications for INTRODUCTORY BUSINESS STATISTICS  READY
 Overview  Data  Files  Processors  Set Up  Exit                   | F1=Help

   This program accompanies the
  INTRODUCTORY BUSINESS STATISTICS
  text by Shiffler and Adams.  It is
  to aid in learning about the appli-
  cation of statistics to business
  problems.

  Press the Enter key to continue or
 F8 to proceed directly to the Menu.
```

Figure 1.3 First Overview Screen

```
≡        Microcomputer Applications for INTRODUCTORY BUSINESS STATISTICS  MENU
 Overview  Data  Files  Processors  Set Up  Exit                   | F1=Help
                         Simple random sample
                         Frequency distributions
                         Descriptive statistics
                         Least squares line
                         Discrete probability distributions
                         Normal probability distribution
                         One sample inference
                         Two sample inference
                         Analysis of variance (ANOVA)
                         Regression
                         Time series
                         X² (chi-square)
```

Figure 1.4 List of Processor Choices

in Figure 1.6. If an error message is received instead, disk 2 may not be in its proper location.

Running the Sample Entity Numbers Program

As the screen in Figure 1.6 suggests, the processor will automatically select a random group of n numbers and display them on the screen. If you input $n = 1$, the processor will select a single number. *(Note:* you could use this number as a starting point in a systematic sample.) This routine does *not* select the sample data; it simply identifies the entity numbers for the sample. Once the entity numbers are displayed, the processor is finished.

```
≡        Microcomputer Applications for INTRODUCTORY BUSINESS STATISTICS   MENU
   Overview  Data  Files  Processors  Set Up  Exit                    | F1=Help
                         Simple random sample
                          Sample entity numbers
                          Sample data points
                                                ibutions
                         Normal probability distribution
                         One sample inference
                         Two sample inference
                         Analysis of variance (ANOVA)
                         Regression
                         Time series
                         X² (chi-square)
```

Figure 1.5 Subprograms Within Simple Random Sample Processor

```
≡                          Sample Entity Numbers                           READY
   Introduction  Data  *eneration  *esults  Exit                      | F1=Help

     This processor will generate a
    simple random sample of size n from
    a finite population of size N.

     Two inputs are required: n and N.
    The output consists of a set of n
    entity numbers, each between 1
    and N, inclusive.
     See Chapter 1 for information.
```

Figure 1.6 Overview Screen for Sample Entity Numbers

To illustrate the processor, we will select a simple random sample of $n = 5$ entities from a population of $N = 274$ entities. (The maximum possible value of N will depend on your machine's unused memory space.) To proceed, move the highlighter to the right, over to Data. By touching Enter, you will get a screen asking for the population size to be specified. Figure 1.7 shows this screen *after* we have entered 274. If you now touch Enter, you will be asked to state the desired sample size. Figure 1.8 shows the screen *after* we have entered 5.

Now touch Enter twice, move the highlighter to Generation, and touch Enter twice more. The routine now chooses your sample entity numbers at random. When the selection is complete, usually within seconds, the screen shown in Figure 1.9 will appear. You are now ready to see your results.

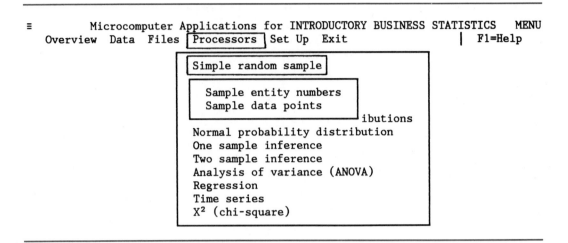

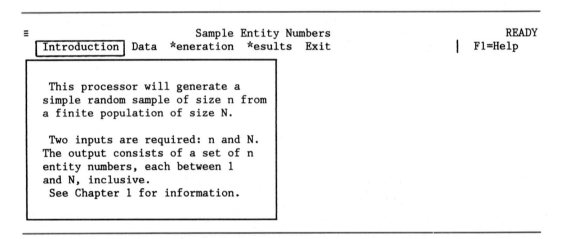

```
≡                       Sample Entity Numbers                          EDIT
   Introduction [Data] *eneration *esults Exit                      | F1=Help
              Population and sample size

               ══════════ Population ══════════
                 Enter population size, N =  274
```

Figure 1.7 Entering *N*, the Population Size

```
≡                       Sample Entity Numbers                          EDIT
   Introduction [Data] *eneration *esults Exit                      | F1=Help
              Population and sample size

               ══════════ Sample ══════════
                  Enter sample size, n =  5
```

Figure 1.8 Entering *n*, the Sample Size

```
≡                       Sample Entity Numbers                          READY
   Introduction *ata [Generation] *esults Exit                      | F1=Help
                   Start process
                      Complete
```

Figure 1.9 End of the Sampling Process

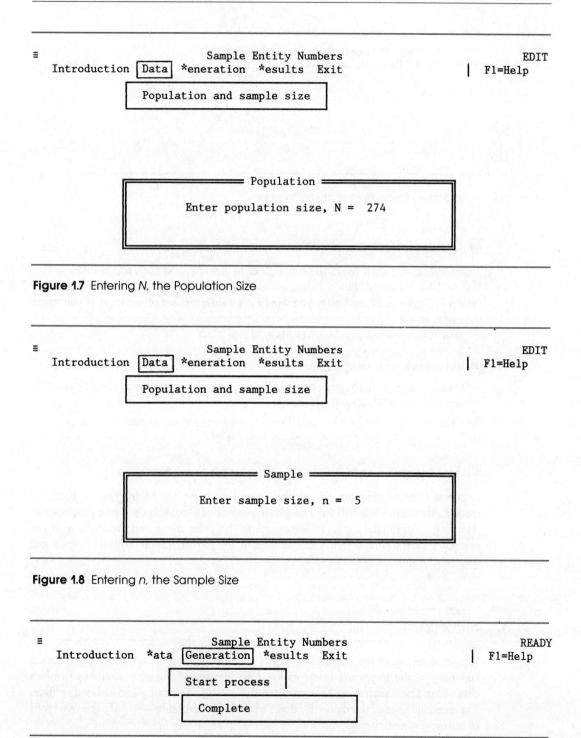

```
≡                            Sample Entity Numbers                              LIST
   Introduction  *ata  Generation [Results] Exit                          | F1=Help

                Entities
    1 74
    2 71
    3 209
    4 219
    5 170
```

Figure 1.10 Results Screen for Sample Entity Numbers

Outputting the Results

Touch Enter, and then move the highlighter to Results, and then touch Enter twice. You will be presented with a listing of your sample entity numbers. Our results are shown in Figure 1.10; naturally you would get a different set of numbers if you made the same inputs.

For future reference to your results, you can

1. Write down your entity numbers.
2. Print the screen, if you are connected to a printer (F6 key or touch the Print Screen key with the Shift Key depressed).
3. Save the results into a file (see Chapter 2 for instructions on saving results).

Exiting the Simple Random Sample Processor

It is now time to move on. Touch the Enter key; move the highlighter to Exit and touch Enter again. You will then be given a choice of repeating the same processor or returning to the main menu. If you return to the main menu and are done with the computer for the time being, you may sign off by moving the highlighter to Exit and then touching the Enter key twice.

1.5 SUMMARY

A careful review of the chapter objectives at the beginning of this chapter provides a summary of the important concepts we have discussed. Business statistics involves data. This first chapter has been devoted to ways of classifying and collecting data. The intended uses of the data—for description or for inference—specify the two types of business statistics procedures.

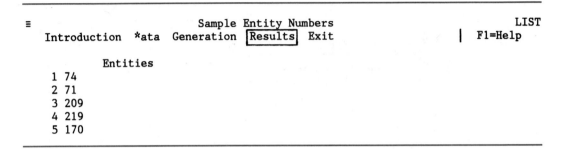

Data, in their original raw form, are a necessary part of the problem-solving process. But to be useful, data must be transformed into information. The function of business statistics is to provide meaningful information from a set of numbers or words.

In the classical theory of information processing, it is argued that the process evolves from data to information to knowledge to wisdom. Too often in business, information remains inert. It is buried in a report or memo, or it remains in the possession of one individual. G. Ray Funkhouser, author of the *Power of Persuasion: A Guide to Moving Ahead in Business and Life*, commented in an interview published in the *Bottom Line /Personal* (May 30, 1987): "In today's world of mass communication we are smothered with useless information. Information you can use becomes precious knowledge—and knowledge is power."

For us to gain knowledge we must *use* the information in a statistical analysis. Otherwise, business statistics will be perceived merely as a collection of data processing techniques. Throughout the book we will point out applications. Some of these will be in the text; some will be in the examples and exercises; and still others will show up in the "To Be Continued" sections. For those who doubt the pervasiveness of statistics in business, read on!

1.6 TO BE CONTINUED . . .

In most chapters there will be a section titled "To Be Continued . . . ." The purpose of these sections is twofold:

1. To show you where a specific business statistics concept from the material in that chapter likely will reappear in a future business course
2. To indicate applications of the chapter material in the "real world" by spotlighting uses in business and/or in the media

We will reference paragraphs from standard textbooks used in other courses and/or mention areas within certain courses where the business statistics material is applicable. Graphs, tables, and passages from the media, as well as vignettes illustrating potential uses of statistics in a job setting, will be featured. We hope you benefit by seeing the interrelationships among business courses, business statistics, and the business world.

. . . In Your College Courses

Although collecting data through sampling is an integral part of every business statistics problem, auditing is one area in business where sampling methods are discussed in detail. Thus, much of what has been presented in our first chapter will be seen again in an auditing course, typically offered through the accounting department.

A popular textbook for such a course is *Auditing: An Integrated Approach* by Arens and Loebbecke. In Chapter 10 (titled "Determining Sample Size Using Attributes Sampling, and Selecting the Items for Testing") of their book, we find these comments: "In the early part of the chapter, the selection of items from the population by the use of *judgmental* and *random sampling* is examined."

Arens and Loebbecke refer to all nonrandom sampling as judgmental sampling and use the phrase "random sampling" interchangeably with the phrase "statistical sampling." Continuing from their Chapter 10,

> *A judgmental sample is the determination of the sample size and the selection of the individual items in the sample on the basis of sound reasoning by the auditor. It differs from statistical sampling primarily in the lack of objectivity in selecting the sample items and the inability to measure sampling risk. The use of judgmental sampling is widespread . . . among . . . auditors. . . . .**

As we mentioned, judgmental sampling is a viable sampling method, though we cannot assess the sampling error. However, our earlier point that random sampling is preferred for inferential statistics problems is supported by the following passage:

> *It is preferable to use random selection methods for selecting samples whenever it is practical. . . . It is improper and a serious breach of due care to use* statistical measurement techniques *if the sample is selected by the haphazard, block or any other judgmental approach. Only* random selection *is acceptable when the auditor intends to evaluate a population statistically.**

Arens and Loebbecke go on to describe methods of randomly selecting a sample using random number tables and computer generators. The remainder of their tenth chapter is devoted to sampling as it relates to the auditing function.

Clearly your encounter with business statistics in general and sampling in particular is not an isolated part of your undergraduate business education. As the material from the auditing text indicates, those planning to major in accounting will need a good foundation in business statistics. We will try to highlight applications from other business areas in later chapters.

. . . In Business/The Media

Perhaps the most familiar application of statistics is the opinion poll. We see it in newspapers and in magazines, and hear about polls on television. They have impressive names: The CBS–*New York Times* Poll, The Gallup Poll, The Roper Poll, The NBC News Poll, and so forth. And they give us a wealth of facts, figures, and (yes) misinformation.

In a story titled "The Numbers Racket: How Polls and Statistics Lie," we find these opening comments:

> *Statistics are an American obsession. In an election year, they become a positive mania. The poll data pour in daily . . . . It's not just polls. Numbers of every description have become the currency of American life.*†

*Arens/Loebbecke, *Auditing: An Integrated Approach,* 2/E, © 1980, pp. 324, 327. Reprinted by permission of Prentice-Hall, Inc., Englewood Cliffs, NJ.

†Adapted from *U.S. News & World Report,* July 11, 1988. With permission.

The article exposes common abuses with poll data and weaknesses of surveys. In addressing the ". . . standard 3-percent-plus-or-minus-warning," the article states,

> *What does it mean when a poll says "margin of error of plus or minus 3 percent"? . . . That standard warning label refers only to the most obvious possible source of error—the statistical chance that a perfect, randomly selected sample doesn't reflect the country as a whole.**

Do you recognize what the article is referring to? Yes—sampling error! Indeed, we might be wise to assign this article as required reading for a review of sampling and nonsampling errors.

Other sources of nonsampling errors are named and illustrated, including "the skewed sample," "the ignorance factor," and "the 'pseudo' poll." The authors also mention that sampling error ". . . doesn't account for the fact people may refuse to answer, may lie or may be influenced by leading questions."

Although the terms *sampling error* and *nonsampling error* were not mentioned specifically, they do apply to the information presented in the article. The lesson here is clear: Our textbook terms may not appear verbatim in the media, but a firm understanding of the textbook concepts is necessary to see the relationship between what we read in this book and what we read and hear in the media.

CHAPTER 1 EXERCISES

1.30 Identify the entities and some possible values of the following variables:

a. number of sales per hour.
b. number of sales per person.
c. daily high temperature in Phoenix.
d. annual inflation rate.
e. daily number of outpatient surgeries per hospital.
f. baseball player's uniform number.

1.31 Refer to Exercise 1.30. For each variable classify the data as qualitative or quantitative.

1.32 Refer to Exercise 1.30. Describe the entities in a possible sample if the entities in the population are as follows:

a. the first ten hours of business on Monday.
b. all sales personnel in District 25.
c. the 31 days in the month of May.
d. the years 1982 through 1991.
e. all the hospitals located in Washington County and the 92 days in the last three months of the year.
f. all baseball players on the Baltimore Orioles' 40-man roster.

1.33 Assuming the sample data are error free, identify the following as an inference or a description by deciding whether the statements are risk free.

a. A business plan for a start-up company projects sales in the first year of $150,000.

*Adapted from *U.S. News & World Report*, July 11, 1988. With permission.

b. From a set of sample data, the Environmental Protection Agency issues a highway rating of 27 miles per gallon to accompany a certain type of new car.
c. A sample of people meters leads the A. C. Nielsen Company to estimate a total viewing audience of 17.4 million people for a popular television show.
d. Based on an analysis of a company's finances relative to the health of the economy, Value Line predicts the company's earnings-per-share to be $1.18 for the year.
e. The market share for a company's soft drink was 13.3 percent for the most recent calendar year.

1.34 What term describes the likelihood that an inference is wrong?

1.35 Describe the difference between an entity and a variable.

1.36 Define a population and a sample.

*1.37 In an article on the randomness of "streaks" ("The Orderly Pursuit of Pure Disorder," *Discover,* January 1987) Kevin McKean writes: "Athletes and fans alike have long supposed that basketball players shoot in streaks: a player who sinks three or four shots in a row has the hot hand and is more likely to make his next shot. But when (Amos) Tversky, Robert Vallone of Stanford, and Thomas Gilovich of Cornell analyzed shooting statistics for four National Basketball Association teams, including the world-champion Boston Celtics, the number of streaks was about what would be expected from chance. . . . The most troubling finding of the basketball study, says Tversky, is that even repeated exposure to the random process doesn't diminish people's belief in the hot hand . . . . In fact, when Tversky and colleagues asked volunteers to judge the randomness of computer-generated patterns of basketball shots . . . , the ones they picked as random were often very far from being so."† Here are 21 consecutive shots, X = hit, 0 = miss, by three basketball players who shoot about 50%. Are any of them *not* scoring at random?
a. 0 X X X 0 X X X 0 X X 0 0 0 X 0 0 X X 0 0
b. X 0 0 X X X 0 0 X 0 X 0 X 0 0 0 X 0 X X X
c. X 0 X 0 X 0 0 0 X X 0 X 0 X 0 0 X X X 0 X
(In Chapter 19, we will explain how to answer this question using a statistical analysis. For now you must use your intuition.)

1.38 Explain the difference between a framed population and a target population.

1.39 The words *dynamic* and *static* are sometimes used in conjunction with descriptive and inferential statistics procedures. Which word is associated with which type of procedure?

1.40 Is the reasoning used in inferential statistics an example of inductive or deductive reasoning?

1.41 Distinguish a sampling error from a nonsampling error.

1.42 *Business First,* a weekly newspaper specializing in business news in the Louisville, Kentucky, area, conducted a census on the commercial and office market in the east end of Louisville and published the results in the *Leasing Directory*. For each building, the data in Table 1.8 were recorded on the available square feet of office space.
a. Select a simple random sample of size $n = 18$ from this population using Table 1.6. Start at row 21, column 1, and proceed across row 21.
b. Comparing the proportion of 0's in your sample and in the population, demonstrate the meaning of the term *sampling error*.
c. Using the population, demonstrate the meaning of the term *nonsampling error*.

*Optional
†From Kevin McKean/© 1987 Discover Publications. Reprinted with permission.

Table 1.8 Data for Exercise 1.42

Building	Available Space	Building	Available Space	Building	Available Space	Building	Available Space
1	1,400	35	0	69	2,600	103	0
2	2,256	36	0	70	1,200	104	7,306
3	2,000	37	8,000	71	672	105	1,100
4	8,800	38	1,500	72	0	106	10,000
5	0	39	2,236	73	15,000	107	5,487
6	0	40	0	74	1,600	108	0
7	0	41	5,400	75	500	109	20,387
8	0	42	0	76	11,000	110	20,000
9	375	43	0	77	3,500	111	NA
10	1,400	44	1,650	78	9,415	112	2,000
11	13,000	45	0	79	4,330	113	0
12	1,100	46	3,325	80	2,488	114	NA
13	2,000	47	0	81	1,800	115	NA
14	1,050	48	25,400	82	10,900	116	1,717
15	0	49	0	83	0	117	460
16	2,500	50	2,000	84	70,000	118	9,500
17	5,000	51	0	85	NA	119	0
18	0	52	2,400	86	1,200	120	8,913
19	3,200	53	0	87	6,000	121	1,450
20	0	54	1,700	88	NA	122	13,500
21	4,199	55	0	89	750	123	11,500
22	0	56	0	90	0	124	14,000
23	0	57	2,400	91	0	125	1,500
24	10,000	58	0	92	0	126	1,300
25	400	59	0	93	9,945	127	66,000
26	2,000	60	1,500	94	1,770	128	27,800
27	17,125	61	850	95	1,400	129	76,830
28	NA	62	50,390	96	70,000	130	165
29	3,459	63	0	97	0	131	12,500
30	0	64	3,779	98	16,000	132	0
31	18,000	65	7,579	99	0	133	NA
32	0	66	37,000	100	4,200	134	1,000
33	0	67	15,000	101	0		
34	37,000	68	20,000	102	0		

NA = no response from the building manager.

Source: Leasing Directory, Business First of Louisville, June 8, 1987. Reprinted with permission.

1.43 In selecting a simple random sample using Tables 1.5 or 1.6, identify the entity numbers to be included in the sample if the following information is given:

a. $N = 572$, $n = 11$, starting point = row 27, column 7, and direction = up column 7.
b. $N = 89$, $n = 9$, starting point = row 1, column 3, and direction = across row 1.
c. $N = 913$, $n = 5$, starting point = row 8, column 24, and direction = down column 24.

1.44 The purity of data is always an issue in business statistics. In Robert J. Samuelson's article "The Joy of Statistics," we read: "If you're going to use a number, you'd better know where it comes from, how reliable it is and whether it means what it seems to mean . . . . Or, as British economist Sir Josiah Stamp once put it: 'The Government [is] very keen on amassing statistics. They collect them, add them, raise them to the *n*th power, take the cube root and prepare wonderful diagrams. But you must never forget that every one of these figures comes in the first instance from the village watchman, who just puts down what he damn well pleases.' "† This passage, though humorous, addresses what kind of error in statistics?

1.45 A systematic sample of 81 sales invoices is to be selected from a population of invoices numbered consecutively from 14,522 to 21,471. Suppose a young auditor chooses a random starting point of 51. Is it possible to use 51 as a starting point or should he start over? If yes, what are the first 3 invoice numbers? If no, explain how he should proceed.

1.46 Identify the goal of descriptive statistics and the goal of inferential statistics.

***1.47** An auditor is preparing to randomly sample vouchers from last year's register. Each voucher is filed according to month and a voucher number, starting with 1 at the beginning of each month. The number of vouchers per month are as follows:

Month	Vouchers	Month	Vouchers
January	407	July	101
February	418	August	186
March	222	September	114
April	275	October	116
May	260	November	106
June	71	December	233

a. What types of random sampling are most amenable to this situation? Why?
b. Suppose the size of the sample is to be $n = 62$. When taking a sample within several groups, a popular way of subdividing n into the different groups is called *proportional allocation*. To illustrate, January accounts for approximately 16.2 percent (407/2509) of all the vouchers; thus, 16.2 percent of the 62 sample entities is allocated to January. Determine the proportional allocation for each month, assuming a stratified sample is to be taken.
c. Select the entity numbers for each month from the appropriate table according to your answer in part b. Use as a starting point for each month, row = the month number and column = the number of letters in each month's name. Proceed across the row for an odd-numbered month, and down the column for an even-numbered month.

†*Newsweek,* November 4, 1985. Reprinted with permission.

1.48 Several years ago advice columnist Ann Landers was asked by a young couple whether having children was worth all the problems involved. As she sometimes does, Ann Landers invited her readers to respond to this question. A few weeks later the results of about 10,000 reader replies were presented in Ann Landers' column. If we call the 10,000 replies a sample of parent opinion, do you think it is a random or nonrandom sample? Justify your answer.

1.49 Refer to Exercise 1.48. Ann's readers' responses were as follows: About 30 percent said children were worth the trouble, while 70 percent said they were not. This result got a lot of attention in the media. *Newsday* subsequently commissioned a professional polling of 1373 parents, using the identical words as appeared in Ann Landers' column. *Newsday* found 91 percent responding that children were worth the trouble versus 9 percent saying they were not. How do you account for the difference in these results?

1.50 In testing a new product, companies try to locate markets that are considered "average" U.S. cities or states. Consider the following news brief: "So you've got a new cereal with as much fiber as a redwood, or a new detergent that gets out those stubborn stains. It still has to play in Peoria—or any of the 44 other 'most average' cities and markets in the USA. The list is compiled by ad agency Saatchi & Saatchi Compton, Inc., for its annual *Guide to Test Marketing*. The five markets most often used to test new products: Minneapolis–St. Paul; Portland, Ore.; Kansas City, Mo.; Syracuse, N.Y.; and Columbus, Ohio. 'All are near-perfect microcosms of the U.S.,' says Saatchi. It may be a dubious honor, but Ohio is the most average state: Dayton, Columbus, Cincinnati, and Cleveland all made it. The Midwest is the most average region, with 22 of 45 markets."* In choosing cities such as Minneapolis or Syracuse as test markets, marketers are selecting what type of random or nonrandom sample? Would a simple random sample be better suited for this specific marketing problem? Explain.

1.51 The editorial cartoon reproduced in Figure 1.11 (page 42) humorously demonstrates a problem in collecting data through surveys. (For historical purposes, the people named in the first and last panels were candidates for the presidential nomination in the 1988 Democratic party primary elections.) The polltaker has misrecorded the respondent's choice. Is this a sampling error or a nonsampling error?

Computer Exercise

C1.1 The purpose of this computer exercise is to demonstrate the Simple Random Sample processor. As discussed in Section 1.4, the processor will select the entity numbers for a simple random sample of size n from a finite population of N.

Refer to Exercise 1.42 in which we selected a simple random sample using Table 1.6. Let us repeat this exercise using the processor instead. Execute the Simple Random Sample processor to identify the entity numbers to be included in a sample of size eighteen. Then produce the set of sample data by writing down the values corresponding to each entity selected.

*"Average states get first crack at new products," Copyright 1987, *USA Today*. Excerpted with permission.

Figure 1.11 A Data Collection Problem (Reprinted with special permission of King Features Syndicate, Inc.)

REFERENCES

Arens, A. A., and J. K. Loebbecke. 1980. *Auditing: An Integrated Approach,* 2nd Edition. Prentice-Hall, Inc., Englewood Cliffs, NJ.

Bottom Line/Personal. 1987. "The Secret of *Real* Power," an interview with G. Ray Funkhouser. May 30, pp. 11–12.

Churchill, Jr., G. A. 1979. *Marketing Research: Methodological Foundations,* 2nd Edition. Dryden Press, Hinsdale, IL.

Conover, W. J., and R. L. Iman. 1983. *Introduction to Modern Business Statistics.* John Wiley & Sons, New York.

Ferber, R., P. Sheatsley, A. Turner, and J. Waksberg. 1980. "What Is a Survey?" American Statistical Association, Washington, D.C.

Kotler, P. 1988. *Marketing Management: Analysis, Planning, Implementation, and Control,* 6th Edition. Prentice-Hall, Inc., Englewood Cliffs, NJ.

Kreitner, R. 1986. *Management,* 3rd Edition. Houghton Mifflin Company, Boston.

Landon, Jr., E. L., and S. K. Banks. 1978. "An Evaluation of Telephone Sampling Designs," *Advances in Consumer Research, 5:* 103–108.

McClave, J. T., and P. G. Benson. 1985. *Statistics for Business and Economics,* 3rd Edition. Dellen Publishing Company, San Francisco.

Neter, J., W. Wasserman, and G. A. Whitmore. 1988. *Applied Statistics,* 3rd Edition. Allyn and Bacon, Boston.

Table of 105,000 Random Decimal Digits. 1949. Interstate Commerce Commission, Bureau of Transportation, Economics, and Statistics, Washington, D. C.

Van Matre, J., and G. Gilbreath. 1983. *Statistics for Business and Economics,* Rev. Edition. Business Publications, Inc., Plano, TX.

Waksberg, J. 1978. "Sampling Methods for Random Digit Dialing," *Journal of the American Statistical Association, 73:* 40–46.

Warner, S. L. 1965. "Randomized Response: A Survey Technique for Eliminating Evasive Answer Bias," *Journal of the American Statistical Association, 60:* 63–69.

Chapter Maxim *Raw, unprocessed data are seldom free of recording errors, missing observations and/or spurious values.*

CHAPTER 2

ORGANIZING AND GRAPHING DATA

*Optional

Objectives

After studying this chapter and working the exercises, you should be able to

1. Describe several data management activities involved in processing data.
2. Distinguish between discrete and continuous data.
3. Explain the Principles of Inclusion and Exclusion.
4. Transform raw, processed data into organized data.
5. Construct histograms from frequency distributions.
6. Compile qualitative data into a one-way tabulation.
7. Construct a pie chart and a bar graph.
*8. Execute the Frequency Distributions processor to verify answers and/or solve problems.
9. Recognize applications of data organization in marketing research.

Our first goal after generating a set of data through sampling is to sort it out. This may take many forms, such as ordering the data from low to high, compiling it into a table, or graphing it to form a visual image. However, we should anticipate that organizing and presenting data from a quantitative variable necessarily will be different from that for a qualitative variable. This is because, in general, qualitative data are a collection of words while quantitative data are a collection of numbers.

The purpose of this chapter is to present methods for organizing and displaying qualitative and quantitative data in tables and graphs. As a result, we hope to achieve clarity and extract information that might not have been apparent about the sample in its raw, unprocessed form.

2.1 PRELIMINARIES TO ORGANIZING DATA

Collecting and recording data is a hands-on activity that is not easily simulated in a textbook such as this. Data that we provide in the examples and exercises of this book are usually neat and free of complications. In reality, data sets are rarely this antiseptic. Before any analysis of the data can begin we must sift through the data and "clean them up."

*Applies to optional section.

Data Management

When we first obtain raw, unprocessed data, a variety of problems often surfaces and must be resolved immediately. For example, a survey of the commuting distances from home to work administered to workers in a downtown area included the question: How far do you commute to work each day? This seemingly simple question generated answers such as 5 miles, 10 miles, and 17½ miles, but it also produced such responses as about 10 miles, 12 to 14 miles, 3 blocks, 205 miles, and zero. What do we do with these nonstandard responses?

We must perform some "processing" activities. First, we must examine the data and decide upon a standard degree of accuracy. For example, the data analysis for the variable X = daily commuting distance from home to work will be easier if each piece of data has the same units and is rounded off to the same number of decimal places. If we choose to record X to the nearest tenth of a mile, then the response "about 10 miles" would be recorded as 10.0, the "12 to 14 mile" answer might be recorded as 13.0, and "3 blocks" might become .2 mile.

Second, we must develop a system for dealing with missing data. For example, in a consumer behavior survey questionnaire consisting of several questions, we might discover that answers to some questions were not provided. A decision must be made about the usefulness of a questionnaire with only partial information: Should it be included in the analysis or deleted? Often, the answer depends on the degree to which the questionnaire has been completed. If there are but a few instances of missing information, two remedies are possible. One is to estimate the missing response (this cannot always be done, and with a couple of exceptions, is not addressed in this book); another and more likely strategy is to ignore the missing data point and reduce the sample size by one for that question.

A third processing activity requires that we check the data for unusual responses. A response of "zero" for the commuting distance seems odd, but it may have a practical explanation (perhaps the individual is self-employed and works out of his/her home). On the other hand, the response of "205 miles" for commuting distance seems absurd. Was this number really 20.5 miles or 2.5 miles mistakenly recorded as 205? Extreme values in a data set that appear to be inconsistent with the others are called **outliers.** (A more detailed discussion of outliers can be found in Chapter 3.) Remedying this problem is generally not as simple as ignoring it. First, we carefully recheck the origin and validity of an unusual response. If we find nothing in error, then a workable solution is to isolate (temporarily) the data point, organize the remaining $n - 1$ observations, and then include the outlier in a separate category.

These and other activities, such as coding the data or creating a data file for later use with a computer program, are considered necessary parts of managing data. We will use the term "processed" to describe the raw data *after* performing the various data management activities. A more complete treatment of data management techniques usually occurs in a marketing research course or in an advanced statistical methods class, so we will not delve into this area in more detail here. We shall proceed on the assumption that our data sets have been "cleaned up" as well as possible.

Before considering ways to organize and visually present processed data, we need to develop some ground rules.

Two Types of Quantitative Data

In Chapter 1 we learned that sampling produces one of two forms of data: qualitative and quantitative. While qualitative data can be thought of as one broad category, we will find it useful to break down quantitative data into two types: discrete and continuous. "Discrete" means separate and distinct; "continuous" means unbroken or extending without interruption. A quantitative variable that has "separate" values at specific points along the number line, with gaps between them, is called a **discrete variable**. A quantitative variable that has a "connected" string of possible values at all points along the number line, with no gaps between them, is called a **continuous variable.**

Deciding whether quantitative data are values of a continuous variable or a discrete variable is sometimes difficult. In Figure 2.1 we have indicated a characteristic of each type of data along the lines connecting the boxes. A discrete variable usually produces data by counting. For example, counting the number of employees in a company yields discrete data. Continuous data usually result from a measurement or computation. For instance, dividing the number of unemployed people by the size of the work force we generate continuous data for the variable "unemployment rate." Continuous variables are also characterized by physical measurements such as size, time, area, and so forth.

To aid us in classifying quantitative data as discrete or continuous, we may resort to the Refinement Test described in the following box.

The Refinement Test

To decide whether quantitative data are discrete or continuous, conduct the following test:

1. Try to identify two potentially consecutive values of the variable.
2. Find the midpoint of the two values by adding the values and dividing by two. Is the computed midpoint a *possible* value for the variable?
   - **a.** If the answer is YES, repeat step 2 using one of the values from step 1 and the midpoint from step 2. If step 2 can be repeated indefinitely, then the data originated from a *continuous* variable.
   - **b.** If the answer is NO, or if step 2 cannot be repeated indefinitely, then the data originated from a *discrete* variable.

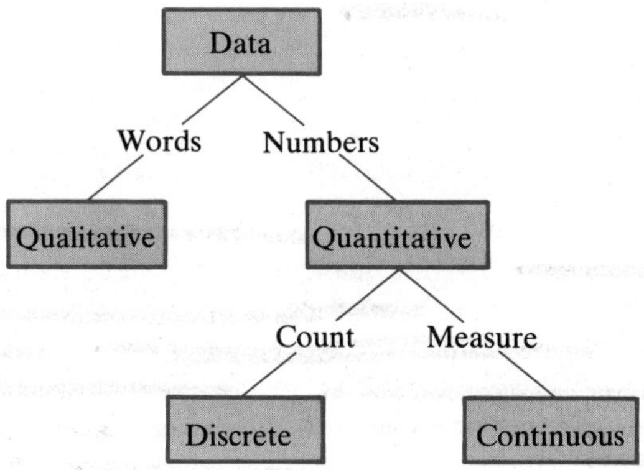

Figure 2.1 Data Classifications and Characteristics

EXAMPLE 2.1

Classify the following data as discrete or continuous:

a. Number of times per week an individual eats the evening meal at a restaurant

b. Interest rate charged on bank credit cards

c. Monthly percentage change in wholesale prices

d. Daily closing price of a share of McDonald's stock on the New York Stock Exchange (NYSE).

Solution:

a. Since this variable produces numbers like 0, 1, 2, 3, . . . , the data are countable and therefore *discrete*.

b. Let us use the Refinement Test here to decide whether we are dealing with discrete or continuous data. Two potentially consecutive values might be 18 percent and 19 percent. The midpoint is 18½ percent, which also could qualify as an interest rate. Another repetition of the Refinement Test—finding the midpoint of 18 percent and 18½ percent—gives 18¼ percent, which also might be the interest rate. Further iterations yield 18 and any fraction as possible interest rates. These data are continuous. Please note that although the data are continuous in theory, they are recorded as discrete numbers in practice. A bank has the prerogative to charge any interest rate it wishes. Usually, a whole number plus a simple fraction is selected to make the computations easier.

c. Possible consecutive values of the variable like 1 percent and 2 percent may be continually refined to any value in between. Practically speaking, the percentage change is usually rounded off and reported to the nearest tenth of a percent. The data from this variable, like the interest rate data in part b, are *continuous* in nature, but treated as discrete in application.

d. A share of McDonald's stock could close at a price (in dollars) of 50 or perhaps 50½. Applying the Refinement Test leads us to ask whether the stock could close at 50¼, 50⅛, 50 1/16, and so on. The sharp reader will recall that prices of stocks listed on the NYSE generally are quoted in increments of ⅛. This means that closing prices of 50¼ and 50⅛ are possible values of the variable, but 50 1/32 is not. The Refinement Test cannot be applied indefinitely; the closing price of McDonald's stock yields *discrete* data.

□

COMMENT You cannot always determine whether quantitative data are discrete or continuous by looking at the numbers. Due to space and accuracy limitations, continuous data always are rounded and therefore look like and are even treated like a set of discrete numbers. For example, the time required to close a sales deal is, in theory, continuous but reported discretely such as 83 minutes, 83.2 minutes, and so on. Although we cannot list all the trailing decimals for continuous data, this does not nullify the continuous nature of the data.

Principles of Inclusion and Exclusion

Another basic prerequisite to organizing data is to employ two straightforward rules.

Principle of Inclusion

There must be a category into which each piece of data can be assigned.

Principle of Exclusion

No piece of data can be assigned to more than one category.

Simply stated, the Principles of Inclusion and Exclusion require each piece of data, whether qualitative or quantitative, to fit into one and only one category. We do not wish to exclude any pieces of data, nor do we wish to "double-count" any data points.

We use the word *category* in a generic sense. For qualitative data, the term is self-explanatory. For quantitative data, a category is either an actual value of the variable or a *class* of consecutive values. To make it easier to organize data, we will adopt a special definition for the term *class*.

Definition

A class that begins at point a and ends at point b is a set of consecutive values that are greater than or equal to a, but less than b.

For example, the class from 45 to 55 for the variable X means $45 \leq X < 55$.

EXAMPLE 2.2

Develop several categories to account for possible values for these variables:

a. Number of times per week an individual eats the evening meal at a restaurant
b. Interest rate charged on bank credit cards

Solution:

a. If we assume that an individual eats only one evening meal per day, then possible categories would be the integers from 0 to 7. In this case the categories are the actual values of the variable.

b. The minimum value would be 0 percent, but the maximum value is unknown. We might expect a ceiling figure to be around 24 percent. With this in mind, here are some potential classes:

Classes (%)
0–6
6–12
12–18
18–24

Applying our definition to the class 0–6 percent, for example, means all interest rates between 0 and 6 percent, except 6 percent, are included in this class. Interest rates of 6, 12, and 18 percent belong to the classes 6–12 percent, 12–18 percent, and 18–24 percent, respectively.

□

COMMENT Although our term *class* may be used interchangeably with the term *interval,* the definition of a class may be different in other disciplines or books. Some consider a class to include both endpoints, while others introduce concepts such as class limits or class boundaries in order to satisfy the Principles of Inclusion and Exclusion.

We have chosen to classify data as qualitative or quantitative. In other disciplines a different classification scheme, called the *scales of measurement,* may be used. The four scales, in order of increasing sophistication, are nominal, ordinal, interval, and ratio.

A **nominal scale** is one that simply lists names as potential categories. As an example, the variable *type of industry* produces nominal responses such as construction, retail, banking, and so on. An **ordinal scale** implies a ranking or ordering of the categories. Various cuts of beef have such labels as prime, choice, good, and so on, that imply a difference in quality. Usually, qualitative data are measured on nominal or ordinal scales.

The interval and ratio scales differ only by the meaning attached to the value of zero on the scale. When the value zero implies a lack of, absence of, or none, the variable is measured on a **ratio scale**. For example, if the variable is revenue, then $0 in revenue is meaningful and suggests no sales; revenue is a ratio scale variable. When the meaning of the value zero is arbitrary, the variable is measured on an **interval scale**. An example of a scale with an arbitrary zero point is the Fahrenheit temperature scale. The only true ratio scale for temperature is the Kelvin or absolute scale, on which zero (about minus 460° Fahrenheit) represents complete lack of heat—the coldest possible temperature. Generally, quantitative data are measured on interval or ratio scales.

Where we go from here depends on the type of data. First, we will consider how to organize and display quantitative data, and then we will discuss the treatment of qualitative data.

SECTION 2.1 EXERCISES

2.1 Use the Refinement Test, if necessary, to identify the data from the following quantitative variables as discrete or continuous:

a. yearly subscription fee to financial newsletters.

b. number of ATM (automated teller machine) transactions in a 15-minute period.

c. proportion of yearly income consumed by federal income tax, for a sample of taxpayers.

2.2 Refer to Exercise 2.1. Develop several categories to account for possible values for each variable listed above.

2.3 Explain the Principles of Inclusion and Exclusion.

2.4 On a survey, the question "How comfortable are you with your current level of debt?" was followed by these categories:

(1) very comfortable	(2) comfortable
(3) slightly uncomfortable	(4) very uncomfortable

How well do these categories adhere to the Principles of Inclusion and Exclusion?

2.5 Use the Refinement Test, if necessary, to identify the data from the following quantitative variables as discrete or continuous:

a. amount of time spent per month managing personal investments.

b. number of bytes of available space per used disk.

c. trunk space in a new car.

2.6 Refer to Exercise 2.5. Develop several categories to account for possible values for each variable listed in that exercise.

2.7 Suppose a survey question is worded "What is your marital status?" The possible choices are as follows:

(1) never married	(2) live with lover
(3) married	(4) divorced or separated
(5) widow or widower	

How well do these categories adhere to the Principles of Inclusion and Exclusion?

2.8 Think of a variable that would produce discrete data, and define it in words. Repeat this instruction for continuous data.

2.9 A professor's syllabus indicated that a student's final average would be computed by adding the student's scores on 3 exams and dividing by 3. The final grade would be awarded based on this scale:

Final Average	Grade
90–100	A
80– 89	B
70– 79	C
60– 69	D
0– 59	F

Comment on this grading scale with respect to the Principles of Inclusion and Exclusion.

2.10 Why isn't the Principle of Inclusion by itself a sufficient guideline with which to organize data?

2.11 Identify three data management activities.

2.2 ORGANIZING QUANTITATIVE DATA

Once a quantitative data set has been processed, we have two options for organizing it. The first is called an ungrouped frequency distribution. This option is preferred whenever the number of different values of the variable is ten or fewer.

Ungrouped Frequency Distribution

To construct an ungrouped frequency distribution, we simply list the different values of the variable in one column and set up a second column to tally the data. After all the data have been tallied, we add the tallies corresponding to each value and report this figure in a third column labeled "frequency." At this point the tallies are superfluous; thus, the first and third columns constitute the *ungrouped frequency distribution*.

> **Definition**
>
> **An ungrouped frequency distribution** is a table consisting of two columns of information: the values of the variable and the frequency with which each value occurs in the data set.

EXAMPLE 2.3

To monitor the mobility of our society, a private consulting group surveyed 47 individuals and asked them to respond to several questions, one of which was "How many times within the last three years have you changed residences?" The responses to this question were as follows:

5	1	0	1	1	1	5	3
0	0	1	1	1	1	3	0
2	4	0	2	4	0	0	3
0	0	3	0	0	6	1	2
4	5	0	2	3	0	4	0
1	4	3	0	7	2	0	

Identify the variable of interest, classify the data, and organize these data into an ungrouped frequency distribution.

Solution:

The variable is X = the number of times within the last three years that an individual has changed residences. It is a quantitative variable that produces discrete data. Since the number of different values of the variable in the sample is 8, an ungrouped frequency distribution is appropriate. The values from 0 to 7 are listed below in the column labeled X, and the 47 data points are tallied in the second column:

X	Tallies
0	卌 卌 卌 1
1	卌 卌
2	卌
3	卌 1
4	卌
5	111
6	1
7	1

By adding the tallies we generate the frequency of each X-value. The ungrouped frequency distribution then appears as

X	f
0	16
1	10
2	5
3	6
4	5
5	3
6	1
7	1
	$n = 47$

Note that f is the symbol for the frequencies, and the total of the frequency column matches the number of individuals surveyed.

□

Organization has been achieved. We have adhered to the Principles of Inclusion and Exclusion. The 47 numbers have been neatly sorted and compiled into a table. Quickly and easily we gain valuable information from the frequency distribution that we could not see so clearly in the raw data. For example, a majority of the individuals surveyed moved once or not at all during the three-year period.

Usually discrete data are candidates for this type of organization, provided that the number of values in the X column is not too large. We would gain less clarity if the ungrouped frequency distribution stretched the length of this page. For this reason, we recommend a cap of ten different values for an ungrouped frequency distribution.

When there are many different values of the variable, a second option for organizing the data called a grouped frequency distribution becomes more viable.

Grouped Frequency Distribution

The idea behind a *grouped frequency distribution* is to segregate neighboring values together into a class or an interval of numbers. Rather than listing all the actual values X can assume, we list the endpoints of each class plus the number of data points falling within the class. The result is a table consisting of a column of classes and a column of frequencies.

Definition

A **grouped frequency distribution** is a table consisting of at least two columns of information: the values of the variable organized into classes and the frequency of the values occurring within each class.

As we learn to develop the classes and the frequency distribution, we will be adding other columns to the table. The purpose of each additional column will be explained as it is introduced. First we need some guidelines to help generate the classes, as there are many different approaches to grouping data. Without such guidelines, grouping tends to be haphazard.

Step 1: Number of Classes Generating classes or intervals poses an immediate problem. That is, how many classes should we use? If we develop too many, then each class will have relatively few observations in it and the efficiency of grouping will be lost. On the other hand, too few classes overly bunches the data together and may hide certain patterns within the data set. As a guideline we recommend basing the *number of classes,* denoted by c, on the size of the data set as indicated in Table 2.1.

COMMENT In Table 2.1 the relationship between n and c is based on powers of 2. For example, if n is between 16 and 32, we recommend 4 to 5 classes since $16 = 2^4$ and $32 = 2^5$. In general, use the smaller value of c for smaller sample sizes and the larger value of c for larger samples within the indicated ranges for n. If n is exactly in the middle of a sample size range, choose the larger value of c.

Step 2: Class Width Once we have chosen c, we must decide how wide to make each class. It is a good practice to maintain a constant value for the *class width.* This assures uniformity and makes it easier to construct the frequency distribution. To satisfy the Principle of Inclusion, we must ensure that the c classes span the data set. Thus, the class width, denoted by w, can be found as follows:

$$w = \frac{\max - \min}{c} \tag{2–1}$$

where max and min refer to the maximum and minimum values in the data set, respectively.

In practice, the division called for in Equation 2–1 usually yields a number with several decimal places. We should round up this number to the accuracy of the original data. For example, if the data are recorded in tenths and if $w = 1.327 \ldots$, then

Table 2.1 Recommended Number of Classes per Sample Size

Sample Size n	Number of Classes c
11– 16	3–4
16– 32	4–5
32– 64	5–6
64–128	6–7
128–256	7–8
256–512	8–9
Over 512	10 or more

round up w to 1.4. When using Equation 2–1 to determine the class width, *always round up*!

COMMENTS

1. For moderately large to large data sets, it is sometimes difficult to pinpoint the maximum and minimum values without considerable effort. To save time, you may need to estimate these values. For instance, if you were grouping a large set of data representing hourly wages, you might be able to use the federal minimum wage as the smallest value and \$18 as the largest if the data set warrants doing so.
2. In the rare case that Equation 2–1 produces a value for w that is of the same degree of accuracy as the raw data, we still recommend rounding-up. For instance, if we had a data set consisting of integers from 17 to 47 and we used $c = 5$, then w would work out to be exactly 6. Even though no rounding is apparently needed, we suggest rounding up to $w = 7$. This ensures that exactly $c = 5$ classes will be sufficient. Otherwise, with $w = 6$, you would need six classes. Exercise 2.23 demonstrates this anomaly.
3. Do not perform step 2 blindly. Look at the data set closely before you automatically put the maximum and minimum values into Equation 2–1. Occasionally, the minimum (or maximum) value is not consistent with the remaining values. In such cases, you may wish to use the second smallest (or second largest) value in Equation 2–1. Then, create an extra, open-ended class to accommodate the unusually small (or large) value. See Exercise 2.22.

EXAMPLE 2.4

Suppose a random sample of $n = 60$ long-distance telephone calls yielded data on the variable X = cost (in dollars and cents) per phone call. Determine the number of classes and the class width for grouping the data. The minimum and maximum charges were \$.30 and \$9.22, respectively.

Solution:

According to Table 2.1 we should use $c = 6$ classes. Evaluating Equation 2–1 produces:

$$w = \frac{\max - \min}{c} = \frac{9.22 - .30}{6} = 1.48666$$

The accuracy of the original data is two decimal places, so the interval width should be rounded up to $w = 1.49$.

□

Step 3: Generating the Classes There are several ways to begin generating the classes; we recommend the following. Starting with the minimum value in the data set, consecutively add the value of the class width. The resulting sequence of numbers forms the classes.

To demonstrate, consider the information in Example 2.4. The minimum value was .30 and $w = 1.49$. Consecutively adding w produces this sequence: .30, 1.79,

3.28, 4.77, 6.26, 7.75, 9.24. We can stop at 9.24 since the maximum value of 9.22 is accounted for. The classes are as follows:

Classes
.30–1.79
1.79–3.28
3.28–4.77
4.77–6.26
6.26–7.75
7.75–9.24

Our definition of a class implies that the first class, for example, is understood to be $.30 \le X < 1.79$.

EXAMPLE 2.5

A sample of 60 long-distance telephone calls produced the following values for X = cost per phone call:

\$3.28	\$ .39	\$1.14	\$6.87	\$7.06	\$6.48	\$1.29	\$1.21	\$3.67	\$2.60
.57	2.17	9.18	1.57	9.22	5.99	2.91	1.34	.39	4.57
.56	.53	4.50	4.87	2.84	1.83	1.91	4.55	3.60	1.78
2.42	1.53	2.06	.80	2.27	1.25	4.94	.93	4.57	2.97
.30	1.30	3.41	1.43	3.83	.31	6.18	3.30	1.78	2.60
.45	3.80	4.40	4.19	5.61	2.95	2.42	5.04	2.97	2.37

Organize these data into a grouped frequency distribution.

Solution:

The minimum and maximum values are \$.30 and \$9.22, respectively, and as determined in Example 2.4, there should be 6 classes with $w = 1.49$. The classes are as follows:

Classes	Tallies	f
\$ .30–1.79	~~1111~~ ~~1111~~ ~~1111~~ ~~1111~~ 1	21
1.79–3.28	~~1111~~ ~~1111~~ ~~1111~~	15
3.28–4.77	~~1111~~ ~~1111~~ 111	13
4.77–6.26	~~1111~~ 1	6
6.26–7.75	111	3
7.75–9.24	11	2
		$n = 60$

Each value of X is tallied into the correct class of values. For example, the first value of \$3.28 is tallied in the class 3.28–4.77. The tallies are added to yield frequencies; these counts are entered in the f column above. Notice that the frequencies add to 60, the sample size.

□

Generating the classes and the frequencies is not the end of our organizational efforts, but rather the beginning. By taking a set of raw, processed data in whatever form—on sheets of paper, on index cards, on questionnaires, in a computer file, and so forth—and reducing it to a concise table of values and frequencies, we gain order from chaos. A frequency distribution gives us a compact, yet global view of the entire data set. Our cost has been a slight loss of accuracy. For instance, in Example 2.5 we note that there are two pieces of data in the class \$7.75 to \$9.24, but we cannot see the exact values of these two data points from the table. If it were important later to know the specific values, we would need to go back to the raw data that generated this table.

COMMENT

Although a constant class width is preferred, there are occasions when the first or last class is left "open" to handle extreme observations. For example, a grading scale might appear as 90–100, A; 80–90, B; 70–80, C; 60–70, D. The letter grade of F is then designated as either 0–60 or below 60. Either way this class docs not have the same width as the other four. Unequal class widths also may be necessary to handle a data set with a very concentrated set of values in a narrow subrange or one with a very thin set of values over a wide subrange. Exercise 2.24 shows a distribution with the former characteristic, while Exercise 2.28 contains a data set with the latter property. Notice in Exercise 2.28 that it might have been more efficient to combine some of the last few classes in the data rather than maintain a constant width. There are no general guidelines for when to use unequal interval widths. If you keep in mind that our objective is to extract information from the data set through grouping, then classes with unusually large (or small) frequencies may require further partitioning (or contraction).

EXAMPLE 2.6

A certain part used in the assembly of several of a firm's different engine models has the design specification that its diameter should be 5.000 millimeters. A random sample of 20 of these parts finds the diameters listed below, in millimeters. Organize these data into a grouped frequency distribution:

4.961	4.982	4.991	4.999	5.001
4.964	4.984	4.992	4.999	5.003
4.975	4.984	4.994	5.000	5.004
4.975	4.987	4.997	5.001	5.007

Solution:

Step 1: With $n = 20$, we should use $c = 4$ classes.

Step 2: The class width is

$$w = \frac{\max - \min}{c} = \frac{5.007 - 4.961}{4} = .0115$$

Since the accuracy of the raw data is thousandths, we round up to $w = .012$.

Step 3: The classes and frequencies form the following grouped frequency distribution:

Classes	Tallies	f
4.961–4.973	11	2
4.973–4.985	~~1111~~	5
4.985–4.997	1111	4
4.997–5.009	~~1111~~ 1111	9
		$n = 20$

□

Additional Columns of Information

We have said a grouped frequency distribution consists of *at least two* columns of information: classes and frequencies. From these columns, we often produce other columns of information and include them in the frequency distribution.

Relative Frequency For comparative and graphical purposes, we introduce a quantity called ***relative frequency*** as a means of determining the proportion of the sample within each class.

Definition

The **relative frequency**, denoted by rf, of a class is the frequency of observations in that class divided by n.

Relative frequency converts each frequency into a number between 0 and 1, making it easy to compare two or more data sets with unequal sample sizes. In addition, when we study probability, we will discover a close relationship between relative frequency and probability.

Cumulative Frequency Another concept that will prove useful is that of accounting for each data point in a natural progression from the minimum to the maximum value. Imagine that each piece of data occupies a position from 1 to n. For an ungrouped frequency distribution, it is easy to match up each value with each position.

This is not the case for a grouped frequency distribution since the exact values of the data points are lost within the classes. For instance, in Example 2.5 we know that the 21 data points in the first class precede the 15 values in the second class. Rather than trying to list the individual values, we account for them by posting the cumulative sum in a column titled *cumulative frequency*. The cumulative frequency tracks the 1-to-n sequence of positions occupied through the classes.

Definition

The **cumulative frequency**, denoted by *cf*, of a class is the sum of the frequencies in that class and all other classes preceding it.

A column of cumulative frequencies is useful when we are trying to locate the class containing the median and other measures of position that we will discuss in Chapter 3.

Midpoint Finally, in our analysis of the organized data we often need to summarize the values within a class for graphical and/or computational purposes. The frequency of each class certainly tells us how many values there are, but we seek a symbolic value to represent or characterize the entire class. The simplest solution is to use the *midpoint* of the class as a proxy for all the values in the class. Of course, some of the actual values may be more than this value and some may be less, but on average, the midpoint should be a reasonable approximation to a typical value in the class.

Definition

The **midpoint** of a class, denoted by M, is the sum of the class endpoints divided by 2.

EXAMPLE 2.7

Generate the relative frequencies, cumulative frequencies, and midpoints of each class in the grouped frequency distribution of Example 2.5. Summarize the values into columns and indicate the positions occupied within each class.

Solution:

To the Classes and frequency (f) columns from Example 2.5, we append four columns labeled *rf*, *cf*, M, and Positions Occupied (pos):

Classes	*f*	*rf*	*cf*	*M*	Positions Occupied (pos)
\$ .30–1.79	21	.350	21	1.045	1 through 21
1.79–3.28	15	.250	36	2.535	22 through 36
3.28–4.77	13	.217	49	4.025	37 through 49
4.77–6.26	6	.100	55	5.515	50 through 55
6.26–7.75	3	.050	58	7.005	56 through 58
7.75–9.24	2	.033	60	8.495	59 through 60
	$n = 60$	1.000			

□

The first entry of .350 in the *rf* column is found by dividing the frequency of the first class, 21, by $n = 60$. The remaining relative frequencies are computed in a similar manner and carried out to three decimal places. The second entry, 36, in the *cf* column represents the total frequency in the first two classes: 21 + 15. The third entry, 4.025, in the *M* column is the result of adding 3.28 and 4.77 and then dividing by 2. Notice that the midpoints are exactly \$1.49 apart; this is the value of the class width. The fifth entry in the positions occupied column is 56 through 58. The cumulative frequency preceding the fifth class is 55, meaning that the first 55 positions are occupied. Thus the three values in the fifth class occupy the next 3 positions: 56, 57, and 58.

COMMENTS

1. The sum of the relative frequencies should be 1, but occasionally rounding errors cause the total to be .999 or 1.001, and so on. Round off as necessary, and report relative frequencies to three decimal places.
2. The cumulative frequency refers to the total observations accumulated by the *end* of the class, not at the beginning or in the middle. It is only at the end of a class that we can be sure all the data points are accounted for. In Example 2.7, the cumulative frequency for the class \$1.79–\$3.28 is 36, which means 36 of our data points must have occurred before \$3.28. The last entry in the *cf* column must be equal to *n*.
3. The midpoint of a class is also referred to as a class mark.

We have introduced many new terms and steps in this section, and it is easy to get lost in the details. *Unprocessed data* becomes *processed data* (hereafter just called "raw data") after doing the various data management functions. Our goal has been to transform raw data into organized data. **Raw data** refers to a bunch of numbers, listed in rows and/or columns, representing the observed values of a variable. **Organized data** refers to a table of numbers, arranged into classes or in sequence, followed by a slate of frequencies. Organized data takes two forms: an ungrouped frequency distribution or a grouped frequency distribution. Figure 2.2 shows the transformation of data as it relates to the execution stage of the problem-solving process.

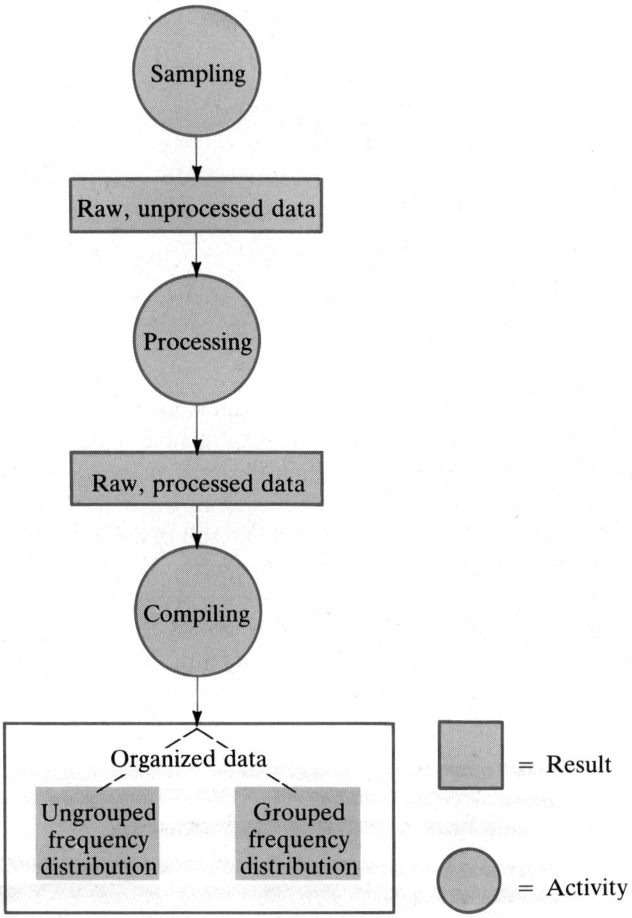

Figure 2.2 Transformation of Data

EXAMPLE 2.8

A large company specializing in insurance and benefits packages surveyed businesses in 27 different industries to gain information on the average projected salary increases for the next year. The percentage increases for each industry were as follows:

5.9	6.1	6.4	5.7	6.4	5.8	6.4	6.4	6.0
6.1	6.3	6.7	5.7	5.9	7.1	6.3	6.1	6.0
5.4	5.7	6.0	6.6	5.2	5.6	6.2	5.8	6.8

Organize the raw data into a grouped frequency distribution, with relative frequency, cumulative frequency, midpoint, and position occupied (pos) columns.

Solution:

Compiling the data will follow the steps outlined earlier in this section.

Step 1: Number of classes: From Table 2.1 with $n = 27$, we elect to use $c = 5$ classes.

Step 2: Class width: Each class will have a width of $w = .4$, since

$$w = \frac{\max - \min}{c} = \frac{7.1 - 5.2}{5} = .38$$

and .38 is rounded up to the accuracy of the raw data (tenths).

Step 3: Generating the classes: Beginning with 5.2, we consecutively add .4 to generate the class endpoints, yielding 5.2, 5.6, 6.0, 6.4, 6.8, and 7.2.

Putting these three steps together and tallying the data produces the following grouped frequency distribution:

Classes	f	rf	cf	M	pos
6.8–7.2	2	.074	27	7.0	26 through 27
6.4–6.8	6	.222	25	6.6	20 through 25
6.0–6.4	9	.333	19	6.2	11 through 19
5.6–6.0	8	.296	10	5.8	3 through 10
5.2–5.6	2	.074	2	5.4	1 through 2
	$n = 27$	.999			

The rf, cf, M, and pos columns are listed above as well. Notice that the rf column totals .999 because of rounding errors. Also, the classes in this example are listed with smaller values at the bottom and larger ones at the top of the column. Consequently, the cumulative frequencies and positions occupied ascend from bottom to top.

□

CLASSROOM EXAMPLE 2.1

Creating an Ungrouped Frequency Distribution

The express checkout lane in a grocery store is intended for customers purchasing ten or fewer items. A random sample of customers from the express line reveals the following numbers of items per customer:

4	3	3	4	10	1
6	3	2	7	6	8
4	3	1	8	9	5
7	8	1	2	3	3
2	7	1	3	4	1
1	4	1	2	6	2

a. Define the variable of interest.

b. What is an entity for this variable?

c. Does the variable produce discrete or continuous data?

d. List the values of the variable X.

e. Tally the sample data.

f. Add the tallies for each value and record its frequency. Determine the value for n.

CLASSROOM EXAMPLE 2.2

Creating a Grouped Frequency Distribution

The proliferation of television commercials prompted a newspaper television critic to study the number of minutes of actual programming per 60-minute show. He randomly sampled 24 one-hour shows and recorded the following seconds of programming:

2491	2625	2576	2746
2505	2535	2615	2613
2555	2671	2838	2637
2580	2686	2937	2880
2627	2732	2637	3060
2550	2505	2815	2968

a. What is the accuracy of the raw data?

b. To group these data how many classes are needed?

c. Determine the class width.

d. Generate the classes.

e. Tally the data and record the frequency values. What is the sum of the entries in this column?

f. Generate and record the relative frequencies.

g. Generate and record the cumulative frequencies.

h. Generate and record the midpoints.

i. List the positions occupied (pos).

SECTION 2.2 EXERCISES

2.12 The human resource department of a company recorded the number of sick days taken by a sample of employees from the main office for the last six months. These data are as follows:

3	7	2	5	3	4	5	1	6
4	0	2	0	9	7	2	0	7
5	3	7	8	2	6	5	2	3
4	0	6	3	4	2	5	7	2
4	3	5	2	1	7	9	2	2
2	3	9	7	3	4	2	3	2
0	4	3	7	4	4	9	0	4
9	6	2	1	2	8	4	4	2
4	2	3	7	3	3	3	3	6
2								

a. Define the variable represented by these figures.
b. Does the variable produce discrete or continuous data?
c. What is an entity in this problem?
d. Should these data be organized into a grouped or ungrouped frequency distribution? Why?
e. Generate the appropriate frequency distribution.

2.13 Consider the following frequency distribution based on a set of sample data:

X	0	1	2	3	4	5
f	10	17	14	9	3	1

a. What size is the sample?
b. What is the relative frequency of $X = 2$?
c. What type of frequency distribution is this?
d. Generate the set of cumulative frequencies.
e. The value $X = 2$ occupies what positions in the ordered data set?

2.14 How many classes are needed to organize a data set consisting of the following observations?
a. $n = 60$ **b.** $n = 200$ **c.** $n = 15$

2.15 A sample of size $n = 63$ is to be organized into a grouped frequency distribution.
a. How many classes should be used?
b. If the minimum and maximum values are 61.1 and 83.7, respectively, what should be the value of the class width?
c. What are the endpoints of the first class?

2.16 Organize these data into an ungrouped frequency distribution:

17	17	13	15	12	11	15
14	15	12	15	12	14	15
12	12	17	11	11	17	13
16	12	13	16	12	12	14
14	16	15	13	14	14	13

2.17 For a frequency distribution with equal class widths, is it possible to determine the class width solely from a column of midpoints? Explain.

2.18 Consider the following grouped frequency distribution:

Class Number	Classes	f
1	47– 58	7
2	58– 69	18
3	69– 80	24
4	80– 91	12
5	91–102	5

a. Determine the value of w.
b. What is the midpoint for class 2?
c. What is the relative frequency for class 3?
d. Generate a cumulative frequency column.
e. Identify the positions occupied by the values in the fourth class.

2.19 If the minimum value is 51 and the maximum value is 96, determine the class width needed to organize a set of $n = 58$ pieces of data into a grouped frequency distribution. What are the endpoints of the first class?

2.20 Organize the following data set into a grouped frequency distribution:

3.4	4.5	3.3	5.2	3.6	2.5	3.6
2.4	2.1	5.3	3.4	6.3	5.7	4.9
4.4	1.4	7.6	3.5	1.4	5.5	3.8
4.2	8.2	6.2	8.5	4.6	5.3	1.6

2.21 Consider the following grouped frequency distribution:

Class Number	Classes	f
1	−4.4– 1.8	2
2	1.8– 8.0	7
3	8.0–14.2	17
4	14.2–20.4	21
5	20.4–26.6	19
6	26.6–32.8	5

a. What is the cumulative frequency at the end of class 4?
b. Generate the relative frequency column.
c. Determine the value of n.
d. What is the midpoint for class 1?

2.22 Automobile dealers' inventories are measured by the number of days supply of vehicles they have on their lots. A survey of several Chevrolet dealerships was taken to determine the number of days supply of Cavaliers, yielding the following data:

13	31	12	16	11
63	18	21	22	20
7	24	15	11	23
10	20	9	30	17
20	32	24	18	23
16	8	19	8	
6	15	29	29	
12	17	16	26	

a. Define the variable represented by these figures. Is the variable discrete or continuous?
b. Identify the minimum and maximum values.
c. The maximum value seems inconsistent with the rest of the data. What do we call extreme values like this?
d. Organize the data set into a grouped frequency distribution, temporarily omitting the maximum value. Compute the class width using the second largest number. After generating the classes, create an open-ended class, such as "36 and above," for the maximum value.

2.23 Organize these raw data into a grouped frequency distribution:

23	20	17	25	31	43
40	27	19	19	30	39
21	19	47	28	21	45
25	29	28	25	24	22
19	33	27	35	32	21

2.24 A company's package of employee benefits is becoming as important in attracting qualified applicants as salary scales traditionally have been. One of the desirable items on a candidate's checklist is a 401(k) plan. A 401(k) plan, named after a section of the tax code, is basically a tax-deferred means of saving for retirement. The Hay Group, a consulting firm, surveyed 618 companies that offer a 401(k) plan as a benefit and asked each firm to indicate the percentage of employees participating in the plan. The results are summarized in the following frequency distribution:

Workers in Plan (%)	Number of Companies
0– 25	49
25– 50	111
50– 65	105
65– 75	130
75– 90	155
90–100	68

Source: "401(k) plans are popular," Copyright 1988, *USA Today*. Adapted with permission.

Notice that the classes do not maintain a constant width. The width for the first two classes is 25, but then alternates between 15 and 10 in the next four classes. One reason for this is the disproportionate number of values in the 50–100% subrange. By partitioning the 50–75% and 75–100% classes, we reveal the concentrations in the 65–75% and 75–90% subranges.

a. Generate the set of relative frequencies for this distribution.

b. What percentage of companies have between 65 percent and 90 percent of their employees participating in a 401(k) plan?

c. Develop a column of midpoints.

2.25 A survey was conducted in 1986 to compare the annual costs (tuition and fees) of attending a state supported college or university by a resident of that state. The following data represent these costs for the state's largest school and main campus:

$1304	$1000	$1996	$1029	$1542
1036	1040	2662	1470	1323
1136	2262	1970	820	863
1030	1760	1727	1092	1326
1347	1390	1567	1704	3198
1526	1290	1330	921	2238
1937	1332	1524	1487	1605
2306	1727	1080	2996	1260
1058	1565	2625	2130	1570
1662	1601	2308	2028	778

Source: "State colleges: What residents and non-residents pay," *USA Today*, November 10, 1986.

a. Organize these data into a grouped frequency distribution.

b. Suppose someone asked you to identify the in-state annual cost at a "typical" state university in the United States. A quick and simple answer could be found by using the midpoint of the interval with the largest frequency. If you used this approach, what would be your response?

c. An alternative way of answering the question in part b is to use a "median" cost. (We will learn more about this in the next chapter.) Since the median is the halfway point in the data set, one way to locate it is to look for an occupied position of 25, in this case. By generating a column of cumulative frequencies or a column of positions occupied, locate the *class* containing the 25th occupied position.

2.26 What is the difference between the following types of data?
a. unprocessed and processed data **b.** raw and organized data

***2.27** Another quantity sometimes generated in a frequency distribution is the *cumulative relative frequency,* denoted by *crf*. Similar to cumulative frequency, the *crf* of a class is the sum of the relative frequency in that class and all others preceding it. Develop a column of cumulative relative frequencies for the frequency distribution in Exercise 2.21.

2.28 Lightning is a major cause of forest fires, especially in times of prolonged dry weather. In the heavily forested province of British Columbia, a network of field detectors was set up to locate lightning strikes and determine their intensity with a measure called the normalized signal strength. A sufficiently intense bolt of lightning should be detected simultaneously by several field detectors, producing potentially different readings. Thus, an average normalized signal is computed for each strike. Between 1983 and 1985, the network recorded over 165,000 lightning strikes. The distribution of the average normalized signals follows:

Average Normalized Signal	Frequency
0– 20	2,702
20– 40	21,528
40– 60	34,315
60– 80	28,790
80–100	20,743
100–120	14,810
120–140	10,667
140–160	7,765
160–180	5,746
180–200	4,210
200–220	3,106
220–240	2,183
240–260	1,733
260–280	1,337
280–300	1,025
300–320	802
320–340	598
340–360	497
360–380	391
380–400	301
400–420	239
420–440	207
440–460	162
460–480	163
480–500	129
500 and up	983
	165,132

Source: B. R. Johnson, "Lightning Detection in British Columbia: An Example of Using System Operation at Unknown Reliability to Estimate Component Reliability," *The Canadian Journal of Statistics, 16* (1988): 105–115. Reprinted with permission.

*Optional

Notice that the author collapsed the classes above 500 into one class, rather than continue the classes up to the maximum average normalized signal.

a. What is the class width of the other classes?

***b.** Refer to Exercise 2.27. Create a column of cumulative relative frequencies, to four decimal places.

***c.** By referring to the cumulative relative frequency column, determine the subrange containing roughly 75 percent of the weakest average normalized signals.

2.3 GRAPHICAL TECHNIQUES FOR QUANTITATIVE DATA

Our descriptive analysis of a set of data does not end with the construction of a frequency distribution. Now that the data have been organized, we may wish to display them in a graph. Quantitative data are graphed in a variety of familiar ways: line graphs, pictographs, and bar graphs. Graphs are popular because they have an immediate visual impact that frequency distributions lack. By graphing a data set we can quickly, easily, and succinctly convey information with a minimum of words.

Time-Related Versus Time-Free Data

Some graphs are appropriate only for data that have been gathered or recorded over regular intervals of time. For instance, if we record the changes in the Consumer Price Index per month, then an entity is a month. The values are chronological and represent *time-related* data. There are two main graphical techniques for time-related data: a line graph and a time bar graph. We will discuss the line graph and the general analysis of time-related data in Chapter 18. Time bar graphs will be covered in Section 2.4 in the subsection on bar graphs.

Data for which the entity is not time are called *time-free* data. This is the type of data with which we've been working. For example, recording the number of days of sick leave taken by a sample of employees over the past six months is time-free data, even though it took six months to generate an observation. The key is that recording the number of sick days is done once for each entity (an employee) included in the sample. In this example, we're interested in the six-month total of sick days, not the month-by-month figures.

Unless we specifically note them as time-related, "data" hereafter will refer to time-free data.

Histograms

To graph data that have been organized into a frequency distribution, we can use a histogram. A **histogram** is a collection of rectangles that displays a frequency distribution. There are two major types of histograms: frequency histograms and relative frequency histograms. They differ by what is plotted on the vertical axis: frequency or relative frequency. The horizontal axis of a histogram plots either the values of the variable (for ungrouped data) or the classes (for grouped data).

Frequency Histogram In Example 2.3, we developed the following ungrouped frequency distribution for the variable X = the number of times within the last three years that an individual has changed residences:

X	f
0	16
1	10
2	5
3	6
4	5
5	3
6	1
7	1

Suppose we wish to present these data as a histogram. Such a graph would plot the eight integer values of X along the horizontal axis and the frequency on the vertical axis in the form of connected rectangles. The frequency of $X = 0$ from the distribution is 16, so the rectangle will be 16 units tall. Where, on the horizontal axis, should the rectangle begin and end? Remember we are trying to depict the frequency for $X = 0$. If we begin at 0 and end at 1, it will be unclear whether the rectangle is associated with $X = 0$ or $X = 1$.

To avoid ambiguity such as this, we choose to begin the rectangle to the left of $X = 0$ and end it to the right of $X = 0$, being careful not to overlap $X = 1$, as illustrated in Figure 2.3. We call the beginning and ending points class boundaries. Numerically, for $X = 0$ they are the numbers $-.5$ and $.5$ on the horizontal axis. Forming the class boundaries for individual values of X by subtracting one-half and adding one-half enables us to construct a rectangle that is centered over the actual value of X. Each distinct X-value has its own pair of class boundaries; for example, the boundaries for $X = 5$ are 4.5 and 5.5.

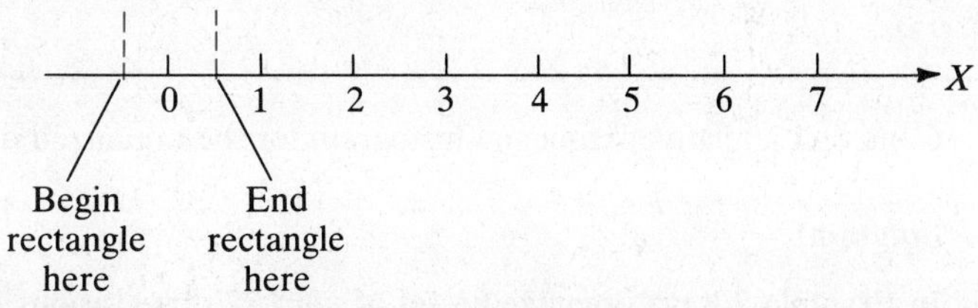

Figure 2.3 Class Boundaries for $X = 0$

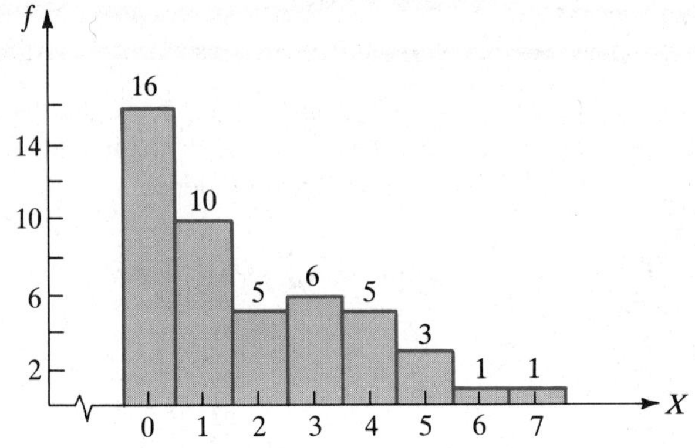

Figure 2.4 Number of Residence Changes Within the Last Three Years

Figure 2.4 illustrates a frequency histogram for the data in the ungrouped frequency distribution. Several points relative to Figure 2.4 should be noted. First, the horizontal axis is labeled X and shows the eight values of the variable; the class boundaries are implied, but not explicitly labeled. Second, the vertical axis is labeled f. Third, the title of the graph does *not* include the phrase "frequency histogram," since the connected rectangles always are characteristic of histograms. Fourth, the numbers resting atop each rectangle, representing the frequencies, are optional. Fifth, along the horizontal axis immediately to the right of the vertical axis is a V-shaped symbol. This symbol indicates that the horizontal axis has been "broken" and shifted to the right so that $X = 0$ no longer is associated with the point of intersection between the two axes. Breaking the horizontal axis is done to center the rectangles in the first quadrant.

Relative Frequency Histogram A **relative frequency histogram** is also a series of connected rectangles drawn above the values or classes on the horizontal axis, with the height of each rectangle corresponding to the relative frequency. Otherwise the construction is identical to that of a frequency histogram.

EXAMPLE 2.9

Construct a relative frequency histogram for the organized data in Example 2.8.

Solution:

In Example 2.8 we organized a set of $n = 27$ observations into a grouped frequency distribution and generated the relative frequencies for each class. These data are reproduced here:

Classes	*rf*
6.8–7.2	.074
6.4–6.8	.222
6.0–6.4	.333
5.6–6.0	.296
5.2–5.6	.074
	.999

The relative frequency histogram will plot the five values appearing in the *rf* column on the vertical axis. Along the horizontal axis, we plot the classes. Figure 2.5 displays the relative frequency histogram.

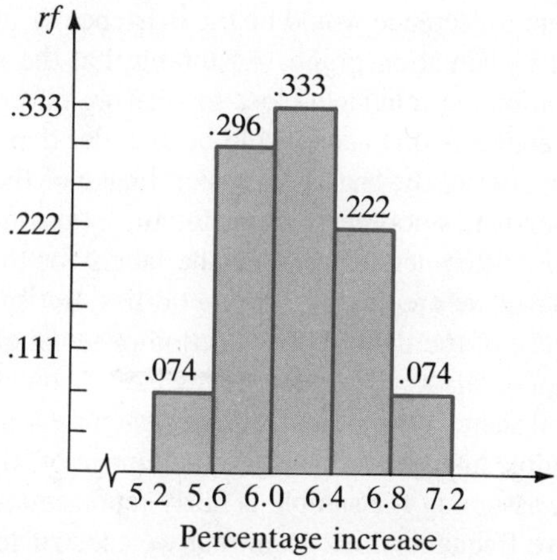

Figure 2.5 Projected Salary Increases for a Sample of 27 Industries

□

The careful reader should note from Figures 2.4 and 2.5 that every graph must be titled and include a label for each axis. The horizontal axis was shortened as indicated by the V-shaped symbol to center the rectangles in the first quadrant. Including the relative frequencies at the top of each rectangle is optional. It is a good idea to include the sample size in the title of a relative frequency histogram. This is not necessary for a frequency histogram.

Graphing data into either a frequency or relative frequency histogram often reveals clusters of values that are not apparent to the eye in the raw data set. Relative frequency histograms are often more popular in business since the relative frequencies are easily converted into percentages, facilitating comparisons. Relative frequency histograms also play an important role in our study of probability.

Histograms are not the only means of graphing quantitative data. There are other techniques with names such as polygons and ogives that also are popular and effective.

Each graph highlights a different aspect of the data set. For this reason, one graph does not dominate, though histograms are very common. No descriptive statistical analysis would be complete without a graph.

Common Shapes of Data Sets

One reason why we advocate the construction of histograms is that the graph enables us to envision the rough shape of the population from which the sample was drawn. As we develop inferential techniques later in the book, we will focus on the shape or form of the population. For now, we argue that the histograms developed from the sample reflect a general view of the population. If we had access to the entire population and if we constructed its histogram, what differences and similarities between the sample and the population histograms would we find?

A major difference would be the existence of many more classes, and thus rectangles, in the population graph. Assuming that the minimum and maximum values in the population are relatively close to what we observed in a sample, then the effect of more rectangles is that each would be skinnier than the ones in the sample histogram. Appearance-wise, the sharp, stair-step image of the sample histogram would contrast with a blended, smoother picture for the population histogram. Figure 2.6 demonstrates this difference. (Note that the labels for the horizontal and vertical axes are omitted since we are dealing in generalities, not specifics.)

Because of the dense concentration of rectangles for the population histogram, a smooth curve, also called a frequency curve, sometimes is drawn instead to represent the general shape. Using this tactic of sketching a smooth curve through the rectangles of the sample histogram forms a crude image of what the population histogram might look like, assuming the sample is fairly representative of the population. For example, the relative frequency histogram that we constructed in Example 2.9 yields the shape illustrated in Figure 2.7.

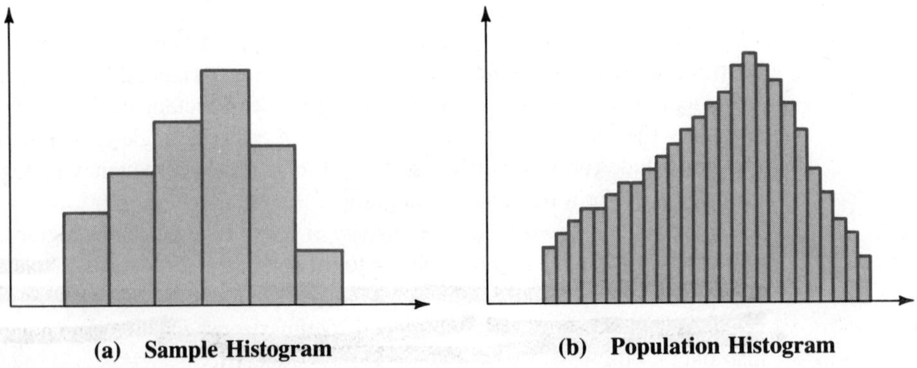

Figure 2.6 Sample Histogram Versus Population Histogram

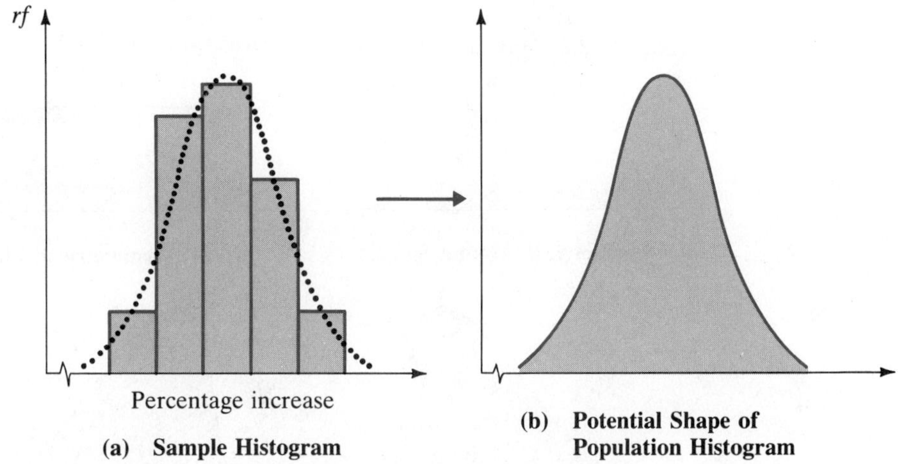

Figure 2.7 Shape of Relative Frequency Histogram from Figure 2.5

Of course we would be foolish to think that the shape of the sample histogram *perfectly* describes the shape of the population histogram. What we hope is that the shape of the population is close to the shape of the sample histogram. As we progress in our study of business statistics, we will define "close to" in a formal sense and develop a way to compare the shape of sample histograms to one or two classic shapes of population histograms.

Most, but not all histograms, resemble one of the common shapes in Figure 2.8. The relative frequency histogram we constructed in Figure 2.5 for the projected percentage salary increases resembles the symmetrical, mound-shaped figure of panel (a). In Figure 2.4 where we graphed the number of residence changes in the last three years, we detect the positive, asymmetrical shape of panel (d). Note that the shape in panel (d) has a long tail to the right.

A graph with a long tail to the left has the negative, asymmetrical contour of panel (c). If we graphed a sample of data for the variable X = the number of days that lapse between the time a company receives a bill and pays the bill, we might see such a graph. Companies usually receive bills with terms such as "net 30," meaning the net amount is due 30 days from the date of the bill. Other bills may be due upon receipt. Thus, the variable X would produce occasional small values and a heavy concentration of values between 20 and 30, for example.

Many states that sponsor lotteries offer a game called Lotto in which six numbers are selected from the first 40 or 44 integers. We would expect each digit to be selected an equal number of times over the course of many drawings. Otherwise, the repetition of one or more digits would cause gamblers to play those numbers more often, creating an unfair gambling proposition. Consequently, numbers drawn in Lotto games often follow the symmetrical, flat shape of panel (b).

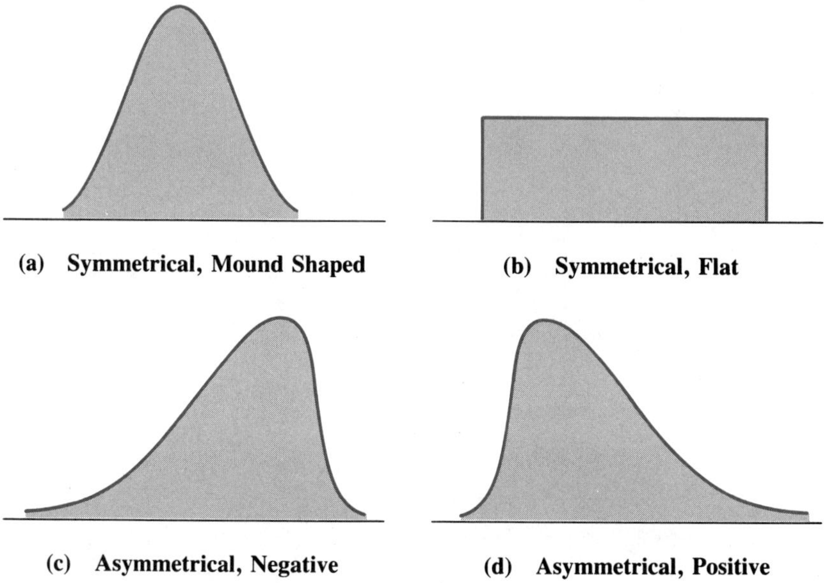

Figure 2.8 Common Shapes of Histograms

CLASSROOM EXAMPLE 2.3

Constructing a Histogram

In Classroom Example 2.2, we organized a data set representing seconds of television programming per 60 minutes into the following grouped frequency distribution:

Classes	*f*	*rf*
2491–2605	8	.333
2605–2719	8	.333
2719–2833	3	.125
2833–2947	3	.125
2947–3061	2	.083

a. Construct a frequency histogram for these data. Label and scale both axes.

b. Construct a relative frequency histogram for these data.

c. Which common shape in Figure 2.8 is most closely approximated by the histograms you have drawn?

SECTION 2.3 EXERCISES

2.29 Construct a frequency histogram from the following ungrouped frequency distribution:

X	0	1	2	3
f	6	5	8	5

2.30 Would the data generated by the following variables be time-related data or time-free data?

a. X = Percentage of operating capacity per factory for the month of June.
b. X = Current popularity (in ratings points) of syndicated TV shows.
c. X = Number of new home sales per year.
d. X = Difference between adjustable mortgage rates and the rates on conventional loans per week.

2.31 Develop a relative frequency histogram for the following grouped frequency distribution:

Classes	f
.45– 8.05	3
8.05–15.65	5
15.65–23.25	8
23.25–30.85	4

2.32 Refer to Exercise 2.31.

a. What is the value of w?
b. What is the midpoint of the class 23.25–30.85?
***c.** What is the cumulative relative frequency (*crf*) at the end of the class 8.05–15.65? (See Exercise 2.27 for the definition of *crf*.)

2.33 A sample of 50 stocks was taken from Standard & Poor's to obtain the following data on the variable X = debt as a percentage of capital (i.e., the "debt ratio"):

17.3%	9.9%	20.2%	6.8%	21.7%
0.0	6.1	38.4	21.9	11.6
1.9	15.6	25.3	4.5	18.0
2.7	30.2	25.5	15.9	12.9
21.8	2.5	4.9	53.9	20.2
24.2	15.4	40.9	34.9	0.0
32.4	34.3	34.7	5.2	12.6
6.4	27.7	2.4	18.2	20.4
4.4	38.0	53.6	9.5	32.4
19.7	0.0	10.3	11.7	32.7

*Optional

a. What type of variable (discrete or continuous) is X?
b. Organize these data into a frequency distribution, assuming none of the observations is an outlier.
c. Construct a frequency histogram.
d. Does the sample histogram resemble a common shape? If so, what is it?

2.34 In the May, 1986, issue of *Changing Times* magazine, the article titled "Cars That Bloom in the Spring" previewed several new cars that were introduced in mid-year. Included in the article were data on the latest available specifications for each model. One of the attributes listed was the miles-per-gallon figure for city driving. Some of these values were as follows:

20	21	19	19
25	17	19	23
17	26	18	18
24	21	17	25
18	17	18	22
18	17	18	18
19	20	19	18

a. Organize these data into a frequency distribution.
b. Construct a relative frequency histogram.
c. From the sample histogram constructed in part b, project a potential shape for the population histogram.

2.35 Propose realistic variables that, when the values are graphed, would resemble each of the shapes in Figure 2.8.

2.36 Construct a frequency histogram for each of the following frequency distributions:

a.

Classes	f
1.77–2.19	7
2.19–2.61	24
2.61–3.03	35
3.03–3.45	21
3.45–3.87	10

b.

X	f
10	1
11	5
12	7
13	0
14	12
15	3

c.

Classes	f
45–50	8
40–45	17
35–40	11
30–35	5
25–30	4
20–25	2

Identify, if possible, the common shape of the sample histogram.

2.37 Into which class should we tally the value $X = 14$, for the following classes?

Classes
10–14
14–18
18–22

2.38 Consider the frequency histogram in Figure 2.9 for a set of organized data.

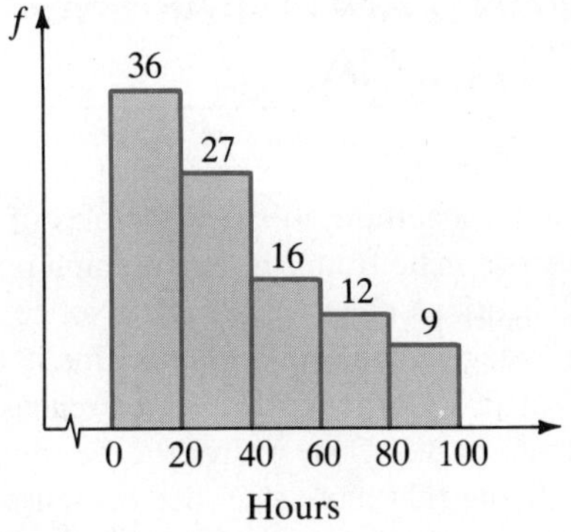

Figure 2.9 Hours Spent Flying per Month by a Sample of Executives

a. Reconstruct the frequency distribution from this graph, including a column of classes.
b. What is the size of the sample?
c. What common shape does the sample histogram exhibit?

2.39 Would the data generated by the following variables be time-related data or time-free data?
a. gross national product per quarter for the current year
b. gross national product of a sample of 6 countries as of the first quarter of the current year
c. amount of time required to complete a banking transaction using an automated teller machine
d. number of runners finishing the New York Marathon per year

2.40 Consider the relative frequency histogram in Figure 2.10 for a set of organized data.
a. What is the class width?
b. Is it possible to determine the frequency of the last class? If so, what is it?

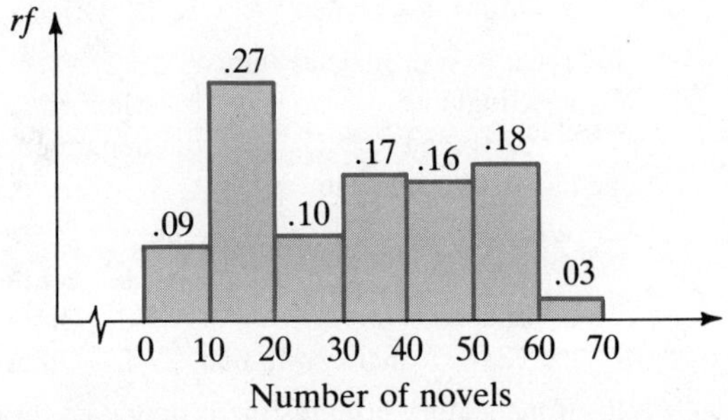

Figure 2.10 Number of Paperback Novels Read in a Calendar Year for a Sample of 231 Adults

2.4 ORGANIZING AND GRAPHING QUALITATIVE DATA

Qualitative data, by definition, are characterized by words or categories, making them easy to organize according to the Principles of Inclusion and Exclusion. Categorical data typically originate from surveys, opinion polling, questionnaires, and so on, such as the brief, fictitious one in Table 2.2. Notice in questions 2, 4, and 5 in the table, the catch-all category labeled "Other." This is necessary to ensure that the Inclusion Principle is satisfied, although an "Other" category is not always needed, as questions 1 and 3 demonstrate. Conversely, the wording of question 5 includes the plural "drugs," implying that more than one response is permitted. This violates the Principle of Exclusion and is not recommended. In this section we will explain how to organize, tabulate, and graph qualitative responses.

Compiling the Data

The process of organizing qualitative data is an exercise in compilation. Given the raw data, we proceed to list the different levels or categories of each variable and then tally the category counts. In effect, we compile the data into a table, called a *one-way tabulation*. Example 2.10 illustrates this process.

Table 2.2 Fictitious Survey

1. What is your gender?
   a. Male b. Female
2. Which of the following characterizes your place of residence?
   a. Detached single home b. Apartment
   c. Condominium d. Trailer
   e. Multiplex home f. Other
3. What is your marital status?
   a. Single b. Married
   c. Divorced d. Separated
   e. Widow or widower
4. What is your ethnic background?
   a. Black b. White
   c. Hispanic d. Oriental
   e. Native American Indian f. Other
5. If mandatory drug testing is approved for all workers, on which drugs should workplace tests focus?
   a. Cocaine b. Marijuana
   c. Heroin d. Alcohol
   e. Rx drugs f. Other

EXAMPLE 2.10

Fifty people were surveyed and asked to name the fast food chain that makes the best french fries. Compile the responses given below into a table:

Burger King	Dairy Queen	Roy Rogers	McDonald's	McDonald's
Wendy's	Wendy's	McDonald's	McDonald's	McDonald's
McDonald's	McDonald's	Hardee's	McDonald's	Arby's
Hardee's	Hardee's	Wendy's	Hardee's	Burger King
McDonald's	Wendy's	Arby's	Hardee's	Burger King
McDonald's	D'Lites	Burger King	Wendy's	Wendy's
Hardee's	Burger King	McDonald's	Wendy's	Wendy's
Wendy's	Burger King	Hardee's	Burger King	Hardee's
McDonald's	Carl's, Jr.	Wendy's	McDonald's	Hardee's
Burger King	Wendy's	Dairy Queen	McDonald's	Burger King

Solution:

The qualitative variable of interest is X = name of the fast food chain that makes the best french fries. To compile this data, we list the names of the chain in one column and then tally the 50 responses in another column.

Chain	Tally	Chain	Tally
Burger King	𝍸 1111	Dairy Queen	11
Wendy's	𝍸 𝍸 1	D'Lites	1
McDonald's	𝍸 𝍸 1111	Carl's, Jr.	1
Hardee's	𝍸 1111	Roy Rogers	1
Arby's	11		

The last five chains listed above have significantly fewer responses than the first four chains. In Table 2.3, we choose to lump them together into a category called "Other" and present the organized data in a one-way tabulation.

Table 2.3 Consumer Survey of Fast Food Chains That Make the Best French Fries

Chain	Number of Responses	Percentage of Responses
Burger King	9	18
Wendy's	11	22
McDonald's	14	28
Hardee's	9	18
Other	7	14
	n = 50	100

□

COMMENTS

1. Table 2.3 depicts qualitative data even though the last two columns are sets of numbers. These figures merely represent the count of each category and do *not* make the data quantitative. The qualitative/quantitative distinction hinges on the nature of the variable X, not on the frequency of the categories/values of X. In this instance, the variable X (fast food chain) is inherently qualitative.
2. As a general guideline we recommend using from four to six categories, unless the data naturally lend themselves to more or less. For example, the qualitative variables X = sex and Y = month in which companies recorded their lowest sales should be reported with 2 and 12 categories, respectively.
3. Quantitative data are sometimes collapsed into verbal categories and treated as qualitative data. For example, ages may be grouped together like 18 through 29, 30 through 49, or 50 through 65 and be given labels such as young adults, middle-aged adults, and mature adults.

One-way tabulations like Table 2.3 are a concise and effective means of organizing a set of raw data. To complement the tabulations we often display the organized data in a graph. For qualitative data, the two most popular graphical techniques are the *pie chart* and *bar graph*.

Pie Chart

A **pie chart** is a circle in which each category of qualitative data is apportioned a slice of the pie. The size of the slice depends on the number of times each category occurred in the sample. Usually the raw frequencies are converted into a percentage. The percentages are used to partition the 360 degrees in a circle into angles that correspond to the relative size of each category. For example, if a category occurred 40 percent of the time in a sample, then the angle for this category would be 40 percent times 360° or 144°. The following box summarizes the guidelines for constructing a pie chart.

Guidelines for Constructing a Pie Chart

1. For each category, determine its percentage of the total.
2. Multiply 360 by the percentages from step 1 to determine the angle to allot to each category.
3. Beginning at the 9 o'clock position create an angle in accordance with the appropriate number of degrees for the category with the largest percentage.
4. Beginning where the first category ended, repeat step 3 for the category with the second largest percentage. Repeat this procedure until all categories have been allocated.
5. Label each slice with the category name and percentage.
6. Title the graph.

Using a compass or protractor, we draw line segments from the center of the circle to its perimeter in accordance with the appropriate angles, starting at the 9 o'clock position and working in a clockwise direction. To demonstrate the construction of a pie chart for the data in Table 2.3, we first determine the angle for each category by multiplying 360° by the category's percentage:

Chain	Percentage	Angle (°)
Burger King	18	65
Wendy's	22	79
McDonald's	28	101
Hardee's	18	65
Other	14	50

Figure 2.11 summarizes the percentages and angles into a pie chart.

COMMENTS

1. Starting at the 9 o'clock position, though recommended, is not strictly enforced. Depending on the space available and the intent, the categories can begin anywhere on the circle. For example, the 12 o'clock position often is used as a starting position. Typically though, the categories are sequenced after the most popular one in descending order of frequency.
2. Most computer software packages automatically determine the angles in the construction of a pie chart.

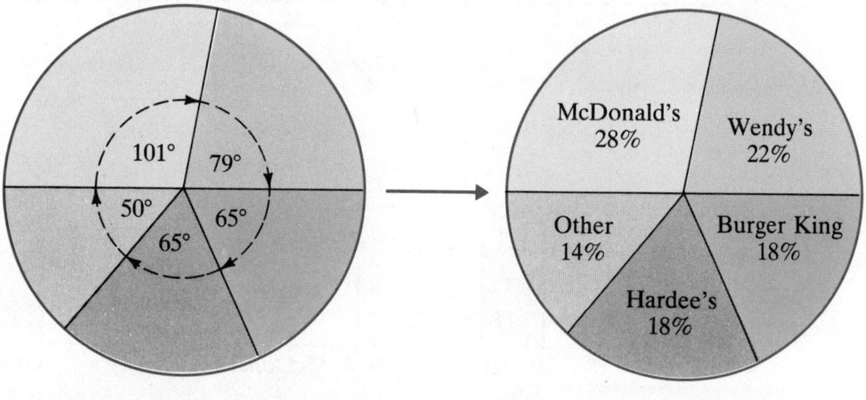

(a) **Apportioning the Degrees** (b) **Pie Chart**

Figure 2.11 Constructing a Pie Chart for Consumer Survey of Fast Food Chains Making the Best French Fries

Bar Graph

Probably the most familiar-looking graph is a bar graph. The traditional form of a **bar graph** is a series of disjointed rectangles, where each rectangle represents a category of a qualitative variable and the height of each rectangle corresponds to the frequency of the category. We recommend keeping the rectangles separated so as to distinguish the appearance of a bar graph from a histogram, which is intended for quantitative data. However, many chartists for newspapers and magazines construct bar graphs with connected rectangles.

In the following box, we suggest guidelines for constructing a bar graph, and in Figure 2.12, we use the guidelines to construct a bar graph for the data in Table 2.3.

Guidelines for Constructing a Bar Graph

1. For each category, determine its frequency of occurrence.
2. Draw a pair of axes perpendicular to one another.
3. Label the vertical axis frequency and scale it accordingly.
4. Label the horizontal axis according to the variable under study, and place the name of each category on this axis, separated by some space.
5. Erect a rectangle above each category name so that the height corresponds to the frequency of the category, keeping the rectangles separate.
6. Title the graph.

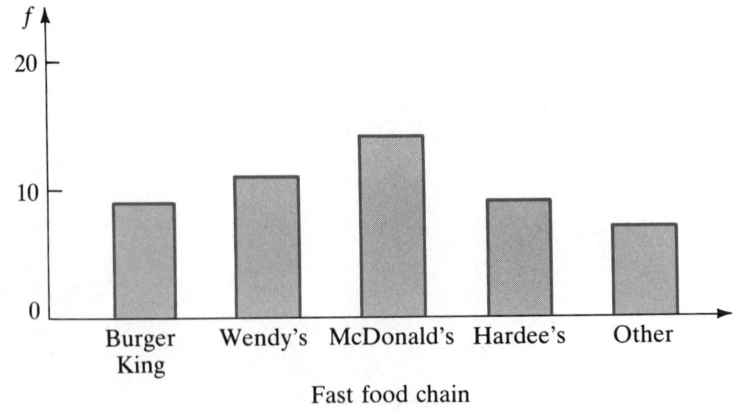

Figure 2.12 Bar Graph for Consumer Survey of Fast Food Chains Making the Best French Fries

COMMENTS

1. The percentage for each category could be substituted in lieu of the frequency and graphed on the vertical axis.
2. For bar graphs, nothing is sacred about the axes; you may put frequency on the horizontal axis and categories on the vertical.
3. Some people prefer to put the category with the largest frequency first and arrange the remaining categories in descending order of frequency thereafter.

Technically, a bar graph is a graphical tool for qualitative data only. Nevertheless, bar graphs also may be used to represent the *amount* of a quantitative variable by plotting the entities on one axis and the amount of the variable on the other axis. The left panel in Figure 2.13 is such a bar graph. Notice that the variable *parking costs* produces quantitative data. On the vertical axis several entities—airports, in this case—are listed, while on the horizontal axis, the value of the variable, *the amount charged* for each entity, is shown as a rectangle.

Bar graphs are common because of their flexibility. For example, one variation in presentation is the side-by-side bar graph in the right panel of Figure 2.13. This graph shows changes in credit forms for two different months. There are many varieties of bar graphs, each of which summarizes and displays the data differently from the

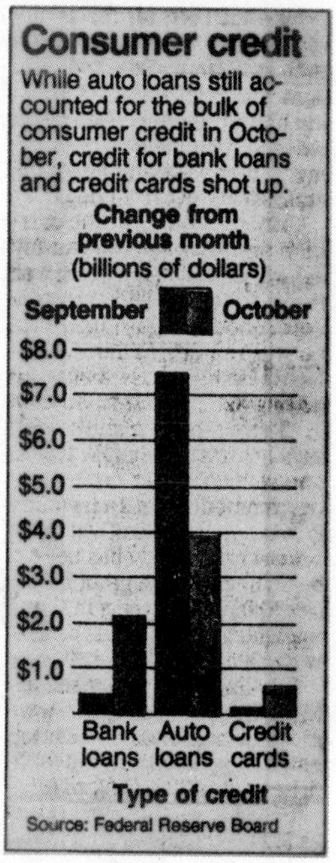

Figure 2.13 Bar Graph Variations (*Left:* Copyright 1985, *USA Today.* Reprinted with permission. *Right:* Copyright 1986, *USA Today.* Reprinted with permission.)

traditional bar graph, but in an appealing manner. For this reason, there is not just one style of bar graph, and each person selects the appropriate form depending on his or her creative instincts.

■ Postscripts

Graphs are indisputably popular as a means of conveying information quickly. Advances in microcomputer graphics make it easy to construct attractive bar graphs and pie charts. Some computer configurations generate graphs that give the illusion of three-dimensions both on the screen and in hard copy. Today's manager cannot afford to ignore the power of a graph as it is practically a standard part of a written report or an oral presentation involving data.

We urge caution in constructing and viewing graphs, since there are ways to present data that are deliberately misleading. One well-documented way of deceiving an audience is to alter the vertical axis by deleting a portion of it. Figure 2.14 shows a time bar graph, in which the bottom portion of the bars are missing. The effect is to accentuate the differing heights of the bars, conveying the implication that the frequency in 1988 was over seven times the frequency in 1989, when actually the ratio is less than 2:1. A small percentage difference in frequencies among the categories is thereby magnified to fool the viewer into thinking a large difference exists. Watch for this and other types of deception!

COMMENT There are additional references on deceptive graphing. Several chapters in *How to Lie with Statistics* (Huff, 1954) deal with this problem, as does the chapter titled "Cheating Charts" in *Flaws and Fallacies in Statistical Thinking* (Campbell, 1974).

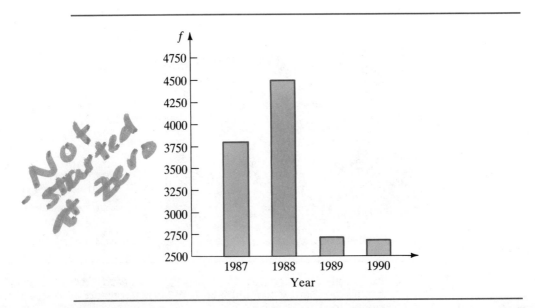

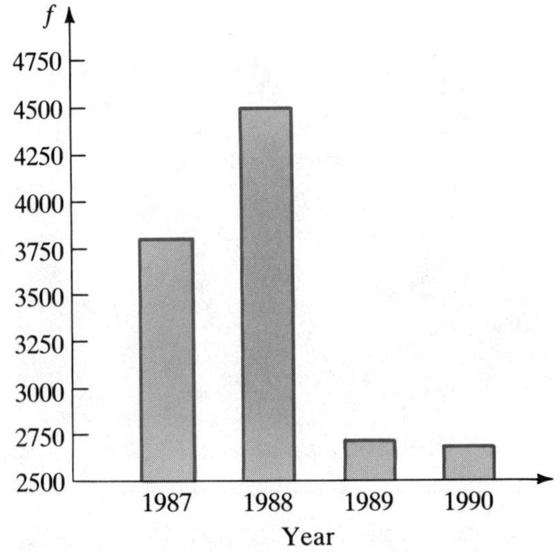

Figure 2.14 Time Bar Graph with Portion of Vertical Axis Omitted

CLASSROOM EXAMPLE 2.4

Presentation of Qualitative Data

A retailer specializing in bedding records each size bed sold during a special, holiday weekend sale. The available sets are single (*S*), extra-long single (*E*), double (*D*), queen (*Q*), and king (*K*). The first forty sales were as follows:

Q	*K*	*D*	*S*	*D*	*D*	*Q*	*Q*
Q	*D*	*D*	*D*	*Q*	*Q*	*S*	*Q*
S	*S*	*D*	*S*	*K*	*Q*	*S*	*S*
D	*D*	*S*	*K*	*E*	*S*	*K*	*S*
D	*D*	*K*	*Q*	*D*	*D*	*Q*	*D*

a. Compile the responses into a one-way tabulation.

b. Construct a bar graph for these data. Label and scale as needed.

c. Suppose a pie chart is desired. Determine the percentage and angle for each category.

d. Construct and label a pie chart for these data.

SECTION 2.4 EXERCISES

2.41 At the end of a management development seminar, the leader handed out an evaluation form that included the following statement: "This seminar will have an immediate impact on my performance as a manager." The seminar participants were asked to react to this statement by checking one of five responses: strongly agree, agree, neutral, disagree, strongly disagree. A list of their responses follows.

Participant	Response	Participant	Response
1	Agree	15	Agree
2	Neutral	16	Agree
3	Neutral	17	Strongly agree
4	Disagree	18	Agree
5	Agree	19	Neutral
6	Agree	20	Neutral
7	Disagree	21	Agree
8	Neutral	22	Agree
9	Neutral	23	Agree
10	Strongly agree	24	Disagree
11	Agree	25	Agree
12	Agree	26	Agree
13	Strongly agree	27	Neutral
14	Agree	28	Neutral

a. Compile these data into a one-way tabulation.
b. If you were the seminar leader, would you consider your seminar successful in terms of making an immediate impact on the participants' performance at their companies, based on their responses? Why?

2.42 A survey was taken to investigate the voting behavior of the American public. One of the questions asked for the respondents' opinion about the following statement: "I feel that one vote really does count." The responses from 1057 people were as follows:

Strongly agree	705
Moderately agree	214
Moderately disagree	64
Strongly disagree	53
Don't know/wouldn't answer	21

Construct a bar graph for these data.

2.43 Refer to Exercise 2.42. In the same survey, respondents were asked to rate the most important race in the upcoming "off-year" election (i.e., no presidential race). The results were summarized into the bar graph in Figure 2.15. Comment on the appropriateness and/or accuracy of this graph.

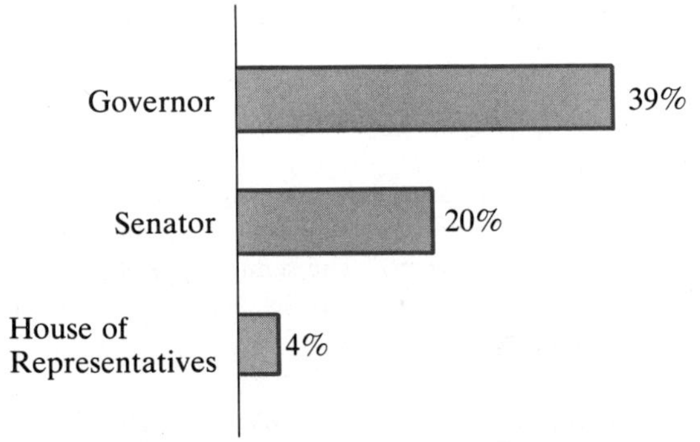

Figure 2.15 What Race We Rate Most Important

2.44 Refer to the fictitious survey in Table 2.2. Suppose a survey of 761 people yielded the following responses to question 4:

Black	179	Oriental	26
White	427	American Indian	2
Hispanic	83	Other	44

a. Determine the percentage for each category relative to the total sample size.
b. Calculate the angle to allot to each category if the data were to appear in a pie chart.
c. Construct a pie chart for these data.

2.45 Sales, in billions of dollars, of personal computers for the second quarter of the year were as follows:

Company	Sales
IBM	4.3
Apple	1.7
Tandon	2.8
Other	1.2

Construct a bar graph for these data.

2.46 Consider the bar graph in Figure 2.16, in which X = type of job held by a sample of 83 minority workers.

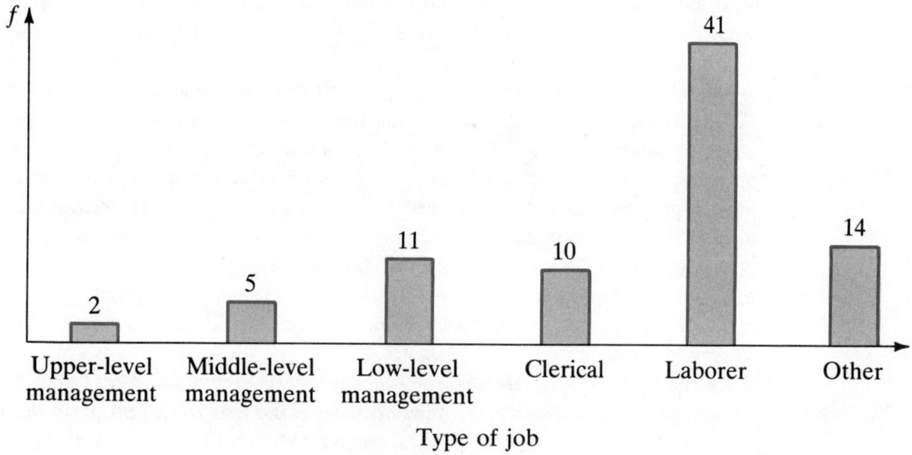

Figure 2.16 Types of Jobs Held by Minority Workers

a. Do the categories satisfy the Principles of Inclusion and Exclusion?
b. Suggest a way to improve the organization and display of the data.

2.47 A restaurant, through computerized billing, tracks the main entrees ordered by its evening customers by the following major categories: beef, chicken, fish, pork, other. The manager listed the following numbers of orders at the end of one particularly busy night.

Entree	Number
Beef	68
Chicken	82
Fish	34
Pork	44
Other	22

Develop a pie chart for these data.

2.48 In 1985 the supermarket sales shares of America's favorite snacks were as follows:

Snack	Share (%)
Potato chips	51.3
Corn/tortilla chips	30.3
Pretzels	8.6
Popcorn	3.5
Fabricated snacks (e.g., Cheese Puffs)	3.0
Other	3.3

Develop a pie chart to depict these data.

2.49 Homeowners traditionally subsidize their local governments through property taxes paid to the county in which they live. The disposition of these monies in a certain county was reported as follows: The state government received 34.09 percent of all tax dollars, the county government kept 17.57 percent, 44.15 percent was allocated to the public schools within the county, and the remaining funds were distributed for fire and sewer obligations. When the county experienced a rapid growth spurt last year, the commissioners proposed a 12% increase in the fire and sewer fund for an additional firehouse and improvements in the water treatment facilities. By law, the commissioners may increase the property tax rate a maximum of one-half cent per $100 of assessed valuation without a voter referendum. The present property tax rate in the county is $.972 per $100 of assessed valuation.

a. Determine the current property tax rate for each category, assuming the percentage of tax dollars is directly related to the tax rate.

b. Find the proposed new tax rate for the fire and sewer allocation.

c. Will a voter referendum be necessary to fund the additional monies needed for fire and sewer service?

d. If the property tax rates for the other categories remain constant and the fire and sewer category is increased by 12 percent, what is the new (total) property tax rate?

e. Suppose the state's share and the county's share of the tax dollars must remain fixed at 34.09 percent and 17.57 percent, respectively, by law. Construct a pie chart for the distribution of tax dollars based on the new (total) property tax rate from part d.

2.50 Sales figures in 1985 from General Foods indicate that consumers purchased 389 million boxes of Jell-O gelatin. The favorite flavor was strawberry with 75.9 million boxes sold. Cherry and raspberry were close behind with sales of 50.6 and 48.6 million boxes, respectively. Orange flavored Jell-O boxes accounted for another 44 million, while a strawberry and banana combination sold 35.8 million boxes.* Construct a bar graph for Jell-O's sales by flavors. After constructing and examining the bar graph, do you find anything unusual about its appearance? If so, suggest a way to improve the appearance of the graph.

2.51 A home furnishings company reported total sales of $1.14 million in the fiscal year just ended. The following figures represent sales by region of the country, plus exports.

*Data taken from "Yelling for Jell-O," October 30, 1986. Copyright 1986, *USA Today*. Reprinted with permission.

Region	Sales ($ millions)
North	.318
East	.368
South	.307
West	.113
Exports	.034

Construct a pie chart to display the company's sales.

2.52 Mutual funds are classified by their investment objective or specialization. Typical categories include long-term growth, income, and sector. A survey of mutual funds produced the following data on the variables *type of mutual fund* and the (most-recent) *one-year return*.

Fund	Type	Return (%)	Fund	Type	Return (%)
1	Long-term growth	5.09	17	Long-term growth	9.05
2	Long-term growth	3.14	18	Aggressive growth	6.04
3	International	71.78	19	International	15.96
4	Income	1.94	20	International	15.64
5	Sector	−.80	21	International	9.74
6	International	10.58	22	Long-term growth	4.46
7	Sector	−.78	23	Income	−.36
8	Sector	5.70	24	Sector	2.65
9	Sector	6.09	25	Long-term growth	15.61
10	Income	1.53	26	Aggressive growth	2.87
11	Long-term growth	12.58	27	Aggressive growth	16.82
12	International	8.50	28	International	14.61
13	Long-term growth	4.60	29	Aggressive growth	6.21
14	Long-term growth	5.81	30	International	13.88
15	Sector	3.24	31	Long-term growth	9.40
16	Aggressive growth	4.81	32	Long-term growth	5.44

Organize the data for the variable *type of mutual fund*.

*2.5 PROCESSOR: FREQUENCY DISTRIBUTIONS

Preliminaries: Creating a Data Set

The Frequency Distributions processor is designed to transform raw, processed data into organized data (see Figure 2.2). Either an ungrouped or a grouped distribution can be created, depending on the wishes of the user. For this and several other processors, the user is required to enter his or her data into a data set before accessing

*Optional

2	4	2	2	4
0	1	3	4	5
2	4	0	4	1

the processor. We will illustrate this processor by transforming the above fifteen-observation sample data set into an ungrouped frequency distribution.

After obtaining the Overview screen (Figure 1.3), touch F8 to clear the screen, and then use the right arrow key to move the highlight to Data. After touching Enter twice, we will be presented with a choice of several types of data sets that we could create. In this case, we wish to use the first type: Multivariate Raw Data. When we touch Enter with the highlight on this option, we will be asked to specify our sample size. Figure 2.17 shows the screen after we have responded 15, but *before* touching Enter. After touching Enter, we will be asked the number of variables: 1, in this case. After we enter 1, a new screen will appear, asking us to name our variable. Let us agree to call the variable *example,* which we now type in. Next we will be asked to supply the 15 individual values, one per row. Figure 2.18 shows the screen after we have entered the fifteenth value (but before touching Enter for this last value). Before touching Enter for this last data value, it may be a good idea to check the correctness of our data entry process by comparing the screen against our original listing. If an error was made, note its location (a problem can be fixed, as we will see momentarily). Now touch Enter twice.

Editing Incorrect Data Entries

Should you need to correct a data entry error, use the View/Edit option within the Data section of the main menu. With the highlight on Data, touch Enter, select the View/Edit choice and then the Data Vectors option. To change a value, position the highlight on the value that is in error. Now touch F5, supply the correct value (use

```
≡        Microcomputer Applications for INTRODUCTORY BUSINESS STATISTICS   EDIT
   Overview [Data] Files  Processors  Set Up  Exit                  | F1=Help

           [Create  ]
            *iew/Edit



                    ═══════════ Data set size ═══════════
                    
                      How many entities in the data set, n =  15
                      
```

Figure 2.17 Preparing to Create a Data Set

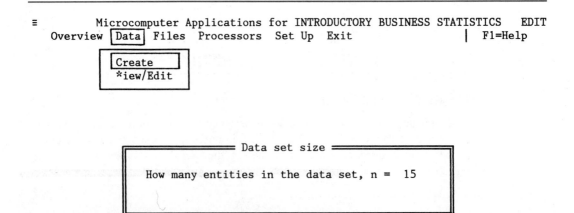

```
≡         Microcomputer Applications for INTRODUCTORY BUSINESS STATISTICS   EDIT
   Overview [Data] Files  Processors  Set Up  Exit                 | F1=Help

            example
    1 2
    2 4
    3 2
    4 2
    5 4
    6 0
    7 1
    8 3
    9 4
   10 5
   11 2
   12 4
   13 0
   14 4
   15▪1
```

Figure 2.18 Screen Appearance After Entering Fifteenth Value

the Delete key if needed to erase any extra digits), and then touch Enter. When all editing is complete, touch Enter to return to the main menu.

■ Saving a Data Set

Our software package has the ability to save data into a file for future reference. To make use of this feature, the user needs to furnish her or his own *formatted data disk* for data storage. (There is very little room for saving files on the two *program disks*, so a *data disk* supplied by the user is necessary for saving data.)

If you wish to save a newly created data set, move the highlight to the Files section of the main menu and touch Enter. Select the Save option on the submenu; Figure 2.19 shows the resulting screen. Of the four available formats, we will use the

```
≡         Microcomputer Applications for INTRODUCTORY BUSINESS STATISTICS   MENU
   Overview  Data [Files] Processors  Set Up  Exit                 | F1=Help
             ┌────────────┐
             │ Retrieve   │
             │[Save]      │
             │ ┌──────────────────────────┐
             │ │[WKS spreadsheet format]  │
             │ │ DBF database format      │
             │ │ TXT fixed length format  │
             │ │ ASCii ", delimited format│
             └─└──────────────────────────┘
```

Figure 2.19 Screen Appearance After Selecting the Save Option on the Files Menu

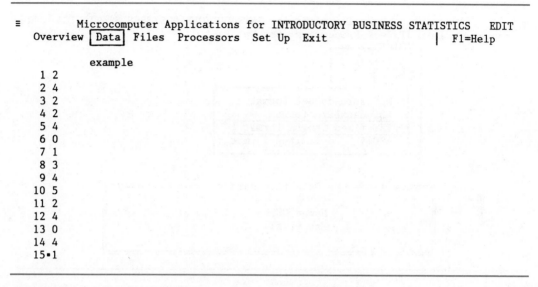

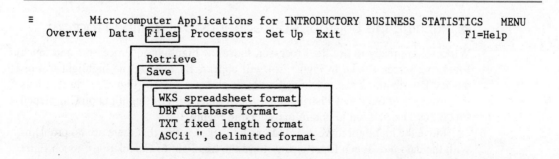

```
≡        Microcomputer Applications for INTRODUCTORY BUSINESS STATISTICS   EDIT
   Overview  Data  [Files]  Processors  Set Up  Exit                   | F1=Help

                 Retrieve
                 [Save]

                   WKS spreadsheet format
                   DBF database format
                   [TXT fixed length format]
                   ASCii ", delimited format

                          ===== Data file name =====

                            Name=[CHP2R    ]
                            Ext =[TXT]
```

Figure 2.20 Naming a Saved Data File

TXT Fixed Length format: Move the highlight to this choice and select it. A box will appear with eight blank spaces inside the brackets next to Name. We choose to name our data file CHP2R (see Figure 2.20).

After we touch Enter, another box will appear, telling us that the default save location is drive A. If you have a computer with two drives, the data disk should be placed in drive B. Correspondingly, we must change the A to B by typing the letter B over the A. Now touch Enter. The computer will proceed to write the data set and file name onto your data disk in drive B for possible retrieval at a later time. Saving these data will not take long. When the "Save Successful" message appears, touch Enter twice to return to the main menu. The data file CHP2R.TXT containing 15 pieces of raw data is not only in the computer's memory awaiting further use, but also is stored safely on a floppy disk.

Running the Ungrouped Frequency Distribution Program

When we are ready to use the Processor, move the highlight to Processors and touch Enter; the screen shown in Figure 1.4 will appear. By moving the highlight to Frequency Distributions and touching Enter, we obtain a screen (Figure 2.21) that lists the two types of frequency distributions. Touch Enter with the highlight on Ungrouped Frequency Distribution to obtain the screen in Figure 2.22.

Move the highlight to Vector, and touch Enter to verify that there are no problems with the data file. Touch Enter again; move the highlight to Calculations; touch Enter twice more. When the Complete message appears, touch Enter and move the highlight to Results. After touching Enter, we will obtain the screen shown in Figure 2.23. If we choose the first option here, we will view a finished product, as shown in Figure

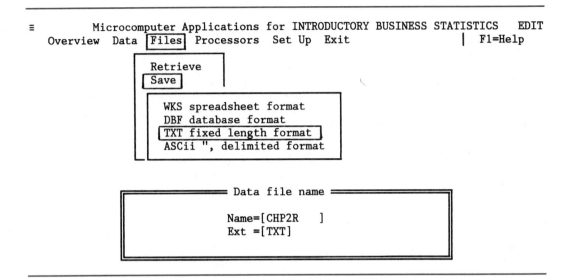

```
≡       Microcomputer Applications for INTRODUCTORY BUSINESS STATISTICS   MENU
   Overview  Data  Files  [Processors]  Set Up  Exit               | F1=Help

                         Simple random sample
                        [Frequency distributions]

                           Ungrouped frequency distribution
                           Grouped frequency distribution

                         One sample inference
                         Two sample inference
                         Analysis of variance (ANOVA)
                         Regression
                         Time series
                         X² (chi-square)
```

Figure 2.21 Options Within the Frequency Distributions Processor

```
≡                    Ungrouped Frequency Distribution                    READY
  [Introduction]  Vector  *alculations  *esults  *ave  Exit         | F1=Help

    This processor generates an
   ungrouped frequency distribution
   from raw data of 10 or fewer
   different X values.

    All data will be treated as
   sample, rather than population,
   data.
    See Chapter 2 for information.
```

Figure 2.22 Introductory Screen for Ungrouped Frequency Distribution Program

```
≡                    Ungrouped Frequency Distribution                     MENU
  Introduction  Vector  Calculations  [Results]  Save  Exit         | F1=Help

                                   Frequency distribution
                                   Intermediate results
                                   Summary measures
```

Figure 2.23 Different Results Options

```
≡                  Ungrouped Frequency Distribution                      READY
   Introduction  Vector  Calculations [Results] Save  Exit          | F1=Help
```

Notation:

f = frequency

rf = relative frequency

cf = cumulative frequency

pos = positions occupied

X	f	rf	cf	pos
0.0000	2	0.133	2	1 to 2
1.0000	2	0.133	4	3 to 4
2.0000	4	0.267	8	5 to 8
3.0000	1	0.067	9	9 to 9
4.0000	5	0.333	14	10 to 14
5.0000	1	0.067	15	15 to 15
SUM	15	1.000		

Figure 2.24 Output from Ungrouped Frequency Distribution

2.24. The other choices in Figure 2.23 provide supplemental items that we will study in Chapter 3; explore them if you wish.

Saving a Frequency Distribution

At this point we have three options:

1. Print the screen shown in Figure 2.24.
2. Write the results down by hand.
3. Use the Save option on Figure 2.24 to place the distribution in a data file.

This Save feature is *not* the same one discussed earlier in this section. We already have discussed saving raw data; now we have the chance to save a set of organized data. If we save the results shown in Figure 2.24, we will be saving the values of X and the corresponding frequencies.

```
≡                  Ungrouped Frequency Distribution                      MENU
   Introduction  Vector  Calculations  Results [Save] Exit          | F1=Help
                                          Save distribution
                                             [WKS spreadsheet format]
                                              DBF database format
                                              TXT fixed length format
                                              ASCii ", delimited format
```

Figure 2.25 Preparing to Save Processor Output

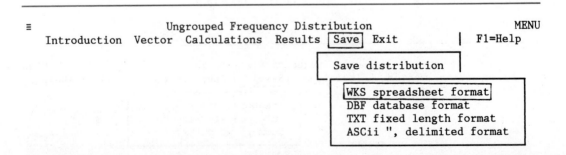

```
≡               Ungrouped Frequency Distribution                         EDIT
   Introduction  Vector  Calculations  Results [Save] Exit        |  F1=Help

                                             Save distribution
                                               WKS spreadsheet format
                                               DBF database format
                                              [TXT fixed length format]
                                               ASCii ", delimited format

                    ======== Data file name ========

                             Name=[CHP2U    ]
                             Ext =[TXT]
```

Figure 2.26 Naming a File to Save Processor Output

With the highlight on Save, touch Enter twice to reproduce Figure 2.25. Again we wish to choose the TXT format; with TXT highlighted, touch Enter. Figure 2.26 shows the screen after we have typed the file name—CHP2U—for storing the organized data. The procedure for saving a frequency distribution will now closely parallel that for saving a data set, as described above. If you would like to see what this data set looks like after it has been saved, return to the main menu and access the View/Edit choice under the Data section.

■ Running the Grouped Frequency Distribution Program

Suppose we wish to develop a grouped frequency distribution with the following data set:

3.4	7.2	9.0	4.3	10.4
10.2	5.5	5.2	7.1	4.7
9.6	11.3	6.9	8.8	3.9

As with Ungrouped Frequency Distribution, we must begin by creating a data set. (*Note:* If you try to create a new data set when a previously created file exists but has not been saved, you will be asked if you are certain that you wish to proceed. Doing so will erase the first data set.)

This subprogram parallels Ungrouped Frequency Distribution with one exception. When you touch Enter with Vector highlighted, an Accuracy Code screen will appear (Figure 2.27). Since this example data set is accurate to one decimal place, we choose the second offering on the list by moving the highlight and touching Enter. By

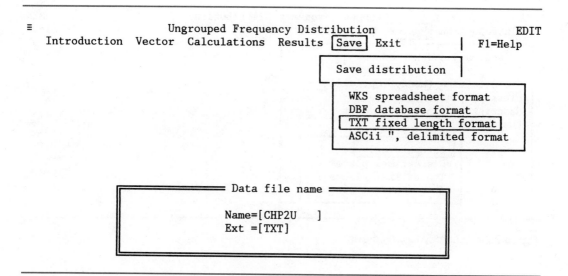

```
≡                  Grouped Frequency Distribution                        MENU
  Introduction [Vector] *alculations *results *ave Exit              | F1=Help

  Accuracy code

  To determine the appropriate
 classes, please select a
 description of the data from
 the menu.

            [Integers only]
             One decimal place
             Two decimal places
             Three decimal places
```

Figure 2.27 Accuracy Code Screen

```
≡                  Grouped Frequency Distribution                        READY
  Introduction *ector Calculations [Results] Save Exit               | F1=Help
```

#	Classes		f	Midpoint	rf	cf	pos
1	3.4	5.4	5	4.40	0.333	5	1 to 5
2	5.4	7.4	4	6.40	0.267	9	6 to 9
3	7.4	9.4	2	8.40	0.133	11	10 to 11
4	9.4	11.4	4	10.40	0.267	15	12 to 15

Figure 2.28 Output for Grouped Frequency Distribution

following the same procedures as for the Ungrouped option, we will be able to arrive at the output shown in Figure 2.28. These results can be saved in a file for future reference if desired. Alternately you may finish your work with this processor by printing the screen or by writing down the results.

2.6 SUMMARY

In this chapter we presented the first step of a descriptive statistical analysis: transforming raw data into organized data. The two main approaches for accomplishing this are creating tabular displays and constructing graphs.

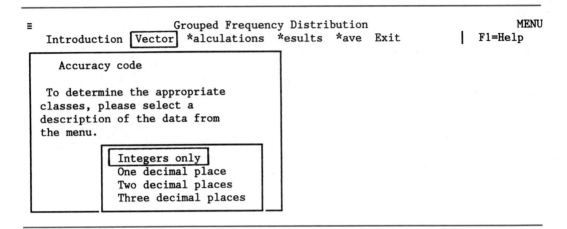

Tabular displays refer to the ungrouped and grouped frequency distributions we introduced in Section 2.2 for quantitative data, and to the one-way tabulations we outlined in Section 2.4 for qualitative data. Graphical techniques for the two types of data include histograms, pie charts, and bar graphs.

Our next challenge will be to further condense processed data into descriptive measures that represent global characteristics of the data. As an analogy, we often read or see capsulized movie reviews. These summaries condense a motion picture that involves hours of direction, production, acting, and so forth, into a rating, perhaps based on a 1- to 4-star system. Similarly, we hope to summarize a data set into one or more informative numbers, called numerical descriptive measures. This is the subject of Chapter 3.

2.7 TO BE CONTINUED . . .

. . . In Your College Courses

Perhaps your next academic encounter with organizing data will occur in a marketing research course. As the name implies, marketing research is an area within marketing that surveys people and explores situations in search of information and relationships. Marketing research may involve a study of consumer behavior, the development of a new product, or a sales analysis with respect to advertising. Each application includes designing a research process, collecting information or data, analyzing it, and summarizing the results into reports, projections, or recommendations. One of the steps in this procedure is the analysis of the data for which some of our Chapter 2 material forms the basis.

To see how business statistics and marketing research dovetail, we will examine two standard textbooks in marketing research. In Chapter 12 of *Marketing Research,* D. S. Tull and D. I. Hawkins discuss ". . . *data reduction,* which refers to *the process of getting the data ready for analysis and the calculation of summarizing or descriptive statistics.*"* This builds on our earlier discussion of data management. Tull and Hawkins cover this aspect and more when obtaining data through surveys and questionnaires. As a standard procedure, they state: ". . . the data for each variable are tabulated as a separate, one-way frequency distribution." Table 12.3 in their text displays an ungrouped frequency distribution with additional columns representing relative frequency and cumulative frequency.

G. A. Churchill, Jr. discusses frequency distributions in Chapter 12 of his textbook *Marketing Research: Methodological Foundations,* although he refers to it as a "one-way tabulation" which we have reserved for qualitative data distributions. He mentions several uses of one-way tabulations:

> *The third use of the one-way frequency tabulation is to determine the* empirical distribution *of the characteristic in question. The distribution often is best visualized through a histogram.*†

*From D. S. Tull and D. I. Hawkins, *Marketing Research: Measurement and Method,* 4th ed. (New York: Macmillan, 1987), © 1987 Macmillan Publishing Company. Reprinted with permission.

†Excerpt from *Marketing Research: Methodological Foundations,* Third Edition, by Gilbert A. Churchill, Jr., copyright © 1983 by Holt, Rinehart, and Winston, Inc. Reprinted by permission of the publisher.

Figure 12.1 in his book shows a frequency histogram representing the distribution of incomes which is ". . . skewed to the right" (synonymous with our phrase "asymmetrical, positive"). A skewed to the right distribution appears in Figure 2.8, panel d. As the passages demonstrate, organizing and graphing data is a much-needed skill in a marketing research class.

. . . *In Business/The Media*

Organizing data is not limited to college courses; managers in all areas of business eventually face this task. The guidelines and examples presented in this chapter should make the job easier and more orderly.

Written reports and oral presentations are responsibilities that every manager must accept. Conciseness and clarity are demanded. Therefore, tables and graphs summarizing data become mandatory supplements. Indeed, these data-reduction tools may have more of an impact than words. If time is short, readers (or viewers) are more inclined to look at a graph or a distribution of numbers than to wade through pages of text.

Think about your own tendencies when you read a magazine or newspaper. Where do your eyes go first? Most people look at the pictures and headlines, and often that's all they see or read of the entire article. For this reason, the media have added color, artwork, very brief titles and labels, and other visual gimmicks to attract the reader's eye. Several publications are known for their accurate, colorful and abundant graphs, including *USA Today,* magazines such as *U.S. News and World Report, Business Week,* and *Forbes,* as well as television shows like "The Nightly Business Report."

Although bar graphs and pie charts dominate, we find other graphs as well. A *statistical map* may be used if the variable relates to geography. To show different categories or levels on a map, we use different colors or shadings. For instance, the way states (or counties within a state) voted in an election—Republican, Democrat, other—is conveyed naturally with a statistical map.

Some graphs appear regularly but do not have a special name. For example, the daily activity of the Dow Jones Industrial Average is published in *The Wall Street Journal.* The graph shows a vertical line extending from the day's low point to the high point with a mark somewhere in between indicating the closing value.

Frequency distributions and histograms also appear in the media, though they may not be labeled as such, or even at all! For example, Figure 2.29 is actually a frequency histogram for an ungrouped frequency distribution. Compare it with Figure 2.4 to see the similarities. The variable is X = number of wins in the first 10 games for the 76 division-winning baseball teams from 1969 through the 1987 season. Eleven values of X—0 through 10—are plotted on the vertical axis and a rectangle is erected to the right of each value, corresponding to its frequency of occurrence. The sum of the frequencies is 76. The main difference in Figures 2.4 and 2.29 is that the axes have been reversed: The values of X appear on the horizontal axis of Figure 2.4.

Rather than diminishing in importance, graphs and tables are exploding in value and prominence. Examples, reflecting our textbook material, abound in the real world,

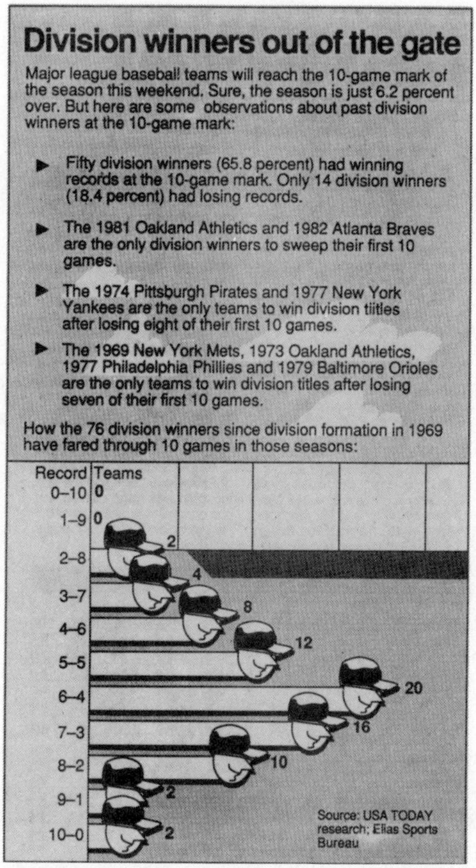

Figure 2.29 Frequency Histogram (Copyright 1988, *USA Today.* Reprinted with permission.)

but you must recognize the subtle differences in construction and in terminology to see the connection. It is almost impossible to go through a business day *without* encountering a graph or table or data.

CHAPTER 2 EXERCISES

2.53 Make up two examples each of a variable producing discrete data and a variable yielding continuous data.

2.54 A telephone survey of 515 eligible voters included the following question: "Which campaign is most important to you this year in your state?" The responses were as follows:

U.S. Senator	253
Governor	211
Congress	41
Other	10

Construct a pie chart for these data.

2.55 In 1985 the soft-drink market accounted for $30 billion in sales. The market share of each company was as follows:

Company	Share (%)	Company	Share (%)
Coca-Cola Co.	38.6	Cadbury Schweppes Inc.	.7
Pepsico Inc.	27.4	Double-Cola Co.	.6
Dr. Pepper Co.	7.1	Faygo Beverages Inc.	.5
Seven-Up Co.	6.3	Frank's Beverages	.4
RJR Nabisco,		Barq's Inc.	.3
(Sunkist, Canada Dry)	4.6	Cragmont	.2
Royal Crown Cola Co.	3.5	Chek	.2
Shasta Beverages Inc.	1.4	Big K	.2
Monarch (Moxie)	1.4	C&C Cola Products	.2
Crush International	1.3	Jos. E. Seagram & Sons Inc.	.2
Squirt Co.	1.2	Cotton Club	.1
A&W Root Beer	1.0	Others	1.7
Dad's Root Beer	.9		

Source: "The soft drink market," *USA Today*, May 27, 1986. Copyright 1986, *USA Today*. Excerpted with permission.

a. Approximately how much revenue did the Squirt Company generate from its soft-drink sales during 1985?
b. What graphical technique(s) is (are) most appropriate for displaying these data?
c. Are these data time-free or time-related?
d. Graph these data. Combine all shares less than 2 percent into an "other" category.

2.56 What is an outlier? Why do outliers present problems when we are organizing data into a frequency distribution?

2.57 A nationwide restaurant chain has thousands of company-owned and franchised stores. The operations department at the home office developed an "efficiency index" that summarizes each store's operation with respect to factors such as budgeting, sales, cleanliness, and so on. A sample of stores yielded the following indices (85.00 is the maximum possible):

73.65	74.01	78.02	63.15	69.73	63.68
58.14	71.25	74.28	73.90	76.26	69.27
73.13	74.20	78.06	71.32	71.88	72.45
70.12	72.35	79.00	56.85	74.53	64.43
70.75	72.11	68.48	70.30	77.53	74.22
78.64	79.01	72.62	78.26	74.89	69.56

70.87	78.63	73.26	79.74	80.20	52.36
66.81	73.53	71.90	81.14	53.88	65.90
79.66	59.58	73.61	78.16	74.08	70.11
60.35	66.92	74.24	59.15	67.92	60.09
51.92	65.70	66.32	54.90	71.54	63.58
60.89	56.56	60.26	53.41	60.03	74.92
71.64	69.43	68.04	69.84	80.14	69.30
67.06	59.56	75.73	74.18	70.20	64.32
73.81	68.80	71.60	71.54	78.81	78.54
57.86	65.35	67.47	73.92	68.22	76.37

a. How many classes should be used to group these data?
b. Determine the value of w.
c. Organize these data into a grouped frequency distribution.
d. Develop the set of cumulative frequencies.
e. Stores with indices below 60.28 are given a warning and placed on probationary status. How many stores in the sample are on probation?

2.58 Refer to Exercise 2.57. Draw a frequency histogram from the grouped frequency distribution.

2.59 How many classes are needed to organize a data set consisting of the following observations:
a. $n = 25$ **b.** $n = 73$ **c.** $n = 116$

2.60 A survey of families living in the western suburbs of a large city was conducted to gain information on the variable X = the number of times in a given week a family ate the evening meal in a restaurant. The data were as follows:

5	0	1	5	4	2	4	2	3	4	1	5	3	3	2	5	1	6
2	4	5	2	3	4	1	5	3	3	3	4	4	2	0	3	3	0
3	2	1	2	4	0	5	2	0	4	2	0	2	4	0	0	5	0
5	1	0	2	6	0	2	4	0	0	3	0	1	5	0	1	2	0
4	3	0	2	1	0	3	1	0	2	0	0	2	4	0	3	1	0
0	7	0	4	2	1	2	1	0	3	1	0	1	1	0	2	1	2

a. The variable X produces what type of data?
b. To organize these data should we use an ungrouped or grouped frequency distribution? Why?
c. Organize these data.
d. What proportion of the families sampled eat out one or fewer times per week?

2.61 Consider the relative frequency histogram in Figure 2.30 for a survey asking people how busy their work week is. Of 1004 people sampled, 3 percent gave no response.
a. Why doesn't the sum of the relative frequencies total 1.00?
b. The last class has no endpoint. Is this merely an oversight or is there another explanation?
c. Approximately how many workers spend less than 20 hours per week on job-related activities?

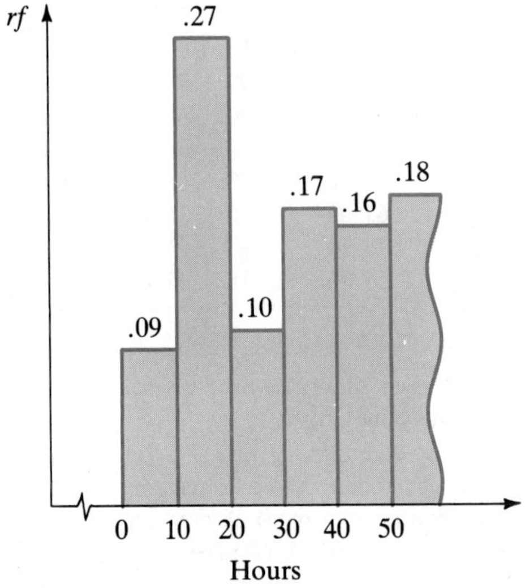

Figure 2.30 Hours Spent on Job-Related Duties

2.62 The owner of a coin-operated laundry business was concerned about repeated complaints from his customers that his washing machines did not have enough hot water. When he purchased his equipment—32 machines and one boiler—from the wholesaler, he was assured that the boiler was of adequate size to provide sufficient hot water to all his machines. The wholesaler's recommendation was based on the national average of approximately 25 percent of all machines in actual use at any one time in a coin laundry. The owner decided to spot check his business, counting the number of washing machines in use at random times. He recorded the following data:

23	0	14	6	14	7	9	22	5	19	4
14	12	4	3	23	8	3	0	14	14	6
3	5	19	25	11	27	18	1	9	19	2
10	24	14	6	6	25	17	8	10	9	12
9	1	9	10	6	16	15	23	11	15	7
12	16	2	11	7	22	4	3	4	1	

a. Organize these data into a frequency distribution.
b. Construct a frequency histogram for these data.
c. Does the national average appear to be a representative figure for the actual equipment use in this laundromat? Explain.

2.63 A survey reported in the October 22, 1986, issue of *USA Today* asked adults in the United States and in Japan how they think people become rich. The results were reported in the pie charts shown in Figure 2.31.

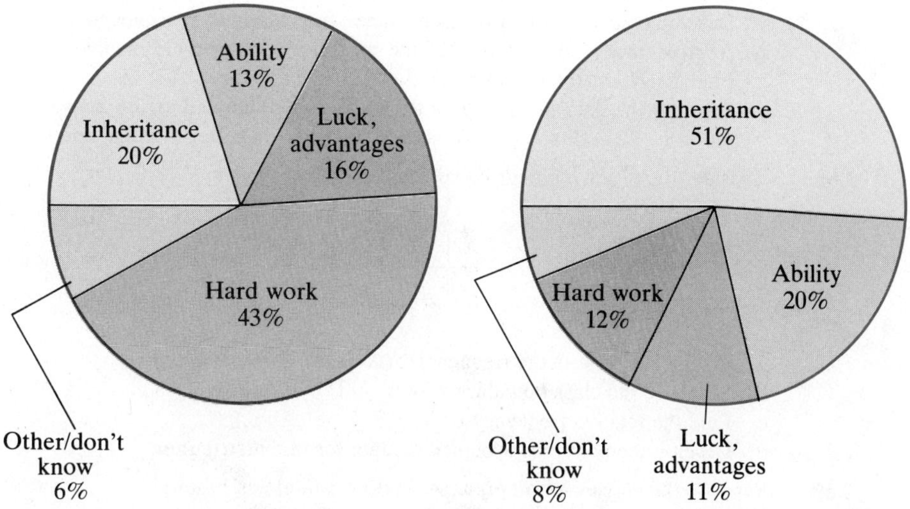

Figure 2.31 Different Countries' Views as to How People Become Rich (From "Views on rich are a world apart," *USA Today*, October 22, 1986.)

a. Comment on how well the five categories satisfy the Principles of Inclusion and Exclusion.

b. The countries corresponding to each graph were purposely omitted. Which graph do you think represents the U.S. and which represents Japan? Why?

2.64 Why isn't the Principle of Exclusion alone a sufficient guideline for organizing data?

2.65 Suppose the minimum value in a data set is 7.7 and the maximum is 24.6. Determine the class width necessary to group the data when n has the following values:

a. $n = 20$ **b.** $n = 45$ **c.** $n = 75$

2.66 Refer to Exercise 2.65. Identify the endpoints of the first class for each part.

2.67 The director of data processing for a local company sought information on the length of time to complete projects within the department. One Monday morning, at the weekly staff meeting, the director asked the manager of applications programming to select a random sample of 50 project requests from the programming log book and organize the data. The variable of interest was the number of hours required for each project. Unfortunately, one of the entries in the log book was indecipherable due to a coffee stain. The remaining data were as follows:

4	16	29	41	46	57	84
4	22	30	42	47	60	88
5	26	31	42	48	61	93
6	26	33	42	48	67	105
7	27	37	43	49	68	126
9	28	38	43	51	73	129
12	28	39	45	56	77	221

a. Create a grouped frequency distribution from these raw data.
b. Develop columns of relative frequencies and cumulative frequencies.
c. Approximately 85.8 percent of the projects in the sample involved less than X hours. Find X. (*Hint:* Refer to part b.)
d. Regroup the data using 129 as the maximum value and create an open-ended class for the value 221. How is this frequency distribution better than the one in a?

2.68 Consider the following ungrouped frequency distribution:

X	1	2	3	4	5	6
f	10	17	15	8	?	2

a. If $n = 59$, what is the frequency of $X = 5$?
b. What are the class boundaries for $X = 1$?
c. Find the relative frequency of $X = 3$.
d. Develop the positions occupied column for this distribution.

2.69 Suppose the classes for an organized set of data are as follows:

.235– .875
.875–1.515
1.515–2.155

a. What are the midpoints for each class?
b. What is the value of w?

2.70 What quantity is usually plotted on the vertical axis of the following types of diagrams?
a. frequency histogram
b. bar graph
c. relative frequency histogram

2.71 A manufacturer of automotive batteries carefully monitors the percentage of lead in its Oxide Manufacturing Area. Oxide (powdered lead), when mixed with water and sulfuric acid, produces a paste that is applied to the lead skeleton (called a "grid") of the battery plate. A high grid weight is very costly since over 60 percent of the total cost of a battery is lead. The desired percentage of lead is between 25.4 percent and 32.2 percent. A random sample of 63 grids revealed the following data on the percentage of lead:

31.6	29.2	24.8	28.0	25.6	30.0	32.8
25.6	30.0	29.2	26.0	28.2	28.0	26.0
33.6	28.0	30.8	25.6	30.0	29.2	25.2
30.0	30.0	29.6	31.2	34.0	32.0	26.8
27.6	28.0	30.0	28.0	30.0	27.6	30.0
28.4	29.6	29.6	32.0	28.0	26.4	27.6
23.6	27.6	27.2	30.4	26.4	31.2	26.4
24.0	28.0	28.8	20.8	25.2	33.2	27.2
26.0	27.6	27.2	30.0	23.2	33.6	27.6

a. Organize these data into a grouped frequency distribution.

b. The manufacturer hopes that only 5% of its grids yield a percentage of lead outside the desired range. Based on this sample data are they achieving this objective? Explain.

2.72 Suppose the classes for an organized set of data are as follows:

11.2–12.9
12.9–14.6
14.6–16.3

Into which class is the value 14.6 assigned?

2.73 Retail sales are known to be affected by the calendar. A shoe boutique showed the following dollars of total revenue for each day indicated:

Day	M	Tu	W	Th	F	Sa	Su
Revenue	$1613	$702	$1477	$1865	$1629	$1944	closed

Develop a bar graph for these data.

2.74 Due to rising costs many medical insurance companies will pay surgical fees for minor operations only if the operation is performed on an outpatient basis. Consequently, hospitals have streamlined their systems and procedures for handling outpatients. One of the variables that is closely monitored is the number of outpatient surgeries scheduled per day. A sample of thirty-eight days yielded the following data:

7	5	3	2	8	3	3	6	7	1
1	3	5	6	5	5	2	1	7	5
3	3	3	7	1	2	2	1	4	4
5	5	1	6	6	5	4	4		

a. Organize these data into a frequency distribution.

b. Construct a frequency histogram.

c. Develop the column of cumulative frequencies.

d. What percentage of the days had 4 or fewer outpatient surgeries scheduled?

e. For each value in the distribution indicate the positions occupied.

2.75 As of September 30, 1986, the total number of U.S. military active duty personnel was 2,169,112, broken down as follows:

Army	780,980
Air Force	608,199
Navy	581,119
Marine Corps	198,814

Graph these data on a bar graph.

2.76 Consider the following grouped frequency distribution:

Classes	f
70.5– 76.5	6
76.5– 82.5	8
82.5– 88.5	11
88.5– 94.5	4
94.5–100.5	1

a. Is it possible to determine the accuracy of the raw data? If so, what is it?
b. Is it possible to determine the numerical value of the minimum value in the data set? If so, what is it?
c. Develop a column of relative frequencies.
d. Find the midpoint for each class.
e. What is the numerical value of w for this frequency distribution?
f. Draw a relative frequency histogram.
g. Develop a column of positions occupied.

2.77 Suppose a data set contains 74 numbers with a minimum value of 13.18 and a maximum value of 19.77.

a. Determine the appropriate number of classes needed to organize these data.
b. Find the value of w.
c. Generate the classes.
d. Develop the midpoints for each class.

2.78 Consider the graph in Figure 2.32 for a set of sample data.

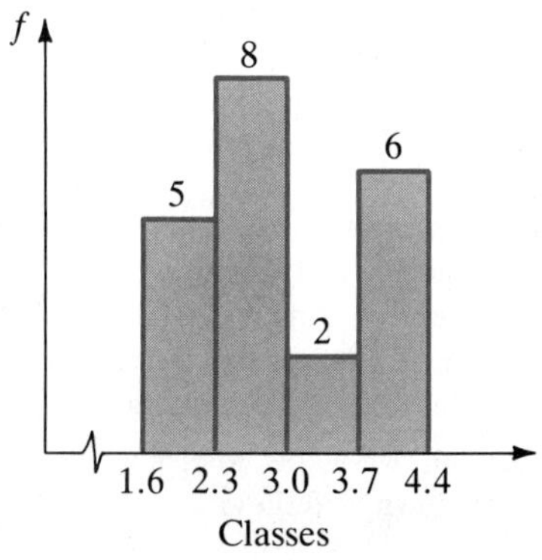

Figure 2.32 Graph of a Grouped Frequency Distribution

a. What is the name of this type of graph?
b. What size was the sample?
c. What is the cumulative frequency at 3.7?

2.79 The U.S. Travel Data Center supplied the pie charts in Figure 2.33 to illustrate the method of travel for a sample of people during the most recent three-month winter season and summer season.

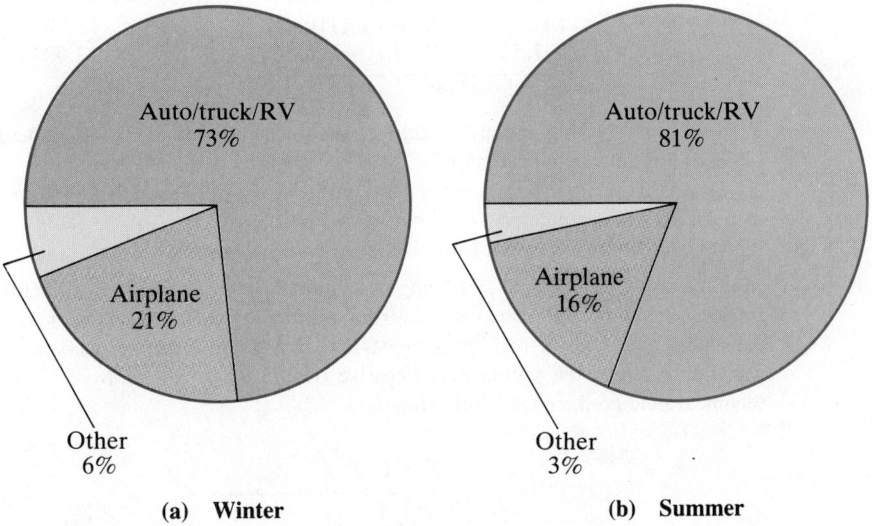

Figure 2.33 Method of Travel (From "USA Snapshots," *USA Today,* November 10, 1986.)

a. Express in words the variable being depicted in these graphs.
b. What type of variable is it?
c. Do the categories in the pie charts satisfy the Principles of Inclusion and Exclusion?
d. Name other modes of travel that fall into the "Other" category.
e. What could have been done in creating the pie charts to provide the viewer with more information about the variable under study?

2.80 Scheduling the number of crew persons per shift in a fast food restaurant is often facilitated with guidelines set forth in a procedures manual from the home office. For example, one company recommends the following crew sizes based on hourly sales:

Sales ($)	Crew
0– 60	3
60–104	4
104–111	5
111–177	6
177–210	7

At a particular franchise, the manager decided to see whether her present staff of four people was sufficient for the morning shift (6:30 a.m. to 10:30 a.m.). For a one-week (weekday) period she recorded the following hourly sales figures:

\$128.75	\$100.25	\$136.75	137.75	\$106.75
112.50	111.25	103.00	158.00	102.00
101.50	128.50	110.25	163.25	115.50
97.50	118.50	106.50	98.25	117.00

a. According to the company guidelines, determine the number of crew people needed for each hour of sales listed above.
b. Organize the crew sizes needed from part a into a frequency distribution.
c. Generate the cumulative frequencies.
d. Do you think her present staff size is sufficient? Explain.

2.81 Monitoring the health and well-being of a company is accomplished, sometimes naively, by tracking the company's performance from quarter to quarter in terms of net profit. Let the variable X represent the number of quarters during the most recent one-year period in which a company's reported profits were negative (i.e., a loss). A sample of 23 firms in the bio-technical field produced the following data:

X	0	1	2	3	4
f	3	11	2	5	2

a. Develop the cumulative frequencies for this distribution.
b. What percentage of the companies sampled operated in the red for at most one quarter?
c. Some analysts form the ratio of the value of X for each company to the industry-wide average to gauge how far out-of-line a company's track record is. If this ratio exceeds 2.1, the general conclusion is that the company is "financially troubled." What would the industry-wide average have to be in order for the two companies with X-values of 4 to be declared troubled?

2.82 Refer to Exercise 2.52. Organize the data for the variable *one-year return*.

2.83 Give several reasons why we wish to organize and graph a set of data.

2.84 Identify the data generated by the following variables as qualitative or quantitative. If the data are quantitative, identify the type as discrete or continuous:
a. number of new residential electrical "hookups" by a utility company per week.
b. amount of pressure, in pounds, a cardboard box can withstand.
c. weight of a new car.
d. type of zoning per parcel of land.
e. monthly amount of barrels of oil imported by the United States.

2.85 The owner of the franchising rights for a fast food business was considering building another restaurant fairly close to an existing one. However, he feared that the proposed site would take business away from his existing operation. To investigate this, he hired a consulting firm to survey customers at the current restaurant. The survey was conducted in two parts: personal, on-site interviews with customers and telephone calls to residents in the surrounding community. The personal interviews were taken in two-hour intervals at the breakfast,

Table 2.4 Survey Data for Exercise 2.85

Meal	Response	Meal	Response	Meal	Response
breakfast	home	dinner	home	lunch	work
breakfast	work	breakfast	home	lunch	home
breakfast	work	breakfast	other	breakfast	work
breakfast	home	dinner	work	lunch	home
breakfast	home	lunch	shopping	lunch	work
lunch	home	breakfast	home	dinner	home
dinner	work	breakfast	work	lunch	work
dinner	shopping	lunch	work	breakfast	home
dinner	home	breakfast	home	breakfast	home
lunch	home	dinner	work	breakfast	other
lunch	home	lunch	home	lunch	work
lunch	work	breakfast	other	breakfast	home
lunch	work	breakfast	home	lunch	other
breakfast	home	breakfast	home	lunch	other
breakfast	other	breakfast	work	lunch	work
lunch	work	breakfast	home	lunch	work
breakfast	home	lunch	work	breakfast	home
breakfast	home	lunch	work	lunch	home
lunch	work	dinner	home	lunch	home
breakfast	home	breakfast	home	lunch	work

lunch, and dinner mealtimes. One of the survey questions was "From where did you come before visiting this restaurant?" Possible answers were work, home, school, shopping, other. Some of the data obtained from the survey are shown in Table 2.4.

a. Organize the data for the variable *meal* into a one-way tabulation.

b. What percent of the survey responses listed involved breakfast customers?

2.86 Refer to Exercise 2.85. Organize the responses to the question "From where did you come before visiting this restaurant?" into a one-way tabulation. Where did most of the customers come from before visiting the restaurant?

REFERENCES

Campbell, S. K. 1974. *Flaws and Fallacies in Statistical Thinking*. Prentice-Hall Inc., Englewood Cliffs, NJ.

Churchill, G. A., Jr. 1983. *Marketing Research: Methodological Foundations,* 3rd Edition. Dryden Press, Hinsdale, IL.

Huff, D., and I. Geis. 1954. *How to Lie with Statistics*. Norton, NY.

Tull, D. S., and D. I. Hawkins. 1987. *Marketing Research: Measurement and Method,* 4th Edition. Macmillan Publishing Company, NY.

Chapter Maxim *The mean and standard deviation are interpreted most easily when a data set is approximately mound shaped and symmetrical.*

CHAPTER 3

NUMERICAL DESCRIPTIVE MEASURES

*Optional

Objectives

After studying this chapter and working the exercises, you should be able to

1. Differentiate between a parameter and a statistic.
2. Associate a symbol with each measure we study.
3. Compute and interpret each measure.
4. Be able to classify each measure by its purpose: central location, spread, shape, or position.
5. Use the relationships between certain measures as reasonability checks on computations.
6. Create a mental image of the distribution of a data set from its measures.
*7. Execute the Descriptive Statistics processor to verify answers and/or solve problems.
8. Recognize applications of descriptive statistics in marketing research.

This may be the most important and most referred to chapter in the book because on the following pages we present the basic building blocks of descriptive statistics.

We will present a variety of numerical ways to summarize quantitative data. The need to do so is self-evident: We simply cannot deal effectively with a large collection of numbers. For example, suppose we had purchased 100 shares of common stock in a well-known company at $35 per share. At the close of each business day, we could record the price of the stock. If we recorded the closing price for three months, we would generate approximately 65 values like 35, 35⅛, 34½, and so on, which could be listed on a sheet of paper. If someone should ask about the progress of our investment, we could whip out our sheet of paper and show them 65 numbers. Would they be impressed? Probably not, because the original, raw data are too voluminous (and boring) to be of much help. More than likely, our inquisitive friend is merely interested in a *summary* of the data, as opposed to the day-by-day details. To describe our investment, we would be more likely to use short, one-word descriptions like up, down, or unchanged. Certainly, "up" may be an oversimplification of 65 days of closing stock prices, but it does convey the message succinctly. This is the goal of descriptive statistics in general and this chapter in particular: to describe or summarize a set of data in concise terms.

*Applies to optional section.

3.1 PRELIMINARIES TO NUMERICAL DESCRIPTION

Think about how we use numbers for description. In everyday life, we do this when we use height and weight to characterize a person. Specifying these two quantities can conjure up an image in our minds of the stature of the individual. Six feet two inches tall and a weight of 220 pounds suggests a "big" person, while a height of five feet two inches and a weight of 100 pounds implies a "petite" person. By the end of this chapter, we hope that certain statistical measures will evoke images of data sets just like height and weight evoke images to describe people.

Along the way we must take care not to confuse a set of sample data with a set of population data. Summary measures of a population are called *parameters*, while summary measures of a sample are called *statistics*. For example, suppose a population consists of the values {1,2,3,4,5} while a sample of size $n = 2$ from this population produced the values {1,5}. One possible descriptive measure is the *total* of all the values in the data set. Since it characterizes a population, the total of 15 is a parameter. The sample total of 6 is a statistic.

Definitions

A **parameter** is a numerical descriptive measure of a population.

A **statistic** is a numerical descriptive measure of a sample.

To distinguish population and sample measures, we often employ letters of the Greek alphabet for parameters and letters of the English alphabet for statistics. Not all descriptive measures have widely agreed upon symbols; thus, some quantities will not be given their own special notation. This presents a problem when it is not clear whether a data set is a sample or a population. We approach this chapter with the belief that you are more likely to encounter sample data sets than population data sets. The numerical descriptive measure presented will be explicitly defined for a sample, not the population.

SECTION 3.1 EXERCISES

3.1 For each of the following situations, indicate whether the numerical descriptive measure (italicized) is a statistic or a parameter.

a. A group of five test scores is selected from a class of 52 students, and the *total* of the five scores is computed.

b. The cost of a one-gallon container of low-fat (2 percent) milk is recorded for a randomly selected group of four major grocery stores in a medium sized city. The *average* of these prices is then computed.

c. To determine the popularity of "car pooling," a person stations himself one day on a bridge over a major commuter route during rush hour into the city and counts the number of people in every tenth car. The *percentage* of cars with more than one occupant is determined.

d. A company, offering a one-time $5 rebate on its top-of-the-line product for a period of three months, prints and distributes to retail stores 100,000 rebate certificates. After the deadline expires, the company determines the *percentage* of certificates redeemed.

3.2 Define a parameter and a statistic.

3.3 In the December 1, 1986, issue of *The Sporting News* there was an article about a survey conducted by the Cleveland *Plain Dealer* newspaper. To gauge the recognition level of sports personalities, the *Plain Dealer* ran side-by-side pictures of Billy Packer and Nelson Burton, Jr., and asked readers to write in and identify them. Four readers correctly identified Packer as a CBS college basketball commentator; 37 people recognized Burton as a commentator for the Professional Bowlers Association telecasts.

a. Is this survey a census or a sample? A random sample?

b. Should we consider the *total* number of readers who recognized each personality as a statistic or a parameter? Why?

3.4 If a nonrandom sample is selected, can we still call a numerical descriptive measure of it a "statistic," or does a "statistic" only refer to a quantity that describes a random sample?

3.5 Salary data on licensed real-estate agents in the state of Louisiana is compiled by a team of newspaper reporters for use in an upcoming feature story on careers in the real-estate industry within the state. Suppose variables X and Y have the following definitions:

X = Salary of licensed agents employed by a specific company operating in Louisiana

Y = Salary of licensed agents in Louisiana

a. Are the following measures statistics or parameters relative to the state-wide focus of the article?

(1) largest value of X

(2) percentage of Y-values exceeding $20,000

(3) average value of Y

b. For each measure in part a that you identified as a statistic, explain how the measure could be considered a parameter by redefining the population.

c. For each measure in part a that you identified as a parameter, explain how the measure could be considered a statistic by redefining the population.

3.2 MEASURES OF CENTRAL LOCATION

Numerical descriptive measures are classified into various categories, depending on the characteristics of the data set they describe. There are two major categories—central location and dispersion—and several minor categories, including position and shape. In this section, we present the common measures of central location.

As the name implies, measures of central location try to define the center of the data set. However, the word *center* has a variety of meanings: the point halfway between the largest and smallest piece of data, the point assuring that half the data lies on its either side, and the point representing the center of gravity. Each definition of "center" produces a different central location measure. We will examine three popular measures—mode, median, and mean—in the text and present two others as optional material in Exercises 3.20 and 3.23 at the end of this section.

Mode

The easiest way to describe central location is to determine which, if any, value occurs most frequently. If one particular number appears again and again, its repetitious behavior will force the data set to pile up at or near this point. For example, one question in a survey at a predominantly commuter university asked respondents to give the distance in miles from their place of residence to the university. A majority of people responded "10 miles," not 10.2 or 9.33, but 10 miles. Curiously, an exact ten-mile radius from the university drawn on a map intersected very few residential communities, yet ten was the most common response. In this example, the value 10 is called the modal response or simply the *mode*.

Definition

The **mode**, denoted by *Mo*, is the value in the data set that occurs most frequently.

There is no formula per se to use in finding the mode. Simply count the number of occurrences of each value and identify the value with the largest frequency. Sometimes the frequencies already exist in the form of a frequency distribution, making identification of the mode a simpler task. Two examples illustrate the procedure for finding the mode.

EXAMPLE 3.1

The following sample data were obtained for the variable X = the number of telephones per residence:

1	4	1	0	2	1	1	3	2	9	1	2

Find the mode.

Solution:

Since the number 1 occurred five times, which is more than any other frequency, the mode is the value 1. Note that the number 5 is *not* the mode. The number 5 represents the mode's frequency of occurrence.

□

EXAMPLE 3.2

A random sample of size 37 produced the following ungrouped frequency distribution:

X	0	1	2	3	4
f	12	1	6	11	7

Find the mode.

Solution:

Since the data are already organized into a frequency distribution, it is easy to locate the largest frequency of 12. The mode is therefore the value 0.

□

COMMENTS

1. If all the values in the data set occur with equal frequency, we say there is no mode.
2. If two of the values in the data set have the same frequency, and this frequency is greater than any of the others, then the data set is said to be *bi-modal*. For example, in Example 3.2, if the frequency for the value 3 had been 12 instead of 11, then the data set would have been bi-modal with the two modes of 0 and 3. Rarely are there extensions to the case of three or more modes.
3. If the data are organized into a grouped frequency distribution, the *modal class* is defined to be the class with the largest frequency. The *mode* is the midpoint of this class.

Obviously, the attractiveness of the mode is its simplicity. Yet, this positive feature also can be its major drawback. As the frequency distribution of Example 3.2 shows, zero has the largest frequency, but zero is hardly the perceived "center" of the data set. In some sense then, it might be more accurate to describe the center as being a point with values above and below it. A second measure of central location—the median—takes into account the positioning of the values rather than just the value with the largest frequency.

Median

Anyone who has driven a vehicle on one of the interstate highways in the United States probably has seen the road sign "Keep Off the Median." In this case the median is that strip of land or concrete that separates the two halves of the highway. The term *median* has a similar purpose in business statistics: to separate a data set into two halves. First, however, the data must be arranged in ascending (or descending) order.

Definition

The **median,** denoted by *Md,* is a number that divides an ordered data set in half.

As with the mode, no specific formulas are necessary to calculate the median for raw or ungrouped data. Actually it might be more realistic in practice to say that usually we *find* the median rather than calculate it.

When the size of the data set is an odd number, like 7 or 13, the median is uniquely defined: It is the number occupying the $(n+1)/2$th position. For $n = 7$, the median is the number in the fourth position; for $n = 13$, the median is the number in the seventh position. In both cases, there is an equal number of values above and below the median.

EXAMPLE 3.3

A random sample of nine savings institutions produced these data for X = percent return on 180-day certificates of deposit:

6.25	6.40	6.40	6.40	6.50	6.25	6.38	6.37	6.40

Find the median.

Solution:

Since $n = 9$, the median is the value occupying the $(n+1)/2$ = fifth position. At present, the value in the fifth position is 6.50, but this value is *not* the median because the data have not been arranged in order of magnitude. The ordered data are as follows:

Pos	1	2	3	4	5	6	7	8	9
Value	6.25	6.25	6.37	6.38	6.40	6.40	6.40	6.40	6.50

The median is 6.40, the value of X in the fifth position.

□

When the size of the data set is an even number, such as 6 or 20, the halfway point is somewhat ambiguous. To maintain consistency with the notion that the median occupies the $(n+1)/2$th position, we define the median as the number halfway between the two middle values. For $n = 8$ the median is the number halfway between the fourth and the fifth ordered values. In Example 3.3, for instance, if the value 6.50 in the ninth position were deleted from the data set, the median of the remaining eight values would be 6.39, the number halfway between 6.38 and 6.40.

In the case of an ungrouped frequency distribution, we find the median by utilizing the positions occupied (pos) column, searching for the $(n+1)/2$th position. As before, if n is odd, there will be exactly one such position. If n is even, we must determine the values at the two consecutive positions on either side of the $(n+1)/2$th position.

EXAMPLE 3.4

Find the median for these organized data:

X	f	cf	pos
0	12	12	1 through 12
1	1	13	13
2	6	19	14 through 19
3	11	30	20 through 30
4	7	37	31 through 37

Solution:

The positions occupied column (pos) reinforces the idea that each value occupies a position in the ordered data set. For example, the six occurrences of the value 2 occupy positions numbered 14 through 19. Since $n = 37$, the median will reside in the $(n+1)/2 =$ nineteenth position. One of the occurrences of $X = 2$ is in the nineteenth position; thus, $Md = 2$.

□

Grouped frequency distributions create more problems in finding the median because we cannot determine which specific value occupies each position. Consider the following grouped frequency distribution:

Classes	f	cf	pos
4– 7	5	5	1 through 5
7–10	8	13	6 through 13
10–13	6	19	14 through 19
13–16	1	20	20

With a sample of 20 observations, we would declare the median to be the value occupying the $(n + 1)/2 =$ 10.5th position. Although the 10.5th position is found in the class 7–10, without seeing the raw data, we do not know how the frequency of 8 for this class is allocated to the values within this class. For this reason, we cannot say exactly which value of X occupies the 10.5th position. Therefore, for grouped data, we define the median as the point that divides the total area of an accompanying histogram in half. This point is *not* necessarily a *position* number; it represents a dividing line on a graph. It is necessary in order to deal with the *distribution* of the grouped data rather than the raw data. Figure 3.1 illustrates this concept.

Finding such a point could be done in a hit-or-miss manner involving areas and geometry, or we could rely on Equation 3–1, which is an interpolation formula. It prorates the median class to accomplish the fifty-fifty division of area pictured in Figure 3.1. An assumption of this formula is that the values are evenly distributed across the median class.

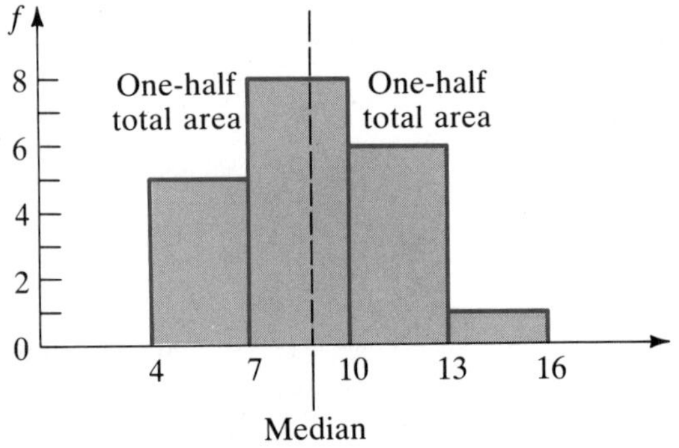

Figure 3.1 Frequency Histogram Showing Location of the Median

Median of a Grouped Frequency Distribution

$$Md = L + \frac{w}{f_m}\left(\frac{n}{2} - cf_p\right) \tag{3–1}$$

where

L = lower class endpoint of *the class containing the median*
w = the width of *the class containing the median*
f_m = frequency of *the class containing the median*
n = the sample size
cf_p = cumulative frequency preceding *the class containing the median*

Of the five quantities defined for Equation 3–1, four of them—L, w, f_m, and cf_p—involve a repeating phrase: the class containing the median. If we can pinpoint this class, the computation of the median will reduce to an arithmetic problem.

Initially a methodical, step-by-step procedure might be helpful in working with Equation 3–1. The following box lists these steps and also demonstrates each step using the data in the preceding grouped frequency distribution.

For the data in the grouped frequency distribution, the median is 8.875 or, rounded off, 8.9. Neither 8.875 nor 8.9 is likely to be the true value of the median of the raw data. Anytime we group the data, some accuracy is lost. Although none was sacrificed in computing the median via Equation 3–1, the answer 8.875 is *exact*

Step-by-Step Procedure for Calculating the Median from a Grouped Frequency Distribution

Step	Example
1. Compute $n/2$	$n/2 = 10$
2. Search the pos column for the value obtained in step 1.	
a. If $n/2$ corresponds to the last position occupied within a class, *stop*. The median is the endpoint of that class.	The number 10 is not the last position occupied within any class.
b. If $n/2$ does not correspond to the last position occupied within a class, *continue*. Find the class that contains the $n/2$th position occupied. This is *the class containing the median*.	The tenth position occupied is in the 7–10 class. *The class containing the median* is 7–10.
3. Identify L.	$L = 7$
4. Identify f_m.	$f_m = 8$
5. Read cf_p.	$cf_p = 5$
6. Determine w.	$w = 3$
7. Evaluate Equation 3–1.	$Md = 7 + (3/8)(10 - 5) = 8.875$

relative to the frequency distribution and histogram but must be regarded as *approximate* relative to the unseen (raw) data set.

COMMENTS

1. For raw data and ungrouped frequency distributions, the median involves no complex calculations; thus, there is no round-off rule needed. For grouped frequency distributions, our suggestion is to round off, if necessary, the computations to one more decimal place accuracy than the classes. In the preceding example, the classes were in integers; the median was reported as 8.9, in tenths.
2. For grouped frequency distributions, if the value of $n/2$ falls between the last position of one class and the first position of the next class, use the latter as the class containing the median. For instance, suppose the class 40–60 includes positions 6 through 13, while the next class 60–80 includes positions 14 through 19. If $n = 27$, the value of $n/2$ is 13.5, or halfway between positions 13 and 14. Therefore we use the class 60–80 as the class containing the median. Exercise 3.15 demonstrates this situation.
3. Because the grouped data formula for the median divides a geometric figure into two equal areas, its formula uses $n/2$ instead of the $(n + 1)/2$ term needed to isolate the middle position of ungrouped data.

The median appears frequently in the print and TV media. It is a popular measure of central location because of its relative stability. To see this, look at the data in Example 3.1. The number 9 is somewhat unusual relative to the other numbers in the data set. If the 9 had been a 90, the median still would have been 1.5. This insensitivity to very large or very small numbers is an attractive feature of the median. Perhaps its biggest limitation is that the data must be ordered. Even with computer programs, rearranging a large data set is time consuming. For most large data sets, the median often is close to the value of another measure of central location: the mean, which is easier to compute and use.

Mean

How many times have you asked an instructor what the *average* score was on a test you took? The information you requested is perhaps the most well-known numerical descriptive measure in business statistics. The average of a data set is synonymous with the mean (or arithmetic mean) of that data set.

Definition

The **mean** is the total of the values in a data set divided by the number of values.

Due to the pervasiveness of the mean throughout statistics, its symbolism is special. (See Table 3.1.) The Greek letter μ (mu) denotes the population mean, while the symbol $\overline{X}$ represents the sample mean.

In computing the mean we must total or add all the values in a data set. To express this process concisely, we use *summation notation*, a widely used method of symbolizing addition. A more complete explanation of summation notation can be found in the accompanying *Student Solutions Manual*. For raw data, the sample mean is computed according to Equation 3–2.

Table 3.1 Symbols for the Mean

Measure	Symbol
Population mean	μ
Sample mean	$\overline{X}$

Mean for Raw Data

$$\overline{X} = \frac{\Sigma X}{n} \tag{3–2}$$

where

X = any value in the data set
n = sample size

EXAMPLE 3.5

A sample of $n = 5$ shoppers produced the following data on X = the amount of money spent on groceries:

\$30.95	\$5.51	\$96.23	\$115.72	\$60.25

Find the average amount of money spent on groceries.

According to Equation 3–2, the mean is

$$\overline{X} = \frac{\Sigma X}{n} = \frac{30.95 + 5.51 + 96.23 + 115.72 + 60.25}{5}$$
$$= \frac{308.66}{5} = \$61.732 \quad \text{or } \$61.73 \text{ (rounded)}$$

□

For organized data the formula for the mean is streamlined to account for multiple occurrences of the values.

Mean for Organized Data

$$\text{Ungrouped: } \overline{X} = \frac{\Sigma fX}{n} \tag{3–3a}$$

$$\text{Grouped: } \overline{X} = \frac{\Sigma fM}{n} \tag{3–3b}$$

Equation 3–3a is appropriate for ungrouped frequency distributions, while Equation 3–3b is to be used for grouped frequency distributions. Since the actual values are melded into classes in a grouped distribution, the midpoint (M) is a proxy for all observations in the class and replaces X in the formula. The following example demonstrates the application of Equation 3–3b.

EXAMPLE 3.6

Find the mean of the following data:

Classes	f
4– 7	5
7–10	8
10–13	6
13–16	1

Solution:

The first step is to generate a column of midpoints M:

Classes	f	M	fM
4– 7	5	5.5	27.5
7–10	8	8.5	68
10–13	6	11.5	69
13–16	1	14.5	14.5
	$n = 20$		$\Sigma fM = 179$

Next, we form the product of f times M and create a fourth column labeled fM. The mean, according to Equation 3–3b, is

$$\overline{X} = \frac{\Sigma fM}{n}$$

$$= \frac{179}{20} = 8.95$$

Rounded off, the mean is 9.0.

□

COMMENTS

1. For raw data and organized data in ungrouped frequency distributions, the mean should be rounded off, if necessary, to one decimal place beyond the accuracy of the raw data. For organized data in grouped frequency distributions, round off, if necessary, to one decimal place beyond the accuracy of the classes.

2. Like the median of a grouped frequency distribution, the mean via Equation 3–3b is only an approximation to the true value of $\overline{X}$ for the raw data.
3. The mean has an appealing physical interpretation, as it represents the center of gravity for the data set. Imagine that the horizontal axis of a histogram for the data set functions as a "seesaw." The histogram (seesaw) is balanced whenever a fulcrum is placed at the value of the mean.

The mean is relatively easy to compute, involves all the values in the data set, and possesses favorable theoretical characteristics for use in inferential statistics. Its worst characteristic is its sensitivity to extreme observations. A few or even one very large or very small number tends to swamp the other numbers in the calculations, producing an artificially large or small value of $\overline{X}$. In these situations, the median is sometimes the preferred measure of central location.

There are other measures of central location that we have not covered explicitly. Two in particular—the midrange and the geometric mean—will be introduced briefly in Exercises 3.20 and 3.23, respectively. For now, familiarize yourself with the symbols and computations for our three primary measures. As this chapter unfolds, we will show you applications and relationships of these measures.

CLASSROOM EXAMPLE 3.1

Central Location Measures for Raw Data

The yearly percentage change in earnings, represented by X, was recorded for an airline over a six-year period:

Year	$X(\%)$
1	11.6
2	7.2
3	−3.1
4	4.6
5	−7.7
6	5.4
	$\Sigma X =$ _____

a. Find the mode.

b. Find the median:

Position	1	2	3	4	5	6
Ordered X's	_____	_____	_____	_____	_____	_____

c. Find the mean: $\overline{X} = \dfrac{\Sigma X}{n} =$ _____.

*d. Find the midrange (see Exercise 3.20): Mg = (min + max)/2 = ______.

*e. Find the geometric mean (see Exercise 3.23).

CLASSROOM EXAMPLE 3.2

Central Location Measures for a Grouped Frequency Distribution

Twenty utilities were surveyed and their net income per share for the previous year was recorded. The organized data are as follows:

Income/Share ($)	f	cf	pos	M	fM
.18–1.10	4				
1.10–2.02	8				
2.02–2.94	5				
2.94–3.86	3				
	n = ______				ΣfM = ______

a. Identify the modal class.

b. What is the value of the mode?

c. Compute the median:

Step 1: $n/2$ = ______.

Step 2: the class containing the median is ______.

Step 3: L = ______.

Step 4: f_m = ______.

Step 5: cf_p = ______.

Step 6: w = ______.

Step 7: $Md = L + (w/f_m)(n/2 - cf_p)$ = ______.

d. Compute the mean: $\overline{X} = \Sigma fM/n$ = ______.

SECTION 3.2 EXERCISES

3.6 Find the mode, median, and mean for these sample data:

a. 12 9 11 12 10 14 10 11 10

b. 2 0 6 3 9 2 0 2

c. 8.9 −5.1 12.3 4.5 8.4 5.0

3.7 Find the mode, median, and mean for these sample data:

a. −4 −6 −1 0 −2

*Optional

b. .08 .03 .05 .03 7.2 .07 .01 .05
c. 0 0 0 0 1 0 0

3.8 Consider the following frequency distribution:

Classes	f
313–369	12
369–425	9
425–481	18
481–537	12
537–593	7

a. Identify the modal class.
b. What is the value of the mode?
c. Compute the sample median.
d. Compute the sample mean.

3.9 Compute the mode, median, and mean for the sample data in the following frequency distribution:

Classes	f
14.3–20.1	5
20.1–25.9	11
25.9–31.7	21
31.7–37.5	7
37.5–43.3	30
43.3–49.1	18

3.10 A sample of six brokerage houses yielded the following values for X = rate of return on a four-year certificate of deposit:

8.5%	8%	8.5%	7.9%	7.7%	7.5%

Find the median and the average rate of return.

3.11 Prospective salespeople with an insurance company often are screened through a battery of tests, from which a "career profile" score is obtained. The maximum possible score is 25; the higher the score, the greater the potential for success. A sample of 27 applicants generated the following distribution of career profile scores:

Scores	f
4– 8	3
8–12	8
12–16	7
16–20	7
20–24	2

a. What is the value of the class width?
b. Identify the mode.
c. Determine the mean career profile score.
d. What number divides the set of scores in half?

3.12 In the mid-1980s, the United States began to experience a shortage of nurses. The American Hospital Association released the following figures on the number of nurses per hospital bed by state as of 1986:

State	Nurses	State	Nurses	State	Nurses
Alabama	0.72	Kentucky	0.75	North Dakota	0.58
Alaska	0.91	Louisiana	0.64	Ohio	0.83
Arizona	0.86	Maine	0.80	Oklahoma	0.70
Arkansas	0.76	Maryland	0.80	Oregon	0.85
California	0.88	Massachusetts	0.82	Pennsylvania	0.81
Colorado	0.77	Michigan	0.86	Rhode Island	0.86
Connecticut	0.79	Minnesota	0.61	South Carolina	0.78
Delaware	0.67	Mississippi	0.61	South Dakota	0.55
D.C.	0.88	Missouri	0.73	Tennessee	0.70
Florida	0.77	Montana	0.59	Texas	0.71
Georgia	0.76	Nebraska	0.61	Utah	0.95
Hawaii	0.80	Nevada	0.81	Vermont	0.76
Idaho	0.74	New Hampshire	0.87	Virginia	0.75
Illinois	0.76	New Jersey	0.74	Washington	0.88
Indiana	0.67	New Mexico	0.77	West Virginia	0.74
Iowa	0.62	New York	0.75	Wisconsin	0.62
Kansas	0.59	North Carolina	0.77	Wyoming	0.54

Source: "Shortage of nurses topic at convention," *USA Today*, December 5, 1986.

Find the mean, median, and modal number of nurses per bed. Assume these 1986 data represent a sample.

3.13 Refer to Exercise 3.12. Organize the data into a grouped frequency distribution. Compute the mean, median, and mode for the resulting distribution and compare these values to the values in Exercise 3.12. Are they identical? Why?

3.14 The number of miles flown in a one-year period by a sample of "frequent flyers" was recorded and organized into the following frequency distribution:

Number of Miles Flown (Thousands)	Number of Frequent Flyers
40–47	3
47–54	9
54–61	17
61–68	14
68–75	10
75–82	6

a. Is this a grouped or an ungrouped frequency distribution?
b. Which class is the modal class?
c. Find the median number of miles flown by the sample of frequent flyers.

3.15 Find the mean and the median of the following organized data:

Classes	f
6.2– 7.8	5
7.8– 9.4	2
9.4–11.0	7
11.0–12.6	3
12.6–14.2	12

3.16 Consider the following ungrouped frequency distribution:

X	−4	−2	−1	0	1	3
f	24	56	33	17	8	2

a. Find the mean.
b. What value is the mode?
c. Determine the median of these data.

3.17 Teenagers are notorious for viewing a motion picture many times. Young people in their teen years were surveyed and asked to indicate the number of times they saw their favorite motion picture within the last year. The responses were organized into the following frequency distribution:

Number of Times Movie Was Seen	Sample Count
1	19
2	44
3	35
4	21
5	11
6	3
7	1

a. Is it possible to determine how many teenagers responded to the survey? If so, how many?
b. What is the mean number of viewings per favorite movie?
c. Determine the median.
d. Is it accurate to say that teenagers are most likely to see their favorite movie twice? Why?

3.18 The owner of a new car recorded the miles per gallon (mpg) achieved with each fill-up. The resulting mpg figures were as follows:

20.909	20.798	23.331	21.847	23.220
19.974	21.111	25.177	20.779	23.492
21.362	20.952	27.627	22.595	23.096
20.503	20.450	22.050	21.495	21.146
20.952	22.177	22.643	23.076	22.495

a. Find the median mpg figure for this sample.
b. Compute the mean mpg.
c. This particular model of car was given an overall mpg rating of 22. Would you agree that the car is achieving the 22 mpg figure? Explain.

3.19 Facing declining revenues, the owner of a video club arbitrarily selected 68 files from his membership list to determine the number of videos rented per household last month. He obtained these data:

X	0	1	2	3	4	5	6	7	8	10	14
f	13	5	9	7	11	9	4	1	3	4	2

a. Define, in words, the variable X.
b. Identify an entity for the variable X.
c. Find the median number of videos rented.
d. The profit per rented video is roughly $.50. If there are a total of 2648 members, estimate the monthly profit using the median number from the survey.

***3.20** A simple measure of central location is the *midrange*, which is defined as the average of the smallest and largest values in a data set. The midrange is used by the U.S. Weather Service to report the average daily temperature of cities. The formula to compute the midrange, denoted by Mg, is

$$Mg = \frac{\text{min} + \text{max}}{2}$$

where min and max refer to the minimum and the maximum values, respectively, in a data set. If the high temperature reached 98° and the low was 72°, what would the Weather Service report as the average (i.e., midrange) temperature?

***3.21** A sample of nine states produced the following data on X = the value in millions of dollars of the yearly burley tobacco crop:

$775.7	$741.3	$821.5
$647.5	$930.6	$1,318.5
$1,487.3	$853.9	$1,373.0

Find the mean and midrange.

***3.22** Find the midrange, median, and mean for these data: 46 13 46 26 38 20 28 15.

*Optional

***3.23** Another commonly used measure of central location, with applications in finance, is the *geometric mean*. The geometric mean is used when dealing with rates of return or percentage yields which involve a time factor. For example, the percentage change of a stock's price depends on the previous year's price as well as the current price—that is, percentage change = (new price − old price)/old price. Thus, the change is a relative measure of gain (or loss), not an absolute measure. To compute the geometric mean, denoted GM, we use

$$GM = \sqrt[n]{(1 + r_1)(1 + r_2)\cdots(1 + r_n)} - 1$$

where the $r_1, r_2, \ldots, r_n$ symbols represent the rates or percentages expressed in *decimal* form (8.1 percent becomes .081, for example). Note that the r_i terms can be negative and that in the worst case, r_i could be -1 (signifying the complete loss of principal). As an example, let's find the geometric mean of these data: 24.4% −9.2% 40.7%.

$$GM = \sqrt[3]{(1 + .244)(1 - .092)(1 + .407)} - 1$$

$$= \sqrt[3]{1.5892797} - 1$$

$$= 1.16699 - 1 = .16699 \quad \text{or } 16.70\%$$

If the above data represented rates of return, then the figure 16.70 percent could be viewed as the constant yearly growth rate for the investment. There are two major features of the geometric mean that make it attractive for use in finance-related problems. First, the geometric mean is a more conservative measure of central location than the average (see Exercise 3.26). Second, since the geometric mean is computed using relative changes, it preserves the meaning of a "zero growth rate" (see Exercise 3.27).

Find the geometric mean of these data: .4% 5.8% −2.3%.

***3.24** Find the geometric mean of these data: 26.4% 25.5% 18.4% 18.9% 21.2%.

***3.25** A stock posted the following yields for the last 4 years:

4.8%	−2.3%	6.7%	11.5%

What is the geometric mean yield?

***3.26** The Standard and Poor's total return for a ten-year period follows:

Year	Return (%)	Year	Return (%)
1	5.1	6	−2.1
2	−8.3	7	27.8
3	18.0	8	17.4
4	20.5	9	15.2
5	19.5	10	23.3

Find the arithmetic and geometric mean returns and compare the values. The more conservative value is the smaller value; which is the more conservative mean?

*3.27 Suppose you invest $2000 in an Individual Retirement Account (IRA). One year later you've earned $200 in interest, giving you an account balance of $2200. The next year your IRA loses $200, returning your balance to $2000.

a. Determine the percentage change in your account balance for the first year and then again for the second year. (See Exercise 3.23.)

b. Compute the arithmetic mean of these percentage changes.

c. What is your geometric mean rate of return?

d. After two years your account has experienced a zero growth rate. Which mean reflects this phenomenon?

*3.28 The speed of computers in processing data promotes the use of looping algorithms to compute many quantities, including the mean. To demonstrate, suppose the mean for a set of n values $X_1, X_2, \ldots, X_n$ has been calculated via Equation 3–2 and denoted $\overline{X}_n$, where the subscript n indicates the mean is based on n values. If another observation, X_{n+1}, is added to the data set, then the mean of the $(n+1)$ values often is found from this relationship

$$\overline{X}_{n+1} = \frac{n\overline{X}_n + X_{n+1}}{n + 1}$$

This equation permits the mean to be updated with each new value instead of processing the entire data set each time. If the mean of a set of $n = 4$ values is $\overline{X}_4 = 5.75$ and a fifth observation, $X_5 = 7$, is added to the data set, what is the mean of the set of five values?

3.3 MEASURES OF DISPERSION

Central location measures do not describe fully all the dimensions of a data set. Although measures like the mean, median, and mode provide us with information about the middle of the data, they yield no clues about the scattering of values from low to high.

Consider the data sets represented by curves A and B in Figure 3.2. Both data sets share identical values for the measures of central location, yet distribution B is more spread out than distribution A. Intuitively, we would expect a measure of dispersion to produce a larger value for B than for A.

In this section we discuss descriptive statistics that gauge the spread or variability of the values in a data set. The first and simplest measure is the range.

Range

A simple way to measure dispersion involves subtracting the smallest value from the largest value.

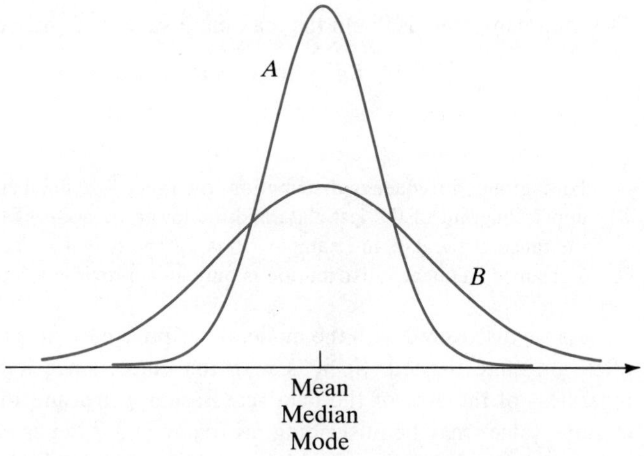

Figure 3.2 Distributions with Identical Central Location Measures but Different Dispersion

Definition

The **range**, denoted by *Rg*, is a single number representing the distance between the minimum and maximum values in a data set.

Equation 3–4 is used to find the range for raw data and for data organized into an ungrouped frequency distribution.

Range

$$Rg = \max - \min \qquad (3\text{–}4)$$

where max and min denote the largest and smallest values of X in the data set.

EXAMPLE 3.7

Find the range for these sample data:

1	4	1	0	2	1	1	3	2	9	1	2

Solution:

The minimum value is 0 and the maximum value is 9; hence, the range is

$$Rg = \max - \min = 9 - 0 = 9$$

□

COMMENT For a grouped frequency distribution, the range is defined as the difference between the upper endpoint of the last class and the lower endpoint of the first class. For example, the range of the data in Example 3.6 is $Rg = 16 - 4 = 12$. The range computed from a grouped frequency distribution is only an approximation to the range of the raw data.

As we discovered with the mode, descriptive measures that are simple to compute often are quite limited. So it is with the range, since it involves only two values, regardless of the size of the data set. Basing a measure of dispersion solely on the extreme values may be misleading, as Example 3.7 indicates. Eleven of the 12 values in the data set are between 0 and 4, yet the range of 9 does not reflect this concentration. A better way to measure dispersion should involve all the data.

Measures Based on Deviations

Although the range may be a nonrepresentative quantity, it does suggest that dispersion can be measured with differences or deviations. Rather than focusing on the deviation between the minimum and maximum values, we consider the deviations between each value in the data set and a common reference point. This is intuitively appealing, especially if we relate this to other disciplines. For example, within the field of psychology, there is an area of study called abnormal psychology that analyzes "deviant behavior"—that is, behavior straying from the "norm."

For descriptive statistics purposes, we must define a norm and then consider deviations or distances from it. There are many candidates for this reference point. Traditionally, the mean has been adopted as the norm so that *deviations from the mean* often are simply called "deviations." Deviations are symbolized $(X - \overline{X})$, where X represents any value in the data set and $\overline{X}$ denotes the sample mean. Each value of X gives rise to a corresponding deviation $(X - \overline{X})$, which can be a positive or negative number or the number zero.

EXAMPLE 3.8

Generate the deviations for each value of this data set:

4	9	5	4	4	10

Solution:

Verify that the mean of these data is 6. Next, subtract $\overline{X} = 6$ from each value in the data set. This is simplified by putting the data into columns:

X	$X-6$	$= (X-\overline{X})$
4	$4-6$	-2
9	$9-6$	3
5	$5-6$	-1
4	$4-6$	-2
4	$4-6$	-2
10	$10-6$	4
		0

The last column, labeled $(X - \overline{X})$, is the collection of deviations.

□

Notice in Example 3.8 that there are positive and negative deviations and that the total of the $(X - \overline{X})$ column is zero:

$$\Sigma(X - \overline{X}) = 0$$

If the positive and negative deviations could be modified so as not to cancel out, we could measure the spread more easily. Two simple solutions to this problem—taking the absolute value of each deviation and squaring each deviation—give rise to the next two measures of dispersion.

Average Deviation This measure is found by averaging the deviations after all the negative signs have been eliminated.

Definition

The **average deviation,** denoted by AD, is the mean of the absolute value of the deviations.

The formula for computing the average deviation for a sample of raw data is given as Equation 3–5.

Average Deviation for Raw Data

$$AD = \frac{\Sigma\,|X - \overline{X}|}{n} \tag{3–5}$$

where

$|X - \overline{X}|$ = the absolute value of the deviation $X - \overline{X}$

EXAMPLE 3.9

Find the average deviation for this sample data set:

4	9	5	4	4	10

Solution:

From Example 3.8 we generated the column of deviations $(X - \overline{X})$. These values and their absolute values are as follows:

X	$(X - \overline{X})$	$\lvert X - \overline{X} \rvert$
4	−2	2
9	3	3
5	−1	1
4	−2	2
4	−2	2
10	4	4
	0	$\Sigma\lvert X - \overline{X}\rvert = 14$

The average deviation is

$$AD = \frac{\Sigma \lvert X - \overline{X} \rvert}{n} = \frac{14}{6} = 2.333\ldots$$

Rounded off to one more decimal place accuracy than the raw data, $AD = 2.3$. This means that the average distance from the mean, without regard to direction, for the collection of values is 2.3.

□

COMMENTS

1. For an ungrouped frequency distribution, the formula for the average deviation is

$$AD = \frac{\Sigma f \lvert X - \overline{X} \rvert}{n}$$

For a grouped frequency distribution,

$$AD = \frac{\Sigma f \lvert M - \overline{X} \rvert}{n}$$

where M is the midpoint of each class.

2. The average deviation is known by other phrases and symbols: average absolute deviation (AAD), mean deviation (MD), and mean absolute deviation (MAD).

Variance An alternative to taking the absolute value of each deviation is to square each deviation. This leads to the measure of dispersion known as the variance.

Definition

The **variance** is the (corrected) mean of the squared deviations.

The parenthetical word "corrected" is our first opportunity to point out the differences between theory and application. Omit the word "corrected" from the definition above, and we have the theoretical definition of variance. Include "corrected" in the definition, and we have the practical or working definition of variance.

In theory, the variance is found by dividing the sum of squared deviations by n. If our entire exposure to business statistics were limited to descriptive statistics, we would divide by n. However, business statistics is both descriptive and inferential. In the latter case, using the theoretical definition of the variance creates bias or error. To alleviate this problem, we "correct" the error by introducing a slightly different definition for the sample variance, one in which the divisor of the sum of squared deviations is $n-1$, rather than n.

Because of its importance in the study of business statistics, the variance demands special notation. As shown in Table 3.2, for a sample variance, we use the notation S^2, where the exponent 2 emphasizes the squared deviations approach. For a population variance, we use the lower case Greek letter sigma, with the exponent 2: σ^2.

In Equation 3–6, we provide two equivalent formulas for S^2: one that arises from the definition above, which we call the "defining formula," and another that sometimes is called the "calculating formula." Though different in appearance, both expressions produce the identical value for S^2.

Variance for Raw Data

$$S^2 = \underbrace{\frac{\Sigma(X - \overline{X})^2}{n - 1}}_{\text{Defining formula}} = \underbrace{\frac{\Sigma X^2 - \dfrac{(\Sigma X)^2}{n}}{n - 1}}_{\text{Calculating formula}} \qquad (3\text{–}6)$$

Both formulas have a denominator of $(n - 1)$, representing the *corrected* mean of the squared deviations. In the calculating formula, the numerator involves two similar

Table 3.2 Symbols for the Variance

Measure	Symbol
Population variance	σ^2
Sample variance	S^2

looking terms: ΣX^2 and $(\Sigma X)^2$. The former term literally says "sum the X^2's." This means we first square each value of X, and then total. The latter term reverses this order: First we obtain the total of the values of X, (ΣX), and then we square that result.

EXAMPLE 3.10

Compute the sample variance of the following raw data:

4	9	5	4	4	10

Solution:

We shall demonstrate the equivalency of the defining formula and the calculating formula by computing S^2 using each. In Example 3.8 we generated the deviations $(X - \overline{X})$ for the raw data:

X	$(X - \overline{X})$	$(X - \overline{X})^2$
4	−2	4
9	3	9
5	−1	1
4	−2	4
4	−2	4
10	4	16
		$\Sigma(X - \overline{X})^2 = 38$

Each number in the column of deviations is squared in the third column and then totaled. According to the defining formula,

$$S^2 = \frac{\Sigma(X - \overline{X})^2}{n-1} = \frac{38}{6 - 1} = 7.6$$

To use the calculating formula, we generate the column X^2 by squaring each value of X:

X	X^2
4	16
9	81
5	25
4	16
4	16
10	100
$\Sigma X = 36$	$\Sigma X^2 = 254$

The totals and the appropriate summation notation appear at the bottom of the two columns. According to the calculating formula,

$$S^2 = \frac{\Sigma X^2 - \frac{(\Sigma X)^2}{n}}{n - 1} = \frac{254 - \frac{(36)^2}{6}}{6 - 1} = \frac{254 - 216}{5} = \frac{38}{5} = 7.6$$

As expected, both formulas yield the same value of 7.6 for S^2.

□

Examine the two measures of dispersion—average deviation and variance—that we calculated in Examples 3.9 and 3.10. For the *same* sample of $n = 6$ values, we found $AD = 2.3$ and $S^2 = 7.6$. Six values deviating from the same point of reference $\overline{X} = 6$ produced two dispersion measures that are not of the same magnitude. However strange this may seem, it does have a simple explanation. The average deviation preserves the units of the original data. For example, if the values of X represent measurements in inches, then the deviations $X - \overline{X}$ are still in inches and the average deviation of 2.3 is in inches. On the other hand, the variance involves squared deviations, meaning that $S^2 = 7.6$ is in units of *square* inches not inches. Comparing square inches to inches is like comparing apples to oranges. Thus, the variance and the average deviation are not directly comparable. Shortly we will resolve this difficulty with another measure of dispersion, the standard deviation.

When the variance is computed for organized data in frequency distributions, Equations 3–6 are not applicable. Equations 3–7 are used instead.

Variance for Organized Data

$$S^2 = \frac{\Sigma f(X - \overline{X})^2}{n - 1} = \frac{\Sigma fX^2 - \frac{(\Sigma fX)^2}{n}}{n - 1} \qquad \text{(3–7a)}$$

↑ Defining formulas ↓   ↑ Calculating formulas ↓

$$S^2 = \frac{\Sigma f(M - \overline{X})^2}{n - 1} = \frac{\Sigma fM^2 - \frac{(\Sigma fM)^2}{n}}{n - 1} \qquad \text{(3–7b)}$$

Equations 3–7a are for data that have been organized into an ungrouped frequency distribution; X refers to an actual value of the variable. Equations 3–7b are for data that have been organized into a grouped frequency distribution; M is the midpoint of each class.

EXAMPLE 3.11

Find S^2 for the following grouped frequency distribution:

Classes	f
4– 7	5
7–10	8
10–13	6
13–16	1

Solution:

This is the data of Example 3.6, for which we already generated the column of midpoints M. To find S^2 using the calculating formula, we need additional columns for fM and fM^2:

Classes	f	M	fM	fM^2
4– 7	5	5.5	27.5	151.25
7–10	8	8.5	68	578
10–13	6	11.5	69	793.5
13–16	1	14.5	14.5	210.25
	$n = 20$		$\Sigma fM = 179$	$\Sigma fM^2 = 1733$

The totals of the appropriate columns are indicated, along with the symbols, immediately beneath the numbers. The variance is

$$S^2 = \frac{\Sigma fM^2 - \dfrac{(\Sigma fM)^2}{n}}{n - 1} = \frac{1733 - \dfrac{(179)^2}{20}}{19} = 6.8921053\ldots$$

Rounded off, we get $S^2 = 6.9$.

□

COMMENTS

1. The defining and calculating formulas for S^2 are presented for use in different situations. The defining formula works best when the computations for $\overline{X}$ produce a number with the same degree of accuracy as the raw data (or at worst, one decimal place more). The calculating formula is preferred when $\overline{X}$ involves many decimal places.
2. The final answer for S^2 for raw data or organized data in an ungrouped frequency distribution should be rounded off to one more decimal place than the raw data. For a grouped frequency distribution, round off to one more decimal place than the accuracy of the classes.
3. The corrected variance, also referred to as the "mean squared error," is *never* a negative number.

Squaring is a necessary part of the computations for the variance. Though this alleviates the earlier problem in which the positive and negative deviations summed to zero, it creates a new problem in that the units of measurement for the raw data are not the same as the units for S^2. The solution to this difficulty is the standard deviation.

Standard deviation Suppose someone asks you how far apart Detroit and Toronto are. If you are an American, you would probably answer about 240 miles. But if you are a Canadian, you might say about 385 kilometers. Which answer is correct? Both answers are correct, though they are different. The difference stems from the *standard* unit of measuring distances in the two countries: a mile versus a kilometer.

We face an analogous problem in business statistics: What standard should we use in measuring distances or deviations from the mean? That is, for every data set, we wish to define a *standard* measure of distance, in terms of the units of the original data. The average deviation could be used, although working with absolute values is not desirable. The variance is less appealing because it measures variability in units *squared*. For purely mathematical ease, we elect neither measure but propose a different one: *standard deviation*.

Definition

The **standard deviation** is the (positive) square root of the variance.

The standard deviation is literally the standard distance from the mean, or the standard against which other distances may be compared. It is specific to a data set. The standard deviation for a data set is like the dollar in our economy or the watt in measuring electrical power. We judge the comparable worth of a commodity by its dollar value, just as we purchase light bulbs that use differing amounts of power (60 watt, 100 watt, and so forth).

Similarly, we view the standard deviation as the baseline deviation and use its value in two ways. First, we measure the distance from any value to the mean relatively, not absolutely, in terms of the number of standard deviations. The second role of the standard deviation is a more global one, representing the spread of the entire data set in one number. This number can be used to compare variability between data sets. For example, if two data sets have the same value for the mean, but different values for the standard deviation, then the one with the larger standard deviation is more spread out or dispersed than the one with the smaller value. A comparison such as this is effective, especially in finance when comparing investment returns, provided the means are equal. When the means are unequal, a simple comparison of standard deviations is not justified. The remedy for this situation also is upcoming in the next section in a measure called the *coefficient of variation*.

Since the standard deviation is a function of the variance, we use a similar notation, without the exponent 2. (See Table 3.3.) Of course, the formula for computing the standard deviation is actually a mathematical operation, taking the square root. Most of the labor involved in finding the standard deviation for a set of data is contained in the computation of the variance.

Table 3.3 Symbols for the Standard Deviation

Measure	Symbol
Population standard deviation	σ
Sample standard deviation	S

Standard Deviation for Raw and Organized Data

$$S = +\sqrt{S^2} \tag{3–8}$$

No separate formula is needed to handle raw data as opposed to organized data; simply take the square root of the variance.

EXAMPLE 3.12

Find the standard deviation for these data:

4	9	5	4	4	10

Solution:

In Example 3.10 we computed $S^2 = 7.6$; hence

$$S = +\sqrt{S^2} = +\sqrt{7.6} = 2.7568\ldots$$

Rounded off, the sample standard deviation is 2.8.

□

COMMENTS

1. The *positive* square root of the variance is used to define the standard deviation, since the square root of a number can be either positive or negative.
2. Do not round off the variance before taking its square root to find the standard deviation. Round off the standard deviation to one more decimal place of accuracy than the original data.

CLASSROOM EXAMPLE 3.3

Dispersion Measures for Raw Data

The yearly percentage change in earnings, represented by X, was recorded for an airline over a six-year period:

Year	$X(\%)$	$X - \bar{X}$	$\lvert X - \bar{X}\rvert$	$(X - \bar{X})^2$	X^2
1	11.6				
2	7.2				
3	−3.1				
4	4.6				
5	−7.7				
6	5.4				

$\Sigma X =$ _____ $\Sigma(X - \bar{X}) =$ _____ $\Sigma\lvert X - \bar{X}\rvert =$ _____ $\Sigma(X - \bar{X})^2 =$ _____ $\Sigma X^2 =$ _____

a. Find the range.

b. Find the average deviation: $AD = \dfrac{\Sigma\lvert X - \bar{X}\rvert}{n} =$ _____.

c. Find the variance using the defining formula: $S^2 = \dfrac{\Sigma(X - \bar{X})^2}{n - 1} =$ _____.

d. Find the variance using the calculating formula:

$$S^2 = \frac{\Sigma X^2 - \dfrac{(\Sigma X)^2}{n}}{n - 1} = \_\_\_\_\_.$$

e. Find the standard deviation: $S = +\sqrt{S^2} =$ _____.

CLASSROOM EXAMPLE 3.4

Dispersion Measures for a Grouped Frequency Distribution

Twenty small businesses were surveyed and their number of full-time employees was recorded. The organized data are as follows

Number of Employees	f	M	fM	$(M - \bar{X})$	$f\lvert M - \bar{X}\rvert$	$f(M - \bar{X})^2$	fM^2
2–12	5						
12–22	7						
22–32	5						
32–42	3						

$\Sigma fM =$ _____ $\Sigma f\lvert M - \bar{X}\rvert =$ _____ $\Sigma f(M - \bar{X})^2 =$ _____ $\Sigma fM^2 =$ _____

a. Find the range.

b. Find the average deviation:

$$AD = \frac{\Sigma f|M - \overline{X}|}{n} = \_\_\_\_\_.$$

c. Find the variance using the defining formula:

$$S^2 = \frac{\Sigma f(M - \overline{X})^2}{n - 1} = \_\_\_\_\_.$$

d. Find the variance using the calculating formula:

$$S^2 = \frac{\Sigma fM^2 - \frac{(\Sigma fM)^2}{n}}{n - 1} = \_\_\_\_\_.$$

e. Find the standard deviation:

$$S = \sqrt{S^2} = \_\_\_\_\_.$$

SECTION 3.3 EXERCISES

3.29 Find the range, average deviation, variance, and standard deviation for the following sample data sets:

a. .201 .407 .639 .657

b. 14.11 14.85 15.47 12.58 12.34 16.63

c. 29 20 17 20 21 28 21

d. −4 −6 −1 0 −2

e. 0 0 0 0 1 0 0 0 0 0

f. 5 9

3.30 Find the standard deviation for these data: 5.784 5.784 5.784.

3.31 Consider the following data set:

3	5	6	0	2	6	9	9	7	3

a. What is the mean?

b. What is the standard deviation?

c. In absolute terms, how far from $\overline{X}$ is the number 9? (*Hint:* Compute $X - \overline{X}$.)

d. In relative terms, how far from $\overline{X}$ is the number 9? (*Hint:* Compute $(X - \overline{X})/S$.)

3.32 Compute the average deviation and standard deviation for the following set of organized data:

X	-4	-2	-1	0	1	3
f	24	56	33	17	8	2

3.33 The number of corporate takeovers per month was monitored for a period of twenty months, producing the following data:

X	5	8	9	10	11	12	14
f	3	6	2	5	1	2	1

a. Find the mean number of takeovers per month.
b. Compute the standard deviation of these data.
c. What is the value of the average deviation?

3.34 A sample of $n = 17$ values produced these sums: $\Sigma X = 9$ and $\Sigma X^2 = 33$. Find the variance of the data set. Express your answer to two decimal places.

***3.35** Consider the following sample data set for the variable X:

54	36	49	57	51	53

a. Determine the sample mean and median.
b. Compute $\Sigma|X - \overline{X}|$ and $\Sigma|X - Md|$. Which value is lower? (*Note:* This result is a general one for all data sets where $\overline{X} \neq Md$.)
c. Compute $\Sigma(X - \overline{X})^2$ and $\Sigma(X - Md)^2$. Which value is lower? (*Note:* This is also a general result when $\overline{X} \neq Md$.)

3.36 If $\Sigma f = 25$, $\Sigma X = 15$, $\Sigma|X - \overline{X}| = 6.2$, and $\Sigma f|X - \overline{X}| = 21.6$ for a set of organized data in an ungrouped frequency distribution, find the average deviation. Find, if possible, the mean.

3.37 Find the range, average deviation, variance, and standard deviation for these data:

X	0	1
f	9	1

3.38 Compute the variance and standard deviation of the following data:

Classes	f
.02– .44	26
.44– .86	6
.86–1.28	7
1.28–1.70	5
1.70–2.12	6
2.12–2.54	2

*Optional

3.39 Miles per gallon figures for a sample of 1990 model year cars (all the same model) were organized into a grouped frequency distribution:

Classes	f
17.12–19.64	4
19.64–22.16	10
22.16–24.68	12
24.68–27.20	2
27.20–29.72	3

a. Find the range for this frequency distribution.
b. Compute the mean.
c. Find S.

***3.40** As we pointed out in Exercise 3.28, computations done by a computer often are facilitated with so-called recursive formulas. In the case of the variance, a recursive formula is generally more accurate, especially for extremely dispersed data sets. Suppose the variance for a set of n values is denoted by S_n^2:

$$S_n^2 = \frac{\Sigma (X - \overline{X}_n)^2}{n - 1}$$

where $\overline{X}_n$ is defined as the mean of the n numbers. Adding another piece of data, X_{n+1}, to the data set creates a new mean $\overline{X}_{n+1}$ and variance S_{n+1}^2. The recursive formula for the variance is

$$S_{n+1}^2 = \left(\frac{n-1}{n}\right) S_n^2 + \left(\frac{1}{n+1}\right) (\overline{X}_n - X_{n+1})^2$$

If the mean and variance of a set of $n = 5$ values are $\overline{X}_5 = 6$ and $S_5^2 = 16$, respectively, and a sixth observation, $X_6 = 10$, is added to the data set, what is the variance of the six numbers?

***3.41** There are many algebraically equivalent expressions for the variance, S^2. Three other formulas that could be used are

$$S^2 = \frac{n\Sigma X^2 - (\Sigma X)^2}{n(n - 1)}$$

$$S^2 = \frac{\Sigma X^2 - n\overline{X}^2}{(n - 1)}$$

$$S^2 = \frac{1}{2n(n - 1)} \sum_{i=1}^{n} \sum_{j=1}^{n} (X_i - X_j)^2$$

The sample variance for the data set 1, 3, 8, 11, and 7 is 16. Verify that each of the above formulas yields this value for the data set.

3.42 "Brokerage stocks can be a great way to play a rally, but when they stumble they can fall a lot further than other stocks." So says Susan Antilla in an article titled "Brokerages Sit Uneasily Atop Bull." She goes on to compare the stock performances of eleven major brokerage houses since the beginning of the calendar year. Some of the data presented in the article are as follows:

Company	Year-to-Date Change in Stock Price	P–E Ratio
Bear Stearns	$+5\frac{1}{8}$	11
A. G. Edwards	$+9\frac{1}{8}$	15
First Boston	$+7$	11
E. F. Hutton	$+2\frac{3}{4}$	28
McDonald & Co.	$+1\frac{1}{8}$	14
Merrill Lynch	$+6\frac{7}{8}$	10
Morgan Stanley	$+8\frac{3}{4}$	9
Paine Webber	$+5\frac{1}{8}$	15
Piper Jaffray	$+1\frac{1}{2}$	9
L. F. Rothschild	$+\frac{3}{4}$	9
Salomon	$-\frac{1}{4}$	10

Source: "Brokerages sit uneasily atop bull," *USA Today,* January 30, 1987.

a. Find the standard deviation for the year-to-date changes in the stock prices.
b. Find the standard deviation for the P–E ratios.
c. Is there more dispersion in the stock price changes or the P–E ratios? Base your answer on the respective values of the standard deviations.

3.43 Refer to Exercise 3.42. Recompute the standard deviation of the P–E ratios for ten companies, eliminating E. F. Hutton. Compare this value of S to the one in Exercise 3.42, part b. What effect does E. F. Hutton's abnormally large P–E ratio have on the standard deviation?

3.44 The length of stay in a hospital for major surgery is recorded for a sample of 161 patients:

Length of Stay (Days)	Patient Count
9	18
10	21
11	29
12	25
13	10
14	22
15	8
16	12
17	5
18	11

a. Find the range in the values of X = length of stay.
b. Compute the average deviation.
c. What is the deviation from the mean for $X = 10$?
d. Find the standard deviation of X.

3.45 A branch bank manager was asked by the home office to monitor the frequency with which the automated teller machine (ATM) was used per month per customer. Let X = number of

ATM transactions per month per customer. The assistant manager took a random sample of customers from a computer tape of transactions and summarized the values of X into the following distribution:

X	0	1	2	3	4	5	6	7	8
f	42	86	77	59	71	35	28	14	12

Find the average deviation and standard deviation for these data.

3.46 Suppose you were asked to measure accurately and record the weights of 10 five-pound sacks of sugar and 10 five-pound bags of apples. If you computed the standard deviation for both sets of 10 numbers, for which set would the value of S likely be larger? Why?

3.4 OTHER SUMMARY MEASURES

Together, measures of central location and measures of dispersion provide a fairly concise portrayal of the data set. In fact, the mean and standard deviation by themselves are sometimes sufficient to describe and summarize an entire sample. However, some data sets need further description, beyond that available in any of the measures presented thus far. Of the many secondary measures in business statistics, we will discuss four in this section and a few others in the exercises.

Coefficient of Variation

The coefficient of variation is classified as a measure of dispersion because it measures the dispersion in a data set relative to its mean. It is useful for comparing two or more sets of data that are calculated in the same units but which have different means and standard deviations.

For example, suppose you are offered the opportunity to invest money in one of the two projects listed in Table 3.4. Both projects involve risk, meaning that the expected return on your investment is *not* guaranteed. Which project is financially more attractive to you?

Project *A* looks better if we examine the projected mean returns, since 7.6 percent is a higher return than 6.8 percent. But, Project *A* also looks riskier or potentially

Table 3.4 Return on Investment

	Projected Return (%)	
Project	Mean	Standard Deviation
A	7.6	3.2
B	6.8	2.5

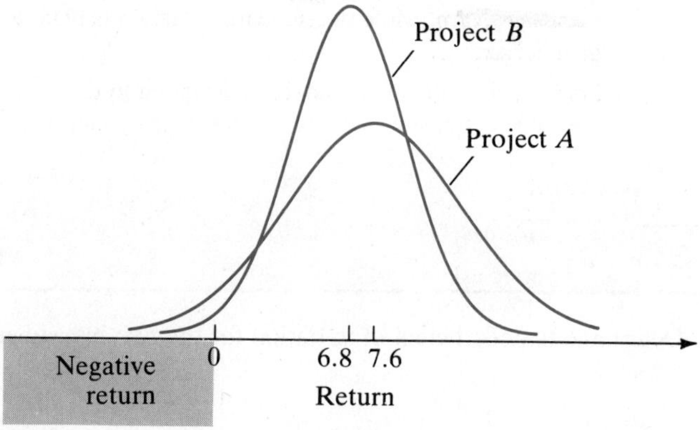

Figure 3.3 Distributions of Returns (Symmetrical Shapes Assumed)

more volatile than Project *B*, since its standard deviation is larger. Should we make our decision based on the higher mean or on the lower standard deviation? Figure 3.3 shows possible distributions of returns for these two investments, including the region labeled "negative return," where we begin to lose some of our investment principal. The financial dilemma we face may be resolved by *combining* the information in the mean and the standard deviation to measure the risk relative to the return.

Definition

The **coefficient of variation,** denoted by *CV*, is the ratio of the standard deviation to the mean.

The formulas for the *CV* are given in the following box.

Coefficient of Variation

$$\text{Sample } CV = \frac{S}{\bar{\bar{X}}} \qquad (3\text{–}9a)$$

$$\text{Population } CV = \frac{\sigma}{\mu} \qquad (3\text{–}9b)$$

Equation 3–9a is applicable for both organized and raw data.

COMMENTS

1. In financial terms, the mean is considered to be a measure of an investment's expected return, while the standard deviation is regarded as a measure of the investment's risk.
2. Forming the ratio of the standard deviation to the mean results in a decimal. Often, the value is transformed into a percentage by multiplying by 100.

EXAMPLE 3.13

Determine the coefficient of variation for the two investment projects in Table 3.4.

Solution:

Technically the values in Table 3.4 are not sample-based, so that $\overline{X}$ and S are not the correct symbols to attach to the numbers. In Chapter 5, we will explain more about the origins of the mean and standard deviation for an investment project that is proposed but not yet realized. For now, let's assume that the figures in Table 3.4 represent population parameters μ and σ. For Project *A*, the population coefficient of variation (*CV*) is, expressed as a percentage,

$$\begin{aligned}\text{Pop } CV &= \frac{\sigma}{\mu} \times 100 \\ &= \frac{3.2}{7.6} \times 100 \\ &= .4210526 \times 100 = 42.11\%\end{aligned}$$

For Project *B*,

$$\begin{aligned}\text{Pop } CV &= \frac{2.5}{6.8} \times 100 \\ &= .367647 \times 100 = 36.76\%\end{aligned}$$

In terms of relative dispersion, Project *B* is less variable than Project *A*. If we are conservative investors who do not like to take risks (risk avoiders), Project *B* is the better choice. If we are risk takers, Project *A* suits us better.

□

■ Z-score

A measure of position we will use throughout this text is called the *Z*-score. A *Z-score* measures the placement of a specific value in terms of the number of standard deviations it is away from the mean. For each value in a data set, there is a corresponding *Z*-score.

Definition

The **Z-score,** denoted by *Z*, corresponding to a value of *X* is the distance between *X* and its mean in units of the standard deviation.

Both population *Z*-scores and sample *Z*-scores share the same notations—namely, the letter *Z*—much like a value in a data set is denoted by *X* regardless if it is a sample value or a population value. The equations for computing population *Z*-scores and sample *Z*-scores involve different symbols.

Z-scores

$$\text{Sample } Z\text{-score: } Z = \frac{X - \bar{X}}{S} \qquad (3\text{–}10a)$$

$$\text{Population } Z\text{-score: } Z = \frac{X - \mu}{\sigma} \qquad (3\text{–}10b)$$

Equation 3–10a is applicable for both organized and raw data.

EXAMPLE 3.14

A sample of seven hospitals in a mid-sized city revealed the following information on X = daily charge for a semi-private room:

\$215	\$180	\$230	\$275	\$325	\$260	\$265

How far from the mean (in terms of standard deviations) is the maximum value? The minimum value?

Solution:

First, we compute the sample mean and standard deviation.

X	$X - \bar{X}$	$(X - \bar{X})^2$
215	−35	1225
180	−70	4900
230	−20	400
275	25	625
325	75	5625
260	10	100
265	15	225
$\Sigma X = 1750$	0	$\Sigma(X - \bar{X})^2 = 13100$

The mean and variance are

$$\overline{X} = \frac{\Sigma X}{n} = \frac{1750}{7} = 250$$

$$S^2 = \frac{\Sigma(X - \overline{X})^2}{n - 1} = \frac{13100}{6} = 2183.3333\ldots$$

Thus, the standard deviation is 46.7, to one decimal place. The maximum value is \$325, which is \$75 larger than the mean, or in terms of standard deviations away from the mean:

$$Z = \frac{X - \overline{X}}{S} = \frac{325 - 250}{46.7} = \frac{75}{46.7} = 1.6059\ldots$$

The maximum value is about 1.61 standard deviations above the mean. The minimum value of 180 is \$70 smaller than the mean, or

$$Z = \frac{X - \overline{X}}{S} = \frac{180 - 250}{46.7} = \frac{-70}{46.7} = -1.4989\ldots$$

approximately 1.50 standard deviations below the mean. By measuring distance in units of the standard deviation, we turn an absolute distance of \$75 dollars into a relative distance of 1.61 standard deviations. The units of the raw data have disappeared, and distances are measured relative to the data set's mean $\overline{X}$ and standard deviation S.

□

In Section 3.5 we will discuss the concentration of data between certain Z-scores. Another encounter with Z-scores will appear in Chapter 7; in fact, we will continue to use the concept of Z-scores for several chapters to come.

COMMENTS

1. A Z-score can be a positive or negative number or the number zero. Positive Z-scores are associated with values of X above the mean and negative Z-scores correspond to those values below the mean. The Z-score of zero occurs only when the value of X coincides with the mean.
2. Use the rounded values of the mean and standard deviation in Equations 3–10a and 3–10b when computing a Z-score. We recommend rounding and reporting Z-scores to two decimal places to the right of the decimal point.

Coefficient of Skewness

The shape of the distribution of a data set, both for samples and populations, is of interest. A common shape is the bell- or mound-shaped, symmetrical shape illustrated in Figure 3.4. The concept of symmetry requires the distribution to be identical in every respect on either side of a reference line.

Graphing a data set reveals the shape of the sample distribution. An alternative to graphing to check the condition of symmetry is to find the *coefficient of skewness*.

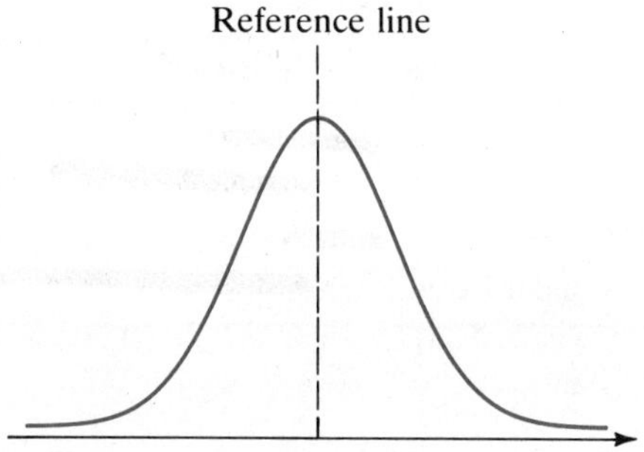

Figure 3.4 A Mound-Shaped, Symmetrical Distribution

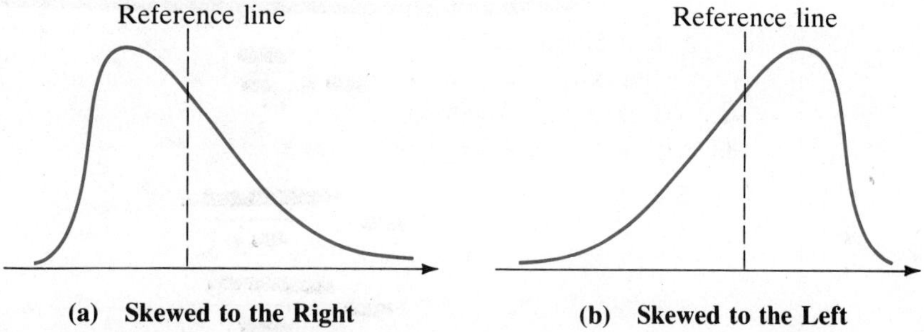

Figure 3.5 Skewed Distributions

It quantifies the asymmetry or lack of symmetry in the data set and represents a measure of shape. The verb *skew* means to move or slant sideways. Relative to Figure 3.4, to skew the distribution means the mound is moved sideways, away from the reference line, causing the curve to be slanted in one direction. Figure 3.5 shows a distribution slanted or skewed to the right (positively asymmetrical) in panel (a) and one slanted or skewed to the left (negatively asymmetrical) in panel (b). Neither distribution is symmetrical, relative to the dashed reference line.

Definition

The **coefficient of skewness** is a measure of a data set's deviation from symmetry.

Table 3.5 Symbols for the Coefficient of Skewness

Measure	Symbol
Population coefficient of skewness	γ_1
Sample coefficient of skewness	g_1

The notation for the population and sample coefficients of skewness is in Table 3.5. To compute the coefficient of skewness, we use Equations 3–11, 3–12a, and 3–12b when dealing with raw data. We omit the corresponding formulas for organized data; Exercise 3.52 asks you to supply them.

Sample Coefficient of Skewness for Raw Data

$$g_1 = \frac{m_3}{m_2^{3/2}} \tag{3–11}$$

where

$$m_3 = \frac{\Sigma(X - \overline{X})^3}{n} \tag{3–12a}$$

$$m_2 = \frac{\Sigma(X - \overline{X})^2}{n} \tag{3–12b}$$

The quantity m_2 is similar to the variance S^2 except for the divisor of n. The quantity m_3 represents the average of the cubed deviations from the mean.

EXAMPLE 3.15

Find the sample coefficient of skewness for these data.

2	0	6	3	9	2	0	2

Solution:

First, you can verify that the sample mean is $\overline{X} = 3$. Next, we need to generate deviations, squared deviations, and cubed deviations:

X	$X - \overline{X}$	$(X - \overline{X})^2$	$(X - \overline{X})^3$
2	−1	1	−1
0	−3	9	−27
6	3	9	27
3	0	0	0
9	6	36	216
2	−1	1	−1
0	−3	9	−27
2	−1	1	−1
	0	$\Sigma(X - \overline{X})^2 = 66$	$\Sigma(X - \overline{X})^3 = 186$

The quantities m_2 and m_3 are found according to Equations 3–12a and 3–12b:

$$m_2 = \frac{\Sigma(X - \overline{X})^2}{n} = \frac{66}{8} = 8.25$$

$$m_3 = \frac{\Sigma(X - \overline{X})^3}{n} = \frac{186}{8} = 23.25$$

The coefficient of skewness is

$$g_1 = \frac{m_3}{m_2^{3/2}} = \frac{23.25}{(8.25)^{3/2}} = \frac{23.25}{23.6963} = .98116\ldots$$

Rounded to three decimal places, $g_1 = .981$.

□

The interpretation of g_1 revolves around the value of zero. If the distribution of the data is symmetrical, $g_1 = 0$. If $g_1 \neq 0$, the distribution is skewed. Thus, when $g_1 > 0$, the distribution is skewed to the right (or positively skewed) as indicated in panel (a) of Figure 3.5. If $g_1 < 0$, the distribution is skewed to the left (or negatively skewed) as in panel (b) of Figure 3.5. Since $g_1 = +.981$ in Example 3.15 we conclude that the distribution of the sample data is not symmetrical but skewed to the right, as shown in Figure 3.6.

Rarely will a set of data be perfectly symmetrical so that $g_1 = 0$. However, if g_1 is close to zero, the degree of skewness in the distribution is very slight. When this happens, we usually say the distribution is *approximately* symmetrical, rather than skewed. We are tolerant of slight imperfections in our distribution but not major flaws. This suggests that the magnitude of the value for g_1 plays a key role in how we describe the distribution in words.

COMMENTS

1. There are other ways to measure skewness. Popular alternatives to g_1 involve comparing the distance between the mean and median or mean and mode, relative to the standard deviation, as in

$$\frac{\overline{X} - Md}{S} \quad \text{or} \quad \frac{\overline{X} - Mo}{S}$$

These measures are less sensitive to extreme observations than is g_1.

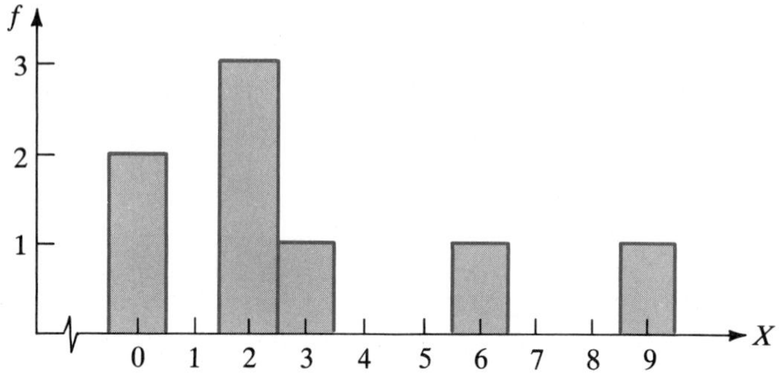

Figure 3.6 Skewed to the Right Sample Data Set

2. If $\overline{X}$ does not work out to a convenient value, the squared and cubed deviations can become messy. We provided a computational alternative to $\Sigma(X - \overline{X})^2$ in Equation 3–6. A similar trick for $\Sigma(X - \overline{X})^3$ is

$$\Sigma(X - \overline{X})^3 = \Sigma X^3 - \frac{3(\Sigma X^2)(\Sigma X)}{n} + \frac{2(\Sigma X)^3}{n^2}$$

3. Use the unrounded values of m_2 and m_3 when computing the coefficient of skewness. We recommend rounding and reporting g_1 to three decimal places to the right of the decimal point.

Percentiles

Like the Z-score, a percentile is a measure of position. Percentiles are numbers that divide an ordered data set into 100 equal parts. Associating a value in a data set with a specific percentile indicates the value's position in the data set. For example, if the one-year return of 16.8 percent for a mutual fund corresponds to the 63rd percentile, then we know that this mutual fund outperformed 63 percent of all mutual funds for that year. Conversely, about 37 percent of the mutual funds had a higher one-year return than 16.8 percent.

Definition

The ***k*th percentile,** denoted by P_k, for a distribution of values is a number that divides the distribution into two regions. The area of the region below P_k *is k* percent of the total area; the area of the region above P_k is $(100 - k)$ percent of the total area.

In our definition, the letter k can be any number between 0 and 100, although usually k is an integer. For a set of ordered, raw data, the definition of P_k changes slightly:

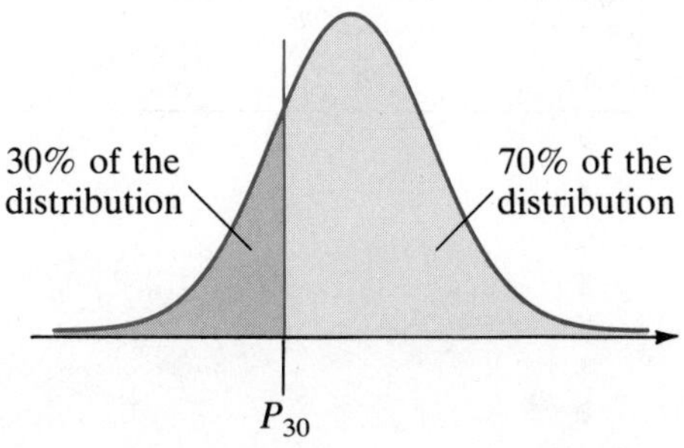

Figure 3.7 The 30th Percentile

The region below P_k—between the minimum value and P_k—contains at least k percent of the values. Figure 3.7 illustrates the position of the 30th percentile in a bell- or mound-shaped, symmetrical distribution and its partitioning of the total area.

Some percentiles—like the 25th, 50th, and 75th—are used more often than others. The difference between the 75th percentile, P_{75}, and the 25th percentile, P_{25}, sometimes is employed as a "modified" range, since between them the middle 50 percent of the distribution is concentrated. The modified range is called the *interquartile range;* see Exercise 3.61. The 25th, 50th, and 75th percentiles also are known as the first, second, and third *quartiles,* respectively. The notation for quartiles is Q_1, Q_2, and Q_3. Finally, the median serves double duty as a measure of central location and a measure of position since $Md = P_{50} = Q_2$.

A formula for computing P_k for a grouped frequency distribution is provided in Exercise 3.62. An algorithm for finding P_k for a set of raw data or for organized data in an ungrouped frequency distribution is included in Exercise 3.60.

Although there are additional measures for data sets, the four measures introduced in this section—coefficient of variation, Z-score, coefficient of skewness, and percentile—provide an adequate supplement to the measures of central location and dispersion.

CLASSROOM EXAMPLE 3.5

Computing the Coefficient of Skewness

Twenty-four-hour "hotline" numbers have become a common service in most major companies. A poll of several companies taken to determine the number of employees staffing the telephones during the "graveyard shift" from 11:00 p.m. to 7:00 a.m. yielded these data:

6	3	10	8	8	4	7	2

Find the sample coefficient of skewness.

X	$X - \overline{X}$	$(X - \overline{X})^2$	$(X - \overline{X})^3$
6			
3			
10			
8			
8			
4			
7			
2			
$\Sigma X =$ ______		$\Sigma(X - \overline{X})^2 =$ ______	$\Sigma(X - \overline{X})^3 =$ ______

$$m_2 = \frac{\Sigma(X - \overline{X})^2}{n} = \_\_\_\_\_\_$$

$$m_3 = \frac{\Sigma(X - \overline{X})^3}{n} = \_\_\_\_\_\_$$

$$g_1 = \frac{m_3}{m_2^{3/2}} = \_\_\_\_\_\_$$

SECTION 3.4 EXERCISES

3.47 Consider the sample data set

95	23	0	25	3	52

and find the following:

a. $\overline{X}$ **b.** S

c. CV **d.** g_1

3.48 A middle level manager recorded the number of long distance calls he made each day for a sample of seven working days. The data are as follows:

5	3	6	2	7	0	8

a. Compute the mean of these data.

b. Find the variance.

c. What is the Z-score for the value 5?

d. A second manager recorded the same information for a period of nine working days: 4 0 2 0 1 9 9 8 7. Which manager has the higher relative variation in daily numbers of long distance calls?

3.49 Compute the coefficient of skewness for these sample data: 1.89 5.66 9.65 3.08.

3.50 A sample of five cash sales during the first hour of business in a certain department of a retail store yielded the following figures:

\$59.50	\$22.90	\$46.55	\$19.19	\$25.00

How far from the mean in terms of standard deviations is the maximum value? The minimum value?

3.51 Find the sample coefficient of skewness for the following data:

20.8	27.6	16.6	19.2	15.8

Indicate whether the distribution would be skewed or symmetrical.

3.52 Edit Equations 3–12a and 3–12b to produce expressions that could be used to find g_1 for a grouped frequency distribution.

3.53 Using the equations from Exercise 3.52, compute g_1 for these organized data:

Classes	f
4– 7	7
7–10	8
10–13	5
13–16	1

3.54 Find the Z-score corresponding to $X = 0$ in a sample data set with a mean of 8 and a variance of 5.

3.55 The return on investment for two alternatives is summarized as follows:

	Projected Return (%)	
Alternative	Mean	Standard Deviation
A	9.2	5.5
B	13.1	7.8

a. Which alternative is riskier in terms of relative dispersion?

b. Practically speaking, why is it a moot point to declare yourself a risk-avoider or a risk-taker when choosing one of these alternatives?

3.56 If the mean and standard deviation for a set of sample data are 31.5 and 12.4, respectively, and if the Z-score corresponding to an unknown value of X is 1.25, find X.

3.57 Two investments are projected to have the following amounts of expected return and risk:

	Projection (%)	
Investment	Expected Return	Risk
1	10	2.5
2	5	1.5

Which investment carries the larger relative risk?

3.58 Three financial opportunities—P, Q, and R—are estimated to have the following expected returns and risks:

	Projection (%)	
Opportunity	Expected Return	Risk
P	10.21	4.68
Q	7.97	3.98
R	8.11	3.95

a. Find the coefficient of variation for each investment opportunity. Which opportunity has the lowest relative risk?

b. Find the Z-score corresponding to a return of 0 percent for each opportunity. Which opportunity has the smallest Z-score for 0 percent? Explain the significance of this in terms of the likelihood of losing money.

3.59 A set of seven values produces the following sums:

$$\Sigma X = 12.91 \qquad \Sigma X^2 = 23.8459 \qquad \Sigma X^3 = 44.116087$$

Find the coefficient of skewness and describe, in words, the shape of the sample data set.

***3.60** To find the kth percentile, P_k, for a set of raw data consisting of n values, we suggest the following procedure:

(i) Compute $nk/100$.
(ii) Find the position occupied by P_k:
If $nk/100$ is an integer, position $= (nk/100) + .5$.
If $nk/100$ is not an integer, position $=$ next integer above $nk/100$.
(iii) Locate the position occupied in the *ordered* data set. (Note: Positions that end in .5 are halfway between adjacent positions.)
(iv) P_k is the value of X in that position.

For example, find P_{30} for this set of data:

1.36	1.82	1.65	1.31	1.30	1.86	1.40	1.55

*Optional

The steps are

(i) $\frac{nk}{100} = \frac{8(30)}{100} = 2.4$

(ii) Position occupied = 3
(iii) Ordered data set:

pos	1	2	3 ↓	4	5	6	7	8
X	1.30	1.31	1.36 ↑	1.40	1.55	1.65	1.82	1.86

(iv) $P_{30} = 1.36$

For the same set of data, find the following:
a. P_{67} **b.** P_{81} **c.** Q_1

***3.61** The *interquartile range*, denoted IR, is the difference between the third and first quartiles:

$$IR = Q_3 - Q_1$$

Find the IR for the data in Exercise 3.47.

***3.62** To find the kth percentile, P_k, for a grouped frequency distribution, we recommend the following formula:

$$P_k = L + \frac{w}{f_k}(k\% \text{ of } n - cf_p)$$

where

L = lower class endpoint of the class containing P_k
w = width of the class containing P_k
f_k = frequency of the class containing P_k
cf_p = cumulative frequency preceding the class containing P_k
n = sample size

As we encountered with the median, the most important step is finding the class containing P_k. To do so, compute k percent of n. Then find the class that contains the position corresponding to k percent of n. If k percent of n falls between the last position occupied of one class and the first position occupied in the succeeding class, use the succeeding class as the class containing P_k.

For example, to find P_{44} for the data in Exercise 3.53, we compute 44 percent of $n = 21$, or 9.24, which is not the last position occupied within any of the four classes. Rather, the class 7–10 contains the 9.24th position. The 44th percentile is

$$P_{44} = 7 + \frac{3}{8}(9.24 - 7) = 7.84$$

For the data in Exercise 3.53, find
a. P_{80} **b.** P_{23} **c.** Q_3

***3.63** In describing the shape of a distribution we often use a second measure to supplement the information from the coefficient of skewness. The *coefficient of kurtosis,* denoted by g_2 for the sample and by γ_2 for the population, is a measure of a data set's peakedness or flatness relative to a bell-shaped distribution. A formula for computing g_2 is

$$g_2 = \frac{m_4}{m_2^2} - 3$$

where

$$m_4 = \frac{\Sigma(X - \overline{X})^4}{n}$$

and m_2 is defined in Equation 3–12b. By subtracting 3 in the formula for g_2, we establish a baseline value of $g_2 = 0$ for the mound-shaped, symmetrical distribution of Figure 3.4. When $g_2 > 0$, the distribution is "taller" than this standard shape, and the distribution is sometimes called *leptokurtic*. When $g_2 < 0$, the distribution is "flatter" than the standard shape, and the distribution is called *platykurtic*. When $g_2 = 0$, the distribution is called *mesokurtic*. As an illustration of computing g_2, let's use the data in Example 3.15. To find m_4 we compute $\Sigma(X - \overline{X})^4 = (-1)^4 + (-3)^4 + (3)^4 + \ldots + (-1)^4 = 1542$; hence $m_4 = 1542/8 = 192.75$. Since we already computed $m_2 = 8.25$,

$$\begin{aligned} g_2 &= \frac{m_4}{m_2^2} - 3 \\ &= \frac{192.75}{(8.25)^2} - 3 \\ &= 2.832 - 3 = -.168 \end{aligned}$$

indicating that the data set is slightly flatter than the standard.

Find g_2 for the data in Exercise 3.47.

***3.64** For these data,

13	7	2	13	5	2	11	0	4	3

compute the following:

a. m_2 **b.** m_3 **c.** m_4 **d.** g_1 **e.** g_2

f. Characterize the distribution in terms of skewness and kurtosis.

3.5 RELATIONSHIPS AMONG MEASURES

Convenient relationships in the form of inequalities exist between and among several of the descriptive measures presented thus far. There are two main benefits to be gained from studying these relationships. First, they provide a common-sense check on the accuracy of your computations. Since most of the descriptive statistics involve calculations, you run a risk of making simple math errors when working the problems in this chapter.

Second, the relationships help to build visual images of the data sets. If we told you that we were both 6 feet, 3 inches tall and weighed 250 pounds, you might imagine us as interior linemen for a football team, not as textbook authors. Alternatively, the image created in your mind of a person 5 feet, 1 inch tall and weighing 105 pounds would be quite different. Similarly, when you compute a positive number for g_1, a picture of a skewed to the right data set with a certain positioning for the measures of central location should emerge in your mind without having to first graph the data.

In the graphs that follow we will draw smooth curves in lieu of histograms. It is important to note that some of these relationships are valid *only* for smooth curves (or histograms that can be approximated by smooth curves) and not raw data. Please refer to the Comments included in each section for the appropriate conditions.

Central Location Relations

When the distribution of the data is bell shaped and symmetrical, as in Figure 3.8, the mean, median, and mode are equal and the coefficient of skewness is zero. When the distribution is skewed to the right, the skewness coefficient is positive, and the mode is less than the median which, in turn, is less than the mean, as Figure 3.9 shows. As we might expect, a skewed to the left distribution reverses the order of the measures of central location: the mean is less than the median which is less than the mode. (See Figure 3.10.) These relationships are summarized in the following box.

Measures of Central Location Inequalities

Distribution	Coefficient of Skewness	Measures of Central Location
Mound shaped, symmetrical	$g_1 = 0$	Mean = median = mode
Skewed to the right	$g_1 > 0$	Mean > median > mode
Skewed to the left	$g_1 < 0$	Mean < median < mode

COMMENT The relationships in the preceding box apply to distributions that are smooth curves and have only one mode. They also apply, generally, to organized data in grouped frequency distributions that contain no empty intervals. The relationships are not always true for small sets of raw data. (See Exercise 3.66.) Sometimes the mode does not exist (see Exercise 3.69), or perhaps two of the three measures are equal.

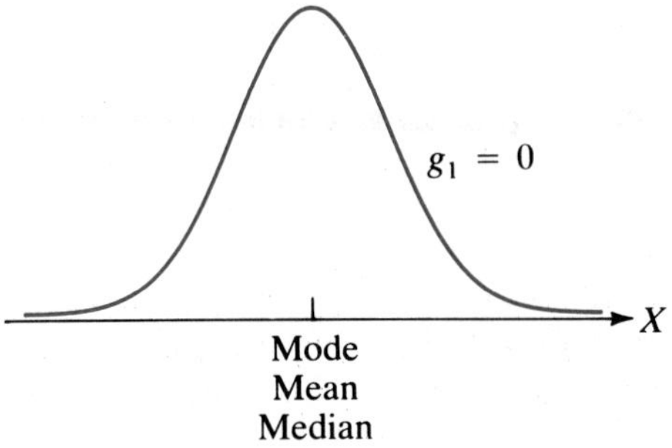

Figure 3.8 Mound-Shaped, Symmetrical Distribution Central Location Relations

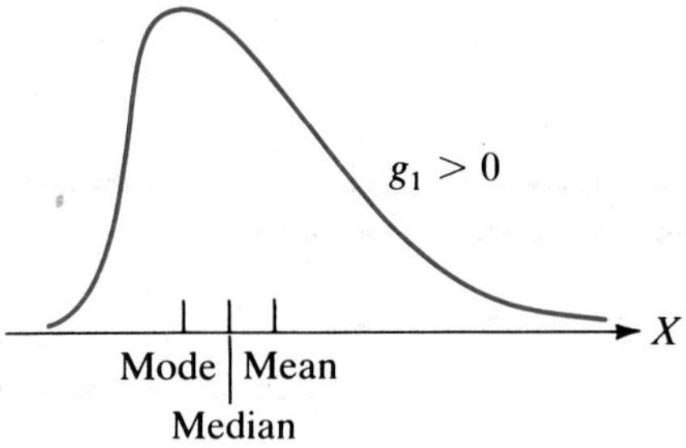

Figure 3.9 Skewed to the Right Distribution Central Location Relations

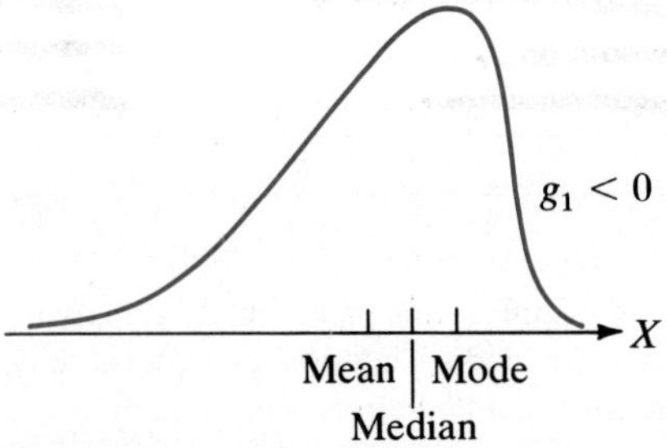

Figure 3.10 Skewed to the Left Distribution Central Location Relations

EXAMPLE 3.16

A production manager was responsible for incorporating the data generated by his quality assurance staff into a monthly report, complete with graphs and summary statistics. One of the products on which data were gathered has a hardness specification of at least 2.90. Of the 640 units manufactured last month, the quality assurance engineers sampled 60 and noted the product's hardness with a scope reading. In his monthly report, the production manager included the frequency histogram shown in Figure 3.11 and the following summary statistics:

Descriptive Statistics

Sample size = 60	Standard deviation = .023
Mean = 2.933	Skewness = −.373
Median = 2.930	

After reviewing his report, the production manager's boss scheduled a meeting ". . . to discuss a potential discrepancy in the monthly report." What raised the boss's suspicions?

Solution:

The graph and the statistics are not consistent. The histogram shows a distribution with a slightly longer tail to the right that suggests a skew in that direction. For such a distribution, the mean value should exceed the median. This is borne out by the reported values of 2.933 and 2.930, respectively. However, the indicated value for the

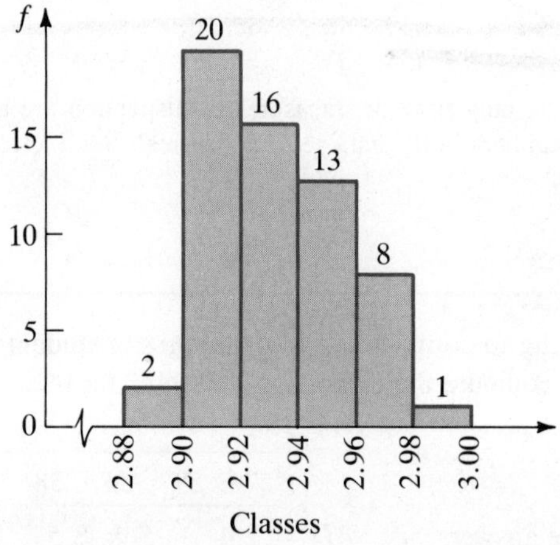

Figure 3.11 Sample of 60 Hardness Readings

coefficient of skewness is negative, which contradicts the previous discussion. Either the graph and the values for the median and mean are correct (which means g_1 is incorrect), or g_1 is correct, and the graph, the median, and the mean are not. More than likely g_1 is in error, and the computations need to be rechecked.

□

Dispersion Relations

Unlike the inequalities relating the measures of central location which depended on the shape of the distribution, the inequality linking the measures of dispersion is valid for *any* set of raw data. Since each measure of dispersion must be a positive number, the number 0 is the smallest possible value for each statistic. The following box summarizes the remaining relationships among the measures.

Measures of Dispersion Inequality for $n \geq 4$

$$0 \leq AD \leq S \leq .6Rg$$

This inequality implies that the standard deviation is always greater than or equal to the average deviation. Also, the standard deviation's value cannot exceed 60 percent of the range for sample sizes of four or more. Actually, this percentage decreases as n increases, but 60 percent qualifies as a general purpose value. Usually, the standard deviation is much less than $.6Rg$.

COMMENTS

1. Shiffler and Harsha (1980) derived the measures of dispersion inequality and showed how it can be modified for $n = 2$ and $n = 3$ by changing $.6Rg$ to $.707Rg$ and $.612Rg$, respectively.
2. A lower bound for the standard deviation involving a function of the range is included in Exercise 3.80.
3. The only time the measures of dispersion are equal and equal zero is when all the numbers in the data set are identical. See Exercise 3.68.

EXAMPLE 3.17

In hurrying to complete an in-class quiz, a student gave the following answers when asked to compute the dispersion measures for these data:

57	59	54	58	52

Student's answers: $AD = 2.4$ $S = 8.5$ $Rg = 7$

Why should she have known immediately that something was in error?

Solution:

One more glance at her answers with the Dispersion Inequality in mind might have revealed the error. The standard deviation must be less than or equal to 60 percent of the range. Her answer of 8.5 for S violates this part of the Dispersion Inequality. She forgot to take the square root to get the standard deviation. The correct answer for S is $\sqrt{8.5}$ or 2.9.

□

Empirical Rule

The Empirical Rule combines a measure of central location—the mean—with a measure of dispersion—the standard deviation—as it relates the concentration of values in a data set to positions within the data set. It is not an equality. It is a rule of thumb and applies to data sets which are mound or bell shaped and symmetrical, or approximately so.

The Empirical Rule

If the distribution of a data set is approximately mound shaped and symmetrical, then

1. The interval from $\overline{X} - S$ to $\overline{X} + S$ will contain approximately 68 percent of the values.
2. The interval from $\overline{X} - 2S$ to $\overline{X} + 2S$ will contain approximately 95 percent of the values.
3. The interval from $\overline{X} - 3S$ to $\overline{X} + 3S$ will contain approximately 99.7 percent of the values.

The Empirical Rule tells us what to expect in terms of percentages of the data inside and outside certain regions of the distribution. Roughly two-thirds of the data reside within one standard deviation of the mean. Since the distribution is symmetrical, the percentage of measurements beyond one standard deviation is split evenly in the tails of the curve, as Figure 3.12 shows. Two standard deviations above and below the mean account for approximately 95 percent of the data, leaving only 5 percent to be distributed in the upper and lower tails of the curve.

Figure 3.13 has a tremendous impact on both descriptive and inferential statistics. It indicates where most of the data will be found. Ninety-five percent is a hefty percentage, the result of answering 19 out of 20 questions correctly on a test, for instance. The two standard deviation interval implicitly labels values that are beyond it as "rare"; only 5 percent of the time should they appear.

As a result, the interval $\overline{X} \pm 2S$ stakes out reasonable limits for the values of a data set. Values falling inside the interval are considered likely or expected, while

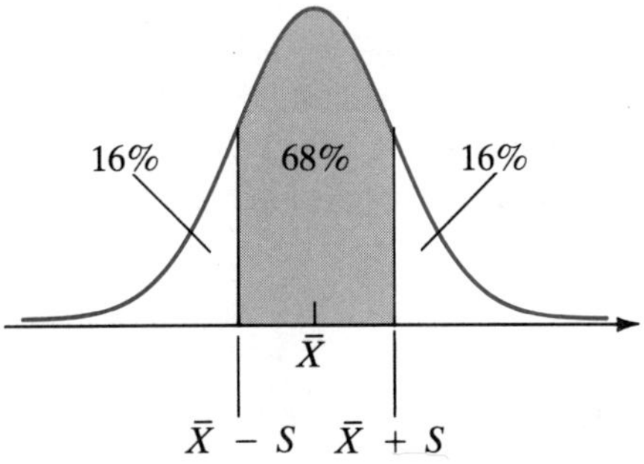

Figure 3.12 Plus and Minus One Standard Deviation Interval

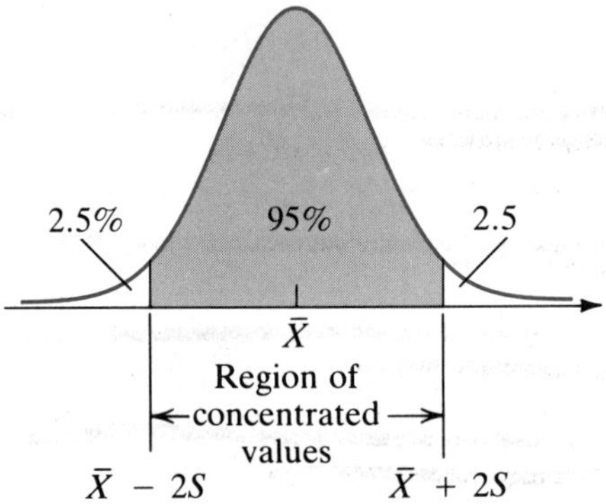

Figure 3.13 Plus and Minus Two Standard Deviation Interval

values outside this region are unusual or unlikely. The following definition formalizes these concepts.

> **Definition**
>
> The **region of concentrated values** extends two standard deviations on either side of the mean and, for mound-shaped distributions, contains 95 percent of the values. For a sample, the region is $\bar{X} \pm 2S$; for a population, it is $\mu \pm 2\sigma$.

EXAMPLE 3.18

Refer to the data in Figure 3.11 of Example 3.16. Find the region of concentrated values and determine the actual percentage of values falling within it. Compare this percentage to the figure provided by the Empirical Rule.

Solution:

The first point to note is that Figure 3.11 is somewhat mound shaped, but not very symmetrical, suggesting that the figure of approximately 95 percent *may not* be accurate for this data set. The mean and standard deviation from Example 3.16 are 2.933 and .023, respectively. The region of concentrated values is therefore

$$\bar{X} \pm 2S = 2.933 \pm 2(.023) = 2.887 \text{ to } 2.979$$

The region of concentrated values spans all values between 2.89 and 2.97. This includes the middle four classes of Figure 3.11, plus part of the first class, representing a frequency of at least 57 of the 60 values. Approximately 95 percent of the data generally should be found in the region of concentrated values; this result is found to hold in this case.

□

The Empirical Rule also describes the concentration of values within three standard deviations of the mean. As we deviate farther from the mean, we account for more of the values, and the percentage of non-included values decreases.

Notice in Figure 3.14 that the distance spanned from $\bar{X} - 3S$ to $\bar{X} + 3S$ is six standard deviations. Since almost all of the data is contained within these limits, the minimum value in a mound-shaped, symmetrical data set is located fairly close to the point $\bar{X} - 3S$, and the maximum value is near $\bar{X} + 3S$:

$$\max \cong \bar{X} + 3S \qquad \min \cong \bar{X} - 3S$$

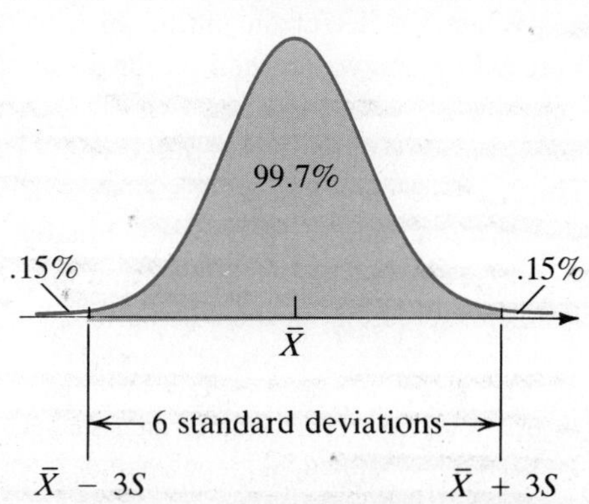

Figure 3.14 Plus and Minus Three Standard Deviation Interval

Recall that the range is the difference between the maximum and minimum values. Thus, for *mound-shaped* distributions, we have a relationship between the standard deviation and the range: Six standard deviations is approximately equal to the range. We can use this information to create a reasonability check for the computation of the standard deviation in large data sets. For sample sizes exceeding, say, 100, Equation 3–13 is a reasonable *approximation* for S:

$$S \cong \frac{Rg}{6} \tag{3–13}$$

In samples containing (maybe) 30 to 100 entities, a better *approximation* for the standard deviation is given by Equation 3–14:

$$S \cong \frac{Rg}{4} \tag{3–14}$$

When n falls below 30, the approximation is not stable and is not recommended.

Values that are more than three or four standard deviations from the mean are obviously unusual. When we discover such an extreme value, we may question whether it is truly consistent with the rest of the data.

An extremely large or small value relative to the other values in the data set greatly influences measures such as the mean, standard deviation, and coefficient of skewness and may mislead us into believing that a distribution is severely skewed or highly variable. An extreme value is called an *outlier*.

Definition

An **outlier** is a value located more than four standard deviations from the mean.

Our definition is a rule of thumb for identifying outliers in data sets with roughly 18 or more values. Equivalently, an outlier is any value with a Z-score below -4 or above $+4$. When we detect an outlier in a sample, we recommend the following remedial procedure. As we learned in our discussion of managing data in Chapter 2, we first recheck the unusual data point for accuracy and reasonability. Second, we recheck the computations for $\overline{X}$ and S and verify that the extreme value qualifies as an outlier. Third, if the value is an outlier, we can hold it out and redo the descriptive analysis on the remaining $n - 1$ values. Finally, we display both sets of descriptive statistics—one set including the outlier and one set excluding it—to show the effect of the extreme observation on the calculations.

COMMENTS

1. The Empirical Rule holds for raw and organized data sets. When n is fairly small or the distribution is other than mound shaped, the Empirical Rule is unreliable and should not be applied. Another, more general, rule called Chebyshev's theorem applies for *all* data sets regardless of their shape—see Exercises 3.133 and 3.134.
2. If the data set is a population rather than a sample, the Empirical Rule still applies with the symbols μ and σ replacing $\overline{X}$ and S, respectively.

3. There are other ways to define outliers. Tukey (1977) identifies potential outliers as those values beyond the "inner fences." The lower inner fence is found by subtracting the quantity 1.5 times the interquartile range (see Exercise 3.61) from Q_1. The upper inner fence is $Q_3 + 1.5$ IR. Shiffler (1988) showed that outliers, as spelled out in our definition above, cannot exist for small data sets in which $n \leq 17$ (see Exercises 3.77 and 3.78). Consequently, Tukey's definition may be more appropriate for small data sets, while ours is more appropriate when $n \geq 18$.

SECTION 3.5 EXERCISES

3.65 Find the mean, median, and mode for these data:

2	0	6	3	9	2	0	2

Based on the values of these three summary statistics, can you infer the sign of g_1? Compare your guess with the value of g_1 in Example 3.15.

3.66 Consider the following set of sample data:

0	6	6	8	10

a. Compute the mean, median, and mode.
b. Based on your answers in part a, infer the sign of value of g_1.
c. Compute g_1.
d. Compare your answers to parts a and c with the measures of central location inequalities listed in the box on page 163. What lesson has this example demonstrated?

3.67 A bag of dog food treats advertises a total weight of 3 oz. To check this, a consumer advocacy group selected ten grocery stores in the area and randomly sampled six bags of this product from each. Data on the variable X = actual weight of a 3 oz. bag of treats is listed below:

3.03	3.02	2.97	2.98	3.01	2.94
2.93	2.93	2.93	2.91	2.98	3.05
2.97	2.95	3.02	2.94	2.98	2.98
2.99	2.97	3.00	2.89	2.98	2.94
2.94	2.95	2.96	2.91	2.95	2.98
2.93	2.98	3.00	2.90	3.04	2.99
2.96	2.92	2.92	2.93	2.96	2.96
2.94	2.93	2.91	3.00	2.93	2.96
2.95	2.95	2.95	2.98	2.97	2.98
2.91	2.97	2.95	3.02	3.02	2.94

a. Organize these data into a grouped frequency distribution.
b. Construct a frequency histogram.

c. Based on the graph of part b, identify the shape of the distribution.
d. Compute the mean, median, and mode for the organized data. Do they line up as suggested by the shape of the distribution?
e. Compute S for the organized data.
f. Generate the intervals $\overline{X} \pm S$ and $\overline{X} \pm 2S$. Using the raw data, count the number of values falling into each interval. Change the counts into percentages and compare with the Empirical Rule percentages.

3.68 Consider the following set of sample data: .3 .3 .3 .3.
a. Find the range.
b. Compute the average deviation and standard deviation.
c. Do the values satisfy the Dispersion Inequality? Explain.

3.69 A sampling of prime-time network TV shows on one network produced the following "shares" (percentage of sets in use):

24	19	16	23	21	29	22
33	18	13	46	44	34	36

a. Treating these as raw data, find the mean, median, and mode.
b. Compute the range, average deviation, and standard deviation.
c. Do the answers in part b satisfy the Dispersion Inequality?
d. Do the answers in part a conform to one of the Central Location Inequalities?
e. Predict the sign of g_1 and then compute g_1 as a check.

3.70 An answer given on a test revealed a value of -17 for the variance. Why is this obviously wrong?

3.71 The mean and standard deviation for a set of sample data are 41 and 4, respectively. If the data set is approximately mound shaped, what percentage of data falls between the following sets of numbers:
a. 37 and 45 b. 33 and 45 c. 49 and 53

3.72 Cholesterol levels of adults are approximately mound shaped with a mean of 220 milligrams/decaliter (mg/dl) and a standard deviation of 30 mg/dl. What approximate percentage of adults' cholesterol readings fall in the following ranges?
a. between 130 and 190 mg/dl
b. above 250 mg/dl
c. between 160 and 250 mg/dl

3.73 Suppose the distribution of a data set is mound shaped and symmetrical. Express the location of the 16th percentile in terms of the mean and standard deviation. Where is the 84th percentile located? (*Hint:* Study Figure 3.12.)

3.74 The mean and standard deviation for a set of sample data are 65 and 12, respectively. If the data set is approximately mound shaped, what percentage of the data falls within the following groupings?
a. below 53 b. between 53 and 89
c. above 41 d. between 65 and 101
e. between 29 and 89

3.75 The percentage increase (or decrease) in building permits from one year to the next was recorded for a sample of 50 cities and organized into the following frequency distribution:

Classes	f
−42.3– −20.6	2
−20.6– 1.1	4
1.1– 22.8	6
22.8– 44.5	9
44.5– 66.2	12
66.2– 87.9	10
87.9– 109.6	7

a. Compute $\overline{X}$ and S.
b. Find the region of concentrated values.
c. Construct a frequency histogram and plot the region of concentrated values.
d. Determine the percentage of actual values in the region of concentrated values and compare with the baseline figure of 95 percent.

3.76 Suppose you were given the following descriptive statistics for a sample data set of size $n = 34$.

$Mo = 38$	$Rg = 32$
$Md = 37$	Min $= 18$
$\overline{X} = 35.4$	$S = 6.6$

Sketch the probable shape of a frequency curve for this data set.

***3.77** For any set of n values there are limits to the magnitude of the associated Z-scores. Shiffler (1987) showed that the largest possible Z-score (associated with the maximum value of X), call it $Z_{\max}$, is constrained by the following inequality

$$\sqrt{1/n} \leq Z_{\max} \leq (n - 1)/\sqrt{n}$$

Similarly, the smallest Z-score, $Z_{\min}$, satisfies this relationship

$$-(n - 1)/\sqrt{n} \leq Z_{\min} \leq -\sqrt{1/n}$$

a. Consider the following data set: 0, 0, 6, 6. Find $\overline{X}$, S, and the Z-score ($Z_{\max}$) corresponding to $X = 6$. Does $Z_{\max}$ satisfy the above inequality?
b. Find $\overline{X}$ and S for the following data set:

0	10	10	10	10	10	10	10	10	10

Compute $Z_{\min}$, the Z-score corresponding to $X = 0$, and compare it to the inequality for $Z_{\min}$ when $n = 10$.

*Optional

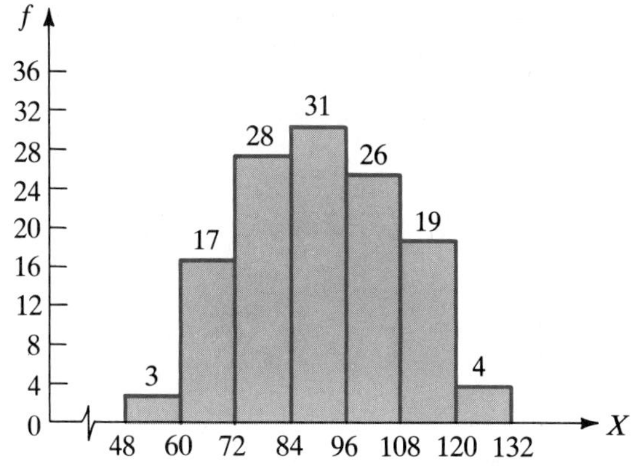

Figure 3.15 Graph of a Grouped Frequency Distribution

*3.78 Compute the mean, standard deviation, and Z_{max} for the following data sets. Identify extreme values that represent outliers.

a. 0 0 0 0 100
b. 0 0 0 0 0 0 0 0 0 1 million
c. 0 0 0 0 0 0 0 0 0 0 0 0 0 0 0 0 0 0 1

3.79 Refer to the frequency histogram in Figure 3.15. Using the Empirical Rule, estimate the values of $\overline{X}$ and S.

*3.80 Macleod and Henderson (1984) developed a tighter set of bounds on the sample standard deviation for a set of raw data:

$$\frac{Rg}{2}\sqrt{\frac{2}{n-1}} \leq S \leq \frac{Rg}{2}\sqrt{\frac{n}{n-1}}$$

where

Rg represents the sample range

a. If $n = 5$ and $Rg = 10$, evaluate the inequality to form bounds on S.
b. Find the standard deviation for these data: 0 1 4 10 4
c. Does the value of S you computed in part b satisfy the inequality you generated in part a?

*3.81 Macleod and Henderson (1984) also showed that the sample standard deviation for organized data in a grouped frequency distribution must satisfy

$$\frac{(c-2)w}{2}\sqrt{\frac{2}{n-1}} \leq S \leq \frac{cw}{2}\sqrt{\frac{n}{n-1}}$$

where c is the number of classes and w is the class width. Verify that the frequency distribution in Exercise 3.67 satisfies this relationship.

***3.82** The sample mean (also known as the *arithmetic* mean) and the geometric mean (see Exercise 3.23) are related via $GM \leq \overline{X}$. This relationship suggests the geometric mean usually will produce a smaller value than the arithmetic mean, supporting the geometric mean's reputation in finance as being a more conservative measure. Find the GM and $\overline{X}$ for these data and verify their relationship:

10.4%	10.1%	10%	10.7%	9.5%	9.4%

*3.6 PROCESSOR: DESCRIPTIVE STATISTICS (RAW DATA)

■ Background

The Descriptive Statistics processor will compute the value of most of the summary statistics discussed in this chapter, for a set of either raw or organized data. We will discuss the Descriptive Statistics processor for raw data in this section and for organized data in Section 3.7. Here is a list of the measures that the processor will determine for raw data: mean, median, mode, midrange, minimum value, maximum value, range, variance, standard deviation, average deviation, coefficient of variation, coefficient of skewness, coefficient of kurtosis, and the geometric mean. The processor requires a data set as input, so we must either create one first (refer to the subsection Creating a Data Set in Section 2.5) or retrieve an existing file. To demonstrate the processor and file retrieval, we will use the following set of raw data:

2	4	2	2	4
0	1	3	4	5
2	4	0	4	1

Recall that we used this data set in Chapter 2 to illustrate the Frequency Distributions Processor and saved it in a data file named CHP2R.TXT.

■ Retrieving a Data File

After obtaining the Overview screen (Figure 1.3), touch F8 to clear the screen and then use the right arrow key to move the highlight along the main menu to Files. Be sure the data disk containing the CHP2R.TXT data file is in the appropriate drive. (For purposes of this discussion, we will assume you have a two-drive computer with the data disk in drive B.)

*Optional

```
≡        Microcomputer Applications for INTRODUCTORY BUSINESS STATISTICS   EDIT
   Overview  Data  [Files]  Processors  Set Up  Exit                 | F1=Help
                  ┌─────────────┐
                  │ [Retrieve]  │
                  │ *ave        │
                  │ List        │
                  │ Rename      │
                  │ Erase       │
                  │ Directory   │
                  │ OS          │
                  └─────────────┘

  ╔══════════════ Data file "[drive:][\]path\]name.ext" ══════════════╗
  ║ Dir A:\*.*                                                         ║
  ╚════════════════════════════════════════════════════════════════════╝
```

Figure 3.16 Retrieving a File

Touch Enter; then, with the highlight on the Retrieve option, touch Enter again. The screen should appear as in Figure 3.16. In the box at the bottom of the screen is the DOS command Dir A:*.*, which means a directory of all files on the disk in drive A will be displayed by touching Enter. Since our data disk is in drive B, we must change the drive designation from A to B *before* touching Enter. Figure 3.17 shows the general appearance of the screen *after* changing the drive designation and touching Enter. (*Note:* Our data disk has several files on it, as Figure 3.17 indicates. Yours may contain only one file—CHP2R.TXT—at this point, so do not expect the screen to resemble Figure 3.17 *exactly*.) Move the highlight using the arrow keys to the file labeled CHP2R.TXT and touch Enter to produce the screen shown in Figure 3.18. The query listed in the box must be answered affirmatively (by touching the letter Y) in order to retrieve the file. After the Retrieve Successful message appears, touch Enter to return to the main menu.

■ Running the Descriptive Statistics for Raw Data Program

Proceed across the main menu with the right arrow key to Processors, then touch Enter and then select the Descriptive Statistics processor. Figure 3.19 illustrates the options available with this processor. Since we wish to demonstrate the raw data option in this section, we select the first option.

```
≡        Microcomputer Applications for INTRODUCTORY BUSINESS STATISTICS   FILE
   Overview  Data [Files] Processors  Set Up  Exit                | F1=Help
 ─────────────────────────── Retrieve file ───────────────────────────────────
  . <CURRENT>     <DIR>                 | .. <PARENT>     <DIR>
   CE161.TXT            733    4-29-85  | CE162.TXT            547     4-29-85
   CE183.TXT            387    8-15-89  | CE183R.TXT           227     4-29-85
   CH13MP.TXT           345    7-29-89  | CH14ONE.TXT          438     7-29-89
   CH14RB.TXT           361    7-29-89  | CH14TWO.TXT          788     7-29-89
   CH15SEC7.TXT         429    7-29-89  | CH18GOFT.TXT         215     9-04-89
   CH30.TXT             261    7-28-89  | CH4.TXT              513     7-28-89
   CH5.TXT              303    7-28-89  | CHP2G.TXT            261     5-10-89
   CHP2R.TXT            403    7-28-89  | CHP2RG.TXT           403     7-28-89
   CHP2RR.TXT           403    4-29-85  | CHP2U.TXT            345     7-28-89
   EXAMPLE.TXT          303    9-04-89  | MRTEST1.TXT          857     4-29-85
   MRTEST2.TXT          733    4-29-85  | MRTEST3.TXT         1773     5-16-89
   SRTEST1.TXT          429    5-23-89  | UNIT5X12.TXT         733     7-16-89
   XM153.TXT            429   11-02-89  | XM161.MTW            272     4-29-85
   XM161.TXT            857    4-29-85  | XM162.TXT           2429     4-29-85
   XM16T.TXT            485    4-29-85  | XM181.TXT            765     6-14-89
   XM192.TXT            299    9-06-89  | XR151.TXT            429     6-12-89
 ─────────────────────────── Directory B:\*.* ────────────────────────────────
```

Figure 3.17 Directory of Files on Drive B

```
≡        Microcomputer Applications for INTRODUCTORY BUSINESS STATISTICS  QUERY
   Overview  Data [Files] Processors  Set Up  Exit                | F1=Help
 ─────────────────────────── Retrieve file ───────────────────────────────────
   . <CURRENT>     <DIR>                | .. <PARENT>     <DIR>
   CE161.TXT            733    4-29-85  | CE162.TXT            547     4-29-85
   CE183.TXT            387    8-15-89  | CE183R.TXT           227     4-29-85
   CH13MP.TXT           345    7-29-89  | CH14ONE.TXT          438     7-29-89
   CH14RB.TXT           361    7-29-89  | CH14TWO.TXT          788     7-29-89
   CH15SEC7.TXT         429    7-29-89  | CH18GOFT.TXT         215     9-04-89
   CH30.TXT             261    7-28-89  | CH4.TXT              513     7-28-89
   CH5.TXT              303    7-28-89  | CHP2G.TXT            261     5-10-89
  [CHP2R.TXT            403    7-28-89] | CHP2RG.TXT           403     7-28-89
   CHP2RR.TXT           403    4-29-85  | CHP2U.TXT            345     7-28-89
   EXAMPLE.TXT ═════════════ Retrieve file ═════════════════════       4-29-85
   MRTEST2.TXT                                                         5-16-89
   SRTEST1.TXT        Retrieve file CHP2R.TXT? (y/n) No                7-16-89
   XM153.TXT                                                           4-29-85
   XM161.TXT                                                           4-29-85
   XM16T.TXT  ═════════════════════════════════════════════════       6-14-89
   XM192.TXT            299    9-06-89  | XR151.TXT            429     6-12-89
 ─────────────────────────── Directory B:\*.* ────────────────────────────────
```

Figure 3.18 Retrieving CHP2R.TXT File

```
≡        Microcomputer Applications for INTRODUCTORY BUSINESS STATISTICS    MENU
   Overview  Data  Files  [Processors]  Set Up  Exit                 | F1=Help

                    Simple random sample
                    Frequency distributions
                   [Descriptive statistics]

                      [Descriptive statistics for raw data]
                       Descriptive statistics for organized data

                    Two sample inference
                    Analysis of variance (ANOVA)
                    Regression
                    Time series
                    X² (chi-square)
```

Figure 3.19 Accessing the Raw Data Option of the Descriptive Statistics Processor

```
≡                  Descriptive Statistics For Raw Data                       MENU
   Introduction  Vector  [Type]  Calculations  *esults  Exit            | F1=Help

                        [Sample    ]
                         Population
```

Figure 3.20 Identifying the Data Set as a Sample

After reading the introductory screen, touch Enter or F8 to clear the screen and move the highlight to the Vector section of the main menu. Touch Enter to verify that there are no problems with the data file. Touch Enter again and move across the main menu to Type. Upon touching Enter, we should see the screen illustrated in Figure 3.20. We must inform the processor as to the type of data set we have: sample or population. With the Sample choice highlighted, touch Enter.

Now move across the main menu to Calculations and touch Enter twice. After the "Complete" message appears, touch Enter again and move to the Results section of the main menu. Touching Enter produces a screen with two choices: Summary Mea-

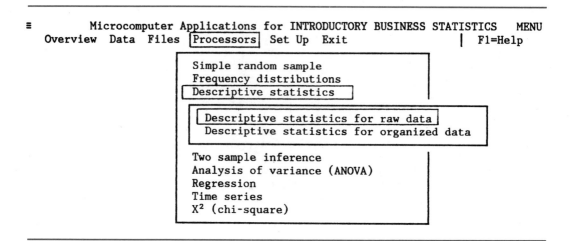

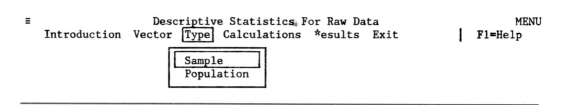

```
≡                    Descriptive Statistics For Raw Data                    READY
  Introduction  Vector  Type  Calculations [Results] Exit              | F1=Help

       Measure                          Value

        Mean..................... 2.5333
        Geometric mean........... 2.1072
        Median................... 2.0000
        Mode..................... 4.0000
        Midrange................. 2.5000
        Range.................... 5.0000
        Minimum value............ 0.0000
        Maximum value............ 5.0000
        Variance................. 2.5524
        Standard deviation....... 1.5976
        Average Deviation........ 1.3689
        Coefficient of Variation.. 63.06%
        Skewness.................-0.1713
        Kurtosis.................-1.1914
```

Figure 3.21 Results Screen for Descriptive Statistics for Raw Data Subprogram

sures or Intermediate Results. Figure 3.21 shows the screen that appears when we select the Summary Measures option. (The Intermediate Results option shows the values of the various sums needed to compute the measures.)

As you can see from Figure 3.21, the values of 14 descriptive statistics are computed and displayed for the data set listed at the beginning of this section. You can write down the results or print the screen at this point.

To run Descriptive Statistics with a set of data that have not been saved previously, create a data set—see Section 2.5—and then go directly to the Processor choice on the main menu, as described above.

*3.7 PROCESSOR: DESCRIPTIVE STATISTICS (ORGANIZED DATA)

■ Background

The Descriptive Statistics processor also can be used to find the mean, median, mode, variance, and standard deviation for a set of organized data in the form of an un-

*Optional

grouped or a grouped frequency distribution. We will demonstrate the processor with the following grouped frequency distribution:

Classes	*f*
2–12	5
12–22	7
22–32	5
32–42	3

This data set appeared in Classroom Example 3.4.

Creating the Data Set

Advance to the Data section of the main menu and select the Create option. For this application, we wish to create an Organized Grouped Frequency type of data set; select this type by touching the letter "G" or by highlighting the phrase and touching Enter. The response from the processor is a question as to the number of classes, c, in the frequency distribution. Since our organized data set has four classes, we type the number 4 in the box; Figure 3.22 illustrates the screen at this point.

To input the frequency distribution, we are required to provide the midpoint, M, and the frequency, f, of each class. Figure 3.23 is a replica of the screen after we have entered the four values of M and f. Upon touching Enter we should receive the "Data Entry Successful" message. Return to the menu by touching Enter. If we anticipate using the data set later, we should save it; otherwise, we proceed directly to the Processor section of the main menu.

Running the Descriptive Statistics for Organized Data Program

With the highlight on the Processor section of the main menu, touch Enter to bring up the choices. Either move the highlight to Descriptive Statistics and touch Enter or simply touch the letter D to access the two options within Descriptive Statistics. Select the second option, Descriptive Statistics for Organized Data.

Move the highlight to Data and touch Enter to verify that the data set is compatible with the processor. Then go to Calculations and depress the Enter key twice to complete the computations. Move across to the Results section of the menu and touch Enter to reproduce the screen shown in Figure 3.24. Choosing the first option—Summary Measures—will produce the results of Figure 3.25.

```
≡        Microcomputer Applications for INTRODUCTORY BUSINESS STATISTICS   EDIT
  Overview [Data] Files  Processors  Set Up  Exit                    | F1=Help

           [Create]
            *iew/Edit



                  ========== Data set size ==========

                    How many classes in the data set, c =  4

```

Figure 3.22 Creating an Organized Grouped Frequency Data Set

```
≡        Microcomputer Applications for INTRODUCTORY BUSINESS STATISTICS   EDIT
  Overview [Data] Files  Processors  Set Up  Exit                    | F1=Help

              M                  f
  1 7                        5
  2 17                       7
  3 27                       5
  4 37                       3
```

Figure 3.23 Inputting the Midpoint and Frequency of Each Class

The Intermediate Results option (refer to Figure 3.24) will display the values of these sums:

$$\Sigma fM \qquad \Sigma fM^2 \qquad \Sigma f(M - \overline{X})^2$$

while the Frequency Distribution option will recreate the grouped frequency distribution from the midpoints and frequencies that make up the data set. You may wish to verify the results given by the processor with the answers you supplied to Classroom Example 3.4.

```
≡                 Descriptive Statistics For Organized Data                MENU
   Introduction  *ata  Calculations [Results] Exit                 |  F1=Help

                                     Summary measures
                                     Intermediate results
                                     Frequency distribution
```

Figure 3.24 Available Output from Descriptive Statistics for Organized Data Program

```
≡                 Descriptive Statistics For Organized Data               READY
   Introduction  *ata  Calculations [Results] Exit                 |  F1=Help

          The number of grouped sample observations is n = 20

                     Measure                Value

                     Mean................ 20.0000
                     Median.............. 19.143
                     Mode................ 17.0000
                     Variance............ 106.3158
                     Standard deviation.. 10.3110
```

Figure 3.25 Summary Measures for Descriptive Statistics for Organized Data Program

To run this processor with an ungrouped frequency distribution, use the Create option within the Data section of the main menu to create an Organized Ungrouped Frequency type of data set, and then execute the processor exactly as described above.

3.8 SUMMARY

Business statistics fills a special role in decision making. It is not concerned strictly with computations, though the material presented so far might suggest otherwise. It should be thought of as a tool that is used much like a hammer or saw is used in construction. Without a human being operating the other end of the tool, the tool would be useful perhaps as a paperweight only. Similarly, statistics does not do anything on its own; rather we use it to help us make decisions and create order from numerical chaos.

Our objective is to describe data by reducing the entire set of numbers into a few descriptors. These measures attempt to portray several characteristics of the data set: central location, spread, shape, and/or position. Although the notation and computation of each measure have been emphasized, we have tried to stress applications too.

To help you assess the sensibility of numerical answers, we presented several reasonability checks in the form of inequalities and figures in Section 3.5. Our descriptive analysis of data continues in Chapter 4 as we describe methods of treating *bivariate data* (two variables defined on the same entity).

3.9 TO BE CONTINUED . . .

. . . *In Your College Courses*

Marketing majors usually are required to take a course in marketing research. For other business majors, such a course may be an elective. Three popular textbooks for the marketing research course are: *Marketing Research* (Aaker and Day, 1986), *Marketing Research* (Parasuraman, 1986), and *Marketing Research: Measurement and Method* (Tull and Hawkins, 1987).

Research in the marketing environment is performed for a variety of purposes: to make decisions about marketing a product; to measure consumer reaction to a new product, service, or mode of advertising; to track consumer buying patterns; to assess brand loyalty; and so forth. Marketing research generates data which must be described and analyzed. Beginning to learn to do so is the objective of our Chapter 3! Indeed, the interrelationship between business statistics and marketing research is described by Tull and Hawkins:

> *There are two major kinds of summarizing statistics. . . . The first . . . is known as* measures of central tendency. *The second . . . is known as* measures of dispersion.*

Tull and Hawkins recommend ". . . three primary measures of central tendency . . . the *arithmetic mean,* the *median,* and the *mode.*" Their term "central tendency" and our term "central location" are interchangeable. Aaker and Day, and Parasuraman, limit their discussion to the mean only.

With regard to measuring dispersion, Parasuraman covers only the standard deviation, while Aaker and Day present both the variance and the standard deviation.

*From D. S. Tull and D. I. Hawkins, *Marketing Research: Measurement and Method,* 4th Ed. (New York: Macmillan, 1987). © 1987 Macmillan Publishing Company. Reprinted with permission.

Tull and Hawkins write: "The *standard deviation, variance,* and *range* are measures of how 'spread out' the data are."

Regardless of the text, it is apparent that the mean and standard deviation have well-established roles in marketing research. Additionally, marketers have chosen to preserve the notation that business statisticians use for these quantities. The symbols $\overline{X}$, S^2, and S appear in all three texts as the notation for the sample mean, variance, and standard deviation, respectively. This is not a trivial point! The lack of standard notation for common statistics is probably the major cause of confusion among students as they take other courses in the business curriculum.

Descriptive statistics have applications in other areas of business as well. As we mentioned, the coefficient of variation and the geometric mean are popular in finance. In economic forecasting, the average deviation and the variance are used as criteria for comparing various forecasts. With the mean and standard deviation common to most business disciplines, we hope you see that numerical descriptive measures form the basis for describing data.

. . . In Business/The Media

We encounter descriptive statistics, especially measures of central location, every day in the business world. In the financial markets, for instance, the Dow Jones Industrial Average (DJIA) is perhaps the most recognized statistic reported in the media. Though a weighted average, the DJIA nonetheless is a summary of the current stock prices for 30 U.S. firms. Often included with a mention of the "closing Dow" is a statement about the effect of the day's trading on a share of stock, such as, "The average price of a share of stock rose three cents."

The average, as we have seen, is a simple measure to compute and interpret, fueling its popularity as a summary statistic. Potential applications include the following:

- A bank that may compute an average checking account balance each month per account and use this value to determine service charges should the average fall below some minimum figure
- A life insurance company that must estimate average life expectancies in order to set premium rates on its policies
- An income tax guide, such as *J. K. Lasser's Your Income Tax,* which lists the average amount of deductions claimed on Tax Schedule A according to category (interest, taxes, and so forth) and adjusted gross income, based on Internal Revenue Service figures

The median is a popular measure as well and often is cited when salary, income, home prices, sales data, and so forth are being described or compared. For instance, the median annual salary and bonus payment for chief executive officers was included in a study of major U.S. companies, as reported in "A Great Leap Forward for Executive Pay" (*Wall Street Journal,* April 24, 1989, page B1). Coldwell Banker frequently quotes the median selling price of houses in *USA Today* articles that focus on the nation's housing industry.

Measures of dispersion appear in subtle ways. Though the range may not be reported per se, we may find the minimum and maximum values listed. For example, the U.S. Weather Service reports daily high and low temperatures for large cities across the country. Many newspapers list the high, low, and closing prices each day for the issues traded on the New York Stock Exchange. From the high and low prices, investors can form their own opinion about the volatility of the market and/or a particular stock.

The Environmental Protection Agency (EPA) tests new cars each year and reports their estimated fuel economy. For instance, here is a quote from the EPA sticker affixed to the window of a 1988 subcompact car: "Results reported to EPA indicate that the majority of vehicles with these estimates will achieve between . . . 24 and 34 mpg on the highway." This statement may be regarded as an illustration of the region of concentrated values defined earlier in this chapter. Therefore we might expect about 95 percent of the vehicles with the EPA sticker quoted above to obtain mpg readings between 24 and 34.

With a renewed emphasis on quality in manufacturing and production, we must understand measures such as the mean, range, and standard deviation in order to compete in the marketplace of the 1990s, since *tolerance limits* and *product specifications* represent combinations of these measures. Quality control (also known as statistical process control) and control charts are based on $\overline{X}$, Rg, and S, as we will see in Chapter 11.

Summary statistics are the rule, not the exception, whenever we wish to describe data. And, as the media and real world applications demonstrate, describing data is a requisite job skill.

SECTION 3.9 EXERCISES

3.83 A sample of college women in an advertising class was divided at random into two groups. Each group was shown two current music videos with a test commercial for a new personal-care product shown in between. Two versions of the test commercial were shown; each student was asked to give an opinion of the commercial she saw, with the following results:

	Response Count	
Coded Response	Test Ad *A*	Test Ad *B*
5 = Liked a lot	29	52
4 = Liked somewhat	31	16

(continues)

	Response Count	
Coded Response	Test Ad *A*	Test Ad *B*
3 = Neither liked nor disliked	33	11
2 = Disliked somewhat	15	12
1 = Disliked a lot	9	26

Let the coded numerical response represent X, the viewer's reaction to the test commercial she saw.

a. Find the median reaction for each commercial.
b. Find the modal reaction for each commercial.
c. Find the mean reaction for each commercial.
d. By inspection, which commercial's reaction has the larger standard deviation?
e. If you were told that one of the two commercials was "controversial," which would you guess it is?

3.84 A sample of size ten yielded the following numbers:

5	5	4	1	3	4	4	5	2	1

Compute the following:
a. mean **b.** variance **c.** standard deviation

3.85 A survey questionnaire contained the following statement with which the respondents were asked to identify their degree of agreement: "My state legislators should enact a law requiring all front seat occupants of a vehicle to use seat belts." The results from a sample of 1213 people were as follows:

Responses	Coded Scale	f
Strongly agree	5	421
Agree	4	378
Neutral	3	59
Disagree	2	251
Strongly disagree	1	104

Using the coded scale as a surrogate for X = response to the statement, find the following:
a. mean **b.** variance **c.** standard deviation

3.86 Parasuraman (1986), in Chapter 12 of his text, discusses estimating the standard deviation: "If the minimum and maximum values of the variable in the population are known, . . .

the standard deviation can be estimated as follows: S = (maximum value − minimum value)/6." This suggestion is consistent with our recommendation in Equation 3–13. Use the result to estimate the standard deviation of wages for the population of autoworkers. Assume the union contract calls for a floor of \$5.16 per hour and a ceiling of \$16.35 per hour. Can you also estimate the average hourly wage?

CHAPTER 3 EXERCISES

3.87 Organize these data into a grouped frequency distribution. Compute the mean and median for the organized data.

200	225	267	199	227
150	104	183	213	209
175	249	202	204	183
164	127	141	199	218
264	152	183	174	238

3.88 A business junior college offers programs specializing in office administration. Graduates from the word processing program are touted as being quick and accurate. A random sample of 20 students' words typed per minute produced the following data:

62	64	63	57	60
56	58	58	60	57
57	60	58	60	58
59	57	57	55	55

Organize these data into an ungrouped frequency distribution. Find the region of concentrated values.

3.89 The mean and standard deviation for a set of sample data are 7.7 and 1.3, respectively. If the data set is approximately mound shaped, what percentage of the data falls in the following groupings?

a. between 6.4 and 11.6 **b.** between 5.1 and 10.3 **c.** below 9

3.90 If the mean and standard deviation for a set of sample data are 6.0 and 1.3, respectively, compute the Z-scores to two decimal places for the following values of X:

a. $X = 10$ **b.** $X = 7$ **c.** $X = 6$ **d.** $X = 0$

3.91 Find the value of X corresponding to a Z-score of −.86 if the standard deviation is 5.00 and the mean is 1.40.

3.92 Find the mode, median, and mean for these data:

1044	976	1313	1117	820	891	1006

3.93 For this data set,

−1	−5	−6	−6	−2	−2	−8	−6

find the following:

a. the average deviation **b.** the variance **c.** the standard deviation

3.94 The Z-score corresponding to $X = .04$ is 2. If the population standard deviation is .003, find μ.

3.95 A recently completed survey on consumer behavior revealed the following data on X = number of times per month a banking customer uses a 24-hour automatic teller:

2	0	2	4	5	1	0	5
6	1	5	3	1	8	0	0
6	3	5	0	0	2	8	2
1	6	5	5	5	2	6	0
7	2	0	2	9	8	0	6
0	8	4	7	5	6	7	0
2	6	7	0	0	5	0	1

a. Find the mean of the raw data.
b. Organize these data into an ungrouped frequency distribution.
c. Find the mean of the organized data. Compare with your answer to part a. Are they the same? Why?
d. Find the median of the organized data.
e. Find the mode.
f. Suppose someone told you that the standard deviation of the raw data was 6.2. Without computing S, is this a believable value? Why?

3.96 Within a certain county last year, the mean selling price of a single family residential home was $108,000, while the median was $86,500. Which measure is more representative of the "typical" home sale? Why? Describe the possible shape of the distribution of house prices.

3.97 At the end of 1986, there were a total of 1338 mutual funds investing in stocks and bonds. A sample of 86 of these funds produced the following data on their 12-month return through December 31:

Return (%)	Number of Funds
56.2– 72.6	1
39.8– 56.2	4
23.4– 39.8	18
7.0– 23.4	46
−9.4– 7.0	14
−25.8– −9.4	3

a. Find the average return of the funds in the sample.
b. Compute the standard deviation.
c. If a mutual fund had a return of 10 percent, what would be its position in the sample in terms of its Z-score?
d. If a mutual fund in the sample had a Z-score of 2.15, what would be its return?
e. For the population of mutual funds, the coefficient of variation was 93.45 percent. Is this sample representative of the population in terms of relative variation? More volatile? More stable?

3.98 A three-day convention drew the following numbers of people per session:

	Day		
Session	1	2	3
Morning	342	313	210
Afternoon	401	287	184

a. Find the average morning and the average afternoon attendance.
b. Compute the standard deviation for both the morning and the afternoon sessions.
c. Find the coefficient of variation for the morning and for the afternoon attendance figures. Which session type had the largest CV?

3.99 The Z-score corresponding to $X = -7$ is .4. If the population standard deviation is 2.0, find μ.

3.100 Since the standard deviation is defined as the square root of the variance, must the variance always be a larger number than the standard deviation? Explain.

3.101 Each new employee of a company is required to take an aptitude test administered by the Personnel Department. The maximum score is 40, and the test covers basic English and math skills. A sample of 14 test scores produces these values:

17	25	18	33	35	14	35
17	24	23	38	28	31	37

a. Treating these as raw data, compute the mean and standard deviation.
b. What percentage of the 14 scores fall in the region of concentrated values? Is this close to the guideline suggested by the Empirical Rule? Why or why not?
c. Find the coefficient of variation for this sample.
d. Suppose the manager of the Personnel Department reported in a company memo that the average score for all new employees starting on or after January 1 of the current calendar year is 21.6, with a standard deviation of 24.9. Is this reasonable? Why or why not?

3.102 The State Fair runs ten days during the summer with the largest attendance usually occurring on the last three days (a weekend). Typically, the average attendance for the first seven days is a harbinger of the final weekend. To meet revenue projections, the fair needs to average 60,466 daily customers for the entire run. The average daily attendance figure for the first week of this year's fair is 57,656. What must be the average attendance for the final three days in order for the fair to meet the projections?

3.103 The mean and median are sometimes called *location* measures because they function as an anchor for the location of the data set.

a. Find the mean and median of these numbers:

1.5	3.7	1.8	2.3	2.6

b. Find the mean and median of these numbers:

2.2	4.4	2.5	3.0	3.3

c. Notice that each number in the data set of part b is exactly .7 larger than the corresponding value in the data set of part a. Verify that the mean and median you found in part b are exactly .7 larger than the mean and median you found in part a.

d. Generalize this phenomenon to an arbitrary data set $X_1, X_2, \ldots, X_n$ in which a constant d is added to each value. Give the mean and median for the following data set:

$$X_1 + d, X_2 + d, \ldots, X_n + d$$

e. Suppose an arbitrary data set $X_1, X_2, \ldots, X_n$ is "relocated" by subtracting the value of $\overline{X}$ from each number. What would be the mean of the relocated data set?

3.104 A baseball player's batting average is computed by dividing his number of hits by his number of "at bats." At midseason a player has a batting average of .248 as a result of getting 63 hits in 254 at bats. If the player wishes to have a batting average of .300 by the time he has batted 500 times, what must be his batting average for his next 246 at bats?

3.105 The sample Z-score corresponding to $X = 34.1$ is -1.68. If $S = 5.25$, find $\overline{X}$.

3.106 The sample Z-score corresponding to $X = 19$ is $-.5$. If $\overline{X} = 24.0$, find S.

*3.107 Find the geometric mean of these data:

−2.70%	−4.11%	13.09%	1.09%	−.79%

3.108 The owner of a small bookstore dealing in used paperback books has learned that the tenant adjacent to her store is not renewing his lease. She has two options: expanding her bookstore by renting this property or maintaining the status quo. Presently her business averages \$340 a week in gross revenue with a standard deviation of \$45. With some help from friends, the owner has projected average weekly gross revenues of \$435 with a standard deviation of \$70 if she expands. The expansion will cost an additional \$140 per month in rent.

a. Find the coefficient of variation for the weekly net (gross revenue − weekly rent) if she maintains the status quo. Her weekly rent is now \$50.

b. Find the *CV* for the weekly net revenues if she expands. Assume four weeks per month.

c. If she bases her decision on the option with the smaller relative variation, what should she do and why?

3.109 Find the mean, variance, and standard deviation using this information:

*Optional

$$n = 33$$

$$\Sigma X = 211.4$$

$$\Sigma X^2 = 1847.69$$

Assume the accuracy of the raw data is in tenths.

3.110 The 52-week average market value index (1973 = 100) for a sample of industry groups appears as follows:

136	355	742	376	37
800	259	313	113	354
263	377	101	210	86
253	125	51	99	174
375	144	163	338	106
71	346	98	80	201
509	865	241	113	347
266	84	88	92	238
370	435	979	189	294
676	124	446	469	384

a. Compute the mean of the raw data.
b. Organize these data into a frequency distribution.
c. Compute the mean of the organized data. Compare with your answer to part a.
d. Find the median of the organized data.

***3.111** The American Stock Exchange is open Monday through Friday, excepting holidays. As of the close of business on Thursday, the average number of shares traded (in millions) for the week was 11.4, with a standard deviation of 1.70. If Friday's volume was 10.6 million shares, find the average number of shares traded and the standard deviation for the week's activity. (*Hint:* See Exercises 3.28 and 3.40.)

3.112 The sample *Z*-score corresponding to $X = .6$ is 2. If the mean is zero, find S.

3.113 If the coefficient of variation is .40 and the mean is 6, what is the value of the standard deviation?

3.114 The prices paid to farmers for milk produced for fluid consumption in a metropolitan area are as follows for several years:

Year	Federal Minimum Price	100 lbs. of Milk Dairymen, Inc., Price	Dairymen, Inc., Premium
1	$14.28	$15.21	$ .93
2	14.18	15.07	.89
3	14.24	15.01	.77
4	14.00	14.67	.67
5	13.45	14.48	1.03
6	12.89	13.70	.81
7	13.09	14.15	1.06

Dairymen, Inc., is a 17-state milk marketing cooperative. The preceding figures could be converted to prices per gallon via 8.6 lbs = 1 gal.

a. Find the mean and median of the federal prices.
b. Find the coefficient of skewness for the Dairymen's premium.
c. Compute the average deviation, variance, and standard deviation for the Dairymen's price.
d. Find the Dairymen's price of a gallon of milk in year seven.

3.115 In order to expand his customer base, the owner of a specialty store decided to experiment with a mail-order catalog. He kept detailed records of each order received from the initial mailing. The sales are summarized in the following table:

Sales ($)	Number
10–25	122
25–40	97
40–55	41
55–70	58
70–85	36
85 and up	49

a. What is the class width for this distribution, excluding the "$85 and up" category?
b. Find the midpoint of each class. For the open-ended category "$85 and up," estimate the midpoint by adding the class width to the midpoint of the immediately preceding class.
c. Find the median sales amount. Does the last interval of sales affect the accuracy of this figure? Explain.
d. Find the average sales per order. Does the last interval of sales affect the accuracy of this figure? Explain.

3.116 Calculate the mean, median, and standard deviation for the following distribution:

f	1	6	17	44	23
X	1	2	3	4	5

3.117 In surveys, the responses to questions or statements often are artificially coded on a 1 to 5 scale (or 1 to 7 scale, depending on the number of responses). A sample of 737 adults was asked to select one of these responses—strongly agree, agree, neither agree nor disagree, disagree, or strongly disagree—to this statement: The economy will be much improved whenever the federal budget is balanced again. If we use a 1 (strongly agree) to 5 (strongly disagree) scale, find the mean and modal responses for these data:

Response	Number
Strongly agree	18
Agree	209
Neither agree nor disagree	134
Disagree	245
Strongly disagree	131

3.118 Refer to Exercise 3.117. What problems in interpretation arise when we use an artificial coding system on the responses to a qualitative variable such as "opinion"?

***3.119** The manager of the auto loan department in a bank was asked to analyze the repayment patterns on its 48-month auto loans. One of the department's employees sampled 147 loans and summarized the data for the number of months until the loan is paid off. In his analysis, the employee reported an average loan life of 36.1 months and a standard deviation of 25.3 months. Explain why the manager was skeptical about the analysis and asked the employee to re-do it. (*Hint:* See Exercise 3.80.)

***3.120** The sum of the deviations from the mean will add to zero for a set of raw data, provided the value of $\overline{X}$ is not rounded off. For organized data, a similar property holds, provided each deviation is weighted by its frequency.

a. For the following ungrouped frequency distribution, show that $\Sigma f(X - \overline{X}) = 0$. Is it true for these data that $\Sigma(X - \overline{X}) = 0$? Why?

X	f
0	5
1	2
2	1
3	4
4	7
8	1

b. For the following grouped frequency distribution show that

$$\Sigma f(M - \overline{X}) = 0 \qquad \Sigma(M - \overline{X}) \neq 0$$

Classes	f
−3–0	2
0–3	3
3–6	7
6–9	8

3.121 The local Emergency Medical Service (EMS) is proud of its reputation for quick responses to emergency calls. A reporter for a local television station interviewed the director of EMS for a short spot on an upcoming newscast. The director claimed that EMS responds to 95 percent of the calls it receives in 1 to 8 minutes. Assuming response times are fairly mound shaped, estimate the average time and the standard deviation it takes EMS to respond to an emergency call.

3.122 Find the coefficient of skewness for the following data set: 2.0 16.6 14.5 4.5 10.9.

3.123 Find the mean and median for these data:

.7	2.6	1.9	1.3	1.8

Suppose the last number had been 11.3 instead of 1.8. Recompute the mean and the median. What changes in these measures of central location occur when one of the values is inconsistent with the others?

3.124 The current rate offered on a money market deposit account was recorded for a sample of federal savings and loan institutions. The data, expressed as percentages, are as follows:

6.00	5.95	6.05	6.00	6.00	6.00	6.00	6.05	6.10

a. Define the variable of interest.
b. Find the mean, median, and mode for these data.
c. If you were reporting this information to the public, which measure of central location would you use to convey the prevailing money market rate? Why?

3.125 Find g_1 for these data:

71	85	144	106	124	77	93

Characterize the sample data set in terms of skewness.

*3.126 Consider the following data set:

7.85	7.80	7.70	7.75	8.20

a. Find the variance and standard deviation for these data.
b. Multiply each number in the data set by two. Recompute the variance and standard deviation.
c. Compare the answers to parts a and b. What changes in these measures have occurred?
d. Add three to each number in the original data set. Recompute the variance and standard deviation. What changes occurred?
e. Generalize the changes in S^2 and S to these situations:
   (i) Each value in a data set is multiplied by a constant
   (ii) A constant value of d is added to each value in the data set.

3.127 Consider the following grouped frequency distribution:

Classes	f
55–63	4
63–71	11
71–79	15
79–87	10
87–95	6

a. Find the mean.
b. Find the median.
c. Find the variance.

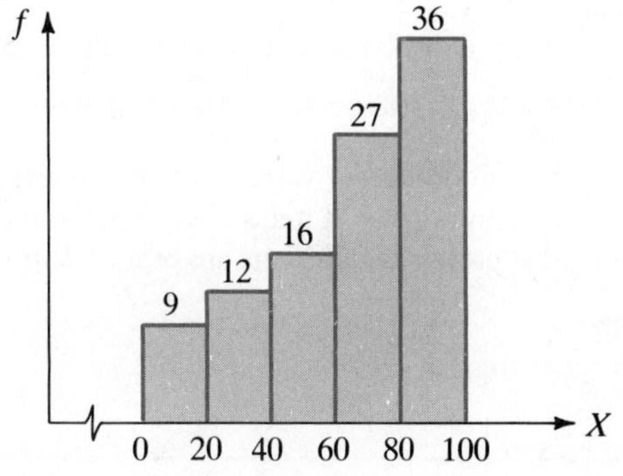

Figure 3.26 Graph of a Grouped Frequency Distribution

3.128 A sample data set produces the frequency histogram in Figure 3.26. Decide whether each of the following statements is correct. Make computations as necessary. If the statement is incorrect, explain why.

a. The sample mean is between 80 and 100.
b. The coefficient of skewness is a negative value.
c. The average deviation is more than the standard deviation.
d. The standard deviation is more than 60.
e. The median is between 60 and 80.

3.129 If $\Sigma X^2 = 2470$, $\overline{X} = 17$, and $n = 8$, what is the sample standard deviation?

3.130 If $\Sigma X = 2.4$, $\Sigma X^2 = 3.89$, and $n = 7$, find S.

3.131 Suppose a sample set of data is fairly bell shaped and symmetrical with $\overline{X} = 8.6$ and $S = .4$. Draw a rough sketch of what a sample histogram for this data set might look like between the values 7.4 and 9.8.

3.132 Suppose a sample data set of size $n = 30$ yielded the following numerical descriptive measures:

$$\overline{X} = 300 \qquad S = 50 \qquad g_1 = 1.22$$

Roughly sketch a histogram for these data. Be sure to indicate the scale on your horizontal axis.

3.133 While the Empirical Rule generally applies only to symmetrical, bell-shaped distributions, another well-known result for *all* distributions is Chebyshev's theorem. This theorem states that the proportion of values more than k standard deviations above or below the mean cannot exceed $1/k^2$. For instance, if $k = 2$, we can state that no more than $1/2^2$, or 1/4, of the items differ from the mean by more than two standard deviations. Conversely, we could also state that at least 3/4 of the items must lie within plus or minus two standard deviations from the mean. Chebyshev's theorem holds regardless of the shape of the distribution. For a distribution of unknown shape, determine the following:

a. At most, what proportion of items are more than three standard deviations from the mean?

b. At least, what proportion of items are within plus or minus 1.5 standard deviations from the mean?

c. At most, what proportion of items are more than 2.5 standard deviations from the mean?

3.134 (See Exercise 3.133.) A skewed-to-the-right distribution has a mean of \$210, with a standard deviation of \$100.

a. At least, what proportion of values will be found from \$50 to \$370?

b. At most, what proportion of items are not in the interval from \$10 to \$410?

c. At most, what proportion of items are beyond four standard deviations from the mean?

REFERENCES

Aaker, D. A., and G. S. Day. 1986, *Marketing Research,* 3rd Edition. John Wiley & Sons, New York.

Hildebrand, D. K., and L. Ott. 1987. *Statistical Thinking for Managers,* 2nd Edition. PWS-KENT, Boston.

J. K. Lasser Institute. 1989. *J. K. Lasser's Your Income Tax, 1989*. Simon & Schuster, Inc., New York.

Macleod, A. J., and G. R. Henderson. 1984. "Bounds for the Sample Standard Deviation," *Teaching Statistics, 6:*72–76.

McClave J. T., and P. G. Benson. 1983. *Statistics for Business and Economics,* 3rd Edition. Dellen Publishing Company, San Francisco.

Parasuraman, A. 1986. *Marketing Research*. Addison-Wesley Publishing Company, Reading, MA.

Shiffler, R. E. 1987. "Bounds for the Maximum Z-Score," *Teaching Statistics, 9:*80–81.

Shiffler, R. E. 1988. "On Maximum Z-Scores and Outliers," *The American Statistician, 42:* 79–80.

Shiffler, R. E., and P. D. Harsha. 1980. "Upper and Lower Bounds for the Sample Standard Deviation," *Teaching Statistics, 2:*85–86.

Tukey, J. W. 1977. *Exploratory Data Analysis*. Addison-Wesley Publishing Company, Reading, MA.

Tull, D. S., and D. I. Hawkins. 1987. *Marketing Research: Measurement and Method,* 4th Edition. Macmillan Publishing Company, New York.

Chapter Maxim *Correlation is not causation.*

CHAPTER 4

ORGANIZING AND DESCRIBING BIVARIATE DATA

*Optional

Objectives

After studying this chapter and working the exercises, you should be able to

1. Represent quantitative bivariate data with a scatter plot.
2. Compute a measure of the strength of relation for quantitative bivariate data.
3. Use sample results to judge whether two variables are related or unrelated to each other.
4. Explain the least-squares criterion.
5. Compute the regression line for a sample data set.
6. Interpret the slope of a regression line.
7. Use a regression equation for prediction purposes.
8. Recognize problem situations that can be analyzed by regression and correlation.

*9. Execute the Least-Squares Line processor to verify answers and/or solve problems.

10. Construct and interpret a contingency table for qualitative bivariate data.

In the previous chapters, we have learned some useful and widespread methods to visually present and to numerically describe data. Among methods with a visual appeal, we considered some tabular and graphical techniques of displaying data, such as frequency distributions, bar graphs, and histograms. To numerically describe important properties of a data set, we developed concepts such as the mean and median to measure central location and the standard deviation to measure dispersion. These approaches are appropriate when we want to be able to summarize and describe data for *one* variable considered by itself, as has been the case through Chapter 3.

In this chapter, we will be interested in looking at *two* variables at the same time. In addition to graphical and descriptive analysis, we will have a new objective in mind when we consider two variables together: to see whether these variables are related. For instance, an automobile industry researcher conducts a study of this year's new car estimated miles per gallon and the car engine size. For a sample of several dozen car models, the researcher records data on two variables: each car's fuel economy rating and its engine size. No doubt you would anticipate that smaller engines would tend to be found on cars that achieve higher mileage ratings. If indeed this proves to be true, how can we quantify the relation between these two variables? How can we describe whether the relation between engine size and miles per gallon is a weak one or a strong one? If we know a car model's engine size, can we roughly predict its

*Applies to optional section.

miles per gallon? In this chapter, we will consider techniques that are intended to answer questions like these and to help us summarize relations between variables.

Most of this chapter is devoted to the case where both variables of interest produce quantitative data. Section 4.5 will discuss the display of qualitative bivariate data. In later chapters, we will consider applications where one of two variables is qualitative and the other is quantitative.

4.1. QUANTITATIVE BIVARIATE DATA

Paired Observations

As the name would suggest, *bivariate data* refer to observations made on two variables, which we will denote as X and Y. The observations are paired in the sense that each value of X has a corresponding value of Y associated with it; a sample of size $n = 10$ would have ten pairs of values for (X, Y). Table 4.1 illustrates bivariate data for a sample of ten observations of new car model engine size and the car's estimated miles per gallon.

Definition

Quantitative bivariate data are paired observations (X, Y) in which the first variable, X, is called the independent, or predictor, variable and the second variable, Y, is called the dependent, or response, variable.

Table 4.1 Sample Data on Engine Size and Estimated Miles per Gallon

Car Model	Engine Size (Cubic Inch Displacement)	Estimated Miles per Gallon
A	263	23
B	242	24
C	148	31
D	130	30
E	180	27
F	318	18
G	160	27
H	202	25
I	305	18
J	216	26

In bivariate analysis, we will let X denote the *independent variable* and let Y denote the *dependent variable*. The *independent* (predictor) variable X is the one that is measured to predict or explain the value of the other variable. The *dependent* (response) variable Y is the one whose values are to be predicted or explained on the basis of its relation to the independent variable. For the sample data set in Table 4.1, if we intend to use engine size to predict or forecast miles per gallon, we want to label engine size as X and miles per gallon as Y. As another example, if someone were studying the relation between the current retail price of coffee and the world supply of coffee, we would say the independent variable is supply and the dependent variable is price. It is not difficult to argue that supply affects price, instead of vice versa; therefore, current price, Y, *depends* in some way on supply, X.

The examples given may have been fairly clear as to which variable should be labeled X and which one Y. This is because a cause-and-effect relation between the two pairs of variables seems apparent. For instance, larger engine sizes (and therefore heavier cars) tend to cause fuel economy to fall. In such cases, the independent variable X represents the "cause," while the dependent variable Y represents the "effect." However, there are many instances where it is not at all clear that the variables have a cause-and-effect relation. There may be in reality no relation at all, but still we must label one variable as X and the other as Y. We must thus acknowledge that the labels X and Y on variables do not by themselves necessarily mean that we believe a cause-and-effect relation exists.

Scatter Plots

The most commonly used method to visually present quantitative bivariate data is by means of a scatter plot, also called a *scatter diagram,* the purpose of which is to reveal graphically any relation between variables X and Y.

Definition

A **scatter plot** is a two-dimensional graph of quantitative bivariate data with the independent variable plotted on the horizontal axis and the dependent variable plotted on the vertical axis.

Each plotted point represents one pair of values for X and Y. The independent variable X always is shown on the horizontal axis with the dependent variable Y on the vertical axis. The visual representation of the data made possible by a scatter plot can convey information very quickly about whether and how X and Y are related. Figure 4.1 shows the ten data points for engine size and miles per gallon on a scatter plot.

In this figure, it is clear that for larger engine sizes, the miles per gallon becomes lower. This is indicative of a negative, or *inverse,* relation between the variables, since increasing values of X tend to imply decreasing values of Y. The relation between

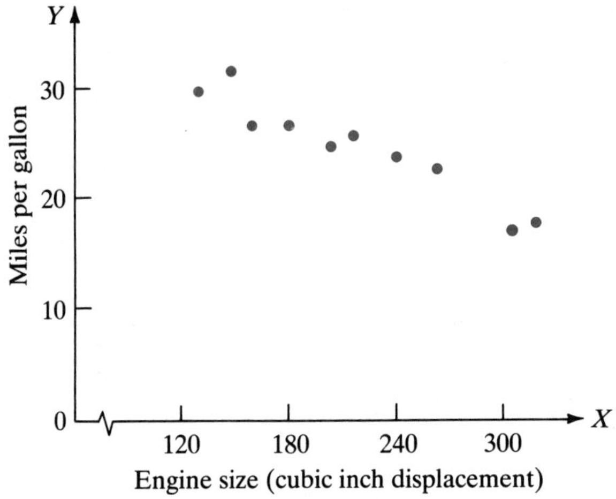

Figure 4.1 Scatter Plot for Data in Table 4.1

engine size and miles per gallon also can be described as *linear*, meaning that the pattern in the scatter plot can be captured by a straight line, as in Figure 4.2. A positive, or *direct*, relation between two variables is represented in Figure 4.3, where increasing values of X are associated with increasing values of Y. Another possibility

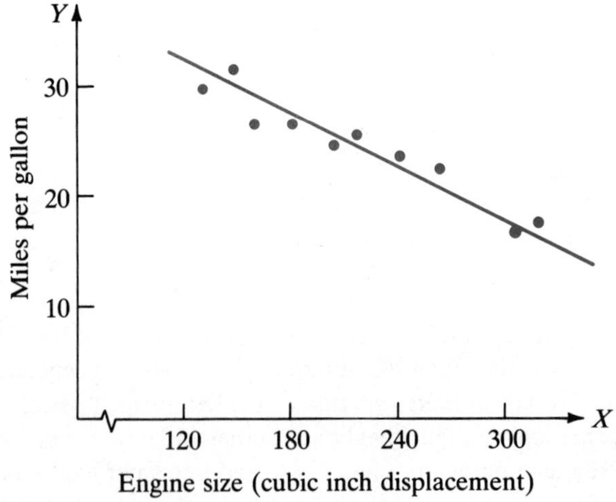

Figure 4.2 Scatter Plot from Figure 4.1 with Straight Line Added

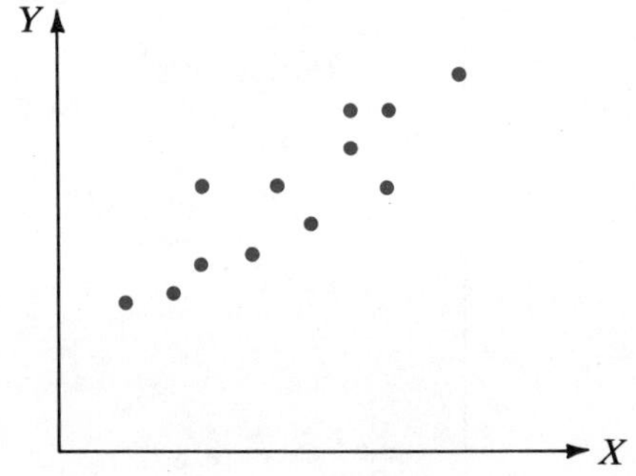

Figure 4.3 Scatter Plot Showing Direct Linear Relation

is that two variables may have *no linear relation* between them at all. Such a situation is suggested in Figure 4.4, where the scatter plot fails to reveal any particular trend or pattern to the data. A final possibility is that X and Y are related, but not linearly. Figure 4.5 shows a nonlinear, or curvilinear, relation between X and Y. In this chapter, we will be considering variables that are linearly related; methods to deal with nonlinear relations will be discussed in Chapter 17.

It is strongly recommended that any bivariate analysis include displaying the data on a scatter plot. This visual representation of information provides quick insight into the nature of the relation that may exist between variables X and Y. Further, this practice also may call attention to any data points that happen to fall far away from the main body of the data. Such an occurrence conceivably could be caused by an error in recording the data.

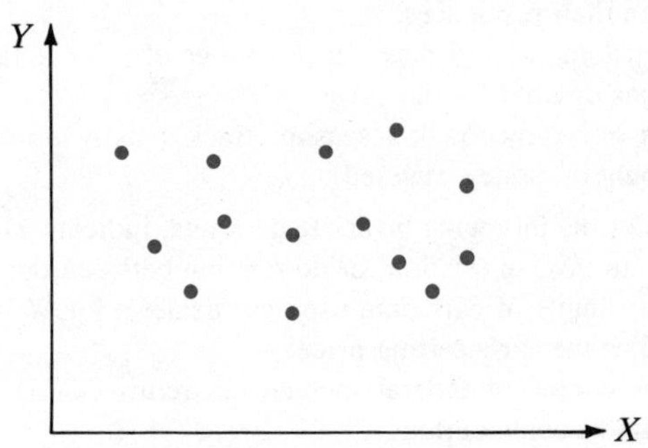

Figure 4.4 Scatter Plot for Unrelated Variables

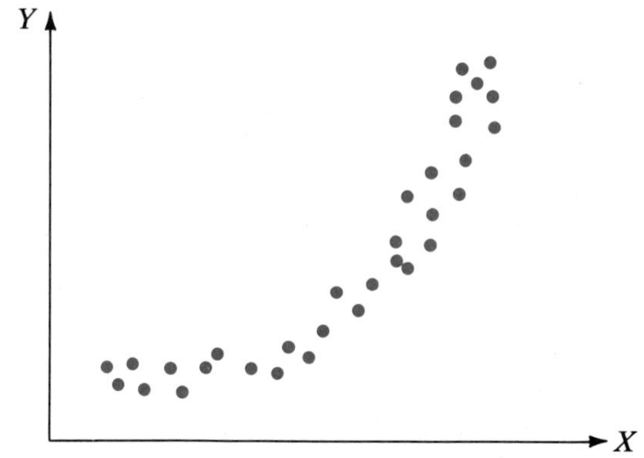

Figure 4.5 Scatter Plot for a Nonlinear Relation

SECTION 4.1 EXERCISES

4.1 The data in Table 4.1 were described as a sample of ten bivariate observations. Suppose you obtain a second sample of the same two variables, also of ten observations, and you plot these points on a scatter diagram.

a. Do you think your second sample's ten points would be identical to those of the first sample (given in Table 4.1)?

b. If you answer no to part a, do you think your second sample's scatter diagram would show a negative relation between the two variables, as does Figure 4.1?

4.2 For each of the following bivariate data sets, indicate which variable should be the independent variable and which should be the dependent variable, and explain the reason for your choices:

a. for residential electric utility customers: annual electricity usage and number of square feet in their residences.

b. for custom-ordered new cars: number of days it takes to get delivery and number of options ordered for the car.

c. for a horse racetrack's season: track's daily attendance and track's daily "handle" (amount of money wagered).

4.3 For each of the following bivariate data sets, indicate whether you anticipate finding a direct relation, an inverse relation, or no relation between the two variables.

a. For a sample of cars on a used car dealer's lot, X is number of miles on the odometer, and Y is the car's selling price.

b. For a sample of federal income tax returns, X is gross income, and Y is amount of charitable contributions.

c. For a sample of college underclassmen, X is height, and Y is grade point average.

d. For a sample of gymnasts competing at a meet, X is rating of expert judge 1, and Y is rating of expert judge 2.

4.4 Refer to Exercise 4.3 where you were asked to express the bivariate relation as direct, inverse, or none. Would any of your answers change if you switched labels, changing each X to Y and vice versa?

4.5 For a random sample of five college sophomores, the scatter plot in Figure 4.6 shows grade point average (GPA) in high school, X, and in college, Y. The points are labeled as J, K, L, M, and N.

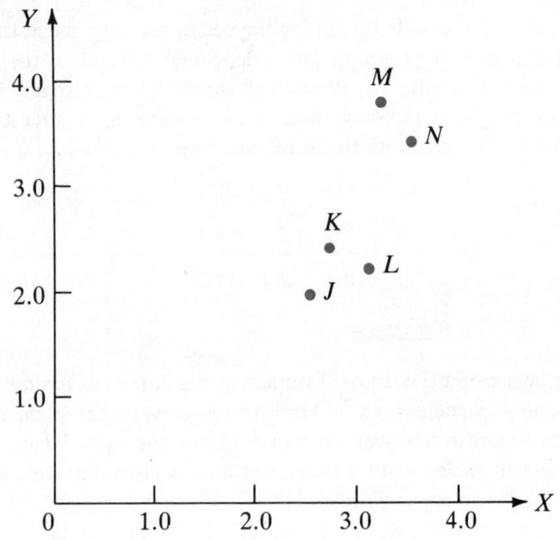

Figure 4.6 Scatter Plot of GPA for Five Students: High School, X, Versus College, Y

The actual values are as follows:

Point	High School GPA	College GPA
J	2.5	2.1
K	2.7	2.5
L	3.1	2.3
M	3.2	3.9
N	3.5	3.5

a. Which of the five points represents the student whose college GPA is lowest relative to his or her high school GPA? Which student is highest relative to high school GPA?
b. By graphical inspection, does X or Y appear to have the larger standard deviation?

c. Does there appear to be a direct relation, inverse relation, or no relation between the variables X and Y?

4.6 Make a scatter plot for the following data:

X	3	5	6	7	4	2
Y	5	2	4	3	6	7

4.7 An employer sampled eight employees to see how much they were using the firm's family dental insurance program. The independent variable is the number of eligible people in the employee's family; the dependent variable is last year's family dental bill, expressed in hundreds of dollars. Show these bivariate data on a scatter diagram. Does your graph suggest a direct, inverse, or no linear relation between X and Y?

X	2	2	3	5	3	4	1	2
Y	0.9	2.1	1.2	3.1	1.7	4.5	0.4	0.5

4.8 The owner of a recreational vehicle dealership went through his records and determined the number of vehicles sold in April for the most recent six years. He also determined what the prime interest rate was on April 15 for the same years. Which variable should be the dependent variable? Plot the points on a scatter diagram. Does your graph reveal a direct, inverse, or no apparent relation?

Number of Vehicles Sold in Month	Prime Interest Rate in Middle of Month
100	10
80	12
86	13
90	14
85	18
75	17

4.9 The data in the table that follows is adapted from an article in *The Sporting News* (July 6, 1987) about Chicago baseball teams. For a twenty-year period beginning in 1967, we have the Chicago Cubs' home attendance, A, and the number of games won, GW. We have omitted the strike-shortened 1981 season from the table.

a. If there is a cause-and-effect relation between GW and A, which variable would you choose as the cause?

b. Construct a scatter plot for the data.

Year	Games Won, GW	Home Attendance, A (Millions)	Year	Games Won, GW	Home Attendance, A (Millions)
1967	87	1.0	1977	81	1.4
1968	84	1.0	1978	79	1.5
1969	92	1.7	1979	80	1.6
1970	84	1.6	1980	64	1.2
1971	83	1.7	1982	73	1.2
1972	85	1.3	1983	71	1.5
1973	77	1.4	1984	96	2.1
1974	66	1.0	1985	77	2.2
1975	75	1.0	1986	70	1.9
1976	75	1.0			

4.2 CORRELATION

The Sample Correlation Coefficient, *r*

Given that we have a bivariate data sample (and perhaps a scatter plot), we are likely to be interested in measuring the strength of the linear relation between X and Y. Let's first consider the two different sets of observations whose scatter plots are shown in Figure 4.7. While the scatter plots differ in appearance, both sets of data do display a direct relation for X and Y. In addition, a straight line with about the same steepness,

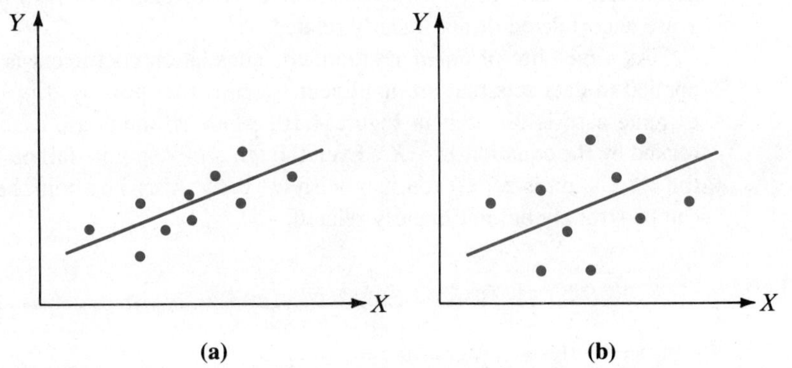

Figure 4.7 Two Different Ten-Observation Scatter Plots

or slope, gives a good representation of the overall pattern in each case. However, it should be apparent that the variables shown in Figure 4.7(a) have a stronger relation than do those in Figure 4.7(b). This is because there is less scatter in panel (a) of the figure—that is, individual points fall closer to the reference line in panel (a) than they do in panel (b). Accordingly, what we want to develop in this section is a measure that quantifies the *strength* of relation between X and Y. We want such a measure to be able to discriminate between the two situations suggested in Figure 4.7. We want the measure to somehow inform us, even if we cannot see the scatter plots, that a stronger relation exists in panel (a) than in panel (b). The measure we develop is called the *correlation coefficient*.

Definition

The sample **correlation coefficient,** denoted by r, is a unitless measure of the strength of the linear relationship for quantitative bivariate data.

The sample correlation coefficient is a *single number* that describes the degree of linear relation between X and Y. Its purpose is to indicate whether our sample data are strongly related, weakly related, or unrelated. As Figure 4.8 suggests, r is constrained to take on values from -1.0 to $+1.0$, inclusively, regardless of the units of measurement for X and Y. Two variables with an inverse relation will have a negative correlation; a direct relation will yield a positive correlation. Strength of the linear (straight-line) relation is indicated by the *absolute* value of r. The stronger the relation, the closer r will be to negative 1.0 or positive 1.0. The plus or minus sign of r indicates nothing more than whether the relation is inverse or direct; a correlation of $-.80$ shows a relation just as strong as does a correlation of $+.80$.

Figure 4.9 presents scatter plots illustrating different degrees of linear correlation. Note that the extreme values of -1.0 and $+1.0$ can only occur when all the plotted points can be connected by a straight line. A correlation of zero indicates that X and Y are uncorrelated or not linearly related.

As a measure of *linear* relation, the correlation coefficient is not intended to be applied to data sets that are nonlinear, such as that portrayed in Figure 4.5. A more extreme case is pictured in Figure 4.10, where X and Y are exactly but nonlinearly related by the equation $Y = X^2$. Even though all five points fall on the line of relation, the correlation is zero. (You may wish to verify this.) The point here is that variables can be strongly but not linearly related.

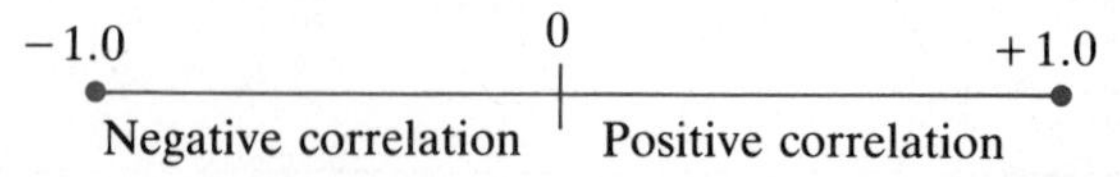

Figure 4.8 Range of Possible Values for r, the Sample Correlation Coefficient

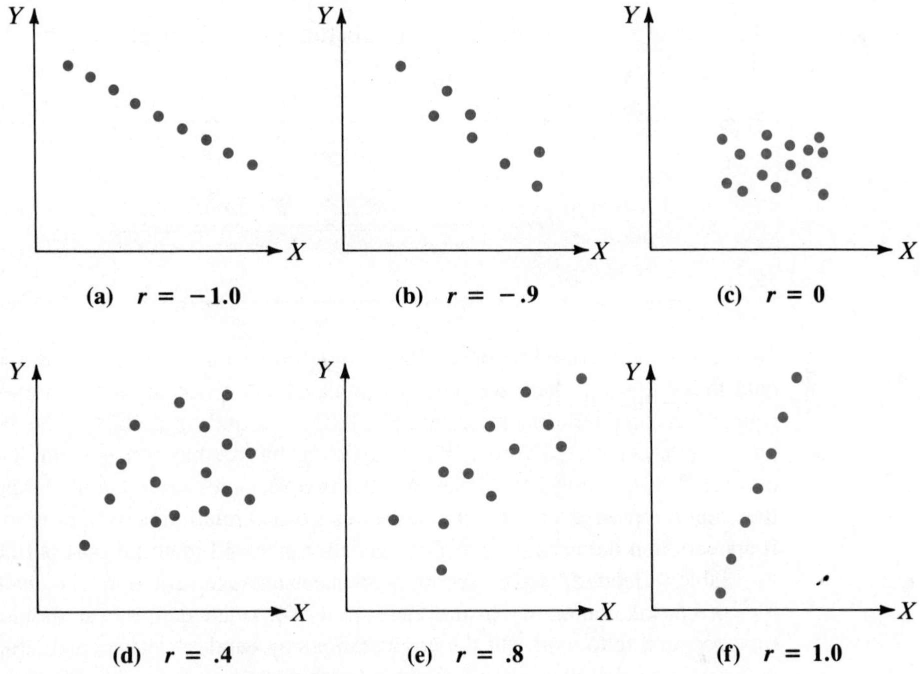

Figure 4.9 Scatter Plots Showing Various Degrees of Correlation

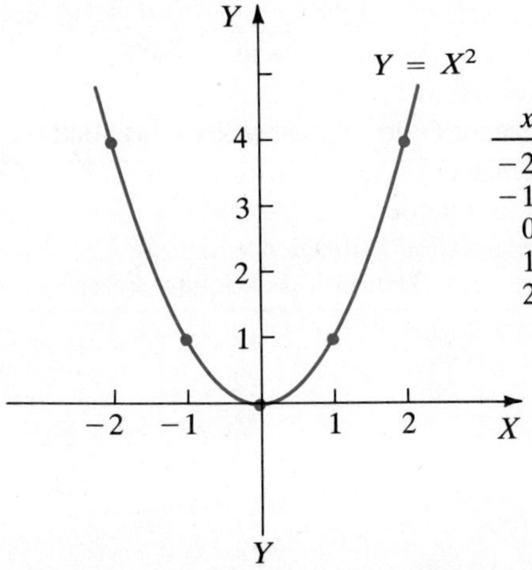

x	y
−2	4
−1	1
0	0
1	1
2	4

Figure 4.10 Exact Nonlinear Relation Having Zero Correlation

Computing the Sample Correlation Coefficient

The value of r, the sample correlation coefficient, is obtained from the formula in the following box.

Sample Correlation Coefficient

$$r = \frac{n\Sigma XY - \Sigma X \Sigma Y}{\sqrt{n\Sigma X^2 - (\Sigma X)^2}\sqrt{n\Sigma Y^2 - (\Sigma Y)^2}} \qquad \text{(4–1)}$$

To illustrate the computation of the sample correlation coefficient, we will use the data in Table 4.2, where we have a sample of ten observations on home size, X (in square feet), and building material cost, Y (in thousands of dollars). The data represent ten recently completed homes that a particular builder has constructed. The builder is interested in learning how strongly these two variables are related. A scatter plot for these data, shown in Figure 4.11, indicates a direct relation between the two variables. It appears that the relation between residence size and material cost is linear.

Table 4.3 presents the type of worksheet arrangement that is convenient to use for correlation and/or regression analysis. For a small data set, it should not be too time consuming to work out the computations by hand or with a calculator. For larger data sets, you will want to use a calculator that has correlation/regression capabilities, an electronic spreadsheet, or a computer program. (See Section 4.4.)

Using the column totals given in Table 4.3, we have

$$r = \frac{10(11072) - 217(496)}{\sqrt{10(4929) - 217(217)}\sqrt{10(25096) - 496(496)}}$$

$$= \frac{3088}{\sqrt{2201}\sqrt{4944}} = \frac{3088}{3298.7} = .936$$

Table 4.2 Sample Data on Home Size and Building Material Cost

Home	Residence Size, X (Hundreds of Square Feet)	Building Material Cost, Y (Thousands of Dollars)
1	17	46
2	29	60
3	18	42
4	19	43
5	21	50
6	21	47
7	14	39
8	24	58
9	26	53
10	28	58

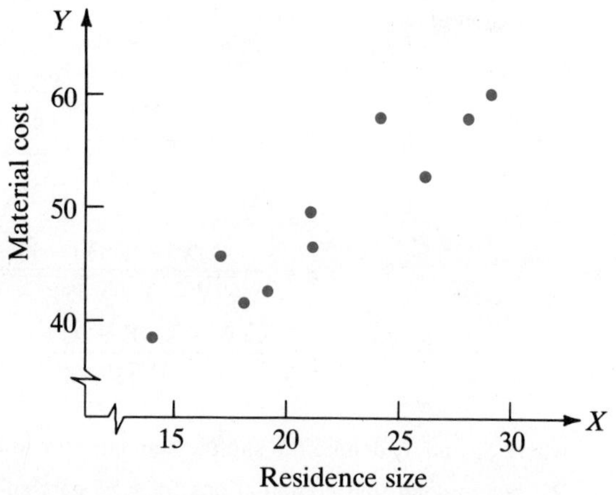

Figure 4.11 Scatter Plot for Data in Table 4.2

Table 4.3 Worksheet for Regression and Correlation Sums

X	Y	X^2	XY	Y^2
17	46	289	782	2116
29	60	841	1740	3600
18	42	324	756	1764
19	43	361	817	1849
21	50	441	1050	2500
21	47	441	987	2209
14	39	196	546	1521
24	58	576	1392	3364
26	53	676	1378	2809
28	58	784	1624	3364
$\Sigma X = 217$	$\Sigma Y = 496$	$\Sigma X^2 = 4929$	$\Sigma XY = 11072$	$\Sigma Y^2 = 25096$

This high correlation shows a very strong linear relation between the size of the residence built and the builder's material cost.

COMMENTS

1. As Exercise 4.4 implies, the labels X and Y do not matter with respect to the value you will obtain for r. However, the labels do matter for finding the equation of the regression line in the next section of this chapter.
2. Any point in a data set that is some distance away from the main body of observations can influence heavily the value of the correlation coefficient (as well as the equation of the regression line). Use of a scatter plot can indicate whether variables have a linear relation as well as alert you to the presence of such points. (See Exercise 4.17.)

3. The following formulas for the sample correlation coefficient are equivalent to the expression given in Equation 4–1:

$$r = \frac{\Sigma XY - \dfrac{\Sigma X(\Sigma Y)}{n}}{\sqrt{\Sigma X^2 - \dfrac{(\Sigma X)^2}{n}}\sqrt{\Sigma Y^2 - \dfrac{(\Sigma Y)^2}{n}}}$$

$$r = \frac{\Sigma(X - \bar{X})(Y - \bar{Y})}{\sqrt{\Sigma(X - \bar{X})^2}\sqrt{\Sigma(Y - \bar{Y})^2}}$$

$$r = \frac{\Sigma(X - \bar{X})(Y - \bar{Y})}{(n - 1)\, S_X S_Y}$$

where S_X and S_Y denote the sample standard deviations of the variables X and Y.

4. We recommend that computations for r be carried out to four decimal places with the final result being rounded back to three decimal places.

EXAMPLE 4.1

An ice cream parlor manager recorded the following sales and average temperatures for his most recent eight-month operating season:

Month	Mean High Temperature (°F)	Sales (Hundreds of Gallons)
April	65	8
May	77	11
June	86	14
July	90	13
August	83	12
September	84	10
October	71	8
November	59	7

A scatter diagram for these data appears in Figure 4.12. The manager is interested in knowing the correlation for these two variables.

Solution:

We can let X represent the mean monthly high temperature and let Y represent sales and then set up a worksheet:

X	Y	X^2	Y^2	XY
65	8	4225	64	520
77	11	5929	121	847
86	14	7396	196	1204
90	13	8100	169	1170
83	12	6889	144	996
84	10	7056	100	840
71	8	5041	64	568
59	7	3481	49	413
$\Sigma X = 615$	$\Sigma Y = 83$	$\Sigma X^2 = 48117$	$\Sigma Y^2 = 907$	$\Sigma XY = 6558$

We can now use Equation 4–1 to determine r.

$$r = \frac{n\Sigma XY - \Sigma X \Sigma Y}{\sqrt{n\Sigma X^2 - (\Sigma X)^2}\sqrt{n\Sigma Y^2 - (\Sigma Y)^2}}$$

$$= \frac{8(6558) - 615(83)}{\sqrt{8(48117) - 615(615)}\sqrt{8(907) - 83(83)}}$$

$$= .904$$

Though the number of observations is small, this sample shows a strong relation between sales and monthly high temperatures.

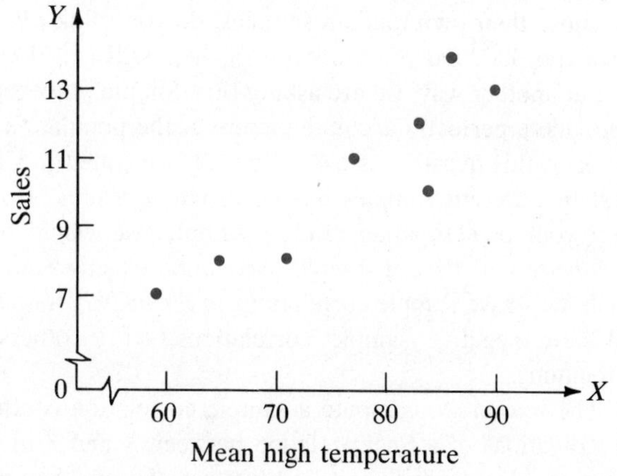

Figure 4.12 Scatter Plot for Example 4.1

□

■ A First Look at Statistical Inference

In Chapter 1, we noted that there are two types of business statistics procedures: description and inference. To this point, we have focused all our attention on descriptive statistics. Activities such as graphing, putting data into a table, or computing measures of central location and dispersion are forms of data description. Although we have not yet covered enough material to begin discussing statistical inference formally, we would like to provide an intuitive look ahead to this major concept.

The Nature of Inference Inference is more complex than description. Description involves summarizing what is *seen*—that is, the sample data that has been gathered. Inference involves making a statement or judgment about the *unseen*—that is, the entire population from which the sample was drawn. Inference also differs from description with regard to risk. Description is a risk-free undertaking; inference is not. Correct application of procedures and formulas will yield descriptions that are accurate, with 100% certainty. But making inferences about an entire population based on seeing only a portion—the sample—obviously cannot be done with 100% certainty. We can make an inference, however, if we are willing to accept a small degree of risk of making an incorrect statement.

Let us now look at a specific example of making a statistical inference. We will use r, the sample correlation coefficient, to judge whether an underlying population of X and Y are correlated or uncorrelated.

Evidence of Correlated Variables Suppose you have a very large bivariate population in which X and Y are *known* to be uncorrelated. That is, if the correlation coefficient were determined with all entities of the population involved in the computation, the answer would be .000. Now imagine you are permitted to sample at random from this population: Let's say your sample size is 50. If you compute your sample's correlation r, do you think it would be exactly .000? If ten of your classmates also chose their own random samples, do you think *all* of them will get a correlation of exactly .000? Do you think *any* of them will get .000?

Put another way, we are asking this: Should you expect any single random sample to provide a perfectly accurate picture of the population from which it is drawn? The answer to this question is clearly no. As we stated in Chapter 1, sampling error will result in different samples having different results. If you and ten classmates each select your own separate random sample, we would be willing to bet money that (a) *nobody* will have a sample correlation exactly equal to zero, (b) most if not all people will have sample correlations *in the neighborhood of zero,* and (c) some people will have negative sample correlations, while others will have positive sample correlations.

The reason we compute a sample correlation coefficient is to see whether there are indications of a linear relation between X and Y in the population. If the sample shows us a large positive or negative correlation, then we have reason to believe that there is a relation in the population. On the other hand, if the sample's r value is near zero, then we doubt that the variables X and Y are linearly related. At some place between 0 and $+1.0$ (as well as between 0 and -1.0), there will be a cutoff point that separates a region for concluding that there likely is a relation from another region that represents concluding that no relation exists between the variables X and Y. As Figure 4.13 suggests, we should not be willing to believe a relation between the variables X and Y exists in the population unless the (absolute) value of the sample

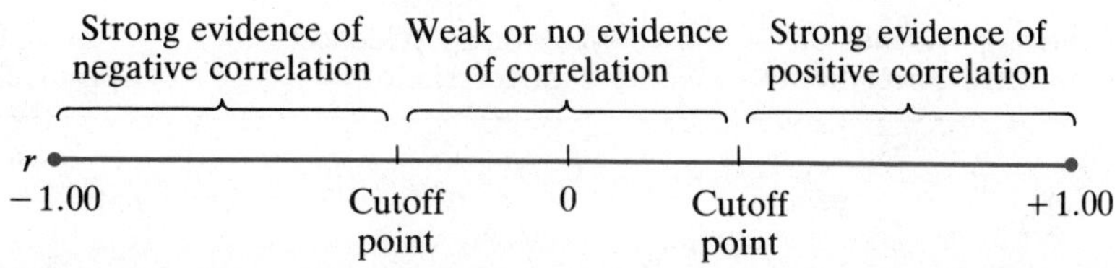

Figure 4.13 Cutoff Points for Drawing Conclusions About the Correlation of Two Variables in a Bivariate Population

Table 4.4 Cutoff Points for r

Sample Size	Cutoff Points	Sample Size	Cutoff Points
5	±.878	18	±.468
6	±.811	20	±.444
7	±.754	25	±.396
8	±.707	30	±.361
9	±.666	40	±.312
10	±.632	50	±.279
12	±.576	60	±.254
14	±.532	80	±.220
16	±.497	100	±.196

correlation r exceeds a certain point. Table 4.4 gives values for these cutoff points. Notice that these points get closer to zero for larger sample sizes. Also note that the same cutoff point is not appropriate for all sample sizes, since a large sample utilizes more information (observations) than does a small sample.

To complete our hypothetical example, suppose your sample size of 50 reveals a sample correlation of $r = .085$. Table 4.4 indicates cutoff points of $\pm.279$ for $n = 50$. Thus you conclude that no strong evidence of a relation between X and Y in the population is found. You now have drawn a conclusion *about an entire data set,* even though you observed only 50 members of that set!

EXAMPLE 4.2

In the previous section, we introduced the calculation of r by considering a data set of ten observations on residence size and building material cost. We found that the sample correlation was $r = .936$. To make an inference using $r = .936$ and Table 4.4, we note that the cutoff points are $\pm.632$ when the sample size is ten. In this case, the positive cutoff of .632 is easily exceeded by the sample value of .936,

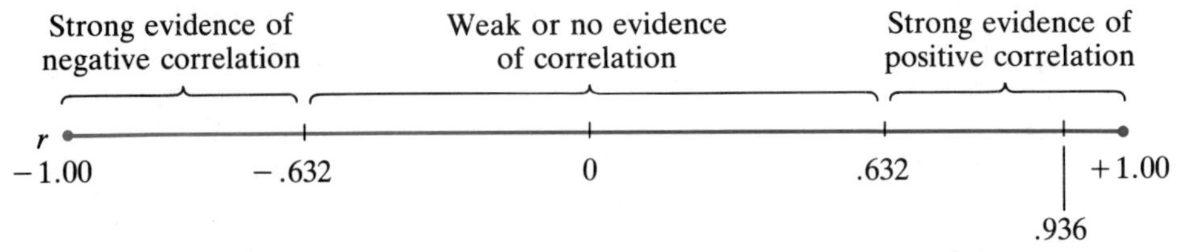

Figure 4.14 Cutoff Points for $n = 10$ in Example 4.2

offering convincing evidence that a direct relation does exist in the bivariate population from which the sample was drawn. (See Figure 4.14.) □

Much of business statistics involves instances of what we have just illustrated: using sample evidence to draw a conclusion about a larger body of data. Keep in mind that there is a small risk that the conclusion is wrong. After studying probability and probability distributions in the next few chapters, we will be ready to fully understand the basis of inference.

COMMENTS

1. The finding of strong evidence of correlated variables does not necessarily mean the existence of a cause-and-effect relation. Two variables may happen to be correlated but not cause-and-effect related. For instance, assume variable A has a strong effect on both variables B and C. In that event, variables B and C could appear highly correlated even if B does not affect C and C does not affect B.
2. Even if causation exists, its direction may not always be clear. For instance, many firms' annual sales, S, will correlate positively with annual advertising expenditures, A. If we believe there is cause and effect here, does A cause S, or does S cause A? While you may think that advertising causes sales, some firms use the rule that advertising should be a certain percent, say 5 percent, of sales. For a firm following such a rule, we might argue that sales is the cause and advertising the effect.
3. We have briefly and informally exposed you to some of the major ideas of statistical thinking when we discussed cutoff points for different conclusions that you might draw about the population from which a sample has been taken. The origin of these cutoff points and establishing the risk of making an incorrect judgment are beyond our scope at this point. For now, we can intuitively think of the cutoff points as being the limits of a "region of concentrated values" when zero correlation exists in the population of X and Y. We hope that when you are reading about hypothesis tests in Chapter 10, you can turn back to this chapter and note the similarities.

CLASSROOM EXAMPLE 4.1

Computing a Sample Correlation Coefficient

A restaurant owner decides to advertise through DMA, a company specializing in direct-mail advertising. The restauranteur plans to promote a "buy one entree, get a second entree free" sale by printing coupons and having DMA mail them to residents in the surrounding neighborhood. Over a period of six months, the restaurant owner plans five mailings. Each time, he plans to print a different number of coupons and record the number redeemed. Let X = the number of coupons printed (in thousands) and Y = the number of coupons redeemed (in hundreds). For instance, if 2000 coupons are printed and if 300 are redeemed, then $X = 2$ and $Y = 3$. Suppose he records the following data:

X	Y	XY	X^2	Y^2
1	1			
2	1			
3	2			
4	2			
5	4			

a. Construct a scatter plot of these data.

b. Supply the data for the columns labeled XY, X^2, and Y^2. Find the sum for each column.

c. Compute the sample correlation coefficient between X and Y.

d. Is there evidence that the variables X and Y are correlated in the population?

SECTION 4.2 EXERCISES

4.10 For each of the following pairs of variables, indicate whether you would expect them to be positively correlated, negatively correlated, or uncorrelated:

a. for a sample of households: monthly water consumption and number of people in the household.

b. for a sample of homes in Minneapolis in January: number of inches of attic insulation and monthly use of natural gas.

c. for a sample of bank customers: number of bad checks written last year and mean checking account balance last year.

d. for a sample of two-career households: age of the wife and age of the husband.

4.11 Compute the sample correlation coefficient for the data in the following exercises:

a. Exercise 4.5. **b.** Exercise 4.6. **c.** Exercise 4.7.

4.12 Compute r for the data in Exercise 4.8. Is the sample correlation strong enough to convince you that a relation exists in the population?

4.13 You have sampled 12 observations from a bivariate population of interest to you. You wish to draw a conclusion about whether X and Y are correlated in this population, based on your sample evidence. What should be said if $r = .44$?

4.14 Answer Exercise 4.13 for $n = 40$, with $r = -.58$.

4.15 Refer to Exercise 4.9, which gives information about the Chicago Cubs' attendance and games won over a 20-year period. Treating these data as a sample, determine the following:

a. Compute the correlation coefficient. (Note that $n = 19$ since one pair of values was not used.)
b. Is the value of r you determined of sufficient magnitude for you to conclude the variables A and GW are related in the population?

4.16 Following is a listing for millions of people unemployed in the United States and the number of new unemployment insurance claims filed in the state of Kentucky. The data available are the January figures for the most recent available 14 years:

Thousands of New Claims Made in Kentucky, *A*	Millions of U.S. Unemployed, *B*
24.6	3.9
31.6	5.9
28.5	5.8
26.8	4.9
32.0	5.1
58.8	8.1
51.2	7.9
50.4	7.5
45.6	6.4
51.3	5.9
65.1	6.3
66.3	7.5
86.7	8.6
89.4	10.4

Source: Kentucky Economic Information System, Center for Business and Economic Research, University of Kentucky, Lexington, KY.

a. Plot the data, with B on the horizontal axis. Does a linear relation appear reasonable?
b. Compute the correlation coefficient.
c. Use Table 4.4 to determine whether it is reasonable to believe that the variables A and B are uncorrelated in the population.

4.17 A real estate agent chooses a random sample of 15 listings from an area multiple listing service booklet. In the following table are the data on home square footage and asking price:

Sample Item	Square Feet (Hundreds)	Asking Price (Thousands of Dollars)
1	28.0	164.9
2	15.5	54.9
3	14.5	69.9
4	12.4	59.0
5	19.0	84.0
6	12.0	48.5
7	32.0	387.5
8	17.5	42.5
9	24.0	103.0

Sample Item	Square Feet (Hundreds)	Asking Price (Thousands of Dollars)
10	20.0	119.0
11	16.0	69.9
12	14.9	58.9
13	20.4	96.0
14	19.5	104.8
15	22.5	116.9

a. Plot the data.

b. Compute r, the sample correlation coefficient.

c. We noted in the text that an unusual observation can have a noticeable impact on the value for r. Remove what appears to be the most unusual sample item and recompute r to assess the influence of the removed item.

4.18 **a.** Find the correlation between the total number of races and the average net purse per race based on the following random sample of six years of data from a Kentucky racetrack:

Number of Races	Average Net Purse
194	\$ 4,731
240	4,894
249	6,842
248	13,432
256	17,289
256	20,248

Source: Data from "How Kentucky Tracks Have Done," Louisville *Courier-Journal,* November 18, 1986, p. A6.

b. Code the purse amounts in thousands (4731 = 4.731, for example) and recompute r. Do you obtain the same result as in part a?

4.19 A random sample of size $n = 16$ produced the following sums of squares:

$$\Sigma X = 562.12 \qquad \Sigma Y = 29.36 \qquad \Sigma XY = 1052.3884$$

$$\Sigma X^2 = 23{,}540.851 \qquad \Sigma Y^2 = 55.8194$$

Determine the sample correlation coefficient.

4.20 Ten randomly selected Indiana hospitals reported the following values for average daily cost, X, and average length of patient stay, Y:

X(\$)	561	710	454	470	461	655	512	629	579	583
Y	5.9	6.5	6.1	6.8	5.9	6.7	4.8	4.8	5.9	5.8

a. Does a scatter plot suggest no relation, a direct relation, or an inverse relation between these variables?

b. Determine the sample correlation coefficient.

4.21 Following are data for variables S and T. Variable S represents expenditures for new plant and equipment; variable T represents new housing starts. Both variables are reported on an annual basis for the United States from 1980 through 1987.

Year	Expenditure for New Plant/Equipment, S (Billions of Dollars)	New Housing Starts, T (Millions)
1980	314	1.3
1981	349	1.1
1982	347	1.1
1983	343	1.7
1984	399	1.8
1985	432	1.7
1986	427	1.8
1987	440	1.6

Source: U.S. Department of Commerce, Bureau of Economic Analysis, Washington, D.C.

a. Plot the data with variable S on the horizontal axis. Does a strong relation between S and T seem apparent?
b. Compute the sample correlation coefficient.
c. Use Table 4.4 to determine whether the sample value for r provides evidence that S and T are correlated in the population.

4.3 SIMPLE LINEAR REGRESSION

Statistical and Functional Relations

Before developing an equation to quantify the relation between two variables X and Y, it is useful to make the distinction between a statistical relation and a functional relation for variables. You are no doubt already familiar with functional relations. As an example, consider the equation that converts degrees Celsius, X, into degrees Fahrenheit, Y:

$$Y = 32 + 1.8X$$

Figure 4.15 shows a scatter plot for four pairs of (X, Y) values. The straight line drawn through these points represents the functional relation. If X and Y are functionally related, we can say that each value of X has one and only one value of Y associated with it. For instance, if $X = 20$, Y can be determined exactly; the equation gives us $Y = 68$. For a *functional relation,* all points will fall on the line of relation.

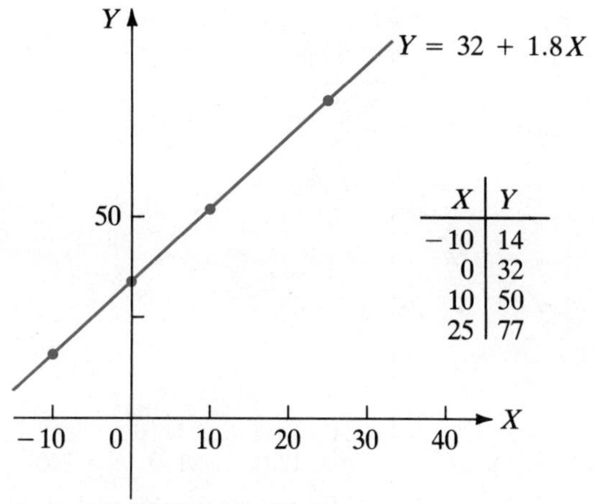

Figure 4.15 Scatter Plot for a Functional Relation

> **Definition**
>
> A **functional relation** between variables X and Y is an exact correspondence, such that each X-value is associated with a unique Y-value.

In contrast, a *statistical relation* is one where the variables are related, but the relation is not perfect or exact. Figure 4.16 illustrates a statistical relation between engine size and miles per gallon for a data set given earlier in this chapter; it is clear that a straight line drawn on the scatter plot will not touch all the points. However, it is also apparent that most points would be close to such a line, indicating that some underlying or average relation between engine size and miles per gallon does exist. In business and economic applications, it would be unusual to have functionally related variables. It is much more likely that we will find statistical relations.

> **Definition**
>
> A **statistical relation** between variables X and Y is an inexact correspondence, such that each X-value is associated with a predicted Y-value.

We are interested in developing the *equation* of a line that describes a statistical relation; we will call this line the **regression line.** One major use of the regression line will be to make forecasts, or predictions. The sample correlation coefficient, in contrast, is a *number* used to measure the strength of the relation.

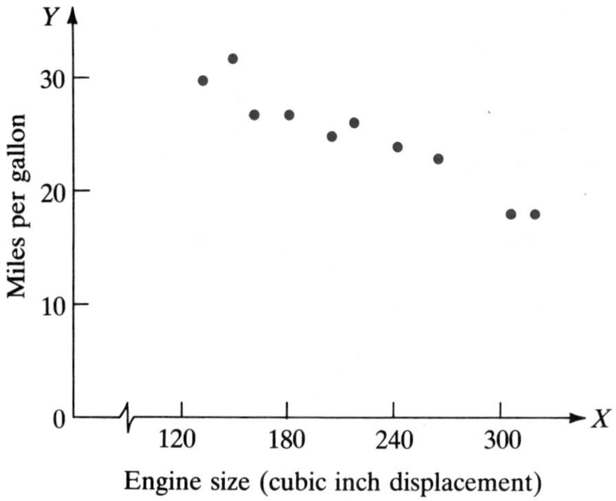

Figure 4.16 Scatter Plot for Data in Table 4.1

Straight Lines and Least-Squares Lines

The general equation for any *straight line* can be written as

$$Y = b_0 + b_1X \tag{4–2}$$

where b_0 is referred to as the Y-intercept and b_1 is the slope of the line. Substituting numerical values for b_0 and b_1 defines a specific line. In our equation for converting Celsius into Fahrenheit ($Y = 32 + 1.8X$), the Y-intercept is 32, and the slope is 1.8. The Y-intercept is the value of Y when $X = 0$; you will note in Figure 4.15 that if $X = 0$, the line crosses or intercepts the Y-axis at 32. The slope represents the *change* in variable Y associated with a *change* of one unit in variable X. In this instance, a slope of 1.8 indicates that a change of one degree on the Celsius (X) scale corresponds to a 1.8 degree change on the Fahrenheit (Y) scale. The slope is sometimes written as

$$b_1 = \frac{\text{change in } Y}{\text{change in } X} = \frac{\Delta Y}{\Delta X}$$

where Δ (delta) is used to denote change. The slope will be a positive number for a direct relation between X and Y; it will be a negative number when X and Y are inversely related. Another way to express the slope is to say b_1 = rise/run. (See Figure 4.17.)

The *regression line* we compute for a given sample data set will be symbolized as follows:

$$\hat{Y} = b_0 + b_1X \tag{4–3}$$

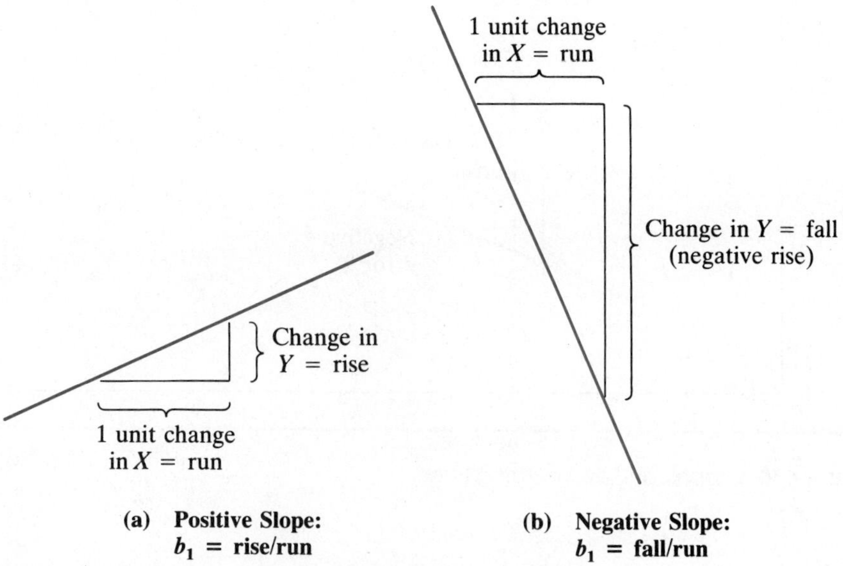

Figure 4.17 Equating the Slope to Rise/Run

Keep in mind that since the regression line is a straight line used to describe an imperfect (statistical) relation, we cannot have it go through each point on the scatter plot. Accordingly, note that we replace Y of Equation 4–2 with $\hat{Y}$ in the above expression. The "hat" symbol placed above a variable denotes an *estimated* or *predicted* value for that variable.

Although we do not expect our straight line to give us a perfect fit, we do wish to find the straight line that *best fits* the sample data. Our method of finding the best fitting line revolves around the idea of a "fitting error," or for short, simply an "error." An error is the vertical (Y-axis) distance that an actual point is away from the regression line $\hat{Y}$. Symbolically, an error is $(Y - \hat{Y})$. Points above the line will have positive errors; points below will have negative errors. (See Figure 4.18.) An error is a deviation—that is, the amount by which a given value differs from a reference point. We already have encountered deviations. Some of our dispersion measures in Chapter 3 are based on deviations and squared deviations from the mean.

How can we measure how well a straight line fits a data set? While more than one way exists to assess the quality of a line's fit, the criterion most often used is this: *Minimize the sum of squared fitting errors*. Accordingly, we will say that the "best" fitting straight line is the one with the smallest possible *sum* of squared errors; it also is called the least-squares line.

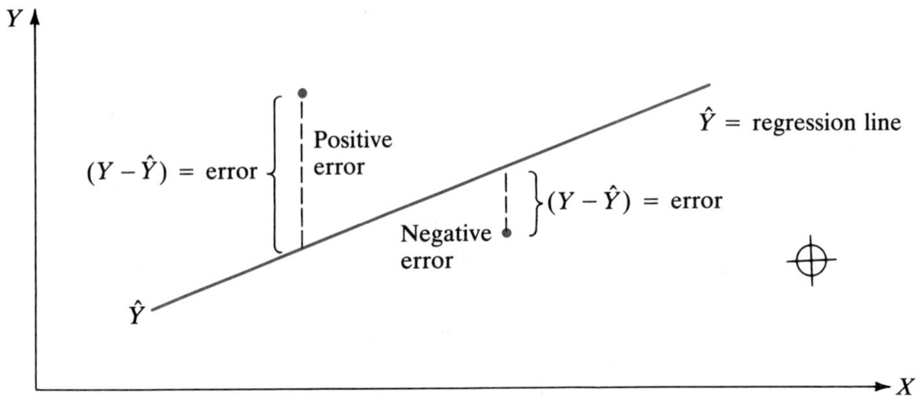

Figure 4.18 Graphical View of Fitting Errors

Definition

The **least-squares criterion** is selection of the line that minimizes the sum of squared fitting errors.

To help understand this criterion, consider as an example the four-point scatter plot shown in Figure 4.19(a). We would like to find the best fitting line for these points; two of many possible straight lines that appear to fit the data are suggested in panel (b) of the figure. Each possible straight line that is applied to our data will have its own fitting errors. Panel (c) shows the fitting errors for line 2. Panel (d) gives numerical values for the four errors of line 2. However, our criterion for best fit involves the *sum of squared errors,* or $\Sigma(Y - \hat{Y})^2$. For line 2, this sum would be $1^2 + (-3)^2 + 4^2 + (-2)^2 = 30$. If in fact line 2 is the best fitting line, then no other line that could be drawn through the data will have a sum of squared errors as low as 30. For any data set, there always will be one unique straight line that achieves the minimum possible value of $\Sigma(Y - \hat{Y})^2$.

How can we determine the intercept and slope of the line that minimizes the sum of squared errors? The answer comes from calculus. We will give the results without proof in this chapter.

Definition

A linear statistical relation satisfying the least-squares criterion is called the **least-squares line** and is of the form

$$\hat{Y} = b_0 + b_1X$$

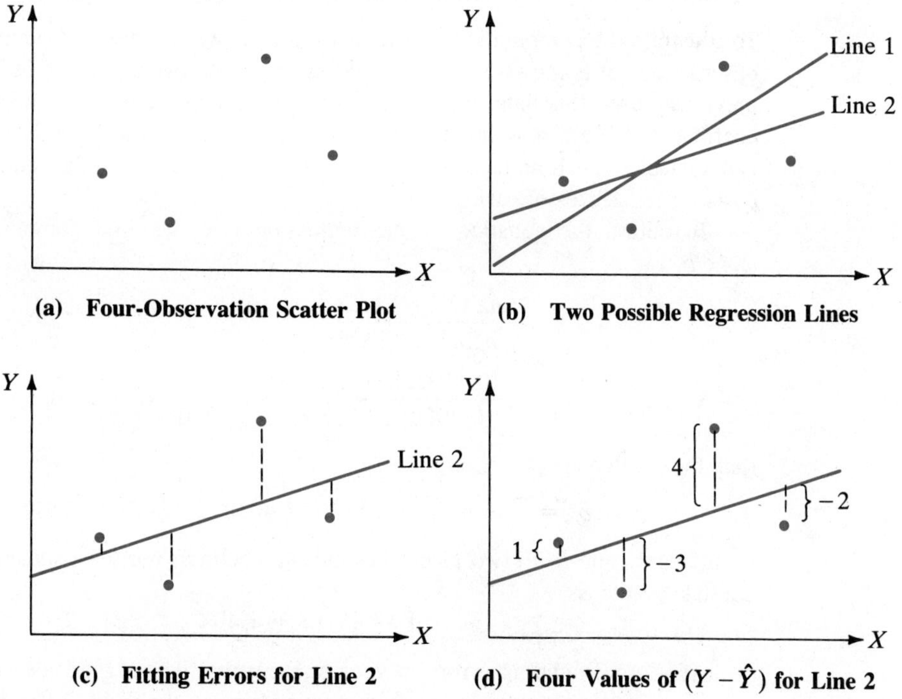

Figure 4.19 Determining Fitting Errors for a Straight Line

where

$$b_1 = \frac{n\Sigma XY - \Sigma X \Sigma Y}{n\Sigma X^2 - (\Sigma X)^2} \quad (4\text{–}4)$$

$$b_0 = \overline{Y} - b_1 \overline{X} \quad (4\text{–}5)$$

and where $\overline{X} = \Sigma X/n$ is the mean value of X, and $\overline{Y} = \Sigma Y/n$ is the mean value of Y.

Since b_1 is part of the solution to b_0, b_1 needs to be determined before computing b_0. Note that the numerator of b_1 is identical to the numerator of r, the correlation coefficient.

COMMENT Algebraically equivalent forms of Equation 4–4 include the following:

$$b_1 = \frac{\Sigma(X - \overline{X})(Y - \overline{Y})}{\Sigma(X - \overline{X})^2} \qquad b_1 = \frac{\Sigma XY - n\overline{X}\overline{Y}}{\Sigma X^2 - n\overline{X}^2} \qquad b_1 = \frac{\Sigma XY - \dfrac{\Sigma X(\Sigma Y)}{n}}{\Sigma X^2 - \dfrac{(\Sigma X)^2}{n}}$$

Computing and Interpreting the Regression Line

To illustrate the computation of a regression line, we will use the sample data of ten observations of home size, X, and building material cost, Y, shown in Table 4.5. We previously used this data set and the sums shown to compute the sample correlation coefficient. We are now interested in developing the equation that best relates these two variables. Such an equation may be useful in predicting future building material costs for residences of various sizes.

To obtain the equation of the regression line, we first compute the sample slope b_1:

$$b_1 = \frac{n\Sigma XY - \Sigma X \Sigma Y}{n\Sigma X^2 - (\Sigma X)^2}$$

$$= \frac{10(11072) - 217(496)}{10(4929) - 217(217)} = \frac{3088}{2201} = 1.4030$$

then the Y-intercept is

$$b_0 = \overline{Y} - b_1\overline{X} = 49.6 - 1.4030\,(21.7) = 19.1549$$

Rounding b_0 and b_1 to two places beyond the decimal, our least-squares line for the ten data points is

$$\hat{Y} = 19.15 + 1.40X$$

To avoid large rounding errors in your final results, it is a good idea to keep terms such as $\overline{Y}$, $\overline{X}$, and b_1 accurate to at least four places to the right of the decimal as you work along.

Table 4.5 Sample Data Set and Quantities Needed for Computing the Regression Line

Home	Residence Size, X (Hundreds of Square Feet)	Building Material Cost, Y (Thousands of Dollars)
1	17	46
2	29	60
3	18	42
4	19	43
5	21	50
6	21	47
7	14	39
8	24	58
9	26	53
10	28	58

$\Sigma X = 217$ $\Sigma Y = 496$ $\Sigma XY = 11072$ $\Sigma X^2 = 4929$

Note that $\overline{X} = 21.7$ and $\overline{Y} = 49.6$.

The slope of 1.40 shows how X and Y change relative to each other. We interpret 1.40 to mean that Y will change by 1.40 thousand dollars ($1400), on average, for a change in X of 1.0 hundred square feet (100 ft^2).

The regression equation often is used for prediction purposes. If two variables are related, then we take advantage of this fact and use one variable to predict the other. For instance, suppose the next contract that the builder signs calls for a house with 2500 square feet ($X = 25$). We can use the regression equation to estimate or predict that the building material cost will be

$$\begin{aligned}\hat{Y} &= 19.15 + 1.40(25)\\ &= 19.15 + 35.0\\ &= 54.15 \quad \text{or about \$54,000}\end{aligned}$$

EXAMPLE 4.3

In a test market of a new toy, a department store offered the toy at six different prices in six of its stores over a two-week period. The stores were comparable in overall sales volume. Compute the regression line for the following data:

Store	Price, X ($)	Mean Daily Toys Sold, Y
1	8	10
2	12	3
3	10	6
4	14	2
5	7	9
6	9	12

Solution:

As we did previously, we should organize the computations—four columns are needed.

X	Y	X^2	XY
8	10	64	80
12	3	144	36
10	6	100	60
14	2	196	28
7	9	49	63
9	12	81	108
$\Sigma X = 60$	$\Sigma Y = 42$	$\Sigma X^2 = 634$	$\Sigma XY = 375$

Note that $\overline{X} = 10$ and $\overline{Y} = 7$.

First,

$$b_1 = \frac{n\Sigma XY - \Sigma X \Sigma Y}{n\Sigma X^2 - (\Sigma X)^2}$$
$$= \frac{6(375) - 60(42)}{6(634) - 60(60)} = \frac{-270}{204}$$
$$= -1.3235$$

then

$$b_0 = \overline{Y} - b_1\overline{X} = 7 - (-1.3235)10 = 20.235$$

Rounding to two places beyond the decimal, we then have

$$\hat{Y} = 20.24 - 1.32x$$

The negative slope reflects an inverse relation between price and number of toys sold. Figure 4.20 plots these data and the regression line.

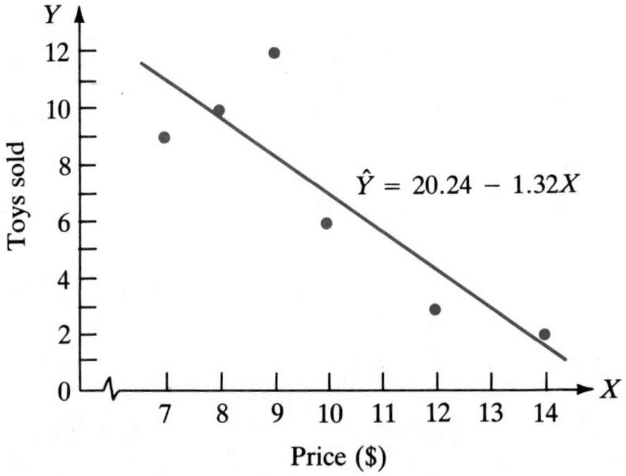

Figure 4.20 Scatter Plot and Regression Line for Example 4.3

□

COMMENTS

1. After computing a regression line, it is a good idea to graph your line on your scatter diagram to verify that you have a reasonable fit. An obviously poor fit may suggest a computational error.
2. *Least-squares line* and *regression line* are interchangeable terms in this and most introductory books.

3. The regression equation should not be used to make predictions outside of the range of X used in developing the line equation. In Example 4.3, for instance, the regression line was computed using values of X ranging from 7 to 14. Making a prediction for Y when $X = 4$ would be dangerous, since we do not know the relation is linear outside of the range of our sample data.
4. Note that ΣY^2 is not needed to compute the least-squares line, although it is needed to determine the correlation coefficient.
5. The regression equation by itself conveys no information about the strength of relation for X and Y; we need the correlation coefficient to do this. As you might expect, having a large (absolute) value for r implies that one variable will be a good predictor of the other variable. A correlation near zero implies that one variable is practically useless in predicting the other.
6. If b_0 or b_1 is less than one, we recommend that you report at least two nonzero digits to the right of the decimal point. For instance, if the computations yield $b_1 = .0001578$, then we would report b_1 as $b_1 = .00016$. If three nonzero digits are desired, we give $b_1 = .000158$.
7. In using the regression equation for prediction, we recommend rounding the predicted value $\hat{Y}$ to the accuracy of the observed Y-values.

CLASSROOM EXAMPLE 4.2

Computing a Sample Regression Line

Refer to the situation in Classroom Example 4.1, where X = the number of coupons printed (in thousands) and Y = the number of coupons redeemed (in hundreds). After ten mailings, suppose the owner has observed these data:

X	Y	XY	X^2
1	1		
2	1		
2	2		
2	3		
3	3		
3	2		
3	4		
4	2		
5	3		
5	4		

a. Supply the data for the XY and X^2 columns, and find the sum of each column.

b. Find the equation of the least-squares line.

c. If the restaurant owner prints and mails 2500 coupons, approximately what number would you predict will be redeemed?

SECTION 4.3 EXERCISES

4.22 Using the equation $\hat{Y} = b_0 + b_1X$, answer the following questions.

a. What property would the line have if the Y-intercept (b_0) had a value of zero?

b. Is it possible for b_0 to be negative?

c. If X and Y were completely unrelated, what would the value of b_1 be? (See Figure 4.4.)

4.23 For a given value of X, what do the symbols (X,Y) and $(X,\hat{Y})$ represent?

4.24 Indicate whether the following statements about the least squares criterion are true or false.

a. The least squares line minimizes the sum of the fitting errors.

b. To apply the least squares criterion, we must find the minimum values for b_0 and b_1 in Equation 4–3.

4.25 Determine the regression equation for the data in Exercise 4.5. Does a straight line appear to offer a reasonable fit to the data?

4.26 Determine the regression equation for the data in Exercise 4.6. Does a straight line appear to offer a reasonable fit to the data?

4.27 **a.** Determine the least-squares line for the data in Exercise 4.7.

b. Use your equation to predict Y for $X = 4$.

c. The sixth data point in this set is (4, 4.5). What is the fitting error for this particular point?

d. What is the meaning of the slope in terms of family size and dental expenditures?

4.28 **a.** Determine the least-squares line for the data in Exercise 4.8.

b. Use your equation to predict Y for $X = 15$.

c. The second (X, Y) pair in this data set is (12, 80). Find the squared fitting error for this point.

d. What does b_1 mean in terms of vehicle sales and the prime interest rate?

4.29 The following table gives the engine horsepower, X, and the weight, Y, for several riding mowers produced by John Deere:

Model Number	Engine Horsepower, X	Weight, Y (Pounds)
R70	8	275
S82	8	335
S92	11	370
130	9	415
160	12.5	470
180	17	500
185	17	525

Source: Data from John Deere Company.

a. Plot the seven observations. Do the data appear to be linearly related?

b. Compute the correlation coefficient for these data.

c. Determine the equation of the regression line.

d. Plot your regression equation on the scatter diagram.

e. What weight would you predict if a mower with 15 horsepower were developed?

4.30 For a sample of 100 observations, the following quantities have been computed:

$$\Sigma X = 600 \qquad \Sigma Y = 1600 \qquad \Sigma X^2 = 5200$$

$$\Sigma XY = 13600 \qquad \Sigma Y^2 = 37700$$

Compute the equation for the straight line that would best fit the data.

4.31 With $n = 22$ and the following quantities

$$\Sigma X = 134 \qquad \Sigma Y = 120 \qquad \Sigma Y^2 = 4804$$

$$\Sigma X^2 = 6161 \qquad \Sigma XY = 5200$$

determine the slope and intercept of the straight line that will minimize the sum of squared fitting errors for the data set.

4.32 Refer to Exercise 4.17, where we have a sample of 15 values of X, home square footage, and Y, home asking price.

a. Develop a prediction equation so that we can use values of X to forecast Y.
b. Should this equation be used to predict Y for $X = 7.5$? Why or why not?
c. Based on your equation, estimate what Y would be for a home with $X = 26$.
d. Eliminate sample item 7 and develop a prediction equation using the remaining 14 points. Use this equation to predict Y when $X = 26$. Which prediction—part c or part d—reflects the more realistic asking price for a 2600 square-foot home?

4.33 Refer to Exercise 4.16, where we have a sample of 14 values for new Kentucky unemployment claims, A, and the number of U.S. unemployed, B.

a. Letting A be the dependent variable, compute the least-squares line.
b. Predict the dependent variable if the value of the independent variable is 5.0.

*4.4 PROCESSOR: LEAST-SQUARES LINE

Background

The processor for this chapter is named Least-Squares Line. It can compute the sample correlation coefficient, the equation of the regression line, and predict Y for a given value of X. To run this processor, the user first must enter the data into a file. We will illustrate the processor with the ten-observation data set on home size, X, and building material cost, Y, given in Table 4.5:

X	17	29	18	19	21	21	14	24	26	28
Y	46	60	42	43	50	47	39	58	53	58

Creating the Data File

File creation is the same here as was discussed in Chapter 2, Section 2.5, with one small difference: The number of variables, v, will be 2 instead of 1. Also, we will

*Optional

```
≡          Microcomputer Applications for INTRODUCTORY BUSINESS STATISTICS    EDIT
   Overview [Data] Files  Processors  Set Up  Exit                   |  F1=Help

                Y                         X
   1 46                      17
   2 60                      29
   3 42                      18
   4 43                      19
   5 50                      21
   6 47                      21
   7 39                      14
   8 58                      24
   9 53                      26
  10 58                     •28
```

Figure 4.21 Creating a Bivariate Data File

wish to place our dependent variable in column 1 and the independent variable in column 2. Figure 4.21 shows the screen after the 10 bivariate data points have been entered. Note that the variable names are *Y* and *X* and that *Y* has been placed in the first column. Now touch Enter twice to return to the main menu.

Running the Least-Squares Line Program

With the highlight on Processors, touch Enter and then choose Least-Squares Line. After reading the introductory screen, move the highlight to Vector and touch Enter to verify which variable is to be treated as *X* and which as *Y*. If Y was placed in the first column during data entry, no further action is needed; touch Enter twice and proceed to Calculations. (If you wish to make the variable in row 2 become the dependent variable, follow this sequence after touching Vector: F5, down arrow, F5, F5, up arrow, F5, Enter twice.) Touch Enter twice with Calculations highlighted and then, after the "Complete" message, move to Results. Touching Enter now gives the user three options:

1. To obtain the correlation coefficient and the line equation
2. To see the intermediate statistics
3. To use the regression line to make predictions

Figure 4.22 shows the screen after choosing the first of these options; you should explore the others. More extensive regression processors appear in Chapters 15 and 16.

```
≡                          Least Squares Line                          READY
  Introduction  Vector  Calculations [Results] Exit                | F1=Help

      Your sample correlation coefficient is r =  0.9361
                                                  ^
  The least squares line is Y = b0 + b1 * X.  For your sample,

                           b0 = 19.154930
                           b1 = 1.402999

  Therefore your regression line equation is, to 4 decimal place accuracy,
                          ^
                          Y =  19.1549 + 1.4030X.
```

Figure 4.22 Correlation and Line Equation for Data in Table 4.5

4.5 QUALITATIVE BIVARIATE DATA

In Chapter 2 we considered methods to present data for variables that are qualitative, or categorical, in nature. At that time we limited ourselves to tabulations and graphs for a single variable. We now want to extend our discussion to bivariate qualitative data.

Contingency Tables

As discussed in the first section of this chapter, *quantitative* bivariate data can be displayed with a scatter plot. *Qualitative* data do not lend themselves well to graphical presentation; instead, the best way to convey information about bivariate qualitative data is in table form. Tables for qualitative data, called **cross-tab** (short for *cross-tabulations*) or contingency tables, are simply two-dimensional formats with one variable for the rows and the other variable for the columns.

As an illustration of developing a cross-tab table, examine the survey data given in Table 4.6. A sample of 50 university students was asked to specify their favorite sport among baseball, basketball, and football. The sex of each respondent also was noted.

These data are bivariate: The two variables are *sex* and *favorite sport*. These data are also qualitative in that *sex* has two possible categories (male and female), while *favorite sport* has three (baseball, basketball, and football). After tallying the individual responses, a contingency table for these data has the appearance shown in Table 4.7.

Table 4.7 is said to be a two-by-three table, meaning that it contains two row categories and three column categories. All the numbers in the table represent fre-

Table 4.6 Raw Data for Sport Preference Survey Used in Section 4.5

Person	Sex	Favorite Sport	Person	Sex	Favorite Sport	Person	Sex	Favorite Sport
1	Female	Basketball	18	Male	Baseball	35	Female	Football
2	Female	Football	19	Male	Football	36	Male	Basketball
3	Male	Basketball	20	Female	Basketball	37	Female	Football
4	Male	Football	21	Male	Baseball	38	Male	Baseball
5	Female	Basketball	22	Female	Basketball	39	Male	Baseball
6	Male	Football	23	Male	Baseball	40	Female	Basketball
7	Female	Football	24	Male	Basketball	41	Male	Basketball
8	Male	Football	25	Female	Basketball	42	Female	Football
9	Male	Basketball	26	Female	Football	43	Male	Basketball
10	Female	Football	27	Female	Baseball	44	Female	Baseball
11	Male	Baseball	28	Female	Basketball	45	Female	Basketball
12	Female	Basketball	29	Male	Football	46	Male	Baseball
13	Male	Basketball	30	Female	Baseball	47	Male	Basketball
14	Female	Baseball	31	Male	Basketball	48	Female	Baseball
15	Male	Basketball	32	Male	Football	49	Female	Basketball
16	Male	Baseball	33	Male	Football	50	Male	Football
17	Female	Basketball	34	Female	Basketball			

Table 4.7 Contingency Table for Sport Preference Survey

	Favorite Sport			
Sex	Baseball	Basketball	Football	Total
Male	8	10	8	26
Female	5	12	7	24
Total	13	22	15	50

quencies, or counts. A two-by-three table possesses six *cells,* where we find the bivariate frequencies for the six combinations of sex and favorite sport. For instance, the 8 in the upper left cell indicates that eight respondents had the two attributes of being male and selecting baseball.

The frequencies given outside the cells are called *marginal* totals, which represent univariate totals for each of the two variables. To illustrate, the number 13 at the bottom of the left column of the table denotes that altogether 13 people chose baseball as the favorite sport; the marginal row total of 26 shows a total count of 26 males in

Table 4.8 Relative Frequency Contingency Table

	Favorite Sport			
Sex	Baseball	Basketball	Football	Total
Male	.16	.20	.16	.52
Female	.10	.24	.14	.48
Total	.26	.44	.30	1.00

the survey. The overall total of 50 in the lower right position of the table represents the total count of each row and column variable.

As was done in Chapter 2, we might want to express our counts as relative frequencies, yielding decimal values from zero to one, inclusive. Table 4.8 provides a relative frequency contingency table for these survey data. The values are obtained by dividing each frequency by the overall number of observations, in this case 50.

COMMENTS

1. For a cross-tabulation, it generally makes no difference as to which variable is placed on the rows and which on the columns.
2. The margins of a cross-tab contain two *one-way* tabulations. It is the cells of the table that provide our *bivariate* counts or relative frequencies.

SECTION 4.5 EXERCISES

4.34 A random sample of 30 corporation stockholders was conducted to collect opinions on a possible merger with another firm. The following bivariate data give each individual's number of shares held and their opinion:

24, opposed	1200, for	100, opposed
100, opposed	5, opposed	275, for
50, for	55, for	40, for
80, no opinion	100, opposed	10, opposed
350, for	200, for	400, opposed
613, for	150, opposed	10, opposed
100, opposed	400, opposed	1500, for
100, opposed	10, opposed	100, opposed
50, for	75, no opinion	100, for
120, for	100, for	700, for

a. Construct a contingency table, letting the shares held variable be categorical with two possibilities: *small* (100 shares or less) and *large* (more than 100).
b. Express your table in relative frequency form.
c. Is there any noticeable pattern in the data? Explain.

4.35 A midwestern-based insurance company was among the first automobile insurers to offer a discount for nonsmoking drivers. A company study of policyholders' smoking behavior against their driving records revealed the following:

	Driving History		
Driver Status	**No Accidents or Violations**	**Median Accident-Violation Group**	**High Accident-Violation Group**
Principal Driver Smokes	93	165	270
Principal Driver Does Not Smoke	407	335	230

a. How many policyholders were surveyed?
b. Compute the relative frequencies.
c. What evidence in the table suggests that nonsmokers may represent a better insurance risk?

4.6 SUMMARY

The primary focus of this chapter has been on graphing and describing relationships that may exist between two quantitative variables. We began our study with two-dimensional diagrams of the data, called scatter plots. We recommend creating scatter plots for all data sets, since this visual representation can inform the observer, at least approximately, about the nature and strength of the relations.

We have developed the sample correlation coefficient, a major descriptive statistic for bivariate data. The purpose of the correlation coefficient is to summarize with a single number the nature (direct or inverse) and strength of linear relation for two variables. Another way to describe the manner in which variables X and Y are statistically related is to develop the equation of a straight line that best fits the data. We introduced the criterion of minimizing the sum of squared errors as our means of obtaining this line: the regression line. We discussed how to use the regression line to make predictions, noting that if two variables are related it is then possible to use values of one variable to help estimate or predict values of the other variable. Regression is likely the most widely used of all statistical procedures, with many applications in economics, finance, forecasting, and general business research. To begin to become truly knowledgeable in this area, a student would need to know considerably more than we have attempted to present in this overview chapter. A more detailed and advanced discussion of correlation and regression with more emphasis on assumptions and problem areas appears in Chapters 15, 16, and 17.

While scatter plots, correlation, and regression are appropriate tools of analysis for quantitative bivariate data, qualitative (categorical) variables are best investigated with a two-dimensional table called a contingency table. In Chapter 19, we will study the counterpart idea to correlation that can be applied to qualitative bivariate data.

4.7 TO BE CONTINUED . . .

. . . In Your College Courses

You are likely to encounter the ideas of correlation and/or regression again in your college course work. Here are a few possibilities:

- Many business schools require a course in management science/operations research. You may use regression in such a course as a sales forecasting technique.
- In an introductory psychology course, correlation is often used to show how strongly two variables are related to each other. For instance, a study of juvenile delinquency may attempt to correlate variables such as *problems at home* with *crime rate*. Correlation research enables psychologists to make predictions about behavior.
- If you take a second term of business statistics, you will certainly study regression and correlation further.
- If you take a course in economic analysis, econometrics, or forecasting, you will see applications of regression to problems of predicting demand, sales, unemployment, tax collections, and so forth.
- If you take a course in cost accounting, you might analyze data represented graphically as in Figure 4.23 or in Figure 4.24. Figure 4.23 suggests a case where there is interest in predicting direct labor costs for any given size batch a firm might produce. Figure 4.24 uses regression analysis in a situation where mixed or semivariable costs are present. The objective may be to predict the total cost of a production run when there is a fixed element, such as overhead, and a variable element that depends on the amount of production.
- If you sit for the CPA exam, you may be asked questions about regression or correlation, perhaps in an accounting setting as suggested above. Questions A and B, which follow, are two examples of questions taken verbatim from previous exams:*

  A. *What is the appropriate range for the coefficient of correlation* (r)?
  a. $0 \leq r \leq 1$ *b.* $-1 \leq r \leq 1$ *c.* $-100 \leq r \leq 100$ *d.* $-\infty \leq r \leq \infty$

  B. *A scatter chart depicting the relationship between sales and salesmen's auto-*

**Uniform CPA Examination,* copyright American Institute of Certified Public Accountants, Inc., New York. Reprinted by permission.

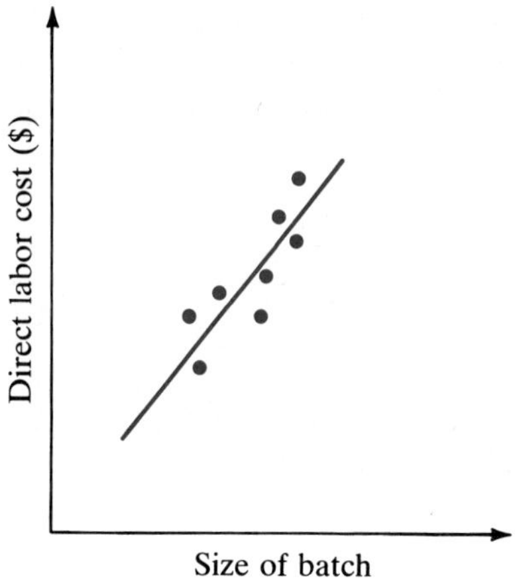

Figure 4.23 Scatter Plot and Regression Line for Relating Direct Labor Costs and Size of Batch

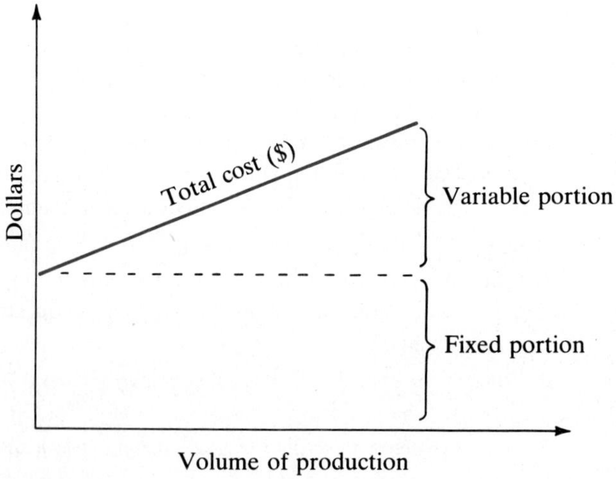

Figure 4.24 Example of Regression Line to Model Total Costs

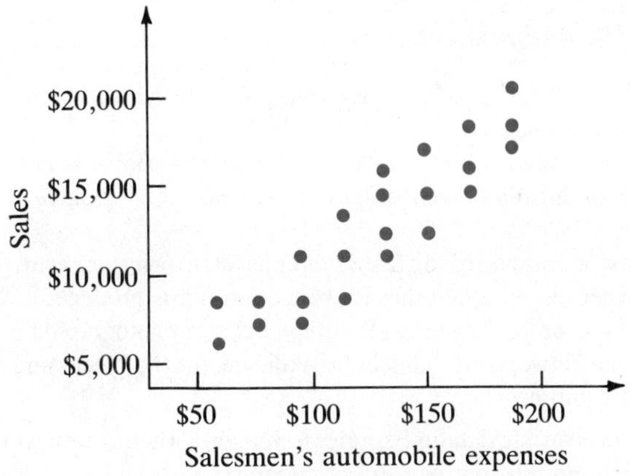

Figure 4.25 Scatter Plot of Automobile Expenses Versus Sales (*Source: Uniform CPA Examination,* copyright American Institute of Certified Public Accountants, Inc. Reprinted by permission.)

mobile expenses is set forth [*in Figure 4.25*]. *What can we deduce from the chart about the relationship between sales and salesmen's automobile expenses?*

a. A high degree of linear correlation.
b. A high degree of nonlinear correlation.
c. No apparent correlation.
d. Both sales and salesmen's automobile expenses are independent variables.

. . . *In Business/The Media*

Regression and correlation are widely used analytical techniques in business practice, and we have tried to suggest a variety of problem settings where it would be appropriate to plot the data to see if and how the variables under consideration are related. Applications exist in economics, finance, production, accounting, and other areas. Many business uses of regression and correlation are more complex than those presented in this chapter. Specifically, many situations exist where there may be more than one independent variable X available to relate to our dependent variable Y. In later chapters, we will consider *multiple regression*—that is, finding the best combination of X's to predict or model our Y variable.

Survey research is a business application that is likely to generate bivariate qualitative data. Marketing research, in particular, often involves consumer studies where questions with categorical response options are asked about individuals' interests, opinions, purchase intentions, lifestyle, income, age, and so forth. Cross-tabulation of survey data is almost always a standard part of the analysis. Methods for establishing the existence of relations in contingency tables will appear in Chapter 19.

CHAPTER 4 EXERCISES

4.36 For the following pairs of variables, specify which should be the independent variable in regression analysis.

a. For a sample of a firm's employees, one variable is last year's medical expense claimed in the firm's benefit program, and the other variable is the number of dependents the employee has.

b. For a sample of different months at a printing firm, one variable is manufacturing expenses, and the other is production (units produced).

c. For a sample of pieces of equipment at a photographic plant, one variable is the average time between mechanical breakdowns for the piece, and the other is the age of the piece of equipment.

4.37 For the bivariate data in Exercise 4.36, state whether you would anticipate finding an inverse relation, no relation, or a direct relation.

4.38 A realtor wants to examine the relation between her appraised value of homes and the actual sale price. She samples five recent transactions for homes she had listed. The figures shown are in thousands of dollars.

Appraised Value, X	Sale Price, Y
89.0	86.5
110.0	109.0
91.5	92.0
101.0	99.5
93.5	88.0

a. Diagram the data with a scatter plot.

b. Determine the equation of the least-squares line.

c. Compute the sample correlation coefficient.

d. If the realtor has appraised a particular home at $90,000, estimate its selling price.

4.39 If you computed the sample correlation for X and Y based on only two pairs of observations of (X, Y), what value for r would result in the following cases:

a. $Y_1 = Y_2$ **b.** $Y_1 \neq Y_2$

(Note: In both cases, assume $X_1 \neq X_2$.)

4.40 A concert promoter wants to relate his concerts' advance ticket sales with the actual concert attendance. He thinks a better understanding could help him plan the proper amount of security, seating, supplies for concessions, and so forth. He has data for the last six concerts. The independent variable is ticket sales up to 48 hours before the concert, X, and the dependent variable is concert attendance, Y. All figures shown are in thousands.

X	7.9	3.9	4.1	5.4	7.7	4.0
Y	11.2	4.8	6.0	8.0	9.1	5.9

a. Compute the regression line for this sample data. Interpret the meaning of the slope in terms of this problem setting.

b. Assess the strength of relation between X and Y by determining the sample correlation coefficient.

c. Predict attendance for a concert whose advance sales are 6500 tickets.

4.41 A storage and transfer company sampled 10 full-time employees' records from last year to see if any relation exists between *age of employee* and *absenteeism*.

Age of Employee, X	Days Absent Last Year, Y
48	1
55	15
21	16
35	4
31	7
20	3
44	5
48	4
60	8
28	7

a. Plot the data on a scatter diagram. Is there any apparent relation?

b. Compute r, the sample correlation coefficient.

c. Given that the sample size is ten, does the correlation you have computed provide strong evidence of a relation between age and absenteeism in the population?

4.42 If a sample of eight data pairs shows a sample correlation coefficient of $r = .30$, is r large enough that you can be reasonably certain that X and Y are correlated in the population? For a correlation of $r = .30$ to suggest that X and Y are related, approximately how large a sample size is required?

4.43 For a representative sample of 40 new car models, X is the estimated highway fuel economy (in miles per gallon) and Y is the estimated overall or combined city–highway estimated mpg.

a. Do you anticipate a direct, inverse, or no relation between X and Y?

b. Given the following data:

$$\Sigma X = 1050 \qquad \Sigma Y = 830 \qquad \Sigma X^2 = 31{,}000$$

$$\Sigma Y^2 = 21{,}500 \qquad \Sigma XY = 24{,}900$$

compute the equation of the regression line.

c. Predict Y for $x = 30$ mpg.

d. Determine the sample correlation coefficient.

4.44 The instructor of a statistics class collected the following data:

$$\Sigma X = 120 \qquad \Sigma Y = 3500 \qquad \Sigma X^2 = 510$$

$$\Sigma Y^2 = 258{,}000 \qquad \Sigma XY = 7850$$

Here, Y = score on the first exam, X = number of student absences prior to the first exam, and $n = 50$.

a. Do you think these variables are likely to be directly or inversely related?

b. Find the sample correlation coefficient.

c. Determine the equation of the least-squares line.
d. Interpret your slope value in terms of scores and absences.
e. Predict Y when $X = 5$.

4.45 Compute the sample correlation using the summary quantities given in Exercise 4.30.

4.46 Compute the sample correlation r using the summary quantities given in Exercise 4.31. Even without $n = 22$ appearing in Table 4.4, can you tell if there is reason to believe that X and Y are correlated in the population?

4.47 *If the coefficient of correlation between two variables is zero, how might a scatter diagram of these variables appear:
a. Random points.
b. A least-squares line that slopes up to the right.
c. A least-squares line that slopes down to the right.
d. Under these conditions, a scatter diagram could not be plotted on a graph.

4.48 Variable X is the annual estimated population of Jefferson County, Kentucky, and Y is the annual residential electricity consumption for the area. The following sample data represent 13 years in the 1970s and 1980s.

Year	County Population, X (Thousands of People)	Residential Consumption, Y (Billions of Kilowatt Hours)
1	706	1.49
2	703	1.62
3	710	1.81
4	712	1.75
5	708	2.03
6	705	2.00
7	698	2.31
8	696	2.27
9	692	2.27
10	685	2.52
11	680	2.37
12	685	2.37
13	685	2.57

Source: Kentucky Economic Information System, Center for Business and Economic Research, University of Kentucky, Lexington, KY.

a. Plot the data. Does a strong relation between X and Y seem obvious?
b. Compute r, the sample correlation coefficient.
c. Use Table 4.4 to arrive at a yes or no conclusion as to whether the variables X and Y are related in the population.
d. Does your correlation coefficient value seem unusual? If so, suggest an explanation for your result.

4.49 A federally sponsored study on smoking by college students was conducted by the University of Michigan's Institute for Social Research. Conducted over the five-year period ending in 1986, the study involved interviews with about 1100 college students each year. One result of this study is given in the following contingency table:

*Exercise 4.47 is reprinted by permission from *Uniform CPA Examination*, copyright American Institute of Certified Public Accountants, Inc., New York.

Sex	Smoke	Do Not Smoke
College Men	275	2475
College Women	495	2255

Source: Data from Louisville *Courier-Journal,* July 8, 1986.

a. What percentage of college men in the survey do not smoke?
b. What is the overall percentage of college students in the survey who smoke?

4.50 The United Way recently sponsored a large-scale survey of employees within a metropolitan area. One of the questions was intended to see whether men and women had the same view of the United Way's efficiency. The survey item read, "Local United Way agencies can provide health and social services more efficiently than state or federal governments can." The responses were cross-tabulated against the sex of the respondent as follows:

	Efficiency Question Response			
Sex	Strongly Disagree	Disagree Somewhat	Agree Somewhat	Strongly Agree
Female	147	241	174	89
Male	81	108	228	123

Source: Data from A. J. Adams and S. C. Lonial, "Metro United Way Contributor Survey," Metro United Way, Louisville, KY, 1981.

a. What is the relative frequency of the category "strongly disagree"?
b. What evidence in the table suggests that the sexes may hold noticeably differing views on this question?

4.51 For a variety of occupations, we have a listing of P, percentage of positions in the occupation held by females, and R, ratio of female pay to male pay in that occupation. Treat the following data as a sample of all occupations:

Occupations	P	R
Registered Nurses	92.7	.91
Bookkeepers	93.0	.74
Elementary Teachers	81.9	.95
Cashiers	79.8	.75
Social Workers	60.0	.73
Office Supervisors	70.3	.64
Editors and Reporters	48.2	.67
Designers	44.2	.55
Real Estate Sales	42.5	.66
Insurance Sales	27.4	.62
Janitors/Cleaners	21.0	.69
Lawyers	15.2	.63

Source: Data from *USA Today,* September 4, 1987.

a. Establish a linear relationship between the variables, and interpret the slope of the line.
b. Measure the strength of the relation, and infer the status (correlated or uncorrelated) of the variables in the population.

REVIEW PROBLEMS CHAPTERS 1–4

R1 The commissioner for a large metropolitan county has been asked by the media to forecast voter turnout for the upcoming county election just five days away. Not wanting simply to rely on her instincts, she decides to look at historical records for clues. Finding data for the most recent eight county elections, she notes the following:

Voter Turnout (Thousands)	Absentee Ballots Cast Prior to Election
46.1	553
37.1	705
51.6	823
29.9	389
31.5	536
42.4	731
40.8	725
34.0	588

a. Treating these data as a random sample of voting outcomes, determine the linear relation between the variables. Is there strong evidence that these variables are related? Why?
b. If you determine that there is strong evidence of a relation, predict this election's voter turnout knowing that 611 absentee ballots were cast.
c. What assumption is necessary for your analysis?
d. Use a graphical method to assess the reasonableness of the assumption referred to in part c.

R2 Over the last 15 years, the National Weather Service in a Tennessee city has recorded the following dates for the fall's first measurable snowfall:

Nov. 17	Nov. 22	Oct. 30
Nov. 9	Nov. 7	Nov. 19
Nov. 21	Nov. 13	Dec. 1
Nov. 20	Nov. 24	Nov. 11
Nov. 10	Nov. 19	Dec. 3

Treating these data as a sample, determine the following:
a. the median date of first snowfall
b. the mean date of first snowfall
c. the dates at the beginning and end of a region of concentrated values for first snowfalls

R3 By inspecting a random sample of grocery shoppers' orders (read into the store's computer by checkout scanner), an assistant manager obtains the following data for the number of dairy items purchased per order, X:

4	2	2	3	0	5	1	1
7	3	2	3	1	1	2	0
0	2	2	2	2	4	1	8
6	4	3	2	5	1	1	7
1	4	0	2	4	4	4	3
0	1	2	4	6	4	5	2

a. Organize these data into a frequency distribution.
b. Generate a set of relative and cumulative frequencies.
c. Which value of X occupies the 24th position in the ordered data set?

R4 A metropolitan county is beginning to control air pollution levels by requiring all vehicles registered in the county to pass an annual inspection. A newspaper reporter covering the opening of the inspection stations wondered how stable the pollution measurement process is. He asked for and received permission to have his own vehicle run through the testing station eight times to see how the readings would fluctuate. The readings, with X = parts per million of hydrocarbons in vehicle exhaust, are as follows:

117	121	124	127	120	131	126	118

a. Compute three different quantities that would offer the reporter a way to summarize what he wishes to measure.
b. Of the three measures you computed, which would the general reading public most likely understand?

R5 A random sample of repair-call history of six copiers serviced by a copier repairman showed the following:

Age of Machine, X (Months)	Number of Repair Calls in Last 30 Days, Y
4	2
7	1
12	3
10	2
2	0
13	4

$\Sigma X = 48 \quad \Sigma X^2 = 482 \quad \Sigma Y = 12 \quad \Sigma Y^2 = 34 \quad \Sigma XY = 123$

a. Graph the data. Does a straight-line relation appear reasonable?
b. Determine the sample correlation coefficient.
c. Is the sample correlation an indication that X and Y are related in the population?

R6 In a study of shoppers' use of store coupons ("The Coupon-Prone Consumer: Some Findings Based on Purchase Behavior Across Product Classes," *Journal of Marketing*, October 1987), Kapil Bawa and Robert Shoemaker report on a sample of 462 households that kept purchase diaries over a one-year period.

a. Consider the variable X, where X = percent of household purchases made with a store coupon in a product class. For the ready-to-eat cereal product class, the authors report that the mean value of X is 13.6, while the median value of X is 7.8. Are the 462 values of X skewed left, skewed right, or symmetrical?

b. For the facial tissue product category, the authors report that the mean value of X is 9.3, and that 47 percent of households made some purchases with a coupon. What is the median value of X for facial tissues?

R7 A rural county treasurer advertised for sealed bids in an auction of a four-acre parcel for which the property taxes had become delinquent. Of 26 bids submitted, the low bid was $7901, and the high bid, $10,351. If a grouped frequency is desired, consider the following questions.

a. How many classes would you recommend?

b. What width would your classes be?

c. What would be the midpoint of the class that contains the bid of $8880?

R8 Over a 40-year period ending in 1988, the time of the winning horse in the Kentucky Derby has averaged 2 minutes, 2.17 seconds, with a standard deviation of 1.13 seconds. (Source: Louisville *Courier-Journal*, May 8, 1988.)

a. The median winning time was 2 minutes, 2.20 seconds. This value was also the mode. Do you think the winning times are nearly symmetrical or noticeably skewed?

b. The best winning time in this period is 1 minute, 59.4 seconds by Secretariat in 1973; the worst is 2 minutes, 5.0 seconds by Tim Tam in 1958. Determine the range.

c. If we view an outlier as a value more than four standard deviations from the mean, is either extreme in part b an outlier?

d. Suppose the data set is mound shaped and symmetrical. Give an interval that will contain about 68 percent of the winning times.

R9 Refer to the data in Review Problem R5.

a. What value of Y would you predict when $X = 8$?

b. What are the mean values for X and for Y? What property of a regression line does your answer in part a suggest relative to the mean values for X and for Y?

R10 In a study involving people's opinions about lawyers who advertise, Robert Hite and Edward Kiser ("Consumers' Attitudes Toward Lawyers with Regard to Advertising Professional Services," *Journal of the Academy of Marketing Science*, Spring 1985) report on a questionnaire given to the 500 households that comprise the Arkansas Household Research Panel. One of the statements in the survey was this: "If a lawyer advertises, his credibility is lowered." The responses of those who returned the questionnaire were as follows:

Strongly agree: 14	Agree: 60	No opinion: 47
Disagree: 307	Strongly disagree: 61	

Let $X = 1$ for strongly agree, $X = 2$ for agree, and so forth up through $X = 5$ for strongly disagree. Treating the responses as a random sample of Arkansas opinions, determine the following:

a. the sample mean, median, and mode.
b. the average deviation and the standard deviation.
c. whether there is much chance for nonresponse error or bias for the 500 households that comprise the panel. (Could the results in parts a and b be noticeably distorted because nonresponders have different opinions than responders?)

R11 How soon after receiving store coupons do consumers redeem them? In an article in the *Journal of Marketing Research,* Bawa and Shoemaker use scanner panel data to study coupon redemption behavior.

The accompanying table summarizes one part of the study. The results are for a frequently purchased product class having an average purchase cycle of less than three weeks.

Pattern of Brand *A* Coupon Redemptions by Purchase Occasion After Coupon Delivery Date

Purchase Occasion After Coupon Delivery Date	Brand *A* Coupon Redemptions			
	Frequency	Percentage	Cumulative Frequency	Cumulative Percentage
1st	110[a]	29.6	110	29.6
2nd	61	16.4	171	46.0
3rd	53	14.3	224	60.3
4th	33	8.9	257	69.2
5th	18	4.9	275	74.1
6th	16	4.3	291	78.4
7th	19	5.1	310	83.5
8th	11	3.0	321	86.5
9th	11	3.0	332	89.5
10th or more	39	10.5	371	100.0
Total	371	100.0		

[a]To be read as: 110 households redeemed the brand *A* coupon on the first purchase occasion after the coupon delivery date.

Source: Reprinted with permission from Kapil Bawa and Robert Shoemaker, "The Effects of Direct Mail Coupons on Brand Choice Behavior," *Journal of Marketing Research, 24* (November 1987). Published by American Marketing Association.

a. Is the table an example of a grouped or ungrouped frequency distribution?
b. What positions are occupied by the data in the fourth purchase occasion?
c. What is the modal purchase occasion?
d. Find the median purchase occasion, if possible.
e. Which purchase occasion corresponds to the 65th percentile?

R12 What effect does scoring first have on a National Football League (NFL) team's chances of winning the game? A check of each team's won–loss record when they scored first and when their opponents scored first showed that 26 of the 28 teams fared better during the 1987 season when they put the first points of the game on the scoreboard. In the accompanying data, in which tie scores count as a .5 win, X_1 = the number of games in which the team scored first, Y_1 = the number of games won when the team scored first, X_2 = the number of games in which the opponent scored first, and Y_2 = the number of games won when the opponent scored first.

X_1	Y_1	X_2	Y_2	X_1	Y_1	X_2	Y_2
4	2	11	1	9	4	6	4
5	4	10	3	7	4	8	4
8	6	7	5	9	6	6	2
12	4	3	0	12	10	3	2
9	7	6	3	9	4	6	2
6	4	9	3	7	4	8	2
7	6	8	4.5	7	4	8	3
8	4	7	0	7	4	8	3
6	3.5	9	2	7	4	8	4
3	2	12	7	8	5	7	3
7	6	8	3	7	7	8	6
6	4	9	0	9	7	6	2
8	4	7	1	8	2	7	2
8	5	7	1	8	7	7	4

Source: Data from "NFL Stats: Scoring First Means Finishing First," *Sport* (August 1988).

a. Find the least-squares line and correlation coefficient between X_1 and Y_1 for the 28 NFL teams, to four decimal places.
b. Find the least-squares line and correlation coefficient between X_2 and Y_2, to four decimal places.
c. Are both correlation coefficients large enough to infer that a relation exists?

R13 A random sample of radio stations from around the country revealed the accompanying data on the following variables:

X_1 = Call letters of the radio station
X_2 = Number of disk jockeys employed by the station
X_3 = AM or FM station
X_4 = Style of music
X_5 = Market share

Radio Station	Call Letters	DJs	AM/FM	Style	Market Share (%)
1	WKRP	10	FM	Easy Listening	11
2	WFPK	8	FM	Classical	6
3	WCII	15	AM	Country	21
4	WQMF	14	FM	Rock	17
5	WWWC	9	FM	Top 40	15
6	WKOK	14	AM	Top 40	14
7	WAMZ	14	AM	Country	23
8	WQXI	12	FM	Rock	25
9	WQSB	16	FM	Rock	21

Radio Station	Call Letters	DJs	AM/FM	Style	Market Share (%)
10	WARX	10	AM	Top 40	12
11	WDRB	11	FM	Rock	19
12	WQUE	12	FM	Country	18
13	WMAK	16	FM	Rock	18
14	WFPL	10	FM	Classical	8
15	WQUT	9	FM	Classical	10
16	WWIT	16	FM	Country	19
17	WAKY	13	AM	Top 40	16
18	WQIP	16	FM	Rock	24
19	WCAN	8	AM	Easy Listening	9
20	WQRC	11	FM	Rock	15
21	WDAY	16	FM	Country	24
22	WFUN	17	FM	Country	19
23	WCUT	13	AM	Rock	21
24	WKAR	14	AM	Top 40	17
25	WQLT	10	FM	Classical	9
26	WFIX	15	FM	Rock	18
27	WDIG	10	FM	Country	12
28	WKQQ	10	AM	Top 40	11
29	WVEZ	9	FM	Easy Listening	8
30	WALK	8	AM	Easy Listening	10

Organize the values of the variable X_2 into an ungrouped frequency distribution.

R14 Refer to R13. Organize the values of the variable X_5 into a grouped frequency distribution. Describe the shape of a histogram that could be constructed from the distribution.

R15 Refer to R13.

a. Organize the values of the variable X_4 into a one-way tabulation.

b. Organize the values of the variables X_3 and X_4 into a cross-tab table. Verify that the one-way tabulation from part a coincides with one of the marginal totals in the cross-tab table.

R16 Refer to R13. Assume the data set is complete and represents the frame for a population. Select a simple random sample of size $n = 4$ from the frame. Use row 4, column 5 in Table 1.5 as a starting point and proceed up column 5. Write out the call letters of the radio stations included in the sample.

R17 Motorists who register at least 0.10 on a breathalyzer test are subject to immediate arrest for drunken driving in most states. A sampling of 254 test readings revealed the following data:

Class Number	Reading	Frequency
5	.20 and up	3
4	.15–.20	14
3	.10–.15	27
2	.05–.10	68
1	.00–.05	142

a. Find the midpoint of each class. (*Hint:* To estimate the midpoint of class 5, add the width of the other classes to the midpoint of class 4.)
b. What is the average breathalyzer reading? What assumption, if any, is necessary to compute this measure?
c. What is the median breathalyzer reading? What assumption, if any, is necessary to compute this measure?

R18 A random sampling of 21 business statistics textbooks generated the following data on the variables *edition number*, X, and *number of text pages*, Y:

X	Y	X	Y	X	Y
2	677	2	680	1	869
1	672	1	770	1	735
1	641	3	799	1	497
3	571	2	660	1	676
1	687	2	872	2	581
2	784	4	830	4	675
3	907	1	711	5	976

a. What type of data do the variables X and Y produce?
b. Find the mean and standard deviation of the raw data for Y.
c. Find the median of the raw data for X.
d. Do you think X and Y have a direct or inverse relation? Substantiate your answer by determining the correlation coefficient.
e. Is there evidence of a relation between X and Y in the population of all business statistics textbooks? Assume the cutoff points for r when $n = 21$ are $\pm .433$. Explain.

R19 A survey of different bank credit cards yielded the following interest rates and annual fees:

Annual Fee ($)	Interest Rate (%)	Annual Fee ($)	Interest Rate (%)
20	18	20	18
15	18.5	20	n/a
20	18	20	13
20	13	20	16.5
none	18	20	16.5
20	15.5	15	18
20	13.5	15	18.875
15	18	20	21
none	18	24	21
15	19.75	12	12.375

Note: n/a means not available.

a. What is an entity in this situation?
b. Apply the principles of data management to the set of interest rates. Allow one decimal place accuracy. For the missing value, use the average interest rate of all remaining credit cards with an annual fee of $20.

c. Organize the annual fees into an ungrouped frequency distribution. What would be the most descriptive measure of central location to characterize the "typical" annual fee?

d. Organize the interest rates into a grouped frequency distribution. Characterize the "typical" interest rate by computing the median of the frequency distribution.

R20 In Joe Ward's "At Work" column (Louisville *Courier-Journal,* June 13, 1988), he cited figures from the Bureau of Labor Statistics indicating that the number of women working in Kentucky increased by 18,000 in 1987, while the number of Kentucky men with jobs declined by 14,000. Ward also quoted economist Julia Lane, who said these figures should be taken with "a grain of salt." She explained that the work force estimates of Kentucky's labor force are based on a sample, and that the state's apparent loss of jobs for men simply could be the result of sampling error. Based on this information, indicate whether the following statements are correct. If incorrect, explain why.

a. Lane's explanation about the work force estimates means the Bureau of Labor Statistics' sample must be flawed because it contains sampling error.

b. It is possible that the Kentucky men might have experienced an *increase* in jobs during 1987.

c. The Bureau of Labor Statistics' report that shows a gain of 18,000 jobs for women is an example of descriptive statistics.

d. Lane's comments imply that the Bureau of Labor Statistics' figures are *not* risk-free statements.

Chapter Maxim *In general, probability has meaning only in a "before the fact" or "prior to sampling" sense.*

CHAPTER 5

PROBABILITY AND PROBABILITY DISTRIBUTIONS

*Optional

Objectives

After studying this chapter and working the exercises, you should be able to

1. State the goal of probability.
2. Define a probability.
*3. Contrast statistical independence with conditional probability.
4. Differentiate between a random variable and the values of a random variable.
5. Identify the two types of random variables.
6. Display a probability distribution for a discrete random variable in tabular, graphical, or formula form.
7. Recognize the difference between point probabilities and interval probabilities.
8. Compute the mean, variance, and standard deviation of a discrete probability distribution.
9. Understand the differences between discrete and continuous probability distributions.
*10. Use the Discrete Probability Distributions processor to find the parameters of any discrete distribution.
11. Recognize applications of discrete probability distributions in finance.

With this chapter, we enter the second phase of our book: dealing with probability. Probability may be viewed as the bridge between descriptive statistics and inferential statistics. In the first four chapters covering descriptive statistics, we learned how to describe sets of data. Our approaches included descriptive measures such as the mean and standard deviation, tabular forms such as frequency distributions and cross-tabulations, and graphs such as histograms and pie charts. Eventually, we will learn how to use sample information to make inferences about the population from which the sample was obtained. To do this, we require a theoretical framework from which to operate. Probability is this basis.

The purpose of this chapter is to discuss basic concepts of probability and to introduce probability models. We will see that the term *probability* generally refers to the chance that one specific outcome occurs, while the collection of the probabilities for all possible outcomes forms a *probability model*. Probability models or distributions are the building blocks for our subsequent study of inferential statistics.

*Applies to optional section.

5.1 PROBABILITY OVERVIEW

We encounter probability almost daily in our lives if we listen to weather forecasts such as "The probability of precipitation tomorrow is 70 percent." What do we do with this number? We base many of our daily activities on its magnitude. A high chance of rain might cause us to take an umbrella or to choose a certain pair of shoes. A large chance of a heavy snowfall may encourage us to procrastinate a homework assignment in hopes of classes being cancelled the next day.

Probability influences our actions; it also influences the business world. Consider the following illustration: Several years ago, as a cold front rapidly was approaching the Deep South, a severe freeze was forecast for the central sections of Florida where a bumper crop of oranges was growing. Shrewd investors acting on this information bought futures in oranges. (Futures are contracts to buy or sell a commodity—oranges, for example—on a future date at a specified price.) The freezing weather descended on Florida, severely damaging the orange groves and cutting the harvest in half. Weeks later the investors holding the "buy" futures reaped a financial profit when the price of orange juice and orange-related products tripled. The extensive damage to the orange crop was not a certain event, but it was "possible" in terms of probability. In this case, the probability of freezing weather was a key factor in the investors' decision to buy orange futures.

The Role of Uncertainty

Probability is applicable only where there is uncertainty. If you went to class yesterday, there is no need to talk about the probability that you attended that class. We know the outcome, so there is no uncertainty. You may have every intention of going to class tomorrow, but something may happen to prevent you from going. In this case there is uncertainty, and it makes sense to assess the "probability" of your attendance. Let us consider the existence of probability for another scenario involving uncertainty.

EXAMPLE 5.1

With mortgage rates falling, a young couple decided to sell their starter house and "move up" to a larger, more expensive residence. They contacted a realtor and signed a 60-day contract authorizing the realtor ". . . to aggressively market the property in order to find a buyer." Does this contract remove the uncertainty about selling the house, or does probability apply to this situation? What effect, if any, does the term of the contract have on whether probability applies?

Solution:

Probability definitely applies to this home-selling situation. The 60-day contract simply gives the realtor certain freedoms in *trying* to sell the house as well as certain monetary rewards if successful. However, the contract does not obligate the realtor or the couple to sell the house within 60 days. The term of the contract has no bearing on whether probability applies. Uncertainty about selling the house persists as long as the house remains on the market. This may extend beyond the 60-day contract period.

□

The Direction of Reasoning

In descriptive statistics, we selected a sample from a population and proceeded to organize and summarize the data. The sample was tangible and known in that we could see the actual observations. We did not see all the data that comprised the population, however, but we knew it existed. Relative to Figure 5.1, we could say that descriptive statistics operated only within the region representing one particular sample. The area outside the sample represented the remaining, unknown entities of the population, with respect to descriptive statistics.

In studying probability, we assume that the composition of the population is known. Interest is then focused on the samples that might arise when selecting from this population. Specifically, we attempt to assess quantitatively the chance of a particular sample occurring; thus, probability reasons from the population to the sample.

Definition

The **goal of probability** is to quantify the uncertainty of sampling from a population.

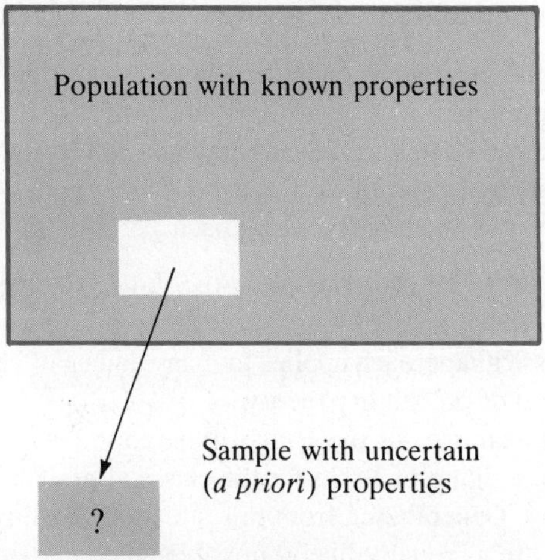

Figure 5.1 Probability Reasons from the Population to the Sample

A simple example illustrates the flow of reasoning. In the game of Bingo, 15 numbers are assigned to each of the 5 letters, *B, I, N, G,* and *O*. The numbers 1 through 15 are associated with the letter *B*, the numbers 16 through 30 with the letter *I*, and so on. The 75 different letter–number pairs like *B*–7, *I*–21, or *O*–74 constitute the population. Selecting letter–numbers represents the sampling process. Suppose we are interested in obtaining the letter *B* on the first draw, an uncertain outcome before sampling. One approach to quantifying the sampling uncertainty is to use the proportion of *B*'s in the population—15 out of 75 (or .20).

In our discussion of probability, we must recognize that the direction of reasoning is different from that of inferential statistics. Probability is deductive, reasoning from the general to the specific—that is, from the population to the sample. As we will see in later chapters, inferential statistics is inductive, reasoning from the specific to the general—that is, from the sample to the population.

Definition of a Probability

Just as a sports judge measures an athlete's performance for a gymnastics routine or diving maneuver, we wish to measure uncertainty. This could be done qualitatively with words such as "very unlikely," "unlikely," "likely," and "very likely." Or we could measure uncertainty quantitatively on a numerical scale. This is exactly what we attempt to do by means of a probability. The generally agreed-upon scale is from 0 to 1 (or 0 to 100 percent), with 1 representing certainty.

Definition

A **probability** is a number between 0 and 1 that quantifies uncertainty.

Philosophies of Generating Probabilities

Although probability quantifies uncertainty, there may be more than one way to generate probabilities in a given situation. We therefore need to understand the philosophies or views of assessing probability: *objective* and *subjective*.

Objective probabilities can be established in two different ways: the classical approach and the empirical approach.

The Classical Approach Also called the theoretical, or *a priori,* approach, this method yields probabilities based on complete knowledge of the sampling situation. The classical approach applies to many games of chance which, incidentally, represent the origin of probability theory.

For example, in the Bingo illustration, we used our knowledge of the game to arrive at a figure of 15/75 for the classical probability of obtaining the letter *B* on the first draw. Generalizing from this situation, we form classical probabilities as follows: If there are n (equally likely) possible outcomes from sampling and if m outcomes are associated with the event of interest, then the **classical probability** of the event is

m/n. In the Bingo example, $n = 75$, $m = 15$, and the event of interest was obtaining the letter B on the first draw.

Classical or Theoretical Probability

$$\text{Event probability} = \frac{m}{n} = \frac{\text{Number of outcomes associated with the event}}{\text{Number of equally likely outcomes possible}}$$

The classical approach is an effective means of developing probabilities provided we can determine the values for m and n. For this purpose, we often use counting rules such as those for counting permutations or combinations, or we simply enumerate all the possible outcomes. Notice that a theoretical probability is developed without actually sampling. We did not need to select the first chip in the Bingo game in order to declare a probability of 15/75 for the letter B; we knew before sampling that m was 15 and that n was 75. We only needed to assume all chips were equally likely to be drawn.

The Empirical Approach There are cases where the classical method is not applicable. For instance, in drawing samples from a population not previously encountered, we would have no a priori way to know m and/or n. We therefore need another objective approach which relies on data, not theory. Such an approach to developing probabilities is called the empirical or relative frequency approach. This method produces approximate probabilities based on the actual outcomes from sampling. For example, if 120 consumers in a random sample were asked to name their favorite soft drink and if 24 of them said Diet Pepsi, then the empirical probability that a randomly selected consumer prefers Diet Pepsi is 24/120 or .20. In general, we form empirical probabilities as follows: If a sample of size n yields m outcomes associated with the event of interest, then the **empirical probability** of the event is m/n.

Empirical or Relative Frequency Probability

$$\text{Event probability} = \frac{m}{n} = \frac{\text{Number of times event has occurred}}{\text{Number of times event could have occurred}}$$

The empirical approach depends on past experience or observed results and is, therefore, quite sensitive to the size and content of the sample. Empirical probabilities are natural approximations to the "true" probabilities. Generally, as n increases, the empirical probability stabilizes and becomes more accurate. For example, if the employee mix at a large company is 60% female and 40% male, then the "true" probability of randomly selecting a female employee is .60. However, in sampling employees, we might observe the following empirical probabilities as the sample size increases:

n	Number of Females	Empirical Probability
10	5	.500
20	11	.550
50	31	.620
100	59	.590
200	117	.585
500	302	.604

Although we may never observe exactly 60% women in a sample, the empirical probability approaches this figure as n gets larger.

The Subjective Approach A third approach to generating probabilities involves our intuitive feelings about the event of interest and is called the subjective approach. As suggested by the name, there is no theoretical or empirical basis for producing subjective probabilities. However, do not imagine that all subjective probabilities are merely hocus-pocus figures. Eliciting a subjective probability may involve examining historical data and/or theoretical models. For example, when a company introduces a new product, a marketing manager may be required to provide subjective probabilities for three potential scenarios: low demand, average demand, and high demand for the product. The manager must combine judgment and expertise with historical perspectives (What happened the last time our company introduced a new product?) and available test-market data to form the subjective probabilities.

A difficulty with using subjective probabilities is the potential for more than one value for the same event, depending on whose intuitive feelings are involved. The theoretical and empirical approaches yield only one probability for a given problem.

Subjective probabilities are common in everyday life. When we make an investment, drop a letter in the mail, or travel by airplane, we tend to assess the chances of getting a return on our investment, having the letter delivered promptly, or arriving safely at our destination with subjective, rather than objective, probabilities. Often applied in unique or ill-structured problems, subjective probability is better treated in a higher level course, such as one on decision analysis. We will not delve further into this subject here.

COMMENTS

1. In Chapter 1, we introduced and defined the concept of simple random sampling. Recall that random sampling ensures an equal chance of selection for each entity of a finite population. This ties in nicely with the classical philosophy of probability.

Table 5.1 Philosophies of Probability

Name	Basis
Classical	Knowledge of possible outcomes
Empirical	Observed data
Subjective	Personal beliefs

Thus, the classical approach is in force *prior to* the selection of a simple random sample.

2. Almost exclusively, we interpret a probability as the approximate proportion of times the event occurs in repeated sampling. For example, the probability of 4/52 attached to the event of selecting an ace from a deck of cards is formulated via the classical approach to probability. We interpret 4/52 as the approximate proportion of times we would pull an ace out of a 52-card deck, assuming we repeated the process many times. We do *not* interpret 4/52 as meaning we will select an ace exactly 4 times in a series of 52 repetitions of selecting a card from a full deck.

In summary, a probability measures uncertainty. Uncertainty exists when we select a random sample from a population, since we cannot describe the composition of the sample in advance. Probability can be used to quantify the various outcomes associated with sampling. Table 5.1 summarizes the three methods of developing probabilities.

SECTION 5.1 EXERCISES

5.1 Define a probability.

5.2 Identify the different philosophies of developing a probability.

5.3 In the following scenarios, determine whether uncertainty exists for the situation described. Then explain whether probability applies.

a. In applying for a home loan, a prospective buyer is offered a choice between a fixed rate mortgage at 10% and an adjustable rate mortgage that will be set at 8 percent for the first two years. Is there uncertainty associated with the interest rate on the mortgage five years from now, if she opts for (1) the fixed rate or (2) the adjustable rate—that is, does probability apply in either case?

b. A store owner offers each buyer of a compact disk (CD) player the option of purchasing a two-year service contract. Suppose a buyer also purchases the service contract. Does probability apply to the following situations:
(1) the CD player will need repairs within two years?
(2) the buyer will have to pay for any repairs within two years?

c. In many companies there is a clear distinction between management and labor. One of the distinguishing features is whether a person receives a salary—management—or an hourly wage—labor. Typically, a management person draws no overtime pay, whereas a labor individual can. Suppose an employee is interested in determining whether his gross pay will exceed $1500 in an upcoming four-week period.

(1) If the employee is a management type, does probability apply?

(2) If the employee is a labor type, does probability apply?

5.4 A 680-room hotel offers four different types of accommodations: single, double, king single, and suite. The single and double rooms have queen-sized beds, while each king single has one king-sized bed. The suite has a king-sized bed plus additional amenities. Half of the king single rooms are designated nonsmoking; none of the 10 suites is. Only 20 of the 300 single and 30 of the 320 double rooms are nonsmoking rooms. Suppose a radio station sponsors a contest with a "getaway weekend" at the hotel as the prize.

Each hotel room number is written on a ball, and the balls are mixed in a large container. The contest winner, however, must select one room number at random to claim the prize. Find the probability that the winner will select one of the following, and tell what philosophy of probability you used to answer:

a. a single room.

b. a room with a king-sized bed.

c. a nonsmoking room.

d. a nonsmoking room with a queen-sized bed.

5.5 During a recent weeknight between 9 P.M. and 10 P.M., a regional independent television rating service claimed the following number of television sets were tuned into these networks:

Network	Number of TV sets (Millions)
ABC	3.1
CBS	6.3
NBC	1.7
All other	2.4

What is the empirical probability that a television set in use was tuned into CBS during this time period?

5.6 Which philosophy of assigning probabilities is used in each of the following situations?

a. To make room for a downtown redevelopment project, a contractor is hired to demolish an existing building without damaging the surrounding property. The existing building is situated on top of a rock fault which presents additional problems. The contractor assesses the probability of successfully demolishing the building at .98.

b. Wall Street analysts have a keen interest in the result of pro football's Super Bowl every year. The Super Bowl Theory holds that when a team from the old National Football League (NFL) wins, stock prices rise. When an old American Football League (AFL) team wins, stock prices fall. Suppose this theory has been correct 12 of the 13 times an NFL team has won the Super Bowl and 10 of the 11 times an AFL team won, measured against the New York Stock Exchange composite index. Then the probability that stock prices rise in the coming year is .92 if an NFL team wins the Super Bowl.

c. On the back of a lottery ticket are the number of prizes of each kind to be awarded as well as the anticipated number of lottery tickets to be sold. Three runner-up prizes of $50,000 will be awarded if two million tickets are sold. If all tickets are sold, then the probability of buying one lottery ticket and winning $50,000 is .0000015.

*5.2 ELEMENTARY PROBABILITY CONCEPTS

Probability applies to sampling situations. When sampling, we are not sure which outcome will be observed, so we attach a probability to each outcome to represent its uncertainty. We will use the term *event* to describe a sampling outcome. In this section, we will consider some elementary probability concepts as well as ways to assign probabilities to simple events.

Marginal (Unconditional) Probability

The simplest type of event probability arises when we wish to assign a probability to the occurrence of a single event; such a probability is called a **marginal,** or **unconditional, probability.** For example, in our earlier Bingo illustration, the event we discussed was selecting the letter *B* on the first draw of the game. The probability of 15/75 that we assigned to this event is considered a marginal probability, since only this one event was under consideration. Many game-related marginal probabilities can be evaluated by the classical approach mentioned in the previous section. That is,

$$\text{Event probability} = \frac{m}{n} = \frac{\begin{array}{c}\text{Number of outcomes}\\ \text{associated with the event}\end{array}}{\begin{array}{c}\text{Number of equally}\\ \text{likely outcomes possible}\end{array}}$$

For instance, you could use the expression above to determine that the probability of picking an ace from a standard deck of playing cards is 4/52 or that the chance of getting a five on a single roll of a die is 1/6.

Contingency, or cross-tab, tables also offer us the opportunity to study elementary probability concepts. Let us refer to the sports preference survey from Chapter 4, where the favorite sport of 50 university students was cross-classified with the sex of the respondent. Table 5.2 displays the responses.

Suppose we wish to know the probability that a student selected randomly from this group preferred football. Since 15 students in the survey stated a preference for

Table 5.2 Sports Preference Survey

	Favorite Sport			
Sex	Baseball	Basketball	Football	Total
Male	8	10	8	26
Female	5	12	7	24
Total	13	22	15	50

*Optional

football, the marginal probability would be 15/50 or .30. We also could use the row variable, *sex,* to specify an event of interest. For instance, to evaluate the marginal probability that a survey respondent was a male, we could determine immediately from the table that the probability is 26/50 or .52.

Recall from the previous chapter that we referred to the various row and column totals of a cross-tab as *marginal* totals. For data in a contingency table, a marginal probability is always the relevant row or column marginal total divided by the overall total (when sampling at random). Most of the probability problems we will encounter in this book will require marginal (unconditional) probabilities. However, some problems may be more complex than those encountered to this point.

Joint Probability

While marginal probability has to do with the chance that a single event occurs, **joint probability** refers to the chance that two or more events occur together. Continuing with our sports preference survey, suppose we are interested in the probability that someone selected at random is a female who prefers basketball. If so, we now require two events, female and basketball, to occur simultaneously. Inspection of Table 5.2 shows that the 12 individuals in the middle cell of the second row are part of the event *female* and at the same time part of the event *basketball.* The joint probability of these two events is therefore 12/50 or .24. As another example, the joint probability of *baseball and male* is 8/50 or .16.

It should be clear that a joint probability for a two-dimensional cross-tab is a cell frequency divided by the overall frequency, whereas a marginal probability is a row or column frequency divided by the overall frequency.

COMMENTS

1. The marginal probability of event A is symbolized as $P(A)$. The joint probability of events A and B is represented as $P(A$ and $B)$. Use of the word *and* denotes a joint probability.
2. Joint probability can be extended to three or more events occurring together. With a two-dimensional cross-tab, however, we can consider only joint probability for two events.
3. On a Venn diagram, joint probability corresponds to shared area, or intersections, of circles. In the same sense, joint probability in a cross-tab is related to the intersection of a particular row with a particular column.

The Addition Rule

We can use our knowledge of marginal and joint probabilities to evaluate the probability that *either* of two events occurs. The addition rule is a simple expression that enables us to find such a probability. Using $P(A$ or $B)$ to symbolize the probability that either event A or event B occurs, we have the following rule.

Addition Rule for Events *A* and *B*

$$P(A \text{ or } B) = P(A) + P(B) - P(A \text{ and } B) \qquad (5\text{–}1)$$

The probability that either of two events occurs is found by adding the two events' marginal probabilities and then subtracting their joint probability.

Referring to Table 5.2, we can determine the probability that someone selected at random is either a male or a baseball enthusiast as

$$\begin{aligned} P(\text{male or baseball}) &= P(\text{male}) + P(\text{baseball}) - P(\text{male and baseball}) \\ &= \frac{26}{50} + \frac{13}{50} - \frac{8}{50} \\ &= \frac{31}{50} \end{aligned}$$

As another example, the probability of someone being either a female or a football fan is $24/50 + 15/50 - 7/50 = 32/50$.

COMMENTS

1. The term *P(A* or *B)* refers to the chance that *any* of the following happen: (a) only *A* occurs, (b) only *B* occurs, (c) *A* and *B* occur together.
2. The subtraction term in Equation 5–1 prevents double counting. If events *A* and *B* do not intersect, this subtraction term will equal zero.
3. The addition rule often is referred to as the *union* operation on a Venn diagram. Use of the word *or* denotes addition or union.

Conditional Probability

Besides marginal and joint, the third type of probability we wish to describe is conditional probability. A **conditional probability** involving two events—one of which is uncertain and a second which is certain—measures the uncertainty of one event, given the knowledge that the other event has occurred already. The occurrence of the second event is a known, or given, condition; hence, the term *conditional* probability.

A conditional probability is symbolized as $P(A|B)$, which can be read as "the probability event *A* occurs, given or subject to the condition that event *B* has occurred." To evaluate a conditional probability, we can make use of Equation 5–2.

Conditional Probability for Event *A*, Given Event *B*

$$P(A|B) = \frac{P(A \text{ and } B)}{P(B)} = \frac{\text{Joint probability}}{\text{Marginal probability}} \qquad (5\text{–}2)$$

Thus, a conditional probability is simply the ratio of the two events' joint probability to the marginal probability of the known condition.

For an example, let us refer again to the sports preference survey in Table 5.2. We will determine the probability that the favorite sport is basketball, given that the sex is known to be female. Following Equation 5–2, we would represent this problem as

$$P(\text{basketball}|\text{female}) = \frac{P(\text{basketball and female})}{P(\text{female})}$$
$$= \frac{12/50}{24/50} = \frac{12}{24} = .50$$

Had we sought the marginal (unconditional) probability of basketball, the answer would be 22/50, or .44. By saying that the sex is known to be female, we specified a condition that narrowed our attention to the female row of the contingency table.

Statistical Independence

The final elementary probability concept that we wish to discuss is statistical independence. In the example above, we saw that the unconditional probability of basketball was .44, but that the conditional probability of basketball, given a female respondent, was .50. In other words, the second event, female, affected the probability of the first event. We classify events as *dependent* when the occurrence of one event affects the probability of another event.

The opposite case is independence. Two events are statistically independent if the occurrence of one event has no effect on the probability of the other event. In symbols, two events A and B are independent if $P(A|B) = P(A)$—that is, if the conditional and unconditional probabilities are equal. Perhaps the simplest way to check for the existence of independent events in a cross-tab is to apply the following test.

Test for Independent Events

Events A and B are independent if
$$P(A \text{ and } B) = P(A)P(B)$$

The expression above states that two events are independent if their joint probability is equal to the product of their marginal probabilities. For an example from Table 5.2, suppose we wish to establish whether the events male and baseball are independent. We will declare them independent *if*

$$P(\text{male and baseball}) = P(\text{male}) \cdot P(\text{baseball})$$

The left side of this equation is 8/50, or .16; the right side is 26/50 times 13/50, or .1352. Since $.16 \neq .1352$, these events are not independent. Instead there is a conditional, or dependent, relation present.

EXAMPLE 5.2

	Days on the Market			
Selling Price	Under 45	45–90	Over 90	Total
Under $75,000	75	77	3	155
$75,000 to $125,000	151	192	41	384
Over $125,000	20	31	10	61
Total	246	300	54	600

The cross-tabulation above provides information for the 600 single-family homes that changed hands in an Ohio city last year. For a sale selected at random, determine the following:

a. Find the probability that the home was priced over $125,000.

b. Find the probability that the home was one that was priced in the $75,000–$125,000 range and was on the market from 45–90 days.

c. Determine the probability that a home was on the market over 90 days, given that it was priced under $75,000.

d. What is the probability that either a home was sold within 45 days or was priced under $75,000?

e. Are the two events "$75,000 to $125,000" and "45–90 days" statistically independent?

Solution:

a. This is a marginal probability—we are interested in a single event, price over $125,000, without regard to time on the market. The probability is 61/600, or .102.

b. Since we are looking for two events to occur together, we need to locate the appropriate cell and divide by the overall total. We obtain 192/600, or .32, as the joint probability for these two events.

c. This is a conditional probability—we know that the home in question was one of the 155 homes in the lowest price category. Using Equation 5–2, we have

$$P(\text{over } 90 | \text{under } \$75{,}000) = \frac{P(\text{over 90 and under } \$75{,}000)}{P(\text{under } \$75{,}000)}$$
$$= 3/155 \quad \text{or } .019$$

d. Using the addition rule given in Equation 5–1,

$$\begin{aligned} P(\text{under 45 days or under \$75,000}) &= P(\text{under 45 days}) + P(\text{under \$75,000}) \\ &\quad - P(\text{under 45 days and under \$75,000}) \\ &= \frac{246}{600} + \frac{155}{600} - \frac{75}{600} = \frac{326}{600} \quad \text{or } .543 \end{aligned}$$

e. To see if these events are independent, we will multiply their marginal probabilities together and compare the result against their joint probability (the joint probability is .32, determined in part b above). The marginal probability of "$75,000 to $125,000" is 384/600, or .64. The marginal probability of "45–90 days" is 300/600, or .50. Since .50 multiplied by .64 does equal the joint probability of .32, these events are independent. No other events in the table are independent, however.

□

Classifying events as independent or dependent is specific to the events under study. In later chapters, we will be interested in extending this idea to two *variables*. Notice in Example 5.2 that the independent events "$75,000 to $125,000" and "45–90 days" are actually distinct values of the variables *selling price* and *days on the market,* respectively. If we change the values to "under $75,000" and "under 45 days," for instance, you can verify that these events are not independent. But the larger idea to be investigated is the dependence or independence of the *variables,* not the events.

Determining whether two variables are independent depends on the structure—descriptive or inferential—of the problem. For the data in Example 5.2, we would declare the variables *selling price* and *days on the market* to be dependent, since all pairs of events are not independent. The business statistics problem in this case involves *description,* not inference.

Alternatively, we might wish to *infer* the relationship between *selling price* and *days on the market* in the population of all real estate transactions involving single-family homes, based on the sample data in Example 5.2. At this point, we do not have the ability to generalize to the population from the sample and say whether the variables are independent or dependent in the unseen population. In Chapter 19, we will learn how to analyze the sample data to make such an inference.

The statistical independence of two variables is a status that is either present or absent. Establishing independence may be the sole objective of an inferential statistics procedure, or confirming independence may be part of a larger set of inferential objectives. The independence or dependence of two variables is rarely of interest in a descriptive statistics framework.

COMMENTS

1. If we solve Equation 5–2 for the joint probability, we obtain $P(A \text{ and } B) = P(A|B)P(B)$. This equality is referred to as the *multiplication rule*.
2. The idea of correlation for two quantitative variables is closely related to the idea of independence for qualitative variables, as discussed above. If the quantitative variables X and Y are statistically independent, then their population correlation coefficient is zero.

CLASSROOM EXAMPLE 5.1

Probabilities for a Cross-Tabulation

A sales clerk tabulates the following for the 250 sales made in the past week at a women's clothing store.

	Method of Payment		
Sales Amount	Check	Cash	Credit Card
Under $35	40	42	18
$35 and up	60	6	84

a. Supply the marginal totals and the overall total for the table above.

b. For the past week, what is the probability a sale was either under $35 or paid in cash?

c. If we know a particular sale was paid for in cash, what is the probability that it was for an amount less than $35?

d. Express all the numbers in the table above in terms of relative frequency to the 250 sales.

e. What is the probability or relative frequency that a sale is for $35 or more and is paid for by check?

f. Find two events that are statistically independent.

SECTION 5.2 EXERCISES

5.7 Identify the following as either a joint probability, an unconditional probability, or a conditional probability:

a. $P(A)$ b. $P(B|A)$ c. $P(A \text{ and } B)$

5.8 The output of a firm that produces small, inexpensive electronic parts is classified as follows:

	Production Plant		
Status	Chicago, C	Atlanta, A	Syracuse, S
Defective, D	.020	.032	.041
Good, G	.400	.328	.179

a. Find $P(G \text{ and } S)$.

b. Express $P(G|C)$ in words and determine this probability.

c. Find $P(S \text{ or } G)$.
d. Find the probability that a part is defective, given that it is made in Syracuse.
e. What is $P(C \text{ and } A)$?

5.9 The success of a particular college basketball conference whose teams have played 180 nonconference games over the past three years is summarized in the following table:

Game Outcome	Game Site		
	Home Court	Opponent Court	Neutral
Win	64	35	18
Loss	16	41	6

For a game chosen at random, find the following:
a. the probability of a win.
b. the probability of a win, given that it was a home game.
c. the probability the game was played on an opponent's court, given the game was a loss.

5.10 In a study of sex roles in television advertising, a researcher sampled 275 commercials broadcast in the United States. In one part of the study, the men and women who appeared in the commercials were categorized by their (estimated) age as shown in the accompanying table. A total of 301 commercial characters were classified—only those appearing on camera for at least three seconds and/or who had at least one line of dialogue were counted.

Character Sex	Estimated Age Category		
	Under 35	35–50	Over 50
Women	68	86	15
Men	26	82	24

Source: Data from Mary C. Gilly, "Sex Roles in Advertising: A Comparison of Television Advertisements in Australia, Mexico, and the United States," *Journal of Marketing* (April 1988).

a. For a character chosen at random from this study, what is the probability of being under 35?
b. If the character is known to be a woman, what is the probability of being under 35?
c. If the character is known to be under 35, what is the probability of being a woman?
d. What is the probability of being a woman and over 50?

5.11 Suppose the following contingency table applies when a physician recommends that a patient undergo a treadmill stress test to evaluate the condition of the heart:

Actual Patient Condition	Stress Test Reading	
	Normal	Not Normal
Normal	.79	.05
Not Normal	.01	.15

a. What is the probability that the patient either is not normal or will test out as not normal?
b. What is the probability that the patient is normal, given that the test result is not normal?

c. A "false-positive" happens when the patient is in fact normal, but the test indicates otherwise. For a stress test subject chosen at random, what is the probability of a false-positive?

5.12 Consider the contingency table in Exercise 4.35, reproduced as follows:

	Driver History		
Driver Status	No Accidents or Violations	Median Accident-Violation Group	High Accident-Violation Group
Principal Driver Smokes	93	165	270
Principal Driver Does Not Smoke	407	335	230

Define the following events:

A: No accidents or violations
B: Principal driver smokes

a. Find the unconditional probability of event A.
b. Find the conditional probability of event A, given event B.
c. Determine whether the two events are independent.

5.13 A survey of 191 managers in a large company was conducted. Three levels of management rank were identified: upper, middle, and lower. Only 5 of the 28 upper-level managers were women, and half of all male managers held a lower-level rank. There were a total of 77 female managers in the company, and 6 were middle-level managers.
a. Find the probability that a randomly selected manager is a man.
b. Find the probability that a randomly selected manager is a woman holding a middle-level rank.
c. What is the probability that a randomly selected manager holds an upper-level rank, given the manager is a woman?
d. What is the probability that a randomly selected manager is a man or holds an upper-level rank?

5.14 Refer to Exercise 5.4. What is the probability that the winner will do the following:
a. sleep in a nonsmoking room, given the room has a king-sized bed.
b. sleep on a queen-sized bed or in a nonsmoking room.
c. sleep in a smoking room, given the winner selected a suite.
d. sleep on a king-sized bed in a double room.

5.3 RANDOM VARIABLES AND PROBABILITY

Random Variable Approach to Probability

A probability problem reasons from the population to the sample. Our objective is to determine the probability of observing specific samples. In relatively uncomplicated situations involving games of chance and cross-tabulations, determining a probability is fairly simple. But as the sampling gets larger and involves complex events, our objective becomes harder to achieve.

For example, suppose the proportion of all brokerages with an international office is .3. If one brokerage firm were to be randomly selected, then it is a simple matter to state that the probability it has an office outside the United States is .3. However, if ten brokerage firms were randomly sampled, then the probability that exactly one has an international office is not as easy to determine. Attacking this problem and others like it is facilitated by *defining a characteristic of interest relative to the sampling situation,* a step we call *defining a random variable*. Once this has been done, the objective then becomes one of developing a probability model to represent the variable.

Definition

A **random variable** is a variable whose numerical values are determined from a random sampling.

As we learned in Chapter 1, a variable may produce either qualitative or quantitative data. For example, a random sample of five long-distance telephone calls will produce quantitative data for the variable *duration of each call, D,* in seconds, and qualitative data for the variable *T, type of billing rate in effect*—day, evening, or night—for each call. Consider the following values:

D	30	415	127	89	366
T	Day	Day	Evening	Evening	Night

In probability, a random variable is a special kind of variable, which assigns a number to an entity or set of entities in a random sample. The variable *D* in our example already does this, so *D* qualifies as a random variable. But the variable *T* is not a random variable. We could create one for the set of entities for *T* by defining *X* to be the number of day-rate phone calls in the sample. For the listed sample, *X* would realize the value of 2. Hence, a random variable may be thought of as a function, accepting entities from a random sampling as inputs and producing numbers as outputs.

As implied by this discussion, there are two types of random variables. The random variable *D* represents a measurement of time; therefore, it is classified as a *continuous* random variable. Alternatively, since the values for *X* are obtained by counting, not measuring, we classify *X* as a *discrete* random variable. The following box lists the types of random variables.

Types of Random Variables

- Discrete
- Continuous

EXAMPLE 5.3

Suppose you have saved $2,000 and wish to put this money into an individual retirement account (IRA) managed by a mutual fund. Your financial objective is growth, so you focus on mutual funds classified as growth funds. There are numerous growth funds, but you eventually whittle the choices down to five:

F = Fidelity Magellan
K = Kemper Growth
P = Putnam Voyager
T = Twentieth Century Growth
V = Value Line Fund

Suppose you decide to pick two from this list and invest $1000 in each. Although you hope to achieve at least a 12 percent return, you have no preferences among the five growth funds and decide to select the two finalists randomly. Identify the possible pairs of funds in which to invest your money.

Solution:

The five funds may be thought of as making up the population in this example. Selecting two finalists from this group generates ten possible pairs of funds; each pair may be viewed as a potential sample. The ten samples are

FK	KT
FP	KV
FT	PT
FV	PV
KP	TV

where TV signifies Twentieth Century and Value Line, for example. Notice that FF is not a possible sample, because you have decided to invest in two *different* funds. □

Now that the potential samples have been identified, how do we find the likelihood of selecting a particular one? Recall from Section 5.1 that we have three principal methods—classical, empirical, or subjective—with which to determine probabilities. Concentrating on objective approaches, we should use the classical method here because we have complete knowledge of the sampling situation—prior to sampling, we can specify ten equally likely outcomes. As we learned in Chapter 1, a random selection process ensures an equal opportunity of occurrence—1/10—for each pair. Uncertainty is associated with the choice of the two funds; the probability figure of 1/10 is a measure of this uncertainty.

Of course, you don't know in advance what the return will be on your investment. Let's suppose that the one-year returns on these five funds turn out as follows:

F: 16.7%	T: 21.9%
K: 18.3%	V: 8.2%
P: 11.4%	

When you invested your \$2000, you hoped that both funds selected would return at least 12 percent. In retrospect, three of these funds exceeded your expectations and two did not, but this information was unknown at the time of your decision. Let us define a characteristic with respect to your expectation of a 12 percent return and the ten possible samples:

$$X = \text{Number of funds in the pair selected with at least a 12\% return}$$

We now have defined a discrete random variable. If we apply our random variable X to each of the ten possible samples, we generate the following values:

Possible Samples	Value of X	Probability
FK	2	.1
FP	1	.1
FT	2	.1
FV	1	.1
KP	1	.1
KT	2	.1
KV	1	.1
PT	1	.1
PV	0	.1
TV	1	.1

Rather than worrying about the probability of selecting a specific sample (.1), we focus our attention on the probability that both funds selected have 12 percent or better returns—3/10, that one of the two funds selected has a 12 percent or better return—6/10, or that neither fund selected has the desired return—1/10.

This example demonstrates our approach to solving probability problems. We begin by defining a characteristic of interest, X. Then we evaluate the characteristic with respect to each possible sample. Finally, we organize the results of these evaluations into a table or graph, perhaps using the organizational techniques we learned in descriptive statistics. In probability language, the characteristic of interest is a *random variable,* and our method of solving probability problems using it is called the random variable approach.

Random Variable Notation

We will use the notation in Table 5.3 to organize the possible outcomes of a random variable. With this notation, we can summarize the results of our mutual fund selection problem from Example 5.3. (See Table 5.4.) Table 5.4 specifically defines our random

Table 5.3 Probability Notation for Random Variables

Notation	Symbolic Meaning	Remarks
Capital letter, such as X	Description of a random variable	The description is always a collection of words.
Lowercase letter, such as x	Numerical values of the random variable	The values are always a collection of numbers.
$P(X = x)$	The probability that the random variable X takes on a particular value	In specific applications, lowercase x will be replaced by a number.

Table 5.4 Values and Probabilities for X = Number of Funds with a 12 Percent (or Better) Return

x	0	1	2
$P(X = x)$	.1	.6	.3

variable X in its title. The lowercase x row lists the possible numerical values of the random variable that can occur, and the $P(X = x)$ row gives the probability for each numerical value. Using this summary table, it is now clear that the probability that both funds in the sample return 12 percent or more is $P(X = 2) = .30$, that only one of the two funds returns 12 percent or more is $P(X = 1) = .60$, and that there is a probability of .10 that neither fund realizes this target.

COMMENTS

1. We recommend that you start every probability problem by defining the random variable in words.
2. The notation $P(X = x)$ may appear as $P(x)$ in other textbooks and/or other disciplines.

SECTION 5.3 EXERCISES

5.15 For each random variable defined in the following cases, classify it as discrete or continuous.

a. V = Number of Big Macs sold per day at a McDonald's franchise for a random sample of different days.

b. W = Number of first-class letters received by a business for a random sample of different business days.

c. X = Actual flying time (liftoff to touchdown) for flight 261 from Denver to Los Angeles for a random sample of different days.

d. Y = The current ratio (current assets ÷ current liabilities) computed for each company in a random sample of ten companies.

5.16 Classify each of the following random variables as discrete or continuous.

a. R = Daily exchange rate of the Canadian dollar with the American dollar for a random sample of different days.

b. S = Number of mid-air plane collisions per year for a random sample of several years.

c. T = Price of a barrel of intermediate crude oil on the spot market per week for a random sample of ten weeks.

d. U = Yearly attendance at the major sight-seeing attractions in Central Florida for a random sample of eight years.

e. Z = Amount of a company's debt as a percentage of its capital for a random sample of companies.

5.17 Define a random variable of potential interest for the following situations.

a. A marketer plans to analyze the results of a consumer preference study. One of the questions on the survey requires the respondents to indicate whether they preferred aspirin-type tablets in round or oblong shapes. The marketer is interested in knowing the number of consumers preferring each shape.

b. The personnel manager of a medium-sized company instructed his staff to prepare an in-house supervisory seminar for all salaried employees. At the end of the seminar, the manager plans to ask each participant to rate the usefulness of the seminar on a one (extremely useless) to five (extremely useful) scale, with three representing neither useful nor useless. The manager is interested in knowing the number of participants who found the seminar to be extremely useless.

c. Each month, domestic airline carriers report the number of complaints filed against them. This figure is then converted into a rate per 100,000 passengers. A reporter for the business section of a newspaper is planning to compare the figures for the month of May for all carriers. She is interested in knowing the current month's rate per 100,000 passengers for each airline.

5.18 Consider the following samples that could arise from randomly selecting $n = 3$ entities without replacement from a large population containing equal numbers of P's and Q's:

PPP	PPQ	PQP	QPP
QQQ	QQP	QPQ	PQQ

Let the random variable X be defined as

$$X = \text{Number of } Q\text{'s in each sample}$$

Identify the possible values for X, find the probability of each value, and summarize your results in a table.

5.19 Consider each of the following random variables and classify it as discrete or continuous.

a. X = Time required to fill out a two-page job application for a random sample of applicants.

b. Y = Amount of sale for each customer checking out through the "express" lane in a grocery store for a random sample of customers.

c. Z = Quick ratio [(current assets − inventory) ÷ (current liabilities)] for each company in a random sample of eight companies.

5.20 When attempting to solve a probability problem, what is the recommended first step?

5.21 Suppose the following samples are equally likely when randomly sampling from a large population:

ab	aj	bg	eg	gj
ae	ak	bj	ej	gk
ag	be	bk	ek	jk

Let the random variable X be defined as

$$X = \text{Number of vowels in each pair of letters}$$

Identify the possible values for X, find the probability of each value, and summarize your results in a table.

5.22 Refer to Exercise 5.21. Let the random variable Y be defined as

$$Y = \text{Number of times the letter } k \text{ appears in each pair of letters}$$

Identify the possible values for Y, find the probability of each value, and summarize your results in a table.

5.23 For next year's inflation rate, an econometrician developed the following table of projected rates (integer values only) and probabilities.

Rate (%)	4	5	6	7	8
Probability	.24	.36	.15	.15	.10

a. Define the random variable X that the economist has in his model. What type of random variable is it?
b. What symbol would you use to denote the entries in the *rate* row?
c. What symbol would you use to denote the entries in the *probability* row?
d. $P(X = 6\%) = ?$
e. $P(X = 3\%) = ?$

5.4 PROBABILITY DISTRIBUTIONS FOR DISCRETE RANDOM VARIABLES

Whether a random variable is discrete or continuous, we will wish to specify its probability distribution. A *probability distribution* includes three components: (1) a random variable, (2) a specification of the values of the random variable, and (3) a listing of the probability of each value or a means of finding these probabilities. Probability distributions represent a *model* for the population; they may appear as graphs, tables, or as formulas.

Forms of the Probability Distribution

Let us consider defining a probability distribution for a discrete random variable. Suppose a population consists of $N = 10$ pieces of data—seven of which are the value 1, three of which are the value 0. In Figure 5.2, the top portion illustrates this population as a set of data. The bottom portion of the figure models the population with three different forms of a discrete probability distribution. The following definition formalizes the models shown in the figure.

Definition

A **probability distribution for a discrete random variable** X is a table, graph, or formula that displays the values of X and the probability associated with each value, subject to the following constraints:

$$\text{(i)}\quad 0 \le P(X = x) \le 1$$

$$\text{(ii)}\quad \sum_{\text{all } x} P(X = x) = 1$$

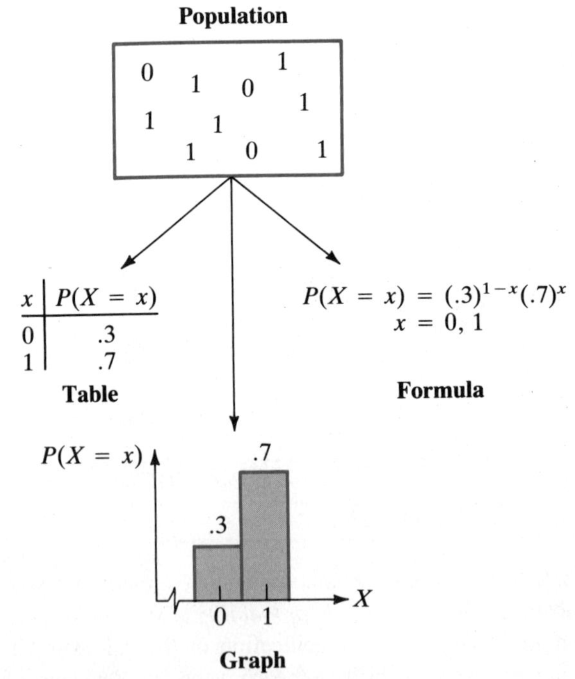

Figure 5.2 Modeling a Population with a Discrete Probability Distribution

Note the difference between a probability and a probability distribution. A *probability,* a number between 0 and 1, represents the likelihood of a single value of the variable X. A *probability distribution* is a collection of all values of the variable and their probabilities in the form of a table, graph, or formula as indicated in Figure 5.2. Note that the expression $x = 0,1$ in the formula shown in Figure 5.2 is shorthand to indicate that the discrete variable X can only take on the value 0 or 1.

An attractive feature of a probability distribution is that the probabilities for *all* values of X are displayed. By reading the table or graph or by evaluating a formula, we can determine any *point probability* of the form $P(X = x)$ that involves only one value of the variable. *Interval probabilities* involving two or more values of the variable such as $P(2 \leq X \leq 5)$ can be determined by adding the appropriate point probabilities. Example 5.4 demonstrates these ideas.

EXAMPLE 5.4

Imagine that you are looking at the daily records of a county emergency medical service. If X is the number of calls responded to per day, then X is a discrete random variable. Suppose the possible values for X are 0, 1, 2, 3, or 4, and the corresponding probabilities are generated according to the following formula:

$$P(X = x) = \frac{9 - 2x}{25} \qquad x = 0, 1, 2, 3, 4$$

a. What are the point probabilities associated with these five values of X?

b. What is $P(X > 2)$?

c. Find $P(X \leq 1)$.

Solution:

a. The chance of one response in a given day is found by substituting the number 1 for the symbol x in the formula

$$P(X = 1) = \frac{9 - 2}{25} = \frac{7}{25} = .28$$

If we substitute the different values for x into this formula, we generate all the possible point probabilities. Table 5.5 summarizes these computations.

b. To find the interval probability $P(X > 2)$, add the appropriate point probabilities and use the *addition rule* of probability.

$$\begin{aligned} P(X > 2) &= P(X = 3, 4) \\ &= P(X = 3) + P(X = 4) \\ &= .12 + .04 = .16 \end{aligned}$$

Table 5.5 Values of X and Their Point Probabilities (X = Number of Calls Responded to per Day)

x	$P(X = x)$
0	.36
1	.28
2	.20
3	.12
4	.04
	1.00

c. The interval $X \leq 1$ includes the points $x = 0$ and $x = 1$:

$$\begin{aligned} P(X \leq 1) &= P(X = 0, 1) \\ &= P(X = 0) + P(X = 1) \\ &= .36 + .28 = .64 \end{aligned}$$

□

Notice that all probabilities in the $P(X = x)$ column of Table 5.5 are between 0 and 1 and that the probabilities total 1. Since both constraints of the probability distribution definition are satisfied, Table 5.5 qualifies as a valid probability distribution.

Mathematically, the formula in Example 5.4 can accommodate *any* value for X such as $x = 5$, $x = 13.7$, or even $x = -2$, but statistically, the probabilities associated with these X-values are meaningless. For instance, if we tried to evaluate the formula when $x = 5$, we would get a negative number, $-.04$, violating the definition of a probability. Therefore, a probability formula is an *incomplete model* without a specific listing, like $x = 0, 1, 2, 3, 4$, of the possible values of the random variable. In the listing, we are making an implicit statement that all values *not* included are understood to have a probability of 0. Relative to the formula in Example 5.4, then, we declare $P(X = 5) = 0$, $P(X = -2) = 0$, and so on.

Developing a Probability Distribution from Data

Deriving a probability model is generally difficult. In practice, probability distributions are developed either empirically from data or they are postulated theoretically. When certain sampling conditions are known to exist, theoretical probability distributions are assumed to be applicable models. We will study some of these special distributions in Chapters 6 and 7.

Alternatively, we sometimes develop a probability distribution from existing data by viewing the observed relative frequency of each value as its empirical probability. An underlying assumption, of course, is that the historical data will accurately model the data from subsequent samplings. Example 5.5 illustrates this procedure.

EXAMPLE 5.5

Suppose you worked 238 days for a car dealership last year. On 60 of those days you sold exactly 1 car, on 52 days you sold 2 cars, on 19 days you sold 3 cars, and you sold 4 cars in one day four times.

a. Develop a probability distribution for X = the number of new cars sold per day.

b. Find $P(X \geq 1)$.

Solution:

a. As a first step, we can organize the information into an ungrouped frequency distribution as depicted in Table 5.6. Since you worked a total of 238 days, you failed to close a deal on $238 - (60 + 52 + 19 + 4) = 103$ days. Thus, the entry in the frequency column for $x = 0$ is 103. The third column in Table 5.6 is the relative frequency column, formed by dividing each f by $n = 238$. If relative frequency is interpreted as probability, then the first and third columns of Table 5.6 form the probability distribution for X. We have rounded off the relative frequencies in Table 5.6 and summarized the results in Table 5.7.

Table 5.6 Ungrouped Frequency Distribution for X = Number of New Cars Sold per Day

x	f	rf
0	103	.433
1	60	.252
2	52	.218
3	19	.080
4	4	.017
	$n = 238$	

Table 5.7 Probability Distribution for X = Number of New Cars Sold per Day

x	$P(X = x)$
0	.43
1	.25
2	.22
3	.08
4	.02
	1.00

b. One way to find $P(X \geq 1)$ would be to identify the individual values equivalent to the interval $X \geq 1$ and add the respective point probabilities. These values are $x = 1, 2, 3,$ or 4. Thus

$$\begin{aligned} P(X \geq 1) &= P(X = 1) + P(X = 2) + P(X = 3) + P(X = 4) \\ &= .25 + .22 + .08 + .02 = .57 \end{aligned}$$

Note that we could approach this problem from an alternate view by saying

$$1 = P(X = 0) + P(X = 1) + P(X = 2) + P(X = 3) + P(X = 4)$$

or

$$1 = P(X = 0) + P(X \geq 1)$$

A bit of simple algebra produces

$$\begin{aligned} P(X \geq 1) &= 1 - P(X = 0) \\ &= 1 - .43 = .57 \end{aligned}$$

The principle behind this alternative solution is the *complementary rule* of probability. Instead of directly solving the probability problem by adding, we indirectly solved it by subtracting, using the complementary collection of values of the random variable ($x = 0$ in this case) and the fact that the total probability is 1. □

In Example 5.5, we easily made the transition from an ungrouped frequency distribution (Table 5.6) to a probability distribution (Table 5.7). A similar connection exists between a relative frequency histogram and a probability distribution in graphical form. Although these descriptive statistics ideas relate nicely to our discussion of discrete probability distributions, we will discover in Section 5.7 that models for continuous random variables are slightly different.

CLASSROOM EXAMPLE 5.2

Verifying Conditions for a Discrete Probability Distribution

For each of the following cases, determine whether the table, graph, or formula qualifies as a valid probability distribution for an arbitrarily defined discrete random variable X.

a.

x	−2	0	6
$P(X = x)$	.42	.17	.41

b.

x	1	2	3	4
$P(X = x)$	.38	.30	.14	.15

c. See Figure 5.3.

d. $P(X = x) = \dfrac{x}{15} \qquad x = 1, 2, 3, 4, 5$

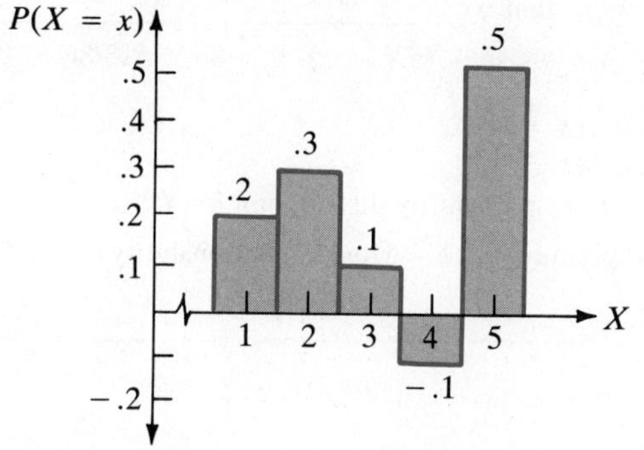

Figure 5.3 Proposed Probability Distribution for *X* in Classroom Example 5.2

SECTION 5.4 EXERCISES

5.24 By evaluating the formula $P(X = x) = \dfrac{3x + 7}{35}$ for $x = -2, -1, 0, 1, 2$, display the probability distribution for X in tabular form.

5.25 a. In order for the following table to satisfy the requirements of a probability distribution, what would be the value of $P(X = 5)$?

x	0	5	10	20
$P(X = x)$	.08		.59	.23

b. Find $P(X > 5)$.

c. What law of probability did you use to find $P(X > 5)$?

5.26 Which of the following are valid probability distributions for X?

a. $P(X = x) = \dfrac{x^2 - 5}{51} \qquad x = 3, 4, 5, 6$

b.

x	0	1
$P(X = x)$	.4	.6

c. $P(X = x) = .25 \qquad x = -1, 0, 1, 2$

d.

x	1	5	10	12
$P(X = x)$	.1	.5	.6	−.2

5.27 The probability distribution for X is given by the following table:

x	2	3	4	5
$P(X = x)$	6/86	15/86	26/86	39/86

a. Find $P(X = 1)$.
b. Find $P(X = 3)$.
c. Graph the probability distribution for X.

5.28 The graph in Figure 5.4 depicts the probability distribution for X.

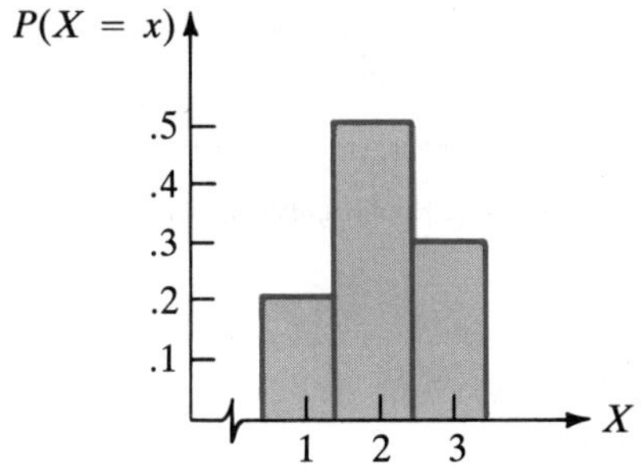

Figure 5.4 Probability Distribution for X in Exercise 5.28

a. What is $P(X = 2)$?
b. Find $P(X = 4)$.
c. Display the probability distribution for X in tabular form.
d. Find $P(X < 3)$.

5.29 The probability distribution for X is given by the following table:

x	0	1	2	3	4
$P(X = x)$	.063	.250	.375	.250	.062

a. Find $P(X > 2)$.
b. Find $P(X \leq 3)$.
c. Find $P(X = 0 \text{ or } X = 1)$.
d. What law of probability did you use to solve part c?

5.30 In a strategic planning meeting with her staff of 10, the supervisor of budgets asked each person to project the number of new full-time positions that the budgeting area will need in the next three years. The most popular response, given by four staff members, was four. Two people thought the area would need three new positions. Everyone agreed on the need for at least two new positions, but only one person thought as many as five positions would be necessary.

a. Define a random variable X as the number of new, full-time positions needed in the next three years. What type of random variable is X?
b. Develop a probability distribution for X.
c. Find $P(X = 1)$.
d. Find $P(X \geq 3)$.
e. The supervisor of budgets must make a recommendation to the vice-president of operations on her future staffing needs. How might the information gained from her staff be summarized to produce a specific staffing recommendation?

5.31 Explain two different ways of developing probability distributions.

5.32 If we equate relative frequency and probability, what assumption must we make?

5.33 Define a probability distribution for a discrete random variable.

5.34 Suppose X is a discrete random variable with possible values 0, 1, 2, 3, 4. Identify the following as point probabilities or interval probabilities:
a. $P(X = 2)$ b. $P(X > 1)$ c. $P(X = 3 \text{ or } X = 4)$ d. $P(X < 1)$

5.35 In *Money* magazine, a summary of activities in the financial markets and the following 12-month forecast appeared:

				Stock	
12-Month Forecast	CPI	T-Bills	Long Bonds	Dow	S&P 500
Projected results	2.0%	6.6%	7.8%	1900	250
High estimate	2.5	7.3	8.3	2000	260
Low estimate	1.5	6.2	7.3	1700	220

Source: "Investment Scorecard: Cloudy Forecasts for Sky-High Markets," *Money* (May 1986):34. Reprinted by permission.

The high and low estimates are self-explanatory, while the projected results are taken to be the most likely event. Percentage estimates are made to the nearest tenth, while the Dow is projected to the nearest hundred and Standard & Poor's to the nearest ten.

a. Define X to be the forecasted Dow. Relative to the textual material in this section, the column of estimates for the Dow does not qualify as a probability distribution. Why?
b. Suppose the high and low Dow estimates each have a probability of .10, and the projection of 1800 has a probability of .30. Complete the probability distribution for X.
c. Assume a probability of .02 is assigned to each of the high and low estimates and .20 is assigned to the projected results for the T-bill forecasts. Assign probabilities between .02 and .20 to all remaining values between 6.2 and 7.3 percent, such that the distribution is a valid probability distribution and is skewed to the right. Graph your distribution.
d. Assume a probability of .02 is assigned to each of the high and low estimates and .20 is assigned to the projected results for the CPI forecasts. Assign probabilities between .02 and .20 to all remaining values between 1.5 and 2.5 percent such that the distribution is valid and is mound shaped and symmetrical.
e. Suppose the high and low projections for the S&P 500 each have a probability of .05, and the most likely projection of 250 has a probability of .45. Develop a probability model for Y = the 12-month forecast of the S&P 500 that would be negatively skewed.

5.36 Over the past five years, a dog breeder has kept detailed records of the number of puppies per litter for the last 30 litters. Only once did she record a litter as small as five, and only once was a litter as large as 10 puppies. The records showed that 50 percent of the litters contained eight puppies, 10 percent contained nine pups, and 20 percent contained seven. With her next litter she wants to run an advertisement in a national sporting magazine. The ad will cost $2800 and must be purchased before the puppies are born in order to meet publication deadlines. However, the ad will enable her to charge $750 per pup. Without the ad, she can sell each puppy for $350. Regardless of the selling price, $80 per puppy always goes toward expenses. Assume all puppies can be sold.

a. Develop a probability distribution for X = the number of puppies per litter based on her historical data.

b. Suppose she does not run the ad in the magazine. Letting Y = profit (defined as selling price − cost of the ad) per litter, develop a probability distribution for Y.

c. If she does run the ad, develop a probability distribution for Z = profit (selling price − expenses − cost of the ad) per litter.

d. At what size litter would both strategies (run the ad/do not run the ad) produce the same profit?

e. What strategy would you recommend for her? Why?

5.37 Which of the following qualify as probability distributions?

a. X = Type of sales transaction (0 = cash, 1 = charge):

x	0	1
$P(X = x)$	.34	.66

b. X = Projected number of years until the next recession starts:

$$P(X = x) = \frac{x}{10}$$

c. X = Number of people in a sample of 20 who use ibuprofen as their favorite nonprescription painkiller.

5.5 PARAMETERS OF PROBABILITY DISTRIBUTIONS

Just as every random variable has a probability distribution, every probability distribution has one or more parameters that characterize it. In Chapter 3 we defined a parameter as a numerical descriptive measure of a population. A probability distribution, though not a population per se, is a model for the population, and therefore the term parameter applies to it as well. General examples of parameters include the important descriptive quantities like the mean, variance, and standard deviation. However, the symbols and formulas that we used in Chapter 3 for the different descriptive statistics must be modified to account for a different representation of the data—that is, a probability distribution instead of raw data.

Mean (Expected Value)

Let us first consider the popular measure of central location: the mean. For a set of sample data, we symbolized the mean with $\overline{X}$ and determined its value for an ungrouped frequency distribution with the formula

$$\overline{X} = \frac{\Sigma fX}{n}$$

If we change the order of operations, dividing each term by n before adding, we can rewrite this formula as

$$\overline{X} = \sum \frac{f}{n} X$$

where f/n is the relative frequency of each value. Recall that the empirical approach to generating probability equates the relative frequency of an event to its probability. Hence, replacing $\overline{X}$ with μ and the relative frequency f/n by the probability $P(X = x)$, we obtain the equation for the mean of a discrete probability distribution.

Mean (Expected Value) of a Discrete Probability Distribution

$$\mu = \Sigma x P(X = x) \qquad (5\text{–}3)$$

EXAMPLE 5.6

If X is a discrete random variable representing the number of personal computers per household with the following probability distribution, find the mean μ:

x	0	1	2
$P(X = x)$	.37	.16	.47

Solution:

From Equation 5–3, we compute

$$\begin{aligned} \mu &= \Sigma x P(X = x) \\ &= 0\,P(X = 0) + 1\,P(X = 1) + 2\,P(X = 2) \\ &= 0(.37) + 1(.16) + 2(.47) \\ &= 1.1 \end{aligned}$$

The average number of personal computers per household is 1.1.

□

All of the hints and interpretations that we provided in the third chapter as we were discussing the sample mean $\overline{X}$ apply here as well. Specifically, remember that the answer you obtain for μ must be between the smallest and largest values for X. Secondly, the mean or average value can yield a number that is not a possible value for the random variable. Referring to Example 5.6, no household can have 1.1 personal computers in it. This does not diminish or invalidate the importance of the mean, because the mean is a global characteristic of *all* the households of interest. It is not a specific characteristic of any one individual household.

Just as we did for a set of sample data, we could consider other measures of central location to describe a discrete probability distribution, such as the median or the mode. However, we choose not to, relying on the mean as a sufficient measure for our purposes in this textbook. Similarly, there are many measures of dispersion that we can apply to probability distributions, but we will focus on only two: the variance and the standard deviation.

Variance

The variance of a discrete probability distribution is derived in an analogous manner to the mean (treating relative frequency as probability). Omitting the details, the defining formula is given in the following box.

Variance of a Discrete Probability Distribution: Defining Formula

$$\sigma^2 = \Sigma(x - \mu)^2 P(X = x) \quad \text{(5–4)}$$

This is consistent with the definition that the variance is an average of squared deviations from the mean. The squared deviations are each weighted by the associated probability—$P(X = x)$—and then totaled.

EXAMPLE 5.7

Find the variance of the probability distribution representing X = the number of personal computers per household.

x	0	1	2
$P(X = x)$	.37	.16	.47

Solution:

In Example 5.6, we computed the value of the mean to be $\mu = 1.1$ for this distribution. To find σ^2, let us perform some intermediate calculations. First, subtract μ from each value of X, then square the result. The columns labeled $(x - \mu)$ and $(x - \mu)^2$, respectively, represent these operations:

x	$P(X = x)$	$(x - \mu)$	$(x - \mu)^2$	$(x - \mu)^2\, P(X = x)$
0	.37	−1.1	1.21	.4477
1	.16	−.1	.01	.0016
2	.47	.9	.81	.3807

$$\Sigma(x - \mu)^2 P(X = x) = .8300$$

Finally, multiply the entries in the $(x - \mu)^2$ column by the respective probabilities. The variance, according to Equation 5–4, is the sum of numbers in the last column. Rounded, $\sigma^2 \approx .8$.

□

As we discovered in descriptive statistics, finding the variance directly from the deviations $(x - \mu)$ can be inaccurate if the mean works out to be a number with many decimal places. Common sense tells us to round off before proceeding, but the process of rounding introduces error into the final answer. Alternatively, another formula for computing the variance provides some relief in much the same way that the calculating formula for S^2 was used in Chapter 3. A different, but equivalent, method of finding σ^2 is presented in Equation 5–5.

Variance of a Discrete Probability Distribution: Calculating Formula

$$\sigma^2 = \Sigma x^2 P(X = x) - \mu^2 \tag{5–5}$$

Unlike the repeated subtractions of μ required by the defining formula, the calculating formula requires only one subtraction, that of μ^2. Example 5.8 illustrates its use.

EXAMPLE 5.8

Find the variance of the probability distribution in Example 5.7 for the random variable X = the number of personal computers per household, using the calculating formula.

Solution:

Using Equation 5–5 with $\mu = 1.1$, we have

$$\begin{aligned}\sigma^2 &= \Sigma x^2 P(X = x) - \mu^2 \\ &= [0^2(.37) + 1^2(.16) + 2^2(.47)] - (1.1)^2 \\ &= [0 + .16 + 1.88] - 1.21 \\ &= .83 \approx .8\end{aligned}$$

Thus, using either Equation 5–4 or 5–5 we have arrived at the same value of .8 for σ^2.

□

Standard Deviation

The final measure of dispersion for discrete probability distributions is the standard deviation, found by taking the positive square root of the variance.

Standard Deviation of a Discrete Probability Distribution

$$\sigma = +\sqrt{\sigma^2} \qquad (5\text{–}6)$$

EXAMPLE 5.9

Find the standard deviation of the probability distribution for X in Example 5.6.

Solution:

In Examples 5.7 and 5.8, we found the variance, before rounding, to be .83. The standard deviation is therefore

$$\begin{aligned}\sigma &= +\sqrt{\sigma^2} \\ &= +\sqrt{.83} \\ &= .9 \text{ personal computers} \qquad \text{(rounded)}\end{aligned}$$

□

COMMENTS

1. The value generated by Equation 5–3 and attached to the symbol μ is referred to interchangeably as the *mean of the probability distribution* or the *expected value of the random variable X*.
2. As a guideline for computational accuracy, we suggest the following round-off rules:
   (a) When the mean is computed as a final answer, round it off to one more decimal place than the accuracy of the X-values, *but* as an intermediate calculation for the variance, do not round it off.
   (b) When the variance is computed as a final answer, round it off to one more

decimal place than the accuracy of the X-values, *but* as an intermediate calculation for the standard deviation, do not round it off.

(c) When the standard deviation is computed as a final answer, round it off to one more decimal place than the accuracy of the X-values.

An interesting example of the application of the mean of a discrete probability distribution follows.

EXAMPLE 5.10

A straight bet on a sporting event at many casinos in Las Vegas and Atlantic City has a payoff of 10 to 11, or \$10 for every \$11 wagered. For example, if you bet \$11 and win, you will receive \$21 in return—your original \$11 wager plus the \$10 you won. Although many bettors believe they are smarter than the odds makers, the fact is that point spreads and 10-to-11 odds favor the bookmakers in the long run. If we assume that winning and losing are equally likely for the bettor (the purpose of a point spread), find the mean for the random variable X representing the bettor's net gain—that is, X = (amount received − amount bet) when the amount bet is \$55.

Solution:

It might be helpful to construct a table representing the bettor's outcomes and probabilities. If we bet \$55 and lose, then the amount we receive is \$0, and the net gain is $X = 0 - 55 = -55$. If we win, we receive \$105, and the net gain is $X = 105 - 55 = +50$.

Outcome	x	$P(X = x)$
Win the bet	+50	.5
Lose the bet	−55	.5

The mean value for X is

$$\begin{aligned}\mu &= \Sigma x\, P(X = x) \\ &= (+50)(.5) + (-55)(.5) = -\$2.50\end{aligned}$$

The important part of the above computation is the negative sign. This type of wager is a losing proposition for gamblers in the long run. It is naturally a profitable proposition for bookmakers. If a total of \$1.1 million were wagered in the form of \$55 bets with the half the bettors taking positions on each side of the point spread, then a bookmaker stands to make \$50,000, regardless of the outcome of the sporting event! From a financial perspective, it is obvious why sports in this country are truly "big business." An estimated \$40 million is *legally* bet on pro football's Super Bowl every year. The only way the casino can lose is if it doesn't balance the dollar amount of its customers' selections on either side of the point spread. This is why the point spread fluctuates prior to the sporting event.

□

CLASSROOM EXAMPLE 5.3

Computing Parameters of a Probability Distribution

Consider the following probability distribution of X:

x	$P(X = x)$	$xP(X = x)$	x^2	$x^2P(X = x)$
1	.05			
2	.10			
3	.20			
4	.25			
5	.40			
		$\Sigma xP(X = x) =$ ______		$\Sigma x^2P(X = x) =$ ______

a. Find the mean (expected value), μ.

b. Find the variance, σ^2.

c. Find the standard deviation, σ.

SECTION 5.5 EXERCISES

5.38 Find the mean, variance, and standard deviation of the following probability distribution for X:

x	−5	0	270
$P(X = x)$	.91	.06	.03

5.39 Find the mean and standard deviation of the following probability distribution for X:

x	0	5	10	20
$P(X = x)$	.08	.10	.59	.23

5.40 Find the mean and standard deviation of the following probability distribution for X:

x	−2	−1	0	1	2
$P(X = x)$	1/35	4/35	7/35	10/35	13/35

5.41 To see the effect of the concentration of probability on the value of σ, consider the probability distributions in Figure 5.5. In each distribution the value of μ is 0. Find σ for each distribution.

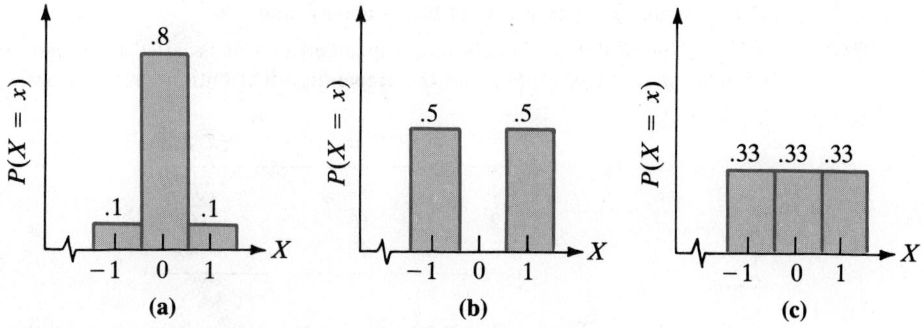

Figure 5.5 Probability Distributions for X in Exercise 5.41

5.42 Find the mean and variance of the following probability distribution for X:

x	0	1	2	3	4
$P(X = x)$	.0016	.0256	.1536	.4096	.4096

5.43 Refer to Exercise 5.41 and Example 5.10. Using the results of Exercise 5.41, guess at the value of σ for the bettor's net gain. Then compute the standard deviation and compare.

5.44 Find the mean and standard deviation of the following probability distribution for X:

x	0	1	2	3	4
$P(X = x)$	.6065	.3033	.0758	.0126	.0018

5.45 Suppose the probability that the closing Dow Jones Industrial Average moves up ($x = 1$), down ($x = -1$), or remains the same ($x = 0$) as the previous day's close is

$$P(X = x) = \frac{4x^2 + x + 1}{11} \qquad x = -1, 0, 1$$

Find the mean of this probability distribution.

5.46 The projected return (in percentages) for U.S. Treasury bills one year from now is summarized as follows:

Return	Probability	Return	Probability
7.3	.02	6.7	.16
7.2	.05	6.6	.20
7.1	.07	6.5	.09
7.0	.07	6.4	.08
6.9	.11	6.3	.03
6.8	.12		

Find the expected return.

5.47 Why are the symbols μ and σ^2 used to denote the mean and variance, respectively, of a probability distribution instead of the symbols $\overline{X}$ and S^2?

*5.48 Totaling the probability at and beyond a specified X-value is called a *decumulative probability* and is denoted $P(X \geq x)$. Suppose the probability distribution for X is as follows:

x	$P(X = x)$	$P(X \geq x)$
1	.37	1.00
2	.16	.63
3	.47	.47

The third column, labeled $P(X \geq x)$ contains the decumulative probabilities; for example $P(X \geq 2) = P(X = 2) + P(X = 3) = .16 + .47 = .63$. One benefit of this column is its usefulness in checking your calculations for the mean, μ. Shiffler (1984) showed that

$$\mu = \Sigma P(X \geq x)$$

provided X takes on positive integer values only. That is, the mean of the above probability distribution can be found simply by adding the numbers in the third column. Verify that 2.10, the total of the third column, is the value of μ using Equation 5–3. Two cautions, however, follow: (1) The values of X must start at 1, and (2) the summation extends over *all* values of X, including any missing values.

5.49 Consider the following probability distribution for X:

x	$P(X = x)$	$P(X \geq x)$
1	.1	1.0
2	.5	.9
3	.2	.4
4	.1	.2
5	.1	.1

a. Find the mean using Equation 5–3.
***b.** Find the mean using the result from Exercise 5.48.

5.50 Consider the following probability distribution for X:

x	1	2	4
$P(X = x)$	.37	.41	.22

a. Generate the decumulative probabilities. (*Hint:* There are four such probabilities, including $P(X \geq 3)$.)
b. Find the mean using Equation 5–3.
***c.** Find the mean using the result from Exercise 5.48.

*Optional

5.51 A manufacturer of microwave ovens offers a service contract on its more expensive models. For $50 a year, the oven purchaser can insure herself or himself from any repairs that are needed in the upcoming 12 months. Suppose the manufacturer estimates that the following table will accurately represent their costs and probabilities of product failures:

Category of Microwave Failure	Average Cost to Correct Failure ($)	Category Probability
Serious	100	.15
Moderate	60	.15
Minor	35	.05
No Failure	0	.65

Let X = Repair cost to the manufacturer per insured microwave.

a. Find the average value of X.

b. Find the standard deviation of X.

c. About what proportion of oven buyers are better off for having purchased the service contract?

*5.6 PROCESSOR: DISCRETE PROBABILITY DISTRIBUTIONS

The Discrete Probability Distributions processor will compute the mean, variance, and standard deviation for a user-supplied distribution. We illustrate its application with the following probability distribution (from Classroom Example 5.3):

x	1	2	3	4	5
$P(X = x)$	.05	.10	.20	.25	.40

Creating the Data Set

When we reach the main menu, we move the highlight to Data and touch Enter twice to obtain the list of data types. For this processor, we require the Probability Distribution option. After we make this selection, we are asked the number of values for X in our distribution; 5 in this case. Figure 5.6 shows the screen after entering 5. After touching Enter, we then provide the values of X and their respective probabilities. After the final entry of .40 as the probability for $X = 5$ has been made, the screen should appear as in Figure 5.7.

*Optional

```
≡        Microcomputer Applications for INTRODUCTORY BUSINESS STATISTICS   EDIT
  Overview [Data] Files  Processors  Set Up  Exit                    | F1=Help
            Create
            *iew/Edit




                    ═══════════ Data set size ═══════════
                    How many values for X are in the dist, k =  5
```

Figure 5.6 Preparing to Create a Probability Distribution Set

```
≡        Microcomputer Applications for INTRODUCTORY BUSINESS STATISTICS   EDIT
  Overview [Data] Files  Processors  Set Up  Exit                    | F1=Help

                 x                  P(X = x)
   1 1                 .05
   2 2                 .10
   3 3                 .20
   4 4                 .25
   5 5                ▪.40
```

Figure 5.7 Screen Appearance After Final Data Item Entry

■ Running Discrete Probability Distributions

With data creation complete, move to Processors on the main menu and choose Discrete Probability Distributions. A submenu of three choices will now appear. Choose the second option, labeled Parameters; the Binomial and Poisson options will be investigated in Chapter 6.

After reading the introductory screen, move to Data and touch Enter to verify that there are no problems with the data set. Then proceed to make the Calculations. When complete, move to Results and touch Enter. A submenu with three choices is

```
≡              Parameters Of A Discrete Probability Dist                    READY
 Introduction  *ata  Calculations [Results] Exit                          | F1=Help

                      For the probability distribution

                       Measure                        Value

                        Mean.................... 3.8500
                        Variance................ 1.4275
                        Standard deviation...... 1.1948
```

Figure 5.8 Summary Results Screen

now presented. If we select the Parameters choice, we will obtain the screen shown in Figure 5.8. The other Results options provide intermediate calculations and a complete listing of the probability distribution.

5.7 PROBABILITY DISTRIBUTIONS FOR CONTINUOUS RANDOM VARIABLES

In Section 5.4, we discussed the probability distribution for a discrete random variable. In this section, we will describe probability distributions for continuous random variables.

Point Probabilities for Continuous Variables

In Example 5.4 the random variable X = the number of new cars sold per day was discrete. Suppose we define a related random variable Y to be the amount of time (in hours) needed to close the sale. Since time is a measurable quantity, the random variable Y is continuous.

The probability model to characterize the selling time involves the usual three components: the random variable, its values, and a method for finding the probabilities. The random variable is Y and its values are all the numbers between 0 and, say, 20; *not* just the integer numbers, but *all* the numbers. Can we list all these values and the probabilities of each? No, we cannot.

Therefore, we simply declare the probability of each point value for Y to be zero. For any continuous random variable like Y, the point probability of each value is zero. This is not true for discrete random variables; it applies only to continuous random variables.

Definition

If X is a continuous random variable, then all **point probabilities** are zero—that is,

$$P(X = \text{any specific number}) = 0$$

Interval Probabilities for Continuous Variables

For discrete random variables, we can find point *and* interval probabilities. A point probability for the random variable X representing the number of calls responded to per day in Example 5.4 could be zero or nonzero: $P(X = 1) = .28$ and $P(X = 5) = 0$. An interval probability for a discrete random variable usually is solved by breaking it up into its component point probabilities. Continuing from Example 5.4, for instance:

$$\begin{aligned} P(1 \le X \le 3) &= P(X = 1, 2, 3) \\ &= P(X = 1) + P(X = 2) + P(X = 3) \\ &= .28 + .20 + .12 = .60 \end{aligned}$$

On the other hand, if X were a continuous random variable, then $P(X = 1)$ and $P(X = 5)$ would both be zero, and the $P(1 \le X \le 3)$ would *not* be found by adding $P(X = 1)$, $P(X = 2)$, and $P(X = 3)$. We can evaluate probabilities for continuous variables only by considering intervals.

Because of this, we adopt a different tack for identifying the values of a continuous random variable. In the formula in Example 5.4, we wrote $x = 0, 1, 2, 3, 4$ to list the values of the discrete variable X. This is a standard convention. For discrete variables, we separate the values by commas and maybe add three dots if the string of values is very long (for example, $x = 0, 1, 2, \ldots, 30$). For continuous variables we do not have this luxury due to the number of values. Instead we use inequality signs and, if possible, the smallest and largest values that X can assume. For the example involving the amount of time needed to close the sale, we specify the values of Y as $0 \le y \le 20$.

Interval Probabilities: The Special Case of Cumulative Probability

Since point probabilities are always 0 for continuous random variables, we turn our attention to interval probabilities of the form $P(X \le a)$, $P(b \le X \le c)$, or $P(X \ge d)$, where a, b, c, d represent numbers. The first form—$P(X \le a)$—is of special interest, as it denotes the cumulative probability at the value a. All the reference tables in this book that give probabilities for continuous variables are set up as cumulative probabilities.

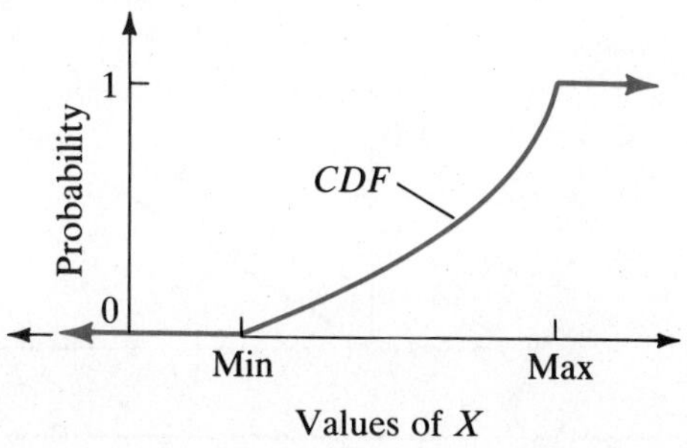

Figure 5.9 Typical Shape of a Cumulative Distribution Function (CDF)

The concept of *cumulative probability* is similar to the notion of cumulative frequency that we studied in Section 2.2: The idea is to accumulate the probability from the lowest value up to the value of interest. At the minimum value and any value below it, the cumulative probability is 0. At the maximum value and any value above it, the cumulative probability is 1. In between these extreme values, the cumulative probability increases as the values increase. Figure 5.9 shows this pattern.

For discrete random variables, we can develop cumulative probabilities merely by adding the appropriate point probabilities. But this is not possible for a continuous random variable. Instead, we rely on a mathematical function called a *cumulative distribution function* (CDF), denoted by $F(x)$, in Equation 5–7, to generate the cumulative probability for every value of X.

Cumulative Distribution Function

$$F(x) = P(X \leq x) \qquad \textbf{(5–7)}$$

For example, a simple cumulative distribution function is

$$F(x) = x/2 \qquad 0 \leq x \leq 2$$

It is understood that $F(x)$, in this case, yields the value 0 for any $x < 0$ and $F(x) = 1$ for $x > 2$. The graph of this function appears in Figure 5.10. If we wanted to find the cumulative probability at the value $x = 1$—that is $P(X \leq 1)$—we evaluate the function $F(x)$ at this value:

$$F(1) = P(X \leq 1) = 1/2$$

This means that 50 percent of the probability occurs in the interval from $x = 0$ through $x = 1$.

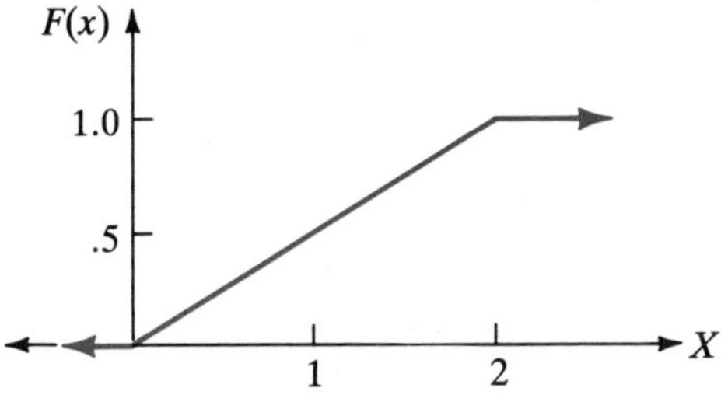

Figure 5.10 Cumulative Distribution Function ($F(x) = x/2, 0 \le x \le 2$)

Most cumulative distribution functions that have applications in business are rather sophisticated. We introduce the concept here for two reasons. First, a probability distribution for a continuous random variable is developed theoretically from a CDF by differentiating the function with respect to x. Since differential calculus is not a prerequisite for this text, we will not pursue this concept. When we need to work with a continuous random variable, we will simply provide the appropriate formula for its probability distribution. Second, for all major continuous probability distributions that we encounter, we will evaluate the appropriate CDF for many values of X and summarize the results into an easy-to-read table of cumulative probabilities. This eliminates the reliance on calculus for dealing with continuous probability distributions and shifts the emphasis to reading the various tables.

Forms of the Probability Distribution

This brings us to the definition of a probability distribution for a continuous random variable. In Section 5.4, we learned that discrete probability distributions are displayed in one of three forms—a table, graph, or formula. For continuous probability distributions, the tabular form detailing the point probabilities does not exist. The graphical approach is still feasible, although the graph usually resembles a smooth curve as opposed to a series of connected rectangles. Unlike the formula for the discrete case, the formula for a continuous probability distribution generates the graph but not the point probabilities, which are always zero. The following definition formalizes the requirements for a continuous probability distribution.

Definition

A **probability distribution for a continuous random variable** X is a formula $f(x)$ or its graph and an identification of the values of X subject to the following constraints:

(i) $f(x) \ge 0$
(ii) Total area $= 1$

The first constraint in the definition states that the formula for a continuous probability distribution cannot yield a negative number when evaluated. Equivalently, this means that the graph of the distribution always must lie on or above the horizontal axis, never below it. This property is analogous to the property that a discrete probability distribution must have probabilities between 0 and 1. Notice that $f(x)$ does not have to be less than 1. This clearly implies that $f(x)$, the formula, does *not* produce probabilities when it is evaluated. It simply produces a point on the graph.

The second constraint requires that the total probability for a graph be 1. In the discrete case, we added the point probabilities and required the total to equal 1. In the continuous case, the means of verifying a total probability of 1 is accomplished by verifying a total graphical *area* of 1. The concept of area in mathematics is parallel to the concept of probability in statistics.

COMMENT The formula $f(x)$ that defines a probability distribution for a continuous random variable is also known as a *probability density function*.

Figure 5.11 illustrates the terms introduced in this section for the continuous probability distribution $f(x) = e^{-x}$, $x \geq 0$. To generate the graph, we plotted the formula $f(x) = e^{-x}$. A calculator with an e^x or e^y button could be used to evaluate $f(x)$ at various X-values. As we know, a table of point probabilities for a continuous random variable is nonexistent. An interval probability for a continuous random variable is represented as an area under the graph. The CDF for $f(x) = e^{-x}$ is $F(x) = 1 - e^{-x}$. (The development of this formula is based on integral calculus and is not presented here.) Given the CDF, we can find any interval probability such as $P(X \leq 2)$ by evaluating the CDF at $x = 2$. For example,

$$F(2) = P(X \leq 2) = 1 - e^{-2} = .8647$$

As we continue our study of continuous probability distributions in Chapter 7, we will be required to find interval probabilities. Unfortunately, many of the continuous probability distributions of interest do not have an easy-to-use CDF such as the one in Figure 5.11. Therefore, we will organize selected values of the CDF into a table of cumulative probabilities as needed. In working with continuous probability distributions, you must learn to read the tables in lieu of finding areas under graphs.

For now, we should recognize that continuous probability distributions create slightly different problems than discrete distributions, and these problems are solved with a more sophisticated level of mathematics—calculus, rather than algebra. Table 5.8 summarizes various attributes about discrete and continuous probability distributions.

Determining Probabilities

Although a cumulative distribution function (or a table of selected values from a CDF) yields probabilities of the form $P(X \leq a)$, we can use the CDF to find interval probabilities such as $P(b \leq X \leq c)$ or $P(X \geq d)$, as Example 5.11 illustrates.

Continuous Probability Distribution Formula:

$$f(x) = e^{-x}, \qquad x \geqslant 0$$

Continuous Probability Distribution Table:
Does not exist.

Continuous Probability Distribution Graph:

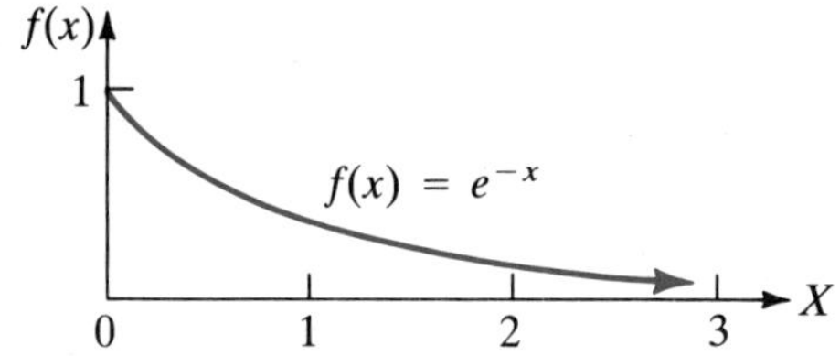

Point Probability:

$$P(X = 2) = 0$$

Interval Probability:

$$P(X \leqslant 2) = \text{area under } f(x) \text{ between } x = 0 \text{ and } x = 2$$

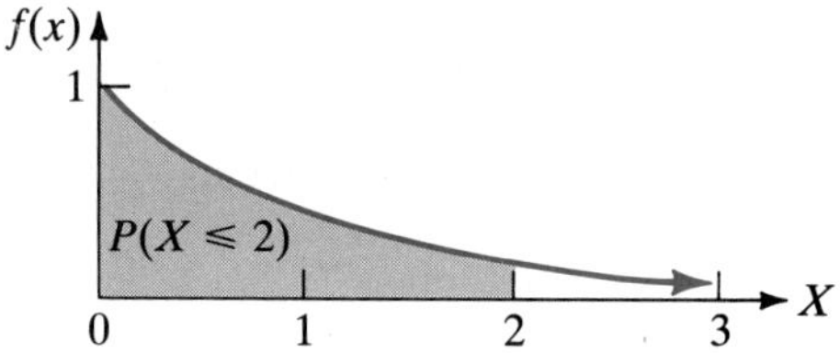

Cumulative Distribution Function for $f(x) = e^{-x}$:

$$F(x) = P(X \leqslant x) = 1 - e^{-x}$$

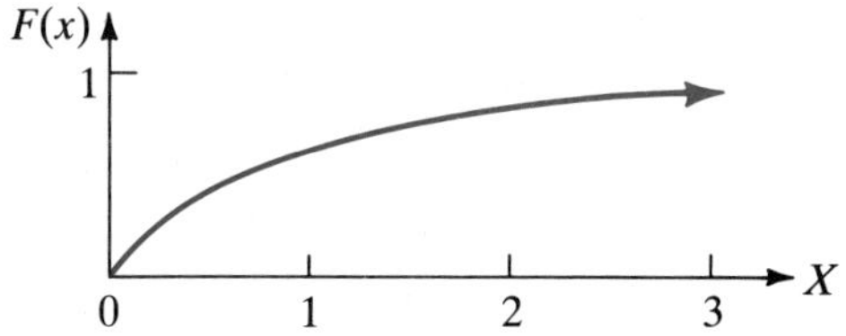

Figure 5.11 Illustration of Terms for a Continuous Random Variable

Table 5.8 Comparison of Probability Distributions for Discrete and Continuous Random Variables

Attribute	Discrete	Continuous
Point probabilities	Zero or nonzero	Zero
Existence of table of point probabilities	Yes	No
Appearance of graph	Connected rectangles	Smooth curve
Notation of formula	$P(X = x)$	$f(x)$
Specification of values	Separated by commas, e.g., $x = 0, 1, 2$	Inequalities linking extreme values, e.g., $0 \leq x \leq 2$

Table 5.8 *(Continued)*

Attribute	Discrete	Continuous
Within a given interval	Only certain values can occur	Any value can occur
Table of interval probabilities	Not used in this book but can be created easily using a point probability table	Cumulative probability tables are the best way to find probabilities
Graphical representation of interval probabilities	An area under the probability distribution	An area under the probability distribution

EXAMPLE 5.11

A fast food company models the random variable X, where X = the elapsed time (in minutes) from placing an order at the drive-through speaker to picking up the order at the drive-through window, with the continuous probability distribution shown in Figure 5.12. The minimum value for X is 30 seconds, or .5 minute, while the maximum service time is thought to be 5 minutes. Suppose some of the cumulative probabilities have been determined and are shown in the following table:

x	.5	1	1.5	2	2.5	3	3.5	4	4.5	5
$P(X \le x)$	0	.398	.639	.786	.874	.928	.961	.981	.993	1

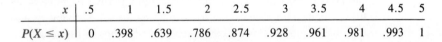

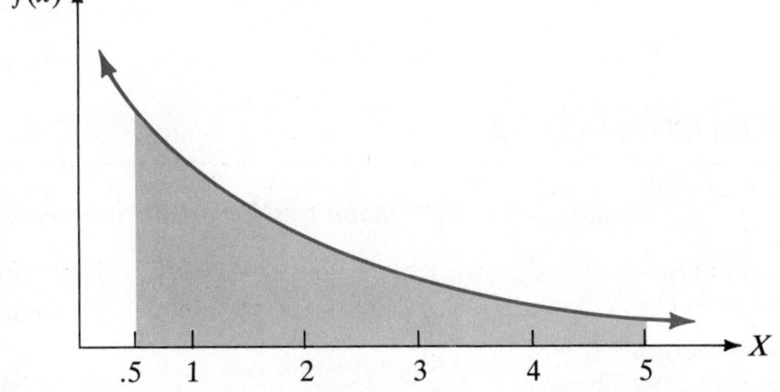

Figure 5.12 Probability Distribution for X in Example 5.11

Find these probabilities:

a. $P(X \leq 3)$ b. $P(X = 2)$ c. $P(X > 1)$ d. $P(1.5 \leq X \leq 4)$

Solution:

a. The cumulative probabilities in the table allow us to declare $P(X \leq 3) = .928$. This value is the area in the graph under the curve from $x = .5$ to $x = 3$ and represents the probability of receiving an order in three minutes or less from the time it was placed.

b. $P(X = 2) = 0$, since this is a point probability.

c. From the complementary law of probability,

$$P(X > 1) = 1 - P(X \leq 1) = 1 - .398 = .602$$

Roughly 40 percent of the time, service is completed within one minute. Conversely, 60 percent of the time service takes longer than one minute. Note that $P(X > 1)$ is equivalent to $P(X \geq 1)$, since X is a continuous variable.

d. We can express $P(1.5 \leq X \leq 4)$ as the difference in two cumulative probabilities:

$$P(1.5 \leq X \leq 4) = P(X \leq 4) - P(X \leq 1.5) = .981 - .639 = .342$$

□

As Example 5.11 demonstrates, the technical details of determining areas for continuous probability distributions are eliminated with a table of cumulative probabilities. Recognizing the capability of relating area to probability will facilitate all of our probability problems for continuous distributions in future chapters.

COMMENT If the graph of a continuous probability distribution takes the form of a common geometrical shape such as a triangle or rectangle, we can use the appropriate formulas for area to find interval probabilities. Exercises 5.56 and 5.77 involve this principle.

CLASSROOM EXAMPLE 5.4

Finding Interval Probabilities for a Continuous Random Variable

A continuous random variable X has a probability distribution given by the formula $f(x) = 3x^2/125$, $0 \leq x \leq 5$. For selected values of X, some cumulative probabilities are as follows:

x	0	1	2	3	4	5
$P(X \leq x)$	0	.008	.064	.216	.512	1.0

Find the following probabilities:

a. $P(X = 1)$ b. $P(X \leq 1)$ c. $P(X > 3)$

d. $P(2 \leq X \leq 5)$ e. $P(-1 \leq X \leq 3)$ f. $F(4)$

g. Suppose the cumulative distribution function is $F(x) = x^3/125$. Find $P(X \leq 2.5)$.

SECTION 5.7 EXERCISES

5.52 By evaluating the function $f(x) = .19 - .018x$, $0 \leq x \leq 10$ for several values of X between 0 and 10, display the probability distribution for X in graphical form.

5.53 In order for the graph in Figure 5.13 to be a valid probability distribution for a continuous random variable X, what should be the height of the graph?

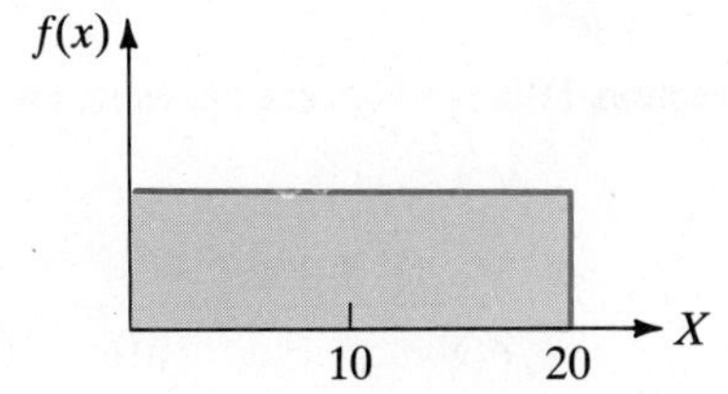

Figure 5.13 Probability Distribution for X in Exercise 5.53

5.54 A continuous random variable X has a probability distribution given by the formula

$$f(x) = \frac{3}{80}(-x^2 + 2x + 8) \qquad 0 \leq x \leq 4$$

For selected values of X, here are some cumulative probabilities:

x	0	1	1.5	2	3.5	4
$P(X \leq x)$	0	.3625	.5766	.6500	.9734	1.0

Find the probabilities in a through e:

a. $P(X > 1.5)$ b. $P(1 \leq X \leq 3.5)$ c. $P(X = 4)$ d. $P(X \geq 5)$ e. $P(X \leq 2)$

f. If $F(x) = \frac{3}{80}(-x^3/3 + x^2 + 8x)$, find $P(X \leq 3)$.

5.55 Based on the specifications of the values for X that follow, indicate whether X is a discrete or continuous random variable:

a. $x > 0$ b. $x = 0, 1, \ldots, 10$ c. $x = 0, 1$

d. $3 < x < 7$ e. $x = 1, 2, \ldots$ f. $0 \leq x \leq 1$

*5.56 Suppose the continuous random variable X has the probability distribution shown in Figure 5.14. Using the formulas for the area of a triangle and rectangle, find the following interval probabilities:

a. $P(X < 2)$ **b.** $P(5 \leq X \leq 8)$ **c.** $P(X > 4)$

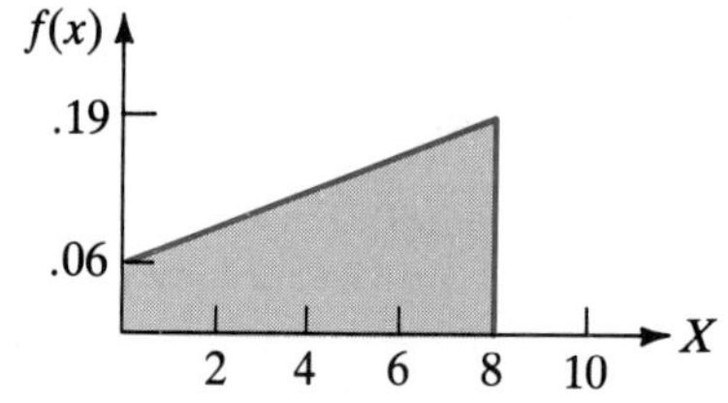

Figure 5.14 Probability Distribution for X in Exercise 5.56

5.57 The cumulative probabilities for a continuous random variable X include some of the following values of X:

x	0	.25	.50	.75	1
$P(X \leq x)$	0	.016	.125	.422	1

Find the following:

a. $P(X \leq -1)$ **b.** $P(X \leq .5)$ **c.** $P(X > .25)$
d. $P(X = .75)$ **e.** $P(X \leq 2)$ **f.** $P(X > 1)$

5.8 SUMMARY

Sampling creates uncertainty about the values of a random variable that might be observed. Probability measures this uncertainty on a zero to one scale. Combining the values and the associated probabilities into a graph, formula, or table are ways to form a probability distribution. Probability distributions can be developed from data or derived from assumptions about the sampling process. As a model for a population, a probability distribution can be characterized by its parameters: μ, σ^2, and σ.

In the next two chapters, we will present several prominent discrete and continuous probability models. These special probability distributions have well-known formulas and/or tables of probabilities that facilitate their use in the social sciences and business.

*Optional

5.9 TO BE CONTINUED . . .

. . . *In Your College Courses*

The subject of this chapter—random variables and probability distributions—likely will reappear in future business courses. Your next encounter with this material may be in a finance course, perhaps titled Introduction to Finance, Corporate Finance, or Principles of Financial Management. Popular textbooks for this course include *Fundamentals of Financial Management* (E. F. Brigham, 1986), *Essentials of Managerial Finance* (J. F. Weston and E. F. Brigham, 1982), and *Contemporary Financial Management* (R. C. Moyer, J. R. McGuigan, and W. J. Kretlow, 1984).

A financial application of our Chapter 5 material can be found in assessing the riskiness of different assets, investment alternatives, capital projects, and the like. Due to the uncertainty that accompanies these situations, financial analysts attach probabilities to the various outcomes and, in the process, develop discrete probability models. Since a typical problem may involve many assets, alternatives, or projects, some means of quantifying the risk or stability of each is needed. As Weston and Brigham (1982) state:

> *Two measures developed from the probability distribution have been used as initial measures of return and risk. These are the mean and the standard deviation of the probability distribution.*

The mean and standard deviation to which they refer are the same parameters we discussed in Section 5.5 of this chapter.

What causes trouble for many students when they encounter this material in finance are the different terms, notations, and formulas that are used in the finance textbook. Although we cannot compare all the finance texts and symbols to ours, we wish to demonstrate some of the differences using the Weston and Brigham and Moyer, McGuigan, and Kretlow texts. In Table 5.9 we outline these differences.

The important point for you to absorb is this: When business statistics concepts are applied in a functional area like finance, the vocabulary and symbols may change, but the theory remains constant. For example, the mean of the probability distribution is called the *expected value,* the *expected rate of return,* or just the *expected return* in finance. Since we deal with rates of return in finance, the variable may no longer be denoted by X but by some variations of the letter R to signify returns. Finding the mean of the probability distribution (or the expected value) remains the same: Multiply each value of the random variable by its probability and then total. For example, notice the differences between this book and Moyer, McGuigan, and Kretlow for computing the mean:

Source	Formula
This textbook, Equation 5–3	Mean $= \mu = \Sigma x P(X = x)$
Moyer et al.	Mean $= \hat{R} = \Sigma R_j P_j$

Moyer et al. use $\hat{R}$ instead of μ, R_j to represent x, and P_j as the symbol for probability.

Table 5.9 Differences Between Business Statistics and Finance Textbooks with Respect to Random Variables and Probability Distributions

	Textbooks		
Term	Shiffler & Adams	Weston & Brigham	Moyer, McGuigan, & Kretlow
Random variable	X	R_a	R
Value of the random variable	x	$\tilde{R}_a$	R_j
Probability of each value	$P(X = x)$	P_s	P_j
Mean of the probability distribution	Mean	Expected return	Expected value
Symbol for the mean	μ	$\bar{R}_a$	$\hat{R}$
Symbol for the standard deviation	σ	σ_a	σ

EXAMPLE 5.12

The rate of return on asset A is related to the performance of the economy as a whole. A financial analyst developed the following probability model for asset A's return:

State of the Economy	Return on Asset A (%)	Probability
Down	−4	.1
Average	7	.7
Up	12	.2

Find the expected rate of return and the standard deviation.

Solution:

We did not denote the columns in the foregoing table with symbols so that we can apply the material in this chapter to this financial application. The discrete random variable is X, the return on asset A, and the *expected rate of return* is μ:

$$\text{Expected rate of return} = \mu = (-4\%)(.1) + 7\%(.7) + 12\%(.2)$$
$$= 6.9\%$$

To find the standard deviation, generate a column of deviations (return–expected return) and squared deviations:

Return	Probability	Deviation = Return − 6.9%	Squared Deviation	Probability × Squared Deviation
−4%	.1	−10.9	118.81	11.881
7%	.7	.1	.01	.007
12%	.2	5.1	26.01	5.202
				17.090

The variance of the returns is 17.09, the sum of the entries in the last column. To find the standard deviation of the returns, we take the square root of the variance: Standard deviation = 4.1 percent.

□

The concepts of a probability distribution and its mean and standard deviation are born in business statistics but apply in many areas, especially finance. By mastering such topics in this course, you develop a theoretical base for future encounters with these ideas.

. . . *In Business/The Media*

As the discussion above indicates, you are likely to encounter probability distributions in a finance course. This suggests their applications in real world financial and investments practice as well. Even at the level of personal finance, we may choose between different stocks, bonds, and other investment opportunities based upon beliefs (possibly subjective) about expected rates of return and riskiness.

The insurance industry represents another area of business practice that uses probability distributions. Life insurance companies, for instance, rely on mortality tables (empirical death probabilities) as an input into setting rates. Automobile and home insurance policies also have rates that are influenced by the company's current or recent past probabilities of having to pay claims.

The gaming industry, most prominently represented by the casinos of Nevada and New Jersey, is based on what we have called theoretical or classical probabilities. The casinos offer the customers games of chance with known probabilities (at least to the casino) of various outcomes. While some players will win money, you can be certain that the expected profit for the casino for each game is a positive number!

Speaking of money, one of the monthly features in *Money* magazine is a column called "The Numbers." Past performances and current indices for stocks, long bonds, and U.S. Treasury bills (T-bills) are summarized in an "Investment Scorecard." A five-year forecast for these three primary investments also is provided in the form of a *projected compound annual return*. For long-term investors, the suggested asset mix among stocks, bonds, and T-bills to achieve a targeted rate of return is given. (See Table 5.10.)

Table 5.10 Projected Returns for Long-Term Investors and Asset Model

Projected Compound Annual Returns (%):

	T-Bills	Long Bonds	Stocks
Five-year forecast	6.1	7.5	12.8

Asset Model:

Annual Target Rate of Return (%)	Lowest-Risk Portfolios Asset Mix (%)		
	T-Bills	Long Bonds	Stocks
12	10	0	90
11	10	20	70
10	5	45	50
9	20	45	35
8	35	45	20

Source: "Investment Scorecard: Cloudy Forecasts for Sky-High Numbers," *Money* (October 1986): 34. Reprinted by permission.

To see the relationship of these data to our Chapter 5 material, let us define a discrete random variable X to be the projected compound annual return. Only three distinct values are given for this variable—the returns in the first row of Table 5.10. The figures in each row of the lower portion of the table are the necessary proportion of the portfolio invested in each type of financial instrument to yield the annual targeted rate of return. The proportions represent probabilities, and the annual targeted rate of return is the expected return.

For instance, consider the 8 percent targeted rate of return. If we form the following probability distribution for X,

Financial Instrument	x	$P(X = x)$
T-bills	6.1	.35
Long bonds	7.5	.45
Stocks	12.8	.20

then you can verify that the mean of this distribution—the expected return—is 8.07 percent, or 8 percent rounded off. Changing the probabilities in our distribution according to the ones given in Table 5.10 changes the expected return to the targeted rate of return.

As you can see, the concept of expected value is more than just a section of material in a textbook. It applies in the media too, as the *Money* magazine feature evidences.

SECTION 5.9 EXERCISES

5.58 The probability distributions of future returns for stocks P and Q are as follows:

P's Return (%)	Q's Return (%)	Probability
−3	0	.15
4	3	.36
10	8	.47
11	11	.02

Find the standard deviation and expected rates of return for stocks P and Q.

5.59 Financial analysts frequently compare two investment alternatives using the *coefficient of variation*. The alternative with the smaller coefficient of variation is judged "less risky." An individual who wishes to minimize his or her risk (known as a risk-avoider or one who is risk averse) therefore would select the less risky alternative. From Exercise 5.58, find the coefficients of variation for both stocks and recommend the better investment alternative for a risk-avoider.

5.60 The expected return for asset C is 18 percent, with a standard deviation of 12 percent. The expected return for asset D is 13 percent, with a standard deviation of 8 percent. Using the coefficient of variation criterion outlined in Exercise 5.59, which asset is riskier?

5.61 Projects J, K, and L have the following expected returns and standard deviation of returns:

Project	Expected Return (%)	Standard Deviation of Returns (%)
J	10	7
K	12	6
L	15	10

Which project is the least risky, according to the coefficient of variation criterion of Exercise 5.59? Most risky?

5.62 A firm is considering two capital investment projects R and S. The projected cash flows and the associated probabilities are indicated as follows: ("Cash flow" can be interpreted loosely as income or revenue on a cash basis.)

Project R Cash Flow	Project S Cash Flow	Probability
\$12,000	\$10,000	.1
\$21,000	\$19,000	.4
\$30,000	\$35,000	.3
\$33,000	\$36,000	.2

a. Find the expected cash flow for each project.

b. Suppose Project R costs \$50,000, and Project S costs \$53,000. Also suppose that the cash flows in the table represent *yearly* revenues with the probabilities and cash flows remaining constant from year to year. If the decision criterion for choosing between R and S is the project with the greater profitability (expected cash flow − costs) after two years, which do you recommend? (Ignore the time value of money over the two-year period.)

5.63 Refer to Table 5.10. Verify that the expected return is 11 percent for a portfolio mix of 10% T-bills, 20% long bonds, and 70% stocks.

CHAPTER 5 EXERCISES

5.64 Name the different types of random variables.

5.65 What is the goal of probability?

5.66 The probability distribution for a random variable X is

$$P(X = x) = \frac{(x + 1)^3 + 2}{44} \qquad x = -1, 0, 1, 2$$

Find the mean and standard deviation of this probability distribution.

5.67 Name the different forms of exhibiting a probability distribution for a discrete random variable.

5.68 To qualify as a valid probability distribution, what constraints must be satisfied for each of the following:

a. discrete random variables.

b. continuous random variables.

5.69 Suppose X is a discrete random variable with the following probability distribution:

x	3.7	3.8	3.9	4.0	4.1	4.2
$P(X = x)$	.17	.24	.28	.19	.11	.01

a. Find $P(X > 4)$.

b. Graph the probability distribution for X.

c. Is the distribution for X symmetric? If not, in which direction is it skewed?

d. Compute μ and indicate its value on the graph in part b.

5.70 In planning an expansion of an existing hotel, a consultant developed the following probability model for the random variable X = anticipated occupancy rate (expressed as a percentage of total rooms available) for the first year of business:

x	50	60	70	80	90
$P(X = x)$	.41	.34	.15	.08	.02

a. What type of random variable is X in the model?

b. What is the most likely occupancy rate?

c. Find the expected occupancy rate.
d. Determine the region of concentrated values for X.

5.71 Suppose X is a discrete random variable with the following probability distribution:

x	−10	−5	0	5	10
$P(X = x)$	.04	.16	.20	.42	.18

Find the mean and standard deviation.

5.72 Financial newsletters are so prolific that there is even a newsletter exclusively devoted to summarizing the information in the other newsletters. Near the end of the calendar year, analysts typically make projections about the coming year, including the percentage gain in consumer spending. A positive projection is usually a sign of optimism. Suppose the eclectic newsletter combined the various forecasts into the following probability distribution for X = percentage change in consumer spending for the next 12 months:

x	−10	−7.5	−5	−2.5	0	2.5	5	7.5	10	20
$P(X = x)$	.03	.08	.12	.07	.09	.12	.24	.13	.07	.05

a. What is the modal projection?
b. Would you classify the analysts as generally optimistic or pessimistic? Why?
c. Find the mean projected change.
d. What is the Z-score associated with the projection of −10 percent?

5.73 The probability distribution for X is given by the formula

$$P(X = x) = \frac{2^x}{14} \qquad x = 1, 2, 3$$

a. Find $P(X = 2)$.
b. Display the probability distribution for X in tabular form.
c. Graph the probability distribution for X.

5.74 Suppose X is a discrete random variable with the following probability model:

x	−3	−2	−1	0	1	2	3
$P(X = x)$	.1	.2	.1	.2	.3	.05	.05

a. What is the most likely value for X?
b. Compute μ.
c. Find $P(X < 1)$.
d. Compute σ.
e. Find the region of concentrated values for X.

5.75 In the article "Financial Planners: How to Pick the Best for You" (*Changing Times,* May 1986, p. 36), one of the recommendations is "You should see worst-case scenarios and middle of the road projections, not just the numbers that would result if everything the planner suggests turns out wonderfully." Suppose a financial planner offers you a potential investment in real estate, costing \$20,000 and yielding the following revenues (return − investment) in five years:

Scenario	Revenues
Worst case	−20,000
Middle of the road	5,000
Best case	50,000

a. Compute the expected revenue if the probabilities of the scenarios are .2, .5, and .3 (worst to best).
b. Repeat part a using probabilities of .3, .5, and .2 (worst to best).
c. Suppose you could invest your $20,000 in a no-risk venture paying 10 percent per year. If the probability of the middle of the road scenario is .50, what would the (minimum) probability of the best case scenario have to be in order for this real estate investment, specifically the expected revenue, to be financially more attractive? (Ignore the time value of money over the five-year period.)

5.76 Suppose X is a continuous random variable with the following cumulative probabilities for selected values of X:

x	0	.1	1	2	3	4	5	6
$P(X \le x)$	0	.0488	.3935	.6321	.7769	.8647	.9179	.9502

Find the following probabilities:
a. $P(X \ge -1)$ **b.** $P(X = 3)$ **c.** $P(X \le 4)$
d. $P(2 \le X \le 5)$ **e.** $P(X < 1 \text{ or } X > 5)$
f. Between which two values of X is there approximately 90% of the total probability?
g. If $F(x) = 1 - e^{-.5x}$, find $P(1.5 < X < 4.2)$.

***5.77** Fast food restaurants have a reputation for quick service. Within the corporate offices of a major chain, the research and development team has developed the following probability model for the variable Y, waiting time until service begins:

$$f(y) = .4 - .08y \qquad 0 < y < 5$$

where Y is measured in minutes.
a. What type of random variable is Y?
b. Graph the probability model.
c. Using geometry, find $P(Y < 3)$.
d. The company wishes to ensure that 95 percent of its customers are served within four minutes of entering the service queue. Are they achieving their goal with this probability model?

5.78 Identify the components of a probability distribution.

5.79 Is the following a valid probability model for a discrete random variable X? Explain.

$$P(X = x) = (\tfrac{1}{2})^x$$

5.80 For the coming year, an economist developed the following probability distribution for the variable X, inflation rate:

*Optional

Rate (%)	4	5	6	7	8
Probability	.24	.36	.15	.15	.10

a. Find the mean of this distribution.
b. Compute the standard deviation.
c. Describe the projections as skewed right, skewed left, or symmetrical.

5.81 Classify each of the following random variables as discrete or continuous:
a. X = Market share of a particular brand of 35mm SLR camera for a random sample of cameras.
b. Y = Number of new employees hired by a company in a randomly sampled three-month period.
c. Z = Waist size on the label of a pair of blue jeans in a random sampling of jeans.

5.82 Does probability reason from the sample to the population or vice versa?

5.83 To gauge the effectiveness of television advertising, a marketing research concern makes random telephone calls to inquire whether viewers remember seeing a certain ad. The variable of interest is X, percentage of viewers in the sample who recall the ad, and some of its cumulative probabilities are shown in the following table:

x	0	10	20	30	40	50	60	80	100
$P(X \leq x)$	0	.0500	.2000	.3875	.5500	.6875	.8000	.9500	1.0

a. What is the probability that at most 20 percent of the viewers recalled the ad?
b. What is the probability that more than 50 percent of the viewers recalled the ad?
c. What is the probability that between 10 percent and 40 percent of the viewers recalled the ad?
d. What value of X corresponds to the 80th percentile?
e. Between what two values of X is there 90 percent of the probability?

5.84 A major league baseball player has an attendance clause in his contract that ties certain bonus money to his team's attendance at home games. Each game in which the attendance is at least 30,000 but no more than 34,999, the player receives $1,000; between 35,000 and 39,999, he receives $2,000; between 40,000 and 44,999, he receives $4,000; and 45,000 and over he receives $8,000. Last season, the home team had only three games in which the attendance was 45,000 or more, as well as the following attendance figures:

Attendance	Number of Games
30,000–34,999	12
35,000–39,999	5
40,000–44,999	3

a. If the team plays 81 home games, construct a probability distribution for the player's bonus money per game this year based on last season's attendance figures.
b. Find the average amount per home game the team can expect to pay the player in bonus money because of the attendance clause.
c. Find the region of concentrated values for the bonus money per home game.
d. Use your answer in part b to estimate the team's *total* bonus obligation to the player for the entire season.

*5.85 The probability distribution for the discrete random variable X is as follows:

x	1	2	3	4	5	6
$P(X = x)$	.02	.17	.11	.35	.06	.29

Use the results of Exercise 5.48 to find the mean of this probability distribution.

5.86 In Section 5.4, we used two laws of probability to solve interval probability problems involving a discrete random variable. Name these laws and explain how they are useful.

5.87 The formula $P(X = x) = (3x - 2)/22$ can be used to find probabilities when $x = 1, 2, 3$, or 4.

a. Find the probability when $X = 2$.
b. What is the probability when $X = 4$?
c. What is wrong with the result if the formula is used to find the probability when $X = 0$?

5.88 A grocery store's delicatessen bakes several large sheet cakes each morning. Customers buying the cakes can have "Happy Birthday" or other messages written on top of the cake at no charge. The deli manager has developed the following distribution of X, daily demand for sheet cakes, over a period of time:

x	2	3	4	5	6	7
$P(X = x)$	.05	.10	.35	.30	.15	.05

a. What is the average value of X?
b. What is the standard deviation of X?
c. Assume cakes sold generate a profit of \$7.00 each but that cakes unsold at the end of the day are thrown away, representing a \$3.60 cost for the ingredients for each cake discarded. Suppose the store would like to bake a fixed number of cakes, C, each day—the most profitable number over the long haul. For instance, if they adopt the policy that C should be 2, daily profit always will be \$14 (\$7 times 2, less a discard loss of 0). On the other hand, if C is fixed at 7 cakes, there is a probability of .05 that 5 cakes are discarded, plus a probability of .10 that 4 cakes are discarded, and so forth for a \$49 profit. Verify that for the fixed rule of baking 7 cakes each day, the expected profit is \$23.03.
d. Determine the daily expected profit for $C = 3, 4, 5$, and 6. What value for C do you recommend the deli adopt?

5.89 What is the difference between a probability and a probability distribution?

5.90 One of the first states to operate a lottery was New York. The early version of their game sold tickets for 50 cents each. After each one million tickets were sold, a drawing was held and prizes were awarded to the lucky ticket holders. At each drawing, there were the following results:

Number of Winners	Amount
900	\$50
90	\$500
9	\$5,000
1	\$50,000

a. If you bought one ticket, what is the probability that you would win a prize?
b. What was the average payoff of a single lottery ticket?
c. What was the average payoff of a winning lottery ticket?
d. From the game player's point of view, what is the "fair" price to pay to play the game (buy one ticket)?

5.91 In Section 5.1, an example was presented involving the selection of one Bingo chip from a collection of 75 similar chips. The probability of drawing the letter *B* on the first selection was given as 15 out of 75. What philosophy of probability was used to arrive at this figure?

5.92 Which of the following are valid probability distributions for X:

a.

x	0	1	2
$P(X = x)$	.49	.42	.09

b. $f(x) = .5 \quad 0 \le x \le 20$
c. $P(X = x) = .2 \quad x = 1, 2, 3, 4, 5$

5.93 What parameters of a probability distribution are used in finance to measure return and risk?

5.94 Suppose Figure 5.15 is an accurate representation of the probability distribution for a discrete random variable X. Also assume that k is a fixed but unknown number. Find, if possible, the value of k.

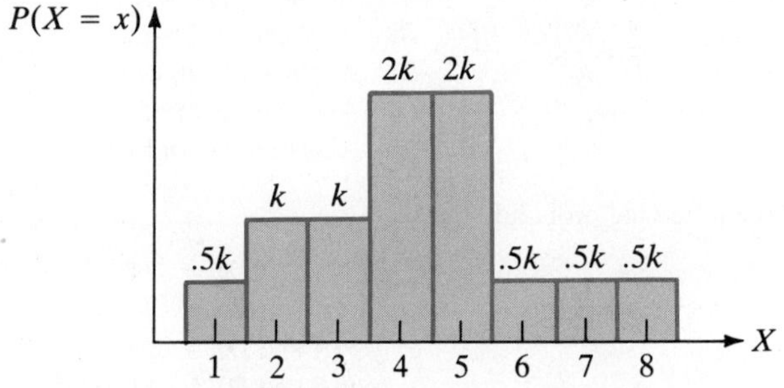

Figure 5.15 Probability Distribution for *X* in Exercise 5.94

5.95 Suppose the probabilities that the closing Dow Jones Industrial Average (DJIA) is up ($x = 1$), down ($x = -1$), or the same ($x = 0$) as the previous day's close are given by the following model:

$$P(X = x) = \frac{4x^2 + x + 1}{11}$$

a. Find the probability that the DJIA closes up on a given day.
b. What is the probability that the DJIA does not close down on a given day?

***5.96** The results of the first *USA Today*/Cable News Network poll, conducted nationwide September 8–13, 1987, were published in the September 30, 1987, edition of *USA Today*. In the accompanying article, "Reagan Is Liked More Than Policies," the following statement was made: "Half the respondents say they're financially better off since Reagan took office. Most likely to say that: those under 25. Least likely to say that: those 65 and older."† Define the following events:

R: Under 25 years of age
S: 65 years of age and older
T: Better

Event T is one of the responses to the question: Are you financially better off under Reagan? Possible responses were: better, worse, same, not sure. Decide whether the following statements are true or false for a randomly selected respondent to this poll.

a. The probability of event T is .5.
b. The probability of event T, given event R is larger than the probability of event T, given any other age group.
c. The probability of event T, given event S is smaller than the probability of event T, given any other age group.

***5.97** Answer true or false to the following statements. If false, explain why.
a. The union of two events, A and B, is denoted by $P(A \text{ and } B)$.
b. If two events are statistically independent, then $P(B|A) = P(A)$.
c. If events A and B do not intersect, then $P(A \text{ or } B) = 0$.

***5.98** Consider the following events for the variables *age* and *applicant status:*

A: Applicant is under 21
B: Applicant is 21–25
C: Applicant is 26 or older
D: Applicant is hired
E: Applicant is not hired

and the following probabilities:

$$P(D) = .34$$
$$P(A) = .36$$
$$P(A \text{ and } D) = .17$$
$$P(B \text{ and } D) = .11$$
$$P(E \text{ and } C) = .21$$

a. Develop a complete contingency table.
b. What proportion of applicants are hired?
c. What is the probability an applicant is hired if you know the applicant is 26 or older?

***5.99** Can a joint probability for two events ever exceed the marginal probability of either event? Why or why not?

***5.100** In "Consumers' Attitudes Toward Lawyers with Regard to Advertising Professional Services" by Robert Hite and Edward Kiser (*Journal of the Academy of Marketing Science*, Spring 1986), the authors surveyed residents of Arkansas about their attitudes toward lawyers advertising. The data in the accompanying table resulted when the respondents' income was cross-tabulated against the answer checked to this statement: "Advertising would help consumers make more intelligent choices between lawyers."

Income	Response		
	Agree or Strongly Agree	No Opinion	Disagree or Strongly Disagree
Low	125	31	25
Middle	105	37	45
High	63	16	40

Source: Journal of the Academy of Marketing Science, Spring 1986. Reprinted by permission.

For someone selected at random from the survey, find the following (to two decimal places):

a. P(low income|agree or strongly agree)

b. the probability that someone is in the high income group and disagrees or strongly disagrees.

c. the probability that someone disagrees or strongly disagrees.

d. the probability of disagree or strongly disagree, given that the respondent is in the high income category.

e. Does knowing that the respondent is in the high income category increase or decrease the (unconditional) probability of disagree or strongly disagree?

5.101 Using the following probability distribution of returns, find the mean, variance, and standard deviation of returns.

Return (%)	Probability
−4	.07
−2	.15
0	.18
3	.44
5	.16

REFERENCES

Brigham, E. F. 1986. *Fundamentals of Financial Management,* 4th Edition. The Dryden Press, Chicago, IL.

Daniel, W. W., and J. C. Terrell. 1986. *Business Statistics: Basic Concepts and Methodology,* 4th Edition. Houghton Mifflin Company, Boston.

Johnson, R. R., and B. R. Siskin. 1985. *Elementary Statistics for Business,* 2nd Edition. PWS-KENT Publishing Company, Boston.

Kendall, M., and A. Stuart. 1977. *The Advanced Theory of Statistics, Vol. 1,* 4th Edition. Macmillan Publishing Company, Inc., New York.

McClave, J. T., and P. G. Benson, 1985. *Statistics for Business and Economics,* 3rd Edition. Dellen Publishing Company, San Francisco.

Moyer, R. C., J. R. McGuigan, and W. J. Kretlow. 1984. *Contemporary Financial Management,* 2nd Edition. West Publishing Company, St. Paul, MN.

Shiffler, R. E. 1984. "Alternative Moment Expressions for Discrete Probability Distributions," *International Journal of Mathematical Education in Science and Technology,* 15: 394–396.

Weston, J. F., and E. F. Brigham. 1982. *Essentials of Managerial Finance,* 6th Edition. The Dryden Press, Chicago.

Chapter Maxim ***To use the binomial and Poisson probability models, define in words an appropriate random variable X and specify the values of n and π, or the value of λ.***

CHAPTER 6

DISCRETE PROBABILITY MODELS

*Optional

Objectives

After studying this chapter and working the exercises, you should be able to

1. Recognize and describe binomial sampling situations.
2. Recognize and describe Poisson sampling situations.
3. Determine the mean and standard deviation for binomial and for Poisson distributions.
4. Use a hand calculator or the reference tables to find binomial and Poisson probabilities.
5. Explain the concept of acceptance sampling as a business application of the binomial distribution.
6. Use the "region of concentrated values" idea to determine binomial and Poisson sampling results that are unlikely to occur.
7. Use the Poisson to approximate the binomial distribution.

*8. Use the Binomial and Poisson Probability Distribution processors to compute probabilities and solve exercises.

In the previous chapter, we introduced the concept of probability distributions for random variables. We noted that while there is a limitless number of possible probability distributions, we can generally classify our random variables as being either *discrete* or *continuous*. In this chapter, we will be concerned with learning to understand and use the most important discrete distributions. In the following chapter, we will focus our attention on the most commonly used continuous distributions.

Although a variety of discrete distributions exists, there are a few that have widespread applications in business which we will study in some detail. These distributions are used so often that special formulas and reference tables for them have been developed to make these distributions easier for us to work with.

6.1 THE BINOMIAL DISTRIBUTION

Without question, the most widely used discrete probability distribution is the binomial distribution. As its name suggests, the binomial is used in sampling situations where two and only two results are possible for each entity sampled, or trial. Consider the following scenarios as illustrations of having two possible results for each item sampled:

- Situation: Listing results of vehicles taking a city's exhaust emissions test.
  Results: Each vehicle passed or failed the test.

*Applies to optional section.

- Situation: An auditor is verifying a firm's accounts payable balances.
  Results: Each account balance will be found to be incorrect or correct.
- Situation: A quality assurance inspector selects finished parts from a production line.
  Results: Each part is determined to be within specifications or not.
- Situation: A credit manager is evaluating applications for her store's charge card.
  Results: Each applicant is issued a charge card or is not issued a charge card.

EXAMPLE 6.1

A manufacturer of tools and supplies for the do-it-yourself automobile owner conducted a market research survey of 650 randomly selected adult auto owners. One question on the survey listed a certain minor maintenance item and asked the respondent to indicate his or her favorite method of handling the item. The results were as follows:

Method of Handling the Problem	Count
Would do the maintenance myself	208
Would take car to service station	260
Would take car to dealer for service	182
	650

On the face of it, this is not a binomial sampling situation since each of the 650 sample entities could result in one of three, not two, possible responses. However, the survey sponsor may be interested in gauging the size of the do-it-yourself market for the maintenance item in question, and for the company's purposes the above data could be viewed as

Method of Handling the Problem	Count
Would do the maintenance myself	208
Would not do the maintenance myself	442
	650

We now have two possible outcomes into which each survey participant could be classified.

□

While all the above situations may be examples of binomial problem settings, not all two-outcome-possible situations qualify. Technically, for the binomial distribution to be applicable in a given set of circumstances, the following conditions must be present.

Binomial Sampling Conditions

1. Each of n sample entities (trials) has two possible outcomes; we will refer to these outcomes generically as "success" and "failure."
2. The probability of success, denoted by Greek letter π, remains the same for all entities. The probability of failure, denoted by $(1 - \pi)$, is constant as well.
3. The outcomes are independent; this means that the result from any one entity does not affect the result from any other entity.

If these conditions are met, then the random variable X, defined as the number of successes observed in the n sample entities, is called the binomial random variable.

Minor Violations of Binomial Conditions

In practice, some two-outcome-possible sampling situations do not *perfectly* meet conditions 2 and 3. However, we will not be concerned if these conditions are still approximately satisfied. For instance, assume that a shipment (population) of 4000 parts is received at an assembly plant and that a random sample of 30 parts is to be selected for inspection. Let's imagine that the shipment contains 100 defective parts. We would then say that the probability of the first part inspected being defective is 100/4000 or .0250. But the probability of the second part being defective will differ from .0250 very slightly, depending on the result of the first part. If the first part tested is defective, then the chance of the second being defective is 99/3,999 or .024756; on the other hand, if the first part is a good one, the chance of the second being defective is 100/3,999 or .025006. Although we would have to carry our work out several decimal places to notice it, the probabilities do shift from one entity or trial to another, and strictly speaking this violates conditions 2 and 3. But as you can imagine, these minor variations have negligible impact on any results we may be interested in; for all practical purposes we can say that the probability of a defective part remains at .025 throughout the 30 trials. A commonly applied rule of thumb is that the binomial distribution can be used when the sample size is small (5 percent or less) compared to the size of the population.

Rule of Thumb

Binomial sampling probabilities are not seriously affected by nonreplacement sampling conditions as long as $n \leq .05N$.

In this example, the sample of 30 is less than 1 percent of the population, so we can reasonably characterize the situation as binomial. In contrast, if a population had only 15 entities and a sample of size 6 were to be drawn, then the probabilities would fluctuate much more dramatically as sampling proceeded and the binomial distribution would not be appropriate. If the sample constitutes more than 5 percent of the population in a two-outcome application, it is best to use the hypergeometric distribution instead of the binomial. As you will see (Exercises 6.51 and 6.73), the hypergeometric is not nearly as convenient to work with as is the binomial.

COMMENTS

1. The above discussion involves sampling *without replacement* (discussed in Section 1.3). If we consider sampling *with replacement* then we do satisfy all the binomial conditions. However, sampling with replacement is usually undesirable and/or impractical. The great majority of business sampling applications are performed without replacement.
2. In *accounting,* two-outcome sampling situations are sometimes referred to as *sampling for attributes:* A sample entity is said to either possess an attribute of interest or not. For instance, an auditor may be checking purchase orders and noting the presence or absence of an authorized signature for each order.

The Binomial Formula

A given binomial sampling situation will involve a certain sample size or number of entities (n) and a certain probability (π) for obtaining a success on each entity sampled. We will have an interest in knowing the chance that some specific number of successes will be found within the n entities sampled.

For instance, suppose that 80 percent of all vehicles pass a city's required exhaust emissions test on their first attempt. We are to choose ten vehicles' records at random. Let's say we want to know the probability that our sample will happen to have six vehicles that passed the test on their first attempt. Note that our interest in the number of successes (six in this case) is overall in nature; we don't specifically care about the second vehicle tested or the seventh vehicle—we want to know the chance that overall a certain number of successes occur, without regard to specific details. The binomial formula will permit us to answer such questions.

Binomial Probability Formula

$$P(X = x) = \frac{n!}{x!(n-x)!}\pi^x(1-\pi)^{n-x} \quad x = 0, 1, \ldots, n \qquad (6\text{–}1)$$

where X represents the number of successes observed in the sample.

This expression may look imposing at first, but it is not difficult to work with. You have likely encountered the factorial symbol ! before. As a quick example of using factorials, recall that $4! = 4 \times 3 \times 2 \times 1 = 24$. You may remember that $0!$ is defined to be equal to 1.

Let's now use the binomial formula to solve the problem of determining the probability that six vehicles in a sample of ten passed the test on the first attempt. We begin by noting the following points:

In General	In This Example
—A "success" is the attribute of interest.	—A success is defined as "passing the test on the first attempt."
—Capital X denotes the binomial random variable. Capital X plays no part in the computations; it denotes a verbal description.	—X denotes the number (count) of vehicles in the sample that pass the test on the first attempt.
—Small x denotes some specific number of successes that we're interested in.	—We want x to be 6.
—n denotes the sample size or number of entities.	—$n = 10$ vehicle records to be randomly selected.
—π denotes the probability of success for each entity.	—$\pi = .80$. Eighty percent in the population pass the test.

Substituting our specific numeric values for x, n, and π into the formula, we have

$$\begin{aligned} P(X = 6) &= \frac{10!}{6!\,4!}(.8)^6(.2)^4 \\ &= 210\,(.262144)(.0016) \\ &= 210\,(.00041943) \\ &= .0881 \end{aligned}$$

That is, we can expect about a 9 percent chance that a random sample of 10 vehicle records would contain 6 successes.

Note that the first term on the right of the equal sign in the binomial formula, $n!/x!(n - x)!$, is a counting term called the number of *combinations*. Its value of 210 in the example above tells us that the six successes could occur in any one of 210 different sequences. The six successes could be the first six vehicles tested, the last 6 tested, or one of the 208 other sequences. The other term, $\pi^x(1 - \pi)^{n-x}$, or .00041943 in the example, is the probability that any specific one of these 210 combinations will occur. We can therefore view our binomial formula as

$$\begin{array}{ccccc} P(X = x) & = & (n!/x!(n - x)!) & \times & (\pi^x(1 - \pi)^{n-x}) \\ P\left(\begin{array}{c}\text{Exactly } x \text{ successes} \\ \text{in } n \text{ entities}\end{array}\right) & = & \left[\begin{array}{c}\text{Number of different} \\ \text{sequences that} \\ \text{yield } x \text{ successes}\end{array}\right] & \times & \left[\begin{array}{c}\text{Probability of any} \\ \text{one such sequence} \\ \text{occurring}\end{array}\right] \end{array}$$

EXAMPLE 6.2

Suppose 30 percent of a store's charge card customers pay off their monthly balance in full. If you sample five charge accounts at random, find the probability that three are paid in full.

Solution:

If a success is defined as an account being paid in full and X = number of successes, then we want

$$\begin{aligned} P(X = 3) &= \frac{5!}{3!\,2!}(.3)^3(.7)^2 \\ &= 10\,(.027)(.49) \\ &= .1323 \end{aligned}$$

That is, the probability is about 13 percent that three of the five accounts would be paid in full.

□

EXAMPLE 6.3

For the sampling conditions given in Example 6.2 ($n = 5$, $\pi = .30$), present a complete probability distribution in both tabular and graphical form.

Solution:

The values for the probability distribution can be determined by repeated use of the binomial formula, starting with $X = 0$ and working up to $X = 5$:

$$P(X = 0) = \frac{5!}{0!\,5!}(.3)^0(.7)^5 = .1681$$

$$P(X = 1) = \frac{5!}{1!\,4!}(.3)^1(.7)^4 = .3602$$

$$P(X = 2) = \frac{5!}{2!\,3!}(.3)^2(.7)^3 = .3087$$

$$P(X = 3) = \frac{5!}{3!\,2!}(.3)^3(.7)^2 = .1323$$

$$P(X = 4) = \frac{5!}{4!\,1!}(.3)^4(.7)^1 = .0283$$

$$P(X = 5) = \frac{5!}{5!\,0!}(.3)^5(.7)^0 = .0024$$

Table 6.1 Binomial Probability Distribution for $n = 5, \pi = .30$

x	$P(X = x)$
0	.1681
1	.3602
2	.3087
3	.1323
4	.0283
5	.0024

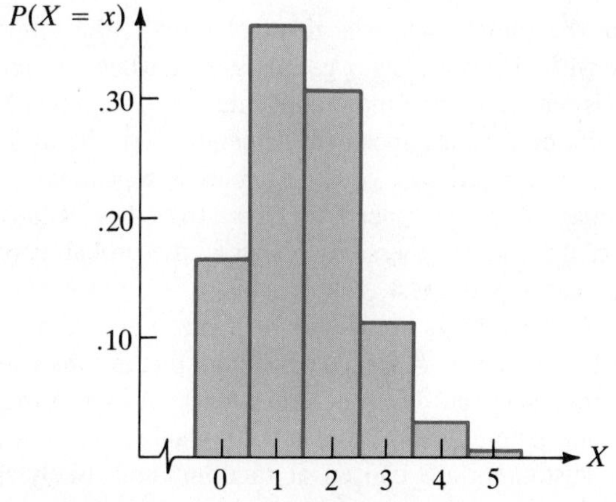

Figure 6.1 Binomial Probability Distribution for $n = 5, \pi = .30$

We can then organize these data into a table similar to those discussed in Chapter 5—see Table 6.1. Notice in this table that we had previously computed $P(X = 3)$. Also note that the probabilities in the table should sum to 1.0000. This table can easily be represented as a graph—see Figure 6.1.

□

COMMENTS

1. Either one of the two possible outcomes can be labeled as "success" with the other being "failure." As long as we match up each outcome with its respective probability of occurrence, it makes no difference how we proceed. To illustrate, refer to Example 6.2 where we found the probability that three of five accounts would be paid in full. The question could just as easily have been worded: Find the chance that two of five accounts would *not* be paid in full. If that were the case, then a success would be finding an account not paid in full, and you would want $P(X = 2)$ when $\pi = .70$. You should verify that the result is .1323, and note that this is the same as the

answer to Example 6.2. In general, a success occurs when the outcome of interest, however defined, occurs.

2. Note that if we want the probability of zero successes in n entities, the binomial formula simplifies to $(1 - \pi)^n$. On the other hand, if we want the chance of all entities being successes, the formula reduces to π^n. Example 6.3 illustrates both these cases.
3. The probability of success (π) is also referred to as the *proportion* of successes. These expressions can be used interchangeably.
4. We recommend rounding and reporting the probabilities obtained using the binomial formula to four decimal places.

Binomial Tables

When the sample size n is relatively small, the binomial formula is not difficult to work with. However, as n gets larger or when the probabilities of several outcomes are desired, it may be more convenient to obtain results from a prepared table.

Binomial tables appear in Appendix A at the back of the book. To help illustrate their use, a small part of these tables is reprinted in Table 6.2. In order to use the binomial tables, you need to know the same values as needed to work with the binomial formula—the sample size n, the probability of success π, and the number of successes of interest, x.

Binomial tables are usually organized according to sample size; Table 6.2 shows probabilities for $n = 8$. The columns in the tables are for various values of π; the rows represent the different values of X. As an example, suppose that 15 percent of cable television customers in a county subscribe to a particular premium channel. If eight customers are chosen at random, how likely is it that four subscribe to the

Table 6.2 Section of Binomial Tables for $n = 8$

		π											
n	x	.05	.10	.15	.20	.25	.30	.35	.40	.45	.50		
8	0	.6634	.4305	.2725	.1678	.1001	.0576	.0319	.0168	.0084	.0039	8	
	1	.2793	.3826	.3847	.3355	.2670	.1977	.1373	.0896	.0548	.0312	7	
	2	.0515	.1488	.2376	.2936	.3115	.2965	.2587	.2090	.1569	.1094	6	
	3	.0054	.0331	.0839	.1468	.2076	.2541	.2786	.2787	.2568	.2188	5	
	4	.0004	.0046	.0185	.0459	.0865	.1361	.1875	.2322	.2627	.2734	4	
	5	.0000	.0004	.0026	.0092	.0231	.0467	.0808	.1239	.1719	.2188	3	
	6	.0000	.0000	.0002	.0011	.0038	.0100	.0217	.0413	.0703	.1094	2	
	7	.0000	.0000	.0000	.0001	.0004	.0012	.0033	.0079	.0164	.0312	1	
	8	.0000	.0000	.0000	.0000	.0000	.0001	.0002	.0007	.0017	.0039	0	8
		.95	**.90**	**.85**	**.80**	**.75**	**.70**	**.65**	**.60**	**.55**	**.50**	x	n
		π											

premium channel? To answer this, let a success be defined as a premium channel subscriber; then locate the $\pi = .15$ column in Table 6.2 and read down until you find the value in the row where x is 4. You should easily locate the probability of .0185, which of course is the same value that the binomial formula would yield.

EXAMPLE 6.4

Twenty-five percent of a firm's accounts receivable are "past due." You are to sample eight accounts selected at random. Letting X represent the number of accounts past due, find these probabilities:

a. $P(X = 0)$

b. $P(X \geq 4)$

c. $P(X = 2 \text{ or less})$

Solution:

a. The probability that none of the accounts will be past due is .1001, or about 10 percent. This is found by intersecting the $x = 0$ row with the $\pi = .25$ column in Table 6.2.

b. Using the additive law of probability, we have

$$\begin{aligned} P(X \geq 4) &= P(X = 4) + P(X = 5) + \ldots + P(X = 8) \\ &= .0865 + .0231 + \ldots + .0000 = .1138 \end{aligned}$$

c. The chance that two or fewer accounts are past due is .6786, obtained in a manner similar to part b above.

□

EXAMPLE 6.5

Suppose that 90 percent of all federal tax returns filed last year are free of arithmetic errors. A random sample of 8 returns is to be drawn. Defining a success as a return being error free, find

a. $P(X = 7)$

b. $P(X \leq 6)$

c. the value of X most likely to occur

Solution:

When $\pi > .50$, you will find the appropriate π value along the bottom of the table. You will also use the x column on the right side of the page. (The colored ink on these bottom and right side headings is to remind you that they are to be used together.)

a. The probability of 7 returns being error free is .3826.

b. The chance of 6 or fewer being error free is

$$.1488 + .0331 + .0046 + .0004 = .1869$$

c. For $n = 8$ with $\pi = .90$, the most probable value of X is 8. The chance of eight successes (.4305) is larger than any other value in the column.

□

COMMENTS

1. All individual table values are accurate to the fourth decimal place. An actual probability of .00002, for instance, will be rounded in the tables to .0000. An actual probability of .00166 would appear as .0017 in the tables. Due to the effects of rounding, each column's entries may not sum exactly to 1.0000.
2. Due to rounding effects and the fact that many problems can be solved in more than one way, you may get an occasional answer which differs by .0001 from what your classmates arrive at or from the answers that appear in the book.
3. Binomial tables in this and other books cannot anticipate all possible combinations of n, π, and x. For instance if you want $P(X = 10)$ when $n = 23$ and $\pi = .332$, there may be no tables in existence to help you. However, you can approach this problem in one of three different ways:
   a. Use the binomial formula (Equation 6–1). Although a bit tedious to use for large n, it will give the proper answer if you avoid arithmetic errors.
   b. Use a computer program to do the work for you. One such program is included in the software accompanying this book and is briefly discussed later in this chapter.
   c. Use an approximation method. One such approach will be illustrated in the next chapter.

Application: Acceptance Sampling

Suppose a manufacturing firm receives a large shipment of parts from a particular supplier. Before accepting the shipment, the firm wants to examine some of the parts. Since a complete inspection of all parts (a census) would be too time-consuming and costly, a sample is used to decide whether the shipment's quality level is acceptable. The firm's procedure is to sample 25 parts chosen at random and then employ this rule:

- If the sample contains three or fewer defectives, accept the shipment.
- If the sample contains four or more defectives, refuse the shipment.

The scenario above illustrates an example of *acceptance sampling,* a widely used quality monitoring procedure. In its simplest form, a sample of predetermined size is drawn from a larger group of interest. The entities sampled are classified as defective or acceptable. If the number of defectives exceeds some set number, the entire lot is refused; otherwise it is accepted. Acceptance sampling can also be applied to work in progress as it moves from one department to another within a production facility.

EXAMPLE 6.6A

Suppose a firm receives a shipment which overall contains 5 percent defective. Using the acceptance sampling plan criteria above ($n = 25$; refuse if 4 or more defectives), how likely is it the shipment will be accepted? What is the probability the shipment will be refused?

Solution:

Let X = the number of defective parts. Using the binomial tables in Appendix A for $n = 25$ and $\pi = .05$, we can find

$$\begin{aligned} P(\text{accept the shipment}) &= P(X \leq 3) \\ &= P(X = 0) + P(X = 1) + P(X = 2) + P(X = 3) \\ &= .2774 + .3650 + .2305 + .0930 \\ &= .9659 \end{aligned}$$

The probability of refusing the shipment is then $1 - .9659$, or .0341.

□

EXAMPLE 6.6B

Rework the problem, assuming that the shipment's defective rate is 35 percent.

Solution:

Proceeding as above but using the $\pi = .35$ column of the tables, we find

$$\begin{aligned} P(\text{accept the shipment}) &= .0000 + .0003 + .0018 + .0076 \\ &= .0097 \end{aligned}$$

Thus, there is about a 1 percent chance the sample would lead to the decision to accept the shipment and about a 99 percent chance of refusal. It is likely that a shipment with 35 percent defective would be considered a very poor quality lot. If so, there is about a 1 percent risk of accepting a low-quality lot with this sampling plan. Risk of an undesired outcome is unavoidable when we make decisions based on sample evidence. The risk of error can usually be reduced to zero only by conducting a census.

□

■ Characteristics of a Binomial Distribution

The binomial tables show us probability distributions for various combinations of n and π. A limitless variety of different binomial distributions exists; a few are shown in graphical form in Figure 6.2. Even though many binomial probability distributions exist, they all share certain common properties.

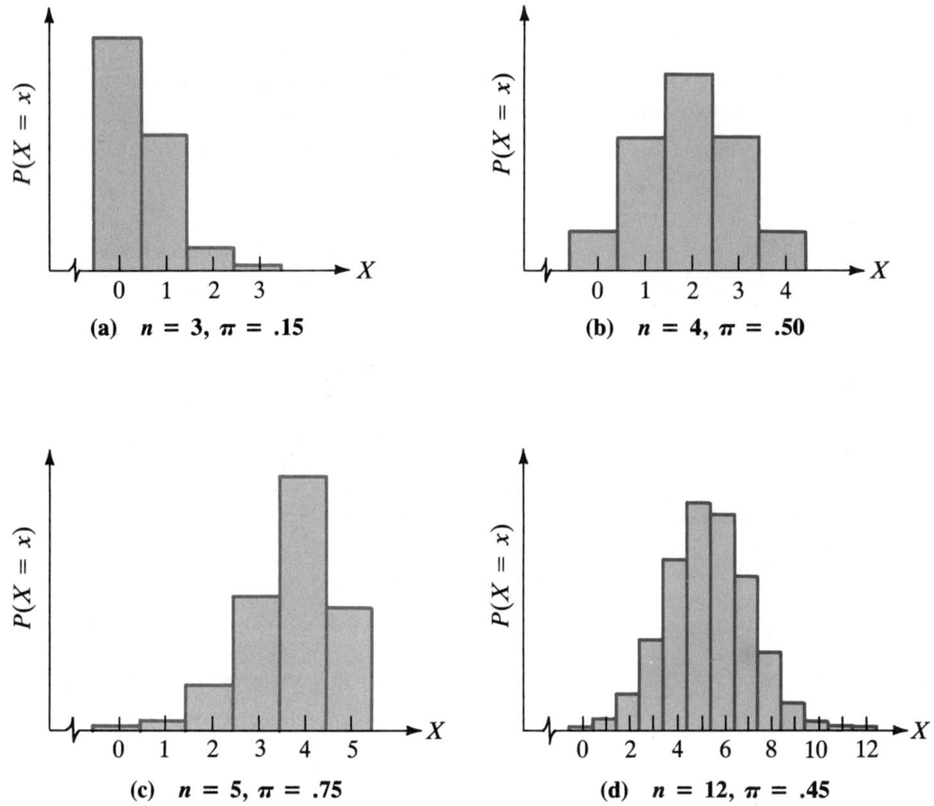

Figure 6.2 Binomial Distributions for Different Combinations of n and π

Mean and Standard Deviation For any binomial sampling application, the mean and standard deviation are easily determined if you know the number of trials or entities, n, and the probability of success, π. The values of n and π are referred to as the *parameters* of the binomial distribution.

Binomial Parameters

Mean: $\mu = n\pi$ (6–2)

Standard Deviation: $\sigma = \sqrt{n\pi(1 - \pi)}$ (6–3)

Shape The binomial distribution will be skewed left, skewed right, or symmetrical depending on the value of π. When $\pi < .50$, the distribution will be skewed right as in panel (a) of Figure 6.2. When $\pi = .50$, the distribution will be symmetrical, as in panel (b). Panel (c) shows that a skewed left distribution results when $\pi > .50$. Panel (d) illustrates that even if $\pi \neq .50$, the binomial distribution becomes nearly symmetrical as n increases.

EXAMPLE 6.7

Suppose that 70 percent of mortgage loan applications at a major savings and loan are approved. For a sample of 15 applications, determine the mean, standard deviation, and shape of the distribution of X, where X = number of applications approved.

Solution:

The expected number of approvals is $\mu = n\pi = 15(.7) = 10.5$ approvals. The standard deviation is $\sigma = \sqrt{n\pi(1 - \pi)} = \sqrt{15(.7)(.3)} = 1.8$. With $\pi > .50$, the distribution will be skewed to the left.

It should be noted that the mean value of 10.5 approvals cannot occur in any one sample of size 15. A given sample of 15 loan applications will have 8 or 12 or some other *integer* number of approvals. To help interpret the mean value of 10.5, you might imagine that a large number of samples, each of size 15, is drawn. The number of approvals in these different samples would vary. The value of 10.5 would then represent the mean number of approvals across *all* the samples; the standard deviation of 1.8 approvals would measure the dispersion among these different samples.

□

EXAMPLE 6.8

For binomial sampling with $n = 2$ and $\pi = .60$, show that the general formula for the mean of any discrete random variable (Equation 5–3) gives the same result as the special case formula for the binomial distribution (Equation 6–2).

Solution:

Equation 6–2 quickly tells us $\mu = n\pi = 2(.6) = 1.2$. Equation 5–3 is more cumbersome to work with here, since it is for *all* discrete distributions, not just the binomial. From the binomial tables in Appendix A, for $n = 2$ with $\pi = .60$, we have

x	0	1	2
$P(X = x)$	.16	.48	.36

Using Equation 5–3, we have

$$\begin{aligned}\mu &= \Sigma x P(X = x) \\ &= 0(.16) + 1(.48) + 2(.36) \\ &= 1.20\end{aligned}$$

□

Example 6.9 is based on the authors' consulting experience and illustrates how knowledge of the region of concentrated values can be helpful in anticipating the results of a particular sampling situation.

EXAMPLE 6.9

A woman planned to sell her relatively small but profitable mail-order catalog business. She had about 12,000 customers, each being represented by a 5 × 8 index card (sorted alphabetically in shoe boxes, by customer last name) upon which she had recorded the customer's address and ordering history.

When one potential buyer of the business asked the owner what proportion of customers had ordered goods within the previous calendar year, the reply was "between 40 and 45 percent." Suppose, at the prospective buyer's suggestion, a sample of 400 customers is chosen by a random selection method to see whether the more conservative number (40 percent) is realistic. Should we be willing to believe that 40 percent of the population has ordered within the previous year when the sample count is found to be 131 customers?

Solution:

Let X = number of customers having ordered within the previous year. Assuming the 40 percent figure to be valid, the expected number in the sample is $n\pi = 400(.4) = 160$ customers; the standard deviation is $\sqrt{n\pi(1 - \pi)} = \sqrt{400(.4)(.6)} = 9.8$ customers. The region of concentrated values (mean ± 2 standard deviations) is therefore 160 ± about 20 customers, or from 140 to 180 customers.

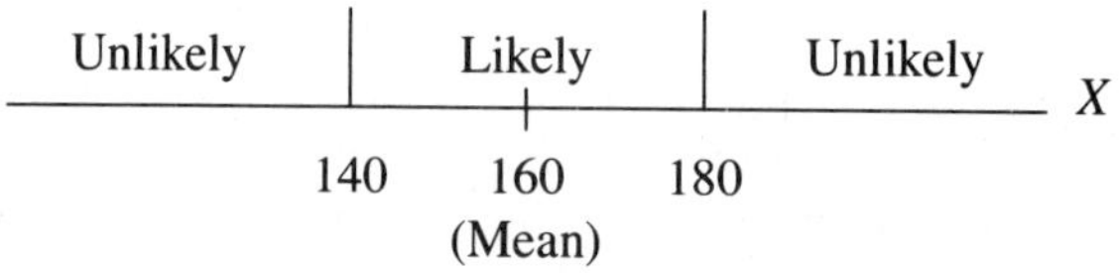

The sketch above suggests that the sample count has a high probability of being between 140 and 180 if it is correct that 40 percent of all customers have ordered in the last year. But the sample count was $X = 131$ customers—what's going on here? Do you think 40 percent is correct and that the *unlikely* result is an example of the capriciousness of random chance? Or do you think the assumption (that $\pi = .40$) that centered our region of concentrated values at $X = 160$ was overly optimistic?

□

COMMENTS

1. There is a subtle distinction between a *parameter in a formula* (a "use" parameter) and a *parameter of a probability distribution* (a "summary" parameter.) In the latter case, quantities such as μ and σ are parameters of a probability distribution since they summarize characteristics of the distribution. In the former case, quantities such as n and π are parameters in a formula since they are used to define the binomial

probability formula of Equation 6–1. A summary parameter may be a function of use parameters as Equation 6–2 demonstrates: μ, a parameter of the probability distribution, is computed by multiplying n and π, which are the parameters in the binomial formula.

2. We recommend rounding and reporting the values of μ and σ to one decimal place.

CLASSROOM EXAMPLE 6.1

Using the Binomial Distribution Tables

Before unveiling a new advertising campaign, a soft drink manufacturer determined that its products were preferred by 20 percent of consumers in a major city. Four weeks after the new ads ran, the company randomly sampled 25 consumers in this city to determine the number of consumers expressing a preference for its products. Suppose the ad campaign has had no effect on consumers' preferences.

a. Define (in words) a binomial random variable of interest.

b. Identify the following values: $n =$ ______ $\pi =$ ______.

c. What is the probability that exactly six people in the sample prefer the company's soft drinks?

d. What is the probability that fewer than two people prefer the company's soft drinks?

e. What value of X, the number of consumers preferring the company's soft drinks, is most likely to occur?

f. What is the probability that exactly 15 people do *not* prefer the company's soft drinks?

g. Determine the mean and standard deviation of X.

h. Determine the region of concentrated values for X.

i. Explain the significance of observing $X = 12$ relative to the assumption that the ad campaign had no effect on consumers' preferences.

SECTION 6.1 EXERCISES

6.1 Cite two examples, other than those in the chapter, of binomial sampling applications.

6.2 Which of the following could be binomial sampling applications:

a. for a sample of employees: how long it takes to perform a particular task.

b. for a sample of sixth graders: height measurements.

c. for a sample of drivers passing through a toll station: whether their seat belts are in use.

6.3 Which of the following could be binomial sampling applications:

a. for a sample of students: how far away their home is from campus.

b. for a sample of a student's courses currently enrolled in: whether the course is taken for a grade or pass-fail.

c. for a sample of 100 letters passing through the post office: the number that do not have a zip code as part of the address.

6.4 Use the binomial formula to find $P(X = 1)$ when $n = 6$ and $\pi = .12$.

6.5 A restaurant has found that 28 percent of its dinner parties request seating in the smoking section. Use the binomial formula to determine the chance that three of the next four dinner parties will request the smoking section.

6.6 If a maternity ward records 20 births on a given day, how likely is it that 15 or more babies are girls? (What assumption do you need to work this problem?)

6.7 For $n = 18$ and $\pi = .20$, use the binomial tables in Appendix A to find the following:
a. $P(X = 4)$ b. $P(2 \le X \le 6)$ c. $P(X \ge 7)$

6.8 Radio station WBRS is thought to have a 12 percent share of the listening audience during the morning "drive time" from 7 A.M. to 9 A.M. In a random sample of 20 people, 12 said they regularly listened to the radio between 7 A.M. and 9 A.M., while 8 did not. Of those who listen, what are the following probabilities:
a. that none tune to WBRS.
b. that one or two listen to WBRS.

6.9 Draw a sketch similar to Figure 6.1 to show the binomial probability distribution for $n = 4$ and $\pi = .35$. Label your axes.

6.10 For $n = 20$ and $\pi = .10$, use the binomial tables in Appendix A to find the following:
a. the chance of no successes.
b. the chance of at least five successes.
c. the chance of four or fewer successes.

6.11 For a sample of 15 items taken from a population where the proportion of successes is .90, use the binomial tables in Appendix A to find the following:
a. $P(X = 14)$ b. $P(X \ge 13)$ c. $P(X \le 10)$

6.12 Prepare a binomial probability distribution table (similar to Table 6.1) for $n = 2$, $\pi = .38$.

6.13 Prepare a binomial probability distribution table (similar to Table 6.1) for $n = 3$, $\pi = .61$.

6.14 A small motel has 20 guest rooms. They have accepted 25 reservations for a particular Sunday night, however, knowing from past experience that 30 percent of their reservations fail to appear. How likely is it the motel will have more guests than available rooms?

6.15 A baseball team has a .250 hitter who has five official at bats in a particular game. Find the following:
a. the chance the player will go hitless for the game.
b. the most likely number of hits the player will get.
c. the chance of getting exactly three hits in five at bats.

6.16 An acceptance sampling plan calls for a sample of 10 items, with the decision being to accept if the sample shows either zero or one defective. If a lot has 5 percent defective, how likely is it to be accepted?

6.17 Refer to the acceptance sampling plan in Exercise 6.16. Suppose the lot being sampled has 25 percent defective. What is the probability the sample result will lead to the decision to refuse the lot?

6.18 Refer to the sampling situation in Exercise 6.5. Determine the expected number of dinner parties requesting the smoking section.

6.19 Determine the mean, standard deviation, and shape of the binomial distribution used in Exercise 6.7.

6.20 Determine the mean, standard deviation, and shape of the binomial distribution used in Exercise 6.8.

6.21 Refer to the baseball player of Exercise 6.15. For games in which he has five official at bats, find the following:
a. the expected number of hits.
b. the standard deviation.
c. the chance he will make four outs.

6.22 Suppose a sample of 1250 is drawn from a binomial population where the proportion of successes is .24.
a. What is the expected number of successes?
b. What is the standard deviation?
c. What is the region of concentrated values ($\mu \pm 2\sigma$) where your sample number of successes is about 95 percent certain to fall?
d. Would a sample count of 350 successes in 1250 trials be in the likely region or the unlikely region when $\pi = .24$?

6.23 A pizza business has sent out 5431 coupons good for three dollars off their large pizza, if redeemed within a month. From past experience, they expect about 6 percent of the coupons to be used.
a. What is the expected number of redeemed coupons?
b. What is the standard deviation?
c. What is the region of concentrated values where the sample count is about 95 percent likely to fall?
d. If the actual sample count showed 315 redeemed coupons, would you consider this as being consistent with past experience?

6.24 For a set sample size, what value of π will make the binomial standard deviation as large as it can be?

*6.2 PROCESSOR: BINOMIAL PROBABILITY DISTRIBUTION

This processor can determine binomial probabilities and parameters for various combinations of n and π, up to a maximum sample size of 25. We will illustrate its capabilities for a specific problem discussed earlier in this chapter: finding $P(X = 10)$ when $n = 23$ and $\pi = .332$. Since no tables exist for this combination of n and π, one way to obtain an answer would be to use a computer to evaluate Equation 6–1.

Processor Selection and Parameter Entry

After obtaining the main menu, proceed directly to Processors and select Discrete Probability Distributions. One of the three submenu choices offered is Binomial

*Optional

```
≡                    Binomial Probability Distribution                          EDIT
  Introduction [Data]   *alculations   *esults   Exit                    |  F1=Help

    To compute probabilities using the binomial formula:

                                      n!          x          n - x
                    P(X = x) =  ───────────  π  (1 - π)
                                 x! (n - x)!

                           number of trials, n = 23

              ═══════════ Success Probability ═══════════
              ║                                           ║
              ║   Probability of success, π =  .332       ║
              ║                                           ║
              ═════════════════════════════════════════════
```

Figure 6.3 Binomial Parameter Entry

Probability Distribution—select it. (*Note:* Unlike other processors we have used thus far, this one does not require a data file.) After reading the introductory screen, move the highlight to Data and touch Enter. At this point, we are to input values for n and π. Touch Enter, then respond 23 when asked for n. Touch Enter again, then respond .332 when asked for π (see Figure 6.3).

Calculations and Results

After touching Enter, move to Calculations and touch Enter twice. When the computations are complete, move the highlight to Results and touch Enter to call up the Results submenu. To address our question (find the probability of 10 successes in 23 trials when $\pi = .332$), select the Individual Probability option. After the value of $X = 10$ is supplied, the correct probability will be presented (see Figure 6.4). Note that while the screen shows $\pi = .33$ rather than .332, the computation is based on $\pi = .332$. Other Results options can provide the user with the complete probability distribution to four decimal place accuracy (Figure 6.5) and the parameters (Figure 6.6). Note that in Figure 6.5, certain individual probabilities are listed as .0000+. This means that these point probabilities, while nonzero, still must be reported as .0000 when rounded to four decimal places.

```
≡                     Binomial Probability Distribution                    QUERY
  Introduction  *ata    Calculations [Results] Exit                    |  F1=Help

   To find P(X = x), replace x with an integer between zero and 23.

                           23!            10              23 - 10
         P(X = 10) = ─────────────── * 0.33  * (1 - 0.33)
                      10! (23 - 10)!

                   = 0.09815805

              ═════════ Probability ═════════
               Calculate another individual prob? (y/n) No
```

Figure 6.4 Calculation of an Individual Binomial Probability

```
≡                     Binomial Probability Distribution                    READY
  Introduction  *ata    Calculations [Results] Exit                    |  F1=Help
```

x	P(X = x)	x	P(X = x)	x	P(X = x)
0	0.0001	9	0.1411	18	0.0000+
1	0.0011	10	0.0982	19	0.0000+
2	0.0058	11	0.0577	20	0.0000+
3	0.0203	12	0.0287	21	0.0000+
4	0.0504	13	0.0121	22	0.0000+
5	0.0952	14	0.0043	23	0.0000+
6	0.1419	15	0.0013		
7	0.1713	16	0.0003		
8	0.1703	17	0.0001		

Figure 6.5 Probability Distribution for $n = 23$, $\pi = .332$

```
≡                    Binomial Probability Distribution                    READY
  Introduction  *ata    Calculations  [Results]  Exit                 |  F1=Help

          For a binomial random variable X, with n = 23 and π = 0.332

                    Measure                        Value

                     Mean.................... 7.6360
                     Variance................ 5.1008
                     Standard deviation...... 2.2585
                     Skewness................ right
```

Figure 6.6 Summary Measures for Binomial Sampling with $n = 23$, $\pi = .332$

6.3 THE POISSON DISTRIBUTION

Another discrete distribution with widespread applications is the *Poisson distribution,* named after a French mathematician. The Poisson is used to model the number of occurrences of some event of interest within a specified unit of measurement such as a period of time, a distance, an area, or a volume. The following examples illustrate conditions in which the random variable would likely be described by the Poisson distribution:

- Number of vehicles per hour arriving at a toll bridge
- Number of times machines in a factory break down per day
- Number of lost-time accidents per month for a given company
- Number of surface blemishes on newly painted refrigerators
- Number of typographical errors per page in a newspaper
- Number of potholes per mile of city street

At first the Poisson may seem somewhat like the binomial, but brief consideration of the circumstances should tell you which distribution is appropriate. Table 6.3 provides a comparison of these two distributions. The Poisson and binomial are alike in that they are discrete distributions where the random variable can take on only positive integer values or zero. One major difference is the absence of the sample size concept in the Poisson distribution. The binomial is used to count occurrences (successes) *for a given sample size,* while the Poisson counts occurrences *within a given interval of time,* area, and so forth. Expressed another way, when using the binomial

Table 6.3 Comparing the Binomial and Poisson Distributions

Similarities	Poisson	Binomial
Type of distribution	Discrete	Discrete
Random variable X is	Number of occurrences	Number of successes
Possible values of X	$X = 0, 1, 2, 3, \ldots$	$X = 0, 1, 2, 3, \ldots$, up to n
Differences		
Number of entities (sample size) concept	Absent	Present
Interval of time (or other measurement) concept	Present	Absent
Countable outcomes	One	Two

we can count either of two opposite outcomes: the *successes* or the *failures* in a given number of entities. But with the Poisson, we can count only one outcome: the number of *occurrences* of some event of interest. Referring to some of the Poisson examples given above, we can easily count the number of vehicles arriving at a toll bridge, but it makes no sense to speak of the opposite (the number that do not arrive). Likewise, we can count lost-time accidents as they happen, but it is meaningless to speak of the number of accidents that did not happen.

Like the binomial distribution, the Poisson requires some specific conditions to be met.

Poisson Sampling Conditions

1. We are counting the number of times some event happens within some defined continuous interval.
2. The number of occurrences in any one interval is independent of the number in any other nonoverlapping interval.
3. The probability of an occurrence is the same in all intervals of the same length.

Notice that the latter two conditions (independence and constant probability) are binomial sampling requirements as well.

The Poisson Formula and Tables

When the random variable X is defined as the number of occurrences and the above conditions are satisfied, the Poisson probability distribution formula follows.

Poisson Probability Formula

$$P(X = x) = \frac{\lambda^x e^{-\lambda}}{x!} \qquad x = 0, 1, 2, \ldots \tag{6–4}$$

where

λ (lambda) = the mean number of occurrences per interval
e = the base of the natural logarithm system (approximately 2.71828)

Unless you have a calculator with an e-power key or a computer program at your disposal, it is advisable to use Poisson tables to obtain probabilities. Such tables are provided in Appendix B at the back of the book. As we did with the binomial tables, we have reprinted a small part of these tables to illustrate their use (see Table 6.4). To use the tables, you need to know the same values as you would need to work with the Poisson formula: the mean number of occurrences per interval, λ, and the value of interest, x.

Poisson tables are organized according to λ. Table 6.4 provides probabilities for Poisson applications where the mean number of occurrences per interval ranges from 2.2 to 2.8. As an example, suppose the mean number of mechanical breakdowns in a fleet of taxicabs is 2.5 per week, and that the pattern of breakdowns follows the Poisson distribution. Use the Poisson tables to find the probability of five or more

Table 6.4 A Portion of the Poisson Probability Tables

	λ						
x	2.2	2.3	2.4	2.5	2.6	2.7	2.8
0	.1108	.1003	.0907	.0821	.0743	.0672	.0608
1	.2438	.2306	.2177	.2052	.1931	.1815	.1703
2	.2681	.2652	.2613	.2565	.2510	.2450	.2384
3	.1966	.2033	.2090	.2138	.2176	.2205	.2225
4	.1082	.1169	.1254	.1336	.1414	.1488	.1557
5	.0476	.0538	.0602	.0668	.0735	.0804	.0872
6	.0174	.0206	.0241	.0278	.0319	.0362	.0407
7	.0055	.0068	.0083	.0099	.0118	.0139	.0163
8	.0015	.0019	.0025	.0031	.0038	.0047	.0057
9	.0004	.0005	.0007	.0009	.0011	.0014	.0018
10	.0001	.0001	.0002	.0002	.0003	.0004	.0005
11	.0000	.0000	.0000	.0000	.0001	.0001	.0001
12	.0000	.0000	.0000	.0000	.0000	.0000	.0000

breakdowns in a given week. To solve this, let X = number of taxicab breakdowns per week. Locate the $\lambda = 2.5$ column and obtain $P(X \geq 5) = .0668 + .0278 + .0099 + .0031 + .0009 + .0002 = .1087$, or about an 11 percent chance of five or more mechanical breakdowns in a week.

EXAMPLE 6.10

Suppose X = the number of imperfections per square yard of carpeting follows the Poisson distribution, with a mean of 2.2. Defining an occurrence as locating an imperfection, determine the following for one square yard selected at random:

a. $P(X = 0)$

b. $P(X \leq 2)$

c. the value of X most likely to occur

Solution:

Using the $\lambda = 2.2$ column of Table 6.4, you should have little problem finding the following:

a. $P(X = 0) = .1108$

b. $P(X \leq 2) = .1108 + .2438 + .2681 = .6227$

c. In a random sample of one square yard, the most probable (highest point probability) value for X is 2.

□

EXAMPLE 6.11

A specially equipped trauma-emergency room at a hospital has been in operation for 40 weeks and has been used a total of 108 times. Assuming the weekly pattern of demand for this facility is Poisson, find the following:

a. the mean demand per week.

b. the probability the room is not used in a given week.

c. the probability the room is used seven or more times in a week.

d. the mean demand per two-week period.

Solution:

Let X = number of times per week the room is used. Then X is a Poisson random variable.

a. With 108 occurrences over 40 weeks, λ is 108/40, or 2.7 occurrences per week.

b. Referring to Table 6.4, $P(X = 0) = .0672$.

c. $P(X \geq 7) = .0205$.

d. The mean demand per two weeks would be *twice* the mean for one week, or 5.4 occurrences. Any probabilities desired for a two-week period would come from the $\lambda = 5.4$ column of the Poisson tables.

□

Characteristics of a Poisson Distribution

Just as is true for the binomial distribution, there is an infinite variety of Poisson distributions possible. A few are shown in graphical form in Figure 6.7. The following properties apply to all Poisson distributions.

Poisson Distribution Properties

Expected value (mean): λ

Standard deviation: $\sqrt{\lambda}$

Shape of distribution: Skewed right

The Poisson distribution has only one parameter, λ, while the binomial distribution has two parameters, n and π. All Poisson distributions are skewed right, with the skewness becoming less prominent as λ becomes larger (see Figure 6.7).

COMMENTS

1. Some textbooks use μ (mu) to symbolize the mean of both the Poisson and binomial distributions. We prefer the more traditional λ for the Poisson mean, since λ specifically denotes Poisson applications.
2. Poisson tables are more compact than binomial tables, since the Poisson distribution has just one parameter. It is the need for different tables for each sample size (n) that makes the binomial tables more cumbersome.
3. When solving Equation 6–4 with a calculator, be advised that some calculators may have an e^x (but no e^{-x}) key. To resolve this problem, either (a) enter a negative value for the mean prior to touching the e^x key, or (b) after entering a positive x and touching e^x, touch the inverse ($1/x$) key, thereby obtaining e^{-x}.
4. If you use your calculator to compute probabilities simultaneously with the memory function to add two or more of them, you may get a sum that differs slightly from some answers at the back of the book. The answers we provide that result from adding a series of numbers were arrived at by summing probabilities that already have been rounded back to the fourth decimal place.

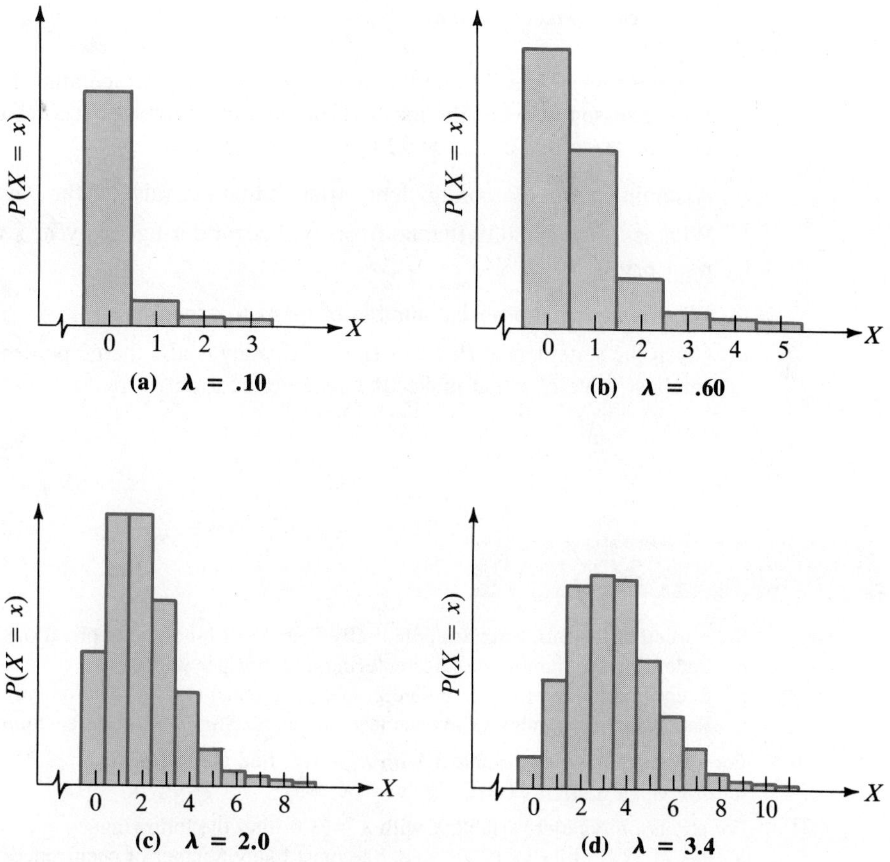

Figure 6.7 Poisson Distributions for Various Values of λ

CLASSROOM EXAMPLE 6.2

Using the Poisson Distribution Tables

"Citrus growers could be better prepared for the frosts that kill the young buds on their trees if they knew the probable number of frosts that would occur during the growing season." So says Dr. Peter Waylen, a Geography Professor at the University of Florida. Dr. Waylen discovered that the timing and number of Florida's frosts followed a statistical curve.

> *When [he] plotted the state's frost occurrences from the past twenty years on a graph, it closely matched the curve called the Poisson distribution. . . . Dr. Waylen*

*generates his probabilities by first calculating the average number of frosts a year in a particular area of Florida.**

Suppose we define X to be the number of times the temperature falls under 32°F per winter season in a specific locale. From historical weather records dating back 20 winters, we learn that a total of 32 frosts have occurred.

a. Assuming X is a Poisson random variable, find the value of the Poisson mean λ.

b. What is the probability that no frosts will occur during next year's winter season?

c. Find $P(X \le 2)$.

d. What is the most probable number of frosts to expect?

e. Given the statement, "$P(X \ge x$ frosts per year) is less than 2 percent," what is the smallest value of x that makes the statement correct?

SECTION 6.3 EXERCISES

6.25 State whether the following are potentially Poisson or binomial applications.
a. Recording the number of ships entering a harbor per week.
b. Recording the number of spoiled grapefruit in each crate of a shipment.
c. Recording the number of fire-damage claims reaching an insurance company per day.

6.26 For a Poisson random variable X with $\lambda = 1.2$, find the following:
a. $P(X = 1)$ **b.** $P(X \ge 3)$ **c.** $P(X \le 3)$

6.27 For a Poisson random variable X with $\lambda = 3.0$, find the following:
a. $P(X \ge 3)$ **b.** $P(X \le 4)$ **c.** the most likely number of occurrences

6.28 If a Poisson sampling situation has 120 occurrences over 50 intervals, find its mean and standard deviation.

6.29 For a Poisson sampling situation with 90 occurrences over 100 intervals, find the mean, standard deviation, and $P(X = 0)$.

6.30 A spray painting process produces a mean of .20 noticeable blemishes per appliance. Using the Poisson distribution, find the following:
a. the probability of no blemishes on a randomly chosen appliance.
b. the probability of at most one blemish.
c. the probability of at least three blemishes.

6.31 The mean number of vehicles that become disabled on Interstate 80 as it crosses the Mississippi River is .40 per week. Assume a Poisson distribution for X = number of disabled vehicles per week, and determine the following:
a. the standard deviation.

*Source: Dorothy Lerman, "Scientists Can Now Predict Freezes," *A Touch of CLAS* (Summer 1987):8. Reprinted by permission.

b. the likelihood that one or more vehicles will become disabled in a given week.
c. the $P(X = 3)$ for a given week.
d. the probability of three disabled vehicles for a two-week period.

6.4 USING THE POISSON TO APPROXIMATE THE BINOMIAL

Under certain binomial sampling conditions (large n, π close to 0 or 1), the Poisson distribution is a very good approximation to the binomial distribution. We can take advantage of this fact and use the Poisson tables to obtain binomial probabilities that would otherwise be unavailable or that would be time-consuming to compute.

Why and When

To demonstrate how to use the Poisson approximation, consider sampling $n = 50$ items out of a population with $\pi = .01$ defective. Let X = number of defectives, and suppose we wish to know the chance that three items in the sample will be defective. This is a binomial sampling situation, but we do not have binomial tables for $n = 50$, and use of the binomial formula to solve the problem would take some time. But we can get a reasonably accurate answer very quickly out of the Poisson tables. Simply set the Poisson mean equal to the binomial mean, and then find $P(X = 3)$ in the Poisson tables. In this case, set $\lambda = n\pi = 50(.01) = .5$ and find, in the $\lambda = .5$ column of the Poisson tables, that $P(X = 3)$ is .0126.

What would you obtain for this problem if you took the time and trouble to work this out exactly by the binomial formula? You would get .0122, different from the Poisson approximation by 4 ten-thousandths. Table 6.5 shows how similar the Poisson tables for $\lambda = .5$ and the exact binomial tables for $n = 50$ with $\pi = .01$ are. This approximation method requires a binomial sampling situation with large n and small

Table 6.5 Comparing Exact Binomial Values with Poisson Approximate Values for $n = 50$ with $\pi = .01$

	$P(X = x)$	
x	Exact Binomial Formula	Poisson Tables with $\lambda = .5$
0	.6050	.6065
1	.3056	.3033
2	.0756	.0758
3	.0122	.0126
4	.0015	.0016
5	.0001	.0002

π (or $1 - \pi$). A frequently used rule of thumb is that the sample size should be at least 20 with the probability of success (or failure) being .05 or smaller.

Rule of Thumb

Binomial sampling probabilities can be reasonably approximated by the Poisson distribution if $n \geq 20$ and either $\pi \leq .05$ or $(1 - \pi) \leq .05$.

EXAMPLE 6.12

Suppose a life insurance company has written policies on 1000 women of age 24. If the chance is .0005 that a woman in this age group will die during a given year, find the following:

a. The probability that none dies in the next year.

b. The probability that two or more die in the next year.

Solution:

With $n = 1000$ and $\pi = .0005$, the conditions for using the Poisson approximation are easily satisfied. We will need $\lambda = n\pi = 1000(.0005) = .50$. If we let X = number of women that die in the next year, the Poisson tables will then give us the following:

a. $P(X = 0) = .6065$, or about a 61% chance.

b. $P(X \geq 2) = .0902$, or about a 9% chance.

□

EXAMPLE 6.13

The owner of a computer supply store is considering stocking a new line of diskettes. The bulk price for an order of 10,000 of these disks is substantially lower than that of other disk brands she carries, but she is concerned about quality as well as price. One measure of a disk's quality is the number that are returned to the store because of a defect. Since the store offers a free exchange for defective disks, the owner wants some assurance that the new variety is not of significantly lower quality than her current offerings. To help her decide whether to purchase 10,000 disks, she first buys 200 to monitor the return rate. The return rate for the other brands she carries has averaged about 1 in 50. Suppose 6 of the 200 new disks get returned. What decision should the owner make about a bulk purchase?

Solution:

The small order of 200 can be viewed as a random sample from a large population of unknown size. Let X = the number of new disks returned. The random variable X is binomial, and if the new line of disks is comparable to the other brands, $\pi = 1/50$, or .02. With large n and small π, we can approximate this binomial sampling situation with a Poisson distribution with $\lambda = n\pi = 4$. The owner therefore knows the mean number returned out of 200 should be 4 disks, if the return rates are comparable. The standard deviation is $\sqrt{\lambda} = 2$ disks. The region of concentrated values (mean ± 2 standard deviations) is then 4 ± 4 disks, or from 0 to 8 disks.

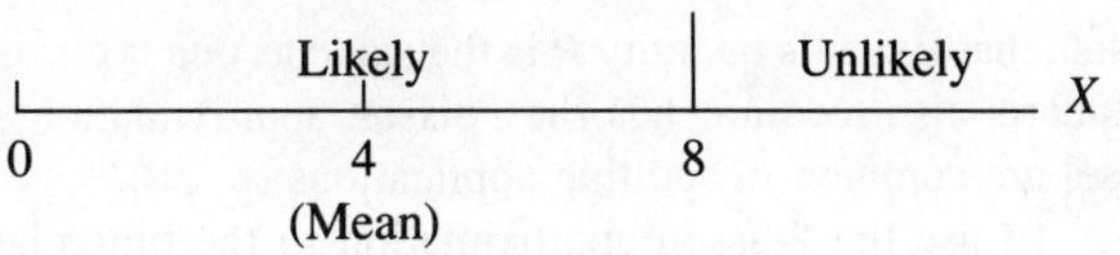

As the above sketch suggests, if the sample count of returned disks falls from 0 to 8, then it would appear that the new line's return rate is comparable to that of the other brands. A sample count above 8 is unlikely to result if the new line is to have the same return rate as the existing brands; a count of 9 or more defectives should warn the owner that the new line's defective rate is very likely too high. Since the sample contained 6 returns, she should conclude that the return rate for the new brand is consistent with that of her other brands. A bulk purchase is justified.

□

Application in Auditing Environments

Although many auditing problems conform to the binomial sampling conditions outlined in Section 6.1, auditors frequently rely on the Poisson distribution to approximate the binomial. For example, an auditor might be interested in the variable X, number of bills in error in the accounts receivable. Billing errors occur in a variety of ways: inaccurate billing quantities or prices, errors in mathematical computations, or incorrect billing addresses. The auditor's objective is to confirm (or dispute) that the proportion of a firm's bills in error is small or immaterial.

To achieve this objective, an auditor takes a random sample of accounts and classifies each as a success—no errors present—or a failure—one or more errors per bill. Even though sampling is conducted without replacement, the binomial conditions are generally in force, provided the set of accounts is fairly large. If the observed number of successes does not deviate greatly from the expected number, then the auditor declares the accounts' condition acceptable. Otherwise, he or she warns that an unacceptably high proportion of erroneous bills may be present.

Situations like the one just described are routine in auditing. Textbooks such as *Statistical Auditing: Review, Concepts and Problems* (Bailey, 1981) and *Statistical Auditing* (Roberts, 1978) describe several "attribute sampling" procedures, including the Poisson approximation to the binomial, as potential avenues for auditors to take. Bailey says,

> *For large values of* n, *the binomial probabilities are time-consuming to compute and difficult to tabulate. Computer assistance reduces these drawbacks significantly. Nevertheless, it is useful to seek more efficient methods. . . . For small values of* P

and large sample sizes n, *the Poisson probability distribution is a good approximation of the binomial probability distribution. . . .*

One of the reasons it is difficult to tabulate and compute with the binomial probability distribution is that it is based on two parameters, n *and* P. *The Poisson distribution is based on only one parameter,* λ = nP. *Numerous binomial distributions have the same* λ *value, so that many binomial computations can be approximated by a single column in a Poisson table* [*p. 106*].

Note that Bailey's quantity P is the same as our π for the binomial success probability. Roberts also recommends the Poisson approximation and notes, "conditions [for its use] are common in auditing applications (p. 30)."

To use the Poisson approximation to the binomial, we match up the centers of both distributions. Setting the mean of the Poisson equal to the mean of the binomial enables us to determine the value of the Poisson parameter from the binomial n and π. Thus, we have $\lambda = n\pi$ in order to determine the required probabilities with the Poisson probability formula. Example 6.14 demonstrates this procedure.

EXAMPLE 6.14

A medium-sized company has monthly billings of 1852 invoices. Although the company tries to maintain current addresses for each account, occasionally bills are returned due to incorrect addresses. The company would like an address error rate of 1 percent (or less). Suppose an auditor randomly samples 80 current invoices and checks the accuracy of each address.

a. What is the probability that two or fewer invoices have incorrect addresses?

b. The auditor finds three invoices with incorrect addresses. Based on this sample, is the company complying with its desired error rate?

Solution:

a. Let X = the number of incorrectly addressed invoices. Then X is a binomial random variable with $n = 80$ and $\pi = .01$ (or less). To find $P(X \leq 2)$, we *could* compute the three point probabilities—$P(X = 0)$, $P(X = 1)$, and $P(X = 2)$—with the binomial probability formula. Or, since $\pi < .05$ and $n > 20$, we could use the simpler Poisson approximation with $\lambda = n\pi = 80(.01) = .8$. Now, $P(X \leq 2)$ is found by adding the first three values in the Poisson tables under $\lambda = .8$:

$$\begin{aligned} P(X \leq 2) &= P(X = 0) + P(X = 1) + P(X = 2) \\ &= .4493 + .3595 + .1438 = .9526 \end{aligned}$$

(*Note:* This is an approximate answer. The exact binomial answer using Equation 6–1 can be shown to be .9534.)

b. Essentially, the values $X = 0$, $X = 1$, and $X = 2$ represent the region of concentrated values, since together they account for about 95 percent of the probability. This means we would "likely" see at most two bills in error in a sample of 80 invoices. If three or more bills are in error, then we have observed an event with probability only .0474, assuming $\pi = .01$. Since this is an "unlikely" event, we report that the error rate in the population may exceed the 1 percent ceiling desired.

□

The judgment about the error rate in the population of billing invoices in Example 6.14 was based on concepts of likely and unlikely regions where approximately 95 percent and 5 percent, respectively, of the probability are found. As we will learn, different amounts of probability (other than 95 percent and 5 percent) can be allocated to each region, depending on the degree of risk desired. For now, our risk is 5 percent of rendering an erroneous judgment about the error rate.

CLASSROOM EXAMPLE 6.3

Approximating Binomial Probabilities with the Poisson Distribution

Suppose you are auditing the accounts payable of a company, and in the quarter just ending the company paid over $675,000 as a result of receiving 612 invoices. You have been asked to focus your interest on the percentage of incorrect invoices, not the total dollar amount. The company has informed you that it can tolerate errors in no more than 3 percent of all the bills. Checking all 612 bills is impossible, so you sample 27 of them and record your observations as shown in Table 6.6.

a. Define a binomial random variable of interest: $X =$ ______.

b. Identify the following quantities: $N =$ ______ $n =$ ______ $\pi =$ ______.

Table 6.6 Recorded Value Versus True Value for a Sample of 27 Invoices

Invoice No.	Recorded Value	True Value	Invoice No.	Recorded Value	True Value
1	$385	$385	15	869	869
2	3,564	3,564	16	847	847
3	1,100	1,100	17	275	275
4	350	350	18	900	900
5	275	275	19	812	812
6	1,122	0	20	928	928
7	500	500	21	1,045	1,045
8	1,350	1,350	22	1,276	1,276
9	2,556	2,556	23	2,970	2,970
10	459	459	24	660	606
11	1,683	1,683	25	3,575	3,575
12	1,023	1,023	26	2,475	2,475
13	430	430	27	990	990
14	480	430	Totals	$32,899	$31,673

c. Verify that the conditions to use the Poisson approximation for the binomial distribution are satisfied.

d. Compute λ.

e. Determine the region of concentrated values, and place it on a number line.

f. Scan the Recorded Value and True Value columns in Table 6.6, and determine the value of X. Locate the value of X on your number line for part e.

g. What is your recommendation to the company regarding the accounts payable and the desired limit for billing errors?

SECTION 6.4 EXERCISES

6.32 Use the Poisson approximation to the binomial to find the following:

a. $P(X = 1)$, when $n = 70$, $\pi = .02$.
b. $P(X = 2)$, when $n = 80$, $\pi = .04$.
c. $P(X = 3)$, when $n = 90$, $\pi = .03$.

6.33 Suppose 600 people have made cruise ship reservations and that typically 1 percent of those making reservations fail to show up at the time of departure. Let X = number of no-shows.

a. Is this a binomial or a Poisson sampling situation?
b. What is the expected number of no-shows?
c. Find $P(X = 0)$ using the Poisson approximation to the binomial.
d. Find the chance of at least ten no-shows.

6.34 To audit a set of 14,000+ accounts, an auditor selected a random sample of 137 accounts. The company's objective is a maximum error rate of 2 percent.

a. What is the (approximate) probability that the sample will produce exactly three errors?
b. Find and interpret the region of concentrated values.
c. Suppose the audit reveals four accounts in error in the sample. What decision relative to the company's desired error rate is warranted?

***6.35** To approximate the binomial with the Poisson, we matched the means of both distributions so that $\lambda = n\pi$. Let us investigate the effect on the dispersion within each distribution by comparing their standard deviations. The standard deviation of the Poisson is $\sqrt{\lambda}$, while the standard deviation of the binomial is $\sqrt{n\pi(1 - \pi)}$. Let n remain fixed at $n = 50$. Compute the standard deviations for both distributions based on the following values of π, carrying computations out five decimal places. Then determine the percentage error, relative to the binomial distribution, to three decimal places.

π	Standard Deviation of Poisson, $\sqrt{\lambda}$	Standard Deviation of Binomial, $\sqrt{n\pi(1 - \pi)}$	Percentage error = $\frac{\sqrt{\lambda} - \sqrt{n\pi(1 - \pi)}}{\sqrt{n\pi(1 - \pi)}} \times 100$
.01			
.05			
.10			
.20			
.50			

*Optional

Is it apparent why we want π to be small in order to use the Poisson approximation?

*6.5 PROCESSOR: POISSON PROBABILITY DISTRIBUTION

The Poisson processor is similar in many respects to the one for the binomial distribution. It will determine point probabilities for any Poisson distribution where the mean is 10 or less occurrences per interval. To illustrate, let us work with a Poisson random variable with a mean of $\lambda = .85$ (a value not available in our Poisson table).

Processor Selection and Parameter Entry

After obtaining the main menu, proceed directly to Processors and select Discrete Probability Distributions. Then choose the Poisson option on the submenu. After reading the introductory screen, move the highlight to Data and touch Enter. At this point, we will be asked to specify a value for the Poisson mean. Figure 6.8 shows the screen after we have touched Enter and given the value of .85 for the mean. Note that since our screen cannot display Greek lambda (λ), our processor uses L to represent lambda.

```
≡                    Poisson Probability Distribution                      EDIT
 Introduction [Data] *alculations  *esults  Exit                      | F1=Help

    To compute probabilities using the Poisson formula:

                                                x    - L
                                               L   · e
                                 P(X = x) =  ----------
                                                  x!

    Where:   e = 2.718281828...  &  L = average number of occurrences.

                ================ Lambda ================
               ||                                      ||
               ||   Enter a value for lambda, L =  .85 ||
               ||                                      ||
                ========================================
```

Figure 6.8 Poisson Parameter Entry

*Optional

```
≡                        Poisson Probability Distribution                         QUERY
   Introduction  *ata  Calculations [Results] Exit                        |  F1=Help

     To find P(X = x), replace x with an integer between zero and 30.

                                                  2     -0.85
                                              0.85    · e
                                P(X =  2) =  ----------------
                                                   2!

                                         = 0.15440364

                   ================= Probability =================
                   ||                                            ||
                   ||  Calculate another individual prob? (y/n) No ||
                   ||                                            ||
                   ================================================
```

Figure 6.9 Computing an Individual Poisson Probability

Calculations and Results

After touching Enter, move to Calculations and touch Enter twice more. When the calculations are complete, move the highlight to Results and touch Enter to call up the Results submenu.

If we would like to know a specific point probability such as $P(X = 2)$, we would choose the Individual Probability option. After the value of $X = 2$ is given, the point probability is displayed—see Figure 6.9. Other results available include a complete probability distribution and summary descriptive information, similar to those generated by the Binomial processor.

6.6 SUMMARY

This chapter examines two discrete probability distributions, the binomial and the Poisson, which have many practical applications. Both are special cases within the class of discrete distributions, as Figure 6.10 illustrates. In the next chapter, we will discuss some special cases in the continuous class.

The binomial distribution is used in many sampling situations where each of n entities being examined can produce one of two possible outcomes. For relatively small sample sizes, you can determine binomial probabilities with the binomial formula or with the prepared tables. For larger sample sizes, you may use a computer program or, if π is small, a Poisson approximation. Formulas for the mean and

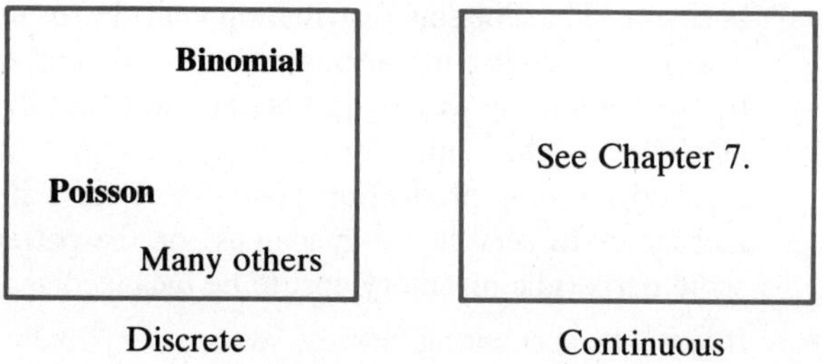

Figure 6.10 Types of Probability Models

standard deviation of the binomial and Poisson distribution are easy to work with, and direction of skewness is quickly ascertained by applying simple rules.

The procedure of acceptance sampling was described and illustrated, showing one type of situation where we might use sample information to help make a judgment about a much larger population. We will use the binomial distribution often in the chapters ahead.

The Poisson distribution is used to count occurrences of a given event over some interval of time, distance, area, or some other continuum. The absence of a fixed number of entities differentiates the Poisson from the binomial. The Poisson has numerous applications, many involving "customers" (people, cars, telephone calls, insurance claims, orders for goods, and so forth) arriving at "service facilities" (checkout counters, intersections, switchboards, claims offices, warehouses, and so forth). Poisson probabilities can also be assessed with tables or with computer programs. Poisson tables also give good approximations for binomial sampling where n is large and π is small. Use of the Poisson approximation for the binomial distribution under these conditions can save tedious computations. Looking ahead to Chapter 7, we will also see that a continuous probability model (the normal distribution) can help us approximate binomial probabilities when the sampling conditions are beyond the limited capacity of the binomial tables.

6.7 TO BE CONTINUED . . .

. . . In Your College Courses

As you enroll in other business courses, you are likely to see the binomial and/or the Poisson distributions again. Here are a few possibilities.

- If you take a course in production management or operations research, you will encounter the Poisson distribution. It is widely used in an area of study called waiting-line theory. Consider a bank where customers arrive to transact their business. The Poisson distribution may be useful to model the number of customers arriving in various time periods and to help manage the trade-off between too many waiting in line (annoyed customers) and too few in line (idle employees). The same sort of analysis might apply to the pattern of machine breakdowns in a production plant (need for scheduling a proper number of mechanics to service the machines) or the pattern of demand for expensive spare parts (idle inventory cost to be balanced against the risk of downtime).

- If you take accounting classes, you are very likely to see examples of binomial sampling (sampling for attributes) as well as instances of acceptance sampling, perhaps in an auditing context. An acceptance sampling plan to inspect invoices or balances for errors may call for a particular sample size to be drawn, to be followed by an inference as to the correctness of the population, based on the sample evidence. As suggested in the chapter, acceptance sampling can also serve a quality control role as well.
- Marketing research is an area where a lot of data is collected from consumers, especially in survey form. Many studies are done in which questions are posed with two answers possible, such as "expect to buy in the next year" and "do not expect to buy." The binomial distribution would be appropriate to analyze results of this type.

. . . In Business/The Media

You can also notice frequent examples of binomial applications in newspapers and magazines. For instance, the Bureau of Labor Statistics samples adults in the United States each month to ascertain whether they have a job. The results are released monthly in the form of the unemployment rate, a widely watched indicator of the economy's performance. Another binomial application is represented by the various polling organizations that survey the public on social and economic issues and, particularly when elections are approaching, on their voting intentions. We will discuss polling in some detail in Chapter 12.

CHAPTER 6 EXERCISES

6.36 A company that makes chain saws is going to have its gas-driven saw appear on four live commercials broadcast during a college football game. In each commercial, the saw will be given one pull. From past experience, the probability that the saw will start on one pull is .98. What is the probability that the saw will be a hero and start on one pull in all four commercials?

6.37 The owner of a minor league baseball franchise states that 20 percent of his customers are female. A random sample of 25 customers shows 11 females, however. If the owner's figure is correct, what is the expected number of females in the sample? How likely is it the sample would contain 11 or more females?

6.38 A student took a 25-question true–false exam, answering the questions randomly without reading them. If a score of 18 correct is the minimum passing grade, how likely is it the random guesser will pass the exam?

6.39 A county school system requires new students entering its schools to provide proof of vaccination against childhood diseases. Suppose that 95 percent of new students can furnish proof on the first day of class. For a random sample of 12 new students chosen on the first day of class, find the following probabilities:

a. all can furnish proof of vaccination.
b. two do *not* have proof.
c. at least one does *not* have proof.

6.40 A small commuter airline states that 90 percent of its flights arrive on time (within 15 minutes of the scheduled arrival time). If you sample 20 arrival records selected at random, what is the probability of each of the following:

a. all flights arrive on time.
b. 18 or more arrive on time.
c. 5 are late.

6.41 Refer to Exercise 6.40. If your sample of 20 records showed that only 10 flights were on time, would you have cause to doubt the statement that 90 percent of flights are on time? Use table values to make your case.

6.42 A baseball team has a .250 hitter and a .350 hitter. Assume each bats eight times (officially) in a doubleheader.

a. Which player is more likely to get exactly two hits in eight at bats?
b. Which player is more likely to get no hits?
c. Find the expected number of hits for each player.
d. Find the most likely number of hits for each player.

6.43 Suppose exactly half the population intends to vote for candidate K. What is the chance that a random sample will show 60 percent or more of voters intend to vote for candidate K?

a. when $n = 5$. **b.** when $n = 15$. **c.** when $n = 25$.

6.44 Refer to Exercise 6.43. What generalization can be drawn about the probability that the sample proportion could exceed the population proportion by 10 points or more as we let n increase?

6.45 In an acceptance sampling plan, the sample size is 20; the decision rule is to accept the lot if two or fewer defectives are found.

a. Find the probability of refusing a lot which is 5 percent defective.
b. Find the probability of accepting a lot whose proportion of defectives is .20.

6.46 Suppose that 30 percent of a firm's accounts receivable are 60 days old or older. An auditor is going to sample 120 accounts chosen at random. The firm in question is a large one, with several thousand accounts receivable.

a. Find the expected number of accounts receivable that are 60 days or older.
b. Find the standard deviation.
c. Find the region of concentrated values, where there is a 95 percent probability the sample count will lie.
d. Would a sample count of 32 accounts receivable in the 60 days or older category be viewed as a likely or unlikely result?

6.47 State whether the following are potential applications of the binomial, the Poisson, or neither:

a. number of houses sold by a real estate agent each day.
b. price–earnings ratio of a sample of utility stocks on the New York Stock Exchange.
c. number of customers in a computer store who make a purchase.

6.48 Suppose the number of chocolate chips in a bakery's cookie dough has a Poisson distribution with a mean of nine.
a. What proportion of cookies have four or fewer chips?
b. What proportion have exactly nine chips?
c. Using the rule of thumb of plus and minus two standard deviations, where is the region of concentrated values?

6.49 Assume the number of vacancies occurring on the U.S. Supreme Court per four-year presidential term is a Poisson distributed random variable with a mean of 1.60.
a. What is the standard deviation?
b. How likely is it a president will appoint three or more Supreme Court justices in a four-year term?
c. How likely is it a president will have no vacancies occur in a four-year term?
d. Suppose a president serves two full terms. What is the probability no vacancies occur on the Supreme Court?

6.50 If a lot of 7500 ball bearings is purchased from a supplier whose defective rate is .002, how likely is it that 20 or more bearings will be defective? Use the Poisson approximation.

***6.51** When you have two-outcome sampling without replacement and the sample size is more than 5 percent of the population, the appropriate distribution is the *hypergeometric*. Let

$$n = \text{sample size; } N = \text{population size}$$
$$s = \text{successes in the sample; } S = \text{population successes}$$
$$f = \text{failures in the sample; } F = \text{population failures}$$

Then the chance of s successes in n trials will be

$$\frac{S!\,F!\,n!\,(N-n)!}{s!\,f!\,(S-s)!\,(F-f)!\,N!}$$

To illustrate, assume a sample of $n = 4$ is to be taken from a population of size $N = 20$. Imagine that 17 items in the population are good; 3 are bad. How likely is it the sample will contain 2 good and 2 bad items? The chance of two good items ($s = 2$) is

$$\frac{17!\,3!\,4!\,16!}{2!\,2!\,15!\,1!\,20!} = .0842$$

Use this same problem setting to find the following:
a. the chance that all 4 items sampled are good.
b. the chance that the sample shows 3 good, 1 bad.
c. the chance that the sample shows 1 good, 3 bad.
(Your answers to parts a, b, and c should sum to 1.000 if you also include the result of .0842 given above.)

6.52 Refer to the survey data in Example 6.1. Assuming you were to sample 15 respondents selected at random, answer the following.
a. What is the expected number who would take their cars to service stations?

*Optional

b. What is the probability that the sample would contain 10 or more who would take their cars to service stations?

6.53 Explain how acceptance sampling is conducted.

6.54 Which of the following could be binomial sampling applications?

a. For a sample of 50 work shifts in a computer-assisted operating room, the number of machine failures is recorded.

b. For a sample of 100 births at a hospital, the sex of the child is recorded.

c. For a sample of fast-food restaurants, daily sales records are obtained.

6.55 Seven percent of a dentist's appointments are no-shows. In a sample of six patients chosen at random, what is the probability that three are no-shows?

6.56 A certain prestigious journal accepts for publication 8 percent of the manuscripts it receives. Suppose the journal receives 22 submissions in January of this year. Let X = the number that will eventually get published in the journal.

a. What is the expected value for X?

b. What value of X is most likely to occur?

c. The probability is at least .90 that the number published will be x or less. Find the lowest value for X that makes the previous sentence true.

6.57 Suppose the number of daily machine breakdowns requiring a certain part follows a Poisson distribution with a mean of three breakdowns per day. What minimum number of spare parts should be available at the start of a day in order to be at least 95 percent sure of not running out? To be at least 99 percent sure?

6.58 A batch production process makes pieces, each of which has a probability of .90 of being good. A customer has placed an order for six pieces. How large a batch should be made in order for the batch to have a 98 percent chance (or better) of yielding six or more good pieces?

6.59 To complete one particular schedule on the Federal Income Tax return, the taxpayer is required to make six separate computations. Assume that the binomial distribution applies to the series of computations, with the probability of an error on each calculation being 2.5 percent. What percent of taxpayers complete the schedule with no errors?

6.60 A retail store's catalog operation experiences customer returns of items on 6 percent of orders. In one particular month, 3300 orders were handled.

a. What is the expected count of orders which have items returned?

b. Should an actual count of 211 orders involving returns in this month be viewed as an unlikely result? Explain why or why not.

6.61 The number of customers arriving per minute in the ordering line for the drive-through window of a fast-food restaurant generally satisfies the Poisson sampling conditions. If the average number of customers arriving per minute is 1.33 in the 5–7 P.M. interval, what is the probability that four customers arrive during any one-minute period?

6.62 A study by Mary C. Gilly ("Sex Roles in Advertising: A Comparison of Television Advertisements in Australia, Mexico, and the United States," *Journal of Marketing*, April 1988) found that 30 percent of U.S. television commercials were for food, snacks, and soda.

a. For a random sample of ten commercials, what is the probability that half or more are for food, snacks, or soda?

b. What is the probability that none of the ten are of this type?

c. Answer parts a and b for a random sample of ten Australian commercials, using the result that about 10 percent of Australian commercials were for food, snacks, and soda.

6.63 Bruce Buchanon, Moshe Givon, and Arieh Goldman report ("Measurement of Discrimination Ability in Taste Tests: An Empirical Investigation," *Journal of Marketing Research,* May 1987) on some experiments that measured subjects' ability to distinguish between two slightly different product formulations. In one experiment, 180 subjects were given four unlabeled cola samples, two of each formulation, to see if they could correctly identify the two pairs which were identical.

a. As the authors note, the probability of a nondiscriminating subject correctly identifying the two identical pairs by guessing is 1/3. How many subjects are expected to make correct identifications if the task is so difficult that people can only guess?

b. Form a region of concentrated values for the number of correct identifications under the conditions described in part a.

c. Suppose you wish to choose between statement A ("The subjects cannot discriminate and are only guessing") and statement B ("A substantial number, but not all subjects, can discriminate"). Which statement do you choose if the experiment were to find 71 correct identifications? Why?

6.64 If X is a Poisson random variable with $\lambda = 1.88$, find the following:

a. $P(X = 0)$ **b.** $P(X \leq 2)$

c. the integer values of X contained in the region of concentrated values

6.65 If X is a Poisson random variable with $\lambda = .53$, find the following:

a. $P(X = 3)$ **b.** $P(X > 1)$ **c.** σ, to two decimal places

6.66 From historical records, the manager of a dry cleaning business has determined that an average of 12.5 orders per month are not retrieved by customers within 180 days. Assume the Poisson distribution approximately describes the random variable X = the number of orders per month that are not claimed by customers within 180 days. Is it unlikely or likely for 20 or more orders from a given month to go unclaimed? Explain.

6.67 To promote sales at a car dealership, the sales manager devised a contest for his sales representatives. The salesperson selling the most cars during the month of February would receive a substantial cash bonus. By the end of the third week, two salespeople—Clara and David—had established a commanding lead over the other representatives. Clara had sold 15 new cars to David's 12. Assume the dealership is open six days a week and both Clara and David had worked every day of the month so far.

a. What is the average number of new cars sold per working day during the month of February for Clara? For David? Round your answers to two decimal places.

b. Assuming the number of new cars sold per day is a Poisson random variable, what is the probability that Clara will sell two new cars the next working day?

c. What is the probability David will sell at most one new car the next working day?

d. Suppose both Clara and David will work all six days during the last week in February. Also suppose that we know that Clara will end up selling only one more new car in the last week. What is the probability that David will overtake her and win the sales contest?

6.68 When the editor of a newspaper decides to replace an old comic strip with a new one, he or she expects a certain amount of irate letters and phone calls from disgruntled fans of the old strip. From past experience, the editor knows that the newspaper receives an average of 6.2 complaints per day during the first two weeks after a comic strip change. Assume the complaint pattern follows the Poisson distribution.

a. What is the probability the newspaper receives only one complaint in any day during the first two weeks after a comic strip change?

b. What is the probability of at least four complaints in a day?

c. Supply the editor with a likely region for the *total* number of complaints the newspaper can expect to receive during the first two weeks after a change.

6.69 A soft drink company considers its bottling plant to be operating "in control" if 99.5 percent of all two-liter bottles are filled to within .01 liter of 2.00 liters. To verify these specifications, the quality manager samples 120 bottles at random per shift and, using a precise scale, determines the amount of beverage in each bottle. Let X = the number of two-liter bottles *not* meeting the fill specifications.

a. What type of discrete random variable is X?

b. Verify that the sampling conditions for using the Poisson approximation are in force.

c. Find λ.

d. Using the appropriate Poisson distribution, find the region of concentrated values for X.

e. If a sample of 120 two-liter bottles shows two bottles not meeting the fill specs, what should we infer as to whether the bottling process is in control?

6.70 A nuclear submarine is about to embark on a three-month voyage. A key component of its propulsion system is subject to failure, partly due to the extreme depths at which it must work. Based on fleet records, the number of failures of this component follows the Poisson distribution with an average of .80 failures per year. Taking spare components along is therefore necessary, although the components are very bulky and storage space is extremely limited. How many spare components should be taken if the captain decides the maximum risk of the submarine having to surface and call for assistance before three months have elapsed should be the following:

a. 10 percent b. 5 percent c. 1 percent d. 0.1 percent

6.71 In a professional basketball game played in March 1989, Craig Hodges of the Chicago Bulls made five three-point baskets in five attempts. Suppose that in his lifetime career, Hodges is successful on 25 percent of shots from the three-point range. Assume you are told before the game that Hodges would attempt five three-pointers.

a. What would you predict as his most likely number of made shots?

b. What probability would you assign to his making all the shots?

6.72 A municipal street department surveys the city's 52 miles of residential streets, mapping 338 potholes large enough to warrant repair. Assume that the variable X = number of potholes per mile follows the Poisson distribution.

a. What is the modal value of X?

b. What is the probability a randomly selected mile of city street has 10 or more potholes?

c. Does the interval "mean plus and minus three standard deviations" capture about 99.7 percent of the probability (as the Empirical Rule states)?

***6.73** (See Example 6.51.) In a sex discrimination case, 19 qualified military officers (15 men, 4 women) applied for four openings that were viewed as desirable. When three of the four openings were granted to women, one of the men claimed that the men were treated unfairly. Assume that by virtue of passing qualifying exams, all applicants had an equal chance at selection. Find the following probabilities, to four places beyond the decimal.

a. What is the probability that three of the four selectees would be women?

b. What is the probability that all four would be women?

c. What is the probability that all four would be men?

6.74 An airline offers two dinner entree choices on its Denver to Honolulu flight: chicken and beef. Based on past experience, 40 percent of passengers ask for chicken, and 60 percent opt for beef. A particular flight is stocked with 300 dinners in the indicated ratios, although only 275 passengers end up boarding the plane. Assume all passengers request dinner.

a. What is the expected number of requests for the chicken entree?

b. What is the standard deviation?

c. Roughly speaking, how likely is the airline to get more requests for chicken than it can handle (choose one)?

(1) Almost impossible to happen: It could happen theoretically, but not realistically.

(2) Very unlikely: It would take a sample count more than two standard deviations above expectation.

(3) Somewhat unlikely: It would take a sample count more than one standard deviation above expectation.

(4) Rather likely: The probability would be at least 50 percent.

6.75 Suppose that 15 percent of businesses with 50 or more employees have one or more facsimile machines. For a random sample of 18 such businesses, let X = the number that do have facsimile machines, and find the following:

a. $P(X = 3 \text{ or } 4)$ **b.** $P(X \geq 1)$

c. the mean and standard deviation for X

6.76 Based on extensive surveys in the previous years, a Missouri amusement park has established that 86 percent of its patrons live within a 30-mile radius. A random sample of 400 patrons is conducted during this season.

a. If the 86 percent figure still holds, define a likely (.95 probability) range of values for X = number of patrons in the sample who live within a 30-mile radius. Report your end points as the smallest and largest integers contained in the range.

b. Would a sample count of X = 366 be viewed as consistent with past experience or not?

REFERENCES

Bailey, Jr., A. D. 1981. *Statistical Auditing: Review, Concepts and Problems*. Harcourt, Brace, Jovanovich, Inc., New York.

Roberts, D. M. 1978. *Statistical Auditing*. American Institute of Certified Public Accountants, New York.

Chapter Maxim *Normal distributions with different averages and spreads fit one common mold after standardizing.*

CHAPTER 7

CONTINUOUS PROBABILITY MODELS

*Optional

Objectives

After studying this chapter and working the exercises, you should be able to

1. Describe the characteristics of a normal probability distribution.
2. Explain the process and purpose of standardizing data.
3. Determine any probability on a normal curve.
4. Find any percentile on a normal curve.
5. Use a normal distribution to approximate binomial probabilities.

*6. Execute the Normal Probability Distribution processor to verify answers and/or solve problems.

7. Describe the characteristics of an exponential probability distribution.
8. Evaluate probabilities under an exponential distribution curve.

We have just finished our study of the major *discrete* probability distributions. In this chapter, we will consider the most useful *continuous* probability distributions, giving special attention to the class called the normal distribution. The normal distribution plays a special role in both the theory and practice of statistics, and we will find many applications of normal distributions in the chapters beyond this one. As we did for the binomial and Poisson distributions, we will learn to use tables that will greatly facilitate our study.

7.1 NORMAL DISTRIBUTIONS

Quite possibly the most useful of all probability models is the one called the *normal distribution*. It has applications in many disciplines in both the natural and social sciences.

■ Appearance and Properties

Figure 7.1 shows the general appearance of a normal probability distribution for the continuous random variable X. Some of its most important features are given in the following box.

*Applies to optional section.

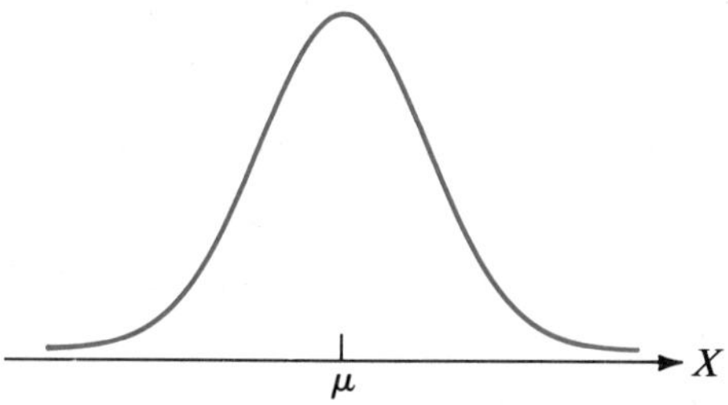

Figure 7.1 A Normal Probability Distribution

Normal Distribution Characteristics

1. The distribution is symmetrical about its mean, μ.
2. Other measures of central location, such as the median and the mode, will equal the mean.
3. The total area (probability) under the curve is 1.
4. The values of the random variable X extend from $-\infty$ to $+\infty$.
5. A normal distribution has two parameters, μ and σ. The value of μ locates the center of the distribution, while the value of σ (the standard deviation) indicates the dispersion within the distribution.
6. The normal distribution is really a *family* of curves, not just a single curve. Different members of this family exist for each possible different pair of values for μ and σ.

Figure 7.2 illustrates three members of the normal distribution family. Distributions R and S have different values for μ but the same standard deviation. Distributions S and T share the same mean, but T has more dispersion than S. All three are normal probability distributions.

Examples

One major reason normal distributions play an important role in business statistics is that they give a good approximation (model) for many data sets observed in practice. Consider the following possibilities:

- In the production of various consumer and industrial goods, the distribution of certain measurable attributes such as weight, thickness, volume, length, specific gravity, breaking strength, and so forth often follows the normal curve. In fact, many firms' process control and inspection procedures are based on the as-

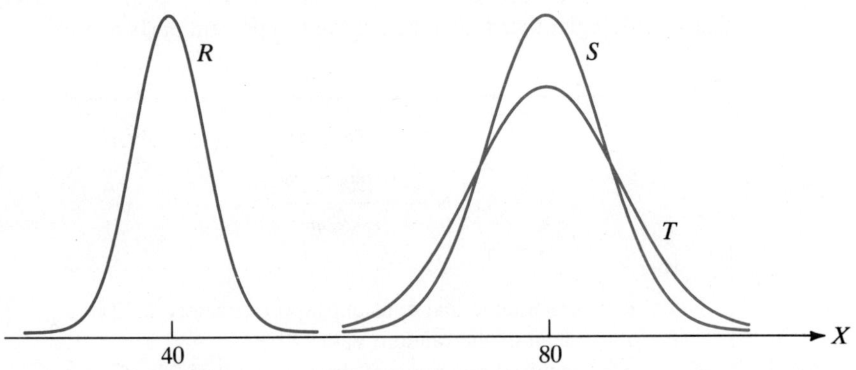

Figure 7.2 Three Members of the Normal Distribution Family

sumption that the production process will generate a normally distributed random variable. For a specific instance from manufacturing, the actual fill level of one brand of soft drink's two-liter bottles has been found to be a continuous random variable having a normal probability distribution with a mean of 2.020 liters and a standard deviation of .016 liter.

- In a similar vein, the distribution of random variables representing the service life of manufactured items such as light bulbs, television picture tubes, disk drives, oil pumps, radio batteries, and so forth often follows a normal distribution. For example, one producer of light bulbs may make a 100-watt bulb whose pattern of service life is a normal distribution with a mean of 960 hours and a standard deviation of 240 hours.
- Many human variables such as scores on aptitude tests, time required to perform a particular task, heights, percent body fat, and so forth have distributions that tend to closely resemble a normal distribution.

The above discussion is meant merely to suggest a few of the possible applications of normal distributions. The fields of economics, business, psychology, engineering, and the natural sciences offer many other examples. However, we should not blindly assume that every continuous variable follows a normal distribution. The preliminary inspection of our data by use of frequency distributions and graphical analysis can give indications as to whether a random variable is normally distributed. For instance, a histogram with obvious skewness should suggest that the random variable being represented is probably not normally distributed.

Mathematical Formula

There is a specific mathematical expression that generates the bell-shaped curve we call a normal distribution. First published in the 1700s, the formula and credit for developing the normal distribution is often attributed to the German scientist Karl Gauss, although other researchers played important parts as well.

Normal Probability Distribution

$$f(x) = \frac{e^{-(1/2)[(x - \mu)/\sigma]^2}}{\sigma\sqrt{2\pi}} \qquad -\infty < x < \infty \qquad (7\text{–}1)$$

where

e = a mathematical constant (approximately 2.72)
μ = mean of the random value X
σ = standard deviation of the random variable X
π = a mathematical constant (approximately 3.14)

In our work with normal distributions, we will rely on tables that summarize integral calculus results as opposed to working directly with the formula. Note that the random variable X in Equation 7–1 has no upper or lower bounds. Strictly speaking, this means that the tails of the normal curve approach but do not touch the horizontal axis as we move away from the mean in either direction. In practice, most variables do not range from negative infinity to positive infinity. However, this technical point does not prevent a normal distribution from being a good model for many continuous random variables.

COMMENTS

1. In Chapter 5, we defined a probability distribution for a continuous random variable as a formula $f(x)$ or its graph. Notice that Equation 7–1 is the formula and Figure 7.1 is the graph of a normal probability distribution.
2. In reality, few continuous random variables perfectly follow the normal probability distribution pattern given by Equation 7–1. Many data sets are approximately normal, however, and the normal distribution offers a practical model in such cases. When we refer to a data set as being "normally distributed," we will mean that its appearance closely matches Figure 7.1 and that Equation 7–1 will, for all practical purposes, enable us to answer questions of interest about the data.

The Empirical Rule Revisited

In Chapter 3, we presented the so-called Empirical Rule for mound-shaped, symmetrical data sets. The Empirical Rule stated that knowledge of the numerical value for the mean and the standard deviation permits us to say *approximately* what percent of the data falls within certain common distances of the mean. As you may now suspect, the basis for the Empirical Rule is in fact a normal probability distribution.

Table 7.1 Normal Distribution Probabilities

Interval	Probability Within Interval
$\mu \pm \sigma$	.68268
$\mu \pm 2\sigma$	.95450
$\mu \pm 3\sigma$	.99730
$\mu \pm 4\sigma$	.99994

In the special case where a distribution is known to follow a normal curve *exactly*, we can be a bit more precise than was the case in Chapter 3. Table 7.1 provides normal curve areas (probabilities) to five decimal place accuracy for selected intervals. Note that less than 3/10 of 1 percent of the normal distribution lies beyond ± 3 standard deviations from the mean.

While it is helpful to remember the Empirical Rule or the values in Table 7.1, we now need to turn our attention to being able to use tables to find *any* area under a normal curve. This is discussed in the next section. For continuous random variables, note that terms such as percent of the data, probability, and area are interchangeable.

SECTION 7.1 EXERCISES

7.1 Indicate whether the following pairs of values for μ and σ are members of the normal distribution family. If not, explain why.

a. $\mu = 2, \sigma = 1$ **b.** $\mu = 2, \sigma = 10$
c. $\mu = -1, \sigma = 3$ **d.** $\mu = 0, \sigma = 1$
e. $\mu = -1, \sigma = -3$ **f.** $\mu = 7.2, \sigma = .35$
g. $\mu = 1, \sigma = 0$ **h.** $\mu = 8, \sigma = 8$

7.2 Refer to the exercise above. Of those pairs of values that are members of the normal family, for which one would the following values most likely and least likely be observed?

a. a value of -8 or less **b.** a value of 24 or more

7.3 How much area is under the normal probability distribution from μ to $+\infty$?

7.4 Use Table 7.1 to find the probability under a normal curve for these intervals:

	Interval Beginning	Interval Ending
a.	μ	$\mu + 2\sigma$
b.	$\mu - 4\sigma$	μ
c.	$\mu - 4\sigma$	$\mu + 2\sigma$
d.	$\mu + \sigma$	$\mu + 3\sigma$
e.	$-\infty$	$\mu + 3\sigma$

7.2 THE STANDARD NORMAL DISTRIBUTION

As mentioned earlier, there is a *family* of normal distributions, one for each possible pair of values for μ and σ. Although there is a limitless number of different normal distributions, we can use *one* reference table to find probabilities for all of them.

Standardizing: One Size Fits All

In order to have one table to accommodate any normal distribution, we need only to standardize our data, using the following formula.

The Z-transformation for X

$$Z = \frac{X - \mu}{\sigma} \qquad (7\text{–}2)$$

This simple equation converts our random variable X into a new random variable Z that has the very convenient properties of being normal with $\mu = 0$ and $\sigma = 1$. Since the one normal curve table we will employ presumes that our normal variable has $\mu = 0$ and $\sigma = 1$, we will need to standardize our data prior to referencing the normal table. Figure 7.3 illustrates the concept of standardizing.

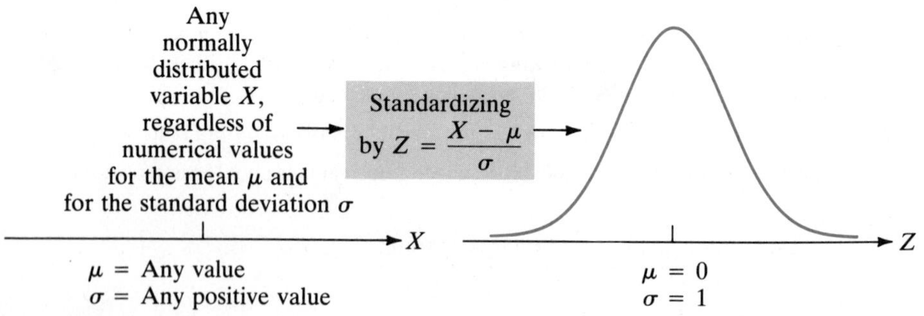

Figure 7.3 Creating the Standard Normal Variable, Z

EXAMPLE 7.1A

For a normal distribution with $\mu = 72$ and $\sigma = 8$, express the point $X = 84$ as a standardized value Z.

Solution:

Simply substitute the known quantities into Equation 7–2 and solve for Z:

$$Z = \frac{X - \mu}{\sigma}$$

$$= \frac{84 - 72}{8} = \frac{12}{8} = 1.50$$

As we mentioned in Chapter 3, the standard score Z represents the distance, expressed as the number of standard deviations, that a given point X is away from the mean. Thus, we can state that the point $X = 84$ is 1.5 standard deviations above the mean.

□

EXAMPLE 7.1B

For the same distribution, convert $X = 54$ into its corresponding standard score Z.

Solution:

As in Example 7.1A, substitute and solve for Z:

$$Z = \frac{54 - 72}{8} = \frac{-18}{8} = -2.25$$

The minus sign indicates that the point of interest is below the mean. In this problem, our value of 54 is 2.25 standard deviations below the mean.

Figure 7.4 offers a graphical view of standardizing for Examples 7.1A and 7.1B. In this figure, the top graph labeled with the X-axis represents the unstandardized data; the Z-axis at the bottom applies after we have standardized using Equation 7–2.

□

We will also find it convenient to be able to work the standardizing formula in the other direction: to change a standard score Z back into its equivalent value of X. This ability will be especially useful when we are interested in determining percentiles on the normal curve.

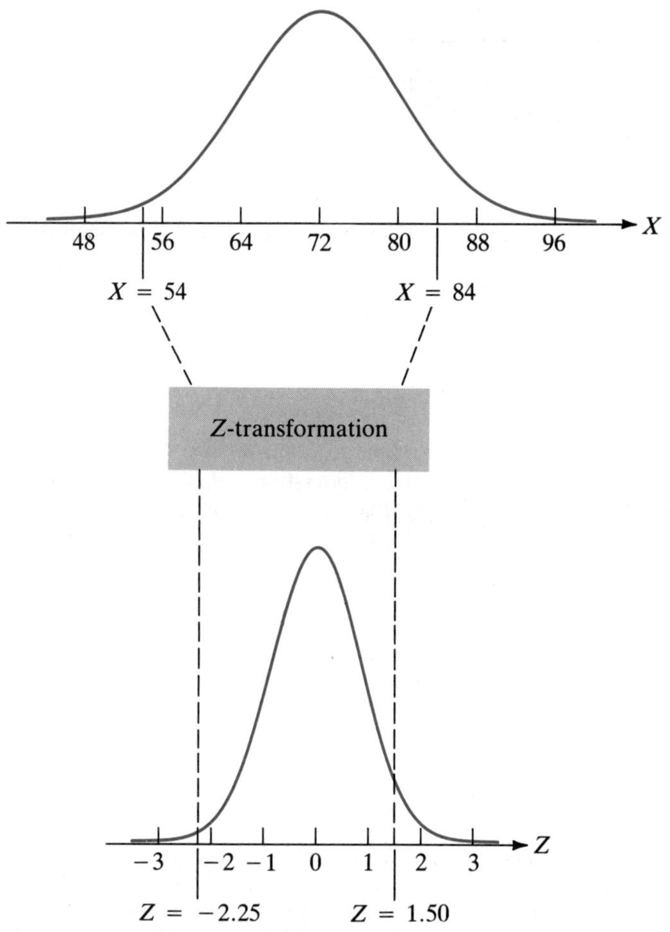

Figure 7.4 Graphical Representation of Example 7.1

EXAMPLE 7.2A

Express $Z = 1.35$ as an unstandardized score from a normal distribution with $\mu = 760$ and $\sigma = 140$.

Solution:

Our Equation 7–2 is needed here, although we will solve for X in this situation since X is the unknown quantity:

$$1.35 = \frac{X - 760}{140}$$

A little algebra should give us the answer of $X = 949$. Alternatively, we could rewrite Equation 7–2 as follows to work a problem of this type:

The *X*-Reversion

$$X = \mu + Z\sigma \tag{7–3}$$

□

EXAMPLE 7.2B

For the same distribution ($\mu = 760$, $\sigma = 140$), change $Z = -.80$ into its corresponding value of X.

Solution:

Using Equation 7–3, we have

$$\begin{aligned} X &= 760 + (-.80)140 \\ &= 648 \end{aligned}$$

□

The *Z*-Table

Our table of the standardized normal distribution appears in Appendix C as well as on the inside front cover of the book. The graph at the beginning of the *Z*-table shows us how the table is set up: Each entry in the table represents the cumulative probability, to four decimal-place accuracy, from negative infinity over to the *Z*-score of interest. The values of Z that we can look up range from $Z = -3.49$ to $Z = +3.49$. We will rarely interpolate between table values. For instance, we will tend to treat a *Z*-score of 1.788 as being $Z = 1.79$ as opposed to interpolating between the table values for $Z = 1.78$ and $Z = 1.79$: Compute and report *Z*-scores to *two* decimal place accuracy.

COMMENTS

1. The normal probability distribution with $\mu = 0$, $\sigma^2 = 1$, and $\sigma = 1$ is called the *standardized normal probability distribution*, or *standard normal*, for short. The random variable associated with the standard normal is traditionally denoted by Z.
2. We should note that the *Z*-table differs from the binomial and Poisson tables of the previous chapter in an important way. Binomial and Poisson tables give the probability that X equals some specific value (a point probability). However, since the normal is a continuous distribution, any point probability is zero. Accordingly, the normal table reads cumulative probabilities instead of point probabilities; in particular, the *Z*-table gives the probability that our variable is *equal to or less than* some value.

3. The software that accompanies this book contains a more extensive Z-table. The disk version lists values of Z up to $Z = \pm 7.00$ and provides up to ten decimal place accuracy.

Finding Probabilities Using the Z-Table

The Z-table can enable us to determine interval probabilities for three types of problem situations:

1. Finding probability to the left of a given point
2. Finding probability to the right of a given point
3. Finding probability between two given points

To illustrate, let us work with a population of disk drives whose service life is a normally distributed random variable with $\mu = 760$, $\sigma = 140$ hours. Throughout our examples, we will include sketches to highlight the areas for which we are looking—we recommend that you follow this practice when working exercises.

Probability to the Left Since our Z-table is specifically set up to read probabilities to the left of various points, this is the easiest situation to have. All we need to do is to standardize the variable and locate the resulting Z-score in the table.

EXAMPLE 7.3A

Suppose a disk drive is randomly chosen from our population ($\mu = 760$, $\sigma = 140$). Find the probability it will have a life of 550 hours or less.

Solution:

In symbols, our problem is to find $P(X \leq 550)$, where X = length of service life. We first need to express $X = 550$ as its equivalent Z-score:

$$Z = \frac{X - \mu}{\sigma}$$

$$= \frac{550 - 760}{140} = \frac{-210}{140} = -1.50$$

This tells us that $P(X \leq 550)$ is the same as $P(Z \leq -1.50)$. When we look up $Z = -1.50$ in the table, we will see .0668. This is the probability that Z is less than or equal to -1.50 as well as the chance that X will take on a value of 550 or less. Nearly 7 percent of the population fail to last longer than 550 hours. Figure 7.5 shows both the unstandardized (X) and the standardized (Z) distributions. Note that since the normal variable is continuous, $P(X \leq 550)$ is the same as $P(X < 550)$, since the point probability $P(X = 550)$ is zero.

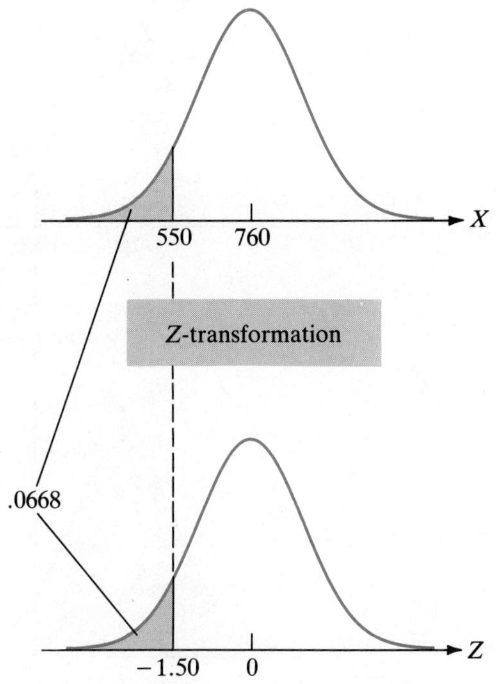

Figure 7.5 Standardizing $X = 550$ ($\mu = 760$, $\sigma = 140$)

□

EXAMPLE 7.3B

What is the probability a disk drive will fail before 1000 hours have elapsed?

Solution:

As before, we need to standardize prior to consulting the table:

$$P(X < 1000) = P\left(Z < \frac{1000 - 760}{140}\right)$$
$$= P\left(Z < \frac{240}{140}\right) = P(Z < 1.71)$$

The table provides our answer, .9564. (See Figure 7.6.) We also can interpret our result in terms of percentiles: The point $X = 1000$ hours is the 95.64th percentile of the distribution of X. Recall that the kth percentile of a distribution is a point such that k percent of the distribution is equal to or less than that point.

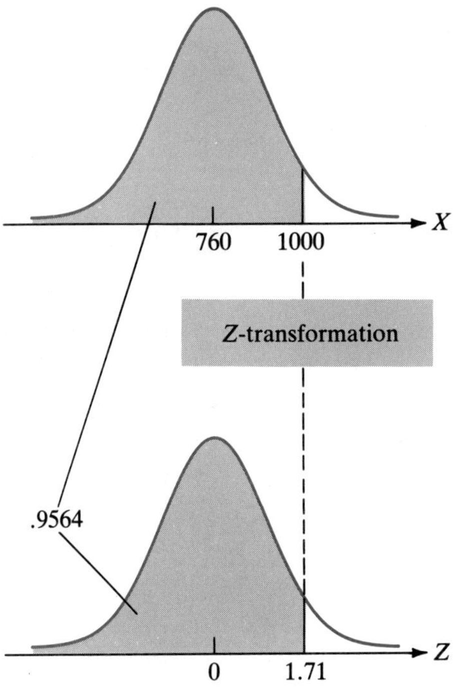

Figure 7.6 Finding $P(X < 1000)$ ($\mu = 760, \sigma = 140$)

□

Probability to the Right Since our table always tells us the probability (area) to the left of a given Z-score, we need to subtract the table value from 1.0 to obtain the probability to the right of a point. This is an application of the *complementary* law of probability discussed in Chapter 5.

EXAMPLE 7.4A

What proportion of the disk drives can be expected to exceed a life of 900 hours?

Solution:

We first express our problem in terms of Z:

$$P(X > 900) = P\left(Z > \frac{900 - 760}{140}\right) = P(Z > 1.00)$$

The table value of .8413 represents $P(Z \leq 1.00)$. To get $P(Z > 1.00)$, we simply subtract: $1 - .8413 = .1587$. About 16 percent of the population will last longer than 900 hours. (See Figure 7.7.) In general, we can say

$$P(Z > \text{point of interest}) = 1 - P(Z \leq \text{point of interest})$$

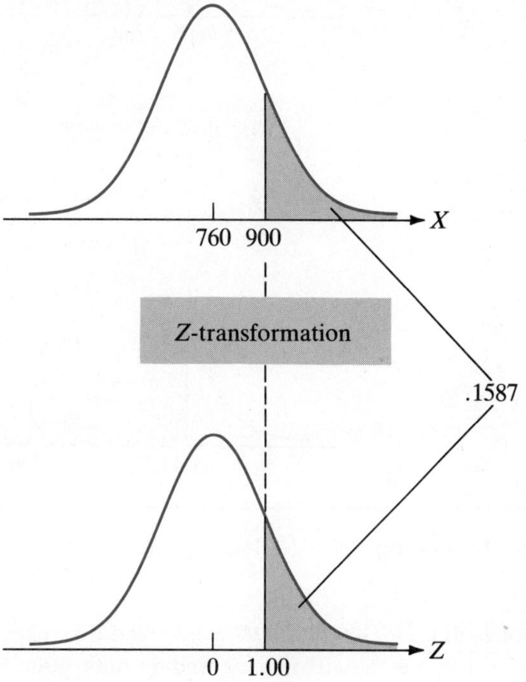

Figure 7.7 Finding $P(X > 900)$ $(\mu = 760, \sigma = 140)$

□

EXAMPLE 7.4B

Find the probability a disk drive will last more than 600 hours.

Solution:

$$P(X > 600) = P\left(Z > \frac{600 - 760}{140}\right)$$

$$= P(Z > -1.14) = 1 - .1271 = .8729 \qquad \text{(See Figure 7.8.)}$$

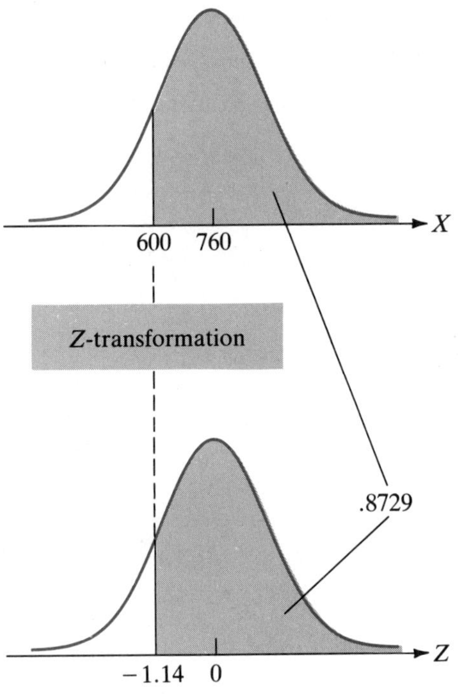

Figure 7.8 Finding $P(X > 600)$ $(\mu = 760, \sigma = 140)$

□

Probability Between Points As you may expect, determining the probability between two points will involve finding both points' Z-scores and then subtracting the smaller table value from the larger.

EXAMPLE 7.5A

What is the probability a disk drive chosen at random will be one whose life is between 700 and 800 hours?

Solution:

In symbols, this problem is $P(700 < X < 800)$. We need to change both 700 and 800 into their standard scores:

$$
\begin{aligned}
P(700 < X < 800) &= P\left(\frac{700 - 760}{140} < Z < \frac{800 - 760}{140}\right) \\
&= P(-.43 < Z < .29) \\
&= P(Z < .29) - P(Z < -.43)
\end{aligned}
$$

Consulting the table, we have .6141 − .3336 = .2805. About 28 percent of the population lies in the interval from 700 to 800 hours. Note that the table value of $Z = .29$ (.6141) is the probability from $-\infty$ to $Z = .29$. Since we do not want all this probability, we must subtract .3336 to exclude the region from $-\infty$ to $Z = -.43$. Figure 7.9 shows the probabilities involved, on the Z-axis.

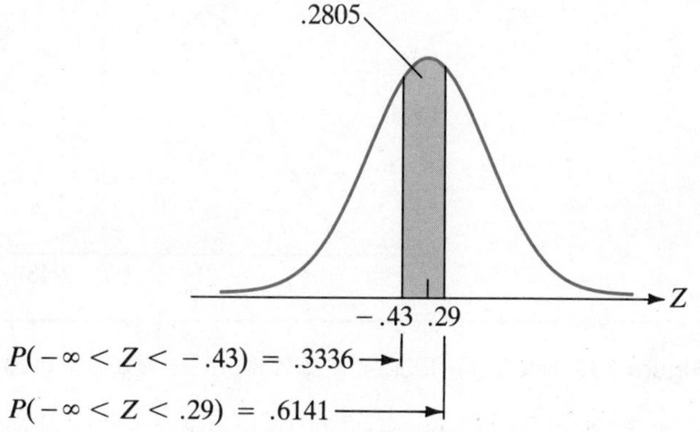

Figure 7.9 Finding Probability Between Two Points

□

EXAMPLE 7.5B

If a disk drive is selected at random, how likely is it that its life will be between 1000 and 1100 hours? (See Figure 7.10.)

Solution:

$$
\begin{aligned}
P(1000 < X < 1100) &= P(1.71 < Z < 2.43) \\
&= P(Z < 2.43) - P(Z < 1.71) \\
&= .9925 - .9564 = .0361
\end{aligned}
$$

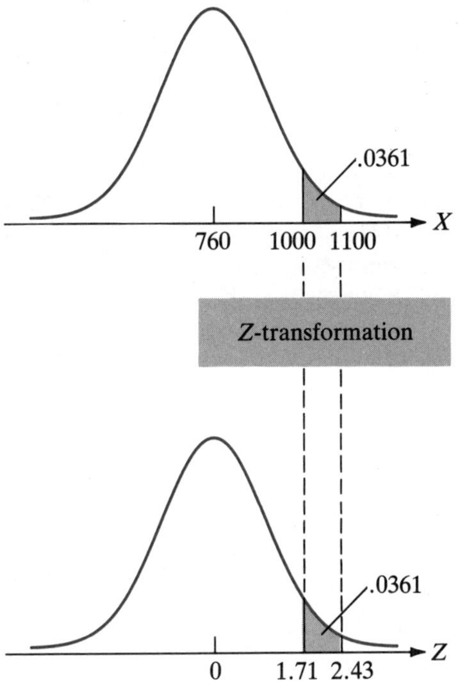

Figure 7.10 Finding $P(1000 < X < 1100)$ ($\mu = 760, \sigma = 140$)

□

Finding Percentiles Using the Z-Table

The Z-table can also help us find any percentile of interest in the distribution. As discussed in Chapter 3, percentiles can help us pinpoint the location of noncentral values in a given data set. Determining a percentile for some random variable X will involve finding the equivalent Z-score first and then converting the Z-score into its corresponding value of X.

EXAMPLE 7.6A

Continuing with our population of disk drives ($\mu = 760$, $\sigma = 140$), find the point X that is the 40th percentile in the distribution.

Solution:

Since the entries in the Z-table are probabilities to the left of various values of Z, we need to look for the probability of interest—in this case, .4000. We cannot find .4000 exactly, so we will use the closer alternative of .4013 and .3974, which is .4013.

Because the area .4013 occurs in the $Z = -.2$ row and in the .05 column, the Z-score of interest is $Z = -.25$. The 40th percentile of any normal distribution occurs one-fourth of a standard deviation below the mean. To complete the problem, we need to express $Z = -.25$ in unstandardized form. This involves converting Z into X using Equation 7–3:

$$X = \mu + Z\sigma$$
$$X = 760 + (-.25)140$$

which can be solved for $X = 725$ hours. We thus find that 40 percent of the disk drives last 725 hours or less; the other 60 percent last beyond that point. Figure 7.11 presents a graphical view of the problem; since we start with a Z-score and convert it into X, the Z-axis graph appears on top.

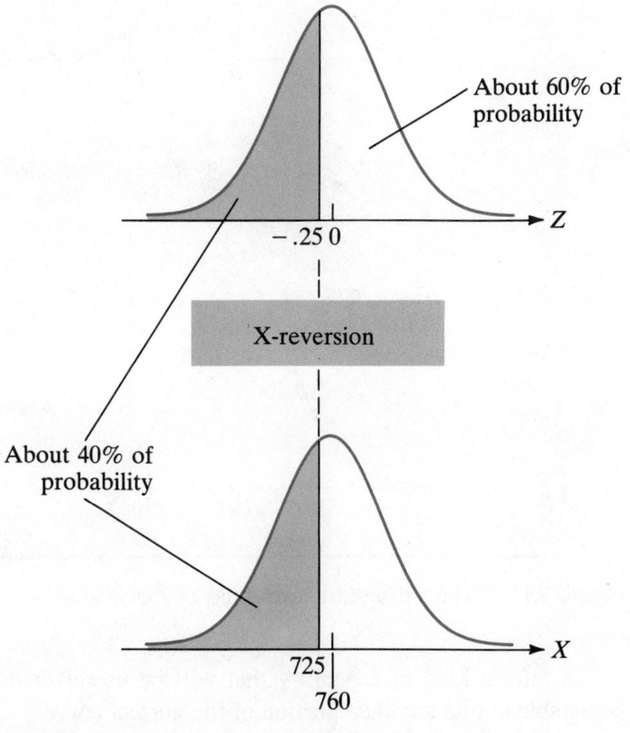

Figure 7.11 Finding the 40th Percentile of Z and X

□

EXAMPLE 7.6B

Find the amount of time that has elapsed when only 2 percent of the population of disk drives are still working.

Solution:

We want the point such that 98 percent of the area is to the left and 2 percent is to the right. We look for .9800 in the Z-table; we find .9798 at $Z = 2.05$. (See Figure 7.12.) This standard score can be converted into its equivalent X-value by solving: $X = 760 + 2.05(140)$. The resulting value of X is 1047 hours. Only 2 percent of the disk drives last longer than 1047 hours.

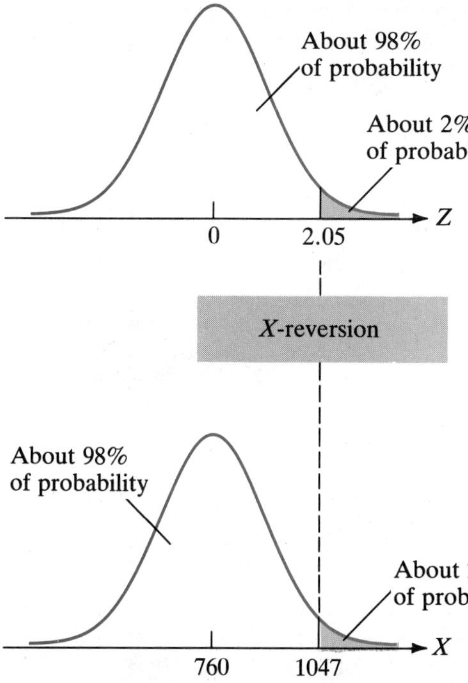

Figure 7.12 Finding the 98th Percentile of Z and X

□

A related kind of capability that will be useful in the chapters ahead involves being able to find a *middle* portion of the normal curve.

EXAMPLE 7.7

For our distribution of disk drives ($\mu = 760$, $\sigma = 140$), determine the points between which the middle 80 percent of the distribution lie.

Solution:

Figure 7.13 shows the area of interest: the middle 80 percent in the central region of the curve with the extreme 20 percent divided evenly, 10 percent in each tail. It should be clear that the middle 80 percent of any normal distribution begins at the 10th percentile and ends at the 90th percentile.

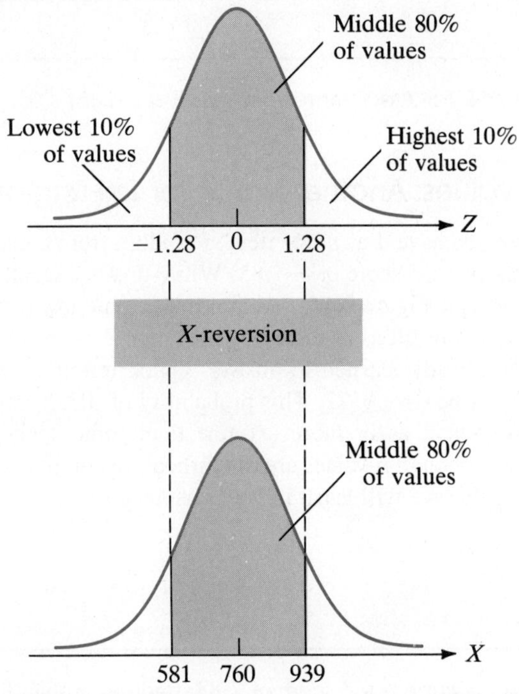

Figure 7.13 Finding the Middle 80 Percent of a Normal Distribution

The 10th percentile is found by looking for .1000 in the Z-table (closest is .1003 at $Z = -1.28$); the 90th percentile is found by seeking .9000 (closest is .8997 at $Z = 1.28$). Converting these two Z-scores into their equivalent values of X, we will arrive at (to the nearest whole hour) $X = 581$ and $X = 939$ hours.

□

Is it coincidence that the two Z's in the solution above have the same absolute magnitude of 1.28? Of course not! Due to the symmetry of the curve, *any middle percentage* will involve going out the same distance in either direction from the mean. With this in mind, we would know when the first Z-score is determined to be $Z = -1.28$ that its partner will have to be $Z = +1.28$.

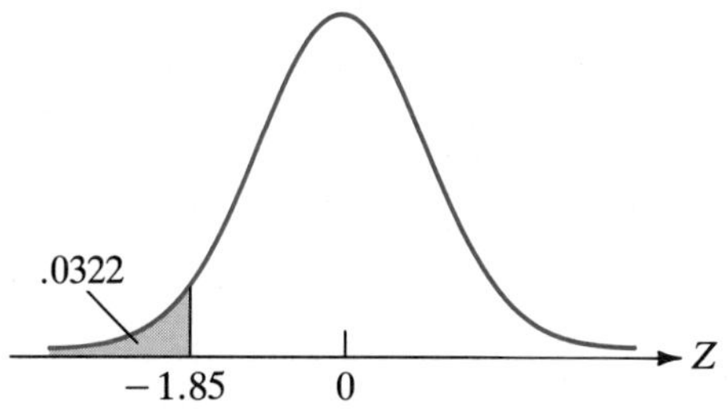

Figure 7.14 Tail Area Representing $P(Z \le -1.85)$

p-values: Another Name for Tail Regions

Suppose you have drawn one item at random from a normal distribution. Assume this one item has a Z-score of -1.85. While this is a result from the relatively low end of the curve (see Figure 7.14), even more extreme low results are possible. What are the chances of a result as or more negative than $Z = -1.85$? This is the type of question we have already learned to answer. Quick reference to the Z-table will tell us that $P(Z \le -1.85) = .0322$. This probability of .0322, representing an area for which Z is of the curve as or more extreme than some specified point, is referred to as a *p*-value. Although *p*-values are not difficult to compute or interpret, they have a special meaning that we will learn in later chapters.

EXAMPLE 7.8

An observation has a Z-score of 2.83; find the probability of a value with the same or a more positive Z-score.

Solution:

We are looking for an upper-tail *p*-value. In symbols, our problem is to find $P(Z \ge 2.83)$. This can easily be determined by

$$\begin{aligned} P(Z \ge 2.83) &= 1 - P(Z < 2.83) \\ &= 1 - .9977 = .0023 \end{aligned}$$

This small *p*-value indicates that there is very little probability beyond $Z = 2.83$.

□

The examples above illustrate finding an extreme region on either the lower end or on the upper end of the curve. If we are interested in the chance of an extreme value without regard to its sign, then we say that we want a two-tailed *p*-value.

EXAMPLE 7.9

What are the chances of an observation falling 2.50 or more standard deviations away from the mean in either direction?

Solution:

We want a two-tailed p-value (see Figure 7.15), which can be symbolized as

$$p\text{-value} = P(Z \leq -2.50) + P(Z \geq 2.50)$$

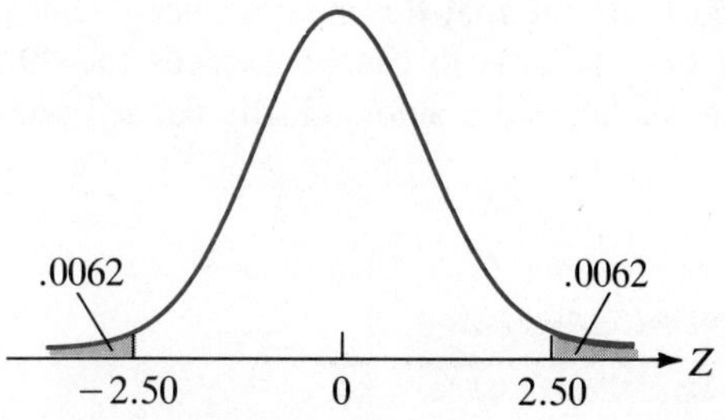

Figure 7.15 Tail Areas Representing $P(|Z| \geq 2.50)$

Due to the symmetry of the normal curve, this can also be written as

$$\begin{aligned} p\text{-value} &= 2 \times P(Z \leq -2.50) \\ &= 2(.0062) = .0124 \end{aligned}$$

A probability of .0062 lies in each tail. If the plus or minus direction an extreme value takes on is unimportant, the p-value here represents $P(|Z| \geq 2.50) = .0062 + .0062 = .0124$, where $|\;|$ is the operation of absolute value.

□

CLASSROOM EXAMPLE 7.1

Using the Z-Table (Part A)

If Z designates the standard normal random variable, find the following probabilities:

a. $P(Z > -1.80)$

b. $P(-.17 < Z < 2.69)$

c. $P(Z < -.45)$

d. $P(1.03 < Z < 2.93)$

e. Find the Z-score that has 1.5 percent of the total area to its right.

CLASSROOM EXAMPLE 7.2

Using the Z-Table (Part B)

Of those federal taxpayers who itemize deductions and have adjusted gross incomes between $30,000 and $40,000, the amount of gifts to charity claimed on Schedule A is approximately normally distributed. The average deduction is $503 in this category, with a standard deviation of $168. Suppose an itemized tax return from this income group is randomly selected.

a. Find the probability that less than $400 in gifts to charity will be claimed.

b. Find the probability that between $475 and $700 will be claimed.

c. Suppose the Internal Revenue Service decides to audit any return for which the deduction for gifts to charity exceeds the 99.5th percentile. Above what dollar figure is a tax return automatically flagged according to this criterion?

SECTION 7.2 EXERCISES

7.5 Given a normal distribution with $\mu = 140$ and $\sigma = 40$, express the following values of X in standardized form:
a. $X = 160$ b. $X = 140$ c. $X = 110$

7.6 Express the following values of X in Z-score form, given that the values of X come from a population with a mean of $45,100 and a standard deviation of $16,500:
a. $X =$ $65,000 b. $X =$ $28,500 c. $X =$ $81,950

7.7 Convert the following Z-scores into their equivalent values of X (the random variable X has $\mu = 50$ and $\sigma = 12$):
a. $Z = -1.75$ b. $Z = -.20$ c. $Z = 1.10$

7.8 A normal distribution has a mean of 175 and a Z-score of .40 when $X = 195$. Find σ.

7.9 A normal distribution has $\sigma = 120$ and a Z-score of -1.20 when $X = 300$. Find the mean value of X.

7.10 Find the following interval probabilities on a standard normal curve:
a. $P(Z > -.44)$ b. $P(Z < 1.75)$ c. $P(-1.89 < Z < 1.89)$

7.11 Find the following probabilities on a standard normal curve:
a. $P(Z < -1.04)$ b. $P(-2.77 < Z < 2.77)$ c. $P(Z > .15)$
d. $P(Z > 4.44)$ (*Suggestion:* Since the table does not extend to $Z = 4.44$, make your answer an educated guess, to four decimal places.)

7.12 Consider the standard normal distribution.
a. What value of Z cuts off the lowest 12 percent of the curve?
b. What value of Z cuts off the highest 4 percent of the curve?
c. Between what values of Z does the middle 98 percent of the area lie?

7.13 Consider the standard normal distribution.
a. What Z-score has 20 percent of the probability to its left?

b. What Z-score has 30 percent of the probability to its right?
c. Between what Z-scores do the middle 20 percent of values lie?

7.14 The manufacturer of a light beer makes a product with a mean (determined by chemical analysis) of 95.0 calories per 12-ounce can. Due to variations in the brewing process as well as in the quality of the ingredients, the standard deviation of calorie count is 4.20. The distribution of calorie content follows the normal curve.
a. Find the proportion of cans that contain between 90 and 100 calories.
b. How likely is it that a can will contain over 100 calories?
c. What percent of cans have a calorie content within $\pm$ 2.0 calories of the mean?
d. What is the probability a can has less than 85 calories?
e. What calorie content is exceeded only 5 percent of the time?

7.15 The fill weight of a certain brand of cereal, X, is a normally distributed random variable with $\mu = 910.0$ grams and $\sigma = 5.0$ grams. Assume you select one box of cereal at random from this population.
a. How likely is it to contain less than 900 grams?
b. How likely is it to weigh within 2 grams of the population mean?
c. Find the probability it will contain more than 895 grams but less than 905 grams.

7.16 Refer to Exercise 7.15, where $\mu = 910.0$ and $\sigma = 5.0$ grams.
a. Only 1 percent of boxes will weigh more than X grams. Find the value of X.
b. Between what two weights does the middle 50 percent of the distribution lie?

7.17 A certain type of battery has an average shelf life of 240 days with a standard deviation of 36 days. Using the normal distribution, find the following:
a. the probability a battery has a shelf life of over 300 days.
b. the proportion of batteries with a shelf life from 200 to 300 days.
c. the percent with a shelf life less than 200 days.
d. the value of X, length of shelf life, at the 33rd percentile of the distribution.

7.18 Suppose that a brand of microwave oven is sold with a one-year guarantee against breakdown. Also suppose that the life of the oven to first breakdown, X, is a normally distributed random variable with $\mu = 2.10$ years and $\sigma = .65$ year.
a. Find the value of X at the 15th percentile of the distribution.
b. What percent of ovens will fail within plus or minus 6 months of the population mean?
c. Find the value of X beyond which the longest-lived 3 percent of ovens last without a breakdown.
d. What percent of ovens experience the first breakdown while still under guarantee?

7.19 The amount of time it takes for a bank's customers to make a simple withdrawal of cash from an automatic teller machine has been shown to be a normally distributed random variable with a mean of 80 seconds and a standard deviation of 18 seconds.
a. What is the probability a customer will make a withdrawal in less than one minute?
b. What proportion of customers require more than two minutes to make a withdrawal?
c. What is the probability a customer will complete the transaction somewhere between 60 and 90 seconds?

7.20 Suppose grade point averages (GPAs) at a large university can be approximated by a normal distribution with $\mu = 2.77$ and $\sigma = .41$. Answer the following to the nearest whole percentage.
a. What percentage of students have a GPA between 2.50 and 3.00?
b. What percentage of students are on the Dean's List (minimum requirement is a GPA of 3.35)?

c. What percentage of students are on probation (GPA below 2.00)?
d. What percentage of students have grade points that differ from the university mean by no more than .25?

7.21 Find the following one-tailed p-values:
a. $P(Z \geq .61)$ **b.** $P(Z \leq -2.19)$ **c.** $P(Z \geq 2.90)$

7.22 Find the following one-tailed p-values:
a. $P(Z \leq -1.61)$
b. $P(Z \leq -3.11)$
c. $P(Z \geq 1.58)$
d. $P(Z \leq -5.17)$

7.23 Find the following two-tailed p-values:
a. $P(|Z| \geq 3.40)$
b. $P(|Z| \geq 1.49)$
c. $P(|Z| \geq 2.22)$

7.24 Between what values of Z does the middle 95 percent of the normal distribution lie?

7.25 Why can't we represent the point probabilities of a normal probability distribution in tabular form?

7.26 If Z is a standard normal variable, find the following:
a. $P(Z > .23)$
b. $P(1.15 < Z < 2.08)$
c. $P(Z > 0)$

7.27 If Z is a standard normal variable, find the following:
a. $P(Z < -2.34 \text{ or } Z > 2.34)$
b. $P(-\infty < Z < \infty)$
c. $P(Z < .76)$

7.28 If Z is a standard normal variable, find the following:
a. $P(Z = 1.49)$
b. $P(-1 < Z < 1)$
c. $P(Z < -.58)$

7.29 If Z is a standard normal random variable, find the Z-score that has the following:
a. 60 percent of total area to its left.
b. 20 percent of total area to its right.
c. 18 percent of total area to its left.

7.30 If Z is a standard normal random variable, find the following:
a. P_{25}
b. P_{62}
c. P_{75}

***7.31** Refer to Exercise 7.30. The distance along the horizontal axis between the 25th and 75th percentiles is called the *interquartile range* and is sometimes used as a measure of dispersion. Find the interquartile range for a standard normal distribution, making sure to attach the unit of measurement to your numerical answer.

*Optional

7.32 The merit pay increases for salaried employees in the services industry is expected to average 5.30 percent next year, with a standard deviation of 1.44 percent, according to a recent survey of companies. Assume the population of merit pay increases is approximately normally distributed.

a. What proportion of salaried employees in the services industry will receive increases of 7 percent or more?

b. Eighty percent of all salaried employees will receive increases of x percent or less. Find x.

c. Suppose the average merit pay increase in the financial institutions industry is going to be 6.2 percent, with a standard deviation of 1.5 percent. Further assume a two-career household has the wife in the financial institutions industry and the husband in the services industry. If the husband receives a 6.8 percent merit increase, what percent merit increase would the wife need to earn to be in the same percentile for her industry as her husband is in his industry?

7.3 THE NORMAL APPROXIMATION TO THE BINOMIAL DISTRIBUTION

Now that we have learned to standardize values and to find probabilities on a normal curve, we will consider extending this capability to large-sample binomial applications.

Binomial Sampling Example (No Tables Available)

Besides being the most useful family of distributions for continuous variables, normal distributions can also be used to give approximate answers to problems involving discrete distributions such as the binomial.

For example, consider the following: State highway officials claim that 40 percent of vehicles on their interstate highways exceed the speed limit by more than five miles per hour. A random sample of $n = 200$ vehicles is obtained. If the percent of speeders (vehicles going 5 miles or more per hour over the limit) in the population is in fact 40 percent, how likely is it the sample will show 50 percent or more speeders (in this case, 100 or more speeders)?

The problem setting described in the preceding paragraph is clearly binomial: We have 200 entities with two possible outcomes: vehicle does or does not speed. However, we have no binomial tables for $n = 200$; in fact, we have no tables beyond $n = 25$. Nevertheless, we can obtain a good approximate answer to this binomial question by using a normal curve. The approach used is called the *normal approximation to the binomial*. As long as $n\pi \geq 5$ and $n(1 - \pi) \geq 5$, we can reasonably approximate the discrete binomial distribution by employing a continuous normal distribution.

Conditions for Using the Normal Approximation to the Binomial

1. Binomial sampling application
2. $n\pi \geq 5$ and $n(1 - \pi) \geq 5$

Binomial Example Solved With a Normal Curve

We now illustrate working this binomial sampling problem (to find the chance of 100 or more successes when $\pi = .40$ and $n = 200$) by employing a normal curve and using the Z-table. We first verify that $n\pi \geq 5$ and that $n(1 - \pi) \geq 5$. In this case $n\pi = 80$ and $n(1 - \pi) = 120$, so these requirements are easily met. If we fail to meet either condition, the approximation is not reliable.

You should recall from the previous chapter that the binomial mean is

$$\mu = n\pi$$

In this problem, we have 200 entities with a probability of success .40, so the mean or expected number of successes is 80. Also recall that the binomial standard deviation is

$$\begin{aligned} \sigma &= \sqrt{n\pi(1 - \pi)} \\ &= \sqrt{200\,(.4)(.6)} \\ &= 6.93 \end{aligned}$$

We can now use a normal curve with $\mu = 80$, $\sigma = 6.93$ in lieu of the binomial histogram involving rectangles. Our problem is to find the probability that X, the number of speeders, is 100 or more. (See Figure 7.16.)

From this point forward, this is an ordinary normal curve problem similar to those we have learned how to solve using the Z-transformation. We want

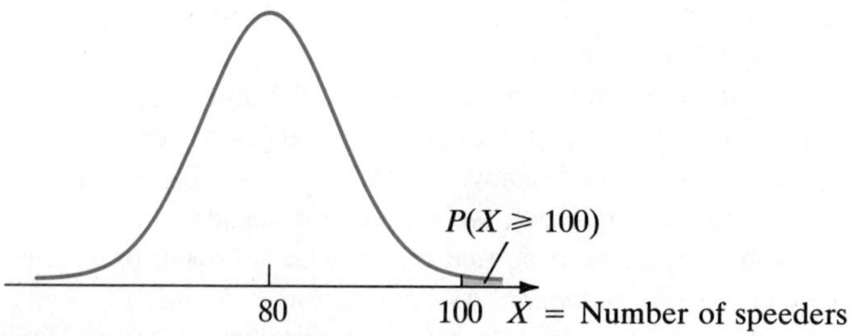

Figure 7.16 Normal Curve with $\mu = 80$, $\sigma = 6.93$

$$P(X \geq 100) = P\left(Z \geq \frac{100 - 80}{6.93}\right)$$
$$= P(Z \geq 2.89) = 1 - .9981 = .0019$$

We have solved a binomial sampling question with the help of a normal curve. The probability of .0019 tells us that we are very unlikely to observe 100 or more speeders if the claim of the highway officials is correct.

COMMENT Unless otherwise instructed, round the binomial standard deviation to two decimal places for use in the normal approximation.

EXAMPLE 7.10

Suppose that, in a certain large city, 30 percent of eligible voters are not registered. A sample of $n = 100$ eligible voters is chosen at random. Find the probability that the sample will show 25 or fewer who are not registered.

Solution:

This is a binomial problem, as we have two outcomes possible on each of 100 voters. With no binomial tables at our disposal, we can approximate this discrete sampling situation with the normal curve.

Letting X = number of eligible voters not registered, we first note that $n\pi = 30$ and that $n(1 - \pi) = 70$, both values exceeding 5. We will use a normal curve with mean $= n\pi = 30$, standard deviation $= \sqrt{100(.3)(.7)} = \sqrt{21} = 4.58$. (See Figure 7.17.) Therefore, our problem becomes

$$P(X \leq 25) = P\left(Z \leq \frac{25 - 30}{4.58}\right)$$
$$= P(Z \leq -1.09) = .1379$$

or about a 14 percent chance.

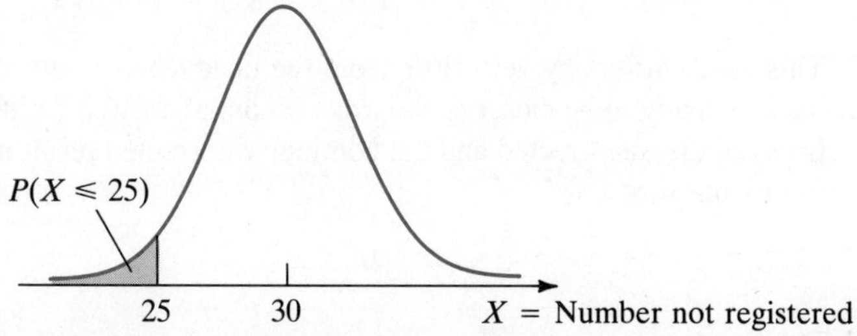

Figure 7.17 Normal Curve with $\mu = 30$, $\sigma = 4.58$

□

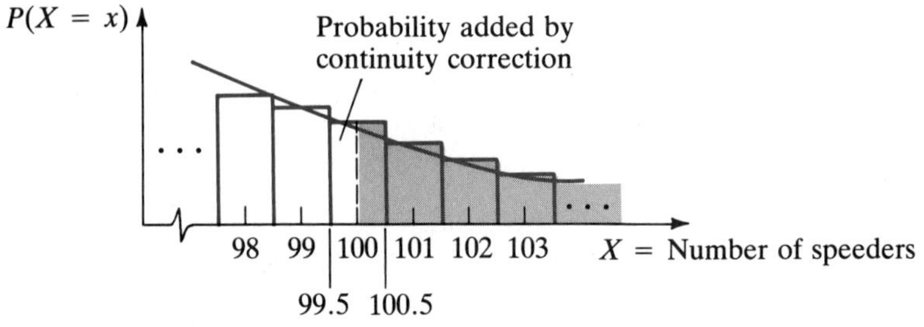

Figure 7.18 Incremental Probability Added by the Continuity Correction

Fine Tuning: Use of a Continuity Correction

The normal approximation to the binomial can be made a bit more accurate by use of an adjustment term (always equal to $-.5$ or $+.5$) called the **continuity correction.** To illustrate, let us refer back to the highway example where we wanted $P(X \geq 100)$ when $\mu = 80$ and $\sigma = 6.93$. Figure 7.18 shows a close-up view of a small part of the right tail of the distribution. As in Chapter 6, we show the graph of a binomial distribution as a set of connected rectangles. The smooth line in the figure represents the normal curve that approximates this binomial distribution. Note that the rectangle that represents $P(X = 100)$ actually takes up space on the number line from $X = 99.5$ to $X = 100.5$, its class boundaries. But when we computed $P(X \geq 100)$ previously, we did not include the half of this rectangle that lies from $X = 99.5$ to $X = 100$. Instead we began at the dashes, missing part of the rectangle that represents $P(X = 100)$. Since we want to include the whole rectangle, we employ the continuity correction and change the problem from $P(X \geq 100)$ to $P(X \geq 99.5)$. We can then determine that

$$P(X \geq 99.5) = P\left(Z \geq \frac{99.5 - 80}{6.93}\right)$$
$$= P(Z \geq 2.81) = 1 - .9975 = .0025$$

This result differs by very little from the unadjusted result of .0019 obtained earlier, but it is likely to be closer to the exact binomial answer. In other cases, the difference between the uncorrected and the continuity-corrected result may be larger than it was in this instance.

EXAMPLE 7.11

Re-do Example 7.10 with the continuity correction.

Solution:

Figure 7.19 shows a portion of the distribution. We want the probability of 25 or fewer not registered. Again we argue that to include the full probability that $X = 25$, we need to change $P(X \leq 25)$ to $P(X \leq 25.5)$. Then

$$P(X \leq 25.5) = P\left(Z \leq \frac{25.5 - 30}{4.58}\right) = P(Z \leq -.98) = .1635$$

a bit different from the unadjusted answer of .1379. Binomial tables for $n = 100$ (not found in this book) give .1632 as the exact answer for this problem. The continuity-corrected approximation is very precise in this instance.

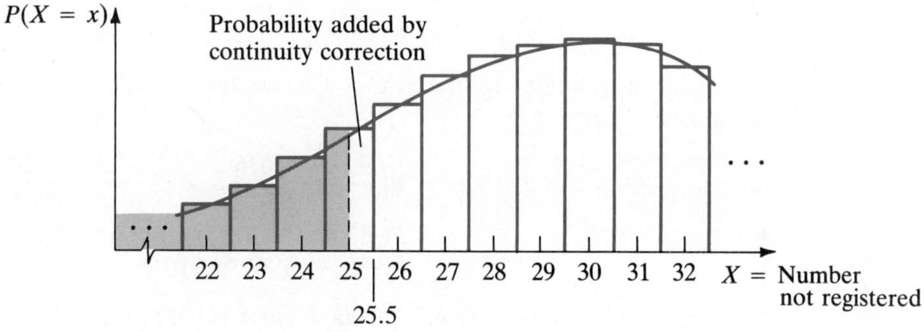

Figure 7.19 Adjusting $P(X \leq 25)$ to $P(X \leq 25.5)$

□

The continuity-corrected answer is often very close to the exact binomial result and is usually more accurate than the unadjusted answer. However, the effect of the continuity correction on the answer obtained becomes smaller as the sample size becomes larger and as Z gets further into the tail of the curve. Table 7.2 offers guidelines for adding or subtracting .5 from X when making the continuity correction.

As a final example, let us refer to the binomial problem we posed (following Example 6.5) in the previous chapter: that of finding $P(X = 10)$ when $n = 23$ and $\pi = .332$. At that time, we stated that this problem could be done by the binomial formula (very time consuming), by computer program (see Figures 6.4 and 6.5), or by an approximation, specifically the normal approximation to the binomial. The approximation will utilize a normal curve with mean $= 23(.332) = 7.636$ and standard deviation $= \sqrt{23(.332)(.668)} = 2.2585$. (For additional accuracy in this example, we will waive the recommendation to round σ to two decimal places.)

Finding $P(X = 10)$ corresponds to case D in Table 7.2. Unlike the other situations, case D *requires* use of the continuity correction. If we did not apply the correction, our answer would be zero, since $P(X = 10)$ is a point probability on a

Table 7.2 Continuity Correction Terms*

Case	Binomial Probability to Be Approximated	Continuity Correction	Example
A	$P(X \geq \text{number})$	Subtract .5 from number	$P(X \geq 10)$ becomes $P(X \geq 9.5)$.
B	$P(X \leq \text{number})$	Add .5 to number	$P(X \leq 20)$ becomes $P(X \leq 20.5)$.
C	$P(\text{number} \leq X \leq \text{number})$	Widen by .5 in both directions	$P(25 \leq X \leq 35)$ becomes $P(24.5 \leq X \leq 35.5)$.
D	$P(X = \text{number})$	Widen by .5 in both directions	$P(X = 40)$ becomes $P(39.5 \leq X \leq 40.5)$.

*Before using this table, convert any binomial inequality into an equality. For example, change the binomial $P(X > 15)$ into $P(X \geq 16)$ before using this table.

continuous distribution. To find $P(X = 10)$, we apply the continuity correction and obtain (see Figure 7.20):

$$\begin{aligned} P(9.5 \leq X \leq 10.5) &= P\left(\frac{9.5 - 7.636}{2.2585} \leq Z \leq \frac{10.5 - 7.636}{2.2585}\right) \\ &= P(.83 \leq Z \leq 1.27) \\ &= .8980 - .7967 = .1013 \end{aligned}$$

Note that this answer is off only about 3 parts in 1000 from the result of .0982 obtained in Figure 6.4.

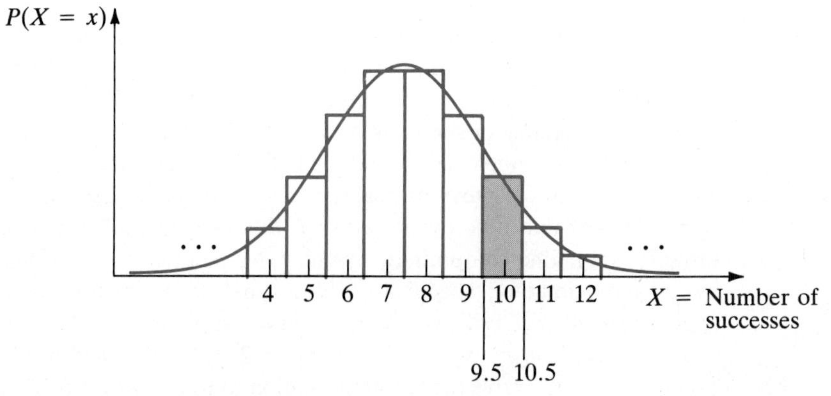

Figure 7.20 $P(X = 10)$ After the Continuity Correction

Table 7.3 Normal Approximation Recommendations for Combinations of n and π

Smaller of $n\pi$ or $n(1 - \pi)$ is	Normal Approximation	Comment
Less than 5	Not advised	See Chapter 6 for alternatives.
From 5 to 10	Advised, with continuity correction	Continuity correction offers fine tuning.
Above 10	Advised; correction term required for case D; otherwise optional	Effect of correction can be very small for cases A, B, and C.

The normal approximation to the binomial distribution can be a tremendous time-saver compared to computing with the binomial formula or accessing computer programs. The accuracy of the approximation improves as the sample size n becomes larger and as the probability of a success π approaches .50. For most practical or managerial purposes, there is little difference between continuity-corrected and unadjusted results. For this reason, we will opt for simplicity and generally omit the correction term when $n\pi$ and $n(1 - \pi)$ both exceed 10, unless we are finding a binomial point probability as just discussed. Table 7.3 gives our recommendations regarding the use of the normal distribution to approximate the binomial.

CLASSROOM EXAMPLE 7.3

Solving a Binomial Problem with a Normal Curve

A clothing retailer has traditionally generated most of its revenues through its mail-order catalogs. However, records indicate that only one of every five customers on the mailing list places an order. Competition in the mail-order catalog business has forced the retailer to consider adding a section of "sale" merchandise in the middle of the catalog. As an experiment, the retailer had 1000 catalogs with a special sale section mailed to a random sample of customers from the mailing list. Let $X =$ the number of orders from the sample of customers receiving the special catalogs. Assume the proportion of orders from the special catalogs will be the same as from the regular catalogs.

a. Verify that the binomial sampling conditions are in force, and that $n\pi \geq 5$ and that $n(1 - \pi) \geq 5$.

b. Determine the values of the parameters μ and σ in order to use the normal approximation to the binomial, and carry out computations to three decimal places: $\mu =$ ______ $\sigma =$ ______.

c. What is the probability that 206 or more orders will be received from the special catalog mailing?

d. What is the probability that from 190 to 220 orders (inclusive) will be received?

e. The retailer decides to include a permanent sale section in all future catalogs if the number of orders is much higher than usual. If the decision criterion is based on the 98th percentile, what value of X forms the dividing line?

SECTION 7.3 EXERCISES

7.33 For binomial sampling with $n = 80$ and $\pi = .40$, use the normal approximation to find $P(X \geq 40)$ in the following situations:
a. without the continuity correction.
b. with the continuity correction.

7.34 For a binomial sampling situation with 880 trials and a probability of success of .225, use the normal approximation without the continuity correction to find the probability of 200 or fewer successes.

7.35 Refer to Exercise 7.34. Recompute your answer, applying the adjustment for continuity.

7.36 A large population of fuses contains 6 percent defective. Find the chance that a sample of 1000 fuses will contain 50 or less defective. Use the normal approximation without the continuity correction.

7.37 Refer to Exercise 7.36. Recompute your answer using the continuity correction.

7.38 Refer to Exercise 7.36. Compute $P(50 \leq X \leq 70)$, where X = number of defective fuses, in the following situations:
a. without the continuity correction.
b. with the continuity correction.

7.39 Within a certain county, 10 percent of registered automobiles are foreign imports. If a sample of 100 automobiles is drawn at random, how likely is it the sample will show six or fewer foreign imports? Use the normal approximation with the continuity correction. (The exact binomial answer is .1172.)

7.40 Refer to Exercise 7.39. Compute $P(X = 15)$ using the normal approximation and the continuity correction. The exact binomial result is .0327. Is the approximation in the ballpark?

7.41 Public opinion polls are regularly conducted in the United States, and the results often appear in the media. Suppose a poll includes a question that probes the sentiment towards one of the United States' foreign trading partners. Suppose 63 percent of the American public has a favorable opinion of the foreign nation in question. Find the probability that 900 or more people in a poll of 1409 respond favorably.

7.42 An entrepreneur plans to start a mail-order business. To obtain financing, he prepares a business plan in which he anticipates getting an order from 35 percent of the catalog recipients.
a. If the response rate is as planned, how likely is a mailing of 1000 catalogs to yield 325 or fewer orders?
b. What is the region of concentrated values for X, where X is the number of orders obtained from a sample of 1000 catalogs mailed?

*7.4 PROCESSOR: NORMAL PROBABILITY DISTRIBUTION

The normal curve processor can do the same two basic operations with normal distributions as were discussed in Section 7.2: Finding interval probabilities for X and finding the value of X for a given percentile of the distribution. To illustrate, let us suppose that we have a normal distribution with $\mu = 462$, $\sigma = 97$ and that we wish to know the answers to two questions:

1. What is the probability that X exceeds 600?
2. What value for X is at the eighth percentile of the distribution?

Before we can address these questions, we need to call up the normal curve processor. Starting on the main menu, move to Processors and touch Enter; then select Normal Probability Distribution. After reading the introductory screen, move the highlight to Table and touch Enter twice—this tells the computer to retrieve the Z-table from a file and read it into memory. After this has been done, move to Data.

Finding Probabilities on a Normal Curve

When we touch Enter, a pair of choices will be presented: the first for finding probabilities, the second for finding percentiles. To address our first question, $P(X > 600)$, we select the first choice. Touching Enter now yields a submenu requesting us to specify whether we want a probability to the left of X, to the right of X, or between two values of X. To evaluate $P(X > 600)$, we choose the second alternative listed. At this point, three successive requests will appear, asking us to enter a value for X, for μ, and for σ. Figure 7.21 shows the screen after we have provided the final quantity, $\sigma = 97$, but before touching Enter. We now move to Calculations, touch Enter twice, then move to Results when the computations are complete. By touching Enter, we can see the requested probability and a rough sketch of the normal curve with the appropriate area shaded in. (See Figure 7.22.)

Finding the X-value for a Given Percentile

To find a given percentile, we get back to Data on the Normal Probability menu, touch Enter, then select the second submenu choice. Again three successive requests will appear, asking for the values of μ, σ, and k, the desired percentile. Figure 7.23 shows

*Optional

```
≡                          Normal Probability Distribution                          EDIT
  Introduction  *able [Data] *alculations  *esults  Exit                    |  F1=Help

     Enter x, μ, and σ to find P(X ≥ x).
   (If you are working directly with Z, then enter the Z-score for x;
   also let μ = 0 and σ = 1.)

                        The value for x = 600
                        The value for μ = 462

                  ═══════════ P(X ≥ x) ═══════════
                    Enter a value for σ, σ =  97
```

Figure 7.21 Data Entry Screen to Find a Probability

the screen after we have responded $k = 8.00$. After touching Enter, we now can proceed to Calculations and then to Results. Figure 7.24 shows the answer to our question—what value of X is at the 8th percentile of this distribution?

```
≡                          Normal Probability Distribution                          READY
  Introduction  *able  Data  Calculations [Results] Exit                    |  F1=Help
```

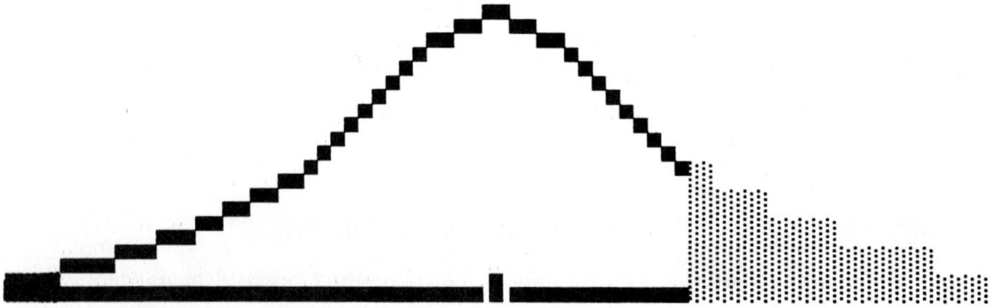

```
          P(X ≥ 600.00) = P(Z ≥ 1.42) = 0.0778
```

Figure 7.22 Finding a Probability

```
≡                        Normal Probability Distribution                        EDIT
  Introduction  *able [Data] Calculations  Results  Exit                | F1=Help

    Enter μ, σ, and a desired percentile k to find P(X ≤ ?) = k.
  Acceptable inputs for k are numbers between 1 and 99.  For example,
  enter k = 60 for the 60th percentile; enter k = 2.5 for the 2½th
  percentile.  If you are working directly with Z, let μ = 0 and σ = 1.

                    The value for μ = 462
                    The value for σ = 97

            ══════════════════ Percentile ══════════════════
               Enter the desired percentile, k =  8.00
```

Figure 7.23 Data Entry Screen for Finding Percentiles

```
≡                        Normal Probability Distribution                       READY
  Introduction  *able  Data  Calculations [Results] Exit                | F1=Help
```

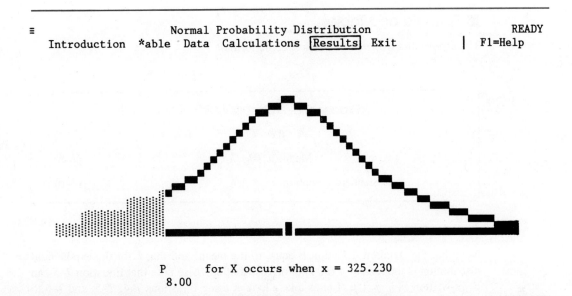

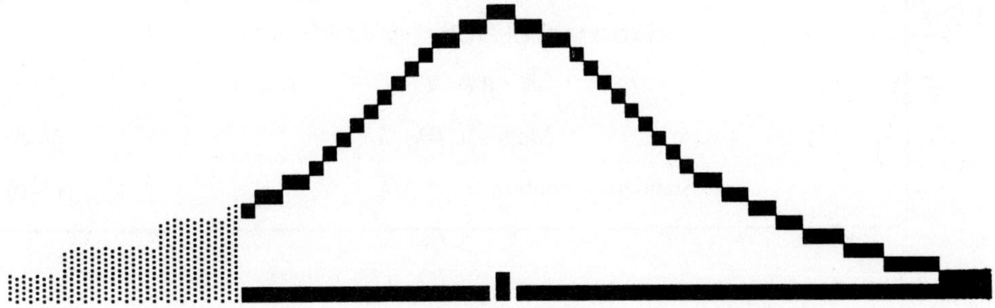

```
                 P        for X occurs when x = 325.230
                  8.00
```

Figure 7.24 Finding a Percentile

7.5 EXPONENTIAL DISTRIBUTIONS

Another continuous model with applications in business is the exponential distribution. As Figure 7.25 suggests, the exponential is a family of curves, with different members resulting depending on the value of the parameter λ (Greek letter lambda). Regardless of the value of λ, the **exponential distribution** is skewed to the right, with steadily decreasing probability as X increases. The total probability under the curve is naturally defined to be 1.

The exponential is related to the Poisson distribution. In Chapter 6, we cited several instances when we used the Poisson to model or describe the *number of arrivals or occurrences* of some event per interval of time (or some other unit of measure). If we do have a Poisson process, then X, the *time between successive arrivals or occurrences,* will form an exponential distribution. Examples where the exponential distribution might apply are

- Describing the time between arrivals of users at a bank branch automatic teller machine
- Modeling the time between mechanical failures of a copying machine
- Describing the time between arrivals of vehicles at a fast-food restaurant drive-through ordering station

Formula and Tables

The *exponential distribution* has a specific formula.

Exponential Probability Distribution

$$f(x) = \lambda e^{-\lambda x} \qquad 0 < x < \infty \tag{7–4}$$

$$\text{Mean: } \mu = 1/\lambda \tag{7–5}$$

$$\text{Standard deviation: } \sigma = 1/\lambda \tag{7–6}$$

Note that the standard deviation is equal to the mean, and that λ for the exponential distribution is the mean for the Poisson distribution. Also note that Equation 7–5 can be re-written as $\lambda = 1/\mu$. Let us take a look at using Equations 7–4, 7–5, and 7–6 to describe an exponential distribution.

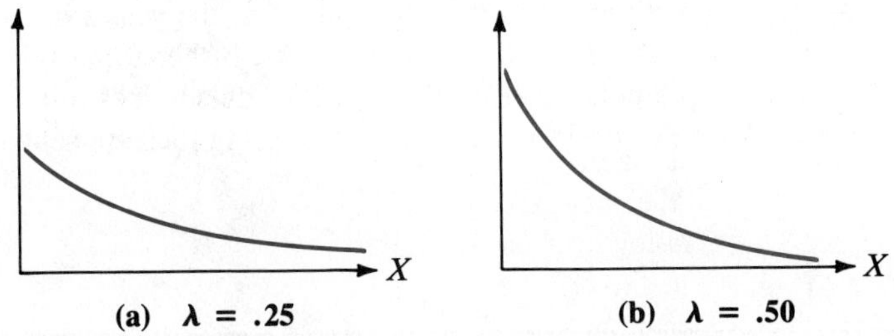

Figure 7.25 Two Members of the Exponential Distribution Family

EXAMPLE 7.12

For the exponential distribution shown in Figure 7.25(a), determine the mean, standard deviation, and the interval $\mu \pm 2\sigma$.

Solution:

Using Equation 7–5, the mean is $\mu = 1/.25 = 4.0$, and the standard deviation is also 4.0. The interval $\mu \pm 2\sigma$ is nominally from -4 to 12, but since X cannot be negative, we would say that in practice our interval would be from $X = 0$ to $X = 12$. This means that approximately 95 percent of the probability occurs between 0 and 12. □

EXAMPLE 7.13

A firm's most sophisticated color copier has a mean number of mechanical failures per week requiring a vendor service call of $\lambda = .80$. The failure pattern follows a Poisson distribution. Find the mean time between failures occurring and the standard deviation of time between failures.

Solution:

If we know the Poisson *mean number* of occurrences λ, it is easily converted into the exponential *mean time between occurrences* by Equation 7–5:

$$\mu = 1/\lambda = 1/.8 = 1.25 \text{ weeks between failures}$$

The standard deviation of time between failures is also 1.25 weeks. □

Our exponential distribution table appears in Appendix D at the back of the book. As the graph at the top of the table indicates, the values in the table are probabilities to the left of a specific value of X. This table is therefore similar to our normal Z-table in that it is cumulative from the left. All table values are accurate to four decimal places. We now consider two examples of using the exponential table.

EXAMPLE 7.14

For the exponential distribution shown in Figure 7.25(a), where $\lambda = .25$, find the following:

a. $P(X \leq 1)$

b. $P(1 \leq X \leq 2)$

c. $P(X > 3)$

Solution:

a. To use the table, we will locate the $\lambda = .25$ row and read over until we are in the column where $X = 1$. The entry of .2212 represents the probability that X is less than or equal to 1.

b. Just as we would in the Z-table, to find probability between two points we will find the two table values and subtract the smaller from the larger. In this problem, we would have:

$$\begin{aligned} P(1 \leq X \leq 2) &= P(X \leq 2) - P(X \leq 1) \\ &= .3935 - .2212 \\ &= .1723 \end{aligned}$$

c. This is a right tail region, so subtraction of a table value from 1 is indicated:

$$\begin{aligned} P(X > 3) &= 1 - P(X \leq 3) \\ &= 1 - .5276 \\ &= .4724 \end{aligned}$$

□

EXAMPLE 7.15

A firm's color copier averages .80 failure per week in a Poisson process. If the machine failed at 9 A.M. on Monday, what is the probability it will fail again before 9 A.M. the following Monday? What is the probability that the copier will run trouble free for more than four weeks?

Solution:

To find the chance the copier will fail during the week, we need to intersect $\lambda = .80$ with $X = 1$ week: The result is a probability of .5507. The chance of operating trouble free beyond four weeks is $1 - .9592 = .0408$.

□

COMMENTS

1. Like the Poisson, the exponential distribution has only one parameter, λ.
2. Not all exponential applications are time between Poisson occurrences. For instance, the pattern of useful service life of a telephone answering machine might follow an exponential distribution.
3. Many inexpensive hand calculators can compute e^y (where y is any number of interest) and can therefore be used to find exponential probabilities. For instance, if you wanted $P(X \leq 2.33)$ when $\lambda = .54$, you could not get an exact answer from our tables since we provide neither a $\lambda = .54$ row nor an $X = 2.33$ column. But if you have a calculator with an e^y key, let $y = \lambda X = 1.2582$. Now touch your $\pm$ key to get -1.2582, and then touch your e^y key, obtaining .284165, or .2842 to four decimal places. Subtracting this result from 1 gives an answer of $P(X \leq 2.33) = .7158$ when $\lambda = .54$. (The values in the exponential distribution table are simply $1 - e^{-y}$, where $y = \lambda X$.)

CLASSROOM EXAMPLE 7.4

Finding Exponential Distribution Probabilities

Along a congested one-half mile stretch of road, the number of accidents per week is Poisson distributed with an average of 2.0.

a. What is the distribution of the random variable X, the time measured in weeks between successive accidents on this one-half mile strip? Specify the distribution's parameters.

b. What is the probability that two or more weeks elapse between successive accidents?

c. What is the probability that no more than 12 hours pass between successive accidents?

SECTION 7.5 EXERCISES

7.43 For an exponential distribution with $\lambda = .30$, find the following:

a. the mean and the standard deviation

b. $P(X \leq 1)$

c. $P(X \geq 6)$

7.44 For an exponential distribution with mean $= 4.0$, find the following:

a. $P(X \leq .50)$
b. $P(.50 \leq X \leq 4.0)$
c. $P(X \geq 5.0)$

7.45 Suppose the number of consumers calling a toll-free question/complaint telephone number is Poisson distributed with a mean of five callers per hour.
a. Find the probability that more than 30 minutes (half of an hour) passes between two successive calls.
b. Find the probability that an hour or more will pass between two calls.

7.46 The amount of time a grocery customer waits in line until a checker begins "ringing up" the items is exponentially distributed with a mean of two minutes.
a. What is the probability a customer waits in line five minutes or more before the check-out process begins?
b. Find the probability that a customer will have a wait of a minute or less.
c. Given that 75 percent of customers wait in line fewer than x minutes before service begins, find x. (Your answer may be approximate.)

*7.47 On the same graph, sketch the exponential probability distributions specified by the following values of λ. Locate the mean for each distribution on the horizontal axis.
a. $\lambda = .20$
b. $\lambda = .30$
c. $\lambda = .40$

*7.48 Refer to Exercise 7.47.
a. What happens to μ as λ increases?
b. What happens to the variability of X as λ increases?
c. Are the distributions skewed left, skewed right, or roughly symmetrical?

*7.49 (*Note:* A computer or calculator with an e^y or e^{-y} key is helpful for this exercise.) If X is a continuous random variable with an exponential probability distribution, find the following probabilities, to four decimal places:
a. $P(X \leq 1.75)$, when $\lambda = .66$
b. $P(X > 1.14)$, when $\lambda = 2.83$
c. $P(2.12 < X < 2.97)$, when $\lambda = 1.68$

7.50 In tracking the performance of stocks, financial analysts sometimes monitor the variable X = the number of days between record high (or low) stock closing prices for the past year. A record price means a new high (or low) closing price. In a bull market, X is assumed to be approximately exponentially distributed with $\lambda = .05$.
a. In a bull market, how likely is it a new record price will occur within five days of the previous record?
b. Find the probability that more than ten days will pass before a new record occurs.

7.6 SUMMARY

Although there are many different continuous distributions, we have focused our attention in this chapter on the exponential and normal distributions. There are three primary reasons why the family of normal distributions is especially important to us:

*Optional

1. It provides a good description or fit for many continuous variables. Many data distributions have the symmetrical, mound-shaped appearance of a normal curve. We mentioned several examples at the beginning of the chapter.
2. It can be used to provide reasonable approximations to discrete distributions. We illustrated the normal approximation to the binomial in problem settings where binomial tables were not available.
3. It is at the heart of the theoretical basis for the process of reasoning we call *inference:* drawing conclusions about populations and making decisions based on sample evidence. This is discussed in the chapters immediately ahead.

A key but simple concept to working with normal distributions is standardizing: converting our random variable X into variable Z. This linear transformation gives us the "standard" normal variable that has the desirable properties of mean $= 0$ and standard deviation $= 1$. We can then use the Z-table to locate interval probabilities and percentiles of interest for any variable that has a normal distribution.

In later chapters we will introduce other continuous distributions (the t-distribution, the F-distribution, and the Chi-square distribution) that have applications under special sampling conditions.

7.7 TO BE CONTINUED . . .

. . . In Your College Courses

It would not be surprising if normal distributions appear in one or more of the courses you take in the future. This is because a normal distribution is often considered as a possible model to fit or describe the distribution of values of continuous and even some discrete variables.

- In courses in psychology or personnel management, a normal distribution is likely to be used to fit the pattern of scores obtained by individuals on intelligence tests, aptitude tests, personality trait measurements, and so on.
- In courses in management science/operations research, a normal distribution is heavily utilized. For instance, the topic of cost–volume–profit (break-even) analysis usually employs a normal distribution as a model for sales volume expected in some future time period. Many colleges and universities require such a course as part of their Business School programs.
- In courses such as marketing research, auditing, and research methods that involve gathering information through sampling procedures and surveys, you will find normal curve applications.

The exponential distribution is extensively used in management science/operations research textbooks to model waiting lines, time required for service, and time between arrivals of customers in a Poisson process.

. . . *In Business/The Media*

As we will see in the next few chapters, the normal curve plays a very important role in business applications where we use sample information to draw inferences about a given population.

A more specific application of the normal curve is represented by the practice of quality control. Most production processes continually monitor relevant physical properties of the units they produce. The underlying model for understanding random variation in manufactured items is the normal curve. Since this is a major business statistics application, special attention is given to this topic in Chapter 11.

CHAPTER 7 EXERCISES

7.51 A normal distribution has $\mu = 39.1$ with $\sigma = 4.74$. Express the following points as Z-scores:
a. $X = 40$ **b.** $X = 35$ **c.** $X = 45.7$

7.52 Refer to the distribution in Exercise 7.51. Change the following into unstandardized form:
a. $Z = -3.50$ **b.** $Z = .83$ **c.** $Z = -1.20$

7.53 A normal distribution has $\sigma = 400$ and a Z-score of .30 when $X = 2820$. Find the mean value of X.

7.54 Find the following probabilities on a standardized normal curve:
a. $P(-3.44 \leq Z \leq -2.44)$
b. $P(Z \geq -.81)$
c. $P(Z \leq -.99)$

7.55 Consider a standard normal curve.
a. What Z-score is exceeded 82 percent of the time?
b. What Z-score has 97.5 percent of the area to its left?
c. Between what Z-scores do the middle 30 percent of values lie?

7.56 A corporation gives an employment exam to its potential production workers at the time of their application for work. Scores obtained on the test are approximately normally distributed with a mean of 150 and a standard deviation of 50. The personnel manager wants to establish a policy that only those who score in the top 20 percent on the test should be hired. Under this policy, what would be the minimum hiring score?

7.57 A two-liter bottle of a soft drink is filled to a mean of 2.020 liters, with a standard deviation of .016 liters. The distribution of fill amounts closely resembles the normal distribution.
a. What percent of bottles get less than 1.98 liters?
b. What fill volume corresponds to the first percentile of the distribution?
c. What percent of bottles get less than the label amount of soft drink?
d. If we want the first percentile to occur at $X = 2.000$ liters (so that only 1 percent of bottles are underfilled), what should be the process mean?

7.58 Suppose that monthly expenditures for food items for a family of two adults and one child in a certain state is a normally distributed random variable with mean \$390 and standard deviation \$72.
a. What percent of such families spend no more than \$400 monthly?

b. What percent get by on \$300 or less per month?
c. What percent spend more than \$500 monthly?
d. What food expenditure, x, is the 75th percentile of the distribution?

7.59 Find the following p-values:
a. $P(Z \geq 3.01)$
b. $P(Z \leq -1.97)$
c. $P(Z \geq .31)$

7.60 A normal distribution has $\mu = \$475$ and $\sigma = \$48$. What percent of the population lies further than \$105 away from the mean (in either direction)?

7.61 Suppose adult weights are normally distributed, with $\mu = 155$ and $\sigma = 20$ pounds for men and with $\mu = 125$ and $\sigma = 16$ for women. For a man who weighs 210 pounds, what woman's weight would place her in the equivalent percentile for her sex?

7.62 For binomial sampling with $n = 385$ and $\pi = .24$, find $P(X \geq 100)$ using the normal approximation without the continuity correction.

7.63 Refer to Exercise 7.62. Use the continuity adjustment in solving $P(X \geq 100)$.

7.64 Suppose that 46 percent of adults in the population favor the death penalty as punishment for certain crimes. If a national polling organization surveys 900 adults chosen by a random selection method, how likely is it the sample will show 50 percent or more (450 or more adults) in favor of the death penalty?

7.65 Refer to Exercise 7.64. Letting X represent the number in the sample favoring the death penalty, find $P(400 \leq X \leq 428)$.

7.66 A marketing research firm studied eating-out patterns of young suburban working couples by having a group of subjects keep diaries of their spending for food items. One result was that the mean amount spent dining outside the home was \$220 per month, with a standard deviation of \$55. The variable of interest is approximately normally distributed.
a. Find the probability a couple spends more than \$300 per month dining outside the home.
b. What percent of couples spend from \$200 to \$300 per month?
c. What percent of couples spend less than \$100 per month?

7.67 A firm that builds prefabricated homes believes that the number of days, X, it takes to complete its most popular model is a normally distributed random variable with mean = 18.5 and standard deviation = 2.0 working days.
a. How likely is it a given home will be completed in 15 or fewer working days?
b. What completion time x is exceeded by 25 percent of the homes built?
c. What proportion of homes have a completion time from 15.0 to 25.0 workdays?

7.68 Suppose in a rural county that 18 percent of single-family residences have a satellite dish. If a random sample of 150 residences is taken, how likely is it the sample will contain 15 or less residences with dishes?

7.69 Suppose the service life of a telephone answering machine is exponentially distributed with a mean of four years. If the warranty on the machine is good for six months, how likely is the machine to fail before the warranty expires?

7.70 An emergency medical team responds to an average of two situations per eight-hour shift. The pattern of number of response situations per shift follows the Poisson distribution.
a. What is the mean time between responses?
b. How likely is it two or more shifts will pass between successive responses?
c. How likely is it that successive calls will be no more than four hours apart?

*7.71 A continuous distribution of somewhat lesser importance is the *uniform distribution*. If all values of a continuous variable are equally likely to occur across a given interval, the variable is then said to be uniformly distributed. Consider the following sketch

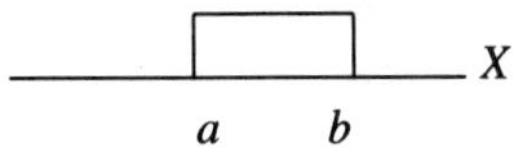

which suggests the rectangular shape of the continuous uniform distribution results in probability that is evenly spread out (quite unlike the normal and exponential) across the range of possible values, from a to b. To ensure that the total enclosed area is 1.00, the height of the rectangle must be $1/(b - a)$. The uniform distribution has mean $(a + b)/2$, with standard deviation $(b - a)/\sqrt{12}$. Although its real-world uses are limited, the uniform distribution has applications in computer simulation and in random number generation.

Suppose arriving passengers at a major airport utilize a monorail train car to get from outlying gates to the terminal–baggage claim area. A car leaves every four minutes. Since passengers arrive randomly at the train departure point, the uniform distribution would apply.

a. What is the mean passenger wait time for a car?
b. What is the standard deviation?
c. What would be the height of the rectangle shown above?
d. How likely is it a given passenger has to wait more than 1.5 minutes for a ride?

7.72 Refer to Example 7.10, where a sample of 100 voters was drawn with $\pi = .30$. Find the region of concentrated values for X, where X is the number of unregistered voters. Report your result to the nearest whole numbers.

7.73 Suppose daily rental car rates for a particular model are approximately normally distributed. If 75 percent of rates are $37 or less, and if 10 percent of rates are $23 or less, find the distribution's mean and standard deviation, to two decimal places.

7.74 The Environmental Protection Agency (EPA) estimates that a particular automobile model will achieve a highway rating of 32 miles per gallon. The EPA label in the car window also says, ". . . the majority of cars will achieve between 25 and 39 mpg." Assume the highway mpg figures are approximately normally distributed.

a. Suppose we interpret "majority of cars" to mean 95 percent. Find the standard deviation for the distribution of mpg figures, to two decimal places.
b. Using your result in a, what is the probability a car will achieve 30 or more highway mpg?
c. Suppose the city driving estimate is 23 mpg, with a majority between 18 and 28. Determine the standard deviation as above and find the chance a car will get 20 mpg or less in city driving.

7.75 A fast-food corporation has set up an end-of-the-year incentive plan for the managers of its company-owned stores. As the following table indicates, managers' bonuses are determined by comparing each store's actual year sales against the projected or target sales:

*Optional

Percent Actual Sales Exceed Target Sales	Manager's Bonus
15 or more	\$10,000
10 to 15	5,000
7.5 to 10	2,000
Less than 7.5	0

Suppose the distribution of actual sales divided by target sales, X, is approximately normal for the stores, with a mean of 1.02 and a standard deviation of .075.

a. Find the percent of store managers that will be in each bonus category. Report category results to the nearest whole percent.

b. Among the managers receiving a bonus, what is the average amount of the bonus?

***7.76** To test the randomness of a random number generator, systems analysts often compare the theoretical number of values in certain ranges with the observed number of values. The uniform and binomial distributions work together in this case. Let

$$Y = \text{a random number between 0 and 1}$$
$$X = \text{the number of } Y\text{-values between 0 and .25}$$

The variable Y is assumed to have a uniform distribution, and variable X has a binomial distribution with $\pi = P(0 \leq Y \leq .25) = .25$.

a. If 10,000 values of Y are generated, find the expected value of X.

b. Find the region of concentrated values of X, to the nearest whole numbers.

c. What would you conclude about the randomness of the random number generator if the observed value of X fell outside the region of concentrated values?

***7.77** (See Exercise 7.71.) A gravel sales company sells various sizes of stones in bulk to its customers. The weight of each purchase is determined through use of a truck scale. Each customer checks in and gets his truck weighed. Then the customer drives to the loading area to get his order. Finally, the customer has his truck reweighed, paying his bill according to the difference in the two weights. The net weight of each customer's order is rounded off to the nearest 50 pounds. Assume that X, round-off error per order (actual weight minus billed weight), is a uniformly distributed random variable.

a. What are the largest and smallest values X can be?

b. Find $P(X \geq 15 \text{ pounds})$.

c. Find the mean value for X.

d. Find the standard deviation for X.

***7.78** Refer to Exercise 7.77. Suppose customer orders are rounded to the nearest 100 pounds instead of 50 pounds.

a. Graph the distribution. What would be the height of the rectangle?

b. Find the mean value for X.

c. Find the standard deviation for X.

7.79 According to a Roper organization survey on the "underground economy" (*Roper Reports*, 5, New York, 1985), 18 percent of U.S. adults surveyed had traded services with someone within the previous six months. An example would be an accountant who does a dentist's books in exchange for dental care—a nontaxable transaction. Suppose this finding remains accurate for the current adult population, and that a random sample of 1500 adults is selected. Let X = number of participants in the underground economy.

a. What are the mean and standard deviation of X?
b. What is the probability that the sample will show no more than 250 participants?
c. There is less than a 1 percent chance the sample will show x or more participants. What is the lowest integer value for x that makes the previous statement true?

Computer Exercise

C7.1 We suggest you solve the problems below by hand, and then run the Normal Probability Distribution program to verify your answers. (Minor discrepancies may occur since the computer's table values are accurate to more decimal places than those in the book.)

Let X be a normally distributed random variable with mean = \$565 and standard deviation \$206. Find the following:

a. $P(\$565 < X < \$900)$
b. $P(X < \$300)$
c. $P(X > \$1000)$
d. the 14th percentile of the X distribution
e. $P(X < \$750)$
f. the value of X with 63 percent of area to the right
g. $P(\$250 < X < \$450)$
h. the value of X with 97.5 percent of area to the left
i. the values of X between which the middle 98 percent of the distribution lies
j. $P(X > \$375)$

Chapter Maxim *A sample mean is likely to be in the neighborhood of the population mean but off by random chance.*

CHAPTER 8

THE SAMPLING DISTRIBUTION OF THE MEAN

Objectives

After studying this chapter and working the exercises, you should be able to

1. Discuss why the sample mean $\bar{X}$ is viewed as a random variable.
2. Describe the concept of a sampling distribution.
3. Be able to compute and explain the standard error of the mean.
4. State the Central Limit Theorem and explain its significance.
5. Find probabilities on $\bar{X}$'s sampling distribution curve using the Z–table.
6. Recognize the conditions when the standard error should be adjusted by a correction term.
7. Describe the relation between sample size and sampling error.

Perhaps the single most important idea in this textbook is the concept of a sampling distribution. We will use this idea in almost every chapter throughout the rest of the book. In order to understand inferential statistics, it is vital to comprehend what a sampling distribution represents. In this chapter, we will focus on one particular sampling distribution: the sampling distribution of the sample mean, $\bar{X}$.

Although this chapter may be somewhat more theoretical than others, your own common sense should help. Some of the major results that we will arrive at should seem readily believable and consistent with your intuition. For instance, let's consider a question we've posed previously in one form or another. Suppose you take a random sample of items from a large population (for instance, a sample of 10 stock prices from the New York Stock Exchange on a given day) and then compute the sample's mean value, $\bar{X}$. Then suppose 50 other people each take their own sample of 10 NYSE stocks and compute their own mean, $\bar{X}$. Now let's imagine comparing the 51 values of $\bar{X}$ that have resulted. Will all 51 values be the same? Your intuition should say "No!" Although taken from the same population, each sample would be made up of a different collection or subset of population items; therefore, we would expect to see variation among the sample means. In fact, it is unlikely that *any* two sample means would be the same. If this seems reasonable to you, then you are well on your way to understanding sampling distributions.

The major objective of this chapter is to be able to understand and describe the variability that sample means possess; we will call this pattern the sampling distribution of $\bar{X}$. In subsequent chapters, we will introduce in less detail the sampling distribution for other statistics, such as the sample proportion and the sample variance.

As we proceed through this chapter and others that follow, we will often use the phrase "random sample." There is more than one kind of random sample, as we discussed in Chapter 1. It should, therefore, be understood that we will be referring specifically to a simple random sample in all instances.

8.1 SAMPLING DISTRIBUTION OF $\overline{X}$

Let's suppose that grade point averages (GPAs), represented by the variable X, at a certain college are approximately normally distributed with $\mu = 2.77$ and $\sigma = .41$. (See Figure 8.1.) There are 860 students at the college. Further suppose that we are going to randomly choose 36 individual student GPAs and then compute the sample mean, $\overline{X}$.

To refresh ourselves on notation, let's refer to Table 8.1 before proceeding. The symbols μ and σ denote *parameters*—that is, numerical properties of a population. The symbols $\overline{X}$ and S denote *statistics*—that is, numerical properties of a sample.

Viewing $\overline{X}$ as a Random Variable

Now getting back to our sample of 36 values of X, do we know in advance what the resulting sample mean $\overline{X}$ will be? Obviously not; our sample may happen to include a majority of stronger students and thus yield a sample mean above 2.77; however, it's just as likely our sample will include a majority of weaker students, causing the sample mean to be below 2.77. Or the sample may be one where the strong and weak students tend to balance one another fairly evenly, yielding a sample mean $\overline{X}$ that is quite close to 2.77. The point here is that *the sample mean $\overline{X}$ is a random variable*. The mean of the population μ is a constant, but the sample mean (as well as the sample standard deviation) is a variable. It is for this reason that we show question marks for the sample results in Table 8.1; we don't know in advance what a particular sample is going to tell us.

Potential Values for $\overline{X}$

We know that the value of $\overline{X}$ obtained from a given sample depends on which population members happen to be selected to make up the sample. The sample mean $\overline{X}$ can assume a variety of values because many different combinations or sets of observations could be selected to constitute the sample. To illustrate, in the sampling

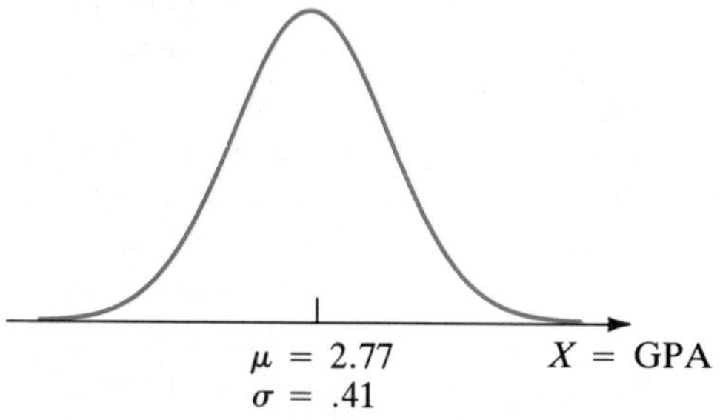

Figure 8.1 Normal Distributrion of Individual GPAs

Table 8.1 Notation for First Sampling Example

Symbol	Meaning	Numerical Value
μ	Population mean GPA	$\mu = 2.77$
σ	Population standard deviation	$\sigma = .41$
$\overline{X}$	Sample mean GPA	$\overline{X} = ?$
S	Sample standard deviation	$S = ?$

situation described above (a sample of size $n = 36$ is to be drawn from a population of size $N = 860$), the number of different possible samples or collections of individual items available to be chosen is, in scientific notation (see the Comment Section on page 414):

$$\frac{860!}{36!\,824!} \approx 5.6 \times 10^{63}$$

which is a huge number. You should be willing to believe that the following statements apply to the 5.6×10^{63} different possible sets of observations of size $n = 36$:

- If simple random sampling is employed, each of the sets has the same chance of being chosen when the sample is drawn.
- There is a value for $\overline{X}$ for each set.

Let's formalize this discussion into a statement called an *axiom*.

Axiom 8.1

Different random samples, each of size n taken from the same population, will generally contain different sets of individual values; therefore, we expect sample statistics such as $\overline{X}$ to vary randomly from one sample to another.

We move now to the main concept of this chapter: describing the distribution that these numerous sample means would form if we could somehow see them all. This distribution of all possible sample means is called the *sampling distribution of* $\overline{X}$.

Definition

The **sampling distribution of $\overline{X}$** is a probability distribution of all the possible values of $\overline{X}$.

Let's now imagine that we can observe all 5.6×10^{63} values for the random variable $\overline{X}$. If we were to graph these sample means onto a frequency curve, what pattern do you think these values of $\overline{X}$ might form? In particular, consider these questions:

1. Where do you think these values of $\overline{X}$ would be centered?
2. Do you think that these sample means would be more variable or less variable than the individual values of X represented in Figure 8.1?

To answer the first question, your intuition should tell you that the distribution of possible sample means should be centered near the actual population mean, $\mu = 2.77$. When sampling from a normal distribution, it should seem reasonable that a sample mean $\overline{X}$ is just as likely to land below the population mean μ as it is to land above. To answer the second question, your common sense should tell you that sample means (which are in between the smallest and largest individual values of X within a given sample) would be less spread out than the various values of X that make up the population. Let's organize these beliefs into a second axiom.

Axiom 8.2

The distribution of possible sample means (*the sampling distribution of* $\overline{X}$) has the same mean as but less dispersion than the underlying population of individual items.

Figure 8.2 illustrates this discussion. The curve at the top of the figure represents the population of individual GPAs; accordingly, its horizontal axis label is X. The arrow indicates that, when a sample is drawn and $\overline{X}$ is computed, we are no longer on the curve for individual GPAs; rather we are on a sampling curve whose horizontal axis label is $\overline{X}$. Axiom 8.2 is illustrated by centering this distribution (representing 5.6×10^{63} means) at 2.77, but showing it with less dispersion than the population pictured above it.

COMMENT *Scientific notation* is a concise way to express large numbers. As an example, several times in this section we referred to 5.6×10^{63}. Rather than write out the full number (it would involve moving the decimal point over 63 places to the right), we use this shorthand. As another example 1,430,000,000,000 can be written as 1.43×10^{12}. Many hand calculators use scientific notation for large numbers; check the owner's manual to see whether yours does and to ascertain how it is displayed. (*Note:* A discussion of scientific notation is included in the *Partial Solutions Manual and Math Review* that accompany this text.)

Sampling from Normal Populations

We can provide more specifics about the sampling distribution of $\overline{X}$. In particular, when we know that our sample is taken from a normal population, we apply the rule given in the following box.

Sampling from a Normal Distribution of X

The sampling distribution of $\overline{X}$ is also a normal distribution with mean (denoted by $\mu_{\overline{X}}$) equal to the population mean μ and with standard deviation (denoted by $\sigma_{\overline{X}}$) equal to $\sigma/\sqrt{n}$.

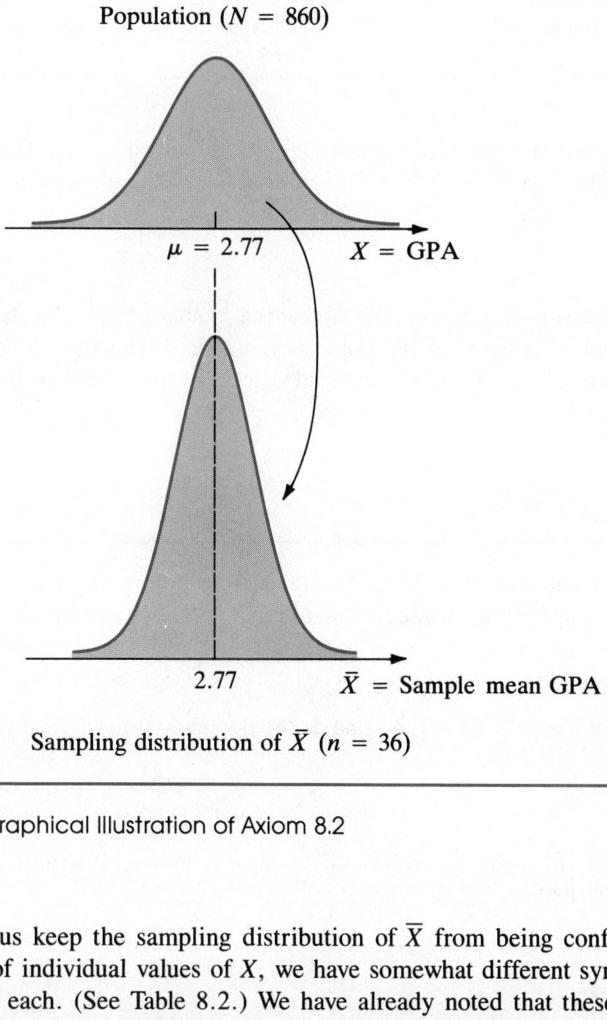

Figure 8.2 Graphical Illustration of Axiom 8.2

To help us keep the sampling distribution of $\overline{X}$ from being confused with the distribution of individual values of X, we have somewhat different symbols and terminology for each. (See Table 8.2.) We have already noted that these distributions have the same average value ($\mu_{\overline{X}} = \mu$), but that sample means are less variable than individual values of X. The standard deviation of the sample means is called the

Table 8.2 Distribution Notation and Use: Individual Values Versus Sample Means

	Distribution	
Characteristic	Individual Values in the Population	Sample Means (Sampling Distribution)
Symbol for mean	μ	$\mu_{\overline{X}}$
Horizontal axis label	X	$\overline{X}$
Measure of dispersion	Standard deviation (of X)	Standard error (of $\overline{X}$)
Symbol for dispersion measure	σ	$\sigma_{\overline{X}}$
Applicability	Interest in individual values of X	Sample is drawn; interest in computed value of $\overline{X}$

standard error of the mean and is symbolized by $\sigma_{\overline{X}}$. The key relation between the dispersion of the individual values of X and the dispersion of means $\overline{X}$ is this:

$$\sigma_{\overline{X}} = \frac{\sigma}{\sqrt{n}} \tag{8–1}$$

This simple equation tells us that the standard error (the standard deviation of sample means) is equal to the population standard deviation divided by the square root of sample size. With this knowledge, we can now describe the sampling distribution in detail.

EXAMPLE 8.1A

For our sample of 36 GPAs drawn from a normal population with $\mu = 2.77$, $\sigma = .41$, specify the mean, standard error, and shape of the sampling distribution of $\overline{X}$.

Solution:

The mean is 2.77, the same as the population mean. The standard error is

$$\sigma_{\overline{X}} = \frac{\sigma}{\sqrt{n}} = \frac{.41}{\sqrt{36}} \approx .0683$$

and since the individual GPAs have a normal distribution, so does the sampling distribution.

□

EXAMPLE 8.1B

Answer the same questions as in Example 8.1A, assuming that the sample size is 15.

Solution:

The mean still will be 2.77 and the sampling distribution of $\overline{X}$ still will be normal, but the standard error now will be larger:

$$\sigma_{\overline{X}} = \frac{\sigma}{\sqrt{n}} = \frac{.41}{\sqrt{15}} \approx .1059$$

Figure 8.3 shows these two sampling distributions; it should also suggest that there is a sampling distribution for each different sample size *n;* the larger *n* becomes, the more compact the curve will be.

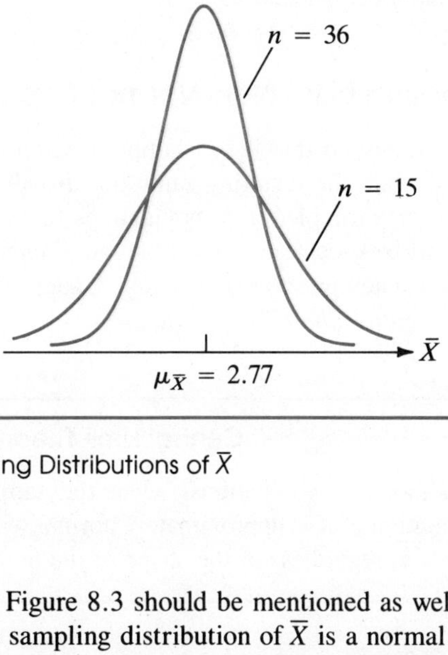

Figure 8.3 Two Sampling Distributions of $\overline{X}$

□

Another aspect of Figure 8.3 should be mentioned as well. Note that regardless of the sample size, the sampling distribution of $\overline{X}$ is a normal curve centered around the value of μ.

When a random sample is drawn, we do not know in advance what its mean $\overline{X}$ will be, since $\overline{X}$ is a random variable. But due to the symmetrical shape of the sampling distribution of $\overline{X}$, we do know this: A sample mean $\overline{X}$ is just as likely to be below μ as it is to be above μ. As a result, we can say that $\overline{X}$ has the property of being *unbiased,* since the average value of $\overline{X}$ is the same as the population mean μ. Dictionaries define *bias* as prejudice or favoritism. If $\overline{X}$ were biased, then it would be more likely to fall below μ than to fall above, or vice versa. Being unbiased is an important and useful statistical property.

The discussion in this chapter also should have made it apparent that when a sample is drawn, $\overline{X}$ most likely will differ from μ. The term *sampling error* often is used to denote this difference.

Definition

Sampling error is the amount by which the value obtained in a random sample (such as $\overline{X}$) differs from the counterpart value in the population (such as μ).

Sampling error is to be expected since $\overline{X}$ is based on a portion of the population and μ is based on the entire population. In real-world applications, we rarely will know the magnitude of the sampling error since this would require having complete knowledge of the population—that is, a *census*. The potential magnitude of sampling error is related to the size of the sample that is drawn. Inspection of Figure 8.3 should suggest that the sampling error is likely, though not certain, to be smaller for a sample of $n = 36$ than it is for a sample of $n = 15$. This is because the sampling distribution becomes more compact as n increases.

Sampling from Non-Normal Populations

We have established that if our sample is taken out of a population that is normally distributed, then the resulting sampling distribution of $\overline{X}$ is also normal. But very often we'll take samples from populations that are skewed or otherwise non-normal. And we will have occasion to sample from populations where we have no idea about shape but cannot presume normality. Under these conditions, an important result called the *Central Limit Theorem* applies.

Central Limit Theorem (CLT)

For almost all populations, when the sample size is large, the sampling distribution of $\overline{X}$ is approximately normal with mean $= \mu$ and standard error $= \sigma/\sqrt{n}$, regardless of the shape of the population being sampled.

The CLT is very important because, for large sample sizes, it tells us that we do not need to know the shape of the population from which we sample in order to declare the sampling distribution of $\overline{X}$ an approximately normal curve. For instance, if we sample from a population of X that is skewed to the left, the CLT informs us that the sampling distribution of $\overline{X}$ will be approximately normal for large n as opposed to retaining the skewed left shape of the underlying population. Figure 8.4 generalizes this phenomenon by implying that a normal curve describes the sampling distribution of $\overline{X}$, regardless of the shape of the population.

In stating the Central Limit Theorem, we referred to a large sample size. There is no universal agreement as to what constitutes a sufficiently large sample size for the normal distribution to be a good model for the sampling distribution of $\overline{X}$. If the underlying population is symmetrical (but not normal), the sampling distribution may be nearly normal for $n = 10$ or 15; if the population is moderately skewed, a sample

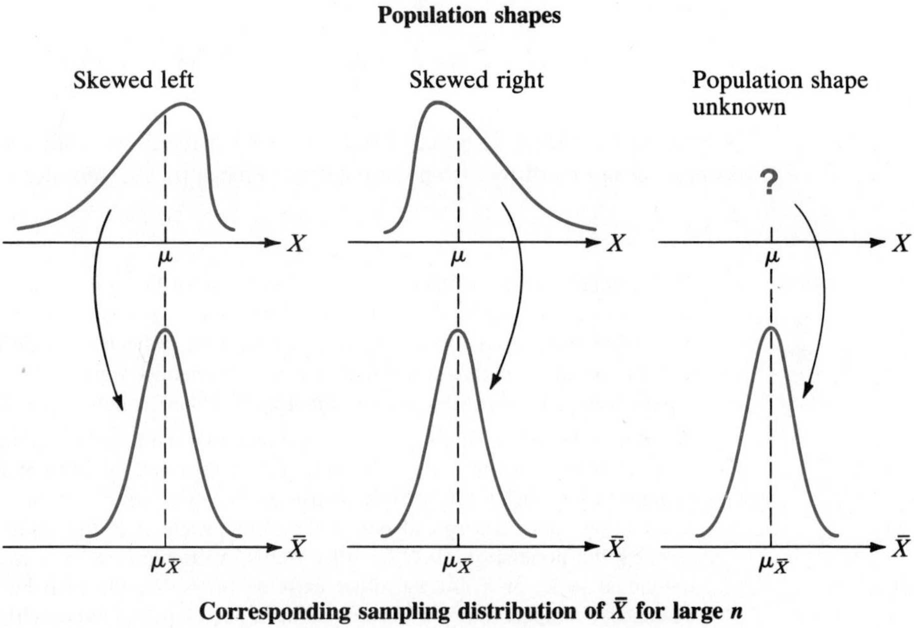

Figure 8.4 The CLT: Normal Sampling Distributions Result from All Populations

of 15 is not large enough to ensure a normal sampling distribution. A commonly used rule of thumb is that the sample size should be at least 30 for the CLT to apply. Distributions with extreme skewness would require larger samples to have approximate normality of the sampling distribution.

EXAMPLE 8.2

Suppose the selling price of single-family residential homes in a medium-sized metropolitan area is noticeably skewed to the right, with a mean of \$114,500 and a standard deviation of \$39,600.

a. Describe the sampling distribution of $\overline{X}$ if a sample of 75 home sales is to be drawn.

b. Specify the region of concentrated values for $\overline{X}$.

Solution:

a. Although the individual selling prices in the population being sampled form a skewed right curve, the CLT applies since our sample is large. Therefore, we can say that the sampling distribution of $\overline{X}$ will be approximately normal with $\mu_{\overline{X}} =$ \$114,500 and a standard error = \$39,600/$\sqrt{75}$, or about \$4,573.

b. For a normal sampling distribution, our region of concentrated values is approximately:

$$\begin{aligned}\text{Mean} \pm 2 \text{ standard errors} &= \$114{,}500 \pm 2(\$4{,}573)\\ &= \$114{,}500 \pm \$9{,}146\end{aligned}$$

A random sample of 75 prices has about a 95 percent probability of yielding a sample mean that differs from the population mean by $9,146, or less.

□

COMMENTS

1. The sample mean is not our only measure of central location for a data set, but it does have certain desirable mathematical properties and is therefore the measure most often used. It is for this reason that we have concentrated on the sampling distribution of $\overline{X}$ in this section. If the sample median were as important as the sample mean, we might develop its sampling distribution here as well.
2. When we referred to Figure 8.3, we mentioned that there is a sampling distribution for each different sample size n. To help picture this, you might think about the two extreme values that n can assume. If n is at its lowest ($n = 1$), then the sampling distribution curve has spread out to the point where it is the same as the curve showing the population of X. In other words, a sample of $n = 1$ is really just an individual value of X. At the other extreme ($n = N$), the distribution of $\overline{X}$ has become so compact that it is a single point; it is no longer a curve with area beneath it. This is because a census (100 percent sample) has been taken, and the sample mean will coincide with the population mean; all variability and uncertainty have been removed.
3. Special procedures exist for situations where small samples ($n < 30$) are drawn. For the time being we will presume large samples; in Chapter 11 the situation of small samples will be discussed.
4. We recommend carrying out computations for $\sigma_{\overline{X}}$ to at least the same number of decimal places as are available for σ.

SECTION 8.1 EXERCISES

8.1 A population is normally distributed with $\mu = 840$ and $\sigma = 110$. Specify the mean, standard error, and shape of the sampling distribution of $\overline{X}$ for a sample of size 50.

8.2 Given a normal population with mean = 4.20 percent and standard deviation = 1.20 percent, describe the sampling distribution of $\overline{X}$ when $n = 100$.

8.3 A population is believed to be strongly skewed to the right. If its mean is $72,300 and its standard deviation is $21,300, specify the standard error and shape of the sampling distribution of $\overline{X}$ for a random sample of 125 items.

8.4 A population of unknown shape has $\mu = \$17{,}300$ and $\sigma = \$7{,}550$. Specify the shape of the sampling distribution of $\overline{X}$ and the value of the standard error for the following size samples:

a. $n = 35$ **b.** $n = 75$ **c.** $n = 450$

8.5 Suppose a population's standard deviation is 11,400.

a. What is the standard error of $\overline{X}$ when the sample size is 40?

b. What sample size would be required to reduce your answer to part a by 50 percent?

8.6 A sample of size 60 is drawn from a large population that has $\mu = 32.12$ and $\sigma = 18.40$.

a. What is the standard error of $\overline{X}$?

b. What is the region of concentrated values (where the sample mean has about a 95 percent probability of landing)?

c. How do your answers to parts a and b change if $n = 200$?

8.7 A population has a mean of 120 seconds with a standard deviation of 25 seconds. A sample of size 300 is to be randomly selected.

a. What is the standard error of $\overline{X}$?

b. What is the region of concentrated values of $\overline{X}$?

c. How do your answers to parts a and b change if $n = 40$?

8.8 Suppose a population of accounts receivable has an unknown mean, but we know $\sigma =$ \$72.50. Assume a sample of 75 accounts receivable is drawn and we compute $\overline{X}$.

a. What is the standard error?

b. The sample mean $\overline{X}$ has about only a 5 percent chance of being different from the population by more than some amount that we will call J. What is the numerical value of J?

8.9 Suppose the mean weight per brick, μ, in a population of bricks is unknown, but we have reason to believe that the standard deviation of brick weights is 50 grams. A sample of 64 bricks is to be chosen at random. The sample mean will be computed.

a. What is the standard deviation of the possible sample means that could result?

b. What value, arbitrarily denoted as J, will make the following a true statement? The probability is about .95 that $\overline{X}$ will fall within J grams above or below μ, the unknown population mean.

8.10 In order to specify the parameters of the sampling distribution of $\overline{X}$, we need to know three quantities. What are these quantities?

***8.11** Suppose we investigated sampling distributions for measures other than $\overline{X}$; for example, the median, range, or standard deviation. Using the notation in Table 8.2 as your guide, write down the symbols that would depict the following quantities:

a. the mean of the sampling distribution of the sample median.

b. the standard deviation of the sampling distribution of the sample range.

c. the mean of the sampling distribution of the sample variance.

d. the standard error of the sample standard deviation.

8.12 Suppose the sampling distribution of $\overline{X}$ is given in tabular form as the following frequency distribution:

$\overline{X}$	−2.0	−1.5	−1.0	−.5	0	.5	1.0	1.5	2.0
f	1	4	1	4	8	4	1	4	1

a. Compute $\mu_{\overline{X}}$. (*Hint:* View the sampling distribution of $\overline{X}$ as an ungrouped frequency distribution and then use Equation 3–3a to determine $\mu_{\overline{X}}$.)

b. Can you state the value of μ, the population mean? If so, what is it?

*Optional

8.13 Characterize as normal, approximately normal, or non-normal the shape of the sampling distribution of $\overline{X}$ in each of the following sampling situations:
a. samples of size $n = 18$, from a normal population with $\sigma = 3$.
b. samples of size $n = 52$, from an exponential population.
c. samples of size $n = 92$, from a binomial population.
d. samples of size $n = 8$, from a Poisson population.
e. samples of size $n = 26$, from a symmetrical population.

8.2 PROBABILITY STATEMENTS ABOUT $\overline{X}$

In Chapter 7, we learned how to evaluate probability statements about individual values of a random variable X. In this chapter, we are interested in taking a random sample and computing $\overline{X}$, the sample mean. Accordingly, we would now like to evaluate probability statements about the behavior of $\overline{X}$. To do so, we will again use the Z-table.

Creating the Standardized Variable Z

Since we know from the CLT that the sampling distribution of $\overline{X}$ follows the normal distribution for large n, we can find probabilities of interest by standardizing $\overline{X}$ and using the familiar Z-table. In Chapter 7, we gave the following formula to change X into its corresponding value of Z:

$$Z = \frac{X - \mu}{\sigma}$$

There are two changes to this expression to keep in mind when we take a sample and determine $\overline{X}$:

1. Instead of some individual X, we are now interested in the value of $\overline{X}$, the sample average or mean.
2. The sample size n affects the standard deviation of the sample mean (the standard error).

Accordingly, the formula for the Z-transformation of $\overline{X}$—for standardizing a value of $\overline{X}$—is given in the following box.

Z-Transformation for $\overline{X}$

$$Z = \frac{\overline{X} - \mu_{\overline{X}}}{\sigma_{\overline{X}}} = \frac{\overline{X} - \mu}{\sigma/\sqrt{n}} \quad (8\text{–}2)$$

Note that $\overline{X}$ replaces X to indicate our interest in the sample mean as opposed to some individual value of X. Also note that $\sigma_{\overline{X}} = \sigma/\sqrt{n}$ replaces σ to indicate the standard deviation of means instead of the standard deviation of individual values of X.

Finding Probabilities for $\overline{X}$

We will now use Equation 8–2 to standardize and we will find probabilities for $\overline{X}$ in exactly the same fashion as in the previous chapter. Again we recommend drawing a graph to visualize the portion of the curve where you wish to find probability.

EXAMPLE 8.3

A sample of size 50 is to be drawn from a roughly symmetrical distribution that has $\mu = 90$, $\sigma = 12$. Evaluate the following probability statements:

a. $P(\overline{X} \geq 95)$

b. $P(\overline{X} < 91)$

c. $P(87 < \overline{X} < 93)$

d. Redo part c for a sample size of $n = 100$.

Solution:

Although the underlying population being sampled is not a normal one, our sample size is large enough for the Central Limit Theorem to apply, thereby giving us a normal sampling distribution of $\overline{X}$.

a. We express $\overline{X}$ as a Z-value and then find the corresponding probability in the Z-table. The value $\overline{X} = 95$ is equivalent to

$$Z = \frac{\overline{X} - \mu}{\sigma/\sqrt{n}}$$
$$= \frac{95 - 90}{12/\sqrt{50}} = 2.95$$

Therefore,

$$P(\overline{X} \geq 95) = P(Z \geq 2.95)$$
$$= 1 - .9984 = .0016$$

A sample mean as large as 95 is highly unlikely. See Figure 8.5 for graphs of all parts of this example.

b. $$P(\overline{X} < 91) = P\left(Z < \frac{91 - 90}{12/\sqrt{50}}\right)$$
$$= P(Z < .59) = .7224$$

c. $P(87 < \overline{X} < 93) = P\left(\frac{87 - 90}{12/\sqrt{50}} < Z < \frac{93 - 90}{12/\sqrt{50}}\right)$

$= P(-1.77 < Z < 1.77) = .9616 - .0384 = .9232$

d. For $n = 100$,

$$P(87 < \overline{X} < 93) = P(-2.50 < Z < 2.50)$$
$$= .9938 - .0062 = .9876$$

Although the interval of interest for $\overline{X}$ extends from 87 to 93 as in part c, the larger sample size has made the sampling distribution more compact allowing a greater probability that $\overline{X}$ will fall in the interval. See Figure 8.5(d).

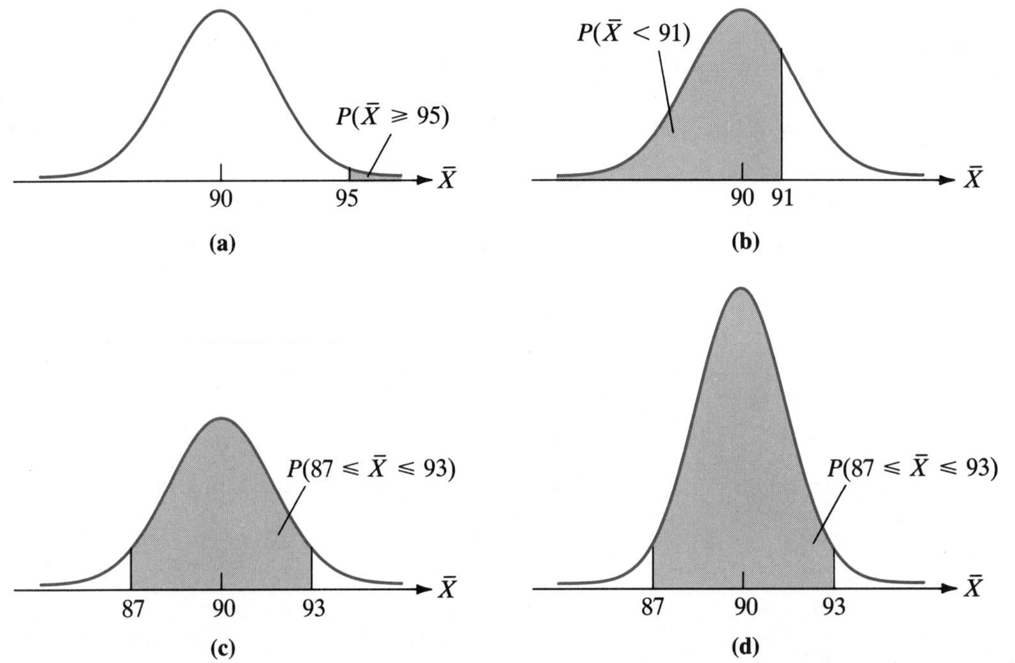

Figure 8.5 Probabilities for Example 8.3

□

EXAMPLE 8.4

A sample of 36 grade point averages is drawn at random from a normal population of GPAs having $\mu = 2.77$ and $\sigma = .41$.

a. What is the region of concentrated values for $\overline{X}$?

b. How likely is the sample average to exceed 3.00?

c. Find the probability the sample mean $\overline{X}$ will differ from the population mean by no more than $\pm .10$.

d. Could we answer part c if the value of the population mean μ were unknown, but we do know $\sigma = .41$?

Solution:

Since the individual GPAs in this population are normally distributed, the sampling distribution of $\overline{X}$ is also normal, with $\mu_{\overline{X}} = 2.77$ and $\sigma_{\overline{X}} = .41/\sqrt{36} = .0683$.

a. The region of concentrated values for $\overline{X}$ is

$$\mu_{\overline{X}} \pm 2\sigma_{\overline{X}} = 2.77 \pm 2(.0683) \approx 2.77 \pm .14$$

The probability is about .95 that the sample mean $\overline{X}$ will fall between a GPA of 2.63 and 2.91.

b. $$P(\overline{X} > 3.00) = P\left(Z > \frac{3.00 - 2.77}{.0683}\right) = P(Z > 3.37)$$
$$= 1 - .9996 = .0004$$

The point $\overline{X} = 3.00$ is at the 99.96th percentile of the sampling distribution, well outside of the region of concentrated values.

c. $$P(2.67 \leq \overline{X} \leq 2.87) = P\left(\frac{2.67 - 2.77}{.0683} \leq Z \leq \frac{2.87 - 2.77}{.0683}\right)$$
$$= P(-1.46 \leq Z \leq 1.46) = .9279 - .0721 = .8558$$

There is about an 86 percent chance that $\overline{X}$ based on a sample of $n = 36$ will land within .10 of the mean of the population.

d. Yes, the question is still answerable. By simply letting the lower and upper limits of our interval of interest be symbolized as $(\mu - .10)$ and $(\mu + .10)$, we would have

$$P[(\mu - .10) \leq \overline{X} \leq (\mu + .10)]$$
$$= P\left[\frac{(\mu - .10) - \mu_{\overline{X}}}{.0683} \leq Z \leq \frac{(\mu + .10) - \mu_{\overline{X}}}{.0683}\right]$$

Since $\mu_{\overline{X}} = \mu$, this expression reduces to

$$P\left(\frac{-.10}{.0683} \leq Z \leq \frac{+.10}{.0683}\right) = P(-1.46 \leq Z \leq 1.46) = .8558$$

as in part c.

□

COMMENT

Written out in words, the Z-transformation can be expressed generally as follows:

$$Z = \frac{\begin{matrix}\text{Variable value} \\ \text{of interest}\end{matrix} - \begin{matrix}\text{Mean of} \\ \text{variable}\end{matrix}}{\text{Standard deviation of variable}} \tag{8–3}$$

We have studied two special cases of this expression. When we are interested in individual values of X as in the previous chapter, the Z-transformation is

$$Z = \frac{X - \mu}{\sigma}$$

In this chapter we have begun to discuss sampling; when we sample and have interest in the value of the mean $\overline{X}$, the Z-transformation is

$$Z = \frac{\overline{X} - \mu_{\overline{X}}}{\sigma_{\overline{X}}} = \frac{\overline{X} - \mu}{\sigma/\sqrt{n}}$$

where $\sigma/\sqrt{n} = \sigma_{\overline{X}}$ is the standard deviation of the variable $\overline{X}$ (also called the *standard error*). We provide Equation 8–3 to suggest that the computer program Normal Probability Distribution introduced in the previous chapter can be used for exercises in this chapter as well. To find probabilities for $\overline{X}$, simply let $\overline{X}$ become the value of interest and let $\sigma/\sqrt{n}$ become the standard deviation of variable; then proceed as before.

CLASSROOM EXAMPLE 8.1

Working with a Sampling Distribution

A nationwide package delivery service handles thousands of parcels each day through its fleet of airplanes and trucks. To monitor its ground transportation system, the company requires each truck driver to log the total number of miles driven per day. Let the random variable X represent the daily mileage. Last year's records indicate that drivers logged an average of 72.4 miles per day with a standard deviation of 23.8 miles.

Suppose a random sample of $n = 62$ logs will be taken from company records and the sample mean computed.

a. Identify, if possible, the shape of the sampling distribution of the mean.

b. What is the value of the mean of the sampling distribution of $\overline{X}$?

c. To two decimal places, find the standard error of $\overline{X}$.

d. What is the probability that the average number of miles driven for the sample of $n = 62$ logs will be 65.0 or less? 75.0 miles or more?

e. Provide an interval where the sample mean will have a probability of about .95 of being found.

SECTION 8.2 EXERCISES

8.14 A normal distribution has mean = 140 and standard deviation = 40. A sample of 50 items is drawn. Express the following values of the sample average $\overline{X}$ in standardized form:
a. $\overline{X} = 130$ b. $\overline{X} = 150$ c. $\overline{X} = 125$

8.15 Given that the following Z-scores are based on samples of size 50 being drawn from a population with mean = 140 and standard deviation = 40, convert each into its equivalent value of $\overline{X}$:
a. $Z = -1.28$ b. $Z = +2.33$ c. $Z = -1.96$

8.16 We select a sample of 64 items from a moderately skewed right distribution with μ = \$51,000 and σ = \$11,200. Evaluate the following probability statements:
a. $P(\overline{X} < \$50{,}000)$ b. $P(\$50{,}000 < \overline{X} < \$52{,}000)$ c. $P(\overline{X} \geq \$52{,}000)$

8.17 Refer to Exercise 8.16.
a. Between what two values of $\overline{X}$ that are equidistant from μ is there a probability of .80 that the sample mean will fall?
b. Find the value of $\overline{X}$ that is the 99th percentile of the sampling distribution.

8.18 A sample of 60 items is taken from a mound-shaped distribution with mean = 28.0 and standard deviation = 3.40. Find the probabilities regarding the following sample means:
a. the sample mean exceeds 29.1.
b. the sample mean will be between 27.5 and 28.5.
c. the sample mean will be below 27.25.

8.19 Tread life, X, for a particular tire model has a mound-shaped distribution with mean = 28,000 miles and standard deviation = 2,900 miles. Suppose a consumer testing agency acquires a sample of 32 tires. Assume the sample is indeed a random one.
a. To the nearest mile, specify the region of concentrated values for $\overline{X}$.
b. Within what region centered about the population mean does the sample mean $\overline{X}$ have a probability of .50 of falling?
c. How do your answers above change if the sample size is increased to 96. Recompute parts a and b.

8.20 A population of graduate school aptitude test scores has an unknown mean, but the standard deviation of the scores can reasonably be presumed to be 100. If a sample of 120 scores is randomly chosen, find the probability the sample mean will differ from the unknown population mean by no more than 15 points.

8.21 A sample of size 36 has been drawn from a population with $\mu = 100$ and $\sigma = 24$.
a. Find $P(\overline{X} < 98.5)$.
b. What is the probability that the sample average is greater than 109?
c. What is the region of concentrated values of $\overline{X}$?
d. Suppose you wanted your region of concentrated values from part c to be narrower; let's say you'd like 100 ± 6.0 instead of the answer obtained in part c. Use algebra or a trial-and-error process to determine what sample size n would be required to do this. (In a later chapter, we will give a formula that tells us the sample size needed.)
e. What sample size would be required to make the region of concentrated values 100 ± 5.0?

8.22 An auditor is about to sample the accounts receivable of a building supply firm. The firm's computer shows that it has 4118 accounts receivable, with a mean balance of $278.41 and a standard deviation of $188.92. A sample of 200 accounts receivable is planned.

a. The auditor employs the computer to graph the frequency distribution of accounts receivable, using classes of width $50. Do you think the graph of the individual account receivable balances would be skewed left, skewed right, or approximately symmetrical?

b. If the firm's records are correct, how likely is the sample average to be less than $250?

c. Assuming the books are correct, what is the region of concentrated values for the sample mean?

d. How likely is the sample mean to land within plus or minus $20 of the population mean?

8.23 Ten-year industrial bonds are issued by firms seeking new or additional capital. Such bonds usually carry a rating to symbolize their safety or riskiness. Bonds with a AAA rating are considered the safest, although some risk is still present. In a particular one-month period, hundreds of ten-year industrial bonds, rated AAA and carrying various yields, were issued. The shape of the distribution of yields is unknown. One investment house states that the month's offerings project an average yield of 7.9 percent with a standard deviation of .62 percent. Suppose a sample of 36 AAA bonds is randomly selected.

a. What is the shape of the sampling distribution of the average yield of these bonds?

b. Is it possible to determine the chance that one particular bond will yield more than 8 percent? If so, find the probability; if not, explain why.

c. For a sample of 36 bonds, find $P(7.5\% \leq \overline{X} \leq 8.0\%)$.

8.3 FINITE POPULATION CORRECTION TERM

To this point in our chapter about sampling distributions, we have used the following expression for the standard error of $\overline{X}$:

$$\sigma_{\overline{X}} = \frac{\sigma}{\sqrt{n}}$$

This equation is, however, a simplification of the exact formula for the standard error. We now present the exact formula and the conditions under which it should be used instead of the more convenient simplification.

Exact Standard Error of $\overline{X}$

$$\sigma_{\overline{X}} = \frac{\sigma}{\sqrt{n}} \cdot \sqrt{\frac{N-n}{N-1}} \qquad (8\text{–}4)$$

where

n = sample size
N = population size

The additional term, $\sqrt{(N - n)/(N - 1)}$, called the **finite population correction** (FPC) term, will be a number less than 1 and will thus make the uncorrected standard error smaller. The FPC should be employed when the following sampling conditions exist.

Sampling Conditions for Applying the FPC

1. The population size N is known,

and

2. The sample size n is 5 percent or more of the population.

In real-world applications of business statistics, these conditions do arise fairly often. When they do, we should use the FPC since its effect is to make the sampling distribution of $\overline{X}$ more compact.

EXAMPLE 8.5

Suppose a regional supervisor of a fast-food chain is responsible for 173 stores. She is interested in employee turnover and is planning to sample 50 of her stores to determine the average time her current store managers have been in that position.

a. Should the FPC be employed?

b. Determine the standard error if the population mean is 19.0 months and the standard deviation is 9.5 months.

Solution:

a. The sample constitutes 50/173 or about 29 percent of the population, so the FPC is needed here.

b. The standard error would be

$$\begin{aligned}\sigma_{\overline{X}} &= \frac{\sigma}{\sqrt{n}} \cdot \sqrt{\frac{N - n}{N - 1}} \\ &= \frac{9.5}{\sqrt{50}} \cdot \sqrt{\frac{173 - 50}{173 - 1}} \\ &= (1.3435)\,(.8456) \\ &= 1.1361\end{aligned}$$

The FPC always reduces the standard error; in this case from 1.3435 to 1.1361. □

EXAMPLE 8.6

Consider the average manufacturing work week estimated monthly by the U.S. government's Bureau of Labor Statistics. Suppose that the Bureau bases its estimate on a sample of about 6,500 individuals in the manufacturing sector in a given month. Should the FPC be used?

Solution:

Although the sample appears large, 6,500 is small compared to the total manufacturing work force, roughly 20 million. Use of the FPC is not required since 6,500/20 million = .000325, which is much less than 1 percent. If we did compute the FPC for this example, we would get .9998, a number so close to 1 that it would not reduce the standard error by any meaningful amount.

□

While we recommend that the sample constitute 5 percent of the population of interest before applying the FPC, this is an arbitrary cutoff point. Some authors employ different rules of thumb—often 10 percent, but extending to as high as 20 percent. Since Equation 8–4 is the exact formula for the standard error, *it is never wrong to use the FPC,* even when the sample is less than 5 percent of the population. As Example 8.6 suggests, however, the FPC is close to 1 and therefore has little effect when the sample is small relative to the size of the population.

COMMENTS

1. The word "finite" in FPC refers to the fact that the correction term applies when the population size N is known and is, as population sizes go, on the small side.
2. In Chapter 6, we mentioned that sampling *with replacement* rarely occurs in business. If we do sample and then put the sampled item back into the population before selecting the next item, the population would be infinite (we'd never run out of items to sample) and the FPC would not apply.
3. The CLT generally applies for very large or infinite populations. When sampling without replacement from a finite population, increasing the sample size does not always guarantee a more normal-shaped sampling distribution. See the article referenced at the end of this chapter for details.

SECTION 8.3 EXERCISES

8.24 For a population that contains 516 items, compute to three decimal places the FPC for the following sample sizes.

a. $n = 10$ **b.** $n = 25$ **c.** $n = 50$ **d.** $n = 103$

8.25 Refer to Exercise 8.24. Suppose the population referred to has a standard deviation of 12.305.

a. Compute the standard error for each sample size, employing the FPC. Keep three decimal-place accuracy.

b. Compute the standard error for each sample size, ignoring the FPC.

c. For each sample size, determine the percentage of lost accuracy when the FPC is ignored. Lost accuracy is found by subtracting the corrected standard error from the uncorrected standard error and then dividing that result by the corrected standard error.

8.26 A medium-sized company employs 243 people. The manager of human resources is interested in analyzing the number of sick days that the employees claimed last year. Historical records indicate that the distribution of X, number of sick days per year per employee, is fairly mound shaped and symmetrical, with a mean of 6.2 days and a standard deviation of 2.1 days. Suppose she samples 27 personnel files to record the value of X from each file.

a. If the figures from the recent past are still valid, describe the shape of the sampling distribution of $\overline{X}$ and specify $\mu_{\overline{X}}$ and $\sigma_{\overline{X}}$.

b. Determine the region of concentrated values for $\overline{X}$.

c. What would you conclude about the validity of the values of $\mu_{\overline{X}}$ and $\sigma_{\overline{X}}$ used in part b if the following were the case:

(1) The sample mean landed inside the region of concentrated values of $\overline{X}$.

(2) The sample mean fell outside the region.

8.27 There are 184 airports in a certain region of the United States with state-of-the-art radar screens and air-traffic equipment. It is thought the new equipment will help reduce the delays in departing flights. The Federal Aviation Administration keeps records on X, the number of delays per 1000 flights at each airport. The distribution of X is skewed right with a mean of 215 and a standard deviation of 66. Suppose a sample of 48 of these airports is planned, and the average value of X is to be determined.

a. Should the FPC be used to find the standard error of $\overline{X}$?

b. Find $\sigma_{\overline{X}}$.

c. Determine the chance that $\overline{X}$ will be between 200 and 240.

d. Would it be unusual if only one of the 48 airports in the sample yielded a value of X larger than 350? Why?

8.28 A population of size $N = 371$ has a mean of 508 and a standard deviation of 95. A sample of 49 items is to be drawn.

a. Should the FPC be employed?

b. Determine the standard error of the mean.

c. Find the region of concentrated values for $\overline{X}$.

d. How likely is $\overline{X}$ to lie within plus or minus 15 of the population mean?

8.29 Consider the two extreme values a sample size can be: 1 and N, where N is larger than 1.

a. If $n = 1$, what is the value of the FPC? What does Equation 8–4 then reduce to?

b. If $n = N$, what percent of the population has been sampled? What would be the value of the FPC? What would be the standard error?

8.30 The assistant editor of a small weekly community newspaper chose 36 issues at random from the last two years to determine the average amount of space devoted to classified advertisements. Suppose in the population the mean is 12.5 columns with a standard deviation of 3.0 columns.

a. What is the value of the FPC?

b. Determine the uncorrected and the corrected standard error.

c. Find $P(12.0 < \overline{X} < 13.0)$.
d. What effect does the FPC have on the mean value of $\overline{X}$?

8.4 THE RELATION OF SAMPLE SIZE TO POTENTIAL SAMPLING ERROR

We have defined *sampling error* as the difference between the population mean μ and the mean $\overline{X}$ computed from a random sample. Since $\overline{X}$ is a random variable, the amount of sampling error that occurs in a given instance is a random variable as well, depending on the luck of the draw. Whenever we take a random sample, we therefore expect some sampling error to be present, and naturally we would like it to be as small as possible. In this section, we will relate potential sampling error to n, the size of the sample drawn.

Figure 8.6 shows two sampling distributions for the same population, based on two different sample sizes. Which sampling distribution is more likely to yield the smaller sampling error? Clearly the smaller sampling error is more likely to result from the more compact sampling distribution, the one based on the larger sample size. The relation of interest to us is simply stated: Increasing the sample size n will tend to make the potential sampling error become smaller, or conversely, a smaller size tends to make potential sampling error larger.

Does this mean that we generally want samples to be as large as possible? We might say "yes" to this if sampling could be done without cost and without taking any time, but such is not the case. Sampling often is needed to acquire information about some population of interest, but the increased accuracy that comes with increased sample size may not be worth the cost.

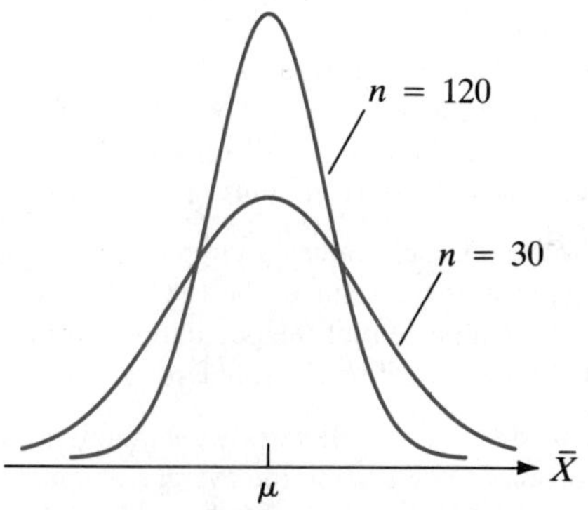

Figure 8.6 Sampling Distributions for $n = 30$ and $n = 120$

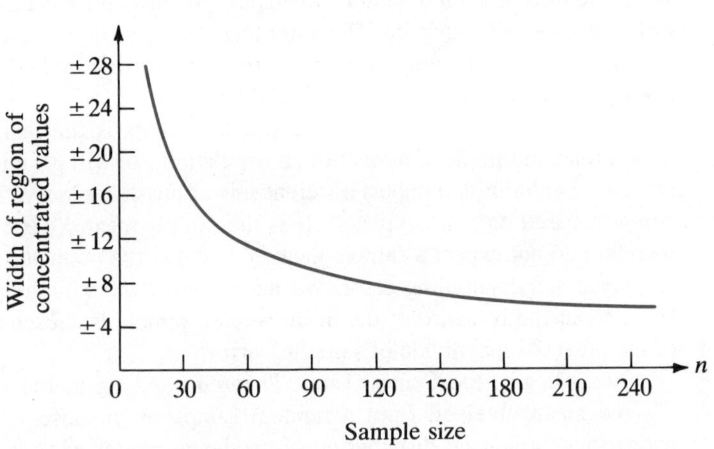

Figure 8.7 Effect of Sample Size on Region of Concentrated Values (σ = 43.8)

Examination of Equation 8–1, restated here, will lead us to another point:

$$\sigma_{\overline{X}} = \frac{\sigma}{\sqrt{n}}$$

We have used this formula throughout the chapter to determine the standard error. If we wish to cut a given standard error in half, we need to multiply the sample size by four, not two, due to the square root sign on n. To illustrate, assume we have a population with $\sigma = 43.8$. For $n = 30$, the standard error is $43.8/\sqrt{30}$, or 8.0. To reduce the standard error in half (down to 4.0) would require a sample four times as large, or $n = 120$.

Figure 8.7 shows the relation between the sample size utilized and the resulting region of concentrated values when sampling from this population. It illustrates that there is a diminishing return of sorts to additional sampling. As n becomes larger, the gain in accuracy (decrease in potential sampling error) from each additional item sampled becomes less. Excessively large samples are not usually needed or worth the cost.

8.5 SUMMARY

A major part of the field of business statistics involves using the information or evidence contained in a sample to help us make decisions or draw conclusions about the population from which the sample was drawn. This process of reasoning from the sample to the population is called *inference*. We have introduced the idea of a sampling distribution in this chapter. The sampling distribution is the concept that links the sample to the population. Sampling distributions will be relied on heavily in the next several chapters as we study statistical inference.

A key to understanding sampling distributions is the realization that when we draw a random sample of items from a population, we will get one of many (sometimes one out of an infinite number) different sets of possible observations. Any measure we compute based on our sample, such as the sample mean $\overline{X}$, is a random variable. We therefore do not expect a sample mean $\overline{X}$ to equal the population mean μ; instead we anticipate some sampling error. We have studied the sampling distribution of $\overline{X}$ to learn about the behavior of the mean when a sample is chosen and to become aware of the potential magnitude of sampling error.

According to the Central Limit Theorem, as long as the sample size is at least 30, the mean obtained from a random sample is an observation drawn out of an approximately normal distribution of means no matter what the shape of the underlying population of individuals. We therefore can talk about a region of concentrated values for $\overline{X}$ as well as evaluate probability statements for $\overline{X}$ using the Z-table. For instance, if we know the standard deviation in the population, we can determine the chance that $\overline{X}$ based on a given sample size will lie within any given plus or minus distance of the actual population mean μ.

The amount of evidence or information a sample contains is related to the sample size n, but this does not necessarily mean we want our samples to be as large as possible. A single sample, even a relatively small one, can tell us quite a bit about the population. In the chapters ahead, you may be surprised at the type and precision of conclusions that can be made with a limited amount of information. On the other hand, any sample large or small is an incomplete look at the whole population, so there is always a risk of an incorrect conclusion. Only a complete census can eliminate risk and sampling error; in the ensuing chapters, we will discuss the chances of error and the extent to which we can control it.

In this chapter, we have chosen to emphasize the concept of sampling distributions by focusing only on the sampling distribution of the sample mean, $\overline{X}$. We will see other sampling distributions in future chapters. For instance, we often will sample from a binomial population where each item can be classified as a success or as a failure. A specific example would be a survey of college students to see what percent or proportion own a compact disc player. We then would be dealing with the sampling distribution of the proportion of successes instead of the sampling distribution for means. The formula for the standard error of a binomial proportion will be somewhat different than that for the standard error of the mean. However, the underlying rationale and development of the sampling distribution for proportions will be very similar to that presented in this chapter.

Two things to remember for all sampling situations: First, the purpose of sampling is to learn about the population without having to observe a substantial portion of the population, and, second, the dispersion of the sampling distribution (the standard error) is always dependent on the size of the sample.

CHAPTER 8 EXERCISES

8.31 A population is normally distributed with $\mu = 5.60$ percent and $\sigma = 1.90$ percent. Specify the mean, standard error, and shape of the sampling distribution of $\overline{X}$ when $n = 12$.

8.32 A population is skewed left with mean \$48,500, standard deviation \$13,600. Specify the standard error and shape of the sampling distribution of $\overline{X}$ when a sample of 225 items is drawn.

8.33 A population of unknown shape has $\mu = 320$ parts per million (ppm) and $\sigma = 48$ ppm. Specify the shape of the sampling distribution of $\overline{X}$ and the value of the standard error for the following:

a. $n = 40$ **b.** $n = 100$ **c.** $n = 363$

8.34 Refer to Exercise 8.33 above. What is the region of concentrated values for $\overline{X}$ when the following are the case:

a. $n = 40$ **b.** $n = 100$ **c.** $n = 363$

8.35 Suppose the mean score achieved on a college entrance exam by students at a state university is unknown, but the standard deviation of scores is known to be 100 points. A sample of 50 student scores is to be selected at random, and the sample mean will be computed.

a. What is the standard error?

b. The sample mean has a 95 percent probability of being different from the unknown population mean by no more than some amount that we will call J. What is the value of J?

8.36 We take a sample of 250 observations from an exponential distribution with mean = 4.00 minutes. Evaluate the following probability statements:

a. $P(\overline{X} < 4.20)$ **b.** $P(\overline{X} > 4.40)$ **c.** $P(3.504 < \overline{X} < 4.496)$

8.37 Suppose you have sampled 45 items from a skewed left distribution with $\mu = \$1640$ and $\sigma = \$402$.

a. To the nearest whole dollar, what is the region of concentrated values of $\overline{X}$?

b. A sample result above \$1800 would seem unlikely. Find the p-value for $\overline{X} \geq \$1800$.

c. How likely is the sample average to land within plus or minus \$75 of the population mean?

d. Answer part c if $n = 90$, instead of $n = 45$.

8.38 A candy bar's label says it contains 240 calories. This number is an average; due to minor variations in size and ingredients, the population standard deviation of calorie count is 8 calories. Assume a sample of 30 bars subjected to a chemical analysis.

a. How likely is the sample mean to differ from the population mean by 2 calories or more?

b. A sample mean below 235 calories is not likely to be obtained. Find the p-value for $\overline{X} = 235$ or fewer calories.

c. To the first place beyond the decimal, find the region of concentrated values for X.

d. If we double the sample size, will the region of concentrated values be half as wide? What if the sample size is multiplied by four?

8.39 A bank manager states that the average time for a customer to make a simple withdrawal at an automated teller machine (ATM) is 90 seconds, with a standard deviation of 20 seconds. Suppose 48 randomly selected ATM withdrawals are monitored. Assume the manager's statement is true.

a. Find $P(88 \leq \overline{X} \leq 92)$.

b. How likely is $\overline{X}$ to fall within plus or minus 5 seconds of the population mean?

c. If the sample were to show $\overline{X} = 100$ seconds, would you believe or doubt the manager's statement?

8.40 Last year, a mail-order catalog business experienced an average mailing cost of \$2.85 per order, with a standard deviation of \$1.52. Assume that these figures are still current and that we sample 64 packages at random.

a. Find the chance the sample mean is less than \$2.50.

b. Determine $P(\overline{X} \geq \$3.00)$.

c. What is the region of concentrated values of the sample mean?

d. Your answer to the above should be wider than $\pm\$.25$. Let's now suppose we would like the region of concentrated values to be $\pm\$.25$. Use a trial-and-error process to find the sample size necessary to make about 95 percent of the sampling distribution within \$.25 of the population mean.

8.41 A state trooper is about to randomly sample speeds of 50 vehicles traveling over a level stretch of interstate highway. Past studies at this location have shown that the standard deviation of vehicle speeds is 3.0 miles per hour. Assume the trooper's radar gun is accurate. When the trooper computes his sample mean $\overline{X}$, how likely is $\overline{X}$ to differ from the actual population mean μ by no more than 1.0 mile per hour? What is the probability that the sampling error will be .50 mph or less?

8.42 Refer to Exercise 8.41. If the sample size were 100 vehicles instead of 50, would the trooper be *guaranteed* a smaller sampling error? Or would he simply be *more likely* to have a smaller sampling error when $n = 100$?

8.43 A video rental store has 674 members in its club. Club members pay a fee in advance and get certain discounts and other privileges. The owner is about to sample 30 members' files to determine the average number of rentals per member in the last six months.

$$\mu = 21 \qquad \text{and} \qquad \sigma = 8$$

a. What is the region of concentrated values for $\overline{X}$?

b. What effect would a sample of size 100 have on your answer to part a? Recompute the region of concentrated values.

8.44 An employee in personnel in a large production plant is doing a study of hourly employees who have retired in the past five years. In particular, she wants to know the average length of service in the plant at the time of retirement. If the standard deviation in the population is 6.2 years and a sample of size 64 retirees is randomly chosen, how likely is the sample mean that results to differ from the unknown population mean by 1.00 year or less?

***8.45** Make a graph similar to Figure 8.7 for samples drawn from a roughly normal population with $\sigma = \$1186$. On the horizontal axis, let each 100 items sampled = one inch; on the vertical, let 1 inch = each $\pm\$100$ for the region of concentrated values. Choose several different sample sizes so that you can plot your curve accurately.

*Optional

***8.46** The sampling distribution of the sample total T is important for making inferences about a population total τ. For example, auditors often face the problem of estimating the total amount outstanding in accounts receivable. A simple relation between the sample mean and T ($T = n\overline{X}$) allows us to use the results of the sampling distribution of $\overline{X}$ to describe the sampling distribution of T. When the population is normal or whenever the sample size is large, the sampling distribution of T is also approximately normally distributed with an average value of

$$\mu_T = n\mu$$

and a standard error of

$$\sigma_T = \sqrt{n}\sigma$$

For instance, if we take a sample of size $n = 22$ from a normal population with $\mu = 3.5$ and $\sigma = .64$, then the sampling distribution of T will be normal with a mean of $\mu_T = 77$ and a standard error of $\sigma_T = 3.002$.

Describe as normal, approximately normal, or non-normal the sampling distribution of T in each of the following sampling situations:

a. samples of size 9 from a normal population.
b. samples of size 49 from a normal population.
c. samples of size 211 from a non-normal population.
d. samples of size 19 from a non-normal population.

***8.47** See Exercise 8.46. Since the sample total T has an approximately normal sampling distribution, there is a Z-transformation that converts values of T into Z-scores:

$$Z = \frac{T - \mu_T}{\sigma_T}$$

$$= \frac{T - n\mu}{\sqrt{n}\sigma}$$

Express the following values of T in standardized form, assuming samples of size 43 are drawn from a normal population with $\mu = 20$ and $\sigma = 8$.

a. $T = 901$ **b.** $T = 792$ **c.** $T = 866$ **d.** $T = 1000$

***8.48** See Exercise 8.47. Suppose we select a sample of 32 accounts from a moderately skewed right distribution of accounts receivable where the population mean is \$2182 and the population standard deviation is \$1186. Evaluate the following probability statements:

a. $P(T > \$50{,}000)$
b. $P(\$60{,}000 < T < \$80{,}000)$
c. $P(T \leq \$75{,}116)$
d. Find the region of concentrated values for T, rounding your result to the nearest hundred dollars.

8.49 Between what two values of Z does the middle 80 percent of the sampling distribution of $\overline{X}$ lie? 95 percent? 99 percent?

***8.50** Using Table 8.2 for reference, express the following symbols in words.

a. μ_S **b.** σ^2_{Md} **c.** σ_{AD} **d.** μ_{g_1}

8.51 The management of a large bank would like to know about the overall friendliness, knowledgeability, and helpfulness of its loan officers when they deal with the public. The bank plans to hire a management consulting firm that will send trained interviewers, posing as customers, to the bank. The interviewers will apply for a loan and record their perceptions of the encounter. An overall rating from 0 to 100 is then derived for each encounter, based on these perceptions. The standard deviation of overall ratings has historically been about 20. For this service, the consulting firm charges \$10,000 in fixed costs plus \$250 per encounter. With a budget of \$25,000 for this project, the bank is reluctant to hire the firm, especially since the sample size the consulting firm would obtain would be, in the bank's opinion, too small. If you worked for the consulting firm, how would you try to convince the bank that the sample of size $n = 60$, which their budget would allow, is enough?

REFERENCE

Plane, D. R., and K. R. Gordon. 1981. "A Common Misconception About the Central Limit Theorem," *Proceedings of American Institute for Decision Sciences*. Boston.

REVIEW PROBLEMS CHAPTERS 5–8

R21 Each year the National Basketball Association (NBA) teams conduct a draft to choose new players who want to play professional basketball. In the table following, X represents the round in which an NBA player was picked when it was his year to be in the draft and $P(X = x)$ is the relative frequency of players chosen in round x, based on the opening day rosters of the 1987–1988 season:

x	1	2	3	4	5	6	7	8
$P(X = x)$	.5770	.2525	.0984	.0426	.0066	.0098	.0033	.0098

Source: Data from *Hoop*, March 1988.

a. Is X continuous or discrete?

b. What value of X is the median? The mode? The mean?

c. What is the standard deviation of X?

d. Define a "success" as meaning a player was a first-round draft choice. For a random sample of five 1987–88 NBA players, what is the probability that all are first-round choices?

e. For a random sample of five 1987–88 NBA players, what is the most probable number of first-round picks to be found?

R22 According to *Bridge Odds Complete* by Frederick H. Frost (George Cofflin, Publisher, Waltham, MA), the probability of a bridge player being dealt a hand with eight cards in the same suit is .00467.

a. If 8024 computer-generated bridge hands are created, what is the expected number of hands with an eight-card suit? Assume the computer program does in fact randomly deal the hands; report to the nearest integer.

b. Let X = number of hands with an eight-card suit. What is the region of concentrated values of X for the 8024 hands?

c. In an article in the November 1981 issue of *The Contract Bridge Bulletin,* Sid Kilsheimer reports on the degree to which computer-generated hands used in bridge tournaments conform to theoretical patterns. A study of 8024 such hands found 36 with eight-card suits. Is this result in the unlikely or the likely region?

d. The chance of a nine-card suit is .00037. What is the expected number of hands with a nine-card suit over 8024 deals (to the nearest integer)?

e. Refer to part d above and your integer answer. Over the 8024 deals, what is the probability of six or more nine-card deals occurring? Use the Poisson approximation method.

f. In part e, are the conditions for using a normal distribution to approximate the binomial satisfied?

R23 A treadmill stress test for the heart is estimated to show a false–positive (indicates a problem or irregularity when in fact the patient is normal) in about 6 percent of tests. A hospital runs 250 stress tests in a year.

a. How many false–positives are expected?

b. Find the probability of six or fewer false–positives occurring during the year.

R24 From each lot manufactured, a producer of tennis balls sets aside several dozen cans of balls in its warehouse to simulate their sitting on store shelves. At regular intervals, a sample of cans is opened to test the balls' bounce. The balls are released from a predetermined height, and an electronic device measures the height of the bounce. Suppose the distribution of bounce heights is approximately normal with a mean of 37.65 inches and a standard deviation of .70 inch.

a. What is the probability an individual ball will bounce 38.0 inches or higher?

b. If a sample of 33 tennis balls is measured, what is the probability the sample average will equal or exceed 38 inches?

c. For the sample of 33 tennis balls, what is the probability that the sample average will differ from the population average by no more than .15 inch?

R25 Let's consider the fortunes of a major league baseball team, such as the Baltimore Orioles. Suppose we wish to model the number of games won by the Orioles over a particular period during the season.

a. Of the binomial, Poisson, normal, or exponential, which distribution would be most appropriate?

b. Even though one of the distributions may offer a reasonable model, the model's assumptions are probably not satisfied exactly. If you are familiar with baseball, explain where the violation might be.

c. In 1988, the Baltimore Orioles began the season by losing their first 21 games, the worst start in the history of major league baseball. Let's suppose that we can obtain accurate results with the distribution chosen in part a. To seven decimal-place accuracy, determine the probability that the Orioles would lose their first 21 games, assuming the following:

(1) that they are a .450 ballclub (will win 45 percent of their games).

(2) that they are a .300 ballclub.

(3) that they are a .414 ballclub (their 1987 figure).

R26 According to the "Odds Chart" accompanying the McDonald's *Monopoly Game Rules,* their 1988 promotion offered these prizes:

X = Prize Level	Number of Prizes
$2,000,000	1
1,000,000	1
300,000	2
150,000	8
20,000	16
2,457*	4,872
25	47,500
5	300,000
1**	60,351,167

*An average value of various prizes ranging from $100 to $16,200.

**A food prize: An average value of $1 was assumed.

a. Most game pieces are losers. But suppose someone has won a prize. What is the expected value of X? What is the median value of X?

b. Assume you have gone to a McDonald's and obtained a single game piece. If, as the complete "Odds Chart" indicates, your probability of winning a prize is .08382 (1 chance in 11.93), what is the expected value of your game piece?

R27 A volunteer Emergency Medical Technician (EMT) keeps a pager with him at all times in order to be able to respond quickly to EMT calls within his rescue unit's area. After 20 weeks of being on call, the volunteer notes that he has been paged 22 times. Let X = the number of EMT calls per week.

a. What is the mean value for X?

b. Which probability distribution would be best suited to model X?

c. For a given week, what is the probability the volunteer will get three or more calls?

d. What is the most probable value of X for a particular week?

e. Suppose the volunteer is about to take a two-week vacation out of state. When he returns, what is the probability that no calls occurred in his absence?

R28 Suppose the average cost to repair a VCR at an appliance service center is $85.50 with a standard deviation of $41.20. For a random sample of 32 repair bills, find the following probabilities.

a. What is the probability the sample mean will exceed $100?

b. What is the probability the sample mean will be at least $78 but not more than $93?

R29 A metalworks firm supplies strips of aluminum to a suitcase manufacturer. One size of suitcase requires strips with a length of 224.400 centimeters. When working properly, the firm's production process creates strips with an average length of 224.400 centimeters and a standard deviation of .055 cm. The distribution of lengths closely approximates a normal distribution.

a. What proportion of strips are within .100 cm of the desired length?

b. Suppose strips that have a length of 224.225 cm or less are too short to be usable. Such strips are melted down and recycled through the manufacturing process. What percent of strips produced have to be recycled?

c. To three decimal places, what length of strip is at the fortieth percentile of the distribution?

R30 A Missouri brewery has tallied the following for the last one million 12-ounce cans from one of its four major can suppliers:

Filled and shipped	999,935
Defective: Returned	
Poor Lithography	18
Dents or Damage	18
Flange Leaker	7
Bulged	22

Assume the next 100,000 cans from this supplier can be viewed as a sample drawn from an infinite population having the defective rate indicated above.

a. What is the probability of a can being returned as a defective?

b. What is the expected number of defectives?

c. For the next 100,000 cans, what is the probability of 10 or more returns?

R31 One of the most efficient mail-order catalogers, L.L. Bean, Inc., takes pride in its speed and accuracy in handling customer orders ("Secret of L.L. Bean's Efficiency? People," Louisville *Courier-Journal,* December 29, 1985). According to the Louisville *Courier Journal,* Bean's internal records in 1985 indicated that 99.89 percent of all orders were shipped correctly. To ensure this high accuracy rate, Bean maintains productivity and accuracy records for each picker (an employee who selects the ordered merchandise from the warehouse shelves) and packer (one who prepares the merchandise for shipping).

a. Suppose we wish to judge the accuracy rate for current orders against the 1985 company figure. If we randomly select 720 orders, what is the probability distribution of the random variable X, number of orders incorrectly shipped?

b. Is it possible to approximate the distribution of X with an alternative probability distribution? Explain.

c. What is the expected value of X?

d. Suppose our sample revealed 718 correct orders. What conclusion about the accuracy rate, relative to the company standard, is warranted? Why?

R32 The number of milligrams of fat per nugget of chicken cooked at a fast food restaurant is approximately normally distributed with a mean of 236 mg and a standard deviation of 28 mg.

a. What is the probability that there is more than 260 mg of fat in one nugget of chicken?

b. What is the probability that exactly three nuggets in a six-piece serving of chicken nuggets will have more than 260 mg of fat?

c. Answer the question in part b for a nine-piece serving.

R33 For the large screen (25–27 inch) TV sets it sells, a large appliance store has kept track of whether the buyer also purchased the optional long-term service contract to protect against any needed repairs. Data for the past six months are given in the following table:

Brand	Number of TV Sets Sold	Number of Service Contracts Sold
Sanyo	52	8
Philco	42	8
Sony	56	12

a. Arrange the preceding information into a two-dimensional cross-tabulation.
b. What is the probability that a sale chosen at random is a Philco?
c. What is the probability that a large screen TV buyer opts not to buy the service contract?
d. What is the probability of a service contract purchase, given that a Philco was purchased?

R34 The time between successive arrivals at a large airport during the 9 A.M. to 5 P.M. period follows the exponential distribution with a mean of 4.0 minutes.
a. What is the probability that 10 or more minutes elapse between successive arriving planes?
b. What is the probability that no more than 90 seconds pass between successive arrivals?

R35 Stocks make up 35 percent of the dollars in an investor's portfolio, while bonds account for 20 percent and a money market account represents the remainder. At year's end, she receives statements indicating that her stocks rose 11.2 percent, the bonds appreciated 3.7 percent, and the money market account returned 5.9 percent. One of her financial objectives for the year was a required rate of return of at least 7 percent.
a. Did she achieve her objective? Explain.
b. Measure the risk of her portfolio by computing the standard deviation.

R36 Warner-Lambert Company, maker of an in-home pregnancy test called e.p.t., included a summary of the accuracy of laboratory and actual tests with an early version of the product. The actual tests, performed at home by individuals, produced the following results:

Actual Clinical State	e.p.t. Reading	
	Positive	Negative
Pregnant	451	36
Not Pregnant	15	183

Source: Warner-Lambert Company, Morris Plains, NJ. (The e.p.t. kit to which the given data apply is no longer marketed; field-test data for the newer version of the e.p.t. kit are currently unavailable.)

A positive reading indicates the presence of the hormone HCG—Human Chorionic Gonadotropin—and implies that the woman is pregnant. A negative reading suggests she is not pregnant.
a. How many at-home tests were performed?
b. What is the probability of a positive reading, given the woman is pregnant?
c. What is the probability the woman is not pregnant, given a positive reading?
d. Describe, in terms of this problem setting, a false–negative reading.
e. The company's brochure included these interpretations of the actual test results: "e.p.t. was 97 percent accurate when a positive result was obtained. e.p.t. was 84 percent accurate when a negative result was obtained." Explain how these figures were determined.

R37 Suppose a cigarette manufacturer claims that its new, low tar cigarette has an average of 4.0 milligrams of tar per cigarette with a standard deviation of 1.0 milligrams. A laboratory test is devised to measure the tar in 100 randomly selected cigarettes, after which, the mean of the tar measurements will be calculated.

a. Assuming the manufacturer's claim is true, which one of the following statements best describes the sampling distribution of the mean tar measurements?
(1) a distribution with mean 4.0, standard deviation .1, and unknown shape.
(2) a distribution with mean 4.0, standard deviation 1.0, and unknown shape.
(3) a distribution with unknown mean and standard deviation, but with an approximately normal shape.
(4) a distribution with mean 4.0, standard deviation .1, and an approximately normal shape.
(5) a distribution with mean 4.0, standard deviation 1.0, and an approximately normal shape.

b. Find the probability the sample mean will exceed 4.1 milligrams of tar.

R38 A fast food company operates a chain of 434 company-owned and franchised stores throughout the western United States. Forty-five stores in one region are targeted to receive a special promotional campaign that will run for three months. Quarterly sales per store average \$250,000 with a standard deviation of \$35,000. The home office anticipates that the campaign will significantly increase sales. Based on the historical averages and the region of concentrated values, what average quarterly sales figure for the stores receiving the promotion should the home office set as a dividing line for determining the success or failure of the campaign?

R39 A company has a lump-sum incentive plan for salespeople, depending on their level of sales. If they sell less than \$100,000 per year, they receive a \$1000 bonus; from \$100,000 to \$200,000, they receive a \$5000 bonus; and above \$200,000, they receive \$10,000. If the annual sales per salesperson is approximately normally distributed with a mean of \$180,000 and a standard deviation of \$50,000, find the average bonus payout per salesperson.

R40 In "Statistical Analysis of Freezing Temperatures in Central and Southern Florida" (*Journal of Climatology*, 1988), Peter Waylen investigates the probability distribution of the variable X, number of frosts per year, at 25 meteorological stations in Florida. At one of the stations in Fort Myers, 42 years of complete annual daily minimum temperatures were available. An ungrouped frequency distribution, reconstructed from a figure in the article, shows the following values for X:

X	0	1	2	3
f	35	6	0	1

a. Find the mean of this distribution, to two decimal places.
b. Generate a column of relative frequencies for the distribution, to four decimal places.
c. Using the value found in part a as the value of λ for a Poisson random variable, generate the probabilities of the four values of X, to four decimal places.
d. Compare the Poisson probabilities to the observed relative frequencies. What is the largest absolute difference between an observed relative frequency and the corresponding Poisson probability? Does the Poisson probability distribution appear to be a good probability model for the variable X?

Chapter Maxim *Evidence that contradicts an assumption generally carries more weight than evidence that supports the assumption.*

CHAPTER 9

INTRODUCTION TO STATISTICAL INFERENCE

Objectives

By the time you complete this chapter, you should be able to

1. Explain what it means to make a statistical inference.
2. Identify the two general approaches to making a statistical inference.
3. Differentiate between point estimates and interval estimates.
4. Describe or define the term *hypotheses.*
5. For a given set of test conditions, formulate the appropriate null and alternative hypotheses.
6. Name and identify the essential elements of hypothesis testing.
7. Explain why we cannot be 100 percent sure of any statistical inference.
8. Define Type I and Type II errors.

This chapter is short, but important. We are about to lay the groundwork for the rest of the book. All the chapters that follow will use the concepts introduced in this chapter.

We are on the threshold of learning inferential statistical procedures. Our book began with a study of descriptive statistics, where we learned how to summarize *sample* data, thereby giving rise to numerical descriptive measures such as the mean, $\overline{X}$, and the standard deviation, S. In Chapters 5 through 8 we studied probability distributions and explained how they are used to model a *population*. We introduced the parameters μ and σ as summary measures for populations. Our goal in this chapter is to synthesize these topics into the study of inferential statistics.

This chapter serves as a transition from probability and descriptive statistics to inference. The concepts, structure, and vocabulary of inferential procedures will be presented in this chapter. Put your calculator aside! Give it a rest as we focus on the *what* and *why* of statistical inference.

9.1 INTRODUCTION

In Chapter 1, we stated that *statistics* involves both a sample and a population, each representing a set of data. By definition, a *sample* is part—perhaps a very small part—of a larger population. In *descriptive* statistics, we concentrated on describing the sample data. Our goal in *inferential* statistics is to generalize about the population from the sample data. To achieve this goal, we will work with summary measures of the population and the sample, called parameters and statistics, respectively. Thus, the population mean μ, variance σ^2, and proportion of successes π are examples of parameters whose unknown values we wish to infer. The corresponding sample statistics $\overline{X}$, S^2, and p, the sample proportion of successes, will provide the basis for the statistical inference.

Definition

The **objective of inferential statistics** is to make a statement about the value of a parameter based on the information found in a sample.

For example, a common objective in inference is to make a statement about the value of the population mean μ, based on the observed sample mean $\overline{X}$. As an illustration, a firm making a wire product might wish to learn about μ, the average breaking strength of a specific type of wire. To do so, they might select some pieces of this wire, subject each piece to sufficient stress to break it, and then determine $\overline{X}$, the average breaking strength. The sample information in this example is $\overline{X}$, the sample mean breaking strength. With this sample evidence, the firm can then make an inference about μ. Note that (unless a complete census is conducted) the exact value of μ, the population mean breaking strength, remains unknown!

There are two ways of making a statistical inference. One is called *estimation,* in which we try to provide a reasonable value or a range of reasonable values for the unknown population parameter. The other is referred to as *hypothesis testing,* in which we decide on one of two opposite statements about potential values of the parameter.

Ways to Make Statistical Inferences

1. Estimation
2. Hypothesis testing

These two methods of making inferences will be discussed in more detail in the next two sections.

Components of an Inferential Statistics Problem

Clearly we cannot use inference to support decision-making in the business world unless we can recognize the components of an inferential statistics problem. We must ensure that we are using the right tools for the job, so to speak. Perhaps the greatest abuse of statistics is its application to business problems that are *not* correctly formulated because one or more of the ingredients listed in the following box is missing.

Components of an Inferential Statistics Problem

1. A specific objective to be achieved
2. An accessible population
3. A random sample

The first component specifies the purpose of the analysis: estimation or hypothesis testing. The second and third components reinforce the fact that statistics problems require a sample and its parent population. Unlike descriptive statistics, an inferential statistics problem involves a population parameter as well as the summary statistics from the sample. Example 9.1 demonstrates how these components could apply in a business problem.

EXAMPLE 9.1

The newly hired manager of a restaurant wishes to see some historical data on the restaurant's daily revenues. He plans to examine the data to familiarize himself with the average daily receipts. Identify the components of this inferential statistics problem.

Solution:

The restaurant manager's *objective* is to estimate the numerical value of the parameter μ, the average daily revenue for all days of operation, using the information found in the random sample. The *population* is the collection of numbers representing each day's receipts since the restaurant opened. A *random sample* would be a group of some of the daily receipts, obtained by using a random selection procedure (see Chapter 1) such as a simple random sample.

□

Inference works when all the components of the problem exist and are well defined. On the other hand, applying statistical procedures to business problems may yield poor results if any of the following situations occur:

1. A manager fails to clearly define his/her objective of the statistical analysis.
2. Some of the population data are inaccessible.
3. The sample data set is biased.

The first problem is more often a communication problem than a statistics problem. Nonetheless, inferential statistics seems to be the scapegoat when a misguided analysis fails to achieve a vague objective. The second problem occurs when some of the data are not available for each entity in the population. When the targeted population and the framed population (from which the sample is drawn) are not the same, the problem of inaccessible data is present. A related problem in survey research is participant nonresponse, which creates a potential bias in the subsequent results. Nonresponse raises this question: Are those who choose to respond inherently different in some way from those who do not respond? For a review of other potential nonsampling errors that create such problems, you may wish to reread portions of Chapter 1.

The third problem—having biased data—can be attributed to nonrandom sampling. As we mentioned in Chapter 1, inferential techniques are invalid for data generated by nonrandom sampling. Care must be taken to ensure that random selection methods are used to gather the data. If we develop the habit of addressing each component of a problem, as given in the preceding box, before applying inferential methods, perhaps the mystique of business statistics will be reduced.

SECTION 9.1 EXERCISES

9.1 What are the different ways to make statistical inferences?

9.2 Identify a potential flaw in the proposed analysis for the following scenario. A survey of the major advertising agencies in a metropolitan area was conducted. Each firm was sent a questionnaire and asked to supply information about the number of full-time employees, gross income, annual billings, services offered, and notable area clients. Two firms declined to participate, and several others refused to disclose some of the information requested. The results of the survey are to be used to estimate the average number of full-time employees, average gross income, and so forth within the metro area.

9.3 Decide whether each of the following is a correct inferential statistics objective:
a. to estimate μ based on $\overline{X}$.
b. to test a hypothesis about σ^2 using S^2.
c. to estimate the sample standard deviation using the population standard deviation.
d. to compute the range.

9.4 The manager of the auto loan department at a commercial bank would like to have a "ballpark" figure for the average actual term for a 36-month loan. She asks one of her assistants to select randomly about 100 loans from the file of expired loans and record the number of months until the loan was paid off. Identify the components of this inferential statistics problem.

9.5 Refer to Exercise 9.4. The manager also would like to compare the standard deviation in loan life for 36-month loans to the standard deviation for 48-month loans. A recent investigation of the 48-month loans revealed a standard deviation of 6.8 months for all expired four-year loans. She wishes to know if the standard deviation of the three-year loans is the same as this figure or different. Identify the components of this inferential statistics problem.

9.2 ESTIMATION

Types of Estimates

When we use sample information to try to establish reasonable values for a population parameter, we are developing an estimate. The fact that we are estimating suggests two things. First, we do not know the value of the parameter of interest (if we did know the value, no estimate would be needed!). For instance, the restaurant manager in Example 9.1 was unfamiliar with the store's revenue patterns and sought information about an average day's receipts. Second, any sample-based estimate of the value of the parameter will be potentially inexact because it involves only a part of the population (see Chapter 8 Maxim!).

There are two ways of estimating: (1) generate a single number called a *point estimate,* which represents a most educated "guess," and/or (2) develop a range of values called an *interval estimate,* within which we are likely to find the value of the parameter.

Types of Estimates

1. Point 2. Interval

EXAMPLE 9.2

By law, automobile manufacturers are required to post a sticker listing the Environmental Protection Agency's (EPA) fuel economy rating on a window of each new car or truck for sale. A sample of such a sticker appears in Figure 9.1. Explain how this sticker depicts estimation in general and point and interval estimates in particular.

Solution:

The sticker represents the fuel economy ratings for 1988 Honda Accords. Let us concentrate on the highway miles-per-gallon (mpg) rating in this discussion. The EPA produces the mpg figures from a sample of vehicles built at the beginning of the model year. Obviously, they could not conduct a census since a majority of the Accords had not been built at the time the EPA administered its tests.

The parameter of interest is μ, the average mileage per gallon of gas for all such Accords. Since the actual value of this parameter is unknown, the EPA uses the statistical technique of *estimation* to provide a reasonable approximation for it. The *point estimate* is the figure 29, which means the EPA estimates a 1988 Honda Accord will average 29 miles per gallon in highway driving. Few Accords will achieve this value exactly.

Reading the fine print reveals this phrase: ". . . the majority of vehicles with these estimates will achieve between . . . 24 and 34 mpg on the highway." The limits of 24 and 34 constitute an *interval estimate*. This means most Accords will average between 24 and 34 miles per gallon in highway driving.

Compare this vehicle to others in the

FREE GAS MILEAGE GUIDE

available at the dealer.

CITY MPG

23

Gas Mileage Information

DOE/EPA

HIGHWAY MPG

29

1988 ACCORD
119 CID, 4-CYL ENGINE
ELECTRONIC FUEL INJECTION
4-SPEED AUTOMATIC TRANSMISSION
FEEDBACK FUEL SYSTEM

Actual mileage will vary with options, driving conditions, driving habits, and vehicle's condition. Results reported to EPA indicate that the majority of vehicles with these estimates will achieve between

19 and 27 mpg in the city, and between 24 and 34 mpg on the highway

Estimated Annual Fuel Cost: $600

For Comparison Shopping, all vehicles classified as COMPACT have been issued mileage ratings ranging from

10 to 33 mpg city and 16 to 42 mpg highway.

Figure 9.1 EPA Fuel Economy Estimates

□

The Vocabulary of Interval Estimates

An interval estimate for a parameter in business statistics is usually called a *confidence interval** and is formed as follows:

$$\text{Point estimate} \pm \text{Potential sampling error}$$

In Example 9.2, the potential sampling error for the highway rating was 5 mpg, the difference between 24 (or 34) and the point estimate of 29. An explanation of how to determine a point estimate and the potential sampling error will be forthcoming in

*Another type of interval estimate is a *prediction interval*.

the next chapter. Other phrases used interchangeably with "potential sampling error" are "error of the estimate" or simply "margin of error." The term *error* is a part of each phrase since we are using a portion of the population to estimate a summary measure of the entire population. We are therefore subject to the randomness of sampling.

It is important to recognize two facts about confidence intervals. First, the numerical value of the parameter being estimated by the confidence interval is fixed, but unknown. Second, an interval estimate is specific to a sample. This means each sample conceivably produces an interval estimate covering different values. For example, if an independent testing firm replicated the EPA's study on highway mpg for Honda Accords, it might produce an interval estimate from 22 to 30 instead of the 24 to 34 interval the EPA developed. While different samples would yield different intervals, the parameter these samples are attempting to estimate (mean mpg for the population) is a constant. It is stationary in its unknown location.

SECTION 9.2 EXERCISES

9.6 Explain the difference between a point estimate and an interval estimate.

9.7 Why is it not possible to be 100 percent certain that a sample-based interval estimate includes the parameter of interest?

9.8 In a copyrighted story by Hedrick Smith titled "Public's Approval of Reagan in Poll Rising But Limited" (*New York Times,* July 3, 1983), we read: "In 1365 telephone interviews conducted June 20–26, 47 percent of those polled said they approved of Reagan's job performance . . . . The latest poll, which has a sampling error margin of three percentage points plus or minus . . . ."*

a. Which form of making inferences is characterized by this opinion poll?
b. What parameter is of interest?
c. What is the value of the point estimate of this parameter?
d. Determine, if possible, the potential sampling error for the poll.
e. Construct a confidence interval for the parameter of interest.

9.9 If a point estimate for the parameter π is .27 and the potential sampling error is given as .11, form an interval estimate for π.

9.10 Suppose a confidence interval extends from 7.1 to 9.4. Determine, if possible, the point estimate and the potential sampling error.

9.11 Refer to Example 9.1. Suppose a random sample of 54 days yields a point estimate of $2178 for the average day's receipts.

a. Is it likely that μ = $2178?
b. If the potential sampling error is $36, is μ guaranteed to be between $2142 and $2214?
c. Is μ a constant, or is μ a variable?

9.12 Why do we *estimate* a population parameter from the sample data and run a risk of being incorrect? Why not conduct a census and *compute* the value of the parameter?

9.3 TESTING HYPOTHESES

Besides estimation, our other method of making a statistical inference is by conducting a hypothesis test. Just what is a hypothesis? A **hypothesis** can be defined as a proposition tentatively put forth to explain certain observations or a situation. For an example from a children's story, when Chicken Little was struck by a falling acorn, she formed the hypothesis "the sky is falling." Sir Isaac Newton, as the legend goes, was hit on the head by a falling apple. To him this suggested that all particles of matter exerted an attraction on each other. Chicken Little's hypothesis was incorrect; Newton's was correct and led to the development of the laws of gravity.

Hypothesis testing is a decision-oriented method of making an inference. Unlike estimation, the bottom line in hypothesis testing is not a range of values; instead it is a yes-or-no judgment about the reasonableness of the hypothesis.

In statistics, a hypothesis explains some aspect of a population. By its suppositional nature, a hypothesis may or may not be true. For a very simple example, we might hypothesize that the maximum value in a population is 72. In making this hypothesis we also are implying that, first, we do not know for certain that 72 is the population's maximum value and, second, because of the first point, we will conduct an investigation to scrutinize the hypothesis. The investigation is called *testing a hypothesis*. For example, to test the hypothesis that 72 is the maximum value in the population, we could take a random sample from the population and note the largest value in the sample. If the sample maximum is less than or equal to 72, then it seems logical that we should retain the hypothesis. Had the largest number in our sample been 75, though, we would have reason to abandon, or reject, our hypothesis that the maximum value in the population was 72.

The rationale for conducting a hypothesis test in statistics is borrowed from a form of argument in logic called the "method of indirect proof," or *reductio ad absurdum*. The idea of that method is to assume as true the opposite of what we wish to prove. If the argument "reduces to an absurdity" by leading to a contradiction, then logic holds that the original assumption must be false. Therefore, its negation—what we set out to prove—must be true.

As an analogy, consider a court trial. Recall that the judicial system is based on the premise that a defendant is assumed innocent until proven guilty. Thus the trial begins with the assumption that the defendant is innocent. Under the method of indirect proof, the prosecutor is responsible for presenting evidence to contradict this assumption. If, in the eyes of the jury, insufficient evidence is presented, the original assumption is retained and the defendant is found not guilty. If the jury sees sufficient evidence, the original assumption of innocence is discarded and the defendant is declared guilty.

Null and Alternative Hypotheses

The starting point of a statistical test is a statement, or a hypothesis. There are two basic forms: a *general hypothesis,* which is made up of words or sentences such as "the mean has not changed" or "the variables are uncorrelated," and a *statistical hypothesis,* which translates a general hypothesis into symbols and numbers.

Definition

A **statistical hypothesis** is a statement about the numerical value of a parameter.

Examples of statistical hypotheses include:

Hypothesis: $\mu = 7$
Hypothesis: $\pi = .4$
Hypothesis: $\sigma^2 = .15$

A statistic *never* appears in a statistical hypothesis; only parameters (usually denoted by Greek letters) are eligible.

Like the court trial, a statistical test involves two opposite statements called the *null hypothesis* and the *alternative hypothesis*. The word *null* means not any, none, or nil, suggesting a zero value. If we think of the position of the number 0 on a number line, we recognize it as a starting point or a point of reference. As applied to a hypothesis, null means the hypothesis represents a starting point in our investigation. The null hypothesis is not a binding proposition, however; we can discard it. If the null hypothesis were binding, we would not be able to change our minds about it in the face of contradictory sample evidence, and there would be no sense testing it. Literally, testing a hypothesis means testing a *null* hypothesis.

The alternative hypothesis embodies an opposite statement from the null hypothesis. Since the null hypothesis is not binding, we may, depending on the evidence gathered, decide to abandon it and embrace the alternative, in effect declaring the alternative hypothesis to be the more reasonable statement. Relative to the court trial, the general hypotheses are

Null hypothesis: The defendant is innocent.
Alternative hypothesis: The defendant is guilty.

In logic and theoretical mathematics, certain propositions are capable of an indirect proof in the strictest sense of the word. However, proof of guilt in a court trial has a looser interpretation since a strict proof may not be possible. While a jury may not be *absolutely* sure, they still can declare the defendant guilty if the evidence convinces them of guilt beyond a reasonable doubt. Likewise in business statistics, we cannot be absolutely sure whether our null hypothesis is false. Accordingly, we will borrow the courtroom's reasonable doubt standard and *avoid* using the word *proof* to describe our conclusion. A census could yield a proof, but a sample cannot.

Our conclusion for a statistical test will be based on the sample evidence we see. If there is sufficient sample evidence to contradict the null hypothesis, we will believe that the alternative hypothesis is the more reasonable statement. If the evidence is insufficient, we will continue to believe that the null hypothesis is a reasonable statement. In neither case do we declare the null statement to be true or false, a declaration that would necessarily follow if our conclusion were interpreted as proof.

Formulating the Hypotheses

Formulating the appropriate null and alternative hypotheses is based on the same fairness principle governing a court trial—that is, the nonbinding hypothesis of innocence is assumed correct until and unless subsequent evidence contradicts it. The null hypothesis is a statement about the value of a parameter that is assumed to be true. In the spirit of the method of indirect proof, we require the null hypothesis to be formulated as a statement of equality about a parameter's value. For instance, here are examples of correctly formulated null hypotheses:

$$\text{Null hypothesis: } \mu = 10$$
$$\text{Null hypothesis: } \sigma^2 = 3$$
$$\text{Null hypothesis: } \pi = .60$$

In each case, the null hypothesis is a specification of the parameter's value. The equality symbol always appears in a (statistical) null hypothesis.

An alternative hypothesis, conversely, is an opposite statement relative to the null hypothesis. The following statements thus qualify as correctly formulated alternative hypotheses:

$$\text{Alternative hypothesis: } \mu \neq 10$$
$$\text{Alternative hypothesis: } \sigma^2 \neq 3$$
$$\text{Alternative hypothesis: } \pi \neq .60$$

However, we also may interpret the notion of an opposite statement rather loosely. There will be problem settings in which interest is restricted to determining whether the parameter's value is strictly greater than or strictly less than the specified value in the null hypothesis. In such cases, the following statements also qualify as examples of alternative hypotheses:

$$\text{Alternative hypothesis: } \mu > 10$$
$$\text{Alternative hypothesis: } \sigma^2 < 3$$
$$\text{Alternative hypothesis: } \pi > .60$$

With respect to the first alternative hypothesis, note that the null hypothesis could be written as

$$\text{Null hypothesis: } \mu \leq 10$$

or simply as

$$\text{Null hypothesis: } \mu = 10$$

Table 9.1 offers further examples of null and alternative hypotheses.

Table 9.1 Examples of Null and Alternative Hypotheses

Hypothesis	Type
$\mu = 30$	Null
$\mu > 30$	Alternative
$\pi < .2$	Alternative
$\pi \neq .5$	Alternative
$\sigma^2 \leq 8$	Null

EXAMPLE 9.3

The Shroud of Turin is thought by many Christians to be the burial cloth of Jesus Christ. Although it contains the image of a man who died by crucifixion, many people doubt the authenticity of the shroud. Scientific testing to date the shroud's linen material was conducted in 1987. Prior to the tests correspondent Lee Dye wrote the following in the copyrighted story:

> *The test, long awaited by scientists and religious leaders alike, could prove the artifact is far too new to have been the burial cloth. However, even if the results show the shroud dates back to the time of Christ, it will not prove that it was indeed the burial cloth. . . . Thus, the seven labs are in the position of being able to prove a negative, but not a positive.**

Use the scientific dating situation to illustrate the method of indirect proof by setting up the general null and alternative hypotheses regarding the age of the shroud.

Solution:

Since the scientific testing can contradict that the shroud is the burial cloth, we set up these hypotheses:

Null hypothesis: The shroud is at least 1950 years old.
Alternative hypothesis: The shroud is less than 1950 years old.

The logic of indirect proof would allow scientists to conclude the shroud cannot be the burial cloth if the scientific tests show the artifact is too new to be authentic.† □

*From Lee Dye, "Shroud of Turin: Tests May Give It an Age," *Los Angeles Times*, October 26, 1986. Copyright, 1986, *Los Angeles Times*. Reprinted by permission.

†In the October 3, 1988, issue of *Newsweek* it was reported that results of three independent laboratories placed the Shroud's manufacture at about 1350 A.D.

Essential Elements of Testing Hypotheses

A statistical test of hypothesis, as we will see in later chapters, contains five essential elements: null hypothesis, alternative hypothesis, test statistic, rejection region, and conclusion. The definition and notation for each term appears in the following box.

Essential Elements of Hypothesis Testing

Element	Notation	Definition
Null hypothesis	H_0	A statement including equality about the value of a parameter
Alternative hypothesis	H_a	A statement of inequality about the value of a parameter
Test statistic	*TS*	A summary of the sample data into a single number that helps us decide the reasonability of H_0
Rejection region	*RR*	A set of unlikely values of the test statistic that imply rejection of H_0
Conclusion	*C*	The decision to accept or reject H_0

The **null** (H_0) and **alternative** (H_a) **hypotheses** represent statements about a population parameter. The **test statistic** (*TS*) summarizes the sample data and may be thought of as the evidence.

In our earlier exposures to inference (see Chapter 4 and Chapter 6), we used the region of concentrated values to help us reach a decision. Those values outside the region of concentrated values formed the unlikely region, which we now term the **rejection region** (*RR*).

The **conclusion** (*C*) is found by determining whether the test statistic falls inside or outside the rejection region. If the *TS* is in the rejection region, we conclude the sample is inconsistent with the null hypothesis and *reject* the hypothesis. Rejecting H_0 often is called a "strong conclusion" because rejection means sufficient evidence has been found to persuade us that H_0 is not a reasonable statement. If the test statistic does not fall into the rejection region, we have evidence that the sample is consistent with the null hypothesis so we *do not reject*—that is, we accept—the hypothesis. Accepting the null hypothesis means we have not found sufficient evidence to reject it. We therefore view our initial assumption as *supported* (but not proved). Acceptance is often referred to as a "weak conclusion."

Possible Errors

Testing the null hypothesis is to decide its reasonability in the light of sample data. In the chapters that follow, we will learn how to use the sample evidence to arrive at

Table 9.2 Possible Outcomes When Testing a Hypothesis

	Actual Status of the Null Hypothesis, H_0	
Conclusion	True	False
Reject H_0	Type I error	Correct conclusion
Accept H_0	Correct conclusion	Type II error

a conclusion to accept or reject the null hypothesis. Before we end our preliminary look at hypothesis testing, we need to discuss the possibility that the conclusion we make is mistaken.

Cut and dried decisions to accept or reject the null hypothesis after seeing part but not all of the population open the door for potential decision-making errors. We cannot be *certain* any decision is the correct one when it is sample based. For instance, it is possible for the null hypothesis to be a true statement, yet the sample evidence indicates (erroneously) that we should reject the null hypothesis. Conversely, the decision to accept the null statement when it is, in fact, a false statement is also an error. These situations, summarized in Table 9.2, are formally defined as follows.

Definitions

A **Type I error** is committed whenever a true null hypothesis is rejected.

A **Type II error** is committed whenever a false null hypothesis is accepted.

In testing statistical hypotheses we risk committing one of these errors with every conclusion we make. If we decide to reject the null hypothesis, we are liable for a Type I error only, just as we are susceptible to a Type II error only with the decision to accept the null hypothesis. As we will discover shortly, we can quantify our potential for committing these errors and design a testing procedure to control for one or both of them. Often, the severity and implication of the errors dictate the design of the test, as the following examples illustrate.

EXAMPLE 9.4

The water treatment plant servicing a city routinely discharges filtered sewage into a nearby river. The plant must filter the waste water to meet minimum government guidelines so as not to endanger the environment. Periodically, an inspector from the

Environmental Protection Agency (EPA) visits the plant and selects water samples to measure the impurities and test for compliance with government regulations.

a. What set of general hypotheses should the EPA official formulate to test for compliance?
b. State, in words, the meaning of both a Type I and Type II error for the hypotheses of part a.
c. Which error is more severe?

Solution:

a. The nonbinding assumption should be that the plant is in compliance with the government regulations. Although a clean water sample is supporting evidence that the waste water is being sufficiently filtered, a clean sample cannot *prove* this assumption. *All* water (a census) would have to be tested to *prove* purity. On the other hand, one impure water sample is sufficient evidence to contradict the nonbinding claim. Thus, the hypotheses are

Null hypothesis:	The waste water is filtered sufficiently to meet government regulations.
Alternative hypothesis:	The waste water is not filtered sufficiently to meet government regulations.

b. A Type I error is committed by concluding that the plant is *not* filtering the waste water sufficiently when, in fact, it is. A Type II error is committed by deciding that the water is pure enough to meet regulations, when actually it is not.
c. Perspective plays an important role in answering this question. From an environmental point of view, the Type II error could be catastrophic since the water being discharged would be impure. This error impacts on the quality of life. Ethically speaking, we would hope the manager of the treatment plant shares this viewpoint. The Type I error possibility forces the plant to spend more time and money filtering already pure-enough water. This error results in an unnecessary cost. In weighing a danger to the environment versus an unnecessary cost, we conclude that the Type II error is more severe in this case, and the test should be designed to minimize it.

□

EXAMPLE 9.5

A pharmaceutical company has developed a new drug for treating migraine headaches. The company believes the drug is superior to currently available medications. However, the company must comply with Federal Drug Administration (FDA) guidelines concerning efficacy and safety before the drug is approved. The company must convince the FDA of the effectiveness of its drug in reducing the pain due to migraines while not endangering the health of the individual.

a. What set of general hypotheses concerning safety should the FDA formulate?
b. State, in words, the meaning of both a Type I and Type II error for the hypotheses of part a.
c. Which error is more severe?

Solution:

a. The pharmaceutical company must provide evidence beyond a reasonable doubt that its drug is safe. The FDA is cautious and conservative; it does not automatically assume new drugs are safe. Before approval, the FDA requires persuasive evidence that the drug should be licensed. The hypotheses are

   Null hypothesis: The new drug is not safe.
   Alternative hypothesis: The new drug is safe.

b. A Type I error is incurred by rejecting a true null hypothesis. For the set of hypotheses in part a, a Type I error means the FDA concludes the new drug is safe when it is not. A Type II error happens when the new drug is erroneously declared unsafe.
c. A Type II error creates a lost opportunity for the pharmaceutical company: The FDA will not approve the drug, but the company may be able to win subsequent approval with further tests. A Type I error has severe consequences for the public. For example, in the 1950s the drug diethylstilbestrol (DES) was widely prescribed for women who were susceptible to miscarriages during pregnancy. Although the drug was tested and approved, it was later found to cause cancer in the first generation of daughters. This drug is still troublesome today, as researchers have discovered its effects in the second generation of daughters. Clearly, a Type I error is more severe for the hypotheses in part a.

□

CLASSROOM EXAMPLE 9.1

Preparing for a Hypothesis Test

An auditor faces the task of rendering an opinion about the account balance for a company's set of accounts receivable. Unable to conduct a census, the auditor must base his or her decision on a sample of accounts. Most companies tolerate small errors in the account balance as being immaterial and are concerned only when the discrepancy becomes material. The auditor's job is to decide whether the stated account balance is essentially correct (the discrepancy is immaterial) or incorrect (the discrepancy is material). Traditionally, an audit evolves from the belief that the firm employs competent people who know how to perform their jobs.

a. Under what set of general hypotheses is the auditor operating?
b. State, in terms of problem setting described above, the meaning of both a Type I and a Type II error.
c. Which error type do you think is more severe for this problem setting?

SECTION 9.3 EXERCISES

9.13 Name the essential elements of hypothesis testing.

9.14 Identify the following as either a null hypothesis or an alternative hypothesis:
a. Hypothesis: $\mu < 0$ b. Hypothesis: $\sigma^2 = 65$
c. Hypothesis: $\mu = 50$ d. Hypothesis: $\pi \geq .9$
e. Hypothesis: $\sigma \neq .7$

9.15 Identify the following as either a null hypothesis or an alternative hypothesis:
a. Hypothesis: $\lambda = 1.5$ b. Hypothesis: $\mu \leq 157$
c. Hypothesis: $\pi < .2$ d. Hypothesis: $\sigma > 9$
e. Hypothesis: $\sigma^2 = .006$

9.16 Why do statistical hypotheses involve parameters but not statistics?

9.17 In the production of plastic wrap for food products, a manufacturer passes a white resin through a hot extruder that transforms it into a clear film to be gathered on rollers. A key characteristic of the resin which must be continually monitored is its moisture content. If the moisture content gets too low, the film will take on a brownish cast; if too high, the risk of holes in the film and a process breakdown increases. Unless spot tests of the moisture content on samples of the film reveal dramatic departures from the ideal moisture content, production continues. What set of general hypotheses about the moisture content does the manufacturer formulate for the spot tests?

9.18 Four months ago a medical plastics firm changed its credit terms, offering a bigger cash discount for early payment, in hopes of reducing the average age of accounts receivable from its historical value of 57 days.
a. What set of general hypotheses about the average age of the accounts receivable should the firm formulate if it wishes to make the case that the change in credit terms has lowered the average?
b. State, in words, the meaning of a Type I error and a Type II error for the hypotheses of part a.

9.19 In exploring sites for the possible location of a convenience store, the parent chain wishes to be as certain as possible that the traffic volume along the road of the prospective site is heavy. The field manager has pinpointed a potential site and sent his recommendation to the vice president of real estate in the home office. However, the vice president is skeptical about the volume of traffic and asks the field manager to collect a sample of traffic counts, analyze the data, and convince him that the site meets their volume requirements. Under what set of general hypotheses about the traffic volume is the vice president of real estate operating?

9.20 Occasionally an automaker will issue a recall for all vehicles of a certain make. The purpose is to eliminate a possibly dangerous problem through repairs or adjustments. A recall is a costly decision in terms of money and prestige and is done only if an intensive investigation of a few cars indicates that a serious problem is likely to exist in the entire fleet.

a. What set of general hypotheses regarding the seriousness of the alleged problem should the automaker form?

b. State, in words, the meaning of a Type I and a Type II error.

c. From the consumer's viewpoint which error is more severe?

9.21 Human resource managers, in the screening of applicants for a position within the firm, often utilize the following set of general hypotheses in their pre-interview assessment of the candidate:

Null hypothesis: The candidate is potentially a good employee.
Alternative hypothesis: The candidate is potentially a bad employee.

If the manager concludes the null statement is true after interviewing the candidate, the candidate is offered a job. If the manager believes the alternative hypothesis, the candidate is not offered a job.

a. Identify a Type I and a Type II error.

b. Which error is more severe to the company?

9.22 One of the biggest problems for new businesses (especially small businesses) is cash flow. A company can be making a profit on paper yet not have enough cash to pay the bills. This unfortunate situation sometimes occurs shortly after the owner has made a decision to increase the company's debt, thinking the company can support it. Increasing the debt outstanding causes one of two consequences. Either the company can support the debt and subsequently grows, or the company cannot support it and subsequently experiences a critical cash flow problem. Suppose the owner of a new business must decide between the following hypotheses:

Null hypothesis: The company cannot support increased debt.
Alternative hypothesis: The company can support increased debt.

a. Identify a Type I and a Type II error.

b. From the owner's perspective which error is more severe?

9.4 SUMMARY

The study of inferential statistics is quite different from other areas of inquiry because it integrates material from logic, descriptive statistics, and probability models. Specific inferential problems will be explained in detail in the next several chapters. All problems will involve three components—a specific objective to be achieved, an accessible population, and a random sample—and will represent either estimation or hypothesis testing. Generally, *estimation* is the preferred avenue for business problems when the mission is *to explore* the population in an information-gathering vein. *Testing hypotheses* is used *to support or to contradict* preconceived ideas about characteristics of the population.

9.5 TO BE CONTINUED . . .

. . . *In Your College Courses*

The two ways of making statistical inferences—estimation and hypothesis testing—reappear in other business disciplines. For students who are accounting majors we present a brief glimpse into auditing, where these inferential tools are used extensively.

In Chapter 12 of Arens and Loebbecke's textbook titled *Auditing: An Integrated Approach,* we read the following:

> . . . *an auditor can choose between calculating confidence intervals or doing hypothesis testing. When the auditor calculates confidence intervals,* the objective is to determine the total value of an account balance or other total . . . .
>
> *When the auditor uses hypothesis testing,* the objective is to determine whether the client's recorded balance is correct. *Hypothesis tests are much more commonly used in auditing than calculation of confidence intervals because the objective in most audits is to evaluate whether an account balance is correctly recorded.**

Notice that auditors use both types of statistical inferences and that the first component of an inferential statistics problem—an objective—is specifically stated.

Arens and Loebbecke go on to discuss the risks associated with the two types of hypothesis testing errors:

- ARIA, the auditor's acronym for the "acceptable risk of incorrect acceptance," which is the chance of committing a Type II error.
- ARIR, the "acceptable risk of incorrect rejection," which is the chance of committing a Type I error.

Clearly a student pursuing a career in auditing or public accounting will encounter statistical hypothesis testing on a regular basis. Hypothesis testing is also an integral part of business courses in marketing research, econometrics, forecasting, portfolio theory, and to a lesser extent in management science and personnel management.

. . . *In Business/The Media*

Estimation and hypothesis testing are found in many areas of the business world, although the applications may not appear as pure as what we have described in this chapter.

Clearly, the EPA's efforts to estimate and report the gas mileage of new cars (see Example 9.2) is a pure application of statistical inference. As another example of the government's use of statistical methods, consider the New Car Assessment Program conducted by the National Highway Traffic Safety Administration (NHTSA). Each

*From Arens/Loebbecke, *Auditing: An Integrated Approach,* 2/E © 1980, pp. 414, 415. Reprinted by permission of Prentice-Hall, Inc., Englewood Cliffs, NJ.

year NHTSA buys about 30 new cars and crashes them into a solid barrier at 35 miles per hour. Two dummies wired with electronic sensors are placed in the front seats of each car crashed. The sensors generate data about the severity of the impact and are used to estimate the so-called HIC (Head Injury Criterion) score and the "chest G's" measurement (the deceleration load at the chest) for each car. NHTSA's annual report is published in the media in July or August.

As the information in the first half of this section indicates, auditors use statistical hypothesis testing in their everyday work. For example, a company's Annual Report includes an Auditor's Report in the form of a letter from the accounting firm that audited the company's financial records. In the letter, the accounting firm renders a judgment about the company's financial state of affairs. This letter summarizes the results of many tests of hypotheses into an *opinion*. Indeed, the word "opinion" suggests some uncertainty, as Needles, Anderson, and Caldwell (1981) note:

> *The use of the word* opinion *is extremely significant because the auditor does not certify or guarantee that the statements are absolutely correct. To do so would exceed the truth, because many such items, such as depreciation, are based on estimates. Instead, the auditor merely gives an opinion as to whether, overall, the financial statements "present fairly" the financial condition of the company* [*p. 237*].

The passage reinforces what we have been stressing: Sample-based inferences always possess a risk of error. An auditor's nightmare, in fact, is committing one of these errors. Exercise 9.32 relates to a real-life example of such a situation.

CHAPTER 9 EXERCISES

9.23 Identify the components of an inferential statistics problem.

9.24 Consider the plight of a financial analyst who must absorb all the current financial information and suggest strategies for clients. Suppose we focus on a particular stock listed on the New York Stock Exchange. Basically, the analyst must choose one of the following recommendations for clients that own stock in the BRS company:

Null hypothesis: It is wise to hold onto BRS stock.
Alternative hypothesis: It is wise to sell BRS stock.

Assume a hold recommendation means the analyst believes the price of the stock will increase. A sell recommendation means the analyst is convinced the price of the stock is headed down. Identify the correct conclusions, a Type I error, and a Type II error, using Table 9.2 as your guide.

9.25 Identify three situations where inferential statistics procedures often break down or yield poor results.

9.26 Suppose the following statistical hypotheses are formulated:

$$H_0\colon \mu = 17$$
$$H_a\colon \mu < 17$$

a. Which hypothesis do we find more reasonable if a strong conclusion results?
b. What is a Type II error for these hypotheses?
c. Suppose the sample data support the null hypothesis. Can we say with certainty that $\mu = 17$? Why?

9.27 A large company tolerates a fraction of errors in the small invoices (total bill under $1000) it receives. A 1 percent ceiling for the error rate is the firm's ideal. Auditors are hired to test this assumption. What set of statistical hypotheses should the auditors form if the objective is to determine whether the sample data are consistent with the firm's stated error rate?

9.28 Suppose you read the following statement in an article concerning the popularity of home computers: "Seven of 10 Americans think that a home computer is a necessity." Later in the article, you read this: "Because of chance variations, these results are subject to a 3 percent margin of error. Although the poll was based on 1500 telephone interviews, if every American owning a telephone were interviewed the chance is only 1 in 50 that the true answer would vary by more than 3 percentage points."
a. What parameter is being estimated?
b. What is the point estimate of this parameter?
c. What is potential sampling error?
d. Is it correct to conclude that a definite majority of Americans think that a home computer is a necessity?

9.29 "Buy low and sell high" is recommended as a strategy for investing in the stock market. Unfortunately, knowing when the time is right to be buying and selling is difficult. Suppose we set up these general hypotheses:

Null hypothesis: Now is the time to sell.
Alternative hypothesis: Now is the time to buy.

Further assume that these are the only two conditions possible and that if one condition or hypothesis is false the other must be true.
a. Identify a Type I error and a Type II error.
b. Which error is more severe to an investor?

9.30 When the father of a child is in question, a blood test sometimes is conducted to compare the blood types of the possible father and the child. A blood test cannot prove the man is the father. It can prove only that he is not. Formulate the general hypotheses in this situation.

9.31 According to the Louisville *Courier-Journal* ("FTC, Brown & Williamson Begin Court Battle over Low-Tar Claims," September 13, 1983), in 1981 Brown & Williamson Tobacco Company introduced a super low-tar cigarette called Barclay. The Federal Trade Commission (FTC) conducted a tar test and reported Barclay had only 1 milligram (mg) of tar. A controversy soon developed when Brown & Williamson used the 1-mg figure from the FTC tests in its advertisements. Competitors complained to the FTC that the cigarette filter had special chambers and air vents that smokers block with their fingers, increasing the tar intake, but which are not blocked by the FTC's testing machines. A second FTC tar test resulted in a reported 3 to 7 mg of tar per cigarette and an order from the FTC to Brown & Williamson to stop using the 1-mg figure in its advertisements. Brown & Williamson then objected and argued it had test data that "proved" Barclay left smokers with only 1 mg of tar. Suppose we set up these hypotheses:

Null hypothesis: Barclay has 1 mg of tar per cigarette.
Alternative hypothesis: Barclay has more than 1 mg of tar per cigarette.

a. Which hypothesis represents Brown & Williamson's position and which represents the FTC's?

b. Whose claim can be contradicted (given sufficient evidence), based on the above hypotheses?

9.32 According to the *New York Times,* Alexander Grant & Company, based in Chicago, was the eleventh-largest accounting firm in the country in 1984. ESM Government Securities, Inc., retained Alexander Grant as its auditor until ESM was closed on March 4, 1985, by a federal court order, obtained by the Securities and Exchange Commission (SEC). The SEC charged ESM with engaging in a pattern of fraud, hiding cumulative losses totaling $196.5 million in an affiliated dummy company. Alexander Grant was in a position to see this but presented a healthy balance sheet of ESM. The audited financial statement and balance sheet of ESM Government Securities were the only independent verification that ESM's customers could use to judge its soundness. Formulate the general hypotheses that Alexander Grant used in auditing ESM and identify the type of error the accounting firm made.

9.33 Refer to Exercise 9.5. Set up a set of statistical hypotheses for the manager.

9.34 In business statistics, is a hypothesis a statement about the sample or is it a statement about the population?

9.35 Suppose you are willfully driving at a high rate of speed (in excess of 65 miles per hour) on an interstate highway when your radar detector starts to sound off. Use the following set of hypotheses to identify each action as a Type I error, a Type II error, or a correct decision.

Null hypothesis: The radar detector is giving a false signal.
Alternative hypothesis: The radar detector is giving a true signal.

a. Maintaining your present rate of speed through a radar trap.

b. Slowing down through a radar trap.

c. Maintaining your present rate of speed when there is no radar trap.

d. Slowing down when there is no radar trap.

9.36 A customer relations manager at a bank is facing a dilemma with respect to a husband and wife's account. According to the couple's version of the story, they paid their Visa bill in full last month before the due date. However, the bank claims the check was not received in time to be credited to their account; thus, a finance charge was assessed. The couple believes the bank failed to process the check promptly and asks the manager to remove the finance charge from their account. The manager must decide between the following statements:

H_0: The check was received in time.
H_a: The check was not received in time.

If the manager believes H_0 is true, she will remove the finance charge. If she believes H_0 is false, she will not remove the finance charge. Identify parts a through d as either a Type I error, a Type II error, or a correct decision.

a. The manager removes the finance charge when, in fact, the check was not received in time.

b. The manager does not remove the finance charge when, in fact, the check was not received in time.

c. The manager does not remove the finance charge when, in fact, the check was received in time.

d. The manager removes the finance charge when, in fact, the check was received in time.

REFERENCES

Arens, A. A., and J. K. Loebbecke. 1980. *Auditing: An Integrated Approach,* 2nd Edition. Prentice-Hall, Inc., Englewood Cliffs, NJ.

Needles, Jr., B. E., H. R. Anderson, and J. C. Caldwell. 1981. *Principles of Accounting,* Houghton Mifflin Company, Boston.

Chapter Maxim *Risk of an incorrect inference is never zero when that inference is based on sample information.*

CHAPTER 10

LARGE SAMPLE INFERENCES ABOUT μ

*Optional

Objectives

After studying this chapter and working the exercises, you should be able to

1. Explain why an interval estimate is preferred over a point estimate.
2. Use sample evidence to construct and interpret a confidence interval for the population mean.
3. List the factors that influence the width of a confidence interval.
4. Determine the sample size necessary to obtain a confidence interval possessing a predetermined precision.
5. Set up and execute a formal statistical test of a hypothesis.
6. Compute and provide an interpretation of the *p*-value for a statistical test.
7. Define Type I and Type II errors and draw a sketch to show how each could happen.
8. Explain the meaning of α (Greek alpha) for confidence intervals and for hypothesis tests.
9. Differentiate the objectives of a hypothesis test and a confidence interval.

*10. Execute the computer processor Large Sample Inferences for μ to verify answers and/or solve problems concerned with large sample inferences about μ.

In business statistics, we often wish to make a statement or draw a conclusion about a population of interest, based on the results of a sample—this process is called *statistical inference*. In this chapter we will investigate statistical inferences in some detail. There are two methods of making an inference: *estimation* and *hypothesis testing*. While these forms of inference have different objectives, we will see that they have several elements in common.

Actually, we have already provided some examples of making an inference with information derived from a sample. For instance,

- In Section 2 of Chapter 4, we discussed how to use the correlation coefficient found in a bivariate sample to draw a conclusion as to whether X and Y are related to each other in the population.
- In Section 1 of Chapter 6, we described acceptance sampling as a business application of the binomial distribution. This procedure involved using sample information (number of defective items found) to indicate a course of action—accept or refuse—about the larger population.

*Applies to optional section.

These examples were presented somewhat informally without delving deeply into the underlying rationale involved. But with an understanding of sampling distributions (Chapter 8) as well as an overview of inference (Chapter 9), we are now ready to proceed more formally and with a better appreciation of the logic employed.

This chapter focuses on inferences about the population mean μ when the sample size is large ($n \geq 30$). Later chapters will consider inference procedures for μ for small samples ($n < 30$) as well as inferences about other parameters. Most of the inference methods of this chapter will reappear throughout the rest of the book; in many cases, we will see that changing symbols in certain formulas is all that is required to adjust from one sampling situation to another. Thus, the structure of inferential techniques remains the same, as does the interpretation.

10.1 ESTIMATION OF THE POPULATION MEAN

One of the most commonly occurring problems in statistical inference is the need for estimating a population's mean value μ. We are now ready to do so, using the available sample evidence to provide two types of estimates.

Point and Interval Estimates

Consider the following scenario: A manufacturer of peanut butter and other peanut-related products purchases its peanuts from a sheller in Georgia. The peanuts are shipped to the manufacturing plant in 50-kilogram packages via railroad; a shipment typically contains several thousand such packages. The manufacturer is billed for the peanuts based on the net weight indicated on the package—50 kilograms (about 110.2 pounds). However, a small but noticeable decline in yields at the plant has management searching for an explanation. One part of the investigation involves checking a sample of incoming package net weights, determined by individually weighing 32 packages selected at random. The plant's scale, recently calibrated and quite accurate, indicates weights in pounds: The sample mean, based on $n = 32$ observations of X (X = package net weight), is $\overline{X} = 108.9$ lbs.

The sample mean of 108.9 lbs. is a *point estimate,* a single numerical value that estimates the unknown population mean μ. However, since any sample is an incomplete look at the population, we cannot expect $\overline{X}$ to equal the population mean μ. Furthermore, a point estimate by itself conveys no information about the potential sampling error involved. In other words, the point estimate cannot tell us how close to μ our $\overline{X}$ might be. Has our sample mean landed within ± 1.0 lb. of μ? Or is the margin for sampling error larger, such as ± 5.0 lbs.? Could we possibly be accurate to within $\pm .25$ lb.? Questions like these can be answered by developing an interval estimate. An *interval estimate* is a range of values (centered at $\overline{X}$) within which the unknown value of μ is thought to lie. An interval estimate, therefore, can be more informative than a point estimate. In the pages ahead, we will learn how to express our degree of faith or confidence that a given interval we have computed does in fact include the population mean μ.

Definitions

A **point estimate** of a parameter is a single numerical value of a statistic.

An **interval estimate** is a range of values within which the unknown parameter value is thought to lie.

COMMENTS

1. A random variable that is used to estimate a population measure is called an *estimator*. The sample mean $\overline{X}$ is therefore an estimator of μ. The specific numerical value of an estimator is called the *point estimate*. In the preceding discussion, the point estimate was 108.9 lbs.
2. Other estimators for μ besides $\overline{X}$ exist. For instance, we could choose the sample median, mode, or midrange to estimate μ. Statisticians have established some criteria for determining which of several possible candidates is the best estimator. Perhaps the most important criterion is unbiasedness, discussed in Chapter 8 as a desirable property of $\overline{X}$. Other criteria also exist, but since the sample mean can be shown to be the best estimator relative to these criteria, it will be the only estimator for μ discussed. In later chapters, any new estimator introduced will be the one generally regarded as the best one for the sampling situation at hand.

Confidence Intervals

In estimation, the mean value of the population (μ) is unknown. The objective is to use sample information to specify a range of values that has a certain probability, or *confidence*, of including μ. The bases for developing this interval estimate, also known as a confidence interval for μ, are the Central Limit Theorem (CLT) and our ability to locate any probability of interest using the Z-table.

When we take a random sample and compute the sample mean, the CLT tells us that our value of $\overline{X}$ (one of perhaps millions of $\overline{X}$'s that could have resulted) comes out of an approximately normal distribution, even if the population of X is not normal. Figure 10.1 illustrates this by showing the distribution of possible values for $\overline{X}$ as a normal curve centered at the population mean μ, and with a standard deviation (standard error) of $\sigma_{\overline{X}} = \sigma/\sqrt{n}$. Our knowledge of the Z-table, in particular our ability to determine a *middle* section of the curve, permits us to place limits on how far from the population mean μ a point estimate such as $\overline{X}$ is likely to fall. (See Example 7.7 if you need a quick review on finding a middle portion of a normal distribution.)

For instance, let's consider how we would construct a 98 percent confidence interval. Figure 10.2 illustrates the situation by showing the middle 98 percent of probability centered at $Z = 0$, with the extreme 2 percent of area being shown as 1 percent in the lower and 1 percent in the upper tail. The middle 98 percent therefore

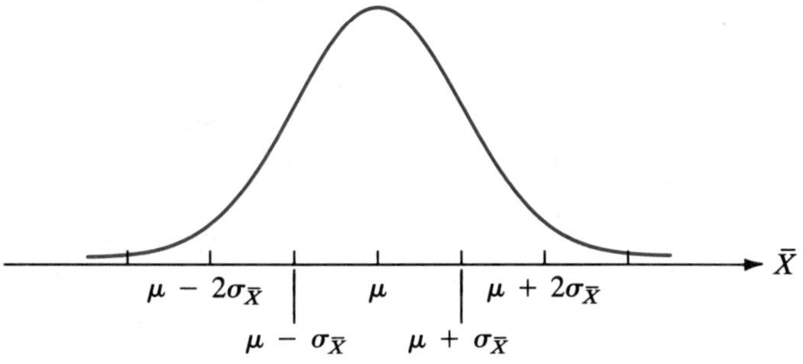

Figure 10.1 Normal Sampling Distribution of $\overline{X}$ for large n

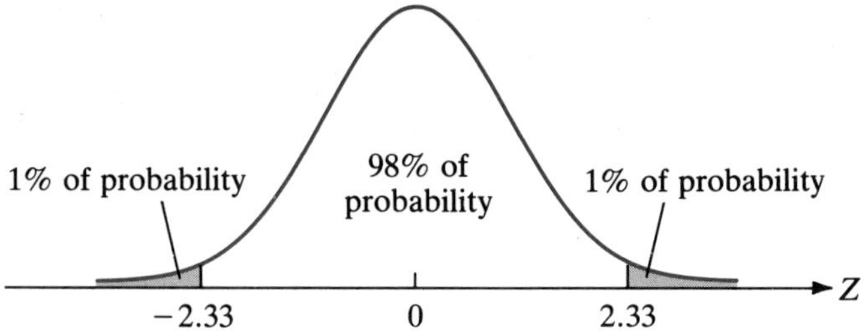

Figure 10.2 *Z*-scores for a 98% Confidence Interval

begins at the first percentile and ends with the 99th percentile. By looking in the body of the *Z*-table for .0100 (or for .9900), we can determine that the *Z*-scores of interest for a 98 percent confidence interval are $Z = \pm 2.33$. In other words,

$$P(-2.33 \leq Z \leq 2.33) = .98$$

When sampling with $n \geq 30$, we know from Chapter 8 that

$$Z = \frac{\overline{X} - \mu}{\sigma/\sqrt{n}}$$

Substituting, we then have

$$P\left(-2.33 \leq \frac{\overline{X} - \mu}{\sigma/\sqrt{n}} \leq 2.33\right) = .98 \qquad \textbf{(10–1)}$$

Applying some algebra and rearranging terms to get μ as the only term between the inequality signs, we obtain

$$P(\overline{X} - 2.33\sigma/\sqrt{n} \leq \mu \leq \overline{X} + 2.33\sigma/\sqrt{n}) = .98 \qquad \textbf{(10–2)}$$

We interpret Equation 10–2 as follows: If we are going to take a random sample, the interval $\bar{X} \pm 2.33\sigma/\sqrt{n}$ will have a probability of .98 of including the unknown value of μ. On the other hand, there is a probability or risk of .02 that the interval $\bar{X} \pm 2.33\sigma/\sqrt{n}$ will fail to include μ. Figure 10.3 illustrates the possibilities. If the sample mean $\bar{X}$ that we will obtain happens to be one from the middle 98 percent of the sampling distribution, then our interval will include μ, since the point estimate would be within $\pm 2.33\sigma/\sqrt{n}$ of μ. Conversely, if the sample mean should happen to be one from either the lower 1 percent or the upper 1 percent of possible means, then the point estimate $\bar{X}$ would miss μ by more than ± 2.33 standard errors.

What will happen when we take one random sample of size n and compute the point estimate $\bar{X}$? Will $\bar{X}$ come from the middle portion of the curve or from the tail regions? We don't know since $\bar{X}$ is a random variable. But we do know that the probability or confidence is .98 that the point estimate $\bar{X}$ will differ from μ by no more than ± 2.33 standard errors. We will use the word *confidence* to refer to the probability that the interval estimate that we compute will include the unknown value of μ.

Definitions

For a confidence interval, the term **confidence** denotes the (prior-to-sampling) probability that the interval estimate will include the unknown value of μ.

For a confidence interval, the term **risk** denotes the (prior-to-sampling) probability that the interval estimate will fail to include the unknown value of μ.

Note that *confidence* and *risk* are terms that apply before the sample is drawn (see Chapter 5 Maxim). Before-the-fact probability applies, since one of two outcomes (interval to be constructed includes μ or does not include μ) is about to happen. But once a sample is taken and specific plus or minus limits are attached to a specific $\bar{X}$, then the resulting interval either does or does not include μ. After the fact, the interval

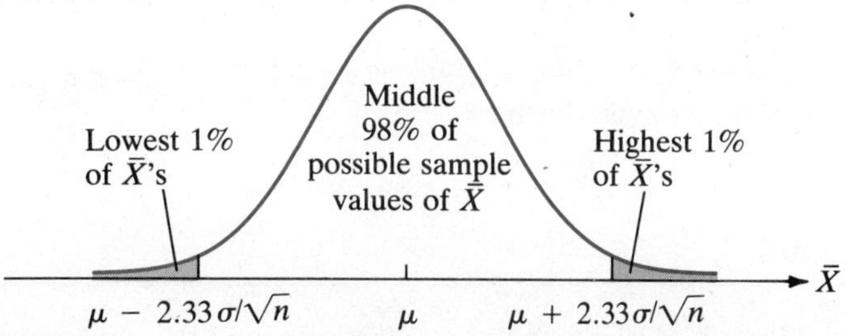

Figure 10.3 A 98% Probability Region for $\bar{X}$

is either correct or it is incorrect, although we do not know which outcome has occurred since μ is still unknown. Therefore we can express only the (prior-to-sampling) degree of confidence that a specific interval includes μ.

In the previous chapter we described the idea of a confidence interval in words as

Point estimate ± Potential sampling error

Changing these words into specific symbols, we now express a confidence interval for μ as

Large Sample Confidence Interval for μ (σ Known)

$$\overline{X} \pm Z\frac{\sigma}{\sqrt{n}} \tag{10–3}$$

where Z is the appropriate Z-score for the level of confidence desired. The most commonly used levels of confidence are 90, 95, and 99 percent. The Z-table on the inside front cover of the book provides the Z-score for these and other frequently used levels of confidence.

It is most often the case that the numerical value for σ in Equation 10–3 will not be known. Our objective is to estimate μ, which is unknown. If we do not know a population's mean, we are not likely to know its standard deviation either. In such cases, we will substitute the computed sample standard deviation S for σ. This approximation works quite well when the sample size is 30 or more; the case of sample size less than 30 is considered in the next chapter. We will use Equation 10–4 more often than Equation 10–3 since in practice σ is generally not known.

Large Sample Confidence Interval for μ (σ Unknown)

$$\overline{X} \pm Z\frac{S}{\sqrt{n}} \tag{10–4}$$

Both expressions for a confidence interval are quite easy to use, as the following examples should illustrate.

EXAMPLE 10.1

Most colleges and universities that offer an MBA program require their applicants to submit their scores on the Graduate Management Aptitude Test (GMAT). The GMAT score is usually a factor in granting admission. Scores on the GMAT can be as low as 200 and as high as 800. GMAT scores are known to have $\sigma = 100$. The director of admissions at a large university has received over three thousand scores from GMAT takers in the past year. He asks that a sample of 75 scores be randomly chosen to develop a 90% confidence interval for the mean GMAT received in the past year. The sample mean is computed to be $\overline{X} = 534$. Construct the desired interval estimate.

Solution:

With σ assumed to be 100, we use Equation 10–3:

$$\overline{X} \pm Z\frac{\sigma}{\sqrt{n}}$$

The sample mean $\overline{X} = 534$ is our point estimate for μ. The Z-values for 90% confidence are ± 1.65, found in the Z-table. With $\sigma = 100$ and $n = 75$, our confidence interval is

$$534 \pm 1.65\frac{100}{\sqrt{75}} \qquad \text{or } 534 \pm 19 \text{ (rounded)}$$

Another way to present our result is to say that our confidence interval extends from 515 to 553. The numbers 515 and 553 represent the lower and upper limits of the confidence interval, respectively.

□

In Example 10.1 does the population mean μ for the 3000-plus GMAT scores lie somewhere between 515 and 553? How should we answer such a question? One way is, "It is likely but we don't know for 100 percent certain." Only a complete census could tell us exactly where μ is, and we've seen only 75 scores. Inferences based on a sample always entail a risk. Here there is a 10% risk that our interval will not include μ; the confidence is 90 percent that our interval will include μ.

EXAMPLE 10.2

A marketing researcher for a paper manufacturer was permitted to sample cash register tapes from a typical supermarket in order to estimate μ, the mean amount spent per customer order on paper products (tissue paper, napkins, paper plates, and so on). The researcher checked 147 tapes and computed $\overline{X} = \$2.12$ and $S = \$1.66$. Develop an interval that would have a risk of only 5 percent of not including μ.

Solution:

Without knowledge of the population's standard deviation but with a large sample available to provide a reasonable approximation for σ, we employ Equation 10–4, using Z-scores of ± 1.96 for a 95% confidence interval.

$$\overline{X} \pm Z\frac{S}{\sqrt{n}} = 2.12 \pm 1.96\frac{1.66}{\sqrt{147}}$$
$$= 2.12 \pm .27$$

or from \$1.85 to \$2.39.

□

EXAMPLE 10.3

A realtor's association in an urban county is interested in seeing how long the average single-family home that sold through its Multiple Listing Service during the last 12 months was on the market before it sold. A random sample of 45 homes shows a sample mean of 58 days and a sample standard deviation of 31 days. Provide a 92% confidence interval for this population's mean.

Solution:

Again, the dispersion observed in the sample is used to approximate σ. The Z-values for a 92% confidence interval are $Z = \pm 1.75$ (see Figure 10.4). We then have

$$\overline{X} \pm Z\frac{S}{\sqrt{n}} = 58 \pm 1.75\frac{31}{\sqrt{45}}$$
$$= 58 \pm 8$$

Our interval encompasses the span from 50 days to 66 days.

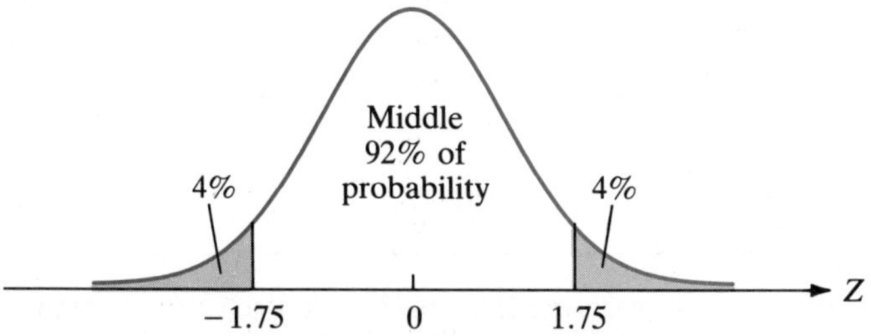

Figure 10.4 Locating Z-Values for a 92% Confidence Interval

□

COMMENTS

1. Confidence and risk are probabilities that sum to 1.0 (or 100 percent). If we desire a .90 (90%) confidence interval, then the risk is .10 (10 percent) that the interval to be computed will not include μ. In developing confidence intervals, we let α (Greek alpha) represent risk and let $(1 - \alpha)$ represent confidence. If we want a 90% confidence interval, for example, then $\alpha = .10$ and $(1 - \alpha) = .90$.

2. A confidence interval for μ or any other parameter will necessarily involve a $\pm$ margin for error. Only a complete census can eliminate all uncertainty. We can think of a census result as being a point estimate that has 100% confidence and 0% risk.
3. Since the point estimate $\overline{X}$ is a random variable, the lower and upper limits of the confidence interval are also random variables. It is for this reason that we *avoid* statements such as, "the probability is .95 that μ will fall within a given interval," since this would imply that μ is the variable, when in fact it is a constant. Since the limits of the interval are the random variables, it is mathematically proper to talk in terms of "a .95 probability that the confidence interval will be one which includes μ," as we have done in this section.
4. With respect to rounding confidence interval results, we recommend reporting the $\pm$ potential sampling error to the same or to one more decimal place accuracy as $\overline{X}$.
5. We will symbolize $\sigma_{\overline{X}}$, the standard error of $\overline{X}$, as $\hat{\sigma}_{\overline{X}}$ when the sample standard deviation S is used in place of σ. The hat symbol over any quantity is used to denote an estimate of that quantity. If the population standard deviation σ is known, then the standard error is $\sigma_{\overline{X}} = \sigma/\sqrt{n}$. More likely σ is not known. In such cases, the (estimated) standard error is $\hat{\sigma}_{\overline{X}} = S/\sqrt{n}$. Some statisticians prefer to write $\hat{\sigma}_{\overline{X}}$ as $S_{\overline{X}}$.

Factors Affecting Confidence Interval Width

A confidence interval for μ has been symbolized as

$$\overline{X} \pm Z\frac{\sigma}{\sqrt{n}}$$

What determines how narrow or wide the range of values above and below the point estimate turns out to be? Let's consider the three terms that affect the potential sampling error:

1. Z reflects the degree of confidence desired. To increase the confidence from 90 to 99 percent, for example, means changing Z from ± 1.65 to ± 2.58. The effect of raising our confidence (lowering the risk) is therefore to widen the interval. You might think that it is better to be 99% certain than 90% certain, but this extra confidence is not free. We pay for this additional confidence by having to report a wider potential sampling error. Other things being equal, we would rather have a narrow than a wide confidence interval. A 90% confidence interval is 1.65/2.58 or about 64 percent as wide as a 99% confidence interval.

2. The symbol σ represents the variability in the population as measured by the standard deviation. When σ is unknown, the sample counterpart S is used to approximate its value. For a given sampling situation, we have no control over σ. We do know, however, that the larger σ is, then the larger is the potential sampling error. More dispersion in the population spreads out the sampling distribution of $\overline{X}$ and therefore widens the confidence interval, other things being equal.

3. The letter n denotes the sample size used in constructing the interval estimate. A large sample provides more information than does a small one. A point estimate $\overline{X}$

from a large sample is therefore more likely to be closer to μ than is an $\overline{X}$ from a small sample. Increasing n narrows the confidence interval. Large samples have less potential error, but recall (see Section 8.4) that there are diminishing returns in accuracy for each additional item sampled.

In Chapter 8 we defined *sampling error* as the amount by which $\overline{X}$ differs from μ. In a confidence interval, the distance above and below our point estimate ($Z\sigma/\sqrt{n}$) reflects the *maximum sampling error* possible for a given level of confidence. To illustrate, in Example 10.1 we computed this 90% confidence interval for the mean GMAT score:

$$534 \pm 19$$

As mentioned earlier, the *actual* sampling error is of course unknown since μ is unknown; the *maximum possible* sampling error for this 90% confidence interval is 19. Therefore, we view a sampling error of 19 as the worst-case value. This is why we state that the point estimate is in error by *no more than* 19, with 90% confidence. Our actual sampling error may be much less—we have no way to know.

We will find it convenient to symbolize our maximum sampling error for a given level of confidence as E.

Maximum Sampling Error for a Given Confidence Level

$$E = \frac{Z\sigma}{\sqrt{n}} \qquad \text{(10–5)}$$

Figure 10.5 illustrates E and its relation to the confidence interval. Note that we generally will treat E as a positive number, letting it denote maximum sampling error without regard to the direction of the error.

Sample Size Requirements

When we sample to determine a confidence interval, is there ever a set sample size that should be taken? The answer to this question is "yes" if we have a predetermined

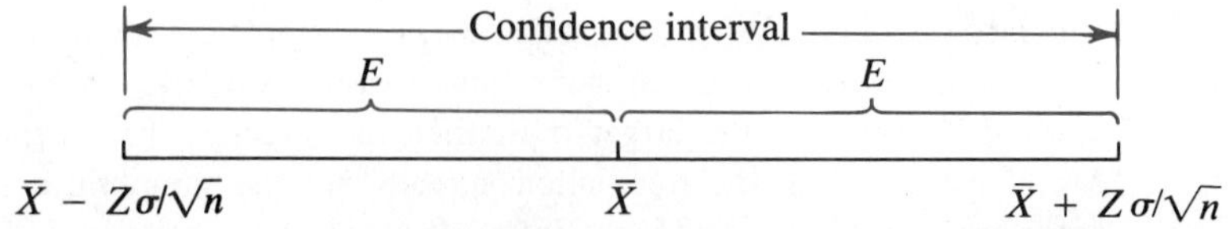

Figure 10.5 Relation of Confidence Interval to E

precision (maximum sampling error E) in mind. If so, the sample size requirement will depend on

1. The specific maximum sampling error E that we are willing to tolerate
2. The dispersion in the population of interest, as measured by the standard deviation σ
3. The level of confidence for our interval, represented by the appropriate value from the Z-table

As an example of choosing a sample size to achieve a desired or target accuracy, suppose Illinois state economists want to estimate annual family nonfarm income in a rural portion of the state. They would like their estimate to be accurate to within ±$500 of the population mean, with 95% confidence. Assume the population standard deviation is $2500. What sample size should be obtained?

The selection of sample size can be quite important. Generally it takes time and costs money to gather a sample, particularly when it is necessary to survey people. Therefore, it would be a waste of resources to take a larger sample than required to get the desired results. In this instance, if the state economists think that an estimate that is accurate to within plus or minus $500 is sufficiently precise for their purposes, why sample 600 families if, for example, a sample of 100 could provide the desired accuracy? On the other hand, if the sample size is too small, then the resulting confidence interval might be so wide as to have little practical use.

A simple rearrangement of Equation 10–5 will give us an expression that allows us to determine in advance the sample size required to meet our objectives.

Sample Size Formula for Estimating μ

$$n = \left(\frac{Z\sigma}{E}\right)^2 \tag{10–6}$$

In our example, Z is 1.96 for 95% confidence, σ is $2500, and the maximum sampling error E is $500. We then have

$$n = \left(\frac{1.96(2500)}{500}\right)^2 = 96.04$$

Since n must be an integer, we would report the required sample size as $n = 97$ families. A sample of this size would produce a point estimate that would be within $500 of the unknown mean, with 95% confidence. (Use of $n = 96$ would cause E to exceed $500; to guarantee that E does not exceed the desired value, we will always *round up* the sample size requirement to the next integer.)

EXAMPLE 10.4

In Example 10.1, a sample of 75 GMAT scores produced a confidence interval of 534 ± 19. Suppose the director of admissions now wants the sampling error to be no more than 15. What sample size is required to produce a margin for error of ±15, retaining the same 90% confidence?

Solution:

For 90% confidence, Z is 1.65. The population standard deviation is 100, and E is equal to 15. We have

$$n = \left(\frac{Z\sigma}{E}\right)^2 = \left(\frac{1.65(100)}{15}\right)^2 = 121 \text{ test scores}$$

Since 75 scores have already been collected, 46 new observations would be needed to tighten the confidence interval to ±15, centered at the $\overline{X}$ based on all 121 GMAT scores.

□

Strictly speaking, Equation 10–6 applies only when σ, the population standard deviation, is known. In practice, σ is usually unknown. However, we still may obtain a reasonable approximation of the sample size requirement if we can estimate σ. Table 10.1 lists possible ways to do so.

The *comparable study* approach can be used if knowledge of dispersion exists from some previous similar sampling situation. For instance, certain state and federal agencies regularly sample employees in certain industries to estimate quantities such as average hourly wage or length of average work week. If the standard deviation observed in previous studies is believed to be reliable, it can be used in our sample size formula to provide an approximate sample size.

If no value from previous experience is available, a *range-based approximation* is possible provided we have some idea about the lowest and highest values in the

Table 10.1 Possible Approximations for σ in the Sample Size Formula

Method	Relies on
Comparable study	Value for σ or S from a previous study
Range-based approximation	Estimate of highest and lowest values of X in the population
Pilot sample estimate	S, the standard deviation found in a preliminary sample

population. Recall from Chapter 3 that for a mound-shaped, symmetrical distribution, the range is equal to about six standard deviations. As an example of using this method, suppose bank officials are considering eliminating certain checking account fees for their senior citizens. To help evaluate the cost of doing so, they would like to estimate the mean number of checks written per month by seniors to within ± 2 checks with 90% confidence. What sample size is needed? The bank treasurer guesses that the population is roughly normal and that almost no values would be less than 2 or more than 50. With a range of 48, we could estimate the standard deviation as 48/6 or 8 checks per month. Equation 10–6 would then yield a sampling requirement of about 44 senior citizen accounts. The range-based approximation is not as reliable when a population is noticeably skewed.

A *pilot sample estimate* is probably the most commonly used method to estimate σ in the sample size formula. The standard deviation that is found in a pilot, or preliminary, sample is used to determine the full sample size needed to reach a specific degree of accuracy, as the following examples illustrate.

EXAMPLE 10.5

Management of a medium-sized theme park would like to know the average amount of money a family (defined as one or two adults, with at least one child) spends per day inside the park. An admission fee good for all rides and attractions is collected at the gate—the goal of management here is to estimate the mean expenditure for other items such as food, gifts, souvenirs, film, and so forth. One week, officials chose 15 families at random as they were leaving the park. The families were offered a choice of a $10 gift certificate in any park shop or a free pass for any future date in exchange for answering questions about how they spent money within the park that day. These are the results from 14 families who chose to answer the questions:

$$\bar{X} = \$23.14$$
$$S = \$\ 8.49$$

Based on this preliminary sample, how many families should be surveyed for $\bar{X}$ to be within $2 of the actual mean, with 90% confidence?

Solution:

Using the pilot sample standard deviation in Equation 10–6,

$$n = \left(\frac{1.65(8.49)}{2.00}\right)^2 = 49.05 \qquad \text{or 50 (rounded)}$$

A full sample of about 50 families (36 additional usable responses) should provide the desired accuracy. The major purpose of the pilot sample is to estimate σ. The computed confidence interval will be centered at the mean of the full sample, not the mean of the pilot sample.

□

EXAMPLE 10.6

The manufacturer of peanut-related products mentioned at the beginning of this section wanted to develop a sampling plan for shipments of packages received from their primary supplier. The nominal weight per package is 110.2 lbs. The firm's controller wants to estimate μ, the mean weight per package in each shipment, to within plus or minus .25 pound, with 90% confidence. With no previous knowledge about package weight variability, a preliminary sampling was done to estimate σ. The pilot study showed $S = .97$ lb, a figure the supplier later said closely matched his own estimate of package weight variability. Using Equation 10–6,

$$n = \left(\frac{1.65(.97)}{.25}\right)^2 = 40.14 \qquad \text{or 41 (rounded)}$$

The firm therefore knows that although each shipment contains several thousand packages, they can estimate the mean package weight to within .25 lb with 90% certainty by using a random sample of about 41 packages.

□

COMMENTS

1. There is an inverse relation between the square root of the sample size n and maximum sampling error E. Making E smaller can be accomplished only by increasing n. Recall (see Section 8.4) there can be a point of diminishing returns with regard to increasing the sample size.
2. When σ is estimated in the sample size formula, the resulting sample size requirement should be viewed as approximate. This is because the estimated σ is unlikely to equal the actual σ. If the estimated $\sigma <$ actual σ, the confidence interval that results will tend to be wider than desired. Conversely, if the estimated $\sigma >$ actual σ, the confidence interval will tend to be narrower than required.
3. When a pilot sample estimate of σ is employed, the required sample size given by Equation 10–6 will more likely than not be too small to achieve the desired accuracy. The nature of this phenomenon and correction factors to adjust for it are described in an article by your text authors in the August 1987 issue of the *Journal of Marketing Research*.

CLASSROOM EXAMPLE 10.1

Constructing a Confidence Interval for the Population Mean

As part of a space utilization study at a downtown parking garage, management took a random sample of 53 customers' parking stubs to estimate the average time the customer parked in their facility. The sample mean was found to be 179 minutes, with a standard deviation of 128 minutes.

a. What is the estimated standard error of $\overline{X}$?

b. What value of Z should be used if we wish to construct a 90% confidence interval for μ?

c. To the nearest whole minutes, specify the lower and upper bounds of a 90% confidence interval for μ.

d. Suppose that the interval determined above is judged to be not precise enough, and that we would like the maximum possible sampling error to be 15 minutes. What sample size would you recommend to accomplish this objective? (Maintain the 90% confidence factor.)

e. If we wanted .05% risk of the interval failing to include μ instead of the 10% risk in part d, how many parking stubs would need to be sampled?

SECTION 10.1 EXERCISES

10.1 Find the Z-scores that would be appropriate to use to construct the following:

a. a 75% confidence interval.

b. an 85% confidence interval.

c. a 99.5% confidence interval.

10.2 Show how to develop Equation 10–2, using Equation 10–1 as your starting point.

10.3 A random sample of 64 items is obtained from a population with $\sigma = 11$. The sample average is computed to be $\overline{X} = 85.1$.

a. Develop a 98% confidence interval for the population mean, reporting your result to one decimal place accuracy.

b. If a second sample of size 64 were drawn, would its mean be 85.1?

c. Is a confidence interval symmetrical around $\overline{X}$, or is it symmetrical around μ?

10.4 We would like to develop, to one decimal place accuracy, an interval that has only a 1% risk of failing to include μ. What should this interval be when a sample of 105 items from a large population shows $\overline{X} = 64.4$ and $S = 21.1$?

10.5 Thirty-two observations from Population H show $S = 4.10$ and $\overline{X} = 8.95$. To two decimal-place accuracy, construct an interval that has a risk of 10 percent of failing to include the mean value of Population H.

10.6 Using Equation 10–4, a 95% confidence interval for μ based on $n = 50$ was computed to be 140.0 ± 7.60. To two decimal-place accuracy, what was the sample standard deviation?

10.7 List the three factors that affect confidence interval width and explain what effect a lowering of each factor has on confidence interval width.

10.8 The owner of a large automobile dealership wants to determine the average age of vehicles being traded in by customers buying a new car or truck at his agency. He asks the office manager to inspect 30 transactions within the last six months to find the average. After checking the records, the office manager computes a mean age of trade-in vehicles as 4.6 years, with a standard deviation of 2.1 years. Develop an 80% confidence interval for the parameter of interest.

10.9 To help summarize library activity to her board of directors, a public library administrator is interested in estimating the average cardholder usage in the summer months. Using a random selection method, she checks 35 cardholders' activity and finds a sample mean of 8.8 books checked out and a sample standard deviation of 5.1. With a risk of 10 percent of being incorrect, provide a confidence interval for average cardholder usage. Report the lower and upper limits of the interval.

10.10 A grocery store surveyed 117 of its nonexpress lane Saturday shoppers. One result of the survey is shown in the following table, where the variable X denotes the number of people living in the shopper's household.

X	1	2	3	4	5	6	7	8
f	12	21	24	30	19	7	1	3

a. Verify that $\overline{X} = 3.54$ and that $S = 1.59$.

b. Assuming this is a random sample of nonexpress lane shoppers, compute a 90% confidence interval for the population mean.

10.11 The survey referred to in Exercise 10.10 also yielded the following table, where the random variable was the shopper's bill for her or his order.

Class ($)	Count
10–30	6
30–50	18
50–70	31
70–90	39
90–110	20
110–130	3

a. Verify that $\overline{X} = \$69.91$ and that $S = \$23.43$.

b. Assuming this is a random sample, compute a 90% confidence interval for the population mean. Give the upper and lower limits of the interval.

10.12 According to Fred Bronson's *The Billboard Book of Number One Hits* (*Billboard,* New York, 1985), *Billboard* began its weekly reporting of the best selling records in America in 1940. Curious to see how long the average number one song holds that position, we chose the 30 number one songs from September 1983 to April 1985, finding a sample mean of 2.9 weeks and a sample standard deviation of 1.3 weeks. If the number one songs of this time period can be viewed as a random sampling of past, present, and future *Billboard* number one songs, compute a 90% confidence interval for the population mean number of weeks atop the charts.

10.13 The state water resources commission regularly monitors the level of a harmful pollutant at various locations in a particular stream. The results of 38 measurements taken last year showed an average level of the pollutant at 77.2 parts per million, with a standard deviation of 21.3 ppm. Viewing these readings as a simple random sample of all possible measurements, compute an interval that has a risk (α) of .01 of not including the mean level of the pollutant in the stream last year.

10.14 A public broadcasting station began a week-long on-air promotional telethon to get viewers to pledge money to help support the station. In the first 90 minutes, 46 callers made pledges averaging \$26.44 with a standard deviation of \$18.13. If these callers represent a simple random sample of pledges that will be made in the next week, compute an 85% confidence interval for the mean pledge amount to be received during the promotion.

10.15 Refer to Exercise 10.14. If your interval had 95 instead of 85% confidence, would it be wider or narrower? Which interval would have the larger chance of missing the population mean?

10.16 Show how to develop Equation 10–6, using Equation 10–5 as a starting point.

10.17 What sample size would be required to estimate μ to within ± 2.00 with 99% confidence, if the population standard deviation is 11.90?

10.18 Determine the sample size necessary to have the point estimate $\overline{X}$ miss μ by no more than \$10 in either direction, using a risk of .10. The population standard deviation is known to be \$48.

10.19 A 95% confidence interval for the population mean is desired. We would like the point estimate to miss the unknown parameter value by no more than \$250. If the standard deviation in the population is \$1800, what sample size do you recommend?

10.20 Refer to Exercise 10.13. What sample size appears necessary to achieve a maximum sampling error of 5 ppm, retaining the same degree of confidence?

10.21 A university registrar would like to estimate the mean GPA of all graduating seniors within the past four years. The GPAs take on values from 2.00 to 4.00. If it is desired that the estimate be accurate to within $\pm .05$ with 95% confidence, what sample size do you recommend? You have no knowledge of σ, but it occurs to you that a range-based approximation could be useful in this instance.

10.22 Refer to Exercise 10.21. Based on the sample size you recommended, your study of transcripts chosen by a random selection method shows $\overline{X} = 2.88$ and $S = .27$.

a. To two decimal-place accuracy, determine a 95% confidence interval for μ, using your sample evidence.

b. Is the maximum sampling error of your interval less than or greater than the desired $\pm .05$ accuracy? Why did it turn out this way?

10.23 An automobile/travel club wants to estimate the mean price of unleaded gas at stations along the interstate systems of a three-state area. A small pilot study shows a sample standard deviation of 6.5¢ per gallon. It is decided to estimate the mean to within 1.5¢ with 90% confidence. The club estimates it costs \$2.25 to sample each station (clerk time plus phone cost). What will it cost to gather enough sample information to ensure the desired accuracy? What would be the cost for 99% confidence?

***10.24** A confidence interval for the population *total* can be expressed as

$$N\left(\overline{X} \pm Z\frac{S}{\sqrt{n}}\right)$$

where N is the number of items in a finite population. For example, a sample of 45 items yields a mean of \$50.50 and a standard deviation of \$13.90. If we know that the population from which the sample came contains 2100 items, then a 95% confidence interval for the population total is

$$2100\left(50.50 \pm 1.96\frac{13.90}{\sqrt{45}}\right) = 2100(50.50 \pm 4.06)$$

$$= 106{,}050 \pm 8{,}526 \quad \text{or} \quad \$97{,}524 \text{ to } \$114{,}576$$

A building supply firm's books show 4118 accounts receivable. An audit sample based on 120 accounts selected at random shows a mean accounts receivable balance of \$259.60

*Optional

and a standard deviation of \$177.98. Use this information to develop a 90% confidence interval for the firm's total dollar value of accounts receivable.

*10.25 A *one-sided confidence interval* is one that has either an upper confidence limit or a lower confidence limit, but not both. For instance, an upper-tail interval would be developed when we wish to place an upper limit on the population parameter and have no concern about the lower limit. An upper-tail confidence interval limit will be $\overline{X} + Z\sigma/\sqrt{n}$; a lower-tail limit will be $\overline{X} - Z\sigma/\sqrt{n}$.

As an example, suppose an ambulance service wants to place an upper limit on its average response time, with 95% confidence (see Figure 10.6). A sample of 50 recent calls

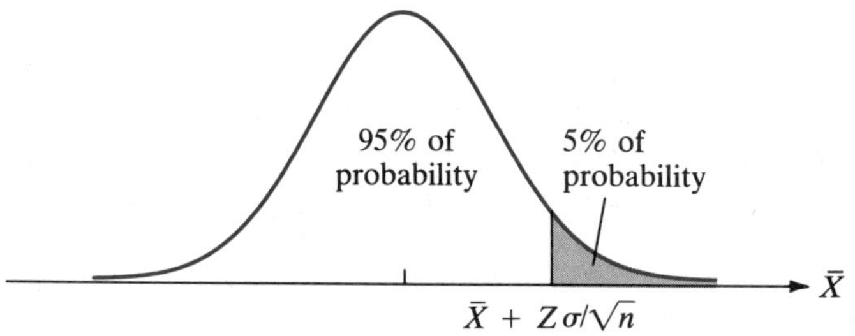

Figure 10.6 An Upper-Tail Confidence Interval

is chosen to estimate the mean amount of time from when the call is received until the ambulance reaches the scene of the emergency. The sample shows $\overline{X}$ = 8.4 minutes, S = 3.4 minutes. The ambulance service can therefore be 95 percent confident that its mean response time does not exceed $8.4 + 1.65(3.4)/\sqrt{50}$, or 9.2 minutes. Note than the Z-value for a one-sided interval is not the same as that for a two-sided interval of the same confidence. This is because the risk is distributed all into one tail for the one-sided interval, whereas a two-sided interval divides the risk evenly into both tails.

A sample of a company's accounts receivable shows a mean balance of \$89.20 with a standard deviation of \$46.12. The sample size is 36. We can be 90 percent certain the population mean balance is no lower than what amount?

10.2 HYPOTHESIS TESTS ABOUT A POPULATION MEAN

One approach to using sample information to make an inference about the larger population is *estimation,* as typified by developing a confidence interval. The other major approach is *hypothesis testing*. Compared to estimation, hypothesis testing is more decision oriented.

The word *hypothesis* has several synonyms, including "assumption," "unproved theory," "supposition," and "educated guess." In business statistics, a hypothesis is an assumption about the numerical value of a parameter, such as the population mean

μ. As the word suggests, a "test" of a hypothesis means that its believability is to be examined using the information or evidence found in a sample taken from the population in question. In light of the sample evidence, we then decide whether the hypothesis has been contradicted or supported. As is the case for confidence intervals, the logic for hypothesis tests relies on our knowledge of sampling distributions.

A Two-Tailed Hypothesis Test: The General Argument

All hypothesis tests begin with a statement, called the *null hypothesis,* which is to be subjected to a statistical analysis. Recall from Chapter 9 that the null hypothesis (H_0) is the assumable or nonbinding statement of equality about a population value; the alternative (H_a) is one of inequality. In symbols, our hypothesis statements will appear as

$$H_0\text{: } \mu = \mu_0 \text{ (some number)}$$
$$H_a\text{: } \mu \neq \mu_0 \text{ (some number)}$$

Using our knowledge of sampling distributions, we argue that if H_0 is true, then our sample evidence will come from a normal sampling distribution for $\overline{X}$ centered at μ_0, the value we specify in our H_0 statement (see Figure 10.7). We also know that since the sample mean $\overline{X}$ is a random variable, we should not expect the value of $\overline{X}$ that we obtain to equal μ_0, even if H_0 is correct. But, we do expect our sample mean to be in the neighborhood of μ_0. This leads us to the idea of rejection and acceptance regions. As the line at the bottom of Figure 10.7 indicates, the acceptance region for H_0 encompasses the great majority of the sampling distribution. Accordingly, the acceptance region can be thought of as an interval of values where $\overline{X}$ is likely to be found,

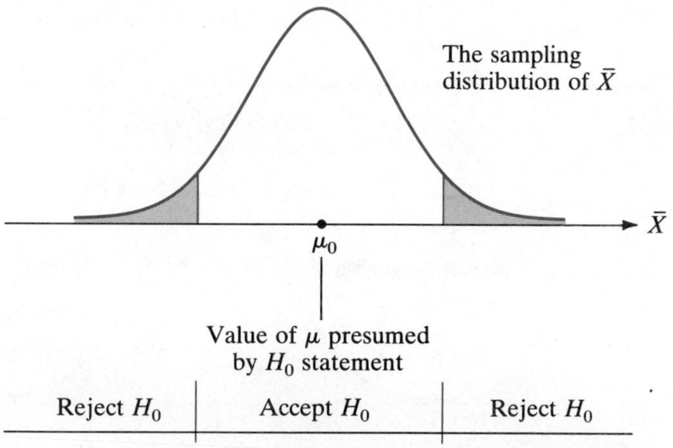

Figure 10.7 Rejection and Acceptance Regions for a Two-Tailed Test

when H_0 is in fact true. The two rejection regions (the basis for the term *two-tailed test*) indicate unlikely values for $\overline{X}$ when H_0 is true.

Therefore, our strategy is this when testing hypotheses: We first presume that H_0 is true, much like the defendant in court who is initially presumed innocent. If our H_0 is true, then we can determine a range of values where the sample evidence is likely to be found—the acceptance region. If our sample evidence lies outside this region, we then conclude that our beginning premise (that H_0 is true) is unlikely to be correct; in the spirit of proof by contradiction, we would then say that H_0 should be rejected. On the other hand, if the sample evidence is found within the acceptance region, we reason that an insufficient case for rejecting H_0 has been made; we would say that we cannot reject or that we accept H_0.

A statistical test is always set up so that we are likely to accept and unlikely to reject when H_0 is true. Perhaps you have heard the expression "guilty beyond a reasonable doubt" to describe the courtroom standard for rejecting the presumption of innocence. Likewise in statistics, we require a convincing case to be made against H_0 before we are willing to conclude that H_0 is not correct. For this reason, "Reject H_0" is sometimes referred to as a strong conclusion, while "Do not reject H_0" or "Accept H_0" is regarded as a weak conclusion.

Now, what constitutes a convincing case that we should reject H_0? Where do we draw the line, as in Figure 10.7, that separates acceptance region from rejection region? The answer to this depends on the probability, or risk, of incorrectly rejecting H_0 (Type I error) that we are willing to tolerate. This risk, denoted by α, also is called the *level of significance* for the test. Commonly used levels of significance are .01, .05, and .10. While the choice of α in a given situation may be somewhat arbitrary, we generally want α to be low. Error risk cannot be zero, however, since we will be basing our conclusion on sample information. Graphically, α represents the percent of the sampling distribution in the rejection region when H_0 is true—see Figure 10.8. For our often-used levels of significance, Table 10.2 provides values of Z that separate acceptance from rejection regions. Such Z-scores are referred to as *cutoff* or *critical*

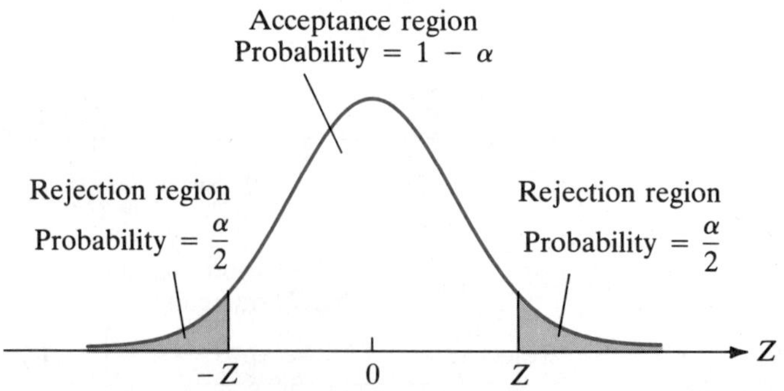

Figure 10.8 Acceptance and Rejection Region Probabilities, Two-Tailed Test

Table 10.2 Table (Cutoff) *Z*-Values for Commonly Used Levels of Significance for Two-Tailed Tests

$\alpha = P$(Type I Error)	*Z*-Table Value
.10	± 1.65
.05	± 1.96
.01	± 2.58

values. Additional cutoff *Z*-values are given in the table at the front of the book. To find any desired cutoff *Z*-score for a two-tailed test, we simply call on our ability to find the middle portion of the normal curve. For $\alpha = .08$, for instance, the acceptance region would be the middle 92 percent of probability, while the 8% risk of error is divided evenly into the two tail regions. The middle 92 percent of the curve therefore lies from $Z = -1.75$ to $Z = +1.75$.

We generally specify H_0, H_a, and α prior to taking our sample. This is desirable—we should not let the sample data influence our specification of hypotheses or our choice of error risk. This also means that the cutoff *Z*-values for acceptance/rejection are established before we see the sample evidence. For instance if we choose a two-sided test with $\alpha = .10$, the cutoff *Z*-scores are predetermined, at $Z = \pm 1.65$.

To see where our sample result lands on Figure 10.8, we use the sample information (mean, standard deviation, and sample size) to compute a number called the *test statistic* (*TS*), denoted by Z^*. The test statistic is simply a *Z*-score that combines the hypothesized value of μ with the sample evidence to produce a number that is used to make the decision to reject or accept H_0. By comparing our computed Z^* against the cutoff *Z*-value, the conclusion for a hypothesis test will be self-evident. When we test a hypothesis about the population mean, we use the test statistic given in the following box.

Large Sample Test Statistic for μ

In words, we have

$$Z^* = \frac{\text{Observed sample mean} - \text{Hypothesized value of } \mu}{\text{Standard error of sample mean}}$$

In symbols, we have

$$Z^* = \frac{\overline{X} - \mu_0}{\sigma/\sqrt{n}} \qquad \text{(10–7)}$$

If σ is unknown in Equation 10–7, the sample standard deviation S can replace it. As mentioned earlier in this chapter, this approximation works well when $n \geq 30$. Note that Equation 10–7 is identical to Equation 8–3 except that μ_0 replaces μ. In Chapter 8, we often presumed μ was known to help us develop the concept of the sampling distribution; in hypothesis testing we are investigating an assumption (H_0), which states the population mean is equal to a specific value, μ_0.

Examples of a Two-Tailed Hypothesis Test

Let us consider as our first example a relatively small mail-order catalog business. Two years ago when the business was just getting started, the shipping department conducted a census that established that the average weight per order shipped was 36.0 ounces. To cut down on handling and paperwork, the company decided at that time to charge its customers a flat fee for shipping and postage regardless of order size. While they think this has been an efficient policy, management knows that the "mix" of items sold has changed somewhat over the past two years. Accordingly, management would like to discover whether the mean package weight is still the same. While the current volume of business precludes conducting another census, we may select a random sample of orders and make a determination based on the sample evidence.

For this statistical test, what should the null and alternative hypotheses be? Since H_0 is the assumable or nonbinding statement of equality about the population mean, our null hypothesis would be, in words, an assertion such as, "the current mean package weight in the population is still 36 ounces" or "there has been no change in mean package weight." The alternative hypothesis would simply be an opposite statement. In symbols, we have

$$H_0\colon \mu = 36.0 \text{ ounces}$$
$$H_a\colon \mu \neq 36.0 \text{ ounces}$$

Assume we choose the .05 level of significance for our test. Table 10.2 shows that the cutoff Z-scores are ± 1.96 for a two-tailed test conducted with 5% risk of Type I error. Suppose that a clerk chooses a random sample of 125 orders and computes these statistics: $\overline{X} = 36.9$ ounces and $S = 15.5$ ounces. Our test statistic then would be

$$Z^* = \frac{\overline{X} - \mu_0}{S/\sqrt{n}} = \frac{36.9 - 36.0}{15.5/\sqrt{125}} = \frac{.90}{1.386} = .65$$

Since $-1.96 \leq Z^* \leq 1.96$, our conclusion is to accept H_0 (see Figure 10.9). Since we have found $\overline{X}$ in its likely region (presuming H_0 to be true), we can say that the evidence supports the presumption of no change in mean package weight and that the .90 ounce discrepancy between $\overline{X}$ and μ_0 is "not statistically significant." Table 10.3 summarizes the elements of a hypothesis test for this example.

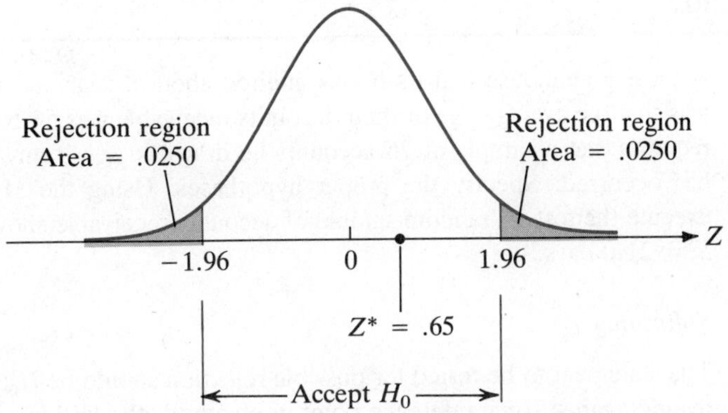

Figure 10.9 Graphical View of Package Weight Example

Table 10.3 Hypothesis Test Elements with Specifics for Package Weight Example

Element, Symbol	General Remarks	In Package Weight Example
Null Hypothesis, H_0	The assumption being subjected to a statistical test, H_0 always contains an equal sign as it specifies a numerical value for the unknown quantity.	H_0: $\mu = 36.0$ ounces, where μ denotes mean package weight in the population
Alternative Hypothesis, H_a	Also called the "research hypothesis," it is the opposite of H_0. H_a always carries an inequality sign. H_0, H_a, and α should be stated prior to taking the sample.	H_a: $\mu \neq 36.0$ ounces
Test Statistic, *TS*	Computed from sample evidence, it is a single number that summarizes the test against H_0.	*TS*: $Z^* = \dfrac{\overline{X} - \mu_0}{S/\sqrt{n}} = \dfrac{36.9 - 36.0}{15.5/\sqrt{125}} = .65$
Rejection Region, *RR*	Always the tail region(s) of the sampling distribution, the *RR* indicates unlikely values for the *TS*, if H_0 is in fact true. Where *RR* begins depends on α and whether H_a is one or two tailed.	*RR*: For $\alpha = .05$, two-tailed, the *RR* is $\lvert Z^*\rvert > 1.96$; the acceptance region is $\lvert Z^*\rvert \leq 1.96$. See Figure 10.9.
Conclusion, *C*	Arrived at by noting whether the *TS* lies in *RR*. *C* must be either "reject H_0" or "accept (do not reject) H_0." For those users of the information who may not know statistics terminology, we recommend phrasing *C* in terms of the situation being investigated.	*C*: Since $\lvert Z^*\rvert = .65 \leq 1.96$, *C* is "accept (do not reject) H_0." The sample evidence is consistent with (supports) the assumption that there has been no change in average package weight.

EXAMPLE 10.7

A local greenhouse had its books audited about a year and a half ago. One finding was that the average age of their accounts receivable was 51.5 days. The manager now requests that a sample of 75 accounts be drawn to see if any change in this quantity has occurred. Specify the proper hypotheses. Using the .10 level of significance, execute the test. A random sample of accounts receivable shows $\overline{X} = 57.3$ days, with $S = 21.0$ days.

Solution:

The statement to be tested for possible rejection should be H_0: $\mu = 51.5$ days. When testing against some reference point or standard, the null hypothesis is often one that presumes no change or no difference. Since the average age could, if it has changed, move in either direction, our alternative should be H_a: $\mu \neq 51.5$ days. For 10% risk of Type I error, the cutoff Z-values are ± 1.65. Our *TS* is

$$Z^* = \frac{\overline{X} - \mu_0}{S/\sqrt{n}} = \frac{57.3 - 51.5}{21/\sqrt{75}} = \frac{5.8}{2.425} = 2.39$$

Since $Z^* = 2.39 > 1.65$, C is "reject H_0" (see Figure 10.10). We conclude that the mean age of accounts receivable has changed. Our logic is that if H_0 is true, the *TS* has a high probability of being found between $Z = -1.65$ and $Z = +1.65$. Since Z^* did not fall in this region, we are persuaded that the mean age has changed. Management may need to investigate the billing/collection process to learn why this change has occurred and to see if any corrective measures can be developed.

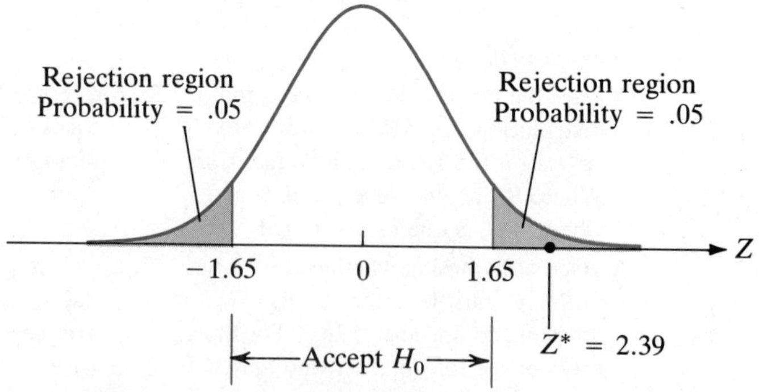

Figure 10.10 Graphical View of Example 10.7

□

Unstandardized Approach to Tests

Up to this point, we have given the dividing line between acceptance and rejection regions as a Z-value from the standard normal table. We will find it useful on occasion to express our cutoff points in an alternative but equivalent form: *unstandardized*.

Unstandardized Acceptance Region

In words, an unstandardized acceptance region for H_0 is

Expected value under H_0 ± Cutoff value (standard error)

In symbols, we have

$$\mu_0 \pm Z\frac{\sigma}{\sqrt{n}} \tag{10–8}$$

For $n \geq 30$, S can replace σ in Equation 10–8. If you are thinking that Equation 10–8 looks like an expression for a confidence interval, you are correct. Centered at μ_0, it is an interval for likely values of $\overline{X}$, presuming H_0 to be true.

To illustrate, consider Example 10.7 above where we tested H_0: $\mu = 51.5$ days. As illustrated in Figure 10.11, the unstandardized acceptance region can be used for Example 10.7:

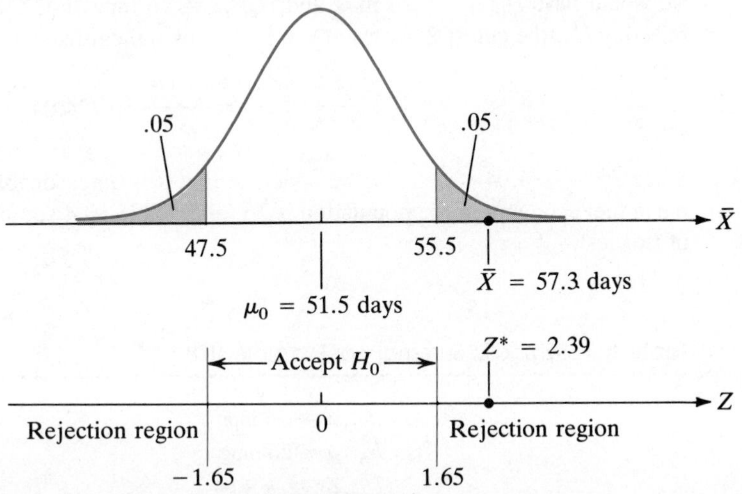

Figure 10.11 Unstandardized and Standardized Cutoffs for Example 10.7

$$\mu_0 \pm Z\frac{S}{\sqrt{n}} = 51.5 \pm 1.65\frac{21}{\sqrt{75}}$$
$$= 51.5 \pm 4.0 \text{ days} \qquad \text{or } 47.5\text{–}55.5 \text{ days}$$

In using the unstandardized cutoffs, our logic would be that if the population mean μ is unchanged from 51.5 days, then the probability is about .90 that our sample mean $\overline{X}$ will be found between 47.5 and 55.5 days. Since $\overline{X} = 57.3$ days falls outside this range, a strong case is made that μ has changed, and we therefore reject H_0. Figure 10.11 illustrates that the same conclusion will result whether the limits of the acceptance region are in terms of $\overline{X}$ (unstandardized) or Z (standardized).

EXAMPLE 10.8

A consumer interest organization obtains 32 models of a new imported car for a variety of performance and safety tests. One of the tests will involve the Environmental Protection Agency (EPA) fuel economy rating. This particular model has been given an EPA average rating of 26 mpg. Suppose we want to determine whether the EPA figure is credible. Specify the null and alternative hypotheses. Using the .05 level of significance and the sample results of $\overline{X} = 24.5$ mpg, $S = 2.1$ mpg, show that the standardized and unstandardized cutoff points yield the same conclusion.

Solution:

The null hypothesis should be one which presumes the EPA figure is correct. Therefore we would have H_0: $\mu = 26$ mpg and H_a: $\mu \neq 26$ mpg. For 5% risk of incorrectly rejecting H_0, the cutoff Z-scores are ± 1.96. Our standardized *TS* is

$$Z^* = \frac{24.5 - 26.0}{2.1/\sqrt{32}} = \frac{-1.5}{.371} = -4.04$$

Since $Z^* = -4.04 < -1.96$, we conclude that it is unreasonable to believe that the mean fuel economy in the population is 26 mpg. Table 10.4 summarizes the elements of this test.

Table 10.4 Symbolic Summary of Example 10.8

H_0: $\mu = 26$ mpg
H_a: $\mu \neq 26$ mpg
TS: $Z^* = -4.04$
RR: Reject H_0 if $Z^* < -1.96$ or if $Z^* > 1.96$
C: Reject H_0

If we choose to express our cutoff points in terms of $\overline{X}$ instead of Z, we have

$$\sigma_0 \pm Z\frac{S}{\sqrt{n}} = 26 \pm 1.96\frac{2.1}{\sqrt{32}}$$
$$= 26 \pm .73 \quad \text{or } 25.27\text{–}26.73 \text{ mpg}$$

We argue that a sample finding between 25.27 and 26.73 mpg is likely if H_0 is true; the actual result of $\overline{X} = 24.5$ mpg is outside of this range and thereby contradicts H_0. Rejection of H_0 means that the discrepancy between $\overline{X}$ and μ_0 (1.5 mpg in this case) is considered to be "statistically significant." Figure 10.12 shows the equivalence of these two approaches.

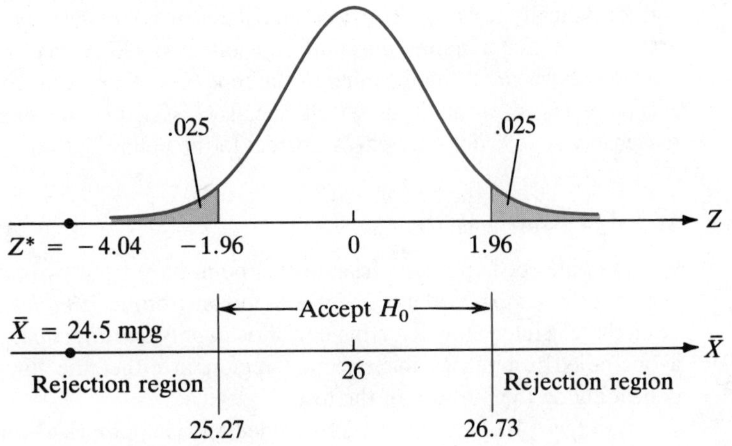

Figure 10.12 Equivalence of Standardized and Unstandardized Approaches for Example 10.8

□

Semantics of Statistical Test Results

When the evidence is not strong against H_0, should we say "accept H_0" or "do not reject H_0"? Strictly speaking, "do not reject H_0" is the preferred statement, although the use of "accept H_0" is widespread. In our book, we will use "accept H_0" most of the time to reflect the business perspective of making a decision. While we will consider the terms interchangeable, it should be understood that acceptance should not be taken as meaning that we believe the value stated in H_0 is the literal truth. In reality, H_0 is seldom *exactly* true. Acceptance does *not* mean that H_0 has been proved; rather, it means that the sample result is not strong enough to reject H_0. In a similar vein, a finding of "reject H_0" does not establish as a guaranteed certainty that H_0 is false; rather, it means that H_0 is *unlikely* to be true. We cannot be 100% certain that a conclusion we make about the population is correct when that conclusion is based on a sample. The risk of an incorrect conclusion may be low, but it is always present.

Another phrase that we have added to our vocabulary is **statistically significant,** an expression we use to describe a difference between $\overline{X}$ and μ_0, which, for a given level of significance, is too large to be consistent with H_0 being true. Finding a significant difference is equivalent to rejection; an insignificant difference equates with acceptance of H_0. Use of either of these expressions is necessarily tied to some specified level of significance, α. A test result that is statistically significant at $\alpha = .05$, for instance, may or may not be statistically significant at the stricter $\alpha = .01$.

Accountants, and auditors in particular, often use the word *material* instead of significant to denote a substantial misstatement of account balance. As Arens and Loebbecke (1981) observe in *Applications of Statistical Sampling to Auditing,* an auditor typically tests H_0: The account balance is not materially misstated. A material error is one that requires an audit adjustment or additional work. Relative to our remarks above about the meaning of "accept H_0," Arens and Loebbecke observe that accepting H_0 in an auditing setting means ". . . that the financial statements are reasonably stated, not necessarily correct to the penny."

One-Tailed Tests

Our examples of hypothesis tests to this point have been two-tailed: Rejection of H_0 occurs if the sample evidence $\overline{X}$ deviates too far from μ_0 *in either direction*. In practice, often there is interest in detecting deviation *in only one direction*. As the name denotes, a one-tailed test places the rejection region in either the lower or the upper tail, depending on the purpose of the test.

As Figure 10.13 suggests, a two-sided test is appropriate when we want to establish whether μ has changed or differs in a positive or a negative direction from some particular value. With two halves of the rejection region, a two-tailed test is therefore sensitive to μ being either above or below some reference point. In general, when it is possible for μ to deviate up or down and we are interested in detecting either occurrence, a two-tailed test is in order.

In contrast, a one-tailed test is employed when we want a procedure that can alert us to a deviation or change in a particular direction. Since a one-tailed test has an H_a that anticipates the direction of possible movement or change, we should have a reason or theory behind us to support the selection of a one-tailed procedure in favor of a two-sided test. We also should keep in mind that the sample evidence should not influence our choice of hypotheses; we generally form hypothesis statements prior to sampling.

An upper-tail test (middle panel of Figure 10.13) has the rejection region in the right tail and therefore can detect that μ has increased or is above some reference point. On the other hand, a lower-tail test is one that is sensitive to μ having decreased or being below a reference point. Like the two-tailed test, acceptance of H_0 for a one-tail test means that the evidence is consistent with the premise of no deviation or no

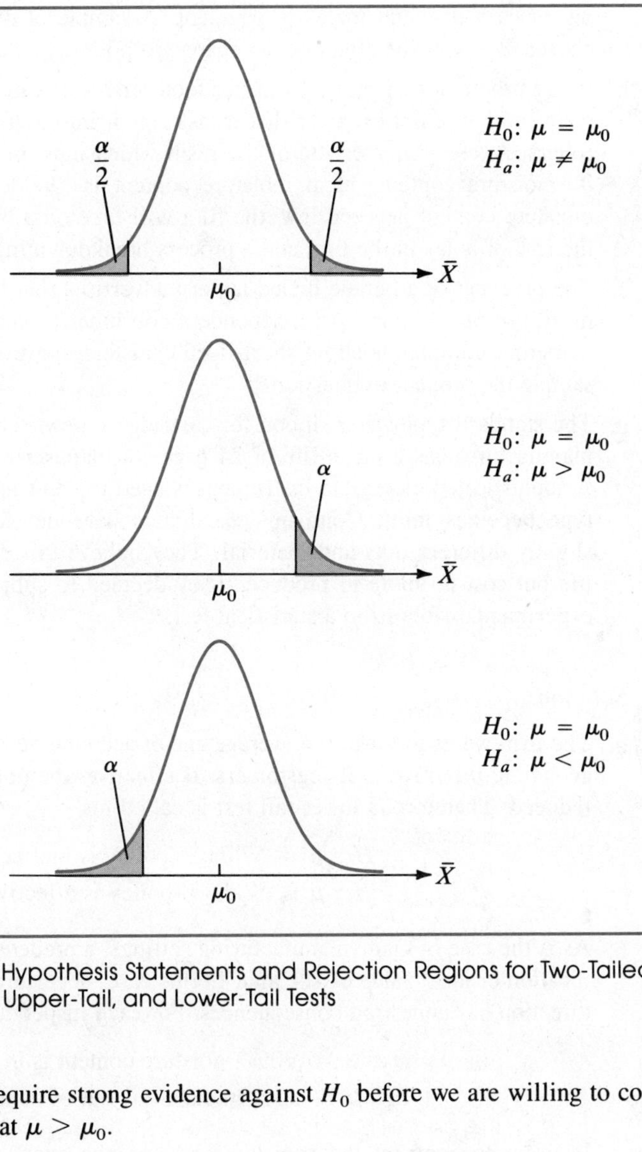

Figure 10.13 Hypothesis Statements and Rejection Regions for Two-Tailed, Upper-Tail, and Lower-Tail Tests

change; we require strong evidence against H_0 before we are willing to conclude that $\mu < \mu_0$ or that $\mu > \mu_0$.

EXAMPLE 10.9

Consider the following sampling situations and indicate whether an upper-tail, lower-tail, or two-tailed test should be used.

a. Because a medical-plastics firm's controller thought the average age of accounts receivable was too high, at 57 days, the firm changed its credit terms, offering a bigger cash discount for early payment. A sample of 40 accounts is about to be chosen to assess the effect of the change in policy.

b. In the production of plastic wrap for food products, a manufacturer passes a white resin through a hot extruder that transforms it into a clear film to be gathered on rollers. A key characteristic of the resin which must be continually monitored is its moisture content. Ideal moisture content is .30 percent of resin weight. If moisture content gets too low, the film will take on a brownish cast; if too high, the risk of holes in the film and a process breakdown increases.

c. The producer of a butane fueled lantern advertises that its fuel canisters provide a mean use of 20 hours. An independent consumer protection agency has received numerous complaints about short-lived canisters, however. The agency decides to sample the product in question.

d. The standard typewriter ribbon for cartridge typewriters made by one particular manufacturer has a mean life of 24 pages of manuscript before it becomes faint. A sophisticated electronic instrument is used to determine the point at which the type becomes faint. Company researchers have developed a new ribbon with slightly different inks and material. They believe this ribbon will have a greater life but cost no more to produce. They decided to subject a sample of 50 of the experimental ribbons to a statistical test.

Solution:

a. The firm wants to lower the average age of accounts receivable and has offered an economic incentive to its customers. If effective, the policy will result in μ being reduced. Therefore, a lower-tail test is called for

$$H_0\colon \mu = 57 \text{ days (policy is ineffective)}$$
$$H_a\colon \mu < 57 \text{ days (policy is effective)}$$

b. As is the case in many manufacturing settings, a predetermined engineering specification or ideal value exists, and, in this case, departure from this value in either direction has undesired consequences. For each inspection we have:

$$H_0\colon \mu = .30\% \text{ (moisture content is in control)}$$
$$H_a\colon \mu \neq .30\% \text{ (moisture is out of control)}$$

c. In order to confront the canister producer, the agency would need convincing evidence to uphold its case. To be confident that the mean is less than 20 hours, the agency would need to reject H_0 in favor of a lower-tail H_a:

$$H_0\colon \mu = 20 \text{ hours (advertised value is correct)}$$
$$H_a\colon \mu < 20 \text{ hours (advertised value is too high)}$$

d. The purpose of the research is to develop an improved product. Before being willing to adopt the experimental ribbon over the current formulation, management would want to be quite certain that it is in fact an improvement. We want a test capable of demonstrating that the mean is above 24.

$$H_0: \mu = 24 \text{ pages (experimental is no better than existing)}$$
$$H_a: \mu > 24 \text{ pages (experimental is superior to existing)}$$

□

A one-tail test follows the same modus operandi as a two-tailed test. After we formulate the hypotheses and establish α, a random sample is selected. We then compute a test statistic and note whether it lies in the acceptance or rejection region. However, it should be apparent (see Figure 10.13) that the cutoff Z-scores for any α are not the same for one-sided as for two-sided tests. This is because the probability which corresponds to α is placed in just one tail; in contrast, a two-tailed test has probability corresponding to $\alpha/2$ in each of two tails. Table 10.5 provides selected cutoff Z-scores for one-tail tests; others are given in the table at the front of the book. Your knowledge of percentiles on the standard normal curve should enable you to quickly determine any one-tail cutoff Z.

Table 10.5 Selected Cutoff *Z*-Scores: One- and Two-Tailed

Risk of Type I Error, α	Type of Test: Lower Tail	Upper Tail	Two Tail
.10	−1.28	1.28	±1.65
.05	−1.65	1.65	±1.96
.01	−2.33	2.33	±2.58

EXAMPLE 10.10

Four months ago, a medical-plastics firm began offering its customers a larger cash discount for early payment in hopes of reducing the average age of accounts receivable (see Example 10.9a). Test H_0: $\mu = 57$ days against H_a: $\mu < 57$ days at the .05 level of significance. A random sample of 40 accounts receivable shows $\overline{X} = 53.5$ with $S = 26$ days. Can we conclude that the change in credit terms has lowered the average age of accounts receivable in the population?

Solution:

For $\alpha = .05$, lower tail, the cutoff Z is -1.65. Our *TS* is

$$Z^* = \frac{\overline{X} - \mu_0}{S/\sqrt{n}} = \frac{53.5 - 57}{26/\sqrt{40}} = \frac{-3.5}{4.111} = -.85$$

Since $Z^* = -.85 > -1.65$ we do not reject H_0 (see Figure 10.14); the sample evidence is consistent with H_0 being true. Keep in mind that if H_0 ($\mu = 57$ days) is still correct, the probability is about .50 that a random sample will give a value of $\overline{X}$ below 57 days. Therefore, not just any $\overline{X}$ below 57 days should lead us to conclude that μ has decreased; we require convincing evidence before we are willing to believe that the new credit conditions have had the desired effect.

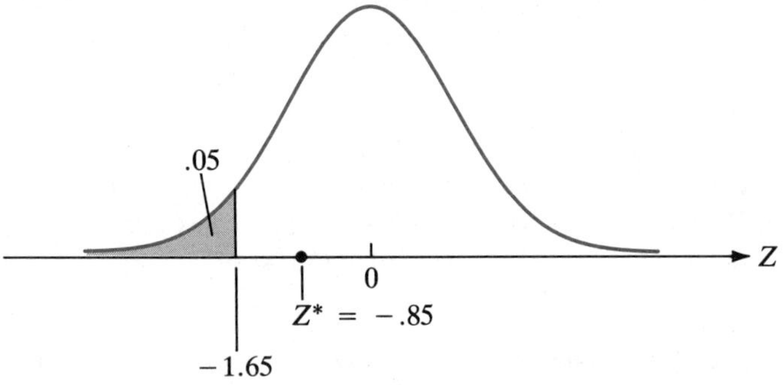

Figure 10.14 Graphical Presentation of Example 10.10

□

EXAMPLE 10.11

A newly formulated typewriter ribbon is to be tested to determine whether its mean life exceeds the current standard (see Example 10.9d). Test H_0: $\mu = 24$ pages versus H_a: $\mu > 24$ pages with 1 chance in 100 of incorrectly rejecting H_0. A random sample of 50 experimental ribbons yields these statistics: $\overline{X} = 24.9$ and $S = 1.5$ pages. Are you persuaded that the mean of the new formulation will exceed 24 pages?

Solution:

For $\alpha = .01$, upper tail, the dividing point between acceptance and rejection is $Z = 2.33$. Our *TS* is

$$Z^* = \frac{\overline{X} - \mu_0}{S/\sqrt{n}} = \frac{24.9 - 24}{1.5/\sqrt{50}} = \frac{.9}{.212} = 4.24$$

Since $4.24 > 2.33$, H_0 is rejected. If it costs no more than the present ribbon, the new formulation should be adopted.

An alternative way to reach the same conclusion is to determine the cutoff point in unstandardized form. For an upper-tail test, the dividing line would be $\mu_0 + ZS/\sqrt{n}$. In this particular example,

$$\mu_0 + Z\frac{S}{\sqrt{n}} = 24 + 2.33(.2121)$$
$$= 24 + .49 \quad \text{or about 24.5 pages}$$

Since $\overline{X} = 24.9$ pages, it is quite likely that the sample did not come out of a population whose mean is 24 pages (see Figure 10.15).

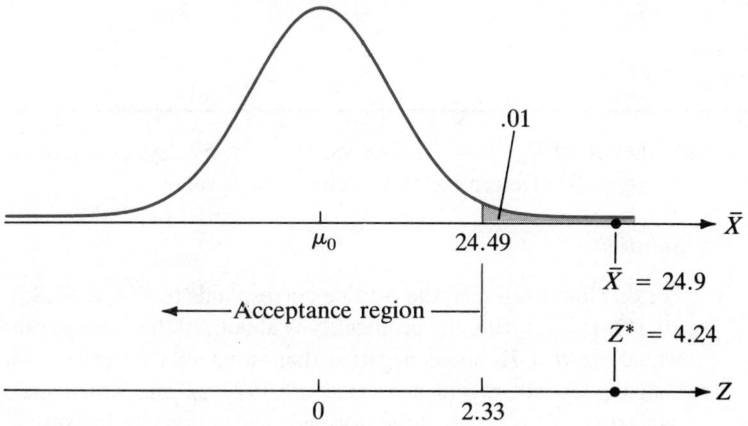

Figure 10.15 Standardized and Unstandardized Solution to Example 10.11

□

COMMENT Some statisticians prefer statements such as H_0: $\mu \geq \mu_0$ and H_0: $\mu \leq \mu_0$ for the null hypothesis when performing lower- and upper-tail tests, as opposed to the format of H_0: $\mu = \mu_0$ which we have employed. For instance, when the alternative is to be H_a: $\mu > 50$, a truly opposite null hypothesis would be H_0: $\mu \leq 50$. However, to test H_0: $\mu \leq 50$, we must assume that the mean of the sampling distribution of $\overline{X}$ is at some exact point, namely 50. For this reason, we choose to write H_0: $\mu \leq 50$ as H_0: $\mu = 50$. This semantic difference in specifying H_0 in no way affects the conclusion of the test. For instance, if H_0 is rejected for $\mu_0 = 50$, it would also be rejected for any possible mean less than 50. Therefore, we will always specify an equality in the null hypothesis; the H_a statement is the one that indicates if the test is to be upper, lower, or two tailed.

p-Values: An Alternative to Fixed Levels of Significance

When discussing the standard normal curve in Chapter 7, we introduced the concept of *p*-values—the area in the tail(s) of the curve beyond a given point (see Examples 7.8 and 7.9 for review). In hypothesis testing, *p*-values represent an alternative means of reporting the results of a test.

Definition

The ***p*-value** for a hypothesis test is the probability (assuming H_0 is true) of observing a test statistic which makes as strong or stronger case against H_0 than what was actually observed.

If we have computed the test statistic, Z^*, then the *p*-value (also called the *observed significance level*) is readily ascertained.

EXAMPLE 10.12

In the test of H_0: $\mu = 57$ days vs. H_a: $\mu < 57$ days (see Example 10.10), the *TS* was $Z^* = -.85$. Determine and interpret the *p*-value.

Solution:

For this lower-tail test, the *p*-value corresponds to $P(Z \leq -.85)$, or .1977 (see Figure 10.16). If H_0 is true, the probability is about .20 that, due to random chance, a sample would show a *TS* more negative than what was observed. Most users of business statistics do not regard a *p*-value of .1977 as one which makes a convincing case against H_0: A relatively large *p*-value such as this one is likely to occur when H_0 is, in fact, true.

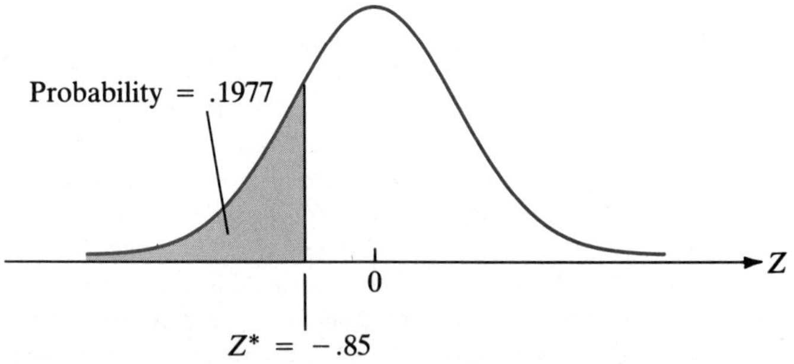

Figure 10.16 p-value for $Z^* = -.85$ and a Lower-Tail H_a

□

EXAMPLE 10.13

In the test of H_0: $\mu = 24$ pages versus H_a: $\mu > 24$ pages (see Example 10.11), we computed a *TS* of $Z^* = 4.24$. Find and discuss the *p*-value for this test.

Solution:

Given the observed *TS* of $Z^* = 4.24$, the *p*-value is $P(Z \geq 4.24)$, or .00001 (found by consulting a *Z*-table more extensive than the one in this book). Random chance is unlikely to produce such a *p*-value. A low *p*-value naturally casts doubt on the assumption that H_0 is true.

□

EXAMPLE 10.14

In the test of H_0: $\mu = 36$ ounces versus H_a: $\mu \neq 36$ ounces (see Table 10.3), the *TS* was $Z^* = .65$. Determine and interpret the *p*-value for this test.

Solution:

Since evidence against H_0 can be found *in either direction* when H_a is two-sided, the *p*-value is $2P(Z \geq .65) = 2(.2578)$, or .5156 (see Figure 10.17). A *p*-value this large does not constitute the evidence "beyond a reasonable doubt" we seek before we reject H_0.

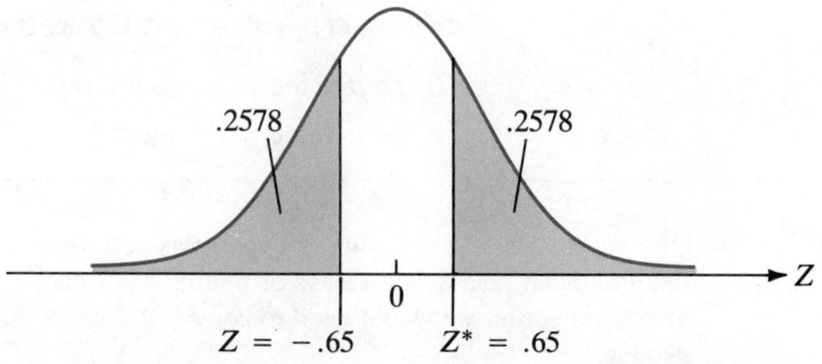

Figure 10.17 Probabilities for Developing *p*-value; Two-Sided H_a with $Z^* = .65$

□

Why do we develop the idea of *p*-values? After all, if we have computed Z^*, why do we want another way to assess the strength of the evidence against H_0? The answer to this lies in the widespread belief that we should try to avoid using arbitrary, predetermined levels of significance such as .10 or .05 as a basis for reporting an accept/reject conclusion. Instead, the argument goes, we should simply report the *p*-value and let the user of the report judge the weight of the evidence against H_0. In other words, the user should be free to draw his or her own conclusion from the sample without being restricted by an arbitrary cutoff point imposed by someone

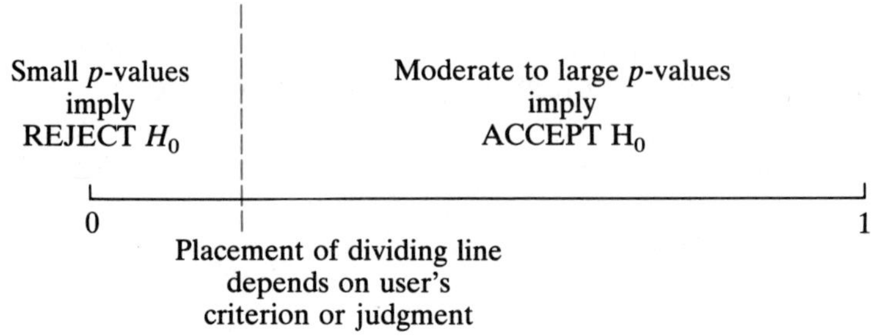

Figure 10.18 Relation of p-value to Conclusion

else. In this spirit, research results published in scientific and other scholarly journals often include the *p*-values for any statistical tests conducted. In such cases, there is often no reject/accept conclusion explicitly stated—this is left to the reader. Likewise, most computer software programs which perform statistical tests routinely print out *p*-values, leaving the interpretation to the user. Figure 10.18 illustrates the spirit of the *p*-value concept.

If we do wish to relate a given *p*-value to some α-risk that we are willing to tolerate, the decision rule is simple.

> **Relating a p-Value to a Desired α**
>
> If the *p*-value $\geq \alpha$, do not reject H_0.
>
> If the *p*-value $< \alpha$, reject H_0.

COMMENTS

1. One advantage of expressing a hypothesis test result in terms of Z^* instead of the unstandardized $\overline{X}$ is the ease of finding a *p*-value. If we do a test using unstandardized cutoffs, we would need to convert $\overline{X}$ into Z^* before looking up the proper *p*-value.
2. For a large sample *TS* with magnitude beyond our normal table maximum of $Z = \pm 3.49$, we can report the *p*-value as *p*-value $< .0002$ or *p*-value $< .0004$, depending on whether H_a is one or two tailed. More exact *p*-values corresponding to Z^*s greater than 4.0 can be found using the One Sample Inference processor.
3. Resist the temptation to think of a *p*-value as the probability H_0 is true, given the sample evidence. A *p*-value is the converse—the probability of observing such evidence, given that H_0 is true.
4. Most users of statistical test results usually want to know more information than simply "reject H_0" or "accept H_0" or the *p*-value associated with the *TS*. Especially when we find evidence against H_0, it is a good idea to report $(\overline{X} - \mu_0)$, the amount by which the sample mean differs from the hypothesized population mean.
5. Relative to the boxed material above, we also could interpret a *p*-value as the smallest level of significance for which we can reject H_0, given the sample data.

Error Possibilities

Whenever we carry out a hypothesis test, the chance for an erroneous conclusion exists. It is important to understand how mistaken conclusions can occur.

Type I Error—Rejecting a True H_0 Type I error probability is subject to our direct control. Accordingly, we typically restrict Type I error to having a small chance of happening. When we set the level of significance α for a test, we are by definition specifying the chance of incorrectly rejecting H_0. Perhaps the most commonly used risk of Type I error in business applications is $\alpha = .05$.

To illustrate how a Type I error could occur, consider this hypothetical example: The registrar at our university has a computer program that equates each student's hometown to a distance, in miles, away from campus. For the entire population ($N = 20{,}000+$ students), the registrar has determined that the mean is $\mu = 73.0$ miles, with a standard deviation of $\sigma = 40.8$ miles. This distance data file is then placed in our computer laboratory.

Now assume that the business statistics instructors give their students the following assignment, to be completed within a week:

Go to the Computer Lab where a random selection program will choose a sample of 100 observations of student hometown distance from campus. Each student will get his or her own unique sample. The population from which you are sampling has $\sigma = 40.8$. Use the mean that results from your sample to test, at $\alpha = .05$,

$$H_0\colon \mu = 73.0 \text{ miles}$$

versus

$$H_a\colon \mu \neq 73.0 \text{ miles}$$

Turn in a short report that documents why you reject or do not reject H_0.

What will the instructors find when the students hand in their results? Unknown to each student, she or he is testing a statement which is *in fact true*. Does this guarantee that each student will report H_0 should be accepted? Unfortunately, the answer to this is "no."

To see how the sample could indicate "reject H_0" when in the population H_0 is actually true, it is helpful to locate the *RR* when $\alpha = .05$. As Figure 10.19 illustrates, the *RR* is $|Z^*| > 1.96$. (You may wish to verify that in unstandardized terms, the acceptance region is 73 ± 8 miles while the *RR* is $\overline{X} < 65$ miles and $\overline{X} > 81$ miles.)

With a large sample size, the sampling distribution is approximately normal. Since about 5 percent of a normal curve lies more than 1.96 standard deviations in either direction away from the mean, then due to no other cause but random chance, about 5 percent of our students will get a sample result that lies in *RR*. The students in our hypothetical example do not *know* that H_0 is in fact true. They do know to reject H_0 if their $\overline{X} < 65$ miles or if $\overline{X} > 81$ miles. Since such results will naturally occur 5 percent of the time, each student has a 5% risk (prior to sampling) of landing in *RR*. If 1000 students participate in our assignment, we would expect about 950 to accept H_0. The others, whose evidence would lead them to reject H_0, would likely be

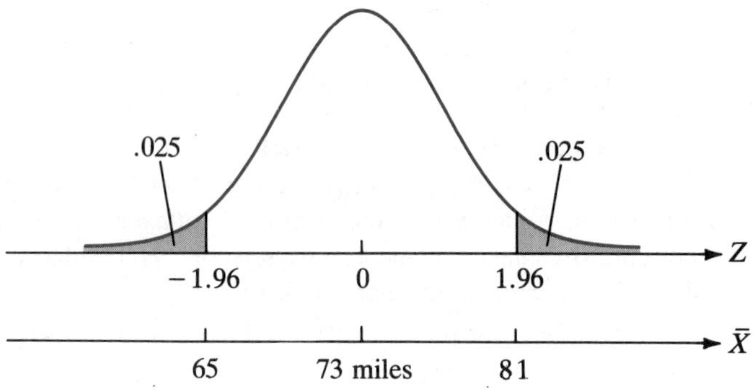

Figure 10.19 Rejection Regions for H_0: $\mu = 73.0$ miles, Two-Tailed, $\alpha = .05$

fairly evenly divided between getting a result in the upper *RR* and getting a result in the lower *RR*. Table 10.6 summarizes this discussion.

Type II Error—Accepting a False H_0 While Type I error can be fixed at some desired level, Type II error is less subject to our control. Its probability of occurrence, denoted by Greek letter β, is usually unknown. We do know that α and β are in a trade-off relation: lowering of α raises β, and vice versa. For a fixed sample size, we cannot simultaneously lower both types of risk. Just as a Type I error is possible only when H_0 is in fact true, a Type II error can occur only when H_0 is in fact false.

To demonstrate how a Type II error could come about, let us reconsider our hypothetical computer project. Imagine that all sampling conditions are the same as described above, except that the hypotheses are now as follows:

$$H_0\text{: } \mu = 60 \text{ miles}$$

versus

$$H_a\text{: } \mu \neq 60 \text{ miles}$$

With the H_0 given above, each student would be asked to test a statement which is *in fact false,* since, in the population, $\mu = 73$ miles. However, even though H_0 is not true, the sample evidence might indicate acceptance of H_0.

Table 10.6 Conclusion Probabilities When H_0 Is True

Conclusion *C*	*P*(*C*)
Accept H_0	$1 - \alpha$
Reject H_0	α
	1.0

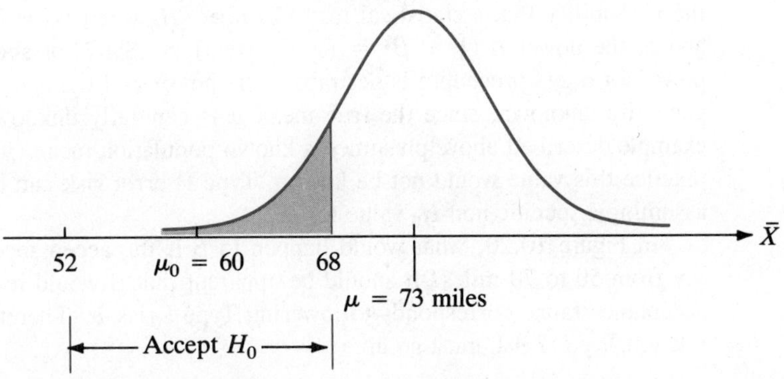

Figure 10.20 Acceptance Region for H_0: $\mu = 60$ miles, Two-Tailed, $\alpha = .05$

Figure 10.20 illustrates the possibilities. With $\mu_0 = 60$, the acceptance range is $\mu_0 \pm Z\sigma/\sqrt{n}$, or from 52 to 68 miles. Our students will know to accept H_0 if $52 \le \bar{X} \le 68$, and to reject H_0 otherwise. The sampling distribution is centered at the true population mean, however, in this case, at $\mu = 73$ miles. Although the center of the sampling distribution is in *RR*, it is possible to obtain a random sample whose mean $\bar{X}$ lies in the acceptance region (see shaded area of Figure 10.20). If this were to occur, Type II error would result. Given the acceptance region and the knowledge that the population mean is 73 miles, we can compute the chance that a sample will signal acceptance of H_0.

$$\begin{aligned}
\beta = P(\text{Type II error}) &= P(52 \le \bar{X} \le 68) \\
&= P(\bar{X} \le 68) - P(\bar{X} < 52) \\
&= P\left(Z \le \frac{68 - 73}{40.8/\sqrt{100}}\right) - P\left(Z < \frac{52 - 73}{40.8/\sqrt{100}}\right) \\
&= P(Z \le -1.23) - P(Z \le -5.15) \\
&= .1093 - .0000 = .1093
\end{aligned}$$

Although H_0: $\mu = 60$ miles is not true, there is about an 11% chance that a random sample will show a result that lies in the acceptance region. Table 10.7 summarizes this discussion.

Table 10.7 Conclusion Probabilities When H_0 Is False

Conclusion C	$P(C)$
Accept H_0	β
Reject H_0	$1 - \beta$
	1.0

The quantity $(1 - \beta)$ in Table 10.7 is called the *power* of the test. It represents the probability that a statistical test will reject H_0 when H_0 is false. In the example above, the power is $(1 - \beta) = (1 - .1093) = .8907$, or about 89 percent. High power for a test procedure is desirable. The power and the error risk β for a test are generally unknown, since the true mean μ is generally unknown. The hypothetical example described above presumed a known population mean ($\mu = 73$ miles), but in practice this value would not be known. Type II error risk can be computed only by assuming a specific non-H_0 value for μ.

In Figure 10.20, what would happen to β if the acceptance region were wider, say from 50 to 70 miles? It should be apparent that β would increase. Widening the acceptance range corresponds to lowering Type I risk α. Therefore, by the trade-off relation Type II risk must go up.

COMMENTS

1. Graphically, we can think of Type II error risk β as being the percentage of the sampling distribution in the acceptance region when H_0 is false—see Figure 10.20. In a similar vein, Type I error risk α is the percentage of the sampling distribution in the *RR* when H_0 is true—see Figure 10.19.
2. Though α and β are in a trade-off relation, it is *not* generally true that $\alpha + \beta = 1$.

CLASSROOM EXAMPLE 10.2

Testing a Hypothesis About the Population Mean

In response to complaints from workers about the noise level along an appliance assembly line, plant management claims that the average sound level is no more than 75 decibels. When the workers contend that 75 decibels is an understatement, a mutually agreed upon third party is brought in to take a random sample of thirty readings.

a. If we wish to test H_0: $\mu = 75$ decibels, should the alternative hypothesis be lower tailed, two tailed, or upper tailed? State the appropriate H_a.

b. What would be the cutoff *Z*-score if we are willing to tolerate 10% risk of Type I error?

c. Compute the test statistic when the sample results are $\overline{X} = 76.38$, $S = 8.06$ decibels.

d. Has management's claim been discredited?

e. The sample average was 1.38 decibels above the tested value of 75 decibels. Should this difference of 1.38 decibels be described as statistically significant?

SECTION 10.2 EXERCISES

10.26 Determine the cutoff *Z*-scores for a two-tailed alternative hypothesis for these conditions:
a. $\alpha = .10$ **b.** $\alpha = .12$ **c.** $\alpha = .001$

10.27 Test H_0: $\mu = 48$ versus a two-sided alternative. Your sample results, based on $n = 60$, are $S = 6.1$ and $\overline{X} = 46.1$. Use a 10% level of significance.

10.28 In a test of the presumption that the population mean equals \$1200, a sample of 76 items shows a sample mean of \$1241 with a sample standard deviation of \$196. Use a two-sided test with 1% risk of Type I error to determine whether the sample result supports your null hypothesis.

10.29 In a government census a year and a half ago, the average rent for a two-bedroom apartment in Louisville, Kentucky, was found to be \$455. The Chamber of Commerce is preparing a brochure about the city and wants to see if this figure still holds. A sample of 65 apartments is selected at random from a complete listing. The sample statistics are a mean of \$484, with a standard deviation of \$89. With a 1 in 10 chance of incorrectly rejecting, test the assumption that the average rent has not changed in the last year and a half.

10.30 One particular brand of smoke detector claims that on average its units sound off when exposed to 375 parts per million (ppm) of smoke in the air. To test this assertion, a sample of three dozen units is exposed to increasing concentrations of smoke in a laboratory. The trigger point (X, in ppm) for each unit is recorded below.

301	319	321	329	341	341
341	343	348	357	360	360
361	369	381	383	384	384
386	388	391	391	392	393
394	401	404	407	407	411
419	435	444	451	475	476

Compute the sample statistics (using the raw data formulas) needed to test the presumption that the manufacturer's statement about the mean trigger point is accurate—not too high and not too low. Using a 5% risk of Type I error, express the cutoff points in both standardized and unstandardized form. Do the data contradict or support the manufacturer's claim?

10.31 If we're so concerned about the chance of rejecting H_0 when H_0 is in fact true, why not lower this controllable risk all the way to zero?

10.32 What, if anything, is improper about the following pairs of hypotheses:

a. H_0: $\overline{X} = 20$
H_a: $\overline{X} \neq 20$

b. H_0: $\mu = 20$
H_a: $\mu \geq 20$

c. H_0: $\mu = 30$
H_a: $\mu < 30$

d. H_0: $\mu < 30$
H_a: $\mu \geq 30$

10.33 A national trade magazine has reported that the average amount of optional equipment sold on small pickup trucks is \$1800. The southeast regional manager for one auto manufacturer requests that a random sample of 100 recent small truck invoices be inspected to see if sales of optional equipment in the southeast region are comparable to the national norm.

a. What should be your hypotheses?

b. The sample results show an average of \$1779 in the southeast region, with a standard deviation of \$560. Choose a commonly used level of significance and perform the statistical test.

c. Does your sample evidence support or contradict your presumption about mean sales in the region?

10.34 The label on a can of chunky beef soup states that the sodium content is 970 milligrams per serving. Suppose we collect a random sample of 40 soup cans and subject them to a chemical analysis that shows: $\overline{X} = 982$ with $S = 47$ mg. At the .05 level of significance, are you willing to accept the label statement as a credible figure for the mean sodium content in the population? Use a test procedure sensitive to the label statement being either an overstatement or an understatement of the sodium content.

10.35 The manager of a branch bank states that the average withdrawal of cash from the branch's automatic teller machine is \$75. Test this statement at the .10 level of significance. A sample of 50 cash withdrawals yields a mean of \$69.20 with a standard deviation of \$18.50. In your conclusion, indicate whether the \$5.80 difference between the observed and hypothesized values is statistically significant.

10.36 Suppose that the average number of total points scored in NCAA Division I basketball games was 142.9 in the five years prior to a major rule change. A random sampling of 150 Division I games chosen since the new rule went into effect shows $\overline{X} = 147.1$ with $S = 17$ points. Select a Type I error risk and conduct a two-tailed test. State whether the discrepancy of 4.2 points should be viewed as evidence that the population mean has changed or as chance variability around the historical value.

10.37 Determine the cutoff Z-scores for these pairs of hypotheses:

a. H_0: $\mu = 50$
   H_a: $\mu \neq 50$ with $\alpha = .01$

b. H_0: $\mu = 60$
   H_a: $\mu < 60$ with $\alpha = .10$

c. H_0: $\mu = 70$
   H_a: $\mu > 70$ with $\alpha = .02$

10.38 Find the value of Z that separates the acceptance from the rejection region(s) for these situations:

a. $\alpha = .06$, two-tailed test

b. $\alpha = .005$, upper-tail test

c. $\alpha = .15$, lower-tail test

10.39 Test H_0: $\mu = \$11{,}600$ against an upper-tail research hypothesis using the .10 level of significance. A sample of 127 items shows a mean of \$11,891 and a standard deviation equal to \$2,886.

10.40 Refer to Exercise 10.39. In terms of dollars, what is the dividing line between the rejection and acceptance regions?

10.41 You are about to test H_0: $\mu = 190$ versus H_a: $\mu < 190$. The risk of a Type I error is to be .005. Report and justify your conclusion when the sample mean is found to be 178, the standard deviation is 24, and the sample size is 88.

10.42 Refer to Exercise 10.41. Determine the cutoff point for rejection/acceptance of H_0 in terms of $\overline{X}$.

10.43 Two years ago, the average level of a certain noxious chemical at the mouth of the Kentucky River was 180 parts per million. Since then, new water treatment facilities have been added upstream and dumping regulations have been strictly enforced, leading to the belief that the average level of the pollutant has been reduced. Select a frequently used α and state the hypotheses you would employ if you want a procedure that can show the mean level has been reduced. Report conclusions if a sampling of 38 river water samples shows a mean of 141 and a standard deviation of 45 parts per million.

10.44 The county clerk motor vehicle office performs a variety of services for the public, including issuing license plate renewals, registering new residents' vehicles, clearing titles on paid-off loans, and collecting personal property taxes on vehicles. When entering the county clerk's office, people take a number and are seated until their number is called. The county clerk (up for re-election in a few months) believes that she should have enough personnel working the counter so that, on average, people do not have to wait more than five minutes before getting attention. Set up the appropriate hypotheses. A sample of 40 patrons chosen at random times reveals a sample mean wait time of 6.05 minutes and a standard deviation of 1.70 minutes. Is it likely that the population mean is in fact five minutes, with this result due to random chance? Or does the clerk definitely need more counter help? Tie your conclusions to some predetermined level of significance.

10.45 A major manufacturer's top-of-the-line camcorder has recently been introduced, carrying a suggested retail price of $999. To management's surprise, the firm's sales representatives are reporting that discounting of the camcorder by retail outlets is prevalent. Management asks selected sales representatives to check the current retail price in a random sampling of stores. The following values of X are the retail prices of the camcorder found at 35 stores.

$899	899	899	899	905	909	909
909	915	919	919	919	924	925
927	929	929	930	930	933	939
939	940	944	949	949	949	949
949	949	969	980	995	999	999

To the nearest whole dollar, provide an interval which has 90% confidence of including the current average selling price for all stores carrying the camcorder. Use the raw data formulas to compute $\overline{X}$ and S.

10.46 Refer to Exercise 10.45. Prior to seeing the sample results, the assistant director of marketing research guessed that the average suggested retail price was below $949. At the .10 level, does the sample evidence back up the assistant director's belief?

10.47 In exploring sites for possible location of a convenience store, the parent chain wants to establish that more than 200 vehicles per hour pass by the location during the 7 A.M. to 6 P.M. time period. At one available site, a sample of 48 hours is gathered over a one-month interval. The sample shows an average of 219 vehicles per hour with a standard deviation of 37. With 1% risk of Type I error, has a convincing case been made that the long-run average for this site exceeds 200 vehicles per hour?

10.48 Refer to Exercise 10.47. Suppose the sample mean had turned out to be 199 vehicles per hour. Could you immediately determine the conclusion for the test without bothering to compute a test statistic or ascertaining the dividing line for reject/acceptance region? Why or why not?

10.49 The USDA has established that to qualify as extra lean, beef should have no more than 5% fat content. A USDA inspector has visited the regional distribution center of a major grocery chain, gathering a sample of cuts of beef labeled "extra lean." The numbers that follow represent the fat content, in percent, as measured by the USDA.

3.9	4.1	4.3	4.3	4.3	4.4	4.4	4.6
4.7	4.8	4.9	5.0	5.0	5.1	5.1	5.1
5.1	5.1	5.2	5.2	5.2	5.3	5.3	5.3
5.3	5.4	5.6	5.7	5.8	6.0	6.4	6.7

a. What should the hypothesis statements be for a test of the correctness of the store's labeling?
b. Verify that this sample has a mean of 5.08, with a standard deviation of .63.
c. Execute the proper test and report your conclusions. Use the .05 level of significance.

10.50 Given the following hypothesis statements and value for the *TS*, determine the *p*-value for the test.

a. H_0: $\mu = 110$ $\quad$ *TS*: $Z^* = -.61$
H_a: $\mu < 110$
b. H_0: $\mu = 120$ $\quad$ *TS*: $Z^* = -1.34$
H_a: $\mu \neq 120$
c. H_0: $\mu = 130$ $\quad$ *TS*: $Z^* = 2.77$
H_a: $\mu > 130$

10.51 Refer to Exercise 10.47 above. Determine the *p*-value.

10.52 In text Example 10.7 we computed $Z^* = 2.39$ using a two-sided test procedure. Find the *p*-value for the test.

10.53 For the following *p*-values and values for α, state whether you would reject or accept H_0.
a. *p*-value $= .1685$, $\alpha = .10$
b. *p*-value $= .0022$, $\alpha = .01$
c. *p*-value $= .0606$, $\alpha = .01$
d. *p*-value $= .0322$, $\alpha = .05$

10.54 In a test of H_0: $\mu = 11{,}000$ against an upper-sided research hypothesis, a sample of 40 observations is found to have $\overline{X} = 11{,}505$ and $S = 2{,}811$. Determine the *p*-value. Are you willing to conclude that H_0 is not true?

10.55 This question appeared on a business statistics quiz: "Construct a 90% confidence interval for the population mean, given that a random sample of 108 items shows a sample mean of 19.43 with a sample standard deviation of 1.89." One student's answer was "$19.13 \le \overline{X} \le 19.73$." In what way is the student's answer flawed?

10.56 For any hypothesis testing situation, what symbol of α, β, $1 - \beta$, or $1 - \alpha$ would correspond to the following:
a. correctly rejecting H_0.
b. incorrectly accepting H_0.

10.57 Refer to Exercise 10.49. In terms of the problem setting described, what would be the meaning of committing a Type I error?

10.58 Refer to Exercise 10.33. How would we describe a Type II error in terms of the variable of interest and the hypotheses employed?

10.59 For a test of H_0: $\mu = 4750$ versus a two-sided H_a, describe in words the meaning of these symbols:
a. $1 - \beta$ **b.** $1 - \alpha$

10.60 Can Type I error have occurred if the conclusion of the test is to not reject H_0?

10.61 Refer to Figure 10.20. With the acceptance region being from 52 to 68 miles, what would be the risk of Type II error if the actual mean in the population were 68 instead of 73 miles? What would be the power of the test?

***10.62** A *power curve* is a graph that shows the probability of rejecting H_0 for different possible values of the population mean μ. For a two-tailed test, the power curve has the general appearance of an upside down normal curve, as shown in Figure 10.21. As an example,

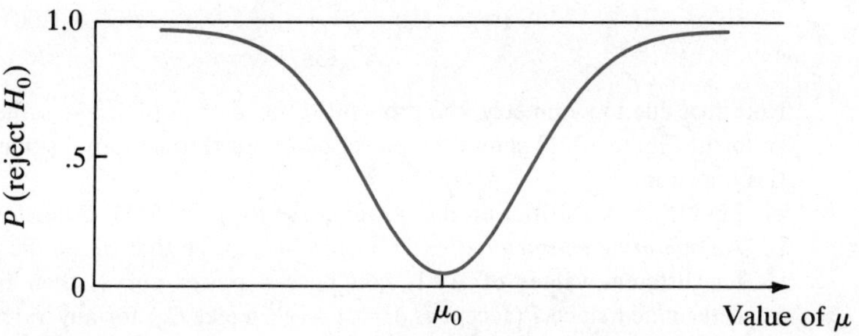

Figure 10.21 A Power Curve

consider a production process that makes ball bearings with an engineering specification that the diameter should be 5.00 centimeters (cm). Periodically a sample of four bearings is drawn, and the sample mean is computed. The purpose of each sample is to test

$$H_0: \mu = 5.00 \text{ cm (process in control)}$$

versus

$$H_a: \mu \neq 5.00 \text{ cm (process out of control)}$$

When the sample mean is computed, this decision rule applies:

If 4.97 cm $\leq \overline{X} \leq$ 5.03 cm, continue production.
If $\overline{X} < 4.97$ or if $\overline{X} > 5.03$ cm, stop production to recalibrate the equipment.

*Optional

Assume the standard deviation of the bearing diameters is $\sigma = .02$ cm. The power curve can be drawn after determining a few points. For instance, if $\mu = 5.00$ cm,

$$\begin{aligned} P(\text{reject } H_0) &= P(\bar{X} < 4.97) + P(\bar{X} > 5.03) \\ &= P\left(Z < \frac{4.97 - 5.00}{.02/\sqrt{4}}\right) + P\left(Z > \frac{5.03 - 5.00}{.02/\sqrt{4}}\right) \\ &= P(Z < -3.00) + P(Z > 3.00) \\ &= .0013 + .0013 = .0026 \end{aligned}$$

If $\mu = 5.01$ cm,

$$\begin{aligned} P(\text{reject } H_0) &= P(\bar{X} < 4.97) + P(\bar{X} > 5.03) \\ &= P\left(Z < \frac{4.97 - 5.01}{.02/\sqrt{4}}\right) + P\left(Z > \frac{5.03 - 5.01}{.02/\sqrt{4}}\right) \\ &= P(Z < -4.00) + P(Z > 2.00) \\ &= .0000 + .0228 = .0228 \end{aligned}$$

If $\mu = 5.02$,

$$\begin{aligned} P(\text{reject } H_0) &= P(Z < -5.00) + P(Z > 1.00) \\ &= .1587 \end{aligned}$$

Note that due to symmetry, the probability for $\mu = 5.01$ is the same as for $\mu = 4.99$, and so forth. Figure 10.22 shows the power curve for this sampling situation with the probabilities drawn in.

a. Find the probabilities on the power curve for $\mu = 5.03, 5.04, 5.05$, and 5.06 cm.

b. An *operating characteristic* (*OC*) curve is a graph that shows the chance of accepting H_0 for different values of μ. If you have a power curve, then the *OC* curve is easily determined since $P(\text{accept } H_0) = 1 - P(\text{reject } H_0)$ for any value of μ. In the example above, for what values of μ is $P(\text{accept } H_0)$ equal to $P(\text{reject } H_0)$?

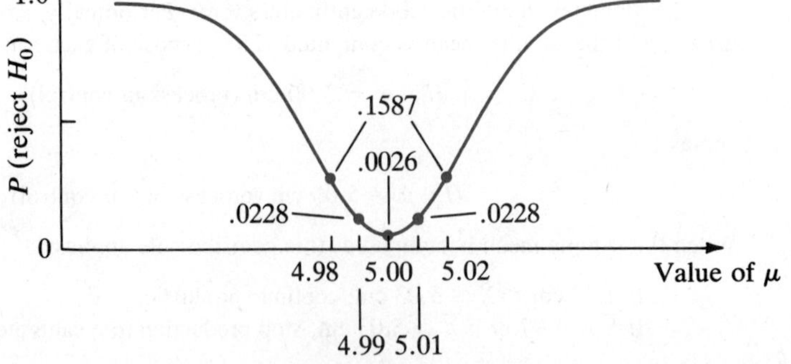

Figure 10.22 Power Curve for Testing H_0: $\mu = 5.00$

***10.63** In a test of H_0: $\mu = 65$ mph with a two-sided alternative, H_0 will be accepted if the sample mean is found to be in the interval from 63.0 to 67.0 mph. Assume that $\sigma = 4.5$ mph and that a sample size of 30 vehicles will be utilized. For integer values from $\mu = 60$ mph to $\mu = 70$ mph, determine and plot the probabilities that H_0 will be rejected.

***10.64** Refer to Exercise 10.63. The power curve for a one-tailed test has a different appearance than the power curve for a two-tailed test. To illustrate, assume we will test H_0: $\mu = 65$ mph against an upper tail alternative, employing the following decision rule:

$$\text{Accept } H_0 \text{ if } \overline{X} \leq 67.0 \text{ mph}$$
$$\text{Reject } H_0 \text{ if } \overline{X} > 67.0 \text{ mph}$$

Plot the power curve for integer values from $\mu = 65$ to $\mu = 70$ mph. Assume $\sigma = 4.5$ mph and a sample size of 30.

10.3 INFERENCES FOR FINITE POPULATIONS

In Section 3 of Chapter 8, we introduced the finite population correction (FPC) term as an adjustment to the standard error. We mentioned that the FPC should be used when the population size N is known and the sample consists of more than 5 percent of the population. When constructing confidence intervals and testing hypotheses, we will want to employ the FPC when these conditions are present, as it will improve our results.

We have given the following formula as the exact standard error of the sample mean:

$$\sigma_{\overline{X}} = \frac{\sigma}{\sqrt{n}} \cdot \sqrt{\frac{N - n}{N - 1}}$$

Recall that the FPC (the term to the right of the multiplication sign) is always a number less than one. When the population standard deviation σ is unknown in the above expression, it can be replaced by the sample standard deviation S when the sample size is at least 30.

EXAMPLE 10.15

A video rental store has 674 members in its club. The owner wants to estimate the average number of titles rented by members in the month just ended. A sample of 75 members' records shows $\overline{X} = 4.7$ with $S = 3.1$ titles. Develop a 95% confidence interval for the average of all 674 members.

Solution:

Since $n/N = 75/674 =$ about 11 percent, the FPC should be used in determining the estimated standard error.

$$\hat{\sigma}_{\overline{X}} = \frac{3.1}{\sqrt{75}} \cdot \sqrt{\frac{674 - 75}{673}}$$
$$= .3580\ (.9434)$$
$$= .3377$$

The 95% confidence interval is then

Point estimate ± Potential sampling error
Point estimate ± Table value (estimated standard error)

$$= \overline{X} \pm Z\hat{\sigma}_{\overline{X}}$$
$$= 4.7 \pm 1.96\ (.3377)$$
$$= 4.7 \pm .66 \text{ titles rented last month}$$

□

Earlier in this chapter, we gave the following formula to establish the sample size necessary to achieve a predetermined margin for potential sampling error:

$$n = \left(\frac{Z\sigma}{E}\right)^2$$

This formula is modified somewhat when we are dealing with finite populations.

Sample Size Formula for Estimating μ: Finite Population Case

$$n = \frac{NZ^2\sigma^2}{E^2(N - 1) + Z^2\sigma^2} \qquad \textbf{(10–9)}$$

EXAMPLE 10.16

Refer to Example 10.15. Suppose the store owner would like to estimate μ to within plus or minus .50 with 95% certainty. What sample size would be required?

Solution:

Using the standard deviation of 3.1 titles found in the first sample, we have

$$n = \frac{NZ^2\sigma^2}{E^2(N - 1) + Z^2\sigma^2} = \frac{674(1.96)^2(3.1)^2}{.5^2(673) + 1.96^2(3.1)^2}$$
$$= 122 \text{ club member records}$$

□

The FPC also should be used in hypothesis testing when the sample constitutes more than 5 percent of the population.

EXAMPLE 10.17

The regional supervisor of a fast-food chain is interested in the average time her store managers have been in that position. Due to increased employee turnover in the industry, she suspects that the mean time for her managers has fallen below 28 months, the figure established in a census three years previously. She decides to sample 50 of her 173 stores to see whether her suspicions are correct. Set up the proper hypothesis statements and report conclusions when the sample results are $\overline{X} = 22$ months with $S = 16$ months.

Solution:

With a belief that μ may have decreased, we should have

$$H_0\text{: } \mu = 28 \text{ months}$$
$$H_a\text{: } \mu < 28 \text{ months}$$

Since $n/N = .29$, the FPC factor should be used in computing the test statistic. The *TS* is

$$Z^* = \frac{\overline{X} - \mu_0}{\dfrac{S}{\sqrt{n}} \cdot \sqrt{\dfrac{N - n}{N - 1}}} = \frac{22 - 28}{\dfrac{16}{\sqrt{50}} \cdot \sqrt{\dfrac{173 - 50}{172}}} = \frac{-6}{1.913} = -3.14$$

The p-value for this test is $P(Z \leq -3.14) = .0008$, thus providing a persuasive demonstration that H_0 is not true. The six-month discrepancy between $\overline{X}$ and μ_0 is unlikely due to chance. For any commonly used α, we conclude that the current mean time of store managers on the job is significantly below 28 months.

□

COMMENT As opposed to ignoring the FPC, use of the FPC has two desirable effects in inference: It makes a confidence interval narrower, and it increases the power of a hypothesis test.

SECTION 10.3 EXERCISES

10.65 A sample of 103 items drawn at random from a population of size 516 reveals a mean of 491.72 with a standard deviation of 101.17. Determine an interval that is 99% certain to include the population's mean.

10.66 Refer to Exercise 10.65. What sample size would be needed to have the point estimate accurate to within plus or minus 15.00, keeping the same degree of confidence?

10.67 The maximum sampling error E is, for finite populations,

$$E = Z\frac{\sigma}{\sqrt{n}} \cdot \sqrt{\frac{N - n}{N - 1}}$$

Solve this expression for n—you should arrive at Equation 10–9.

10.68 Test H_0: $\mu = 1400$ versus a two-sided alternative. Your sample statistics are $\overline{X} = 1369$ and $S = 299$, based on 36 observations taken from a population containing 181 items. Use 5% risk of a Type I error.

10.69 A random sample of 37 employees at a medium-sized company shows a mean of 6.6 days absent from work last year, with a standard deviation of 4.1 days. To one decimal place, compute an interval that you are 90% sure includes μ, the mean number of days absent from work for all 243 employees in the company.

10.70 Refer to Exercise 10.69. Could H_0: $\mu = 8.0$ days be rejected in favor of a lower-tail H_a at the .05 level of significance?

10.71 The assistant editor of a small weekly community newspaper chose 36 issues at random from the last two years to estimate the average amount of space devoted to classified advertisements. The sample showed a mean of 11.6 columns with a standard deviation of 2.8 columns. To one decimal place accuracy, develop an 80% confidence interval for the parameter of interest.

10.72 What sample size is required to estimate μ with sampling error no more than plus or minus \$10 with 90% certainty? The population consists of 149 items; an educated guess for the population standard deviation is \$40.

*10.4 PROCESSOR: LARGE SAMPLE INFERENCES FOR μ

The processor for this chapter can compute confidence intervals for μ and/or the test statistic Z^* for a hypothesis test about the value of μ. To access this program, move the highlight to Processors on the main menu and touch Enter. Then select the One Sample Inference processor; a submenu of four choices will appear. Choose the first offering—Large Sample Inferences For μ ($n \geq 30$). After reading the introductory screen, move the highlight to Table and touch Enter twice; this reads the Z-table into memory. (*Note:* This may take several seconds; be patient!) When ready, move the highlight to Data and touch Enter.

Data Entry

We now are asked if we intend to call for raw data previously entered or if we are going to supply a data set's summary statistics. In the illustrative example that follows, we will use the second option—entering summary statistics—to construct the 95% confidence interval that was developed in Example 10.2. Upon selecting the summary statistics option, we view a screen telling us that the sample size, mean, and standard deviation are needed. Figure 10.23 shows this screen after we have supplied these values: $n = 147$, $\overline{X} = 2.12$, and $S = 1.66$ (but before touching Enter for the final time).

*Optional

```
≡                   Large Sample Inferences For μ (n ≥ 30)                    EDIT
   Introduction  *able [Data] *esults  Exit                              | F1=Help


         To compute the confidence interval or test statistic Z*,
       the sample size, mean, and standard deviation are needed.

                     number of entities, n = 147

                     sample mean, X-bar = 2.12

            ═════════════ Standard deviation ═════════════
             Enter standard deviation, S (or σ) =  1.66
```

Figure 10.23 Entry of Summary Statistics

■ Constructing a Confidence Interval

After touching Enter twice to complete data entry, move to Results and touch Enter. A screen will appear asking which type of inference we wish to make: a Confidence Interval or a Hypothesis Test. After choosing Confidence Interval, we will be asked to furnish a confidence level. Once this is done, a screen similar to Figure 10.24 will provide the desired results. Note that we selected 95.0 for our confidence level in this example.

■ Hypothesis Testing

Data entry is the same for hypothesis testing as it is for developing a confidence interval. To demonstrate the hypothesis-testing option, let us repeat the test given in Example 10.7: a two-tailed test of H_0: $\mu = 51.5$ days. After entering the summary data ($n = 75$, $\overline{X} = 57.3$, $S = 21.0$), we choose Hypothesis Test from the Results menu. We then will be asked whether we wish a two-tailed, lower tail, or upper tail alternative hypothesis. After making the proper choice (two-tailed in this case), we will be asked to state the test value—the value for μ that is to appear in the null hypothesis. To test H_0: $\mu = 51.5$, we enter 51.5 at this time. Figure 10.25 shows the screen that summarizes the hypothesis test results. Note that the p-value for the test is furnished. It is left to the user to decide whether the p-value is low enough to warrant rejection of the null hypothesis. (The p-value for the test is .0168. In Example 10.7 we wished to use $\alpha = .10$. Since the p-value is less than the desired risk of Type I error, we would reject H_0 in this instance.)

```
≡                 Large Sample Inferences For μ (n ≥ 30)                 QUERY
   Introduction  *able  Data [Results] Exit                        | F1=Help

              For 95.0% confidence, the interval is
                   point estimate ± Z * (standard error)

                   =  2.120 ± Z * (1.660 / √n)
                   =  2.120 ± 1.960 * (1.660 / 12.124)
                   =  2.120 ± 0.268

              The lower limit of the confidence interval is  1.852
              The upper limit of the confidence interval is  2.388

              ═══════════════ Confidence ═══════════════
                   Compute another interval? (y/n) No
```

Figure 10.24 Confidence Interval Results for Data in Example 10.2

```
≡                 Large Sample Inferences For μ (n ≥ 30)                 READY
   Introduction  *able  Data [Results] Exit                        | F1=Help

     Hypothesis test      Ho: μ =  51.500
                          Ha: μ not =  51.500

                      sample mean - hypothesized value of μ
            TS:  Z* = ---------------------------------------
                               standard error

                           57.300 -  51.500
                      Z* = ----------------- =  2.392
                           21.000 /  8.660

            The p-value for this test is 0.016800000000

      The decision to "reject Ho" or "do not reject Ho" depends on your
      judgment or tolerable risk of Type I error (see text Figure 10.18).
```

Figure 10.25 Hypothesis Test Results for Data in Example 10.7

10.5 SUMMARY

One major objective in the study of statistics is to be able to make inferences or decisions about a population using the information available in a random sample. We have presented two related methods of making an inference about μ: estimation with confidence intervals and hypothesis testing. Table 10.8 summarizes the similarities and differences of these two approaches.

Table 10.8 Comparison of Estimation and Hypothesis Testing

	Form of Question Intended to Answer	Form of Answer Provided	Meaning of α	Intervals and Tests: Common Elements
Estimation	What is a reasonable value to believe μ is?	An interval	The risk that the interval does not contain μ	Sample evidence, standard error, and risk (α)
Hypothesis Testing	Is it reasonable to believe that μ = (some specific value)?	"Yes" or "no" (accept H_0 or reject H_0)	The risk that the sample indicates "no" when, in fact, in the population, the answer is "yes"	

Note: In later chapters and different problem settings, μ may be replaced by π or some other symbol.

In the chapters ahead, we will study other inference situations, such as binomial sampling, small sample conditions, comparing two populations, and finding relations between two or more variables. The major ideas of this chapter—developing interval estimates and subjecting assumptions to a statistical test—will be used often in subsequent chapters. The sampling conditions and formulas may change in the chapters that follow, but the underlying logic developed in this chapter will remain the same.

10.6 TO BE CONTINUED . . .

. . . *In Your College Courses*

An understanding of the theory underlying point estimates, interval estimates, and hypothesis tests may serve you well in other business and economics courses.

- If you take a course in operations management, you may study statistical quality control (also called quality assurance or process control). Particularly in a manufacturing plant where items are sampled at regular intervals to verify conformity with standards, quality control is a classic example of hypothesis testing (H_0 being that the process being sampled is "in control").

- In marketing research, you may read about a newly developed product going into a test market as the last step prior to national introduction. A test market serves to evaluate different marketing strategies as well as to alert the marketer to a product that has a substantial chance of being a failure if distributed nationwide. The test market (sample cities and stores) is supposed to simulate the national market (the population) in miniature. Relevant estimates found at the test market stage (share of market, unit sales, and so on) are used to help make the go/no-go decision for national introduction. The effort made in choosing the test cities and stores, as well as monitoring the test for unusual abnormalities, are in the spirit of attempting to have a representative sample upon which to base decisions.
- In courses in econometrics and operations management, you must often deal with forecasting future values of demand, tax collections, product sales, stock and bond prices, and so on. The concept of developing confidence intervals for point predictions is an integral part of the forecasting process.

. . . In Business/The Media

- A general knowledge of random sampling and interpretation of sample evidence is essential to an auditor. For example, the American Institute of Certified Public Accountants states, in reference to using point and interval estimates to establish the value of a client's inventory that, ". . . if statistical sampling methods are used by the client in the taking of the physical inventory, the auditor must be satisfied that the sampling plan has statistical validity, that it has been properly applied, and that the resulting precision and reliability (confidence), as defined statistically, are reasonable in the circumstances" (Statement on Auditing Standards 1, Section 331.11, American Institute of Certified Public Accountants, New York, 1972).
- The media regularly reports results of opinion polls. The most prominent numbers in these stories are almost always the point estimates; however, most such articles also provide information about margin for error and degree of confidence for the results. Polls are discussed in more detail in Chapter 12.
- The media often report results of scientific studies or tests, particularly for health-related items. When you see a term such as *significantly different* or *significantly better,* it is often in reference to a formal statistical test where it was possible to reject the null hypothesis. On the other hand, labeling a finding as *a statistically insignificant difference* implies that the evidence was not strong enough to reject the tested hypothesis.

CHAPTER 10 EXERCISES

10.73 Find the Z-score that would be appropriate for constructing the following:

a. a 70% confidence interval.
b. an 88% confidence interval.
c. a 97.5% confidence interval.

10.74 A sample of 161 items has been drawn from a population, yielding a sample mean of 347.52 and a sample standard deviation of 185.85. Construct an interval that has a confidence level of .90 of capturing the population mean.

10.75 An automobile insurance company received 82 claims involving collisions last week. The mean and standard deviation for amount claimed are \$1191 and \$629. Assuming the 82 claims are a representative sampling, construct an interval that has a confidence level of .95 of including the unknown population mean.

10.76 A manufacturer of a nonprescription pain reliever has asked its sales representatives who make calls on drug/grocery/discount stores to pick one box of the product from the front row of its shelf, for each store they call on. The sales representative is to record the date of the sales call as well as certain coded information on the package. When sent to the company headquarters, these codes can tell the exact date the product was packaged. In all, 189 boxes from different stores were checked. The sample evidence shows a mean age of 53 days and a standard deviation of 29 days. To the nearest whole number, provide an interval that will have a confidence level of .98 of including μ (mean age of all boxes on store shelves). Assume the stores selected constitute a random sample of stores in which the product is sold. State the lower and upper limits of the interval you compute.

10.77 Workers at a small appliance manufacturer's warranty/repair center fill out time slips to indicate how long it takes to process (evaluate, repair or replace, complete the paperwork) each appliance received from customers. Now that the company's newest model of hand-held hair dryer has been on the market for several months and more of these units are arriving at the center, management wants to estimate the average processing time. Over a one-week period, 112 time slips indicate work on this model. Compute to the nearest whole minute a 98 percent confidence interval for the parameter of interest, given that $S = 9$ minutes and $\overline{X} = 26$ minutes.

10.78 We want to estimate μ with a sampling error of no more than \$25. The standard deviation in the population is presumed to be \$90. For 98 percent confidence, what sample size do you recommend?

10.79 A school district is planning to estimate the average amount of time its junior high school students watch television in the period Monday through Thursday. In a pilot study, 20 students kept diaries of their viewing habits. The pilot showed $\overline{X} = 9.61$ hours and $S = 2.40$ hours. If the school district wants its estimate to be within 30 minutes (.50 hour) of the unknown population mean with a confidence level of .90, how large a sample should be undertaken?

10.80 Refer to Exercise 10.79. Using the required sample size indicated by the pilot study, the sample of junior high students yielded a sample mean of 10.20 hours and a sample standard deviation of 2.70 hours.

a. Use the full sample evidence to compute the 90% confidence interval for μ. Report to two-decimal place accuracy.

b. Is the maximum sampling error of your interval less than or greater than the desired $\pm .50$ hour accuracy? Why isn't the interval's maximum sampling error exactly .50 hour as originally wanted?

10.81 Refer to Exercise 10.76. What sample size is necessary to estimate mean age of boxes on store shelves to within ± 3 days, assuming we keep the confidence level at .98?

10.82 An urban university has many full-time students who work part-time jobs. The director of student life wants to determine the average number of hours per week worked by full-time students. A preliminary sample of 20 full-time students shows a mean of 10.4 hours and a standard deviation of 5.2 hours. For a 90% confidence interval to have a maximum sampling error of 1 hour, what sample size should be obtained?

10.83 A large auto-parts supplier has a plant in a medium-sized city but employs many workers from nearby communities. The personnel director wants to estimate the average commute distance for its employees to within ± 2 miles, with 5% risk of missing the population mean. The director knows that no one has a shorter commute than 1 mile and doubts if anyone commutes farther than 40 miles.

a. Use the range-based approach to recommend a sample size that should be close to the actual sample size needed.

b. Suppose the confidence interval which resulted from using the sample size you recommended above turned out to be 14.0 ± 1.4 miles. Was your range-based estimate for σ larger or smaller than the actual population standard deviation, σ?

10.84 To demonstrate the error inherent in measurement, a college physics instructor brought a child's plastic ball to a lab session one day and told the students that their task was to ascertain the diameter of the ball by whatever means they chose. Forty-one students measured the ball, generating these results: $\overline{X} = 23.59$ cm and $S = .37$ cm. Treating these measurements as a random sample, compute to two decimal-place accuracy the 99% confidence interval for μ, the actual diameter of the ball.

10.85 For the type of test and error risk indicated, find the cutoff Z-scores.

a. Lower-tail test, $\alpha = 8\%$

b. Two-tail test, $\alpha = 8\%$

c. Upper-tail test, $\alpha = .25\%$

10.86 Test H_0: $\mu = \$45.00$ versus H_a: $\mu \neq \$45.00$. Your sample evidence is a mean of \$46.12 with a standard deviation of \$3.84, based on 33 observations. Choose a frequently used level of significance in your test against H_0.

10.87 A grocery store manager states that the average nonexpress customer redeems 5 coupons per order. A sample of 85 cash register tapes chosen at random shows $\overline{X} = 6.6$ with $S = 3.9$ coupons. Test the manager's claim at the 2% level of significance. Is the discrepancy of 1.6 coupons per order statistically significant?

10.88 The amount of active ingredient in a time release capsule is supposed to average 20 grains. If the dosage is too low, the medicine does not produce the desired effect; if the dosage is too high, the chance of undesired side effects increases. Suppose we are to sample 60 capsules packaged by a newly licensed producer in order to perform a statistical test. What should our hypotheses be? Use a 1% risk of Type I error to draw a conclusion about the population of capsules when the sample mean is found to be 19.71 grains with a standard deviation of 1.51 grains.

10.89 An auditor is about to sample the accounts receivable of a building supply firm. The firm's computer shows that it has 4118 accounts receivable, with a mean balance of \$278.41. A sample of 200 accounts is randomly selected in order to investigate the presumption that the firm's balances are reasonably stated. With a 5% risk of incorrect rejection, report the proper conclusion if the sample mean is \$246.73 with a standard deviation of \$169.97.

10.90 In a test of H_0: $\mu = \$45{,}000$ with H_a: $\mu > \$45{,}000$, a sample of size 67 shows a standard deviation of \$11,137 and a mean of \$48,614. For the 5% level of significance, report the following:

a. The standardized cutoff point and your conclusion.

b. The unstandardized cutoff point and your conclusion.

10.91 The numbers below represent miles per gallon obtained on the first 30 tankfuls of a new economy car. Assuming these figures to be representative of fuel economy to be realized over the life of the car, determine a 95% confidence interval for the long-run average. Report results to two places beyond the decimal. Use the raw data formulas to compute $\overline{X}$ and S.

The sorted values of X = miles per gallon between fill-ups are given in the following table:

27.64	27.88	27.94	28.18	28.19	28.57	28.61
28.64	29.12	29.12	29.12	29.60	29.63	29.63
29.73	29.91	29.95	30.12	30.12	30.18	30.22
30.40	30.42	30.44	30.44	30.48	30.64	30.71
31.06	31.21					

10.92 In many parts of the country, the Bell System offers "measured service," an option for local call billing. Users choosing this plan pay a low base rate plus a fixed charge for each local call made. For customers who use their phones infrequently, the plan should result in a lower monthly bill than the traditional plan with its fixed charge for unlimited use (*source*: Bell-South Advertising and Publishing Corporation, Atlanta, GA.). Suppose a South Central Bell study investigated adopters of this plan to see if it reduced their telephone usage. A study of 125 adopters showed an average of 12.6 calls placed per month in the first four months on measured service, with a standard deviation of 7.1 calls. Company records indicate these customers had averaged 16.6 calls per month in the three years prior to adoption. Assuming these four months are a random sampling of post-adoption behavior, is it possible to conclude that the mean number of calls placed per month has decreased for users of measured service? Use $\alpha = .025$.

10.93 Compute the p-value for Exercise 10.90 above.

10.94 Given the following hypothesis statements and values for the TS, determine the p-value for the test:

a. H_0: $\mu = 170$ $\quad$ TS: $Z^* = 1.01$
H_a: $\mu > 170$

b. H_0: $\mu = 180$ $\quad$ TS: $Z^* = -.35$
H_a: $\mu < 180$

c. H_0: $\mu = 190$ $\quad$ TS: $Z^* = 4.11$
H_a: $\mu \neq 190$

10.95 For the following p-values and levels of significance, state whether the conclusion should be to reject or not reject the null hypothesis:

a. $\alpha = .05$, p-value $= .0872$
b. $\alpha = .02$, p-value $= .0137$
c. $\alpha = .001$, p-value $= .1162$

10.96 In "Management Judgment Forecasts, Composite Forecasting Models, and Conditional Efficiency" (*Journal of Marketing Research,* August 1984) Mark Moriarty and Arthur Adams compare the accuracy of forecasts for appliance sales from a sophisticated econometric model against the subjective forecasts made by the firm's marketing department. Without mentioning Z^*, the authors report that the subjective forecasts were superior and parenthetically remark "p-value $< .005$." Assume the 72 forecast periods over which the comparisons were made constitute a random sample of possible comparisons. Specify a risk factor of your choosing and state whether or not the results make a convincing case that the firm's judgmental predictions will be more accurate than the quantitative model's predictions in the long run.

10.97 Refer to Exercise 10.95 above. For each conclusion stated, specify whether a Type I or Type II error is possible.

10.98 In testing H_0: $\mu = 6.0$ minutes versus an upper-tail H_a, describe in words the meaning of these symbolic expressions:
a. $1 - \beta$
b. α

10.99 Refer to Exercise 10.88 above. In terms of the problem setting, what would be the meaning of committing a Type II error?

10.100 If a researcher has concluded that insufficient evidence to reject has been found against H_0, is a Type I error possible?

10.101 In what way does the FPC affect the computation of the sample mean?

10.102 A random sample of 30 passengers' carry-on luggage on a commuter flight from New York City to Boston showed an average weight of 12.6 pounds with a standard deviation of 5.7 pounds. Determine a 90% confidence interval for the mean if we view the population as being each of the following:
a. the 74 passengers on this particular flight.
b. all passengers on commuter flights between New York and Boston.
Report to the nearest tenth of a pound.

10.103 The owner of a mail-order business recently stated that the average price per item in the catalog should be no more than \$5.00. Set up a null hypothesis such that rejection would indicate that the average price is higher than desired. Use the .05 level of significance. A random sampling of 125 items in the most recent catalog shows a mean of \$6.13 with a standard deviation of \$3.83. The catalog lists approximately 700 different items. State your conclusion in terms of the problem setting.

10.104 An auditor has sampled 60 of a firm's 1312 accounts payable, finding an average balance of \$185.60 with a standard deviation of \$80.81. Provide a 90% confidence interval for the following:
a. the mean balance of accounts payable in the population.
***b.** the total dollar value of all the firm's accounts payable.

***10.105** The Food and Drug Administration (FDA) examines the caffeine content of many different beverages on a regular basis. In March of 1984 they reported (Associated Press, March 18)

*Optional

that an average 12-ounce serving of Coca-Cola contains 45.6 milligrams (mg) of caffeine. Suppose this result was based on a sample of 42 different containers of Coca-Cola and that a 95% confidence interval for the population mean was 45.6 ± 1.3 mg. Assuming a normal distribution, estimate the highest and lowest values of X, mg of caffeine in a single 12-ounce serving, found in the sample of 42 servings. (Hint: Consider using Equation 10–4 with n equal to one.)

10.106 A county water company randomly sampled customer accounts in order to estimate the average number of days from billing the user until payment is received. With X = number of days from billing to payment, the sample results are as follows

27	11	5	13	13	3	7	38	17	9	3
12	27	10	5	14	27	19	37	20	8	29
30	46	3	4	22	7	9	27	5	3	30

a. A 90% confidence interval is desired. Use the data to develop the interval, to one decimal-place accuracy. (Use the raw data formulas to compute $\bar{X}$ and S.)

b. If management wanted a maximum potential sampling error of ±2.5 days, what sample size would you recommend?

10.107 In a test of shelf life of a particular nonprescription drug, a pharmaceutical firm set aside several thousand pills under typical storage conditions for eighteen months. A random sample was then drawn and chemically tested to assess potency. With X = tablet percent of original label strength, the results are as follows:

94.1	93.6	97.2	96.1	89.8
95.1	97.7	90.9	93.4	96.7
94.0	95.7	92.6	91.8	90.3
93.7	95.1	93.7	94.9	97.9
97.1	90.5	91.9	93.6	82.7
96.4	93.0	92.6	97.6	94.2
94.4	95.4	93.1	98.1	96.0
92.8	92.0	94.8	96.0	95.4

a. Compute $\bar{X}$ and S, using the raw data formulas.

b. To two decimal place accuracy, determine a 95% confidence interval for the mean value of the population from which this sample was obtained.

c. Is it possible to reject H_0: $\mu = 92.5$ at $\alpha = .05$ when the research hypothesis is upper-tail?

d. What is the p-value for the test?

10.108 You desire to evaluate the reasonableness of the book value of the inventory of your client, Draper, Inc. You satisfied yourself earlier as to inventory quantities. During the examination of the pricing and extension of the inventory, the following data were gathered using appropriate unrestricted random sampling with replacement procedures.

Total items in the inventory (N)	12,700
Total items in the sample (n)	400
Total audited value of items in the sample	\$38,400
$\sum_{j=1}^{400} (X_j - \overline{X})^2$	312,816
Formula for estimated population standard deviation	$S_{X_j} = \sqrt{\dfrac{\sum_{j=1}^{j=n} (X_j - \overline{X})^2}{n - 1}}$
Formula for estimated standard error of the mean	$SE = \dfrac{S_{X_j}}{\sqrt{n}}$
Confidence level coefficient of the standard error of the mean at a 95% confidence (reliability) level	± 1.96

a. Based on the sample results, what is the estimate of the total value of the inventory? Show computations in good form where appropriate.

b. What statistical conclusion can be reached regarding the estimated total inventory value calculated in part a above at the confidence level of 95 percent? Present computations in good form where appropriate.*

REFERENCES

Arens, A. A., and J. K. Loebbecke. 1981. *Applications of Statistical Sampling to Auditing*. Prentice-Hall, Englewood Cliffs, NJ.

Shiffler, R. E., and A. J. Adams. 1987. "A Correction for Biasing Effects of Pilot Sample Size on Sample Size Determination." *Journal of Marketing Research*, 24: 319–321.

*Problem 10.108 is reprinted by permission from *Uniform CPA Examination*, by the American Institute of Certified Public Accountants, New York.

Chapter Maxim *The merit of a statistical test is not necessarily dependent on sample size—a conclusion to "reject H_0" based on a small sample is just as valid as "reject H_0" based on a large sample.*

CHAPTER 11

INFERENCES FOR SMALL SAMPLES

*Optional

Objectives

After studying this chapter and working the exercises, you should be able to

1. Recognize the conditions under which the *t*-distribution should be used instead of the *Z*-distribution.
2. Find cutoff points on the *t*-distribution for confidence intervals and hypothesis tests.
3. Construct confidence intervals for μ and perform hypothesis tests about μ, using the *t*-distribution.

*4. Learn to use the Small Sample Inferences processor for developing confidence intervals and/or hypothesis testing for small samples.

*5. Explain the advantages the signed ranks procedure has over the *t*-test.

*6. Perform hypothesis tests using the signed ranks procedure.

7. Distinguish between chance variation and assignable variation in manufacturing.
8. Relate the notion of testing a hypothesis to the practice of quality control.

In Chapter 10, we learned how to estimate the unknown value of μ with a confidence interval and how to assess the reasonableness of a proposed value for μ with a hypothesis test. Both forms of making a statistical inference relied on the Central Limit Theorem to describe the sampling distribution of $\overline{X}$ for large samples. As a result, we were able to reference the *Z*-distribution for cutoff values to use in the confidence interval or to separate the acceptance and rejection regions of a hypothesis test.

However, all sampling situations do not involve large sample sizes. For small samples, the CLT does not apply, and the inferential procedures based on the *Z*-distribution developed in Chapter 10 are not theoretically valid. Making inferences for μ with small sample sizes therefore requires additional information about the sampling distribution of $\overline{X}$. In this chapter, we will learn how $\overline{X}$ behaves when n is small and certain sampling conditions are in effect. A different continuous probability distribution called the *t*-distribution will be presented as the model for the small sample behavior of $\overline{X}$. We will use the *t*-distribution as the basis for small sample confidence interval construction and hypothesis testing about μ in a parallel manner to our development in Chapter 10.

*Applies to optional sections.

11.1 INFERENCES FOR μ USING THE t-DISTRIBUTION

In Chapter 10, we discussed inferences (confidence intervals and hypothesis tests) for the population mean μ. In the great majority of these sampling situations, the numerical value for the population standard deviation σ will be unknown to us. But we *need* to know σ in order to make an inference! To illustrate, recall that our expression for a confidence interval contains σ:

$$\overline{X} \pm Z\sigma/\sqrt{n}$$

Should we want to perform a hypothesis test, σ appears again, this time as part of our test statistic:

$$Z^* = \frac{\overline{X} - \mu_0}{\sigma/\sqrt{n}}$$

Our two forms of inferences, therefore, seem to rely on knowing the value of σ. How can we proceed if σ is not known? In Chapter 10 (where we limited ourselves to large sample size applications), our strategy was this: Substitute S, the sample standard deviation, into the expressions above in place of σ. Our expressions for a confidence interval and for a test statistic then become

$$\overline{X} \pm ZS/\sqrt{n} \qquad \textbf{(11–1)}$$

and

$$Z^* = \frac{\overline{X} - \mu_0}{S/\sqrt{n}} \qquad \textbf{(11–2)}$$

To solve our problem of not knowing σ, we use the current sample information to compute S, the sample standard deviation. We then state that S is a reasonable approximation of σ *when the sample size is 30 or more*.

However, samples are sometimes small. The two major constraints on sampling (time and cost) may necessitate a limited number of observations. Or perhaps we have access to only a few population members. Or perhaps the sampling situation at hand involves destructive sampling, forcing a restricted view of the population. For these and other reasons, a sample may be less than 30 observations. If σ is unknown and n is small, we no longer have the luxury of assuming that S can be treated as equal to σ. If we must estimate σ by computing S, we would rather do so with a big sample than with a small sample. This is because the value which we compute for S is more likely to approximate σ closely when n is large, say $n = 100$, as opposed to when n is small, such as $n = 3$.

When n is large and S can be treated as if it equals σ, then expressions (11–1) and (11–2) each have a single random variable, $\overline{X}$, as the other terms are nonrandom. We already have learned that the sampling distribution of $\overline{X}$ is approximately normal when n is large. But when n is small, there is uncertainty about how close S is to σ, and expressions (11–1) and (11–2) now each have two random variables, $\overline{X}$ and S.

This extra uncertainty means that the sampling distribution of $\overline{X}$ is no longer approximately normal. The exact form of the sampling distribution is known for samples drawn from a normal population when σ is unknown—it is called the ***t*-distribution.**

t-Distribution and *t*-Table

We will employ the t-distribution to make inferences about μ under the following sampling conditions.

t-Distribution Sampling Conditions

1. The population standard deviation σ is unknown but estimated by the sample standard deviation S,

and

2. The sample size is less than 30,

and

3. The underlying population of X can be assumed to be normally distributed.

The reason we have the t-distribution is this: It gives us a way to develop inferences about the mean of a normal distribution when σ is not known. Figure 11.1 shows the general shape of a t-distribution. Some of its most important features are given in the next box.

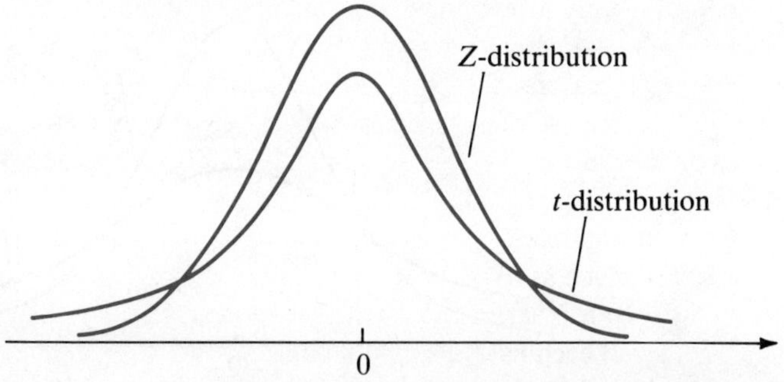

Figure 11.1 General Shape of a t-Distribution Relative to the Z-Distribution

t-Distribution Characteristics

1. The curve is mound shaped and symmetrical around its mean value of 0.
2. A *t*-distribution is a continuous probability distribution.
3. The total area (probability) under the curve is 1.
4. The values for the random variable *t* extend from $-\infty$ to $+\infty$.
5. The standard deviation of a *t*-distribution is always greater than 1.
6. The *t*-distribution is really a *family* of curves, not just a single curve. Different members of this family exist according to the degrees of freedom (*df*) present in the sampling situation.

The term **degrees of freedom** refers to the number of values of a variable that are free to vary, given that there is some restriction on the data. For instance, in computing

$$S^2 = \frac{\Sigma(X - \overline{X})^2}{n - 1}$$

all but one of the values of X can be any number, but the last value has to be one that satisfies the restriction (mentioned in Section 3.3) that $\Sigma(X - \overline{X}) = 0$. For a quick numerical example, if $n = 3$ and $\overline{X} = 20$, perhaps one value of X is 30 and perhaps another is 18; these first two values of X are free to be any numbers we choose. But the third X is restricted—in this example, it must be 12 in order to satisfy the condition $\Sigma(X - \overline{X}) = 0$, or that the sum of the deviations about the mean is zero. (Equivalently, the last X must be 12 in order to have $\overline{X} = 20$.) In this example, there are two *df:* 2 values of X are free to vary. In all instances in this chapter, the degrees of freedom when using the *t*-distribution will be $n - 1$, the sample size less one.

Figure 11.2 illustrates *t*-distributions for two different degrees of freedom. The figure suggests that the *t*-distribution is most spread out when *df* is small; as *df*

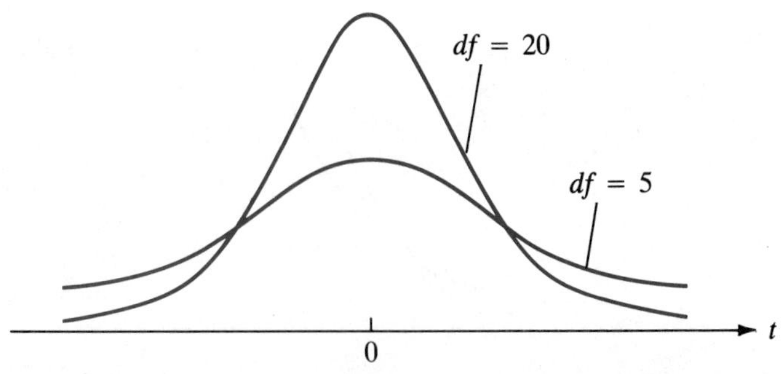

Figure 11.2 Two Members of the *t*-Distribution Family

increases, the t-distribution will become more compact and normal in appearance. The t-distribution has a specific mathematical expression which is rather complicated. However, the t-table makes use of this expression unnecessary. A table of cutoff values for the t-distribution appears in Appendix E. We will work a few examples to familiarize ourselves with the t-table.

EXAMPLE 11.1

Find the t-table values that would be used to develop the following:

a. a 95% confidence interval when $n = 8$
b. a 99% confidence interval when $n = 15$
c. a 90% confidence interval when $n = 5$

Solution:

a. For 95% confidence, our risk will be 5 percent, or .05, divided evenly into the two tails. We therefore locate the $\alpha = .05$ column near the top center of the table. With $n = 8$ the df will be $n - 1$, or 7. We will then find our t-value at the intersection of the $df = 7$ row and the $\alpha = .05$ column. Our result is $t = \pm 2.365$ (see Figure 11.3). If our sample had been large and we used the Z-table, the 95% confidence cutoffs would be the familiar $Z = \pm 1.96$. Since the t-distribution has more dispersion than the Z-distribution, t-values always will be greater in (absolute) magnitude than the Z-values for the same level of risk.
b. For 99% confidence, we locate the .01 risk column near the top of the table and then read down to the $df = n - 1 = 14$ row. We find $t = \pm 2.977$.
c. For $\alpha = .10$ and $df = 4$, the appropriate values for t are ± 2.132.

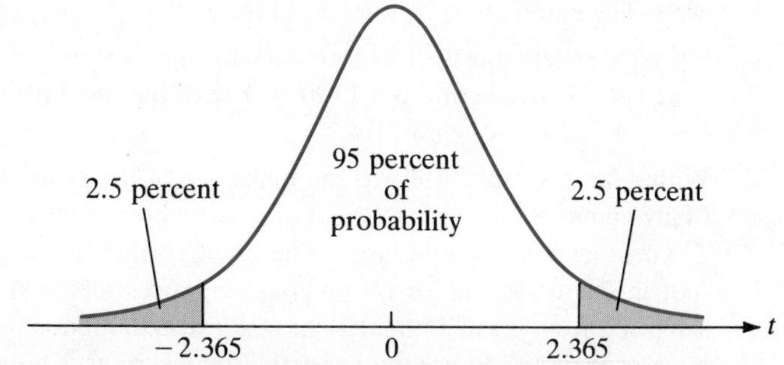

Figure 11.3 Cutoff Values for 95% Confidence When $df = 7$

□

Note that the bottom row of the *t*-table is labeled as "$df \geq 30$"; it can be used for any sample size above 30. In reality, there is a *t*-score for each integer above 30, but these values are little different from the Z-score counterparts. For instance, for two-tailed $\alpha = .05$ with $df = 45$, the actual *t*-value (not shown in our table) is $t = \pm 2.014$; for $df = 75$ the exact *t*-value is ± 1.992. These are close enough to the Z-score of ± 1.96 that some statisticians adopt the convention that for large samples ($n \geq 30$) where σ is unknown, we can use the Z-table for cutoff points instead of the *t*-table. Although it is theoretically correct to use *t*-table values when σ is unknown, regardless of sample size, in practice we will approximate the *t*-distribution with the Z-distribution when the sample size is 30 or more. The cutoff values in the bottom row of the *t*-table are in fact Z-table values expressed to three places beyond the decimal—to see this, simply compare them against your Z-table.

EXAMPLE 11.2

For a hypothesis test under the following conditions, find the correct cutoff values:

a. $n = 6$, upper-tailed H_a, $\alpha = .10$

b. $n = 12$, two-tailed H_a, $\alpha = .10$

c. $n = 14$, lower-tailed H_a, $\alpha = .05$

d. $n = 149$, lower-tailed H_a, $\alpha = .05$

Solution:

a. As the sketch and example at the bottom of the *t*-table indicate, we are to use the risk column headings at the bottom of the table when H_a is one tailed. For $\alpha = .10$ and 5 degrees of freedom, the cutoff for an upper-tailed H_a is $t = 1.476$.

b. Since a two-tailed test is desired, the .10 risk column near the top of the table is used. The cutoffs will be $t = \pm 1.796$.

c. For a lower-tailed test, it is necessary for the user to provide a minus sign for the table value. Intersecting the 13 *df* row with the one-tail .05 column, we arrive at $t = -1.771$ (see Figure 11.4).

d. With a large sample size, we can make use of the $df \geq 30$ value of $t = -1.645$. Many commercial software programs will automatically report *t*-scores instead of Z-scores even when n is large. The exact *t*-value for this instance, $t = -1.655$ (not in our book), is for all practical purposes identical to the Z-table value. In sampling from normal populations, the *t*-distribution is in fact the general case; it is never incorrect to use the *t*-distribution when (as is almost always the situation) σ is unknown. The special case of $n \geq 30$ is where we choose to use, for reasons of convenience and simplicity, the more familiar Z-distribution to approximate the *t*-distribution.

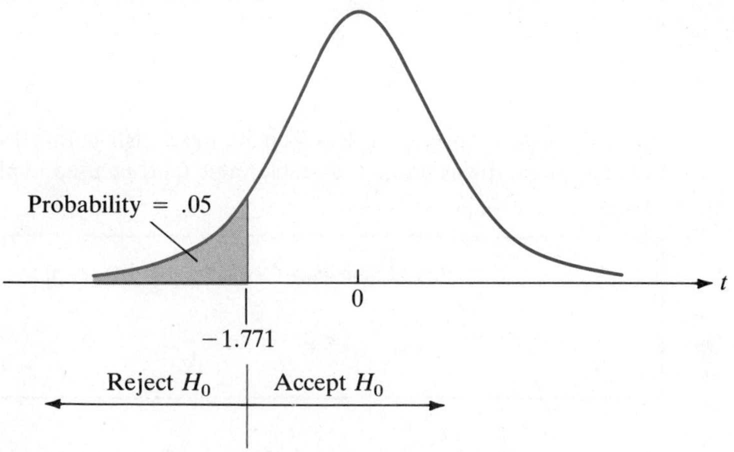

Figure 11.4 Cutoff Point for Reject H_0/Accept H_0 in Example 11.2

□

COMMENTS

1. The t-distribution is also known as *student's t*. In the early 1900s, Englishman W. S. Gosset realized that S was not a reliable estimator of σ when n was small. Under his pen name "Student," Gosset published the derivation of the sampling distribution that he called *student's t*. The t-distribution has been important in small sample inference ever since.
2. The standard normal, or Z-distribution, has $\mu = 0$ with $\sigma = 1$; the t-distribution has $\mu = 0$ with $\sigma = \sqrt{df/(df - 2)}$ (σ is undefined when df is less than 3). Since the t-distribution's standard deviation exceeds 1.0 for any finite sample size, it is more spread out than is the Z-distribution.
3. In one way or another, both our large and small sample inference procedures for μ presented so far require the normal distribution. For large samples ($n \geq 30$), we used the Central Limit Theorem that assures us of an approximately normal sampling distribution of $\overline{X}$ even if the individual X's in the population were skewed or otherwise abnormal. For small samples, we introduce the strong assumption (not needed for large n where the CLT applies) that the *individual X's* in the population are normally distributed or approximately so.
4. Later in this chapter, we present the signed ranks test procedure for hypothesis tests involving small samples. This alternative method is becoming more widely used, one reason being that it does not rely on the assumption that the population of X's is normally distributed.

Confidence Intervals for μ

In previous chapters, we have used these words to express the concept of a confidence interval:

$$\text{Point estimate} \pm \text{Potential sampling error}$$

For large sample estimates of μ where σ is estimated by S, these words are symbolized as

$$\overline{X} \pm Z\frac{S}{\sqrt{n}}$$

But as discussed above, if n is less than 30, the Z-distribution does not apply; assuming the population of X is normal, a t-table value must be used in place of a Z-table value.

Small Sample Confidence Interval for μ

$$\overline{X} \pm t\frac{S}{\sqrt{n}} \tag{11–3}$$

Our logic and procedures for developing a confidence interval for μ when n is small are similar to those of the large sample case. The differences are that: A t-score is used instead of a Z-score, and we must assume that the population of individual values of X is normally distributed. We now consider two examples of small sample size confidence intervals.

EXAMPLE 11.3

A bank manager wants to know how many customers are using the bank's extended business hours (4 to 7 P.M.) on Fridays. Over the next six weeks, a count of customers is obtained. Let X = the number of customers between 4 and 7 P.M. on Friday evenings. The observed values of X are 74, 73, 82, 86, 81, and 78. Assuming that X is normally distributed, determine a 90% confidence interval for the mean number of customers utilizing the extended hours service.

Solution:

Since a large sample would take more than six months to collect, the estimate will be based on a rather small sample size. As usual, the sample mean and standard deviation must be calculated.

X		$X - \overline{X}$	$(X - \overline{X})^2$	
74		-5	25	
73		-6	36	
82	$\overline{X} = \frac{\Sigma X}{n}$	3	9	$S = \sqrt{\frac{\Sigma(X - \overline{X})^2}{n - 1}}$
86		7	49	
81	$= \frac{474}{6}$	2	4	$= \sqrt{\frac{124}{5}}$
78		-1	1	
$\Sigma X = 474$	$= 79$		$\Sigma(X - \overline{X})^2 = 124$	$= 4.98$

For $n - 1 = 5$ *df*, the t-value for $\alpha = .10$ is $t = \pm 2.015$. The confidence interval is then

$$\bar{X} \pm t\frac{S}{\sqrt{n}} = 79 \pm 2.015\,\frac{4.98}{\sqrt{6}}$$
$$= 79 \pm 4.10$$

The bank manager can be 90% certain that the average demand for Friday extended hours is between 75 and 83 customers.

□

In Example 11.3, we would have used $Z = \pm 1.65$ instead of $t = \pm 2.015$ had the sample been 30 or more observations. The cutoff value from the t-table is always larger than that for the Z-table for the same level of confidence. The wider interval that results from using the t-table reflects the fact that there is uncertainty about how well S approximates σ when the sample size is small.

EXAMPLE 11.4

A new snack food undergoes a test market in comparable retail outlets in 12 different cities. After a month, the number of packages sold is recorded for each store. The sample results show an average of 230 packages sold with a standard deviation of 19. Presuming the test outlets to be a random sample of stores where the product would experience wide distribution, develop a 95% confidence interval for the average or expected sales in similar stores.

Solution:

For 5% risk with 11 degrees of freedom, we require $t = \pm 2.201$. Our confidence interval is

$$\bar{X} \pm t\frac{S}{\sqrt{n}} = 230 \pm 2.201\,\frac{19}{\sqrt{12}}$$
$$= 230 \pm 12.1 \quad \text{or } 218\text{–}242 \text{ packages}$$

Whether this level of average monthly sales makes national introduction of the product attractive will be a management decision. As before, our use of the t-distribution presumes a normally distributed population.

□

■ Hypothesis Tests About μ

Besides estimation (development of a confidence interval), our other major approach to making an inference or decision about some population is a hypothesis test. As with confidence intervals, our methods for conducting a statistical test when the sample size is small will closely parallel those of the large sample case.

To assess the strength of the evidence against H_0, we compute a number called a test statistic (*TS*). In words, we have expressed our *TS* previously as

$$\frac{\text{Sample mean} - \text{Hypothesized value of } \mu}{\text{Standard error}}$$

When the sample size is large, the *TS* is a *Z*-value; when n is small, the *TS* is a *t*-value. In symbols, we will have the following formula.

Small Sample Test Statistic for μ

$$t^* = \frac{\overline{X} - \mu_0}{S/\sqrt{n}} \qquad (11\text{–}4)$$

A small sample test about the value of μ differs from the large sample procedure in that: The *TS* is to be compared to a *t*-table cutoff point instead of a *Z*-table cutoff, and as with small sample confidence intervals, we are assuming that the population of *X*'s is normally distributed.

EXAMPLE 11.5

A chemical firm has constructed a state-of-the-art plant to produce a new plastic product. Previous experience with related products as well as in a small-scale pilot plant indicated that an average yield of 93.5 percent can be expected. (Yield is the actual weight of product obtained from a given amount of raw material, expressed as a percentage of the theoretically obtainable weight.) After the plant has come on stream, the first 12 production batches are analyzed. Assume that these batches will be a random sample of the plant's production. Using the .05 level of significance, test the assumption that the average yield of the plant will be 93.5 percent, against a two-tailed alternative. The sample results for X = yield per batch, in percentages, are

91.41	94.47	92.80	92.93	94.19	93.37
92.92	93.16	93.24	92.70	93.66	93.86

Solution:

We wish to test H_0: $\mu = 93.5$ against H_a: $\mu \neq 93.5$ at the .05 level. For a two-tailed test with $n - 1 = 11$ *df*, the cutoff values from the *t*-table are $t = \pm 2.201$. Using the 12 values of X, we can compute that $\overline{X} = 93.23$ percent and that $S = .80$ percent. Our *TS* is then

$$t^* = \frac{\overline{X} - \mu_0}{S/\sqrt{n}} = \frac{93.23 - 93.5}{.80/\sqrt{12}} = \frac{-.27}{.23} = -1.17$$

Since $|t^*| \leq 2.201$, our conclusion is that H_0 has been supported by the sample evidence (see Figure 11.5). Table 11.1 summarizes the elements of this statistical test.

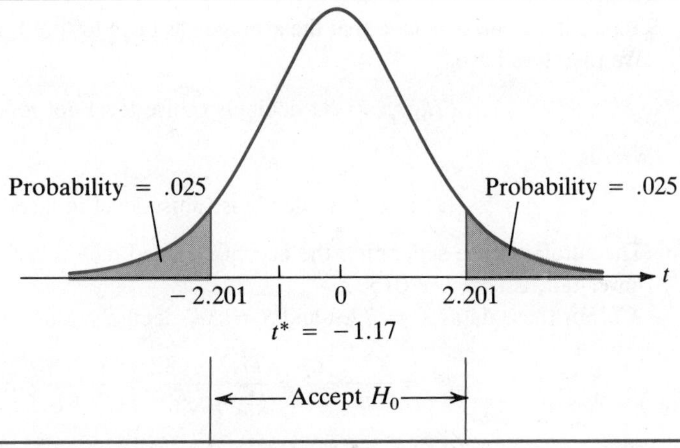

Figure 11.5 Graphical View of Example 11.5

Table 11.1 Symbolic Summary of Example 11.5

H_0: $\mu = 93.5\%$
H_a: $\mu \neq 93.5\%$
TS: $t^* = -1.17$
RR: Reject H_0 if $t^* < -2.201$ or if $t^* > 2.201$
C: Accept H_0

□

In Example 11.5, if we wished to show the lower and upper limits of the acceptance region in terms of the sample average $\overline{X}$, we would compute $\mu_0 \pm tS/\sqrt{n}$, or 92.99 and 94.01 in this example. The sample average of $\overline{X} = 93.23$ lies inside the acceptance region.

EXAMPLE 11.6

The noise rating of one company's most popular push-type lawn mower, a 3.5 horsepower model, is 82 decibels. The rating is based on measurements taken at a point behind the machine about where the operator's ear would be. Company engineers have been asked to try to develop a motor of equal power but with a lower noise level. Six prototype models were built and yielded these noise levels: 75, 77, 80, 83, 74, and 74. Assuming these are a random sample of noise levels which would result from the new motor, can we be at least 95% certain that the average level would be reduced?

Solution:

To see if we can conclude that the average has been lowered, a lower-tail H_a is needed. We therefore have

$$H_0\text{: } \mu = 82 \text{ decibels (noise level not reduced)}$$

versus

$$H_a\text{: } \mu < 82 \text{ decibels (noise level reduced)}$$

The cutoff t-score separating the acceptance and rejection regions for 5 *df*, $\alpha = .05$, lower-tail, is $t = -2.015$.

For these data, $\overline{X} = 77.2$ and $S = 3.7$ decibels. Our *TS* is then

$$t^* = \frac{\overline{X} - \mu_0}{S/\sqrt{n}} = \frac{77.2 - 82}{3.7/\sqrt{6}} = \frac{-4.8}{1.51} = -3.18$$

Since $t^* = -3.18 < -2.015$, H_0 is rejected; it appears that the new model engine will obtain a lower noise rating. This and other hypothesis tests using the t-distribution with $n < 30$ assume that the population of Xs is normally distributed.

□

The p-value concept also applies to tests using the t-distribution, but due to the way the t-table is constructed, we cannot state exact p-values. Instead we can report upper and lower limits for the p-value. For instance, in Example 11.6 we obtained $t^* = -3.18$ in a one-tail test with $df = 5$. Looking at the 5 *df* row, we can see that 3.18 is in between 2.571 (the .025 level cutoff, one tail) and 3.365 (the .01 level cutoff). We therefore can state that $.01 < p\text{-value} < .025$, meaning that our evidence is strong enough to reject H_0 when $\alpha = .025$ but not strong enough to reject H_0 at $\alpha = .01$. Our example involved a one-tail H_a; had H_a been two tailed, the p-value limits would be doubled, yielding $.02 < p\text{-value} < .05$. The software program that coincides with this chapter can compute and report exact p-values for the small sample test statistic, t^*.

CLASSROOM EXAMPLE 11.1

Using the t-Distribution in a Confidence Interval

As a public service to motorists over a three-day holiday weekend, the newspaper in a medium-sized city reports the average price of self-serve unleaded gasoline within a 25-mile radius of the city, based on a random sample of gas stations. Although 25 stations were surveyed by telephone, only 23 agreed to participate. The results were an average price of \$.96 and a standard deviation of \$.06 per gallon.

a. Define the random variable X.

b. Define, in words, the parameter of interest. Give the value of the point estimate for the parameter.

c. If we wished to construct a 90% confidence interval for the parameter of part b, what would be the appropriate t-scores?

d. Construct a 90% confidence interval for μ.

e. What assumption must we make about the probability distribution of the random variable you defined in part a in order for the confidence interval to be valid?

CLASSROOM EXAMPLE 11.2

Using the t-Distribution in Hypothesis Testing

The rate department at a utility company publishes a list of average monthly usage in kilowatt hours (kwh) for electrical appliances. For instance, a frost-free refrigerator uses an average of 108 kwh per month, according to the utility. Meters can be attached to a refrigerator to monitor its exact consumption, but this is awkward and expensive. Consequently, an independent research firm was chosen to hook up meters to eight frost-free refrigerators in a study to test the utility's figure. After one month the firm recorded the following consumption numbers:

126	108	90	104
90	89	163	159

Assume the distribution of monthly kwh consumption for frost-free refrigerators is approximately normally distributed.

a. Set up a null and alternative hypothesis about μ that would enable the independent research firm to agree or disagree with the utility's figure.

b. Calculate the sample mean and standard deviation.

c. Determine the value of the test statistic for testing the hypothesis in part b.

d. Set up a rejection region based on an alpha risk of 10 percent.

e. Does the sample evidence support or refute the utility's figure?

SECTION 11.1 EXERCISES

11.1 Under what conditions should the t-distribution be used instead of the Z-distribution?

11.2 Find the t-table cutoff values that would be required to compute the following:

a. a 90% confidence interval when $n = 6$.

b. a 95% confidence interval when $n = 12$.

c. a 99% confidence interval when $n = 21$.

11.3 Report the t-score(s) that would separate acceptance from rejection region(s) under the following conditions:
a. sample size = 15, α = .10, and two-tailed H_a.
b. sample size = 10, α = .05, and lower-tailed H_a.
c. sample size = 8, α = .025, and upper-tailed H_a.

11.4 What are the numerical values for μ and σ for the Z-distribution? For a t-distribution with 4 degrees of freedom?

11.5 What t-score(s) will, for 7 degrees of freedom, provide the following results?
a. place ½ of 1 percent of the total probability in the lower tail
b. divide 5 percent of the total probability evenly into the tails
c. divide 1 percent of the total probability evenly into the tails

11.6 An automobile manufacturer wants to estimate μ, the average amount of damage that one of its new models will sustain when colliding head-on with a stationary object at 7.5 miles per hour. Due to the destructive nature of obtaining observations, a small sample is employed. A test of eight cars chosen at random yields $\overline{X}$ = average repair cost = \$281 and S = \$84. Assume repair costs are normally distributed.
a. What is the value of the point estimate for μ?
b. If we wished to construct a 95% confidence interval for μ, what would be the appropriate t-score?
c. Construct a 95% confidence interval for μ.

11.7 Refer to Exercise 11.6 and answer each statement following as "true" or "false."
a. Before the sample is drawn, the probability is .95 that the resulting interval will include μ, the population mean repair cost.
b. The population mean, μ, is a random variable.
c. After the interval is constructed, the probability of .95 referred to in part a no longer applies.
d. After the confidence interval is constructed, the population mean μ is either included or not—we don't know which.
e. Had we computed a 99 percent confidence interval, it would be narrower than the 95 percent interval but with a lower risk of failing to include μ.

11.8 A prestigious subdivision has 151 residences. During the last 12 months, seven residences were sold, at the following prices, where X = sale price, to nearest thousand dollars:

214	199	209	214	225	204	219

a. Verify that $\overline{X}$ = 212 and S = 8.8.
b. Assuming that the sample can be viewed as a random sample of subdivision residence values, compute a 90% confidence interval for the subdivision residence mean.
c. What assumption is necessary in developing the interval in part b?
d. If you applied the finite population correction term, would the confidence interval be narrower or wider? By how much?

11.9 An employee at the city water department draws a sample of treated water each weekday to monitor the hardness (level of calcium carbonate) in the local water system. The 21 readings obtained over the past month show an average of 12.3 parts per million of calcium carbonate, with a standard deviation of 1.2 parts per million. Assuming the readings are taken from a normal population, provide a 99% confidence interval for the population mean.

11.10 The National Highway Traffic Safety Administration of the United States Department of Transportation requires the automobile manufacturers to provide information on stopping distances for each car model. Further breakdowns are required for various option and accessory packages as well as for different load conditions. Suppose 24 measurements are obtained randomly for a new car model with a maximum load. If the sample data yield $\overline{X}$ = average stopping distance from 60 mph = 209.0 feet with S = 7.1 feet, develop a 90% confidence interval for the average stopping distance for this car model under the maximum load condition.

11.11 Twenty samples of fishing line are tested from a recent production run. The variable of interest, X, is breaking strength of the line, measured in pounds. The samples show the following (sorted) X-values:

46	47	48	48	48	49	49	49	49	49
50	50	50	50	51	51	51	52	53	55

a. Demonstrate that $\overline{X}$ = 49.75 lbs. and S = 2.07 lbs.

b. Assuming that the population from which this sample was obtained is approximately normally distributed, establish a 95% confidence interval for the population mean.

11.12 In a test of H_0: μ = \$27.50 against a lower-tail alternative, a sample of 14 items shows a sample mean of \$26.10 with a sample standard deviation of \$1.96. Assuming α = .10, consider the following.

a. What value from the t-table separates the acceptance region from the rejection region?

b. Perform the statistical test indicated.

c. Determine, if possible, the upper and lower limits of the p-value for this test.

d. Find the value of $\overline{X}$ that separates the acceptance region from the rejection region and show that the unstandardized approach to reaching a conclusion yields the same result as obtained above.

11.13 A not-for-profit consumer interest group tested several food products for sodium content. One product analyzed was a national brand of chocolate chip cookie that, according to the package label, contains 80 milligrams of sodium per 1-ounce serving. Let X = sodium content in milligrams per 1-ounce serving. A dozen packages are obtained, with the following results of chemical analysis for a 1-ounce serving from each package:

88	86	81	76	72	82
82	74	84	80	87	84

a. What is the entity in this problem?

b. Verify that S = 5.09 and $\overline{X}$ = 81.33.

c. Do you believe the value stated on the package is accurate? Test the appropriate hypothesis, using 1% risk of incorrectly rejecting H_0.

11.14 In the past, workers at a production plant have put in an extra 8-hour shift in weeks where the backlog of unprocessed orders exceeds a certain level. The plant manager is considering other options such as temporary help to handle work in peak periods. The manager suspects that productivity declines substantially when an extra workday is added to the week. If so, the premium pay given for this overtime work may not be a wise expense. Over the past year, the plant production index (a measure of pieces produced per labor hour) has averaged 820 on shifts in weeks where there was no overtime. A total of 13 overtime shifts were worked in the previous year. Assume that these 13 shifts constitute a random sample of past and future overtime experience.

a. Set up hypotheses so that rejecting H_0 would confirm the plant manager's belief.
b. The production index on the 13 overtime shifts is

581	590	618	642	674	701	747	787	811	839	849	855	861

Test your H_0 with 5% risk of Type I error.

11.15 A large real estate company offers a course for adults interested in obtaining a broker's license. The course has been taught by a combination of classroom instruction and student reading assignments (programmed learning). The company is interested in trying a new method of instruction—one that replaces some of the reading and classroom work with prepared videotape lessons that the student can watch at home as many times as desired. The company will adopt the new instructional technique if a convincing case is made that scores obtained on a standardized test taken at the end of the course are higher under the new method. In the past, scores obtained on this test have averaged 78. A random sample of 23 adults is taught by the new approach.
a. To perform a statistical test, what should be the null and alternative hypotheses?
b. If a 5% risk level is employed, what is the table cutoff value that separates the two different conclusion regions?
c. Report the results of the test when $\overline{X} = 81.8$ and $S = 6.9$. Has a convincing case been made for the adoption of the new teaching method?

11.16 Refer to Exercise 11.15.
a. In terms of the problem setting, how would a Type I error be described?
b. Determine upper and lower limits of the p-value for this test.

*11.2 PROCESSOR: SMALL SAMPLE INFERENCES FOR μ

This processor can compute confidence intervals for μ and/or the test statistic t^* for a hypothesis test about the value of μ. This processor is intended for sampling situations where the sample size is less than 30 and where we assume that the underlying distribution of X follows a normal curve. If the user specifies a sample size of 30 or more for a confidence interval, a Z-score will be used in the computations instead of a t-score. However, the hypothesis test p-values reported are for the t-distribution regardless of the sample size.

Processor selection, table reading, data entry, and menu choices for this processor closely parallel those for the large sample processor discussed in Section 10.4. To illustrate confidence interval construction, let us refer to Example 11.4 where we wished to develop a 95% confidence interval for the population mean. After entering the summary statistics ($n = 12$, $\overline{X} = 230$, $S = 19$) and advancing through portions of the Results option, we will obtain a screen offering five different confidence levels for interval construction: 80, 90, 95, 98, and 99% confidence. If we select 95% confidence, we will generate a results screen similar to Figure 11.6.

*Optional

```
≡                    Small Sample Inferences For μ (n < 30)                    QUERY
  Introduction  *able  Data  |Results|  Exit                              | F1=Help

            For 95% confidence, the interval is
                 point estimate ± t * (standard error)

                 =  230.000 ± t * (19.000 / √n)
                 =  230.000 ± 2.201 * (19.000 / 3.464)
                 =  230.000 ± 12.072

            The lower limit of the confidence interval is  217.928
            The upper limit of the confidence interval is  242.072

            ══════════════════ Confidence ══════════════════

                    Compute another interval? (y/n) No
```

Figure 11.6 Confidence Interval Results for Data in Example 11.4

To demonstrate the Hypothesis Test feature, let us reconsider Example 11.6, a test of H_0: $\mu = 82$ against a lower-tail alternative. After furnishing the summary statistics ($n = 6$, $\overline{X} = 77.2$, and $S = 3.7$) and hypothesized value of μ, we will be presented with results, as in Figure 11.7. Note that the test statistic is t^* and that the p-value for our "t-test" is reported to twelve decimal-place accuracy.

```
≡                    Small Sample Inferences For μ (n < 30)                    READY
  Introduction  *able  Data  |Results|  Exit                              | F1=Help

   Hypothesis test       Ho: μ =  82.000
                         Ha: μ <  82.000

                         sample mean - hypothesized value of μ
          TS:   t* = ---------------------------------------
                                   standard error

                              77.200 -  82.000
                    t* = ------------------ = -3.178
                              3.700 /  2.449

            The p-value for this test is 0.012300551108


   The decision to "reject Ho" or "do not reject Ho" depends on your
   judgment or tolerable risk of Type I error (see text Figure 10.18).
```

Figure 11.7 Hypothesis Test Results for Data in Example 11.6

*11.3 THE SIGNED RANKS PROCEDURE FOR HYPOTHESIS TESTS

In the first section of this chapter, we learned how to use the t-distribution to test a hypothesis about the value of μ when we have a small sample. Such a procedure often is called a *t-test*. We now introduce the *signed ranks test* as an alternative to the t-test. The signed ranks procedure is becoming more widely used and has these advantages over the t-test:

1. Its computations are often quicker and easier.
2. It has less restrictive assumptions. The t-test assumes the population of interest is normally distributed. In other words, use of the t-distribution assumes a mound-shaped, symmetrical population that follows the normal curve formula given in Equation 7–1. If the normality condition is not met, the results of using the t-distribution may not be reliable. In contrast, the signed ranks test assumes only population symmetry, a condition easier to satisfy than normality.
3. Its power (probability of detecting a false H_0) is higher than that of the t-test when the sample size is relatively small.

Unlike the tests where we compute Z^* or t^*, the signed ranks test permits us to test a hypothesis about the value of μ *without* computing the sample mean or standard deviation.

We will illustrate the signed ranks test by reconsidering Example 11.5, where we wished to test, for a new plant,

$$H_0\text{: } \mu = 93.5\% \text{ yield}$$
$$H_a\text{: } \mu \neq 93.5\% \text{ yield}$$

We were to use $\alpha = .05$ and a sample size of 12 values of X = yield per batch to ascertain whether H_0 was supported or contradicted. The 12 values of X from Example 11.5 are listed in column (1) of Table 11.2. To begin the computations, we subtract the hypothesized value of μ ($\mu_0 = 93.5$ in this case) from each X, as indicated in column (2). If any X equals μ_0, thus resulting in a 0 in column (2), that item will be dropped from any further computations with the sample size being reduced accordingly. Column (3) shows the absolute values of the column (2) entries. We are about to order, or rank, the numbers in column (3) from low to high. The ranking process could be done without creating column (3), but we are less likely to make an error if we set up this column which temporarily ignores the sign of each difference. Column (4) shows the (absolute) ranks, obtained by assigning the rank of 1 to the smallest value in column (3) and then working up to 12 for the largest value. We then restore the plus or minus sign of each difference onto the ranks, as shown in column (5). The final calculation needed is to determine T_+, the sum of the ranks with positive signs,

*Optional

Table 11.2 Calculations for Signed Ranks Test for H_0: $\mu = 93.5$

(1) Batch % Yield, X	(2) $X - \mu_0$	(3) $\lvert X - \mu_0 \rvert$	(4) Absolute Rank	(5) Signed Rank
91.41	−2.09	2.09	12	−12
94.47	.97	.97	11	+11
92.80	−.70	.70	9	−9
92.93	−.57	.57	6	−6
94.19	.69	.69	8	+8
93.37	−.13	.13	1	−1
92.92	−.58	.58	7	−7
93.16	−.34	.34	4	−4
93.24	−.26	.26	3	−3
92.70	−.80	.80	10	−10
93.66	.16	.16	2	+2
93.86	.36	.36	5	+5

$T_+ = 11 + 8 + 2 + 5 = 26$

$T_- = 12 + 9 + 6 + 1 + 7 + 4 + 3 + 10 = 52$

and T_-, the sum of the ranks with negative signs. These two values are shown at the bottom of Table 11.2. (Although adding the ranks with negative signs produces a negative sum, we define T_- to be the absolute value of this sum.) An arithmetic check to keep in mind is that the sum of T_+ and T_- must equal $n(n + 1)/2$; in this case, $26 + 52 = 12(13)/2 = 78$.

The signed ranks test employs the following rationale: If H_0 is in fact a true statement, the differences in column (2) should vary randomly around 0. In other words, a positive difference of a given magnitude is just as likely to occur as a negative difference of the same magnitude. We would therefore anticipate that T_+ and T_- should be roughly equal when H_0 is true. On the other hand, a strong case against H_0 will be made when T_+ and T_- are dramatically out of balance.

Included in Appendix F is a table of cutoff points for the signed ranks test. We will now refer to this table to see if our example's results ($T_+ = 26$, $T_- = 52$) constitute a strong case against H_0. As the information at the top of the table indicates, we are to use the risk column headings at the top of the table when H_a is two-tailed. We then intersect the α-column of interest with the appropriate sample size row. In this example, we find the cutoff point of "13" at the intersection of the .05 column and the row for $n = 12$. For two-tailed tests, we reject H_0 if the smaller of T_+ or T_- is *equal to or less than* the cutoff point. In this example, the smaller sum is $T_+ = 26$. Since T_+ is greater than the cutoff ($26 > 13$), H_0 is accepted. In order for us to conclude beyond a reasonable doubt ($\alpha = .05$) that the mean yield is not 93.5 percent, we would require the smaller total to be 13 or less. Note that we have no need for the larger total ($T_- = 52$) other than its value serves as the computational check in this expression: $T_+ + T_- = n(n + 1)/2$. The conclusion to accept H_0 was also obtained for these data when we worked Example 11.5 using the t-distribution. The signed ranks test arrives at its conclusion by utilizing ranks instead of sample statistics such as the mean and standard deviation. In addition, the signed ranks test does not assume a normal population as does the t-test.

COMMENTS

1. The signed ranks procedure is also known as the "Wilcoxon signed ranks test," after the American scientist Frank Wilcoxon, who introduced it in 1945.
2. Since the signed ranks test makes only a minimal assumption (symmetry) about the underlying population and does not utilize parameter estimates such as $\overline{X}$ and S, it is called a *nonparametric* procedure. The signed ranks test is the nonparametric counterpart of the t-test, which is a *parametric* procedure.
3. Recall that the median and the mean are the same in a symmetrical distribution. Since the signed ranks test assumes population symmetry, it is simultaneously a test for the location of both the median and the mean.
4. Often a preliminary test is conducted on the sample data to shed light on the assumption of a symmetrical population. Several different tests are available for this purpose. In Exercise 11.26 we outline one such test which is based on the coefficient of skewness, first encountered in Chapter 3.
5. In the process of determining ranks, it is possible to have "ties," where two or more items are equidistant from the H_0 value. This problem is resolved by assigning an average rank to the tied items. This situation arises in the following example.

EXAMPLE 11.7

Because of higher interest rates, the manager of a real estate office believes that the average time on the market before a house is sold probably has increased over the last 12 months. To put her suspicions to the test, a random sample of 15 recent transactions is selected and the time on market for each is established. It is known that a year ago the average time from listing to sale was 84 days. The sample results for the variable X, time on market prior to sale, in days, are

170	119	38	87	84	177	99	141	14	70	110	133	209	131	58

The variable X is believed to have an approximately symmetrical distribution. Conduct the proper statistical test with 10% risk of incorrect rejection.

Solution

We wish to use the signed ranks procedure to test

$$H_0\colon\ \mu = 84 \text{ days (mean has not changed)}$$
$$H_a\colon\ \mu > 84 \text{ days (mean has increased)}$$

Our arithmetic is shown in the following table. We have sorted the X's prior to listing them in the first column since this may make the ranking process easier. Compared to the product yield example, we note two differences. First, one value ($X = 84$) is equal to the value stated in the null hypothesis. This value is now ignored, with the sample size being reduced to 14 items. Second, two values ($X = 58$ and $X = 110$) have the same absolute difference from the H_0 value. We must therefore assign the same absolute ranks to these tied differences. Since these two values have the fourth and fifth lowest absolute difference, we will give each the absolute rank of $(4 + 5)/2$, or 4.5. To continue the ranking process, the next smallest difference will be assigned the rank 6. Assigning the average rank to tied observations ensures that the sum of T_+ and T_- will remain equal to $n(n + 1)/2$.

X	$X - \mu_0$	**Absolute Rank**	**Signed Rank**
14	−70	11	−11
38	−46	7	−7
58	−26	4.5	−4.5
70	−14	2	−2
84	0	dropped from analysis	
87	3	1	+1
99	15	3	+3
110	26	4.5	+4.5
119	35	6	+6
131	47	8	+8
133	49	9	+9
141	57	10	+10
170	86	12	+12
177	93	13	+13
209	125	14	+14

Using the signed rank column, we obtain $T_+ = 80.5$, $T_- = 24.5$ and show that $80.5 + 24.5 = 105 = 14(15)/2$. We are now ready to use the table of cutoff values. Since the H_a is one tailed, we will use the .10 risk column near the bottom of the table. Our directions tell us that when H_a is upper tailed, we will reject H_0 when T_- is equal to or less than the table value. In this case, T_- is 24.5, and the table value is 31 when $n = 14$. Since $24.5 < 31$, we believe H_0 is unlikely to be true—a strong case has been made that the average time of houses on the market has increased.

□

The box on page 552 summarizes the computations involved in carrying out the signed ranks test.

Summary of Computational Steps for the Signed Ranks Test

1. Create a listing of differences by subtracting μ_0 from each value of X.
2. Discard any zero differences and reduce the sample size accordingly.
3. Temporarily ignore the plus or minus signs on the differences.
4. Rank these absolute differences by assigning 1 to the smallest, 2 to the next smallest, and so on.
5. Resolve any ties in the ranking process by giving an average rank to the tied items.
6. Assign the original plus or minus sign to each rank.
7. Obtain T_+ by summing the plus ranks; obtain T_- by determining the (absolute) sum of the minus ranks.

COMMENTS

1. The concept of p-values also applies to signed ranks tests. For instance, in Example 11.7 above, our sum T_- would cause rejection of H_0 when $\alpha = .05$ (since $24.5 < 25$) but not at $\alpha = .025$ (since $24.5 > 21$). The p-value is therefore between .025 and .05 for this test.
2. The values for α (column headings) in the signed ranks table are not necessarily exact. For instance, consider a two-tailed test with a desired risk of $\alpha = .05$ and a sample size of 9 observations. As you can verify from the table, H_0 will be rejected if the smaller sum is 5 or less. However, when H_0 is true the chance that the smaller sum will be 5 or less is in reality about .040 (the computation of this value is too detailed to discuss at this point), somewhat smaller than the desired $\alpha = .05$. But if we were to raise the cutoff point up to 6, the exact α would be .054, a slightly larger risk than desired. Unlike a Z-test or a t-test, the risk of exactly .05 is simply not available here. Some texts will give the table value as 6 in this case, since 6 is the cutoff point whose α is *nearest* the .05 level. We adopt the more conservative policy that the cutoff value should be one so that the column's risk factor cannot be exceeded.
3. Should there be a three-way tie in the rankings, the average rank is given to all 3 observations. For instance if there is a three-way tie for the sixth lowest difference, each observation would receive the rank of 7, since the three observations would occupy the sixth, seventh, and eighth positions. The average rank of $(6 + 7 + 8)/3 = 7$ is given to each.

SECTION 11.3 EXERCISES

11.17 Find the cutoff value from the signed ranks table to use when we have the following:

a. a lower-tailed test with $n = 11$ and $\alpha = .05$.
b. a two-tailed test with $n = 17$ and $\alpha = .10$.
c. an upper-tailed test with $n = 7$ and $\alpha = .10$.

11.18 A sample of eight items is drawn from a population in order to test the presumption that the population's mean is not different from 240. Assuming the population is symmetric, conduct the test with $\alpha = .10$. The sample observations are 249, 245, 260, 253, 232, 236, 268, 257.

11.19 A city water department states that the average hardness of water entering the lines to customers is less than 15 parts per million of calcium carbonate. Assume you wish to test H_0: $\mu = 15$ ppm versus a lower-tailed alternative, using a random sample of 15 readings. Use the signed ranks test with 5% risk of a Type I error. Can H_0 be rejected, thus providing evidence that the claim is correct? The water samples give these results for the variable X, ppm of calcium carbonate:

15.5	14.9	13.6	13.1	13.8
14.2	14.7	15.9	13.7	15.2
12.9	15.1	12.6	13.6	13.2

11.20 A real estate agent has stated that the average property value in a certain subdivision is about \$215,000. Suppose in the last 12 months that seven residences in the subdivision have changed hands at the following prices (to the nearest thousand dollars):

214	199	209	214	225	204	219

Assuming these data for X = sale price can be viewed as a random sample of subdivision home values, test the statement that the average value is \$215,000 against a two-sided alternative, using the signed ranks procedure. Choose a commonly used level of significance and indicate whether the agent's statement is supported or contradicted.

11.21 The numbers in the following table represent X = miles per gallon on the first 30 tankfuls of a new economy car. Assume that these figures have been sorted and will be a random sampling of the fuel economy to be realized over the life of the car. Using the signed ranks test and a risk factor of your own choosing, determine whether or not it is reasonable to believe that the average miles per gallon over the long run will be 30 miles per gallon. Let your alternative hypothesis be that 30 miles per gallon is an overstatement of the fuel economy.

27.64	27.88	27.94	28.18	28.19	28.57
28.61	28.64	29.12	29.12	29.12	29.60
29.63	29.63	29.73	29.91	29.95	30.12
30.12	30.18	30.22	30.40	30.42	30.44
30.44	30.48	30.64	30.71	31.06	31.21

11.22 What boundaries can be placed on the p-value for the test in Exercise 11.21?

11.23 A downtown hotel has recently added a computer "video checkout" service by which guests can check out from their rooms as an alternative to performing this task at the front desk.

The checkout system permits the guest to view an itemized bill on the television screen and, by pushing a button on the TV set, approve its being charged to the guest's credit card. The company which set up the system stated that the average time from requesting the checkout program to completing the checkout process should be no more than 3 minutes. Suppose you are to obtain a random sample of 20 video checkout times. Set up hypotheses so that rejection would indicate that the average checkout time is more than 3 minutes. Use the signed ranks test with 5% risk of Type I error. The sample results for X = time in minutes to complete a video checkout are

1.12	1.28	1.91	2.07	2.74
2.76	3.04	3.17	3.44	3.79
3.81	3.99	4.07	4.19	4.54
4.91	5.06	5.13	5.67	6.00

11.24 Refer to Example 11.23 above.

a. Consider the last four values of X listed above. If you were told only that these values were all above 5.00 minutes, but you were not permitted to see the exact figures, could you still perform the desired statistical test?

b. What boundaries can be placed on the p-value for the test?

11.25 As part of her job, a supervisor for an interstate carrier prepares estimates of moving cost for individual households. A key factor in the total bill estimated is the weight of the household goods. The supervisor monitors her assessments by computing X = the ratio of estimated weight to actual scale weight for each job her company receives. Below are her most recent 15 ratios. Suppose these can be considered as a random sample of her performance. With $\alpha = .10$, test the assumption that her average estimate is correct, using the signed ranks test.

.960	.983	.904	.993	1.088
1.021	.966	1.042	.954	1.065
.921	.969	.987	.957	1.058

***11.26** In Chapter 3, we introduced a measure of symmetry called the *coefficient of skewness*. When this quantity is equal to zero, the data set is perfectly symmetrical. We denoted the sample coefficient of skewness by g_1 and the population coefficient by γ_1. Using the computed value of g_1 and the hypothesis testing principles we developed in Chapter 10, we can set up a test about γ_1 that would enable us to infer whether a population from which we have obtained data is symmetrical. The first 3 elements of the test are:

H_0: $\gamma_1 = 0$ (population symmetrical)

H_a: $\gamma_1 \neq 0$ (population skewed)

TS: $g_1 = m_3/m_2^{3/2}$

*Optional

where m_3 and m_2 are defined by Equations 3–12a and 3–12b, respectively. Intuitively, if g_1 is close to the value 0, we infer the population is symmetrical. A sample result close to 0 is synonymous with being inside the cutoff points of the acceptance region for H_0. The following table provides the cutoff points for the absolute value of g_1 that separate the acceptance and rejection regions for an α-risk of .05:

Cutoff Points for g_1

n	Cutoff Point	n	Cutoff Point
6	1.239	16	1.018
7	1.230	17	.997
8	1.208	18	.978
9	1.184	19	.960
10	1.159	20	.942
11	1.134	21	.925
12	1.109	22	.909
13	1.085	23	.894
14	1.061	24	.880
15	1.039	25	.866

Source: Excerpted, by permission of the Biometrika Trustees, from H. P. Holland, "On the Null Distribution of $\sqrt{b_1}$ for Samples of Size at most 25, with Tables," *Biometrika,* Vol. 64, 1977, p. 408.

For example when $n = 6$, if $-1.239 \leq g_1 \leq 1.239$, we would accept H_0 and infer the population is symmetrical. But if $g_1 < -1.239$ or if $g_1 > 1.239$, we would reject H_0 and declare the population to be nonsymmetrical. When we reject H_0, we should *not* perform the Wilcoxon signed ranks test because we have strong evidence that a key Wilcoxon assumption does not hold.

Suppose a random sample of size $n = 15$ generated a sample skewness coefficient of $g_1 = 1.21$. Make an inference about the symmetry of the underlying population. Should the Wilcoxon signed ranks test be conducted using the sample data set?

***11.27** For each sample size and sample skewness coefficient following, determine the cutoff points for inferring the symmetry of the population and make the appropriate inference:

a. $n = 12$; $g_1 = -.873$
b. $n = 20$; $g_1 = -1.116$
c. $n = 24$; $g_1 = .209$

***11.28** Make a decision about the symmetry of the population that generated the following random sample:

200	225	267	199	227
150	104	183	213	209
175	249	202	204	183
164	127	141	199	218
264	152	183	174	238

(*Hint:* Compute g_1 first.)

*11.29 A business junior college offers programs specializing in office administration. Graduates from the word processing program are touted as being quick and accurate. A random sample of 20 students' words typed per minute produced the following data:

62	65	63	57	60
56	51	53	60	57
57	60	58	60	58
59	57	57	55	55

Can we conclude that the distribution of words typed per minute for all office administration students at this junior college is symmetrical? Explain your reasoning.

*11.30 Refer to Exercise 11.29. The junior college advertises that its graduates type at least 62 words per minute on average. Set up a hypothesis test that assumes the junior college advertisement is truthful. Do the sample data support or discredit this claim? Use the signed ranks test and $\alpha = .10$.

11.4 QUALITY CONTROL AND CONTROL CHARTS

One specialized area of application of statistics in business is quality control. Firms which have a quality control program usually utilize statistical techniques such as taking sample measurements, computing averages and measures of variability, graphing the data, and drawing conclusions based on sample evidence. Individuals who work in quality control also may have responsibilities in establishing product specifications and standards.

Many people think of quality control as a final check or inspection of a finished product, and certainly many such tests are performed at the very end of the production process. However, quality control procedures also may be applied to incoming raw materials and supplies as well as to items at various stages of completion in the manufacturing process.

There is a variety of quality control techniques, and different firms will employ those best suited to their needs. Perhaps the most common quality control concept is the process *control chart*. We will limit our attention to some different types of control charts in this section. We will see that these charts can be viewed as a graphical means of choosing between two opposite statements: The process is "in control," or the process is "out of control."

COMMENTS

1. Firms may call their quality control functions by names such as process control, statistical process control (*SPC*), statistical quality control, or quality assurance.
2. According to the 1987 American Society for Quality Control/Gallup survey (reported in the December 1987 issue of *Quality Progress*), small businesses and service companies are less aware of quality as a business strategy and are less likely than larger firms and industrial companies to have quality control procedures in place.

Process Variability

Suppose we obtain a random sample of several bags of corn chips that have been packaged within the last hour at a food processing plant. Let's consider three questions about our sample that might be of interest to management:

1. What are the net weights of the chip packages? The packages are labelled as containing 16 ounces (454 grams); management does not want to either overfill or underfill the packages, on average.
2. What is the sodium content per ounce? Some consumers make purchase decisions based on the nutritional information printed on the package. If the label states that the sodium content is 240 milligrams per one-ounce serving, management wants this to be accurate.
3. What proportion of the chips are broken? Although broken chips may taste the same as perfect ones, consumers clearly prefer whole chips.

The questions above involve different measures that contribute to the overall quality of the product. The first two questions relate to continuous variables (net weight and sodium content); the third question deals with a binomial variable (each chip is either broken or unbroken).

Regardless of the characteristic of interest, a fact of life in manufacturing is that not all units produced will be identical; measurable variation is unavoidable and to be expected. In quality control, we take the view that there are two types of variation possible in the measurements or counts we make: random, or *chance,* variation and *assignable* variation.

Chance variation is the result of naturally occurring phenomena. Examples of causes of chance variation could include such things as the following:

- Fluctuations in the temperature or humidity in the production facility
- Slight variations from one piece of production equipment to another
- Minor differences in raw materials obtained from different suppliers (For instance, our corn chip producer above has more than one supplier of corn, corn oil, cheeses, and so on. Even for the same supplier, there may be small differences in the same raw material from one shipment to the next.)

Many causes of chance variation are not traceable and therefore remain unknown. Though usually small in magnitude, chance variation cannot be avoided or controlled to any appreciable degree. Therefore we will anticipate random variation for any process. Since chance causes cannot be eliminated, we will want the quality control system to ignore this type of variation.

Assignable variation, on the other hand, is due to a specific cause such as operator error, a faulty or miscalibrated piece of equipment, or nonstandard raw materials. Assignable variation usually can be traced or attributed to its origin and, unlike chance variation, can be eliminated by making adjustments or taking some corrective action.

Since variation due to an assignable cause may result in lower grade quality or units which fail to meet engineering or customer specifications, we will want the quality control system to alert us to such a condition.

Control Charts

In order to keep track of key properties of a manufacturing process, we rely on a graphical tool called a control chart. Many control charts have the appearance of Figure 11.8 (we have placed some hypothetical data points on this chart). The center line indicates the process average, a number which often is based on historical data or engineering specifications. The lines above and below the center line represent the upper and lower *control limits* for the process. To monitor the process, we draw random samples, usually of the same size and at regular intervals of time. Each dot in Figure 11.8 therefore denotes a separate sample. The points are plotted from left to right over time and may or may not be connected.

As long as the points on the chart stay within the upper and lower control limits, there is reason to believe that the process is stable or in control. In other words, the variation observed is attributed to random, or chance, causes. If a point should happen to fall outside the control limits (see the last observation plotted on Figure 11.8), we take this as evidence that some nonrandom or assignable cause is present. Until the cause of this large deviation is located and corrected, the process may remain out of control and may be generating a product which is of unacceptable quality.

A quality control chart exemplifies "management by exception." This is because we set up probability controls so that chance variation is ignored. However, should some major change in the process occur, the next sample drawn is likely to fall outside the control limits. By its placement on the control chart, such an observation should alert management that the probability is high that some nonstandard condition exists.

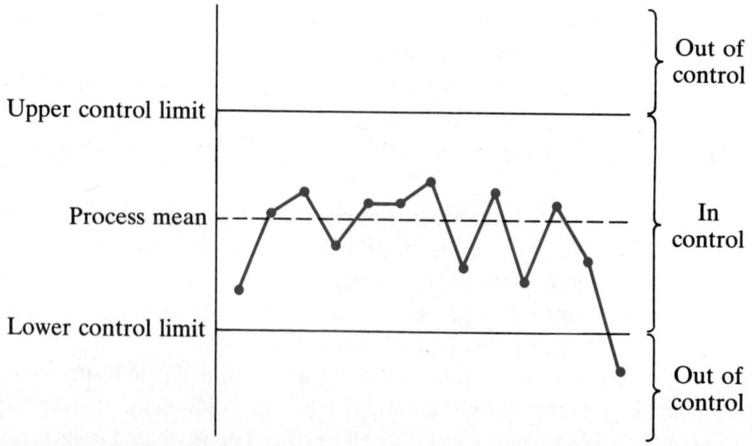

Figure 11.8 General Appearance of a Control Chart

Relationship to Hypothesis Testing Process control is an application of the hypothesis testing principles that we have been discussing in the last three chapters. The test is based on these general hypotheses:

H_0: The process is in control.

H_a: The process is out of control.

Depending on the characteristics to be monitored, the test statistic might be the sample mean $\overline{X}$, the sample range, or the binomial sample proportion p. In order to determine the rejection region we need to know the sampling distribution of these sample quantities. In Chapter 8, we studied the sampling distribution of $\overline{X}$. A quality program monitoring the average will utilize the results of Chapter 8. (We will not present the sampling distribution of the range; instead we will direct the interested reader to a text devoted exclusively to quality control. The sampling distribution of the sample proportion p will be discussed briefly in this section, with a more thorough treatment in Chapter 12.)

From Chapter 8, we learned that $\overline{X}$ has an approximately normal sampling distribution under certain conditions. Thus, rejection regions for testing whether the process is in control can be found in the tails of the normal distribution, as indicated in Figure 11.9. The conclusion about the condition of the process is either "accept H_0," implying that the process is in control, or "reject H_0," implying the process is out of control.

To facilitate the use of this statistical test in a manufacturing or plant setting, we add another dimension to this figure to represent the collection of different samples over time—see Figure 11.10. In Figure 11.10 the normal curve is positioned above the page. The front view of this figure is identical to Figure 11.9. A top view of this figure is identical to Figure 11.8. Therefore, the control chart in Figure 11.8 is merely the top view of the acceptance and rejection regions for the test.

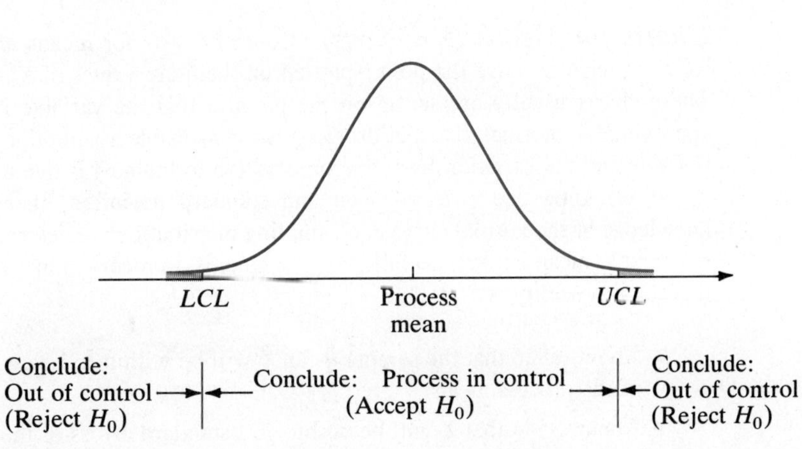

Figure 11.9 Accept and Reject Regions for Quality Control

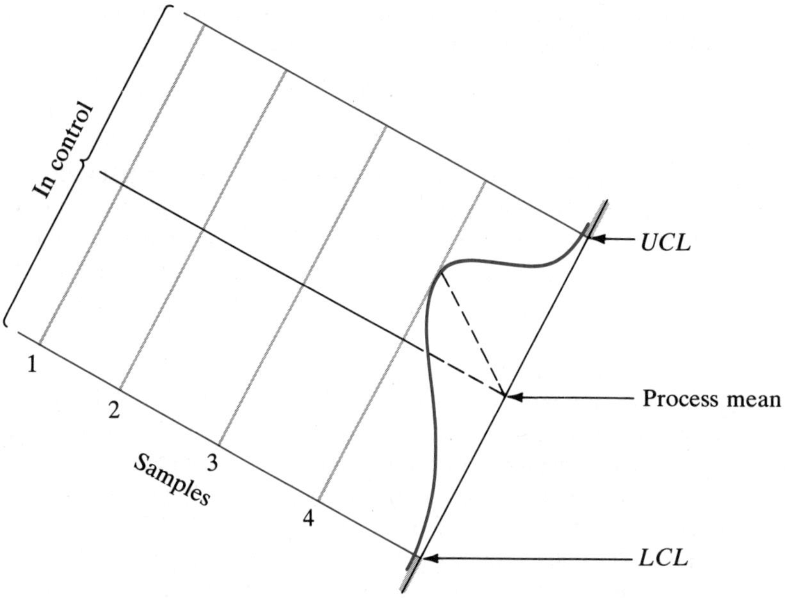

Figure 11.10 Creating a Control Chart by Adding a Dimension to Figure 11.9

By downplaying the formal structure of the hypothesis test—null and alternative hypotheses, the sampling distribution, and the rejection region—we create a graphical tool that can be applied with a minimum of training. It is management's responsibility to set up the control charts properly and to educate the appropriate employees about how to use them.

Charts for Means ($\overline{X}$-Charts) Control charts for means are called $\overline{X}$-charts (or X-bar charts) since the points plotted on them are values of $\overline{X}$, the sample mean. These charts usually are set up on the premise that the variable in question has an approximately normal distribution. This is a reasonable assumption when the process is stable and the variation from one observation to the next is due to chance causes.

If we know the process mean and standard deviation, then we can use our knowledge of the normal curve and sampling distributions to determine regions where the sample mean is likely to fall. For instance, if a process is in control we can state that the probability is

- About .6826 that the sample mean $\overline{X}$ will be within ± 1 standard error ($\sigma/\sqrt{n}$) of the process mean
- About .9544 that $\overline{X}$ will be within ± 2 standard errors of the process mean
- About .9974 that $\overline{X}$ will be within ± 3 standard errors of the process mean

In the United States and many other countries, it is common practice to set the upper and lower control limits of the control charts at ± 3 standard errors. This means that when a process is in fact in control, a random sample has a probability of .9974 of correctly indicating so, with a complementary probability of .0026 of signaling that we are not in control. This low risk of a "false alarm" for a process that is in control is desirable. We want the probability to be very small that a sample would tell us to look for an assignable cause of variation when in fact there is only chance variation present.

Figure 11.11 shows a portion of an $\overline{X}$-chart used by a manufacturer of paper products when producing tubs, or buckets, for use by a fried chicken retailer. The variable of interest is the rim diameter of these containers as measured by a digital

Date		11/28								
Time		7am	8	9	10	11	12pm	1		
Indiv.	1	90	95	85	94	89	94	91		
	2	88	90	95	90	93	86	86		
	3	95	87	85	91	93	93	88		
	4	87	86	87	92	91	99	86		
	5	87	95	85	91	93	95	94		
Total		447	453	437	458	459	467	445		
Average		89.4	90.6	87.4	91.6	91.8	93.4	89.0		
Range		8	9	10	4	4	13	8		

Code:
90 = 8.790

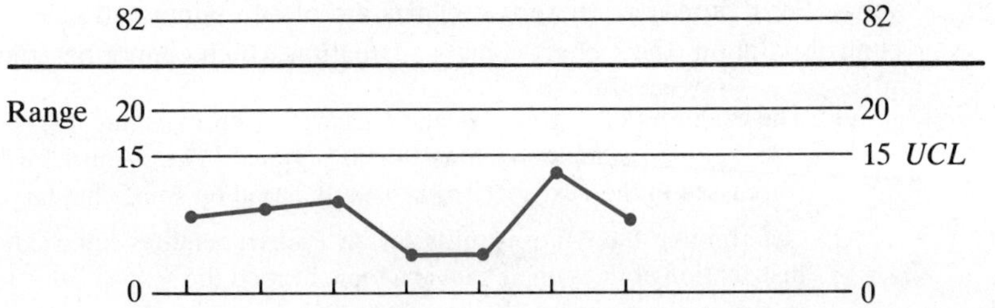

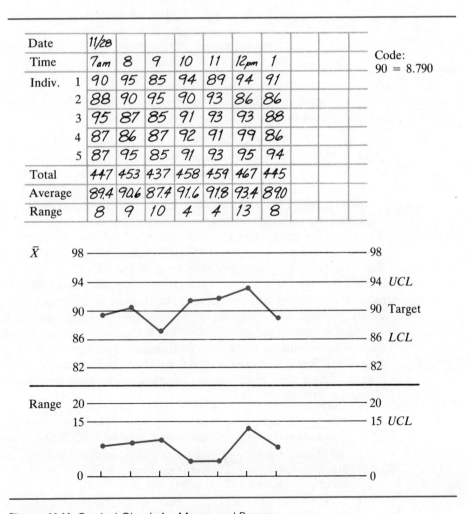

Figure 11.11 Control Charts for Mean and Range

electronic gauge. Once per hour, a random sample of five newly produced tubs is selected, and their individual measurements are recorded. Summary measures such as the sample average and the range are then computed and plotted.

In this case, the specified ideal, or target, diameter is 8.790 inches. Note on the chart that an actual reading of 8.790 would be "coded" as 90: Just the last two numbers of each reading are recorded to avoid needless writing of the entire number. From past experience, the firm knows that the standard deviation of rim diameters is .003 inch. Using the relation that the control limits will be the process mean ± 3 standard errors, we can verify the limits shown for the $\overline{X}$-chart as follows:

$$\begin{aligned}\text{Upper Control Limit} &= \text{Process mean} + 3 \text{ Standard errors}\\ UCL &= 8.790 + 3(.003)\sqrt{5}\\ &= 8.790 + .004\\ &= 8.794 \text{ (coded as 94)}\end{aligned}$$

$$\begin{aligned}\text{Lower Control Limit} &= \text{Process mean} - 3 \text{ Standard errors}\\ LCL &= 8.790 - .004\\ &= 8.786 \text{ (coded as 86)}\end{aligned}$$

A quick inspection of the $\overline{X}$-chart portion of Figure 11.11 shows that none of the points is outside the control limits.

Charts for Ranges (R-Charts) As the bottom portion of Figure 11.11 suggests, an $\overline{X}$-chart is often accompanied by a second chart called an R-chart (or range chart). While the $\overline{X}$-chart is used to monitor the process mean, the R-chart tracks the dispersion of the process.

In Chapter 3 of the text, the first measure of dispersion we discussed was the range. Defined as the difference between the largest and smallest values in a data set, the range is an easily computed and understood measure. For this reason, control charts for the sample range are much more commonly used to monitor process variability than are charts for another measure of dispersion—the standard deviation.

For the example shown in Figure 11.11 the upper control limit for the sample range is 15. In many applications of R-charts, there is no lower control limit, since we are primarily concerned with too much variability, not too little. It is possible for a process to become out of control on the R-chart (the individual values becoming more variable) while at the same time remaining in control on the $\overline{X}$-chart, or vice versa. For this reason, these two charts are often maintained side by side—an out of control point on either chart signals a situation which cannot be ignored.

COMMENTS

1. The examples of a firm's $\overline{X}$- and R-charts given in Figure 11.11 utilize samples of size 5. Small samples such as this are typical. The p-chart for binomial variables (discussed in the next section) is usually based on somewhat larger samples.
2. Determining the control limits for an R-chart requires knowledge of the sampling distribution of the sample range, a topic beyond the scope of this text. The interested reader is directed to A. J. Duncan (1986).
3. In the tub rim diameter example above, the process standard deviation of .003 inch may seem small. However, another supplier furnishes the retailer with plastic lids for these tubs. The lid supplier is also required to have a small process standard

deviation to ensure that the tubs can be closed tightly; too much variability in either part would result in wasted materials. Many manufactured parts demand even stricter uniformity than required in this example.

4. Commercial software that constructs control charts and plots each sample is widely available. Typically, the user needs to specify the control limits for each chart and then input the sample data as they are gathered.
5. Use of electronic robots to inspect each item produced is beginning to result in 100% sampling, or screening, in a limited number of applications.
6. In Chapter 3, we denoted the range by Rg. However, most quality control applications use just the symbol R to represent the range.

Charts for Proportions (p-Charts) Charts that monitor the process percent or proportion defective are called charts for attributes, or p-charts. These charts are set up and maintained much like $\overline{X}$- and R-charts. They differ in that $\overline{X}$- and R-charts involve measurements while p-charts involve a two-outcome-possible (binomial) attribute for each item inspected—defective or not defective. Figure 11.12 shows a

Date	12/1								
Time	1pm	2	3	4	5	6	7		
Sample size	25								
Defect									
Missing print				1					
Smudged print									
Grease in tub			1			1			
Dirt in tub									
Cracked rim							2		
Creased rim			1				4		
Other	1								
Total	1	0	2	1	0	1	6		
Percent	4	0	8	4	0	4	24		

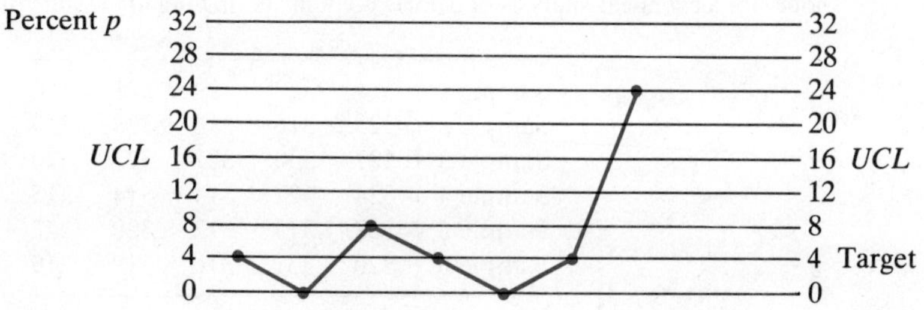

Figure 11.12 Percent Defective Chart

p-chart used by the tub supplier mentioned above. Quality control personnel draw periodic samples of size 25 and inspect for several types of defects that make the tub unacceptable to their customer. Some p-charts such as this one provide details or counts on the frequency of each type of defect found. Like the range chart, this example of a p-chart has no lower control limit. The upper control limit is 16 percent (equivalent to 4 defectives in a sample of 25). Any sample with 5 or more defectives would alert management to an out of control situation. The 7 P.M. sample on this chart was out of control; an investigation found that a forming machine had become out of adjustment, causing jamming of the tubs and resulting in cracked or creased rims.

SECTION 11.4 EXERCISES

11.31 In quality control, what hypothesis is tested each time a sample is drawn? Using the phrases "in control" and "out of control," explain the meaning of a Type I error in a quality control setting.

11.32 Why should points on control charts generally be plotted as soon as possible after the sample has been gathered?

11.33 A manufacturer of electric clothes dryers performs many tests on the machines before they leave the plant. One such inspection involves measuring the air temperature as it enters the drying drum, for a given fabric setting. When in control, the process mean is at the engineering specification of 118° Fahrenheit with a standard deviation of .60°. If a sample of six dryers are checked every 90 minutes for their air temperature, what would be the upper and lower limits of the $\overline{X}$-chart?

11.34 A metal-works firm supplies strips of aluminum to a suitcase manufacturer. The strips serve to hold the two halves of the suitcase together as well as to anchor the locking mechanism. One size of suitcase requires metal strips with a length of 224.400 centimeters. When in control, the aluminum cutting process is on target with a standard deviation of .055 cm. During production, periodic samples of nine strips are measured and the resulting sample means are plotted on an $\overline{X}$-chart. What should be the chart's control limits? Is a sample reading of $\overline{X} = 224.52$ cm indicative of a process in or out of control?

11.35 A candy bar producer knows from past history that the calorie content of its two-ounce size bar is very nearly normally distributed with a mean of 320 calories and a standard deviation of six calories. Suppose a random sample of five newly wrapped bars is chosen every two hours for a chemical analysis of calorie content. Following are results of six separate samples.

Sample 1	314	322	329	321	320
Sample 2	325	316	318	324	319
Sample 3	327	330	322	323	320
Sample 4	324	327	314	314	318
Sample 5	328	319	312	320	322
Sample 6	320	331	316	319	319

a. Construct an X-bar chart by computing the control limits and plotting the sample means. Are any points beyond the control limits?

b. Assume that the corresponding R-chart has no lower control limit, but the upper control limit is 28 calories. Develop an R-chart and determine whether any observations contradict the assumption that the process is in control.

11.36 A packaging process is considered in control when the average weight of product placed in packages is 913.50 grams with a process standard deviation of 2.50 grams. Following are four recent samples of X = net package weight, as determined by a quality assurance technician:

Sample 1	914.1	909.3	914.7	910.6	912.4	911.6
Sample 2	913.0	912.4	910.3	917.1	917.2	913.8
Sample 3	912.4	914.4	916.0	909.1	911.2	910.1
Sample 4	903.1	910.2	914.7	910.4	909.1	909.2

a. Develop an $\overline{X}$-chart by computing the control limits and plotting the four sample means. Is the process in control?

b. Suppose the R-chart for this process has an upper control limit at 14.2 grams. Do any of the R-chart points signal too much process variability?

11.5 SUMMARY

In the first section of this chapter, we considered constructing confidence intervals and testing hypotheses about μ when the sample size is small. We learned that having a small sample poses a new problem. The difficulty is that S, the sample standard deviation, is not a reliable estimator for σ when n is small. However, if we are willing to assume that our sample is drawn from a normal population, use of the t-distribution instead of the Z-distribution takes this problem into account and enables us to conduct proper inferences. (Even when the sample size is large and σ is unknown, the t-distribution is correct—the Z-distribution is simply a reasonable and convenient approximation to the t-distribution.)

For testing hypotheses about the value of μ, a strong alternative to the more traditional but also more restrictive t-test is available in the Wilcoxon signed ranks test. This procedure presumes population symmetry, a weaker condition than the normality assumption of the t-test. Unlike the t- and Z-tests for the population mean, the signed ranks test relies on computations involving sorted or ranked observations instead of computing a mean and standard deviation.

The last section of the chapter discussed control charts, a common element in the quality control program of many firms. Control charts are intended to detect assignable causes of variation while ignoring causes attributable to random chance. The sample size in quality control applications often is quite small. The ongoing null hypothesis in quality control work is that the process in question is "in control" and needs no adjustments. Only when an out of control reading occurs do we reject our hypothesis and look for an assignable cause to explain the nontypical value.

CHAPTER 11 EXERCISES

11.37 For the following conditions, specify the correct t-distribution table cutoff value(s):

a. $\alpha = .05$, $df = 17$, and two tailed.
b. $\alpha = .01$, $df = 11$, and lower tailed.
c. $\alpha = .02$, $df = 19$, and two tailed.
d. $\alpha = .025$, $df = 201$, and upper tailed.

11.38 Find the following probabilities:

a. $P(t > 2.485)$ when $df = 25$.
b. $P(t < -2.131)$ when $df = 15$.
c. $P(t > -2.086)$ when $df = 20$.
d. $P(|t| < 1.337)$ when $df = 16$.
e. $P(t > 2.00)$ when $n = 10$. (*Hint:* Either estimate by interpolation or set upper and lower limits for this probability.)

11.39 In preparing a story about single working parents, a newspaper feature writer contacted 10 licensed day care centers chosen at random in a metropolitan area. One piece of information collected was the weekly charge for a child in the three- or four-year-old category. Following are these sample results for X = weekly fee:

\$77.50	\$79.00	\$79.50	\$82.00	\$81.50
\$79.00	\$81.00	\$78.00	\$77.00	\$79.50

Assuming that X is a normally distributed random variable, determine an interval that, prior to sampling, has 1 chance in 50 of not including the area's average fee.

11.40 A random sample of 16 claim amounts listed in the plaintiff's pretrial papers at a county's small-claims court (damages under \$500) shows an average claim of \$229. The sample standard deviation is \$73. Assuming normality, develop a 90% confidence interval for the average damage amount sought in this small-claims court.

11.41 A medium-sized city has begun taking air quality readings once a week. Following are the measurements for X = suspended particulates, in micrograms per cubic meter, for the first 26 weeks of operation:

19	20	17	20	21
18	22	28	21	19
18	13	27	17	23
22	14	14	18	24
21	16	16	22	22
22				

a. Verify that the sample standard deviation is 3.71 and that the sample mean is 19.77 micrograms per cubic meter.
b. What is the point estimate for μ, the average level of suspended particulates in the city's air over the 26-week period?

c. What t-score would we use if we wished to construct a 90% confidence interval for μ?
d. Construct a 90% confidence interval for μ.

11.42 Refer to Exercise 11.41 above. Is it possible to reject H_0: $\mu = 20$ micrograms in favor of a two-tailed alternative hypothesis? Use $\alpha = .10$ in your test.

11.43 A sample of size 11 is drawn from a population that is believed to be normally distributed. The sample average is found to be 13.61, with a sample standard deviation of 4.92. At the .10 level of significance, can we reject the assumption that the sample came from a population whose mean is 15.0? Use a two-tailed alternative hypothesis.

11.44 Due to an extended dry spell, local water company officials have declared an emergency and have asked their customers to cut back voluntarily on water consumption. According to company records, normal residential usage of water at this time of year is 260 gallons per day. Choosing a random sample of 25 residences, officials obtain two readings (24 hours apart) to see whether customers are complying with the request to conserve water. The 25 measurements of daily water consumption are, in number of gallons per day:

78	121	136	151	157
158	161	177	183	183
193	211	221	228	251
267	279	290	291	299
307	313	319	331	401

a. Compute $\overline{X}$ and S.
b. Test the appropriate hypothesis with 5% risk of Type I error, using the t-distribution. Are customers helping to conserve or not?

11.45 Refer to Exercise 11.44.
a. Explain in terms of the problem setting what a Type II error would be.
b. Determine the upper and lower limits of the p-value for this statistical test.

11.46 A midwestern city is looking for ways to increase income from parking meters on its downtown streets. Since an increase in meter rates would be very unpopular, officials have decided to test a newly produced meter. The new meter contains an electronic "eye" that can detect when its parking place is unoccupied. If unused time remains on the meter, it then resets to zero. By eliminating free parking to vehicles that come along after a space has been vacated, the new meter is supposed to generate more revenue per day than a regular meter. City officials are able to obtain six of the new meters for a two-week trial period. The meters are placed in representative downtown locations where an average meter now generates $4.15 per day. Following are the test meter collections:

Meter Number	Daily Average Over Trial Period, X($)
1	4.98
2	5.40
3	6.01
4	5.77
5	4.77
6	5.83
	($\overline{X} = \$5.46$ and $S = \$.50$)

Due to the higher purchase price and maintenance expected for the new meters, city officials estimate that the new meters need to bring in an extra 75¢ per meter per day just to cover their additional expense. Revenue above that amount would make the meters an attractive option.

a. Set up the appropriate statements for a statistical test.

b. If you are willing to tolerate 10% risk of Type I error, what recommendation do you make? Is a large-scale adoption of the new meters warranted or not? Use the t-distribution in making your decision.

11.47 Refer to Exercise 11.46 above.

a. What assumption about the population does a t-distribution test require?

b. Determine upper and lower limits for the p-value for your test statistic.

11.48 A desirable property of a forecasting model is that it should generate unbiased forecasts. Loosely speaking, this means that on average a model's predictions should be correct as opposed to being regularly too low (or too high). The accompanying table gives the most recent 15 forecasting errors made by a university professor's forecasting model. The model was predicting monthly unemployment for a particular state. Let X = forecast error, in thousands of people.

−8.4	2.2	2.4	−11.3	6.0
4.1	−.9	7.1	6.4	8.0
2.7	11.7	2.9	−7.7	2.5

Negative values indicate that the month's forecast was too low; positive, too high.

a. Verify that $\overline{X} = 1.85$, $S = 6.46$.

b. Set up a null and alternative hypothesis about μ that would enable you to conclude that the model's forecasts are either unbiased or biased.

c. Determine the value of the t-test statistic for testing your null statement in part b.

d. Set up the rejection region, using a 10% risk of incorrect rejection.

e. On average, does the model generate biased or unbiased forecasts?

11.49 What advantages does the signed ranks test procedure have over the t-test when the sample size is small?

11.50 Refer to Exercise 11.48. Test the hypothesis that the average forecast error is zero, using the signed ranks test with $\alpha = .10$.

11.51 Refer to Exercise 11.46. Use the data given to test H_0: $\mu = \$4.90$ against an upper tail alternative, using the signed ranks test. Report results for $\alpha = .10$.

11.52 Refer to Exercise 11.51. Between what two probabilities does the test's p-value lie?

11.53 Prior to its first season of business, management of a new theme park stated that they expected weekday attendance to average 1500 people and weekend attendance to average 3500 people. The accompanying table gives a random sample of 15 weekday attendance figures taken from the first two months of operation (X = weekday attendance). Is it reasonable to believe that these figures are from an infinite population of figures whose average will be 1500? Use the signed ranks test with a lower-tailed alternative hypothesis. Use a level of significance of your own choosing.

591	712	787	1014	1184
1236	1319	1352	1440	1490
1494	1562	1587	1743	1912

11.54 When he took a part-time job delivering orders for a franchized pizza outlet, a statistics student was told that the average customer lived within a mile of the store. The first night on the job, the student delivered 13 pizzas, noting with his odometer the distance of each customer from the store. Assume the data from the first night are a random sample of values of X = distance, in miles. Does this evidence support or contradict the statement that the average customer lives no more than a mile from the store? Use the signed ranks test with 10% risk of Type I error. The data are as follows:

1.7	0.4	0.6	1.0	3.1	0.7	0.2	0.9	0.2	1.9	0.5	1.2	1.9

11.55 A manufacturer of ball bearings has a contract with a customer to supply a product with a mean of 1.5000 centimeters and a standard deviation of no more than .0020 centimeter. What would be the control limits for an $\overline{X}$-chart if the sample size is to be 16 bearings? Would a sample mean of 1.4982 centimeters be an indication of assignable variation?

11.56 When operating properly, a brick making operation produces bricks of an average weight of 1640 grams, with a standard deviation of 30 grams. The distribution of weights is approximately normal.

a. Establish the control limits for an $\overline{X}$-chart, assuming a sample of 5 bricks is to be weighed every 30 minutes.

b. Plot the following observations ($\overline{X}$) from one particular eight-hour work shift on your chart. Was the process in control at all times during the shift?

1611	1646	1620	1647	1674	1598	1646	1662
1650	1615	1648	1649	1670	1606	1621	1668

11.57 The control limits for an operation that fills containers labelled "one liter" are 998 milliliters and 1006 milliliters. If these limits were established with the knowledge that the periodic inspections would be of size 5, what is the process standard deviation of fill weights?

11.58 In quality control, match the terms "undetected problem" and "false alarm" to the two types of hypothesis testing errors.

***11.59** Consider the following data set:

2.0	16.6	14.5	4.5	10.9	11.5

a. Compute the coefficient of skewness.

*Optional

b. Did these data come from a symmetrical population? Explain the justification for your answer.

***11.60** Suppose the null hypothesis H_0: $\gamma_1 = 0$ is rejected. Is a subsequent signed ranks test using the sample data justified? Why?

***11.61** (*Note:* To solve this problem, you need to have studied sections 11.1, 11.3, and Exercise 11.26.) An independent insurance institute conducted a study to determine repair costs for small cars damaged in a five mile per hour (mph) crash. The firm crash-tested several small, two-door cars using four separate 5 mph crashes: head-on into a flat barricade, backed into a pole, backed into a barricade, and into a barrier at a 30-degree angle. An estimate of the total damage was then determined. Ultimately, the institute wishes to infer whether the damage over these types of crashes averages $1000. A computer analysis of the sample data generated these figures:

$n = 29$	$T_+ = 340$
$\bar{X} = \$1316.34$	$T_- = 95$
$S = \$570.56$	$g_1 = .271$

In each of the following situations, make the appropriate inference for the institute, based on $\alpha = .05$. Explain what analysis (or analyses) you used.

a. You believe the population is normally distributed, but σ is unknown.
b. You believe the population is not normally distributed but that it is symmetrical.
c. You know nothing more than the computer results given above. (*Hint:* Cutoff value for g_1 is .818 for $n = 29$.)

***11.62** Is it reasonable to conclude that the following sample data were drawn from a symmetrical population?

71	85	144	106	124	77	93

Why?

11.63 Financial analysts often miss the mark when estimating corporate earnings, but many investors still base their stock decisions on the analysts' predictions. It has been hypothesized that there is a natural tendency among analysts to be too optimistic. One of the large brokerage houses has agreed to test this phenomenon among its analyst trainees. Using one-year-old financial data from an anonymous company, the firm asked a random sample of 12 analysts to predict the company's earnings per share. Suppose we would like to choose between believing that on average

H_0: The population of analysts are not optimistic.
H_a: The population of analysts are optimistic.

Here are the 12 analyst predictions for the company's earnings per share:

$1.97	2.08	2.11	2.20	2.28	2.45	2.45	2.59	2.66	2.70	2.82	2.85

a. Choose a statistical technique presented in this chapter that is capable of determining whether analysts are overly optimistic.
b. Execute the test with $\alpha = .025$. The company in question turned out to have actual earnings of \$2.25.
c. What assumption(s) are necessary to perform your statistical test?

REFERENCES

Duncan, A. J. 1986. *Quality Control and Industrial Statistics,* 5th Edition. Irwin, Homewood, IL.

Mulholland, H. P. 1977. "On the Null Distribution of $\sqrt{b_1}$ for Samples of Size at Most 25, With Tables." *Biometrika, 64:* 401–409.

Ryan, J. 1987. "1987 ASQC/Gallup Survey." *Quality Progress* (December):12–17.

Chapter Maxim *As consumers of information obtained from polls, we should want to know more about the results than simply the winner, the loser, and the score.*

CHAPTER 12

INFERENCES ABOUT π AND σ

*Optional

Objectives

After studying this chapter and working the exercises, you should be able to

1. Draw a graph of a binomial population and a graph of the sampling distribution of the sample proportion.
2. Use sample data to develop a confidence interval for π, the population proportion.
3. Determine the sample size needed to estimate π to within a specified precision.
4. Set up and perform hypothesis tests for π.
5. Recognize the signs of a well-conducted and well-reported opinion poll.
*6. Execute the portion of the One-Sample Inference processor that deals with confidence intervals and hypothesis tests for π.
7. Describe the sampling distribution of S^2, the sample variance.
8. Locate table values for variables that follow a chi-square distribution.
9. Use sample data to construct a confidence interval for σ^2 or σ, our primary measures of population dispersion.
10. Set up and carry out hypothesis tests for σ^2 and σ.
*11. Execute the portion of the One-Sample Inference processor that can do confidence intervals and hypothesis tests for σ^2 and σ.

In Chapters 10 and 11 we learned the general principles of statistical inference and applied them to sampling situations where there was an interest in ascertaining the value of μ, the population mean. In this chapter, we will extend our study of estimation and hypothesis testing to binomial sampling applications where we focus on π, the proportion of items in the population having some particular attribute. We will also consider inference procedures about σ, the population standard deviation.

12.1 INFERENCES ABOUT π, THE POPULATION PROPORTION

We often encounter sampling situations where each unit in the population to be sampled can be labeled as fitting into one of two possible categories. Consider the following examples:

*Applies to optional section.

- A national banking association is contacting member banks to learn how widespread the policy of charging a service fee to customers for each use of automatic teller machines has become.
- *Money* magazine plans to survey upper-level managers in several metropolitan areas to estimate what proportion have a car telephone.
- The faculty of the Marketing Department at our university has polled students in several sections of the introductory marketing course to see what percent have ever made a purchase through a home shopping television show.
- Suppose the IRS states that its "assistors" (employees who respond to taxpayers' questions via toll-free calls) give correct advice at least 90 percent of the time. A consumer interest group wants to test this claim by making several hundred anonymous calls for tax help over a four-week period.

Each of the sampling situations above is binomial in nature (assuming we meet the conditions of independent trials with a constant probability of success). On each trial or unit sampled, the outcome can be described as a "success" or as a "failure." In binomial sampling our interest will be in determining π, the proportion of outcomes in the population that are "successes." With only two potential outcomes per item sampled, binomial sampling is in contrast with the settings discussed in the previous chapters. In Chapters 10 and 11, we considered situations where each unit sampled could take on one of numerous values, and our interest focused on μ, the average value of the variable in the population.

Before proceeding, we need to introduce a new symbol, p, to denote the *percent, or proportion, of successes* found in a sample.

Sample Proportion of Successes

$$p = \frac{X}{n} = \frac{\text{Number of successes in the sample}}{\text{Number of items in the sample}}$$

As is the case for π, p necessarily will be a number between zero and one, inclusive. The symbols π and p are the binomial counterparts for μ and $\overline{X}$ (see Table 12.1). Note that in both instances the population values are symbolized by Greek letters.

Sampling Distribution of p

Before considering the properties of the sampling distribution of p, let us first focus our attention on how we present a binomial population graphically. Figure 12.1 shows the appearance of three different binomial populations. Since each item in any binomial population can have only one of two different outcomes, the pictures of these

TABLE 12.1 Symbols: Means Versus Proportions

Measure	Symbol for Population Parameter	Symbol for Sample Statistic
Mean	μ	$\overline{X}$
Proportion	π	p

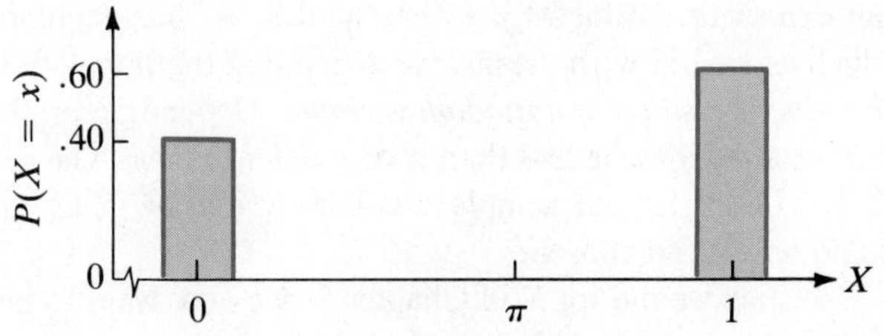

(a) $\pi = .60$

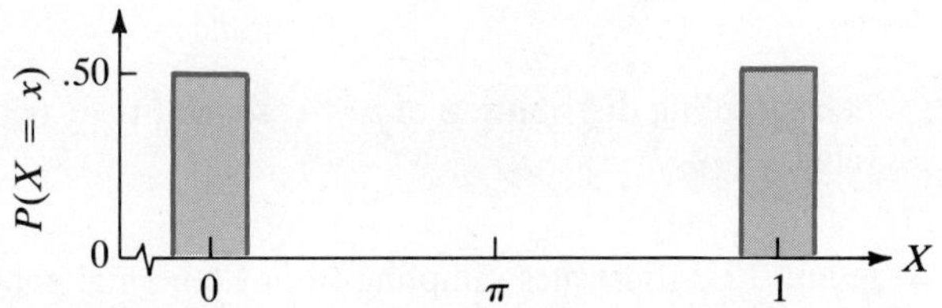

(b) $\pi = .50$

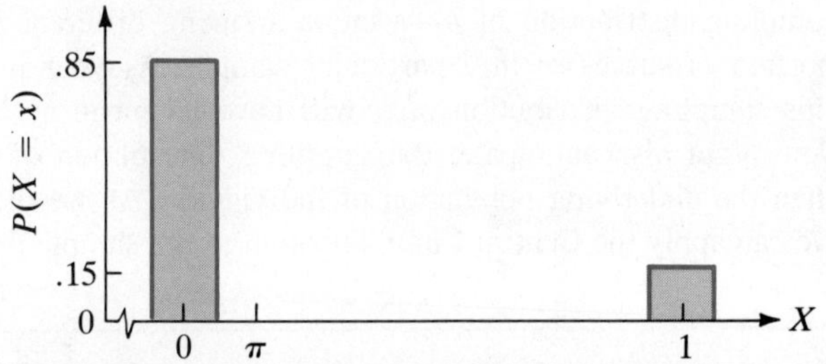

(c) $\pi = .15$

FIGURE 12.1 Binomial Populations

populations show probability only at two locations. By convention we let $X = 0$ represent the outcome failure, while $X = 1$ represents success. As we did in Chapter 6, we use Greek π to denote the proportion of successes (observations where $X = 1$) in the population. In each panel of Figure 12.1, note that π corresponds to the height of the rectangle for $X = 1$ and that the location of π between 0 and 1 corresponds to the population center of gravity.

With this brief review of the nature of binomial populations, we now wish to develop the sampling distribution of the sample proportion, p. As we do so, it may be helpful to make some analogies to the sampling distribution of $\overline{X}$ presented in Chapter 8.

Let us begin by supposing we are about to draw a random sample of $n = 200$ items from a large binomial population. If we find 133 successes in our sample, then $p = X/n = 133/200 = .665$. But due to the random nature of sampling, we would not expect this value of p to be equal to π, the population proportion of successes. Much as we did with means, we argue that the population value π is a constant while *the sample result p is a random variable*. Depending on the luck of the draw, our value of $p = .665$ may be less than π or greater than π. The set of 200 items that happened to be chosen for our sample was simply one of many different sets of size 200 that could have been chosen.

Just as we did for $\overline{X}$ in Chapter 8, we now want to be able to describe the pattern that the various possible values of p (for a given sample size) would form if we could somehow see them all. This distribution is called the *sampling distribution of p*.

Definition

The **sampling distribution of *p*** is a probability distribution of all the possible values of p.

Figure 12.2 illustrates sampling from a binomial population where $\pi = .60$. The top panel in the figure represents the population of failures ($X = 0$) and successes ($X = 1$). The population diagram has X as its horizontal axis label, and its mean (center of gravity) is at $\pi = .60$. The lower panel in Figure 12.2 represents the sampling distribution of p—a curve showing different possible values for the proportion of successes in a particular sample. As your intuition is likely to suggest, this sampling distribution of p will have its mean at π (in this case, at .60). As you might also anticipate, the sampling distribution of p will have less dispersion than the underlying population of individuals. As we did for the sample average $\overline{X}$, we can apply the Central Limit Theorem to the sample proportion p.

Central Limit Theorem—Binomial Sampling

When the sample size is large, the sampling distribution of p is approximately normal with mean $= \pi$ and standard error $= \sigma_p = \sqrt{\pi(1 - \pi)/n}$.

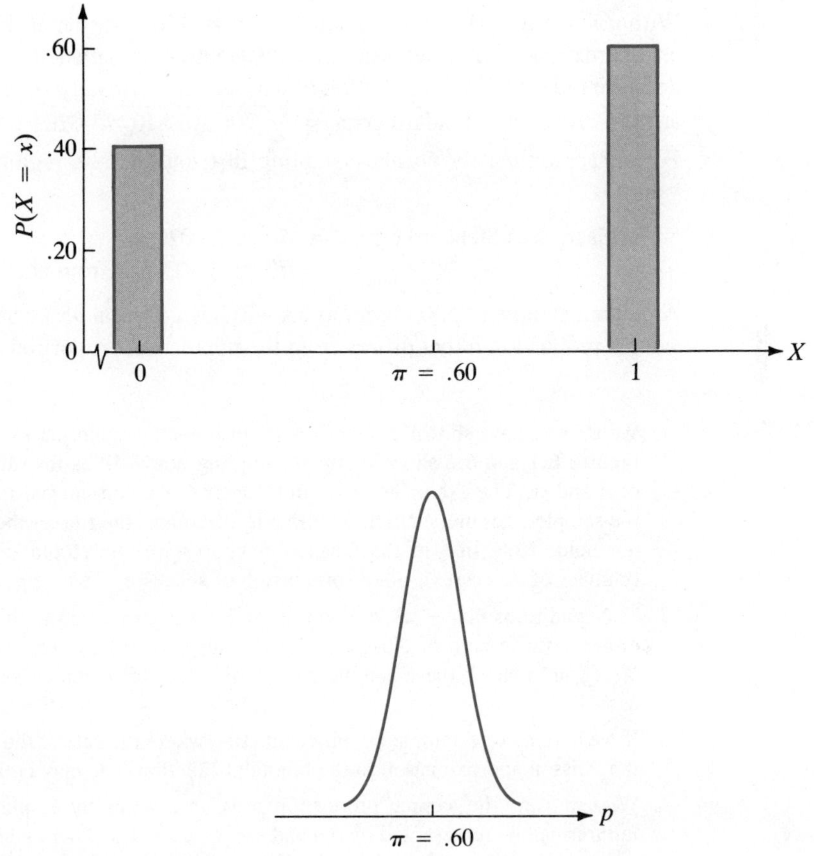

FIGURE 12.2 A Binomial Population (π = .60) and the Sampling Distribution for p

What is a large binomial sample? A generally applied rule-of-thumb is that both $n\pi$ and $n(1 - \pi)$ should be at least 5.0. In practice, samples drawn from binomial populations usually exceed this minimum by a large margin.

EXAMPLE 12.1

Suppose a sample of size 520 is to be drawn from a binomial population where $\pi = .70$.

a. Describe the sampling distribution of p.

b. Specify the region of concentrated values for p.

Solution:

a. With $n\pi = 520(.70) = 364$ and $n(1 - \pi) = 156$, the rule of thumb for attaining an approximately normal sampling distribution is satisfied. Therefore we can describe the sampling distribution of p as an approximately normal curve centered at $\pi = .70$ with a standard error $= \sqrt{.70(.30)/520} = .0201$.

b. For an approximately normal sampling distribution, the region of concentrated values is

$$\text{Mean} \pm 2 \text{ Standard errors} = .70 \pm 2(.0201)$$
$$= .70 \pm .0402 \quad \text{or from about } .66 \text{ to } .74$$

A random sample of 520 observations will have about a 95% chance of yielding a sample proportion p that differs from the population proportion π by .04 or less.

COMMENTS

1. We already have shown some examples of binomial sampling in a previous chapter. Figures 6.1 and 6.2 show binomial sampling possibilities for various combinations of n and π. The examples given in Chapter 6 are non-normal in appearance since the sample sizes are small and, with one exception, the rule of thumb ($n\pi \geq 5$) does not hold. Note that in the Chapter 6 figures, the horizontal axis was labeled X (number of successes), not p (proportion of successes), where $p = X/n$.
2. The conditions $n(1 - \pi) \geq 5$ and $n\pi \geq 5$ are in danger of not being met when π is either close to zero or close to one. The suspicion of extreme values for π should alert you to check these conditions. Small values for n can cause the same problem as well.
3. If we have a large-sample binomial situation where the rule of thumb is not satisfied, the Poisson approximation to the binomial (Section 6.4) may be an alternative.
4. We can view the sample proportion p as an *average* by employing the idea that failures can be represented by 0's and successes as 1's. Then, a given sample of size n will typically contain a mixture of 0's and 1's, so when the total of the 0's and 1's (equivalently, the total number of sample successes) is divided by n, an average, denoted by p, will result.

Probability Statements About p

Since the CLT tells us that the sampling distribution of p follows the normal distribution for large n, we can determine probabilities of interest by standardizing p and using the Z-table. The equation for standardizing binomial sampling results is given in the following box.

Z-Transformation for p

$$Z = \frac{p - \pi}{\sigma_p} = \frac{p - \pi}{\sqrt{\dfrac{\pi(1 - \pi)}{n}}} \tag{12–1}$$

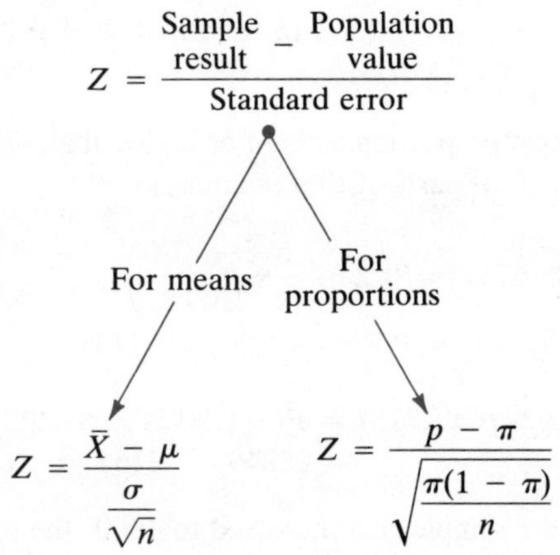

FIGURE 12.3 *Z*-Transformation for Sampling: Means Versus Proportions

Figure 12.3 compares this formula to its counterpart expression for means, given in Chapter 8. The following examples illustrate the use of this expression.

EXAMPLE 12.2

A sample of size 480 is to be drawn from a binomial population. If in the population $\pi = .70$, evaluate the following probability statements:

a. $P(p \geq .75)$
b. $P(p < .675)$
c. $P(.675 \leq p \leq .725)$
d. Re-do part c for a sample size of 1000.

Solution:

With both $n\pi$ and $n(1 - \pi)$ well in excess of 5, the CLT applies, giving us an approximately normal sampling distribution for p.

a. We express the sample proportion p as a Z-value and then find the corresponding probability as usual in the Z-table. The value $p = .75$ is equivalent to

$$Z = \frac{p - \pi}{\sqrt{\dfrac{\pi(1 - \pi)}{n}}}$$

$$= \frac{.75 - .70}{\sqrt{\dfrac{.7(.3)}{480}}} = \frac{.05}{.0209} = 2.39$$

Therefore

$$\begin{aligned} P(p \geq .75) &= P(Z \geq 2.39) \\ &= 1 - .9916 = .0084 \end{aligned}$$

A sample proportion as high or higher than .75 is unlikely. (See Figure 12.4 for a sketch of all parts of this example.)

b. $$\begin{aligned} P(p < .675) &= P\left(Z < \frac{.675 - .70}{.0209}\right) \\ &= P(Z < -1.20) = .1151 \end{aligned}$$

c. $$\begin{aligned} P(.675 \leq p \leq .725) &= P(-1.20 \leq Z \leq 1.20) \\ &= .8849 - .1151 = .7698 \end{aligned}$$

d. With the sample size increased to 1000, the sampling distribution becomes more compact, making it more likely that p will land close to π. For $n = 1000$,

$$\begin{aligned} \sigma_p &= \sqrt{.7(.3)/1000} = .0145 \\ P(.675 \leq p \leq .725) &= P(-1.73 \leq Z \leq 1.73) \\ &= .9582 - .0418 = .9164 \end{aligned}$$

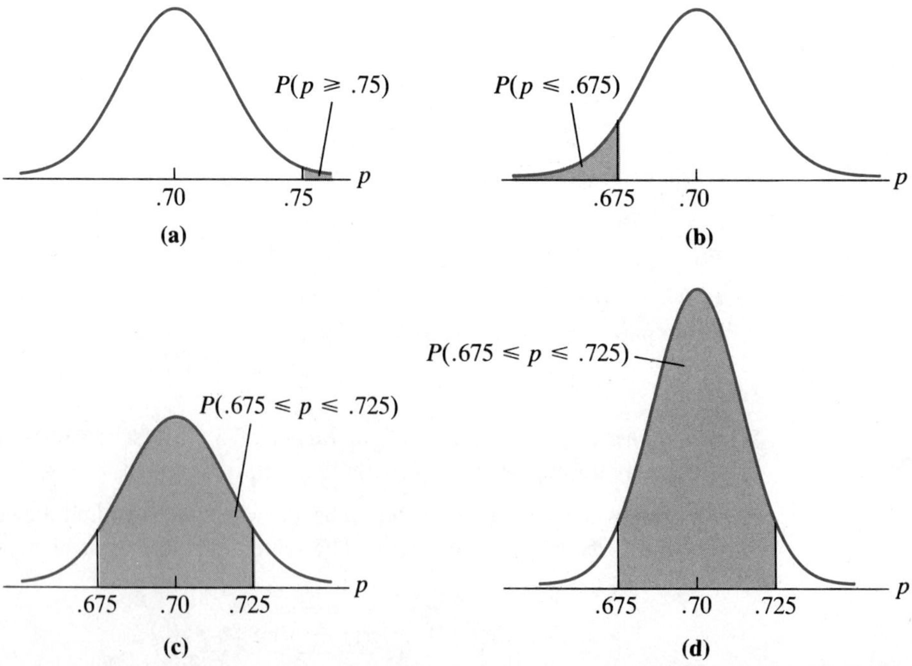

FIGURE 12.4 Probabilities for Example 12.2

□

EXAMPLE 12.3

Suppose an opinion poll is about to be conducted. A total of 1100 adults will be asked whether they are in favor of certain proposed legislation.

a. If in the population 40 percent of adults support the legislation in question, how likely is it the sample will land within plus or minus two percentage points from 40 percent?

b. If 60 percent in the population favor the legislation, how likely is the sample to land within two percentage points on either side of 60 percent?

c. If 20 percent in the population favor the legislation, how likely is the sample to land within two percentage points on either side of 20 percent?

Solution:

a. If $\pi = .40$, the standard error is $\sqrt{.4(.6)/1100} = .01477$. We then have

$$P(.38 \leq p \leq .42) = P\left(\frac{.38 - .40}{.01477} \leq Z \leq \frac{.42 - .40}{.01477}\right)$$
$$= P(-1.35 \leq Z \leq 1.35) = .8230$$

A sample result based on 1100 observations has better than four chances in five of landing within two percentage points of $\pi = .40$.

b. If $\pi = .60$, the standard error is $\sqrt{.6(.4)/1100} = .01477$. Note that this is the same result as in part a since the same numbers, .6 and .4, appear in the numerator of the standard error. Therefore $P(.58 \leq p \leq .62) = P(-1.35 \leq Z \leq 1.35) = .8230$, as in part a.

c. When $\pi = .20$, the standard error becomes $\sqrt{.2(.8)/1100} = .01206$. We then have

$$P(.18 \leq p \leq .22) = P\left(\frac{.18 - .20}{.01206} \leq Z \leq \frac{.22 - .20}{.01206}\right)$$
$$= P(-1.66 \leq Z \leq 1.66) = .9030$$

If π is equal to .20, there is only about one chance in 10 that the sample result p will be in error by more than two percentage points.

□

This example illustrates that for a given sample size, the dispersion (standard error) of a sampling distribution for p depends on the value of π. A binomial sampling distribution for p has its greatest dispersion when $\pi = (1 - \pi) = .50$ (see Exercise 6.24 of Chapter 6).

COMMENTS

1. Public opinion/political polls often are based on a few hundred to over a thousand respondents. Samples of this size generally will provide the sponsor an acceptable margin for potential sampling error, typically from ± 2 to ± 5 percentage points.

2. Standard errors for p involve small decimal numbers, especially prior to taking the square root. We recommend that you *not* round off these numbers until after you have completed the standard error computation. Even then, we recommend keeping results accurate to at least three non-zero numbers. For instance, $\sqrt{.61(.39)/950}$ should be reported and used as .0158 or .01582. If we were to round off .0158 to .02, for instance, we would be introducing a large error into any subsequent confidence interval computation or test statistic value.
3. If more than 5 percent of a binomial population is sampled, the finite population correction term $\sqrt{(N - n)/(N - 1)}$ should be applied to the standard error.

Confidence Intervals for π

We have discussed two ways to make a statistical inference in the previous chapters: by estimation or by a hypothesis test. In this section, we consider confidence interval estimation for binomial sampling applications.

Our approach is comparable to that of estimating μ. In the binomial case, the value of the population percent or proportion (π) is unknown. Our objective is to use the information found in a random sample to specify a range of values that will have a predetermined probability or confidence of including the unknown value of π. As is the case for confidence intervals for the mean, the confidence interval for π will be centered about the point estimate. For binomial sampling the best estimator is p, the proportion of successes found in the sample.

In words, we like to think of a confidence interval as

Point estimate ± Potential sampling error

For large sample binomial applications, this phrase becomes, in symbols, $p \pm Z\sigma_p$.

Now, we have defined σ_p as $\sigma_p = \sqrt{\pi(1 - \pi)/n}$. In confidence interval estimation, however, the term π in this formula is unknown, so σ_p is also unknown. Therefore we will replace π with its sample estimate p, giving us an *estimated* standard error, $\hat{\sigma}_p = \sqrt{p(1 - p)/n}$. (Our use of $\hat{\sigma}_p$ is the direct counterpart to using $\hat{\sigma}_{\bar{X}}$ in the previous chapters when we used S to replace σ in confidence interval estimation.) We then have

Confidence Interval for the Population Proportion

$$p \pm Z\sqrt{\frac{p(1 - p)}{n}} \tag{12–2}$$

Note that the sample result p appears in the standard error as well as serving as the point estimate. Figure 12.5 compares a confidence interval for π with a confidence interval for μ. We are now ready to consider some examples.

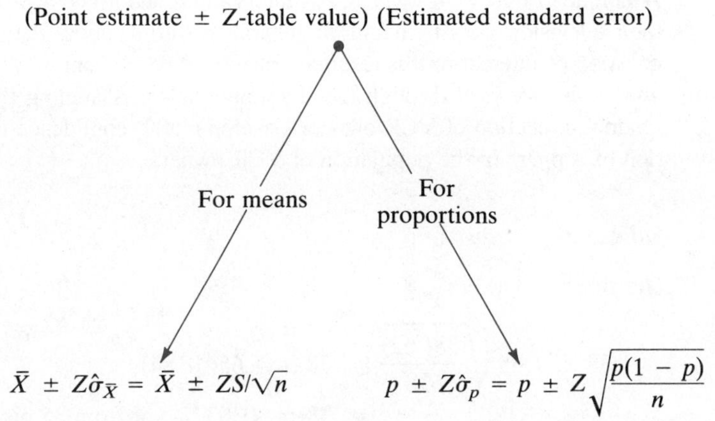

FIGURE 12.5 Large-Sample Confidence Interval Expressions: Means and Proportions

EXAMPLE 12.4

A poll of students enrolled in the introductory marketing course at our university showed that 63 of the 373 respondents have made a purchase at one time or another through a home shopping television show. Assume that these students constitute a random sample of university students with respect to their shopping experience and develop a 99% confidence interval for π, the proportion of all university students who have bought goods via home shopping.

Solution:

The point estimate is $p = X/n = 63/373 = .169$. The Z-table value for 99% confidence is ± 2.58. Our confidence interval therefore is

$$\begin{aligned} p \pm Z\,\hat{\sigma}_p &= .169 \pm 2.58\sqrt{\frac{.169(.831)}{373}} \\ &= .169 \pm 2.58(.0194) \\ &= .169 \pm .0501 \qquad \text{or } .17 \pm .05 \end{aligned}$$

□

EXAMPLE 12.5

At the request of several of its larger clients, one of the major U.S. advertising agencies conducted a research study to assess the extent of "zapping"—the practice by VCR owners of fast-forwarding through commercials when viewing taped television shows. A sample of 698 VCR owners was interviewed and asked a variety of questions about their television viewing habits in general as well as about the use of their VCR. The question of interest in this example found that 38 percent of VCR owners "usually or always fast-forward through taped commercials." Assuming the sample represents a random collection of VCR owners, develop a 90% confidence interval for the proportion of zappers in the population of VCR owners.

Solution:

Our interval will be

$$.38 \pm 1.65\sqrt{\frac{.38(.62)}{698}} = .38 \pm 1.65(.0184)$$
$$= .38 \pm .0303 \qquad \text{or from 35 percent to 41 percent}$$

□

EXAMPLE 12.6

There are about 1075 households in two closely situated towns in a predominantly rural county. A survey of 447 households chosen at random in the two locales was sponsored by a cable television company to estimate the area's degree of interest in cable television. The cable company believes that the subscription rate would need to exceed 50 percent to make the establishment of service in the towns economically attractive. The sample of households found 201 who indicated that they would subscribe if service were to become available. Determine a 95% confidence interval.

Solution:

The point estimate is 201/447, or about .450. With the sample making up 447/1075 or almost 42 percent of the total households, the estimated standard error should be adjusted by the *FPC*. The confidence interval for the proportion interested in cable service will be

$$.45 \pm 1.96\sqrt{\frac{.45(.55)}{447}}\sqrt{\frac{1075 - 447}{1075 - 1}} = .45 \pm 1.96(.02353)(.7647)$$
$$= .45 \pm 1.96(.0180)$$
$$= .45 \pm .0353 \qquad \text{or } 45\% \pm 3.5\%$$

Use of the *FPC* always reduces the standard error, in this case from about .0235 to about .0180. Since the confidence interval ranges from 41.5 percent to 48.5 percent interest, prospects for a profitable cable system in the towns are not encouraging.

□

Sample Size Requirements In the previous chapter, we encountered a formula that can tell us the sample size needed to estimate μ to within a specific margin for error E. As you might expect, a similar expression can be used to establish the sampling requirements for estimating π to a predetermined precision.

Sample Size Formula for Estimating π

$$n = \frac{Z^2\pi(1 - \pi)}{E^2} \qquad \text{(12–3)}$$

This expression is the same as the one for the mean given in Chapter 10, except that here $\pi(1 - \pi)$ replaces σ^2. The binomial sample size required for a targeted degree of accuracy depends on the following:

a. The specific maximum sampling error allowable, E
b. The level of confidence desired, represented by the appropriate Z-table value
c. An *estimated* value for π. Since our objective in sampling is to estimate π, its true value is of course not known to us—if it were known, no need for sampling would exist! Therefore, in Equation 12–3 we must provide an estimate for π. Table 12.2 lists possible ways to do so.

TABLE 12.2 Possible Approximations for π in Binomial Sample Size Formula

Method	Relies on
Comparable study	Value for π or p from a previous study
Pilot sample estimate	p, the proportion found in a preliminary sample
Conservative assumption	Use of $\pi = .5$ when no information about value of π is available

EXAMPLE 12.7

An automobile industry research group is interested in estimating π, the proportion of households in a metropolitan area whose primary vehicle is three years old or older. The estimate is desired to be accurate to within $\pm.03$ with 90% confidence. No reliable information exists from prior studies as to what value π might be. What sample size will achieve the desired precision?

Solution:

With no estimate of π at hand, we should presume $\pi = .50$. For 90% confidence, Z is 1.65. We want the sampling error to be no more than three percentage points, so E is .03. We then have

$$n = \frac{Z^2\pi(1 - \pi)}{E^2} = \frac{1.65^2(.5)(.5)}{.03^2} = 757 \text{ households}$$

The use of $\pi = (1 - \pi) = .50$ is conservative in the sense that any other pairs of values (such as .7 and .3, or .8 and .2, and so on) will indicate smaller sample size requirements. To be certain of attaining our target accuracy, we therefore choose to let $\pi = .50$ when no previous knowledge of π exists.

□

EXAMPLE 12.8

A new drug is being tested for shelf life. Ten thousand capsules are stored at room temperature for six months, after which a pilot sample of 50 capsules is chosen at random. It is found (via an expensive chemical analysis) that 42 capsules have retained an acceptable level of their original strength. How many capsules should be tested to develop a 95% confidence interval for π (π = percent in population retaining potency) that is in error by no more than four percentage points?

Solution:

With the results of a preliminary study giving us an estimate for π as $42/50 = .84$, we have

$$n = \frac{1.96^2(.84)(.16)}{.04^2} = 323 \text{ capsules}$$

If we had no pilot sample and simply used $\pi = .50$, our formula would tell us that we need to test 601 capsules. The conservative value for π always guarantees enough sample size, but to the extent that the actual value for π differs from .50, the conservative approach will indicate a larger sample size than necessary. When available, use of an estimated value for π is warranted.

□

In sampling from finite binomial populations, the sample size formula is given in the following box.

Sample Size Formula for Estimating π: Finite Population Case

$$n = \frac{NZ^2\pi(1 - \pi)}{E^2(N - 1) + Z^2\pi(1 - \pi)} \qquad (12\text{–}4)$$

As an example, suppose management of a credit union wants to estimate the proportion of members that would support the idea of the credit union buying, at member expense, a relatively low cost life insurance policy for all members. An estimate that would be accurate to within plus or minus five percentage points with 90% confidence is deemed suitable. How many members should be questioned? The membership printout contains 655 names.

With no preliminary estimate of π, we presume π to be .50:

$$n = \frac{655(1.65)^2\,(.5)(.5)}{.05^2(654) + 1.65^2(.5)(.5)} = 193 \text{ members}$$

Had we used Equation 12–3 here instead, it would tell us to survey 273 credit union members.

COMMENT As is the case for our other sample size formulas, we will round up any fractional value for n to the next highest integer. We also must view any sample size figure as approximate since π in our equations is estimated, not known.

Hypothesis Tests for π

The logic and method of statistical tests for binomial sampling applications closely parallels that for Z- and t-tests for μ presented in the previous chapters. The procedures bear close resemblance since both rely on normal sampling distributions. While the binomial symbols differ from those for tests on μ, the underlying concepts are identical. The test statistic for the binomial proportion π is given in the following box.

Test Statistic for a Population Proportion

$$Z^* = \frac{p - \pi_0}{\sqrt{\dfrac{\pi_0\,(1 - \pi_0)}{n}}} \qquad (12\text{–}5)$$

Figure 12.6 compares the binomial TS to its counterpart for the tests about μ. We are now ready to look at some examples.

COMMENT Note that the (estimated) standard error for computing a confidence interval is $\sqrt{p(1 - p)/n}$ while the standard error for a hypothesis test is $\sqrt{\pi_0(1 - \pi_0)/n}$ (compare Equations 12–2 and 12–5). A test of H_0 presumes that the sampling distribution is centered at a specific point, the π_0 value. If H_0 is true, then the standard error is indeed $\sqrt{\pi_0(1 - \pi_0)/n}$. On the other hand, a confidence interval carries no presumption as to the value of π; accordingly only the sample result p appears in the standard error for a confidence interval.

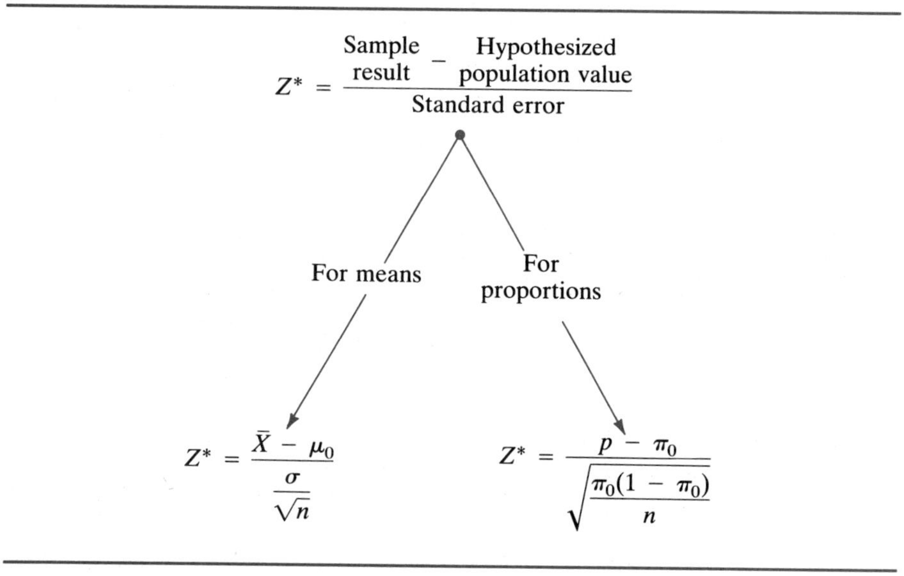

FIGURE 12.6 Large-Sample Test Statistics: Means Versus Proportions

EXAMPLE 12.9

In a study of its credit approval process 18 months ago, a large department store with outlets in several cities learned that customer requests for its store credit card were granted to 65 percent of applicants. One of the company's financial officers wants to see if the approval rate is still 65 percent. Accordingly a random sample of 315 credit applications acted upon in the last 90 days is selected. Test the presumption that no change in the approval rate has occurred, letting $\alpha = .10$. The sample is found to contain 101 denials of credit and 214 approvals.

Solution:

Since the proportion of customers obtaining credit could conceivably deviate in either direction from its past value, a two-sided test should be conducted. Letting π denote the proportion of all applications getting approval, we have

$$H_0\colon \pi = .65$$
$$H_a\colon \pi \neq .65$$

For $\alpha = .10$, the cutoff Z-table value is ± 1.65. Our sample evidence is $p = X/n = 214/315 = .6794$. Therefore, our TS will be

$$Z^* = \frac{p - \pi_0}{\sqrt{\dfrac{\pi_0(1 - \pi_0)}{n}}} = \frac{.6794 - .65}{\sqrt{\dfrac{.65(.35)}{315}}} = \frac{.0294}{.0269} = 1.09$$

Since $-1.65 \leq Z^* \leq 1.65$, we cannot reject H_0; we can describe the difference between p and π_0 (.0294) as not statistically significant. Table 12.3 summarizes this example.

TABLE 12.3 Symbolic Summary of Example 12.9

H_0: $\pi = .65$

H_a: $\pi \neq .65$

TS: $Z^* = 1.09$

RR: Reject H_0 if $Z^* < -1.65$ or if $Z^* > 1.65$

C: Do not reject H_0

□

Unstandardized Approach An alternative and equivalent way to approach Example 12.9 would be to state the cutoff points in unstandardized form. As discussed in Chapter 10, an unstandardized acceptance region is, in words,

$$\begin{matrix}\text{Expected value}\\ \text{under } H_0\end{matrix} \pm \begin{matrix}\text{Cutoff}\\ \text{value}\end{matrix}\begin{pmatrix}\text{Standard}\\ \text{error}\end{pmatrix}$$

For a binomial inference, the formula is given in the following box.

Unstandardized Acceptance Region—Binomial

$$\pi_0 \pm Z\sqrt{\frac{\pi_0(1 - \pi_0)}{n}} \qquad \textbf{(12–6)}$$

For Example 12.9 above, we have the acceptance region as being

$$.65 \pm 1.65\sqrt{\frac{.65(.35)}{315}} = .65 \pm 1.65(.0269)$$

$$= .65 \pm .0443 \qquad \text{or from .606 to .694}$$

If H_0 is true, the sample proportion of credit approvals is likely to be found in this interval. Since $p = .6794$ lies within this region, the sample findings are deemed

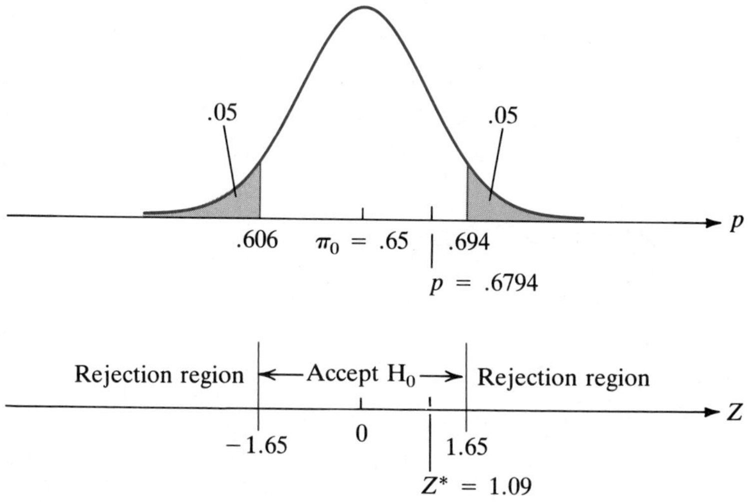

FIGURE 12.7 Unstandardized and Standardized Cutoffs for Example 12.9

consistent with the assumption that the approval rate has not changed. Figure 12.7 illustrates the comparability of the unstandardized and the standardized Z^* approach for this binomial example.

EXAMPLE 12.10

The manager of a large travel agency in a midwestern city suspects, for its business involving travel outside the continental United States, that the agency's bookings to Europe are down. She wants to sample the firm's records for the past 12 months to see if she is correct. She knows that four years ago European destinations accounted for 32 percent of the firm's business outside the continental United States. A clerk is assigned to note the primary destination for a random sample of 300 clients traveling outside the continental United States in the past year. The sample shows 76 European bookings. Specify the hypothesis statements for this sampling situation. Test H_0 with 5% chance of a Type I error.

Solution:

With a belief that π, the population percent of bookings to Europe, may have moved in a downward direction, a lower tail test is in order:

$$H_0\colon\ \pi = .32 \text{ (percent going to Europe has not changed)}$$
$$H_a\colon\ \pi < .32 \text{ (percent going to Europe has declined)}$$

For $\alpha = .05$, the cutoff Z-score is -1.65. Our *TS* is

$$Z^* = \frac{p - \pi_0}{\sqrt{\dfrac{\pi_0\,(1 - \pi_0)}{n}}} = \frac{\dfrac{76}{300} - .32}{\sqrt{\dfrac{.32(.68)}{300}}} = \frac{-.0667}{.0269} = -2.48$$

Since $Z^* = -2.48 < -1.65$, we can reject H_0. The assumption of no change in the proportion bound for Europe has been discredited. The p-value for this test is $P(Z \le -2.48) = .0066$, indicating that our sample result is very unlikely to be one from a sampling distribution centered at $\pi_0 = .32$.

□

EXAMPLE 12.11

As time runs out on the current labor contract, management and the hourly employees' union seem at an impasse in their negotiations for a new agreement. Several union leaders believe that there is a strong sentiment for a strike rather than accepting management's "final" offer. The union leadership tentatively agrees to strike if no better offer is forthcoming from the management side and if more than half the union members would support a strike. A polling of 400 members is conducted to see if the majority are willing to strike. The results are 214 in favor, 162 opposed, and 24 undecided. Union membership is 1217. Use 10% risk of a Type I error in determining whether there is conclusive evidence of strike support.

Solution:

Since a decision to strike may hinge on a finding that more than half would support such action, an upper tail test for π (π = percent of membership in favor of a strike) is called for

$$H_0\colon\ \pi = .50$$
$$H_a\colon\ \pi > .50$$

For $\alpha = .10$, the cutoff Z-table value is 1.28. Since the sample constitutes nearly one-third of the population, we should compute the *FPC* and multiply it times the (uncorrected) standard error.

$$FPC = \sqrt{\frac{N - n}{N - 1}} = \sqrt{\frac{1217 - 400}{1217 - 1}} = .8197$$

Our *TS* is then

$$Z^* = \frac{\frac{214}{400} - .50}{\sqrt{\frac{.5(.5)}{400}}(.8197)} = \frac{.535 - .50}{.025(.8197)} = \frac{.035}{.0205} = 1.71$$

The *TS* $Z^* = 1.71$ exceeds the reject/accept cutoff point of $Z = 1.28$—H_0 is rejected in favor of H_a. A persuasive case has been made that more than half of all members are willing to strike. The sample result of 53.5 percent is said to be significantly above 50 percent.

□

SECTION 12.1 EXERCISES

12.1 Which of the following are not examples of binomial sampling?

a. For a sample of long distance interstate phone calls, a listing of the length of time of the connection is made.

b. For a sample of grocery customers, a count is gathered of those who say they would be likely to fill their prescriptions at the store if a pharmacy were opened inside the grocery.

c. For a sample of service industry employees, X (number of hours worked last week) is recorded.

12.2 Are the following sampling situations designed to estimate μ or to estimate π?

a. A sample of adults is asked if they would consider buying a disposable camera if their regular camera were for any reason unavailable.

b. A sample of adults is asked what they think would be a reasonable price to pay for a disposable camera and its one roll of film.

c. For a sample of fifteen communications stocks on the New York Stock Exchange, the price–earnings ratio is noted.

d. A sample of convenience store customers is asked if they would "occasionally or often" use a new service being contemplated for introduction by the store: overnight laundry and dry cleaning service.

12.3 A sample of size 620 is drawn from a population that contains 80% successes. The sample contains 472 successes. What is the numerical value of π? What is the numerical value of p?

12.4 For a fixed sample size, which binomial sampling distribution has more variability: one where $\pi = .55$ or one where $\pi = .65$?

12.5 For which of the following binomial sampling situations would the rule of thumb indicate that the sampling distribution for p could be viewed as approximately normal?

a. $n = 480, \pi = .21$

b. $n = 48, \pi = .955$

c. $n = 30, \pi = .480$

d. $n = 220, \pi = .005$

***12.6** **a.** For $n = 10, 20, 50$, and 100, solve $n\pi = 5$ for the value of π and then solve $n(1 - \pi) = 5$ for the value of π.

b. Using graph paper or a carefully measured page, plot the points determined in part a and connect them with a curve. Let n be on the horizontal axis, π on the vertical axis.

c. Label the two regions of your curve to show where the CLT-binomial sampling conditions hold and do not hold.

d. When the conditions do not hold, what distribution besides the normal curve might be used to approximate any binomial probabilities of interest?

12.7 A random sample of 180 items is to be drawn from a population that contains 6% defectives.

a. Is the rule of thumb satisfied for the CLT to be applicable?

b. What is the standard error of p?

c. What is the region of concentrated values for p?

d. How likely is the sample to show 3 percent or fewer defectives?

12.8 A large population of subassemblies has been warehoused for several months under unfavorable storage conditions. A sample of 250 is to be drawn to estimate the proportion that are in working condition. Suppose that $\pi =$ the population proportion in working condition $= .70$.

a. Specify the mean and the standard error of the sampling distribution of p.

b. How likely is p to be found within $\pm .025$ of π?

c. How likely is the sample to show that 75 percent or more are in working condition?

d. Answer part b if $n = 400$ instead of $n = 250$.

12.9 A candidate for mayor in the upcoming election commissions a poll of 300 registered voters. Suppose that if a complete census were taken, it would reveal that 52 percent intend to vote for our candidate. How likely is it a randomly conducted poll of 300 voters will show our candidate has less than half of the vote?

12.10 A large retail chain offers maintenance agreements on its major appliances. The appliance manager wants to know what fraction of buyers of the company's microwave ovens is buying the maintenance contract along with the oven. She assigns a clerk to look at the sales slips for the last 200 ovens sold and note whether the customer bought the agreement. The clerk reports 31 service contracts were sold. Specify an interval that has 90% confidence of including the parameter of interest to the manager.

12.11 The 2981 entering freshmen at a state university were given a survey form by the Director of Student Affairs when they enrolled for fall classes. Completion and return of the form were optional. A total of 1312 forms were turned in by the freshmen. One of the questions posed on the survey asked about the student's intended major. Of the 1247 forms where this particular question was answered, 151 students checked one of the areas of the Natural Sciences Department. Use this result to generate a 90% confidence interval for the proportion of all freshmen intending to major in the natural sciences.

12.12 A pizza and pasta restaurant mailed out 300 coupons to postal patrons in its zip code area. The coupons were good for \$3.00 toward the purchase of a large pizza if they were redeemed within two weeks. After the passage of two weeks, a total of 24 coupons were used. The manager of the restaurant is now planning a mass mailing of about 8400 additional coupons. Assuming the preliminary mailing to be a random sample, perform the following.

a. Compute a point estimate for π, the population redemption rate.

*Optional

b. Compute a 98% confidence interval for the response rate to be found in the mass mailing.
c. Develop a 98% confidence interval for the count X, the number of coupons redeemed, in the mass mailing.

12.13 The two major universities in Kentucky, the University of Louisville and the University of Kentucky, do not schedule each other in football. However, interest in such a rivalry is increasing. The Louisville *Courier-Journal* newspaper published the results of its Bluegrass State Poll ("True to Blue: UK Outshoots U of L Among Ranks of Fans," December 21, 1986). This particular poll was based on 737 interviews with Kentuckians 18 years of age or older and was balanced to represent all regions of the state. The question of interest was, "Do you think that UK and U of L should play each other in football or not?" Construct a 95% confidence interval for the proportion of all Kentuckians who would say "yes" to this question. The poll reported that 69 percent of the sample replied "yes."

12.14 The last new episode of the popular television series M★A★S★H was broadcast on February 28, 1983. As of this writing, it remains the single most-watched episode of any series. The following week, the A. C. Nielsen Company reported that 60.3 percent of households in their sample watched all or part of the show ("CBS Clinches Lead with Finale of 'M★A★S★H'," Louisville *Courier-Journal,* March 5, 1983). If at this time the Nielsen survey was based on 1171 households and if we can assume that Nielsen has the equivalent of a random sample of all U.S. television households, develop a 90% confidence interval for the proportion of television households tuned in for the farewell episode.

12.15 There are two hospitals in a particular city. At the smaller hospital, an average of about 15 babies are born per day; the larger one averages about 27 births per day. On any given day, which hospital is more likely to record 2/3 or more of their births as girls? You should assume that the probability of each sex is .50.

12.16 The Louisville *Courier-Journal* reported the following news items in an article titled "Wrong Numbers" on April 9, 1987. The government's General Accounting Office (GAO) placed a series of telephone calls to the Internal Revenue Service's toll-free system (set up to assist the general public in filling out tax forms) to ask questions about how to complete a 1986 tax return. From February 17 to March 20, GAO investigators placed 918 calls, asking questions from a prepared list of 21 typical tax questions. The GAO reported that their questions were answered correctly 78 percent of the time. Generate an 80% confidence interval for π, the accuracy rate of the IRS tax help service.

12.17 A sociologist has sampled the most recent 240 divorce records in the county to estimate the proportion of divorces that involve no children under 18 years of age. The sample reveals 99 such cases. Provide a point estimate and a 95% confidence interval for the parameter of interest.

12.18 Suppose CBS Evening News reports one night that a new CBS News/*New York Times* poll finds that 61 percent of Americans approve of the way the President of the United States has dealt with a recent foreign policy issue. Not mentioned on the air but appearing at the bottom of the screen is "$\pm$4 percent." Assume that a 95% confidence interval is being referred to. About how many Americans were sampled for this poll?

12.19 Suppose you are going to pick 120 stocks at random from the New York Stock Exchange listings. A month later you are going to see how many have outperformed the overall market average during that time. Assume that the chance a stock chosen at random will beat the market average is .50.

a. Develop a 90% confidence interval (centered at .50) for the proportion of winners you will have in your sample.

b. Express your result in part a as an interval for the count of X, the number of market beaters in the sample.

c. Suppose a stock analyst who claims to be able to pick stocks that will outperform the market chooses his own set of 120 stocks to watch. If 65 of the analyst's stocks do beat the market, are you convinced that his long-run success rate is better than that of a random guesser?

12.20 Refer to Exercise 12.12. If your interval had 90% instead of 98% confidence, would it be wider or narrower? Which interval would have the greater chance of missing π?

12.21 What sample size would be needed to estimate π to within $\pm .025$ with 95% certainty if the value of π happened to be each of the following:

a. .50 **b.** .20 **c.** .80 **d.** .025

12.22 How many items from a population of 543 items need to be inspected to estimate the population proportion to within ± 3 percent? Assume that 90% confidence in the resulting interval is desired and that at least two-thirds of the population have the attribute in question.

12.23 Refer to Exercise 12.14. What sample size would have been necessary to estimate the M★A★S★H audience to within 1.5 percentage points on the evening in question? (Nielsen had a secondary group of panelists at that time who kept a written diary of what they watched—these written records were processed several weeks after the "overnight" results reported in the media were available.)

12.24 A regional manager of a grocery chain wants to know the percent of customers who are playing a heavily promoted store bingo game for prizes and groceries. He hires an individual to randomly survey shoppers exiting one store. The researcher finds 39 players and 71 nonplayers. Use this result to approximate the sample size requirement to estimate π, the population proportion of bingo players, to within 5 percentage points with 95% confidence.

12.25 Refer to Exercise 12.17. If the sociologist later decided that enough court records should be studied to end up with his 95% confidence interval being in error by no more than .05, what sample size is needed?

12.26 Sales personnel in an electronics store are required to record information (model number, price, tax, serial number, and so forth) from each transaction onto a sales ticket. A recent audit of 100 sales tickets sampled at random found nine tickets with errors. To estimate the error rate in the population of tickets to within ± 2.5 percent with 90% confidence, what sample size is needed?

12.27 Which of the following pairs of hypotheses are not stated in a correct form? Specify any needed corrections.

a. H_0: $\pi = 20$	**b.** H_0: $\pi = .50$	**c.** H_0: $\pi \neq .40$	**d.** H_0: $p = .60$
H_a: $\pi \neq 20$	H_a: $\pi \geq .50$	H_a: $\pi = .40$	H_a: $p \neq .60$

12.28 In a test of H_0: $\pi = .35$ versus a two-sided alternative, a sample shows 149 successes and 301 failures.

a. What is your conclusion if the test were to be conducted with 10% risk of incorrectly rejecting H_0?

b. What is the test's p-value?

12.29 In an upper-tail test where the null statement is H_0: $\pi = .75$, a sample of 425 items showed $p = .80$. Is the 5 percentage point discrepancy between observed and hypothesized proportions statistically significant at the .05 level? At the .01 level? At the .001 level?

12.30 A researcher is testing the assumption that 40 percent of a certain population possess a particular attribute. The alternative states that 40 percent is an overstatement of the proportion with the attribute. A random sampling of 850 items from the population in question shows 308 items with the attribute, 542 without. Select a commonly used level of significance, execute the proper test procedure, and state your conclusion for the test.

12.31 Determine the p-value for the test in Exercise 12.30.

12.32 A sample of 470 items is drawn from a population containing 2240 items in order to test H_0: $\pi = .20$ against a two-sided alternative. The test is to be conducted with 1 chance in 10 of Type I error. The sample reveals 113 items with the attribute in question. Can you reject H_0?

12.33 A large hardware/building supply store has historically had 39 percent of its store charge card users pay off their monthly balances in full. However, due in part to a profit squeeze, the store instituted two new policies six months ago. These were a raise in the interest rate on unpaid monthly balances from 15 to 18.5 percent and the imposition of a 50¢ per month handling fee on accounts that are not paid in full.

a. What effect on the proportion of accounts paying off monthly balances in full are the changes in policy intended to have?

b. State H_0 and H_a for a test that could determine whether the desired effect was realized.

c. A sample of 275 active accounts will be selected at random. Letting $\alpha = .05$, make the proper inference when the sample shows 124 current account balances are paid in full.

12.34 When smoking on short flights was banned by the FAA in 1988, a southern-based airline routinely had been setting aside 80 percent of seats for nonsmokers. Now that smoking is permitted only on flights over two hours, the airline is planning to sample 400 recent seating requests to see if the 80% figure is appropriate for long flights. A random sample is found to have 326 requests for nonsmoking. Test the presumption that the current demand for nonsmoking seating on long flights is 80 percent, using a 5% risk of Type I error.

12.35 Find the p-value for the test in Exercise 12.34.

12.36 In a large city, the turnout for the mayoral election is expected to be between 80,000 and 90,000 voters. Two candidates are on the ballot, Smith and Barney. Consider the following scenarios shortly after the polls have closed on election day:

Situation	Number of Votes Tallied	Percent of Vote for Smith	Percent of Vote for Barney
A	500	56	44
B	750	54	46
C	1500	52	48
D	3000	51½	48½

Assume the votes tallied above can be viewed as a random sample. For which situation is it safest to project Smith as the winner? For which situation would you be sticking your neck out farthest to proclaim Smith as the victor? (*Hint:* Think of testing H_0: Smith's proportion $= .50$ versus an upper-sided H_a.)

12.37 Refer to Exercise 12.36 above. Answer the same two questions, now assuming the total voter turnout is expected to be about 4000.

***12.38** By multiplying the numerator and denominator of Equation 12–5 by n, we will obtain

$$Z^* = \frac{np - n\pi_0}{\sqrt{n\pi_0(1 - \pi_0)}} = \frac{X - n\pi_0}{\sqrt{n\pi_0(1 - \pi_0)}}$$

This is an alternative way to express our test statistic Z^*. The letter X denotes the binomial random variable, the number of sample successes. To illustrate this version of the *TS*, text Example 12.9 would be

$$Z^* = \frac{214 - 315(.65)}{\sqrt{315(.65)(.35)}} = \frac{9.25}{8.4654} = 1.09$$

Use the above expression for Z^* to verify the value for Z^* worked out in the following text examples:

a. text Example 12.10

b. text Example 12.11 (don't omit the FPC)

12.39 Sponsors of a bill in the state House of Representatives to make a state-run lottery possible claim that two-thirds of the general public support their legislation. Test the claim against a lower-sided alternative at the .05 level. Can H_0 be rejected when a newspaper's random sample shows 545 of 841 adults in favor of the lottery?

12.40 Compute the p-value for the test in Exercise 12.39.

12.2 RELATING OPINION POLLS TO INFERENTIAL STATISTICS

Opinion polls such as the Roper Poll, the Gallup Poll, and the Associated Press–NBC News Poll share the same objective: to take the pulse of the public with respect to beliefs, intentions, lifestyle, and opinions about current events and issues. To do so, a sample, typically between 500 and 1500 Americans, is selected and questioned.

Many students do not make a connection between polling and inferential statistics. However, conducting a poll and analyzing the responses is a real-life business statistics application involving inferences about the population proportion π. To see the relationship, let us go through the polling process from beginning to end.

The Polling Process

First, a pollster must define a *target population*—that is, a population that consists of all people having a certain characteristic. For example, a poll focusing on concerns of senior citizens might have a target population of individuals over age 60. A poll to gauge the popularity of Democratic candidates in the weeks prior to a state's primary election would target registered Democrats within that state.

Next, the pollster must determine the method of sampling. Here is the key for the pollster and his or her poll. The sampling method must ensure that the sample is random in order for the results of the poll to be valid. History has no shortage of biased election polls forecasting "sure" victory for an eventual losing candidate. In 1936, for instance, the highly regarded *Literary Digest* Poll blundered in predicting the outcome of the presidential election. Their now infamous prediction was that Republican Alfred Landon would handily defeat Democrat Franklin Roosevelt. The problem was not sample size—the *Digest* had a sample of more than two million responses! Instead the fatal flaw was that the sampling method was not random; the sample was obtained in a manner that grossly over-represented Republicans. A small random sample is much preferred to a large, potentially biased one.

Classical sampling methods like simple random or stratified (discussed in Chapter 1) involve selecting people from a *frame,* or list of everyone in the population. In selecting a simple random sample of eligible voters, we would need a massive data base for the frame consisting of all voters. An alternative to working from a frame is random digit dialing. This approach utilizes a computerized random number generator which produces a list of potential telephone numbers. The pollster using such numbers then will ask some qualifying, or filtering, questions over the phone to ascertain that the respondent is in fact in the target population.

Telephoning is much quicker than mailing questionnaires or conducting in-person interviews, although both of the latter methods are appropriate in certain situations. No one method of sampling is perfect, and new sampling concepts continue to appear. For instance, some research companies that used to rely on selected panel families keeping a written diary of their grocery purchases now employ optical scanning machines and computers that record package Uniform Product Codes at the checkout counter as a means to monitor panel families' buying behavior. To refine its television rating system, the A. C. Nielsen Company also introduced a new method in selected homes—where individual household members punch an assigned button on a meter by the TV set to indicate their choices. Electric utilities are experimenting with end-use metering, in which meters are attached to certain major appliances and equipment in randomly selected homes to estimate and project components of total consumption. Pollsters and others are guided by the need to design a random sampling process.

A third step involves establishing the size of the sample. Equation 12–3 can be used to determine n. The size of the sample will depend on the level of confidence. Usually, though not always, 95% confidence is selected, yielding $Z = 1.96$, which is sometimes rounded to $Z = 2.0$. In the planning stages of a poll, the sponsor of course does not know the value of π. To be conservative, he or she often selects $\pi = .5$ regardless of the situation. This may generate a sample size larger than needed, but a pollster would rather err by sampling too many rather than by selecting too few. When cost and time factors are critical, a more accurate estimate of π may be needed at this stage.

Once the pollster finishes the design stage, execution can begin. This tedious process of gathering data must be done in a consistent manner so as not to bias the responses. Typically the data will be compiled into tables and/or frequency distributions as outlined in Chapter 2. Clearly this step in the polling process is one of collection and organization.

Fifth, the pollster analyzes the organized data, deriving point estimates and standard errors. When combined, these two pieces of information help to form a confidence interval for π. Finally, percentages for and against each candidate, proposition, and so on are reported along with the margin of potential error (in most cases) and the sample size. The margin for potential error is simply the plus or minus part of Equation 12–2. Newspaper articles and television clips usually do not combine the point estimate and the margin for error into what we call a confidence interval. Since the general public does not know its meaning, the term *confidence interval* is often avoided completely!

In spite of this omission, it is a simple matter for the knowledgeable reader to construct a confidence interval from the information given, as the following example illustrates.

EXAMPLE 12.12

Construct a confidence interval for π, the proportion of voters favoring Ronald Reagan over Walter Mondale, based on the following news item.

> *President Reagan has surged into a 19-point lead over prospective Democratic presidential nominee Walter F. Mondale, according to a Gallup Poll released yesterday.*
>
> *The survey of 908 registered voters, taken June 22–25, gave Reagan 56 percent and Mondale 37 percent, with 7 percent undecided or supporting other candidates. The theoretical sampling error is plus or minus 4 points. . . .**

Solution:

The proportion of the sample voters favoring Reagan is $p = .56$ as reported in the first sentence of the second paragraph. Since p is the sample proportion of "successes"—voters favoring Reagan—then to qualify as a binomial experiment, $1 - p$ will be the sample proportion of the "failures"—voters not favoring Reagan. Thus, $1 - p = .44$, combining those in favor of Mondale with those undecided.

Later in the second paragraph, we read that the theoretical sampling error (or margin of error) for this poll is plus or minus 4 points (or .04). A confidence interval for π is

$$\text{Point estimate} \pm \text{Potential sampling error} = .56 \pm .04$$
$$= .52 \text{ to } .60$$

This result suggests that President Reagan would defeat Walter Mondale in the popular vote, since the confidence interval excludes the critical value of $\pi = .50$.

□

*From David S. Broder, "Poll Shows Reagan Ahead of Mondale by 19 Points," *The Washington Post*.

Complete Reporting of Results

Can you determine in Example 12.12 the level of confidence in the confidence interval? No; the sad fact is the newspaper article (not all shown) omitted this important information. Nowhere does the writer mention risk or confidence. This article is an example of a *poorly reported poll*. As consumers of statistics we seek the missing information.

If the margin of error, denoted E, and the sample size are reported, we can "back into" the level of confidence by solving Equation 12–3 for Z, using the conservative estimate of $\pi = .50$. Algebraic rearrangement yields

$$Z = 2E\sqrt{n} \qquad \textbf{(12–7)}$$

Equation 12–7 gives the Z-score needed to achieve the reported margin of error based on a sample size of n people. The probability between the plus and minus values of the Z-score is the level of confidence associated with the poll.

EXAMPLE 12.13

Determine the level of confidence in the poll of Example 12.12.

Solution:

From Example 12.12 we know that $n = 908$ and $E = .04$. Equation 12–7 becomes

$$Z = 2(.04)\sqrt{908} = 2.41$$

Consulting our Z–table, we find that the probability between $Z = -2.41$ and $Z = 2.41$ corresponds to a level of confidence of 98.4 percent.

□

The 98.4% level of confidence in the Reagan/Mondale poll of Example 12.13 is somewhat unusual and peculiar to explain. But it is just this type of analysis that gives added insight into the scientific methods and accuracy of the pollster. Perhaps the pollster initially sought 99% confidence, which requires a sample size larger than 908, but due to the nonrespondents had only 908 valid responses to analyze. Or maybe the pollster used an initial estimate of π other than .50. Perhaps most likely, the actual sampling error might have been more accurate than the reported .04, such as .037, but a newspaper editor chose to round off and report only .04.

A *well-reported poll* includes not only n and E but also a statement about the risk or confidence involved, as Example 12.14 illustrates.

EXAMPLE 12.14

Wilkerson & Associates, a marketing research company, conducted the Commonwealth of Kentucky Poll to obtain perceptions from Kentucky residents about issues pertinent to the state. The results were reported in the following newspaper article from the Louisville *Courier-Journal:*

> *Lexington, the home of the University of Kentucky Wildcats, Keeneland Race Course and last year's NCAA Final Four basketball tournament, is the state's premier place to live, according to the statewide poll conducted by the two marketing research and public-relation firms. . . .*
>
> *Tom Wilkerson, president of Wilkerson & Associates marketing research company, . . . announced the poll results yesterday. . . .*
>
> *The poll . . . included a question about whether respondents favored the creation of a "flagship university." Wilkerson said 44.9 percent favor one, while 42.7 percent oppose it and 12.4 percent are undecided. Of those who favor the flagship suggestion, 67.2 percent believe it should be the University of Kentucky; the University of Louisville was second with 9.7 percent. . . .*
>
> *Wilkerson said 500 Kentuckians from every region in Kentucky were questioned from Jan. 3 to 5 in the random-sample telephone poll, which was conducted as a public service by the two companies.*
>
> *The margin for error was 4.5 percent on statewide results, 10 percent on regional breakdowns, he said.*
>
> *In theory, that means that statewide, in 95.5 cases out of 100 the survey results would differ by no more than 4.5 percentage points from what would have been obtained by interviewing all adults in the state.**

a. What is the target population?

b. Construct a confidence interval for π, the proportion of Kentuckians who favor the creation of a flagship university.

c. Verify the accuracy of the reported level of confidence.

Solution:

a. The target population for this poll is the adult residents of the state of Kentucky.

b. The point estimate of the percentage of Kentuckians in favor of creating a flagship university is 44.9 percent with a margin of error of 4.5 percent. A confidence interval for the proportion in favor is:

$$\begin{aligned}\text{Point estimate} \pm \text{Potential sampling error} &= .449 \pm .045 \\ &= .404 \text{ to } .494\end{aligned}$$

From this we are able to infer that those Kentuckians in favor of a flagship university do not constitute a majority of the target population.

c. The last paragraph of the article addresses the issue of confidence. The reported level of confidence is 95.5 percent. To verify this, let us substitute $E = .045$ and $n = 500$ into Equation 12–7 and solve for Z:

$$Z = 2E\sqrt{n} = 2(.045)\sqrt{500} = 2.01$$

*From John Voskuhl, "Poll Says Lexington Is State's Finest City," Louisville *Courier-Journal*. Copyright ©, Louisville *Courier-Journal*, January 30, 1986. Reprinted with permission.

The probability between $Z = -2.01$ and $Z = +2.01$ is .9556, or roughly 95.6 percent, which is almost exactly the figure reported. Both the pollster and the newspaper reporter are commended for accurately analyzing and reporting the poll to the public.

□

The next time you encounter a poll in the media, look for these valuable results of the statistical analysis:

1. Point estimates
2. Sample size
3. Method of sampling
4. Potential sampling error
5. Statement of confidence

A well-reported and conducted poll generally will provide most or all of this information. A poorly reported and/or poorly conducted poll often will omit the confidence level, margin for error, or other useful information.

COMMENTS

1. Many commercial organizations that regularly conduct and report polls attempt to follow the standards of minimal disclosure of the National Council on Public Polls (New York, N.Y.). Besides the statistical results listed above, the Council recommends that the poll report should include such things as the identity of the sponsor, definition of the population sampled, exact wording of questions used, and dates of interviews. The American Association for Public Opinion Research (Princeton, N.J.) lists a similar set of guidelines ("Standard for Minimal Disclosure") in its membership directory.
2. A young unknown college graduate named George Gallup obtained recognition in 1936 by, prior to election day, claiming that the *Literary Digest* poll was unreliable and by correctly foretelling a big victory margin for Roosevelt. The *Digest* predicted Roosevelt would get 43 percent of the vote; the actual result turned out to be 62 percent for Roosevelt. (Source: Maurice Bryson, "The Literary Digest Poll: Making of a Statistical Myth," *The American Statistician,* November 1976.)

CLASSROOM EXAMPLE 12.1

Analyzing an Opinion Poll

The Bluegrass State Poll ("Kentuckians Loyalties Split When it Comes to Baseball Season," Louisville *Courier-Journal,* May 15, 1988) surveyed adult Kentuckians with respect to their interest in major league baseball. Respondents were contacted by phone, with the telephone numbers called being randomly selected by computer. A total of 767 interviews were obtained.

a. One question in the poll was, "How much do you follow major league baseball?" Response categories were "a lot," "some," "not very much," and "not at all." Of the respondents, 44 percent answered either "a lot" or "some." Develop a 95% confidence interval for the proportion of all Kentuckians who would answer "a lot" or "some" to this question.

b. The newspaper reported that ". . . in theory, in 19 of 20 cases, the poll results would be no more than 3.5 points above or below" the (unknown) population results. Does your interval above agree with this?

c. Those who responded "a lot" or "some" were asked to name their favorite team. Fifty percent picked the Cincinnati Reds. Verify the article's reported margin for error of 5.4 percentage points for this question.

d. Suppose you had a preconceived notion that 1 adult in 4 would be opposed to allowing women to umpire major league games. Would this idea be supported or refuted by the sample result of 128 adults being opposed, 639 not opposed? (*Note:* In the poll, women and men were found to be equally receptive to women umpires!) Find the *p*-value of the test statistic and relate it to a level of significance of your choosing.

e. How would you define the target population for the question about naming one's favorite team?

f. What segment of Kentuckians had no chance to be in this poll? Do you think their exclusion is likely to have distorted the results?

SECTION 12.2 EXERCISES

12.41 Identify the six steps of a polling process.

12.42 What distinguishes a poorly reported poll from a well-reported poll?

12.43 Use the following poll to answer parts a through c. "U.S. Sen. Walter 'Dee' Huddleston today released a statewide poll showing him with a 67–23 percent lead over County Judge Mitch McConnell in their battle for Huddleston's Senate seat. . . .

"The pollsters—Hamilton & Staff of Washington—asked voters, 'If the election were held today, who would you vote for,' and then read the two candidates' names.

"The poll was paid for by the Huddleston campaign and was conducted by telephone between June 20 and 29. Harrison Hickman, a project director for the firm, said that the statewide results are based on interviews of 1,200 registered voters . . . . The statewide findings could be off by as much as 2½ points in either direction, and the county showings by five points each way, he said."*

a. Determine whether this is a poorly or well-reported poll.

b. Find the level of confidence associated with the poll, using Equation 12–7.

c. Construct a 95% confidence interval for π, the proportion of voters in the state favoring Walter Huddleston (who may have been ahead in June but who lost in November).

*From Lonnie Rosenwald, "Huddleston Poll Shows Big Lead over McConnell," Louisville *Times*.

12.44 Use the following poll to answer parts a through d. "A poll by a Maryland research company indicates that a majority of Kentucky Democrats rate Gov. Martha Layne Collins' performance in office as fair to poor.

"The poll, commissioned by a Cincinnati television station and several Kentucky news organizations, surveyed 808 Democrats between Jan. 30 and Feb. 7. Of those, 3.5 percent rated the governor's job performance as excellent, 37.1 percent as good, 36.5 percent as fair and 17.5 percent as poor.

"Mason-Dixon Opinion Research Inc. of Columbia, Md., said the margin of error was no more than plus or minus 3.5 percent. That margin indicates a 95 percent probability that the sample result would hold true if the entire population of Democratic voters were polled."*

a. Determine whether it is a poorly or well-reported poll.

b. Find the level of confidence associated with the poll, using Equation 12–7.

c. Does your answer in part b agree with the reported level of confidence?

d. Do you think the last sentence in the article is a good interpretation of 95% confidence? How could it be improved?

12.45 **a.** Find and verify the level of confidence in the following poll. "Nearly half the fans at the Tiger's Opening Day game were supposed to be someplace else, according to a *Free Press* survey of 1,278 fans entering the ballpark Monday. . . .

"Asked 'Are you playing hooky from work or school today,' 45 percent of the 1,278 fans interviewed said yes. Fifty-four percent said no, and one percent declined to answer.

"In an effort to gauge opinion among Monday's Opening Day crowd of 51,180 on a range of topics, the *Free Press* placed 25 interviewers—mostly journalism students hired for the day from area universities—outside Tiger Stadium's 21 gates. . . .

"The poll was conducted by the *Free Press* with the help of journalism students from Wayne State University, Oakland University, and others from the Detroit area.

"The findings were based on 1,278 random interviews as fans entered Tiger Stadium on Monday morning and early afternoon.

"Sampling error on a survey of this size is three percentage points. This means that, 95 times out of 100, the results would differ no more than three points either way from the results if everybody who attended the game Monday had been interviewed."[†]

b. Construct a 95% confidence interval for π, the proportion of fans attending the game who admitted to playing hooky from work or school. Use the sample result based on all fans interviewed.

12.46 Use the following poll to answer parts a through c. "Seven of 10 Americans think the space shuttle program is a good investment for the country, according to . . . 1,583 adults in the nationwide telephone poll conducted early last week. . . .

"As with all sample surveys, the results of AP-NBC News polls can vary from the opinions of all Americans because of chance variations in the sample.

"For a poll based on about 1,500 interviews, the results are subject to an error margin of 3 percentage points either way because of chance variations. That is, if one could have talked to all Americans with telephones, there is only 1 chance in 20 that the findings would vary by more than 3 percentage points.

*From "Governor Gets Low Marks," *The Kentucky Post*, February 17, 1987. Copyright © Associated Press. Reprinted with permission.

[†]From Tim Kiska and David Ashenfelter, "Hooky: Name of Game for Many in the Crowd," Detroit *Free Press*, April 9, 1985. Copyright © Detroit *Free Press*. Reprinted with permission.

"Of course, the results could differ from other polls for a number of reasons. Differences in the exact wording of questions, differences in when the interviews were conducted and different methods of interviewing could also cause variations."†

a. Determine whether it is a poorly or well-reported poll.
b. Find the level of confidence associated with the poll, using Equation 12–7.
c. Does your answer in part b agree with the reported level of confidence? If not, mention possible reasons for the discrepancy.

*12.3 PROCESSOR: INFERENCES FOR THE PROPORTION, π

This processor can compute confidence intervals for π and/or the test statistic Z^* for a hypothesis test about the value of π. Selecting the processor and reading the Z-table follow in a manner similar to that for the processors in the previous two chapters. After reading the Z-table and activating the Data section of the menu, we encounter a different data entry screen, as shown in Figure 12.8.

At this point, the processor wishes to know the sample proportion of successes, p. We must either input p directly or provide information so that the processor can compute p. If we select "Integer Value For X," the processor will ask us to enter the number of successes X and then it will compute the sample proportion of successes as $p = X/n$. If we already know the value for p, we would select the option, Decimal Value For p.

After data entry, we are ready to construct a confidence interval or test a hypothesis. To illustrate the former, we have chosen the problem in Example 12.5. A 90% confidence interval for π, the proportion of VCR owners who usually or always bypass recorded commercials, is desired. The sample proportion is .38, based on $n = 698$. After using the second data entry format of Figure 12.8, advancing to Results, and specifying the 90.0 level of confidence, we arrive at the results shown in Figure 12.9.

```
≡                  Inferences For The Proportion, π                      MENU
   Introduction  *able [Data] Results  Exit                          | F1=Help

                        Integer value for X
                        Decimal value for p
```

FIGURE 12.8 Data Entry Options for Inferences About π

†From Timothy Harper, "Most Feel Space Shuttle Is a Good Investment, Poll Shows." Copyright © Associated Press. Reprinted with permission.
*Optional

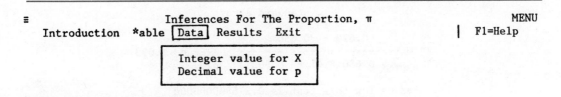

```
≡                    Inferences For The Proportion, π                    QUERY
  Introduction  *able  Data  [Results]  Exit                          | F1=Help

              For 90.0% confidence, the interval is
                   point estimate ± Z * (standard error)

                   =  0.380 ± Z * √((0.380 * (1 - 0.380)) / n)
                   =  0.380 ± 1.645 * √(0.236 / 698)
                   =  0.380 ± 0.030

              The lower limit of the confidence interval is  0.350
              The upper limit of the confidence interval is  0.410

              ══════════════ Confidence ══════════════
              ║                                       ║
              ║   Compute another interval? (y/n) No  ║
              ║                                       ║
              ═════════════════════════════════════════
```

FIGURE 12.9 Confidence Interval Results for Data in Example 12.5

```
≡                    Inferences For The Proportion, π                    READY
  Introduction  *able  Data  [Results]  Exit                          | F1=Help

    Hypothesis test     Ho: π =  0.650
                        Ha: π not =  0.650

                       sample proportion - hypothesized value of π
             TS:  Z* = -------------------------------------------
                                     standard error

                           0.679 -  0.650
                  Z* = --------------------------------- =  1.093
                       √(( 0.650 * (1 -  0.650)) / 315)

              The p-value for this test is 0.275800000000


    The decision to "reject Ho" or "do not reject Ho" depends on your
  judgment or tolerable risk of Type I error (see text Figure 10.18).
```

FIGURE 12.10 Hypothesis Test Results for Data in Example 12.9

We use Example 12.9 to illustrate the output for a hypothesis test about the value of π. We employed the "Integer Value For X" option to enter $X = 214$ and $n = 315$. After advancing to Results, selecting a two-tailed hypothesis test, and entering the test value ($\pi = .65$), we generate the results shown in Figure 12.10.

12.4 INFERENCES ABOUT σ^2 AND σ

To this point, our study of statistical inference has focused on applications involving μ and π. We now turn our attention to confidence intervals/hypothesis tests for the dispersion or variability within a population, as measured by the variance σ^2 or the standard deviation σ. Consider the following situations where knowledge of population variability would be important:

- Many applications in science, engineering, and manufacturing demand precision tools and instruments that not only must perform correctly on average but also must exhibit a minimum amount of variability in their performance.
- A pharmaceutical firm has had a newly approved prescription drug for hypertension on the market for about a year. They would like to estimate the dispersion in the price of the drug that exists at the retail level.
- Mass production and the use of interchangeable parts require uniformity. In many instances, the variability of components and parts has to be monitored and controlled. Produced pieces that have dimensions or physical properties outside of a narrow tolerance for variability may end up as scrap.
- As mentioned in Chapters 3 and 5, in finance and investment applications the notion of riskiness or uncertainty of competing capital projects or portfolio alternatives is quantified by the estimated standard deviation of returns.

In our study of inferences for σ^2 and σ, we will find it more convenient for mathematical reasons to work with the variance, σ^2, rather than the standard deviation, σ. However, due to the square root relation between these two measures, any results for the variance can be converted easily into results for the standard deviation. Our inference procedures about σ^2 and σ will rely on the assumption that the population being sampled is normally or approximately normally distributed.

Review of Variance and Standard Deviation Computation

In Chapter 3 we provided several different formulas for computing the sample variance, S^2. Two variations, repeated in the following formula, provide often-used expressions for S^2 (for raw data):

$$S^2 = \frac{\Sigma(X - \overline{X})^2}{n - 1} = \frac{\Sigma X^2 - (\Sigma X)^2/n}{n - 1}$$

EXAMPLE 12.15

Use both formulas for S^2 to compute the sample variance of quarterly rates of return of a mutual fund over the most recent six quarters. The quarterly rates of return, X, are

4.2 7.9 9.5 11.2 8.1 7.7

Solution:

Using the defining formula $S^2 = \dfrac{\Sigma(X - \overline{X})^2}{n - 1}$, we have

$$\overline{X} = \frac{\Sigma X}{n} = \frac{4.2 + 7.9 + \cdots + 7.7}{6} = \frac{48.6}{6} = 8.1$$

Then

X	$\overline{X}$	$(X - \overline{X})$	$(X - \overline{X})^2$
4.2	8.1	−3.9	15.21
7.9	8.1	−0.2	.04
9.5	8.1	1.4	1.96
11.2	8.1	3.1	9.61
8.1	8.1	0	0
7.7	8.1	−0.4	.16
			26.98

$$S^2 = \frac{\Sigma(X - \overline{X})^2}{n - 1} = \frac{26.98}{5} = 5.396$$

Using the calculating formula, we have

X	X^2
4.2	17.64
7.9	62.41
9.5	90.25
11.2	125.44
8.1	65.61
7.7	59.29
48.6	420.64

$$S^2 = \frac{\Sigma X^2 - (\Sigma X)^2/n}{n - 1}$$

$$= \frac{420.64 - (48.6)^2/6}{5} = \frac{26.98}{5} = 5.396$$

The standard deviation is $\sqrt{5.396}$, or about 2.32. The defining formula emphasizes that the variance and standard deviation measure variability by employing *squared deviations*. The calculating formula is simply an algebraic manipulation of the defining formula to a form that is sometimes easier to work with. Other formulas for S^2 and S for organized data appear in Section 3 of Chapter 3.

□

The Sampling Distribution of S^2

Our outlook in developing the sampling distribution of the sample variance S^2 is much the same as it was for developing the sampling distribution of $\overline{X}$ (Chapter 8) and the sampling distribution of p (Section 12.1). We begin by recognizing that there is a constant population value, in this case, σ^2, whose value is unknown to us. A random sample will yield a point estimate of that value, in this case S^2. The sample result S^2 is unlikely to be equal to the population value σ^2; however, S^2 will generally be either above or below the true value of σ^2. If we knew the pattern that all possible sample results S^2 would form, we would have the sampling distribution of S^2. As we are about to see, the sampling distribution of S^2 differs from the previous sampling distributions encountered in that it does not follow the familiar normal distribution. This is because S^2 involves *squared* deviations.

The sampling distribution of S^2 was derived many years ago. Unfortunately, there is a different sampling distribution for S^2 for each different combination of σ^2 and sample size n. Fortunately, we can simplify matters by standardizing, a process we used with normal distributions. Recall that many different normal distributions exist, depending on the values of μ and σ. By standardizing to create the standard normal variable Z, we can employ one table for all normal distributions. We now apply a somewhat similar procedure for the sampling distribution of S^2. We standardize to take into account the different possible values of σ^2 and n. The standardized variable is called the chi-square (χ^2) random variable, and is given by

$$\chi^2 = \frac{(n - 1)S^2}{\sigma^2} \qquad \text{(12–8)}$$

As Figure 12.11 indicates, the chi-square variable has a continuous distribution with an asymmetric (skewed right) appearance. The chi-square variable has a lower limit of zero but no upper limit.

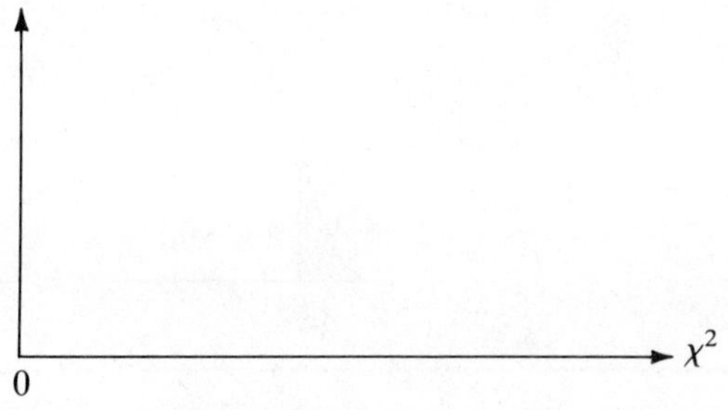

FIGURE 12.11 A Chi-Square Distribution

Use of the Chi-Square Tables

When we develop confidence intervals or conduct hypothesis tests for σ^2, we will need cutoff values from the chi-square tables. These values will depend on the sampling situation's degrees of freedom (*df*). For inferences about a population variance or standard deviation, *df* will be $n - 1$.

Our chi-square tables are given in Appendix G at the back of the book. They provide cutoff values for commonly used risk levels and for various degrees of freedom. Refer to these tables as we consider some examples that illustrate finding cutoff points on χ^2 distributions.

EXAMPLE 12.16

Between what two values of the chi-square distribution does the middle 95 percent of the total probability lie for the following:

a. Five degrees of freedom

b. Ten degrees of freedom

c. Twenty degrees of freedom

Solution:

For a 95% confidence interval, we would need the table values which place 2.5 percent of the total probability in each of the two tails. We find the cutoff point to place 2.5 percent in the left tail on the first page of the χ^2 tables; the right tail cutoff is on the second page. The correct values are

a. 0.831 and 12.833

b. 3.247 and 20.483

c. 9.591 and 34.170 (see Figure 12.12)

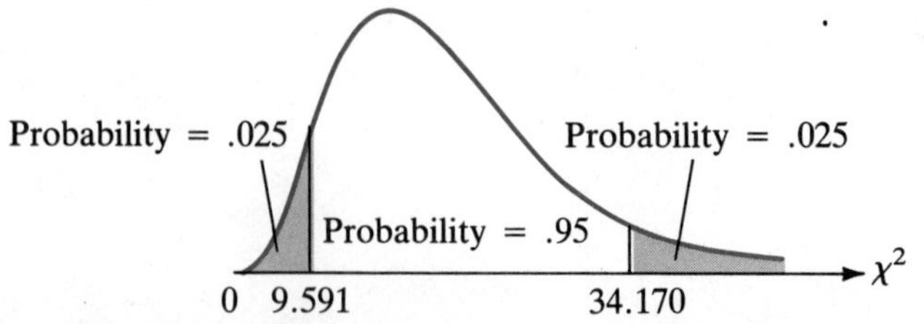

FIGURE 12.12 Middle 95 percent of χ^2 Distribution, with df = 20

To express these results in terms of probability, we could report

a. For $df = 5$, $P(0.831 \leq \chi^2 \leq 12.833) = .95$
b. For $df = 10$, $P(3.247 \leq \chi^2 \leq 20.483) = .95$
c. For $df = 20$, $P(9.591 \leq \chi^2 \leq 34.170) = .95$

□

EXAMPLE 12.17

Find the χ^2 cutoff value(s) for

a. the middle 90 percent of probability, $df = 7$
b. 10 percent of probability in the right tail, $df = 17$
c. 1 percent of probability in the left tail, $df = 11$

Solution:

a. The cutoff points for a 90% confidence interval will be found in the two .05 columns. For seven degrees of freedom, $P(2.167 \leq \chi^2 \leq 14.067) = .90$.
b. For 10% probability in the right tail, simply use the .10 column of the right tail page. You should be able to locate $\chi^2 = 24.769$ (see Figure 12.13).
c. For 1% probability in the left tail, use the .01 column of the left tail page: The cutoff is $\chi^2 = 3.053$.

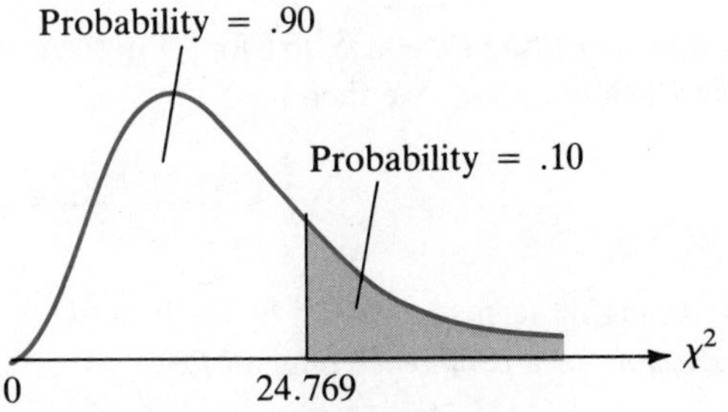

FIGURE 12.13 Upper 10 percent of χ^2 Distribution, with $df = 17$

□

COMMENTS

1. The cutoff points in the chi-square tables are not our end result when we make an inference about σ^2. As we are about to see, these table values are intermediate numbers we need in order to accomplish our objective—to develop a confidence interval and/or perform a hypothesis test for σ^2.

2. Degrees of freedom often are expressed as $n - k$, where n = sample size and k = the number of restrictions on the data. For inferences that involve the computation of S^2, *df* always will be $n - 1$. Note that $n - 1$ is the denominator in the formula for S^2. We referred to the $n - 1$ term in the denominator (as opposed to just n) as a correction in Chapter 3, as we did not want to get into a discussion of *df* at that time. In general, we can think of a variance as being any sum of squared deviations divided by its degrees of freedom.
3. As the tables indicate, there is not just one χ^2 distribution. Rather, there is a family of distributions whose members differ according to their degrees of freedom. The mean of a χ^2 distribution is its *df;* its standard deviation is $\sqrt{2df}$. For instance, a χ^2 distribution with 11 degrees of freedom has a mean or expected value = 11, with a standard deviation = $\sqrt{22}$, or about 4.69. As *df* increase, the mean and standard deviation of χ^2 distributions become larger while the appearance becomes less skewed and more like a normal distribution.

Confidence Intervals for σ^2

Examples 12.16 and 12.17 show that a general expression for a confidence interval involving a chi-square variable can be written as

$$\begin{matrix}\text{Left tail } \chi^2 \\ \text{table value}\end{matrix} \leq \chi^2 \leq \begin{matrix}\text{Right tail } \chi^2 \\ \text{table value}\end{matrix}$$

where the χ^2 table value depends on the desired confidence. Let us denote the lower and upper limits of the interval as χ_L^2 and χ_R^2 (L = left, R = right), respectively, giving us

$$\chi_L^2 \leq \chi^2 \leq \chi_R^2$$

Let us then substitute $(n - 1)S^2/\sigma^2$ for χ^2 in the expression above; this is the equality given in Equation 12–8. We then have

$$\chi_L^2 \leq \frac{(n - 1)S^2}{\sigma^2} \leq \chi_R^2 \qquad \text{(12–9)}$$

Now rearranging terms to get σ^2 by itself between the inequality signs, we arrive at an expression for a *confidence interval for* σ^2.

Confidence Interval for the Population Variance σ^2

$$\frac{(n - 1)S^2}{\chi_R^2} \leq \sigma^2 \leq \frac{(n - 1)S^2}{\chi_L^2} \qquad \text{(12–10)}$$

We are now ready to work some examples that illustrate confidence interval estimation of σ^2 or σ.

EXAMPLE 12.18

The quality control group for a motorcycle manufacturer chooses a random sample of 20 newly assembled bikes for a series of inspection and performance tests. Additional measurements are sought—estimates of odometer accuracy and variability. These measures are taken to begin a performance record with a new company which has recently become their supplier of odometer parts. The bikes are driven over a course known to be exactly 200 miles. The sample of odometer readings yields $\overline{X} = 200.21$ miles, $S^2 = .23$, and $S = .48$. Develop a 95% confidence interval for σ, the population standard deviation over the course. Assume that odometer readings form an approximately normal distribution.

Solution:

For $n = 20$, our $df = n - 1 = 19$. Equation 12–10 gives us

$$\frac{(n-1)S^2}{\chi_R^2} \leq \sigma^2 \leq \frac{(n-1)S^2}{\chi_L^2}$$

The value for χ_R^2 is 32.852, found in the table at the intersection of the $df = 19$ row and the .025 column of the right tail. The χ_L^2 value of 8.907 is located where the .025 column of the left tail intersects the 19 df row. Substituting the numerical values for $n - 1$, S^2, χ_R^2, and χ_L^2, we obtain

$$\frac{19(.23)}{32.852} \leq \sigma^2 \leq \frac{19(.23)}{8.907} \qquad \text{or } .13 \leq \sigma^2 \leq .49$$

The interval given is for σ^2. The interval for σ is developed by taking the square roots of the limits for σ^2, producing .36 miles $\leq \sigma \leq$ .70 miles. Future tests will monitor both the mean and the standard deviation.

Note that the upper and lower limits of the confidence interval are not equidistant from the point estimate. This is due to the skewed right shape of the chi-square sampling distribution. The point estimate is always closer to the lower limit than the upper limit.

□

EXAMPLE 12.19

Fluctuations in the level of the Great Lakes are a natural phenomenon. The current high levels are causing extensive erosion and property damage along their shores. The following readings where X is number of feet above sea level (Department of the Interior, U.S. Geological Survey, Station 4052) are a random sampling of Lake Michigan levels from the 1860s through the 1980s. Assume the population of readings is normally distributed, and develop a 90% confidence interval for σ over this time period.

581.0	580.5	580.1	581.4
580.4	580.0	579.1	578.6
579.3	578.8	578.0	578.7
577.2	579.2	579.2	578.9

Solution:

For these data, $S^2 = 1.223$ and $S = 1.106$ feet. The sample size of 16 tells us to locate chi-square values in the 15 *df* row and, for 90% confidence, in the two .05 columns. The interval for σ^2 is then

$$\frac{15(1.223)}{24.996} \leq \sigma^2 \leq \frac{15(1.223)}{7.261} \qquad \text{or } .734 \leq \sigma^2 \leq 2.527$$

The interval for σ is then .86 feet $\leq \sigma \leq$ 1.59 feet.

□

Hypothesis Tests for σ^2

Examples 12.18 and 12.19 have illustrated making an inference about σ or σ^2 via a confidence interval. We now consider our other inference format—a hypothesis test for σ^2. Hypothesis tests for σ^2 differ from those for μ and π in that the test statistic is a chi-square variable instead of a normal distribution Z variable. The χ^2 test statistic is given in the following box.

Test Statistic for a Population Variance σ^2

$$\chi^{2*} = \frac{(n - 1)S^2}{\sigma_0^2} \qquad \textbf{(12–11)}$$

Consistent with other applications, the zero subscript in the *TS* denotes the value for σ^2 presumed by the null hypothesis.

EXAMPLE 12.20

A prescription drug manufacturer makes time-release capsules that are supposed to contain 40 milligrams per capsule of the active ingredient. The process is also supposed to be uniform—the firm's process control people regularly sample packaged capsules to check variability. The process standard deviation of capsule dosage is not supposed to exceed 1.25 milligrams. A sample of 25 capsules shows $\overline{X} = 40.33$ mg, $S = 1.34$ mg. Is this convincing evidence that dosage has become too variable? The data appear approximately normal. Test at $\alpha = .01$ the presumption that process variability is in control.

Solution:

Many, but not all, tests for variability are upper tail tests. This example is typical—there is interest in detecting if the process fluctuates too much. The packager will need to take action only if the sample reveals excessive variability; no action is required if the process appears in control. Our hypotheses will be

$$H_0\text{: } \sigma = 1.25 \text{ mg (equivalently, } H_0\text{: } \sigma^2 = 1.5625)$$
$$H_a\text{: } \sigma > 1.25 \text{ mg (equivalently, } H_a\text{: } \sigma^2 > 1.5625)$$

With $n = 25$ and an upper tail test with 1% α-risk, our cutoff χ^2 value is found at the intersection of the $df = 24$ row and the right tail .01 column. The table value is $\chi^2 = 42.980$. Our test statistic is

$$\chi^{2*} = \frac{(n-1)S^2}{\sigma_0^2} = \frac{24(1.34)^2}{1.25^2} = 27.580$$

Since the $TS <$ cutoff table value, we cannot reject H_0 (see Figure 12.14). The sample evidence is consistent with H_0 being true; we can declare that the process is "in control." Table 12.4 summarizes this test.

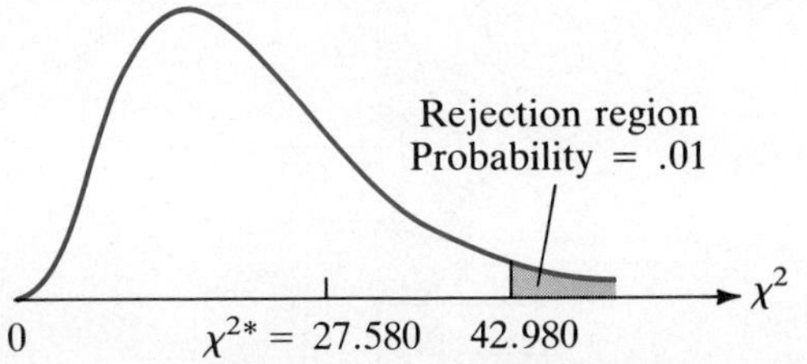

FIGURE 12.14 Test Statistic and Rejection Region for Example 12.20

TABLE 12.4 Symbolic Summary of Example 12.20

H_0: $\sigma = 1.25$ milligrams
H_a: $\sigma > 1.25$ milligrams
TS: $\chi^{2*} = 27.580$
RR: Reject H_0 if $\chi^{2*} > 42.980$
C: Do not reject H_0

□

EXAMPLE 12.21

A midwestern firm creates and administers standardized tests that are taken by college seniors considering going to graduate school. The firm is revising certain tests to reflect changes in undergraduate curricula as well as new knowledge in various fields. Possible scores on the exams range from 200 to 800. The tests historically have been constructed to have $\mu = 500$, $\sigma = 100$. Firm officials think any revised exams must preserve these figures. A new version of the chemistry exam has been prepared and given to a random sample of 81 seniors. Results are $\overline{X} = 497$ and $S = 83$. Test H_0: $\sigma = 100$, using $\alpha = .05$.

Solution:

Since a deviation from the traditional value of $\sigma = 100$ in either direction is undesirable, a two-tailed H_a should be established:

H_0: $\sigma = 100$ (no change in historical value)
H_a: $\sigma \neq 100$ (too much or too little dispersion)

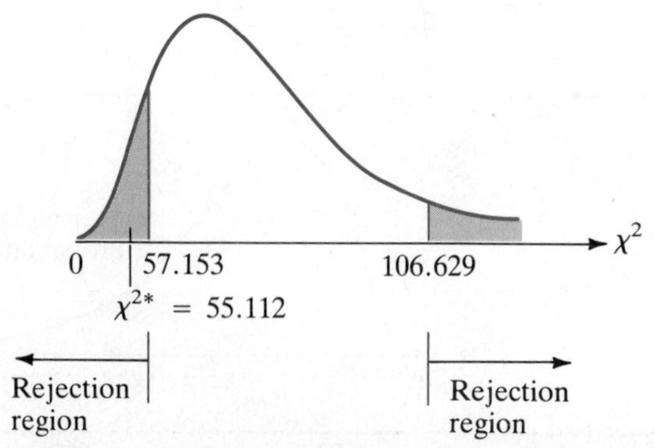

FIGURE 12.15 Test Statistic and Rejection Regions for Example 12.21

The cutoff χ^2 values are 57.153 and 106.629 for 80 *df* and 5% risk of a Type I error. The *TS* is

$$\chi^{2*} = \frac{(n-1)S^2}{\sigma_0^2} = \frac{80(83)^2}{100^2} = 55.112$$

Since the *TS* lies in *RR* (see Figure 12.15), a case is made that σ differs substantially from 100. While this version of the chemistry exam may have the desired mean, it needs changes to spread out the scores further.

□

COMMENTS

1. The inferences for σ^2 and σ assume the population of interest is normally distributed. If this is in doubt, a graphical inspection of the data and/or computation of the coefficient of skewness (Section 3.6) can give indications as to whether the normality assumption is warranted. If a population is noticeably skewed, inferences about σ^2 and σ can be developed using a procedure called the *jackknife,* a method much more computationally involved than the χ^2 method but much less sensitive to departures from normality. See Neter, Wasserman, and Whitmore (1980) or other advanced texts for details.
2. The *p*-value concept applies to tests using the χ^2 distribution, but due to the way χ^2 tables are set up, we can report *p*-values in interval form only. For instance, assume we have an upper-tail test with 5 *df*. If the *TS* were 13.50, we would report $.01 < p\text{-value} < .025$ since the *TS* is in between the .01 and .025 column values. If instead the *TS* were 20.40, we would report $p\text{-value} < .005$. As usual, a two-sided test doubles the one-tail *p*-value result. For instance, assume a two-sided test with 10 *df* yields a *TS* of 3.50. If so, the *p*-value limits of $(.025 < p\text{-value} < .05)$ would be doubled, yielding $.05 < p\text{-value} < .10$. The software program for this section can compute and report chi-square *p*-values in an exact instead of interval form.

CLASSROOM EXAMPLE 12.2

Developing a Confidence Interval for the Population Standard Deviation

According to *Penny Power* ("McDonald's Fast Food Salads," June/July, 1988) a McDonald's chef salad is supposed to have a uniform weight. A random sample of six chef salads was purchased from different McDonald's restaurants by *Penny Power*. Each was weighed—the heaviest was 11.25 ounces; the lightest was 8.25 ounces. Suppose we would like to learn about the variation in weights of McDonald's chef salads and the six weights, *X* (in ounces), obtained were as follows:

10.80	9.60	8.25	9.80	11.25	9.10

a. Determine S^2, the sample variance.

b. If we are to develop a 90% confidence interval, what chi-square table values would be used?

c. Construct the 90% confidence interval for the variance of the population from which the sample was drawn.

d. Give a 90% confidence interval for σ.

e. What assumption is necessary to develop the confidence intervals above?

SECTION 12.4 EXERCISES

12.47 Compute the sample variance for X, miles per gallon per tankful of gas for a new pickup truck:

24.3	25.6	24.1	23.9	25.7	26.4	25.7

12.48 Following are the scores a professional golfer has achieved in the opening round of his last ten tournaments. Compute the sample variance and the sample standard deviation.

74	78	78	77	71	70	75	71	73	76

12.49 Find the cutoff points for the middle 90 percent of a chi-square distribution that has the following:

a. 6 *df*
b. 11 *df*
c. 15 *df*

12.50 For a chi-square random variable with 8 *df*, determine the following:

a. What cutoff value places 5 percent of the probability in the right tail?
b. What cutoff value places 1 percent of the probability in the left tail?
c. What cutoff values would be used for a 99% confidence interval?

12.51 Demonstrate the algebra required to get from Equation 12–9 to Equation 12–10.

12.52 A sample of eight items from a normal population reveals $\overline{X} = 9.64$ and $S = 1.04$. Develop a 90% confidence interval for σ, the population standard deviation.

12.53 A sample of $n = 100$ is drawn from a population believed to be approximately normal. The sample statistics computed are $\overline{X} = 121$ lb, $S^2 = 144$ lb^2, $S = 12$ lb. Construct an 80% confidence interval for σ^2 and for σ, ignoring the small error created by use of the table values for 100 *df* instead of 99 *df*.

12.54 A firm that produces basketball backboards and rims has taken a random sample of 25 rim diameters to estimate the variability of its manufacturing process. The sample variance is $S^2 = .0081$ inch squared. Compute a 90% confidence interval for the population's variance and standard deviation. Assume the basketball rim diameters are approximately normally distributed.

12.55 A packager of 50-pound bags of salt pellets for use in water softeners has taken a sample of 30 bags which have been filled in the past 24 hours. Each bag is carefully weighed, and the following statistics are computed (to two decimal-place accuracy):

$$\overline{X} = 50.22 \text{ lb}$$
$$S^2 = .24 \text{ lb}^2$$
$$S = .49 \text{ lb}$$

a. Construct a 99% confidence interval for σ^2, the filling process variance.
b. Provide a 99% confidence interval for σ.
c. What assumption is made to compute these intervals?
d. Why are the point estimates given above not exactly at the center of your confidence intervals?

12.56 Following is a random sample of 24 stores' monthly profits in thousands of dollars. The stores are established outlets of a franchised fast-food restaurant.

29	42	47	50	64	68	74	77
80	86	88	91	92	93	95	99
106	111	112	119	121	121	126	135

a. Demonstrate that $\overline{X}$ = \$88.6 and S = \$28.4 using the raw data formulas.
b. Compute a 95% confidence interval for σ, the population standard deviation of franchise monthly store profits. Assume the population is approximately normal.

12.57 A financial adviser wants to estimate the dispersion in yields of the various money-market fund offerings. She selects a random sample of 15 such funds, finding a sample standard deviation of .55%. If we assume yields are nearly normally distributed, estimate a 90% confidence interval for the population standard deviation.

12.58 If a normal population is presumed, is a test of H_0: $\sigma^2 = 9$ equivalent to a test of H_0: about 95 percent of values in the population are within ± 6 of the population mean? Why or why not?

12.59 Test H_0: $\sigma^2 = 10$ against an upper-tail H_a at the .05 level. Your sample of 20 items shows $S^2 = 11.6$. Can H_0 be rejected?

12.60 In a two-sided test of H_0: $\sigma = .40$, the sample standard deviation, based on 30 observations, is $S = .66$. Can you be 99% certain (or more) that H_0 is not true?

12.61 State the p-value for Exercise 12.60 in inequality form.

12.62 Suppose state highway officials in a southern state established through frequent sampling in the early and middle 1980s that the standard deviation of vehicle speeds on their interstate highways was 3.1 mph. In 1987 the speed limit was raised from 55 to 65 mph. Test H_0: $\sigma = 3.1$ mph after the limit change against an alternative which states that variability of speeds increases as the speed limit increases. A random sample of 51 vehicle speeds is selected for the test. If $S = 4.4$ mph, should H_0 be rejected? Use $\alpha = .025$.

12.63 Report the p-value for the test in Exercise 12.62.

12.64 An asphalt manufacturer wants the drum-filling process for its 5-gallon drums of driveway sealant to have a standard deviation of .08 gallons or less. A sample of two dozen drums is to be drawn to check for consistency of fill level. What should H_0 and H_a be? Use the .10 level of significance to report your conclusion when the sample standard deviation is .077 gallons.

12.65 Refer to Exercise 12.64. In terms of the problem setting given, how would you express the probability $(1 - \alpha)$ in words?

*12.5 PROCESSOR: INFERENCES FOR σ OR σ^2

This processor can construct confidence intervals for σ and for σ^2 and can compute the chi-square test statistic for a hypothesized value of σ or σ^2. After selecting this processor from the One Sample Inference menu, we view an introductory screen and then read two tables, chi-square and Z. (The Z-table is needed as part of the computation for chi-square p-values.)

We are now ready to enter data for either form of inference—confidence interval or hypothesis test. Let us use the data in Example 12.19 to illustrate confidence interval construction. The objective in this problem is to establish a 90% confidence interval for σ, the standard deviation for X, where X = Lake Michigan level. We move to Data and furnish the summary statistics $n = 16$ and $S = 1.106$. We then move to Results, specify Confidence Interval, select 90 percent, and obtain the output shown in Figure 12.16. This screen provides the confidence interval for σ^2; touch Enter again to view Figure 12.17—results in terms of σ. (*Note:* The computer cannot display Greek letter chi (χ); it uses capital X as a substitute.)

To demonstrate the hypothesis test display, consider Example 12.20, where we wish to test H_0: $\sigma = 1.25$ mg against an upper-tail alternative. After entering our summary statistics ($n = 25$, $S = 1.34$), we choose the Upper-Tail hypothesis test option and specify the H_0 value of 1.25 for σ. The results will appear on a screen similar to Figure 12.18.

```
≡                         Inferences For σ Or σ²                          READY
  Introduction  X² *able  * Table  Data [Results] Exit                 | F1=Help

  The 90% confidence interval for the population variance, σ², is

                     Lower Limit ≤ σ² ≤ Upper Limit

                  (n - 1) * S²          15 * 1.223236
          LL = ------------------ = --------------- = 0.734059
               Upper tail X² value        24.996

                  (n - 1) * S²          15 * 1.223236
          UL = ------------------ = --------------- = 2.526999
               Lower tail X² value        7.261
```

FIGURE 12.16 Confidence Interval Results in Terms of σ^2 for Example 12.19

*Optional

```
≡                              Inferences For σ Or σ²                                    QUERY
  Introduction  X² *able  * Table  Data [Results] Exit                              | F1=Help

    For the population standard deviation, σ, the confidence
   interval is obtained by taking the square root of the limits
   for the variance shown on the previous screen.

          Confidence interval for σ:

                         √LL ≤ σ ≤ √UL

                    0.8568 ≤ σ ≤ 1.5897

              ════════════ Confidence ════════════
                 Compute another interval? (y/n) No
```

FIGURE 12.17 Confidence Interval Results in Terms of σ

```
≡                              Inferences For σ Or σ²                                    READY
  Introduction  X² *able  * Table  Data [Results] Exit                              | F1=Help

   Hypothesis test         Ho: σ =  1.250
                           Ha: σ >  1.250

                                (n - 1) * S²
                     TS:  X²* = ------------
                                     σ²

                                 24 * 1.79560
                          X²* = --------------- =  27.580
                                   1.562500

          The p-value for this test is 0.278133949069

   The decision to "reject Ho" or "do not reject Ho" depends on your
   judgment or tolerable risk of Type I error (see text Figure 10.18).
```

FIGURE 12.18 Results for Hypothesis Test in Example 12.20

12.6 SUMMARY

In the previous chapters, we introduced the logic and process of confidence interval estimation and tests of hypotheses for the population mean. The purpose of this chapter has been simply to take these general concepts and apply them to other settings—for inferences about a binomial proportion and for inferences about population dispersion.

We have learned that the normal or Z-distribution does not apply to all sampling conditions. Small sample inference requires the t-distribution; inferences about dispersion require us to utilize the chi-square distribution. We will have further uses for knowing the sampling distribution of the sample variance as well as the chi-square distribution in subsequent chapters. Overall, our goal will remain the same—to see what we can infer about some underlying population or populations by utilizing information acquired by sampling.

CHAPTER 12 EXERCISES

12.66 The following are three of many questions that appeared on a questionnaire mailed to people who had graduated from a private university 20 years ago. For each, state whether the variable of interest appears to be binomial.

a. "Have you completed any graduate degrees?"

b. "How many Homecomings have you attended since your graduation?"

c. "Do you have any children nearing college age?"

12.67 Suppose 68 percent of all grocery shoppers at a certain store redeem coupons. If a sample of 200 shoppers is monitored, what is the region of concentrated values for p? How likely is the sample to show 75% or more coupon redeemers?

12.68 Management of a minor league baseball franchise wants to estimate the proportion of paying customers that is female. A random sample of 150 customers will be taken. If in fact the true proportion of females is .41, how likely is the sample proportion to land within plus or minus two percentage points of the population figure? If the sample were to be 500 customers, what would be the probability?

12.69 A local utility company plans to visually inspect a sample of 350 of its customers' electric meters to estimate the proportion that will need replacement in the next two years. Suppose that in the population of electric meters, 13 percent have visible indications of upcoming need for replacement.

a. Specify the mean and standard error of the sampling distribution of p.

b. What are the chances that the sample result will miss the true percent in the population by no more than $\pm .02$?

c. Find the probability that the sample percentage will be 10 percent or less.

12.70 A lumber company is about to receive a shipment of over a million boards. They will choose a random sample of 400 boards as part of an acceptance sampling procedure. If the overall percent defective (excessive warpage) in the shipment is 4.5 percent, how likely is it the sample will show 7 percent or more defective?

12.71 Two teenage brothers are about to play a game of Ping-Pong; the winner is to get the family car for the evening. The older one offers his younger brother a choice of a game to 21 or a game to 11. Based on past experience, the younger brother knows he wins about 45 percent of the points against his older brother. What choice should the younger player make?

12.72 Included in an article on religion in the United States ("How the Bible Made America," *Newsweek,* December 27, 1982) were results of a Gallup poll reporting that 44 percent of U.S. Catholics could state correctly four or more of the Ten Commandments. If this result was based on a sample of 770 U.S. Catholics, provide a 90% confidence interval for π, the proportion of all U.S. Catholics who can state four or more commandments.

12.73 Property owners in most localities can appeal their property tax assessment if they think that it is out of line. In one large city, 782 property owners filed appeals after a new assessment evaluation program was implemented. In the first three weeks of hearings, the appeals board heard 214 of these cases and granted relief in 79 instances. Construct a point estimate and a 90% confidence interval for the percent of all appeals that will be successful.

12.74 In his annual report to the readership, the editor of a scholarly journal reports that he accepted 48 manuscripts submitted for publication and rejected 246 in the year just ended. If this year can be viewed as a random sample of the journal's past and future acceptance behavior, develop a 95% confidence interval for the journal's long-run acceptance rate.

12.75 Suppose NBC News conducted a poll of American adults and found that 31 percent had an optimistic outlook for the economy for the next year. The sample was based on 912 adults.

a. For 90% confidence, what is the potential margin for sampling error here?

b. Suppose NBC had asked the same question to a similar number of adults six months earlier, finding 32 percent with an optimistic outlook for the year ahead. Do you think it would be reasonable for NBC to state that Americans are less optimistic now than they were six months ago? Why or why not?

***12.76** A one-sided confidence interval for the binomial parameter is computed in much the same fashion as a one-sided interval for μ (see Exercise 10.25).

An overnight courier service sampled 400 recent deliveries chosen at random to see how often they meet their goal of delivery to the addressee by 10:30 a.m. the next business day. The sample showed 368 on-time deliveries. Compute a point estimate for the proportion of on-time deliveries. The courier service can be 90% certain that its overall on-time delivery rate is no lower than what figure?

12.77 The Louisville *Courier-Journal* reported ("Airports Fail to Detect 20% of Weapons in FAA Tests," June 18, 1987) on random tests of security systems conducted by the Federal Aviation Administration at 28 major U.S. airports. Testers from the FAA attempted to pass through security checkpoints with hidden weapons such as grenades, bombs, and guns. According to the FAA, the tests were done over a four month period and resulted in 1923 weapon discoveries and 496 unchallenged cases. Provide a 99.5% confidence interval for the 28 major airports' success rate in screening out concealed weapons.

12.78 Refer to Exercise 12.77. Suppose the FAA were to repeat their tests of airport security again this year. How many checkpoint encounters would be necessary to estimate the success rate to within four percentage points with 90% confidence? Treat the 1987 study as your "pilot."

12.79 Refer to Exercise 12.78 above. Suppose the cost to the FAA per checkpoint encounter can

*Optional

be assumed to be roughly \$12. What would be the incremental cost in wanting 99% instead of 90% certainty for the resulting interval estimate?

12.80 Refer to Exercise 12.74. How many additional observations from other years would be required to tighten the interval to $\pm .03$, maintaining the same risk factor?

12.81 A local discount department store wants to estimate to within 6 percentage points the proportion of its sales over \$20 that are charged by the customer to a credit card. The store manager believes the proportion is at least .25, but unlikely over .40. How many sales transactions for amounts over \$20 do you recommend be checked? Assume that 90% faith in the interval will be suitable.

12.82 Union leaders at a production plant want to sample some of the membership (a population of about 1330 individuals) in regard to Option *A* and Option *B*, two alternative ways of computing future cost-of-living adjustments being considered by their negotiators. If they want to estimate the proportion favoring Option *A*, what sample size is advised? Assume that a 90% confidence interval accurate to $\pm 5\%$ will be sufficiently precise.

12.83 Most manufacturers of blank VCR tapes state that better reproduction results when the taped program is recorded at the fastest speed. Many consumers prefer the slowest speed simply because more shows can be saved on a given tape. A local television station recently made two tapes of a movie—one on a fresh tape at slow speed, another at fast speed and took the tapes to a shopping mall for an experiment of viewers' ability to discriminate. Adults passing by the booth were given a soft drink and invited to view a few minutes of each tape. They then were asked to choose the one with the better picture.

Although the viewers were apprised of the purpose of the project, they were unaware of which speed they saw first. Further, to avoid "ordering effects," about half of the raters saw the slow speed tape first, while the others saw the fast speed first. Altogether 138 opinions were gathered, 75 for the fast speed, 63 for the slow speed. Test H_0: the viewers cannot discriminate between the two recording speeds, against the appropriate one-sided research hypothesis. Tie your conclusions to the .10 level of significance.

12.84 Refer to Exercise 12.83. State the cutoff point for rejection/acceptance in unstandardized form, retaining the same risk of Type I error.

12.85 A 99% confidence interval estimate for π was given as $.31 \leq \pi \leq .35$.

a. What was the numerical value for p?

b. About what sample size was taken in developing this estimate?

12.86 A Massachusetts firm manufactures the Lo-Jack, a small radio transmitter that can be hidden in a car's door panel, wheel well, and so forth. If the car is stolen, police enter the owner's identification number into a computer that activates the Lo-Jack transmitter. Police can then pick up the signal and locate the car. According to newspaper accounts ("Radio Tracking Device Lets Police Home in Quickly on Stolen Cars," Louisville *Courier-Journal,* June 27, 1987), 56 transmitter-equipped cars were stolen in the system's first year of operation. All of these cars were found; the average recovery time was under an hour. Suppose 40 of the stolen cars were recovered within an hour of the owner's report to the police. Assume the first year's experience with this system can be viewed as a random sample of future recovery results. Calculate a 90% confidence interval for the long-run stolen car recovery rate for the first hour after the theft is reported.

12.87 Illinois passed a vehicle safety law a few years ago requiring front-seat occupants to wear seat belts at all times. An Illinois tollway official recently was quoted as saying "At best, only 3 out of 5 automobile drivers on the Tollway are using their safety belts." A sample of tollway drivers is obtained by having toll booth attendants at various stations note, for 30 minutes of their shift, whether each driver had his or her belt on. On the day of the study, 1848 drivers were observed, and 1316 had their belts on.

a. Test H_0: 3 out of 5 drivers on the tollway use their seat belts. Use $\alpha = .05$. Choose H_a to be a statement that could refute the tollway official's claim.

b. This test involved only drivers passing through stations operated by attendants. This was because the attendants stand above the driver and have an unobstructed view of the front seat. Those cars passing through the exact change/automatic lane stations (the majority, in fact) were ignored in this study. Do you think that basing such a study only on cars passing through the attendant lanes would invalidate any inference about all tollway drivers?

12.88 Refer to Exercise 12.87. In terms of the problem setting being studied, what would the occurrence of a Type II error mean?

12.89 Suppose a presumably fair coin is flipped 500 times, yielding 265 heads and 235 tails. Is this an unusual result? Would you reject the assumption that the coin is fair, if you were willing to tolerate 10% risk of incorrectly concluding that the coin is not fair?

12.90 Nationwide, about 20 percent of beginning freshmen graduate four years later from the college or university of original enrollment. A large state university wants to sample its freshmen of four years ago to see if their experience compares with the national figure. State what your null and alternative hypotheses should be. Four years ago, 3411 freshmen entered the university. From an alphabetical listing of these names, every sixth name is selected and checked by hand against a list of this year's graduating seniors. Altogether 120 names were on both lists, 448 were not.

a. What type of sampling procedure was employed here?

b. Using $\alpha = .10$, state whether we should be willing to believe that this particular freshman class had the same rate as the national norm.

12.91 A popular disc jockey who hosted a certain radio station's weekday morning drive-time program left three months ago for a better job in another city. While with the station, he had consistently achieved about 22% share of the audience. However, when the next local Arbitron radio ratings were done, his replacement was found to have 17.9% share. If based on a sample of 585 drive-time listeners, is this evidence beyond a reasonable doubt that the station has lost market share during the profitable drive-time period?

12.92 The following sample data were obtained from an approximately normal distribution. Construct a 90% confidence interval for σ, the standard deviation of the population.

\$23.75	\$21.95	\$24.95	\$25.15
\$23.95	\$25.10	\$22.95	\$24.20

12.93 The following numbers represent the number of new admissions at a suburban hospital over a three-week period:

19	14	16	17	18	9	10
21	19	12	13	14	8	9
20	20	16	11	11	11	7

a. Verify that $\overline{X} = 14.05$, $S = 4.40$ admissions.

b. If our numbers can be viewed as a random sample, develop a 95% confidence interval for σ, the population standard deviation. Assume population normality.

12.94 A state-wide newspaper ran a feature on various types of tires in a special automobile section supplement. One article in the section discussed a survey of 12 different dealers in the state. The newspaper had obtained quotes from these dealers on a given day for a set of four steel-belted tires of a well-known brand. From a listing of the prices, a sample standard deviation of $18.40 was computed. Use this result to estimate the standard deviation of prices for all dealers in the state offering the tires. Employ an 80% confidence interval.

12.95 How can the null hypothesis that 95 percent of the members of a normal population have values within plus or minus $10.00 of the mean be most compactly stated?

12.96 Using $\alpha = .10$, test H_0: $\sigma = 24$ against an upper-tail alternative. The sample evidence is $S = 33$, based on 28 observations.

12.97 Refer to Exercise 12.93. Can the H_0: $\sigma = 8$ admissions be rejected when employing a lower-tail H_a? Use the .05 level of significance.

12.98 A coffee vending machine is supposed to dispense a consistent amount of coffee per serving. An acceptable amount of variation is a population standard deviation of no more than .40 ounce. In a test of a new machine, a sample of 12 cups yields $S = .44$ ounce. Does this constitute strong evidence that there is too much variability? Or is the discrepancy between the criterion and the sample result not statistically meaningful? Tie your conclusion to any commonly used level of significance.

12.99 A certain part used in the assembly of a firm's different engine models has the specification that its diameter should be 5.000 millimeters, with a standard deviation of no more than .015 millimeter. A sample of 30 of these parts finds the following diameters:

4.961	4.984	4.999	5.004	5.021
4.964	4.987	4.999	5.007	5.024
4.975	4.991	5.000	5.010	5.029
4.975	4.992	5.001	5.013	5.031
4.982	4.994	5.001	5.016	5.032
4.984	4.997	5.003	5.019	5.035

a. Verify that $\bar{X} = 5.001$, $S^2 = .0003884$, and $S = .01971$ using the raw data formulas.
b. Test the hypothesis that the part diameter variability is within specifications, using a 1% risk of incorrectly rejecting H_0.
c. Would a test of H_0: $\mu = 5.000$ millimeters be one tailed or two tailed?
d. Test the hypothesis that the part mean diameter is in control, using the same risk factor.

(Exercises 12.100 through 12.102 presume that you have studied the Quality Control section of Chapter 11. They all involve *p*-charts—control charts for the proportion defective in a binomial process.)

***12.100** If a percent defective (*p*-chart) is based on samples large enough so that $n\pi \geq 5$ and $n(1 - \pi) \geq 5$, then the upper and lower limits of the *p*-chart are $\pi \pm 3\sigma_p$, where $\sigma_p = \sqrt{\pi(1 - \pi)n}$. For instance, if a process produces 6 percent defective when in control, the control limits for samples of size 100 would be

$$.06 \pm 3\sqrt{\frac{.06(.94)}{100}} = .06 \pm .07$$

In practice, this p-chart would have only an upper control limit, at .13 (13 percent defective). Find the control limits for a p-chart for $n = 75$, assuming the process generates 10% defectives when in control.

***12.101** Refer to Exercise 12.100. In tracking a binomial variable on a p-chart, a random sample of 80 items drawn at 8:30 A.M. shows 11 defectives. If the process is deemed in control when the defective rate is 7 percent, does this sample provide convincing evidence that the process is out of control?

***12.102** Refer to Figure 11.12 (a p-chart with $n = 25$), where the process target is $\pi = .04$ defective and a sample with 5 or more defectives will cause the process to be judged as out of control. Using either a computer program or the formula for the binomial distribution, determine the chance that 5 or more defectives will be found in a random sample of size 25 when $\pi = .04$.

12.103 In which case is the evidence strongest against H_0: $\pi = .70$?

a. Case A: $n = 10$, $X = 3$ successes
b. Case B: $n = 100$, $X = 60$ successes
c. Case C: $n = 1000$, $X = 666$ successes

12.104 When President Reagan left office in 1989, the *New York Times*–CBS News Poll showed an overall job performance approval rate of 68 percent. Such polls were begun in FDR's time, but no departing president had previously topped 60% approval.

a. The result was based on interviews with 1533 adults one week before Reagan's term ended. Construct a 99% confidence interval for the population approval rate for President Reagan as he left office.
b. The lowest approval rate at the end of his tenure was for President Nixon: 24 percent. Suppose that a 95% confidence interval for this figure has a margin for sampling error of plus or minus two and one-half percentage points. If so, about how many adults were included in the poll when Nixon left office?

REFERENCE

Neter, J., W. Wasserman, and G. A. Whitmore. 1988. *Applied Statistics,* 3rd Edition. Allyn and Bacon, Boston.

REVIEW PROBLEMS CHAPTERS 9–12

R41 A market researcher for a local television station wants to determine the average age of regular viewers of a late night sports news program broadcast in the area. A preliminary sample of 226 households finds 17 viewers who report watching the program two or more times per week. Their ages are given as follows:

24	29	38	41	48	26	26	22	44
31	34	25	22	37	30	28	39	

Average = 32; standard deviation = 7.9.

a. Suppose the researcher wants to estimate the average age of all local people who watch the sports program at least twice a week to within plus or minus two years. Based on the preliminary sample, how many viewers should be reached? Assume you are willing to tolerate a 10% risk for your interval estimate.

b. Suppose the full sample based on the sample size determined in part a shows an average age of 31.5 years with a standard deviation of 8.9 years. Compute the desired confidence interval. Is your interval wider or narrower than the target margin for sampling error?

c. Why doesn't your interval have a plus and minus factor of exactly two years as planned?

R42 How common is jet lag among people flying across multiple time zones? The Louisville *Courier-Journal* ("9 of 10 Suffer Jet-Lag Woes, Survey Reports," June 10, 1988) reported on a survey of 784 long-distance jet travelers, conducted by the medical director of the United Nations. For travel across three or more time zones, 45 percent reported severe problems (inability to sleep, daytime sleepiness, fatigue). Assuming the survey respondents were a random sample of multiple time zone travelers, establish an interval that has a confidence factor of .98 of including the proportion of all such travelers who experience severe problems.

R43 Can consumers discriminate between different cola formulations in a blind taste test? In "Measurement of Discrimination Ability in Taste Tests: An Empirical Investigation" (by Bruce Buchanan, Moshe Givon, and Arieh Goldman in the May 1987 issue of the *Journal of Marketing Research*), the authors describe an experiment where subjects were given four unlabeled cola samples, two of each formulation. After tasting, each subject was asked to identify the pairs that were the same. The authors note that a subject who simply guesses at random has a probability of being correct of 1/3. Suppose we wish to have a test capable of providing statistical evidence as to whether the subjects have an ability to discriminate.

a. Which pair of statements would we employ, and why?

(1) H_0: The subjects are only guessing.
H_a: The subjects are not guessing.

or

(2) H_0: The subjects are not guessing.
H_a: The subjects are only guessing.

b. For your answer to part a, express your null and alternative hypotheses in symbolic form involving a parameter and a value.

c. For your answer to part a, how could Type I error occur? Express your answer in terms of the problem setting.

R44 The regional manager of a tax-preparation company has two staff members who visit individual company stores, posing as customers. The purpose of such calls is to monitor such things as employee courtesy and knowledge of tax procedures, store cleanliness, amounts charged for services, and so on. One week, a staff member takes the same information (wage and tax statements, deductions, expenses, investment income) to 12 different company stores to have a tax return prepared. The amounts charged for this service are as follows:

$49	46	52	52	49	52
45	51	45	49	49	50

Mean = $49.08; standard deviation = $2.57.

a. Assuming the 12 returns can be treated as a simple random sample of region stores, state the lower and upper limits for a 90% confidence interval for the region mean charge for this type of return.

b. State any additional assumptions required to determine your interval.

R45 In R44 the sample standard deviation for amount charged was \$2.57. Provide an interval that has a confidence factor of .90 of including σ, the population standard deviation, to two decimal places.

R46 For a five-day period, a city employee counts the number of vehicles passing through an intersection near the downtown area. The counts are made during the weekday rush periods (7:30–8:00 A.M. and 5:00–5:30 P.M.). The following observations are recorded: 97, 92, 101, 89, 80, 94, 96, 103, 98, 75 (mean = 92.5; standard deviation = 8.96). The intersection in question currently has stop signs. In the past, the city has installed traffic lights at intersections where studies show that the intersection average count exceeds 90 vehicles per half hour.

a. For $\alpha = .10$, is the case made that traffic lights should be installed? Assume that traffic counts are normally distributed and perform the appropriate test.

***b.** Relax the assumption of a normal population and use a nonparametric procedure to test the same assumption as above. Does the same conclusion result (using $\alpha = .10$)?

R47 Over the past several weeks, a county real estate assessor in a north central state has inspected a random sample of 34 residences built in the county within the past 10 years. Among the statistics generated are these: $\overline{X}$ = mean residence heating/cooling area = 1666 square feet, and S = 349 square feet. Determine a 95% confidence interval for the mean residence heating/cooling area for all county residences built in the last 10 years.

R48 Refer to the residence sampling situation described in R47. Another finding by the assessor was that eight of the sample homes had central air conditioning. The assessor wants to know what percent of all residences built in the past ten years have central air conditioning. Suppose that a point estimate which is within ±8 percentage points of the actual figure will be acceptable. Use the sample evidence already gathered to establish the sample size necessary to achieve the desired precision, with 95% confidence.

R49 A metropolitan cable-TV system is contemplating paying \$40,000 for the local broadcast rights to an upcoming boxing title fight. The system would then offer the fight to its home subscribers for \$24.95 on a pay-per-view basis. Besides the rights fee, the cable system would also have to pay a certain percentage of their gross income to the fight promoters and the fighters themselves. The cable system believes that they will make a profit if more than 3 percent of their subscribers respond to their offer. To assess subscriber interest, a random sample of households is about to be contacted.

a. State an appropriate null and alternative hypothesis.

b. The sample finds 11 households out of 248 contacted would definitely be interested. What decision should be made? Assume cable management wants a risk of no more than 10% of buying the pay-per-view rights when they should not.

R50 An audit during the previous fiscal year showed that the mean gross profit per invoice for a wholesale distributor was \$101.50. Two months into the current year, the company comptroller chooses a random sample of 42 invoices to see if this gross profit figure still holds.

a. Specify the appropriate statement for H_0 and H_a.

b. The sample average is found to be \$98.97, with a sample standard deviation of \$31.01. Choose a commonly used risk level and determine whether the sample data contradict or support your null hypothesis.

R51 Milford Recreation Center is seeking a grant from the state to rewire lights with defective wiring in their youth baseball park. To be eligible to receive the grant, the center must show that more than 25 percent of the lights have defective wiring. There are 11 diamonds in the park; each diamond has 4 light standards with 20 lights per standard.

*Optional

a. The recreation center plans to conduct a sampling of lights to determine the percentage with defective wiring. What method(s) of random sampling is most efficient for this situation? Which method(s) seems to be least efficient?

b. Suppose that a preliminary sample of lights revealed 30 percent with defective wiring. If Milford Recreation Center wishes to be 80% confident, how large a sample should be taken to ensure a 5% error margin? (*Hint:* Be sure to include the finite population correction factor.)

c. Assume that in a full sample of 120 lights, 38 needed to be rewired. Construct an 80% confidence interval for π, the proportion of all lights with defective wiring, to three decimal places.

d. Explain whether these data enable the Center to qualify for the grant, assuming a 20% risk factor.

R52 In 1975 the Financial Accounting Standards Board (FASB) was assigned the task of developing a uniform accounting method in the oil and gas industries. Subsequently, the FASB produced three documents: a Discussion Memorandum (DM), an Exposure Draft (ED), and the Statement of Financial Accounting Standard (SFAS) No. 19, which spelled out their position. The oil and gas companies promptly appealed SFAS No. 19 to the Securities and Exchange Commission (SEC). Before each FASB directive was announced, oil and gas companies lobbied against the proposed changes. E. B. Deakin ["Rational Economic Behavior and Lobbying on Accounting Issues: Evidence from the Oil and Gas Industries," *Accounting Review, LXIV,* no. 1 (1989): 49–68] developed three models, one for each of the three FASB documents, to predict whether a company would engage in lobbying. In testing the predictive abilities of the models, he used a holdout set of $n = 94$ firms, classifying each into one of two groups: lobbying or nonlobbying. After each FASB directive, Deakin compared his model's classification rate with a "chance criterion" that classifies all companies as belonging to the larger group represented in the population, i.e., the nonlobbying group in this case. Consider the following data:

Deakin's Model	Proportion of Correct Classifications	
	By Chance	**By Deakin's Model**
DM	.713	.798
ED	.597	.822
SFAS No. 19	.568	.763

Is there sufficient evidence to conclude that π, the proportion of correct classifications by each of Deakin's models, is better than the classification rate by chance? Use $\alpha = .025$.

R53 Retail price stickers on new cars often are used as a starting point toward an equitable selling price in negotiations between the dealer and the consumer. The difference between the posted retail price and the actual selling price (assuming it is below the posted price) is called the discount. For the present model year, which started October 1, a dealer's discount has averaged \$893 on the sale of 37 cars of a certain model. The dealer has determined that his discount must average at most \$850 on all cars sold of this model in order to maintain the desired profit margin. Assume that 37 cars sold to date represent a random sample of car sales and that the dealer will receive only 512 cars of this model under his yearly allocation from the manufacturer. If the sample standard deviation is \$113, is there sufficient evidence to indicate that the desired profit margin is shrinking? Use $\alpha = .01$.

R54 An investment firm plans to conduct a nationwide poll to determine current trends in investment attitudes. One of the key issues to be investigated is the "program trading" concept that has been linked to stock market volatility.

a. The firm wishes to estimate π, the proportion of investors who are unfamiliar with program trading, with a maximum error of 4 percent. If 99% confidence is specified, how many investors should be sampled?

b. Suppose 1001 investors agreed to participate in the survey and of those, 481 indicated they were unfamiliar with program trading. Construct a 99% confidence interval for π to two decimal places.

c. Explain how the potential sampling error in the confidence interval of part b can agree with the maximum error specified in part a, yet the sample size requirement from part a does not coincide with the actual number of investors included in the survey.

R55 From the mid-1970s to the mid-1980s, the Motorola company's market share in televisions, stereos, and microcomputer memory chips eroded because of the products' poor quality ("Top Quality Is Behind Comeback," *USA Today*, March 28, 1989). A turnaround began in the early 1980s—when, in making cellular telephones, the firm decided to adopt the Japanese approach to quality—and culminated in 1988 with the Malcolm Baldrige National Quality Award. One of the cornerstones of Motorola's revival was a plan called "Six Sigma Quality," in which control limits were set six standard deviations above and below the norm. In order for the Six Sigma Quality plan to help in reducing the number of defective cellular telephones, engineers had to alter the product design to allow more leeway in the manufacturing process. Based on such a criterion, decide whether the following statements are true or false.

a. Six Sigma Quality allows for less chance variation and more assignable variation.

b. The probability is almost zero that a random sample will fall outside the control limits when the process is in control.

c. The new product design allowed less variation in the size of the parts.

R56 In a 1981 article ("Perception, Preference, and Patriotism: An Exploratory Analysis of the 1980 Winter Olympics," *The American Statistician, 35,* No. 3, pp. 170–173), I. Fenwick and S. Chatterjee analyzed scores given by judges of figure skating. One question they investigated was whether judges tended to give higher scores to contestants from their own countries. Suppose we view this as a two-outcome-possible sampling situation: either a judge rates a compatriot higher or lower than expected.

a. If a judge is unaffected by patriotism in evaluating a contestant from his or her own country, what is the probability of a higher than expected (based on the contestant's overall standing) rating?

b. In the article the authors report that in 53 cases where a judge and the contestant were of the same nationality, 51 higher than expected ratings occurred. Treating this as a random sample of such judging opportunities, test the hypothesis ($\alpha = .01$) that judges' marks are unaffected by patriotism.

R57 Car dealerships typically sell cars, parts, and service. At Downey Toyota in Downey, California, parts manager Ron Koontz took a different tack. He determined that the parts department services about 150 people daily and that the average purchase is \$37 ("Dealerships Offer More Than Cars and Parts," Louisville *Courier-Journal,* June 24, 1988). To cater to his customers, Koontz started selling Haagen-Dazs ice cream, leather jackets, and about 2700 other items that have no relation to Toyota!

Suppose a dealership in the north-central section of the country is considering a similar venture into non-auto-related merchandising. The manager believes the average purchase at his parts department is similar to Koontz'. An average significantly less would persuade him to stick to selling cars, parts, and service. He selects a random sample of 72 purchases and finds an average of \$35.81 with a standard deviation of \$14.34. Since his average is less than \$37, should he abandon his non-auto-related merchandising plans? Explain, supporting your answer with a p-value and its interpretation.

R58 The number of calories per serving (one cookie) of Do-Si-Do cookies is thought to be normally distributed with a mean of 50 and a standard deviation of 2. Two boxes containing 22 cookies each were randomly selected and the calorie content of all the cookies was measured, producing a mean of 48.7 and a standard deviation of 2.3. Are the parameters of the normal distribution reasonably stated? Conduct two tests in proper sequence using $\alpha = .05$ for each to support your answer. Assume the right-tail χ^2 cutoff value for $df = 43$ is 62.990 and the left-tail cutoff value is 26.785.

R59 During the 1988 presidential election, there was a school board race in the Second District of Bullitt County, Kentucky, between James A. Matthews III and Thomas Whitt ("Error in 121 Ballots Didn't Taint Election in Bullitt, Court Rules," Louisville *Courier-Journal,* January 21, 1989). Because of a printing error, 121 voters used ballots that omitted the school board race. Although Matthews defeated Whitt by 35 votes, the race was contested and eventually decided by the Kentucky Court of Appeals. The facts are these: 3123 people in the Second District voted in the presidential election; of these, only 1099 voted in the school board race.

a. Determine the proportion of these Second District voters receiving ballots containing the names of Matthews and Whitt who voted in the school board race. Round off your answer to two decimal places.

b. Let X represent the number of the 121 voters receiving tainted ballots who would have voted in the school board race. Develop a region of concentrated values for X using the known proportion of voters who voted in the Second District school board race.

c. Find the largest integer contained inside the region of concentrated values and use it as an (best case for Whitt) estimate of the sample size for those 121 voters who would have voted in the school board race. Partition the sample size into k votes for Whitt and $n - k$ votes for Matthews such that the k votes for Whitt *exceed* the $n - k$ plus 35 votes for Matthews. In other words, what is the minimum number of votes Whitt would have needed to defeat Matthews?

d. Now assume that π, the probability a voter would select either candidate, is .5. Based on the sample size and value of k determined in part c, find the approximate probability that Whitt would have defeated Matthews.

e. Suppose you are a member of the Kentucky Court of Appeals. Based on the probability computed in part d, what decision about the contested race would you render: Uphold the election result as is or throw out the election result and have a revote in the Second District?

R60 In 1976 statisticians Ronald Thisted and Bradley Efron analyzed the 884,647 total words in the Shakespearean canon ["Estimating the Number of Unseen Species; How Many Words Did Shakespeare Know?" *Biometrika, 63* (1976): 435–437] and discovered that Shakespeare's total published vocabulary consisted of 31,534 distinct words. At that time, they attempted to answer the question, "If a new work by Shakespeare were discovered, how many new words would it contain?"

Never expecting to be able to test their theory, the authors (and the literary world) were astonished when, in 1985, Shakespearean scholar Gary Taylor found an untitled poem (hereafter called the "Taylor poem") in a seventeenth century Shakespearean anthology in an Oxford University library. Few experts believed that Shakespeare could have been the poem's author because of obsessive rhyming that was uncharacteristic of Shakespeare's works. Nevertheless, Thisted and Efron's subsequent analysis ["Did Shakespeare Write a Newly Discovered Poem?" *Biometrika, 74* (1987): 445–455] revealed, that, indeed, Shakespeare could have written this poem!

Although we cannot provide all the details of their analysis here, let us consider one of the many tests they conducted. A key variable was m_x = number of distinct words in the poem that occurred exactly x times in the 884,647 total words of the Shakespearean canon. Of particular interest is m_0 which is the number of new, never-before-used words in the 429-word Taylor poem. There were nine such words. (*Note:* Of the 429 words in the poem, 258 were distinct. Thisted and Efron focused on words that were used "rarely," not "commonly." Rare usage meant 99 times or less; common words were used 100 times or more in the canon. With these definitions, only 118 of the 258 distinct words in the Taylor poem were rare, while 140 were common.) Based on rare words only, they claimed nine years before the poem came to light that π, the proportion of new words, should be .073407. Conduct a test of the null hypothesis that the proportion of new words in the Taylor poem is consistent with Shakespearean authorship. Use $\alpha = .05$.

Chapter Maxim *For a statistical test to provide evidence that two populations differ, it is necessary to test and reject a hypothesis of equality.*

CHAPTER 13

COMPARING TWO POPULATIONS

Objectives

After reading this chapter and working the exercises, you should be able to

1. Make inferences about $\mu_1 - \mu_2$, the difference between two population means.
2. Explain what it means to find the number zero inside a two-sample confidence interval.
3. Distinguish independent from dependent sampling situations.
4. Explain the motivation for having a matched pairs sampling plan.
5. Determine confidence intervals and conduct hypothesis tests for matched pairs data.
6. Make inferences about $\pi_1 - \pi_2$, the difference between two population proportions.

*7. Use the Two Sample Inference processors to verify answers and/or solve problems concerned with comparing two populations.

In the three preceding chapters, we have studied statistical inference for the one-sample case. The procedures for developing confidence intervals and conducting hypothesis tests to this point have involved taking a single sample from a given population. We now wish to expand our statistical repertoire by considering the two-sample case: applications where we sample from each of two different populations. The purpose of a two-sample procedure is *comparison*—we will want to make an inference about the difference in means (or proportions) between two populations.

Consider the following scenarios:

- An audience researcher for a local television station conducts a survey of late night television watchers: those who regularly watch the station's reruns of a situation comedy and those who watch a national news program on another channel. The objective of the survey is to ascertain any demographic differences (average age, average income level, sex, and so on) between the two audiences.
- A fried chicken retailer purchases buckets from two different paper companies. While the overall defective rate of buckets is between 2 and 3 percent, no current information exists on individual supplier defective rates. A random sample of 1000 units from each supplier is tagged upon receipt. A count of the number found defective for each supplier will be made to see if they have the same or different defective rates.
- The admissions officer for a university's MBA program wants to learn if there is any difference in the qualifications of applicants whose undergraduate degree

*Applies to optional section.

is in business versus nonbusiness applicants. A sample of admission test scores from each group is obtained and compared.

- Many health-related studies involving new drugs or treatments are set up as two-sample tests. Often one group of subjects receives the new treatment, while another group of subjects (the control group) receives a dummy treatment, or a placebo. The control group then becomes a reference point against which the treatment group is compared. For instance, *Newsweek* (February 1, 1988) reported on such a test for Retin-A, a cream that appeared promising for removing wrinkles. In the study, a group of 15 patients who applied Retin-A to their faces were compared to another group of 15 who used a placebo cream.

Each situation above requires two samples so that we can make an inference about whether the two populations are the same or different with regard to some particular characteristic.

We will examine several different types of two-sample inferences in this chapter. However, the necessary groundwork has already been established and should be familiar. For instance, if we want a confidence interval for a two-sample case, we will express it in words in exactly the same way as we did for the one-sample case:

$$\text{Point estimate} \pm (Z \text{ or } t)\,(\text{standard error})$$

If we wish to conduct a two-sample statistical test, we will express our test statistic as

$$Z^* \text{ (or } t^*) = \frac{\begin{matrix}\text{Sample} \\ \text{result}\end{matrix} - \begin{matrix}\text{Value assumed} \\ \text{under } H_0\end{matrix}}{\text{Standard error}}$$

which is also identical to that of the one-sample case. Basically, the framework for making inferences remains unchanged from previous chapters—only the formulas are different to reflect the fact that two samples are drawn instead of one.

13.1 INFERENCES ABOUT $\mu_1 - \mu_2$: INDEPENDENT SAMPLES

In comparing two populations, the samples we obtain are said to be either independent or dependent. When two separate random samples are drawn from the two populations, we have *independent* samples. Figure 13.1 illustrates that the two-sample procedure for independent samples is simply an extension of the one-sample case: Instead of one sample and one set of summary statistics ($\overline{X}$ and S), we now have two. Therefore we put subscripts on our symbols to help us keep our terms properly identified. The sample sizes n_1 and n_2 may or may not be the same.

Briefly stated, having independent samples means that the particular items selected from the first population have no bearing on which items will be selected from

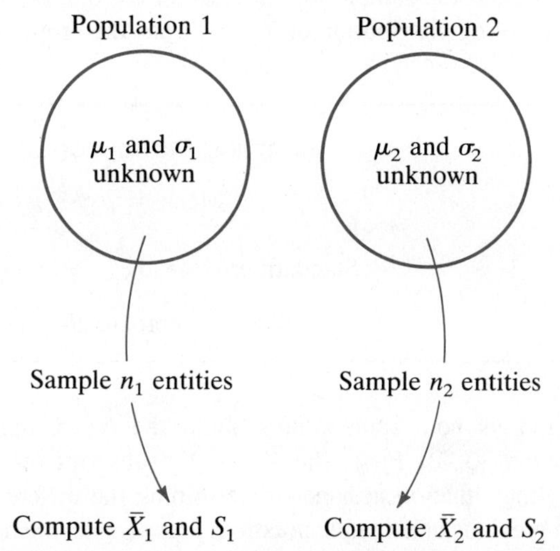

Figure 13.1 Schematic Representation of Independent Samples

the second population: Each sample stands on its own as a separate collection. This is in contrast to dependent sampling (discussed in Section 13.2) where each item in the first sample is compared to a specific, predetermined counterpart item in the second sample.

We will now consider the case of independent samples where n_1 and n_2 are both "large"—at least 30 observations.

Large Sample Inferences About $\mu_1 - \mu_2$

In the one-sample case, we learned that the sample mean is an estimator of the population mean: The observed value of $\overline{X}$ is a point estimate of μ. Now that we are taking two samples with the objective being to compare means, we are interested in $\mu_1 - \mu_2$, the *difference* in population means. It therefore should seem reasonable that the logical estimator for $\mu_1 - \mu_2$ should be $\overline{X}_1 - \overline{X}_2$, the difference in sample means.

Point Estimate of $\mu_1 - \mu_2$

$$\overline{X}_1 - \overline{X}_2$$

In Chapter 8, we developed the sampling distribution of $\overline{X}$. We learned that if the sample size is sufficiently large, the sampling distribution of $\overline{X}$ is a normal distribution centered at μ. We now need to know the sampling distribution of the random variable

$\overline{X}_1 - \overline{X}_2$. If the samples are at least 30 (so that the Central Limit Theorem applies), the sampling distribution of $\overline{X}_1 - \overline{X}_2$ has the properties given in the following box.

Sampling Distribution of $\overline{X}_1 - \overline{X}_2$

$$\text{Mean} = \mu_1 - \mu_2 \tag{13–1}$$

$$\text{Standard error} = \sigma_{\overline{X}_1 - \overline{X}_2} = \sqrt{\frac{\sigma_1^2}{n_1} + \frac{\sigma_2^2}{n_2}} \tag{13–2}$$

Shape: Approximately normal

Let us note some points about the two-sample standard error, $\sigma_{\overline{X}_1 - \overline{X}_2}$ (see Equation 13–2). First, the $\overline{X}_1 - \overline{X}_2$ subscript on σ is there to remind us of the sampling situation at hand—determining the difference in means. This is consistent with previous one-sample notation; for example, $\sigma_{\overline{X}}$ denotes the (one-sample) standard error of the mean. Second, population variances (standard deviations squared) appear in the standard error. Third, most likely in practice the population variances (σ_1^2 and σ_2^2) will *not* be known. If so, the sample counterparts (S_1^2 and S_2^2) will replace them, giving us an estimated standard error as in the following box.

Estimated Standard Error of $\overline{X}_1 - \overline{X}_2$, Large Sample Sizes

$$\hat{\sigma}_{\overline{X}_1 - \overline{X}_2} = \sqrt{\frac{S_1^2}{n_1} + \frac{S_2^2}{n_2}} \tag{13–3}$$

We are now ready to look at large-sample examples of confidence intervals and hypothesis tests for $\mu_1 - \mu_2$, the difference between two population means.

Confidence Interval Estimation To construct a confidence interval for the difference in population means, we will rely on this familiar expression:

Point estimate ± (Z or t) (Standard error)

Equation 13–4 translates these words into symbols.

Large Sample Confidence Interval for $\mu_1 - \mu_2$

$$\overline{X}_1 - \overline{X}_2 \pm Z\sqrt{\frac{S_1^2}{n_1} + \frac{S_2^2}{n_2}} \tag{13–4}$$

EXAMPLE 13.1

The manager of a moderately priced women's clothing store wishes to estimate the difference in average amount charged between shoppers who make their purchases with a Discover card and those customers who charge with a Visa card. Obtaining a random sample of charge slips from the last three months of operation, she determines the following results:

Charge Card	Sample Size	Sample Average, $\overline{X}$	Sample Standard Deviation, S
Discover	59	\$98.33	\$34.03
Visa	66	81.88	30.11

Construct a 90% confidence interval for the mean difference in the populations of charge amounts.

Solution:

It doesn't matter which charge card is labeled as sample 1 or sample 2. We might choose the one with the larger mean, Discover, to be sample 1 so that the point estimate is a positive number. We then have, in words,

Point estimate $\pm$ (Z or t) (Standard error)

and in symbols,

$$\overline{X}_1 - \overline{X}_2 \pm Z\sqrt{\frac{S_1^2}{n_1} + \frac{S_2^2}{n_2}} = \$98.33 - \$81.88 \pm 1.65\sqrt{\frac{34.03^2}{59} + \frac{30.11^2}{66}}$$
$$= \$16.45 \pm 1.65(5.776)$$
$$= \$16.45 \pm \$9.53$$

We are 90% confident that our point estimate (Discover charges average \$16.45 more than Visa) is within \$9.53 of the unknown true difference. Alternatively, we could say that we are 90% certain that $\mu_1 - \mu_2$ lies between \$6.92 and \$25.98. Had we chosen Visa to be denoted as sample 1, the interval would appear as $-\$16.45 \pm \$9.53$. The point estimate of minus \$16.45 would then be interpreted as indicating that the Visa sample average is \$16.45 less than the Discover average.

□

Tests of Hypotheses A two-sample test is usually for the purpose of investigating the following null hypothesis:

$$H_0\text{: } \mu_1 = \mu_2$$

or equivalently,

$$H_0\text{: } \mu_1 - \mu_2 = 0$$

In words, the null hypothesis given implies that there is no difference between two populations' means. In order to conclude that a difference exists, it is necessary to reject a null statement of equality. The alternative hypothesis can be one tailed or two tailed.

The two-sample test with large sample sizes relies on our customary standardized test statistic:

$$Z^* = \frac{\begin{matrix}\text{Sample} \\ \text{result}\end{matrix} - \begin{matrix}\text{Value assumed} \\ \text{under } H_0\end{matrix}}{\text{Standard error}}$$

If our null hypothesis is that two populations have the same mean, then the "value assumed under H_0" in the given expression is 0, indicating zero difference. Equation 13–5 is the symbolic form of the test statistic.

Large Sample Test Statistic for Testing H_0: $\mu_1 = \mu_2$

$$Z^* = \frac{(\overline{X}_1 - \overline{X}_2) - 0}{\sqrt{\dfrac{S_1^2}{n_1} + \dfrac{S_2^2}{n_2}}} \qquad (13\text{–}5)$$

EXAMPLE 13.2

In a test market of a new cereal, two different advertising campaigns are to be evaluated: one campaign positions the product as a nutritional cereal for children and the other has an adult–natural cereal appeal. The children–nutritional campaign is comprised of 32 stores in five different cities, while a different set of 32 stores in six cities represents the adult–natural campaign. The stores are chosen to be of similar sales volume, and comparable advertising expenditures are allocated to the two test campaigns. After a six-week period, the following sales data have been collected. Use these data to make a decision about which, if either campaign, appears more promising. Allow 5% risk of Type I error for your analysis.

Advertising Campaign	Number of Test Stores	Average Cases Sold per Week	Standard Deviation
1. Adult–natural	32	13.9	2.3
2. Children–nutritional	32	12.1	2.0

Solution:

With no a priori reason to anticipate that one campaign should be better than the other, we will employ a two-sided test of H_0: $\mu_1 = \mu_2$. For 5% risk, our cutoff Z-values are ± 1.96. Our test statistic from Equation 13–5 is

$$Z^* = \frac{(\bar{X}_1 - \bar{X}_2) - 0}{\sqrt{\frac{S_1^2}{n_1} + \frac{S_2^2}{n_2}}}$$

$$= \frac{(13.9 - 12.1) - 0}{\sqrt{\frac{2.3^2}{32} + \frac{2.0^2}{32}}}$$

$$= \frac{1.8}{.539} = 3.34$$

Since the *TS* exceeds the cutoff value (see Figure 13.2), we have strong evidence that H_0 is not true: The adult–natural campaign appears to be the more attractive advertising campaign. Table 13.1 summarizes this statistical test.

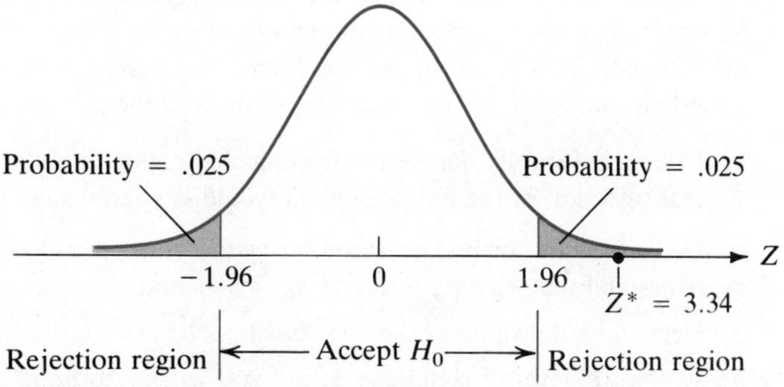

Figure 13.2 Graphical View of Example 13.2

Table 13.1 Symbolic Summary of Example 13.2

H_0: $\mu_1 = \mu_2$
H_a: $\mu_1 \neq \mu_2$
TS: $Z^* = 3.34$
RR: Reject H_0 if $Z^* < -1.96$ or if $Z^* > 1.96$
C: Reject H_0

□

COMMENTS

1. Some texts denote $\hat{\sigma}_{\bar{X}_1 - \bar{X}_2}$, the estimated standard error, as $S_{\bar{X}_1 - \bar{X}_2}$ or as $S(\bar{X}_1 - \bar{X}_2)$.
2. It is possible to test a null hypothesis of a nonzero difference. See Exercise 13.9.
3. Simply by looking at a confidence interval for the difference in means, we can tell whether a formal test of H_0: $\mu_1 = \mu_2$ would be rejected or accepted. The key question is whether zero is included in the interval: Inclusion of zero implies that an H_0 of equality would be accepted, while exclusion of zero implies H_0 would be rejected. For instance, suppose you have determined a 95% confidence interval for $\mu_1 - \mu_2$ is 20.4 ± 6.0. Zero is not included; the interval ranges from 14.4 to 26.4. A test for equal means therefore would be rejected at $\alpha = .05$. On the other hand, if an interval for $\mu_1 - \mu_2$ is determined to be 1.7 ± 8.0, zero is included since the interval extends from -6.3 to 9.7. A test for equal means would result in acceptance in such a case. Said another way, if our measurements have detected a difference of 1.7 but the potential sampling error is 8.0, we cannot be certain we have found a difference at all. It is when the observed difference exceeds the potential sampling error that we can be confident a true difference really exists.

CLASSROOM EXAMPLE 13.1

Comparing Means for Large Samples

A manufacturer of batteries for use in portable radio/cassette players claims in its advertisements that, under conditions of intermittent use (30 minutes on, 60 minutes off), its batteries outlast the leading brand. A private consumer interest group obtains a random sample of 32 fresh batteries of each of the two brands in question for a test.

a. Letting μ_L and μ_C denote the leading challenging brands, formulate hypotheses so that rejection of the null statement would lend credence to the advertised claim.

b. Determine the estimated standard error, given that $\bar{X}_L = 12.97$ and $S_L = .88$ hours and that $\bar{X}_C = 13.20$ and $S_C = .91$ hours.

c. Determine the value of the test statistic Z^*.

d. At the .05 level of significance, are you willing to believe the challenger's advertising claim?

e. Find the p-value for this test. Would the conclusion have been different if a less stringent risk level, such as $\alpha = .10$, were employed?

*■ Processor: Comparing Means (Large Independent Samples)

One of the choices on the processor's menu is Two Sample Inference. If we select this choice, a submenu of four two-sample processors appears. The first option on the submenu—Comparing Means (Large Independent Samples)—can develop confidence intervals and perform hypothesis tests for $\mu_1 - \mu_2$ when the sample sizes are large.

*Optional

```
≡              Comparing Means (large Independent Samples)              EDIT
   Introduction  *able [Data] *esults  Exit                         | F1=Help

        To compute the confidence interval or test statistic Z*,
      the two sample sizes, means, and standard deviations are
      needed.  Enter sample #1 information first.

                 number of entities, n1 = 59
                 sample mean, X-bar #1 = 98.33
                 standard deviation, S1 = 34.03

                 number of entities, n2 = 66
                 sample mean, X-bar #2 = 81.88
            ============ Standard deviation ============

                Enter standard deviation, S2 =  30.11
```

Figure 13.3 Entering Summary Statistics for Two Samples

To use this processor, select it from the Two Sample submenu, read the introductory screen, read the *Z*-table into memory, and then move to Data. You now will be asked to input the summary statistics for your two samples. Let us use the data in Example 13.1 to illustrate the confidence interval feature. Figure 13.3 shows the screen appearance after we have provided the final piece of sample information. After data entry is complete, we can move to Results and choose either Confidence Interval or Hypothesis Test.

Following Example 13.1, we request a 90% confidence interval, thereby obtaining Figure 13.4—information about the point estimate for $\mu_1 - \mu_2$ and the value of the standard error. By touching Enter again, we view Figure 13.5, a final results screen.

```
≡              Comparing Means (large Independent Samples)              READY
   Introduction  *able  Data [Results] Exit                         | F1=Help

       The point estimate for (μ1 - μ2) is (X̄1 - X̄2) =  16.450

            The standard error is the square root of

                    S²   S²
                     1    2
                    -- + --,  which is  5.776
                    n1   n2
```

Figure 13.4 Point Estimate and Standard Error for Data in Example 13.1

```
≡                 Comparing Means (large Independent Samples)                    QUERY
  Introduction  *able  Data [Results] Exit                                   |  F1=Help

              For 90.0% confidence, the interval is
                  point estimate ± Z * (standard error)

                  =  16.450 ± 1.645 * 5.776

                  =  16.450 ± 9.502

              The lower limit of the confidence interval is  6.948
              The upper limit of the confidence interval is  25.952

              ═══════════════════ Confidence ═══════════════════
                  Compute another interval? (y/n) No
```

Figure 13.5 Confidence Interval for $\mu_1 - \mu_2$ for Data in Example 13.1

■ Small Sample Inferences About $\mu_1 - \mu_2$

In Chapter 11, we learned that our large sample procedures for confidence interval estimation and hypothesis testing could be extended to the case of small samples—that is, less than 30 observations. The small sample case differed from the large sample case in that we used cutoff values from the t-table instead of the Z-table and that we found it necessary to assume that the sample came from a normal population.

What if we have a comparative situation with two small samples? In order to make inferences about $\mu_1 - \mu_2$, one additional condition appears: an assumption that the two populations have the same (unknown) variance.

The Pooled Sample Variance, S_p^2 Small sample inference assumes that the two populations share a common variance, σ^2. But when we obtain our two sets of sample statistics, we will have two separate estimates of this unknown value: S_1^2 and S_2^2. Rather than choose one or the other of these values to be our estimate of σ^2, we will weight each variance by its degrees of freedom and then average together into a *pooled estimate* of the population variance. Using S_p^2 to denote this pooled estimate of σ^2, we have the following equation.

Pooled Sample Variance

$$S_p^2 = \frac{S_1^2(n_1 - 1) + S_2^2(n_2 - 1)}{n_1 + n_2 - 2} \qquad (13\text{–}6)$$

To illustrate computing S_p^2, suppose we have taken samples from each of two populations and determined the following:

$$n_1 = 8 \qquad n_2 = 12$$
$$S_1 = 20.8 \qquad S_2 = 24.2$$
$$\overline{X}_1 = 88.3 \qquad \overline{X}_2 = 85.1$$

If we wish to make an inference about $\mu_1 - \mu_2$, we will assume that the populations have the same variance, σ^2. We have two estimates of σ^2: $S_1^2 = 20.8^2 = 432.64$ and $S_2^2 = 24.2^2 = 585.64$. By weighting and averaging these values, we obtain S_p^2, the point estimate of σ^2:

$$\begin{aligned} S_p^2 &= \frac{S_1^2(n_1 - 1) + S_2^2(n_2 - 1)}{n_1 + n_2 - 2} \\ &= \frac{432.64(8 - 1) + 585.64(12 - 1)}{8 + 12 - 2} \\ &= \frac{9470.52}{18} = 526.14 \end{aligned}$$

This pooled estimate would then be used in place of S_1^2 and S_2^2 in computing the estimated standard error of $\overline{X}_1 - \overline{X}_2$, given in the following box.

Estimated Standard Error of $\overline{X}_1 - \overline{X}_2$, Small Sample Sizes

$$\hat{\sigma}_{\overline{X}_1 - \overline{X}_2} = \sqrt{\frac{S_p^2}{n_1} + \frac{S_p^2}{n_2}} \qquad \textbf{(13–7)}$$

The expression in Equation 13–7 is identical to the large sample standard error, Equation 13–3, except that S_p^2 appears twice, replacing S_1^2 and S_2^2.

Having small samples implies use of the t-distribution for cutoff values. For one sample, recall that the row entry to the t-table, degrees of freedom, was $n - 1$. For the two sample case, df will be $n_1 - 1$ plus $n_2 - 1$, more conveniently written as $n_1 + n_2 - 2$. The first sample's degrees of freedom are used in estimating S_1^2; the second sample's df are used in estimating S_2^2.

Prior to looking at some small sample examples, let us summarize the adjustments we make relative to the case of two large samples:

- We assume normal populations with equal variances.
- We will compute a pooled sample variance to use in the estimated standard error.
- We will use t-scores with $df = n_1 + n_2 - 2$.

COMMENTS

1. A rough arithmetic check in computing the pooled sample variance is that S_p^2 must be between S_1^2 and S_2^2 (unless $S_1^2 = S_2^2$). In our preceding example, $S_1^2 = 432.64$ and $S_2^2 = 585.64$; when combined into an averaged estimate, we must get S_p^2 between these values. Compute S_p^2 to at least two decimal-place accuracy.
2. The right side of Equation 13–7 may also be written as $S_p\sqrt{(1/n_1) + (1/n_2)}$, where S_p is the pooled standard deviation—the positive square root of S_p^2.
3. Despite the fact that the sample variances S_1^2 and S_2^2 are not equal in a given situation, the assumption of equal population variances still may be reasonable. Strictly speaking, a formal test of equal population variances should be done to verify the equality assumption. Such a test requires knowledge of the F-distribution and will be introduced in Section 6 of Chapter 14.
4. If $n_1 + n_2 - 2$ equals 30 or more, we may use the Z-distribution to approximate the t-distribution.

Confidence Interval Estimation Except for the adjustments detailed above, small sample estimation for $\mu_1 - \mu_2$ is no different from the large sample case. Equation 13–8 gives the symbolic form of the confidence interval.

Small Sample Confidence Interval for $\mu_1 - \mu_2$

$$\bar{X}_1 - \bar{X}_2 \pm t\sqrt{\frac{S_p^2}{n_1} + \frac{S_p^2}{n_2}} \qquad (13\text{–}8)$$

EXAMPLE 13.3

An amateur investor wants to see what difference in yields may exist between public utility bonds and industrial bonds. Choosing a random sample of 12 from each category, she determines the yield of each bond over the past two years.

Bond Issuer Type	Average Yield (%)	Standard Deviation
Public Utility	7.02	.85
Industrial	7.86	.95

Determine a 90% confidence interval for the difference in average yields in the populations of industrial and public utility bonds.

Solution:

Since the sample sizes are small, we need to compute S_p^2, the pooled sample variance. Letting the industrial bonds be sample 1, we have

$$S_p^2 = \frac{S_1^2(n_1 - 1) + S_2^2(n_2 - 1)}{n_1 + n_2 - 2}$$
$$= \frac{.95^2(11) + .85^2(11)}{12 + 12 - 2}$$
$$= \frac{17.875}{22} = .8125$$

The denominator in the pooled variance computation, 22, represents the degrees of freedom in computing S_p^2. For 90% confidence with $df = 22$, the t-table values are ± 1.717. The confidence interval for $\mu_1 - \mu_2$ can be expressed as

$$\text{Point estimate} \pm t\text{-score (Standard error)}$$
$$= (\bar{X}_1 - \bar{X}_2) \pm t\sqrt{\frac{S_p^2}{n_1} + \frac{S_p^2}{n_2}}$$
$$= (7.86 - 7.02) \pm 1.717\sqrt{\frac{.8125}{12} + \frac{.8125}{12}}$$
$$= .84 \pm 1.717(.368)$$
$$= .84 \pm .63$$

Our point estimate for the difference in average yields is .84% in favor of the industrial bonds, with a potential sampling error of .63 percent. An equivalent expression would be .21 percent $\leq (\mu_1 - \mu_2) \leq$ 1.47 percent.

□

Tests of Hypotheses A formal statistical test about $\mu_1 - \mu_2$ for small samples closely parallels the large sample counterpart. Again, however, there is the assumption of normal populations having equal dispersion. Equation 13–9 is the corresponding test statistic.

Small Sample Test Statistic for Testing H_0: $\mu_1 = \mu_2$

$$t^* = \frac{(\bar{X}_1 - \bar{X}_2) - 0}{\sqrt{\frac{S_p^2}{n_1} + \frac{S_p^2}{n_2}}} \qquad (13\text{–}9)$$

EXAMPLE 13.4

A doctors' office participates in two local insurance plans, MedCare and Partners. When the office manager states that reimbursement of claims from MedCare seems to take longer than from Partners, it is decided to track a random sample of 10 claim submissions to each insurer. A few weeks later, the data in the following table have been gathered. With 5% Type I error risk, does the sample evidence permit concluding that the population mean time to reimbursement of MedCare claims exceeds that of Partners?

Type of Insurance	Number of Claims	Average Time to Repayment (Days)	Standard Deviation
MedCare	10	39.3	6.1
Partners	10	28.0	5.1

Solution:

With a reason to suspect that MedCare takes more time to repay its doctors than does Partners, we will want to test

$$H_0\colon\ \mu_{\text{Med}} = \mu_{\text{Part}}$$

against

$$H_a\colon\ \mu_{\text{Med}} > \mu_{\text{Part}}$$

The cutoff t-score for $10 + 10 - 2 = 18$ *df*, upper tailed, $\alpha = .05$, is $t = 1.734$. The pooled variance will be

$$S_p^2 = \frac{6.1^2(9) + 5.1^2(9)}{18} = \frac{568.98}{18} = 31.61$$

The *TS* is then

$$t^* = \frac{(\overline{X}_{\text{Med}} - \overline{X}_{\text{Part}}) - 0}{\sqrt{\dfrac{S_p^2}{n_{\text{Med}}} + \dfrac{S_p^2}{n_{\text{Part}}}}}$$

$$= \frac{(39.3 - 28.0) - 0}{\sqrt{\dfrac{31.61}{10} + \dfrac{31.61}{10}}} = \frac{11.3}{2.514} = 4.49$$

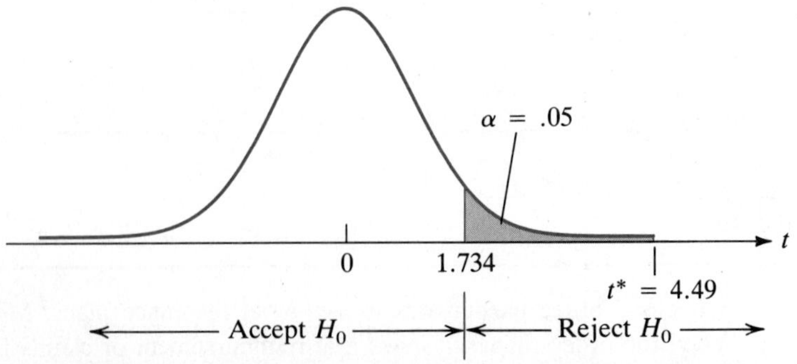

Figure 13.6 Graphical View of Example 13.4

Since $t^* = 4.49 > 1.734$, there is strong evidence against H_0. The 11.3 day difference in average time from filing to repayment is statistically significant—H_0 is rejected in favor of the one-tailed alternative. Figure 13.6 summarizes this statistical test.

□

COMMENTS

1. The test executed in Example 13.4, based on Equation 13–9, is sometimes called the *two-sample t-test.*
2. In Chapter 19, we will present an alternative two-sample hypothesis testing procedure called the *rank-sum* test. Like the Wilcoxon *signed ranks* test described in Chapter 11, this nonparametric procedure has less restrictive assumptions than does the two-sample t-test.

CLASSROOM EXAMPLE 13.2

Comparing Means for Small Samples

Two plastic supply firms have quoted a company's purchasing agent the same price for high-impact plastic. To investigate strength of the plastic, the agent acquires a random sample of eight pieces from each supplier. The company's laboratory then determines X, the breaking point (in pounds per square inch) for each piece. The sample data are

Supplier 1	2480	2660	2520	2630	2580	2880	2750	2700
	(mean = 2650; standard deviation = 129 lb/in.^{2})							
Supplier 2	2740	2610	2720	2380	2680	2590	2540	2500
	(mean = 2595; standard deviation = 121 lb/in.^{2})							

a. Compute S_p^2, the pooled sample variance.

b. Compute the estimated standard error that can be used for making an inference about $\mu_1 - \mu_2$.

c. What table value would be used to establish a 90% confidence interval for the difference in population means?

d. Specify the upper and lower limits of the 90% confidence interval for $\mu_1 - \mu_2$.

e. Use your confidence interval to state whether a two-tailed test of H_0: $\mu_1 = \mu_2$ would be accepted or rejected using a $\alpha = .10$.

f. What assumptions about the populations need to be made for small sample comparisons?

* ■ Processor: Comparing Means (Small Independent Samples)

This processor closely parallels its counterpart for large independent samples. It differs in minor respects: Samples less than size 30 are accepted, a pooled variance will be

*Optional

```
≡            Comparing Means (small Independent Samples)              READY
  Introduction  *able  Data [Results] Exit                          | F1=Help

   Hypothesis test      Ho: μ1 - μ2 =  0.000
                        Ha: μ1 - μ2 >  0.000

                          (X̄1 - X̄2) - (μ1 - μ2)
              TS:   t* = ---------------------
                             standard error

                           11.300 -  0.000
                    t* = ------------------ =  4.494
                                2.514

              The p-value for this test is 0.000159850247


     The decision to "reject Ho" or "do not reject Ho" depends on your
     judgment or tolerable risk of Type I error (see text Figure 10.18).
```

Figure 13.7 Hypothesis Test Results for Data in Example 13.4

computed for use in the standard error, and the *p*-value for hypothesis tests will be based on the *t*-distribution.

Data entry follows the same pattern as seen in Figure 13.3. We will demonstrate this processor using the hypothesis testing problem posed in Example 13.4: Testing the equality of two population means versus a one-tailed alternative. After letting the Medcare data be sample 1 in the data entry step, we chose an upper-tail alternative hypothesis and supplied the null value of zero for $\mu_1 - \mu_2$. Figure 13.7 shows the summary of the hypothesis test.

SECTION 13.1 EXERCISES

13.1 Specify the lower and upper limit of an interval which has a .99 confidence factor of including $\mu_1 - \mu_2$. The sample results are:

Sample	Number of Observations	Mean	Standard Deviation
1	35	71.1	6.5
2	51	64.1	5.7

13.2 Two separate random samples have been drawn from populations *F* and *G*. Use these results to estimate, with 95% confidence, the difference in population means: $\overline{X}_F = \$36.79$, $\overline{X}_G = \$38.12$; $S_F = \$4.78$, $S_G = \$4.01$; and $n_F = n_G = 60$.

13.3 An appliance manufacturer purchases water pumps for washing machines and dishwashers from two different suppliers. Samples from inventory were randomly selected and tested under standard conditions to compare the average service lives of the pumps. Sample results are as follows:

Supplier	Sample Size	Average Life Until Failure*	Standard Deviation*
B	48	5.72	.94
E	48	5.36	1.19

*In thousands of hours.

Develop an interval that has 90% confidence of including the difference in mean service life between the populations of *B* and *E* pumps.

13.4 Do business managers and business school teachers have different views of the merits of various journals? Elaine Fry, C. Glenn Walters, and Lawrence Scheuermann investigated this question in their article, "Perceived Quality of Fifty Selected Journals: Academicians and Practitioners" (*Journal of the Academy of Marketing Science,* Winter/Spring 1985). The authors surveyed business teachers and business managers with respect to the quality of various journals. Respondents were asked to rate each journal's quality on a four-point scale from very low (1) to very high (4). The highest rated of the 50 journals in the study was *Harvard Business Review* (*HBR*). Business managers gave an average rating of 3.685 to *HBR* (S = .543, n = 54); academicians gave *HBR* an average rating of 3.467 (S = .695, n = 240). Assume the respondents are random samples of their populations and compute an interval that has a 90% confidence factor of including the difference in population means.

13.5 A 99% confidence interval for $\mu_M - \mu_K$ has been computed to be 4.19 ± 13.91. Would a hypothesis test whose null statement was $\mu_M = \mu_K$ be accepted or rejected (assume $\alpha = .01$)?

13.6 The MBA program director at a large state university wishes to see whether nonbusiness undergraduate degree applicants differ from business degree applicants with respect to their "index." The index is a number used in the admission decision process, representing a combination of the applicant's undergraduate GPA and his/her score on a standardized aptitude test. The director obtains a random sample of 35 applications received from each group within the past year and determines the following results:

Category of Applicant	Average Index	Standard Deviation
Nonbusiness undergraduate	1141	98
Business undergraduate	1088	91

With 10% risk of Type I error, test the assumption that there is no difference in average index scores in the populations.

13.7 Refer to Exercise 13.6 above. Determine the p-value for the test statistic Z^*.

13.8 A large upscale department store located in the downtown region of a metropolitan area is thinking about opening a second store in the suburbs. The store has options to buy two different parcels of land, about 16 miles apart. To see whether household incomes differ near the two possible store sites, store management commissions a private research company to obtain information from residents within five miles of each site. Accordingly, a random sample of 50 residences within a five mile radius of each store site is obtained. Test the assumption that average household income should not be a factor in store site selection, using $\alpha = .10$.

Area Sampled	Average Household Income ($)	Standard Deviation ($)
Near site 1	33,417	4,016
Near site 2	34,166	3,806

13.9 A company that recycles aluminum products has been utilizing a single supplier but has received a proposal from a second supplier. The new supplier is asking a higher price per ton of raw material but promises to deliver a cleaner product—one with less impurities and more usable aluminum per ton than the competition. Company officials decide to compare the suppliers by analyzing random samples from each. Prior to sampling, a cost analysis is done which reveals that the higher priced raw material would need to yield an average of .75 lb more aluminum per 100 lb to offset its higher cost. Set up statements so that rejection of the null hypothesis would imply that the second supplier is more attractive. Use $\alpha = .10$. Sample results are as follows:

Source	Number of Samples Drawn	Average Yield per 100 Pounds	Standard Deviation
Current supplier	36	96.61	1.22
New supplier	36	97.82	.83

13.10 A family owns two adjacent fishing lakes, Crystal and Mirror. In recent weeks, regular customers have been telling the owners that larger fish are being caught in Crystal Lake. To see if this is true, the owners catch and weigh samples from each lake—data are shown in the table that follows. Set up the appropriate test with respect to average weights, using the 5% level of significance. Assume the fish catches can be viewed as random samples. Does the statistical evidence confirm the customer reports?

Lake	Number of Fish Caught	Average Weight (lb)	Standard Deviation
Crystal	41	5.96	1.54
Mirror	33	5.18	1.33

13.11 Refer to Exercise 13.10 above. If, in fact, the average fish weights in the two lakes are the same, what is the probability that a sample weight discrepancy as large or larger than was observed (.78 lb) would occur?

13.12 To two decimal-place accuracy, determine the weighted average variance, given the following: $S_1 = 10.82$ based on 9 observations; $S_2 = 13.30$ based on 18 observations.

13.13 What assumptions are required to make small sample inferences about the difference in population means?

13.14 Construct a 99% confidence interval for $\mu_A - \mu_B$ using these sample statistics:

Sample	Number of Items	Average Value ($)	Standard Deviation ($)
A	7	69.90	12.81
B	11	53.40	11.18

13.15 A nutritionist wants to estimate the difference in average fat content between patties of ground chuck and patties of ground round. Suppose he has access to the data in the following table, representing laboratory measurements of the amount of fat retained after cooking three-ounce patties of both types. Develop a 90% confidence interval for the difference in the populations' average fat content.

Type of Ground Beef	Sample Average*	Standard Deviation	Sample Size
Ground Chuck, 80% Lean	14.52	.38	9
Ground Round, 85% Lean	11.81	.35	9

*Source: In grams of fat, from "Composition of Foods: Beef Products," U.S. Department of Agriculture Handbook 8–13, Washington, D.C., 1986.

13.16 A research chemist doing pilot plant work on a new batch-process latex polymer has made 12 preliminary runs to compare yields using different catalysts for the reaction. Use the following data to test the assumption that percent yield for the reaction would be the same in the long run under each catalyst condition, using the .10 level of significance.

Catalyst *H* Percent Yield	71 68 65 78 69 63 (mean = 69.0; standard deviation = 5.25)
Catalyst *J* Percent Yield	68 78 80 81 69 83 (mean = 76.5; standard deviation = 6.41)

13.17 A county health department office has been receiving complaints about mosquitoes, many from the area near a small town named Burton. To assess the magnitude of the mosquito population, a county employee places mosquito traps in residential areas of Burton and in Bartlett, a town of the same size about 15 miles from Burton. The next day, the following counts of mosquitoes trapped are recorded:

Burton	10	14	13	19	15
Bartlett	13	9	3	6	7

a. Verify that the Burton data have a mean of 14.2, with a standard deviation of 3.27 and that the corresponding numbers for the Bartlett data are 7.6 and 3.71.

b. Assume that the readings can be viewed as random samples taken from normal populations with equal dispersion. Is the evidence from these small samples strong enough to convince you that the mosquito problem is significantly worse in Burton than it is in Bartlett? Use a Type I error risk factor of 5 percent.

13.18 A large retail home building supply store purchases its goods from various wholesalers. One such product is an 8″ × 8″ × 16″ concrete block used in outer walls of garages and basements. As part of a quality control program, a store clerk randomly chose and weighed 10 such blocks from each of two wholesalers that furnish the blocks. Sample results are as follows:

Wholesaler	Average Weight of Block (lb)	Standard Deviation (lb)
1	35.1	.49
2	36.8	.59

a. Develop a 95% confidence interval for the difference in mean weights of the populations from which these samples were drawn.

b. Use your interval to describe the observed sample difference as statistically significant or not statistically significant.

13.2 INFERENCES ABOUT $\mu_1 - \mu_2$: DEPENDENT SAMPLES

Dependent samples, also called matched pairs, represent a special application of two-sample inference. Dependent sample studies are generally more efficient and informative than are independent studies of equal sample size. If we have a choice about which way to set up a study to learn about population differences, we will prefer the matched pairs approach.

Figure 13.8 is a graphical overview of the dependent samples concept. Using some appropriate matching criteria, we create a population of paired observations where each item in one group is naturally connected to a specific item in the other group. Our sample observations then are *individual pair differences,* symbolized by d's. The computations made are to find $\bar{d}$, the mean of the paired differences, and S_d, the standard deviation of the paired differences.

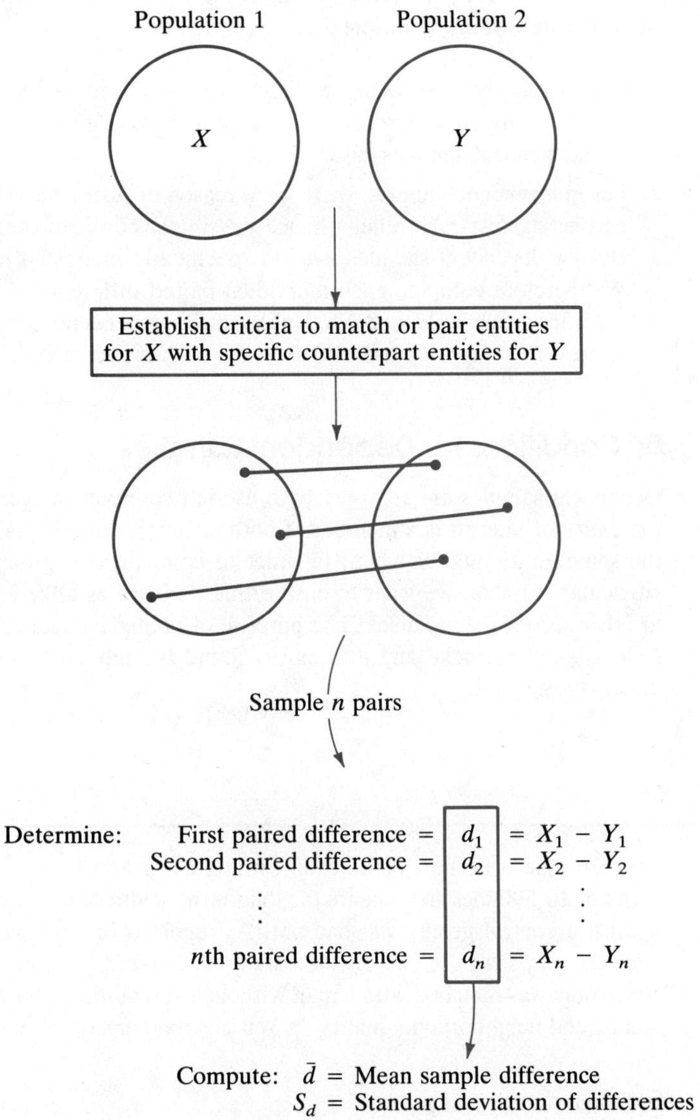

Figure 13.8 Schematic Representation of Dependent Samples

If we compare this figure to that of the independent samples case (Figure 13.1), we will note two major differences:

1. The dependent samples figure looks more involved, reflecting the fact that it is a more sophisticated approach—more of an "experiment"—than is the unmatched or independent samples case.
2. For independent samples, there is no reason or justification for comparing specific entities against each other—hence we simply compare overall means $(\overline{X}_1 - \overline{X}_2)$. But for dependent samples, we are specifically interested in pairwise differences. We therefore compute each individual paired difference d and the corresponding statistics, such as $\overline{d}$ and S_d. Unlike the independent samples case, we have no interest in computing overall means for each group, and do not do so.

Conditions for Dependent Samples

Dependent samples are appropriate to use in two broad categories of situations: when the entity of interest is a member of both samples, thereby giving us two readings on the same entity, and/or when, in order to compare two groups with respect to one particular variable, we want to ensure that they are as alike as possible with respect to other secondary variables. The purpose of taking secondary variables into account is to attempt to make any differences found be due solely to the primary variable under study.

EXAMPLE 13.5

In 1976, the National Institute of Child Health and Human Development began a study of 10,590 men to compare the long-term health of men who had had a vasectomy against a control group who had not. As reported in the Louisville *Courier-Journal* (February 2, 1984), "The institute study was carefully controlled. Each man with a vasectomy was matched with a man without a vasectomy of the same age, race, marital status, and neighborhood quality."* What is the purpose of matching here?

Solution:

The reason the men are matched is so that we can compare "apples with apples" as much as possible. We want to compare, for instance, a 39-year-old white, single male in the $35,000 income category who has had a vasectomy to a counterpart single in the same age, race, and income category but who has not had a vasectomy. The matching variables (age, race, income, marital status) conceivably could bear a relation

 (Note: The newspaper article does not expand on the meaning of "neighborhood quality." We will take it to be a proxy for income/access to health services. . . .)

to one's overall well-being. Had these variables been ignored (as simple independent samples would do) instead of being taken into account, the study would likely have been dismissed as meaningless by the scientific community.

□

To ascertain whether a study involves matched pairs, ask yourself this question: Is there a natural relation or pairing between specific members of each sample?

EXAMPLE 13.6

For each inference situation that follows, determine whether the sampling conditions suggest independent or dependent sampling.

a. A savings and loan firm wants to see whether their two real estate appraisers provide the same average appraisal values. For a sample of 10 homes, the appraisers are each asked to estimate the value of the homes.

b. A company is contemplating a purchase of 30 to 40 desktop copiers. Having narrowed its choices down to two brands, it obtains six of each for a four-week trial period. The company plans to compare the copiers with respect to the number of service calls needed.

c. A university testing service wants to compare scores on two different versions of a test that has recently been prepared to exempt freshmen from a computer programming requirement. A group of 80 freshmen take the test; the first 40 alphabetically get assigned to version 1 and the rest get version 2.

Solution:

a. This is a matched pairs study since we will obtain two readings on each home. We will therefore have an interest in each of the 10 pairwise differences as we proceed to make an inference about $\mu_1 - \mu_2$. Would it make any sense to have independent samples here—to have one appraiser rate one set of homes while the other appraiser rates a different set? We hope you think not, since one sample could be worth more than the other one to begin with, thus yielding results incapable of really determining whether the appraisers rate homes the same or differently. This situation does call for dependent samples—the natural connection between pair members being that the same home generates two observations.

b. This is an unmatched or independent samples study. We have two separate random samples, and no reason exists to compare any particular pair of copiers. In the absence of any matching criteria, we can compare only the overall mean of one brand against that of the other as was done in Section 13.1.

c. Again, this will be independent samples since there is no justification for pairing any two particular students. If, due to random chance, one group of students has more programming knowledge than the other, the results could be misleading. On the other hand, what if the study were set up so that a week later the same students came back and took the opposite version from their first test? Then we would have 80 matched pairs—two readings on the same test-takers— and therefore any previous programming skill differences would be taken into account. What if 3 of the 80 students were no-shows the second week? We would then have 77 pairs; the 3 unmatched scores would have to be discarded.

□

Confidence Intervals and Hypothesis Tests

Dependent samples inference follows the same patterns as those of independent samples. We will utilize some new notation to help us keep our dependent examples separate from our independent examples (see Table 13.2).

In Table 13.2, note that n (the number of matched pairs) is not really a new symbol. But we do wish to point out that, for dependent samples, n denotes the number of pairs not the combined number of observations. This is because it takes two observations, an X and its paired Y, to yield one value of d (see Figure 13.8). Since our inference is about population matched pair differences, n must represent the number of matched pairs in the sample. Two dependent samples yield a single sample of differences. As a result, our standard error ($S_d/\sqrt{n}$) is really the same as the one-sample standard error already encountered in Chapters 10 and 11. The only difference is notational—we put the d subscript on S to remind us of the problem setting.

The formulas needed for a matched pairs analysis therefore should look familiar. Equations 13–10 and 13–11 represent the confidence interval for estimating μ_d and the test statistic for testing H_0: $\mu_d = 0$, respectively. In both cases, the t-statistic is based on $df = n - 1$.

Table 13.2 Notation for Dependent Samples

Symbols	Description
n	The number of matched pairs
d	An individual paired difference
$\bar{d}$	The sample mean difference, computed as $\Sigma d/n$ (the point estimate of μ_d)
μ_d	The (unknown) population mean difference
S_d	The standard deviation of the d's
$S_d/\sqrt{n}$	The (estimated) standard error of $\bar{d}$

Confidence Interval for μ_d

$$\bar{d} \pm t \frac{S_d}{\sqrt{n}} \qquad (13\text{–}10)$$

Test Statistic for Testing H_0: $\mu_d = 0$

$$t^* = \frac{\bar{d} - 0}{S_d/\sqrt{n}} \qquad (13\text{–}11)$$

We are now ready to consider some specific examples.

EXAMPLE 13.7

A new drug treatment for patients with elevated cholesterol levels is administered to six subjects for a four-month test period. Treating the subjects as a random sample of patients with a cholesterol count above 250, estimate with 95% confidence the population mean reduction in cholesterol count.

	Cholesterol Count*	
Patient	Before Treatment	After Treatment
1	252	211
2	311	251
3	280	241
4	293	248
5	312	256
6	327	268

*Milligrams per one-tenth liter of blood.

Solution:

We have a natural pairing between the samples—before and after readings on the same subjects. Matched pairs analysis is therefore appropriate. Our attention focuses on the six paired differences.

Patient	Before–After Difference, d
1	41
2	60
3	39
4	45
5	56
6	59

The average reduction in cholesterol is $\bar{d} = \Sigma d/n = (41 + \ldots + 59)/6 = 50$. (You may wish to verify that S_d, the standard deviation of the six differences, is 9.42.) With a sample size of 6, our table value will be a t-score with 5 df. Our interval estimate for μ_d, the population mean reduction in cholesterol, can be expressed as Equation 13–10:

$$\bar{d} \pm t\frac{S_d}{\sqrt{n}} = 50 \pm 2.571\,\frac{9.42}{\sqrt{6}}$$
$$= 50 \pm 10 \text{ (to the nearest integer)}$$

We are 95% confident that the population mean reduction in cholesterol count over a four-month period is between 40 and 60.

□

As with other small samples, use of the t-distribution in Example 13.7 assumes a normal population; in the case of matched samples, a normal population of paired differences. In a matched pairs study involving large samples, we may use Z-table values to approximate t-table values.

EXAMPLE 13.8

A savings and loan business wants to see if their two real estate appraisers provide the same average appraisal values. For a random sample of 10 homes on the market, each is asked to furnish a valuation. Test the assumption that there is no difference in appraiser means, using 10% risk of Type I error. The numbers given are expressed in thousands of dollars.

Home	Appraiser 1	Appraiser 2	App. 1 − App. 2 Difference, d
A	89.8	88.4	1.4
B	94.5	95.5	−1.0
C	100.8	102.6	−1.8
D	70.3	70.1	0.2
E	88.5	89.1	−0.6
F	106.4	109.8	−3.4
G	77.5	79.0	−1.5
H	110.3	108.5	1.8
I	94.0	95.2	−1.2
J	82.6	81.9	0.7

Solution:

We wish to test

$$H_0\!: \mu_d = 0 \quad \text{and} \quad H_a\!: \mu_d \neq 0$$

The cutoff t-scores for 9 df, two tailed, $\alpha = .10$, are $t = \pm 1.833$. Note that unlike the previous example, some numbers in the difference column are negative while others are positive. Since the sign of each difference is relevant, we must be consistent in subtracting one column entry from the other column as we develop our values of d.

Using our column of paired differences, we can compute that $\bar{d} = -.54$ and that $S_d = 1.59$. The *TS* is then, following Equation 13–11,

$$t^* = \frac{\bar{d} - 0}{\frac{S_d}{\sqrt{n}}} = \frac{-.54 - 0}{\frac{1.59}{\sqrt{10}}} = \frac{-.54}{.503} = -1.07$$

Since $|t^*| \leq 1.833$, we accept H_0 (see Figure 13.9). The sample data are consistent with our null hypothesis that there is no difference in appraiser means.

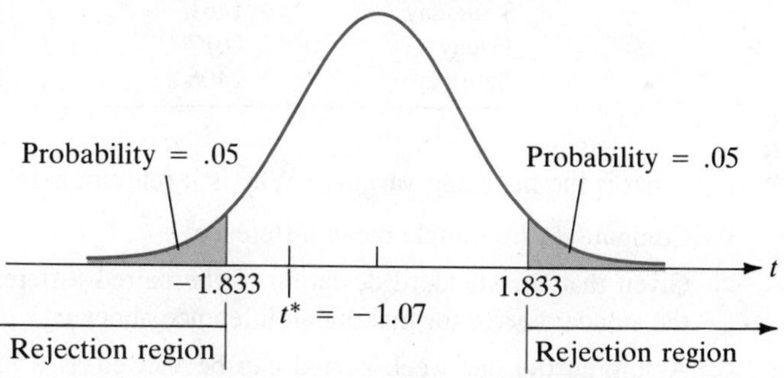

Figure 13.9 Graphical View of Example 13.8

□

COMMENTS

1. If you mistakenly treated a matched pairs data set as if it were independent samples, you would obtain a larger standard error. For instance, if you approached Example 13.7 above as if it were a test involving independent samples, you would ignore the paired differences and compute $\bar{X}_1$, $\bar{X}_2$, S_1, and S_2 instead. While the numerator of your *TS* would be unaffected (since $\bar{d} = \bar{X}_1 - \bar{X}_2$), the independent samples standard error would be much larger than $S_d/\sqrt{n}$. A larger standard error would make it more difficult to reject H_0. The matched pairs standard error is generally smaller, since the influence of other variables is removed. In the case of the appraisals, matching removes the fluctuating levels of home value from the analysis, letting us focus on only the paired differences. Independent samples cannot do this.
2. Undesired dispersion in the variable of interest that is due to other, uncontrolled variables is sometimes called *noise*. Matching or pairing observations is intended to reduce noise.

3. Since matched pair differences are in effect a single sample, the signed ranks test introduced in Chapter 11 can be used to test H_0: $\mu_d = 0$. As mentioned previously, this alternative procedure has fewer and less restrictive assumptions than the one-sample t-test.

CLASSROOM EXAMPLE 13.3

Comparing Matched Pairs

A restaurant manager wants to compare mean daily sales of two of his restaurants located several miles apart in the same city. For a one-week period, the following sales data are recorded:

Day	Restaurant *A*	Restaurant *B*
Sunday	\$1916	\$1771
Monday	991	904
Tuesday	1112	996
Wednesday	1212	1258
Thursday	1461	1385
Friday	2017	1867
Saturday	2406	2192

a. What is the matching variable? Why is it relevant here?

b. Compute $\bar{d}$, the sample mean difference.

c. Given that the standard deviation of the paired differences is \$81.23, determine the standard error for making an inference about μ_d.

d. Assuming the one-week period can be viewed as a random sample, establish a 90% confidence interval for μ_d. Based on the sample, is it reasonable to believe that there is no difference in population means?

e. Suppose the manager picks seven daily sales at random from Restaurant *A* from the last month, and a separate random sample of size seven from Restaurant *B*. Would the samples still be dependent? Why or why not?

*■ Processor: Dependent Samples (Matched Pairs)

This processor can compute a confidence interval for μ_d or test a hypothesis about the value of μ_d. The method of data entry for this processor depends on whether you have only the raw data available or you already know the values for $\bar{d}$ and S_d. The latter case is much easier; we simply supply this information when asked for it within the Dependent Samples processor. The former case requires that we create a data set prior to advancing to Processors on the main menu.

*Optional

Let us replicate the 95% confidence interval for μ_d that was developed in Example 13.7. We will use only the raw data: the six pairs of patient before-and-after cholesterol readings. Since we have the first case mentioned above, we would begin by selecting Data on the main menu, then use the Create option to enter a Multivariate Raw Data set with $n = 6$ entities on $v = 2$ variables. After the data entry is complete, we select the Dependent Samples option from the Two Sample Inference menu.

When the processor asks us whether our data is to be in the form of raw data or summary statistics, we specify raw data. At this point, Figure 13.10 will appear,

```
≡                    Dependent Samples (matched Pairs)                        READY
  Introduction  *able [Data] *esults  Exit                                 | F1=Help

                     Data variable OK



                        number of pairs, n = 6

                        mean difference, d-bar = 50

                        standard deviation, Sd = 9.42338
```

Figure 13.10 Dependent Samples Statistics from a Data Set

```
≡                    Dependent Samples (matched Pairs)                        QUERY
  Introduction  *able  Data [Results] Exit                                 | F1=Help

          For 95% confidence, the interval is
               point estimate ± t * (standard error)

               =   50.000 ± t * (9.423 / √n)
               =   50.000 ± 2.571 * (9.423 / 2.449)
               =   50.000 ± 9.891

          The lower limit of the confidence interval is  40.109
          The upper limit of the confidence interval is  59.891

               ═══════════════ Confidence ═══════════════

                    Compute another interval? (y/n) No
```

Figure 13.11 Confidence Interval Results for Matched Pairs Data in Example 13.7

giving us results for our raw data. When we proceed to request a 95% confidence level, we will view a screen similar to Figure 13.11.

SECTION 13.2 EXERCISES

13.19 Using the data in the following table, determine a 99% confidence interval for μ_d. Assume a normal population of differences.

Observation	Sample H	Sample J
A	10.9	8.1
B	6.1	5.9
C	7.7	7.2
D	14.1	12.7
E	10.8	8.9
F	7.6	7.8

13.20 A sample of 21 matched pairs shows $S_d = 7.7$ and $\bar{d} = 1.30$. At the .05 level of significance, would you accept or reject the assumption of no mean difference in the populations? Use a two-tailed alternative hypothesis.

13.21 Two car wax formulations were compared by putting each onto a sample of six cars of various ages. Wax *T* was applied to half of each car, and Wax *V*, to the other half. After two months of ordinary use, each car was washed and then rated on each side (on a 0 to 100 scale) for remaining wax protection.

Brand of	Protection Index, by Car					
Wax	1	2	3	4	5	6
T	65	72	47	75	91	81
V	60	67	45	68	92	72

a. Compute the sample mean difference of protection index scores.
b. Given that the standard deviation of the differences is 3.56, determine whether it is reasonable to conclude that $\mu_d = 0$. Use a risk factor of your choosing.

13.22 At a major university, students accepted into the MBA program are required to write an essay within a 45-minute test period. The MBAs are given a choice of four topics to write about. Their written work is then graded on a 0–100 scale with point categories for sentence structure, grammar, conciseness, organization, and so forth. Students who obtain a failing score are required to take a graduate writing course during their first semester in the MBA program; those passing are exempted. Two doctoral students from the English Department serve as graders for this somewhat subjective task. To check for grader consistency, a random sample of nine essays was scored by each grader.

	Score, by Essay								
Grader	1	2	3	4	5	6	7	8	9
1	54	83	87	64	97	91	71	93	71
2	46	83	85	59	93	94	75	89	76

a. Verify that the sample mean difference is 1.22, with a standard deviation of 4.49.
b. Use the results from part a to make an inference as to whether the graders have the same or different standards (let $\alpha = .10$).

***13.23** Refer to Exercise 13.21 above. Use the signed ranks test to make the inference required. What assumption inherent in the matched pairs test is not required in the signed ranks test?

13.24 For each inference situation that follows, determine whether the sampling conditions described suggest independent or dependent sampling.
a. In a study to assess the difference in average electricity usage between the summer and winter months, a consumer's group planned to monitor the kilowatt-hour (kwh) consumption for 12 randomly selected apartments. Each apartment had 1200 to 1400 square feet of living space. The total kwh consumption for the three summer months and the three winter months was recorded for each apartment.
b. An editor for a monthly newsletter catering to the gasoline industry wishes to estimate the average difference in cost between regular unleaded and premium unleaded gas. She randomly samples 95 gas stations and records the current price per gallon for each type of gas.
c. An investor would like to compare the average annual return for no-load and load mutual funds. He plans to take a random sample of both types of funds and test the null hypothesis of no difference in average return.

13.25 An automobile insurance agent suspects that one local repair shop, *A*, provides higher estimates for repair costs of cars that have sustained accident damage than does another shop, *B*. Over the next month, the agent arranges to have the next 10 customer cars needing repairs taken to both shops for estimates. Assume that the next 10 cars can be viewed as a random sample and conduct the proper one-tailed test, using a 10% level of significance. Do the data make a convincing case that shop *A* provides estimates whose mean is significantly higher than those of shop *B*?

	Estimate			Estimate	
Car	Shop *A*	Shop *B*	Car	Shop *A*	Shop *B*
1	$ 465	$ 475	6	$1350	$1280
2	920	845	7	1015	965
3	780	715	8	745	725
4	1425	1285	9	860	800
5	690	745	10	1220	1105

13.26 The regional manager of six electronics stores is buying local television advertising time to promote a one-week sale at her stores. To attempt to see if her first venture into this medium has justified the expense, the manager determines the following sales figures:

*Optional

Store	Weekly Sales Prior to the Media Buy*	Weekly Sales After the Media Buy*
1	9.1	11.8
2	10.4	11.9
3	11.6	12.8
4	14.2	14.9
5	15.2	17.8
6	12.2	13.1

*In thousands of dollars.

a. Compute the sample average paired difference and the standard deviation of paired differences.

b. Assume the outcomes above are a random sample of results from such a media buy. Develop a 90% confidence interval for the mean change in store sales the week after television advertising.

13.3 INFERENCES ABOUT $\pi_1 - \pi_2$

Comparative studies also may involve binomial populations—see Figure 13.12 for a schematic representation of the two-sample binomial procedure. Our objective here will be to make an inference about $\pi_1 - \pi_2$, the difference in proportions between two populations, based on the information found in our two samples.

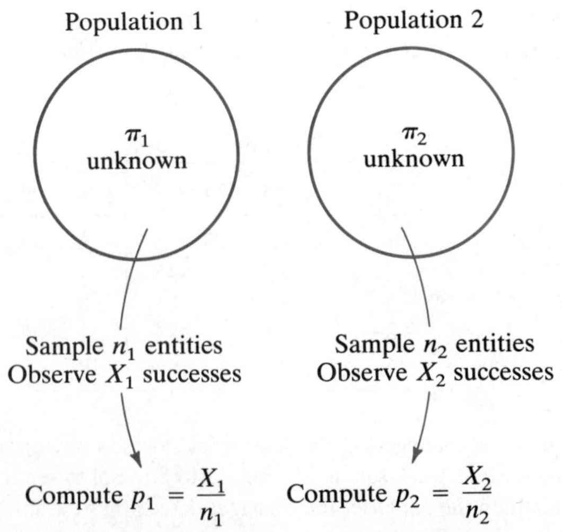

Figure 13.12 Two-Sample Procedure for Binomial Populations

We learned in the previous chapter that the estimator of the population proportion π is the sample proportion p. As you would then anticipate, the estimator of $\pi_1 - \pi_2$ will be $p_1 - p_2$. The discussion which follows presumes that sampling conditions exist ($n_1\pi_1$, $n_2\pi_2$, $n_1(1 - \pi_1)$, $n_2(1 - \pi_2)$ all being at least 5) so that the Central Limit Theorem can assure us of an approximately normal sampling distribution. We also will assume that our two binomial samples are independent.

Confidence Interval Estimation

The sampling distribution of $p_1 - p_2$ has the properties given in the following box.

Sampling Distribution of $p_1 - p_2$

$$\text{Mean} = \pi_1 - \pi_2 \tag{13–12}$$

$$\text{Standard error} = \sigma_{p_1 - p_2} = \sqrt{\frac{\pi_1(1 - \pi_1)}{n_1} + \frac{\pi_2(1 - \pi_2)}{n_2}} \tag{13–13}$$

Shape: Approximately normal

However, we now run into an obstacle similar to what we saw in the one-sample binomial confidence interval: π_1 and π_2 in the formula for the standard error (Equation 13–13) are unknown. Our solution to this is the same as before—we will replace π_1 and π_2 with their sample estimates p_1 and p_2, thereby giving us an estimated standard error.

Estimated Standard Error of $p_1 - p_2$

$$\hat{\sigma}_{p_1 - p_2} = \sqrt{\frac{p_1(1 - p_1)}{n_1} + \frac{p_2(1 - p_2)}{n_2}} \tag{13–14}$$

We use this expression in equation 13–15 to determine confidence intervals for $\pi_1 - \pi_2$, the difference of two binomial proportions.

Confidence Interval for $\pi_1 - \pi_2$

$$p_1 - p_2 \pm Z\sqrt{\frac{p_1(1 - p_1)}{n_1} + \frac{p_2(1 - p_2)}{n_2}} \tag{13–15}$$

EXAMPLE 13.9

An orchard owner has two fields of winesap apple trees that were treated with different insecticides at the start of the season. When the apples matured, a random sample was taken from each field to check for infestation. Use the following sample results to estimate the difference in infestation rates in the apple populations, with 90% confidence:

Insecticide	Sample Size	Number Infested
1	400	44
2	400	24

Solution:

Our problem setting is clearly binomial, since each apple inspected can be classified into one of two mutually exclusive categories: infested or not infested. Letting X = the number of infested apples, our point estimates are

$$p_1 = \frac{X_1}{n_1} = \frac{44}{400} = .11 \qquad p_2 = \frac{X_2}{n_2} = \frac{24}{400} = .06$$

Our desired confidence interval will be

$$p_1 - p_2 \pm Z\sqrt{\frac{p_1(1 - p_1)}{n_1} + \frac{p_2(1 - p_2)}{n_2}}$$

$$= .11 - .06 \pm 1.65\sqrt{\frac{.11(.89)}{400} + \frac{.06(.94)}{400}}$$

$$= .05 \pm 1.65(.0196) = .05 \pm .032$$

With 90% confidence, we believe the population infestation rate is higher where Insecticide 1 was used, by at least .018 (1.8 percent) and perhaps by as much as 8.2 percent.

□

Tests of Hypotheses

Like its counterpart procedure for means, the two-sample binomial test is usually conducted to see whether it is reasonable to believe that two populations do not differ. The null hypothesis therefore will be

$$H_0: \pi_1 = \pi_2$$

or equivalently,

$$H_0: \pi_1 - \pi_2 = 0$$

Depending on the intent of the study, H_a can be one tailed or two tailed.

The standard error for hypothesis tests is slightly different than that for confidence intervals. If we compare Equation 13–16

Standard Error for Testing H_0: $\pi_1 = \pi_2$

$$\hat{\sigma}_{p_1-p_2} = \sqrt{\frac{\bar{p}(1-\bar{p})}{n_1} + \frac{\bar{p}(1-\bar{p})}{n_2}} \qquad (13\text{–}16)$$

to Equation 13–14 we see that $\bar{p}$, denoting a weighted average or pooled proportion, replaces the terms p_1 and p_2. As an average of p_1 and p_2, $\bar{p}$ is computed as

$$\bar{p} = \frac{\text{Total number of sample successes}}{\text{Total sample size}} = \frac{X_1 + X_2}{n_1 + n_2}$$

or as

$$\bar{p} = \frac{p_1 n_1 + p_2 n_2}{n_1 + n_2}$$

an equivalent expression as long as p_1 and p_2 are not rounded off excessively.

There is a reason why this $\bar{p}$ term is computed to replace p_1 and p_2 when we conduct a hypothesis test. Our null hypothesis assumes that $\pi_1 = \pi_2$. If H_0 is true, then a common value of π should appear in each numerator term in the standard error (Equation 13–16). However, we do not know this value of π; instead, we have two estimates for it, p_1 and p_2. Rather than choose one or the other to appear in each numerator term, we average p_1 and p_2 to obtain $\bar{p}$, the best estimate of π. The corresponding test statistic is given by Equation 13–17.

Test Statistic for Testing H_0: $\pi_1 = \pi_2$

$$Z^* = \frac{(p_1 - p_2) - 0}{\sqrt{\frac{\bar{p}(1-\bar{p})}{n_1} + \frac{\bar{p}(1-\bar{p})}{n_2}}} \qquad (13\text{–}17)$$

EXAMPLE 13.10

An automobile financing corporation wishes to learn whether marital status has any bearing on whether a new car loan becomes delinquent within the first year. A random sample of 950 approved financing applications is summarized in the following table. Test the assumption that married and unmarried borrowers are equally likely to become delinquent, using $\alpha = .10$.

Marital Status	Number of Loans Sampled	Number Delinquent in Year 1
Unmarried	413	29
Married	537	47

Solution:

In the absence of a reason to anticipate that one group's rate should be higher than the other, we can form these statements:

$$H_0\text{: } \pi_U = \pi_M \qquad H_a\text{: } \pi_U \neq \pi_M$$

The sample delinquent rate for unmarrieds is 29/413, or .0702; for marrieds it is 47/537 = .0875. Our pooled proportion $\bar{p}$ is determined as (letting X = number of delinquent loans)

$$\bar{p} = \frac{X_U + X_M}{n_U + n_M} = \frac{29 + 47}{413 + 537} = \frac{76}{950} = .08$$

The *TS* is then

$$Z^* = \frac{(p_U - p_M) - 0}{\sqrt{\dfrac{\bar{p}(1 - \bar{p})}{n_U} + \dfrac{\bar{p}(1 - \bar{p})}{n_M}}}$$

$$= \frac{(.0702 - .0875) - 0}{\sqrt{\dfrac{.08(.92)}{413} + \dfrac{.08(.92)}{537}}}$$

$$= \frac{-.0173}{.0177} = -.98$$

For $\alpha = .10$, we would not reject H_0 unless the absolute value of the *TS* exceeded 1.65. Therefore, we would declare that the difference in delinquent rates found in the sample is not statistically significant.

□

COMMENTS

1. Some textbooks denote $\hat{\sigma}_{p_1 - p_2}$ as $S_{p_1 - p_2}$ or as $S(p_1 - p_2)$.
2. The value of the pooled proportion p must be in between p_1 and p_2. It will be exactly equidistant when $n_1 = n_2$.
3. The pooled proportion $\bar{p}$ does not appear in the standard error for confidence intervals since, unlike the case of a statistical test, there is no assumption of equal proportions.
4. The pooling concept for hypothesis tests is theoretically correct, although pooling makes little difference in the value of the standard error when p_1 and p_2 are close to each other.
5. A two-sample procedure was part of a case that recently reached the U.S. Supreme Court. In 1988 the Court ruled 8–0 that employees could use statistical evidence to prove that subjective decisions made by their employers treated them unfairly. The

case in question (*Watson* v. *Fort Worth Bank and Trust*) involved a bank teller who offered evidence that white supervisors who hired only 3.5 percent of black applicants hired 14.8 percent of white applicants. Further evidence showed that for those individuals hired, the same supervisors' average evaluations of blacks were substantially lower than those of whites.

CLASSROOM EXAMPLE 13.4

Comparing Binomial Populations

In a study conducted at the University of Southern California and reported in *Newsweek* (June 29, 1987), researchers looked at 162 nonsmoking males who had coronary artery disease and who had undergone heart bypass surgery. In order to see if diet and niacin played any role in the buildup of plaques on the subjects' new coronary arteries, two groups were formed. One group, the treatment group, t, was put on a diet where fat and cholesterol were strictly limited. They also received niacin, megadoses of which may reduce blood cholesterol. The control group, c, received placebo drugs and followed a less restrictive diet. After two years, 39 percent of the treatment group had formed additional plaques, compared to 61 percent in the control group.

a. Let π_t and π_c represent the proportions of the treatment and control populations that form additional plaques. Use these symbols to specify hypotheses for a test capable of providing convincing evidence that the treatment is effective in suppressing formation of new plaques.
b. Assuming the two sample groups are of the same size, what would be the value of $\bar{p}$, the average, or pooled proportion?
c. Compute the standard error for the test, using $\bar{p}$ from part b.
d. What is the numerical value of the test statistic (TS)?
e. What table value would you compare your TS against for a test with 1% risk of Type I error? At this risk level, do you find that the diet/drug treatment was effective or ineffective?

*■ Processor: Comparing Proportions

Our two-sample binomial processor can provide a confidence interval for $\pi_1 - \pi_2$ or execute a hypothesis test about the value of $\pi_1 - \pi_2$. We illustrate its use with the problem setting of Example 13.9—developing a 90% confidence interval for the difference in infestation rates for two populations of apple fields.

Data entry is similar to that for the one-sample binomial processor. We first specify whether we intend to enter integer values for the X's or decimal values for each p. Figure 13.13 shows the screen after we have supplied each n and each value of X for this data set. After selecting the Confidence Interval option, we view two results screens, the first of which is presented in Figure 13.14.

*Optional

```
≡                           Comparing Proportions                           EDIT
  Introduction  *able [Data] *esults  Exit                              | F1=Help

                      Parameter entry OK

                       X2    number of successes
                 p2 =  --  = -------------------
                       n2         sample size

                      number of successes, X1 = 400
                      proportion of successes, p1 = 44
                      number of successes 0.1100

                      number of successes, X2 = 400
                      proportion of successes, p2 = 24
                      number of successes 0.0600
```

Figure 13.13 Summary of User-Supplied Information for Comparing Proportions

```
≡                           Comparing Proportions                           READY
  Introduction  *able  Data [Results] Exit                              | F1=Help

            The point estimate for π1 - π2 is p1 - p2 = 0.050

            The standard error for computing a confidence
                 interval is the square root of

                 p1 * (1 - p1)    p2 * (1 - p2)
                 ------------- +  -------------,  which is 0.020
                      n1               n2
```

Figure 13.14 Point Estimate and Standard Error for Confidence Interval Construction

SECTION 13.3 EXERCISES

13.27 Estimate $\pi_1 - \pi_2$ with 99% confidence, given the following data:

Population	Sample Size	Number of Successes
1	600	384
2	500	295

13.28 Samples drawn from two binomial populations have been summarized:

Population	Sample Size	Number of Successes
C	650	286
D	650	234

a. Determine the pooled proportion that would be needed as part of a hypothesis test for equal population proportions.
b. With 5% risk of Type I error, do you find that the proportions of success in the populations of C and D are significantly different?

13.29 A firm samples raw material purchased from each of two suppliers—results are as follows:

Supplier	Sample Size	Percentage Defective
S	200	6.5
T	250	9.2

a. Determine the pooled proportion needed to test for equal population proportions.
b. Conduct a two-tailed hypothesis test at the .10 level of significance. Is there persuasive evidence that the suppliers have different defective rates?
c. What is the p-value for this statistical test?

13.30 In "Promised Incentives in Media Research: A Look at Data Quality, Sample Representativeness, and Response Rate" (Edward Goetz, Tom Tyler, and Fay Cook in the May 1984 issue of the *Journal of Marketing Research*), a sample of Chicago area residents were called using randomly chosen telephone numbers. The residents were asked to watch a particular television program; half were offered \$10 to do so and half were offered no reward. Follow-up calls after the program was broadcast were made to check on the rate of task performance for those individuals who agreed to watch an assigned show.

Payment	Sample Size	Number Watching Assigned Program
\$10	132	85
None	118	46

If we obtain the table above, develop a 90% confidence interval for the difference in response rates for the two payment conditions.

13.31 What group has a higher percentage of smokers—college men or college women? According to a study done by the University of Michigan's Institute for Social Research ("College Women More Likely Than Men to Smoke," Louisville *Courier-Journal*, July 8, 1986), the proportion of college women who smoke daily is 18 percent; for men, 10 percent. The sample size is reported as 1100 college students. Assume that men and women were equally represented in the sample and that random sampling methods were employed. With 95% confidence, establish lower and upper bounds for the difference in smoking rates of the populations.

13.32 Refer to your interval in Exercise 13.31. Would a formal hypothesis that the percentage of smokers on the nation's college campuses is the same for both sexes be accepted or rejected?

13.33 A local tax preparation firm has two offices. The firm's owner plans to compare accuracy of returns completed by his employees at the two offices. Accordingly, he takes a random sample of 150 returns from each office. These returns are carefully inspected for any arithmetic or procedural errors. The inspection shows that Office 1 has 15 returns with errors; Office 2 has 6 returns with errors. Test the assumption of no difference in overall office error rates, using the 5% level of significance.

13.34 In a survey of regular viewers of local, late evening news, researchers employed by a particular station showed a photograph of their channel's weather reporter to interviewees and asked them to identify him by first and last name. Of the 200 women surveyed, 61 percent could name the weather reporter; of the 200 men questioned, 46 percent were able to do so. Test the hypothesis that the weather reporter's recognition is the same between all men and women who are regular watchers of local, late evening news. Use a two-tailed alternative and a 10% risk factor of incorrectly rejecting your null statement.

13.4 SUMMARY

The major purpose of this chapter has been to extend our knowledge of inference—developing confidence intervals and conducting hypothesis tests—to situations where we compare two populations.

We first considered inferences for $\mu_1 - \mu_2$ for independent samples and then introduced the idea of collecting data in such a way that the observations are dependent, or paired. Due to the nature of the specific inference problem at hand, often the only avenue available is to employ independent samples. If a choice is available, however, a dependent sampling plan generally is preferred because of its ability to control unwanted variation that might otherwise make it difficult to detect a true difference in means.

We will see this idea of separating variation and creating more of a true experiment developed further in the next chapter as we consider tests involving more than two populations.

CHAPTER 13 EXERCISES

13.35 The owner of a seafood restaurant wishes to compare the average bills of charge card customers versus cash customers for families of size four. She arranges to have random samples of both types of sales collected over a one month period. Use the following information to develop an 80% confidence interval for $\mu_{\text{Chge}} - \mu_{\text{Cash}}$.

Payment Method	$\overline{X}$	S	n
Charge	\$29.22	\$4.71	45
Cash	27.78	4.40	45

13.36 In "The Effect of Problem Recognition Style on Information Seeking" (*Journal of the Academy of Marketing Science*, Winter, 1987), G. C. Bruner develops a problem recognition (*PR*) style score with regard to how consumers recognize clothing problems and search for solutions. Scores on subjects' *PR* were generated for a sample of 439 upperclass students. For the two subcategories of each demographic variable listed, indicate whether a statistically significant difference in *PR* scale scores exists in the populations, using 1% risk of Type I error. Assume random samples.

Demographic Variable	Subcategories	Mean *PR* Score	n	Standard Deviation
Age	Under 25	4.65	399	1.26
	Over 25	4.26	40	1.33
Sex	Male	4.20	281	1.22
	Female	5.36	158	.99

13.37 A 90% confidence interval for $\mu_1 - \mu_2$ has been computed as $-21.3 \leq (\mu_1 - \mu_2) \leq -8.8$. Does this indicate that (with 10% risk) the populations should be viewed as having the same mean?

13.38 Two sections of Principles of Economics were taught by a conventional lecture format while another two sections received a new approach: a mixture of lecture and computer lab time with an interactive "economics teacher" software program. At the end of the term, all students took the same comprehensive final exam. Choose your hypotheses about the average final exam scores so that it would be possible for a convincing case to be made favoring the new method of instruction, using $\alpha = .025$. Assume the students in the study can be viewed as random samples under each instruction method. Do you recommend a switch over to the new approach?

Instruction Method	Average Final Score	Sample Size	Standard Deviation
Conventional	78.9	58	10.8
New	80.1	53	9.8

13.39 Find the p-value for the test in Exercise 13.38.

13.40 Suppose we obtain two sets of predictions for the inflation rate for next year: one from a random sample of Fortune 500 firms, another from a random sample of university economists. Choose your own level of significance and determine whether the populations from which these samples were drawn have the same or different means.

Fortune 500 Firms	4.3 3.8 6.0 4.4 5.1 5.6 4.2 6.1 4.5 4.0 ($\bar{X}$ = 4.80%; S = .84%)
University Economists	4.4 5.9 7.0 5.1 5.9 6.0 6.3 ($\bar{X}$ = 5.80%; S = .84%)

13.41 A large insurance company is planning to build a new data processing facility. It has narrowed the location to two medium-sized cities within the same state. Since many current employees will be transferred to the new locale, the company is investigating housing costs. To try to make a reasonable comparison, a multiple listing service book of homes on the market is obtained from each city, and a random sample of 15 houses in the \$80,000–\$90,000 price range is selected. For the sample houses, the number of square feet is then ascertained. The results are as follows:

City	Average Home Square Feet	Standard Deviation
1	1871	138
2	2002	157

a. With 5% risk, should the company declare that mean square footage (and therefore housing costs) is significantly different between the two cities?

b. Do you think it is reasonable to assume that your conclusion for the \$80–90 thousand price range would be applicable to other nearby price ranges? Why or why not?

13.42 Refer to Exercise 13.41 above. Suppose the null hypothesis tested is in fact a true statement. If this is so, what is the probability that a sample difference as large or larger than 131 square feet would be found? (Since your table cannot read an exact probability, determine upper and lower limits for this probability.)

13.43 A top-of-the-line VCR of a certain brand was priced by phone for a random sample of nine stores each in Boston and in New York City. The Boston average price was \$418, with a standard deviation of \$12. The New York City average was \$405, with a standard deviation of \$13. Estimate the difference in average price for the populations with 80% confidence.

13.44 A nationwide fast-food chain is testing a new breakfast product in 17 selected markets. The participating stores are generally similar with respect to their sales volume. To introduce the new product, two different television advertising campaigns have been developed by the chain's ad agency. Nine stores were chosen at random to represent one proposed campaign; eight stores represent the other campaign. Equivalent television expenditures promoting the new product are made in all test cities, and at the end of the test period the following unit sales have been recorded at each test store:

Campaign 1	1212	989	1312	1146	1333	1460	1051	1196	1218
Campaign 2	1132	1049	1262	1334	1465	1229	1501	1160	

Test the hypothesis that the promotion plans are equally effective. Use a risk factor of your choosing to summarize the results of this test market.

13.45 Two grocery stores within the same town both claim that their prices are lower. A newspaper reporter decides to do a story on this. He makes up a list of 35 food items that a typical shopper might buy. He then prices each item (same brand, same quantity) at the stores. Four items were scratched from the list when it was found that the stores did not carry the same brand. The sample showed an average difference of 2.0 cents, with a standard deviation of 7.2 cents. At the .10 level of significance, is the two-cent difference statistically significant?

13.46 Thirty-eight overweight people have voluntarily entered a health clinic and agreed to follow one of two diet/exercise regimens. In order to compare the regimens, clinic workers created matched pairs of subjects. The two individuals most overweight became a pair, with one randomly assigned to regimen 1 and the other, to regimen 2. This process was continued until the two least overweight people were assigned. Assume the volunteers can be treated as a random sample of regimen followers.

a. What variable is being controlled by the matching process? Why might matching be beneficial in assessing any difference that may exist between regimen weight losses?

b. Suppose the regimen 2 volunteers have lost a mean of 6.7 pounds more than those on regimen 1 at the end of the study period. If the standard deviation of paired differences is 8.9 pounds, estimate the mean amount by which regimen 2 followers' weight loss exceeds that of regimen 1 followers. Use 95% confidence.

13.47 In cooperation with several grocery stores, a local distributor of a sausage product set up an in-store promotion. For several hours on a particular Saturday, store employees cooked little sausage patties at the end of one aisle and offered free samples and cents-off coupons to shoppers. In the following table are product sales, in number of packages, for the promotion Saturday period and the prior Saturday period.

	Sales, by Store							
Time Period	1	2	3	4	5	6	7	8
Prior Saturday	41	12	10	19	64	35	11	47
Promotion Saturday	59	26	21	29	92	51	24	57

a. Verify that $\bar{d} = 15.0$ and that $S_d = 5.98$.

b. Treating the data as a random sample, estimate the increase in average Saturday sales per store for all such promotions, with 80% confidence.

c. How would you describe the relation that connects the paired items?

13.48 A consumer interest organization tests a new premium motor oil that is advertised as increasing fuel economy. A sample of 12 cars of various sizes are driven at a steady pace for 1000 miles on a test track with the new oil. Each car also is driven with a substantially less expensive brand that also has earned the American Petroleum Institute's "energy conserving" designation. To make sure no ordering effects influenced the study, half the cars used the new oil first while half did not. Fuel economy results, in miles per gallon, are as follows:

Car	New Oil Brand	Existing Brand	Car	New Oil Brand	Existing Brand
1	24.5	23.7	7	26.1	25.2
2	27.9	27.2	8	35.7	36.7
3	29.3	28.9	9	23.0	22.5
4	22.6	22.0	10	29.6	28.8
5	18.4	17.4	11	21.4	21.1
6	28.0	28.3	12	30.4	29.5

$\bar{d} = .47$; $S_d = .58$ mpg.

a. Perform a hypothesis test with an alternative statement that the new product will provide superior mean fuel economy relative to the less expensive brand. Is the sample evidence strong enough (α = .05) to permit rejection of your null statement?

b. What assumption about the paired differences is required to use the t-distribution?

13.49 A marketing research survey of couples with children contained this statement: "In a typical workweek, you will eat your evening meal at a fast-food or pizza restaurant at least once." Of 300 couples where both worked, 204 responded by checking the answer, "Always or almost always." Of 300 couples where just one person worked, 138 checked this answer. Assuming random samples, estimate the difference in proportions between these two populations of couples with children. Use a 90% confidence factor.

13.50 A university's alumni giving officer randomly sampled graduates from the past five years to see if donations might be related to degree type. The sample revealed the following:

Type of Degree	Sample Size	Number Making Donation(s)
Undergraduate	312	84
Graduate	118	17

Use a 99% confidence interval to estimate the difference in giving rates for the university's two types of graduates, based on this sample.

13.51 A mail-order firm is about to introduce a new, high-markup item in its next catalog. In a pilot study, the firm sent a preliminary version of the catalog to 1000 regular customers. Half received a catalog with the new item featured on the inside front cover, and half, with the new item featured on the back cover. The catalogs were otherwise identical. After a month, 72 orders that included the new item arrived from inside front cover placement catalogs; 62 orders that included the new item came from back cover versions. At the .10 level of significance, is there clear evidence that sales of the new item will differ according to its placement in the catalog?

13.52 In "Management Judgment Forecasts, Composite Forecasting Models, and Conditional Efficiency" (*Journal of Marketing Research,* August, 1984) Mark Moriarty and Arthur Adams report on a test of forecasting accuracy of two different approaches to predicting sales of a firm's major consumer durable product. In the table below, forecast accuracy is expressed as absolute percent error—the lower this number, the better the forecast.

Prediction Month	Forecast Error (Absolute %)	
	Method 1	Method 2
Jan	6.50	8.10
Feb	3.86	5.75
Mar	13.39	10.12
Apr	0.54	5.20
May	17.35	12.30
Jun	8.39	9.18

Prediction Month	Forecast Error (Absolute %)	
	Method 1	Method 2
Jul	2.88	9.80
Aug	25.60	18.60
Sep	3.75	4.63
Oct	4.51	5.77
Nov	.66	3.11
Dec	8.13	2.00

a. What is the natural relation that connects pairs of observations?

b. Test the null hypothesis that the population of forecasts generated under these methods have the same average accuracy. Use a two-sided t-test and a level of significance of your own choosing to decide whether to reject or to accept your null hypothesis. (Given: $S_d = 4.342$.)

***c.** The article authors chose the signed ranks test over the t-test to make their inference "because of its less restrictive assumptions and its insensitivity to outliers." Using the same level of significance as in part b, what conclusion follows if the smaller sum of ranks is 38, as was reported? If the null statement were true, what is the expected value of the smaller sum?

13.53 On January 1, 1990, the United States Golf Association (USGA) banned square-groove golf irons from use in USGA-sanctioned events. The relatively new square-grooved irons had been gaining popularity over the more traditional V-groove iron. Square grooves on the club face give the ball more spin, resulting in better control of the shot as well as a better chance of getting the ball to stay on the putting green. Many professional golfers advocated the ban, saying that the square-grooved clubs made play too easy. The USGA ran tests comparing the two types of clubs. A golf ball hitting machine was used so that each ball was struck at a constant speed and angle. Use the following data to estimate the difference in average spin for the club face types. Use $\alpha = .10$.

Type of Golf Club	Sample Size	Mean (Rev/Sec)	Standard Deviation
V-groove	48	49.8	4.1
Square-groove	60	79.7	5.1

13.54 A local United Way conducted a lengthy survey of area employees (reported in "Investigation of Giving Behavior to United Way Using Log-Linear Modeling and Discriminant Analysis: An Empirical Study," by Arthur Adams and Subhash Lonial in the Summer 1984 issue of the *Journal of the Academy of Marketing Science*). The goal of the survey was to see what variables (such as age, income, union membership, sex, various attitudes, and so on) were related to the amount of money that employees contributed to the United Way campaign. A large body of data was gathered, and two different statistical models were used to analyze

*Optional

it. Another major purpose of the analysis was to see which model could best predict whether an employee made no pledge, a small (less than $75) pledge, or a large pledge. Treat the data in the table as a sample and test the hypothesis that there is no difference in the percent correct predictions for the statistical models. Let $\alpha = .05$.

Prediction Model	Number of Pledge Category Predictions	Number of Correct Predictions
Discriminant Analysis	274	191
Log-Linear Analysis	274	212

13.55 Undoubtedly you have received a vaccine against polio. Up until the 1950s, however, there was no vaccine available to protect people from this dreaded disease. The number of new cases fluctuated unpredictably from year to year, and there always seemed to be "pockets" of high polio incidence scattered about—some small geographical regions would have many more cases per capita than adjacent communities. Individuals most likely to contract polio were children, especially those from kindergarten through the third grade. A vaccine developed by Jonas Salk in the early 1950s looked promising, as it appeared to work well in the laboratory. Consequently, the U.S. Public Health Service organized a large-scale field trial of the Salk vaccine in 1954.

a. Given the pattern of disease outbreak, would a test with treatment and control groups work better if treatment and control groups were defined within selected school districts or if whole school districts or even larger regions were defined as the treatment and control groups? Why?

b. Would the null hypothesis be that the vaccine is effective or that the vaccine is ineffective? Why?

c. A substantial number of children assigned to the treatment groups were dropped from the study when their parents refused permission for the vaccine to be given. Why do you think some parents withdrew their children?

13.56 A local cable TV system has the ability to send different signals to their subscribers. Working with an advertising agency, the cable system broadcasts two different advertisements for the same product during a popular evening cable show. Some subscribers receive Test Ad 1; others receive Test Ad 2. The next day, the ad agency contacts cable subscribers and ascertains whether they viewed the show in question. Those subscribers who stated that they had watched the show were asked to describe the test commercial. Using an aided recall format, ad agency researchers gave each respondent a "playback" score (from 0 to 100) according to his or her ability to recall the test advertisement. Use the following sample results to develop a 95% confidence interval for the population difference in average recall scores.

Test Ad	Average Recall Score	Standard Deviation	Number of Viewers
1	44.1	14.4	47
2	56.2	19.9	57

Chapter Maxim *In an experimental or controlled study involving two or more populations, correct conclusions about the treatments are most likely to result when the entities are as alike as possible, except for the treatments.*

CHAPTER 14

ANALYSIS OF VARIANCE

*Optional

Objectives

After studying this chapter and working the exercises, you should be able to

1. Find cutoff values on the F-distribution.
2. Explain the terminology of ANOVA experimentation.
3. Conduct one-factor ANOVA tests to compare three or more means.
4. Describe how the ANOVA test statistic is affected when the null hypothesis is false.
5. Perform follow-up analyses to establish which of several means are not equal.
6. Explain the motivation for and be able to carry out a randomized block study.
7. Describe the concept of interaction.
8. Perform two-factor ANOVA tests.

*9. Test to see if two populations have equal variances.

10. Relate the business practice of "test marketing" to ideas found in this chapter.

*11. Use the ANOVA processors to analyze problems.

In Chapter 13, we studied applications where we used sample information to make inferences about the difference in two populations' means. In such cases, there was usually an interest in whether the two populations have the same mean. It is our goal in this chapter to extend this type of analysis to situations where we compare and evaluate differences among three or more sample means. For instance, we may need a procedure to compare weathering ability of five different paint formulations, or to assess the fuel economy of four pickup truck models, or to measure viewers' ability to recall the message in three versions of a test commercial for a new product.

In the previous chapter, when we wished to evaluate an assumption about two means, we tested H_0: $\mu_1 = \mu_2$. If we had large, independent samples, our test statistic was

$$Z^* = \frac{\overline{X}_1 - \overline{X}_2}{\sqrt{\dfrac{S_1^2}{n_1} + \dfrac{S_2^2}{n_2}}}$$

Note that the difference observed between the sample means, $\overline{X}_1 - \overline{X}_2$, constituted the numerator of the test statistic. In this chapter, we may test a statement about four population means such as H_0: $\mu_1 = \mu_2 = \mu_3 = \mu_4$. As we will see, the test statistic

*Applies to optional section.

for comparing multiple means will look considerably different from the expression above, since we would need to consider four sample means simultaneously, instead of just two.

The statistical technique for comparing three or more sample means is called ANOVA, short for analysis of variance. The name ANOVA results from the fact that the test statistic for comparing three or more populations contains no sample means, as does Z^*. Instead, our test statistic (we will call it F^*) is a *ratio of variances*. Although differences in sample means will not appear explicitly in the test statistic, we will see that these differences do determine its numerical value.

ANOVA concepts were originally developed in agricultural research. Today ANOVA is a basic part of planned experiments in business, the social sciences, and the natural sciences. Its strengths include the capacity to consider several samples simultaneously as well as the ability to attribute sample differences to the effects of one or more experimental factors.

14.1 THE *F*-DISTRIBUTION

A Different Test Statistic, F^*

In ANOVA, we will test statements such as H_0: $\mu_1 = \mu_2 = \mu_3$. Our indicator as to whether it is reasonable to believe that H_0 is true will be a *ratio of variances*, F^*. Therefore, F^* will have the following appearance:

$$F^* = \frac{\text{First estimate of } \sigma^2}{\text{Second estimate of } \sigma^2}$$

The test statistic F^* will involve sample evidence collected from three or more populations. Here is the crucial point: The two sample estimates that make up F^* are obtained in such a way that the following are true:

- The expected value of the numerator will equal the expected value of the denominator *only if H_0 is true*. The numerator and denominator in F^* should be roughly equal when H_0 is true. Thus F^* should not differ significantly from 1.00.
- The expected value of the numerator is greater than the expected value of the denominator when H_0 is false.

We will see shortly how the two sample variance estimates that comprise F^* are obtained. For now, it is important to understand that when H_0 is true, the ANOVA test statistic F^* is likely to not differ substantially from 1.00. A false H_0, however, tends to impact the numerator of F^* while having no effect on the denominator. Therefore, we will view a large value of F^* as evidence that H_0 is not true.

COMMENTS

1. Recall that expected value is synonymous with average value. For example, the sample variance S^2 takes on different values from sample to sample, but its *expected value* (or average value) over all possible samples of the same size is σ^2.
2. Another term for a variance is *mean square*. We will use this expression often in this and subsequent chapters.
3. Some textbooks and computer programs refer to the test statistic F^* as the "F-ratio."
4. Recall that when we do a two-tailed Z-test (or t-test), we take extreme positive or negative values of Z^* or t^* as evidence against H_0. F^* differs from Z^* or t^* in that F^* cannot be negative since variances cannot be negative, and F^* can be extreme in only one direction—too large. (We are using "extreme" here to mean being of sufficient magnitude to be statistically significant at some specified risk level, α.)

The Sampling Distribution of F^*

The sampling distribution of F^* was developed in the 1920s by the English statistician R. A. Fisher and later was named the F-distribution in his honor. The F-distribution is similar to the t-distribution and the chi-square distribution in that there is actually a family of distributions. The F-distribution differs in that it takes a *pair* of values for degrees of freedom (*df*) to identify each family member. This is because F^* is based on two sample variance estimates, not just one; hence, use of the F-distribution will require two *df* values: one for the numerator in the F-ratio and one for the denominator.

Figure 14.1 shows the general shape of an F-distribution. While different family members have somewhat different appearances, all are skewed to the right, have one mode, and can take on values from zero to infinity.

Unlike the case of the t- and chi-square distributions, we will not use the F-distribution to construct confidence intervals. Instead, the F-distribution will be used to test hypotheses, such as H_0: $\mu_1 = \mu_2 = \mu_3$. Since a false null hypoth-

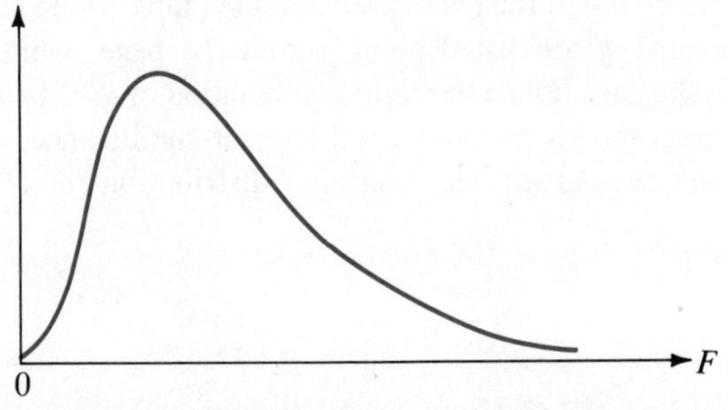

Figure 14.1 An F-Distribution

Table 14.1 Abbreviated *F*-Table

Denominator df	α	Numerator df 1	2	3	4	5
6	.10	3.78	3.46	3.29	3.18	3.11
	.05	5.99	5.14	4.76	4.53	4.39
	.01	13.75	10.92	9.78	9.15	8.75
7	.10	3.59	3.26	3.07	2.96	2.88
	.05	5.59	4.74	4.35	4.12	3.97
	.01	12.25	9.55	8.45	7.85	7.46
8	.10	3.46	3.11	2.92	2.81	2.73
	.05	5.32	4.46	4.07	3.84	3.69
	.01	11.26	8.65	7.59	7.01	6.63
9	.10	3.36	3.01	2.81	2.69	2.61
	.05	5.12	4.26	3.86	3.63	3.48
	.01	10.56	8.02	6.99	6.42	6.06

Note: Entries in this table are cutoff values to place probability α in the rejection region. For example, to have $\alpha = .05$ when numerator $df = 2$ and denominator $df = 7$, the table value is $F = 4.74$.

esis in ANOVA will be indicated by a large value of F^*, we will be interested in cutoff values that place a known probability or risk (α) in the upper tail (rejection region) of the F-distribution.

Using the *F*-Table

Our F-table is given in Appendix H at the back of the book—a brief excerpt appears in this section to illustrate its use. Cutoff values for the F-distribution depend on the problem settings' two values for *df*. The F-table therefore lists cutoff values for commonly used risk levels together with various combinations of *df*.

Referring to the condensed F-table (Table 14.1) in this section, you will note that numerator *df* are listed going across the page, while denominator *df* appear going down the page. Once the right combination of *df* is located, simply find the table value that corresponds to the desired level of significance, α. Please refer to Table 14.1 to familiarize yourself with finding F-distribution cutoff scores.

EXAMPLE 14.1

Use the F-table to determine the following:

a. The cutoff value for numerator $df = 4$, denominator $df = 9$, with $\alpha = .01$.
b. The cutoff value for F with 2 and 8 *df*, $\alpha = .10$.

c. The .05 cutoff for F with 1 and 6 df.

d. Should an ANOVA null hypothesis be accepted or rejected if $F^* = 5.11$ when $\alpha = .05$ and the df are 2 and 9?

e. What would be the p-value for $F^* = 5.11$ in part d?

Solution:

a. After intersecting the numerator $df = 4$ column with the denominator $df = 9$ row, we locate the cutoff value of $F = 6.42$ on the $\alpha = .01$ line.

b. By convention, the first df number always refers to the numerator in the F-ratio, the second to the denominator. For $\alpha = .10$ with 2 and 8 df, we find the cutoff value of $F = 3.11$.

c. The dividing line between the acceptance and rejection region occurs at 5.99 when $\alpha = .05$ with 1 and 6 df.

d. For 2 and 9 df with $\alpha = .05$, the table value is $F = 4.26$. Since $F^* = 5.11 > 4.26$, the null hypothesis should be rejected.

e. In part d, H_0 was rejected at $\alpha = .05$ since $F^* = 5.11 > 4.26$, but it would not be rejected at $\alpha = .01$, where $F^* = 5.11 < 8.02$. The p-value is therefore somewhere between .05 and .01. We can express this as: $.01 < p\text{-value} < .05$. The computer software that accompanies this book can determine exact p-values; using the F-table we can give p-values only in the form of an inequality.

□

COMMENT Since Appendix H does not list cutoff values for F for every possible combination of df, we can use straight-line interpolation to approximate missing values. For instance, if a sampling situation has 3 and 17 df and we need the .05 risk value, we can reasonably use $F = 3.20$, since it is halfway between 3.24 (the 3 and 16 df value) and 3.16 (the 3 and 18 df value). Likewise we would approximate the table value for 3 and 32 df, $\alpha = .10$, as $F = 2.27$. Since 32 is 20 percent of the distance from 30 to 40, we want a table value 20 percent of the distance from $F = 2.28$ to $F = 2.23$.

In the next section we will consider the simplest type of ANOVA: one-way or one-factor ANOVA.

SECTION 14.1 EXERCISES

14.1 What effect does H_0 being false tend to have on the ANOVA test statistic?

14.2 Find the cutoff values from the F-table for these sampling conditions:

a. $\alpha = .10$; $df = 2$ and 18
b. $\alpha = .05$; $df = 3$ and 11
c. $\alpha = .01$; $df = 4$ and 24

14.3 If H_0 is true for an ANOVA test, find the following probabilities:

a. $P(F^* > 2.81)$ for 2 and 12 df

b. $P(F^* < 4.72)$ for 3 and 24 *df*
c. $P(F^* > 3.15)$ for 2 and 60 *df*

14.4 State whether H_0 should be accepted or rejected for $\alpha = .05$, given the following:
a. $F^* = 2.34$; $df = 2$ and 11
b. $F^* = 2.52$; $df = 4$ and 20
c. $F^* = 4.29$; $df = 3$ and 24

14.5 Express the p-value in inequality form for the following:
a. $F^* = 3.51$; $df = 3$ and 8
b. $F^* = 1.37$; $df = 2$ and 10
c. $F^* = 5.01$; $df = 3$ and 22

***14.6** Use interpolation to determine an approximate F-table value for the following:
a. 3 and 21 *df*; $\alpha = .01$
b. 2 and 45 *df*; $\alpha = .01$
c. 2 and 99 *df*; $\alpha = .05$

14.2 ONE-FACTOR ANOVA

We begin our study of ANOVA with its simplest form: one-way, or one-factor ANOVA. Figure 14.2 suggests that we can view one-factor ANOVA as an extension of the independent, two-sample case of the previous chapter. Two differences from the two-

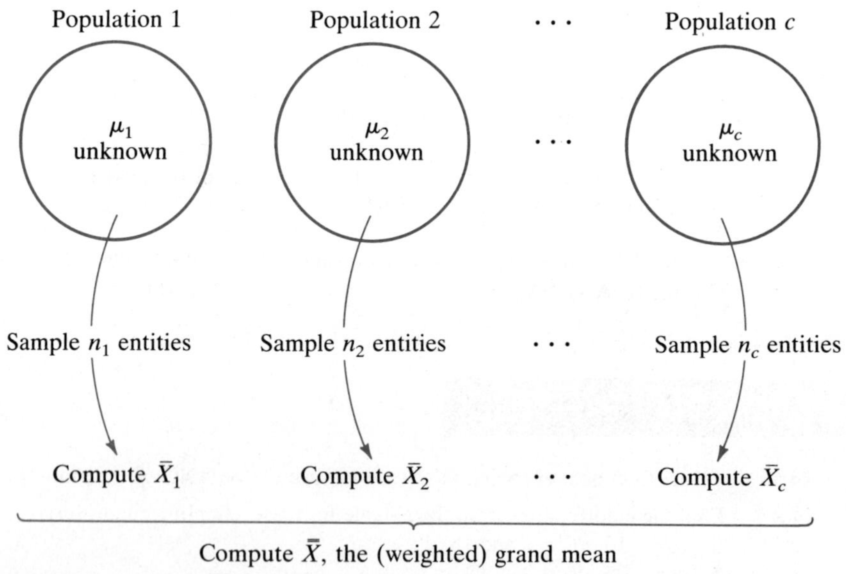

Figure 14.2 Schematic Representation of One-Factor ANOVA

*Optional

sample case are apparent, however: We now will have c samples drawn from c populations (where $c \geq 3$), and we now will have an interest in the observed mean of all the individual X's. This mean, referred to as the **grand mean** and denoted by an X with two bars over it, is really the mean of the means.

Terminology and Notation

ANOVA (as well as the subject of the next few chapters—regression) has some terminology and notation that we have not encountered before. We will introduce some of these terms as we consider a first example of one-factor ANOVA.

Our problem setting: A national automobile club obtains random samples of three comparably priced brands of emergency road flares to assess their average burning time. The following table shows burning time for each flare, recorded to the nearest whole minute.

	Brand 1	Brand 2	Brand 3
	21	23	26
	26	26	29
	25	25	26
	20	18	23
Sample Mean	23	23	26
	Grand Mean $\bar{\bar{X}} = 24$		

This study is investigating one variable, or *factor:* flare burning time. Later in this chapter we will see how to investigate two factors simultaneously. *Treatments* (also called factor levels) are particular forms, or conditions, of the factor. We have three treatments in this example—each of the three brands constitutes a treatment. When we test the assumption that $\mu_1 = \mu_2 = \mu_3$ we will be trying to ascertain whether the different treatments have different means. For this example, we will use the .10 level of significance. Let us now consider some notation that will be useful in analyzing one-factor ANOVA problems. Table 14.2 introduces the necessary symbols with a specific example of each from the flare burning time data.

COMMENTS

1. A special case in computing $\bar{\bar{X}}$ occurs in the example above where each treatment sample size is the same. A shortcut in computing the grand mean in such a case involves an unweighted average of the treatment means: $\bar{\bar{X}} = (23 + 23 + 26)/3 = 24$. This shortcut is not valid when the n_j are not equal.
2. The notation in Table 14.2 assumes that treatments are column headings. In the event that treatments are row headings (see Exercise 14.9, for instance), then c becomes the number of rows and the meanings of i and j are reversed.

Developing the Test Statistic, F^*

We already have mentioned that the ANOVA test statistic F^* is a ratio of two sample variances. Now we will use our road flare data set to develop this ratio. A preliminary step in doing so is to identify and partition the data set's total sum of squares.

Table 14.2 One-Factor ANOVA Notation

Symbol	Meaning	Burning Time Example
c	Number of columns (treatments)	3 brands are sampled: $c = 3$.
i	Row identifier, or index	With 4 rows of data, i can take on values from 1 to 4, inclusive.
j	Column (treatment) identifier, or index	j can take on values from 1 to 3.
X_{ij}	Identifier of the value of the variable in row i of treatment j	For instance: $X_{42} = 18$; $X_{23} = 29$.
n_j	Sample size within treatment j	$n_1 = n_2 = n_3 = 4$. Equality of all n_j is not required, however.
n	Total sample size	$n = \Sigma n_j = 4 + 4 + 4 = 12$.
$\overline{X}_j$	Column (treatment) mean	$\overline{X}_1 = 23$; $\overline{X}_2 = 23$; and $\overline{X}_3 = 26$.
$\overline{\overline{X}}$	Overall, or grand, mean	$\overline{\overline{X}}$ is the sum of all X_{ij}'s divided by n; that is, $\overline{\overline{X}} = (\Sigma\Sigma X_{ij}/n) = 288/12 = 24$. (Use of two Σ's occurs because we are summing in two directions—down columns and across rows—to get $\overline{\overline{X}}$.)

Total Sum of Squares, SST Under the ANOVA null hypothesis, the mean burn times of the three brands of flares are equal. The point estimate of this common mean (μ) is the grand mean, $\overline{\overline{X}} = 24$. We want to measure the variation around this grand mean and then separate it into two parts: the dispersion due to the treatments (brands), and the dispersion due to random variation in burn times.

This variation around $\overline{\overline{X}}$ is called the total sum of squares, or *SST*. It is simply the sum of squared deviations of individual values around the grand mean.

$$SST = \text{Total sum of squares} = \Sigma\Sigma(X_{ij} - \overline{\overline{X}})^2 \qquad \textbf{(14–1)}$$

For our flare burn times, *SST* would be computed as

$$\begin{aligned} SST &= \Sigma\Sigma(X_{ij} - 24)^2 \\ &= (21 - 24)^2 + (26 - 24)^2 + (25 - 24)^2 + (20 - 24)^2 \\ &\quad + (23 - 24)^2 + (26 - 24)^2 + (25 - 24)^2 + (18 - 24)^2 \\ &\quad + (26 - 24)^2 + (29 - 24)^2 + (26 - 24)^2 + (23 - 24)^2 \\ &= 106 \end{aligned}$$

We now want to separate, or partition, the total sum of squares *SST* into its two component parts. As suggested by Figure 14.3, *SST* is made up of *SSTR* and *SSE*. The component that is called the treatment sum of squares, or *SSTR*, is defined as follows:

$$SSTR = \text{Treatment sum of squares} = \Sigma n_j(\overline{X}_j - \overline{\overline{X}})^2 \qquad \textbf{(14–2)}$$

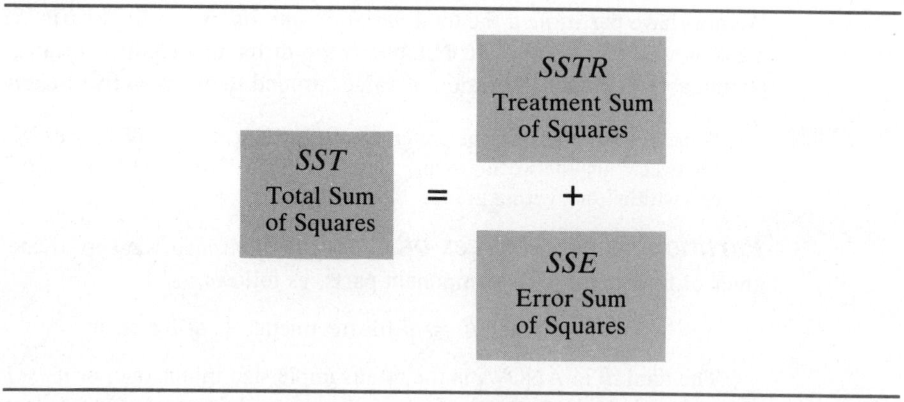

Figure 14.3 Partitioning Total Sum of Squares for a One-Factor ANOVA

If we focus on the terms inside the parentheses in Equation 14–2, it should be apparent that *SSTR* reflects the variation in treatment (column) means around the grand mean. *SSTR* would be zero if each treatment mean $\overline{X}_j$ were equal to the grand mean. On the other hand, the more the various treatment means differ from the grand mean, the larger *SSTR* will be. For the current example,

$$\begin{aligned} SSTR &= \Sigma n_j(\overline{X}_j - 24)^2 \\ &= 4(23 - 24)^2 + 4(23 - 24)^2 + 4(26 - 24)^2 \\ &= 4(-1)^2 + 4(-1)^2 + 4(2)^2 \\ &= 4 + 4 + 16 = 24 \end{aligned}$$

The other component of *SST* is the error sum of squares, or *SSE*, defined as follows:

$$SSE = \text{Error sum of squares} = \Sigma\Sigma(X_{ij} - \overline{X}_j)^2 \tag{14–3}$$

SSE measures the random variation of values about their own treatment mean. If we compare Equation 14–3 versus Equation 14–1, we see that *SSE* differs from *SST* in that we square deviations around the treatment mean $\overline{X}_j$ instead of around the grand mean $\overline{\overline{X}}$.

If you have complete faith in your computations of *SST* and *SSTR*, you can announce that *SSE* is the difference between these numbers. We recommend that you verify the equality by computing *SSE*. For this data set, we have

$$\begin{aligned} SSE &= \Sigma\Sigma(X_{ij} - \overline{X}_j)^2 \\ &= (21 - 23)^2 + (26 - 23)^2 + (25 - 23)^2 + (20 - 23)^2 \\ &\quad + (23 - 23)^2 + (26 - 23)^2 + (25 - 23)^2 + (18 - 23)^2 \\ &\quad + (26 - 26)^2 + (29 - 26)^2 + (26 - 26)^2 + (23 - 26)^2 \\ &= 82 \end{aligned}$$

Summarizing the partitioning, we have

$$SST = SSTR + SSE \quad \text{or} \quad 106 = 24 + 82$$

We now have partitioned the total variation into its two distinct parts, *SSTR* and *SSE*. *SSTR* is viewed as being attributable to the different treatment means, while *SSE* is attributable to random variation of values around their respective treatment means.

COMMENT Some business statisticians prefer to call "treatment" sum of squares by names such as "between" or "between-groups" sum of squares; "error" sum of squares is also known as "within" or "within groups" sum of squares.

Partitioning the Degrees of Freedom We also want to divide the total degrees of freedom into its component parts, as follows:

$$\text{Total } df = df \text{ for treatments} + df \text{ for error} \quad \textbf{(14–4)}$$

The total *df* in ANOVA is the total sample size minus one, or $n - 1$. One degree of freedom is lost when 12 values of X_{ij} are used to compute *SST*. In the burn time example, the total *df* will be $n - 1$, or 11.

The *df* for treatments is the number of treatments minus one, or $c - 1$. One degree of freedom is lost when three values of $\overline{X}_j$ are used to compute *SSTR*. In this example, the *df* for treatments will be $c - 1$, or 2.

The *df* for error is the total sample size minus the number of treatments, or $n - c$. One degree of freedom is lost per treatment (column) in computing *SSE*, since each $\overline{X}_j$ has to be determined. In this example, the *df* for error will be $n - c$, or 9.

Expressing Equation 14–4 in symbols, we have the following equation.

One-Factor ANOVA: Total *df* = Treatment *df* + Error *df*

$$n - 1 = (c - 1) + (n - c) \quad \textbf{(14–5)}$$

In our burning time example, Equation 14–5 becomes $11 = 2 + 9$.

Two Independent Estimates of the Variance As mentioned at the start of this section, the ANOVA test statistic is a ratio:

$$F^* = \frac{\text{First estimate of } \sigma^2}{\text{Second estimate of } \sigma^2}$$

Both numerator and denominator of F^* are sample variances, or mean squares. You may recall that a variance, or mean square, is a sum of squares divided by its degrees of freedom. We now have two sums of squares (*SSTR* and *SSE*) and their respective *df*. We are therefore ready to obtain the numerator and denominator of F^*.

By stipulating that the numerator of F^* should be based on the treatment variance and the denominator on the error variance, we have in words

$$F^* = \frac{\text{Treatment sum of squares}/df \text{ for treatments}}{\text{Error sum of squares}/df \text{ for error}}$$

and in symbols,

$$F^* = \frac{SSTR/(c-1)}{SSE/(n-c)} = \frac{MSTR}{MSE} \tag{14–6}$$

where *MSTR* denotes treatment mean square, and *MSE* denotes error mean square. For our burn time example, our test statistic would be

$$F^* = \frac{24/2}{82/9} = \frac{12}{9.111} = 1.32$$

Since the F-table value for 2 and 9 *df* with $\alpha = .10$ is $F = 3.01$, we would conclude that no significant difference exists among the three population mean burn times—see Figure 14.4. Table 14.3 summarizes this first example of one-factor ANOVA. Notice that the alternative hypothesis is written as "not ($\mu_1 = \mu_2 = \mu_3$)." There are several ways for the null statement to be false, such as $\mu_1 \neq \mu_2$, yet $\mu_2 = \mu_3$; or as $\mu_1 \neq \mu_2 \neq \mu_3$. Rather than list all the possibilities, we simply state H_a as "not H_0."

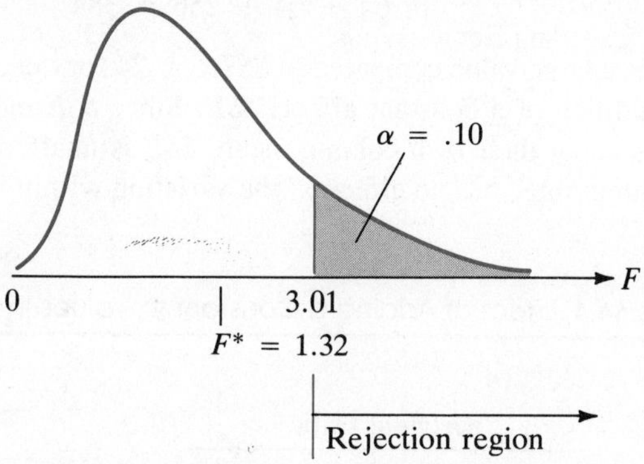

Figure 14.4 Graphical Representation of Rejection Region for Flare Burning Time Example

Table 14.3 Symbolic Summary of Flare Burning Time Example

H_0: $\mu_1 = \mu_2 = \mu_3$
H_a: not ($\mu_1 = \mu_2 = \mu_3$)
TS: $F^* = 1.32$
RR: Reject H_0 if $F^* > 3.01$
C: Do not reject H_0

Understanding the Behavior of F^* Now that we have an example under our belts, let us reconsider how ANOVA works. Assuming three treatments, the null hypothesis for one-factor ANOVA is H_0: $\mu_1 = \mu_2 = \mu_3$. If H_0 is in fact true, then all three treatments have a common mean, and the dispersion measured by *SST* is all due simply to random variation. *SST*'s component parts, *SSTR* and *SSE*, will represent random variation as well. As a result, *MSTR* and *MSE* are likely to be similar in value; that is, the ratio *MSTR*/*MSE* is likely to be near 1.00. In other words, the variance between columns (*MSTR*) is likely to be of the same magnitude as the variance within columns (*MSE*) when H_0 is true. F^*, therefore, is not likely to be significantly greater than 1.00.

Let's now suppose the opposite—that H_0 is false. Refer to Table 14.4, where we have reproduced, on the left side of the table, the flare burning time data and the computations leading to F^*. On the right side, the same data appear again except that we have added nine minutes to each Brand 3 observation, for illustrative purposes. This of course raises $\overline{X}_3$ by nine, to $\overline{X}_3 = 35$; the grand mean increases to 27. How does this addition of a constant to one treatment affect F^*?

Looking first at *SSTR*, we realize that since it measures the spread among treatment means, *SSTR* must be increased. To be specific, *SSTR* is now computed as

$$\begin{aligned} SSTR &= 4(23 - 27)^2 + 4(23 - 27)^2 + 4(35 - 27)^2 \\ &= 64 + 64 + 256 = 384 \end{aligned}$$

This is a large value compared to $SSTR = 24$ for the original data. Now consider how the addition of a constant affects *SSE*. Since *SSE* measures the spread of individual values about their own column mean, *SSE* is unaffected. The addition of a constant to column three has no effect on the variation within that column. With *SSTR* becom-

Table 14.4 Effect of Adding a Constant to Values in a Column

	Original Data				New Data: 9 Added to Brand 3		
	Brand 1	Brand 2	Brand 3		Brand 1	Brand 2	Brand 3
	21	23	26		21	23	35
	26	26	29		26	26	38
	25	25	26		25	25	35
	20	18	23		20	18	32
$\overline{X}_j$	23	23	26	$\overline{X}_j$	23	23	35
	$\overline{\overline{X}} = 24$				$\overline{\overline{X}} = 27$		
	$SSTR = 24$				$SSTR = 384$		
	$SSE = 82$				$SSE = 82$		
	$F^* = \dfrac{24/2}{82/9} = 1.32$				$F^* = \dfrac{384/2}{82/9} = 21.07$		

ing much larger and *SSE* remaining unchanged, the resulting F^* for the right side of Table 14.4 is now substantially larger, at $F^* = 21.07$. In summary, when H_0 is false and the treatment means are significantly different, the numerator of the F statistic will become large relative to the denominator.

COMMENTS

1. One-factor ANOVA is also called the *completely randomized design*.
2. The assumptions of one-factor ANOVA are the same as for a two-sample t-test in the previous chapter. The samples must be independent random samples, selected from populations with normal distributions having equal variances. ANOVA is said to be robust, which means that moderate violations of the normality and equal dispersion assumptions do not seriously affect the results, especially when treatment sample sizes are about equal.
3. The computation of *SSE* is a simple extension of the pooled sample variance computation introduced in the previous chapter for the small sample t-test (see Equation 13–6). In the t-test, we averaged, or pooled the variance from two samples. When we compute *SSE* in ANOVA, we are doing the same thing except that the ANOVA pooled variance is based on three or more samples.
4. The two-sample t-test with a pooled sample variance (see Section 13.1) referred to in Comment 3 is a special case of ANOVA. While we have presented one-factor ANOVA as a technique to test the equality of three or more means, it can be used to test two means.

   To test H_0: $\mu_1 = \mu_2$ versus H_a: $\mu_1 \neq \mu_2$ for independent samples, we have a choice: Approach the data as we did in Chapter 13 and compute t^*, or view the situation as ANOVA with two treatments and compute F^*. Either approach is correct; a data set will yield the same conclusion regardless of the approach used.
5. There are short-cut, or calculating formulas for *SST* and *SSTR* (*SSE* is found by subtraction) that may be quicker to use for hand computation than the defining formulas we have given. These are

$$SST = \Sigma\Sigma X_{ij}^2 - \frac{T^2}{n}$$

$$SSTR = \Sigma\frac{T_j^2}{n_j} - \frac{T^2}{n}$$

$$SSE = SST - SSTR$$

In these formulas, T is the total of all the values in the sample (that is, $T = \Sigma\Sigma X_{ij}$); T_j is the total of the values within each treatment; n and n_j are as already defined. Exercise 14.19 demonstrates these formulas.

The ANOVA Summary Table

Since there are several quantities to be determined in a one-factor ANOVA, we will find it helpful to organize them in a standard format. Such a format is called an ANOVA summary table. Most computer software packages routinely present ANOVA results in the manner shown in Table 14.5.

Table 14.5 General Format of One-Factor ANOVA Summary Table

Source	*SS*	*df*	*MS*	*F*
Treatments	*SSTR*	$c - 1$	*MSTR*	*MSTR/MSE*
Error	*SSE*	$n - c$	*MSE*	
Total	*SST*	$n - 1$		

Table 14.6 ANOVA Summary Table for Flare Burning Times

Source	*SS*	*df*	*MS*	*F*
Treatments	24	2	12	1.32
Error	82	9	9.11	
Total	106	11		

In this table, "source" refers to the source of variation; the sources are treatment, error, and total. The *SS* column denotes the various sums of squares, again, according to treatment, error, and total. The *MS* column refers to mean squares and reports the results of *SSTR* and *SSE* being divided by their respective *df*. The sole entry in the *F* column is F^*, the ratio of the two sample variances.

Table 14.5 is general—in specific ANOVA problems, we want to insert the actual numbers in place of the symbolic expressions. Table 14.6 gives the summary table entries for our original flare burning time example.

The purpose of an ANOVA summary table is to present all the computational results in a compact fashion. To see whether it is reasonable to conclude that the null hypothesis is true, we need to compare only the entry in the *F* column ($F^* = 1.32$ in this case) to the *F*-table value for the appropriate *df* and the desired risk level.

CLASSROOM EXAMPLE 14.1

Conducting a One-Factor ANOVA

An outdoor advertising firm has options to buy three different parcels of land for possible billboard placement. To help evaluate the locations, company employees conduct a vehicle count at each. For four days in one week, employees simultaneously record traffic volume passing each location for a ten-minute period, beginning at an agreed upon time. The data given in the following table are the vehicle counts observed; Location 2 has one less value than the others because the assigned employee was delayed on one day and could not get to the location on time.

Location 1	48	52	44	52
Location 2	44	33	40	
Location 3	46	40	35	39

a. What is the factor (variable) being studied? What are the treatments?

b. Specify the null hypothesis to be tested.

c. Determine the grand mean.

d. Determine the three values for $\overline{X}_{j}$.

e. Compute *SST*, the total sum of squares.

f. Compute the treatment sum of squares, *SSTR*.

g. Compute *SSE* and verify that $SSE = SST - SSTR$.

h. Create an ANOVA summary table.

i. At the .10 level of significance, what is the cutoff value from the *F*-table? Should H_0 be accepted or rejected?

j. Is the evidence against the null hypothesis strong enough so that we can be 99% certain that H_0 is not true?

*■ Processor: One-Way ANOVA

There are three ANOVA processors in our software program. All require that the user create a data set prior to calling for the ANOVA processor on the main menu.

To illustrate the One-Way ANOVA processor, assume we have the following data set:

Treatment 1	47	63	50	48	67
Treatment 2	71	52	58	63	
Treatment 3	54	52	62	48	49

Our first step is to select Data from the main menu, then touch Create. For one-way ANOVA, we select One-Way ANOVA Variables on the submenu. After specifying the number of treatments and the sample size for each, we will be asked to enter all the values within one treatment before entering the next treatment's values. Figure 14.5 shows the screen after we have entered 54, the first value in Treatment 3 in the data set above.

When data entry is complete and correct, move to Processors and select Analysis of Variance (ANOVA). A submenu with three choices will appear—choose the first. After viewing the introductory test, retrieve the *F*-table, verify via the Vector option that the proper data type is on hand, then select Calculations. When the computations

*Optional

```
≡        Microcomputer Applications for INTRODUCTORY BUSINESS STATISTICS   EDIT
  Overview [Data] Files  Processors  Set Up  Exit                    | F1=Help

              TR-1                     TR-2                    TR-3
  1 47                      71                    ▪54
  2 63                      52
  3 50                      58
  4 48                      63
  5 67
```

Figure 14.5 Creating a Data File for the One-Way ANOVA Processor

```
≡                             One Way Anova                              READY
  Introduction  *ata  Calculations [Results] Exit                    | F1=Help
```

ANOVA table

Source	SS	df	MS	F	p-value
Treatments	150.0000	2	75.0000	1.242	0.3262046966
Error	664.0000	11	60.3636		
Total	814.0000	13			

Figure 14.6 One-Way ANOVA Table

are done, we are ready to advance to Results and view the analysis. One of the two results screens that are available is shown in Figure 14.6. It provides the ANOVA summary table and the p-value for the test of equal treatment means.

SECTION 14.2 EXERCISES

14.7 a. Set up an ANOVA summary table for the following data:

Sample 1	16.01	16.89	17.14	14.84
Sample 2	18.18	17.34	16.16	17.72
Sample 3	17.99	17.10	17.08	

b. For a test of H_0: $\mu_1 = \mu_2 = \mu_3$, would you reject H_0 at the .10 level of significance?

14.8 What assumptions are required for one-factor ANOVA?

14.9 For the following data, should we conclude that the samples came from populations that have the same mean? Let $\alpha = .10$.

Sample 1	65	71	70	62
Sample 2	61	63	68	64
Sample 3	69	63	69	75
Sample 4	66	72	64	70

14.10 After entering data into a computer, a student obtained the following table:

Source	*SS*	*df*	*MS*	*F*
Treatments	346.2	3	115.4	20.79
Error	88.8	16	5.55	
Total	435.0	19		

a. How many treatments were present?
b. Should H_0 be rejected at $\alpha = .01$? Why or why not?

14.11 Given the following, complete the ANOVA table and make the correct inference at the .10 level of significance.

Source	*SS*	*df*	*MS*	*F*
Treatments		2	3.24	
Error		17		
Total	40.98			

14.12 Set up an ANOVA table, given the following: $SSTR = 712$, $SSE = 1416$, $c = 4$, and all $n_j = 7$. Then use your test statistic to determine whether the null hypothesis should be rejected at $\alpha = .05$.

14.13 A one-factor ANOVA was conducted with five observations per treatment. Letting MSE_j = sample variance within treatment j, use the following information to develop the ANOVA summary table:

Treatment	$\overline{X}_j$	MSE_j
1	16	8.0
2	22	9.0
3	14	6.0

14.14 A consumer organization wished to see whether the cash offer for a used car would vary according to the appearance of the owner of the car. The test vehicle was a four-year-old car for which cash offers were solicited from a total of 18 dealers in a large metropolitan area.

Three "owners" each took the test car to six randomly chosen dealers. The owners were all men, ranging from a well-dressed professional to a young laborer. Specify the hypothesis to be tested, and then perform the test at the .05 level of significance, using the data given in the following table. Prices are reported to the nearest hundred dollars.

Owner	Price ($)	Owner	Price ($)
1	40	2	47
1	43	2	43
1	44	2	48
1	46	3	44
1	39	3	49
1	40	3	47
2	45	3	48
2	44	3	49
2	46	3	48

14.15 In a test of paint wear, four test strips of five different formulations were applied to a blacktop surface. Six months later, each application was scored on a 0–30 scale for resistance to weathering, with a larger score denoting better wear.

Formulation	Sample Results				Sample Mean
1	23	26	27	24	25.0
2	24	21	26	25	24.0
3	26	20	22	26	23.5
4	27	27	28	24	26.5
5	27	28	26	23	26.0

a. Given that the total sum of squares is 100, develop the ANOVA summary table.
b. At $\alpha = .10$, do you conclude that statistically significant differences in formulation means exist?

14.16 A random sample of eight grocery stores in each of three adjacent suburban counties was selected to compare the price of a 16-ounce box of a particular brand of cereal. The data are as follows:

County	Price ($)							
K	2.09	2.15	2.22	2.19	2.17	2.09	2.11	2.26
D	2.17	2.29	2.25	2.29	2.19	2.15	2.19	2.23
L	2.09	2.10	2.15	2.05	2.17	2.13	2.19	2.16

a. What are the treatments?
b. Set up the ANOVA summary table.

c. If we are willing to tolerate 10% risk of Type I error, should we accept or reject the test hypothesis?

14.17 A college mathematics department has put together an exam for students who have had calculus in high school. Those students passing the test will be given three hours of college credit. For security purposes, the department wishes to develop and rotate three different versions or forms of the exam. A prime consideration is to have an equivalent degree of difficulty for all tests. If 120 randomly assigned students now have taken the test versions with the results given here, is it reasonable ($\alpha = .10$) to conclude that the objective of equivalent forms has been achieved? Results:

$$\bar{X}_1 = 67.6, \bar{X}_2 = 65.8, \bar{X}_3 = 68.2$$

$$n_1 = n_2 = n_3 = 40, SSE = 6074.20$$

14.18 The personnel director of a large manufacturing plant took a random sample of 15 shift production reports and recorded the number of employees per report who did not appear as scheduled to work.

Day Shift	Afternoon Shift	Night Shift
12	11	11
10	8	10
7	9	16
14	6	17
12	6	16

a. What is the factor for this study?

b. What is the value of the sample item that is symbolized X_{32}?

c. Test the appropriate hypothesis at the .05 level.

***14.19** A survey was conducted to compare next year's projected salary increases for five industries: computer services and software, utilities, banking, manufacturing, and retail. Five companies within each industry were contacted and asked to report the average anticipated raise for its employees. The data are as follows:

	Computer	Utilities	Banking	Manufacturing	Retail
	8.0%	6.2%	6.7%	5.7%	4.9%
	7.5	6.1	6.6	5.3	5.9
	7.2	5.7	6.8	5.9	6.4
	7.3	6.0		6.2	6.8
	7.0			6.4	
Total	37.0%	24.0%	20.1%	29.5%	24.0%

Several companies refused to divulge their raise information and were deleted from the study.

*Optional

a. Use the following calculating formulas to find *SST* and *SSTR:*

$$SST = \Sigma\Sigma X_{ij}^2 - \frac{T^2}{n} \qquad SSTR = \Sigma\frac{T_j^2}{n_j} - \frac{T^2}{n}$$

where T is the total of all the values in the sample, and T_j is the total of the values within each treatment.

b. Test the null hypothesis that the average percentage salary increase is the same among the five industries, using $\alpha = .01$. Show all parts of the test in a symbolic summary table like Table 14.3.

14.3 FOLLOW-UP TESTS: THE TUKEY *T*-METHOD

Suppose were are testing four populations in a one-factor ANOVA study. Our null and alternative hypotheses then would be as follows:

$$H_0\text{: } \mu_1 = \mu_2 = \mu_3 = \mu_4 \qquad H_a\text{: not } (\mu_1 = \mu_2 = \mu_3 = \mu_4)$$

An equivalent statement of the alternative hypothesis is H_a: at least two means are different. If the sample evidence leads to our accepting H_0, then there is generally no need for further analysis. In other words, the problem is solved if we have concluded that all treatment means are the same. On the other hand, if we reject H_0, then we are saying that there is at least one inequality in the null statement. But where? There are many ways for H_0 to be false: perhaps because $\mu_1 \neq \mu_4$, or/and perhaps because $\mu_2 \neq \mu_3$, and so forth. By itself, simply rejecting H_0 does not tell us where any inequalities are. There may be, however, an interest in establishing which means are different.

An American statistician named John Tukey has developed a procedure to ascertain which pair(s) of means are not equal, given that the null hypothesis has been rejected. It is intended for use when the treatment sample sizes (the n_j's) are the same. Tukey's method involves determining a cutoff range, called the *T-range,* and then comparing each possible pair of sample mean differences against the *T*-range. The number of such comparisons to be made will always be $c(c - 1)/2$. For any (absolute) difference in treatment means that exceeds the *T*-range, we conclude that the two respective populations have different means. The *T*-range is determined by Equation 14–7.

$$T\text{-range} = Q\sqrt{\frac{MSE}{n_j}} \qquad \textbf{(14–7)}$$

The value of Q will depend on the desired risk factor α as well as on the *df* values associated with the sampling situation. *MSE* and n_j will, of course, be known from the ANOVA computations. Appendix I at the back of the book gives values for Q for various levels of significance and *df* combinations. An abbreviated version of this table appears in Table 14.7—note that it is set up much like the *F*-table.

Table 14.7 Abbreviated Table of Values for Q

Denominator df	α	Numerator df 1	2	3	4	5
	.10	2.59	3.32	3.76	4.08	4.34
9	.05	3.20	3.95	4.41	4.76	5.02
	.01	4.60	5.43	5.96	6.35	6.66
	.10	2.56	3.27	3.70	4.02	4.26
10	.05	3.15	3.88	4.33	4.65	4.91
	.01	4.48	5.27	5.77	6.14	6.43
	.10	2.54	3.23	3.66	3.96	4.20
11	.05	3.11	3.82	4.26	4.57	4.82
	.01	4.39	5.14	5.62	5.97	6.25
	.10	2.52	3.20	3.62	3.92	4.16
12	.05	3.08	3.77	4.20	4.51	4.75
	.01	4.32	5.04	5.50	5.84	6.10

$$T\text{-range} = Q\sqrt{\frac{MSE}{n_j}}$$

Note: Entries are values of Q to place in the equation at left to determine the T-range in ANOVA follow-up tests. For example, for a follow-up analysis with $\alpha = .05$ when the F-statistic df are 2 and 9, the table value is $Q = 3.95$.

EXAMPLE 14.2

In a survey to determine child care costs for working parents, the local Chamber of Commerce randomly samples licensed child care centers in four regions of the metropolitan area. The purpose of the survey is to see whether and how average child care expense varies according to region. The following figures are weekly cost of child care for a two-year-old.

Suburban–East, E	Downtown Area, D	Suburban–West, W	Suburban–South, S
\$90.00	\$94.00	\$82.00	\$86.50
87.50	97.50	84.50	88.00
89.50	94.00	88.00	89.50
90.00	92.50	85.50	85.00

The ANOVA summary table for these data is as follows:

Source	SS	df	MS	F
Treatments	197.50	3	65.83	16.63
Error	47.50	12	3.96	
Total	245.00	15		

a. Establish whether a null hypothesis of equal means across the four regions would be accepted or rejected, using $\alpha = .10$. Is a follow-up analysis appropriate?

b. Compute the four sample treatment means and all possible pairs of differences in treatment means.

c. Compute the T-range for a follow-up analysis, using $\alpha = .10$.

d. Ascertain which pairs of differences are significantly different at the .10 level of significance.

Solution:

a. The hypothesis that $\mu_E = \mu_D = \mu_W = \mu_S$ would be rejected since the test statistic ($F^* = 16.63$) exceeds the table value ($F = 2.61$). Therefore follow-up tests are appropriate to see specifically which means are not equal.

b. The sample treatment means are $\overline{X}_E = 89.25$, $\overline{X}_D = 94.50$, $\overline{X}_W = 85.00$, and $\overline{X}_S = 87.25$. There are always $c(c - 1)/2$ comparisons to be made. In this case, we have $c = 4$ treatments, which will result in 4(3)/2, or 6 pairwise comparisons, as follows:

$$
\begin{aligned}
|\overline{X}_E - \overline{X}_D| &= |89.25 - 94.50| = 5.25 \\
|\overline{X}_E - \overline{X}_W| &= |89.25 - 85.00| = 4.25 \\
|\overline{X}_E - \overline{X}_S| &= |89.25 - 87.25| = 2.00 \\
|\overline{X}_D - \overline{X}_W| &= |94.50 - 85.00| = 9.50 \\
|\overline{X}_D - \overline{X}_S| &= |94.50 - 87.25| = 7.25 \\
|\overline{X}_W - \overline{X}_S| &= |85.00 - 87.25| = 2.25
\end{aligned}
$$

Note that it is the magnitude of the difference in means that is important—the sign does not matter.

c. The T-range is equal to $Q\sqrt{MSE/n_j}$. The raw data tell us that n_j is 4; the summary table specifies that MSE is 3.96. From Table 14.7, we obtain $Q = 3.62$ for 3 and 12 df with $\alpha = .10$. The T-range is then as follows:

$$
\begin{aligned}
T\text{-range} &= Q\sqrt{\frac{MSE}{n_j}} = 3.62\sqrt{\frac{3.96}{4}} \\
&= 3.62(.995) = 3.60
\end{aligned}
$$

Any difference in treatments means that exceeds 3.60 will constitute a statistically significant difference.

d. All comparisons involving the downtown region yield differences greater than 3.60—we conclude that the downtown mean is different from each separate region mean. One other difference—the one involving the east and west regions—is large enough for us to conclude that $\mu_E \neq \mu_W$. The other two differences are not significant.

□

COMMENTS

1. The risk of a Type I error (α) for the Tukey T-method applies to the overall follow-up analysis process. In Example 14.2, the 10% risk of a Type I error applies to the entire follow-up process (which consisted of six separate comparisons).
2. Although the Tukey T-method assumes equal sample sizes, it often is used when the sample sizes are relatively balanced. Under these conditions, the actual α is somewhat smaller than the desired α that equal sample sizes would provide.
3. When F^* is statistically significant, Tukey's method usually will reveal significant treatment differences; however, it is possible that it may fail to do so. Other approaches to follow-up analysis besides the Tukey T-method exist, and some specifically address the possibility of unequal treatment sample sizes. However, these are beyond the scope of this text. The interested reader is referred to an advanced text such as *Applied Linear Statistical Models* by Neter, Wasserman, and Kutner (1985).

SECTION 14.3 EXERCISES

14.20 If the ANOVA test statistic is less than the F-table value, why would there be little or no interest in performing any follow-up analysis?

14.21 Suppose an ANOVA summary table shows $MSE = 37.14$. If the study were based on 3 treatments each of size 7, determine the following:

a. What would be the table value for Q for a follow-up analysis at the .05 level of significance?

b. What would be the T-range?

14.22 Refer to Exercise 14.14.

a. How many pairwise comparisons are to be made for a follow-up analysis?

b. Determine the T-range for a follow-up test at the .05 level of significance.

c. Which, if any, pairwise comparisons do not result in finding a significant difference?

14.23 Refer to Exercise 14.16.

a. Determine Q and the T-range for an $\alpha = .10$ pairwise comparison of all sample means.

b. Report the result of your follow-up tests.

14.24 Refer to Exercise 14.18. At the .05 level of significance, which shifts differ in average attendance?

14.4 THE RANDOMIZED BLOCK DESIGN

In the previous chapter, we learned how to compare means of two populations. We studied the situation of independent samples; then we considered the case of dependent samples, or matched pairs.

We already have mentioned that one-factor ANOVA is an extension or generalization of the independent samples t-test from the previous chapter. We now want to consider extending the dependent samples case to three or more populations—analyzing such a situation requires a *randomized block design*.

Let us restate the conditions and motivation for a dependent samples test, since these same circumstances will call for a randomized block design when there are three or more groups to be compared. Simply stated, a dependent samples test is in order when there is a natural connection among specific members of each sample. In the matched pairs case, we compute each pairwise difference. By analyzing these differences (instead of the raw values), we can "tune out" the dispersion of the raw values, thereby focusing our attention on what matters—the differences between connected observations. For instance, in Example 13.8 we compared two real estate appraisers who each evaluated a sample of 10 homes. By analyzing only the pairwise differences in appraisal value, we tuned out or eliminated from consideration the spread in home prices; in other words, we devoted our attention to what was important (the difference in each home appraisal) while ignoring the "clutter," or uninformative dispersion (the spread among home values). We are now ready to extend this concept to the case of three or more populations.

Terminology, Notation, and Partitioning *SST*

To help illustrate how the randomized block design differs from one-way ANOVA, we will consider an example where a bank's three commercial real estate appraisers each have evaluated four business properties. The following table shows the sample results.

	Appraisal (Thousands $)			
Property	Appraiser 1	Appraiser 2	Appraiser 3	$\overline{X}_i$ = Row Mean
1	202	210	194	202
2	78	82	77	79
3	112	110	105	109
4	248	246	232	242
$\overline{X}_j$ = Column Mean	160	162	152	Grand Mean = $\overline{\overline{X}}$ = 158

The purpose of the analysis is to see whether the bank's appraisers have the same mean; H_0 will be that $\mu_1 = \mu_2 = \mu_3$, as in ordinary one-way ANOVA. Let us assume a Type I error risk of $\alpha = .05$. This problem differs from one-way ANOVA in that we have matched or connected observations: Each piece of property creates three observations. The groups of matched observations are called *blocks*. In this example, the different properties are the blocks—notice that each treatment (appraiser) is found once within each block (property).

Much of the notation for the randomized block design is the same as for one-way ANOVA. Table 14.8 lists the new notation that applies when we have a randomized block design.

Table 14.8 Additional Notation for the Randomized Block Design

Symbol	Meaning	Appraisal Example
k	Number of treatments	$k = 3$
b	Number of blocks	$b = 4$
$\overline{X}_i$	Block mean	$\overline{X}_i$'s are 202, 79, 109, 242
SSB	Block sum of squares	To be computed

The objective is to test the assumption that the appraisers have the same mean. We will employ the blocks to isolate the uninformative property-to-property variation. We then can direct our attention to the matched observations that can be found within each block. We proceed by partitioning SST into its component parts, as indicated by Figure 14.7. Block sum of squares (SSB) denotes the variation that the blocks can isolate in the analysis.

SST and $SSTR$ are computed in exactly the same fashion as for one-way ANOVA. For our appraisal example, we have

$$\begin{aligned} SST &= \Sigma\Sigma(X_{ij} - \overline{\overline{X}})^2 \\ &= (202 - 158)^2 + \ldots + (232 - 158)^2 \\ &= 53222 \end{aligned}$$

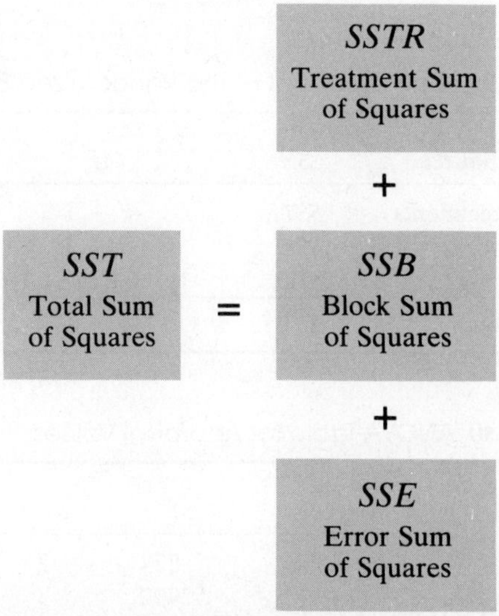

Figure 14.7 Partitioning Total Sum of Squares for a Randomized Block Design

and

$$\begin{aligned} SSTR &= \Sigma n_j (\overline{X}_j - \overline{\overline{X}})^2 \\ &= 4(160 - 158)^2 + 4(162 - 158)^2 + 4(152 - 158)^2 \\ &= 16 + 64 + 144 = 224 \end{aligned}$$

For a randomized block design, n_j will be the same for all treatments.

While *SSTR* involves deviations of treatment means around the grand mean, *SSB* involves deviations of block means around the grand mean:

$$\begin{aligned} SSB &= k\Sigma(\overline{X}_i - \overline{\overline{X}})^2 \\ &= 3[(202 - 158)^2 + (79 - 158)^2 + (109 - 158)^2 + (242 - 158)^2] \\ &= 3(1936 + 6241 + 2401 + 7056) = 52902 \end{aligned}$$

The final term in the partitioning is *SSE*, the error sum of squares. *SSE* is the residual, or remaining, random variation among observations after treatment and block variation have been determined. A defining formula for *SSE* is $\Sigma\Sigma(X_{ij} - \overline{X}_i - \overline{X}_j + \overline{\overline{X}})^2$; we generally will compute it as $SSE = SST - SSTR - SSB$. For this data set, $SSE = 53222 - 224 - 52902 = 96$. We now will organize our efforts into an ANOVA table.

The Randomized Block ANOVA Table

Table 14.9 illustrates the standard format for the randomized block ANOVA table. The table is set up much like the one-way ANOVA table, although it contains an additional line of information about the blocks. Table 14.10 shows the ANOVA table results for our appraisal illustration.

Table 14.9 General Format of the Randomized Block ANOVA Table

Source	*SS*	*df*	*MS*	*F*
Treatments	*SSTR*	$k - 1$	*MSTR*	*MSTR*/*MSE*
Blocks	*SSB*	$b - 1$	*MSB*	
Error	*SSE*	$(k - 1)(b - 1)$	*MSE*	
Total	*SST*	$n - 1$		

Table 14.10 ANOVA Table for Appraisal Values

Source	*SS*	*df*	*MS*	*F*
Treatments	224	2	112	7.00
Blocks	52902	3	17634	
Error	96	6	16	
Total	53222	11		

In this example, our test statistic F^* has a value of 7.00. To see whether the null hypothesis of equal appraisal means has been supported or contradicted, we compare our F-ratio ($F^* = 7.00$) to the .05 level table value for 2 and 6 *df* ($F = 5.14$). Since the test statistic exceeds the cutoff value, we conclude that H_0 should be rejected.

The large value for *SSB* suggests that the blocks represented a great deal of property-to-property variation. Notice that the blocking mean square ($MSB = 17634$) and its associated *df* play no role in developing F^* for our test of equal treatment means. (Sometimes researchers do want to check on whether the blocks have performed their intended function—see Comment 2 in the following.)

COMMENTS

1. If we mistakenly paid no attention to the fact that we had blocks, the problem would then reduce to ordinary one-way ANOVA, where we compute *SST*, *SSTR*, and *SSE*. Since *SST* and *SSTR* are computed the same way for one-way ANOVA and randomized blocks, the difference between the two methods centers on computing *SSE*. With no blocking factor, *SSE* is simply the difference between *SST* and *SSTR*. A blocking factor permits some of this difference between *SST* and *SSTR* to be assigned to the blocks, resulting in a smaller *SSE* and a more powerful test.
2. If a researcher wants to verify that the blocking factor is effective in isolating unwanted variation, she or he can compute $F^* = MSB/MSE$ to test H_0: the blocking factor is ineffective. If F^* were not large enough to reject H_0, it would suggest that the blocks were not relevant to the problem at hand. Rejection of the blocking factor hypothesis indicates that blocking was desirable.
3. Follow-up analysis for a randomized block design is conducted in the same fashion as for one-way ANOVA.
4. A calculating formula for the block sum of squares is available. If B_i represents the total of the values within each block, then

$$SSB = \frac{\Sigma B_i^2}{k} - \frac{T^2}{bk}$$

where

k = the number of treatments
b = the number of blocks
T = the total of all the values in the sample

Exercise 14.28 demonstrates the calculating formulas.

CLASSROOM EXAMPLE 14.2

Analyzing a Randomized Block Design

In a test of three different brands of riding lawn mower, the machines cut test fields of equal size under various conditions. The amount of time to mow each field, in minutes, is recorded in the following table. Assume the .05 level of significance is to be employed.

Mowing Condition	Mower *H*	Mower *I*	Mower *J*	Condition Mean
3″ high grass, level	9.0	8.4	11.1	9.5
7″ high grass, level	14.0	12.6	18.1	14.9
5″ high grass, 10 degree slope	10.6	10.2	14.0	11.6
Mower Mean	11.2	10.4	14.4	$\bar{\bar{X}} = 12.0$

a. What are the blocks? What are the treatments? State the hypothesis to be tested.

b. Given $SST = 73.54$ and $SSTR = 26.88$, determine SSB and SSE.

c. Create an ANOVA table.

d. What is the table value against which the test statistic should be compared? Is the hypothesis accepted or rejected?

*■ Processor: Randomized Block

Like the previous processor, this one requires that we begin by creating a data set. Select the Randomized Block Variables option from the Create submenu and supply the needed information. For illustrative purposes, we will use the appraisal data given earlier in Section 14.4.

When data entry is complete, choose the ANOVA processor and then the Randomized Block option. Proceed from left to right across the Randomized Block menu

```
≡                         Randomized Blocks                              READY
  Introduction  *ata  Calculations [Results] Exit                   |  F1=Help

    The treatment means are:
         160.000             162.000            152.000

    The block means are:
         202.000             79.000             109.000             242.000




        The overall mean, X-double-bar, is  158.000

        Ho: all treatment means are equal
        Ha: not Ho
```

Figure 14.8 Reporting Means for a Randomized Block Design

*Optional

as was done for one-way ANOVA. When we call for Results, two screens can be viewed. The first, showing various means and a statement of the hypotheses, is given in Figure 14.8. The other screen available is much like Figure 14.6; it provides the ANOVA summary table and p-value.

SECTION 14.4 EXERCISES

14.25 Set up an ANOVA table, given that three blocks were used, $SST = 368.4$, $SSB = 211.2$, $SSE = 102.2$ and that the total degrees of freedom is 14. Use your test statistic to ascertain whether it is reasonable to believe that all treatments have the same mean. Use $\alpha = .10$.

14.26 Given the following table, state whether the treatments have equal means, using 10% risk of a Type I error.

Source	*SS*	*df*	*MS*	*F*
Treatments	300			
Blocks	2048	8		
Error		24		
Total	3752			

14.27 Consider the following data:

	Treatments		
Blocks	**1**	**2**	**3**
A	42	44	49
B	37	38	39
C	53	50	50
D	34	30	38

a. Set up an ANOVA table for these data.

b. For a test of H_0: $\mu_1 = \mu_2 = \mu_3$, would you accept or reject the null hypothesis, using a risk factor of $\alpha = .05$?

***14.28** To compare the processing speeds of four different models of lap-top computers, three statistical analyses of varying degrees of complexity were run on each. The times, in thousandths of a second, to complete each analysis are listed as follows:

	Analysis		
Computer Model	**1**	**2**	**3**
A	5	123	34
B	12	141	41
C	3	88	29
D	8	104	31

*Optional

a. Identify the blocks and the treatments. What are the values of b and k?

b. Determine the grand total T using the formula

$$T = \Sigma\Sigma X_{ij}$$

c. Compute the total sum of squares using the formula

$$SST = \Sigma\Sigma X_{ij}^2 - \frac{T^2}{bk}$$

d. Compute the treatment sum of squares using the formula

$$SSTR = \frac{\Sigma T_j^2}{b} - \frac{T^2}{bk}$$

where T_j is the total of the values within each treatment.

e. Compute the block sum of squares using the formula

$$SSB = \frac{\Sigma B_i^2}{k} - \frac{T^2}{bk}$$

where B_i is the total of the values within each block.

f. Set up the ANOVA table.

g. If $\alpha = .10$, is there sufficient evidence to conclude that the average processing times are different for the various models of lap-top computers under study? Why?

14.29 Three employees of a county's public health service recently completed training to be restaurant inspectors. At the end of their instruction, each employee inspected the same randomly chosen set of five restaurants for kitchen cleanliness, storage procedures, food preparation methods, and so on. With $\alpha = .10$, determine whether the three inspectors provide the same average ratings, using the data in the following table. The numbers shown are total ratings points per restaurant on a standard rating form—higher scores denote more sanitary conditions. Each restaurant's ratings were obtained within a 20-minute period on the inspection day.

	Inspector		
Restaurant	1	2	3
A	69	65	70
B	81	80	82
C	98	97	93
D	96	92	94
E	86	86	86

14.30 An economic research group sampled food prices in three different Indiana cities to see if there were differences in price for the same items. The prices shown in the following table were all obtained on the same day from outlets of the same regional grocery retailer.

Product	Price by Store Location		
	Evansville	Ft. Wayne	Indianapolis
Skim milk (gallon)	\$1.85	\$1.97	\$1.79
Sliced cheese (pound)	2.09	2.14	2.07
Sugar (5 pounds)	1.89	1.95	1.89
White bread (pound)	.69	.65	.67
Bananas (pound)	.49	.54	.44
Instant coffee (8 ounces)	5.29	5.89	5.44
Ground beef, 70% lean	1.79	1.79	1.73
Mixed chicken parts (pound)	.87	.99	.99

a. Identify the treatments and the blocks for this study.

b. Would there be any interest as to whether the block means are equal?

c. Construct an ANOVA table, given $SST = 53.36385$, $SSB = 53.12925$, and $SSE = .1623$.

d. Treating the data as a random sample of grocery prices in the three cities, test the assumption that there is no difference in average grocery prices at the .10 level of significance.

e. Place bounds on the p-value for your test statistic.

14.31 A chain of convenience stores tested a display for a new snack product by placing the display in four different locations in various stores: at the entrance, in the snack section, by the cash register, and with the soft drinks. Each display location was utilized in three stores over a one-week test period. Since the 12 stores used to test the display differ somewhat in overall sales volume, they were divided into three categories that served as blocks. Within blocks, assignment of stores to display method was random. Units sold are shown in the following table.

Store Sales Volume	Unit Sales, by Location of Display			
	Entrance	Snacks	Register	Drinks
Below average	46	38	57	54
Average	62	50	67	67
Above average	75	62	89	77

a. If store sales volume were not taken into account and the 12 stores were randomly assigned to the display locations, what possibly could occur?

b. Explain how the use of blocks deals with the potential problem you identified in part a.

c. Given $SST = 2238$, set up the ANOVA table.

d. Is it possible to be 99% certain that the display locations' mean sales are not equal?

14.32 Refer to Exercise 14.31. Conduct the appropriate follow-up analysis (at $\alpha = .01$) to establish which means are significantly different.

14.5 TWO-FACTOR ANOVA

Consider the following scenario: An agricultural researcher wishes to compare soybean yields, using three different fertilizers. Suppose 18 test plots of land are available. The crop will be grown on each plot, and the yield will be recorded at the end of the season. The researcher will use one-factor ANOVA to test the hypothesis that all three fertilizer means are equal. Schematically, we can think of the experiment in this way:

H_0: $\mu_{F1} = \mu_{F2} = \mu_{F3}$
H_a: not $(\mu_{F1} = \mu_{F2} = \mu_{F3})$

Treatment		
Fertilizer 1	Fertilizer 2	Fertilizer 3
↑ 6 plots ↓	↑ 6 plots ↓	↑ 6 plots ↓
$\overline{X}_{F1}$	$\overline{X}_{F2}$	$\overline{X}_{F3}$

Now suppose the same researcher also wishes to compare soybean yields when using three different hybrids of the plant. Another one-factor ANOVA can be employed to investigate this second factor:

H_0: $\mu_{H1} = \mu_{H2} = \mu_{H3}$
H_a: not $(\mu_{H1} = \mu_{H2} = \mu_{H3})$

Treatment		
Hybrid 1	Hybrid 2	Hybrid 3
↑ 6 plots ↓	↑ 6 plots ↓	↑ 6 plots ↓
$\overline{X}_{H1}$	$\overline{X}_{H2}$	$\overline{X}_{H3}$

You may wonder whether our researcher can conduct a *single* experiment instead of two separate ones and still test both hypotheses. The answer to this is "yes"; it is possible to investigate two factors at the same time. Such a procedure is called a two-factor ANOVA and can be set up as follows:

	Fertilizer			
Hybrid	1	2	3	
1	2 plots	2 plots	2 plots	$\overline{X}_{H1}$
2	2 plots	2 plots	2 plots	$\overline{X}_{H2}$
3	2 plots	2 plots	2 plots	$\overline{X}_{H3}$
	$\overline{X}_{F1}$	$\overline{X}_{F2}$	$\overline{X}_{F3}$	$\overline{\overline{X}}$

The different factor combinations of two-way ANOVA are referred to as treatments—in the example above there are nine different fertilizer-hybrid combinations, or treatments.

To test for equal fertilizer means, the researcher can compare the three column means—in effect a one-factor ANOVA that ignores the presence of the row factor. To test for equal hybrid means, the researcher can then compare the three row means—a second one-factor ANOVA, this one without regard to the columns.

A two-factor study has these advantages relative to conducting two separate one-way studies:

1. It is less expensive and more easily done. For time and cost reasons, conducting two tests simultaneously is preferred to running the same two tests separately.
2. It can alert us to the existence of interaction. Interaction occurs when some or all treatment means are unexpectedly lower or higher than one-way ANOVA might predict. For instance, imagine that fertilizer 1 has the highest yield of the three fertilizers, and that hybrid 1 has the best yield of the three hybrids. We might then expect to find the highest treatment mean in the upper left location, where we have the combination of fertilizer 1 and hybrid 1. This outcome is not guaranteed, however. It is conceivable that, for instance, the best treatment mean occurs under the fertilizer 2–hybrid 3 combination. If so, we can say that the interaction of fertilizer 2 with hybrid 3 has a beneficial effect on yield. Interaction can be thought of as the joint effect of particular row–column combinations—it may occur in some locations but not in others. Since it is a phenomenon of two combined factors, interaction cannot be detected by one-factor ANOVA.

COMMENTS

1. We will limit our two-factor ANOVA study to considering examples where all treatments have the same sample size. The case of unequal sample sizes is beyond the scope of this book.
2. Two-way ANOVA may appear similar at first to the randomized block design of the previous section. One difference is that the blocks of randomized blocks are not considered factors in their own right. Another difference is that with the randomized block approach, there is only one observation for each row-column combination. However, interaction cannot be analyzed unless there are two or more observations per row-column combination. Since this is not the case for randomized blocks, it follows that interaction cannot be detected in the randomized block design. Some authors refer to the randomized block design as two-way ANOVA without interaction.

Interpreting Interaction

Perhaps the best way to illustrate the presence or absence of interaction is with some graphs. Let's first suppose that our two-factor ANOVA with fertilizers and hybrids results in the treatment means (yield, in bushels) shown in Table 14.11.

If we graph these means, placing yield on the vertical axis and placing the row factor (hybrids) on the horizontal axis, we obtain a picture such as Figure 14.9. The three line segments represent each column factor (fertilizer). It is easy to see that, regardless of hybrid, fertilizer 1 outperforms fertilizer 2 by an average of two bushels per plot. Fertilizer 2 in turn outperforms fertilizer 3 by an average of three bushels

Table 14.11 Hypothetical Cell Means for Two-Factor ANOVA

	Fertilizer		
Hybrid	1	2	3
1	16.0	14.0	11.0
2	17.0	15.0	12.0
3	15.0	13.0	10.0

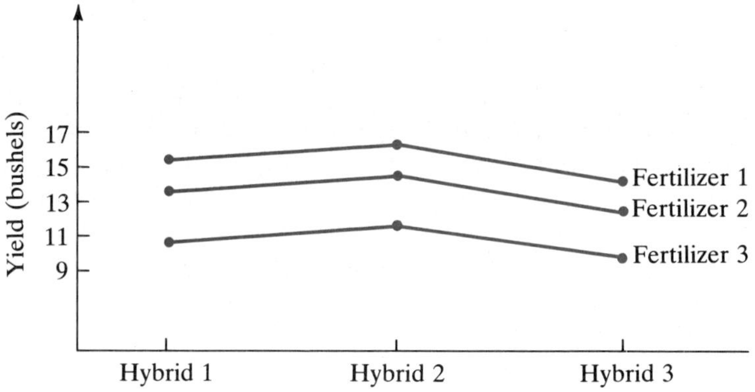

Figure 14.9 Plot of Hypothetical Treatment Means Exhibiting No Interaction (from Table 14.11)

under each hybrid condition. A graph with parallel line segments such as this one will occur when there is no interaction between the two factors. When there is no interaction, we say that we have *additivity*, or additive effects. In this example, relative to fertilizer 3, use of fertilizer 2 adds a constant three bushels to the mean for all hybrids. Likewise, relative to fertilizer 2, use of fertilizer 1 adds two bushels to the mean across all hybrid conditions.

Interaction is present when the difference between two levels of one factor is not constant across the levels of the other factor. Let's now suppose our table of treatment means has the results shown in Table 14.12 and graphed in Figure 14.10. Unlike the case of additivity with its parallel lines, this graph shows that differences between fertilizers depend on the hybrid level. For instance, fertilizer 1 varies from being best, middle, or worst depending on which hybrid it is combined with. Fertilizer 2 is always better than fertilizer 3, but the difference is 3, 2, or 1 bushel, depending on hybrid.

The possibility of having interaction between two factors makes two-factor ANOVA more complex than running two separate one-factor studies. There will be three null hypotheses to be investigated in a two-factor ANOVA. These are

Table 14.12 Hypothetical Cell Means to Illustrate Interaction

	Fertilizer		
Hybrid	1	2	3
1	11.0	13.0	10.0
2	17.0	15.0	13.0
3	13.0	16.0	15.0

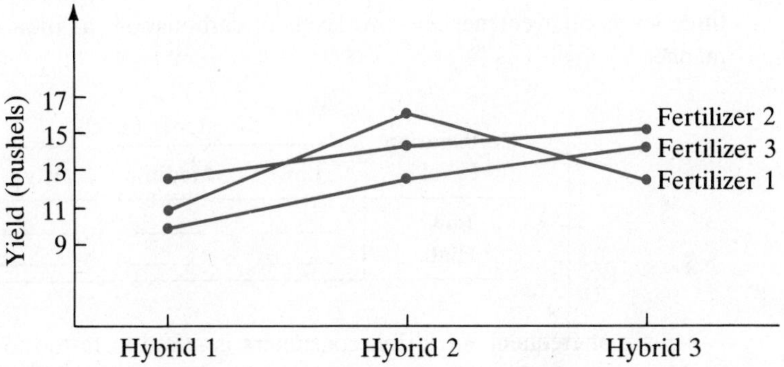

Figure 14.10 Plot of Hypothetical Treatment Means Exhibiting Interaction (from Table 14.12)

H_0: The column means are all equal.
H_0: The row means are all equal.
H_0: There is no interaction present.

As usual, the alternative hypothesis for each H_0 will be an opposite statement. The first two hypotheses listed above often are called the *tests for main effects*. Each main effect hypothesis is essentially an ordinary one-factor ANOVA. It does not matter which factor is labeled as the column factor or row factor. The interaction hypothesis often is called the *test for interaction effects*. When significant interaction is found, it is usually a good idea to graph the treatment means so that an understanding of the interaction pattern can be gained.

COMMENTS

1. We have said that having additive effects equates to parallel lines on a plot of treatment means. However, this does not mean the lines must be *perfectly* parallel to accept the hypothesis of no interaction. As an analogy, if we test the hypothesis that $\pi = .75$ in a binomial population, we do not require that the sample proportion, p, be exactly .75 in order to accept the hypothesis—values of p close to .75 are viewed as supporting the hypothesis. In the same spirit, a graph of sample treatment means with approximately parallel lines can support the hypothesis of no interaction.

2. Graphs of treatment means can be drawn in two different ways. Our graphs show levels of the row factor on the horizontal axis, with each line segment representing a level of the column factor. The graphs are just as informative if the horizontal axis has levels of the column factor with the segments becoming levels of the row factor. In either case, the magnitude of the treatment means is displayed on the vertical axis.

Developing the Two-Factor ANOVA Table

Let us consider the following situation as we develop the ANOVA table for a two-factor study. Suppose a soft drink manufacturer is developing a new fruit-flavored drink and is interested in experimenting with different levels of sweetener and carbonation in the drink before arriving at a final formulation. The study is set up to try three levels of sweetener and two levels of carbonation, as indicated in the following manner.

Carbonation Level	Sweetener Levels		
	Low	Medium	High
Low	___	___	___
High	___	___	___

A sample of frequent soft drink consumers is asked to taste the new drink and then rate it on a scale from 1 = very poor to 7 = very good. Twenty consumers receive each treatment combination; the total sample size is 120. Suppose the following set of treatment, row, and column means is determined:

Carbonation Level	Sweetener Levels			
	Low	Medium	High	
Low	4.0	4.9	4.3	4.4
High	3.8	4.1	4.7	4.2
	3.9	4.5	4.5	$4.3 = \overline{\overline{X}}$

To analyze this two-factor study, we proceed by partitioning *SST* into its component parts, as suggested by Figure 14.11. Notice that unlike the ANOVA settings previously studied, we do not use the label *SSTR*. This is because we no longer have a single factor—we now have two. The column factor is therefore represented by *SSC*, the row factor by *SSR*. Once these two sums are computed, we will organize them into an ANOVA table. The general format that we will follow is given in Table 14.13.

While we will place less emphasis on computation for two-factor sums of squares, it is important to know how the various quantities are determined.

- *SST* is computed exactly the same as for the other ANOVA problem settings: as $\Sigma(\text{each value} - \overline{\overline{X}})^2$. In the soft drink rating example, we are not showing

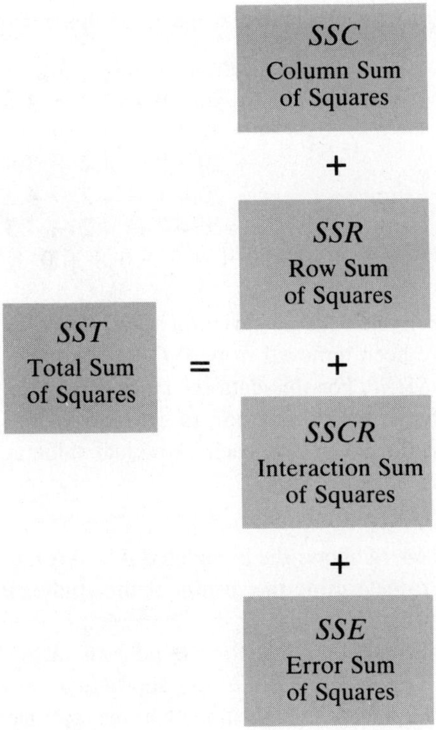

Figure 14.11 Partitioning Total Sum of Squares for a Two-Factor ANOVA

Table 14.13 General Format of the Two-Factor ANOVA Table

Source	*SS*	*df*	*MS*	*F*
Column Factor	*SSC*	$c - 1$	*MSC*	*MSC/MSE*
Row Factor	*SSR*	$r - 1$	*MSR*	*MSR/MSE*
Interaction	*SSCR*	$(c - 1)(r - 1)$	*MSCR*	*MSCR/MSE*
Error	*SSE*	$n - rc$	*MSE*	
Total	*SST*	$n - 1$		

each of the 120 individual ratings, so we cannot show this computation. From entering the data into a computer, we have determined that $SST = 113.76$.

- *SSC* is computed as $\Sigma n_j (\overline{X}_j - \overline{\overline{X}})^2$, which is identical to computing *SSTR* for one-factor ANOVA. Here $n_j = 40$ per column and $SSC = 40(3.9 - 4.3)^2 + 40(4.5 - 4.3)^2 + 40(4.5 - 4.3)^2 = 9.6$.
- *SSR* is computed as $\Sigma n_i(\overline{X}_i - \overline{\overline{X}})^2$, where n_i = number of row observations and $\overline{X}_i$ denotes row mean. $SSR = 60(4.4 - 4.3)^2 + 60(4.2 - 4.3)^2 = 1.2$.

- *SSCR* is computed as $\Sigma n_{ij}(\overline{X}_{ij} - \overline{X}_j - \overline{X}_i + \overline{\overline{X}})^2$, where n_{ij} is treatment sample size and $\overline{X}_{ij}$ is treatment mean. In this example,

$$\begin{aligned} SSCR &= 20(4.0 - 4.4 - 3.9 + 4.3)^2 \\ &\quad + 20(4.9 - 4.4 - 4.5 + 4.3)^2 \\ &\quad + 20(4.3 - 4.4 - 4.5 + 4.3)^2 \\ &\quad + 20(3.8 - 4.2 - 3.9 + 4.3)^2 \\ &\quad + 20(4.1 - 4.2 - 4.5 + 4.3)^2 \\ &\quad + 20(4.7 - 4.2 - 4.5 + 4.3)^2 \\ &= 0 + 1.8 + 1.8 + 0 + 1.8 + 1.8 = 7.2 \end{aligned}$$

- *SSE* is the residual variation after row, column, and interaction sums of squares have been removed from *SST*. We compute *SSE* as $SSE = SST - SSC - SSR - SSCR$. For this data set, $SSE = 113.76 - 9.6 - 1.2 - 7.2 = 95.76$. A defining formula for *SSE* is Σ(each value $-$ each value's treatment mean)^2; we would need to have each individual value at our disposal to compute *SSE* if we didn't already know *SST*.

Table 14.14 shows the completed ANOVA table for this data set—we will refer to this table to determine the results of the study. Assume that we wish a Type I error risk of $\alpha = .05$ to apply.

The main effects test for the column factor (sweetener) has a test statistic of $F^* = 5.71$. For 2 and 120 *df* (an approximation to 2 and 114 *df*), the table *F* value is 3.07. We therefore conclude that the average ratings are not the same for all three sweetener conditions. The table of treatment means indicates that the lowest ratings are associated with the low level of sweetener.

The main effects test for the row factor (carbonation) has $F^* = 1.43$. For 1 and 120 *df* (an approximation to 1 and 114 *df*) the cutoff value is $F = 3.92$. This one-way test therefore finds no significant difference in mean ratings between the two levels of carbonation.

For the interaction test, $F^* = 4.29$, which exceeds the .05 level table value of 3.07. Since the null hypothesis declares that there is no interaction, we conclude that significant carbonation-sweetener interaction is present. Figure 14.12 shows a graph of treatment means. The highest ratings occur under the low carbonation–

Table 14.14 ANOVA Table for Sweetener–Carbonation Example

Source	*SS*	*df*	*MS*	*F*
Columns (Sweetener)	9.6	2	4.8	5.71
Rows (Carbonation)	1.2	1	1.2	1.43
Interaction	7.2	2	3.6	4.29
Error	95.76	114	.84	
Total	113.76	119		

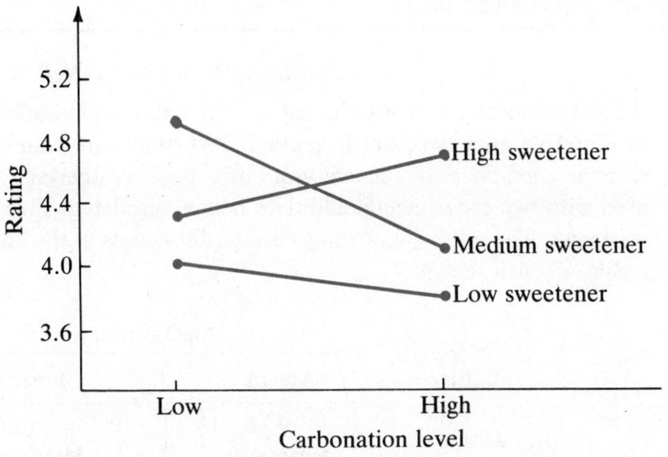

Figure 14.12 Plot of Treatment Means for Carbonation–Sweetener Example

medium sweetener combination and under the high carbonation–high sweetener combination.

COMMENT Calculating formulas for the two-factor study, if needed, are based on the following totals and subtotals:

T = Total of all the values in the sample
n = Number of values in the entire sample
C_j = Total of the values in each column
c = Number of columns
R_i = Total of the values in each row
r = Number of rows
CR_{ij} = Total of the values in each cell
h = Number of values in each cell

The sums of squares are

$$SST = \Sigma\Sigma\Sigma X_{ijm}^2 - \frac{T^2}{n}$$

$$SSC = \frac{\Sigma C_j^2}{rh} - \frac{T^2}{n}$$

$$SSR = \frac{\Sigma R_i^2}{ch} - \frac{T^2}{n}$$

$$SSCR = \frac{\Sigma\Sigma CR_{ij}^2}{h} - \frac{T^2}{n} - SSC - SSR$$

$$SSE = SST - (SSC + SSR + SSCR)$$

The subscripts i, j, and m in x_{ijm} refer to the mth value in the ith row and the jth column. Exercise 14.37 demonstrates these formulas.

CLASSROOM EXAMPLE 14.3

Conducting a Two-Factor ANOVA

A paint manufacturer is experimenting with additives in hopes of reducing the amount of blistering after the paint is applied. Test panels are painted and then subjected to extreme conditions to simulate years of natural weathering. A basic paint formula is used with two experimental additives being considered. Five test panels receive each treatment. The following readings are blister counts at the end of the tests; treatment means are also shown.

SO_2 Additive	MgO Additive: Absent	MgO Additive: Present
Absent	8 9 9 6 13 Mean = 9	10 14 10 10 11 Mean = 11
Present	6 6 5 8 10 Mean = 7	17 14 11 18 15 Mean = 15

a. Complete the ANOVA table.

Source	*SS*	*df*	*MS*	*F*
Column factor	125			
Row factor	5			
Interaction	45			
Error	84			
Total	259			

b. Report the conclusion for each hypothesis test, using $\alpha = .01$.

c. Graph the treatment means to show the interaction pattern, placing the SO_2 conditions on the horizontal axis. Scale and label your graph.

d. What recommendations do you make with regard to using the additives?

e. Determine the contribution to *SSCR* made by the upper left treatment by computing n_{ij} (treatment mean − row mean − column mean + overall mean)^2. Use the following fact to check your answer: When there are four treatments in a two-factor ANOVA, each treatment contributes equally to *SSCR*.

*■ Processor: Two-Way ANOVA

We will use the data given in the following table to illustrate this processor:

*Optional

	Column 1	Column 2	Column 3
Row 1	34 19 24 36 25	16 35 18 16 12	30 23 39 29 24
Row 2	27 24 19 33 37	41 30 17 27 22	28 34 40 27 42

We begin by creating a data set, choosing Two-Way ANOVA variables from the Data–Create submenu. Data entry is accomplished by entering a cell at a time into a labeled column. Figure 14.13 shows the screen as we begin to enter values for the bottom right cell (column 3, row 2). Note that we can see only three of the six cells at a time; pressing the left and right arrow keys as needed can bring unseen values into view.

After data entry has been accomplished, we proceed through the Two-Way ANOVA processor much as we did for One-Way and Randomized Block. When we reach Results, three different screens are available. One screen displays the various cell means; another provides row and column means along with the two-factor ANOVA hypothesis statements. The screen shown in Figure 14.14 summarizes the three hypothesis tests.

```
≡      Microcomputer Applications for INTRODUCTORY BUSINESS STATISTICS   EDIT
   Overview [Data] Files  Processors  Set Up  Exit                  | F1=Help

        Col 2, Row 2          Col 3, Row 1          Col 3, Row 2
   1 41                           30                    28
   2 30                           23                    34
   3 17                           39                    ▪
   4 27                           29
   5 22                           24
```

Figure 14.13 Entering Data for the Two-Way ANOVA Processor

```
≡                           Two Way Anova                                READY
   Introduction  *ata  Calculations [Results] Exit                  | F1=Help

    ANOVA table
```

Source	SS	df	MS	F	p-value
Column factor	336.8000	2	168.4000	2.870	0.0763079427
Row factor	154.1333	1	154.1333	2.627	0.1181577053
Interaction	73.8667	2	36.9333	0.629	0.5414970432
Error	1408.4000	24	58.6833		
Total	1973.2000	29			

Figure 14.14 Two-Way ANOVA Summary Table

SECTION 14.5 EXERCISES

14.33 Women who take oral contraceptives do not have a risk of heart disease substantially higher than those who do not take oral contraceptives. However, women who smoke do have a higher risk of heart disease than those who do not smoke. According to an article in the health section of *Newsweek* (November 25, 1985), "Heart disease may turn out to exact an even higher toll on women smokers, especially those who take oral contraceptives. The combination increases the risk of circulatory problems tenfold." What term found in this chapter is used to describe this "combination effect"?

14.34 Refer to the partially completed ANOVA table shown.

Source	*SS*	*df*	*MS*	*F*
Columns	10.8	2		
Rows	6.6	2		
Interaction	39.6	4		
Error	81.0	27		
Total	138.0	35		

a. How many levels of the row factor were employed?
b. How many observations were in each treatment?
c. At $\alpha = .05$, is significant interaction present or absent?
d. At $\alpha = .05$, which, if either, main effects test finds significant results?

14.35 Refer to the partially completed ANOVA table shown. The table gives results for a 3 × 2 (3 rows, 2 columns) study.

Source	*SS*	*df*	*MS*	*F*
Columns	40.8			
Rows	183.6			
Interaction	61.2			
Error	612.0			
Total		29		

At the .10 level of significance, test the following hypotheses:
a. All row means are equal.
b. All column means are equal.
c. The row and column factors do not interact.

14.36 The following table of treatment means has resulted from a two-factor ANOVA with four observations receiving each treatment.

30	25
34	35
38	45

Source	SS	df	MS	F
Columns				
Rows	784			
Interaction	144			
Error	135			
Total				

a. Compute the column means, row means, and grand mean.
b. Compute SSC and then fill in the ANOVA table.
c. Conduct the hypothesis tests at $\alpha = .01$.
d. If significant interaction is detected, make a graph of treatment means.

*14.37 An insurance company planned to conduct a study of agents' persistency rates for three types of insurance policies—accident and health, traditional life, and universal life—to determine whether the mean persistency rates differ. (The persistency rate is the percent of policies in force that remain in force, excluding maturing policies or policies for which the policyholder has died.) In addition, interest is focused on the difference in persistency rates between two major cities in which the company has offices. A total of 12 agents were randomly sampled in each city to yield four persistency rates per policy. The data are as follows:

	Type of Policy					
City	Accident/ Health		Traditional Life		Universal Life	
A	.90	.93	.91	.69	.87	.89
	.71	.86	.86	.91	.96	.78
B	.97	.96	.74	.91	.90	.91
	.82	.69	.74	.71	.89	.95

a. Identify the values of n, r, c, and h.
b. Compute T using the formula

$$T = \Sigma\Sigma\Sigma X_{ijm}$$

c. Compute SST using the formula

$$SST = \Sigma\Sigma\Sigma X_{ijm}^2 - \frac{T^2}{n}$$

d. Compute SSR, SSC, and $SSCR$ using the formulas

$$SSR = \frac{\Sigma R_i^2}{ch} - \frac{T^2}{n} \qquad SSC = \frac{\Sigma C_j^2}{rh} - \frac{T^2}{n} \qquad SSCR = \frac{\Sigma\Sigma CR_{ij}^2}{h} - \frac{T^2}{n} - SSC - SSR$$

e. Find SSE by subtraction.
f. Set up the ANOVA table for this two-factor study.
g. Are the main effects significant at a 10% risk level?
h. Is there a significant interaction effect at $\alpha = .10$?

*Optional

14.38 In "Perception, Preference, and Patriotism: An Exploratory Analysis of the 1980 Winter Olympics" (*The American Statistician,* August, 1981), Ian Fenwick and Sangit Chatterjee analyzed instances in the Olympic Games where judges of figure skating had the opportunity to rate contestants from their own country. Two-way ANOVA was performed for various events, with the summary table for the contestants' scores having the form shown here.

Source	*SS*	*df*	*MS*	*F*
Contestants				*a*
Judges				*b*
Interaction				*c*
Error				
Total				

a. Would you anticipate the main effect test statistic ("*a*" in the *F* column above) to be significant or nonsignificant? Why?
b. In the interest of common standards of judging, would a significant value for "*b*" in the table above be viewed as desirable or undesirable? Explain.
c. Would a significant value for "*c*" indicate "patriotic" judging or "neutral" judging of one's fellow countrymen?

14.39 Refer to the problem setting described in Exercise 14.38. The ANOVA table for pairs' figure skating in the 1980 Olympics is shown as follows:

Source	*SS*	*df*	*MS*	*F*
Contestants	40.22	10	4.02	205.90
Judges	.23	8	.03	1.45
Interaction	1.58	80	.02	1.01
Error	5.80	297	.02	
Total	47.83	395		

a. Is the significant F^* for contestants a surprising result? Why?
b. How many judges evaluated contestants from their own country?
c. To test ($\alpha = .10$) for equal means across the panel of judges, we need a table value for 8 and 297 *df*—it is not available. Can we still make the appropriate inference? If so, what is that inference?
d. Was significant judge-contestant interaction found for the pairs' figure skating event ($\alpha = .10$)?

14.40 A wine distributor wished to study the effect of wine bottle label on the perception of quality for restaurant patrons who often have wine with dinner. Two different wines were used: one generally known to be of average quality, the other being of high quality. Forty people participated in the test over a two-week period. Each subject was given a bottle of wine with dinner and asked to rate it afterwards on a 20-point scale (0 = extremely poor, 20 = outstanding). Half the people had bottles with the correct label; for the other half the distributor had switched the labels. The participants were assigned at random to the treatment combinations. Individual ratings are given here.

	Wine Quality									
Label	**Average**					**High**				
Correct	9	13	11	10	11	13	18	15	19	18
	14	12	11	9	10	16	14	14	15	18
Switched	14	9	15	15	14	16	12	13	14	17
	10	19	16	16	12	12	17	15	17	17

a. Compute the overall mean as well as the row means, column means, and treatment means.
b. Complete the ANOVA table given $SST = 324$.
c. Conduct the proper hypothesis tests, using $\alpha = .10$.
d. If significant interaction is found, make a graph of treatment means.
e. Interpret your graph in terms of the factors under study.

14.41 A bakery is experimenting with two types of cake mix and three baking temperatures. Specifically, cake texture (the measurement of which is not important here) is being investigated. From two batter mixes, 30 individual cakes are baked and rated. Results are given in the following table; larger numbers are desirable.

	Baking Temperatures								
Mix	**300°**			**325°**			**350°**		
S	18	17	17	17	16	17	13	12	14
	19	16		18	16		12	15	
T	14	12	14	16	15	17	17	16	16
	13	12		16	18		19	18	

a. Compute the overall mean as well as the row means, column means, and treatment means.
b. Set up and fill in the entries for an ANOVA table (Given: $SSE = 30.8$).
c. Conduct the two main effects tests, using $\alpha = .05$.
d. Judging by the main effects tests alone, which baking temperature appears best? Why?
e. At $\alpha = .05$, is significant mix-temperature interaction found? If so, make a graph of the treatment means.
f. What two mix-temperature combinations have the highest means? Could we have learned this from two separate one-way ANOVAs?

14.42 Refer to Exercise 14.41 above. Place bounds on the p-value for the following:
a. the test for equal mix means.
b. the test for equal temperature means.
c. the test for the existence of interaction.

14.43 Two similarly priced brands of watch battery were tested for length of life under conditions of extreme cold and heat. A total of 36 fresh batteries were sampled. The following numbers represent days of continuous operation before battery failure.

Treatment Combination	Life (Days)				
Brand *A*–Heat	17.0	18.5	17.5	16.5	16.5
	16.0	17.0	17.5	17.0	
Brand *A*–Cold	13.0	12.0	10.5	13.5	13.5
	12.0	11.5	11.0	12.5	
Brand *B*–Heat	16.5	18.5	19.0	18.0	19.5
	17.5	16.5	18.5	17.5	
Brand *B*–Cold	13.5	12.5	13.0	14.0	11.5
	12.0	12.5	12.5	12.5	

a. Set up and determine all the entries for an ANOVA table (given $SST = 263.7$).
b. Do the two factors interact, using $\alpha = .10$? If you find significant interaction, prepare a graph of treatment means.
c. Is there a main effect due to the temperature condition ($\alpha = .10$)? Due to brand?
d. Given your answer to part b, would we describe any main effects as being additive or nonadditive?

14.44 A new food product is being test marketed in carefully selected cities and stores. Part of the test concerns finding the best in-store placement for the product. Since it is a bread-related snack, it is placed in the bread section in half the test stores; it is placed in the snack-cookie section in the other half. Another variable under study is the advertising campaign to be used to introduce and support the product. One-third of the test cities had a television promotion featuring kids as users of the product; one-third had Campaign "A" aimed at a target audience of people in their twenties; one-third had Campaign "B," also aimed at consumers in their twenties. Equivalent media expenditures were made in all test cities. The test market lasted six weeks. Unit sales for each test store were recorded, and the following ANOVA table has resulted.

Source	*SS*	*df*	*MS*	*F*
Columns	2,842	2		
Rows	622	1		
Interaction	917	2		
Error	12,615	30		
Total	16,996	35		

a. How many test stores were in each cell?
b. Assume that the overall level of sales in the test market was encouraging enough to warrant a national introduction of the product. What information has been learned in the test market? Relate your findings to the .10 level of significance.

*14.6 THE F-TEST FOR EQUAL VARIANCES

In certain situations, there may be an interest in determining whether two normal populations have the same dispersion, as measured by the standard deviation or the variance. For instance, in Section 1 of the previous chapter, we learned how to perform a t-test to investigate the hypothesis that $\mu_1 = \mu_2$ for the case of small, independent samples. Such a test requires the assumption that the two populations have the same variance. While we did not verify the assumption of equal variances at that time, we can do so now.

We now know that the ratio of two sample variances follows an F-distribution. Therefore we can use our F-table to test this hypothesis:

$$H_0\colon \sigma_1^2 = \sigma_2^2 \quad \text{or equivalently} \quad H_0\colon \frac{\sigma_1^2}{\sigma_2^2} = 1.00$$

versus

$$H_a\colon \sigma_1^2 \neq \sigma_2^2$$

The test statistic F^* therefore will be S_1^2/S_2^2, with df being $n_1 - 1$ and $n_2 - 1$.

Due to the way our F-table is set up (only upper-tail cutoff values are given), we always will need to form our F-ratio as

$$F^* = \frac{\text{Larger sample variance}}{\text{Smaller sample variance}} = \frac{S_1^2}{S_2^2}$$

With no lower-tail F-values available, we must make it possible for H_0 to be rejected by defining the larger sample variance to be the one from population 1. This convention also affects the stated value for α in the F-table. The values for α all must be doubled when we wish to compare variances drawn at random from two independent populations. In other words, our stated table risk factors ($\alpha = .10, .05, .01$) will become $\alpha = .20, .10$, and $.02$ when we test H_0: two populations have the same variance versus H_a: the two population variances are different.

EXAMPLE 14.3

In Exercise 13.16 from the previous chapter, we wished to test $H_0\colon \mu_H = \mu_J$. Our sample results showed the following:

Sample	Mean	Standard Deviation	Size
H	69.0	5.25	6
J	76.5	6.41	6

*Optional

Prior to testing for equal means, we may wish to verify that it is reasonable to believe that the samples came from populations with equal variances. We will assume $\alpha = .10$ for our inference.

Solution:

Our test statistic will be $F^* = S_J^2/S_H^2 = 6.41^2/5.25^2 = 1.49$. (Note that we need to convert our standard deviations into variances.) We will compare $F^* = 1.49$ against the (stated) .05 level F-table value for 5 and 5 *df:* The result is $F = 5.05$. Since F^* does not exceed the table value, we accept the assumption of equal population variances (standard deviations). Summing up symbolically, we have

$$H_0\colon \sigma_H^2 = \sigma_J^2$$
$$H_a\colon \sigma_H^2 \neq \sigma_J^2$$
$$TS\colon F^* = S_J^2/S_H^2 = 1.49$$
$$RR\colon \text{Reject } H_0 \text{ if } F^* > 5.05$$
$$C\colon \text{Accept } H$$

□

COMMENTS

1. We double the stated level of α for this F-test, since defining the larger sample variance as the numerator in F^* doubles the chance that F^* will be found in the rejection region.
2. If the alternative hypothesis is to be upper tail when we test for equal variances, the stated table values for α are not doubled.

SECTION 14.6 EXERCISES

14.45 Given $n_1 = 9$, $n_2 = 9$, $S_1^2 = 2209$, $S_2^2 = 1444$, test the hypothesis of equal population variances, using $\alpha = .20$. Assume the samples are from normal populations.

14.46 If two sample variances happen to be equal, can we immediately accept the null hypothesis of equal population variances regardless of α?

14.47 In Exercise 13.17, we wished to test for equality of means based on two rather small samples ($n_1 = n_2 = 5$). We observed sample standard deviations of 3.71 and 3.27. At $\alpha = .10$, does the sample evidence support or contradict the assumption of equal population variances?

14.48 Suppose a random sample of 120 automobile speeds on Interstate 90 in 1986 showed $S = 4.1$ miles per hour. Another sample was collected along the same portion of the highway in 1989, after the speed limit was raised from 55 to 65 mph. This sample, also of size 120, showed $S = 5.9$ mph. At $\alpha = .02$, can you conclude that the samples came from populations with different variances?

14.49 A random sample of large office supply outlets in different cities compared current retail prices of facsimile machines. The standard deviation of price for Brand M ($n = 7$) was \$44.51, for Brand W ($n = 8$), $S = \$27.78$. At $\alpha = .10$, is it reasonable to believe retail prices of the two brands are equally variable?

14.50 On an automobile test track, a new model car was checked for stopping distance from a speed of 50 mph. Eight measurements were taken using one brand of tire (S = 14.0 feet); another eight measurements were recorded using another brand (S = 20.5 feet). Prior to testing for equal means, we wish to verify whether the assumption of equal variances holds. Execute this test with 10% risk of a Type I error.

14.7 SUMMARY

The purpose of this chapter has been to extend our knowledge of hypothesis testing to applications where we simultaneously compare three or more populations' means. To do so, we became familiar with the F-distribution, a very useful tool that will have further applications in the next few chapters as we consider regression analysis.

We opened the chapter by learning the basic building block—one-factor ANOVA. We developed the idea of partitioning, seeing how to assign total variation into component parts that are due to treatments and to random variation. We then learned how we could form a ratio of two variances that would have the potential of indicating whether the populations have the same mean.

We also developed more advanced procedures. Included here are methods such as the randomized block design and the two-factor study, with its ability to detect interacting variables. As you might expect, experimental studies can involve three or more factors and can be considerably more complex than the examples seen in this chapter.

14.8 TO BE CONTINUED . . .

. . . *In Your College Courses*

You may encounter results of ANOVA studies in a psychology course or in a course in personnel/human resources management. As you might expect, human beings and their behavior under different conditions are the focus of many of these studies.

. . . *In Business/The Media*

Test Marketing If you take a course in Marketing or Marketing Research, you will likely study some examples of test marketing, a common business practice for new products and services. A *test market* is a chance to try out marketing the product on a small scale—typically in more than one city but usually involving less than 3% coverage of the whole U.S. market. In contrast to human behavior studies which are often laboratory experiments, a test market is a "field" experiment that takes place in the real world. The units of analysis in a test market are usually stores that have received different treatments. There are two general purposes for a test market:

1. To estimate the level of sales that would result if a full national introduction, or rollout, followed. Many products are judged to be failures at the test market stage. If the level of sales generated in the test market does not project to a profitable future, the product in question is unlikely to get national introduction. A test market therefore can reduce the risk of an unsuccessful national rollout. The estimation techniques studied in Chapters 10 through 12 can help estimate the unit sales or brand share that would be realized in a national rollout.
2. To try out different marketing strategies for use in national introduction. There are many possible variables involved in marketing a new product. For instance, depending on the nature of the product, market researchers may wish to know the following:

   - Which of three package designs is more likely to catch the shopper's eye?
   - Which of three alternative advertising campaigns has the greater appeal?
   - Should the retail price be \$2.59, \$2.89, or \$3.09?
   - What level of advertising dollars should be spent on the product?
   - Should the media budget be spent mostly on print or television or some combination?
   - Will free samples or couponing generate more first-time buyers?

Most test markets consider levels of at least two different variables (two-factor ANOVA); some test markets are more complex. We should note that there is usually little interest in main effects due to price or advertising expenditure. This is because sales will usually if not always be higher in test stores where the price is lower; likewise, sales will generally be higher in test cities that receive the largest expenditures for advertising. However, price and advertising are known often to have strong interactive effects that may point to the optimum combination of these strategy variables.

Test Cities and Stores There was a movie in the 1940s called "Magic Town." In it, the hero located a city that exactly represented the attitudes, beliefs, demographics, and so on of the entire U.S. population. No such city exists, of course, though test marketers would certainly like one to exist. There are a fair number of middle and larger size cities, however, that many marketing researchers and advertising agencies believe are more or less typical and that often are included on lists of cities for possible test market experimentation. A few of these cities are Buffalo, Des Moines, Columbus, Kansas City, Omaha, and Syracuse.

Within test cities it is important to have test stores that are similar. The most common criterion for store comparability is store sales volume. In order to compare stores that receive different treatments, we do not want the test to be biased by having stores with unequal sales volume involved.

Test markets can help choose the optimum marketing plan for a national rollout, and they can alert the marketer to a product that is a candidate for failure. Individuals who work in marketing research are likely to be familiar with interpreting ANOVA results.

The downside of test marketing is that it is expensive to set up and monitor; it takes time and may give the competition a chance to react; and, like any field experiment, there is the possibility of unforeseen disruptions beyond the control of the experimenter. Despite these challenges, test marketing continues to be a widely used tool in new product and service introduction.

CHAPTER 14 EXERCISES

14.51 State whether F^* is of sufficient magnitude to convince you (at $\alpha = .05$) that H_0 is false:
a. $F^* = 4.81$, with $df = 2$ and 21.
b. $F^* = 3.01$, with $df = 5$ and 60.
c. $F^* = 1.88$, with $df = 3$ and 16.

14.52 Conduct an analysis of variance to establish whether there are differences among the treatment means, at the .05 level of significance.

Treatments		
1	2	3
39	36	40
45	39	32
39	41	36
46	34	35
41	35	32

14.53 Set up a one-factor ANOVA table, given the following: $SSTR = 88.12$, $SSE = 388.65$, $n_1 = 8$, $n_2 = 9$, $n_3 = 11$. Use your test statistic to decide whether your null hypothesis should be accepted at the $\alpha = .05$ risk level.

14.54 A chemical company has developed a new product. Four different catalysts for the chemical reaction have been used to date in the research laboratory. Prior to scaling up the process to an existing plant, further work is to be done in a pilot plant to see if yields vary according to the catalyst. Five batches of the product are made with each catalyst as shown in the following table. The numbers are percent yield per batch.

Catalyst Used			
1	2	3	4
81	80	88	78
86	87	80	84
75	79	89	73
79	85	89	85
84	84	93	81

a. What are the treatments?
b. What is the factor?
c. What is the value of X_{34}?
d. Set up the summary table.
e. At the .10 level, is it reasonable to believe that the observed differences in catalyst means are simply random variation?

14.55 In "Marketing Strategy Implications of the Miles and Snow Strategic Typology" (by Stephen McDaniel and James Kolari in the October, 1987, issue of *Journal of Marketing*), the authors classify 279 banks into three organizational or strategic types: defenders, prospectors, and analyzers. These categories represent the banks' response to changing environmental conditions as reported by the banks' marketing officer or marketing director. Each bank in the survey responded to several questions. One was, "Over the next five years, how important do you think computerized customer information services will be in carrying out the bank's overall strategy?" Respondents answered on a scale where 1 = not important and 5 = extremely important. The following was reported:

Type	n	$\overline{X}$	F^*
Defender	57	3.32	8.16
Prospector	67	4.18	
Analyzer	155	3.94	

a. Specify what the null hypothesis would be.
b. What would be the *df* values for the test statistic?
c. What table value can you use for a test at the .01 level of significance?
d. Can H_0 be rejected?

14.56 Refer to Exercise 14.55 above, where some information was provided. Although the ANOVA summary table was not given, it can be constructed. Develop the table to two decimal place accuracy.

14.57 A one-way ANOVA based on four treatments, each of size eight, has resulted in $F^* = 4.11$.
a. Verify that H_0 would be rejected at the .10 level.
b. What table value of Q would be used to compute the T-range for a follow-up analysis, maintaining $\alpha = .10$?
c. If the summary table shows $MSE = 18.08$, would we judge a difference of 4.0 between two treatment means to be statistically significant?

14.58 Refer to Exercise 14.52.
a. Determine the T-range for an $\alpha = .05$ follow-up analysis of all sample means.
b. Which treatment means are significantly different?

14.59 Refer to Exercise 14.54. Which pair(s) of catalysts can be viewed as significantly different at the .10 level?

14.60 Many college-bound high school students who take the ACT exam will, for a variety of reasons, repeat the test in an effort to bring up their scores. The company that makes the test generally discourages this practice, saying that average scores tend to remain the same over repeated testings. Following is a random sample of scores of individuals who took the exam three times.

	Test Administration		
Individual	1	2	3
1	26	25	27
2	14	14	15
3	28	27	28
4	17	18	19
5	27	29	29
6	30	31	30
7	26	28	29
8	24	24	23

a. What are the blocks?
b. Compute *SSTR*.
c. Given $SST = 666$ and $SSB = 652$, develop the ANOVA table.
d. Does the sample evidence support or contradict the statement that average scores do not change in subsequent testings? Use $\alpha = .01$.

14.61 A store owner is considering three downtown sites as candidates for a second store location. To help in the decision, pedestrian traffic counts passing each location are made for five intervals during a given day—see the following data.

	Location		
Time Period	1	2	3
10–10:15	29	18	22
12–12:15	55	39	53
3– 3:15	23	22	15
5– 5:15	63	41	52
7– 7:15	20	20	23

a. Given $SST = 3590$, construct the ANOVA table.
b. Can a hypothesis of equal location means be rejected at the .10 level of significance?

14.62 Refer to Exercise 14.61.
a. Determine which pairs of means are unequal (at $\alpha = .10$).
b. What would be the value of F^* if the time periods' role as a blocking factor was ignored, resulting in ordinary one-way ANOVA?
c. Would a different conclusion result if the blocks were ignored, as suggested in part b?

14.63 Refer to the following partially complete ANOVA table.

Source	*SS*	*df*	*MS*	*F*
Columns	272.6			
Rows	301.0	2		
Interaction	252.8	4		
Error				
Total	2248.4	44		

a. What was the treatment sample size?
b. Conduct the main effect hypothesis tests, using $\alpha = .05$.
c. At $\alpha = .05$, are the main effects additive or nonadditive?

14.64 The following table of cell means has been computed after obtaining measures for a two-factor ANOVA with six observations per cell.

	A	*B*
C	42	40
D	44	41

a. Make a graph of the treatment means. Do the line segments intersect? Are the line segments parallel?
b. Complete the following ANOVA table.

Source	*SS*	*df*	*MS*	*F*
Columns	37.5			
Rows	13.5			
Interaction	1.5			
Error				
Total	123.5			

c. At $\alpha = .10$, is either one-way hypothesis rejected?
d. At $\alpha = .10$, do the row and column factors have additive or nonadditive effects?

***14.65** Refer to Exercise 14.64.
a. Compute the portion of the interaction sum of squares that is due to the combination of factors *B* and *C* (the upper right cell).
b. Suppose the six values for *X* in the upper right cell are 38, 38, 39, 40, 41, and 44. Compute the contribution that this cell makes to the error sum of squares.

14.66 Some over-the-counter medicines come in packages that have a drug interaction precaution statement printed on the package. The word "interaction" on these packages has a similar meaning to that found in this chapter. In terms of taking medicine, about what potential problem does such a warning alert the would-be user?

14.67 What's in a name? The July 1978 issue of *Decision Sciences* journal reported on a study ("The Effect of Masculine and Feminine Names on the Perceived Taste of a Cigarette," by Hershey Friedman and William Dipple) whose purpose was to determine whether a product's name can affect the user's evaluation of the product. The sample consisted of 200 smokers: 100 males and 100 females. The subjects were told that they were testing a new cigarette brand. Half of each sex group were informed that the new brand was named "Frontiersman"; the other half were told that the cigarette was named "April." A total of 50 subjects were in each sex–brand name treatment, or cell. After trying the cigarette, subjects completed a questionnaire to rate the cigarette on various attributes, using an eight-point scale. One item

*Optional

on the questionnaire asked the smoker to rate her or his cigarette on a scale from 1 = bland flavor to 8 = rich flavor. The cell means were as follows:

	Sex	
Brand	**Male**	**Female**
Frontiersman	4.42	3.30
April	3.84	4.26

The ANOVA results were as follows:

Source	*SS*	*df*	*MS*	*F*
Sex	6.13	1	6.13	1.77
Brand	1.80	1	1.80	.52
Interaction	29.64	1	29.64	8.54
Error	680.12	196	3.47	
Total	717.69	199		

a. At α = .05, is there a significant difference in the males' and females' mean ratings of the cigarette for the flavor question?
b. At α = .05, test the hypothesis that the mean rating under the two names is the same.
c. At α = .05, is there significant sex–brand interaction? If so, describe its apparent pattern.

14.68 Refer to the cigarette rating problem of Exercise 14.67. Another rating variable that the subjects responded to was hot/cool, using a scale of 1 = hot tasting to 8 = cool tasting. The cell means were

"Frontiersman" males = 5.00
"Frontiersman" females = 3.64
"April" males = 4.20
"April" females = 4.28

a. Given SST = 805.24, SSE = 758.52, $SSRC$ = 25.92, set up the ANOVA table.
b. Conduct the appropriate hypothesis tests at α = .05.
c. If a significant interaction is found, prepare a graphical illustration of the pattern.

14.69 Refer to Exercise 14.68 above. Find boundaries for the p-value for the following:
a. the sex hypothesis test.
b. the brand name hypothesis test.
c. the interaction hypothesis test. (*Hint:* The F-value for 1 and 196 df is not in the F-table but is approximately 6.78.)

14.70 A glass manufacturing firm is experimenting in producing a glass formulation for use in commercial buildings. Interest centers on having a product that lets as much light as possible pass through but that at the same time is resistant to heat flow. One factor in the study is formulation; the other is heat source. The panels used in this study are alike with respect to their ability to let light pass through. The readings shown in the following table are taken on the inside surface of the test panels of glass—they represent amount of heat transmitted through the glass.

Heat Source	Formulation		
	A	B	C
Direct	52 50 55 57 54	46 54 50 50 47	48 49 52 48 49
Indirect	45 45 50 51 46	53 51 56 56 53	57 49 47 46 46

a. Compute the grand mean, the row and column means, and the cell means.
b. Set up and fill in the entries for an ANOVA table.
c. Conduct the two one-way hypothesis tests, using $\alpha = .10$.
d. At the .10 level, is interaction between formulation and heat source present or absent? If present, make a graph of treatment means and describe the pattern of interaction.
e. For any of the hypotheses that were rejected at the .10 level, state whether your conclusion would still be the same at $\alpha = .01$.

14.71 In a test marketing experiment conducted in shopping malls in large cities, a total of 480 women (screened beforehand to meet certain demographic objectives) were shown a 12-minute videotape presentation that showed how the special effects in a popular movie had been made. Interspersed in the presentation were five commercials—one was the test commercial while the other four served as clutter. After the presentation, the viewers were interviewed and each given a score (0 to 100) according to their ability to describe or "play back" the test commercial, which was for a new personal care product. As the following table suggests, there were two factors.

Commercial Appeal	Product Name	
	C	*D*
A	______	______
B	______	______

One was the commercial: The company's advertising agency had prepared two commercials with different appeals. The second factor was the name given to the new product: Two different proposed names were utilized. Therefore there were four different test commercials prepared. After entering the recall scores into our computer, we obtained the following table:

Source	*SS*	*df*	*MS*	*F*
Columns	920			
Rows	288			
Interaction	652			
Error	106,310			
Total				

Using the .10 level of significance, determine the following.

a. Do average recall scores vary according to the appeal made in the commercial?

b. Is average message recall the same or different for the two product names employed?

c. Is there appeal-name interaction?

14.72 In "Response to Commercials as a Function of Program Content" (*Journal of Advertising Research*, April 1981), Gary Soldow and Victor Principe showed television commercials to three groups of subjects to see if the subjects' recall of the commercials was affected by their involvement in the program they were watching. The subjects were told that they would view a videotape and then answer questions about it. One group saw an "involving" action-adventure show ("Baretta") while a second group saw a "less-involving" comedy ("The Brady Bunch"). The third group served as controls, viewing the same commercials as the other groups but no program. When brand recall was measured later, these means resulted (higher scores indicate better recall):

Recall Score	Group
1.21	More involved group
2.24	Less involved group
2.28	Control group

a. Reconstruct the ANOVA summary table, given that there were 29 subjects in each group and that $SSE = 85.86$.

b. The F-table does not have values for the df combination in this problem. Determine a usable .01 level value by interpolation.

c. Compare the test statistic to the value you determined in part b. Is there convincing evidence that the samples came from populations with different means?

14.73 After performing a randomized block test, a student noticed that MSB was a smaller number than MSE. Does this suggest that the blocking factor was useful or not useful in screening out dispersion that is unrelated to testing for equal treatment means?

REFERENCE

Neter, J., W. Wasserman, and M. H. Kutner. 1985. *Applied Linear Statistical Models*, 2nd Edition. Irwin, Homewood, IL.

Chapter Maxim *Do not use X to predict Y unless you have established (via hypothesis testing) a relationship between them.*

CHAPTER 15

SIMPLE REGRESSION AND CORRELATION

*Optional

Objectives

After studying this chapter and working the exercises, you should be able to

1. State the goal and objectives of a simple regression analysis.
2. Differentiate between a deterministic and a statistical model.
3. Specify the simple regression model and its assumptions.
4. Explain the least-squares criterion.
5. For any bivariate data set, find the equation of the least-squares line.
6. Graph a set of bivariate data and the least-squares line.
7. Explain the implications when we accept the hypothesis that $\beta_1 = 0$.
8. Identify and execute three ways of testing the utility of a fitted regression model.
9. Construct and interpret confidence intervals for an average value of *Y*, given a value of *X*.
10. Construct and interpret prediction intervals for a specific value of *Y*, given a value of *X*.
11. Contrast the simple correlation model with the simple regression model.

*12. Execute the Regression processor to verify answers and/or solve problems.

13. Recognize applications of regression in accounting and finance.

Regression analysis is probably the most popular statistical technique in the business world. Accountants use it to develop cost functions. Financial analysts employ regression to relate the movement of a stock's price with the movement of the market as a whole. Regression analysis is popular in marketing where store sales are predicted based on factors such as residential population within a certain radius of the store or traffic volume along nearby roads. Economists build regression models to predict tax collections, unemployment, and other variables.

Simple linear regression is the study of the straight-line relationship between *two* variables *X* and *Y*. Hereafter our references to "simple regression" automatically will imply only two variables and a linear relation. In Chapter 4 we introduced bivariate data and mentioned several descriptive aspects of regression—scatter plots, the least-squares criterion, and the regression line. We will expand on that discussion here to include the inferential procedures associated with the line $\hat{Y} = b_0 + b_1X$. We also will explain how to use the equation for prediction.

*Applies to optional section.

When we only wish to know whether two quantitative variables are related rather than predicting one variable from the other, we focus on their correlation. Simple correlation analysis is the study of the nature and degree of the relationship. Nature refers to the direct or inverse behavior between X and Y, while degree measures the strength. For example, marketers target their advertising by studying the correlation between consumers' disposable income and their income levels. Economists correlate variables like the number of building permits or the number of jobless benefits claims with the inflation rate to develop economic indicators.

Regression and correlation are used in planning, budgeting, forecasting, and research. Since they are practical, powerful, and effective analytical techniques, we will study them in much detail in the next three chapters. Simple linear regression and correlation are the subjects of this chapter; multiple regression and correlation are covered in the next. Our last look at regression in Chapter 17 examines a finer point of the analyses: building a model.

15.1 INTRODUCTION

Goal of Simple Regression

In Chapter 4, we learned that bivariate data are paired observations (X,Y) where X is the independent variable and Y is the dependent variable. We wish to study the joint behavior of the variables X and Y for which the relationship is believed to be a straight line.

Determining an equation for the linear relation between X and Y is naturally a part of the regression analysis. But the analysis does not stop there. We also seek answers to questions like, "Can we be reasonably certain that X is useful in explaining Y?" or, "If we use X to predict Y, how accurate can we expect to be?" The following definition summarizes the overall goal of regression.

Definition

The **goal of simple regression** is to analyze the straight-line relationship between X and Y.

To meet this goal we create several achievable objectives that are detailed in the next section. First however, we examine the structural foundation of regression analysis: a model.

Models

Before any analysis takes place, we must understand that regression is based on a model and certain assumptions. In general, a model is a representation of the relationship between variables. There are two ways for variables to be linearly related—exactly or approximately. For instance, the relationship between F = temperature in degrees Fahrenheit and C = temperature in degrees Celsius is

$$F = \frac{9}{5}C + 32$$

This is an exact or *functional relationship*. The equation relating F and C qualifies as a *deterministic model*.

Definition

A **deterministic model** is a functional relationship between variables.

On the other hand, the *statistical relationship*

$$\hat{F} = 2C + 32$$

is only an approximate relationship between F and C. Recall that the "hat" symbol over the F denotes an estimated or predicted value. Slight errors are introduced when converting from C to F using the statistical relationship: 5°C corresponds to a predicted temperature of 42°F. The exact Fahrenheit temperature is of course 41°, so we have incurred an error of 1°F with the $\hat{F}$ expression. Notice that the error changes depending on the value of C: For 10°C, the statistical relationship is off by 2°F.

Statistical relationships can be viewed as exact relationships provided we include a term to account for the error. The equation

$$\begin{aligned} F &= \hat{F} + \text{error} \\ &= 2C + 32 + \text{error} \end{aligned}$$

is called a *statistical model*. In this case, the error term is a variable representing $F - \hat{F}$, the difference between the actual Fahrenheit temperature and the predicted Fahrenheit temperature.

Definition

A **statistical model** is a statistical relationship between variables plus a random error term.

Before leaving the previous discussion, we point out that the deterministic model $F = \frac{9}{5}C + 32$ is preferred over the statistical model. Certain applications in the natural sciences employ exact or deterministic models. Unfortunately, most business problems do not involve functional relationships between variables. Statistical models therefore are useful alternatives for these inexact relations.

15.2 SIMPLE REGRESSION MODEL

Statement of the Model

The basis of simple regression is summarized in the statistical model given in the following box.

Simple Regression Model

$$Y = \beta_0 + \beta_1 X + \epsilon \tag{15–1}$$

Symbol	Name	Classification
Y	Dependent variable	Random variable
X	Independent variable	Fixed variable
β_0	Y-intercept	Parameter
β_1	Slope	Parameter
ϵ	Random error term	Random variable

Beta, the second letter of the Greek alphabet, is used to symbolize the parameters of the model. Both X and Y are quantitative variables. However, Y is considered a random variable, while X is said to be a **fixed variable**—that is, X can assume different values (the notion of a variable), but these values are fixed or predetermined.

In Equation 15–1, we view each value of Y as having three components. The first component is the Y-intercept denoted by β_0. This term represents a constant effect independent of the value of X. The second component is the product of the slope, denoted by β_1, and X. The term $\beta_1 X$ represents the effect of the value of X on the value of Y. We regard β_0 and β_1 as parameters whose values are constant, but usually unknown to us. This is similar to the estimation problem we faced in Chapter 10 where the value of μ was constant and unknown. The third component is the error term denoted by the Greek letter epsilon, ϵ. The error term, a random variable, is included to account for discrepancies between Y and $(\beta_0 + \beta_1 X)$.

Equation 15–1 tells us how to describe or model every piece of bivariate data (X,Y). For example, to see how the point (3,3) "fits" into the linear regression model $Y = 8 - 2X + \epsilon$, refer to Figure 15.1. In panel (a), the point (3,3) is plotted in the (X,Y) plane. For the model $Y = 8 - 2X + \epsilon$, the Y-intercept is $\beta_0 = 8$ and the slope (rise over run) is $\beta_1 = -2$. The constant component of Y, $\beta_0 = 8$, is illustrated in Figure 15.1(b). Figure 15.1(c) shows the proportional effect of X on Y: A "run" of 3 in X produces a "rise" in Y of -6. Finally, in Figure 15.1(d) we see the error of 1, the difference between $Y = 3$ and $\beta_0 + \beta_1 X = 2$.

If we had a sample of four data points as indicated in Figure 15.2, then each point (X,Y) could be modeled in a similar fashion by $Y = 8 - 2X + \epsilon$. In each case,

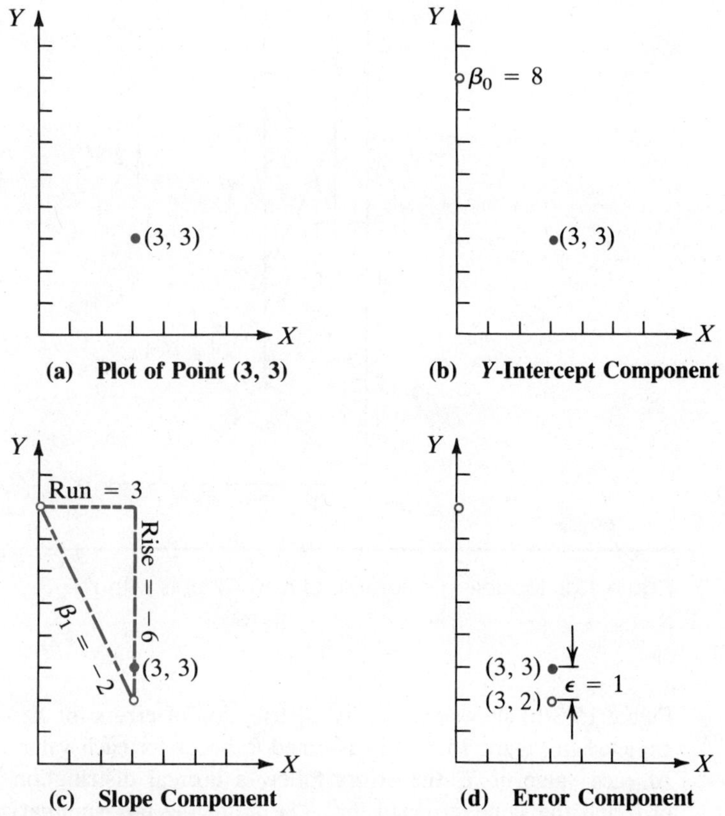

Figure 15.1 Modeling the Point (3,3) with $Y = 8 - 2X + \epsilon$

ϵ denotes the vertical distance from the individual point to the regression line. The contributions of the first two components of the model, $\beta_0 + \beta_1 X$, are the open circles in Figure 15.2; the line connecting them is the regression line. The actual values of Y are the closed circles, with the corresponding errors indicated by the vertical lines. When the actual values of Y are above the line, ϵ is a positive value; and when the Y-values are below the line, ϵ is negative.

Although we can see the *sample* of closed circles, we must imagine that the *population* consists of many points which are unseen in Figure 15.2. For instance, picture all the different values of Y associated with $X = 2$. In Figure 15.3(a) we have represented this collection with many dots along the vertical line through $X = 2$. Each dot (point) has an error associated with it; some errors are positive, some negative, and some zero—whenever the dot corresponds to the point (2,4). The collection of these values of ϵ forms a distribution.

The usual assumption in the simple regression model is that at each X-value ϵ has a normal probability distribution with an average value of 0 and a variance of σ^2.

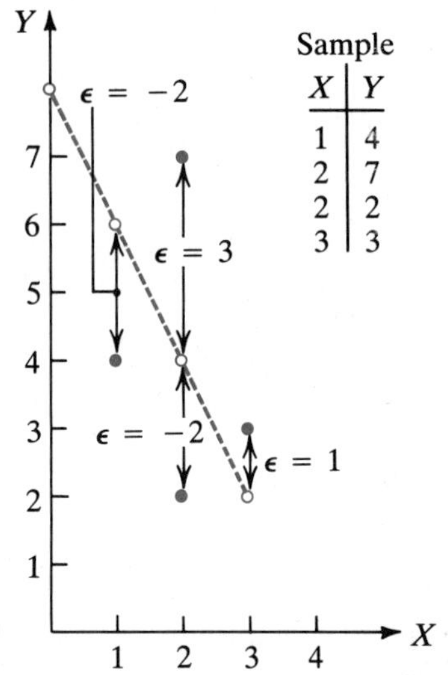

Figure 15.2 Modeling a Sample of $n = 4$ Points with $Y = 8 - 2X + \epsilon$

Figure 15.3(b) shows the normal distribution of errors for $X = 2$. The phenomenon depicted in Figure 15.3(b) is assumed to occur for each value of X, not just $X = 2$. At each value of X, the errors follow a normal distribution with the same mean of 0 and the same variance, σ^2. The following box summarizes these assumptions about ϵ.

Assumptions About ϵ in the Simple Regression Model

1. At each value for X, ϵ is a random variable.
2. At each value for X, ϵ has a normal probability distribution with

$$\text{Mean} = 0 \quad \text{and} \quad \text{Variance} = \sigma^2$$

3. For each pair of values for X, the error random variables are independent.

COMMENTS

1. Classifying X as a fixed variable implies that its values are exactly recorded or measured. This, in turn, leads to the implicit assumption that X and ϵ are independent. If the values of X occur randomly, the general simple regression model is still applicable provided we assume the observed values of X are recorded without error. Other names for X are the *regressor, explanatory variable, predictor variable,* or *design variable*.

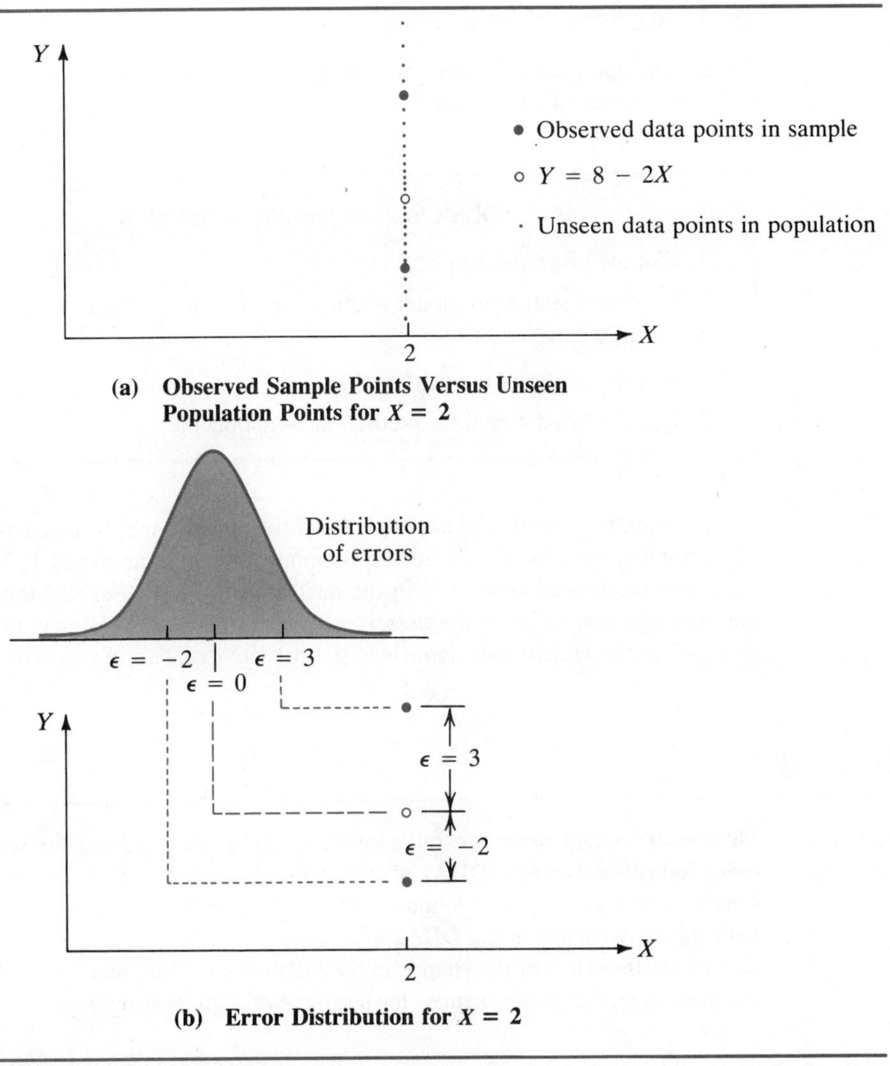

Figure 15.3 Illustrating the Error Distribution

2. The dependent variable Y is a random variable whose distribution depends only on the distribution of ϵ. As a result, Y has a normal probability distribution with a mean $= \beta_0 + \beta_1 X$ and a variance $= \sigma^2$. Other names for Y are the *regressand, explained variable, criterion variable,* or *response variable.*

Our ability to determine the error for each sample point in Figure 15.2 depended on our knowledge of $\beta_0 = 8$ and $\beta_1 = -2$. In general, the values of β_0 and β_1 will be unknown. Therefore, we will have to estimate these values based on the sample data. In Section 15.3, we outline the procedure for developing these estimates which we denote by b_0 and b_1, respectively. Also, we will estimate the error variance, σ^2.

Objectives of Regression

To achieve our goal of analyzing the relationship between X and Y, we list five objectives in the following box.

Objectives of Simple Regression

1. Plot the bivariate data.
2. Propose a statistical model relating the variables.
3. Fit the model.
4. Test the utility of the fitted model.
5. Use the fitted model for prediction, if applicable.

A sixth objective—verify the assumptions of the model—will be discussed in Chapter 17. Achieving the first two objectives is demonstrated in Example 15.1. The remaining objectives are treated separately in the next three sections. For illustrative purposes, our examples and some of the exercises involve relatively small sets of data. Nevertheless, simple regression is applicable to bivariate data sets of any size.

EXAMPLE 15.1

Most stock mutual funds are fully invested in the stock market for which the Dow Jones Industrial Average (DJIA) often is viewed as a barometer. The daily change in a fund's share price can be found in the business section of most newspapers, along with the daily change in the DJIA.

To see the effect of the change in the DJIA on a mutual fund, an analyst recorded the following percentage changes for a sample of eight trading days:

Percentage Change in DJIA, X	Percentage Change in Share Price of Mutual Fund, Y
−.4	−.4
−.1	−.3
+.1	−.1
+.3	+.8
−.2	−.2
0	+.1
+.2	+.4
+.3	+.5

Plot the data and propose a model (statistical or deterministic) to describe the relationship between X and Y.

Solution:

The eight data points are shown in Figure 15.4. Notice that X and Y appear to have a direct, but not exact, relationship. Thus, a statistical model is appropriate. The simple regression model is $Y = \beta_0 + \beta_1 X + \epsilon$. The values of the parameters β_0 and β_1 are unknown and must be estimated. The error term is assumed to follow a normal distribution.

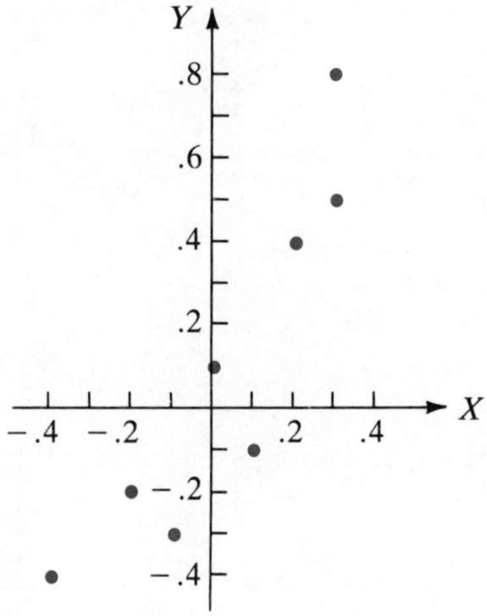

Figure 15.4 Scatter Plot for Data of Example 15.1

□

Analyzing the data in Figure 15.4 is done for two related reasons. First, the historical movements in share prices of mutual funds vis-à-vis the general stock market provide valuable information about the risk level of the mutual fund. Second, knowing the statistical relationship, if any, between variables like X and Y in Example 15.1 enables an investor or analyst to anticipate the vicissitudes of the future.

In the next section, we will provide the necessary tools for the part of the regression analysis we call *fitting the model*.

15.3 FITTING THE MODEL

■ Least-Squares Criterion

Applying the simple regression model, $Y = \beta_0 + \beta_1 X + \epsilon$, to a set of sample data requires a criterion of *fit*. This means we must decide how we will estimate the unknown betas (β_0 and β_1) in Equation 15–1. There are many ways to do this. A

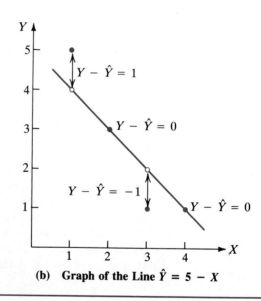

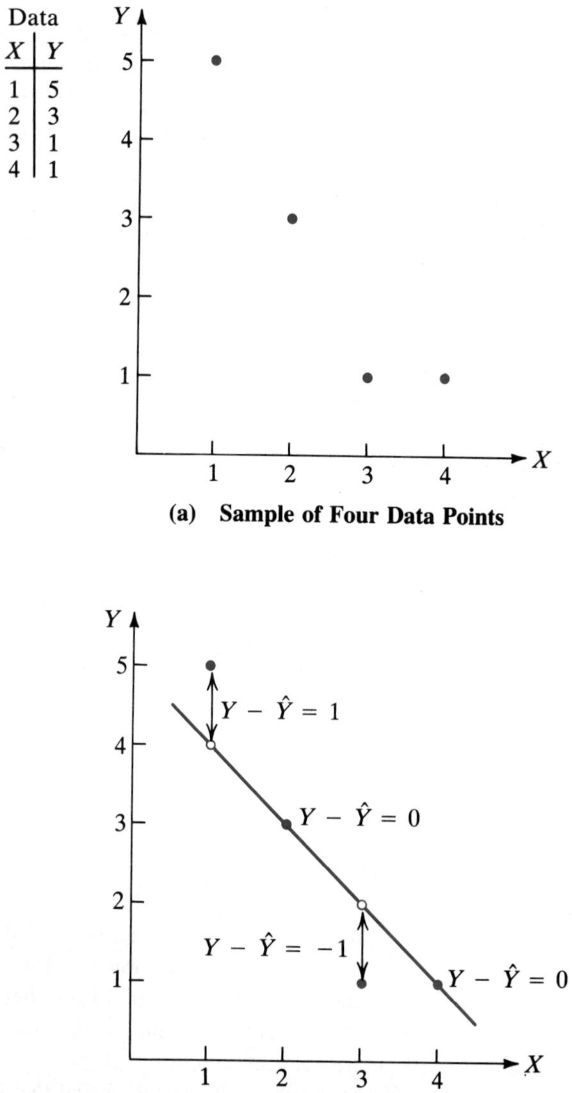

(a) Sample of Four Data Points

(b) Graph of the Line $\hat{Y} = 5 - X$

Figure 15.5 Three Possible Regression Lines for the Data Set in Figure 15.4

simplistic solution would be to find the equation of the line that connects the greatest number of sample data points. Alternatively, we could find the line that minimizes the sum of the absolute errors, $\Sigma|Y - \hat{Y}|$, or the line that goes through the origin and the point $(\overline{X}, \overline{Y})$, if X and Y were directly related.

Each of these criteria has attractive features and could be adopted, but the traditional approach has been to minimize the sum of the squared errors, $\Sigma(Y - \hat{Y})^2$,

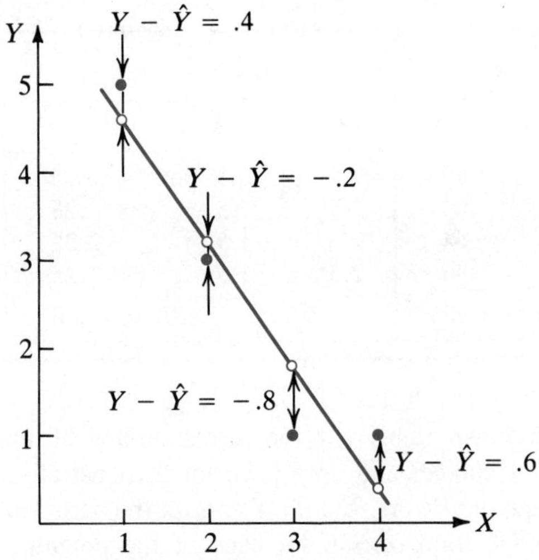

(c) Graph of the Line $\hat{Y} = 6 - 1.4X$

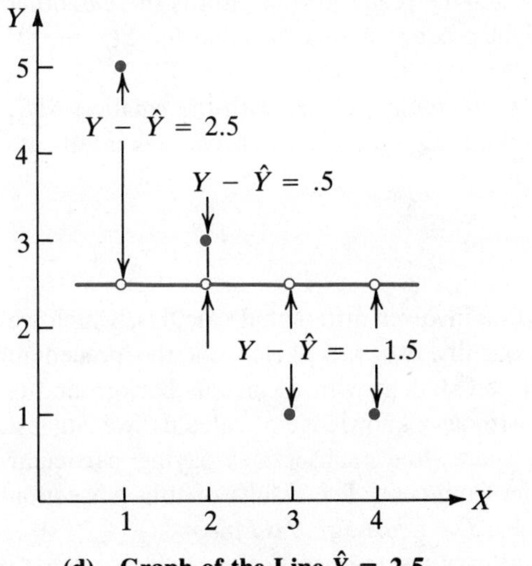

(d) Graph of the Line $\hat{Y} = 2.5$

between the observed Y values and the fitted Y values. As we learned in Chapter 4, this is called the *least-squares criterion*. In short, the line $\hat{Y} = b_0 + b_1X$ is the *least-squares line* provided the quantity $\Sigma(Y - \hat{Y})^2$ is as small as possible, given the configuration of the sample data points.

To illustrate the least-squares notation, consider the graph of the four sample data points in Figure 15.5(a). In Figure 15.5(b)(c) and (d) the lines $\hat{Y} = 5 - X$, $\hat{Y} = 6$

Table 15.1 Computation of *SSE* from Figure 15.5

		(b) $\hat{Y} = 5 - X$			(c) $\hat{Y} = 6 - 1.4X$			(d) $\hat{Y} = 2.5$		
X	Y	$\hat{Y}$	$Y - \hat{Y}$	$(Y - \hat{Y})^2$	$\hat{Y}$	$Y - \hat{Y}$	$(Y - \hat{Y})^2$	$\hat{Y}$	$Y - \hat{Y}$	$(Y - \hat{Y})^2$
1	5	4	1	1	4.6	.4	.16	2.5	2.5	6.25
2	3	3	0	0	3.2	−.2	.04	2.5	.5	.25
3	1	2	−1	1	1.8	−.8	.64	2.5	−1.5	2.25
4	1	1	0	0	.4	.6	.36	2.5	−1.5	2.25
				$SSE = 2$			$SSE = 1.20$			$SSE = 11.00$

$- 1.4X$, and $\hat{Y} = 2.5$, respectively, are drawn along with the corresponding fitting errors $(Y - \hat{Y})$. These lines are arbitrary choices and are drawn for demonstrative purposes. Let us represent the Sum of Squared Errors, $\Sigma(Y - \hat{Y})^2$, with the acronym *SSE*. In Table 15.1, we outline the computation of *SSE* for each of the potential regression lines indicated in Figure 15.5. The $\hat{Y}$ values are obtained by inserting the various values of X into each $\hat{Y}$ equation and performing the indicated arithmetic.

Of the three lines shown in Figure 15.5, the smallest value of *SSE* from Table 15.1 is 1.20, making the line $\hat{Y} = 6 - 1.4X$ the preferred line. But is there another line, which we have not shown, that might produce a smaller value for $\Sigma(Y - \hat{Y})^2$ than 1.20?

Our criterion of least squares tells us to *choose* the line with the smallest *SSE;* however, our choice should not be made on a trial-and-error basis. We seek a fail-safe way to pinpoint *the* least-squares line.

Normal Equations

The method of finding *the* least-squares line involves differential calculus, which we do not assume as a prerequisite. Consequently, we will sketch out the procedure without supplying the details. The interested student with a calculus background is encouraged to fill in the gaps. For those without a knowledge of calculus, we suggest you skim this "how-to-derive-the-least-squares-line" subsection, paying particular attention to the notation introduced in the Comment. The results of this subsection are summarized in the next subsection titled *The Least-Squares Line.*

Let us begin by substituting the equation of the regression line, $\hat{Y} = b_0 + b_1X$, into the expression for *SSE* to produce

$$SSE = \Sigma[Y - \hat{Y}]^2 = \Sigma[Y - (b_0 + b_1X)]^2 \tag{15–2}$$

Equation 15–2 represents the quantity *SSE* as a function of the given data (X, Y) and two unknowns, b_0 and b_1. Choosing the line that minimizes the sum of squared errors is equivalent to determining the values of b_0 and b_1 in Equation 15–2 that yield the smallest *SSE* with respect to the observed values of X and Y.

To find the minimum for *SSE,* we take its partial derivative with respect to b_0 and its partial derivative with respect to b_1. The partial derivatives are set equal to 0, yielding the following pair of *normal equations*.

Normal Equations

$$nb_0 + b_1\Sigma X = \Sigma Y \quad (15\text{–}3)$$

$$b_0\Sigma X + b_1\Sigma X^2 = \Sigma XY \quad (15\text{–}4)$$

The values of b_0 and b_1 that minimize SSE are the simultaneous solutions to these two equations.

COMMENT Throughout our study of regression, we will encounter many quantities involving the sum of squared deviations. These often are designated with acronyms. To represent the sum of squared deviations between the values of X and the mean $\overline{X}$, that is, $\Sigma(X - \overline{X})^2$, we propose the symbol SSX. Similarly, we define SSY to be the sum of the squared deviations between the values of Y and the mean $\overline{Y}$. The following expressions are equivalent ways of computing these quantities:

$$SSX = \Sigma(X - \overline{X})^2 = \Sigma X^2 - \frac{(\Sigma X)^2}{n} = \Sigma X^2 - n\overline{X}^2 \quad (15\text{–}5)$$

$$SSY = \Sigma(Y - \overline{Y})^2 = \Sigma Y^2 - \frac{(\Sigma Y)^2}{n} = \Sigma Y^2 - n\overline{Y}^2 \quad (15\text{–}6)$$

The symbol $SSXY$ represents the sum of the (cross) product of the deviations between the values of X and $\overline{X}$ and the values of Y and $\overline{Y}$:

$$SSXY = \Sigma(X - \overline{X})(Y - \overline{Y}) = \Sigma XY - \frac{(\Sigma X)(\Sigma Y)}{n} = \Sigma XY - n\overline{XY} \quad (15\text{–}7)$$

The Least-Squares Line

With or without the ability to derive the normal equations (15–3 and 15–4), we need only basic algebra to solve them for b_0 and b_1. The results are summarized in the following box.

Estimated Regression Coefficients

The equation of the least squares line is

$$\hat{Y} = b_0 + b_1X$$

where

$$b_0 = \overline{Y} - b_1\overline{X} \quad (15\text{–}8)$$

$$b_1 = \frac{SSXY}{SSX} \quad (15\text{–}9)$$

The equations for b_0 and b_1 are identical to those given in Chapter 4, keeping in mind the meanings of the newly introduced symbols *SSXY* and *SSX*. Example 15.2 demonstrates their use.

EXAMPLE 15.2

Find the equation of the least-squares line relating Y to X based on the following set of bivariate data. Graph the data and the least-squares line.

X	1	2	3	4
Y	5	3	1	1

Solution:

Let's set up a worksheet to generate the four sums needed for Equations 15–8 and 15–9:

X	Y	XY	X^2
1	5	5	1
2	3	6	4
3	1	3	9
4	1	4	16
$\Sigma X = 10$	$\Sigma Y = 10$	$\Sigma XY = 18$	$\Sigma X^2 = 30$

Intermediate calculations are

$$\overline{X} = \frac{\Sigma X}{n} = \frac{10}{4} = 2.5$$

$$\overline{Y} = \frac{\Sigma Y}{n} = \frac{10}{4} = 2.5$$

$$SSXY = \Sigma XY - n\overline{X}\,\overline{Y} = 18 - 4(2.5)(2.5) = -7$$

$$SSX = \Sigma X^2 - n\overline{X}^2 = 30 - 4(2.5)^2 = 5$$

The final calculations for the intercept, b_0, and the slope, b_1, are

$$b_1 = \frac{SSXY}{SSX} = \frac{-7}{5} = -1.4$$

$$b_0 = \overline{Y} - b_1\overline{X} = 2.5 - (-1.4)(2.5) = 6$$

The equation of the least-squares line is $\hat{Y} = 6 - 1.4X$; it is graphed in Figure 15.6.

We recognize $\hat{Y} = 6 - 1.4X$ as the line in Figure 15.5(c). Our earlier question as to whether another line that is not shown could have a smaller value for $\Sigma(Y - \hat{Y})^2$ than 1.20 now is resolved. Any other straight line drawn among the four sample data points of Figure 15.5(a) *must* yield a value for *SSE* that is greater than 1.20. Hence,

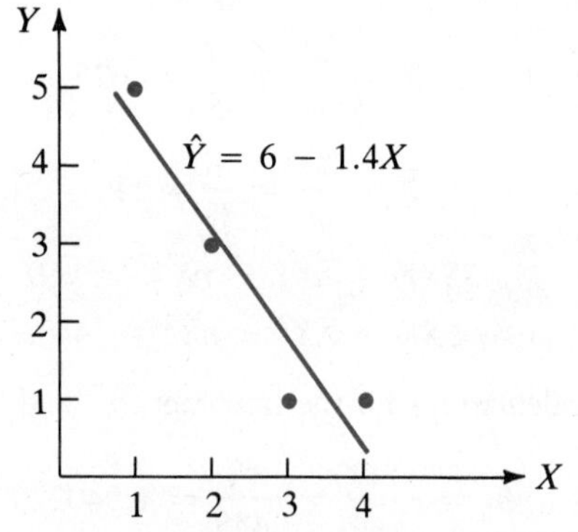

Figure 15.6 Least-Squares Line for Data of Example 15.2

the least-squares line $\hat{Y} = 6 - 1.4X$ is the unique line that minimizes the quantity $\Sigma(Y - \hat{Y})^2$, given the configuration of the four sample data points.

□

EXAMPLE 15.3

Find the equation of the least-squares line relating Y, the percentage change in the share price of a mutual fund, to X, the percentage change in the DJIA based on the following sample. Graph the data and the least-squares line.

X	−.4%	−.1	+.1	+.3	−.2	0	+.2	+.3
Y	−.4%	−.3	−.1	+.8	−.2	+.1	+.4	+.5

Solution:

Equations 15–8 and 15–9 involve four sums: ΣX, ΣY, ΣXY, and ΣX^2. Let's set up a worksheet to generate these quantities:

X	Y	XY	X^2
−.4	−.4	.16	.16
−.1	−.3	.03	.01
.1	−.1	−.01	.01
.3	.8	.24	.09
−.2	−.2	.04	.04
0	.1	0	0
.2	.4	.08	.04
.3	.5	.15	.09
$\Sigma X = .2$	$\Sigma Y = .8$	$\Sigma XY = .69$	$\Sigma X^2 = .44$

Intermediate calculations are

$$\overline{X} = \frac{\Sigma X}{n} = \frac{.2}{8} = .025$$

$$\overline{Y} = \frac{\Sigma Y}{n} = \frac{.8}{8} = .1$$

$$SSXY = \Sigma XY - n\overline{X}\,\overline{Y} = .69 - 8(.025)(.1) = .67$$

$$SSX = \Sigma X^2 - n\overline{X}^2 = .44 - 8(.025)^2 = .435$$

The final calculations for the intercept, b_0, and the slope, b_1, are:

$$b_1 = \frac{SSXY}{SSX} = \frac{.67}{.435} = 1.5402299$$

$$b_0 = \overline{Y} - b_1\overline{X} = .1 - (1.5402299)(.025) = .0614943$$

To three decimal places, the equation of the least-squares line is $\hat{Y} = .061 + 1.540X$; it is graphed in Figure 15.7.

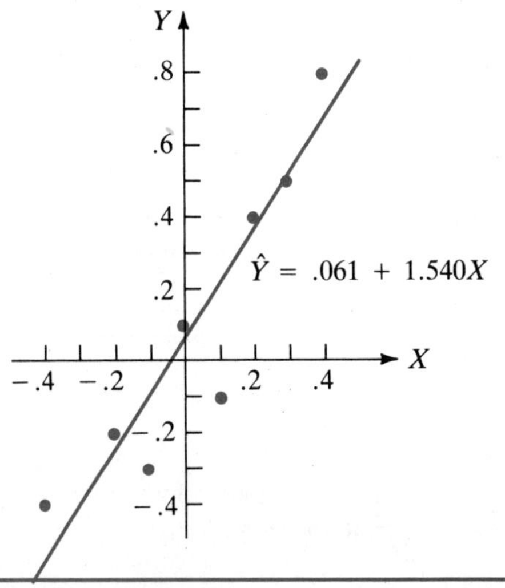

Figure 15.7 Least-Squares Line for Data of Example 15.3

□

COMMENTS

1. The numerical values produced by Equations 15–8 and 15–9 for a set of data also are referred to as the *fitted regression coefficients*, the *estimated betas*, or the *partial regression coefficients*. Other names for the least-squares line are the *fitted regression model*, the *regression line*, the *best fitting line*, or the *prediction equation*.
2. Mathematically, b_0 represents the Y-intercept of the line relating X and Y. But, practically speaking, it oftentimes is not interpretable. Only if the value of 0 is

within the *relevant range* of the X-values (as in Example 15.3) can we interpret b_0 as the average value of Y when $X = 0$. Relevant range refers to the region between the minimum and maximum X-values observed in the sample.

3. Where possible, we recommend reporting the values of b_0 and b_1 to at least three decimal places. Also if possible, use unrounded values of $SSXY$ and SSX in computing b_1 and the unrounded value of b_1 in computing b_0.

Estimating the Variance of ϵ

In Figure 15.5, we see that the positioning of a straight line through a set of bivariate data points creates vertical deviations between the points and the line. Though the regression line minimizes the sum of the squared deviations, the deviations are not all equal to zero. Each deviation is an estimated value of ϵ, the error term in the simple regression model. Recall that at each X-value, we assumed a normal distribution for ϵ with a mean of 0 and a constant, but unknown, variance σ^2.

The sum of the squared deviations can be used to estimate σ^2, the variability of the points around the line. This approach to measuring dispersion is analogous to that used in Chapter 3 when we computed the sample variance by summing the squared deviations from the mean. Using the "hat" notation to signify an estimator, we define

$$\hat{\sigma}^2 = \frac{\Sigma(Y - \hat{Y})^2}{n - 2} = \frac{SSE}{n - 2} \qquad \textbf{(15–10)}$$

Equation 15–10 is similar to the defining formula for S^2 in Equation 3–6. The denominator of $n - 2$ reflects the two degrees of freedom expended to estimate the two parameters of the regression model—β_0 and β_1. Taking the square root of $\hat{\sigma}^2$ yields $\hat{\sigma}$, which often is called the estimated *standard error of the regression*.

Estimated Standard Error of the Regression

$$\hat{\sigma} = +\sqrt{\hat{\sigma}^2} \qquad \textbf{(15–11)}$$

The magnitude of $\hat{\sigma}$ gives us an idea of the precision with which the regression model fits the sample data: The smaller the value of $\hat{\sigma}$, the better the fit.

COMMENTS

1. A more convenient way of computing SSE, instead of generating each $\hat{Y}$ and $Y - \hat{Y}$, is to use the formula

$$SSE = SSY - b_1 SSXY \qquad \textbf{(15–12)}$$

If possible, use unrounded values of SSY, $SSXY$, and b_1.

2. The quantity $\hat{\sigma}^2$ is the same as the quantity "mean square error (MSE)" in an analysis of variance summary table.

3. The fitting errors $Y - \hat{Y}$ are also known as *residuals*. Techniques for analyzing the residuals appear in Chapter 17.
4. Since most inferential procedures in regression depend on $\hat{\sigma}$, computational accuracy is critical. We recommend carrying out the computations and reporting the value of $\hat{\sigma}$ to *at least* five decimal places.

EXAMPLE 15.4

Find the estimated standard error of the regression for the data in the following examples:

a. Example 15.2
b. Example 15.3

Solution:

a. In Table 15.1, we computed $SSE = 1.2$. Our estimate of σ is

$$\hat{\sigma} = \sqrt{\frac{SSE}{n-2}} = \sqrt{\frac{1.2}{4-2}} = \sqrt{.6} = .7745967 \qquad \text{or } .77460 \text{ (rounded)}$$

b. In Example 15.3, we determined that $b_1 = 1.5402299$ (unrounded) and $SSXY = .67$. With $\Sigma Y^2 = 1.36$, we have

$$SSY = \Sigma Y^2 - n\bar{Y}^2 = 1.36 - 8(.1)^2 = 1.28$$

Thus, from Equation 15–12,

$$SSE = SSY - b_1 SSXY$$
$$= 1.28 - (1.5402299)(.67) = .248046$$

and

$$\hat{\sigma} = \sqrt{\frac{SSE}{n-2}}$$
$$= \sqrt{\frac{.248046}{6}} = \sqrt{.041341} = .2033249 \qquad \text{or } .20332 \text{ (rounded)}$$

□

The estimated standard error of the regression is important for three major reasons. First, it enables us to completely specify the probability distribution of ϵ. For example, using the value $\hat{\sigma}$ from Example 15.3, we declare the error term ϵ to be normally distributed with a mean of 0 and an estimated standard deviation of .20332. Note that declaring ϵ to be normally distributed is an assumption and must be clarified. We discuss the procedure for doing so in Chapter 17. Second, $\hat{\sigma}$ helps to confirm or repudiate the utility of the least-squares line as a prediction tool—Section 15.4 explains how. Third, any subsequent confidence interval forecasts with the least-squares line are functions of $\hat{\sigma}$. The smaller the value of $\hat{\sigma}$, the more precise are the predictions. Section 15.5 addresses this issue.

CLASSROOM EXAMPLE 15.1

Computing and Graphing a Least-Squares Line

A manager in the human resources department randomly selected five employee files and recorded the following data on X = number of weeks of paid vacation (annually) and Y = number of sick days claimed by the employee in the previous year. An analysis of the linear relationship between X and Y is desired.

X	Y	XY	X^2	Y^2
1	3			
2	1			
3	1			
4	0			
5	0			
$\Sigma X =$ _____	$\Sigma Y =$ _____	$\Sigma XY =$ _____	$\Sigma X^2 =$ _____	$\Sigma Y^2 =$ _____

a. Plot the data on a graph.

b. Find the equation of the least-squares line.

$$\bar{X} = \_\_\_\_\_ \qquad b_1 = \frac{SSXY}{SSX} = \_\_\_\_\_$$

$$SSX = \_\_\_\_\_$$

$$\bar{Y} = \_\_\_\_\_ \qquad b_0 = \bar{Y} - b_1\bar{X} = \_\_\_\_\_$$

$$SSXY = \_\_\_\_\_$$

The equation of the least-squares line is _____.

c. Plot the least-squares line on the graph made for part a.

d. Find the estimated standard error of the regression.

$$SSY = \_\_\_\_\_$$

$$SSE = SSY - b_1 SSXY = \_\_\_\_\_$$

$$\hat{\sigma} = \sqrt{\frac{SSE}{n-2}} = \_\_\_\_\_$$

SECTION 15.3 EXERCISES

15.1 There is believed to be a relationship between the prevailing 30-year fixed mortgage interest rate and the rate at which consumers default on their mortgage loans. A sample of eight pairs of values for X, mortgage interest rate, and Y, number of mortgage loan defaults per 1000, is indicated as follows:

X	10.9%	10.3	7.3	9.6	11.6	13.0	9.9	10.1
Y	19%	16	7	15	18	25	10	12

The equation of the least-squares line for these data is, to three decimal place accuracy, $\hat{Y} = -17.267 + 3.146X$.

a. Graph the sample data and the least-squares line. Find the fitting errors $(Y - \hat{Y})$ for each data point.

b. Find the average value of the fitting errors. Is it zero, or close to zero? Explain the significance of this value relative to the assumption given earlier that the mean of ϵ is 0.

c. Compute SSE.

d. Determine the value of $\hat{\sigma}$.

e. Describe the observed relationship between the mortgage interest rate and the default rate as direct, inverse, or neither.

15.2 A sample of size $n = 5$ yielded the following data and sums of squares:

X	Y
10	62
12	68
13	65
14	67
15	70

$SSX = 14.8$

$SSXY = 19.4$

a. Graph the data.

b. Find the equation of the least-squares line.

c. Find the five fitting errors, $(Y - \hat{Y})$.

15.3 For each bivariate data set described as follows, indicate which variable is likely to be considered the independent variable and which is to be the dependent variable.

a. for used cars, the age of the vehicle and the percentage of original value retained.

b. for a recreational theme park, the daily attendance and the predicted daily high temperature.

c. for a retail sales company, sales revenue and national business activity as measured by the current gross national product (GNP).

15.4 A "load" for a mutual fund is a fee assessed at the time money is invested to cover administrative costs. Most of the time the load is expressed as a percentage. A sample of the one-year returns for a group of mutual funds is shown in the following table in order to relate Y = the fund's annual return to X = the fund's load. The X-value of 0 percent signifies a no-load fund.

Load, X	0%	6	2	0	8.5	8.5	4	0	0	4.5
Return, Y	10%	7	15	22	18	12	24	19	13	7

a. Graph the data.
b. Find the equation of the least-squares line. Plot it. Give a practical interpretation, if possible, to b_0 and b_1.
c. Find the fitting errors, $Y - \hat{Y}$, for each data point.
d. Determine the value of the estimated standard error of the regression.
e. Describe the observed relationship between a mutual fund's return and its load. What does this suggest about load versus no-load funds?

15.5 Suppose a sample of $n = 5$ data points yielded a value of $SSE = 44$. Find the value of the estimated standard error of the regression.

15.6 Find the equation of the least-squares line for the following data:

X	−2	−1	0	1	2
Y	5	7	8	10	12

15.7 Suppose a sample of $n = 10$ data points yielded $\Sigma(Y - \hat{Y})^2 = 61.3$. Find $\hat{\sigma}$.

15.8 Find the value of SSE from the following information:
a. $n = 10$, $\hat{\sigma} = 2.383$
b. $n = 15$, $\hat{\sigma}^2 = 31.73865$
c. $df = 19$, $\hat{\sigma} = 5.40118$

15.9 A sample of size $n = 6$ yielded the following data:

X	1	2	3	4	5	6
Y	10	8	5	6	3	1

a. Plot the data.
b. Propose a linear model for the variables X and Y.
c. Find the equation of the least-squares line.
d. Plot the least-squares line.
e. Find the estimated standard error of the regression.

15.10 Does the least-squares line have to go through the point $(\overline{X},\overline{Y})$? Explain. (*Hint:* Determine the value of $\hat{Y}$ when you substitute $X = \overline{X}$ into the least-squares equation.)

15.11 Plot the following data:

X	−2	−1	0	1	2
Y	1	3	4	4	2

Find the equation of the least-squares line and plot it. What relationship exists between X and Y that is reflected in the graph, but not in the least-squares line?

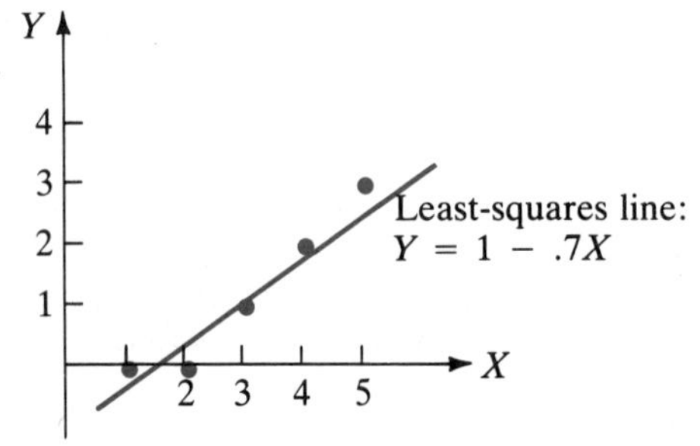

Figure 15.8 Scatter Plot and Least-Squares Line

15.12 A sample of size seven generated the following sums:

$$\Sigma X = 20 \qquad \Sigma X^2 = 58.42 \qquad \Sigma Y = 21 \qquad \Sigma XY = 62.4$$

Find the equation of the least-squares line.

15.13 A sample of size $n = 5$ generated the following sums:

$$\Sigma X = 0 \qquad \Sigma X^2 = 10 \qquad \Sigma Y = 14 \qquad \Sigma X(Y - \bar{Y}) = 3.7$$

Find the equation of the least-squares line.

15.14 Refer to Figure 15.8. There are four errors in the figure. Find and correct them. Assume the sample data is listed correctly as follows:

X	1	2	3	4	5
Y	0	0	1	1	3

15.15 The U.S. trade deficit has widened and remained a sore subject for years. One component of our country's trade with foreign governments that, until recently, has resisted the red ink is the trade balance in high-tech goods. Goods classified as high-tech include televisions, video cassette recorders, computers, and the like. The high-tech trade balance between 1980 and 1986 was as follows:

Year	Trade Balance ($ Billions)
1980	26.7
1981	26.6
1982	23.6
1983	18.8
1984	6.0
1985	3.6
1986	−2.6

Source: "USA's High-Tech Trade Deficit," *USA Today*, February 26, 1987. Copyright 1987, *USA Today*. Adapted with permission.

a. Create a quantitative variable X by assigning the first seven integers to the years 1980 through 1986, respectively.
b. Plot X versus Y = trade balance (in billions) and propose a linear regression model for the variables.
c. Do you expect b_1 to have a positive or negative sign? Why?
d. Find the equation of the least-squares line.
e. What is the fitting error associated with the year 1986?

15.16 A sample of size $n = 5$ produced the following data:

X	.1	.3	.1	.2	.4
Y	12	5	8	9	2

Find the equation of the least-squares line.

***15.17** Complete the missing steps outlined in the text between Equation 15–2 and Equations 15–3 and 15–4 by taking the partial derivatives of SSE with respect to b_0 and b_1.

***15.18** Solve Equations 15–3 and 15–4 for b_0 and b_1.

***15.19** The linear regression model is said to be *invariant* by a change of location for the independent variable X. This means that moving the entire set of values for X by a constant amount while maintaining the corresponding values for Y yields the same value for b_1, the slope, and for SSE. To see this, consider the following set of data:

X	1	2	4	9
Y	0	0	1	3

a. Verify that the equation of the least-squares line is

$$\hat{Y} = -.5789 + .3947X$$

and that

$$SSE = .0789473$$

b. Create a new variable W by subtracting the value of $\overline{X}$ from each value of X, that is, $W = X - \overline{X}$. Verify that this generates the following set of data:

W	−3	−2	0	5
Y	0	0	1	3

c. Verify that the equation of the least-squares line for the variables W and Y of part b is $\hat{Y} = 1 + .3947W$ and that $SSE = .0789473$.
d. The only difference between the equations of parts a and c, besides the representation of the independent variable, is the constant term: −.5789 versus 1. In part c, note that b_0 was computed as

$$1 = \overline{Y} - b_1\overline{W}$$

and in part a, b_0 was computed as

$$-.5789 = \overline{Y} - b_1\overline{X}$$

*Optional

By subtraction we see that the Y-intercepts are 1.5789 units apart. In general, how far apart are the two Y-intercepts?

The preservation of the value of b_1 despite changing the values of X by a constant amount is an important property. Frequently, the variable X in real-life applications assumes very large or very messy values. Converting the original values via $W = X - \overline{X}$ or by some other linear transformation simplifies the computations and difficulty of the problem. (For example, see Exercise 15.15.) Remember:

1. Do *not* transform the values of Y in this manner (the linear regression model is *not invariant* for a change in location of Y).
2. If you subsequently use the equation $\hat{Y} = b_0 + b_1W$ to predict values for Y based on X, be sure to put the correct value for W (*not* X) in the equation.

15.4 TESTING THE UTILITY OF THE MODEL

Are *X* and *Y* Related?

A least-squares line $\hat{Y} = b_0 + b_1X$ exists for every set of bivariate data regardless of the degree of association between the variables X and Y. This is both good news and bad news. It's good since we know the line minimizes the sum of the squared fitting errors and, in this sense, represents an optimal model. But, it is also bad because the variable X might have little or no relation to Y, diminishing the predictive capability of the regression model.

Consequently, failing to examine the strength of the relationship between X and Y before using the regression equation for prediction would be foolish. We seek a means of determining the usefulness of the fitted model. In other words, we must answer the question: Are X and Y related?

Recall that the second component of the regression model—β_1X—is the proportional effect of X on Y. Algebraically, β_1 represents the slope of the line relating X and Y in the population. A line with a slope of 0 is horizontal and indicates that the value of Y remains constant for any value of X. Such a situation renders X useless in predicting Y. On the other hand, if the slope were nonzero, then each one-unit change in the value of X is associated with some change in the value of Y. This suggests a relationship between X and Y.

We now have an answer for our earlier question as to whether X and Y are related. If the slope of the regression model, β_1, is zero the answer is no, X and Y are not related; if the slope is nonzero, the answer is yes, X and Y are related. However, implementing this decision rule is possible only if we know the value of β_1, which we do not, without conducting a census. But we do have a proxy value for β_1 in our sample estimator b_1.

Therefore, our question is answerable provided we are willing to incur the risk involved in making an inference about β_1's value based on the sample information contained in b_1. To determine if X and Y are related, we shall test the null hypothesis that $\beta_1 = 0$.

Sampling Distribution of b_1

In order to test the assumption that $\beta_1 = 0$, we need to know the theoretical properties of b_1. The sample-based estimator b_1 is a random variable with a sampling distribution. As long as the error term ϵ is normally distributed, then so is the sampling distribution for b_1, as illustrated in Figure 15.9. Normal distributions are characterized by their mean and standard deviation. For the sampling distribution of b_1, these quantities are listed in the following box.

Mean and Standard Deviation for the Sampling Distribution of b_1

$$\text{Mean: } \mu_{b_1} = \beta_1 \qquad \textbf{(15–13)}$$

$$\text{Standard Deviation: } \sigma_{b_1} = \frac{\sigma}{\sqrt{SSX}} \qquad \textbf{(15–14)}$$

The symbol μ_{b_1} represents the mean of the sampling distribution of b_1. Its value, β_1, is located in the center of the distribution in Figure 15.9. The symbol σ_{b_1} is shorthand for the standard deviation (standard error) of the sampling distribution of b_1. Its value is related to σ, the standard error of the regression, and to SSX, the dispersion in the values of X. Notice that the normal sampling distribution for b_1, centered at the parameter β_1, is similar to the sampling distribution of $\overline{X}$, centered at the population mean μ, that we developed in Chapter 8.

When the value of σ is unknown, as is usually the case, the standard error of b_1 is estimated by inserting $\hat{\sigma}$ for σ. The *estimated standard error of* b_1, denoted as S_{b_1}, is given in Equation 15–15.

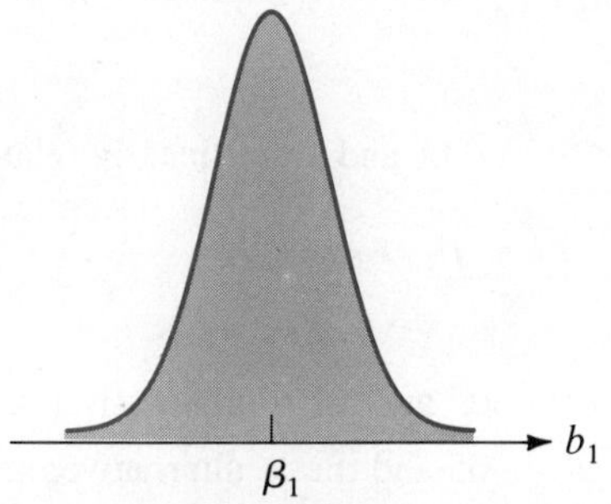

$$\sigma_{b_1} = \frac{\sigma}{\sqrt{SSX}}$$

Figure 15.9 Sampling Distribution of b_1

Estimated Standard Error of b_1

$$S_{b_1} = \frac{\hat{\sigma}}{\sqrt{SSX}} \quad (15\text{–}15)$$

COMMENT If possible, use unrounded values of $\hat{\sigma}$ and SSX in computing the estimated standard error of b_1, via Equation 15–15. We recommend rounding off S_{b_1} to five decimal places for subsequent use.

Testing Hypotheses About β_1

The test procedure to examine the independent variable's relationship to Y is a test of the model's utility and is based on the sampling distribution of b_1. For small samples and unknown σ, the test statistic is a t-statistic with $n - 2$ degrees of freedom. When the sample size is large, a Z-test statistic can be used instead. The null hypothesis states the slope of the population regression line is 0, indicating that the line is horizontal and therefore X and Y are not linearly related. The following box and Figures 15.10 through 15.12 summarize the elements of the test.

t-Test of Model Utility

H_0: $\beta_1 = 0$ (X and Y are not linearly related.)

H_a: $\beta_1 \neq 0$ (X and Y are linearly related.)

or

$\beta_1 > 0$ (X and Y are linearly related in a direct manner.)

or

$\beta_1 < 0$ (X and Y are linearly related in an inverse manner.)

$$TS: \ t^* = \frac{b_1 - \beta_1}{S_{b_1}} \quad (15\text{–}16)$$

RR: For $\alpha = .05$ and the $\neq$ alternative, see Figure 15.10.

For $\alpha = .05$ and the $>$ alternative, see Figure 15.11.

For $\alpha = .05$ and the $<$ alternative, see Figure 15.12.

C: Reject H_0 or accept H_0.

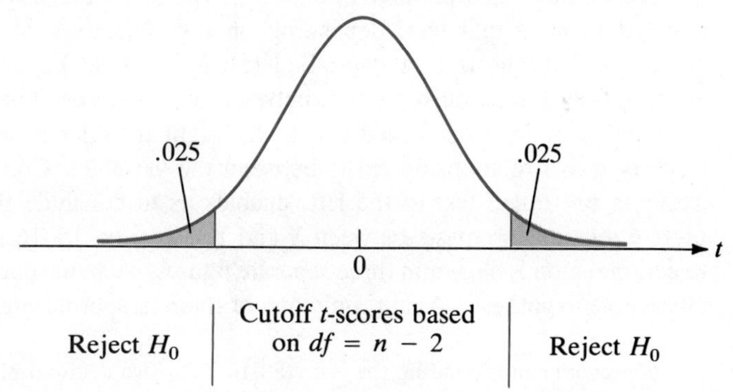

Figure 15.10 Rejection Region for Two-Tailed t-Test of Model Utility, $\alpha = .05$

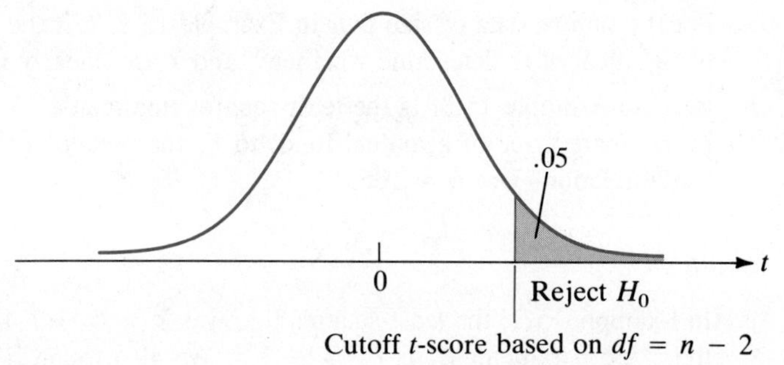

Figure 15.11 Rejection Region for Upper-Tailed t-Test of Model Utility, $\alpha = .05$

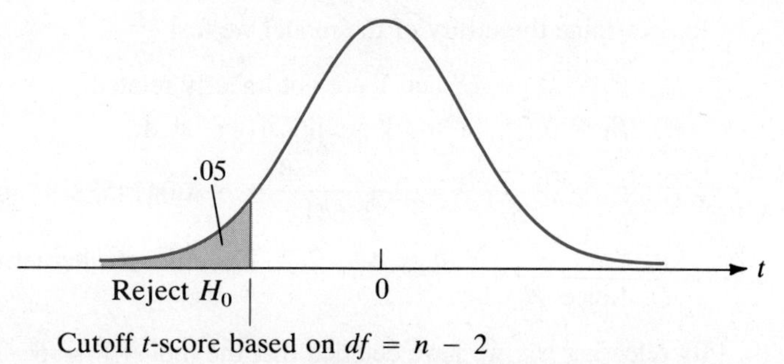

Figure 15.12 Rejection Region for Lower-Tailed t-Test of Model Utility, $\alpha = .05$

Notice that three possible alternative hypotheses are listed. Only one of them is selected in an actual test, depending on the objectives of the analysis. The first alternative hypothesis is a two-tailed test and should be used if we wish only to establish that a relationship exists between the variables. The second alternative hypothesis specifies a one-tailed test to the right in order to conclude that a positive slope or direct relationship exists between the variables. Conversely, the third alternative, a one-tailed test to the left, enables us to conclude that a negative slope or inverse relationship exists between X and Y. Equation 15–16 is the test statistic. The rejection region is shown in three separate figures, each one corresponding to a specific alternative hypothesis. Again, only one of them is appropriate for an actual test.

COMMENT We recommend rounding the test statistic t^* to two decimal places.

EXAMPLE 15.5

a. For the sample data of size four in Example 15.2, test the model at the 10% level of significance to determine whether X and Y are linearly related.

b. Refer to Example 15.3. Is the least-squares line relating Y, the percentage change in the share price of a mutual fund, to X, the percentage change in the DJIA, a useful relation? Use $\alpha = .05$.

Solution:

a. In Example 15.2, the least-squares line was $\hat{Y} = 6 - 1.4X$; thus, the estimate of the slope parameter β_1 is $b_1 = -1.4$. We also found $SSX = 5$ and computed $\hat{\sigma} = .7745967$ (unrounded) in Example 15.4. The estimated standard error of b_1 is

$$S_{b_1} = \frac{\hat{\sigma}}{\sqrt{SSX}} = \frac{.7745967}{\sqrt{5}} = .3464102 \qquad \text{or } .34641 \text{ (rounded)}$$

To determine the utility of the model we test

H_0: $\beta_1 = 0$ (X and Y are not linearly related.)
H_a: $\beta_1 \neq 0$ (X and Y are linearly related.)

$$TS\text{: } t^* = \frac{b_1 - \beta_1}{S_{b_1}} = \frac{-1.4 - 0}{.34641} = -4.0414538 \qquad \text{or } -4.04 \text{ (rounded)}$$

RR: For $\alpha = .10$, $df = 4 - 2 = 2$ and the $\neq$ alternative, see Figure 15.13.
C: Reject H_0.

By rejecting H_0, we have decided that the model $\hat{Y} = 6 - 1.4X$ is a useful linear relation between X and Y.

b. Before testing H_0: $\beta_1 = 0$, we need to compute the estimated standard error of b_1:

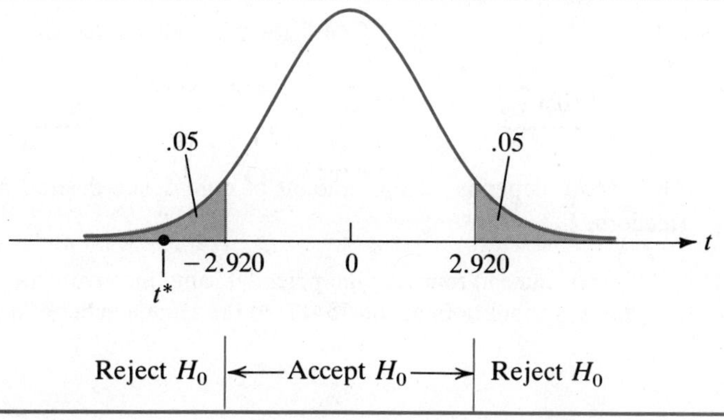

Figure 15.13 Rejection Region for Two-Tailed *t*-Test of Model Utility in Example 15.5a

$$S_{b_1} = \frac{\hat{\sigma}}{\sqrt{SSX}} = \frac{.2033249}{\sqrt{.435}} = .3082804 \qquad \text{or } .30828 \text{ (rounded)}$$

The elements of the test are

H_0: $\beta_1 = 0$

H_a: $\beta_1 \neq 0$

$$TS: t^* = \frac{b_1 - \beta_1}{S_{b_1}} = \frac{1.540 - 0}{.30828} = 4.9954587 \qquad \text{or } 5.00 \text{ (rounded)}$$

RR: For $\alpha = .05$, $df = 8 - 2 = 6$ and the $\neq$ alternative, the cutoff t-scores are ± 2.447.

C: Since $t^* > 2.447$, we reject H_0.

Thus, we have established the least-squares line as a useful relation between X and Y in the population.

□

Confidence Interval for β_1

The model utility test merely establishes the presence or absence of a linear relationship between the independent and dependent variables. Often, we also wish to know the extent of the direct or inverse movement of Y with X. For instance, if we increase or decrease the value of X by one unit, approximately how much will Y change? Our best guess would be the observed value of b_1, but as we remember from Chapter 10, a point estimate like this is unreliable. A confidence interval estimate is preferred due to its reliability. Equation 15–17 is the general expression for a confidence interval for the slope parameter β_1.

Confidence Interval for β_1

$$b_1 \pm (t\text{-score})\, S_{b_1} \qquad \textbf{(15–17)}$$

The t-score depends on the amount of confidence desired and on $n - 2$ degrees of freedom.

COMMENT We recommend rounding the potential sampling error (the computation to the right of the $\pm$ symbol in Equation 15–17) to the same number of decimal places as b_1.

EXAMPLE 15.6

Estimate β_1 from the data of Example 15.3 using a 95% confidence interval.

Solution:

From Example 15.3, we found $b_1 = 1.540$ and in Example 15.5b we computed $S_{b_1} = .30828$. With $df = 6$ and 95% confidence, we use a t-score of 2.447. A 95% confidence interval for β_1 is

$$b_1 \pm (t\text{-score})\, S_{b_1} = 1.540 \pm (2.447)(.30828)$$
$$= 1.540 \pm .754 \qquad \text{or } .786 \text{ to } 2.294$$

□

Notice that the interval in Example 15.6 excludes the value of 0. This reinforces the conclusion to reject H_0: $\beta_1 = 0$ we arrived at in Example 15.5b. Hence, testing the utility of the regression model also can be achieved by examining the confidence interval for β_1 for the presence of the number 0. If 0 is in the interval, we accept H_0: $\beta_1 = 0$ (that X and Y are not linearly related); if 0 is not in the interval, we reject H_0: $\beta_1 = 0$. To use this rule, make certain that the amount of confidence in the interval estimate is complementary to the risk level of the test—that is, 95% confidence ⇔ 5% risk.

■ ANOVA Table

Testing the regression model's effectiveness can be accomplished in a third way with the results from an analysis of variance table. The ANOVA approach to summarizing the test of the relation between X and Y is equivalent to the t-test of model utility. We will use the ANOVA table in multiple regression analysis and find that it is a standard part of the computer output in most regression programs.

In setting up an ANOVA table for a regression analysis, we identify three sources of variability. First we have the overall variability of the Y-values about their mean $\overline{Y}$. We refer to this variability as the *total* variability. Second, there is the discrepancy between $\overline{Y}$ and $\hat{Y}$, the predicted value of Y, that is explainable by the presence of the

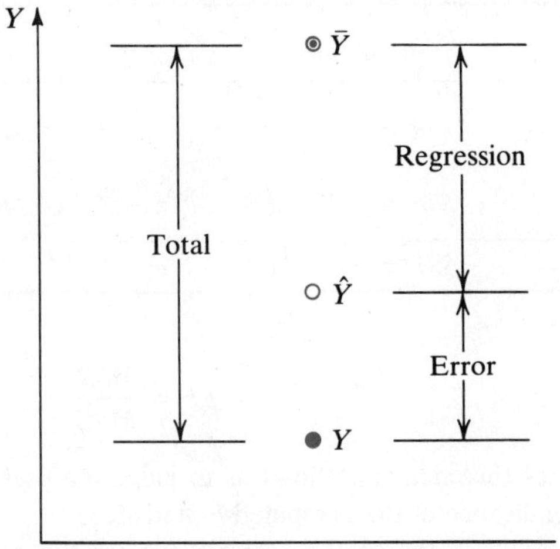

Figure 15.14 Partitioning the *Total* Variability Into *Regression* Plus *Error* Variability

component $\beta_1 X$ in the regression model. We call this *regression* variation. Third, we have the unexplainable variation between the observed value of Y and $\hat{Y}$. Consistent with our earlier discussions, we call this the *error* variability, since it represents the variation of the points about the fitted line. Theoretically, we can show from the relationship

$$\underset{\text{Total variability}}{(Y - \bar{Y})} = \underset{\text{Regression variability}}{(\hat{Y} - \bar{Y})} + \underset{\text{Error variability}}{(Y - \hat{Y})}$$

that the *total* sum of squares, denoted *SST*, can be partitioned as

$$\underset{SST}{\Sigma(Y - \bar{Y})^2} = \underset{SSR}{\Sigma(\hat{Y} - \bar{Y})^2} + \underset{SSE}{\Sigma(Y - \hat{Y})^2}$$

where *SSR* is the *regression* sum of squares and *SSE* is the *error* sum of squares. See Figure 15.14. Furthermore, the degrees of freedom are also additive:

$$df_{\text{Total}} = df_{\text{Regr}} + df_{\text{Error}}$$

Since there are always $n - 1$ degrees of freedom associated with the *total* variability and in simple regression we have $n - 2$ degrees of freedom for the error component, then we must have one degree of freedom corresponding to the regression source.

The mean squares are formed, as we learned in Chapter 14, by dividing each sum of squares by its number of degrees of freedom. Large values of the mean square for regression, *MSR*, relative to the *MSE* suggest the regression model accounts for a large portion of the variability in the Y-values. This would signal a useful model. The F-statistic, defined as

Table 15.2 ANOVA Table for a Simple Regression Analysis

Source	Sum of Squares	df	Mean Square	F
Regression	$SSR = \Sigma(\hat{Y} - \overline{Y})^2$	1	$MSR = \frac{SSR}{1}$	$F^* = \frac{MSR}{MSE}$
Error	$SSE = \Sigma(Y - \hat{Y})^2$	$n - 2$	$MSE = \frac{SSE}{n-2}$	
Total	$SST = \Sigma(Y - \overline{Y})^2$	$n - 1$		

$$F^* = \frac{MSR}{MSE}$$

incorporates this idea and allows us to judge the usefulness of the regression model by the significance of the computed F-statistic.

The results of the analysis of variance are summarized in an ANOVA table like the one in Table 15.2. From the ANOVA table we can reconstruct the statistical test for the regression model's utility. The test is summarized in the following box.

F-Test of Model Utility

H_0: $\beta_1 = 0$ (X and Y are not linearly related.)
H_a: $\beta_1 \neq 0$ (X and Y are linearly related.)
TS: $F^* = \frac{MSR}{MSE}$ **(15–18)**
RR: For $\alpha = .05$ and the $\neq$ alternative, see Figure 15.15.
C: Reject H_0 or accept H_0

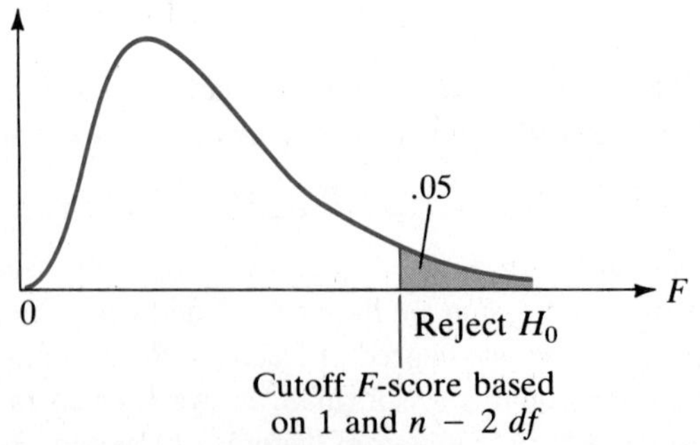

Figure 15.15 Rejection Region for F-test of Model Utility, $\alpha = .05$

COMMENT We recommend rounding the test statistic F^* to two decimal places.

EXAMPLE 15.7

Conduct an F-test of the utility of the regression model $\hat{Y} = .061 + 1.540X$ based on the following ANOVA table. Use $\alpha = .05$.

Source	*SS*	*df*	*MS*	*F*
Regression	1.031954	1	1.031954	24.96
Error	.248046	6	.041341	
Total	1.28	7		

Solution:

The F-test of the model's utility is

H_0: $\beta_1 = 0$ (X and Y are not linearly related.)

H_a: $\beta_1 \neq 0$ (X and Y are linearly related.)

$$TS: F^* = \frac{MSR}{MSE} = \frac{1.031954}{.041341} = 24.96$$

RR: For $\alpha = .05$, numerator $df = 1$, denominator $df = 6$ and the $\neq$ alternative, see Figure 15.16.

C: Reject H_0.

At the 5% level of significance the model is judged useful. This is consistent with our analysis of the 95% confidence interval for β_1 in Example 15.6. Then, as now, we have established that X and Y are linearly related.

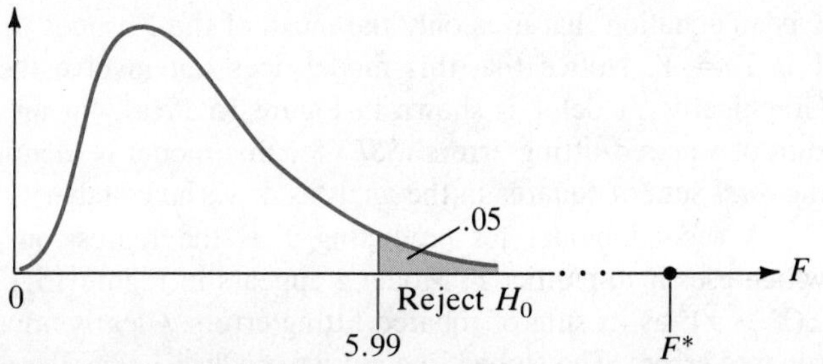

Figure 15.16 Rejection Region for F-Test of Model Utility in Example 15.7

□

COMMENTS

1. To construct an ANOVA table from the sample data, compute the sums of squares as follows:

$$SSR = b_1 SSXY \qquad SSE = SSY - b_1 SSXY \qquad SST = SSY$$

Then fill in the remainder of Table 15.2 as indicated.

2. The mean square for error, *MSE*, is the same value as $\hat{\sigma}^2$.

3. Sources of variability have different labels depending on the discipline and/or computer package. Synonyms for *regression* are "model," "treatment," or "explained." For *error*, you might see "residual" or "unexplained." *Total* is a standard term with perhaps no exceptions.

4. The F-test for model utility is more limited than the t-test because it tests the $\neq$ alternative only. If you wish to conclude that the slope is strictly positive or negative, use the more versatile t-test.

5. Since the F-test for H_0: $\beta_1 = 0$ is equivalent to the t-test, it always must yield the same conclusion for a given level of significance. Note also that $(t^*)^2 = F^*$ for these tests (see Exercise 15.31).

6. In one-way ANOVA we partitioned *SST* into two components, which we called *SSTR* and *SSE*. In regression we call the treatment sum of squares (*SSTR*) by a new name, regression sum of squares (*SSR*). In effect, the "treatment" is now the "regression" of Y on X.

7. In subsequent examples and exercises we may use the notation df_n and df_d to denote the numerator and denominator degrees of freedom, respectively, of the F-statistic.

Coefficient of Determination

There is a descriptive statistic called the *coefficient of determination* that often is used to summarize a linear regression model. Unfortunately, some practitioners (erroneously) use the value of the coefficient of determination by itself as an indicator of a model's utility, in lieu of the estimation or testing procedures for β_1 that we have just discussed.

The development of the coefficient of determination stems from a comparison of two potential models for predicting the dependent variable Y. What we will call model 1 is an equation that uses only the mean of the Y-values to predict Y—that is, model 1 is $\hat{Y} = \overline{Y}$. Notice that this model does not involve the independent variable X. Graphically, model 1 is shown in Figure 15.17(a), for an arbitrary set of data. The sum of squared fitting errors (*SSE*) for this model is identical to $SST = \Sigma(Y - \overline{Y})^2$, the *total* sum of squares in the analysis of variance table.

A second model for predicting Y is the regression model, $\hat{Y} = b_0 + b_1X$, which uses X to predict Y. Model 2 appears in Figure 15.17(b) and generates $SSE = \Sigma(Y - \hat{Y})^2$ as its sum of squared fitting errors. Clearly, model 2 has a smaller sum of squared errors. The sloped line always produces a smaller sum of squared errors than does the horizontal line, no matter how gentle the slant.

Figure 15.18 provides a way to compare model 1 to model 2. Both models have the same *SST*. For model 1 however, since $SST = SSE$, the *SSR* component must be zero. Model 2 has a nonzero value for *SSR;* thus, its value of *SSE* is necessarily

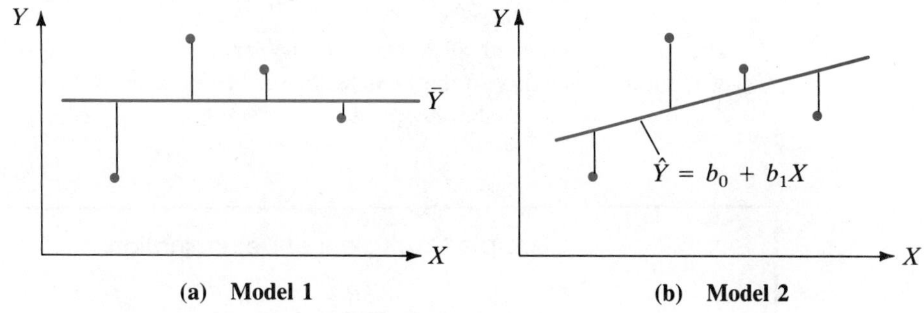

Figure 15.17 Comparison of Model 1 and Model 2

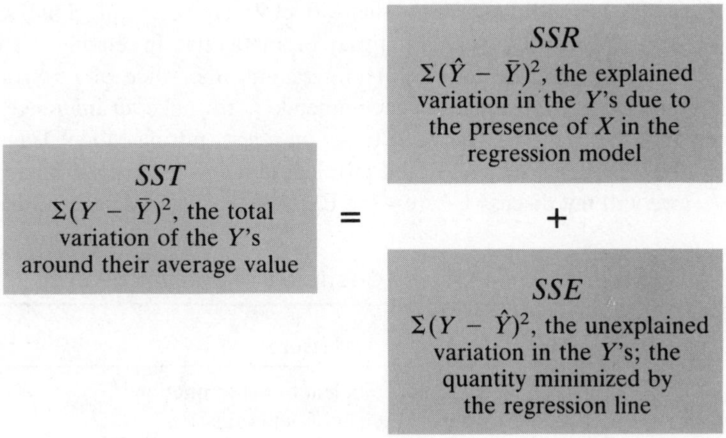

Figure 15.18 Interpreting the Partitioning of a Regression's *SST*

smaller. The better a sloped line fits a data set, the larger *SSR* will become relative to *SSE*. Translated into sums of squares, we have this relationship

$$0 \leq SSR \leq SST$$

or, in relative terms,

$$0 \leq \frac{SSR}{SST} \leq 1$$

The ratio *SSR*/*SST* is called the *sample coefficient of determination* and is symbolized as r^2. Equation 15–19 shows two ways of computing r^2.

Definition

The **sample coefficient of determination,** denoted by r^2, is the percentage of the total variability in the Y-values that is attributable to the influence of X in the regression model.

Sample Coefficient of Determination

$$r^2 = \frac{SSR}{SST} = 1 - \frac{SSE}{SST} \quad (15\text{–}19)$$

In Table 15.3 we present the symbols that are commonly used to distinguish the sample and population coefficients of determination. Clearly, r^2 is a number between 0 and 1. A large value of r^2 means the least-squares line accounts for a large proportion of Y's variability and hints that Model 2 may be useful. Small values of r^2 imply the opposite, that the regression may be ineffective in relating X and Y. Merely judging the regression model's utility by gauging the value of r^2 on a zero-to-one scale is highly subjective and not recommended. To make an inference about the regression model based on r^2 we need to set up a test, pitting Model 1 against Model 2. It can be shown that such a test is equivalent to the F-test given earlier. To avoid redundancy, we will not discuss it here—see Exercise 15.37 for details of the procedure.

Table 15.3 Symbols for the Coefficient of Determination

Measure	Symbol
Population coefficient of determination	ρ^2
Sample coefficient of determination	r^2

EXAMPLE 15.8

Compute the coefficient of determination from the ANOVA table in Example 15.7 and interpret its value.

Solution:

From the ANOVA table, we find $SSR = 1.031954$ and $SST = 1.28$. The coefficient of determination is

$$r^2 = \frac{SSR}{SST} = \frac{1.031954}{1.28} = .8062141 \quad \text{or } .8062 \text{ (rounded)}$$

Of the total variability in the observed Y-values (1.28), approximately 80.62 percent is explained by the regression of Y on X; the remaining 19.38 percent is unexplainable. □

COMMENTS

1. Be careful in interpreting r^2. Its value alone is not indicative of a model's utility. In Exercise 15.39, we provide examples where $r^2 = .877$, yet Model 2 is not declared useful at a 1% level of significance, and where the regression model is significant at a 1% level with $r^2 = .213$. As you might expect, the conclusion depends on the sample size as well as on the magnitude of r^2.
2. Model 1 ($\bar{Y}$) is also known as the *reduced model*. Model 2 ($\hat{Y} = b_0 + b_1X$) is called a *complete model*. These terms will be used extensively in Chapter 17.
3. Where possible, use unrounded values of SST and SSR or SSE in computing the coefficient of determination via Equation 15–19. We recommend rounding r^2 to four decimal places.

In this section we have addressed the question: Are X and Y related? By answering "yes" we imply that the regression model is an effective means of summarizing the linear relationship. A negative response, on the other hand, means the contribution of the second component in the regression model, β_1X, essentially is of no value in describing Y. We presented three methods of testing the model's utility: a t-test, a confidence interval for β_1, and an F-test. For the same level of risk, all three methods will produce the same conclusion based on the same set of data.

CLASSROOM EXAMPLE 15.2

Establishing Whether X and Y Are Related

From the data set of size five in Classroom Example 15.1, here are the summary statistics:

$$SSX = 10 \qquad SSE = 1.1$$
$$SSY = 6 \qquad \hat{\sigma} = .6055301$$
$$SSXY = -7 \qquad b_1 = -.7$$

a. Find the estimated standard error for the sampling distribution of b_1:

$$S_{b_1} = \frac{\hat{\sigma}}{\sqrt{SSX}} = \_\_\_\_\_\_$$

b. Conduct a t-test for the utility of the model with $\alpha = .10$.

c. Construct a 90% confidence interval for the population slope: $b_1 \pm (t\text{-score})S_{b_1} = \_\_\_\_\_\_.$

d. Set up an ANOVA table for this regression analysis.

e. State the hypotheses and conduct an F-test of the model's utility using $\alpha = .10$.

$$H_0: \_\_\_\_\_\_ \qquad H_a: \_\_\_\_\_\_$$

f. Find the sample coefficient of determination and interpret its value: $r^2 = \_\_\_\_\_\_.$

SECTION 15.4 EXERCISES

15.20 Consider the following bivariate data set of $n = 8$ observations:

X	6	3	7	3	9	5	7	1
Y	8	1	9	−1	12	9	6	0

a. Estimate the value of the slope parameter β_1 in the simple regression model.
b. Find the estimated standard error of b_1. Assume $\hat{\sigma} = 2.35979$.
c. Conduct a one-tailed test that would enable us to conclude the existence of a direct relationship between X and Y with $\alpha = .025$.
d. Since the last data point includes the value of 0, is it possible to give a practical interpretation to b_0? Explain.

15.21 A random sample of six observations yielded the following summary statistics:

$$SSX = 4.099 \qquad SSY = 2.098 \qquad SSXY = .6$$

Construct a 90% confidence interval for β_1. Is the regression model useful? Why?

15.22 A simple regression analysis based on a random sample of size $n = 30$ generated the following least-squares line and summary statistics:

$$\hat{Y} = 99.776 + 51.918X$$
$$\hat{\sigma} = 12.1965$$
$$SSX = 4.754$$

Test the utility of the model at the 10% level of significance.

15.23 Complete the following ANOVA table, constructed from a simple regression analysis, and conduct an F-test of the model's utility. Use $\alpha = .05$. Find and interpret r^2.

Source	*SS*	*df*	*MS*	*F*
Regression	9.1			
Error		9		
Total	12.3			

15.24 Explain the difference between β_1 and b_1.

15.25 A simple regression analysis based on fifteen data points produced the following summary statistics:

$$b_1 = -.78 \qquad S_{b_1} = .29$$

Test the utility of the regression model with an alternative hypothesis that implies an inverse relationship between X and Y. Use $\alpha = .01$.

15.26 The Environmental Protection Agency (EPA) estimates the miles per gallon for all models of new cars sold in the United States. In addition to a city and highway rating, the EPA reports a combined rating. Suppose the following data represent the combined rating and the number of cylinders for a sample of new cars:

Number of Cylinders	mpg Rating	Number of Cylinders	mpg Rating
3	56	4	54
3	46	4	30
3	39	4	34
3	39	4	30
4	27	4	24
4	27	4	29
4	39	4	30
4	32	4	33
4	29	4	29
4	29	4	34
4	28	4	29
4	26	6	24

a. Find the equation of the least-squares line relating Y, mpg rating, to X, number of cylinders.
b. Test the fitted regression model's utility with $\alpha = .10$.
c. Interpret the values obtained for b_0 and b_1 in part a.

15.27 The relationship between the variables X, number of years of schooling beyond high school, and Y, annual salary (in thousands of dollars), is studied for a randomly selected group of 25 adults from the same high school graduating class of 1965. The sample data produced the following summary statistics:

$$SSX = 74 \qquad SSY = 1168 \qquad SSXY = 187$$

a. Determine b_1 and the standard error of b_1.
b. Construct a 98% confidence interval for β_1.
c. Each additional year of schooling beyond high school is worth (roughly) how much in average annual salary?
d. Compute and interpret r^2.

15.28 Examine Equation 15–14 closely. For a fixed value of σ, what happens to the value of σ_{b_1} in the following cases?
a. We increase the spread in the values of X.
b. We decrease the spread in the values of X.
c. All the values of X are the same number.

15.29 A random sample of 10 weeks generated bivariate data (in dollars) on Y = average price of a gallon of unleaded gas for a one-week period and X = average price of a barrel of crude oil during the same period. Summary statistics are

$$b_0 = .36 \qquad SSX = 183.0609$$
$$b_1 = .027 \qquad SSY = .1681$$
$$SSE = .034639 \qquad SSXY = 4.943$$

a. What is the equation of the least-squares line?
b. Construct an ANOVA table for this regression analysis.
c. Test the model's utility at the 1% level of significance.
d. Suppose the price of a barrel of oil jumped \$5 in one week. Construct an interval estimate which has 90% confidence for the expected change in the price of a gallon of unleaded gas.

15.30 Find the standard error of b_1 in each of the following situations:
a. $SSE = 886.54$; $n = 10$; $SSX = 10530.1$
b. $n = 15$; $\overline{X} = 92.333$; $\overline{Y} = 5.573$; $\Sigma X^2 = 179{,}661$; $\Sigma Y^2 = 681.32$; $\Sigma XY = 10{,}917.6$
c. $n = 12$; $\hat{\sigma} = 6.021$; $\Sigma X = 438$; $\Sigma X^2 = 16414$
d. $b_1 = .0109$; $\hat{\sigma}^2 = 3.886$; $SSXY = 236.5$

15.31 Refer to the two-tailed t-test of the hypothesis H_0: $\beta_1 = 0$ in Example 15.5, part b, where we computed $t^* = 4.9954587$ (unrounded). Verify that the square of t^* agrees (to one decimal place) with the value of the F-test statistic computed in Example 15.7.

15.32 Indicate whether the linear regression model would be judged useful at the $\alpha = .05$ level of significance in each of the following situations:
a. A 95% confidence interval for β_1 is -1.12 to 5.07.
b. A 90% confidence interval for β_1 is -8.2 to .3.
c. A 99% confidence interval for β_1 is $-.5$ to 9.5, based on $n = 24$.
d. A p-value of .063 for the two-tailed t-test of H_0: $\beta_1 = 0$.
e. A test statistic of $F^* = 6.14$ for the F-test of H_0: $\beta_1 = 0$, based on $n = 16$.

15.33 The American Hotel and Motel Association monitors the \$46 billion-a-year (revenues) hotel industry and annually reports on the average occupancy rate and average room rate. For the six-year period 1981–86 these figures were as follows:

Year	Occupancy Rate	Average Room Rate
1981	70.1%	\$39.14
1982	66.0	43.55
1983	66.1	46.67
1984	67.9	52.13
1985	65.5	53.81
1986	65.0	54.90

Source: American Hotel & Motel Association as printed in *USA Today*, February 23, 1987. Excerpted with permission.

a. Fit a linear model to these data, treating the occupancy rate as the independent variable.
b. Construct an 80% confidence interval for β_1.
c. Is there sufficient evidence at the 20% level to conclude that demand for hotel rooms—that is, occupancy rate—and price are inversely related?
d. From your knowledge of economics, does the relationship between price and demand in this problem conform to the Law of Demand? Can you state a third factor that might also affect the occupancy rate?
e. What percentage of the variability in room rates is explainable by the occupancy rate?

15.34 A random sample of several banks and thrifts produced the following data on X, the interest rate charged on a 30-year fixed, conventional mortgage, and Y, the discount points on the loan (expressed as a percentage of the amount borrowed):

Bank/Thrift	Interest Rate	Discount Points
First Home Federal	9%	4 ¾
	9 ½	3
	10	½
First Federal of the Carolinas	9 ⅞	2
	10	1
	10 ⅛	0
Southeastern	9	5 ½
	9 ½	3 ½
	10	1
American Federal Savings and Loan	10	2
Wachovia	9 ½	2 ¾
	10	0
Old Stone	10	1
Triad	9 ½	2 ¾

Source: Greensboro News & Record, May 12, 1986, p. 23.

a. Plot these 14 data points and find the equation of the fitted simple regression model. Plot the line.
b. Construct an ANOVA table for the regression analysis of these data.
c. Test the utility of the model at $\alpha = .01$.
d. For the discount points to drop 1%, on average, what must happen to the interest rate?
e. Determine and interpret the coefficient of determination.

15.35 Find r^2 in each of the following situations:
a. $SSE = .0346$; $SST = .1681$
b. $SSR = 471.7$; $SST = 1168$
c. $SSY = 2.098$; $SSXY = .6$; $b_1 = .1464$

15.36 If the regression model (model 2) always reduces the sum of squared fitting errors, why shouldn't we *always* use it instead of model 1? Identify some drawbacks to this pragmatic line of thinking.

***15.37** To test model 1 versus model 2 on the basis of r^2, we use the following procedure:

H_0: $\rho^2 = 0$ (model 1)

H_a: $\rho^2 > 0$ (model 2)

$$TS: F^* = \frac{r^2}{\dfrac{(1 - r^2)}{(n - 2)}}$$

RR: For $\alpha = .05$ and the > alternative, see Figure 15.19.

C: Reject H_0 or accept H_0.

*Optional

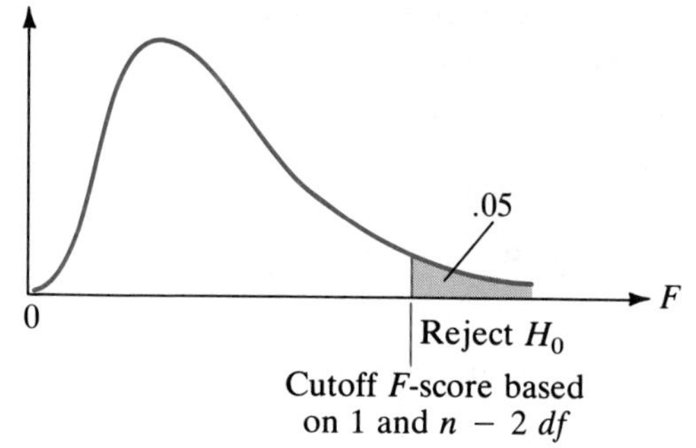

Figure 15.19 Rejection Region for F-Test of Model Utility, $\alpha = .05$

Refer to Example 15.8. Verify that the above testing procedure based on $r^2 = .8062$ produces the same test statistic and conclusion as in Example 15.7.

*15.38 Refer to Exercise 15.37. Determine whether model 1 ($\bar{Y}$) or model 2 (the simple regression model) is preferred if an analysis of $n = 17$ data points yielded $r^2 = .63$. Use $\alpha = .05$.

*15.39 Refer to Exercise 15.37. Determine the utility of the model ($\alpha = .01$) in each of the following situations:

a. $n = 12$; $F^* = 2.77$; $r^2 = .217$
b. $n = 32$; $F^* = 8.12$; $r^2 = .213$
c. $n = 5$; $F^* = 21.39$; $r^2 = .877$

15.40 The sample coefficient of determination is defined in Equation 15–19 as $r^2 = 1 - (SSE/SST)$.

a. What condition would cause r^2 to be equal to 1.0? What is the graphical representation of this condition?

b. What condition would cause r^2 to be equal to 0? What is the graphical representation of this condition?

15.5 USING THE MODEL FOR PREDICTION

Deciding that the regression model is a useful relation between X and Y justifies using X to predict Y. If, on the other hand, the inferential procedures of Section 15.4 had led us to believe that $\beta_1 = 0$, then using X to explain or predict Y is unwarranted. Hence, the fourth objective in our regression analysis—testing the utility of the model—must be accomplished before tackling the final objective of prediction.

There are two forecasting situations in which the fitted regression equation is applicable. To illustrate, consider the linear relationship between X, the percentage change in the DJIA, and Y, the percentage change in the share price of a mutual fund from Examples 15.1 and 15.3. The fitted regression model, $\hat{Y} = .061 + 1.540X$, was deemed useful in Example 15.5, part b. Suppose we knew that the Dow had lost ground today, closing down .2 percent. Although the change in the share price of our mutual fund will be published in tomorrow's newspapers, can we predict it today based on our knowledge of X? The answer is "yes," and this situation is referred to in a regression analysis as *predicting a specific value of Y given a value of X*.

Another forecasting possibility is the long run effect on Y. If we tracked our mutual fund for an extended time period, and we examined all trading days in which the DJIA declined .2 percent, then we might be interested in the average effect on Y rather than a specific one-day fluctuation. This is called *estimating an average value of Y given a value of X*.

In both cases, we rely on a known value for X. If X did not influence Y, then this information would be unnecessary and we would simply use the value of $\overline{Y}$ as our prediction for Y. Since Y is influenced by X, we use the model $\hat{Y} = b_0 + b_1X$ as the basis for our predictions.

Predicting a Specific Value of *Y*

Predicting a specific value of Y is achieved with a single number, called a *point prediction,* or with a range of numbers, called a *prediction interval*. A point prediction for Y is found by substituting the given value of X and solving the regression equation for $\hat{Y}$. Unfortunately, this point prediction has a 100% risk of being wrong. To reduce the risk, we construct a prediction interval that is very similar in form to a confidence interval. Equation 15–20 is a general expression for a prediction interval for a specific value of Y.

Prediction Interval for a Specific Value of *Y* Given *X* = *x*

$$\hat{Y} \pm (t\text{-score}) \cdot \hat{\sigma} \cdot \sqrt{1 + \frac{1}{n} + \frac{(x - \overline{X})^2}{SSX}} \qquad (15\text{–}20)$$

The symbol $\hat{Y}$ is the point prediction; the t-score depends on the amount of risk and on $df = n - 2$; and $(x - \overline{X})^2$ is the squared deviation between the given value of X and the sample mean $\overline{X}$.

COMMENT If possible, use unrounded values of $\hat{\sigma}$, $\overline{X}$, and SSX in computing the prediction interval via Equation 15–20. We recommend rounding off the $\pm$ part of the interval to the same number of decimal places as $\hat{Y}$.

EXAMPLE 15.9

Predict the change in the share price of the mutual fund for the data in Example 15.3 when the Dow Jones Industrial Average declines by .2 percent. Use a 10% level of risk.

Solution:

From Example 15.3, we determined the equation of the least-squares line to be $\hat{Y} = .061 + 1.540X$. The point prediction of Y when $X = -.2$ percent is

$$\hat{Y} = .061 + 1.540(-.2) = -.247$$

The t-score for a 90% prediction interval based on $df = 6$ is $t = 1.943$. With $\hat{\sigma} = .2033249$, $SSX = .435$, $n = 8$, and $\overline{X} = .025$, we compute

$$\begin{aligned}\hat{Y} &\pm (t\text{-score}) \cdot \hat{\sigma} \cdot \sqrt{1 + \frac{1}{n} + \frac{(x - \overline{X})^2}{SSX}} \\ &= -.247 \pm (1.943)(.2033249)\sqrt{1 + \frac{1}{8} + \frac{(-.2 - .025)^2}{.435}} \\ &= -.247 \pm .440 \quad \text{or} \quad -.687 \text{ to } .193\end{aligned}$$

We project a .25% drop, roughly, in the share price when the Dow loses .2 percent in value. Rarely will this hold true exactly. More probably—with 90% confidence—the change in the share price may decline as much as .69 percent or gain as much as .19 percent.

□

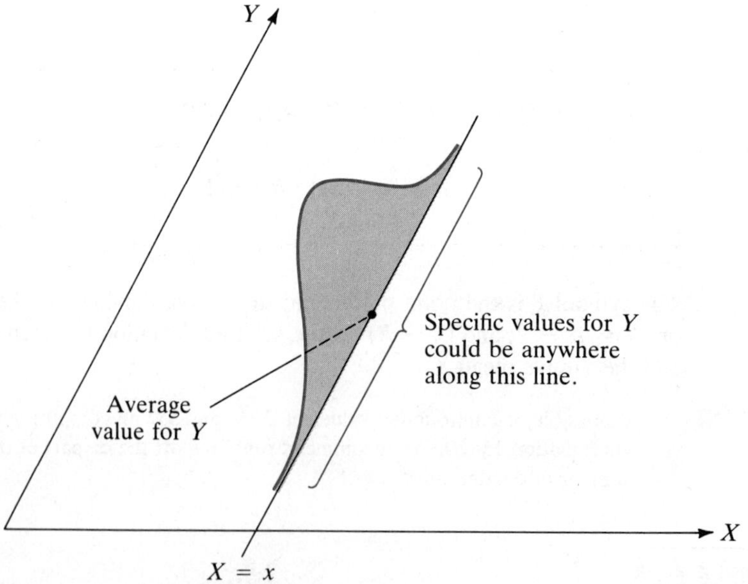

Figure 15.20 Difference Between a Specific Value for Y and an Average Value for Y

The prediction interval in Example 15.9 is rather wide due in part to the difficulty predicting a specific value of a random variable. A **prediction interval** is an interval of potential values of a *variable*. A **confidence interval** is an interval of potential values of a *parameter*. A parameter is a fixed quantity; obviously, a variable is not. We draw the analogy here to a hunter aiming at a moving target (the variable) as opposed to a stationary target (the parameter). Intuitively, we should expect more room for error in the former case.

Figure 15.20 shows the difference in these two forecasting situations. The normal distribution of the Y-values for a given value of X is depicted in the third dimension above the line $X = x$. A particular value of Y could turn out to be *any* point along the line, whereas the average value of Y is the one in the middle of the distribution.

Estimating an Average Value of *Y*

Developing the *point estimate* of the average value of Y follows the same procedure as that for the point prediction for a specific value of Y: Evaluate the regression equation $\hat{Y} = b_0 + b_1X$, using the known value of X. As is true in most estimation problems, the point estimate has a confidence factor of 0 percent. We prefer an interval estimate, or *confidence interval*, summarized in Equation 15–21.

Confidence Interval for an Average Value of *Y* Given $X = x$

$$\hat{Y} \pm (t\text{-score}) \cdot \hat{\sigma} \cdot \sqrt{\frac{1}{n} + \frac{(x - \overline{X})^2}{SSX}} \qquad (15\text{–}21)$$

Compared to Equation 15–20, Equation 15–21 is almost identical, except for the "1" underneath the square root. Its omission in Equation 15–21 results in a narrower interval and is our symbolic "reward" for estimating a parameter rather than predicting a value for a random variable.

COMMENT If possible, use unrounded values of $\hat{\sigma}$, $\overline{X}$, and SSX in computing the confidence interval via Equation 15–21. We recommend rounding off the $\pm$ part of the interval to the same number of decimal places as $\hat{Y}$.

EXAMPLE 15.10

Estimate the average change in the share price for the data in Example 15.3 when the Dow declines by .2 percent. Use a 10% level of risk.

Solution:

A 90% confidence interval for the average value of Y when $X = -.2$ percent is

$$\hat{Y} \pm (t\text{-score}) \cdot \hat{\sigma} \cdot \sqrt{\frac{1}{n} + \frac{(x - \overline{X})^2}{SSX}}$$

$$= -.247 \pm (1.943)(.2033249) \sqrt{\frac{1}{8} + \frac{(-.2 - .025)^2}{.435}}$$

$$= -.247 \pm .194 \quad \text{or} \quad -.441 \text{ to } -.053$$

When the DJIA loses .2 percent in value, we project an average decline of .25 percent (roughly) in the share price. Though the average change is not likely to be exactly −.25 percent, we are 90 percent confident that the share price, on average, could drop as little as .05 percent or as much as .44 percent.

□

Discussion

Both forecasting situations—predicting a specific value and estimating an average response—are common in business, though the former is perhaps a more popular application. However, a general discussion of the finer points of using a regression model for prediction is appropriate before we leave this subject.

Point estimates and point predictions are unstable; yet, there is a tendency to expect them to coincide with the eventual, realized values. Instead, we should rely on interval estimates. For example, the Federal Reserve Board regularly forecasts a range of acceptable values for $M1$, a measure of the supply of money in our economy, rather than a single, target value. No drastic changes in policy occur whenever $M1$ falls in the forecasted interval. Following the Fed's example, we should get in the habit of developing interval forecasts, not point forecasts.

The width of interval estimates reflects the precision of the forecast. Frequently when a regression analysis is performed on a small set of data, like the data set in Example 15.2, the prediction and confidence intervals are wider than desired, almost independent of the strength of the relationship between X and Y. For this reason, forecasts tend to be more precise when based on larger sample sizes.

The precision of a forecast also wanes when the given value of X is relatively far from $\overline{X}$. In fact, for a constant level of confidence, the width of interval estimates increases the further $X = x$ is from $\overline{X}$. Exercise 15.43 demonstrates this phenomenon. An extreme situation—selecting X beyond the observed range of sample values—exacerbates the problem and is not recommended.

To use the fitted regression model effectively for prediction, we must recognize the factors that affect the width of the interval—the size of the sample, the distance $X = x$ is from the mean, and the strength of the relationship between X and Y—and understand their effects. The last factor, the strength of the linear relationship, is the subject of the next section.

CLASSROOM EXAMPLE 15.3

Prediction and Confidence Intervals

Consider the following summary statistics from Classroom Example 15.1:

$$n = 5 \qquad SSX = 10 \qquad b_0 = 3.1$$
$$\bar{X} = 3 \qquad \hat{\sigma} = .6055301 \qquad b_1 = -.7$$

a. Develop a 95% prediction interval for the number of sick days claimed by an employee with two weeks of annual paid vacation.

$$\hat{Y} = b_0 + b_1X = \_\_\_\_\_\_$$

$$\hat{Y} \pm (t\text{-score}) \cdot \hat{\sigma} \cdot \sqrt{1 + \frac{1}{n} + \frac{(x - \bar{X})^2}{SSX}} = \_\_\_\_\_\_$$

b. Develop a 95% confidence interval for the average number of sick days claimed by all employees with two weeks of annual paid vacation.

$$\hat{Y} \pm (t\text{-score}) \cdot \hat{\sigma} \cdot \sqrt{\frac{1}{n} + \frac{(x - \bar{X})^2}{SSX}} = \_\_\_\_\_\_$$

SECTION 15.5 EXERCISES

15.41 A simple regression analysis produced the following summary statistics:

$$n = 11 \qquad SSE = 14{,}741.6$$
$$\bar{X} = 136.364 \qquad SSX = 21{,}704.545$$
$$\hat{Y} = 50.723 + .4867X$$

a. Construct a 95% prediction interval for Y when $X = 100$. Interpret the interval.
b. Construct a 95% confidence interval for the average value of Y when $X = 100$. Interpret the interval.
c. Compare the widths of the two intervals. Roughly, what percentage smaller is the width of the confidence interval compared to the width of the prediction interval?

15.42 A random sample of bivariate data points yielded the following:

X	1	4	5	10	15
Y	55	52	48	32	25

a. Determine the equation of the least-squares line.
b. Test the utility of the model with $\alpha = .10$.
c. What is the point prediction for Y when $X = 1$?

d. Develop an 80% prediction interval for Y when $X = 1$.
e. Develop an 80% confidence interval for the average value of Y when $X = 15$.

15.43 Refer to the data in Example 15.2 that yielded a least-squares line of $\hat{Y} = 6 - 1.4X$ and these summary statistics:

$$n = 4 \qquad \hat{\sigma} = .7745967$$
$$\bar{X} = 2.5 \qquad SSX = 5$$

a. Develop a 95% confidence interval for the average value of Y when $X = 1, 2, 2.5, 3,$ and 4.
b. Imagine that we plotted the lower limit and upper limit of each confidence interval. Describe the shapes formed by connecting the lower limits together and then connecting the upper limits together.
c. Is there a constant width between the lower limits and upper limits? If not, where is the thinnest width? Thickest width?

15.44 Is it ever possible for the width of the prediction interval and the width of the confidence interval to be equal (for the same value of $X = x$)? Explain.

15.45 Summary statistics for a simple regression analysis based on a random sample of size $n = 48$ are as follows:

$$\bar{X} = 395.4167 \qquad SSX = 6{,}156{,}193$$
$$\hat{Y} = 9.793 + .0503X \qquad \hat{\sigma} = 4.16994$$

a. Construct a 99% prediction interval for Y when $X = 250$.
b. Construct a 98% confidence interval for the average value of Y if $X = 400$.

15.46 What factors affect the width of confidence intervals and prediction intervals?

15.47 The cost of television commercials intuitively has a direct relationship to the anticipated "rating" of the program. Selling TV advertising for new shows, sporting events, specials, and so forth is difficult since advertising executives must project a rating for the program before it airs in order to entice advertising. A sample of 10 network television programs revealed the following data on X, cost of a 30-second commercial (in hundreds of thousands of dollars), and Y, rating (1 ratings point represents approximately 874,000 viewers):

X	1	1	1.5	2	2	2.5	2.5	2.5	3	5
Y	5.1	7.0	7.2	6.8	7.2	7.9	7.5	8.3	9.0	11.6

a. Identify and carry out the first four objectives of a regression analysis for these data. Use $\alpha = .05$.
b. Suppose a 30-second commercial for a new show is sold for $250,000. What ratings can the buyer anticipate? Generate a 98% prediction interval. Interpret the interval.
c. Explain the significance of a positive value for $(Y - \hat{Y})$ to the commercial buyer. A negative value?
d. With 95% confidence, within what limits is the average rating for commercials costing $400,000?

15.48 Relatively speaking, describe (wide, narrow, or indeterminate) the width of a confidence or prediction interval in the following situations:
a. Sample size is small.

b. Sample size is large and X and Y are strongly related.
c. The given value of X is three standard deviations from $\overline{X}$.

15.49 An airline plans to initiate service at an airport in a city of approximately one-half million people. To determine the staffing requirements, officials for the airline collected internal data on similar operations in other cities it serves as well as external data from competitors already servicing the airport. The data are as follows:

Flights Departing Weekly	Total Employees
104	88
89	71
94	85
63	40
49	43
48	8
42	23
50	10
21	6

a. Fit a simple regression model to these data. Plot the data and the least-squares line.
b. Industrywide, it is a commonly held axiom that staffing needs increase as the number of weekly flights increases. Test this claim against the null hypothesis that the size of the staff and the number of departing flights are unrelated. Use $\alpha = .01$.
c. The airline plans to offer six flights per weekday, five on Saturday, and eight on Sunday. Develop lower and upper limits on the number of employees needed. Use $\alpha = .05$. Is a prediction interval or a confidence interval more appropriate for this problem? Why? Interpret the interval.
d. What is the point prediction of staffing needs for an airline planning only 22 flights per week, according to the model of part a? Explain the apparent contradiction.

15.50 Most insurance agents are paid on commissions: If they write no policies, they earn no money. An insurance company has developed and tested the following model relating X, the number of whole life insurance policies written per month, and Y, amount of commission (in thousands of dollars): $\hat{Y} = .857X$. With 98% confidence, predict the commissions earned by an agent who writes five policies in one month. Assume $\hat{\sigma} = .535$, $SSX = 17.5$, $n = 16$, and $\overline{X} = 3.5$.

*15.51 When the model utility test dictates accepting H_0: $\beta_1 = 0$, predictions are no longer based on $\hat{Y} = b_0 + b_1X$. Instead, we use the value of $\overline{Y}$ as the point prediction and the point estimate. In this situation, a prediction interval for a particular value of Y is

$$\overline{Y} \pm (t\text{-score}) \cdot S_Y \cdot \sqrt{1 + \frac{1}{n}}$$

where S_Y is the standard deviation of the Y-values. That is,

$$S_Y = \sqrt{\frac{SSY}{n - 1}}$$

*Optional

Thus, the t-score is based on $n - 1$ degrees of freedom, since we are no longer estimating the slope β_1. Similarly, a confidence interval for the average value of Y is

$$\overline{Y} \pm (t\text{-score}) \cdot S_Y \cdot \sqrt{\frac{1}{n}}$$

In both cases, any information about X is ignored since we decided $\beta_1 = 0$.

A random sample produced the following bivariate data and summary statistics:

X	Y
4	3
7	14
9	21
12	10
14	4

$\hat{Y} = 10.898 - .054X$

$SSY = 221.2$

$S_{b_1} = 1.08311$

a. Test the model for utility with a 20% level of risk.

b. Relative to your answer in part a, predict Y when $X = 10$ with a 20% level of risk.

*15.52 Refer to Exercise 15.51. A regression analysis based on a random sample of size 48 produced the following:

Model: $\hat{Y} = 9.58 - .177X$
Relevant range: $.9 \leq X \leq 15.1$
Estimated standard error of the regression: $\hat{\sigma} = 2.49371$
Sums of squares: $SSX = 525.6666$
$SSY = 302.5231$
$SSXY = -93.043$
Means: $\overline{X} = 6.88, \overline{Y} = 8.25$
F-statistic for H_0: $\beta_1 = 0$ $F^* = 2.64$

Estimate the average value of Y when $X = 4.5$ with a 95% confidence interval. Assume a 5% risk level for all inferences.

15.53 Stocks that consistently pay dividends are popular among investors seeking income. However, the fact that a stock's earnings per share is high is not necessarily indicative of a large dividend. A random sample of 27 stocks that have consistently paid dividends yielded the following current data on earnings per share (EPS) and amount of dividend:

Stock	EPS	Dividend	Stock	EPS	Dividend
1	$1.85	$ .32	10	$1.74	$1.68
2	5.55	1.52	11	4.26	2.96
3	3.69	1.00	12	3.58	1.40
4	4.97	1.68	13	7.42	3.60
5	3.83	2.40	14	4.12	1.68
6	4.00	1.84	15	2.68	1.00
7	6.63	2.16	16	2.58	.84
8	6.34	2.80	17	4.27	2.00
9	3.38	2.16	18	2.93	1.12

(continues)

Stock	EPS	Dividend	Stock	EPS	Dividend
19	$8.80	$3.28	24	$1.86	$ .76
20	2.87	1.40	25	1.97	2.04
21	4.44	1.52	26	2.92	1.04
22	3.45	1.76	27	2.29	1.80
23	5.07	2.04			

a. Fit and test ($\alpha = .10$) a simple regression model.
b. Estimate the average dividend for stocks earning $2 per share with an 80% confidence interval.
***c.** Examine the 25th data point closely: The dividend exceeds the EPS. Is this a spurious data point? (A first course in accounting may be needed to answer this.)

15.54 To model the relationship between X, a bank's total assets, and Y, its return on average equity, a random sample of 22 banks yielded the following data:

Total Assets ($ Billions)	Return on Average Equity (%)	Total Assets ($ Billions)	Return on Average Equity (%)
$5.13	11.5	.60	15.9
4.74	12.0	.38	11.4
2.60	9.5	.34	14.1
.68	12.8	.09	6.5
.14	10.0	.06	14.6
.07	8.3	.06	24.0
.04	−4.6	.29	0.1
.01	2.3	.10	8.4
1.46	10.2	.02	11.3
.36	0.5	.04	17.6
.46	15.0	.04	11.2

a. Find the simple regression model that minimizes *SSE*.
b. Construct a 90% confidence interval for β_1. Determine the model's utility from this interval.
***c.** Estimate (98% confidence) the return on average equity for banks with a half-billion dollars in total assets. (*Hint:* Refer to Exercise 15.51.)

15.55 Employee absenteeism can be a serious problem for a company. One of the ways to measure absenteeism is to determine the total time lost (in hours or days) per employee. A study involving the personnel in a chain of retail drugstores provided the data in Table 15.4 on this variable. The purpose of the study was to compare absenteeism with the size of the store's staff.
a. Fit a linear model relating Y = total time lost to X = number of employees.
b. Is Y significantly ($\alpha = .10$) related to X?
***c.** Refer to Exercise 15.51. Estimate the average value of Y for stores with a dozen employees with a 98% confidence interval.

Table 15.4 Employee Absenteeism Data for Exercise 15.55

Number of Employees	Total Time Lost per Employee (Hours)	Number of Employees	Total Time Lost per Employee (Hours)
18	49.5	22	29.1
17	19.4	18	36.7
19	22.8	12	22.8
15	56	16	28.2
15	27	13	62.9
17	4.4	13	38
18	33	14	50.6
18	34.5	19	34.2
12	47.7	15	30.2
19	17.4	13	41.4
14	61.1	14	46.7
12	32.5	16	38.6
22	44.6	11	36.9
19	36.2	11	59.7
14	58.5	9	11.7
11	38.6	12	63
14	34.2	21	39.5
18	28.8	13	37.7
16	61.6	13	32.5
22	50.1	11	19.1
15	33.2	23	36.9
18	48.9	15	29.3
14	43.6	15	42.3
15	48.0	17	38.3

Source: F. Kuzmits, "The Relationship Between Absenteeism and Productivity: An Empirical Assessment," *Proceedings of the Southern Management Association,* Atlanta, Georgia, 1983. Reprinted with permission of the author.

15.6 SIMPLE CORRELATION ANALYSIS

Goal of Correlation

The terms *regression* and *correlation* are linked together so often that we sometimes fail to differentiate them. They share similar computational quantities, but the assumptions and objectives of the two are different.

Definition

The ***goal of simple correlation*** is to analyze the strength of the relationship between X and Y.

A first difference between regression and correlation is seen when we compare definitions. Regression presumes a straight-line relationship between X and Y and sets out to analyze the relation. Correlation is not directly concerned with lines relating X and Y; its mission is simply to characterize the degree to which X and Y behave in a direct or an inverse fashion. In other words, correlation looks at the strength and direction of the relationship, while regression seeks the equation of the best fitting line to describe the relationship.

COMMENT Simple correlation refers to the correlation between two variables. Synonyms are linear correlation, simple linear correlation, or (just) correlation.

Simple Correlation Model

Earlier we stated the simple regression model as $Y = \beta_0 + \beta_1 X + \epsilon$ and identified each term. This equation explicitly relates the variables X and Y in a linear fashion and implicitly defines a hierarchy between X and Y, as evidenced by the labels for the variables: independent and dependent.

The simple correlation model is not an equation per se, but a characterization of the mutual variation between two variables. Mutual is an important word because it reminds us that the observations occur in pairs of X and Y. Since neither variable is labeled the dependent (nor the independent) variable, the variables are said to have a symmetrical role.

As a result, a second major difference between regression and correlation is the status of the variables. In regression, Y is a random variable and X is a fixed variable. In correlation, both X and Y are viewed as random variables.

To describe the mutual variation of X and Y, we use a probability distribution that involves both variables simultaneously and that is called a *bivariate distribution.* Bivariate distributions are necessarily three-dimensional figures as opposed to the two-dimensional shapes we studied in Chapters 6 and 7 for univariate probability distributions like the normal or exponential.

The simple correlation model is merely a specification of an appropriate bivariate distribution to describe the joint behavior of X and Y.

Simple Correlation Model

Both X and Y vary together according to a bivariate probability distribution.

Symbol	Name	Classification
X	Variable	Random variable
Y	Variable	Random variable
ρ	Coefficient of correlation	Parameter

Table 15.5 Symbols for the Coefficient of Correlation

Measure	Symbol
Population coefficient of correlation	ρ
Sample coefficient of correlation	r

Note that a new parameter—the *coefficient of correlation* ρ—is introduced in the correlation model. The *coefficient of correlation* is a measure of the strength of the association between the random variables X and Y. Table 15.5 displays the conventional symbols used to represent the sample and population correlation coefficients. Unlike the coefficient of determination, which varies between 0 and 1, the coefficient of correlation takes on values between -1 and $+1$, inclusive.

Bivariate Normal Distribution

A third major difference between regression and correlation is the assumptions necessary for each analysis. Regression assumes a (univariate) normal distribution for the error term ϵ. In turn, this assumption creates a normal distribution of Y-values for each value of X. No specific assumptions are necessary about the distribution of X.

The key assumption in simple correlation analysis is that the bivariate distribution relating X and Y is normal. In turn, this creates not only a normal distribution of Y-values for each X, but also a normal distribution of X-values for each Y. A bivariate normal distribution is analogous to the (univariate) normal distribution of Chapter 7, with one exception—we add a third dimension. Figure 15.21 is a picture of a bivariate normal distribution.

Assumption About the Bivariate Probability Distribution in the Simple Correlation Model

Variables X and Y have a bivariate normal probability distribution.

The theory is a bit messy here. Yet we must understand the basics of a bivariate normal distribution—and the effect that the correlation coefficient has on it—if we wish to understand correlation analysis. It may be helpful to think of a bivariate normal distribution as a tangible figure—such as a sombrero or a floppy beach hat—rather than a theoretical formula.

All bivariate normal distributions are naturally bell shaped and form a family of distributions (for example, an inventory of different sized sombreros). An individual bivariate normal distribution is characterized by five parameters, one of which is the

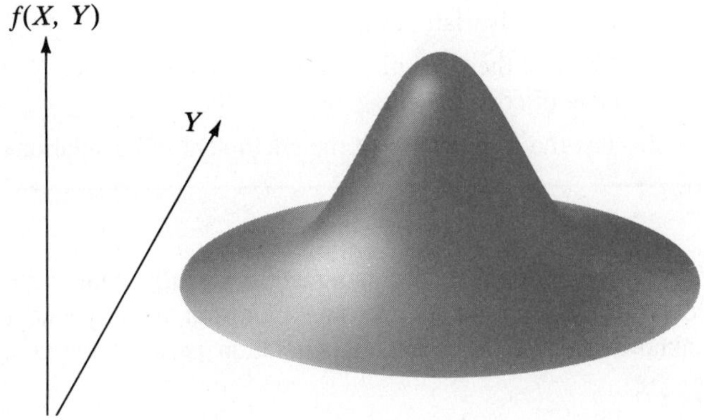

Figure 15.21 A Bivariate Normal Distribution

correlation coefficient ρ. The other four parameters are the mean and standard deviation for both X and Y. While the means of X and Y locate the position of the sombrero in the XY-plane and the standard deviations describe its length and breadth, the population correlation coefficient indicates its shape.

If $\rho = 0$, the sombrero is perfectly symmetrical from a front view and a side view and suggests that X and Y have no association. See Figure 15.22(a). If $\rho > 0$, the sombrero will appear elongated or "stretched" from the lower left hand portion of the XY-plane to the upper right, as indicated in Figure 15.22(b). This implies X and Y are directly related. If $\rho < 0$, X and Y are inversely related and the sombrero is elongated from the upper left corner to the lower right (see Figure 15.22(c)).

The greater the value of ρ, the tighter the sombrero is stretched in a diagonal direction. If $\rho = +1$ or if $\rho = -1$, the sombrero would be stretched to its limit and would appear as a univariate normal curve above a sloped line in the XY-plane.

Objectives of Correlation

As we indicated earlier, regression analysis and correlation analysis have different goals. For a given set of bivariate data as in Figure 15.23(a), regression analysis attempts to describe the population by estimating the parameters in the straight line $Y = \beta_0 + \beta_1 X$ (see Figure 15.23(b)). Correlation analysis attempts to find the appropriate sombrero to "cover" the observed data points (see Figure 15.23(c)) by estimating the value of ρ. In correlation analysis we break down our goal into three objectives, listed in the following box.

Objectives of Simple Correlation

1. Plot the bivariate data.
2. Measure the strength of the association between X and Y with the coefficient of correlation.
3. Test the significance of the coefficient of correlation.

Although it takes five parameters to fully characterize the correct sombrero covering the data points, we are interested in only one of them—ρ. Its value is unknown since it represents a population parameter and we only have access to a

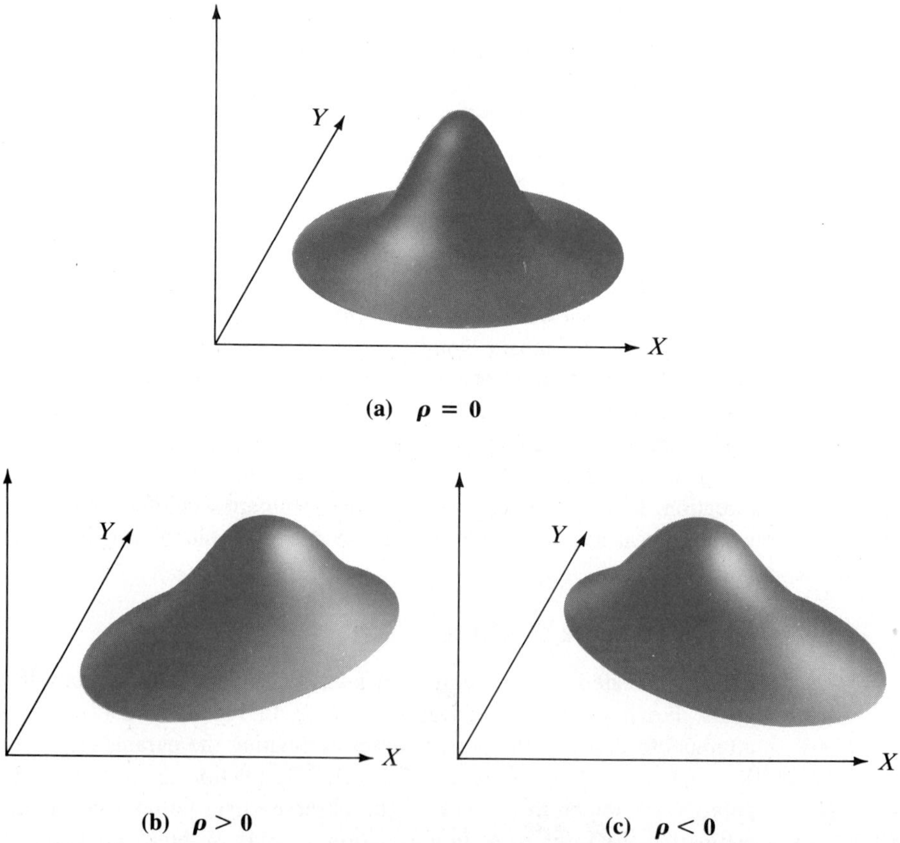

Figure 15.22 Effect of ρ on the Shape of a Bivariate Normal Distribution

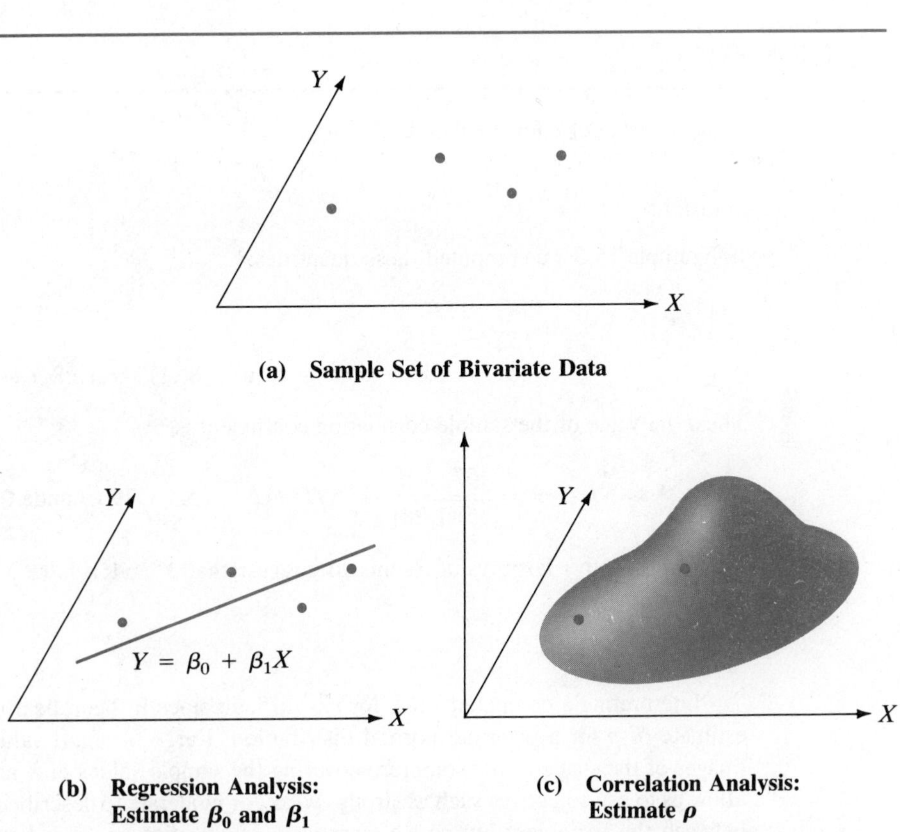

Figure 15.23 Comparison of Regression Analysis and Correlation Analysis

random sample from the population. Thus, we concentrate on estimating ρ with the sample correlation coefficient r. To compute r, we use Equation 15–22.

Sample Coefficient of Correlation Between X and Y

$$r = \frac{SSXY}{\sqrt{(SSX)(SSY)}} \tag{15–22}$$

The sums of squares needed to compute r are the same ones we used in the regression analysis.

COMMENT If possible, use unrounded values of SSX, SSY, and $SSXY$ in computing the correlation coefficient via Equation 15–22. We recommend rounding off and reporting r to four decimal places.

EXAMPLE 15.11

Find the value of r for the data in Example 15.3.

Solution:

In Example 15.3 we computed these quantities:

$$SSXY = .67$$
$$SSX = .435$$
$$SSY = \Sigma Y^2 - n\overline{Y}^2 = 1.36 - 8(.1)^2 = 1.28$$

Thus, the value of the sample correlation coefficient is

$$r = \frac{.67}{\sqrt{(.435)(1.28)}} = .8978942 \qquad \text{or } .8979 \text{ (rounded)}$$

This (apparently) large value seems to suggest that X and Y have a strong, direct relationship.

□

Interpreting a computed value for r is difficult since it theoretically represents an estimate of ρ for a bivariate normal distribution. Large or small values for r create images of the shape of the sombrero covering the sample values of X and Y as well as allow us to use adjectives such as strong, weak, or moderate to describe the association between the variables. But as we learned in interpreting r^2, it is better to attach a meaning to this statistic by testing its significance relative to the value of 0. Depending on the sample size, large values of r may not be statistically significant while small values might be.

Testing Hypotheses About ρ

The imagery of the sombrero helps to explain the purpose of testing hypotheses about ρ. For a given set of sample data points (as in Figure 15.23(a)), many different sized sombreros adequately cover the data. Included in this collection is a one-size-fits-all sombrero (the one when $\rho = 0$) that is the simplest description of the mutual relationship between X and Y. Stretching the sombrero in a diagonal direction ($\rho > 0$ or $\rho < 0$) is a more difficult fit and must be statistically justified before it is selected over the simplest one.

The concepts described above are consistent with our discussion titled "Evidence of Correlated Variables" in Chapter 4. Then, as now, we attempted to answer the question, "Is there a relation between X and Y in the population?" In Chapter 4 we had not yet introduced the elements of hypothesis testing, so the answer was found by comparing r against the cutoff points in Table 4.4. The following box and Figures 15.24 through 15.26 formalize this procedure into a t-test about ρ. The three possible alternative hypotheses are listed under H_a, but as we know, only one is selected and used in a particular test.

t-Test for the Correlation Coefficient

H_0: $\rho = 0$ (X and Y are not related.)

H_a: $\rho \neq 0$ (X and Y are related.)

$\rho > 0$ (X and Y are directly related.)

$\rho < 0$ (X and Y are inversely related.)

$$TS: \; t^* = \frac{r}{\sqrt{\dfrac{1 - r^2}{n - 2}}} \tag{15–23}$$

RR: For $\alpha = .10$ and the $\neq$ alternative, see Figure 15.24.

For $\alpha = .10$ and the $>$ alternative, see Figure 15.25.

For $\alpha = .10$ and the $<$ alternative, see Figure 15.26.

C: Reject H_0 or accept H_0.

COMMENT We recommend rounding t^* to two decimal places.

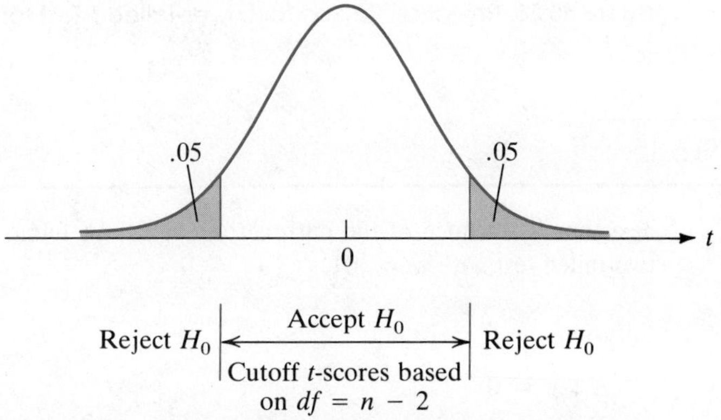

Figure 15.24 Rejection Region for Two-Tailed t-Test for ρ, $\alpha = .10$

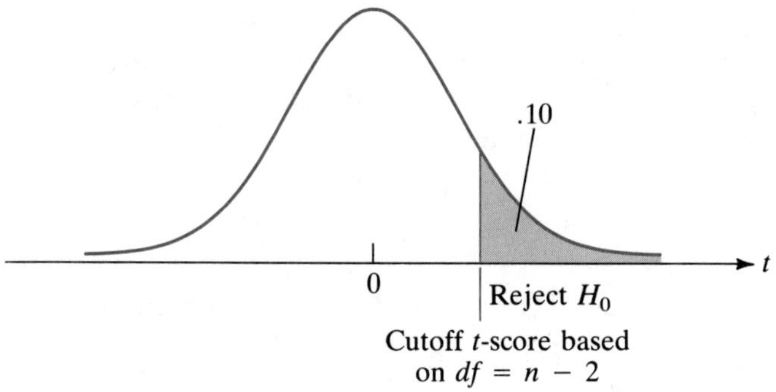

Figure 15.25 Rejection Region for Upper-Tailed t-Test for ρ, $\alpha = .10$

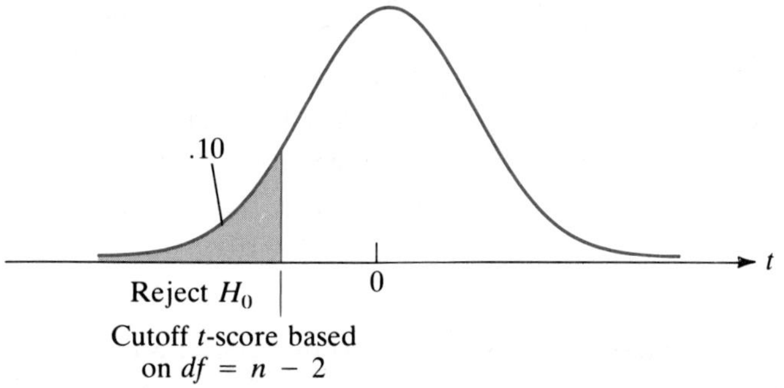

Figure 15.26 Rejection Region for Lower-Tailed t-Test for ρ, $\alpha = .10$

EXAMPLE 15.12

Test the significance of the correlation coefficient found in Example 15.11, using a two-tailed test with $\alpha = .01$.

Solution:

H_0: $\rho = 0$

H_a: $\rho \neq 0$

$$TS: t^* = \frac{r}{\sqrt{\frac{1 - r^2}{n - 2}}} = \frac{.8979}{\sqrt{\frac{1 - (.8979)^2}{8 - 2}}} = 4.9963639 \qquad \text{or } 5.00 \text{ (rounded)}$$

RR: For $\alpha = .01$, $df = 6$ and the $\neq$ alternative, see Figure 15.27.
C: Reject H_0.

The variables X and Y are directly related, and the association between them is strong at the 1% level of significance.

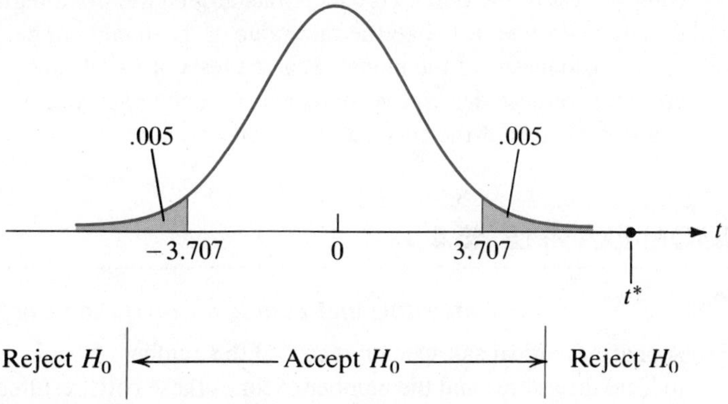

Figure 15.27 Rejection Region for Two-Tailed *t*-Test for ρ in Example 15.12

□

To complete the sombrero analogy with respect to Example 15.12, note that the conclusion to reject H_0 means the (allegedly) "one-size-fits-all" sombrero does *not* fit this particular configuration of data points. Instead we prefer one as pictured in Figure 15.22(b) that is stretched upward to the right.

Although estimates for the mean and standard deviation of X and Y are needed to completely characterize the sombrero, most practitioners stop at this point. Measuring and testing the degree of association between the two variables achieves our correlation analysis objectives.

COMMENTS

1. Algebraically, Equation 15–22 is equivalent to Equation 4–1.
2. The sign (+ or −) of r is the same as the sign of b_1 in the regression equation. Unlike b_1 though, r is a unitless quantity, independent of the X and Y units of measurement.
3. The relationship between the coefficient of correlation and the coefficient of determination is algebraically obvious:

$$(r)^2 = r^2 \quad \text{and} \quad \sqrt{r^2} = \pm r$$

4. Although a regression analysis assumes that X is a fixed variable, we can use regression when X is a random variable as well, under some general restrictions. (The probability distribution for X cannot involve the regression parameters, and we must assume the X values are recorded without error.)
5. In Chapter 4 when we compared r to the cutoff points in Table 4.4, we were, in effect, testing the null hypothesis $\rho = 0$. Now that we understand the process of testing a hypothesis, you should realize that the cutoff values in Table 4.4 are valid

only for the two-tailed alternative hypothesis $\rho \neq 0$ with $\alpha = .05$. In Exercise 15.65 you are asked to verify the alpha-risk, and Exercise 15.66 demonstrates how the values in Table 4.4 were generated.

Statistical practice has led to the reporting of r, the sample correlation coefficient, as a conventional part of the regression analysis. And in a sense, r does measure the compactness of the observed data points around the fitted regression line; but, so does $\hat{\sigma}$. The raw value of r, like the raw value of $\hat{\sigma}$, should *not* be considered as a measure of the "goodness" of the model. It is the tests of significance about ρ and β_1 that give credence to these descriptive statistics. We encourage you not to misinterpret the role and significance of the correlation coefficient.

CLASSROOM EXAMPLE 15.4

Computing and Testing a Correlation Coefficient

A pharmaceutical salesperson recorded the number of "cold calls" she made each day to local drugstores and the number of times these calls resulted in a sale. For a random sample of four days, the data are as follows:

X	Y	XY	Y^2	X^2
1	0			
2	1			
3	1			
4	3			
$\Sigma X =$ ____	$\Sigma Y =$ ____	$\Sigma XY =$ ____	$\Sigma Y^2 =$ ____	$\Sigma X^2 =$ ____

a. Compute the sample coefficient of correlation.

$$SSX = \_\_\_\_$$
$$SSY = \_\_\_\_$$
$$SSXY = \_\_\_\_$$
$$r = \frac{SSXY}{\sqrt{(SSX)(SSY)}} = \_\_\_\_$$

b. Is there sufficient evidence to conclude that the variables are directly related? State the appropriate hypotheses and conduct a test, using $\alpha = .10$.

SECTION 15.6 EXERCISES

15.56 Identify three major differences between regression analysis and correlation analysis.

15.57 Find the value of r given the following information:

a. $SSX = .172$; $SSY = 30.8$; $SSXY = 1.94$
b. $SSY = 2.098$; $SSXY = .6$; $b_1 = .1464$
c. $b_1 = -.47$; $r^2 = .622$

15.58 A random sample of size 10 yielded the following data. Find the sample correlation coefficient.

X	40.7	30.0	53.2	54.1	56.0	43.8	61.6	70.0	81.7	94.0
Y	2068	1933	1158	1026	899	1040	696	917	503	763

15.59 Indicate whether the following situations call for a regression analysis or a correlation analysis by determining the status (fixed or random) of the variables.

a. A businessman budgets quarterly expenditures of \$1000, \$1500, \$1750, and \$1250 for newspaper advertising in the coming year and wishes to study the effect on revenues.
b. A standard skills test is given to all job applicants at a company. The results are compared with each applicant's high school grade point average.
c. A salesperson's monthly expense account and his/her level of sales are examined.
d. The sales volume is recorded for a product at each of three different preset pricing strategies.
e. A bank records Y, the number of months that pass before a 48-month auto loan is retired and X, the amount of the original loan.
f. A rate analyst tracks the monthly average residential kilowatt-hour consumption and the average temperature for the month.

15.60 High-level executives in companies often trade large blocks of stock in their firm. This "insider activity" is thought to be indicative of impending good times—if they buy—and subsequent bad news—if they sell. Figures on the insider activity during a five-day period for a selected group of six companies within the same industry appear in the following table. Each entry represents the number of company executives who have bought or sold stock in their company.

	Stock	
Company	Bought	Sold
A	7	3
B	5	7
C	5	0
D	6	6
E	5	2
F	1	2

a. Would you expect to find a positive or negative correlation or no correlation between these variables? Explain.
b. Find the correlation coefficient between the number of buyers and sellers.

15.61 The yearly percentage returns for two mutual funds were recorded over a five-year period producing the data in the following table:

	Percentage Return	
Year	Mutual Fund 1	Mutual Fund 2
1	31.2%	29.1%
2	41.5	25.8
3	10.5	11.3
4	25.4	29.3
5	20.0	21.6

a. Find the coefficient of correlation for these data.

b. Explain the importance, if any, the following information would have on your decision to invest in one or the other (but not both) of these mutual funds.
(1) The hypothesis H_0: $\rho = 0$ is rejected in favor of H_a: $\rho < 0$.
(2) The hypothesis H_0: $\rho = 0$ is rejected in favor of H_a: $\rho > 0$.

15.62 Office vacancy rates from one year to the next for a sample of 35 cities in the United States are listed in Table 15.6.

a. Determine the correlation coefficient between the 1985 and 1986 vacancy rates.

b. Is this a significant correlation? Use $\alpha = .10$.

c. If Detroit, Michigan, had a high office vacancy rate in 1985, what, if anything, can be said about its office vacancy rate in 1986?

Table 15.6 Office Vacancy Rate Data for Exercise 15.62

	Office Vacancy Rate			Office Vacancy Rate	
Metropolitan Area	Dec. 1985	Dec. 1986	Metropolitan Area	Dec. 1985	Dec. 1986
Atlanta	18.6%	18.3%	New Orleans	22.1%	25.7%
Baltimore	9.5	16.4	Oakland–East Bay, CA	24.9	28.3
Boston	14.1	13.8	Oklahoma City	22.3	26.2
Charlotte, NC	13.8	17.0	Orange County, CA	19.3	22.7
Chicago	14.5	17.3	Orlando	24.5	26.3
Cincinnati	19.0	18.3	Philadelphia	14.5	14.4
Columbus, OH	15.9	15.6	Phoenix	25.4	26.4
Dallas	24.3	27.7	Portland, ME	20.4	19.2
Denver	26.1	26.5	Sacramento, CA	26.1	21.3
Fort Lauderdale	29.4	29.1	St. Louis	10.3	17.3
Houston	28.2	29.9	San Antonio	24.2	29.7
Indianapolis	13.8	18.0	San Diego	22.9	23.1
Greater Kansas City	16.0	18.4	San Francisco	16.6	19.4
Long Island, NY	14.8	13.1	San Jose, CA	24.5	26.6
Los Angeles	16.9	17.3	Seattle	17.4	17.6
Miami	20.3	21.1	Tampa	26.3	22.8
Minneapolis–St. Paul	17.7	17.4	Washington, DC	14.0	14.8
Nashville, TN	20.1	20.9			

Source: Reprinted with permission from Coldwell Banker Commercial Group, Inc., Rosemont, Ill.

15.63 A random sample of $n = 12$ yielded $r = -.942$. Test for evidence that the variables are inversely related. Find the p-value.

15.64 The planned merit pay raises in 1987 for salaried employees in 11 different industries are listed in the following table along with the actual percentage raises in 1986. Find the correlation coefficient for these data.

Industry	1987 Planned	1986 Actual
Insurance	5.6%	6.4%
Financial institutions	5.5	6.2
Services	5.1	6.0
High technology	5.7	5.9
Nondurable goods	5.2	5.7
Consumer home prod.	5.4	5.6
Retail/wholesale	5.4	5.5
Transportation	5.3	5.4
Durable goods	5.0	5.4
Utilities	4.4	5.3
Energy	5.2	4.5

Source: Sibson & Co., Inc., as printed in *USA Today*, Nov. 5, 1986. Excerpted with permission.

15.65 To verify that the t-test for ρ as outlined in the earlier box gives the same conclusion about the relation between X and Y as the informal procedure ("Evidence of Correlated Variables") proposed in Chapter 4, complete the following steps.

a. Refer to Table 4.4. Find the cutoff points for $n = 10$. (These cutoff points represent values of r.)

b. Substitute the cutoff points and $n = 10$ into Equation 15–23 and solve for t^*. This should produce two t-values.

c. Refer to the t-table. Find the probability beyond the t-values obtained in part b.

d. What level of significance and alternative hypothesis do the cutoff points in Table 4.4 represent in terms of a t-test for ρ?

15.66 To see how the values in Table 4.4 were generated, follow these steps:

a. Find the cutoff t-score with $df = 18$ and (a two-tailed) $\alpha = .05$ in Appendix table E.

b. Substitute $n = 20$ and the t-score from part a into Equation 15–23.

c. Solve Equation 15–23 for r.

d. Verify that the r-value(s) from part c correspond to the $n = 20$ cutoff points in Table 4.4.

15.67 The earnings per share (EPS) for a random sample of 30 companies for a two-year period produced the following data:

First Year	Second Year	First Year	Second Year	First Year	Second Year
\$6.15	\$4.12	\$1.29	\$1.65	\$ − 3.76	\$ − 1.94
.56	2.19	.34	.58	3.64	5.14
5.25	2.63	1.61	.54	−.67	.02

(continues)

(continued)

First Year	Second Year	First Year	Second Year	First Year	Second Year
\$2.85	\$3.19	\$1.96	\$−2.17	\$2.31	\$2.04
4.02	3.43	1.41	1.19	5.02	1.21
3.26	3.15	1.27	1.26	1.09	1.01
.23	.95	1.15	1.38	−2.33	2.95
1.29	1.06	.84	.77	4.07	3.46
2.85	2.84	.47	.67	.57	.49
5.61	6.19	5.98	1.84	.04	−.16

a. Find r.
b. Determine whether the two-year EPS figures are directly related at a .025 level of significance.

15.68 Coldwell Banker's Home Price Comparison Index tracks the prices of homes across the U.S.A. for a typical corporate transferee. Homes in the survey must meet several criteria, including: 2,000 square feet, 3 bedrooms, 2 bathrooms, a dining area, a family room, a two-car garage, and an upscale neighborhood. The 1986 data in Table 15.7 represent the costs of homes meeting these guidelines and the average number of days it took to sell the house. Find the correlation coefficient relating these variables and test for a direct relationship ($\alpha = .01$).

***15.69** The coefficient of correlation is unaffected by a change in location for either one or both of the variables. This means the correlation between the variables X and Y is the same value as the correlation between the variables X and $(Y - \overline{Y})$ or the variables $(X - \overline{X})$ and Y or even $(X - \overline{X})$ and $(Y - \overline{Y})$. To illustrate, consider the following data set:

X	1	2	3
Y	5	2	2

a. Find the correlation coefficient between X and Y.
b. Create a new variable $T = X - \overline{X}$ by subtracting the value of $\overline{X}$ from each value of X. Find the correlation coefficient between T and Y.
c. Create another variable $U = Y - \overline{Y}$ by subtracting $\overline{Y}$ from each value of Y. Find the correlation coefficient between X and U.
d. Find the correlation coefficient between T and U.
e. The value of r for each of the previous parts should be the same. Is it?

***15.70** The sample variance, $S^2 = \Sigma(X - \overline{X})^2/(n - 1)$ measures the deviations of the values of X from the mean $\overline{X}$ along one dimension (the X-axis). Analogously, we gauge the deviations of bivariate data (X,Y) in two dimensions from the means $\overline{X}$ and $\overline{Y}$ with the covariance. As the name implies, *covariance* is a measure of the simultaneous variability in X and Y relative to the point $(\overline{X},\overline{Y})$. Each bivariate piece of data (X,Y) generates 2 deviations: one from $\overline{X}$, $(X - \overline{X})$, and the second from $\overline{Y}$, $(Y - \overline{Y})$. The *sample covariance*, denoted as *cov*, is the (corrected) average of the product of these deviations:

$$cov = \frac{\Sigma(X - \overline{X})(Y - \overline{Y})}{n - 1} = \frac{SSXY}{n - 1}$$

*Optional

Table 15.7 House Sales Comparison Data for Exercise 15.68

City	Price ($ Thousands)	Average Days on Market	City	Price ($ Thousands)	Average Days on Market
Birmingham, AL	95	88	Omaha, NE	88	100
Anchorage, AK	152.5	150	Las Vegas, NV	116.8	105
Phoenix, AZ	105	98	Reno, NV	120	75
Tucson, AZ	131.3	106	Manchester, NH	177	45
Beverly Hills, CA	775	167	Bergen County, NJ	320	40
Chula Vista, CA	132	85	Essex County, NJ	228	60
Torrance, CA	254	10	Albuquerque, NM	124	75
San Francisco, CA	385	100	N. Long Island, NY	300	75
Denver, CO	144.8	125	S. Long Island, NY	200	75
Hartford, CN	183	45	Rochester, NY	132	30
Wilmington, DE	108.9	30	Charlotte, NC	105.5	90
Washington, DC	148.6	40	Fargo, ND	119	89
Miami, FL	125.5	75	Cincinnati, OH	103	91
Orlando, FL	119	103	Cleveland, OH	96	96
Tampa, FL	125	90	Tulsa, OK	74.8	78
Atlanta, GA	100	106	Portland, OR	104.8	107
Honolulu, HA	215	90	Philadelphia, PA	162	30
Boise, ID	81	94	Pittsburgh, PA	109	105
Chicago, IL	140	60	Providence, RI	160	65
Indianapolis, IN	90	75	Charleston, SC	100	105
Ames, IA	106	82	Rapid City, SD	95	90
Wichita, KS	100	120	Memphis, TN	106	30
Louisville, KY	100	90	Nashville, TN	135	37
New Orleans, LA	101	145	Austin, TX	161.5	120
Brunswick, ME	140	60	Dallas, TX	125	107
Baltimore, MD	145.5	75	Houston, TX	81.5	170
Boston, MA	213	45	Salt Lake City, UT	125	110
Detroit, MI	125	30	Burlington, VT	125	33
Minneapolis, MN	140	70	Richmond, VA	81.5	125
St. Paul, MN	126	80	Seattle, WA	110.9	73
Jackson, MS	135	181	Charleston, WV	120	130
Kansas City, MO	90	64	Milwaukee, WI	101	41
St. Louis, MO	155	40	Cheyenne, WY	125.5	70
Billings, MT	106	85			

Source: Coldwell Banker, as printed in *USA Today,* May 11, 1987. Excerpted with permission.

The product of the deviations, $(X - \overline{X})(Y - \overline{Y})$, is a positive number in the first and third quadrants and is a negative number in the second and fourth quadrants, where the quadrants are formed from lines intersecting at $(\overline{X},\overline{Y})$ (see Figure 15.28). Intuitively, the covariance will be a positive number if a majority of the data are in the first and third quadrants, and a

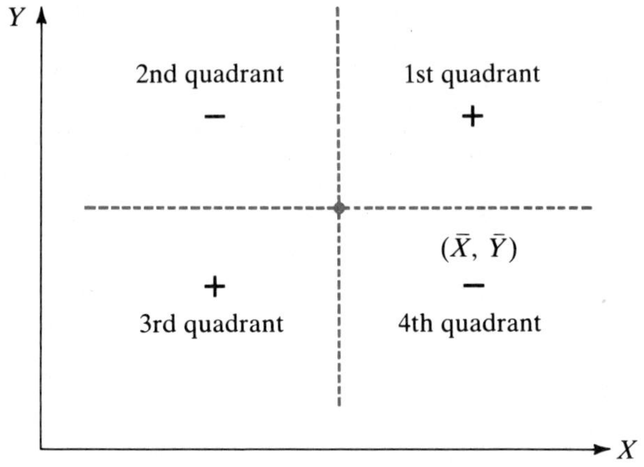

Figure 15.28 Quadrants Formed from Perpendicular Lines Intersecting at $(\bar{X},\bar{Y})$

negative number when the data fall predominantly in the second and fourth quadrants. Should there be a balance of points in all four quadrants, the covariance will be near zero.

Sometimes, the correlation coefficient is defined in terms of the covariance between X and Y, the standard deviation of X (denoted by S_X), and the standard deviation of Y (denoted by S_Y) rather than in terms of the sums of squares:

$$r = \frac{cov}{S_X S_Y}$$

This formulation for r is algebraically equivalent to Equation 15–22, since $S_X^2 = SSX/(n - 1)$ and $S_Y^2 = SSY/(n - 1)$.

Consider the following set of bivariate data:

X	4	7	3	4	8
Y	2	−2	2	0	−4

a. Compute $\bar{X}$ and $\bar{Y}$. Plot the data and locate the point $(\bar{X},\bar{Y})$. Draw a set of axes through this point. Do you expect the covariance to be a positive or a negative number?
b. Find the covariance between X and Y. Is the sign of *cov* as you anticipated?
c. Find the standard deviation of X and the standard deviation of Y.
d. Find r using the covariance formulation.
e. Find r using Equation 15–22. Does your answer agree with part d?

*15.71 The coefficient of correlation, r, and the slope of the least-squares line, b_1, are related measures. Theoretically, it can be shown that

$$r = b_1 \frac{S_X}{S_Y}$$

where S_X is the standard deviation of the X-values, and S_Y is the standard deviation of the Y-values. This formula provides us with an alternative means of computing r from the regression statistics. It also justifies an earlier comment linking the sign (+ or −) of b_1 to the sign of r, since both S_X and S_Y are always positive quantities. To demonstrate the above formula, refer to Classroom Example 15.4 which produces the following summary statistics:

$$SSX = 5 \qquad SSXY = 4.5$$
$$SSY = 4.75 \qquad r = .9233805$$

a. Find b_1 using Equation 15–9.

b. Compute the standard deviation of X and of Y from the following equations:

$$S_X = \sqrt{\frac{SSX}{n-1}} \qquad S_Y = \sqrt{\frac{SSY}{n-1}}$$

c. Verify the (above) relationship between r and b_1.

*15.7 PROCESSOR: REGRESSION

To accompany this chapter (and the next) is a processor titled Regression. Most of the descriptive statistics and inferential procedures associated with a simple regression analysis are available as output. We will demonstrate the processor with the data set from Example 15.3 involving the variables X = the percentage change in the Dow Jones Industrial Average and Y = the percentage change in the share price of a mutual fund.

Getting Started

The first thing to do is to create a data set, unless the data currently reside in a file on your data disk. Activate the Data section on the main menu and then select the Create option. Using the following data, we will create a Multivariate Raw Data set with $n = 8$ entities and $v = 2$ variables:

Y	−.4	−.3	−.1	.8	−.2	.1	.4	.5
X	−.4	−.1	.1	.3	−.2	0	.2	.3

Note: In creating the data set, it is more efficient (for later processing) to put the values of Y in the first column. After data entry, we recommend saving the data set (as a TXT file) using the Save feature under the Files section of the main menu.

Next, move the highlight to Processors and access the Regression option. Of the two choices on the Regression submenu—Simple Regression and Multiple Regression—we will select the former. View the Introduction screen, read in the t-table, then move across the menu to Data and touch Enter.

At this point, we must identify the independent and dependent variables before processing the data. The processor automatically classifies the first variable in the data

*Optional

set as the dependent variable and the second variable as the independent variable. (This is the reason we suggested entering Y first in creating the data set.) If you need to change designations, use the F5 key to re-mark the variables. When the variables are marked properly, touch Enter. The ensuing screen tells us the status of each variable. If the statuses are incorrect, touch Escape and re-mark the variables; otherwise, touch Enter to confirm the designations.

Proceed to Calculations and touch Enter twice to initiate the computations. After the message "Complete" appears, touch Enter, move to the Results section and touch Enter again to generate Figure 15.29 where seven choices of output are displayed.

```
≡                          Simple Regression                            MENU
  Introduction  *able  Data  Calculations [Results] Exit           |  F1=Help

                                     Fitted model
                                     t-test of model utility
                                     Estimated coefficients
                                     ANOVA table
                                     Prediction
                                     Correlation
                                     Intermediate results
```

Figure 15.29 Available Results from Simple Regression Processor

```
≡                          Simple Regression                            READY
  Introduction  *able  Data  Calculations [Results] Exit           |  F1=Help

  The fitted simple regression model, to 5 decimal places, is:

                      ^
                      Y =  0.06149  +1.54023X


     The sum of the squared errors (SSE), to 10 decimal places, is:
                                    ^
                      SSE = Σ(Y - Y)² = 0.2480459770

                                                          ^
     The estimated standard error of the regression (σ), to 5 decimal
     places, is:
                      ^
                      σ = √0.04134  =  0.20332
```

Figure 15.30 The Fitted Model Screen

```
≡                          Simple Regression                          READY
   Introduction  *able  Data  Calculations  [Results]  Exit       | F1=Help

     Summary of estimated regression coefficients
```

Beta	Estimate	Standard Error	t*	p-value
β0	0.061494	0.07229815	0.851	0.4276439809
β1	1.540230	0.30828035	4.996	0.0024616432

Figure 15.31 The Estimated Coefficients Screen

```
≡                          Simple Regression                          READY
   Introduction  *able  Data  Calculations  [Results]  Exit       | F1=Help

        ΣX......... 0.20000

        ΣY......... 0.80000

        ΣXY........ 0.69000

        ΣX²........ 0.44000

        ΣY²........ 1.36000

        SSX........ 0.43500

        SSY........ 1.28000

        SSXY....... 0.67000
```

Figure 15.32 The Intermediate Results Screen

■ Results

The first option—Fitted Model—shows the equation of the least-squares line, and the values of *SSE* and $\hat{\sigma}$. Refer to Figure 15.30.

The *t*-Test of Model Utility will reproduce the first three elements of the test for β_1 outlined earlier. Rather than construct a rejection region and make a conclusion, the processor reports the *p*-value and leaves the decision to you.

The third choice in the Results submenu is labelled Estimated Coefficients. Figure 15.31 displays the screen for our data set. The t^*-values are the test statistics (estimate divided by standard error) for testing H_0: $\beta_0 = 0$ and H_0: $\beta_1 = 0$ against two-tailed alternatives. Correspondingly, the p-values are two-tailed.

Selecting the ANOVA Table option will result in a table similar to the one in Table 15.2. The only difference is that the processor has an additional column for the p-value for the test.

The fifth option—Prediction—allows us to enter a value for X and generate confidence intervals for the average value of Y, denoted by $E(Y)$, and prediction intervals for a specific value of Y as described in Section 15.5.

The Correlation choice gives the values of r and r^2. Finally, the last option on Intermediate Results may prove to be beneficial as you check your answers to some of the exercises. Figure 15.32 shows the relevant sums for our data set.

15.8 SUMMARY

Regression analysis is an efficient and useful way of relating two variables through an equation of a straight line. In this chapter, we explained the rationale, reasons, and method for executing a regression analysis. We gave a summary of the objectives. Perhaps lost in the detail of equations and symbols is the crucial first step: plotting the sample data. Admittedly, hauling out the graph paper and ruler to plot data points is not exciting, but it is very necessary.

For example, Anscombe (1973) dramatically illustrates the need for "seeing" the graphed data before and after the analysis. In Figure 15.33 we have reproduced his four scatterplots. The sample data in each scatterplot have the *same summary statistics* and *equation* of the least-squares line, yet the data sets are not remotely identical. The lesson here is to not depend solely on the fitted model to describe the relationship between X and Y; graph the data too.

Fitting and testing the model are preliminary to making predictions. There are two cautions. First, make interval estimates or predictions rather than point estimates or predictions. Second, recognize that the model is an appropriate description of X and Y over the *relevant range* of observed X-values. Predictions should be limited to values of X in this domain. Predicting outside the relevant range negates the biggest asset of inferential techniques: a probabilistic assessment of risk. We no longer can be assured of 95% confidence, for example, when we extrapolate beyond the minimum or maximum X-values.

Correlation analysis, closely related to regression analysis but with different objectives, provides a quantitative measure of the strength of the association between two variables, X and Y. Restraint is urged in interpreting a correlation coefficient. Although it is tempting to link a large value for r to a cause (X) and an effect (Y) relationship, this practice is not justified. Much has been written about the danger of doing this. Correlation is not synonymous with causation.

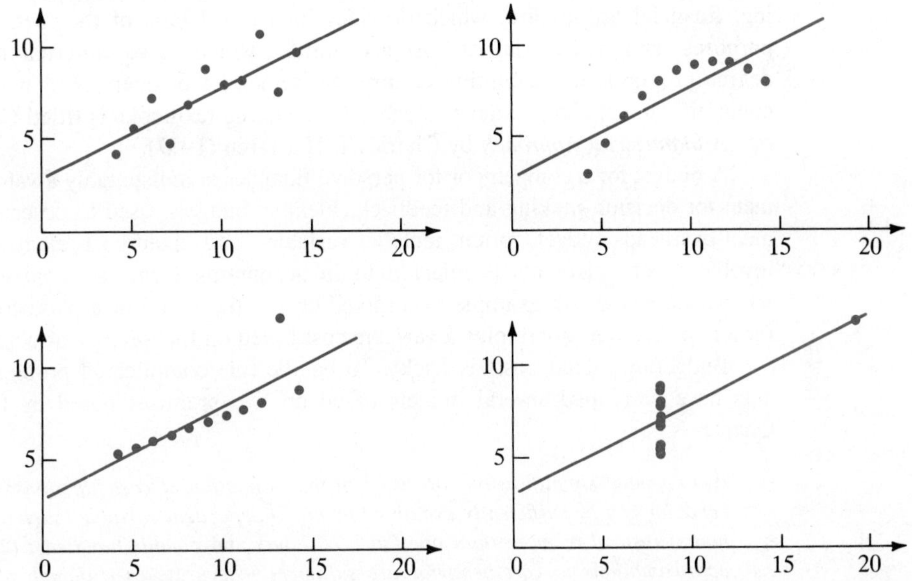

Source: Anscombe, F. J., "Graphs in Statistical Analysis," *American Statistician*, Volume 27, 1973, pp. 17–21. Reproduced with permission.

Figure 15.33 Anscombe's Quartet

All of the inference procedures discussed in this chapter hinged on an assumption of normally distributed errors or residuals. Our study of simple regression is *not* complete until we analyze the merits of this assumption in Chapter 17. Sandwiched between these chapters is a slightly more advanced type of regression—multiple regression—that relates Y to two or more independent variables. The objectives will stay the same, but the mathematics will change. Mastering the ideas of simple regression is a stepping stone to the next chapter.

15.9 TO BE CONTINUED . . .

. . . *In Your College Courses*

At the beginning of this chapter, we indicated that regression analysis pervades all areas of business, including marketing, finance, economic forecasting, and accounting. In this section we would like to concentrate on one of these areas—accounting—and show that regression will reappear in other accounting courses.

There are two major parts recognized in a contemporary examination of accounting: financial accounting, which provides financial details of the firm for external purposes, and managerial or cost accounting, which serves internal needs. Most degree programs in accounting require two semesters or quarters of managerial accounting. One of the popular managerial accounting textbooks is titled *Cost Accounting: A Managerial Approach* by Charles T. Horngren (1977).

A budget for a company or for personal finances is indisputably a valuable instrument for decision-making and feedback. Flexible budgets, used to describe a department-overhead budget, often include variable- and fixed-cost elements. A cost involving both elements is referred to in accounting terms as a *mixed cost* or a *semivariable cost*. An example of a mixed cost is the rental of a photocopy machine for a fixed fee per month plus a variable cost based on the number of copies made.

Budgeting mixed costs is tricky. To handle this complicated problem, accountants have developed several models based on two premises noted by Horngren in Chapter 8:

> *Two common simplifications are used in the estimation of cost functions: (1) cost behavior can be sufficiently explained by one independent variable (such as labor-hours) instead of more than one (such as hours* and *pounds handled); (2) linear approximations to cost behavior are sufficiently accurate even though nonlinear behavior is widespread.*

As you might anticipate, these factors facilitate the use of simple regression in developing cost functions. More on this from Horngren:

> *The estimation of mixed-cost behavior patterns should preferably begin with a scatter chart of past cost levels . . . A line is fitted to the points . . . by the statistical method of least squares . . .*

These steps outlined by Horngren are perfectly consistent with the first three objectives of a regression analysis that we listed on page 748. Continuing from Horngren's text:

> *. . . the major purpose of determining cost behavior patterns is to estimate how costs behave as volume changes . . . Thus, the emphasis is on determining the slope of the line . . .* the behavior of the variable-cost component over the relevant range.

In this passage Horngren states that the slope of the regression line is the key. This ties in with our fourth objective in determining the utility of the fitted model by testing the significance of the slope.

Later, in Chapter 25, Horngren demonstrates the application of regression analysis with an example relating direct-labor costs to the various sizes of batches of output. Included in his discussion are the normal equations (same as our Equations 15–3 and 15–4), the equations for the slope and the Y-intercept (same as our Equations 15–8 and 15–9), the standard error of the estimate (what we call the estimated standard error of the regression, Equation 15–11), the coefficients of determination and correlation, and both the t-test and the confidence interval for β_1. In short, our entire chapter on simple regression and correlation is *briefly summarized* in part of Horngren's Chapters 8 and 25!

For those planning to major in accounting, we submit that you will encounter regression and correlation in core and/or elective courses in your program of study. Regression is not for business statisticians only!

. . . In Business/The Media

Applications of regression and correlation analyses abound in business and in the media.

Projecting Demand for Electricity A utility company, for example, might project the daily demand for electricity during the summer based on the anticipated high temperature. If conditions warrant, the company can make prior arrangements to purchase additional generating capacity from neighboring utilities in order to meet unusually high demand.

Sequencing Computer Runs The operations department at a major insurance company is responsible for updating files and preparing reports at night when access to tapes and data files is not needed. However, by 8 A.M. the next morning the department must deliver the reports and release the data files for use. The operations manager estimates the time to complete his responsibilities by relating the number of procedures called in the job control language for each application to the amount of run time per application. In this way he can determine which jobs require special attention as well as set up the sequence of job runs.

Assessing the Volatility of a Stock A stock's or mutual fund's "beta" (see Example 15.3) is simply the slope of the regression line relating $Y =$ the change in the stock's price to $X =$ the change in a market index like the Dow Jones Industrial Average or the Standard & Poor's 500 stock index. *Changing Times* magazine in the article "Small Companies: Can Investors Still Win Big?" reported:

> *Security analysts measure that price volatility with a statistical device they call a beta, which gauges a stock's fluctuation in relation to a corresponding change in the Standard & Poor's 500-stock index. The S & P change becomes the benchmark statistically equal to 1.00. A stock that on the average rises or falls 10% more than the S & P would have a beta of 1.10 . . . Small company stocks are considered more risky—they usually have higher betas . . . than the giant corporations.**

Indeed a stock's beta is becoming as prominent in the financial world as the traditional price-to-earnings ratio has been.

Predicting Telephone Lines Immediate telephone service is expected when a family moves into a house in a new subdivision. Clearly, the telephone company must plan ahead to provide enough telephone cables to meet the subscribers' demands. The

*Excerpted with permission from *Changing Times* Magazine, © 1984 Kiplinger Washington Editors, Inc., June 1984.

phone company relates the number of houses to the number of residential phone lines and then uses this relation to predict cable needs for future developments.

As these illustrations indicate, regression and correlation analyses are appropriate when we wish to model two quantitative variables having a statistical—that is, inexact—relationship. Depending on the strength of the relationship, we subsequently can generate reliable forecasts from the regression equation to aid in planning and in decision-making.

SECTION 15.9 EXERCISES

15.72 Identify two simplifications that are used in developing cost functions.

15.73 Suppose a company is interested in modeling the relationship between Y, total repair cost per month, and X, number of miles driven per month, for its fleet of automobiles. The transportation manager fits a simple regression model to 18 months of data and obtains the equation $\hat{Y} = \$75{,}073 - \$.74X$. Noting a slope of $\$-.74$, the manager interprets this to mean the repair costs *decrease* as total mileage increases. As this runs counter to his intuition, he decides regression is impractical in this situation. Provide a plausible explanation for this apparent inverse relationship between cost and mileage.

15.74 Develop a budget formula for predicting mixed costs based on the number of hours of labor for the following data.

Hours of Labor	12	16	28	19	20	28	14	36	25
Mixed Cost ($)	31	49	80	61	52	88	46	96	74

15.75 Annual sales for a grocery store are recorded for a six-year period in order to develop a regression model. The data are as follows:

Year	Sales ($ Millions)
1	5
2	4
3	9
4	11
5	10
6	13

Perform a regression analysis on these data and point predict the sales in year 7. Let X = the year number.

15.76 The cost accountant for a small company specializing in engines for lawn mowers is responsible for preparing a quarterly flexible budget for the first quarter of the calendar year. In order to assess the semivariable expenses associated with the production of the engines, the accountant examines five years (20 quarters) of historical data and records the information given in Table 15.8.

Table 15.8 Budget Data for Exercise 15.76

Year	Quarter	Units Produced (Thousands)	Semivariable Manufacturing Overhead ($ Thousands)	Observation Number
1	1	.8	33	1
	2	.6	29	2
	3	.7	24	3
	4	.4	15	4
2	1	1.3	31	5
	2	.8	30	6
	3	.5	20	7
	4	.6	25	8
3	1	1.6	37	9
	2	1.5	38	10
	3	1.2	24	11
	4	1.0	23	12
4	1	2.4	50	13
	2	1.9	33	14
	3	1.4	29	15
	4	.8	24	16
5	1	2.9	53	17
	2	2.2	42	18
	3	1.2	22	19
	4	1.2	20	20

a. Initially, the accountant fit a regression model relating Y, semivariable manufacturing overhead, to the independent variable, *observation number*. What equation did she generate?

b. Perplexed by the almost negligible slope of her first model, she decided to try another model, using *quarter* as the independent variable. Determine the least-squares line in this case.

c. After realizing that both of her first two models ignored the valuable information provided by the independent variable *number of units produced,* she created a third regression equation. Find this equation.

d. In the course of all this analysis, she came upon another angle to analyze these data. Since she was preparing a flexible budget for only the first quarter, why not fit a regression line to just the 5 first-quarter data points relating Y and the *number of units produced?* Help her find this model.

e. Which model would you use to predict the semivariable expense for the first quarter of the sixth year? Are there any models you definitely would not use? What else is needed to use the models of parts c and d for prediction purposes?

f. Use all four models to generate a point prediction and an 80% prediction interval for the first quarter of year 6. You may assume that 2000 engines are planned for production during this quarter.

CHAPTER 15 EXERCISES

15.77 What is (are) the assumption(s) of the simple correlation model?

15.78 Identify the different types of models. The regression model and the correlation model are examples of which type of model?

15.79 Letting the independent variable be the year, find the simple regression equation to model the house winnings profit in gambling casinos in Nevada and Atlantic City over a five-year period.

Year	Profit ($ Millions per Day)
1	10.9
2	12.6
3	13.7
4	15.1
5	15.9

15.80 A life insurance company requires job applicants to submit a sample of their handwriting which is then analyzed by a graphologist. The company believes there is a relationship between the applicant's handwriting score and his/her potential first-year sales. No one with a handwriting score below 70 is offered a job. A random sample of former and current employees of the company revealed the following data:

Handwriting Score	First-Year Sales ($ Thousands)
72	16.4
80	28.1
75	12.8
70	5.2
75	17.7
78	15.9
85	28.4

a. Construct a scatter plot for these data.

b. Find the simple regression equation relating Y = first-year sales to X = handwriting score. Is it a useful model? Let $\alpha = .01$.

c. Plot the least-squares line on the scatter diagram of part a.

d. Suppose an applicant's handwriting score were 76. What would you predict (with 90% confidence) for his/her first-year sales?

15.81 Suppose a sample of size $n = 10$ produced a value of 0 for b_1. What value does b_0 become? What would the graph of the least-squares line look like? What does $b_1 = 0$ imply about the relationship between X and Y?

15.82 Apply the first 3 objectives of simple regression to these data:

X	−5	−3	−1	0	1	3	5
Y	8.8	7.2	7.3	7.0	6.7	6.7	6.0

***15.83** Consider the following equation of a least-squares line, summary statistics, and p-value for the two-tailed test of H_0: $\beta_1 = 0$:

$$\hat{Y} = 2 - .036X \qquad SSX = 28 \qquad p\text{-value} = .933$$
$$n = 7 \qquad \hat{\sigma} = 2.189 \qquad \overline{Y} = 2$$
$$\overline{X} = 0 \qquad SSY = 24$$

a. If $\alpha = .05$ would we judge the model useful? Explain.
b. Relative to your answer in part a, what would be the point prediction of Y when $X = 1$?
c. Relative to your answer in part a, develop a 98% prediction interval for Y when $X = 1$. (*Hint:* Refer to Exercise 15.51.)

15.84 The regression model $\hat{Y} = -15.2 + 3.4X$ was fit to a set of 27 data points and found to be a useful relation. Construct and interpret a 90% confidence interval for the average value of Y when $X = 12$. Assume $\hat{\sigma} = 1.142$, $SSX = 162.5$, and $\overline{X} = 9.8$.

15.85 Describe the effect (increase, decrease, no effect) on the width of a confidence interval in each of the following situations, all other factors remaining essentially unchanged.
a. The level of risk is increased.
b. The distance between $X = x$ and $\overline{X}$ is decreased.
c. The confidence level is increased.
d. The sample size is increased.
e. $\hat{Y}$ approaches $\overline{Y}$.

15.86 A random sample of 18 bivariate data points yielded the following:

X	Y	X	Y	X	Y
7	6	5	15	5	12
6	10	16	5	10	6
6	5	15	5	7	7
18	4	14	6	2	20
6	12	13	9	9	10
5	13	4	17	8	12

a. Plot the data.
b. Propose a linear model to relate X and Y.
c. Fit a least-squares line to these data.
d. Test the model for utility, using $\alpha = .05$.
e. Predict Y when $X = 10$, with a 98% prediction interval.
f. Estimate the average value of Y when $X = 4$, with a 90% confidence interval.

*Optional

15.87 The Boyd Company, a Princeton, N.J., firm, conducted a survey of cities ("What It Costs to Operate a Factory," *USA Today*, February 25, 1987) to compare the yearly costs of running a durable-goods manufacturing plant employing 750 hourly workers, occupying 300,000 square feet, and making annual shipments of 33 million pounds to a national market. The data are as follows:

Metropolitan Area	Labor Cost per Hour	Power Cost (Millions)	Occupancy Cost (Millions)	Total Cost (Millions)
San Francisco	\$13.34	\$705,000	\$3.3	\$32.2
Peoria, IL	13.45	654,000	3.4	31.9
Cleveland	11.99	554,000	3.4	29.1
Baltimore	11.43	639,000	3.2	28.1
Minneapolis	10.96	472,000	3.5	26.9
Los Angeles	10.47	733,000	3.2	26.3
Chicago	10.45	645,000	3.7	26.2
Boston	9.93	861,000	3.6	26.1
Mobile, AL	10.49	501,000	2.8	25.6
Phoenix	10.18	590,000	3.1	25.4
Atlanta	10.16	500,000	2.9	25.0
Houston	10.00	590,000	3.0	24.8
Denver	10.05	531,000	3.3	24.8
Burlington, IA	8.78	478,000	3.2	22.2

a. Plot X = labor cost versus Y = total cost.
b. Find the simple regression equation relating X and Y.
c. Is Y related to X at the 5% level?
d. Suppose a city has an estimated labor cost of \$9 per hour. What is the predicted total cost, with 80% confidence?

15.88 Refer to Exercise 15.87.
a. Find the correlation between power cost and occupancy cost.
b. Is it significant (α = .10)?

15.89 Find the correlation coefficient for the following data.

X	Y
13.430	6.4
8.449	6.1
8.130	9.0
5.760	8.3
5.633	6.9
5.540	6.1
5.158	8.5
4.369	8.9
3.936	4.7
3.195	5.4
3.083	4.2

15.90 Why can't we develop a least-squares regression equation $\hat{Y} = b_0 + b_1X$ to model the relationship between the variables X, type of household heating (oil, gas, electricity, and so forth), and Y, direction the front of the house is facing (north, east, south, west)?

15.91 A random sample of 25 days of trading activity on the New York Stock Exchange generated data on the change in the Dow Jones Industrial Average (DJIA) and the changes in the price of three mutual funds, as given in Table 15.9.

a. Fit a simple regression model relating Y, the change in the price of mutual fund 3, to X, the change in the DJIA, and test its utility at the 2.5% level of significance with a one-tailed test that would imply a direct relation if we were to reject H_0.

b. What would you estimate the average change in mutual fund 3 to be on days when the Dow lost 10 points? Use 90% confidence.

c. Find the correlation between the changes in mutual fund 2 and mutual fund 3. Is it significantly ($\alpha = .01$) different from zero?

15.92 Refer to Exercise 15.91. Repeat parts a and b using mutual fund 1 in place of mutual fund 3.

Table 15.9 Stock Exchange Data for Exercise 15.91

	Mutual Fund Price Changes		
DJIA	Fund 1	Fund 2	Fund 3
−11.11	+.02	+.07	−.04
−1.64	+.04	+.06	−.11
−8.68	−.01	−.17	−.29
+2.13	+.06	+.13	−.01
+33.95	+.17	+.50	+.50
+30.26	+.17	+.19	+.47
+5.40	−.01	−.01	+.11
−5.69	−.03	−.13	−.27
+9.10	+.06	+.10	+.09
−36.79	−.17	−.78	−.74
−57.39	−.39	−1.30	−1.14
+26.28	+.11	+.48	+.44
+4.40	+.10	+.50	+.17
+69.89	+.24	+.99	+1.09
+15.20	+.09	+.32	+.44
−44.60	−.14	−.62	−.68
−32.96	−.29	−.71	−.52
−.42	−.11	−.26	−.19
−51.71	−.41	−1.20	−1.09
−34.09	−.58	−1.58	−.85
+29.97	+.24	+.84	+.56
−5.39	−.06	−.31	−.17
+66.47	+.24	+.79	+.89
−51.13	−.19	−.45	−.62
−4.97	−.03	−.14	−.17

15.93 A simple regression analysis based on $n = 10$ data points generated the following model and summary statistics:

$$\text{Fitted model:} \quad \hat{Y} = 2304.7459 - 20.5853X$$
$$SSX = 3339.229$$
$$SSY = 2350436.1$$
$$SSXY = -68739.03$$

a. Set up an ANOVA table.
b. Conduct an F-test of the model utility using $\alpha = .01$.
c. Find the sample coefficient of determination and interpret its value.

15.94 *Bank Rate Monitor*, a weekly banking periodical, recorded the following average rates for a 15-year fixed rate mortgage and a 1-year adjustable rate mortgage (ARM) for a sample of 8 weeks in 1987 ("A Rundown on Mortgages," *USA Today*, May 22, 1987):

15-Year Fixed	1-Year ARM
8.95%	7.48%
9.11	7.49
9.88	7.61
10.04	7.63
10.19	7.78
10.23	7.82
10.24	7.84
10.46	7.90

Source: Copyright 1987, *USA Today*. Reprinted by permission.

Find the correlation coefficient for these rates.

15.95 Explain the meaning of the symbols Y, $\overline{Y}$, and $\hat{Y}$.

15.96 A real estate company has developed a model to assess the desirability of a marketable home. By factoring in attributes such as location within the city, square feet of living space, number of bedrooms, condition of the house, and so on, the company comes up with a desirability score on a 0 to 100 scale. (The higher the score, the more desirable the house.) In addition, the company keeps track of the number of days the house is listed until it is sold, defined as the time an agreeable contract is signed between the buyer and seller. A random sample of these data are as follows:

Desirability Score	Number of Days Listed
82	22
46	126
90	8
62	39
27	176
44	96
60	60
84	28

(continues)

Desirability Score	Number of Days Listed
58	54
73	22
68	29
71	35

The company would like to be able to tell a prospective client about how long it will take to sell his/her house. Develop a simple regression model from these data. Test the model (use $\alpha = .05$). Then predict with 90% confidence the number of days a house with a desirability score of 75 will remain listed before it is sold.

15.97 A simple regression analysis based on a random sample of size $n = 17$ produced a least-squares line of $\hat{Y} = 62.51 + 1.97X$. Test the utility of this model at the 10% level of significance, using these summary statistics: $\hat{\sigma}^2 = 6.41$ and $SSX = 7.4$.

15.98 In the $\pm$ computation for a prediction interval, what happens to the value of the term $(x - \overline{X})^2/SSX$ as the given value for X gets farther from $\overline{X}$? What is the effect on the width of the prediction interval?

15.99 As word processors became standard equipment in most offices, the classic typewriter seemed headed for oblivion. But a computer chip here and a daisy wheel there and the low-tech typewriter was suddenly hi-tech. In the article "Those Bleeping High-Tech Typewriters," the features, prices, and capabilities of several new electronic typewriters were examined and summarized in a chart listing the following data:

Manufacturer/Model	Speed (Characters/Second)	Price
AT&T 5300	8	$180
Brother AX-20	10	400
Brother EM-401	15	499
Epson Elite 100	12	310
Olympia XL 120	13	299
Panasonic Jetwriter 1	13	519
Sharp PA-3100 E	12	200
Silver Reed EZ20	12	179
Silver Reed EX36	10	349
Smith Corona XL 1000	10	229
Smith Corona XD 8000	15	559

Condensed with permission from *Changing Times* Magazine, © 1987 Kiplinger Washington Editors, Inc., June 1987.

Find the correlation coefficient between speed and price. Is r significantly different from zero to conclude that a direct relationship exists between the variables? Use $\alpha = .025$.

15.100 A simple regression analysis was performed on a random sample, generating the following summary statistics:

$$\Sigma X = 788 \qquad \Sigma Y = 2{,}368 \qquad \Sigma XY = 128{,}592$$
$$\Sigma X^2 = 44{,}162 \qquad \Sigma Y^2 = 380{,}858 \qquad n = 15$$

a. Develop the equation of the least-squares line.
b. Test for a direct relation using $\alpha = .05$.
c. Predict Y when $X = 60$. Assume a 10% risk.

15.101 Find the correlation coefficient for the following data set of size $n = 29$:

X	Y	X	Y	X	Y
19	4	13	15	14	14
16	10	8	18	11	13
16	11	13	13	11	14
13	14	7	17	8	18
18	7	19	8	7	19
16	12	11	14	6	18
17	11	3	20	8	18
22	9	14	9	6	18
16	12	12	15	0	22
14	13	15	10		

Can we conclude the variables are inversely related at the 2.5% level of significance?

15.102 A random sample of 85 utility stocks was taken to compare the amount of their annual dividends to the annualized yields of the dividends. The data are given in Table 15.10.
a. Find the correlation coefficient.
b. Can we conclude that the variables are positively associated within the population of all utilities? Use $\alpha = .01$.

15.103 The economic law of demand holds that the lower the price of a good, the larger the quantity of the good demanded by consumers. Often, a regression analysis helps to model the relationship between price and demand. Consider the following historical data based on the sales of a jar of mustard for a chain of supermarkets.

Retail Price	Monthly Demand (In Number of Cases)
\$ .85	728
.89	775
.95	764
.99	725
1.09	681
1.19	694

a. Find the equation of the line relating Y = demand to X = price.
b. Using the regression model, (point) predict the total revenue in a month when the mustard is priced at \$.99 a jar. There are 12 jars to a case.

*15.104 Some simple regression analyses require no intercept term. For example, a sales position offering compensation based only on commissions means: no sales, no earnings. Thus, the relation between X = sales and Y = earnings must satisfy this requirement: when $X = 0$, $Y = 0$. This forces $\beta_0 = 0$ in the general regression model $Y = \beta_0 + \beta_1 X + \epsilon$. The resulting problem, called *regression through the origin,* is based on the model $Y = \beta_1 X + \epsilon$, where

Table 15.10 Utility Stock Data for Exercise 15.102

Ticker Symbol	Annual Dividend	Annualized Yield	Ticker Symbol	Annual Dividend	Annualized Yield
AEP	$2.26	7.878%	MWE	$1.48	6.172%
ATE	2.62	6.288	NES	1.92	6.115
AYP	2.92	6.057	NGE	2.64	7.611
AZP	2.72	9.117	NMK	2.08	9.151
BGE	1.80	5.125	NPI	1.50	6.963
BKH	1.14	4.500	NSP	1.90	5.235
BSE	3.44	6.589	NU	1.68	6.817
CER	2.28	5.909	NVP	1.44	6.292
CES	2.72	6.502	OEC	1.92	9.531
CIN	2.16	7.855	OGE	2.08	5.887
CIP	1.68	6.004	ORU	2.18	5.997
CNH	2.96	8.014	PCG	1.92	7.706
CNL	2.08	5.869	PE	2.20	9.842
CPL	2.68	6.995	PEG	2.96	6.975
CSR	2.14	6.286	PGN	1.96	5.970
CTP	1.40	7.559	PNM	2.92	8.333
CV	1.90	7.092	POM	2.36	4.658
CWE	3.00	9.213	PPL	2.60	6.761
D	2.84	6.019	PPW	2.40	6.784
DEW	2.02	5.730	PSD	1.76	7.687
DPL	2.00	7.541	PSR	2.00	10.213
DTE	1.68	9.763	RGS	2.20	8.061
DUK	2.68	5.627	SAJ	1.82	5.175
ED	2.68	5.528	SAV	.88	4.704
EDE	1.88	5.697	SCE	2.28	6.549
EUA	2.18	6.195	SCG	2.24	5.716
FPC	2.28	5.402	SDO	2.38	6.162
FPL	2.04	6.018	SIG	1.96	5.124
GMP	1.80	6.325	SO	2.04	8.133
HE	1.72	5.239	SRP	1.72	6.674
HOU	2.80	8.271	TE	2.52	5.138
IDA	1.80	6.496	TEP	3.30	5.443
IEL	1.94	7.571	TNP	1.32	5.944
IOR	1.60	6.352	TXU	2.68	7.921
IPC	2.64	9.250	UCU	1.48	4.528
IPL	3.04	5.709	UEP	1.84	6.480
IPW	1.96	6.783	UIL	2.32	7.044
IWG	2.90	6.622	UTP	2.32	6.995
KAN	3.16	5.411	WPC	2.68	4.604
KU	2.52	5.895	WPL	2.96	5.413
LOU	2.60	6.403	WPS	3.00	5.322
MPL	1.52	5.063	WWP	2.48	8.360
MTP	2.48	6.232			

Source: Federal Register, Vol. 51, No. 203, October 21, 1986, pp. 37269–37270.

the errors are assumed to be normally distributed as in the boxed material on page 746. To fit the "no-intercept" model, we estimate β_1 with

$$b_1 = \frac{\Sigma XY}{\Sigma X^2}$$

and call $\hat{Y} = b_1X$ the least-squares line, based on $n - 1$ degrees of freedom.

Consider the following data representing life and disability insurance premiums relative to the amount of commissions earned, both recorded in thousands of dollars.

Premiums	Commission
16.7	5.8
47.0	23.9
100.5	36.8
93.0	29.0
9.7	7.7
43.7	24.8
25.5	12.6
20.3	10.7
36.1	15.3

Fit a no-intercept model to these data.

***15.105** Fit a no-intercept model to the following set of data:

X	−2	−1	0	1	2
Y	−1	−1	1	1	3

15.106 Refer to Figure 15.34 in responding to parts a through g.

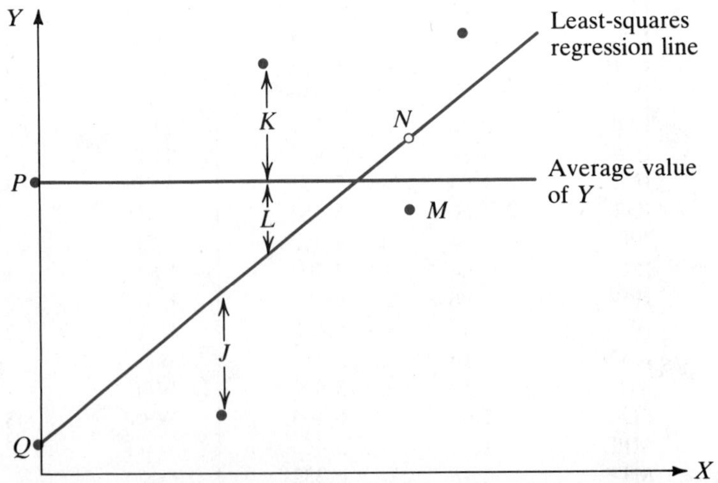

Figure 15.34 Generic Scatter Plot Illustrating Regression Concepts

Identify the point—J through Q—that corresponds to each of the following terms:

a. observed value of Y
b. error variability
c. b_0
d. predicted value of Y
e. regression variability
f. total variability
g. $\overline{Y}$

REFERENCES

Anscombe, F. J. 1973. "Graphs in Statistical Analysis," *American Statistician, 27*: 17–21.

Horngren, C. T. 1977. *Cost Accounting: A Managerial Emphasis,* 4th Edition. Prentice-Hall, Inc., Englewood Cliffs, NJ.

Kleinbaum, D. G., and L. L. Kupper. 1978. *Applied Regression Analysis and Other Multivariate Methods*. PWS-KENT, Boston.

Kuzmits, F. 1983. "The Relationship Between Absenteeism and Productivity: An Empirical Assessment." *Proceedings of the Southern Management Association*.

Mendenhall, W., J. E. Reinmuth, R. Beaver, and D. Duhan. 1986. *Statistics for Management and Economics,* 5th Edition. PWS-KENT, Boston.

Neter, J., W. Wasserman, and M. H. Kutner. 1985. *Applied Linear Statistical Models,* 2nd Edition. Richard D. Irwin, Inc., Homewood, IL.

Younger, M. S. 1979. *A Handbook for Linear Regression*. PWS-KENT, Boston.

Chapter Maxim *Multiple regression is an analysis of the dependencies among variables, while multiple correlation is an analysis of the interdependencies among variables.*

CHAPTER 16

MULTIPLE REGRESSION

Objectives

After studying this chapter and working the exercises, you should be able to

1. State the goal and objectives of a multiple regression analysis.
2. State the assumptions of the multiple regression model.
3. Fit the model to a set of data using the Regression processor.
4. Test the utility of the fitted model with the *F*-statistic.
5. Summarize the multiple regression computations into an ANOVA table.
6. Construct and interpret confidence intervals for an average value of *Y*.
7. Construct and interpret prediction intervals for a specific value of *Y*.
8. Contrast the assumptions of a multiple correlation analysis with those of a multiple regression analysis.
9. Compute and interpret the coefficients of multiple correlation and determination.
10. Interpret each entry in a correlation matrix.

*11. Use matrix algebra to execute objectives 3, 6, and 7.

12. Recognize applications of multiple regression in other business courses and in business settings.

In Chapter 15, we learned how to analyze the straight line relationship between a dependent variable and one independent variable. We will extend our discussion in this chapter to the situation where the dependent variable Y is related to more than one independent variable.

For example, the unit sales of a grocery item are related to several factors such as price, inches of shelf space for displaying the product, the distance the item is displayed from the floor, and the presence or absence of an advertising campaign. Thus, the dependent variable *unit sales* is a function of four independent variables in this case. As another example, financial analysts attempt to relate a company's earnings estimate to variables such as growth of the economy, year-end interest rates, the company's interest expenses, volume of sales, and/or manufacturing expenses. In both examples, we have mentioned only a few of the potential variables that *could* be related to the dependent variable.

The study of the relationship between a quantitative dependent variable Y and several independent variables is called a *multiple regression* analysis. In this chapter we present the basic concepts of multiple regression, patterning the discussion after the simple regression material in the previous chapter. However, a major difference between simple and multiple regression will be evident immediately. We used arithmetic and algebra to solve simple regression problems, but for multiple regression we

*Applies to optional section.

will rely on computer software to perform the calculations. Alternatively, in starred (optional) subsections we will present the matrix algebra formulation of multiple regression for those familiar with this area of mathematics.

16.1 THE MULTIPLE REGRESSION MODEL

Goal and Objectives of Multiple Regression

Studies have shown that regression analysis is one of the most often used statistical techniques in business. Modeling the linear relationship between Y and X, as we did in Chapter 15, sets the stage for an extension of that work: linking two or more independent variables together to model and predict a dependent variable. The use of "multiple" independent variables, rather than just one, suggests the name of this analysis: multiple regression. Accordingly, we now will label our independent variables with subscripts, such as X_1, X_2, X_3, and so on.

Definition

The **goal of multiple regression** is to analyze the relationship between the dependent variable and two or more independent variables.

Our goal is achieved by meeting several objectives, as summarized in the following box.

Objectives of Multiple Regression

1. Identify the dependent variable and several potential independent variables.
2. Propose a statistical model relating the variables.
3. Fit the model.
4. Test the utility of the fitted model.
5. Use the fitted model for prediction, if applicable.

Note that the last four objectives here are identical to the simple regression objectives given in Chapter 15. Hence, much of our discussion in this chapter will parallel that of Chapter 15. Notice that our first step in the multiple regression analysis, though, is different from the recommendation to plot the bivariate data in a simple regression analysis. Since multiple regression may involve several independent variables, graphing the data is more difficult or impossible. For instance, a graph of X_1, X_2, X_3, and Y requires four dimensions. To plot just two independent variables and Y, we need a three-dimensional figure. It would not be as easy to detect patterns on a three-dimensional scatter plot as it would for a simple regression (two-dimensional) scatter plot.

In multiple regression, we first must identify potential variables in the analysis. This forces us to think critically about the problem and pose questions such as the following.

1. What independent variables *might be* related to the dependent variable? If the dependent variable were Y, kilowatt hours of electricity used per month by a residential customer, what independent variables would you consider including in the analysis? There are a couple of obvious ones such as X_1, average temperature during the month, and X_2, square feet of living area in the residence. But there are also some not so obvious ones such as X_3, number of people living at the residence (generally, electrical consumption increases as the number of people at the residence increases), or X_4, temperature setting of the refrigerator (refrigerators are one of the biggest consumers of electricity in a household).
2. Are data available for all the variables specified? For example, including the variables X_1, age, and X_2, annual income, in a multiple regression analysis to predict Y, consumer debt, may be ill-advised if we expect to obtain this information from surveys or applications—that is, job or credit card. An applicant is not required by law to list his or her age, and many people decline to reveal their annual income. As a result, we likely would encounter many missing observations for these variables. A related problem is the accuracy of data collected from surveys, since people are not always truthful in their responses, especially to sensitive questions.
3. Have we proposed pairs of independent variables that are closely or even perfectly related? Suppose the independent variables X_1, the number of long distance phone calls per month, and X_2, the cost of long distance phone calls per month, were included in an analysis to model Y, monthly expenses of sales representatives in a company. These independent variables would be strongly (positively) correlated with each other and would be contributing essentially the same information in the prediction of the dependent variable Y. Both variables may not be needed in the analysis; in fact, highly correlated independent variables are generally undesirable.

Statement of the Model

Once the brainstorming and defining of the variables is completed, we combine them into a statistical model. The following box summarizes the multiple regression model relating the dependent variable Y to several independent variables.

Multiple Regression Model

$$Y = \beta_0 + \beta_1 X_1 + \beta_2 X_2 + \ldots + \beta_k X_k + \epsilon \qquad (16\text{–}1)$$

Symbol	Name	Classification
Y	Dependent variable	Random variable
$X_1, X_2, \ldots, X_k$	Independent variables	"Fixed" variables
$\beta_0, \beta_1, \ldots, \beta_k$	Regression coefficients	Parameters
ϵ	Error term	Random variable

For general purposes, we set up the model with k independent variables, where k is an integer greater than or equal to 2, and designate them as $X_1, X_2, \ldots, X_k$.

Equation 16–1 involves variables—Y, the X_i's and ϵ—and parameters—the β_i's. The values of the variables Y and $X_1, X_2, \ldots, X_k$ result from sampling and therefore are known. However, the values of the β-parameters are unknown and must be estimated. This is a similar problem to the one we faced in Chapter 15 where we needed to estimate the Y-intercept, β_0, and the slope, β_1. Then we used the criterion of least squares to produce the estimators b_0 and b_1, respectively. Now, though we have more betas to estimate, the criterion of least squares still can be used to generate the estimators. This will be discussed in Section 16.3.

The error term in Equation 16–1 represents the difference between Y and the deterministic portion of the model, $\beta_0 + \beta_1 X_1 + \ldots + \beta_k X_k$:

$$\epsilon = Y - (\beta_0 + \beta_1 X_1 + \beta_2 X_2 + \ldots + \beta_k X_k)$$

Conceptually, ϵ represents the unpredictable factors that affect Y and any (as yet) unidentified independent variables related to Y.

In repeated sampling at each setting of the independent variables (for example, at $X_1 = 2$, $X_2 = 6$, and so on up to $X_k = 7$), we will experience positive and negative

Assumptions About ϵ in the Multiple Regression Model

1. At each setting of the values for $X_1, X_2, \ldots, X_k$, ϵ is a random variable.
2. At each setting of the values for $X_1, X_2, \ldots, X_k$, ϵ has a normal probability distribution with:

$$\text{Mean} = 0 \quad \text{and} \quad \text{Variance} = \sigma^2$$

3. For each different pair of settings of the values for $X_1, X_2, \ldots, X_k$, the error random variables are independent.

errors as well as error values of zero. In multiple regression analysis, we assume these errors occur randomly and follow a normal distribution with a mean of 0 and a variance of σ^2. Further, since errors exist at each setting of the X_i's, we require the errors to be independent, as in simple regression. The preceding box summarizes these assumptions.

This model is a general one. The variables and the regression coefficients will have specific meanings and interpretations in the context of the problem. Fitting the model—one of our objectives—logically would follow our present discussion. Before doing so, however, we preview in the next section the math tools needed to achieve our objectives.

COMMENTS

1. As in simple regression, the values of the independent variables are assumed to be fixed and exactly recorded, while the values of the regression coefficients are unknown.
2. The variable Y also is known as the *response* or *criterion variable*. The independent variables also are known as the *design, predictor,* or *explanatory variables* or as the *regressors*. The regression coefficients, β_0 through β_k, also are called the *beta parameters* or *regression parameters*.
3. Equation 16–1 plus the assumptions listed in the preceding box are called the *general linear model* in regression analysis. Linear refers to the understood exponent of 1 for every beta. That is, we can include a squared independent variable, like X_1^2 for example, but not a squared beta parameter, like β_1^2, in this linear model.

16.2 TOOLS FOR A MULTIPLE REGRESSION ANALYSIS

■ Computer

Multiple regression is intuitively appealing as an extension of simple regression, but it is mathematically more complex. For each independent variable added to the model, we add another equation to the set of normal equations. This dramatically increases the difficulty level of the problem. There are at least two alternative ways to deal with the increased complexity: Use a computer program or use matrix algebra.

Relying on a computer program to perform the number crunching for us is a two-edged sword. Although turning the job over to a computer facilitates the analysis, it tends to reduce our involvement in the problem. Creating a data file or inputting the data directly into a program may give us a false sense of believing the analysis is complete (and correct). The program processes the data and displays the output; we print the results and turn off the computer, symbolizing completion.

What is dangerous about this scenario is our dependence on the computer. The computer performs the computation, period. It does not propose the model, gather the data, check the assumptions, make the decisions, and so on. If we merely "run the program" without critically assessing the inputs *and* outputs, then we are assigning all our managerial functions to a machine or program. We must maintain the proper perspective about packaged computer programs. Our responsibility in using the computer as a tool is twofold: *Prepare* the data for input and *evaluate* the resulting output.

Although there are many statistical software packages that perform a multiple regression analysis, we will explain the input and output associated with the Regression processor on the disks that accompany this book. In Section 16.7, we will compare our output to that from other packages.

Preparing the data for input is a data management activity. In Chapter 2, we discussed several procedures for managing the data: maintaining a standard degree of accuracy for recording the values of each variable; checking for unusual values; assuring that there are no missing values; and creating a data file. We can use the accompanying software to create the data file once we have "housecleaned" the data. Example 16.1 contains data for a multiple regression problem involving Y and two independent variables X_1 and X_2 and illustrates the data preparation function of our analysis.

EXAMPLE 16.1

A multiple regression analysis is planned to relate the dependent variable Y, annual number of passengers arriving and departing an airport per airline, to two independent variables for a sample of $n = 12$ airlines. The independent variables are

X_1 = Average number of flights departing weekly

X_2 = Total number of local airline employees

The following data were recorded with values of Y representing units of 10,000 (for example, $Y = 5.1$ means 51,000).

Airline	Y	X_1	X_2
1	19.3	63	40
2	2.8	21	7
3	30.5	94	87
4	6.59	50	10
5	11.7	42	23
6	6.1	21	6
7	7.6	36	11
8	13.4	48	8
9	50.4	104	88
10	14.9	49	43
11	35.4	89	71
12	6.3	25	8

Create a Multivariate raw data set using the accompanying software; save the data set in a file named XM161 on a separate data disk; label the variables as follows:

$$Y = \text{Passengers}$$
$$X_1 = \text{Flights}$$
$$X_2 = \text{Employees}$$

To facilitate later processing, we recommend that you enter the values of Y in the first column of the file.

Solution:

Notice that the value of Y for airline 4 is carried out to 2 decimal places. To maintain a standard degree of accuracy, we will round the value of 6.59 to 6.6. As there are no missing or unusual values, we can proceed to set up the data file.

□

The data in Example 16.1 will be used to fit the model $Y = \beta_0 + \beta_1X_1 + \beta_2X_2 + \epsilon$ in Section 16.3, at which time we will explain the Regression processor. For now, our objective is to refamiliarize ourselves with creating and saving data sets.

*■ Matrix Algebra

A multiple regression problem also can be solved using matrix algebra. In this subsection and subsequent starred subsections in this chapter, we assume the reader has a background in matrix algebra. Specific skills that are necessary to comprehend this material include the matrix operations of addition, multiplication, transposition, and inversion. If you have not had a course in matrix algebra, skip the starred subsections and follow the discussion of the computer solution to multiple regression problems.

Recall that a *matrix* is a rectangular arrangement of numbers into rows and columns. The number of rows and columns is called the *dimension* of the matrix. For example, a 3 × 4 matrix means there are 3 rows and 4 columns. When the array has only one row or one column, we call it a *vector* rather than a matrix. Thus, a 3 × 1 array denotes a vector with 3 rows and 1 column. We use boldface letters like **A** and **B** to represent matrices and vectors.

Suppose we plan to relate two independent variables X_1 and X_2 to a dependent variable Y through the model $Y = \beta_0 + \beta_1X_1 + \beta_2X_2 + \epsilon$, based on the following set of sample data:

Y	X_1	X_2
10	1	1
8	2	0
5	3	1
6	4	1
3	5	1
1	6	0

*Optional

Each data point is modeled by the equation, $Y = \beta_0 + \beta_1 X_1 + \beta_2 X_2 + \epsilon$ as follows:

$$\begin{aligned} 10 &= \beta_0 + 1\beta_1 + 1\beta_2 + \epsilon_1 \\ 8 &= \beta_0 + 2\beta_1 + 0\beta_2 + \epsilon_2 \\ 5 &= \beta_0 + 3\beta_1 + 1\beta_2 + \epsilon_3 \\ 6 &= \beta_0 + 4\beta_1 + 1\beta_2 + \epsilon_4 \\ 3 &= \beta_0 + 5\beta_1 + 1\beta_2 + \epsilon_5 \\ 1 &= \beta_0 + 6\beta_1 + 0\beta_2 + \epsilon_6 \end{aligned} \tag{16–2}$$

The subscripts on the error term—epsilon—were added to identify the different data points or settings of the independent variables. For reference, in Section 15.2 we demonstrated this modeling phenomenon for the simple regression model $Y = 8 - 2X + \epsilon$. Figure 15.1 graphically showed how the point $X = 3$, $Y = 3$ fit into that model, for instance.

The set of six equations in 16–2 can be formulated into one matrix equation.

Multiple Regression Model in Matrix Form

$$\mathbf{Y} = \mathbf{X}\boldsymbol{\beta} + \boldsymbol{\epsilon} \tag{16–3}$$

In applying Equation 16–3 to the equations in 16–2, we set up the Y-values into a 6×1 vector denoted by $\mathbf{Y}$:

$$\mathbf{Y} = \begin{bmatrix} 10 \\ 8 \\ 5 \\ 6 \\ 3 \\ 1 \end{bmatrix}$$

Similarly, we put the three regression coefficients of the model—β_0, β_1, and β_2—into a 3×1 vector denoted by $\boldsymbol{\beta}$:

$$\boldsymbol{\beta} = \begin{bmatrix} \beta_0 \\ \beta_1 \\ \beta_2 \end{bmatrix}$$

The error terms ϵ_1 through ϵ_6 are listed in a 6×1 vector labeled $\boldsymbol{\epsilon}$:

$$\boldsymbol{\epsilon} = \begin{bmatrix} \epsilon_1 \\ \epsilon_2 \\ \epsilon_3 \\ \epsilon_4 \\ \epsilon_5 \\ \epsilon_6 \end{bmatrix}$$

Finally, **X** is a 6 × 3 matrix of coefficients of the betas. From the equations in 16–2, we notice that the coefficient of β_0 is always the number 1, while the coefficients of β_1 and β_2 are the sample X-values for the variables X_1 and X_2, respectively. Matrix **X** is, therefore,

$$\mathbf{X} = \begin{bmatrix} 1 & 1 & 1 \\ 1 & 2 & 0 \\ 1 & 3 & 1 \\ 1 & 4 & 1 \\ 1 & 5 & 1 \\ 1 & 6 & 0 \end{bmatrix}$$

With these definitions of **Y**, **X**, $\boldsymbol{\beta}$, and $\boldsymbol{\epsilon}$, the set of six linear equations in 16–2 are formulated into one matrix equation $\mathbf{Y} = \mathbf{X}\boldsymbol{\beta} + \boldsymbol{\epsilon}$. To see this, perform the matrix multiplication $\mathbf{X}\boldsymbol{\beta}$. Recall that matrix multiplication is achieved by multiplying each element in a row of **X** by the corresponding element in a column of $\boldsymbol{\beta}$. The multiplication produces a 6 × 1 vector, since the product of a 6 × 3 matrix **X** and a 3 × 1 vector $\boldsymbol{\beta}$ takes the row dimension of **X** and the column dimension of $\boldsymbol{\beta}$:

$$\mathbf{X}\boldsymbol{\beta} = \underset{(6 \times 3)}{\begin{bmatrix} 1 & 1 & 1 \\ 1 & 2 & 0 \\ 1 & 3 & 1 \\ 1 & 4 & 1 \\ 1 & 5 & 1 \\ 1 & 6 & 0 \end{bmatrix}} \underset{(3 \times 1)}{\begin{bmatrix} \beta_0 \\ \beta_1 \\ \beta_2 \end{bmatrix}} = \underset{(6 \times 1)}{\begin{bmatrix} 1\beta_0 + 1\beta_1 + 1\beta_2 \\ 1\beta_0 + 2\beta_1 + 0\beta_2 \\ 1\beta_0 + 3\beta_1 + 1\beta_2 \\ 1\beta_0 + 4\beta_1 + 1\beta_2 \\ 1\beta_0 + 5\beta_1 + 1\beta_2 \\ 1\beta_0 + 6\beta_1 + 0\beta_2 \end{bmatrix}}$$

Adding the $\boldsymbol{\epsilon}$ vector to $\mathbf{X}\boldsymbol{\beta}$ yields the right-hand side of each equation in 16–2. Thus, $\mathbf{Y} = \mathbf{X}\boldsymbol{\beta} + \boldsymbol{\epsilon}$ is the matrix representation of the six equations in 16–2.

Most computer routines that perform a multiple regression analysis are based on the matrix formulation we described in this section. We need to concentrate on setting up the vectors and the **X** matrix, given a set of sample data. In the next section, we will learn how to use these arrays to find the estimates of the unknown β-parameters.

*CLASSROOM EXAMPLE 16.1

Identifying Matrices for Regression

A restaurant owner decides to advertise through DMA, a company specializing in direct-mail advertising. The restaurateur plans to promote a "buy one entree, get a second entree free" sale by printing coupons and having DMA mail them to residents in the surrounding neighborhood. Over a period of a year, the owner plans 10 mailings with coupons expiring in approximately one month. Define the variables as follows:

*Optional

Y = Number of coupons redeemed (in hundreds)
X_1 = Number of coupons printed (in thousands)
X_2 = Percentage of coupons mailed to apartment dwellers

The complete set of data for the 10 mailings is

Y	X_1	X_2
1	1	8
1	2	10
2	2	10
3	2	15
3	3	18
2	3	20
4	3	24
2	4	18
3	5	20
4	5	27

Suppose the multiple regression model $Y = \beta_0 + \beta_1X_1 + \beta_2X_2 + \epsilon$ is proposed.

a. Set up the **Y**, $\boldsymbol{\beta}$, and $\boldsymbol{\epsilon}$ vectors.
b. Set up the **X** matrix.
c. Identify the dimension of each vector in part a and the matrix in part b.

SECTION 16.2 EXERCISES

16.1 State the general multiple regression model and identify all terms in the model.

16.2 A multiple regression analysis involving two independent variables is based on the following data. Create a data set and name it XR162.

Y	X_1	X_2
7	−4	9
5	−3	5
4	−2	6
1	−1	2
3	0	2
1	1	2
−1	2	1
−2	3	4
−1	4	2

*16.3 Refer to the data in Exercise 16.2. Suppose the model $Y = \beta_0 + \beta_1 X_1 + \beta_2 X_2 + \epsilon$ is proposed.

a. Set up the $\mathbf{Y}$, $\boldsymbol{\beta}$, and $\boldsymbol{\epsilon}$ vectors.
b. Set up the $\mathbf{X}$ matrix.
c. Identify the dimension of each vector in part a and the matrix in part b.

16.4 Identify the objectives of a multiple regression analysis.

16.5 Nursing homes are the object of a study conducted by a health-care provider in order to model the daily semiprivate room rates. Three independent variables are proposed:

X_1 = The percentage of beds set up for skilled nursing care

X_2 = The total number of licensed beds

X_3 = The total number of registered nurses (RNs) and licensed practical nurses (LPNs) on the staff

Skilled nursing care is defined as nursing service available on a 24-hour basis for convalescent patients. The sample data for the proposed model $Y = \beta_0 + \beta_1 X_1 + \beta_2 X_2 + \beta_3 X_3 + \epsilon$ are

Y	X_1	X_2	X_3
$56	67	252	48
59.75	14	213	32
64	23	223	32
67	8	168	21
61	18	165	25
63	50	200	37
65.75	48	252	42
64	45	195	21
55	57	169	18
70	13	150	18

a. Define the dependent variable.
b. Are any data management activities needed before creating a data set? If so, perform the activity.
c. Create a data set and name it XR165.

16.6 Refer to Exercise 15.4, in which a simple regression model was developed relating Y, one-year return for a mutual fund, to X_1, the administrative fee or "load." Suppose a second independent variable is proposed, defined as X_2 = the percentage of the portfolio made up of blue chip stocks. The sample data are

Y	X_1	X_2
10%	0%	50%
7	6	60
15	2	40

(*continues*)

*Optional

(continued)

Y	X_1	X_2
22	0	30
18	8.5	33
12	8.5	50
24	4	25
19	0	25
13	0	50
7	4.5	45

a. The values of the variable representing the mutual fund's load are recorded in differing degrees of accuracy. What action, if any, would you recommend? Why?

b. Create a data set for these data and name it XR166. Do not round the values of X_1.

16.7 Suppose you wish to develop a multiple regression model to predict Y, the sale price of residential homes in your county last year. Specify four different independent variables that you think would be of use in predicting Y.

16.8 Engineers at a research and development plant for a large automaker hope to develop a model to predict Y, miles per gallon achieved in city driving for a sample of cars. They propose the following independent variables:

X_1 = Number of cylinders
X_2 = Engine size
X_3 = Suggested retail price of the car
X_4 = Weight of the car

Engine sizes are recorded in liters; retail price is coded in thousands of dollars ($X_3 = 5.6$ means \$5600); and weight is coded in thousands of pounds ($X_4 = 2$ means 2000 pounds). A random sample of data from 28 vehicles tested by an independent testing agency revealed the following data:

Car	Y	X_1	X_2	X_3	X_4
1	25	4	1.6	5.6	2.1
2	24	4	2.0	11.0	2.5
3	22	4	2.0	9.5	2.7
4	24	4	1.7	7.3	2.0
5	20	4	1.8	9.8	2.4
6	23	6	2.5	10.0	2.5
7	25	4	1.9	8.1	2.4
8	19	6	3.0	9.9	2.6
9	20	4	2.0	16.6	3.1
10	19	4	2.0	12.3	2.7
11	18	6	3.0	13.7	3.1
12	19	4	2.0	12.3	2.7
13	17	8	3.8	14.3	3.2
14	24	4	2.2	9.5	2.6
15	19	6	3.8	10.4	3.0
16	21	4	2.3	9.5	2.9

(continues)

Car	Y	X_1	X_2	X_3	X_4
17	23	4	2.5	7.8	2.5
18	17	8	5.0	11.6	3.3
19	18	6	4.3	10.6	3.6
20	17	8	4.1	19.7	3.3
21	18	8	5.0	24.6	3.8
22	21	4	2.3	34.8	3.0
23	19	6	3.8	16.8	3.2
24	19	6	2.8	8.9	2.6
25	17	8	5.0	10.7	3.2
26	17	6	3.0	16.8	3.1
27	25	4	2.0	23.4	2.7
28	17	8	4.1	32.1	3.4

Create a data set for these data. Name it XR168.

***16.9** Refer to the data in Exercise 16.8. If we were to set up the data into arrays, identify the dimension of the following vectors or matrices:

a. **Y**

b. $\boldsymbol{\beta}$

c. **X**

16.10 To analyze the pension funds of companies in a metropolitan area, a marketing research firm randomly selected several companies and interviewed their investment managers. The following variables were identified as having a potential relation to Y, the net assets of the company's pension funds (in millions of dollars):

X_1 = Assumed rate of return

X_2 = Number of plans

X_3 = Number of vested employees

X_4 = Number of firms managing the pension funds

Sixteen companies were selected; the available data are

Firm	Y	X_1	X_2	X_3	X_4
1	15.0	7.5	1	373	2
2	29.5	7.0	1	914	1
3	33.3	7.0	1	587	3
4	53.5	8.5	2	2084	3
5	51.2	n/a	1	3078	3
6	13.0	8.0	1	440	1
7	147.2	9.5	9	3530	3
8	6.7	8.0	1	303	1
9	13.4	7.0	2	641	2
10	30.0	8.0	1	605	1
11	50.8	5.5	3	720	2
12	33.7	8.95	1	605	1
13	8.8	7.0	3	321	4

(*continues*)

(*continued*)

Firm	Y	X_1	X_2	X_3	X_4
14	13.4	8.0	1	273	1
15	9.2	9.0	3	426	1
16	Refused to disclose figures				

Note: n/a means not available.

a. The data do not conform to sound data management practices. Identify the violation(s).
b. To correct the problem for the variable X_1 with firm 12, round up.
c. There is also a problem for the variable X_1 with firm 5. Is this a sampling error or a nonsampling error? (*Hint:* See Section 1.3.) Correct the problem by computing $\overline{X}_1$, based on the fourteen available values, and assigning that value to firm 5.
d. Create a data set named XR1610 for these data.

***16.11** Refer to the data in Exercise 16.6. Set up the **X** matrix and the **Y**, $\boldsymbol{\beta}$, and $\boldsymbol{\epsilon}$ vectors for fitting the model $Y = \beta_0 + \beta_1 X_1 + \beta_2 X_2 + \epsilon$ to the data. Identify the dimension of each array.

***16.12** A random sample of $n = 13$ architectural firms was selected in order to model the relationship among these variables:

Y = Total value of construction projects initiated in the last year (in millions of dollars)
X_1 = Percentage of projects classified as commercial
X_2 = Number of registered architects

The sample data are

Y	X_1	X_2
58.0	12	11
9.6	75	3
60.0	10	3
28.0	0	8
10.0	8	5
10.0	67	5
8.0	40	3
60.0	50	4
18.0	55	5
9.9	75	4
14.0	60	4
7.0	60	4
6.4	30	4

a. Set up the **Y**, $\boldsymbol{\beta}$, and $\boldsymbol{\epsilon}$ vectors for the model $Y = \beta_0 + \beta_1 X_1 + \beta_2 X_2 + \epsilon$.
b. Set up the **X** matrix.
c. Identify the dimension of each vector in part a and the **X** matrix in part b.

***16.13** Refer to Classroom Example 16.1. The restaurant owner was considering a potential third independent variable, X_3 = cost of printing the coupons. Why might you discourage him or her from adding this variable to the analysis?

16.3 FITTING THE MODEL

Least-Squares Criterion

The multiple regression model $Y = \beta_0 + \beta_1X_1 + \beta_2X_2 + \ldots + \beta_kX_k + \epsilon$ contains $(k + 1)$ unknown regression coefficients—β_1 through β_k as well as β_0—that must be estimated. Although there are many criteria for developing estimators, we prefer the *least-squares criterion,* as we discussed in the simple regression case.

Briefly, this criterion minimizes the sum of the squared errors between the observed Y-values and the fitted Y-values. In multiple regression, it is harder to visualize this phenomenon because the number of variables involved often exceeds the three dimensions in which we can graph data. The beauty of the mathematics is that we can extend the principle of least squares into $(k + 1)$ dimensions theoretically, even though practically we cannot see it work.

Let us adopt similar notation to that in Chapter 15 and denote the estimated regression coefficients by b_0, b_1, b_2, and so on up to b_k. The fitted multiple regression equation is as follows.

Fitted Multiple Regression Model

$$\hat{Y} = b_0 + b_1X_1 + b_2X_2 + \ldots + b_kX_k \tag{16–4}$$

The fitting errors are $Y - \hat{Y}$, where Y is the observed value of the dependent variable. The sum of the squared errors, $\Sigma(Y - \hat{Y})^2$, is denoted by SSE. The following definition formalizes the least-squares criterion.

Definition

The **criterion of least squares** for fitting the multiple regression model in Equation 16–4 is: Select estimated regression coefficients $b_0, b_1, \ldots, b_k$ that minimize $SSE = \Sigma(Y - \hat{Y})^2$.

Implementing the criterion of least squares is difficult without the aid of the computer or the matrix algebra tools we introduced in Section 16.2. To minimize SSE, we would have to generate a set of $(k + 1)$ normal equations through partial differentiation and then solve these equations simultaneously for $b_0, b_1, \ldots,$ and b_k. Both the computer and matrix algebra approaches can solve the set of normal equations more efficiently than we can by hand.

COMMENT The statistics b_0, b_1, . . . , b_k have many names, including *estimated regression coefficients, estimated beta parameters, estimates of the model parameters,* or *least-squares estimates*. Some books use a hat notation instead of the lower case letters: $\hat{\beta}_0$ in lieu of b_0, $\hat{\beta}_1$ for b_1, and so on.

Processor: Regression

In Chapter 15, we used the Simple Regression option from the Regression processor to help us analyze the relationship between Y and X. The Regression processor also has a Multiple Regression option. Note that the Multiple Regression option is limited to a maximum of ten independent variables.

To demonstrate how to use the processor to fit a multiple regression model, we will use the data in Example 16.1. Earlier, we created and saved a data file called XM161 consisting of one dependent variable, Y = passengers, and two independent variables, X_1 = flights and X_2 = employees. Suppose we wish to fit the multiple regression model $Y = \beta_0 + \beta_1 X_1 + \beta_2 X_2 + \epsilon$ to the data in the XM161 file.

Move the highlight to the Files section of the main menu and touch Enter. With your data disk in drive B and with the Retrieve option highlighted, touch Enter again. Be sure to change the directory from A to B, if you have not already done so, before proceeding. Highlight the XM161 file in the listing and touch Enter to retrieve it.

Go to Processors on the main menu, select the Regression processor, and then select the Multiple Regression option from the submenu. Read the Introduction screen and read in the t-table. Proceed to the Data section and confirm the variable designations before continuing on to Calculations.

When the calculations are complete, move to the Results section of the menu and select the first option, Fitted Model. The Fitted Model screen, shown in Figure 16.1, contains not only the estimated regression coefficients but also several other important measures that will be discussed shortly. As you can see, the fitted multiple regression model is

$$\hat{Y} = -4.86857 + .30669X_1 + .16550X_2$$

The error sum of squares, *SSE*, is 202.4730853627. This is the smallest possible value for *SSE*, given the data set of n = 12 observations in Example 16.1. This means that *any* other fitted model relating Y to X_1 and X_2 in a linear fashion (such as $\hat{Y} = -5 + .4X_1 + .2X_2$, for example) will produce a value for *SSE* larger than 202.47309, to five decimal places. Therefore, the model in Figure 16.1 is called the *best fitting model*.

The estimated standard error of the regression that appears in Figure 16.1 will be explained next, while a discussion of the coefficient of determination will appear in Section 16.6.

COMMENT The Regression processor is programmed to display the estimated regression coefficients to five decimal places and *SSE* to 10 decimal places. In general, we recommend rounding the values of b_0, b_1, . . . , b_k to three decimal places. If the first two or three decimal places of an estimated regression coefficient contain zeroes, then retain four or five digits to the right of the decimal.

```
≡                         Multiple Regression                          READY
  Introduction  Data  Calculations |Results| Exit              |  F1=Help

  The fitted multiple regression model, to 5 decimal places, is:
     ^
     Y = -4.86857  +0.30669X1  +0.16550X2

  The sum of the squared errors (SSE), to 10 decimal places, is:
                                      ^
                    SSE = Σ(Y - Y)² = 202.4730853627
                                                    ^
  The estimated standard error of the regression (σ), to 5 decimal
  places, is:
                    ^
                    σ = √ 22.49701   =   4.74310

  The coefficient of determination (R²), to 5 decimal places, is:

                              R² =  0.91267
```

Figure 16.1 Fitted Model Screen for Example 16.1 Data

Estimating σ

In the general multiple regression model $Y = \beta_0 + \beta_1 X_1 + \ldots + \beta_k X_k + \epsilon$, we assume that the error term, ϵ, is a normally distributed random variable with a mean of 0 and a variance of σ^2. Since the β-values are unknown, the values of ϵ and its variance are unknown too. Using the least-squares criterion though, we have developed estimates of the betas. The domino effect produces estimates of ϵ and σ^2 from the fitting errors, the $(Y - \hat{Y})$s. Letting e represent the estimate of ϵ, we have:

$$e = Y - \hat{Y} = Y - (b_0 + b_1 X_1 + \ldots + b_k X_k)$$

The variance of the observed e's generates our estimate of σ^2.

Notice that $\hat{Y}$ involves $(k + 1)$ estimated parameters—$b_0, b_1, b_2, \ldots, b_k$. The set of b_i's is the simultaneous solution to the $(k + 1)$ normal equations and represents $(k + 1)$ constraints on the n sample data points. Thus, there are $n - (k + 1)$ nonconstrained values, or *degrees of freedom, df*. For the general multiple regression model of Equation 16–1, our estimate of σ^2 is

$$\hat{\sigma}^2 = \frac{\Sigma(Y - \hat{Y})^2}{df} = \frac{SSE}{n - (k + 1)} \qquad \textbf{(16–5)}$$

Equation 16–5 is the variance of the observed fitting errors. The square root of $\hat{\sigma}^2$ is the *estimated standard error of the regression,* see Equation 16–6, and gives a measure of the goodness of the model's fit. Although small values of $\hat{\sigma}$ intuitively indicate a tight fit, we will defer to Section 16.4 and define "small" (or "large") in terms of the F-statistic in a test of model utility.

Estimated Standard Error of the Regression

$$\hat{\sigma} = \sqrt{\hat{\sigma}^2} = \sqrt{\frac{SSE}{n - (k + 1)}} \quad (16\text{–}6)$$

COMMENT Note that the least-squares criterion of minimizing $\Sigma(Y - \hat{Y})^2$ is equivalent to minimizing Σe^2. We also refer to the fitting errors as *residuals*. In Chapter 17, we will analyze the residuals in more detail.

EXAMPLE 16.2

A study of the lodging industry was conducted to develop a model for predicting occupancy rates. Three independent variables and the model $Y = \beta_0 + \beta_1 X_1 + \beta_2 X_2 + \beta_3 X_3 + \epsilon$ were proposed

where

Y = Occupancy rate, expressed as a percentage of maximum capacity

X_1 = Number of competitors within a one-mile radius

X_2 = Nightly room rate

X_3 = Advertising expense, as a percentage of annual gross revenues

A random sampling of 28 hotels and motels yielded the following data:

Entity	Y	X_1	X_2	X_3
1	74	2	$23	9
2	54	1	26	0
3	48	3	25	1
4	69	0	26	5
5	73	0	25	3
6	75	4	34	9
7	54	3	32	7
8	53	1	39	6
9	51	4	40	3
10	65	0	37	5
11	56	4	42	4
12	38	3	34	1
13	73	1	37	10
14	60	3	42	6
15	59	0	36	5
16	65	4	80	2
17	68	4	75	9
18	48	2	73	3
19	67	1	61	2

(*continues*)

Entity	Y	X_1	X_2	X_3
20	60	2	$70	4
21	60	1	58	8
22	53	4	75	3
23	60	1	65	1
24	66	3	52	6
25	76	0	50	5
26	75	2	92	8
27	51	2	87	6
28	63	1	104	9

a. Create and save a data file named XM162.

b. Use the Regression processor to fit the proposed model to these data.

c. What is the equation of the fitted model?

d. Report the values of *SSE* and of $\hat{\sigma}$.

e. How many degrees of freedom are there?

Solution:

a. No further data management activities are necessary before entering the data in a data set. Be sure to save the data in a file on your data disk.

b. Verify the Fitted Model screen shown in Figure 16.2.

```
≡                        Multiple Regression                              READY
  Introduction  Data  Calculations [Results] Exit                   | F1=Help

  The fitted multiple regression model, to 5 decimal places, is:
  ^
    Y =  56.31857  -1.94724X1  -0.00801X2  +1.84047X3

  The sum of the squared errors (SSE), to 10 decimal places, is:
                                  ^
                    SSE = Σ(Y - Y)² = 1633.3318733458

  The estimated standard error of the regression (σ), to 5 decimal
  places, is:
                  ^
                  σ = √ 68.05549    =  8.24958

  The coefficient of determination (R²), to 5 decimal places, is:

                         R² =  0.37389
```

Figure 16.2 Fitted Model Screen for Example 16.2

c. The fitted model is $\hat{Y} = 56.31857 - 1.94724X_1 - .00801X_2 + 1.84047X_3$.

d. To five decimal places, $SSE = 1633.33187$ and $\hat{\sigma} = 8.24958$.

e. There are $n - (k + 1) = 28 - (3 + 1) = 24$ degrees of freedom.

□

Relative to the simple regression models we encountered in Chapter 15, a multiple regression model such as $\hat{Y} = 56.31857 - 1.94724X_1 - .00801X_2 + 1.84047X_3$ possesses the following similarities and differences.

Slopes The interpretation of slope in simple regression is the change in $E(Y)$ per unit change in X. In multiple regression, a slope is the change in $E(Y)$ per unit change in X, *given that all other X's are held constant*. In the equation above, the coefficient of -1.94724 for X_1 indicates that Y, the occupancy rate, will decrease by about 2 percent (on average) if X_1 is increased by 1.0—that is, if another competitor appears within a one-mile radius—given that X_2 and X_3 remain constant.

Y-Intercept The value of $b_0 = 56.31857$ in the above equation is the model's Y-intercept, as in simple regression. If the relevant range—that is, the interval between the smallest and largest X-values in the sample—for each independent variable includes the number 0, then we can interpret b_0 as more than just the Y-intercept. In such a case, b_0 would represent the predicted Y-value when all independent variables have a value of 0. For the data set in Example 16.2, notice that 0 is not included in the relevant range for the variable X_2; thus, b_0 is not interpretable as a predicted Y-value, but only as the Y-intercept.

Types of Relations In simple regression, we could describe the relation between Y and X as direct or inverse, depending on the sign of the slope. In multiple regression, we have more than one slope coefficient and, therefore, can have a mixture of direct and inverse relationships. In the previous equation, X_3 has a direct relation to Y while X_1 and X_2 each have an inverse relation with Y when these three predictors are used.

Fitting Errors Fitting errors can be determined easily in multiple regression, just as they were in simple regression. For instance, let us choose entity 1, where $X_1 = 2$, $X_2 = 23$, and $X_3 = 9$. Placing these values into our fitted model, we obtain $\hat{Y} = 56.31857 - 1.94724(2) - .00801(23) + 1.84047(9) = 68.80409$, or 68.8 to one decimal place. Since the actual Y-value for entity 1 is 74, the fitting error is

$$(Y - \hat{Y}) = (74 - 68.8) = 5.2$$

As Example 16.2 has demonstrated, the process of fitting a multiple regression model to a set of data is greatly simplified with the aid of a computer program. If you are using a different program than the one supplied with the text, you might wish to match our Fitted Model screen to your output. Indeed, in Section 16.7, we will do this for several other packages.

In the next subsection, we outline the matrix algebra solution to a multiple regression problem. Many computer routines utilize this matrix approach in processing the data, and some present it as output.

*■ Matrix Solution

In the general matrix equation $\mathbf{Y} = \mathbf{X}\boldsymbol{\beta} + \boldsymbol{\epsilon}$, we defined $\mathbf{Y}$ as an $n \times 1$ vector of observed Y-values, $\mathbf{X}$ as an $n \times (k + 1)$ matrix of observed settings for the independent variables plus a column of 1's, and $\boldsymbol{\beta}$ as a $(k + 1) \times 1$ vector of unknown regression parameters. To fit this model to a set of data, we must estimate the regression parameters. Let us denote the estimated beta parameters by the $(k + 1) \times 1$ vector $\mathbf{b}$.

$$\mathbf{b} = \begin{bmatrix} b_0 \\ b_1 \\ b_2 \\ \cdot \\ \cdot \\ \cdot \\ b_k \end{bmatrix}$$

It can be shown that the $k + 1$ normal equations can be expressed into matrix notation as follows.

Normal Equations in Matrix Notation

$$(\mathbf{X'X})\,\mathbf{b} = \mathbf{X'Y} \qquad (16\text{–}7)$$

Recall that the symbol ′ (prime) means *transpose,* and that transposing a matrix interchanges rows and columns; thus, while $\mathbf{X}$ has dimension $n \times (k + 1)$, $\mathbf{X'}$ is a $(k + 1) \times n$ matrix. Therefore, the product $\mathbf{X'X}$ is a square matrix with dimension $(k + 1) \times (k + 1)$.

To solve Equation 16–7 for $\mathbf{b}$, we must invert the $\mathbf{X'X}$ matrix. As long as $\mathbf{X'X}$ is nonsingular, $(\mathbf{X'X})^{-1}$ exists, and the matrix solution to Equation 16–7 is as follows.

Estimated Regression Coefficients via Matrix Algebra

$$\mathbf{b} = (\mathbf{X'X})^{-1}\,\mathbf{X'Y} \qquad (16\text{–}8)$$

For a multiple regression analysis with two or three independent variables, the inverse of $\mathbf{X'X}$ can be found easily by hand. However, the problem becomes more difficult as the size of the $\mathbf{X'X}$ matrix increases. This is when computer programs achieve their

*Optional

efficiency. Special subroutines have been developed for the expressed purpose of inverting a square, nonsingular matrix. The computational accuracy of a multiple regression computer program is directly linked to the accuracy of the matrix inversion subroutine. Without qualification, we can say that the $(\mathbf{X}'\mathbf{X})^{-1}$ matrix is the linchpin of the multiple regression analysis.

COMMENT The $\mathbf{X}'\mathbf{X}$ matrix will be nonsingular whenever none of the independent variables is linearly related to one (or more) of the other independent variables. For example, if racy in the $(\mathbf{X}'\mathbf{X})^{-1}$ matrix. For details, the interested reader is directed to Chapter 7 will be singular and the $(\mathbf{X}'\mathbf{X})^{-1}$ matrix will not exist. If X_3 is almost, but not exactly, linearly related to X_1 and X_2, for example, we may have trouble finding the inverse of the $\mathbf{X}'\mathbf{X}$ matrix. Special computing techniques are necessary to ensure accuracy in the $(\mathbf{X}'\mathbf{X})^{-1}$ matrix. For details, the interested reader is directed to Chapter 7 of *Theory and Application of the Linear Model* by F. A. Graybill (1976).

EXAMPLE 16.3

In Section 16.2, we set up the following $\mathbf{Y}$ vector and $\mathbf{X}$ matrix based on the model $Y = \beta_0 + \beta_1 X_1 + \beta_2 X_2 + \epsilon$:

$$\mathbf{Y} = \begin{bmatrix} 10 \\ 8 \\ 5 \\ 6 \\ 3 \\ 1 \end{bmatrix} \qquad \mathbf{X} = \begin{bmatrix} 1 & 1 & 1 \\ 1 & 2 & 0 \\ 1 & 3 & 1 \\ 1 & 4 & 1 \\ 1 & 5 & 1 \\ 1 & 6 & 0 \end{bmatrix}$$

Let the least-squares estimates of the regression parameters be denoted by $\mathbf{b}$.

a. Find the $\mathbf{X}'\mathbf{X}$ matrix and the $\mathbf{X}'\mathbf{Y}$ vector.

b. Find $\mathbf{b}$ using Equation 16–8.

Solution:

a. The transpose of matrix $\mathbf{X}$ is

$$\mathbf{X}' = \begin{bmatrix} 1 & 1 & 1 & 1 & 1 & 1 \\ 1 & 2 & 3 & 4 & 5 & 6 \\ 1 & 0 & 1 & 1 & 1 & 0 \end{bmatrix}$$

Thus, the products $\mathbf{X}'\mathbf{X}$ and $\mathbf{X}'\mathbf{Y}$ are

$$\mathbf{X}'\mathbf{X} = \begin{bmatrix} 1 & 1 & 1 & 1 & 1 & 1 \\ 1 & 2 & 3 & 4 & 5 & 6 \\ 1 & 0 & 1 & 1 & 1 & 0 \end{bmatrix} \begin{bmatrix} 1 & 1 & 1 \\ 1 & 2 & 0 \\ 1 & 3 & 1 \\ 1 & 4 & 1 \\ 1 & 5 & 1 \\ 1 & 6 & 0 \end{bmatrix} = \begin{bmatrix} 6 & 21 & 4 \\ 21 & 91 & 13 \\ 4 & 13 & 4 \end{bmatrix}.$$

and

$$\mathbf{X'Y} = \begin{bmatrix} 1 & 1 & 1 & 1 & 1 & 1 \\ 1 & 2 & 3 & 4 & 5 & 6 \\ 1 & 0 & 1 & 1 & 1 & 0 \end{bmatrix} \begin{bmatrix} 10 \\ 8 \\ 5 \\ 6 \\ 3 \\ 1 \end{bmatrix} = \begin{bmatrix} 33 \\ 86 \\ 24 \end{bmatrix}$$

b. The inverse of $\mathbf{X'X}$ is

$$(\mathbf{X'X})^{-1} = \frac{1}{134}\begin{bmatrix} 195 & -32 & -91 \\ -32 & 8 & 6 \\ -91 & 6 & 105 \end{bmatrix}$$

(Refer to the *Math Review and Student's Partial Solutions Manual* for details about inverting a matrix.) The fraction 1/134 in front of the matrix is a multiplier of each element in the matrix. Thus, the element in row 3, column 3 is 105/134, for example. Using Equation 16–8, we have

$$\mathbf{b} = (\mathbf{X'X})^{-1}\,\mathbf{X'Y} = \frac{1}{134}\begin{bmatrix} 195 & -32 & -91 \\ -32 & 8 & 6 \\ -91 & 6 & 105 \end{bmatrix}\begin{bmatrix} 33 \\ 86 \\ 24 \end{bmatrix}$$

$$= \frac{1}{134}\begin{bmatrix} 1499 \\ -224 \\ 33 \end{bmatrix} = \begin{bmatrix} 11.186567 \\ -1.671642 \\ .246269 \end{bmatrix}$$

(fractional) (decimal)

Both the fractional and decimal forms of **b** are presented. In subsequent calculations, we prefer the fractional form, but in terms of the fitted model we use the decimal form. Thus, the fitted multiple regression equation is $\hat{Y} = 11.187 - 1.672X_1 + .246X_2$.

□

COMMENTS

1. We recommend inverting the $\mathbf{X'X}$ matrix using determinants if possible. This method separates the fractional and integer parts of each element of the matrix and retains a higher degree of accuracy over the row operation method.
2. Matrix algebra also facilitates the computation of the estimated standard error of the regression, $\hat{\sigma}$. In matrix notation, *SSE* is expressed as

$$SSE = \mathbf{Y'Y} - \mathbf{b'X'Y} \qquad (16\text{–}9)$$

where $\mathbf{Y'}$ is the transpose of vector $\mathbf{Y}$, and $\mathbf{b'}$ is the transpose of vector $\mathbf{b}$. Equation 16–6 then can be used to find $\hat{\sigma}$.

EXAMPLE 16.4

Refer to Example 16.3. Find $\hat{\sigma}$ using Equation 16–9 for *SSE*.

Solution:

From Equation 16–9, we compute

$$
\begin{aligned}
SSE &= \mathbf{Y}'\mathbf{Y} - \mathbf{b}'\mathbf{X}'\mathbf{Y} \\
&= [10 \quad 8 \quad 5 \quad 6 \quad 3 \quad 1]\begin{bmatrix} 10 \\ 8 \\ 5 \\ 6 \\ 3 \\ 1 \end{bmatrix} - \frac{1}{134}[1499 \quad -224 \quad 33]\begin{bmatrix} 33 \\ 86 \\ 24 \end{bmatrix} \\
&= 235 - \frac{1}{134}[30995] \\
&= 235 - 231.30597 \qquad = 3.69403
\end{aligned}
$$

With $n = 6$ and $k = 2$, from Equation 16–6, we have

$$\hat{\sigma} = \sqrt{\frac{SSE}{n - (k + 1)}} = \sqrt{\frac{3.69403}{6 - 3}} = \sqrt{1.2313433} = 1.1096591$$

□

COMMENT Carry out the computations of $\hat{\sigma}$ to at least five decimal place accuracy, if possible. When computing *SSE* via Equation 16–9, use the fractional form of the vector **b** rather than the decimal form to maintain as much accuracy as possible.

Equation 16–8 is an all-purpose equation. Regardless of the sample size or the number of independent variables in a multiple regression analysis, the single matrix equation $\mathbf{b} = (\mathbf{X}'\mathbf{X})^{-1}\mathbf{X}'\mathbf{Y}$ will generate the least-squares estimates of the regression coefficients. In future starred subsections, we will encounter the $(\mathbf{X}'\mathbf{X})^{-1}$ matrix in other formulas for the multiple regression analysis. Take time and care in obtaining it.

*CLASSROOM EXAMPLE 16.2

Fitting a Multiple Regression Model with Matrix Algebra

We wish to fit the model $Y = \beta_0 + \beta_1 X_1 + \beta_2 X_2 + \epsilon$ to the following data set:

Y	5	4	1	3	1	−1	−2
X_1	−3	−2	−1	0	1	2	3
X_2	9	4	1	0	1	4	9

*Optional

a. Set up the **X** matrix and the **Y** vector. Indicate the dimension of each array.
b. Find **X′X** and **X′Y**. Indicate the dimension of each product.
c. Compute $(\mathbf{X'X})^{-1}$.
d. What would be the dimension of the product $(\mathbf{X'X})^{-1}$ times **X′Y**?
e. Use Equation 16–8 to find the estimates of the regression coefficients. Write the equation of the fitted model.
f. Determine $\hat{\sigma}$, the estimated standard error of the regression using Equations 16–9 and 16–6.

SECTION 16.3 EXERCISES

16.14 State the criterion of least squares for fitting a multiple regression model.

16.15 Are the following statements true or false?
a. The error term in the multiple regression model is a random variable assumed to follow a normal probability distribution.
b. A residual is the difference between Y and $\hat{Y}$.
c. The more fitting errors that are negative numbers, the smaller SSE is.
d. We call $\hat{\sigma}$ the estimated standard error of the regression.
e. If $\hat{\sigma}$ were equal to 0, then all the fitting errors would be zero.

16.16 Suppose the fitted model in a multiple regression analysis is $\hat{Y} = -4 + 2X_1 - X_2 - 1.5X_3$ based on $n = 24$ entities.
a. Find the fitting error associated with the sample data point $Y = 8$, $X_1 = 4$, $X_2 = 2$, and $X_3 = -6$.
b. Repeat the instructions to part a for the sample data point $Y = -4$, $X_1 = 1$, $X_2 = -5$, and $X_3 = 5$.
c. If both of the data points in parts a and b were used to develop the fitted model, explain why it is impossible for SSE to equal the value 5 for this model.
d. How many degrees of freedom for error are there?
e. Interpret the coefficient of X_2 in the fitted model.

16.17 Find the estimated standard error of the regression in the following situations.
a. $SSE = 4.2319$; $n = 18$; and $k = 2$.
b. $\Sigma e^2 = 157.28$; $n = 41$; and $k = 6$.
c. $SSE = 98.1111$; and $df = 12$.

16.18 Refer to Exercise 16.2. Use the Regression processor and the data set named XR162 to fit the model $Y = \beta_0 + \beta_1 X_1 + \beta_2 X_2 + \epsilon$. Give the equation of the fitted model as your answer. Indicate whether the relation between X_1 and Y is direct or inverse. Repeat for X_2 and Y.

16.19 The model $Y = \beta_0 + \beta_1 X_1 + \beta_2 X_2 + \beta_3 X_3 + \epsilon$ is to be fit to the following sample of data:

Y	X_1	X_2	X_3
15	15	35	2.4
21	22	20	5.3
7	9	30	2.9
16	5	20	2.3

(continues)

(continued)

Y	X_1	X_2	X_3
12	16	45	5.6
24	17	55	5.0
29	8	25	4.1
8	7	46	4.4
17	3	16	2.4
13	6	43	3.1
19	8	38	3.8
19	20	50	5.8

a. Use the Regression processor to fit the model to the data. What is the equation of the fitted model?
b. What is the numerical value of the estimate of σ?
c. For the Y-value of 15, what is the fitting error? (Round $\hat{Y}$ to two decimal places.)
d. Save the data in a file named XR1619.
e. Is b_0 interpretable as a predicted Y-value? If so, give its interpretation. If not, explain why.

16.20 A multiple regression analysis based on $n = 24$ data points yielded the following fitted model:

$$\hat{Y} = 130.0093 - 3.5017X_1 + .0034X_2$$

The error sum of squares was 465.1348. Each of the following statements is incorrect; your job is to correct each.
a. For each one-unit change in X_1, we expect Y to decrease by -3.5017 units.
b. The estimated standard error of the regression is 22.1493.
c. There is a direct relation between b_0 and Y.
d. The fitting error for the data point $Y = 50$, $X_1 = 32$, and $X_2 = 1024$ is 52.1565.
e. If we reduce the value of b_2 from .0034 to .0014, the error sum of squares will be less than 465.1348.

16.21 Refer to Exercise 16.8.
a. Fit the model $Y = \beta_0 + \beta_1X_1 + \beta_2X_2 + \beta_3X_3 + \beta_4X_4 + \epsilon$ to the data set in the file XR168 using the Regression processor. What is the equation of the fitted model?
b. What is the value of $\hat{\sigma}$?

*16.22 A sales manager for a pharmaceutical company wishes to develop a regression model relating Y = sales in thousands of dollars ($Y = 16.3$ means \$16,300, for example) to two independent variables: X_1 = years of experience and X_2 = number of dependent children per sales representative. The manager collected sales data from the first quarter of this year on these variables for 16 sales representatives. The relevant ranges for each variable are as follows:

Variable	Minimum	Maximum
Y	7.8	50.9
X_1	0	6
X_2	0	3

*Optional

Some of the intermediate matrix calculations are

$$\mathbf{X'Y} = \begin{bmatrix} 368.5 \\ 1168.6 \\ 465.4 \end{bmatrix} \qquad (\mathbf{X'X})^{-1} = \frac{1}{9134}\begin{bmatrix} 1769 & -446 & -111 \\ -446 & 288 & -256 \\ -111 & -256 & 735 \end{bmatrix} \quad \mathbf{Y'Y} = 11142.29$$

a. How many degrees of freedom for error are there?
b. Find the equation of the fitted model.
c. What is the interpretation, if any, of b_0?
d. Does the variable *years of experience* have a direct or inverse relation with Y?
e. What sales level does the model predict for a sales representative with five years of experience and one dependent child?
f. Determine $\hat{\sigma}$.

***16.23** If a multiple regression analysis with three independent variables is performed on a data set with $n = 24$ entities, give the dimensions of the following vectors or matrices:

a. **Y** b. **X**
c. **X′** d. **X′Y**
e. **X′X** f. $(\mathbf{X'X})^{-1}\mathbf{X'Y}$

***16.24** Refer to Classroom Example 16.1. The proposed model is $Y = \beta_0 + \beta_1 X_1 + \beta_2 X_2 + \epsilon$, where

Y = Number of coupons redeemed (in hundreds)
X_1 = Number of coupons printed (in thousands)
X_2 = Percentage of coupons mailed to apartment dwellers

a. Find **X′X** and **X′Y**.
b. Compute $(\mathbf{X'X})^{-1}$.
c. Use Equation 16–8 to find the equation of the best fitting model.
d. Determine $\hat{\sigma}$.

***16.25** Refer to Exercise 16.12.

a. Find **X′Y** and verify the following:

$$\mathbf{X'X} = \begin{bmatrix} 13 & 542 & 63 \\ 542 & 31272 & 2257 \\ 63 & 2257 & 367 \end{bmatrix}$$

b. Verify that the inverse of the **X′X** matrix is the following:

$$(\mathbf{X'X})^{-1} = \frac{1}{5181163}\begin{bmatrix} 6382775 & -56723 & -746842 \\ -56723 & 802 & 4805 \\ -746842 & 4805 & 112772 \end{bmatrix}$$

c. Determine the least-squares estimates of the parameters. Write out the regression equation.

***16.26** A survey of weekly city business newspapers was conducted to collect data on several variables, including the number of paid subscribers as a percentage of the population, the size of the city's population, the number of years the weekly newspaper has been in operation, and the number of major daily newspapers published in the city. Some of these data for a sample of weeklies are as follows:

Weekly Newspaper	Paid Subscriptions (%)	Years in Operation	Number of Major Dailies
1	.9	7	1
2	.3	1	2
3	.8	5	2
4	.2	8	3
5	.6	1	1
6	.8	8	2
7	.8	4	2
8	.7	8	1
9	.9	8	1
10	1.1	4	1

a. Set up the **X** matrix and the **Y** vector for fitting the model $Y = \beta_0 + \beta_1X_1 + \beta_2X_2 + \epsilon$, where Y = the percentage of paid subscriptions, X_1 = the number of years in operation, and X_2 = the number of major daily newspapers.

b. Verify that the inverse of the **X′X** matrix is as follows:

$$(\mathbf{X'X})^{-1} = \frac{1}{3160}\begin{bmatrix} 3176 & -212 & -1072 \\ -212 & 44 & -16 \\ -1072 & -16 & 724 \end{bmatrix}$$

c. Fit the model specified in part a.

d. What percentage of paid subscriptions does the fitted model predict for a three-year-old weekly newspaper competing in a city with one major daily newspaper?

e. Compute SSE.

16.4 THE *F*-TEST OF MODEL UTILITY

Background

The fitted multiple regression model is a relation between Y and the set of independent variables $X_1, X_2, \ldots, X_k$. Figure 16.3 depicts this relation as a bridge linking Y to the independent variables.

Using the Regression processor or the matrix equation 16–8, we can generate the fitted model $\hat{Y} = b_0 + b_1X_1 + \ldots + b_kX_k$ (that is, build a bridge). Accomplishing this, we turn our attention to the structural soundness of the bridge. Is the model strong enough to support our objective of prediction?

We might think of the fitted regression coefficients—$b_1, b_2, \ldots, b_k$—as representing the bridge supports. Certainly these sample-based statistics are optimal, in the sense of least squares, for the sample data set, but are they durable enough for the entire population? The answer lies in a test about the set of population coefficients β_1

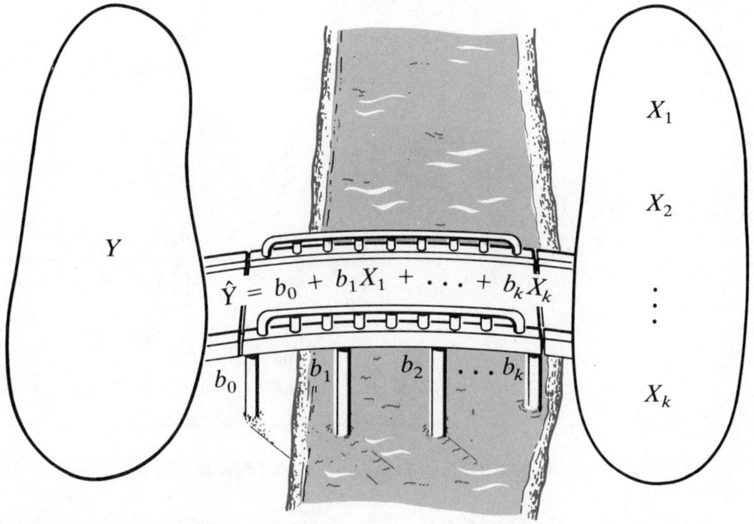

Figure 16.3 Fitted Regression Model as a Bridge Linking Dependent and Independent Variables

through β_k. For now, we view the bridge supports functioning as a team in testing their collective strength in relating Y and a set of X's. In Chapter 17, we will present a method for testing the separate relation of an independent variable with Y.

General Testing Procedure

We have a general procedure for testing a model's utility for predictive purposes.

F-Test of Multiple Regression Model Utility

H_0: $\beta_1 = \beta_2 = \ldots = \beta_k = 0$ (The model is not useful.)

H_a: not H_0 (The model is useful.)

TS: $F^* = \dfrac{MSR}{MSE}$ **(16–10)**

RR: For $\alpha = .05$; numerator $df = k$; denominator $df = n - (k + 1)$, see Figure 16.4.

C: Reject H_0 or accept H_0.

Unlike the simple regression test involving only one parameter, this model utility test involves many parameters. The null hypothesis states that the bridge is not strong

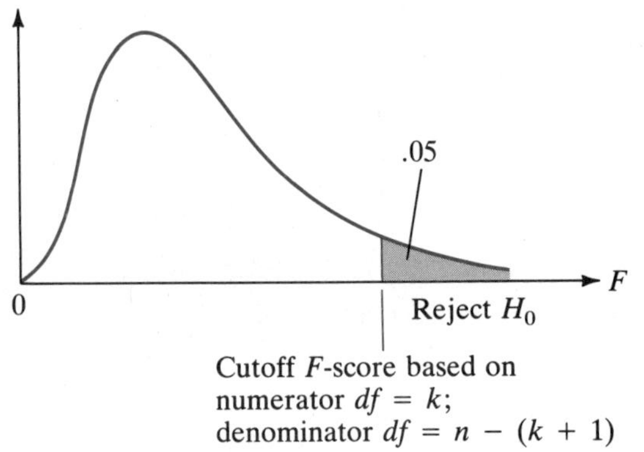

Figure 16.4 Rejection Region for *F*-Test of Multiple Regression Model Utility, $\alpha = .05$

enough to support the burden of prediction, implying that the team of independent variables is not related to Y. The alternative hypothesis, as an opposite statement of H_0, encompasses all the ways of negating the null hypothesis. For example, one way that the null hypothesis H_0: $\beta_1 = \beta_2 = \beta_3 = 0$ could be false is when $\beta_1 = 0$, $\beta_2 \neq 0$ and $\beta_3 \neq 0$. Another way is for $\beta_1 = \beta_2 = 0$ and $\beta_3 \neq 0$. Rather than try to list all the possibilities, we simply say that the alternative hypothesis is "not H_0," much as we did in testing for equal means in Chapter 14.

The test statistic is an F-statistic, defined in Equation 16–10 as the ratio of the mean square for regression (MSR) to the mean square for error (MSE). To find MSR, we use Equation 16–11.

Mean Square for Regression

$$MSR = \frac{SSR}{k} = \frac{SSY - SSE}{k} \qquad \textbf{(16–11)}$$

In this equation, SSR stands for the sum of squares for regression. The error sum of squares (SSE) was defined previously in Equations 16–5 and 16–9, while SSY is the total amount of variability in the observed Y-values.

Total Sum of Squares

$$SSY = \Sigma(Y - \overline{Y})^2 = \Sigma Y^2 - \frac{(\Sigma Y)^2}{n} \qquad \textbf{(16–12)}$$

The F-statistic has a pair of degrees of freedom: k for the numerator—where k is the number of beta parameters listed in the null hypothesis—and $n - (k + 1)$ for the denominator.

COMMENTS

1. Continuing the bridge-model analogy, we imagine the bridge supports b_1 through b_k to be positioned in the water, while b_0 is sunk into the ground along Y's shoreline. The test of model utility concerns only the part of the bridge over the water.
2. Some textbook authors write the alternative hypothesis in the F-test as "at least one β is nonzero," or "not all β's $= 0$," or "some $\beta_i \neq 0$." Also, the null and alternative hypotheses in the F-test box could be written as follows:

$$H_0\text{: No regression effect}$$
$$H_a\text{: A regression effect}$$

3. MSE is just another name for $\hat{\sigma}^2$ in this application.

EXAMPLE 16.5

In Example 16.2, the model $Y = \beta_0 + \beta_1 X_1 + \beta_2 X_2 + \beta_3 X_3 + \epsilon$ was fit to a set of $n = 28$ data points. Test the utility of the fitted model $\hat{Y} = 56.31857 - 1.94724X_1 - .00801X_2 + 1.84047X_3$, with $\alpha = .05$.

Solution:

To five decimal places, we found $SSE = 1633.33187$ and $MSE = 68.05549$ in Example 16.2. From Equation 16–12, we compute:

$$SSY = \Sigma Y^2 - \frac{(\Sigma Y)^2}{n} = 107530 - \frac{(1714)^2}{28} = 2608.714$$

Thus, from Equation 16–11,

$$MSR = \frac{SSY - SSE}{k} = \frac{2608.714 - 1633.33187}{3} = 325.127$$

The test of model utility is

$$H_0\text{: } \beta_1 = \beta_2 = \beta_3 = 0$$
$$H_a\text{: not } H_0$$
$$TS\text{: } F^* = \frac{MSR}{MSE} = \frac{325.127}{68.055} = 4.78$$

RR: For $\alpha = .05$; numerator $df = k = 3$; denominator $df = n - (k + 1) = 28 - 4 = 24$, see Figure 16.5.

C: Reject H_0.

Therefore, the fitted model is a useful relation between Y and X_1, X_2, and X_3.

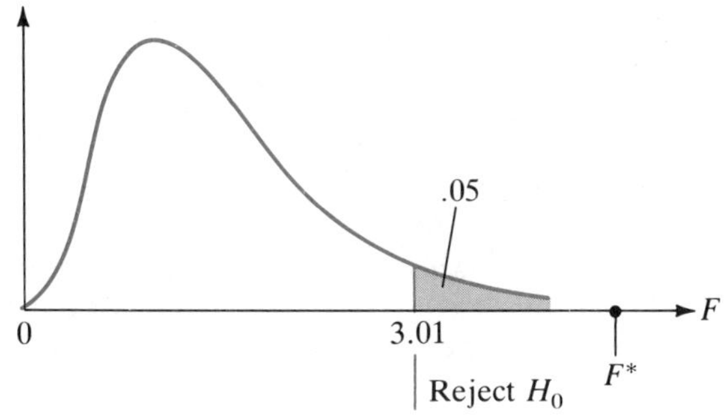

Figure 16.5 Rejection Region for F-Test of Model Utility in Example 16.5

□

Processor: Regression

The F-test of model utility also is available as output in the Regression processor. In the Results submenu, there is an option labeled F-Test of Model Utility. For a given data set, the processor generates the F-statistic and summarizes the information in a format similar to our F-test given in the preceding box. The main difference between the computer output and our test occurs at the rejection region step. Instead of drawing the F-distribution and indicating the cutoff F-value, the program computes the p-value for the test and leaves the conclusion to the user. Example 16.6 demonstrates this option.

EXAMPLE 16.6

Refer to Example 16.2. Run the Regression processor, using the XM162 data file to fit the model

$$Y = \beta_0 + \beta_1 X_1 + \beta_2 X_2 + \beta_3 X_3 + \epsilon$$

At the Results submenu, select the F-Test of Model Utility option. If $\alpha = .05$, are we able to establish the utility of the fitted regression model?

Solution:

Verify the screen shown in Figure 16.6. Notice that the p-value is .0095. Since α is greater than the p-value, we reject the null hypothesis and conclude that the fitted regression model is strong enough to be used for prediction.

```
≡                        Multiple Regression                               READY
  Introduction  Data  Calculations [Results] Exit                     | F1=Help

    F-test of model utility

                   Ho: β1 = β2 = β3 = 0

                   Ha: not Ho

                            MSR   325.127
                   TS: F* = --- = ---------- = 4.777
                            MSE   68.055

                  p-value = 0.0094873722, based on 3 and 24 df.


     The decision to "reject Ho" or "do not reject Ho" depends on your
    judgment or tolerable risk of Type I error (see text Figure 10.18).
```

Figure 16.6 *F*-Test of Model Utility Screen

□

■ ANOVA Table

Using the lodging industry data in Example 16.2 to achieve the third and fourth objectives—fitting the model and testing the utility of the model—of a multiple regression analysis, we needed the following computations:

$SSR = 975.38213$	num $df = 3$	$MSR = 325.127$
$SSE = 1633.33187$	den $df = 24$	$MSE = 68.055$
$SSY = 2608.714$	$F^* = 4.78$	

These figures often are summarized in an analysis of variance (ANOVA) table, as we saw in the simple regression chapter. The Regression processor includes an ANOVA Table option within the Results section of the program. With the XM162 data file we used in Examples 16.2 and 16.6, the processor generates the ANOVA table in Figure 16.7.

There is no separate subsection on matrix algebra here, since the only new computation was *MSR*. As Equation 16–11 indicates, *MSR* is found with basic algebra. We will return to the matrix formulation in the next section when we discuss the predictive features of the model.

COMMENT Although the *F*-test may indicate that the whole set of independent variables is useful, we should not conclude automatically that each one is individually useful as well. Subsequently, we may discover that one (or more) of the independent variable's contribution is slight and that the variable can be removed without seriously affecting the usefulness of the relation between *Y* and the rest of the *X*'s. We will explain how to examine the relative importance of each *X* in Chapter 17.

```
≡                      Multiple Regression                                READY
   Introduction  *able  Data  Calculations  Results  Exit               | F1=Help
```

ANOVA table

Source	SS	df	MS	F	p-value
Regression	975.38241	3	325.127	4.777	0.0094873722
Error	1633.33187	24	68.055		
Total	2608.71429	27			

Figure 16.7 ANOVA Table Screen

CLASSROOM EXAMPLE 16.3

Assessing the Usefulness of a Multiple Regression Model

We fit the model $Y = \beta_0 + \beta_1 X_1 + \beta_2 X_2 + \epsilon$ to the following data in Classroom Example 16.2:

Y	X_1	X_2
5	−3	9
4	−2	4
1	−1	1
3	0	0
1	1	1
−1	2	4
−2	3	9

We obtained $SSE = 5.2857142857$.

a. Compute SSY:

$$SSY = \Sigma Y^2 - \frac{(\Sigma Y)^2}{n} = \_\_\_\_\_$$

b. Compute MSR and MSE, assuming $k = 2$:

$$MSR = \frac{SSR}{k} = \frac{SSY - SSE}{k} = \_\_\_\_\_ \qquad MSE = \frac{SSE}{n - (k + 1)} = \_\_\_\_\_$$

c. Complete the following test of model utility. Let $\alpha = .05$.

$$H_0\colon \beta_1 = \beta_2 = 0 \qquad H_a\colon \text{not } H_0$$

d. Is the model useful?

SECTION 16.4 EXERCISES

16.27 In terms of model utility, what do the following conditions imply?

a. $SSE = 0$

b. $MSR = 0$

16.28 Find the cutoff F-value for the following testing situations:

a. $H_0\colon \beta_1 = \beta_2 = \beta_3 = 0$; $n = 28$; and $\alpha = .05$.

b. $H_0\colon \beta_1 = \beta_2 = \beta_3 = \beta_4 = \beta_5 = \beta_6 = 0$; $n = 35$; and $\alpha = .10$.

c. $H_0\colon \beta_1 = \beta_2 = 0$; $n = 15$; and $\alpha = .01$.

16.29 Construct the ANOVA table from the following F-test of the utility of a multiple regression fitted model. Assume $n = 10$.

$$H_0\colon \beta_1 = \beta_2 = 0$$
$$H_a\colon \text{not } H_0$$
$$TS\colon F^* = \frac{18.031}{1.989} = 9.07$$
$$p\text{-value} = .0114$$

16.30 Refer to Exercise 16.5. Fit the proposed model. Is there sufficient evidence at the 1% level of significance to conclude that the model relating Y to X_1, X_2, and X_3 is a useful relation?

16.31 The model $Y = \beta_0 + \beta_1X_1 + \beta_2X_2 + \epsilon$ was fit to a data set of size $n = 25$, yielding these summary statistics:

$$SSE = .14685 \qquad SSY = 7.53125$$

Test the utility of the fitted model with $\alpha = .01$.

16.32 Colleges often solicit their alumni for contributions. The total amount donated per year per graduating class is thought to be related to two variables:

$$X_1 = \text{Number of years since graduation}$$
$$X_2 = \text{Number of contributors per class}$$

A sample of donations revealed the following figures, where Y is recorded in thousands of dollars ($Y = 5$ means \$5000):

Y	X_1	X_2
\$ 4.0	1	145
7.2	2	174
7.7	3	206
8.6	4	226

(continues)

(*continued*)

Y	X_1	X_2
$ 8.7	5	236
9.3	6	225
12.6	7	249
12.8	8	278
16.2	9	281
14.8	10	254
19.6	15	297
18.4	20	257
26.2	25	204

a. Fit the model $Y = \beta_0 + \beta_1 X_1 + \beta_2 X_2 + \epsilon$ to the data with the Regression processor. Save the data in a file named XR1632.

b. Test the utility of the fitted model, using $\alpha = .05$ and the F-test option. Is the model useful?

c. Obtain the ANOVA table summary. What is the p-value for the model utility test?

16.33 Refer to Exercise 16.6. Fit the model $Y = \beta_0 + \beta_1 X_1 + \beta_2 X_2 + \epsilon$. Conduct an F-test of the null hypothesis H_0: $\beta_1 = \beta_2 = 0$. Use $\alpha = .05$. Is the model useful?

16.34 Generate an ANOVA table for fitting the multiple regression model $Y = \beta_0 + \beta_1 X_1 + \beta_2 X_2 + \epsilon$ to these data using the Regression processor:

Y	X_1	X_2
98	1	4
72	5	9
69	1	4
62	2	2
56	0	5
53	7	2
51	6	8
50	3	0
46	1	2
44	5	5
44	2	8

Save the data in a file named XR1634. What is the p-value?

16.35 A multiple regression analysis produced the following ANOVA table:

Source	SS	df	MS	F
Regression	36.06	2	18.03	9.06
Error	13.92	7	1.99	
Total	49.98	9		

Use the figures in this table to write out all parts of the F-test of model utility, similar to the boxed procedure on page 857. Use $\alpha = .01$.

16.36 Refer to Exercise 16.19. Generate an ANOVA table for the regression analysis. With $\alpha = .05$, is there evidence to indicate the fitted model is useful? Give the p-value for the test.

16.37 Refer to Exercise 16.26. Assuming $SSE = .3437974$ and $SSY = .689$, test the strength of the model's relation. Use $\alpha = .05$.

16.38 Construct an ANOVA table from the following F-test. Assume the sample size is 18.

$$H_0\colon \beta_1 = \beta_2 = 0$$
$$H_a\colon \text{not } H_0$$
$$TS\colon F^* = 14.660/3.997 = 3.67$$
$$RR\colon \text{Reject } H_0 \text{ if } F^* > 3.68.$$
$$C\colon \text{Accept } H_0.$$

What α-risk was used in the test?

16.39 "Your tax dollars at work: Thirty or so times a year, the National Highway Traffic Safety Administration buys a spanking new car, outfits it with an array of measuring devices, straps a couple of dummies into the front seat and smashes the car into a concrete barrier at 35 miles per hour." As reported in the article "Crash Course in Crash Times" (*Changing Times,* June 1987), these crash tests measure the impact on the head, chest, and thighbones of the driver and passenger. A vehicle is considered potentially dangerous when its "head injury criteria" (hic) score is above 1000. Results from 52 vehicles over a two-year period are shown in Table 16.1.

Table 16.1 Crash Test Data for Exercise 16.39

Vehicle	Driver hic Score	Chest Score	Left Thigh Femur Load	Right Thigh Femur Load	Weight of Vehicle
1	584	37	120	346	3520
2	688	50	476	978	3710
3	733	39	736	961	3524
4	908	39	879	1181	3560
5	1237	48	1039	1780	3570
6	405	70	1085	1737	3120
7	479	42	580	1589	3343
8	647	37	945	740	3360
9	699	40	740	240	3250
10	743	64	585	1277	3180
11	773	71	484	—	3390
12	791	62	1008	1863	3320
13	823	36	1019	740	3150
14	846	52	820	1300	3040
15	871	51	1380	1200	3250
16	873	52	717	832	3050
17	909	—	228	112	3062
18	952	48	1260	1752	3450
19	1209	53	824	1485	3460
20	1488	49	1003	748	3000

(continues)

Table 16.1 Continued

Vehicle	Driver hic Score	Chest Score	Left Thigh Femur Load	Right Thigh Femur Load	Weight of Vehicle
21	1831	60	300	546	3360
22	551	42	1146	1053	2740
23	552	44	2176	745	2580
24	576	48	306	1967	2827
25	599	35	791	387	2780
26	603	36	428	1079	2961
27	627	42	382	721	2950
28	716	55	790	345	2610
29	744	38	978	185	2759
30	757	54	2408	1794	2660
31	769	46	863	695	2920
32	784	42	79	658	2999
33	802	66	425	440	2510
34	810	45	5048	1297	2980
35	846	44	879	1848	2990
36	887	52	394	481	2580
37	896	48	1695	1482	2620
38	984	48	1230	110	2666
39	1728	51	838	376	2712
40	1801	52	484	723	2610
41	1809	53	351	747	2570
42	611	42	529	1256	2110
43	1415	59	870	86	2320
44	1855	45	—	585	2320
45	2172	43	486	176	2380
46	758	44	1100	758	3620
47	903	43	776	1595	3660
48	985	44	324	213	3640
49	1052	62	925	923	3554
50	1568	49	286	518	3618
51	1434	65	620	97	3080
52	1764	65	274	279	3349

Condensed with permission from *Changing Times* Magazine, © 1987, Kiplinger Washington Editors, Inc., June 1987.

a. Examine the data closely and note that several values are missing. For example, the chest score for vehicle 17 is missing. Rather than omit the entire seventeenth data point because of this, let's estimate the missing value. Assume that chest scores are fairly constant within a narrow range of hic scores. Locate all hic scores within 10 percent of 909—that is, within 91 points of 909. Find the average chest score corresponding to these hic scores and assign that value as the chest score for vehicle 17. Perform a similar procedure for other missing values. Maintain the proper degree of accuracy.

b. Create a data file named XR1639 for these data. Let the dependent variable be Y = driver hic score, and the independent variables be as follows:

$$X_1 = \text{Chest score}$$
$$X_2 = \text{Left thigh femur load}$$
$$X_3 = \text{Right thigh femur load}$$
$$X_4 = \text{Weight of vehicle}$$

c. Fit the model $Y = \beta_0 + \beta_1 X_1 + \beta_2 X_2 + \beta_3 X_3 + \beta_4 X_4 + \epsilon$. Write out the equation of the fitted model.

d. Test the utility of the fitted model with $\alpha = .10$. Is it useful?

16.40 Suppose we wish to test the utility of a fitted multiple regression model that has 12 estimated regression coefficients (b_0 through b_{11}) based on a sample of 12 entities.

a. How many degrees of freedom for error are there?

b. What is the resulting effect on *MSE*?

c. State a general rule about the sample size needed to estimate σ in a multiple regression analysis, relative to the number of betas in the model.

16.5 USING THE MODEL FOR PREDICTION

Prediction Options

In Section 16.1, we identified five objectives in multiple regression analysis. The first four objectives concerned the formation and development of the model, while the fifth dealt with using the model for prediction. As the simple regression case demonstrated, the model can be used to predict a specific value of Y or to estimate an average value of Y, denoted by $E(Y)$, given a value for each of the independent variables. Table 16.2 summarizes these options.

In the first option, we wish to *predict a specific value of* Y for one new observation or entity at a given setting of the X's. For example, an accountant at the Internal

Table 16.2 Predicting with the Fitted Multiple Regression Model

Option	Quantity to Be Forecast	Classification of Quantity	Information Given	Forecast Intent
1	Specific value of Y	Random variable	Values of $X_1, X_2, \ldots, X_k$	One new entity
2	Average value of Y	Parameter	Values of $X_1, X_2, \ldots, X_k$	Population of entities

Revenue Service (IRS) might develop a multiple regression equation relating Y, the deduction for interest expense on Schedule A of the federal income tax form, to X_1, adjusted gross income (AGI), and X_2, the number of dependents claimed. In examining tax returns, the accountant might use the regression model to predict a taxpayer's interest deduction based on her AGI of \$20,000 and three dependents, for example.

In the second option, we wish to *estimate an average value of Y* for the population of Y-values at a given setting of the X's. For instance, consider a developer who builds energy efficient two-story apartment complexes. To develop promotional material for the apartments, he might construct a regression model to relate Y, a unit's annual utility bill, to three independent variables: X_1 = the number of square feet of living space per unit; X_2 = the location of the unit—that is, first floor, second floor—and X_3 = the number of south-facing windows in the unit. Based on data from existing apartment complexes, the developer might use the fitted regression model to estimate the average utility bill for all first-floor apartments with two south-facing windows and 1500 square feet of living space, for example.

The two prediction options are aimed at different quantities. A specific value of Y is a value of a random variable: a quantity that is not stationary. Thus, predicting a specific value of Y is equivalent to shooting at a moving target. An average value of Y is a parameter: a quantity that is stationary. Hence, estimating an average value of Y is similar to shooting at a fixed target. In the analogy of shooting at a target, the first situation contains two sources of variability: the quivers in our hand and body when aiming (the variability of the estimator) and the movement in the target (the variability in Y). The second situation contains only the source of variability due to aiming (the variability of the estimator). Therefore, the margin for error naturally will be larger for the prediction problem than it will be for the estimation problem. For instance, the IRS accountant would have a wider range of possible values when predicting one taxpayer's interest deduction than he or she would have for estimating the average interest deduction for all taxpayers with the same AGI and number of dependents. Similarly, the developer could estimate the average utility bill more precisely than he could an individual unit's utility bill, given the same location and size of the units.

The type of prediction depends on the problem setting. In general, if we wish to forecast the value of an individual entity, Option 1 applies. If the objective is to forecast the average value of a population of entities, Option 2 applies. A few examples of the two forecasting options follow:

1. Predicting a specific value of Y:

   - Predict the monthly sales of a sales representative named Tyler, based on auto mileage and phone usage.
   - Predict the percentage of cheese in one 2-ounce package of a snack food, based on certain machine settings in the production process.
   - Predict the attendance at a leadership seminar based on the length (in hours) of the seminar, the number of preregistrants, and the cost of the seminar.

2. Estimating an average value of Y:

- Estimate the average monthly sales for all sales representatives, based on auto mileage and phone usage.
- Estimate the average percentage of cheese in all 2-ounce packages of a snack food, based on certain machine settings in the production process.
- Estimate the average attendance for all leadership seminars based on the length (in hours) of the seminar, the number of preregistrants, and the cost of the seminar.

In both types of forecasting situations, we substitute the given values of the independent variables into the fitted regression equation to generate a "best guess" value, which is denoted by $\hat{Y}$. For predicting a specific value of Y, we refer to this value as a *point prediction;* for estimating an average value of Y, we call the value a *point estimate*. In addition to these point forecasts, we seek to generate an interval estimate, called a *prediction interval* for Option 1 and a *confidence interval* for Option 2, with a specified degree of confidence.

In both cases, the intervals are formed as follows:

$$\text{Point forecast} \pm (t\text{-score})(\text{estimated standard error})$$

However, the estimated standard error for multiple regression problems is too complicated to write in a concise algebraic form. If we use matrix algebra, though, we can express the estimated standard errors succinctly; the next subsection provides details. (In particular, refer to the right-hand side of the $\pm$ symbol in Equations 16–15 and 16–16 for the estimated standard errors for prediction and confidence intervals, respectively.)

To demonstrate the construction of such an interval, let us use the data in Example 16.1. Suppose we wish to find a 95% confidence interval for $E(Y)$, the average number of passengers for an airline with $X_1 = 50$ flights per week and $X_2 = 10$ local employees. The point estimate is found by substituting the values of X_1 and X_2 into the fitted model $\hat{Y} = -4.869 + .307X_1 + .166X_2$:

$$\hat{Y} = -4.869 + .307(50) + .166(10) = 12.141$$

Now assume the estimated standard error is 2.984. With $df = 9$, the t-score is 2.262 and the prediction interval becomes

$$\begin{aligned}\text{Point forecast} &\pm (t\text{-score})(\text{Estimated standard error})\\ &= 12.141 \pm (2.262)(2.984)\\ &= 12.141 \pm 6.750 \qquad \text{or } 5.391 \text{ to } 18.891\end{aligned}$$

We are 95% confident that the airline will service on average between 53,910 and 188,910 passengers annually at this airport.

COMMENTS

1. Some authors refer to predicting a specific value of Y as "predicting a particular value of Y" or as "predicting a new response."

2. If the F-test fails to establish a model's utility, we should use $\overline{Y}$ as the point prediction of Y or as the point estimate of $E(Y)$.
3. Exercise 16.83 explains how to use the fitted regression model to perform a third forecasting option: Predict the average value of a sample of m entities.

CLASSROOM EXAMPLE 16.4

Point and Interval Predictions for Y

Refer to Example 16.1. Run the Regression processor, using the data file XM161. Suppose we wish to predict with 95% confidence the annual number of passengers for an airline employing 43 people who service an average of 49 weekly flights.

a. What is the value of the point prediction of Y?

b. Give the upper and lower limits of the prediction interval. Assume the estimated standard error is 2.357.

*■ Matrix Computations

To forecast either Y or $E(Y)$ using the fitted regression model, we need to specify a value for each independent variable $X_1, X_2, \ldots, X_k$. Let us denote the given values as $X_1 = x_1$, $X_2 = x_2$ and so on up to $X_k = x_k$. These values are organized into a column vector, denoted by $\mathbf{X}_s$, as follows.

Given Setting of the *X*'s

$$\mathbf{X}_s = \begin{bmatrix} 1 \\ x_1 \\ x_2 \\ \cdot \\ \cdot \\ \cdot \\ x_k \end{bmatrix} \qquad (16\text{–}13)$$

The initial value of 1 corresponds to the coefficient of b_0 in the regression equations; the remaining X-values match up to the corresponding b_i's in the fitted model.

Point predictions and point estimates are computed by substituting the given values of X into the regression equation:

$$\hat{Y} = b_0 + b_1x_1 + b_2x_2 + \ldots + b_kx_k$$

*Optional

The equivalent matrix operation is shown in Equation 16–14.

Point Prediction of Y or Point Estimate of $E(Y)$

$$\hat{Y} = \mathbf{X}_s' \mathbf{b} \qquad (16\text{–}14)$$

In Equation 16–14, $\mathbf{X}_s'$ is the transpose of the column vector $\mathbf{X}_s$ into a row vector, and $\mathbf{b}$ is the column vector of estimated regression coefficients. Equation 16–15 expands the point prediction into a prediction interval for a specific value of Y.

Prediction Interval for a Specific Value of Y

$$\hat{Y} \pm (t\text{-score})\, \hat{\sigma} \cdot \sqrt{1 + \mathbf{X}_s'(\mathbf{X}'\mathbf{X})^{-1}\mathbf{X}_s} \qquad (16\text{–}15)$$

The cutoff t-score is found in the t-table with $df = n - (k + 1)$ and the desired level of confidence. Equation 16–15 also involves $\hat{\sigma}$, the estimated standard error of the regression, and the $(\mathbf{X}'\mathbf{X})^{-1}$ matrix. The product $\mathbf{X}_s' (\mathbf{X}'\mathbf{X})^{-1}\mathbf{X}_s$ reduces to a single number.

To estimate the average value of Y with a confidence interval, we use Equation 16–16.

Confidence Interval for an Average Value of Y

$$\hat{Y} \pm (t\text{-score})\, \hat{\sigma} \cdot \sqrt{\mathbf{X}_s' (\mathbf{X}'\mathbf{X})^{-1}\mathbf{X}_s} \qquad (16\text{–}16)$$

Notice in Equation 16–15 that the potential sampling error will be larger than the corresponding potential sampling error in Equation 16–16, due to the "1" under the square root. This reinforces our earlier remarks that Option 1 is more variable (and less precise) than Option 2. Example 16.7 demonstrates the use of Equations 16–15 and 16–16.

EXAMPLE 16.7

Using the data in Example 16.3, determine the following:

a. Develop a prediction interval for a specific value of Y with 90% confidence when

$$X_1 = 3 \quad \text{and} \quad X_2 = 1$$

b. Construct a confidence interval for the average value of Y with 90% confidence when

$$X_1 = 3 \quad \text{and} \quad X_2 = 1$$

Solution:

a. In Example 16.3, we fit the model $Y = \beta_0 + \beta_1 X_1 + \beta_2 X_2 + \epsilon$ to a sample of $n = 6$ data points, yielding

$$(\mathbf{X}'\mathbf{X})^{-1} = \frac{1}{134}\begin{bmatrix} 195 & -32 & -91 \\ -32 & 8 & 6 \\ -91 & 6 & 105 \end{bmatrix}$$

and

$$\mathbf{b} = \frac{1}{134}\begin{bmatrix} 1499 \\ -224 \\ 33 \end{bmatrix}$$

The given values of the independent variables, $X_1 = 3$ and $X_2 = 1$, are combined into the $\mathbf{X}_s$ vector, as indicated in Equation 16–13:

$$\mathbf{X}_s = \begin{bmatrix} 1 \\ 3 \\ 1 \end{bmatrix}$$

From Equation 16–14, the point prediction of a specific value for Y, as well as the point estimate of the average value of Y, is

$$\begin{aligned} \hat{Y} = \mathbf{X}_s' \mathbf{b} &= [1 \quad 3 \quad 1]\left(\frac{1}{134}\right)\begin{bmatrix} 1499 \\ -224 \\ 33 \end{bmatrix} \\ &= \frac{860}{134} = 6.42 \end{aligned}$$

To find a 90% prediction interval for a specific value of Y, we must compute $\mathbf{X}_s'(\mathbf{X}'\mathbf{X})^{-1}\mathbf{X}_s$:

$$\begin{aligned} \mathbf{X}_s'(\mathbf{X}'\mathbf{X})^{-1}\mathbf{X}_s &= [1 \ 3 \ 1]\left(\frac{1}{134}\right)\begin{bmatrix} 195 & -32 & -91 \\ -32 & 8 & 6 \\ -91 & 6 & 105 \end{bmatrix}\begin{bmatrix} 1 \\ 3 \\ 1 \end{bmatrix} \\ &= \frac{1}{134}[8 \quad -2 \quad 32]\begin{bmatrix} 1 \\ 3 \\ 1 \end{bmatrix} \\ &= \frac{1}{134}(34) = .2537313 \end{aligned}$$

The estimated standard error of the regression was computed in Example 16.4 to be $\hat{\sigma} = 1.1096591$. For a 90% level of confidence and $df = n - (k + 1) =$

$6 - (2 + 1) = 3$, the t-score is 2.353. Thus, a 90% prediction interval for a specific value of Y when $X_1 = 3$ and $X_2 = 1$ is

$$\begin{aligned}\hat{Y} \pm (t\text{-score})\, \hat{\sigma} \cdot \sqrt{1 + \mathbf{X}_s' (\mathbf{X}'\mathbf{X})^{-1}\mathbf{X}_s} &= 6.42 \pm (2.353)(1.1096591)\sqrt{1 + .2537313} \\ &= 6.42 \pm 2.92 \\ &= 3.50 \text{ to } 9.34\end{aligned}$$

If we set $X_1 = 3$ and $X_2 = 1$, then we are 90% confident that the observed value for Y will be between 3.50 and 9.34.

b. All of the intermediate computations for estimating the average value of Y are already available from part a. A 90% confidence interval for the average value of Y when $X_1 = 3$ and $X_2 = 1$ is

$$\begin{aligned}\hat{Y} \pm (t\text{-score})\, \hat{\sigma} \cdot \sqrt{\mathbf{X}_s' (\mathbf{X}'\mathbf{X})^{-1}\mathbf{X}_s} &= 6.42 \pm (2.353)(1.1096591)\sqrt{.2537313} \\ &= 6.42 \pm 1.32 \\ &= 5.10 \text{ to } 7.74\end{aligned}$$

When $X_1 = 3$ and $X_2 = 1$, we are 90% confident that the average value for Y will be between 5.10 and 7.74. Notice that the confidence interval for $E(Y)$ is more precise (narrower) than the prediction interval of part a for the same level of confidence and the same settings of the independent variables. □

Matrix algebra is useful for constructing confidence and prediction intervals by hand, provided the $(\mathbf{X}'\mathbf{X})^{-1}$ matrix is available. Without it, we must rely on a computer program for the interval estimates.

*CLASSROOM EXAMPLE 16.5

Predicting Y Using Matrix Algebra

Refer to Classroom Example 16.2, where the model $Y = \beta_0 + \beta_1 X_1 + \beta_2 X_2 + \epsilon$ was fit to a sample of seven data points. The vector of estimated regression coefficients was

$$\mathbf{b} = \frac{1}{16464}\begin{bmatrix} 28224 \\ -18228 \\ -588 \end{bmatrix}$$

and the inverse of the $\mathbf{X}'\mathbf{X}$ matrix was

$$(\mathbf{X}'\mathbf{X})^{-1} = \frac{1}{16464}\begin{bmatrix} 5488 & 0 & -784 \\ 0 & 588 & 0 \\ -784 & 0 & 196 \end{bmatrix}$$

*Optional

We also computed $\hat{\sigma} = 1.1495341$ and confirmed the utility of the fitted model in classroom Example 16.3, part d.

a. Set up the $\mathbf{X}_s$ vector if we wish to predict a specific value of Y when $X_1 = 1$ and $X_2 = 1$.
b. Find the point prediction by computing $\hat{Y} = \mathbf{X}_s'\,\mathbf{b}$.
c. Compute $\mathbf{X}_s'(\mathbf{X}'\mathbf{X})^{-1}\mathbf{X}_s$.
d. What cutoff t-value is required for a 95% prediction interval?
e. Determine the 95% prediction interval for a specific value of Y when $X_1 = 1$ and $X_2 = 1$.

SECTION 16.5 EXERCISES

16.41 A coffee salesman services many grocery stores in his territory. At the request of his manager, he occasionally has collected data on weekly unit sales per store, price, average weekly temperature, and square feet of selling area (for coffee) per store. Suppose the manager used the data to develop a multiple regression model for the dependent variable Y = weekly unit sales as a function of the other variables. Relate the two forecasting options to this model by explaining what a prediction interval pertains to and what a confidence interval pertains to.

16.42 For the proposed model $Y = \beta_0 + \beta_1 X_1 + \beta_2 X_2 + \beta_3 X_3 + \epsilon$, the following set of data applies:

Y	X_1	X_2	X_3
3.1	21.2	10.7	14.2
2.6	19.6	12.3	20.4
2.2	26.0	12.8	16.2
2.2	20.9	11.0	15.1
−.2	9.9	7.0	9.0
2.9	12.2	14.8	10.8
3.5	12.2	15.9	12.3
3.8	8.7	11.4	12.3
1.7	3.9	11.5	8.8
.8	.2	11.9	11.8
−1.4	.9	3.5	2.1

Fit the proposed model to the data.
Test the utility of the fitted model with $\alpha = .10$.
Predict Y with 80% confidence when $X_1 = 15.3$, $X_2 = 10.7$, and $X_3 = 4.9$. Assume the estimated standard error is 1.4128.
Estimate $E(Y)$ with 90% confidence when $X_1 = 15.3$, $X_2 = 10.7$, and $X_3 = 4.9$. Assume the estimated standard error is 1.0243.

16.43 Refer to Exercises 16.8 and 16.21. Predict with 98% confidence the miles per gallon achieved in city driving for a new car with a suggested retail price of \$10,500. Assume the vehicle has a four-cylinder, 2.4-liter engine, and weighs 2,700 pounds. Use 1.7286 as the value of the estimated standard error. Would a city rating of 25-miles per gallon be considered unusual for such a vehicle? Why?

16.44 Refer to Exercise 16.10. Predict with 90% confidence the net assets of the pension funds for a company with two plans, two money managing firms, and 484 vested employees. Assume a 7% rate of return, and use 14.870 as the value of the estimated standard error.

16.45 Refer to Exercise 16.44. Do you think the width of the prediction interval would increase or decrease if the company's rate of return was 8 percent instead? Why? If necessary, assume the value of the estimated standard error is 14.846.

16.46 Traditionally, bonds have been necessary hedges against a sluggish economy in most investors' portfolios. However, bond returns are inversely related to interest rates, in general, and tend to hurt the portfolio's return when the economy heats up. A study was conducted to relate Y = the total bond return over a one-year period based on the following variables:

X_1 = Bond term, in years
X_2 = Current yield of the bond
X_3 = Interest rate movement, in points—that is, percents

The following data were gathered:

Y	X_1	X_2	X_3
21.5%	12	16.1%	−1%
27.2	12	16.1	−2
11.3	12	16.1	1
6.7	12	16.1	2
17.5	14	9.7	−1
26.1	14	9.7	−2
2.6	14	9.7	1
−3.7	14	9.7	2
19.9	25	9.3	−1
32.2	25	9.3	−2
.5	25	9.3	1
−7.2	25	9.3	2
20.0	30	9.0	−1
33.4	30	9.0	−2
−.4	30	9.0	1
−8.4	30	9.0	2
15.4	10	8.9	−1
22.3	10	8.9	−2
3.2	10	8.9	1
−2.3	10	8.9	2
11.5	5	8.9	−1
15.0	5	8.9	−2
4.9	5	8.9	1
1.8	5	8.9	2

Source: "Bonds: Boom—or Bomb?" *USA Today*, November 11, 1988. Copyright 1988, *USA Today*. Adapted with permission.

a. Fit the model $Y = \beta_0 + \beta_1X_1 + \beta_2X_2 + \beta_3X_3 + \epsilon$ using the Regression processor. Save the data in a file named XR1646.
b. Test the utility of the fitted model with an α-risk of 10 percent.
c. Predict the one-year total return for a 20-year bond currently yielding 9 percent if interest rates rise 2 points during the year. Use 95% confidence and 4.7039 as the value of the estimated standard error.

16.47 Refer to Exercises 16.2 and 16.18.
a. Develop a 90% prediction interval for Y when $X_1 = 0$ and $X_2 = 3$. Use 1.1895 as the value of the estimated standard error.
b. Develop a 95% confidence interval for $E(Y)$ when $X_1 = 0$ and $X_2 = 3$. Use .4010 as the value of the estimated standard error.

***16.48** Refer to Exercise 16.26. Give a point prediction of the percentage of paid subscriptions for a three-year-old weekly business newspaper competing in a market with one major daily newspaper. (*Hint:* Study the results of Exercise 16.37 closely.)

***16.49** Give the dimensions of the following matrices or vectors if the fitted multiple regression model has 5 independent variables in it and the data set has 20 observations.
a. $\mathbf{X}$ b. $\mathbf{X}_s$
c. $\mathbf{X}_s'$ d. $\mathbf{X}_s'(\mathbf{X}'\mathbf{X})^{-1}\mathbf{X}_s$

***16.50** Refer to Classroom Example 16.5. Construct a 95% confidence interval for $E(Y)$ when $X_1 = 1$ and $X_2 = 1$.

***16.51** A fitted multiple regression model based on $n = 20$ entities is $\hat{Y} = 10.05 + 12.83X_1 - .06X_2$. Assume that the $(\mathbf{X}'\mathbf{X})^{-1}$ matrix is

$$(\mathbf{X}'\mathbf{X})^{-1} = \begin{bmatrix} .67164 & -.16301 & -.01375 \\ -.16301 & .07448 & .00065 \\ -.01375 & .00065 & .00058 \end{bmatrix}$$

If $\hat{\sigma} = 2.5742$, find the following:
a. a 90% prediction interval for Y, when $X_1 = 3$ and $X_2 = 10$.
b. a 95% confidence interval for $E(Y)$, when $X_1 = 3$ and $X_2 = 10$.

16.6 MULTIPLE CORRELATION

Multiple Correlation Model

As we learned in the previous chapter, a correlation analysis often accompanies the regression analysis to shed light on the strength and direction of the relation. In simple correlation, we introduced r as a measure of the association between two random variables X and Y. In multiple correlation, there are more than two variables; for instance, in Example 16.1, there were three variables: Y, X_1, and X_2. To analyze the correlations among the variables we need a theoretical basis from which to operate.

*Optional

In modeling multiple correlations we make two assumptions. The first assumption is that all variables are random variables. The second concerns the probability distribution that characterizes all the variables in the analysis. In simple correlation, we assumed X and Y jointly possessed a bivariate normal distribution. In multiple correlation, we assume that all the variables behave according to a *multivariate normal distribution*. While we cannot show you a multivariate normal distribution (because it would require more than three dimensions), one consequence of such a distribution is that each pair of variables will have a bivariate normal distribution.

COMMENTS The main differences between multiple regression analysis and multiple correlation analysis are:

1. Regression analysis divides variables into two types, independent or dependent; correlation analysis views all variables as "equals."
2. A variable can have "fixed" values in regression analysis, but in correlation analysis all variables take on values randomly.
3. A multivariate normal distribution is assumed in correlation analysis to describe the behavior of all variables jointly, while a (univariate) normal distribution is believed to describe the behavior of Y at each setting of the X's in regression analysis.

Correlation Matrix

Since each pair of variables in correlation analysis has a bivariate normal distribution in theory, there will be a correlation coefficient to measure the association between them. For three variables, Y, X_1, and X_2, we would have three separate correlation coefficients: one describing the mutual variation between Y and X_1, a second for Y and X_2, and a third for X_1 and X_2. In general, for k variables, there will be $k(k - 1)/2$ different pairings, and thus $k(k - 1)/2$ different correlation coefficients.

With multiple correlation coefficients, we encounter a notational problem. Previously, we used r to represent the sample coefficient of correlation between two variables. Now we will add subscripts to r to identify which pair of variables is being correlated. For instance, r_{12} will denote the correlation between X_1 and X_2. Notice that r_{21} will be the same as r_{12} since, in correlation analysis, variables play a symmetrical role. Thus, the correlation between X_1 and X_2 is the same as the correlation between X_2 and X_1. To symbolize the correlation between Y and X_1 we use r_{Y1}. Similarly, r_{Y2} is the correlation between Y and X_2.

The $k(k - 1)/2$ separate correlation coefficients may be arranged in a *correlation matrix*, which we will denote by **C**. For the case of the three variables Y, X_1, and X_2, the correlation matrix would look like this:

$$\mathbf{C} = \begin{bmatrix} r_{YY} & r_{Y1} & r_{Y2} \\ r_{1Y} & r_{11} & r_{12} \\ r_{2Y} & r_{21} & r_{22} \end{bmatrix}$$

The three correlation coefficients down the main diagonal—r_{YY}, r_{11}, and r_{22}—each will be 1.00, since they represent the correlation of a variable with itself. In general, for any correlation matrix, the entries down the main diagonal always will be 1.00. The three correlation coefficients to the right of the main diagonal—r_{Y1}, r_{Y2}, and r_{12}—are repeated to the left of the main diagonal in symmetric positions.

Let us consider a small data set with three variables to demonstrate the computation of the correlation coefficients and the formation of **C**. Suppose the data set is as given in Table 16.3.

Table 16.3 Hypothetical Data Set

	Y	X_1	X_2
	0	−2	4
	0	−1	1
	1	0	0
	1	1	1
	3	2	4
Means	$\bar{Y} = 1$	$\bar{X}_1 = 0$	$\bar{X}_2 = 2$

Table 16.4 Computation of Correlation Coefficients

Y	$Y - \bar{Y}$	$(Y - \bar{Y})^2$	X_1	$X_1 - \bar{X}_1$	$(X_1 - \bar{X}_1)^2$	X_2	$X_2 - \bar{X}_2$	$(X_2 - \bar{X}_2)^2$
0	−1	1	−2	−2	4	4	2	4
0	−1	1	−1	−1	1	1	−1	1
1	0	0	0	0	0	0	−2	4
1	0	0	1	1	1	1	−1	1
3	2	4	2	2	4	4	2	4
		$SSY = 6$			$SSX_1 = 10$			$SSX_2 = 14$

$(Y - \bar{Y})(X_1 - \bar{X}_1)$	$(Y - \bar{Y})(X_2 - \bar{X}_2)$	$(X_1 - \bar{X}_1)(X_2 - \bar{X}_2)$
2	−2	−4
1	1	1
0	0	0
0	0	−1
4	4	4
$SSYX_1 = 7$	$SSYX_2 = 3$	$SSX_1X_2 = 0$

$$r_{Y1} = \frac{7}{\sqrt{6 \cdot 10}} = .904$$

$$r_{Y2} = \frac{3}{\sqrt{6 \cdot 14}} = .327$$

$$r_{12} = \frac{0}{\sqrt{10 \cdot 14}} = 0$$

The correlation between Y and X_1, r_{Y1}, is found according to Equation 15–22 as

$$r_{Y1} = \frac{SSYX_1}{\sqrt{SSY \cdot SSX_1}}$$

where

$$SSYX_1 = \Sigma(Y - \bar{Y})(X_1 - \bar{X}_1)$$
$$SSY = \Sigma(Y - \bar{Y})^2$$
$$SSX_1 = \Sigma(X_1 - \bar{X}_1)^2$$

The correlations r_{Y2} and r_{12} would be computed in a similar manner. Table 16.4 summarizes the computations of each correlation coefficient.

Therefore, the correlation matrix is

$$\mathbf{C} = \begin{bmatrix} 1.00 & .904 & .327 \\ .904 & 1.00 & 0 \\ .327 & 0 & 1.00 \end{bmatrix}$$

A general correlation matrix involving the variables $Y, X_1, X_2, \ldots, X_k$ is given in Equation 16–17.

Correlation Matrix

$$\mathbf{C} = \begin{bmatrix} 1.00 & r_{Y1} & r_{Y2} & r_{Y3} & \cdots & r_{Yk} \\ r_{1Y} & 1.00 & r_{12} & r_{13} & \cdots & r_{1k} \\ r_{2Y} & r_{21} & 1.00 & r_{23} & \cdots & r_{2k} \\ r_{3Y} & r_{31} & r_{32} & 1.00 & \cdots & r_{3k} \\ \vdots & \vdots & \vdots & \vdots & & \vdots \\ r_{kY} & r_{k1} & r_{k2} & r_{k3} & \cdots & 1.00 \end{bmatrix} \quad (16\text{–}17)$$

Ideally, in a regression analysis, we would like to see zeroes in all of the off-diagonal positions below the first row—that is, r_{12}, r_{13}, r_{23}, and so forth—in Equation 16–17. That was the case in our previous three-variable example where $r_{12} = 0$. Zeroes would indicate that all the independent variables in a regression analysis were unrelated to each other. Hence, the correlation coefficients in the first row would measure separately the strength of the relation between each independent variable X_i and the dependent variable Y. For instance, $r_{Y2} = .327$ in Table 16.4 measures the mutual variation between Y and X_2 as if X_1 did not exist in the analysis.

When the off-diagonal elements are nonzero, though, the correlation coefficient r_{Yi} (or equivalently, r_{iY}) measures not only the mutual variation between Y and X_i but also the effects of the other variables that are correlated with X_i. The larger the off-diagonal correlation coefficients, the more the other variables affect r_{Yi}. When r_{ij}

becomes large in absolute value, we cannot identify the separate relations between Y and X_i and between Y and X_j. This condition is called *multicollinearity* and will be addressed in more detail in Chapter 17.

Many software packages including our Regression processor produce a correlation matrix like the one in Equation 16–17 as part of the output of a regression analysis. Though the correlation matrix is a summary of the analysis of the interdependencies among the variables, we use it in regression analysis in two ways: to assess which independent variables have strong and weak relations with Y (by examining the entries in the first row) and to spot confounding relations between independent variables (by examining the off-diagonal entries below the first row).

Coefficient of Multiple Determination

In simple regression, when we square the correlation coefficient r, we generate the coefficient of determination r^2. The value of r^2 represents the proportion of the total variability in the observed Y-values that is explained by the simple regression of Y on X. In multiple regression, there is a corresponding measure of the proportion of explained variability called the *coefficient of multiple determination,* which is denoted by the symbol R^2. We use a capital R^2 here to distinguish the coefficient of multiple determination from the coefficient of simple determination r^2. Equation 16–18 shows how to compute R^2 using the total sum of squares, SSY, and either the regression sum of squares, SSR, or the error sum of squares, SSE.

Coefficient of Multiple Determination

$$R^2 = \frac{SSR}{SSY} = 1 - \frac{SSE}{SSY} \qquad \textbf{(16–18)}$$

The coefficient is a number between 0 and 1—that is, $0 \leq R^2 \leq 1$. If $R^2 = 1$, we have a perfectly fitting model that explains all of the variability in Y. If $R^2 = 0$, the model explains none of the variability and suggests that the independent variables are completely unrelated to Y in their present linear form.

To demonstrate Equation 16–18, consider the small data set listed in Table 16.3. In Table 16.4, we found $SSY = 6$. Suppose the model $Y = \beta_0 + \beta_1 X_1 + \beta_2 X_2 + \epsilon$ were fit to the data set, yielding $SSE = .4571$. According to Equation 16–18, the coefficient of multiple determination is $R^2 = 1 - (.4571/6) = .9238$. This means that 92.38 percent of the six units of variability in the observed Y-values is attributed to the relation between Y and the independent variables X_1 and X_2. Conversely, 7.62 percent of Y's variability is unexplainable or unaccounted for by the fitted regression model.

EXAMPLE 16.8

a. Find the coefficient of multiple determination for the data in Example 16.1, assuming $SSE = 202.47309$, and interpret R^2.

b. Run the Regression processor using the XM161 data file and obtain the correlation matrix. What condition exists between the independent variables?

Solution:

a. From Example 16.1, we compute

$$SSY = \Sigma Y^2 - \frac{(\Sigma Y)^2}{n} = 5820.58 - \frac{(205)^2}{12} = 2318.49667$$

Thus,

$$R^2 = 1 - \frac{SSE}{SSY} = 1 - \frac{202.47309}{2318.49667} = .9126705$$

We conclude that approximately 91.3% of the observed variation in Y-values can be attributed to the variables X_1 and X_2.

b. The processor should produce the correlation matrix displayed in Figure 16.8. Notice that the independent variables are strongly correlated—$r_{12} = .94261$. This makes it practically impossible to assess the separate relation of each variable with Y. A better model, one which eliminates this condition of *multicollinearity*, might include only one of the independent variables. We will learn how to build models like this in the next chapter.

```
≡                        Multiple Regression                          LIST
  Introduction  *able  Data  Calculations [Results] Exit          | F1=Help

    Correlation matrix
    1.00000               0.94752               0.93387
    0.94752               1.00000               0.94261
    0.93387               0.94261               1.00000
```

Figure 16.8 Correlation Matrix Screen

□

COMMENTS 1. A large value for R^2 is a good sign that the set of independent variables is useful in explaining a portion of the variability in Y. However, we should not judge the adequacy of a fitted multiple regression model solely on the R^2 value. A fitted model can be nonsignificant by the ANOVA F-test, yet have a large R^2. Also, a model can be significant by the ANOVA F-test, yet have a small R^2. The value of R^2 is no substitute for a formal test of model utility. As Exercise 16.57 illustrates, the sample size is a major factor affecting the significance of a fitted model.

2. Sometimes the adjective "multiple" is omitted in describing R^2 as the coefficient of determination.

Coefficient of Multiple Correlation

Although we think of correlation as measuring the strength of the relation between two variables, we also can measure the strength of the relation between one variable and a set of variables, provided there is a statistical relation linking them. The fitted multiple regression model is such a relation, tying Y to the set of independent variables $X_1, X_2, \ldots, X_k$. Thus, the *coefficient of multiple correlation,* denoted by R, will be used to measure the linear dependence of Y on the set of X's. Equation 16–19 shows that R is found by taking the positive square root of R^2.

Coefficient of Multiple Correlation

$$R = +\sqrt{R^2} \tag{16–19}$$

In simple correlation, the coefficient of correlation r could be a positive or a negative number. This is not the case in multiple correlation, since we are correlating Y with a best fitting linear function of variables, not with just one variable. While a multiple regression equation may have a mixture of direct and inverse relations, it can be shown that the multiple correlation R is a positive number only.

As with R^2, the coefficient of multiple correlation is a descriptive statistic and should not be the sole determinant of a model's utility. Many computer packages include R as part of the output.

COMMENT Another way of viewing R is to regard it as the correlation between the variables Y and $\hat{Y}$, where $\hat{Y}$ is the value generated by the fitted regression model. Intuitively, we would expect Y and $\hat{Y}$ to behave similarly and to have a positive correlation coefficient R.

CLASSROOM EXAMPLE 16.6

Setting Up a Correlation Matrix

Refer to the following data that have been reproduced from Classroom Example 16.1:

Y	X_1	X_2
1	1	8
1	2	10
2	2	10

Y	X_1	X_2
3	2	15
3	3	18
2	3	20
4	3	24
2	4	18
3	5	20
4	5	27

The variables are defined as

Y = Number of coupons redeemed, in hundreds
X_1 = Number of coupons printed, in thousands
X_2 = Percentage of coupons mailed to apartment dwellers

a. Compute the three correlation coefficients r_{Y1}, r_{Y2}, and r_{12}.
b. Organize the correlation coefficients into a correlation matrix:

$$\mathbf{C} = \begin{bmatrix} & & \\ & & \\ & & \end{bmatrix}$$

c. Which independent variable is, by itself, the stronger predictor of the dependent variable?

SECTION 16.6 EXERCISES

16.52 A sample of 48 observations on Y, X_1, and X_2 yielded the following correlation matrix:

$$\mathbf{C} = \begin{bmatrix} 1.00 & .59 & .38 \\ & 1.00 & .19 \\ & & 1.00 \end{bmatrix}$$

a. Is it possible to fill in the three values in the matrix below the main diagonal? What are the missing values?
b. Which independent variable is, by itself, the stronger predictor of the dependent variable?

16.53 What assumptions do we make in a multiple correlation analysis?

16.54 Develop the correlation matrix for the following data set:

Y	X_1	X_2
4	2	5
8	2	1
4	2	3
12	2	0
10	0	2
9	0	1
7	0	4
10	0	0

16.55 Meteorologists developed a "wind-chill index" to relate the combined effects of temperature and wind to a perceived temperature on exposed skin. Following are some values for the variables Y = perceived temperature; X_1 = wind speed, in mph; and X_2 = thermometer reading, in degrees Fahrenheit.

Y	X_1	X_2
40	0	40
28	10	40
27	5	30
9	15	30
4	10	20
−10	20	20
−18	15	10
−25	20	10
−5	5	0
−21	10	0

Use the Regression processor to do the following:

a. Find R^2 for the model $Y = \beta_0 + \beta_1 X_1 + \beta_2 X_2 + \epsilon$.

b. Generate the correlation matrix.

c. Determine which independent variable is more strongly correlated with Y.

16.56 Refer to the data in Exercise 16.10. Use the Regression processor to do the following:

a. Find the correlation matrix.

b. Determine which independent variable is most strongly correlated with Y.

c. Determine which independent variable is most weakly correlated with Y.

16.57 Consider the following ANOVA tables:

1.

Source	SS	df	MS	F
Regression	5.5429	2	2.7714	12.12
Error	.4571	2	.2286	
Total	6	4		

2.

Source	*SS*	*df*	*MS*	*F*
Regression	11397.03	3	3799.01	18.02
Error	226366.51	1074	210.77	
Total	237763.54	1077		

a. For both ANOVA tables, test the utility of the fitted regression model using $\alpha = .05$.
b. For both ANOVA tables, compute the coefficient of multiple determination.
c. What is the lesson to be learned from parts a and b? (*Hint:* Reread the comments after Example 16.8.)

16.58 Refer to Exercise 16.39. Using the Regression processor, find R^2 and R. Interpret the values of these measures.

16.59 A study of the top graduate programs in business ("The Best B-Schools," *Business Week*, November 28, 1988) produced data on these variables:

$$Y = \text{Number of 1988 graduates}$$
$$X_1 = \text{Annual tuition, in thousands of dollars}$$
$$X_2 = \text{Percent of applicants accepted}$$

Given the following sums of squares, develop the correlation matrix:

$$SSYX_1 = 4587.814 \qquad SSY = 948261.2$$
$$SSYX_2 = -2454 \qquad SSX_1 = 209.935$$
$$SSX_1X_2 = -87.31 \qquad SSX_2 = 1025$$

16.60 Refer to Exercise 16.35. Find and interpret R^2.

16.61 Refer to Exercise 16.46. Find the correlation matrix. Which independent variable is most strongly correlated with Y?

16.62 Suppose, in one problem setting, your *simple* regression analysis yields a correlation of .93 between X and Y. In another simple regression problem setting with different X and Y variables, you obtain $r = .49$. For which situation might you be more likely to consider a multiple regression analysis? Why?

16.63 You have data from 28 outlets in different cities for a retailer that specializes in stereo equipment and supplies. As a first step in examining the data, you have had a computer program generate the following correlation matrix

where

Y = Outlet gross earnings (annual)
X_1 = City population
X_2 = Number of competing stores in city
X_3 = Annual advertising expenditures

Use the following correlation matrix to suggest which predictor variable is most strongly related to earnings. Which independent variable seems unrelated to earnings?

$$\mathbf{C} = \begin{bmatrix} 1.00 & -.09 & -.55 & .39 \\ -.09 & 1.00 & .70 & .80 \\ -.55 & .70 & 1.00 & .32 \\ .39 & .80 & .32 & 1.00 \end{bmatrix}$$

16.7 OTHER COMPUTER PACKAGES

Computer routines facilitate the many calculations needed in a multiple regression analysis, as our Regression processor has demonstrated. There are several commercial computer packages that could be used as well to develop a multiple regression model. Three of the more popular ones are Minitab, the Statistical Analysis System (SAS), and the Statistical Package for the Social Sciences ($SPSS^X$). In this section, we compare the output from these packages to that of our Regression processor.

To facilitate comparisons, we will use the data in Example 16.1 on the variables Y, the annual number of passengers arriving and departing an airport per airline, X_1, the average number of flights departing weekly, and X_2, the total number of local airline employees. The data are listed as follows:

Airline	Y	X_1	X_2
1	19.3	63	40
2	2.8	21	7
3	30.5	94	87
4	6.6	50	10
5	11.7	42	23
6	6.1	21	6
7	7.6	36	11
8	13.4	48	8
9	50.4	104	88
10	14.9	49	43
11	35.4	89	71
12	6.3	25	8

These data were processed by the regression option of each of the three commercial packages. The output for Minitab appears in Figure 16.9; the SAS output is in Figure 16.10; and the $SPSS^X$ output is in Figure 16.11. As you can see, the commercial packages are somewhat different, yet succinct, in their display of the calculations.

We would like to focus on five major summaries in the multiple regression analysis:

A. The equation of the fitted model

B. The estimated standard error of the regression

C. The coefficient of multiple determination and/or multiple correlation

D. The ANOVA table

E. The F-test of model utility

In Figure 16.12—our Fitted Model screen—you find the first three items, each indicated by the appropriate capital letter. The equation of the fitted model is explicitly shown in the Minitab output, but not in the other two. However, by picking off the

```
The regression equation is
PASSNGRS = - 4.87 + 0.307 FLIGHTS + 0.165 EMPLOYEE        (A)

Predictor        Coef        Stdev     t-ratio        p
Constant       -4.869        4.312       -1.13    0.288
FLIGHTS        0.3067       0.1500        2.04    0.071
EMPLOYEE       0.1655       0.1336        1.24    0.247

(B) s = 4.743     R-sq = 91.3%  (C)  R-sq(adj) = 89.3%

Analysis of Variance

SOURCE        DF          SS          MS         F        p
Regression     2      2116.0      1058.0     47.03    0.000
Error          9       202.5        22.5       (D)
Total         11      2318.5
```

Figure 16.9 Minitab Output for Data in Example 16.1

```
                              EXAMPLE 16.1

                          ANALYSIS OF VARIANCE                     (D)

                          SUM OF          MEAN
SOURCE         DF        SQUARES        SQUARE        F VALUE       PROB>F

MODEL           2     2116.02358    1058.01179         47.029       0.0001
ERROR           9      202.47309   22.49700948
C TOTAL        11     2318.49667

(B) ROOT MSE          4.743101      R-SQUARE          0.9127  (C)
    DEP MEAN          17.08333      ADJ R-SQ          0.8933
    C.V.               27.7645

                         PARAMETER ESTIMATES

                     PARAMETER       STANDARD       T FOR H0:
VARIABLE    DF        ESTIMATE          ERROR     PARAMETER=0     PROB > |T|

INTERCEP     1     -4.86856723     4.31187485          -1.129         0.2880
FLTS         1      0.30668650     0.15001002           2.044         0.0713
EMP          1      0.16549770     0.13363612           1.238         0.2469
                       (A)
```

Figure 16.10 SAS Output for Data in Example 16.1

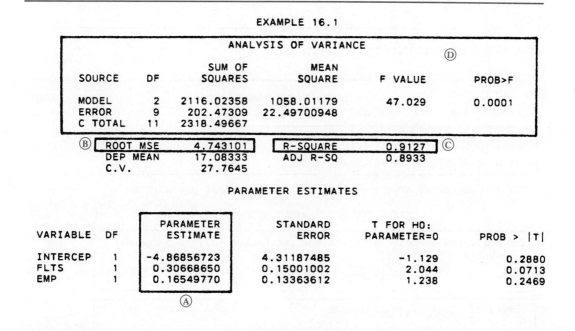

```
Equation Number 1     Dependent Variable..    PASS
Beginning Block Number  1. Method:  Enter
Variable(s) Entered on Step Number  1..    EMP
                                    2..    FLIS

Multiple R              .95534   (C)
R Square                .91267

Adjusted R Square       .89326

Standard Error         4.74310   (B)

ANALYSIS OF VARIANCE                                                      (D)
                        DF          Sum of Squares       Mean Square
Regression               2              2116.02358        1058.01179
Residual                 9               202.47309          22.49701
F =            47.02900        Signif F =   .0000

                    Variables in the Equation

Variable              B  (A)        SE B        Beta           T      Sig T

EMP             .165498          .133636     .365350       1.238      .2469
FLTS            .306686          .150010     .603136       2.044      .0713
(Constant)    -4.868567         4.311875                  -1.129      .2880
```

Figure 16.11 SPSS^X Output for Data in Example 16.1

```
≡                       Multiple Regression                           READY
   Introduction  Data  Calculations [Results] Exit                  | F1=Help

   The fitted multiple regression model, to 5 decimal places, is:
       ^                                                              (A)
     Y = -4.86857  +0.30669X1  +0.16550X2

   The sum of the squared errors (SSE), to 10 decimal places, is:
                                                   ^
                      SSE = Σ(Y - Y)² = 202.4730853627

   The estimated standard error of the regression (σ), to 5 decimal
   places, is:                                                         (B)
                      ^
                      σ = √ 22.49701    =  4.74310

   The coefficient of determination (R²), to 5 decimal places, is:
                                                                       (C)
                             R² =  0.91267
```

Figure 16.12 Fitted Model Screen

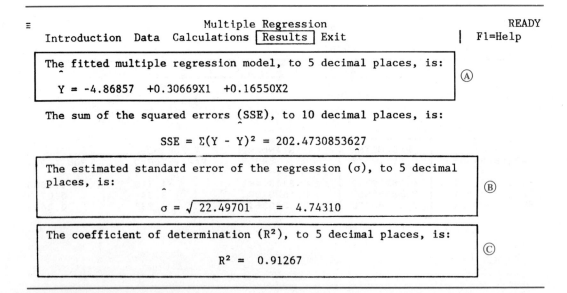

values in the appropriate column of the SAS and SPSS^X outputs, you can construct the equation. Refer to the circled letter A in each figure.

Be prepared to see different names and/or symbols for $\hat{\sigma}$. In Minitab, it is labeled "s"; in SAS, it is called "ROOT MSE"; and in SPSS^X, it is "Standard Error." Look at the circled letter B in each figure.

```
≡                     Multiple Regression                              READY
  Introduction  Data  Calculations [Results] Exit                 |  F1=Help


    ANOVA table
```

Source	SS	df	MS	F	p-value
Regression	2116.02358	2	1058.012	47.029	0.0000171879
Error	202.47309	9	22.497		
Total	2318.49667	11	Ⓓ		

Figure 16.13 ANOVA Table Screen

```
≡                     Multiple Regression                              READY
  Introduction  Data  Calculations [Results] Exit                 |  F1=Help


   F-test of model utility

              Ho: β1 = β2 = 0

              Ha: not Ho

      Ⓔ                 MSR    1058.012
              TS: F* = --- = ---------- = 47.029
                        MSE    22.497

             p-value = 0.0000171879, based on 2 and 9 df.


     The decision to "reject Ho" or "do not reject Ho" depends on your
   judgment or tolerable risk of Type I error (see text Figure 10.18).
```

Figure 16.14 *F*-Test of Model Utility Screen

The coefficients R and R^2 are fairly easy to locate and identify on all the printouts; you can reference them with the circled letter C.

Item D also is obvious from the printouts of the three packages and looks similar to our ANOVA table, displayed in Figure 16.13. Note, though, that our term "*p*-value" is called (just) "p" in Minitab; "PROB > F" in SAS; and "Signif F" in SPSS^X. Also notice that the commercial packages allow only three or four decimal places for the *p*-value, whereas we allocate 10 decimal places.

Lastly, our F-test screen pictured in Figure 16.14 is not available on any of the packages. Experienced users do not need it, but we believe that those who only occasionally run a multiple regression analysis can benefit from it.

We hope this brief exposure to other forms of multiple regression output demonstrates that the calculations all may be the same, but the way the numbers are displayed by a computer package may be quite different.

16.8 SUMMARY

Analyzing the relationship between a dependent variable Y and a set of independent variables is our goal in multiple regression. The analysis can be broken down into steps or objectives. First, we identify several potential independent variables. Second, we propose a linear model linking Y to the independent variables. Third, we fit the proposed model to a sample data set using the criterion of least squares. Fourth, we test the fitted model for its predictive capabilities. If we conclude that there is a relation between Y and the X's, then we can use the fitted model to predict a specific Y-value or to estimate an average Y-value.

Several correlational statistics fall out of our regression analysis, including the coefficient of multiple determination R^2, the coefficient of multiple correlation R, and the correlation matrix C. These descriptive measures are used to quantify the strength of the relation, to locate important and not-so-important regressors, to identify redundant regressors, and to explain the proportion of variation attributed to the model.

In the next chapter, we will make improvements to our multiple regression model by deleting insignificant independent variables or by adding cross-product or interaction terms. In addition, we will examine closely, and in some cases test, the assumptions we made in the regression analysis. These activities are part of a larger process called *model building*.

16.9 TO BE CONTINUED . . .

. . . In Your College Courses

As this chapter occurs late in our book, it suggests that multiple regression is a more advanced statistical tool, requiring a background in sampling distributions, hypothesis testing, and estimation. Similarly, multiple regression rarely is found in the introductory or principles courses in a business core. But it definitely surfaces in the more specialized junior/senior courses in some disciplines.

Marketing Research Research in marketing depends on multiple regression analyses to investigate factors influencing consumer behavior, price sensitivity, or advertising effectiveness, for example. Gilbert A. Churchill, Jr., in his text, *Marketing Research: Methodological Foundations* (1979), devotes an entire chapter to regression analysis, both simple and multiple. For multiple regression, he reinforces the goal we stated at the beginning of this chapter:

> *The purpose will remain the same. We still want to construct an equation that will enable us to estimate values of the criterion variable, but now from given values of* several *predictor variables.*

He discusses the least-squares approach for fitting a model and reiterates the need to test the utility of the fitted model:

> *In the multiple-regression case, it is mandatory that the significance of the overall regression be examined using an* F*-test.*

Churchill mentions the coefficients of multiple correlation and determination as descriptive measures of strength, as we have stated, and presents an example relating Y = the level of sales for ballpoint pens to X_1 = the number of TV advertising spots per month and X_2 = the number of sales representatives.

Auditing Auditors are encouraged to conduct *analytical tests* to find relationships within a client's business that might reveal an unknown problem or signal pending trouble. Alvin A. Arens and James K. Loebbecke discuss analytical tests in a chapter titled "Use of Statistical Techniques in Analytical Tests" in their text, *Applications of Statistical Sampling to Auditing* (1981).

After establishing the need for analytical tests, they argue that

> *Several factors in recent years have motivated auditors to seek more effective analytical testing techniques: 1. Audit Costs . . . 2. Lawsuits . . . 3. Quantitative Approach . . . 4. Quarterly Reviews.*

One of the quantitative approaches referred to in item 3 is regression analysis:

> *Regression analysis is used to make inferences of what financial balances* should be *for comparison to* recorded *balances. By making this comparison, the auditor can identify those balances that, based on the analysis, appear to be out of line.*

Arens and Loebbecke note that relationships within data can be characterized in terms of variables, such that

> *The identification of variables allows the auditor . . . to define the relationships that exist among them. These relationships will never be exact . . . (they) will be* statistical *in nature.*

As an illustration, they suggest relating selling expense to sales and the number of salesmen.

Notice the different perspectives in marketing and accounting. Though both textbooks used the number of sales representatives as an independent variable, Churchill used "sales" as the dependent variable, while Arens and Loebbecke were concerned with sales *expense* and used "sales" as an independent variable!

Econometrics An economist's job is to spot trends and forecast the future. Multiple regression is an often-used technique for doing just that. A major area in economics where it is useful is in demand theory estimation. James L. Pappas, Eugene F. Brigham, and Mark Hirschey in their text, *Managerial Economics* (1983), present a study involving a food processing firm. For one of its products, frozen fruit pies, the company wished to develop a demand equation to predict unit sales (Q) per quarter. Using historical data, management related Q to six independent variables, including retail price, competitors' prices, dollars spent on advertising and promotion, per capita disposable income in the market, population, and a trend factor. An analysis similar to our model utility test revealed that the model could be used "with a great deal of confidence for decision-making purposes."

Multiple regression and correlation may appear in other business courses as well; the three areas mentioned here are not exhaustive by any means. Providing the assumptions are reasonably met, a regression analysis is a powerful predictive tool.

. . . *In Business/The Media*

A fitted multiple regression model is not front page news, nor will one likely make an appearance in a firm's annual report. Nonetheless, multiple regression models are present in many businesses and in many forms.

Real Estate Property Valuation If you own a house or condominium, you probably pay property taxes on the dwelling. Your local tax office needs a means of assessing property value without conducting an on-site examination of every house in the region. No doubt a multiple regression analysis is used to develop a property valuation model. In it are independent variables such as square feet, age of the property, lot size, number of bedrooms, number of bathrooms, central air conditioning (present or absent), attached garage (present or absent), and perhaps several other variables. Using the selling price of other houses in each neighborhood, the tax assessor builds a fairly accurate model to predict property value and thus have a basis for the figures that appear on your tax bill.

Forecasting Peak Electrical Demand Consider the situation facing your local electrical utility company. We, as consumers, expect heat in the winter and air conditioning in the summer whenever we nudge the thermostat in our house or apartment. Of course on very cold or very hot days, we demand more of that invisible power called electricity. Unfortunately, the electrical utility cannot make electricity instantly as it is needed. It must be prepared to meet the *peak demands* of its customers. This requires a great deal of planning and forecasting, usually by personnel in a rate or research department.

The peak demand forecast typically is an evolutionary process with many potential independent variables examined, included in the model, and perhaps later deleted from the model as housing or environmental conditions change. There are some stable variables that remain in the model, such as temperature, an index of growth in the residential sector, and average home size. Appliances are big consumers of electricity, so they too must be factored into the model.

If we were to spend a day with a utility forecaster, we would see many multiple regression analyses and marvel at the complexities of forecasting.

Financial Analysis Buying and selling stocks, bonds, options, and/or futures sounds glamorous and lucrative. However, intensive analysis, sweat, a bit of luck, and more analysis are needed to be successful in the long run in the financial markets. You may hear on television or read an article about a technical analysis. Guess what tool is popular in such analyses? That is correct—multiple regression and correlation.

For an illustration, we refer you to a publication dubbed "The Trader's Magazine" called the *Technical Analysis of Stocks and Commodities*. In any monthly issue, you likely will find correlation matrices, multiple regression models, descriptive statistics like standard deviation and coefficient of variation, and other statistical topics. Clearly, the financial markets are very fertile grounds for statistical analyses such as multiple regression.

CHAPTER 16 EXERCISES

16.64 A commercial developer plans to build a ten-story office building along a major suburban artery. The developer would like to predict the additional traffic volume that the building would produce. If the dependent variable is Y, number of cars using the road per hour, propose several independent variables that might be related to Y.

16.65 Identify several independent variables that might be related to Y, fuel efficiency as measured by miles per gallon, for automobiles.

16.66 A bank will often develop a regression model to help predict the amount of money it will loan a borrower toward the purchase of a used car. Identify several independent variables that might be included in such a model.

16.67 The owners of a theme park wish to predict daily attendance during the summer months in order to schedule employees effectively and to maintain sufficient inventories of supplies. What independent variables might be related to the dependent variable Y, daily attendance?

16.68 Explain the differences between a multiple regression analysis and a multiple correlation analysis.

16.69 Name the assumptions about the error term ϵ in a multiple regression analysis.

16.70 What is the goal of a multiple regression analysis?

***16.71** Suppose we fit the model $Y = \beta_0 + \beta_1 X_1 + \beta_2 X_2 + \epsilon$ to the following set of data:

*Optional

Entity	Y	X_1	X_2
1	3	1	0
2	3	16	3
3	0	32	6
4	4	3	0
5	4	2	0
6	3	6	2
7	18	4	0
8	15	10	0
9	1	15	5
10	29	16	1
11	11	10	1
12	3	3	2

a. Set up the **X** matrix and the **Y** vector.
b. Write out **X′**.
c. Compute **X′Y** and **X′X**.
d. Find $(\mathbf{X'X})^{-1}$.
e. Using Equation 16–8, find **b**.
f. Write the equation of the fitted regression model.

16.72 Refer to Exercise 16.71. Suppose a third independent variable is to be added to the analysis; its values are as follows:

Entity	1	2	3	4	5	6	7	8	9	10	11	12
X_3	9	4	15	21	30	5	7	9	10	2	4	11

Use the Regression processor to fit the model $Y = \beta_0 + \beta_1 X_1 + \beta_2 X_2 + \beta_3 X_3 + \epsilon$.

16.73 An analyst planned to use a multiple regression model to relate the following variables:

Y = Stock price
X_1 = Price-to-earnings ratio
X_2 = Earnings estimate

What relationships are imbedded in the variables that potentially could cause problems in the analysis?

16.74 Refer to Exercise 16.54. Using the Regression processor, fit the model $Y = \beta_0 + \beta_1 X_1 + \beta_2 X_2 + \epsilon$ and test it for utility using $\alpha = .10$.

16.75 A study was conducted by the Center for Science in the Public Interest to assess the nutritional value of fast foods. After analyzing each food's caloric content (X_1), number of teaspoons of fat (X_2), and number of milligrams of sodium (X_3), the Center derived a "Gloom rating." In general, the lower the Gloom rating the better the food is from a nutritional standpoint. A sample of 54 items available at various fast-food restaurants produced the data shown in Table 16.5.

Table 16.5 Fast-Food Data for Exercise 16.75

Company/Product	Chicken			
	Calories	Fat (Tsp.)	Sodium (Mg.)	Gloom Rating
Roy Rogers drumstick	117	2	162	8
Kentucky Fried Chicken original recipe drumstick	147	2	269	11
Church's fried chicken leg	147	2	286	12
KFC drumstick (extra crispy)	173	3	346	15
Wendy's chicken sandwich, multi-grain bun	320	2	500	15
Burger King Chicken Tenders (6)	204	2	636	16
D'Lites chicken filet sandwich	280	3	760	19
McDonald's chicken McNuggets (6)	323	5	512	26
KFC Kentucky Nuggets (6)	276	4	840	27
Arthur Treacher's chicken sandwich	413	4	708	28
Carl's Jr. chicken sandwich	450	3	1,380	28
Dairy Queen chicken sandwich, fried	670	9	870	53
Burger King chicken sandwich	688	9	1,423	58

Company/Product	Hamburgers			
	Calories	Fat (Tsp.)	Sodium (Mg.)	Gloom Rating
McDonald's hamburger	263	3	506	16
Burger King hamburger	275	3	509	17
Wendy's hamburger, white bun	350	4	410	22
Hardee's hamburger	276	3	589	22
Jack in the Box cheeseburger	323	3	749	22
McDonald's cheeseburger	318	4	743	23
Dairy Queen Single with cheese	410	5	790	28
McDonald's Quarter Pounder	427	5	718	31
Roy Rogers hamburger	456	6	495	34
Burger King double cheeseburger	478	6	827	35
Hardee's Big Deluxe	503	7	903	38
Dairy Queen Double with Cheese	650	8	980	46
Burger King Whopper	626	9	842	47
Wendy's Double Cheeseburger, white bun	630	9	835	48
Dairy Queen Triple Hamburger	710	10	690	51
McDonald's McD.L.T.	680	10	1,030	54
Roy Rogers RR Bar Burger	611	9	1,826	57
Burger King Double Beef Whopper with Cheese	970	15	1,206	76
Wendy's Triple Cheeseburger	1,040	15	1,848	85

(continues)

Table 16.5 Continued

Company/Product	Calories	Fat (Tsp.)	Sodium (Mg.)	Gloom Rating
	French Fries			
Arby's	211	2	30	9
Dairy Queen	200	2	115	11
McDonald's	220	3	109	13
D'Lites	260	3	100	14
Long John Silver's	247	3	6	14
KFC Kentucky Fries	268	3	81	15
Jack in the Box	221	3	164	15
Arthur Treacher's Chips	276	3	39	15
Hardee's	239	3	180	15
Wendy's	280	3	95	16
Roy Rogers	268	3	165	16
Burger King	227	3	160	17
Carl's Jr.	250	3	460	21

Company/Product	Calories	Fat (Tsp.)	Sodium (Mg.)	Gloom Rating
	Shakes and Malts			
Jack in the Box, strawberry	320	2	240	13
McDonald's, vanilla	352	2	201	15
McDonald's, strawberry	362	2	207	15
Arby's, vanilla	295	2	245	16
Burger King, vanilla	321	2	205	16
Roy Rogers, vanilla	306	2	282	17
McDonald's, chocolate	383	2	300	17
Burger King, chocolate	374	3	225	18
DQ chocolate malt	520	3	180	21

Source: Dinah Eng, "Quick eating guide," *USA Today*, October 6, 1986. Copyright 1986, *USA Today*. Adapted with permission.

a. Develop a multiple regression model to relate Y = Gloom rating to X_1, X_2, and X_3.
b. What is the estimated standard error of the regression?
c. Is this model a useful relation among the variables? Use $\alpha = .05$.
d. Construct a 95% prediction interval for the Gloom rating for a Big Mac hamburger from McDonald's. Assume the caloric content is 570 with 8 teaspoons of fat and 979 milligrams of sodium. Use 1.6019 as the value of the estimated standard error.
e. Refer to part d. The Center's Gloom rating for a Big Mac was 45. Did your prediction interval correctly predict this possible outcome?

f. Find the correlation matrix and examine the correlations among the independent variables. Based on these values, do the correlations in the first row give an accurate indication of the strength of the separate relations between Y and each X_i? Explain.

16.76 Refer to Exercise 16.32. Estimate the average amount donated for a class celebrating its twelfth year reunion if there are 260 contributors. Use 98% confidence, and .53966 as the value of the estimated standard error.

***16.77** One of the assumptions in a multiple regression analysis is that the random error term ϵ is normally distributed. It can be shown that this assumption, in turn, implies that the (conditional) distribution of Y at a given setting of the X's is normal, with a mean of $\beta_0 + \beta_1X_1 + \ldots + \beta_kX_k$ and a standard deviation of σ. The sample data provide estimates of the mean and standard deviation; namely, $b_0 + b_1X_1 + \ldots + b_kX_k$ and $\hat{\sigma}$, respectively. Using these parameter estimates and recalling Equation 7–2 that we applied in Chapter 7 to use the Z-table, we can find interval probabilities about Y.

For instance in Example 16.2, the fitted model was $\hat{Y} = 56.31857 - 1.94724X_1 - .00801X_2 + 1.84047X_3$ with $\hat{\sigma} = 8.24958$. Determine the probability that the occupancy rate would exceed 70 percent when $X_1 = 2$, $X_2 = 50$, and $X_3 = 5$ percent.

***16.78** Refer to Exercise 16.77 and Exercise 16.39. What is the probability that the driver hic score is less than 1000 if the chest score is 42, the left thigh femur load is 825, the right thigh femur load is 1012, and the weight of the vehicle is 2744 pounds?

***16.79** Consider the general multiple regression model involving k independent variables: $Y = \beta_0 + \beta_1X_1 + \ldots + \beta_kX_k + \epsilon$. Give the dimensions of the following matrices or vectors, based on a data set with n observations:

a. $\mathbf{X}$ b. $\mathbf{X'X}$
c. $\mathbf{X}_s$ d. $\mathbf{X}_s'(\mathbf{X'X})^{-1}\mathbf{X}_s$

***16.80** If a multiple regression analysis with four independent variables is performed on a set of $n = 15$ entities, give the dimensions of the following vectors or matrices:

a. $\mathbf{Y}$ b. $\mathbf{X}$
c. $\mathbf{X}'$ d. $\mathbf{X'Y}$
e. $\mathbf{X'X}$ f. $\mathbf{b}$
g. $\mathbf{X}_s$ h. $\mathbf{X}_s'(\mathbf{X'X})^{-1}\mathbf{X}_s$

***16.81** Refer to Exercise 16.80. Repeat the instructions assuming there are two independent variables and 12 entities.

***16.82** Refer to Exercise 16.12. Construct the ANOVA table for the analysis.

***16.83** Suppose we wish to predict $\overline{Y}_m$, the average value of Y for a set of m new entities, each with the same values for the independent variables. A prediction interval for $\overline{Y}_m$ is as follows:

$$\hat{Y} \pm (t\text{-score})\ \hat{\sigma} \cdot \sqrt{\frac{1}{m} + \mathbf{X}_s'\,(\mathbf{X'X})^{-1}\mathbf{X}_s}$$

where the t-score is based on $df = n - (k + 1)$ and $\mathbf{X}_s$ is the column vector of given values for the independent variables.

Refer to Exercises 16.6 and 16.33 where we fit and tested the model $Y = \beta_0 + \beta_1X_1 + \beta_2X_2 + \epsilon$. The fitted model was $\hat{Y} = 32.787 - .0192X_1 - .4417X_2$ with $\hat{\sigma} = 2.9658$. The inverse of the $\mathbf{X'X}$ matrix was as follows:

$$(\mathbf{X'X})^{-1} = \frac{1}{1363991}\begin{bmatrix} 1868849 & -11010 & -41558 \\ -11010 & 13176 & -812 \\ -41558 & -812 & 1085.25 \end{bmatrix}$$

Predict with 80% confidence the average one-year return for a group of five no-load mutual funds, each of which invests 50 percent of their portfolio in blue-chip stocks.

***16.84** Refer to Exercises 16.83 and 16.24 and to Classroom Example 16.1. Use a 99% prediction interval to estimate $\bar{Y}_4$, the average number of redeemed coupons, if the restaurant owner prints 3000 coupons and mails 25 percent of them to apartment dwellers in each of the next four direct-mail advertisements.

16.85 Construct the correlation matrix from the following regression data:

Y	X_1	X_2
−4	2	5
−2	9	3
7	2	1
−12	5	2
1	8	5
1	6	4
1	1	5
−4	4	6
0	2	9
−8	1	0

16.86 Suppose a multiple regression analysis involving two independent variables produced the following correlation matrix:

$$\mathbf{C} = \begin{bmatrix} 1.00 & 0 & 0 \\ 0 & 1.00 & .357 \\ 0 & .357 & 1.00 \end{bmatrix}$$

Explain whether it is possible to determine the value of R^2 (or R) from this matrix.

***16.87** The matrix formulation of a multiple regression problem also holds for a simple regression problem. Again, the matrix equation is $\mathbf{Y} = \mathbf{X}\boldsymbol{\beta} + \boldsymbol{\epsilon}$, and the method of least squares yields the vector of estimated regression coefficients

$$\boldsymbol{b} = \begin{bmatrix} b_0 \\ b_1 \end{bmatrix}$$

from $\mathbf{b} = (\mathbf{X'X})^{-1}\mathbf{X'Y}$. The dimension of the column vector $\mathbf{Y}$ is $n \times 1$. The matrix $\mathbf{X}$ has dimension $n \times 2$.

a. Set up the matrix **X** and the vector **Y** for a simple regression analysis based on the following data:

X	−2	−1	0	1	2
Y	5	7	8	10	12

b. Find the equation of the least-squares line, using $\mathbf{b} = (\mathbf{X}'\mathbf{X})^{-1}\mathbf{X}'\mathbf{Y}$.
c. Compare your answer to the answer for Exercise 15.6.

***16.88** Refer to Exercise 16.87. Repeat the instructions in Exercise 16.87, using the data in Exercise 15.9.

16.89 A food products company plans to introduce a new muffin mix. In developing the mix, the company set up a taste test involving other brands of muffins, several of the proposed recipes for the muffins, and muffins made from scratch. An independent panel of consumers was solicited and asked to rate the flavor of the muffins on a 1-to-10 scale with 10 the best rating. The main ingredients in muffins are flour, baking powder or soda, sugar, salt, eggs, flavorings, oil, and milk. It is the proportion of each ingredient that defines the muffin's flavor. Following are the ingredients of the muffins tested and the average rating of the panelists:

Flour (Cups)	Baking Soda (Tsp.)	Salt (Tsp.)	Sugar (Cups)	Eggs	Milk (Cups)	Oil (Tbs.)	Rating
2	4	.5	2	1	1	3	9.5
2	2	.5	1	1	.5	4	5.9
1	4	1	1	2	1	3	6.8
1	2	.5	.33	1	1	4	3.2
3	2	1	.33	1	1	4	4.2
1	3	.5	1	2	1.5	3	6.0
1	3	.5	1	1	.5	3	5.7
2	3	.5	1	1	1	4	3.5
2	4	1	1.5	1	2	3	7.2
2	4	1	2	2	1.5	3	6.6
3	4	2	.33	1	1	4	4.8
1	3	1	2	1	1	3	8.0
3	2	1.5	.5	1	.5	3	5.5
3	4	.5	1	1	.33	3	5.7
3	2	1	2	2	1.5	3	8.2
3	2	.5	.5	2	.5	4	4.7

a. The data for some of the variables, cups of sugar for instance, are recorded in differing degrees of accuracy. According to sound data management practices, they all should have the same accuracy. Would you be more inclined to round off all values to one decimal place, to integers (no decimal places), or to leave the data as is and declare all observations to be accurate to two decimal places? Why?
b. Create a data file and name it XR1689.
c. Relate the dependent variable Y = rating to the set of independent variables by fitting the appropriate multiple regression model. Write out the equation of the fitted model.
d. Summarize the calculations into an ANOVA table.
e. Test the fitted model for utility with an α-risk of 10 percent.
f. Estimate the average rating for muffins with these ingredients: 2 cups of flour, 2 teaspoons of baking soda, 1½ teaspoons of salt, 2 cups of sugar, 2 eggs, one-half cup of milk, and 3 tablespoons of oil. Use 98% confidence. Assume the value of the estimated standard error is 1.3116.
g. Obtain the correlation matrix. Which independent variable is most strongly correlated with Y?

h. Which independent variable is most weakly correlated with Y?

i. Do you think the company should include more of the strongly correlated variable—that is, more of that ingredient—and less of the most weakly correlated variable (ingredient) in its formula to obtain a higher rating? Explain the circumstance under which such a strategy would be statistically based.

16.90 A correlation matrix based on a large sample of observations is shown as follows:

$$C = \begin{bmatrix} 1.00 & .32 & .53 & .09 & -.29 \\ .32 & 1.00 & .11 & .31 & -.12 \\ .53 & .11 & 1.00 & .07 & -.09 \\ .09 & .31 & .07 & 1.00 & -.21 \\ -.29 & -.12 & -.09 & -.21 & 1.00 \end{bmatrix}$$

a. Which independent variable appears to be most strongly related to the dependent variable?

b. Which seems to have the weakest relation?

16.91 Suppose a multiple regression analysis produced the following results:

$$SSE = 1000 \qquad b_0 = -1.6$$
$$R = .73 \qquad b_1 = 3.1$$
$$n = 25 \qquad b_2 = 1.9$$

a. Write out the fitted multiple regression model.

b. What is the point prediction of Y when $X_1 = 5$ and $X_2 = 11$?

c. Suppose the actual value of Y is 32 when $X_1 = 5$ and $X_2 = 11$. What is the squared error associated with your estimate in part b?

d. Is the squared error you determined in part c probably less than or more than the squared error of the majority of other points in this data set? Why?

16.92 A survey on laptop personal computers revealed the following data:

Weight (Lb)	Battery Life (Min)	Recharge Time (Hr)	Floppy Disk Drives	Hard Disk Drives	Price ($ Thousands)
14.0	180	2	1	1	5.4
6.3	122	8	1	0	1.2
12.9	43	4	2	0	4.7
12.2	420	8	2	0	1.5
14.3	115	8	1	1	3.7
15.9	218	8	2	0	1.8
4.4	150	8	0	1	3.0
12.4	130	8	1	1	3.0
14.2	101	7	2	0	1.9
15.9	210	8	1	1	3.6
18.3	401	10	2	0	2.5
14.8	310	8	1	1	3.3
10.0	150	10	1	1	3.5
8.5	96	4	2	0	2.8
14.2	184	6	2	0	1.9

a. If the dependent variable is price, relate the remaining variables to it with a fitted multiple regression model.
b. What proportion of the variability in price does the model explain?
c. Is the fitted model useful? Use $\alpha = .10$.
d. Predict the price of a new laptop personal computer about to be introduced by a major computer manufacturer if it has the following specifications: 9.5 pounds, batteries that last for 6 hours and need only half that time to recharge, 1 floppy and 1 hard disk drive. Use 95% confidence, and 1.1429 as the value of the estimated standard error.
e. Which independent variable is the best predictor of price?

16.93 Cereal sales in grocery stores account for roughly 1.30 percent of a store's gross revenues. However, cereal's contribution to revenue is not a function of sales alone, but is thought to be influenced by the available selling space. In a random sample of 15 grocery stores, an accountant for the chain recorded the following data on the variables X_1, the average weekly volume in cases (one case = 24 boxes), X_2 = the square feet of selling area for cereals, and Y, the percentage of a store's gross revenues:

Y	X_1	X_2
1.30	192	252
1.26	251	238
1.73	394	409
1.18	126	151
1.33	347	210
1.29	250	219
1.44	291	253
1.36	317	231
1.27	248	201
1.42	383	325
1.31	269	271
1.27	250	245
1.20	95	188
1.30	274	230
1.29	255	270

a. Fit the multiple regression model $Y = \beta_0 + \beta_1 X_1 + \beta_2 X_2 + \epsilon$ to these data and test its utility.
b. Predict the percentage of gross revenues from cereal sales for a store with 228 square feet of selling area allocated to cereals and average weekly sales of 4,800 boxes of cereal. Use a 90% level of confidence, and .0646 as the value of the estimated standard error.
c. What happens to the prediction interval—wider, narrower, the same—if the store's weekly sales increase to 6,240 boxes while the square feet of selling area for cereals remains the same? Explain why, if possible. Use 90% level of confidence and .0621 as the value of the estimated standard error.

16.94 Develop the correlation matrix for the following data set:

Y	X_1	X_2
6.37	4	.8
5.85	8	1.6

(continues)

(*continued*)

Y	X_1	X_2
6.27	5	.8
6.70	0	2.0
6.90	4	3.0
8.71	5	3.0
8.50	3	2.8
8.50	7	3.2
6.23	6	3.0
6.80	9	3.3
5.10	0	.5
4.69	5	3.0

16.95 Find the estimated standard error of the regression in the following situations:

a. $SSE = 16.8087$; $n = 40$; and $k = 4$.

b. $\Sigma e^2 = 38{,}637$; $n = 53$; and $k = 5$.

c. $SSE = 4.5$; and df for error $= 18$.

16.96 A multiple regression analysis based on 23 data points yielded the following fitted model:

$$\hat{Y} = .5 - 1.50X_1 + 3.00X_2 + .25X_3 + 7.00X_4$$

The error sum of squares was 4.5 and $R^2 = .48$. Each of the following statements is incorrect; your job is to improve or correct each.

a. For each one-unit increase in X_2, we expect Y to be three times as large.

b. There are 17 degrees of freedom available to estimate the standard error of the regression.

c. The total sum of squares is 9.375.

d. The coefficient of multiple correlation is $-.693$.

e. Forty-eight percent of the variability in Y is unexplainable by the fitted regression model.

***16.97** A fitted multiple regression model based on 12 entities is $\hat{Y} = 5.5099 - .0404X_1 + .6208X_2$. Assume that $\hat{\sigma} = 1.2582$ and that the inverse of the $\mathbf{X'X}$ matrix is

$$(\mathbf{X'X})^{-1} = \frac{1}{9983.96}\begin{bmatrix} 5538.51 & -297.86 & -1474 \\ -297.86 & 147.72 & -174 \\ -1474 & -174 & 1016 \end{bmatrix}$$

a. Find an 80% prediction interval for Y when $X_1 = 2$ and $X_2 = 2.5$.

b. Find a 90% confidence interval for $E(Y)$ when $X_1 = 4.5$ and $X_2 = 2$.

***16.98** Firewood burns better after drying than when it is freshly cut. In a study to measure the available heat (in BTUs) per cord, researchers varied the length of drying time, in weeks. The average relative humidity during the drying time also was recorded. For instance, if the drying time were four weeks, the 28 daily relative humidity readings would be averaged to produce the value of X_2. Although many varieties of trees were available, only white oaks, 12 of them, were used in this study. The relevant ranges for each variable were as follows:

Variable	Min.	Max.
Y, heat	17.3	24.6
X_1, drying time	0	11
X_2, relative humidity	50%	90%

A drying time of 0 means the wood was burned within one day of cutting. Some of the intermediate matrix calculations are

$$\mathbf{X'Y} = \begin{bmatrix} 243.8 \\ 1402.2 \\ 16015 \end{bmatrix}$$

$$(\mathbf{X'X})^{-1} = \frac{1}{1459600}\begin{bmatrix} 12531100 & -509600 & -144100 \\ -509600 & 27200 & 5400 \\ -144100 & 5400 & 1716 \end{bmatrix}$$

$$\mathbf{Y'Y} = 5011.44$$

a. How many degrees of freedom for error are there?
b. Find the equation of the fitted model.
c. What is the interpretation, if any, of b_0?
d. Does the variable X_2 have a direct or inverse relation with Y? What happens to the variability of X_2 as X_1 increases? How does this affect the model?
e. Determine $\hat{\sigma}$.

16.99 For a chain of drugstores, you would like to develop a multiple regression equation to predict store monthly sales, Y. List two or more potentially useful independent variables.

16.100 You are interested in predicting grade point averages for a group of college students. What independent variables might prove helpful?

16.101 Construct an ANOVA table from the following F-test. Assume that the sample size was 29.

H_0: $\beta_1 = \beta_2 = \beta_3 = \beta_4 = 0$
H_a: not H_0
TS: $F^* = \dfrac{48.75}{6.33} = 7.70$
RR: Reject H_0 if $F^* > 4.22$.
C: Reject H_0.

What α-risk was used in the test?

16.102 A multiple regression analysis produced the following (partial) ANOVA table:

Source	df	SS
Regression	2	28.70412
Error	9	29.53255
Total	11	58.23667

Use the figures in this table to write out all parts of the F-test for model utility, similar to the box on page 857. Use $\alpha = .05$.

16.103 The model $Y = \beta_0 + \beta_1X_1 + \beta_2X_2 + \beta_3X_3 + \beta_4X_4 + \epsilon$ was fit to a data set of size $n = 24$, yielding these summary statistics:

$$SSE = 2.01758 \qquad SSY = 18.34167$$

Test the utility of the fitted model with $\alpha = .10$.

16.104 A realtor wishes to develop a prediction equation for the selling price (Y) of homes that are listed in the local real estate market. Believing that there are three main influences on selling price besides location—number of square feet of living space, size of the lot, and age of the house—she collects the following data:

Selling Price ($ Thousands) Y	Number of Square Feet (Hundreds) X_1	Number of Acres X_2	Age of Home (Years) X_3
47	14	.2	37
89	22	.4	7
98	29	.5	14
45	15	.4	50
61	17	.3	66
73	20	1.0	8
84	22	.5	3
85	19	.4	1
83	23	1.2	24
59	17	.5	11
91	27	.7	17
77	21	.4	6
105	24	2.4	28
73	20	.5	10
70	19	1.3	12
88	24	.6	4

a. Fit the model $Y = \beta_0 + \beta_1X_1 + \beta_2X_2 + \beta_3X_3 + \epsilon$ to these data.

b. Test the utility of the fitted model with $\alpha = .10$.

c. Predict with 95% confidence the selling price of a newly listed 12-year-old house with 2700 square feet of living space on a one-acre lot. Assume the value of the estimated standard error is 7.5388.

d. Obtain the correlation matrix and determine the following:

1. Which independent variable is most strongly related to Y?
2. Which pair of independent variables are most strongly related to each other?
3. Which pair of independent variables are most weakly related to each other?

16.105 The following correlation matrix of recall and recognition scores appeared in an article by Singh, Rothschild, and Churchill. The subjects in this study were undergraduate students who viewed experimental commercials and later were asked to recall (with no clues available) the product category, brand name, and claim(s) for as many of the commercials as they could.

The subjects then were given verbal recognition tests for the product categories, brand names, and claims. The numbers shown in the matrix are correlations of each subject's aggregate score across all commercials viewed.

	Recall			Recognition		
	Product	**Brand**	**Claim**	**Product**	**Brand**	**Claim**
Recall						
Product	1.00					
Brand	.44	1.00				
Claim	.72	.37	1.00			
Recognition						
Product	.52	.27	.33	1.00		
Brand	.35	.24	.31	.41	1.00	
Claim	.22	.07	.25	.23	.33	1.00

Source: Surendra Singh, Michael Rothschild, and Gilbert Churchill, "Recognition Versus Recall as Measures of Television Commercial Forgetting." Reprinted by permission from *Journal of Marketing*, published by the American Marketing Association, Vol. 25, February 1988.

a. Which two categories of scores are most strongly related?
b. Which two categories of scores appear least related?
c. Assume the sample size is 30 subjects. If so, which correlations are significantly different from zero in a two-tailed test of H_0: $\rho = 0$? (*Hint:* Refer to Section 15.6 and boxed text on page 799.

REFERENCES

Arens, A. A., and J. K. Loebbecke. 1981. *Applications of Statistical Sampling to Auditing*. Prentice-Hall, Inc., Englewood Cliffs, NJ.

Bowerman, B. L., R. T. O'Connell, and D. A. Dickey. 1986. *Linear Statistical Models: An Applied Approach*. PWS-KENT, Boston.

Churchill, Jr., G. A. 1979. *Marketing Research: Methodological Foundations*, 2nd Edition. The Dryden Press, Hinsdale, IL.

Draper, N., and H. Smith. 1981. *Applied Regression Analysis*, 2nd Edition. John Wiley & Sons, New York.

Graybill, F. A. 1976. *Theory and Application of the Linear Model*. PWS-KENT, Boston.

Mendenhall, W., and J. T. McClave. 1981. *A Second Course in Business Statistics: Regression Analysis*. Dellen Publishing Company, San Francisco.

Neter, J., W. Wasserman, and M. H. Kutner. 1985. *Applied Linear Statistical Models*, 2nd Edition. Richard D. Irwin, Inc., Homewood, IL.

Pappas, J. L., E. F. Brigham, and M. Hirschey. 1983. *Managerial Economics*, 4th Edition. The Dryden Press, Chicago.

Younger, M. S. 1979. *A Handbook for Linear Regression*. PWS-KENT, Boston.

REVIEW PROBLEMS CHAPTERS 13–16

R61 Imagine that you are a cardiologist in 1967. A new procedure called "coronary-bypass" has been proposed to prolong the life of patients with coronary disease.

a. Suggest how a two-sample study conceivably could be set up that would have the potential to demonstrate that coronary-bypass prolongs life. (According to "Coronary-Bypass Surgery: Remedy or Racket?", an article in the December, 1984, issue of *The Atlantic*, no such tests took place for years because surgeons viewed it as unethical to withhold this new surgery from candidate patients so that comparative studies could be made—they "knew" these patients would die without the operation.)

b. What null hypothesis would your two-sample study test: that there is no benefit to bypass surgery, or that there are benefits to bypass surgery?

R62 In "Single and Multiple Person Household Shoppers: A Focus on Grocery Store Selection Criteria and Grocery Shopping Attitudes and Behavior" (*Journal of the Academy of Marketing Science*, Winter, 1985), H. F. Ezell and G. D. Russell compared single-person household shoppers versus multiple-person household shoppers. They interviewed shoppers at random in six different grocery stores to determine the relative importance of various supermarket choice criteria. Respondents were given a list of 22 criteria and asked to rate them from 1 (very important) to 5 (very unimportant). Following are results for two of the 22 criteria used in this study. For $\alpha = .10$, indicate for each criterion whether you would accept or reject the assumption that single- and multiple-person households give the same average importance rating.

Store Selection Criteria	Group*	n	$\overline{X}$	Standard Deviation
Fast checkout	S	84	1.619	1.147
	M	286	1.497	1.322
Cleanliness of the store	S	84	1.762	1.134
	M	286	1.511	1.034

*S = single-person household, and M = multiple-person household.

R63 A mail-order photo processing lab has recorded data for 100 working days for X, the number of sacks of mail received, and Y, the number of prints ordered, in thousands. It is anticipated that a prediction equation relating X and Y might be helpful in scheduling overtime and/or hiring temporary workers. A data summary shows:

$$\Sigma X = 600 \qquad \Sigma Y^2 = 37700$$
$$\Sigma Y = 1600 \qquad \Sigma X^2 = 5200$$
$$\Sigma XY = 13600$$

a. Compute the equation of the least-squares line.

b. Develop a 95% confidence interval for β_1, to two decimal places.

c. Determine the sample correlation coefficient.

R64 In a gasoline mileage test, three different makes of compact cars were run on a test track. Each car had an engine of comparable size, and each had a standard transmission, four-cylinder engine. Fuel economy readings were obtained at three different speeds for each car. The figures shown are miles per gallon over a 100-mile drive.

	Car		
Speed	1	2	3
45 mph	37.8	39.8	38.5
55 mph	36.2	37.9	35.7
65 mph	33.1	35.7	33.8

Determine whether the three car makes have the same average fuel economy, at the .05 level of significance.

R65 A residential building contractor wishes to relate home size (X, in hundreds of square feet) to his material cost (Y, in thousands of dollars). The following ANOVA table results after a random sample of completed jobs are analyzed.

Source	*SS*	*df*	*MS*	*F*
Regression	2049.23	1		
Error	91.24	20		
Total	2140.47	21		

a. How many homes were in the sample?
b. Compute the test statistic that would be used to evaluate the hypothesis that X and Y are not linearly related.
c. Can we be 99% certain that a linear relation does exist? Why?
d. What percentage of the total variation in material cost is due to the influence of home size in the regression model?

R66 The business school at a western university offers a noncredit minicourse that is billed as helping a student improve her or his performance on the GMAT, a standardized test required for admission into the school's MBA program. For a $49 fee, students receive three hours of instruction in algebra review, reading graphs, working sample questions, and so on. Recently, 14 people attended this class and took the GMAT. Of these, nine had taken the GMAT within the previous six months. Assume we have a random sample of people who would take such a course. Use the following data to develop a 90% confidence interval for the average increase in GMAT score that results from taking this course.

Student	Exam Score	Previous Score	Student	Exam Score	Previous Score
1	442	410	8	610	—
2	436	423	9	448	404
3	485	490	10	627	—
4	444	370	11	399	—
5	581	—	12	398	366
6	512	495	13	631	—
7	412	380	14	461	430

R67 Twelve stores of comparable sales volume in a large chain carried promotions for a luxury personal product. Four stores offered mail rebates with purchase; four offered a related product attached on-pack; and four utilized free samples to encourage purchase. Following are unit sales over a two-week period. State and test the appropriate hypothesis, using $\alpha = .05$.

Refund by Mail	On-Pack Gift	Mini-Sample
78	91	69
66	75	57
70	74	71
74	92	71

R68 a. You have 40 observations on each of four variables: A, B, C, and D. You are interested in predicting the behavior of the dependent variable D. Set up a correlation matrix and enter possible numerical values into the matrix to simultaneously satisfy these conditions:
1. The dependent variable is strongly and directly related to variables B and A.
2. Variable C is apparently unrelated to variables A and B.
3. The dependent variable is inversely related to variable C.
4. Variables A and B are highly colinear.

b. The following multiple correlation matrix could never occur. However, if each .80 were .70 instead, then the matrix could occur. Why is this?

$$C = \begin{bmatrix} 1.0 & .80 & .80 \\ .80 & 1.0 & 0 \\ .80 & 0 & 1.0 \end{bmatrix}$$

R69 In "Sex Roles in Advertising: A Comparison of Television Advertisements in Australia, Mexico, and the United States" (*Journal of Marketing*, April 1988), Mary C. Gilly sampled commercials broadcast in different countries. The source of the commercials was the major network with the highest viewer ratings at the time of the study (for example, CBS in the United States).

a. One finding was that 15.7 percent of Mexican ads were for personal and beauty care products, while only 4.3 percent of Australian ads were in this category. Is this an indication of a significant difference in the populations? Use any desired level of significance. The sample sizes were 204 Mexican commercials and 138 Australian commercials. Assume these are random samples.

b. In the United States, 10.9 percent of commercials (30 out of 275) were for household cleaning agents; for Australia, the figure was 8.0 percent. Is the 2.9% difference statistically significant? Tie your conclusion to a risk factor of your choosing.

R70 (This problem is based on results reported in "Assessing the Impact of Short-Term Supermarket Strategy Variables" by J. B. Wilkerson, J. Barry Mason, and Christie Paksoy, in the *Journal of Marketing Research*, February, 1982.) To see what effect on product sales resulted from different product displays and different prices for the product, a field experiment was conducted in one store of the Piggly Wiggly supermarket chain. Conceptually, the design included four observations of X, unit sales over a five-day test period, in each of the nine treatment combinations, as follows:

a. Complete the ANOVA table.

b. At $\alpha = .01$, did sales of Camay soap vary according to price level?

Display Type	Price Level		
	Regular Retail	Halfway Between Retail and Cost	At Store Cost
Regular Shelf Space			
Doubled Shelf Space			
Regular Plus Special Display Nearby			

For one product included in this study, Camay soap, the following ANOVA table resulted:

Source	*SS*	*df*	*MS*	*F*
Display Type	21,579			
Price level	5,219			
Price-Display Interaction	4,310			
Error	12,643			
Total	43,751			

c. At $\alpha = .01$, did display type affect sales?
d. Which experimental factor, display type or price level, seemed to have the larger impact on Camay sales?
e. At $\alpha = .01$, was a significant price-display interaction present or absent?

R71 A mutual fund's expense ratio is defined as the percentage of assets taken annually to cover management fees, operating expenses, and 12b–1 marketing fees. In 1988, Jonathan Pond, president of Financial Planning Information, Inc. in Cambridge, Massachusetts, studied stock and bond funds to determine the average expense ratio of funds with different investment objectives ("Mutual Fund Expenses Can Trim Returns for You," *USA Today*, June 28, 1988). Suppose some of Pond's data looked like the following:

Fund Type	Sample 1987 Expense Ratios (in percentages)	1987 Total Return
Balanced	1.1, .9, 1.0, 1.0, .6, .9, .6, 1.3, 1.0, .8, 1.1, 1.2	2.50
Gold	1.5, 1.2, 1.2, 1.4	36.80
Govt. securities	.8, .5, 1.0, 1.1, .9, 1.2, 1.3, 1.4	1.40
Growth & income	1.5, .6, .7, 1.0, .9, 1.2, .7, .8, 1.0, 1.1, 1.5, 1.4	1.20

Fund Type	Sample 1987 Expense Ratios (in percentages)	1987 Total Return
International	1.4, 1.0, 2.2, 1.3, 1.2, 1.0	9.60
Long-term growth	1.0, 1.2, 1.4, .7, 1.5, .9, 1.1, 1.4, 1.6, 1.1	1.60
Max. capital gains	.9, 1.2, 1.2, 1.5, 1.0, 1.3, 1.2, 1.0, 1.2, 1.3	.01
Specialized industry	1.2, 1.1, .7, .8, 1.4, 1.3, 1.6, .8, 1.9	−.60
Tax-exempt bonds	.9, 1.2, .8, .7, .7, 1.0, .5	−.50
Technology	1.0, .8, 1.5, 1.5, 1.2, 1.6	−3.30

a. Do these data indicate that the average expense ratio is constant across fund types? Use $\alpha = .01$. Assume the cutoff F-score for 9 and 74 degrees of freedom is approximately 2.70.

b. Excluding Gold from the analysis, fit a simple regression model to relate Y, 1987 total return, to X, *average* 1987 expense ratio for the $n = 9$ fund types. If necessary, round the values of X to three decimal places.

c. Excluding Gold from the analysis, is there sufficient evidence ($\alpha = .05$) to infer that the correlation between Y and X is different from 0?

R72 A manager in the human resource area of a large firm wishes to study the relationship between an employee's score on a personality index test and the number of rings that they would program a telephone answering machine to wait before responding. From a prior psychological screening, each employee has been classified as having a Type A or Type B personality. A sample of 10 employees was given a personality index test and then asked to indicate the number of rings they would allow a telephone to ring before an answering machine receives the call. Here are the data:

Number of Rings Before Answering	Personality Type	Personality Index Test Score
2	A	72
1	A	84
2	A	67
5	B	43
3	A	70
4	B	61
1	A	75
5	A	55
4	B	58
4	B	64

a. Test the null hypothesis ($\alpha = .10$) that there is no difference in the average index test score for Type A and Type B personalities. Set up an alternative hypothesis that would enable you to conclude that average scores on the personality index test are higher for Type A's.

b. Find the correlation coefficient between the index test score and the number of rings.
c. Is the correlation significantly different from zero at a 10% level of significance?
d. Based on your answers in parts a and c, is it reasonable to conclude that, on average, Type *A* personalities tend to program an answering machine to permit fewer rings than Type *B* personalities?
e. To answer part d directly, we might be tempted to conduct a small sample *t*-test of H_0: $\mu_1 = \mu_2$, where μ_1 is the average number of rings for Type *A* personalities and μ_2 is defined similarly for Type *B*'s. Give a reason why it may not be advisable to perform this test.

R73 The value of a used car is based on many factors, including the age of the vehicle and the odometer reading. To build a valuation model for used cars, the manager of the used car department at a dealership randomly sampled 30 recent sales and recorded data on these variables:

$$X_1 = \text{Age of vehicle, in years}$$
$$X_2 = \text{Odometer reading, in thousands of miles}$$
$$Y = \text{Resale price, in thousands of dollars}$$

A subsequent multiple regression analysis revealed the following statistics and correlation matrix:

$$\hat{Y} = 10.762 - .351X_1 - .048X_2$$

$$SSE = 83.327 \qquad SSY = 318.36$$

$$\mathbf{C} = \begin{bmatrix} 1.0000 & -.8302 & -.8424 \\ -.8302 & 1.0000 & .8967 \\ -.8424 & .8967 & 1.0000 \end{bmatrix}$$

a. Test the utility of the fitted model, using $\alpha = .01$.
b. What is the correlation between Y and X_1? Between X_1 and X_2?
c. Give a point prediction of the resale price of a six-year-old vehicle with 75,000 miles on the odometer.

R74 In presidential elections, there is an adage that states: "As Maine goes, so goes the nation." The implication is that the results from the state of Maine foretell the eventual outcome of the election nationwide. Stock market analysts may have an analogous bellwether of the market in its January performance. Yale Hirsch, editor of *Stock Trader's Almanac,* has proposed a "January Barometer" theory for the market. Hirsch's research reveals that the January performance in the Standard and Poor's (S&P) 500 index, expressed as a percentage change in the index from its value as of the previous December 31, is a good predictor of the market's performance for the ensuing year. In fact, excluding January changes of less than 1 percent as being inconclusive, the first month's activity (up or down) has correctly predicted the full year's activity (up or down) 80 percent of the time since 1950. The following table shows 20 years worth of data, representing the percentage changes in the S&P 500 index.

Year	January's Performance	Year's Performance	Year	January's Performance	Year's Performance
1969	−.8%	−11.4%	1979	+4.0%	+12.3%
1970	−7.6	+0.1	1980	+5.8	+25.8
1971	+4.0	+10.8	1981	−4.6	−9.7
1972	+1.8	+15.6	1982	−1.8	+14.8
1973	−1.7	−17.4	1983	+3.3	+17.3
1974	−1.0	−29.7	1984	−0.9	+1.4
1975	+12.3	+31.5	1985	+7.4	+26.3
1976	+11.8	+19.1	1986	−0.1	+14.6
1977	−5.1	−11.5	1987	+13.2	+2.0
1978	−6.2	+1.1	1988	+4.0	+12.4

Source: " 'January rule' predicts bull market," *USA Today*, January 13, 1989. Copyright 1989, *USA Today*. Adapted with permission.

a. Analyze these data via simple regression and test Hirsch's January Barometer Theory that, excluding January changes of less than 1%, "as January goes (up or down), so goes the year (up or down, respectively)." Let $\alpha = .01$.

b. If in 1989, the S&P 500 index rose 7.1 percent during January, develop a 95% prediction interval for the change in the S&P 500 index for the year 1989.

R75 State government jobs are classified into pay grades in many states. In a study of similar jobs in three states, an analyst compared hourly wages per pay grade. Three workers at each pay grade in each state were randomly sampled and their hourly wages recorded, as indicated in the following table:

	States				States		
Pay Grade	A	B	C	Pay Grade	A	B	C
1	$4.72	$4.13	$5.05	6	$6.38	$6.62	$7.35
	4.90	4.50	5.27		5.97	7.43	6.41
	4.86	4.91	5.31		5.84	6.59	6.52
2	5.17	4.97	4.87	7	6.55	7.65	8.77
	4.69	5.03	5.66		7.14	7.60	7.21
	5.02	5.00	6.59		7.45	6.80	8.41
3	5.13	5.50	5.66	8	7.61	8.78	8.32
	5.45	5.90	5.41		7.25	7.27	9.62
	4.76	5.46	5.70		7.43	8.12	8.84
4	5.51	5.74	5.71	9	8.18	8.71	7.70
	5.30	6.30	5.93		8.09	9.20	9.87
	5.19	6.30	6.00		8.19	8.78	9.69
5	5.81	6.03	6.91	10	8.37	8.84	10.65
	5.71	6.40	6.06		9.30	9.51	10.20
	5.65	6.30	6.77		8.29	9.03	9.53

With a 5% α-risk, analyze the data to answer the following questions:

a. Do average hourly wages differ across the states?

b. Do average hourly wages differ across the pay grades?

R76 According to Jerri Bullard and William Snizek ("Factors Affecting the Acceptability of Advertisements among Professionals," *Journal of the Academy of Marketing Science*, Summer 1988), a random sample of lawyers, dentists, and certified public accountants (CPAs) were asked questions about their attitudes toward members of their profession advertising their services. Those surveyed were picked at random using the telephone directory Yellow Pages in three major metropolitan areas. The respondents were shown five sample print advertisements for their profession, ranging from a very conservative one to a less professional one that encouraged the reader to "check the monthly specials." The subjects were asked to rate the advertisements for appropriateness. The table summarizes some of the results:

Profession	Sample Size	Average Rating, $\overline{X}$
CPAs	137	.95
Dentists	134	.85
Lawyers	111	.64

The higher the average rating, $\overline{X}$, the more inclined the profession was to look favorably on advertising as a means of promoting its practice.

a. To test the null hypothesis of no difference in means across the three professions, what would be the *df* values for the test statistic?

b. The test statistic that resulted in this study was 3.19. Would you reject the null hypothesis at the .05 level of significance?

c. Is the sample evidence strong enough to reject the null hypothesis at the .01 level of significance?

R77 A 15-year performance study was conducted to compare the average returns of no-load and load mutual funds. Over the 15-year period, growth no-load funds averaged 11.2% annual return versus 8.8% for growth load funds. The standard deviations were 3.8% and 3.1%, respectively.

a. Assuming the 15 years under study could be considered a random sample, conduct a test that could enable you to conclude that no-load funds outperform load funds, on average. Use $\alpha = .10$.

b. In the growth and income fund category, the average annual return was 11.5% for no-loads and 11.0% for loads. If the standard deviations were 2.4% and 2.3%, respectively, estimate the average difference in returns with a 95% confidence interval.

c. Based on your answer in part b, can we conclude with 5% risk that the average returns differ for the two types of growth and income funds? Why?

R78 A hotel devised an experiment to monitor its consumption of water. Over a nine-month period, management designated 11 days, chosen at random, to serve as test days. On test days, the hotel's water consumption, Y, was measured by having a representative of the water company read the water meter twice, 24 hours apart. The average 24-hour ambient temperature, X_1, was recorded, as was the number of occupied rooms, X_2, in the hotel. Suppose the experiment yielded the following statistics:

$$\hat{Y} = 1.2433 + .225X_1 + .625X_2$$

$$SSE = 381.35 \qquad SSY = 6154.3$$

a. Test the utility of the fitted model, with $\alpha = .01$.
b. Find and interpret R^2.

R79 In 1986, the Pentagon discovered that test scores on the timed section of the Armed Services Vocational Aptitude Battery had decreased mysteriously during a year and a half period. One possible explanation was the size of the print on the test. During the period in question, the test was prepared with smaller print. To test this theory, two versions of the same test were prepared—one with small print and one with large print. Eight hundred recruits arbitrarily were split into two equal-sized groups, and each group was given a different version of the test. The results were

Test Version	Mean	Standard Deviation
Small print	58.4	18.7
Large print	59.6	19.1

Do these sample results provide sufficient evidence to infer that the average score is lower for the small-print version? Use $\alpha = .10$.

R80 For a random sample of large cities, let X = the population, in millions, and let Y = the number of radio stations.

a. Would you anticipate that X and Y would be inversely related, directly related, or unrelated?
b. Prepare a scatter plot for the following data:

City	Population (Millions)	Radio Stations
Albuquerque, NM	.37	30
Denver, CO	.50	31
Jacksonville, FL	.61	25
Kansas City, MO	.44	27
Memphis, TN	.65	20
San Diego, CA	1.00	36
Phoenix, AZ	.97	41
Seattle, WA	.49	48
Houston, TX	1.73	47

Source: Data from *The 1989 Information Please Almanac*, Houghton-Mifflin Company, Boston.

c. Compute the correlation coefficient for X and Y.
d. Fit the simple regression model $Y = \beta_0 + \beta_1 X + \epsilon$ and, if useful, predict the number of radio stations for a city with a population of 750,000 people. Use an α-risk of 20 percent for all inferential procedures.

Chapter Maxim *Building a multiple regression model is a sequential procedure of fitting and testing various models in order to find a concise set of significant predictors for Y.*

CHAPTER 17

DEVELOPING REGRESSION MODELS

Objectives

After studying this chapter and working the exercises, you should be able to

1. Set up and execute a test about an individual β_i in a multiple regression model.
2. Explain the relation of X_i to Y by interpreting b_i in a multiple regression analysis.
3. Describe the condition of multicollinearity.
4. Test simultaneously the significance of several β's.
5. Analyze the residuals by testing the assumptions of their normality and independence.
6. Justify the inclusion of second order variables in a multiple regression model.
7. Create indicator variables to represent the different levels of a qualitative independent variable.
8. Interpret the betas associated with indicator and interaction variables in a multiple regression model.
9. Describe the principles of and criteria for the variable selection procedures known as forward selection, stepwise, and backward elimination.

In Chapter 16, we learned how to fit a multiple regression model to a set of sample data and then test the significance of the overall model. In such a test of model utility, we are facing an all-or-nothing decision: keep or discard the entire model. As our multiple regression analysis continues in this chapter, we will develop procedures for retaining one or more of the terms in the model while deleting others, depending on each term's relative contribution in explaining Y. Our objective still is to find a useful model for prediction, but one that contains only those independent variables that are strongly related to the dependent variable. A principle of parsimony prevails.

In addition, we will explain how to include qualitative independent variables in a multiple regression model. For instance, the dependent variable Y, the selling price of a house, certainly is related to quantitative independent variables like X_1, number of square feet of living space, and X_2, number of bedrooms, but it also might be related to the qualitative variable, geographical location, which indicates the quadrant—north, east, south, or west—of the house within the city.

Finally, we will analyze the residuals to test the validity of the assumptions of normality and independence for the random error term, ϵ, in the general model.

The diagnostic methods presented in this chapter are part of the process called *model building*. A regression analysis is more than just fitting a model to a set of data. We must address the finer points of the analysis and strive to develop a model that is explanatory, usable, concise, and valid.

17.1 ANALYZING INDIVIDUAL BETAS

In Example 16.2 of Chapter 16, we fit the model $Y = \beta_0 + \beta_1X_1 + \beta_2X_2 + \beta_3X_3 + \epsilon$ to a set of $n = 28$ entities and obtained the equation $\hat{Y} = 56.31857 - 1.94724X_1 - .00801X_2 + 1.84047X_3$. The variables were Y = hotel/motel occupancy rate; X_1 = number of competitors within a one-mile radius; X_2 = nightly room rate; and X_3 = advertising expense, as a percentage of gross revenues. We tested the fitted model via the F-test in Example 16.5 and found it to be useful. As a result, we concluded that the fitted model adequately related Y to X_1, X_2, and X_3 and that it could be used to predict Y, given values of the independent variables.

However, the F-test for model utility applies to the collection of predictors *as a whole,* not to each independent variable. It is possible, therefore, that the whole set of predictors could be judged useful, but that one or more of the individual variables do not merit being in the model. We seek a methodology for analyzing the singular contribution of X_i to Y. Knowledge of the sampling distribution of b_i can help us determine whether an individual predictor is useful in the model.

Sampling Distribution of the b_i's

We also know that, had we selected a different set of $n = 28$ entities, the coefficients of X_1, X_2, and X_3 likely would be different from the ones we obtained. The reason is that the b_i's are random variables that take on specific values depending on the sample selected. As we learned in simple regression, the behavior of the slope coefficient, b_1, in repeated sampling was described by a normal distribution, provided the random error term in the model was normally distributed (see Figure 15.9).

This theoretical result extends to multiple regression as well. That is, the sampling distribution of each estimated regression coefficient b_i will be a normal probability distribution, as depicted in Figure 17.1. Notice in Figure 17.1 that β_i, the beta parameter associated with the independent variable X_i in the general model, corre-

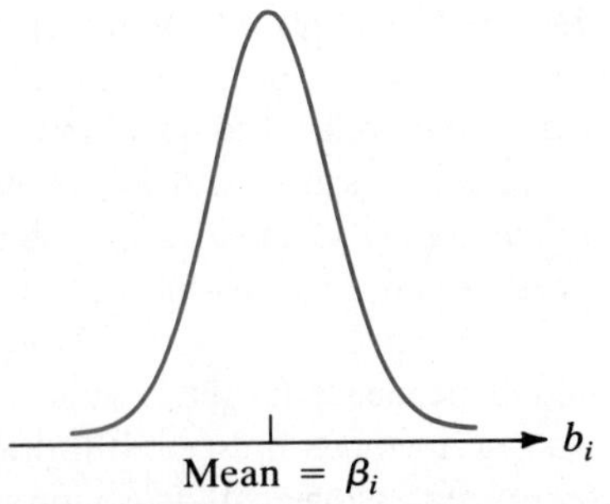

Figure 17.1 Sampling Distribution of b_i

sponds to the mean of the sampling distribution. This suggests that, in repeated sampling, the average value of b_i will be β_i. For one random sample though, we expect that b_i will not coincide with β_i, due to sampling error.

To complete the specification of the sampling distribution of b_i, we need to measure the variability in Figure 17.1. As before, we will use the term *standard error* to describe the standard deviation of the normal curve representing the sampling distribution of b_i. Unlike the formula for the estimated standard error of b_1 that we presented in simple regression (see Equation 15–15), there is not a simple equation to express the standard error of b_i in a multiple regression analysis. Thus, we must rely on either matrix algebra (see ensuing *Comments*) or the output from a multiple regression computer routine to provide the estimated standard errors.

COMMENTS

1. Our notation for the standard error of b_i will be σ_{b_i}. Since this quantity, in theory, involves the unknown population standard deviation σ, we will estimate the standard deviation with $\hat{\sigma}$ and denote the resulting standard error by S_{b_i}. In some commercially available computer packages, the estimated standard error may be denoted by Std Error, SE, or the words Standard Error.
2. For those utilizing the matrix algebra approach to multiple regression, there is an equation for finding the estimated standard error of b_i, using the $(\mathbf{X'X})^{-1}$ matrix:

$$S_{b_i} = \hat{\sigma} \text{ times } \sqrt{c_{ii}}$$

The notation c_{ii} refers to the element along the main diagonal of the $(\mathbf{X'X})^{-1}$ matrix in row i and column i. *Caution:* In the c_{ii} notation, the first row (and column) of the $(\mathbf{X'X})^{-1}$ matrix is numbered 0 to tie it to the b_0 term; the second row is numbered 1 for b_1, and so forth. For example, if $\hat{\sigma} = 1.11$ and the $(\mathbf{X'X})^{-1}$ matrix is

$$(\mathbf{X'X})^{-1} = \frac{1}{134}\begin{bmatrix} 195 & -32 & -91 \\ -32 & 8 & 6 \\ -91 & 6 & 105 \end{bmatrix}$$

then the estimated standard errors of b_0, b_1, and b_2 are

$$S_{b_0} = (1.11) \cdot \sqrt{\frac{195}{134}} = 1.339$$

$$S_{b_1} = (1.11) \cdot \sqrt{\frac{8}{134}} = .271$$

$$S_{b_2} = (1.11) \cdot \sqrt{\frac{105}{134}} = .983$$

3. If the F-test for model utility is nonsignificant, we would have little interest in analyzing individual betas.

Testing a Hypothesis About β_i

In the general multiple regression model, β_i represents the additional contribution of the independent variable X_i in explaining the dependent variable Y, given the presence of the other independent variables in the model. If β_i were 0, then X_i would contribute nothing to the prediction of Y and would automatically drop out of the model. Given the standard error of each b_i, we can proceed to answer the question: Which, if any, of the original set of independent variables are unrelated to Y? Our approach will be to assess how close the observed value of b_i is to a hypothesized value of its mean, β_i. Specifically, we can use the information in the sample estimator, b_i, to test the null hypothesis that $\beta_i = 0$. Accepting the null hypothesis suggests that the associated independent variable X_i has no additional predictive value and therefore could be deleted from the fitted model. Rejecting the hypothesis is a signal that X_i contributes important information about Y and should remain in the model.

For instance, consider the variable X_2, nightly room rate, in the fitted model $\hat{Y} = 56.31857 - 1.94724X_1 - .00801X_2 + 1.84047X_3$. The estimated regression coefficient, $b_2 = -.00801$, is so close to 0 that it seems as if X_2 is explaining a negligent portion of Y and could be removed. However, we *cannot* make this decision on such a judgmental basis. Instead, we must determine how many standard errors b_2 is away from the value of 0. For the data set of $n = 28$ hotels and motels, we can use the Regression processor to determine that the standard error of b_2 is .07063. Thus, $b_2 = -.00801$ is less than one standard error away from 0, which leads to the conclusion that X_2 is not a significant predictor of Y in the presence of X_1 and X_3. The following box and Figures 17.2 through 17.4 formalize this testing procedure about an individual beta.

t-Test for an Individual β

H_0: $\beta_i = 0$ (X_i is unrelated to Y, given the presence of the other independent variables in the model.)

H_a: $\beta_i \neq 0$ (X_i is related to Y, given the presence of the other independent variables in the model.)

or

$\beta_i > 0$ (X_i is directly related to Y, given the presence of the other independent variables in the model.)

or

$\beta_i < 0$ (X_i is inversely related to Y, given the presence of the other independent variables in the model.)

$$TS: \quad t^* = \frac{b_i - \beta_i}{S_{b_i}} \qquad (17\text{–}1)$$

RR: For $\alpha = .05$ and the $\neq$ alternative, see Figure 17.2.
For $\alpha = .05$ and the $>$ alternative, see Figure 17.3.
For $\alpha = .05$ and the $<$ alternative, see Figure 17.4.

C: Reject H_0 or accept H_0.

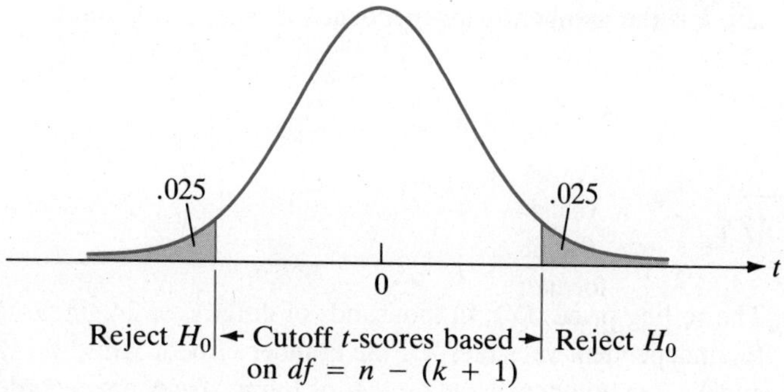

Figure 17.2 Rejection Region for Two-Tailed t-Test of H_0: $\beta_i = 0$, $\alpha = .05$

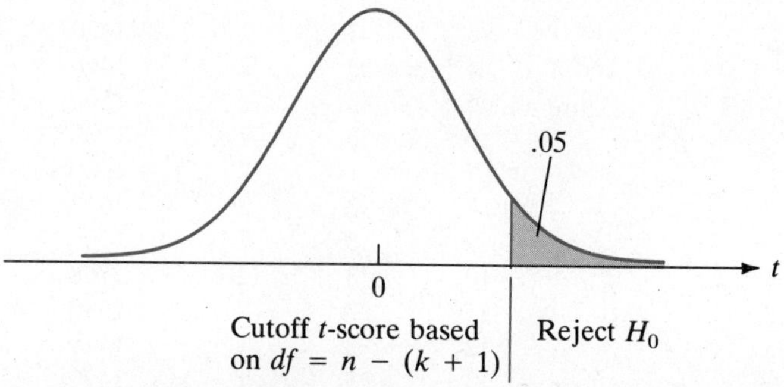

Figure 17.3 Rejection Region for Upper-Tailed t-Test of H_0: $\beta_i = 0$, $\alpha = .05$

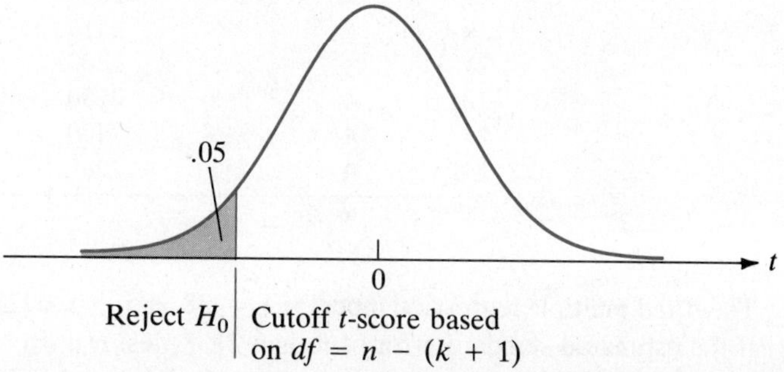

Figure 17.4 Rejection Region for Lower-Tailed t-Test of H_0: $\beta_i = 0$, $\alpha = .05$

This t-test is based on $n - (k + 1)$ degrees of freedom, where n is the sample size and k is the number of independent variables in the model.

EXAMPLE 17.1

The selling prices (Y), in thousands of dollars, of 24 single-family homes and data on the independent variables X_1, the number of bedrooms, X_2, the number of bathrooms, and X_3, the number of square feet of living space, are recorded in the following table:

House	X_1	X_2	X_3	Y
1	3	2	1570	62.0
2	3	2	1500	57.5
3	3	2	1450	62.0
4	3	2	1466	60.0
5	3	2	1500	70.4
6	3	2	1708	68.0
7	3	2	1600	64.0
8	3	2	1617	70.0
9	3	2	1700	71.5
10	3	2	1768	74.5
11	3	2	1990	72.0
12	3	1½	1470	63.9
13	3	1½	1470	56.7
14	3	1½	1580	61.0
15	4	2	2000	61.7
16	4	2	2000	67.9
17	4	2	2000	75.0
18	4	2	2050	81.0
19	4	2½	2346	83.0
20	4	2½	2415	81.9
21	4	2½	2200	75.0
22	4	2½	2150	70.0
23	4	2½	2400	66.5
24	4	2½	2085	78.5

The fitted multiple regression model is $\hat{Y} = 38.285 - 3.092X_1 + 3.513X_2 + .019X_3$. If the estimated standard error of b_2 is 6.2198, test H_0: $\beta_2 = 0$, using $\alpha = .10$ with a two-sided alternative.

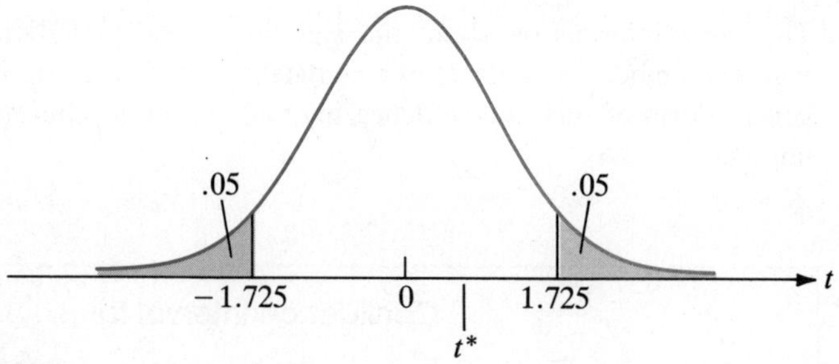

Figure 17.5 Rejection Region for Two-Tailed t-Test of H_0: $\beta_2 = 0$ in Example 17.1

Solution:

The elements of the test are

H_0: $\beta_2 = 0$
H_a: $\beta_2 \neq 0$

TS: $t^* = \dfrac{b_2 - \beta_2}{S_{b_2}} = \dfrac{3.513 - 0}{6.2198} = .56$ (rounded)

RR: The degrees of freedom are $df = n - (k + 1) = 24 - 4 = 20$. For $\alpha = .10$ and the $\neq$ alternative, the critical t-scores are ± 1.725; see Figure 17.5.

C: Accept H_0.

By accepting H_0, we conclude that X_2 contributes no useful information for predicting Y in the presence of X_1 and X_3, and it could be removed from the model.

□

COMMENTS

1. If we remove an independent variable from the model, we should fit the "reduced" model and test it to ensure it still is useful for prediction. The estimated regression coefficients will change when eliminating a variable, so we cannot simply drop out the nonsignificant variable and use the original b_i's. For instance, by deleting X_2 from the model in Example 17.1, the reduced model, when fit to the $n = 24$ data points, becomes $\hat{Y} = 40.918 - 3.258X_1 + .021X_3$. The associated F-test of model utility produces $F = 11.26$, with a p-value of .0005.
2. In Example 17.1, we illustrated the t-test for β_2 only. We could continue the analysis and test H_0: $\beta_1 = 0$ and then H_0: $\beta_3 = 0$. However, conducting a sequence of t-tests on all the betas is not recommended, as we will learn in the next section. We should use t-tests sparingly, applying them to selected variables in which we have a particular interest.

Confidence Interval for β_i

The same information about the significance of an individual beta in a multiple regression model is available in a confidence interval for β_i. Equation 17–2 shows the general form of such a confidence interval, assuming the sampling distribution b_i is normal.

Confidence Interval for β_i

$$b_i \pm t \cdot S_{b_i} \quad \text{(17–2)}$$

The cutoff-t-score in Equation 17–2 is based on $df = n - (k + 1)$.

If the confidence interval for β_i includes the value 0, then we can conclude that the independent variable X_i has no predictive value in the presence of the other variables, and, therefore, could be removed from the model. We would be convinced to retain X_i in the model if the confidence interval excluded zero.

For example, a 90% confidence interval for β_2 in Example 17.1 is formed as follows, using a t-score of 1.725:

$$b_2 \pm t \cdot S_{b_2}$$
$$3.513 \pm (1.725)(6.2198)$$
$$3.513 \pm 10.729 \qquad \text{or } -7.216 \le \beta_2 \le 14.242$$

The interval from -7.216 to 14.242 represents a range of possible values for β_2, with 90% confidence. Since the number 0 is included in this range, we would accept a null hypothesis that states $\beta_2 = 0$ and be justified in removing X_2 from the model. This is necessarily the same result as the formal hypothesis test of Example 17.1 gives us.

Abbreviated Form of Summarizing a Multiple Regression Equation

Most computer programs that are designed to fit a multiple regression model provide complete details of the analysis, including a table of the estimated regression coefficients and their standard errors, an ANOVA table, a correlation matrix, and so on. However, in textbooks for other business courses and in the research literature, we may find an abbreviated summary of the fitted model.

Typically in an abbreviated version, the estimated standard error of each b_i is listed in parentheses immediately beneath the coefficient in the fitted model. For example, in Example 17.1, an abbreviated summary of the fitted model would be

$$\hat{Y} = \underset{(9.3869)}{38.285} - \underset{(4.9303)}{3.092X_1} + \underset{(6.2198)}{3.513X_2} + \underset{(.0091)}{.019X_3}$$

The first line contains the equation of the fitted model, while the second line shows the estimated standard errors. From this format, we know that the estimated standard errors are

$$S_{b_0} = 9.3869$$
$$S_{b_1} = 4.9303$$
$$S_{b_2} = 6.2198$$
$$S_{b_3} = .0091$$

Making inferences about individual predictors is facilitated by the placement of the standard errors beneath their respective regression coefficients.

COMMENT There may be a third line of summary statistics in an abbreviated form, containing the values of R^2, $\hat{\sigma}$, SSE, F^* and/or the p-value. For instance, in Exercise 5.6 of *Managerial Economics*, 4th Edition, by Pappas, Brigham, and Hirschey (1983), we are given the following summary:

$$Q = \underset{(1.2)}{0.5} - \underset{(0.5)}{1.5P} + \underset{(0.5)}{3.0A} + \underset{(0.3)}{0.25I} + \underset{(2.7)}{7.0N}$$
$$R^2 = .48 \qquad \text{Standard Error of the Estimate} = 0.5$$

where the variables are denoted with capital letters. Notice the values of R^2 and $\hat{\sigma}$ (same as standard error of the estimate) in the third line.

Interpreting the Regression Coefficients

In the simple regression model $Y = \beta_0 + \beta_1 X + \epsilon$ we interpreted β_1 as the slope of the population regression line: For each one unit change in X, the average value of Y would change by an amount equal to β_1. A similar explanation holds for each beta in the multiple regression model $Y = \beta_0 + \beta_1 X_1 + \ldots + \beta_k X_k + \epsilon$, provided we assume that the other variables in the model remain constant. Since the true value of each β_i is unknown, we use the sample coefficients, b_i, to estimate the relation between X_i and Y.

For instance in Example 17.1, the fitted model was $\hat{Y} = 38.285 - 3.092X_1 + 3.513X_2 + .019X_3$. Let us interpret the value $b_3 = .019$. Recall that Y represented the selling price of a home, in thousands of dollars, and X_3 was the number of square feet of living space. Thus, the population parameter β_3 would be the average change in the selling price of a house for each additional square foot of living space, provided X_1 and X_2 remain constant. With $b_3 = .019$, we estimate that the average selling price increases by \$19 for each additional square foot of living space (or \$1900 for each additional 100 square feet of living space), provided the number of bedrooms and bathrooms does not change.

Now examine the coefficient of X_1: $b_1 = -3.092$. The negative sign appears somewhat contradictory at first, since it seems to suggest that average selling price decreases as the number of bedrooms increases! This highlights the fallacy of incorrectly interpreting a beta coefficient in a multiple regression analysis as the isolated contribution of X_i to Y. We must take into consideration the presence of the other variables in the model and the interrelationships among the independent variables. In this case, $b_1 = -3.092$ means the average selling price will decline by an estimated \$3092 for each additional bedroom, given that the number of bathrooms and the total square feet of living space remain constant. One possible explanation for this result is that, for the same amount of living space, the extra bedroom would create smaller-sized rooms, making the residence perhaps less desirable.

Another explanation could be developed if we examine the correlation matrix, indicated as follows:

$$\mathbf{C} = \begin{bmatrix} 1.000 & .577 & .615 & .712 \\ .577 & 1.000 & .670 & .302 \\ .615 & .670 & 1.000 & .782 \\ .712 & .302 & .782 & 1.000 \end{bmatrix}$$

Recall that the correlation between Y and each independent variable is found in the first row (or first column). Hence, we see that the correlation between X_1 (the number of bedrooms) and Y is .577, separate of the relations of the other independent variables. We also notice that the correlation between X_1 and X_2 (the number of bathrooms) is relatively high, .670, suggesting that X_1 and X_2 have a fairly strong, direct relation. This reinforces our intuition that more bathrooms in a house are associated with more bedrooms in the house. Further, the separate correlation between X_2 and Y, .615, also implies that additional bathrooms tend to increase the selling price.

Summarizing, we know that, separately, as X_1 increases, Y tends to increase; separately, as X_2 increases, Y tends to increase; and as X_1 increases, so does X_2 (and vice-versa). However, when we include X_1 and X_2 in the same model with Y, the separate contribution of X_1 to Y can become the opposite of what we expect. The reason for this peculiar phenomenon is that X_1 and X_2 are contributing redundant information about Y. We call this situation multicollinearity. **Multicollinearity** is a term describing high correlations between independent variables. One of the effects of multicollinearity is that the apparent relation between X_i and Y, as represented by b_i in the model, may be the opposite of the "true" relation between X_i and Y, as represented by the correlation coefficient between X_i and Y.

The effects of multicollinearity are simultaneously bad and not too bad. The bad news we already have seen: The true relation of an independent variable with Y is masked by the overlapping information from another correlated variable. As a result, the sign of b_i may be contrary to reason and/or the t-test may show that the independent variable is insignificant in the presence of the other variables. In addition, the standard error of b_i may vary widely in the presence and absence of its correlated mate. The not-too-bad news is that multicollinearity, in general, does not affect predictions about Y. If we wish to predict Y or to estimate $E(Y)$, then using a model with correlated independent variables in it yields almost as precise an estimate or prediction as would be obtained using a model without the correlated variables.

If necessary, an obvious remedy for the condition of multicollinearity is to get rid of one of the redundant independent variables. As a general guideline, we recommend that the independent variable having the smaller (absolute) correlation coefficient with Y be removed. For instance, in the real estate model of Example 17.1, variables X_1 and X_2 are redundant. We can remove the variable X_1 from the model, since the correlation between X_1 and Y, .577, is smaller than the correlation between X_2 and Y, .615.

COMMENTS

1. We declare multicollinearity to exist whenever the independent variables are highly correlated, positively or negatively. Recall in Chapter 15 that we presented a procedure for testing H_0: $\rho = 0$ versus H_a: $\rho \neq 0$. We could use it to establish the presence or absence of correlation among pairs of X_i's for any chosen α-risk. Or, we could use Table 4.4 from Chapter 4 for $\alpha = .05$. For instance, with $n = 24$ the cutoff values from Table 4.4 are about $\pm.406$. Since $r_{12} = .670$ in the real estate example, we would reject H_0: $\rho = 0$ and conclude that X_1 and X_2 are positively correlated in the population.
2. Because β_i is interpreted as the contribution of X_i to Y, given the other independent variables in the model remain constant, we sometimes call the β_i parameters *partial regression coefficients*.

Analyzing the contribution of an individual variable in a multiple regression model is tricky because of the interrelationships with the other independent variables in the model. An inference about β_i or an interpretation of b_i does not reflect the separate relationship between X_i and Y, rather it represents the relation of X_i to Y, given the presence of the remaining independent variables in the model. In the next section, we shall investigate the simultaneous effect of some, but not all, of the independent variables on Y.

CLASSROOM EXAMPLE 17.1

Analyzing Individual Betas in a Multiple Regression Model

In Example 16.1, we fit a multiple regression model to a data set of size $n = 12$ to relate Y, the annual number of passengers arriving and departing an airport, to X_1, the average number of flights departing weekly, and X_2, the total number of local airline employees. An abbreviated summary of the analysis is

$$\hat{Y} = \underset{(4.3119)}{-4.869} + \underset{(.1500)}{.307X_1} + \underset{(.1336)}{.166X_2}$$

a. Construct a 90% confidence interval for β_1 and use it to decide if X_1 should remain in or be removed from the model.

b. Interpret the value of b_1.

c. Test whether X_2 is unrelated to Y against an alternative that implies a direct relation between the variables. Use $\alpha = .10$.

d. The correlation matrix for the analysis is

$$\mathbf{C} = \begin{bmatrix} 1.000 & .948 & .934 \\ .948 & 1.000 & .943 \\ .934 & .943 & 1.000 \end{bmatrix}$$

Describe the separate relations of X_1 and X_2 to Y. Are X_1 and X_2 contributing redundant information in the prediction of Y? Explain.

SECTION 17.1 EXERCISES

17.1 An analysis of 25 top money winners playing on the Professional Golfers Association (PGA) tour produced the following multiple regression equation:

$$\underset{}{\hat{Y}} = \underset{(7443.3)}{5867.7} + \underset{(.093)}{.116X_1} + \underset{(2993)}{1902.1X_2} - \underset{(130)}{59.609X_3} - \underset{(141.36)}{92.873X_4}$$

where

Y = Money earned, in thousands of dollars
X_1 = Number of holes played
X_2 = Percentage of greens reached in regulation
X_3 = Average score
X_4 = Average number of putts per round

Test whether the independent variable X_1 is directly related to Y, with $\alpha = .10$.

17.2 Refer to Exercise 17.1. Construct a 95% confidence interval for β_4.

17.3 Refer to Exercise 17.1. Interpret $b_3 = -59.609$ in the fitted model.

17.4 (Use of computer processor recommended.) Fit the model $Y = \beta_0 + \beta_1 X_1 + \beta_2 X_2 + \beta_3 X_3 + \epsilon$ to the following set of data:

Y	X_1	X_2	X_3
2.827	7	75.7	123
25.182	30	74.5	55
60.449	29	73.6	24
37.967	27	74.1	27
70.979	26	73.9	25
67.123	25	73.1	29
200.648	26	72.3	8
197.722	27	72.2	8
25.064	13	73.5	39

Test H_0: $\beta_2 = 0$, using a 5% α-risk. If you decide to accept H_0, delete X_2 from the model and refit the model.

***17.5** (Background in matrix algebra required.) In fitting the model $Y = \beta_0 + \beta_1 X_1 + \beta_2 X_2 + \epsilon$ to a data set of $n = 7$ entities, we obtained the following $(\mathbf{X'X})^{-1}$ matrix:

$$(\mathbf{X'X})^{-1} = \frac{1}{16464}\begin{bmatrix} 5488 & 0 & -784 \\ 0 & 588 & 0 \\ -784 & 0 & 196 \end{bmatrix}$$

If $\hat{\sigma} = 1.1495341$ and $\hat{Y} = 1.714 - 1.107X_1 - .036X_2$, perform the following:

a. Test β_2 for significance at the 10% level.

b. Find S_{b_1}.

c. Construct a 95% confidence interval for β_1. Based on this interval, decide whether X_1 should be retained in the model.

17.6 Based on the following fitted model, test whether X_2 is unrelated to Y.

$$\hat{Y} = \underset{(33.343)}{-38.463} - \underset{(.73755)}{.902X_1} + \underset{(.62698)}{3.602X_2}$$

Use $\alpha = .01$ and assume $n = 25$.

17.7 Refer to Exercise 17.6. Construct an 80% confidence interval for β_1. Does your result suggest X_1 merits inclusion in the model?

***17.8** (Background in matrix algebra required.) The $(\mathbf{X'X})^{-1}$ matrix for a multiple regression analysis involving two independent variables and $n = 16$ data points is

$$(\mathbf{X'X})^{-1} = \frac{1}{9134}\begin{bmatrix} 1769 & -446 & -111 \\ -446 & 288 & -256 \\ -111 & -256 & 735 \end{bmatrix}$$

Assume $SSE = 1063.2527$ and the fitted model is $\hat{Y} = 8.651 + 5.809X_1 + .219X_2$. Conduct a test of the significance for β_2, using a 10% level of significance.

17.9 Using the following correlation matrix, discuss the separate relations of X_1 and X_2 to Y in the fitted multiple regression model.

$$\mathbf{C} = \begin{bmatrix} 1.000 & -.9296 & -.0519 \\ -.9296 & 1.000 & .0000 \\ -.0519 & .0000 & 1.0000 \end{bmatrix}$$

Are X_1 and X_2 contributing redundant information in the prediction of Y? Explain.

17.10 Refer to Exercise 17.1. Suppose the correlation matrix is

$$\mathbf{C} = \begin{bmatrix} 1.0000 & -.0274 & .3248 & -.5438 & -.1946 \\ -.0274 & 1.0000 & -.1719 & .3852 & .2890 \\ .3248 & -.1719 & 1.0000 & -.5657 & .5140 \\ -.5438 & .3852 & -.5657 & 1.0000 & .3635 \\ -.1946 & .2890 & .5140 & .3635 & 1.0000 \end{bmatrix}$$

*Optional

Use Table 4.4 to flag pairs of independent variables that evidence strong positive or negative correlation. Which variable(s), if any, would you recommend deleting from the model because of the redundancy?

*17.11 To see the effects of multicollinearity on b_i, on its standard error, and on a predicted value for Y, consider the real estate data in Example 17.1 and the following three models:

$$\text{Model 1:}\quad Y = \beta_0 + \beta_1 X_1 + \beta_2 X_2 + \epsilon$$
$$\text{Model 2:}\quad Y = \beta_0 + \beta_1 X_1 + \epsilon$$
$$\text{Model 3:}\quad Y = \beta_0 + \beta_2 X_2 + \epsilon$$

a. Fit all three models to the $n = 24$ data points in Example 17.1. (Notice that we are ignoring the independent variable X_3 in this exercise.)

b. Compare the estimated β_1 regression coefficients and the standard errors for models 1 and 2.

c. Repeat instruction b for β_2 in models 1 and 3.

d. Use all three models to predict Y when $X_1 = 3$ and $X_2 = 2$. Compare the point predictions.

17.2 TESTING SEVERAL BETAS SIMULTANEOUSLY

Drawbacks of a Sequence of One-at-a-Time *t*-Tests

In a multiple regression analysis, the dependent variable Y is expressed as a function of several independent variables $X_1, X_2, \ldots$ and so on. We might expect that all the independent variables do not make a significant contribution in explaining the variability in Y. Indeed, in the last section, we learned how to test the contribution of an individual term. While it might seem effective to repeat the t-test outlined in the box on page 920 on each independent variable in the model to flesh out the unimportant ones, such a "one-at-a-time" procedure has at least two major flaws.

First, remember that the inference from each test is conditional on the other independent variables being in the model. For instance, in Example 17.1, we decided that X_2 was a nonsignificant predictor of *Y in the presence of* X_1 *and* X_3. Does it sound like X_2 and Y are unrelated? Not necessarily, since it can be shown that a simple regression involving just Y and X_2 is highly significant (p-value $\approx .0014$). Thus, a one-at-a-time t-test approach is, in a sense, dependent on which independent variables are already in the model.

Second, conducting consecutive t-tests on the independent variables creates a sequence of inferences with an overall α-risk larger than the nominal α-risk associated with each test. For instance, suppose there were five independent variables to be tested, and we specified an α-risk of 5% for each test. Prior to conducting the tests, we know that each test will result in one of two possible outcomes—accept H_0 or reject H_0—and that, given H_0: $\beta_i = 0$ is true, the probability of rejecting H_0 is .05. Thus, for the collection of five tests, our chances of committing *at least* one Type I error is, if all null statements are true,

$$P(\text{at least one Type I error}) = 1 - P(\text{no Type I errors}) = 1 - (.95)^5 = .2262$$

Whereas our nominal α-risk per test is 5%, our overall α-risk for the sequence of five tests inflates to almost 23%. This phenomenon worsens, that is, the overall α-risk increases as the number of individual t-tests increases.

These two drawbacks—the need for a conditional inference interpretation and an inflated overall α-risk—preclude us from relying on multiple t-tests to help thin out the model to its most important contributors. Intuitively, we seek a procedure that simultaneously tests the significance of several independent variables.

Complete and Reduced Models

Suppose we fit a multiple regression model with five independent variables to a data set of $n = 25$ entities and obtain the following model, called the *complete model,*

$$\hat{Y} = -127.72 + 1.513X_1 - .022X_2 - .052X_3 + .014X_4 + 12.137X_5$$

with $SSE_c = 186{,}618$. (The symbol SSE_c denotes the sum of squared fitting errors for the complete model.) Let us further suppose that we have doubts about the explanatory effects of the independent variables X_1, X_2, and X_4. If these variables are nonsignificant, then we would expect their beta coefficients to be zero. Working on this assumption, we could eliminate them from the analysis and fit a *reduced model* containing only X_3 and X_5 to the data set of $n = 25$ entities. Now suppose the resulting model is

$$\hat{Y} = -61.093 - .039X_3 + 12.355X_5$$

with $SSE_r = 192{,}676$. (The symbol SSE_r represents the sum of the squared fitting errors for the reduced model.)

At this point, we make the following observations: SSE increases (or possibly stays the same) as the number of independent variables in the multiple regression model decreases. For instance, as the number of independent variables decreased from five in the complete model to two in the reduced model, SSE increased from 186,618 for the complete model to 192,676 for the reduced model. To decide on the predictive capabilities of a set of independent variables, such as X_1, X_2, and X_4, we will test whether the exclusion of the variables in a reduced model significantly increases SSE in comparison to the complete model. A marginal increase in SSE suggests that the set of additional variables make such a weak contribution in the prediction of Y that they could be dropped from the complete model without dire consequences.

General Testing Procedure

Relative to the complete and reduced models that we have been discussing, we wish to test these hypotheses:

H_0: $\beta_1 = \beta_2 = \beta_4 = 0$

H_a: At least one β among β_1, β_2, and β_4 is nonzero.

The null hypothesis embodies our earlier suspicion that the independent variables X_1, X_2, and X_4 collectively are unrelated to Y. The alternative hypothesis allows for one, or two, or all three of the variables to be related to Y.

For testing a null hypothesis involving several betas set equal to 0, we use the F-statistic defined in Equation 17–3.

F-Test Statistic for a Reduced Model

$$F^* = \frac{\dfrac{SSE_r - SSE_c}{h}}{\dfrac{SSE_c}{n - (k + 1)}} \qquad (17\text{–}3)$$

The SSE_r and SSE_c refer to the sum of the squared fitting errors for the reduced and the complete models, respectively. As usual, n is the sample size, k is the number of independent variables in the complete model, and h is the number of betas in the null hypothesis. The degrees of freedom for this test statistic are h and $n - (k + 1)$.

To demonstrate the computation of the F-statistic for testing H_0: $\beta_1 = \beta_2 = \beta_4 = 0$, we have reproduced the necessary statistics:

$$SSE_r = 192{,}676 \qquad h = 3 \qquad n = 25$$
$$SSE_c = 186{,}618 \qquad k = 5$$

Thus,

$$F^* = \frac{\dfrac{192676 - 186618}{3}}{\dfrac{186618}{25 - (5 + 1)}} = \frac{2019.333}{9822} = .21 \text{ (rounded)}$$

The rejection region for our test is found in the upper tail (always) of the corresponding F probability distribution. Since we used 3 degrees of freedom in the numerator and 19 degrees of freedom in the denominator of the test statistic, we refer to the F-table with 3 and 19 *df*. For $\alpha = .10$, the F-score is (about) 2.40. As the computed F-value of .21 is less than 2.40, we may accept H_0. The F-test confirms our suspicion that the set of independent variables X_1, X_2, and X_4 explain such an insignificant portion of the behavior of Y, relative to the explanatory effects of X_3 and X_5, that they could be eliminated from the complete model.

The general testing procedure is outlined in the following box.

F-Test for Several Betas

H_0: $\beta_1 = \beta_2 = \ldots = \beta_h = 0$ (The subset of independent variables $X_1, X_2, \ldots, X_h$ is unrelated to Y, given the presence of the other independent variables in the model.)

H_a: not H_0 (The subset of independent variables $X_1, X_2, \ldots, X_h$ is related to Y, given the presence of the other independent variables in the model.)

$$TS:\ F^* = \frac{\dfrac{SSE_r - SSE_c}{h}}{\dfrac{SSE_c}{n - (k + 1)}}$$

RR: For $\alpha = .05$, see Figure 17.6.

C: Reject H_0 or accept H_0.

Please note that the null hypothesis is written as if all the betas being tested were numbered consecutively from 1 through h. As we observed in the previous discussion when we tested β_1, β_2, and β_4, this does not have to be the case.

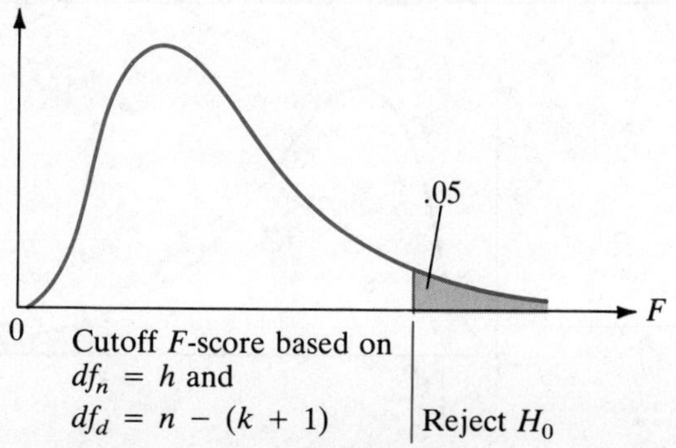

Figure 17.6 Rejection Region for F-Test for Several Betas, $\alpha = .05$

EXAMPLE 17.2

In Example 17.1, we fit the complete model $Y = \beta_0 + \beta_1 X_1 + \beta_2 X_2 + \beta_3 X_3 + \epsilon$ to a data set of $n = 24$ homes and obtained the equation

$$\hat{Y} = 38.285 - 3.092X_1 + 3.513X_2 + .019X_3$$

with $SSE_c = 644.586$. Test the null hypothesis that, in the presence of X_3 = the number of square feet of living space, the two independent variables X_1 = the number of bedrooms and X_2 = the number of bathrooms are insignificant predictors of Y. Assume the reduced model is

$$\hat{Y} = 37.947 + .017X_3$$

with $\text{SSE}_r = 668.99933$. Use $\alpha = .05$.

Solution:

The elements of the test are

H_0: $\beta_1 = \beta_2 = 0$

H_a: not H_0

$$TS: F^* = \frac{\dfrac{668.99933 - 644.586}{2}}{\dfrac{644.586}{24 - (3 + 1)}} = \frac{12.206665}{32.2293} = .38 \text{ (rounded)}$$

RR: For $\alpha = .05$; $df_n = 2$; $df_d = 20$, see Figure 17.7.

C: Accept H_0.

With X_3 already in the model, we gain little additional information about Y from X_1 and X_2. They could be dropped from the model if desired.

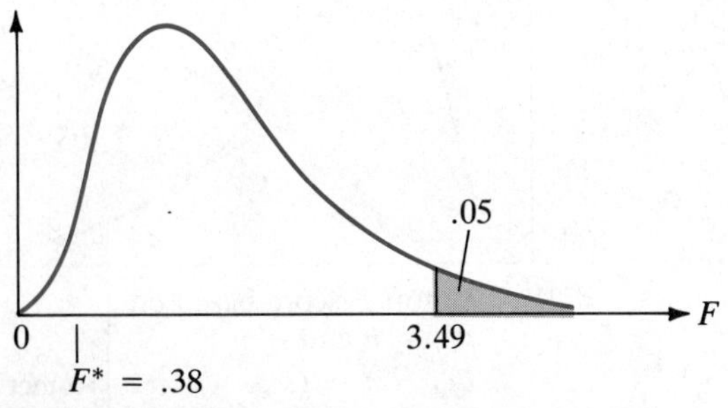

Figure 17.7 Graphical Summary of Example 17.2

□

COMMENTS

1. The decision to drop nonsignificant independent variables from the complete model, based on an accept H_0 conclusion from the F-test (or from a t-test about an individual beta) is left to the user. In some applications, it may be harder to purge the variables from the data file than it would be to just leave them in the model. Though they add little in the prediction of Y if left in, they do not harm the prediction. The main drawback to leaving them in is the waste of several degrees of freedom for estimating σ. However, we strongly recommend that nonsignificant variables be deleted from the model.
2. The F-test of model utility presented on page 857 is a special case of the F-test outlined on page 933. To see this, notice that the denominators of both F-test statistics are the same. When *all* the betas in the complete model are zero, SSE_r is equal to SST, and the numerator of Equation 17–3 becomes $SST - SSE$, also known as the regression sum of squares, SSR. Hence, Equation 17–3 is $F^* = MSR/MSE$.
3. The t-test of an individual beta presented on page 920 also is a special case of the F-test outlined on page 933. Exercise 17.25 demonstrates this equivalence for a set of data.
4. The Fitted Model screen of the Regression processor is designed to list the independent variables in sequence: X_1, X_2, X_3, and so on. Thus, when we fit a reduced model, be sure to reconcile the actual independent variables in the model to the X_1, X_2, and so on variables appearing in the Fitted Model screen. For instance, in Example 17.2 we dropped X_1 and X_2 from the model and fit $Y = \beta_0 + \beta_3 X_3 + \epsilon$ to the data. Had we used the Regression processor, the Fitted Model screen would have shown $\hat{Y} = 37.947 + .017X_1$. In this case, X_1 from the processor corresponds to X_3 in the reduced model.

CLASSROOM EXAMPLE 17.2

Testing Two Betas Simultaneously in a Multiple Regression Model

In Example 16.2, we sampled 28 hotels and motels to relate the variables:

Y = Hotel/motel occupancy rate
X_1 = Number of competitors within a one-mile radius
X_2 = Nightly room rate
X_3 = Advertising expense, as a percentage of annual gross revenues

The model was $\hat{Y} = 56.319 - 1.947X_1 - .008X_2 + 1.840X_3$, and the sum of squared fitting errors was 1633.33187. Test the null hypothesis that X_1 and X_2 are insignificant predictors of Y in the presence of X_3, based on the reduced model $\hat{Y} = 52.018 + 1.839X_3$. Use $\alpha = .10$ and assume the sum of squared fitting errors for the reduced model is 1850.92857.

SECTION 17.2 EXERCISES

17.12 Explain two drawbacks to conducting a sequence of t-tests on the individual betas in a multiple regression model.

17.13 Determine the rejection region for the following tests. (Interpolate if necessary.)
a. H_0: $\beta_1 = \beta_2 = \beta_3 = 0$, $n = 25$, $k = 5$, and $\alpha = .10$.
b. H_0: $\beta_1 = \beta_2 = 0$; $n = 20$, $k = 4$, and $\alpha = .05$.
c. H_0: $\beta_1 = \beta_2 = 0$; $n = 30$, $k = 6$, and $\alpha = .01$.

17.14 Refer to Exercise 17.13. Use the following values of SSE for the complete and reduced models to execute the three tests in Exercise 17.13.
a. $SSE_c = 448.678$, and $SSE_r = 477.826$.
b. $SSE_c = 2.8206$, and $SSE_r = 22.023$.
c. $SSE_c = 125.6717$, and $SSE_r = 134.0934$.

17.15 A multiple regression analysis based on a sample of size $n = 16$ entities yielded a sum of squared fitting errors of 11.135 for the complete model, involving seven independent variables. Conduct a test of H_0: $\beta_4 = \beta_7 = 0$, if $SSE_r = 34.718$. Use $\alpha = .05$.

17.16 Determine the rejection region for the following tests. (Interpolate if necessary.)
a. H_0: $\beta_2 = \beta_3 = 0$, $n = 15$, $k = 3$, and $\alpha = .05$.
b. H_0: $\beta_1 = \beta_3 = 0$, $n = 25$, $k = 3$, and $\alpha = .01$.
c. H_0: $\beta_2 = \beta_3 = \beta_4 = 0$, $n = 49$, $k = 4$, and $\alpha = .10$.

17.17 Refer to Exercise 17.16. Use the following values of SSE for the complete and reduced models to execute the three tests in Exercise 17.16.
a. $SSE_c = 2300.9$, and $SSE_r = 2455.7$.
b. $SSE_r = 30399.32$, and $SSE_c = 21871.7$.
c. $SSE_c = 104030$, and $SSE_r = 123030$.

17.18 Refer to Exercise 17.15. Conduct a test to determine whether the set of independent variables—X_1, X_3, X_5, and X_6—could be dropped from the complete model. Assume $SSE_r = 12.717$ and select an α-risk of your choosing.

17.19 (Use of computer processor recommended.) Closed-end bond funds operate differently from traditional open-ended funds. Open-ended funds sell new shares and redeem existing ones at net asset value (NAV). Closed-end funds have a fixed number of shares that are sold out at the initial public offering. From then on, the shares are bought or sold like shares of stock—for an amount either more or less than the value of the security. When the selling price of a bond fund share is below the NAV, it is said to sell at a discount. When the selling price is above the NAV, it is said to sell at a premium. A comparison of closed-end bond funds, reported in the June 1988 issue of *Changing Times* ("Bond Funds: It Pays To Wait," pp. 71–75), produced the following data on the variables:

Y_1 = Current premium or discount (−) to NAV
Y_2 = Current yield
X_1 = 1986 premium or discount (−) to NAV
X_2 = 1985 premium or discount (−) to NAV
X_3 = 1984 premium or discount (−) to NAV
X_4 = 1983 premium or discount (−) to NAV
X_5 = 1982 premium or discount (−) to NAV
X_6 = 1981 premium or discount (−) to NAV
X_7 = 5-year total return
X_8 = Current expense ratio
X_9 = Number of shares (in millions)

Entity	Y_1	Y_2	X_1	X_2	X_3	X_4	X_5	X_6	X_7	X_8	X_9
1	4.4%	10.2%	7%	−4%	0%	−3%	−1%	−6%	84.61%	0.70	9.9
2	0.0	10.0	7	−2	2	−2	−3	−4	66.97	0.83	1
3	6.9	9.5	24	12	−9	−1	−3	−3	77.65	1.12	2.7
4	2.0	10.5	14	−11	8	10	10	−2	74.08	1.32	2.6
5	0.3	10.9	8	1	−6	−6	−5	−6	83.39	0.90	7.1
6	−2.4	9.2	−3	−7	−1	−9	−8	−4	101.83	1.00	3.7
7	4.1	9.0	12	8	1	1	2	−1	83.82	1.04	2.6
8	−7.7	8.4	1	1	−7	−4	−7	−5	67.13	0.96	2.2
9	−1.3	9.5	−5	−8	−4	−7	−8	−6	92.45	0.80	6.9
10	10.4	9.5	14	2	6	2	0	−2	81.90	0.95	3.1
11	−6.5	10.0	−3	0	−1	−2	−3	−6	76.47	0.88	4.8
12	−3.7	10.0	−3	−4	0	3	2	−7	76.36	0.96	1.8
13	10.9	9.5	12	3	2	3	0	−4	84.21	0.69	10.9
14	−1.2	9.4	0	−3	−9	−9	−9	−7	80.85	0.68	9.3
15	3.6	8.6	4	2	−6	−7	−11	−13	84.16	0.67	6.8
16	−9.2	5.1	−6	−6	−15	−11	−15	−23	88.21	0.93	2.5
17	22.8	9.8	29	−1	−8	−14	−25	−20	100.35	1.95	1.2
18	9.0	10.5	10	7	−3	−5	−6	−5	84.39	0.74	8
19	11.2	9.2	18	11	−3	−3	−5	−7	77.43	0.83	6.7
20	−2.6	9.6	−3	−9	−11	−11	−13	−16	88.15	0.92	5.7
21	−8.9	14.3	2	−19	−20	−1	−5	−20	39.86	1.24	2.6
22	6.1	10.2	9	−4	−7	−9	−7	−11	73.30	0.81	8.3
23	11.2	9.2	3	−1	−3	−4	−6	−2	95.74	0.72	5.3
24	−5.3	10.4	12	0	−2	1	−4	−10	69.70	1.34	5.4
25	−5.5	9.0	−5	−9	−9	−10	−8	−5	96.20	1.00	6.4

Fit the complete model $Y_1 = \beta_0 + \beta_1X_1 + \beta_2X_2 + \ldots + \beta_9X_9 + \epsilon$ to the data. Then fit a reduced model to test the significance of the set of independent variables X_3, X_4, X_6, and X_9, given the presence of the remaining variables, in the prediction of Y_1. Use $\alpha = .05$.

17.20 Refer to Exercise 17.19. Test H_0: $\beta_1 = \beta_2 = \beta_3 = \beta_4 = \beta_5 = \beta_6 = 0$ using $\alpha = .10$, the dependent variable Y_1, and the complete model from Exercise 17.19.

17.21 Refer to Exercise 17.19. Fit the complete model $Y_2 = \beta_0 + \beta_1X_1 + \beta_2X_2 + \ldots + \beta_9X_9 + \epsilon$ to the data. Are the variables X_1, X_3, X_4, X_5, and X_8 a significant set of predictors, given the presence of the remaining variables in the model? Let $\alpha = .01$.

17.22 Refer to Exercise 17.21. Test H_0: $\beta_1 = \beta_3 = \beta_4 = \beta_5 = \beta_6 = \beta_8 = \beta_9 = 0$ using the dependent variable Y_2 and the complete model from Exercise 17.21. Let $\alpha = .05$.

17.23 Suppose a multiple regression analysis involving four independent variables and $n = 49$ entities produced the following correlation matrix:

$$\mathbf{C} = \begin{bmatrix} 1.0000 & .8624 & .6034 & .2253 & -.0641 \\ .8624 & 1.0000 & .6987 & .0343 & .0433 \\ .6034 & .6987 & 1.0000 & -.0159 & .0949 \\ .2253 & .0343 & -.0159 & 1.0000 & -.3643 \\ -.0641 & .0433 & .0949 & -.3643 & 1.0000 \end{bmatrix}$$

What reduced model, if any, would you fit at this point, and why?

*17.24 In Section 17.2, we remarked that it is possible for *SSE* to remain unchanged as we add additional independent variables to the model. The purpose of this exercise is to demonstrate this effect.

a. Fit the complete model $Y = \beta_0 + \beta_1 X_1 + \beta_2 X_2 + \epsilon$ to the following set of data and determine *SSE*.

Entity	Y	X_1	X_2
1	4	2	5
2	8	2	1
3	4	2	3
4	12	2	0
5	10	0	2
6	10	0	1
7	6	0	4
8	10	0	0

b. Add a third independent variable, X_3, to the data set. Let its values be -1, -1, 1, 1, -1, 1, 1, and -1 respectively, for Entity 1 through Entity 8. Now fit the model $Y = \beta_0 + \beta_1 X_1 + \beta_2 X_2 + \beta_3 X_3 + \epsilon$ to the data set and determine *SSE* again. Compare the value with that obtained in part a.

c. Find the correlation matrix for the model in part b. Notice that the new independent variable, X_3, is uncorrelated with X_1 and X_2, as well as with Y. In such a situation, the value of *SSE* will be unchanged when X_3 is added to the model.

*17.25 The t-test of H_0: $\beta_i = 0$ is a special case of the F-test outlined on page 933. To demonstrate this equivalence for a specific set of data, consider the following complete and reduced models:

$$\text{Complete model: } Y = \beta_0 + \beta_1 X_1 + \beta_2 X_2 + \beta_3 X_3 + \epsilon$$
$$\text{Reduced model: } Y = \beta_0 + \beta_1 X_1 + \beta_2 X_2 + \epsilon$$

a. Suppose a sample of size $n = 25$ produced $SSE_c = 21871.7$ and $SSE_r = 23282.4$. With $\alpha = .05$, use the F-test for several betas to test H_0: $\beta_3 = 0$.

b. Now suppose that the estimated regression coefficient for β_3 in the complete model was $b_3 = -8.7233$ with $S_{b_3} = 7.4955$. Use the t-test for an individual beta with $\alpha = .05$ to test H_0: $\beta_3 = 0$.

c. Verify that the square of the t-test statistic in part b is equal to the F-test statistic from part a, to two decimal places. That is, verify $(t^*)^2 = F^*$.

17.3 RESIDUAL ANALYSIS

In Section 16.1, we specified the general multiple regression model as $Y = \beta_0 + \beta_1 X_1 + \beta_2 X_2 + \ldots \beta_k X_k + \epsilon$, where the error term, ϵ, is assumed to be a random variable. No distributional assumptions about ϵ are necessary to use the method of least squares to fit the general model to a set of data, since model fitting is an exercise

*Optional

in descriptive statistics. However, to conduct a test of the model's utility, to construct interval estimates for Y and $E(Y)$, and to test individual and several betas in the model, we require ϵ to satisfy the distributional assumptions laid out on page 832. Specifically, these assumptions are given in the following box.

Assumptions About ϵ in the Multiple Regression Model

1. At each setting of the values for $X_1, X_2, \ldots, X_k$, ϵ is a random variable.
2. At each setting of the values for $X_1, X_2, \ldots, X_k$, ϵ has a normal probability distribution with

$$\text{Mean} = 0 \qquad \text{Variance} = \sigma^2$$

3. For each different pair of settings of the values for $X_1, X_2, \ldots, X_k$, the error random variables are independent.

Let us try to visualize the error random variables, so that we can interpret these assumptions. Suppose we have a multiple regression analysis with two independent variables, X_1 and X_2, and a dependent variable Y. A graph of the data would require three axes: one for Y, one for X_1, and a third for X_2, as suggested in Figure 17.8(a). Figure 17.8(b) shows a particular setting of the independent variables; in this case, the setting is $X_1 = 2$ and $X_2 = 1$. In theory, at this setting there are many Y-values. Figure 17.8(c) shows some of these Y-values as dots along a vertical line through the point $X_1 = 2$ and $X_2 = 1$. For illustrative purposes, only positive Y-values are shown.

The average of all the Y-values along the vertical line is indicated by the point in Figure 17.8(d). At this point, $\epsilon = 0$. For points above the mean, ϵ, which is the difference between a point and the mean, is a positive number, and for points below the mean, ϵ is a negative number. The distribution of the values of ϵ is shown as a normal curve in Figure 17.8(e), per our second assumption. Now, extend this argument to another setting of X_1 and X_2, for instance, $X_1 = 4$ and $X_2 = 1$. Figure 17.8(f) shows that, according to our assumptions about ϵ, a normal distribution describes the error random variable at $X_1 = 4$ and $X_2 = 1$ as well. The third assumption states that the two error random variables—the one at $X_1 = 2$, $X_2 = 1$, and the second at $X_1 = 4$, $X_2 = 1$—are independent.

Although Figure 17.8 applies to a situation involving only two independent variables, the notion of an error random variable at each setting of the X_i's applies to a multiple regression analysis with more than two independent variables as well. When there are more than two X_i's, however, we cannot represent ϵ graphically because it would require more than three dimensions. In spite of our spatial limitations, we hope the development in Figure 17.8 is sufficient to extend the interpretation of the assumptions about ϵ to the general case.

At this point in the multiple regression analysis, we would like to examine the validity of each assumption. As with all other statistical inferences, it will not be possible for us to declare with certainty that an assumption is or is not true. This is

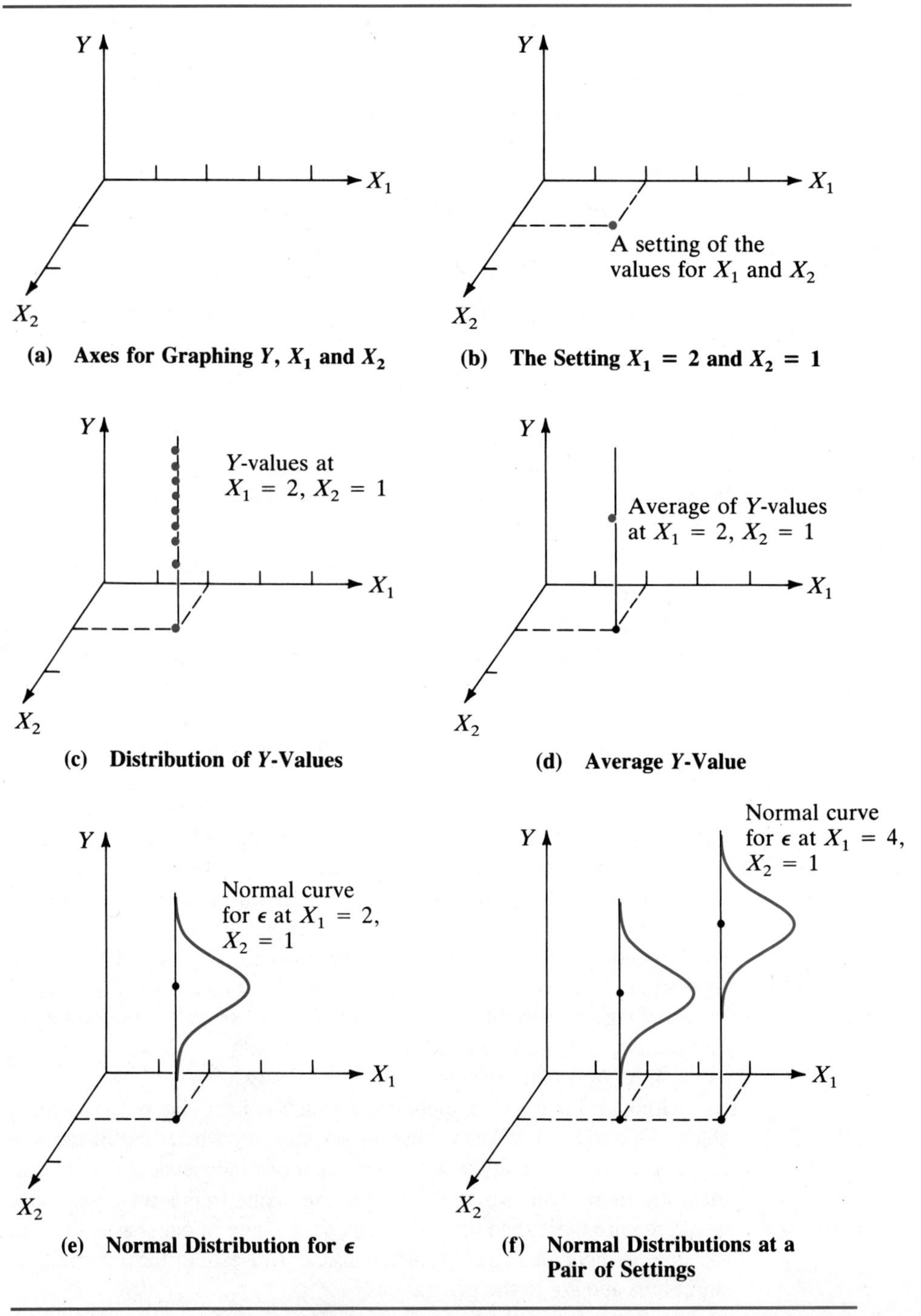

Figure 17.8 Graphical Representation of the Assumptions About ϵ for the Special Case of Two Independent Variables

possible only if we know the true values of β_0, β_1, β_2, . . . , and β_k. What we hope is that the observed fitting errors, or residuals, will offer evidence to support the assumptions, rather than contradict them. Our pending analysis of the assumptions about ϵ, therefore, is called a *residual analysis*.

Assessing the validity of the assumptions is easier to do when there are several residuals at each setting of the X_i's. Unfortunately, we do not encounter this sampling situation very often; indeed, we usually have only one residual per setting. In this case, we will approach the task of analyzing the residuals from a different perspective.

When there are n different settings of the X_i's, in theory there are n error random variables and n normal curves. If assumptions one and two hold, then all n normal distributions look alike—same mean of 0 and same variance of σ^2. Instead of treating each of the n residuals as a sample of size one from its respective normal curve, we will view the n residuals as a sample of size n from one normal curve that has a mean of 0 and a variance of σ^2. In this way, we "pool" the information in each residual to facilitate the residual analysis. This technique is not new; we used the pooling idea to estimate σ^2 in both simple and multiple regression. From this framework, we can analyze the first two assumptions about ϵ.

Residual Plots

The first assumption that ϵ is a random variable is practically indisputable and needs no confirmation. Recall that a random variable is a special kind of variable whose values are quantitative and determined by chance. The observed fitting errors, which represent values of ϵ and are denoted by e_i, satisfy both conditions.

We can see the randomness and the variability in the observed fitting errors by graphing them in a residual plot. A **residual plot** is a scatter plot with the values of e_i on the vertical axis and the values of $\hat{Y}$ on the horizontal axis. (Residual plots also can be constructed with one of the independent variables plotted on the horizontal axis in lieu of $\hat{Y}$.) For instance, for the real estate data in Example 17.1, the following are the observed and predicted values of Y, and the fitting errors, $e_i = Y - \hat{Y}$. The $\hat{Y}$-values were calculated using unrounded b_i's in the fitted model: $\hat{Y} = 38.28506 - 3.09150X_1 + 3.51290X_2 + .01850X_3$.

House	Y	$\hat{Y}$	e_i	House	Y	$\hat{Y}$	e_i
1	62.0	65.09	−3.09	13	56.7	61.48	−4.78
2	57.5	63.79	−6.29	14	61.0	63.51	−2.51
3	62.0	62.86	−.86	15	61.7	69.95	−8.25
4	60.0	63.16	−3.16	16	67.9	69.95	−2.05
5	70.4	63.79	6.61	17	75.0	69.95	5.05
6	68.0	67.64	.36	18	81.0	70.88	10.12
7	64.0	65.64	−1.64	19	83.0	78.11	4.89
8	70.0	65.95	4.05	20	81.9	79.39	2.51
9	71.5	67.49	4.01	21	75.0	75.41	−.41
10	74.5	68.75	5.75	22	70.0	74.48	−4.48
11	72.0	72.86	−.86	23	66.5	79.11	−12.61
12	63.9	61.48	2.42	24	78.5	73.28	5.22

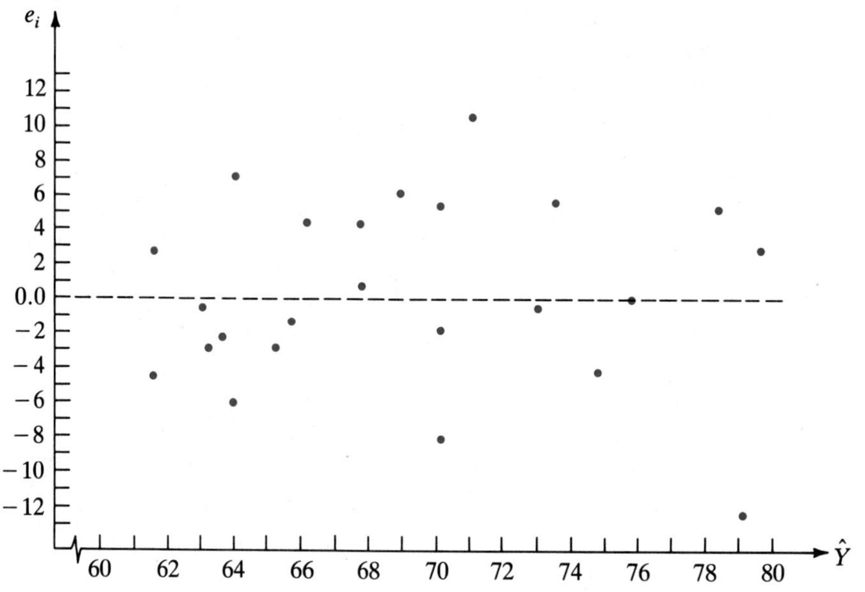

Figure 17.9 Residual Plot of Real Estate Data

A residual plot of e_i versus $\hat{Y}$ appears in Figure 17.9. Note the uniform scatter of the residuals about the horizontal dashed line through $e_i = 0.0$.

Plotting the residuals is recommended so that we can identify any obvious "bad" patterns in the data. For example, Figure 17.10 illustrates three residual patterns that signal trouble with respect to the assumptions about ϵ. In Figure 17.10(a), the residuals are plotted against one of the independent variables, X_i. Notice that the fitting errors exhibit differing amounts of spread about the $e_i = 0$ dashed line. Such a pattern is indicative of a variance for ϵ that does not remain constant at all settings of the X_i's. Figure 17.10(b) shows a pattern of correlated residuals. This pattern suggests that the error random variables are not independent and frequently occurs in time series data when the values of Y are recorded sequentially over time. Note that the independent variable plotted on the horizontal axis represents time. Figure 17.10(c) also could signal correlated errors, over a longer period of time than the time frame in Figure 17.10(b) perhaps. Or, the pattern could arise in a simple regression analysis when the relation between Y and X is a curved one, not a linear one.

The main advantage of a residual plot is to detect outliers (see definition in Chapter 3) that distort the scatter with a very large positive or negative value. Procedures for identifying and dealing with outliers are discussed in Younger (1979) and in Neter, Wasserman, and Kutner (1985).

One of the major disadvantages of a residual plot is that it may be hard for us to see troublesome patterns, unless the effect is pronounced. Further, small sample sizes

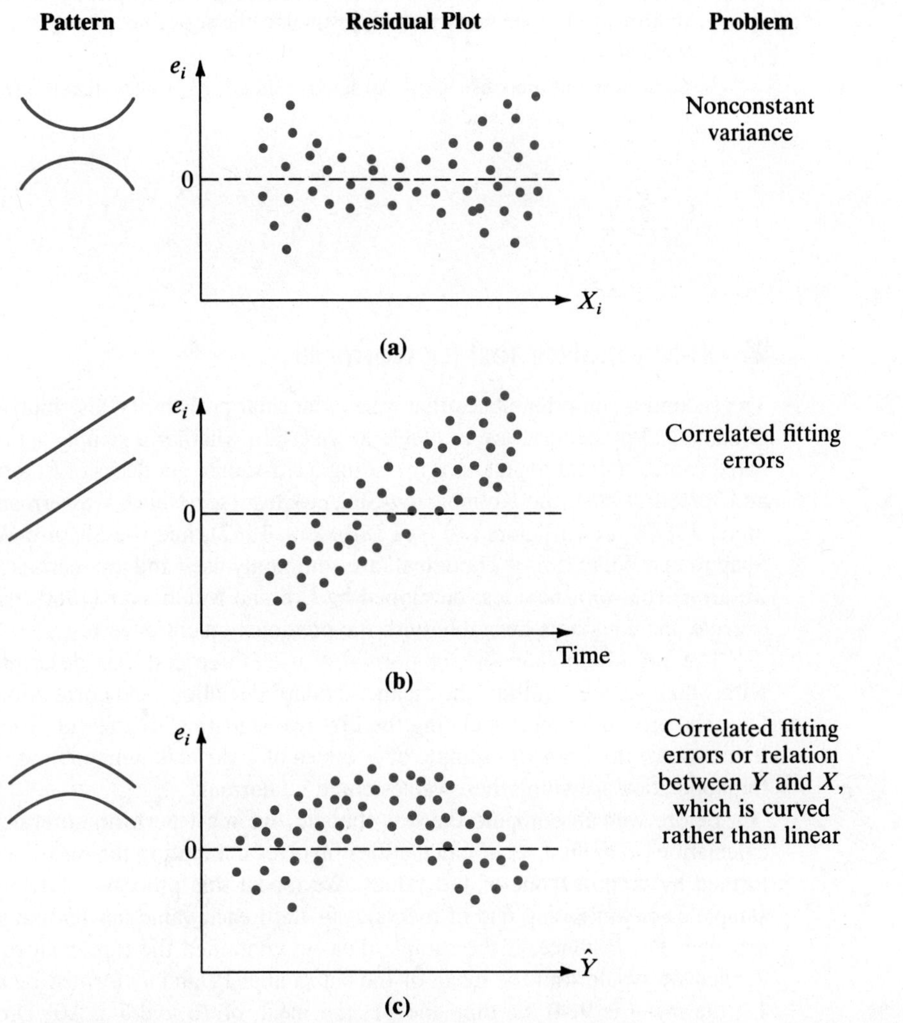

Figure 17.10 Troublesome Patterns in a Residual Plot

tend to create sparse residual plots, making pattern identification even more difficult. Residual plots are very useful in simple regression or when there are many residuals from the same settings of the X_i's. Judging the adequacy of the random error term assumptions only on the basis of a residual plot, however, is a bit risky. We seek more definitive means of examining the assumptions.

COMMENTS

1. Most commercial statistical software packages include residual plots as an option in their multiple regression programs, though the resulting output is not a graphical work of art.
2. Some residual plots use a standardized residual, e_i^*, rather than the (raw) residuals, e_i. A standardized residual is obtained via

$$e_i^* = \frac{e_i}{\sqrt{\hat{\sigma}}}$$

Lin-Mudholkar Test for Normality

The second assumption states that ϵ has a normal probability distribution. A variety of statistical procedures are available to ascertain whether a sample set of data originated from a normal population, including a chi-square goodness of fit test (see Daniel and Terrell, 1986), the Kolmogorov-Smirnov test (see Neter, Wasserman, and Whitmore, 1988), the Lilliefors test (see Lilliefors, 1967), and the Shapiro-Wilk test (see Shapiro and Wilk, 1965). Though all are commonly used and are reasonably powerful, we are partial to a new test developed by Lin and Mudholkar (1980) that is easy to execute and compares favorably with the previously mentioned tests.

The Lin-Mudholkar test for normality uses three common descriptive statistics with which we are familiar: the mean, standard deviation, and correlation coefficient. Only the procedure for calculating the first two statistics is different. To set the stage, let us start with a random sample of n values of a variable. Our job is to test whether the population spawning these values could be normal.

Before we can compute the test statistic, we must perform several intermediate calculations. The first set of calculations involves computing the mean of a subsample formed by dropping one of the values. We repeat this process—starting with a full sample and eliminating one of the values—until each value has had its turn at being dropped. For instance, if the sample data set contained the three values 13, 11, and 7, then we would find the mean of the subsample 11 and 7, formed by dropping the 13; the mean is 9. If we drop the 11, the mean of 13 and 7 is 10. Dropping the 7 yields a mean of 12 for the subsample 13 and 11. Let us denote these means by the symbol m_i.

The second set of calculations involves computing S_i, the standard deviation of each subsample. For 11 and 7, the standard deviation is 2.83; for the subsample 13 and 7, $S_i = 4.24$; and for the subsample 13 and 11, $S_i = 1.41$.

In the third set of calculations, we raise each standard deviation to the power ⅔. Let us denote the resulting values by l_i; that is, $l_i = (S_i)^{2/3}$. For example, for the 11 and 7 subsample, $l_i = (2.83)^{2/3} = 2.00$, to two decimal places. It will be convenient to arrange the data, the l_i, and the m_i values into columns. The entries beside each data value in the m_i and l_i columns are the mean and $(S_i)^{2/3}$ for the subsample formed by dropping that data point. For the sample 13, 11, and 7, following is such a table:

Data	m_i	l_i
13	9	2.00
11	10	2.62
7	12	1.26

Finally, we calculate the correlation coefficient, r, for the m_i and l_i values. (Refer to Equations 4–1 or 15–22 to compute r.) In the above example, $r = -.692$. The resulting value of r is the value of the test statistic in the Lin-Mudholkar test of normality. However, the cutoff values of r presented in Table 4.4 *cannot* be used in this test to separate the acceptance and rejection regions, because r is not calculated from n independent values. Instead, a new table of cutoff values of r for this test is presented in Appendix J. Note that the Lin-Mudholkar test assumes a minimum sample size of $n = 10$.

To summarize, here is the procedure to follow to determine the test statistic for the Lin-Mudholkar test of normality:

From the full sample, do the following.

1. Form a subsample by deleting one of the sample values.
2. Find m_i, the mean of the subsample.
3. Calculate S_i, the standard deviation of the subsample.
4. Find $l_i = (S_i)^{2/3}$.
5. Repeat the first four steps until each of the n sample values has had an opportunity to be deleted.
6. Compute r, the correlation coefficient between m_i and l_i.

The following box outlines the elements of the test.

Lin-Mudholkar Test for Normality

H_0: The residuals are normally distributed.

H_a: The residuals are not normally distributed.

TS: r

RR: Use cutoff values of r from the table in Appendix J.

C: Reject H_0 or accept H_0.

Example 17.3 illustrates the use of the test in deciding whether the residuals in a multiple regression analysis come from a normal population. If we accept H_0, we are stating that the population could be characterized by a normal probability distribution. If we reject H_0, we believe the population of residuals is not normal.

EXAMPLE 17.3

In Example 16.1, we fit the multiple regression model $Y = \beta_0 + \beta_1X_1 + \beta_2X_2 + \epsilon$ to a set of $n = 12$ entities, relating Y = the annual number of passengers arriving and departing an airport to X_1 = the average number of flights departing weekly and X_2 = the total number of local airline employees. The fitted model was $\hat{Y} = -4.86857 + .30669X_1 + .16550X_2$. Conduct the Lin-Mudholkar test to determine if the residuals could have come from a normal population. Use $\alpha = .10$.

Solution:

The observed and predicted Y-values, the fitting errors (e_i), and the values of m_i, S_i, and l_i are summarized in the following table.

Y	$\hat{Y}$	e_i	m_i	S_i	l_i
19.3	21.07	−1.77	.161	4.4623	2.7104
2.8	2.73	.07	−.006	4.5003	2.7258
30.5	38.36	−7.86	.715	3.6761	2.3819
6.6	12.12	−5.52	.502	4.1145	2.5677
11.7	11.82	−.12	.011	4.5002	2.7258
6.1	2.56	3.54	−.322	4.3459	2.6631
7.6	7.99	−.39	.035	4.4986	2.7251
13.4	11.18	2.22	−.202	4.4403	2.7015
50.4	41.59	8.81	−.801	3.4331	2.2758
14.9	17.28	−2.38	.216	4.4312	2.6978
35.4	34.18	1.22	−.111	4.4823	2.7185
6.3	4.12	2.18	−.198	4.4424	2.7024

To compute the correlation coefficient with $n = 12$, we need the following sums:

$$\Sigma m_i = 0 \qquad \Sigma l_i = 31.5958$$
$$\Sigma m_i^2 = 1.674802 \qquad \Sigma l_i^2 = 83.439742$$
$$\Sigma m_i l_i = .0571849$$

According to Equation 4–1 (with $m_i = X$ and $l_i = Y$), the correlation coefficient is

$$r = \frac{12(.0571849) - (0)(31.5958)}{\sqrt{12(1.674802) - 0^2}\ \sqrt{12(83.439742) - (31.5958)^2}}$$
$$= .0886364$$
$$= .089 \quad \text{(to three decimal places)}$$

The test for normality is

H_0: The residuals are normally distributed.
H_a: The residuals are not normally distributed.

TS: $r = .089$

RR: For $\alpha = .10$ and $n = 12$, the cutoff values are $\pm .661$.

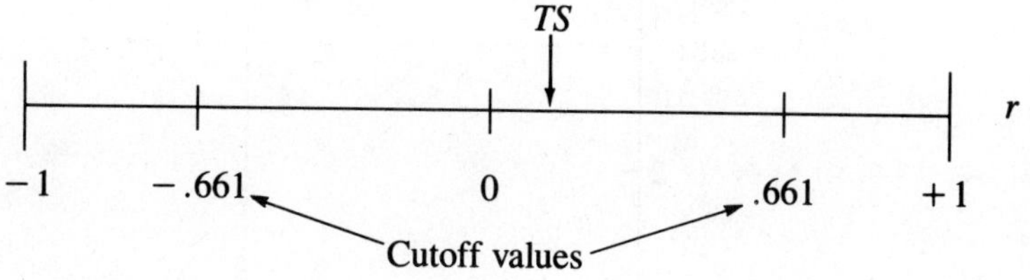

C: Accept H_0.

We conclude that the sample evidence supports the assumption that the residuals could have come from a normal population.

□

COMMENTS

1. The process of successively dropping a sample value and computing a summary statistic like the mean on the remaining values is called a *jackknife procedure*.
2. When generating the $\hat{Y}$-values, try to use the *unrounded* values $b_0, b_1, \ldots, b_k$ rather than the rounded ones to ensure as much accuracy as possible.
3. The repetitive computations of m_i and S_i are not as tedious as you might think. First, make sure that the residuals add to zero. Unless there is a round-off error, this should be the case automatically. Second, compute Σe_i^2. Third, to compute m_i and S_i^2 quickly, use these formulas:

$$m_i = \frac{-e_i}{(n-1)}$$

$$S_i^2 = \frac{\Sigma e_i^2 - \dfrac{ne_i^2}{n-1}}{n-2}$$

To demonstrate, consider the residuals in Example 17.3. You should verify that $\Sigma e_i = 0$ and $\Sigma e_i^2 = 202.5356$. To find m_1 and S_1, for example, we compute:

$$m_1 = \frac{-e_1}{(n-1)} = \frac{-(-1.77)}{11} = .160909\ldots$$

$$S_1^2 = \frac{202.5356 - \dfrac{12(-1.77)^2}{11}}{10} = 19.911789$$

Thus,

$$S_1 = \sqrt{S_1^2} = 4.4622628$$

4. To get an intuitive grasp of how the Lin-Mudholkar test works, imagine a scatterplot of the m_i values on the horizontal axis versus the S_i values (or l_i values) on the vertical axis. For instance, Figure 17.11 illustrates such a plot for the data set -5, 1, 1, 1, and 2. The dotted lines represent the mean (0) and standard deviation (2.83)

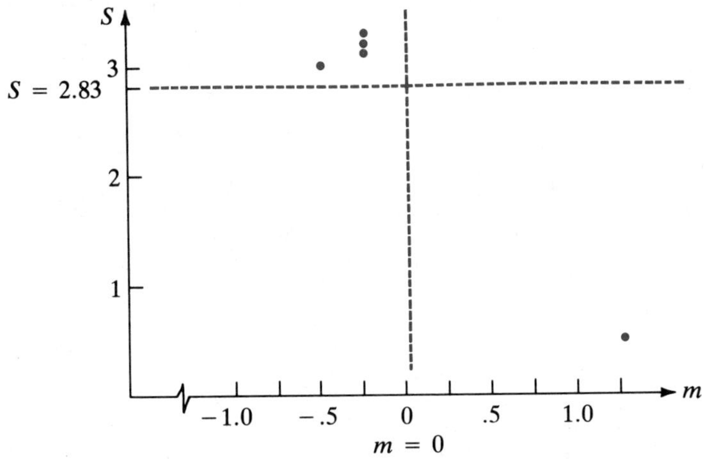

Figure 17.11 Graph of m_i and S_i Values

of the five numbers. The m_i and S_i values for the subsamples are listed in the following table and plotted on Figure 17.11.

Data	m_i	S_i
−5	1.25	.50
1	−.25	3.20
1	−.25	3.20
1	−.25	3.20
2	−.50	3.00

Notice the concentration of points in the second and fourth quadrants, relative to the dotted lines. The data set is skewed to the left with a coefficient of skewness of $g_1 = -1.408$. When the distribution is skewed left, the scatterplot will show an inverse relation between m_i and S_i (and between m_i and l_i); hence, the correlation coefficient will be negative. When the data set is skewed right, the opposite effect is revealed, with a concentration of points in the first and third quadrants. For a normal distribution, the scatterplot should have no discernible pattern; for instance, construct a scatterplot of m_i and S_i for the residuals in Example 17.3 to see this.

CLASSROOM EXAMPLE 17.3

Testing Residuals for Normality

The multiple regression model $Y = \beta_0 + \beta_1 X_1 + \beta_2 X_2 + \epsilon$ was fit to a sample of $n = 10$ entities. The observed and predicted Y-values are listed in the following table, along with the fitting errors (e_i), m_i, S_i, and l_i:

Entity	Y	$\hat{Y}$	e_i	m_i	S_i	l_i
1	10	10.7	−.7		2.753	1.964
2	7	6.2	.8	−.09	2.749	1.962
3	15	15.1	−.1	.01	2.765	1.970
4	22	19.5	2.5	−.28	2.603	1.892
5	18	18.1	−.1	.01	2.765	1.970
6	12	10.5	1.5	−.17	2.708	1.943
7	24	21.7	2.3	−.26	2.629	1.905
8	19	21.7	−2.7	.30	2.575	1.879
9	13	10.7	2.3	−.26	2.629	1.905
10	7	12.8	−5.8	.64		

a. Find the value of m_i to two decimal places for the first entity, assuming $\Sigma e_i = 0$.
b. Find the values of S_i and l_i to three decimal places for the tenth entity, assuming $\Sigma e_i^2 = 61.16$.
c. Compute the correlation coefficient, r, between m_i and l_i. Assume $\Sigma m_i = -.02$, $\Sigma m_i^2 = .7568$, $\Sigma l_i = 18.828$, $\Sigma l_i^2 = 35.679988$, and $\Sigma m_i l_i = -.34671$.
d. Test the assumption that ϵ is normally distributed. Use $\alpha = .10$.

Durbin-Watson Test for Autocorrelation

The third assumption about ϵ states that the error random variables are independent. This assumption is inherently false by virtue of the constraint that the residuals add to zero: $\Sigma e_i = 0$. Hence, the set of error random variables cannot be independent when their values are so constrained. There is a prevailing attitude in the literature, though, that the violation of the independence assumption does not severely affect the analysis, provided the number of residuals is large relative to the number of independent variables. See Kleinbaum and Kupper (1978) or Anscombe and Tukey (1963) for details.

When the data used for a multiple regression analysis are time-dependent though, a more serious threat to the independence assumption arises. Time-dependent or *time series* data means that the data have been collected and recorded sequentially over time. For instance, suppose a realty agency wishes to model Y = number of houses sold per month as a function of three independent variables: X_1 = number of full-time agents; X_2 = average monthly temperature; X_3 = monthly 1-year Treasury bill auction rate. Since we record a value for the variables as each month expires, the resulting data would be time-dependent. We will study time-dependent data in detail in Chapter 18. Time series data tend to produce consecutive residuals that are similar in sign and/or magnitude (see Figure 17.10(b)). This pattern of residuals being related is called *serial correlation* or *autocorrelation*.

There is a test developed by J. Durbin and G. S. Watson (1951) to evaluate consecutive residuals for autocorrelation. If we let ρ (Greek letter rho) signify the true serial correlation coefficient in the population of residuals, then we wish to test H_0: $\rho = 0$ (the residuals are not autocorrelated) versus H_a: $\rho > 0$ (the residuals are positively autocorrelated). Naturally enough, the test statistic, denoted by d, is based in part on the consecutive differences in the residuals, $(e_i - e_{i-1})$.

Durbin-Watson Test Statistic

$$d = \frac{\Sigma(e_i - e_{i-1})^2}{\Sigma e_i^2} \tag{17–4}$$

There are n residuals, e_i, but there can be only $n - 1$ differences in the residuals, $e_i - e_{i-1}$. Thus, the sum in the numerator starts at the second residual ($i = 2$) and involves $n - 1$ terms.

It can be shown that d ranges in value from 0 to 4 and that, when H_0 is true, d should be close to 2. Unlike our previous statistical tests, the Durbin-Watson test may result in one of three decisions: accept H_0, reject H_0, or inconclusive. Figure 17.12 depicts these possibilities based on the cutoff values, denoted by d_L and d_U, of the test statistic d. The values d_L and d_U are listed in Appendix K.

To reference the d_L and d_U values, we need three quantities: n, the sample size, α, the Type I error probability, and k, the number of independent variables in the regression analysis. The table allows only two values for α: .05 and .01. For instance, if we were testing for autocorrelation in an analysis with $k = 4$ independent variables, a sample size of $n = 30$, and $\alpha = .01$, the cutoff values of d would be $d_L = 0.94$ and $d_U = 1.51$.

The following box summarizes the elements of the Durbin-Watson test.

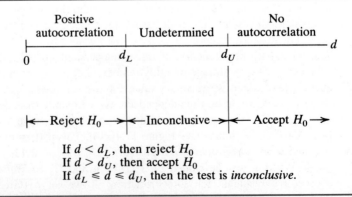

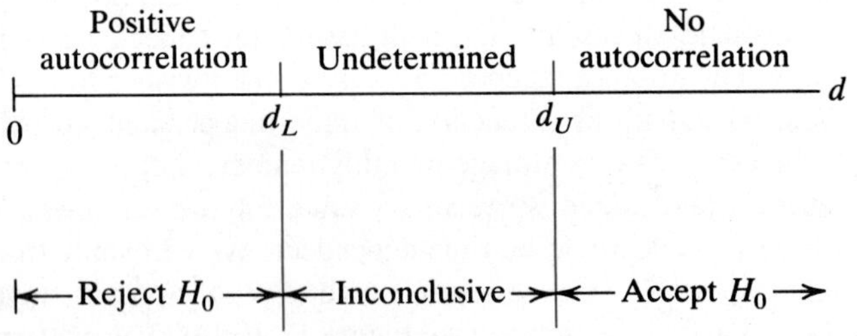

If $d < d_L$, then reject H_0
If $d > d_U$, then accept H_0
If $d_L \leq d \leq d_U$, then the test is *inconclusive*.

Figure 17.12 Rejection Region for Test of H_0: $\rho = 0$ Versus H_a: $\rho > 0$

Durbin-Watson Test for Autocorrelation

H_0: $\rho = 0$ (The residuals are not autocorrelated.)

H_a: $\rho > 0$ (The residuals are positively autocorrelated.)

$$TS: \ d = \frac{\Sigma(e_i - e_{i-1})^2}{\Sigma e_i^2}$$

RR: Use cutoff values, d_L and d_U, from table in Appendix K.

C: Reject H_0 or accept H_0 or inconclusive.

Example 17.4 demonstrates the calculation of d and the subsequent inference about autocorrelation using the Durbin-Watson test.

EXAMPLE 17.4

Churchill Downs race track in Louisville, Kentucky, monitors its attendance and handle (amount of money wagered) during the racing season. Yearly data on the variables Y, average daily handle in millions of dollars, X_1, year of racing season, and X_2, average daily attendance in thousands of people, are listed in the following table. For the variable X_1, years are numbered consecutively with the integers 1, 2, 3, and so on, starting with 1971.

Year	X_1	X_2	Y
1971	1	9.623	$.734
1972	2	8.740	.719
1973	3	8.358	.736
1974	4	9.050	.833
1975	5	8.067	.728
1976	6	9.001	.850
1977	7	8.887	.850
1978	8	9.387	.982
1979	9	9.975	1.107
1980	10	9.644	1.080
1981	11	9.532	1.095
1982	12	9.425	1.060
1983	13	7.105	.861
1984	14	6.909	.806
1985	15	8.950	1.077
1986	16	9.810	1.139
1987	17	10.385	1.240

Source: Copyright 1987, Louisville *Courier-Journal*, June 30, 1987. Reprinted with permission.

Use the fitted model, $\hat{Y} = -.25151 + .02468X_1 + .10728X_2$, to generate the residuals and test for autocorrelation. Round the $\hat{Y}$-values to three decimal places.

Solution:

The predicted values of Y and the residuals, $e_i = Y - \hat{Y}$, are listed in the following table:

Year	Y	$\hat{Y}$	$e_i = Y - \hat{Y}$	e_{i-1}	$(e_i - e_{i-1})$
1971	.734	.805	−.071	—	—
1972	.719	.735	−.016	−.071	.055
1973	.736	.719	.017	−.016	.033
1974	.833	.818	.015	.017	−.002
1975	.728	.737	−.009	.015	−.024
1976	.850	.862	−.012	−.009	−.003
1977	.850	.875	−.025	−.012	−.013
1978	.982	.953	.029	−.025	.054
1979	1.107	1.041	.066	.029	.037
1980	1.080	1.030	.050	.066	−.016
1981	1.095	1.042	.053	.050	.003
1982	1.060	1.056	.004	.053	−.049
1983	.861	.832	.029	.004	.025
1984	.806	.835	−.029	.029	−.058
1985	1.077	1.079	−.002	−.029	.027
1986	1.139	1.196	−.057	−.002	−.055
1987	1.240	1.282	−.042	−.057	.015

From the e_i and $(e_i - e_{i-1})$ columns we compute:

$$\Sigma e_i^2 = .023882 \qquad \Sigma(e_i - e_{i-1})^2 = .019791$$

The Durbin-Watson test for autocorrelation is:

H_0: $\rho = 0$

H_a: $\rho > 0$

TS: $d = \dfrac{\Sigma(e_i - e_{i-1})^2}{\Sigma e_i^2} = \dfrac{.019791}{.023882} = .829$ (rounded)

RR: For $n = 17$, $k = 2$, and $\alpha = .05$, the cutoff values of d are $d_L = 1.02$ and $d_U = 1.54$:

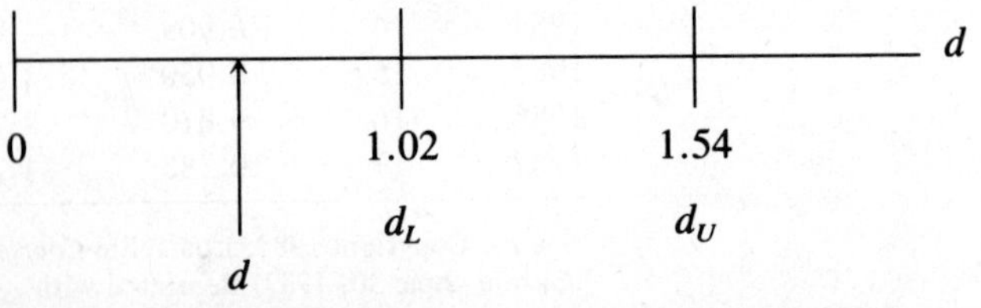

C: Since $d < d_L$, we reject H_0.

In rejecting H_0 we are concluding that the observed fitting errors are so positively autocorrelated that the assumption of independent errors is contradicted. Though this finding does not affect the fitted model, it does preclude us from conducting *F*-tests or *t*-tests about the β-parameters and from constructing interval estimates for Y or $E(Y)$.

□

COMMENTS

1. As a check on the computation for the difference in residuals, $(e_i - e_{i-1})$, be sure to verify that this column of values sums to $e_n - e_1$; that is, $\Sigma(e_i - e_{i-1}) = e_n - e_1$. For example, in Example 17.4, $\Sigma(e_i - e_{i-1}) = .029$ and $e_n - e_1 = -.042 - (-.071) = .029$.
2. In Section 19.4, we will learn about another test called the *runs test* that could be used to test for the presence of autocorrelation in the residuals.

CLASSROOM EXAMPLE 17.4

Testing Residuals for Autocorrelation

Execute the Durbin-Watson test for autocorrelation on the following set of residuals. Assume there are three independent variables in the multiple regression model, and use $\alpha = .01$.

Residual Number	e_i	e_{i-1}	$e_i - e_{i-1}$	$(e_i - e_{i-1})^2$
1	−3.1			
2	−6.8			
3	−0.8			
4	−3.1			
5	6.7			
6	−0.4			
7	−1.1			
8	4.0			
9	3.9			
10	6.1			
11	−1.3			
12	2.7			
13	−3.9			
14	−2.9			
15	0.0			
$\Sigma e_i^2 = 213.38$				

SECTION 17.3 EXERCISES

17.26 What is the purpose of constructing a residual plot?

17.27 Refer to the data in Example 17.1. How many different settings of the independent variables are there?

17.28 Construct a residual plot of $\hat{Y}$ versus e_i for the following $n = 16$ residuals. Examine the plot for any troublesome patterns.

Entity	$\hat{Y}$	e_i	Entity	$\hat{Y}$	e_i
1	7.6	1.9	9	7.6	−.4
2	6.1	−.2	10	8.1	−1.5
3	6.5	.3	11	4.2	.6
4	3.3	−.1	12	7.2	.8
5	4.8	−.6	13	6.3	−.8
6	6.5	−.5	14	6.5	−.8
7	5.9	−.2	15	7.0	1.2
8	3.8	−.3	16	4.1	.6

17.29 Construct a residual plot of $\hat{Y}$ versus e_i for the following $n = 10$ residuals. Examine the plot for any troublesome patterns.

Entity	$\hat{Y}$	e_i
1	−24.5	27.3
2	49.6	−24.4
3	73.5	−13.1
4	22.1	15.9
5	34.1	36.9
6	117.5	−50.4
7	159.2	41.5
8	173.0	24.7
9	50.9	−25.8
10	74.2	−32.6

17.30 Refer to Exercise 17.28. Conduct the Durbin-Watson test for autocorrelation among the residuals, with a 5% risk of making a Type I error. Assume there are five independent variables in the analysis.

17.31 Refer to Exercise 17.28. Suppose the residuals produced a test statistic of $r = .388$ for the Lin-Mudholkar test for normality. With a 5% error risk, what conclusion is warranted?

17.32 Refer to Exercise 17.29. Conduct the Lin-Mudholkar test for normality of the residuals, with $\alpha = .10$. Round off the m_i values to two decimal places and the l_i values to three decimal places.

17.33 Determine the rejection region for the Durbin-Watson test for autocorrelation, based on the following information.

a. $\alpha = .05$; $n = 24$, and $k = 3$.
b. $\alpha = .05$; $n = 40$, and $k = 4$.
c. $\alpha = .01$; $n = 17$, and $k = 2$.

17.34 Refer to Exercise 17.33. Set up the rejection region for the Lin-Mudholkar test for normality for each part.

17.35 What are the effects, if any, on the descriptive and inferential parts of a regression analysis if the Durbin-Watson test for autocorrelation yields a conclusion of accept H_0?

17.36 Complete the Lin-Mudholkar test for normality using the following values for m_i and l_i. Let $\alpha = .05$.

i	m_i	l_i
1	−1.213	7.1148
2	.026	7.2496
3	−2.631	6.5641
4	1.085	7.1423
5	1.908	6.9064
6	.507	7.2265
7	.990	7.1605
8	−3.418	5.9934
9	.125	7.2483
10	.163	7.2473
11	.177	7.2469
12	.761	7.1973
13	1.523	7.0348

17.37 A multiple regression analysis with three independent variables and 45 data points produced the following sums of squares involving the residuals:

$$\Sigma e_i^2 = .8147 \qquad \Sigma(e_i - e_{i-1})^2 = 1.2383$$

Is there sufficient evidence to infer that there is positive autocorrelation in the population of residuals? Assume $\alpha = .05$.

17.38 (Use of computer processor recommended). Road service offered by an automobile association in a city in the northeastern part of the United States is sensitive to the weather. As the temperature turns colder, the association usually experiences an increase in the number of service calls. For a 15-day period in February, the association recorded data on the following variables:

$$Y = \text{Number of service calls per day}$$
$$X_1 = \text{Average daily temperature}$$
$$X_2 = \text{Number of inches of snow per day}$$

Date		Number of Service Calls	Average Daily Temperature	Inches of Snowfall
Feb.	1	43	28	0
	2	36	30	.80
	3	55	28	1.17
	4	57	25	1.52
	5	62	20	0
	6	101	18	0
	7	99	15	0
	8	57	30	0
	9	12	35	0
	10	55	32	.61

(continues)

(continued)

Date	Number of Service Calls	Average Daily Temperature	Inches of Snowfall
11	87	31	4.08
12	24	35	0
13	42	25	0
14	73	24	2.33
15	31	27	.15

The model $Y = \beta_0 + \beta_1X_1 + \beta_2X_2 + \epsilon$ was fit to the data by the method of least squares, producing the equation $\hat{Y} = 143.30 - 3.541X_1 + 10.466X_2$. Test the assumption that the fitting errors came from a population of normally distributed residuals. Use $\alpha = .01$. (*Suggestion:* Round off $\hat{Y}_i$ to one decimal place, and m_i and l_i to two decimal places.)

17.39 Refer to Exercise 17.38. Test the assumption that the fitting errors are not serially correlated, with $\alpha = .01$.

17.4 OTHER TYPES OF INDEPENDENT VARIABLES

In the general multiple regression model $Y = \beta_0 + \beta_1X_1 + \beta_2X_2 + \ldots + \beta_kX_k + \epsilon$, we have considered problem settings where the X_i's have been separately defined quantitative variables. In this section, we will explain how additional independent variables—quantitative and qualitative—can be created and why we might wish to do so.

Second-Order Variables

The general regression model, with each X_i raised to the first power, dictates a linear relationship between X_i and Y. But it may be the case that X_i and Y have a curvilinear relation that is unaccounted for by the term β_iX_i in the model. To describe a curved relation between an independent variable and Y, we must express Y as a function of X_i^2, as well as X_i.

As an example, let us consider the simple regression problem setting of Classroom Example 15.1. The variables are Y = number of sick days claimed by an employee in the previous year and X = the number of weeks of annual paid vacation. Figure 17.13(a) shows the data and a scatter plot. In Figure 17.13(b), notice that an arc sweeps through the scatter plot to describe a curvilinear relation between X and Y. To capture such a relation, we would fit the model $Y = \beta_0 + \beta_1X + \beta_2X^2 + \epsilon$ to the data. The quadratic term X^2 is just the square of the variable X and is called a **second-order variable.**

The fitted model

$$\hat{Y} = \underset{(1.0254)}{4.6} - \underset{(.7814)}{1.9857X} + \underset{(.1278)}{.2143X^2}$$

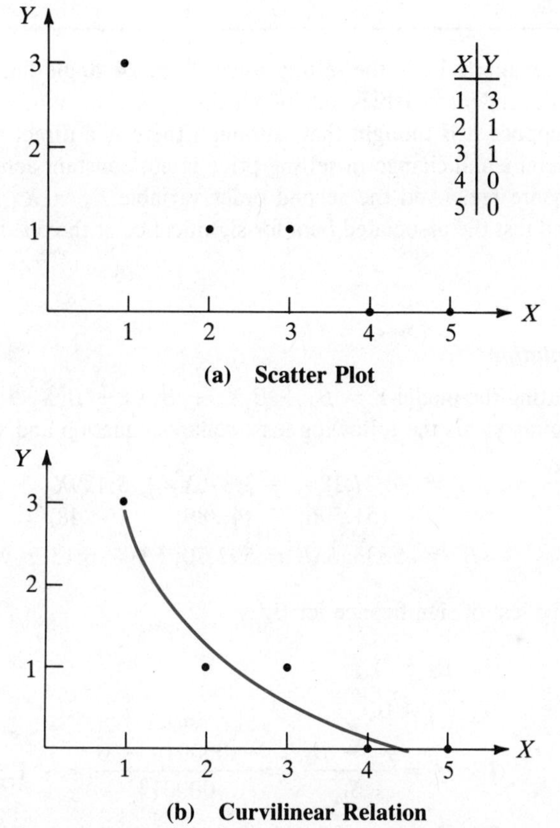

Figure 17.13 Scatter Plot and Possible Curvilinear Relation

characterizes the arc depicted in Figure 17.13(b). With an F-test p-value of .0771, we would declare the model useful for prediction at a 10% level of significance. However, when adding a second-order variable to a model, it is wise to test the significance of the associated beta to assess the wisdom of including the term. Thus, we should test the hypothesis H_0: $\beta_2 = 0$. The test statistic, $t^* = .2143/.1278 \approx 1.68$, translates to a p-value of .2350, which is too large, relative to $\alpha = .10$, for us to reject H_0. Including the second-order variable in the model is not justified. Certainly the small sample size, $n = 5$, and small number of degrees of freedom, $n - (k + 1) = 2$, contribute to our inability to establish X^2 as an important predictor of Y in this example.

Second-order variables can be included in multiple regression analyses too, as Example 17.5 demonstrates.

EXAMPLE 17.5

In Example 17.1, the selling price, Y, of 24 single-family homes was related to three independent variables, one of which was X_3, the number of square feet of living space. Suppose it is thought that, although there is a direct relation between Y and X_3, the incremental change in selling price is not constant across all size houses, in terms of square feet. Add the second order variable $X_4 = X_3^2$ to the model, refit the model, and test the associated beta for significance at the 5% level.

Solution:

Fitting the model $Y = \beta_0 + \beta_1 X_1 + \beta_2 X_2 + \beta_3 X_3 + \beta_4 X_3^2 + \epsilon$ to the $n = 24$ data points yields the following least-squares equation and summary statistics:

$$\begin{array}{l} \hat{Y} = -\,27.186 \quad - \; 3.876X_1 \; + \; 5.139X_2 \; + \; .089X_3 \; - \; .000019X_3^2 \\ \qquad\quad (51.596) \quad\;\; (4.889) \quad\;\; (6.248) \quad\;\; (.055) \quad\;\; (.000014) \end{array}$$

$$R^2 = .5633;\ SSE = 592.70;\ F^* = 6.13;\ p\text{-value} = .0015$$

The test of significance for β_4 is

H_0: $\beta_4 = 0$

H_a: $\beta_4 \neq 0$

TS: $t^* = \dfrac{b_4 - \beta_4}{S_{b_4}} = \dfrac{-.000019 - 0}{.000014} = -1.36$ (rounded)

RR: For $\alpha = .05$ and $df = n - (k + 1) = 24 - 5 = 19$, the critical t-scores are ± 2.093.

C: Since t^* is between -2.093 and $+2.093$, we accept H_0.

The inclusion of the variable $X_4 = X_3^2$ is not justified for these data; the second-order variable should be removed from the model.

□

COMMENTS

1. When a second-order variable is included for all independent variables, the resulting model is called a *second-order model*. For example, the model $Y = \beta_0 + \beta_1 X_1 + \beta_2 X_2 + \beta_3 X_1^2 + \beta_4 X_2^2 + \epsilon$ is a second-order model involving the independent variables X_1 and X_2. Extensions to third-order and higher order models are possible but not practical.
2. Often, in a multiple regression analysis with several independent variables, we can determine the need for including second-order variables by graphing X_i versus Y for those independent variables with potential curvilinear relations with Y.

CLASSROOM EXAMPLE 17.5

Testing the Significance of Second-Order Variables

Refer to Classroom Example 17.1. Upon fitting the model $Y = \beta_0 + \beta_1 X_1 + \beta_2 X_2 + \epsilon$ to the sample of $n = 12$ data points, we obtained the equation

$$\hat{Y} = \underset{(4.3119)}{-4.869} + \underset{(.1500)}{.307X_1} + \underset{(.1336)}{.166X_2}$$

$$R^2 = .9127;\ SSE = 202.47$$

Suppose the second-order model $Y = \beta_0 + \beta_1 X_1 + \beta_2 X_2 + \beta_3 X_1^2 + \beta_4 X_2^2 + \epsilon$ also was fitted to the data set, producing

$$\hat{Y} = \underset{(5.665)}{12.198} - \underset{(.306)}{.671X_1} + \underset{(.207)}{.457X_2} + \underset{(.0034)}{.0113X_1^2} - \underset{(.0031)}{.0074X_2^2}$$

$$R^2 = .9689;\ SSE = 72.885$$

a. What test(s) would you conduct to determine whether the second-order variables are significant predictors of Y?

b. With $\alpha = .05$, execute the test(s) you specified in part a.

Indicator Variables

Qualitative independent variables also may be included in a regression analysis if we adopt a system for assigning numerical codes to each level of the variable. The most popular system uses *indicator,* or *dummy, variables*.

Suppose we planned a simple regression analysis to model Y, the one-year return for mutual funds, with X_1, the proportion of the fund's assets invested in small capitalization companies. In general, if a fund's objective is to obtain large capital gains, then it will invest more heavily in small, fast-growing companies. Suppose we also are curious about the effect, if any, that a load (or sales charge) has on the fund's performance. A no-load fund has no sales charge.

The independent variable *type of mutual fund* is a qualitative variable with two values: load and no-load. To incorporate this variable into the analysis, we introduce an *indicator variable,* X_2, which takes on the numbers 0 or 1, depending on the type of mutual fund. Following is a common way of expressing its values:

$$X_2 = \begin{cases} 1 & \text{if the fund is a load fund} \\ 0 & \text{if the fund is a no-load fund} \end{cases}$$

Although we could have chosen any numbers as codes for load and no-load funds, our choice of 1 and 0, respectively, is a judicious one because it facilitates the interpretation of the associated beta parameter.

To see how the values of the indicator variable are assigned, consider the following set of data for our analysis:

Fund Number	Type of Fund	Y	X_1	X_2
1	Load	19.1%	.603	1
2	No-load	5.3	.117	0
3	No-load	8.4	.116	0
4	No-load	9.8	.105	0
5	No-load	5.1	.146	0
6	Load	21.7	.271	1
7	No-load	29.7	.296	0
8	Load	7.1	.085	1
9	Load	21.5	.569	1
10	Load	20.5	.323	1
11	No-load	12.8	.275	0
12	Load	14.8	.230	1
13	No-load	31.6	.251	0
14	Load	28.1	.311	1
15	Load	2.3	.036	1
16	No-load	17.4	.195	0
17	Load	15.7	.179	1
18	Load	19.0	.204	1
19	Load	14.5	.146	1
20	No-load	4.8	.107	0

Notice that the type of fund description appears in the second column and is translated into the numerical code of 0 or 1 in the last column, according to the definition of X_2. Using the last three columns of data, we can fit the model $Y = \beta_0 + \beta_1 X_1 + \beta_2 X_2 + \epsilon$.

The resulting equation is

$$\hat{Y} = \underset{(3.187)}{7.360} + \underset{(11.781)}{36.482X_1} - \underset{(3.397)}{.412X_2}$$

$$F^* = 5.19;\ p\text{-value} = .0099;\ \hat{\sigma} = 7.178$$

If we wished to use the model to predict the one-year return for a no-load mutual fund with 40 percent of its investments in small capitalization companies, we would substitute the values of $X_1 = .400$ and $X_2 = 0$ into the equation and solve for $\hat{Y}$.

Let us satisfy our curiosity about the relation of a load to a fund's performance. In the population of funds, the average one-year return for load funds, denoted by $E(Y|X_2 = 1)$, is found by substituting $X_2 = 1$ into the general model (not the fitted model):

$$E(Y|X_2 = 1) = \beta_0 + \beta_1 X_1 + \beta_2$$

For no-loads funds ($X_2 = 0$), the average return, denoted by $E(Y|X_2 = 0)$, is

$$E(Y|X_2 = 0) = \beta_0 + \beta_1 X_1$$

If we were to subtract these average returns, we would end up with just β_2:

$$E(Y|X_2 = 1) - E(Y|X_2 = 0) = (\beta_0 + \beta_1 X_1 + \beta_2) - (\beta_0 + \beta_1 X_1)$$
$$= \beta_2$$

Notice that the 0–1 coding system enables us to isolate and interpret β_2: β_2 is the difference in average one-year returns between load and no-load funds with the same investment proportion in small companies. Our sample of $n = 20$ funds generates an estimate of β_2 in the form of b_2. In this case, $b_2 = -.412$ means: For all mutual funds investing the same percent in small capitalization companies, we *estimate* that load funds, on average, will return about .4% less than no-load funds.

Is this a significant difference in performance between load and no-load funds? By testing β_2, we may be able to answer the question. The elements of the test are

H_0: $\beta_2 = 0$

H_a: $\beta_2 \neq 0$

TS: $t^* = \dfrac{b_2 - \beta_2}{s_{b_2}} = \dfrac{-.412 - 0}{3.397} = -.12$ (rounded)

RR: For $\alpha = .05$ and $df = n - (k + 1) = 20 - 3 = 17$, the cutoff values of t are ± 2.110.

C: Accept H_0.

Our sample evidence suggests the one-year performance of mutual funds is unaffected by the type—load or no-load—of fund.

COMMENTS

1. Be careful not to interpret β_2 as the difference in average return between *all* load and no-load funds. In obtaining β_2 by subtracting $E(Y|X_2 = 0)$ from $E(Y|X_2 = 1)$, we note that the term $\beta_1 X_1$ is eliminated *only if* X_1 remains constant. Thus, if a load fund invests 40% ($X_1 = .400$) in small capitalization companies and a no-load fund invests 50% ($X_1 = .500$), the term $\beta_1 X_1$ will not subtract to zero. For this reason, we must interpret β_2 as the difference in average returns, *given the same value of* X_1.
2. There is no such thing as a "second order indicator variable." For instance, if X_2 is a 0–1 indicator variable for the type of mutual fund, we cannot create a second order variable X_2^2, since X_2^2 would be redundant ($X_2^2 = 1$ whenever $X_2 = 1$ and vice versa).

The indicator variable system also is applicable when a qualitative variable has more than two values. For example, the independent variable "season of the year" has four values: winter, spring, summer, and fall. If we wished to relate the dependent variable Y = sales to the season of the year, we would introduce three indicator variables:

$$X_1 = \begin{cases} 1 & \text{if the season is winter} \\ 0 & \text{for any other season} \end{cases}$$

$$X_2 = \begin{cases} 1 & \text{if the season is spring} \\ 0 & \text{for any other season} \end{cases}$$

$$X_3 = \begin{cases} 1 & \text{if the season is summer} \\ 0 & \text{for any other season} \end{cases}$$

Although it may seem as if we are forgetting to include a fourth indicator variable for the fall season, in actuality fall is implicitly included in a special combination of values for X_1, X_2, and X_3. Examine the values in the following table closely:

Season	X_1	X_2	X_3
Winter	1	0	0
Spring	0	1	0
Summer	0	0	1
Fall	0	0	0

When each of the three indicator variables is assigned the value 0, by default the fall season is represented. With respect to a multiple regression model, we would incorporate the seasonal information by including three terms: $\beta_1 X_1$, $\beta_2 X_2$, and $\beta_3 X_3$.

In general, to account for p different values or levels of a qualitative independent variable, we must define $p - 1$ indicator variables and introduce $p - 1$ additional terms into the regression model. Some examples are listed in the following table:

Qualitative Variable	Values or Levels	Number of Indicator Variables Needed
Sex	Male, female	1
Type of business	Corporation, partnership, sole proprietorship	2
Region of the country	North, east, south, west	3
Month	January through December	11

EXAMPLE 17.6

The sales manager of the perfume department in a retail department store plans to monitor the time it takes her clerks to complete a sales transaction. A timing device is installed next to one of the cash registers, and each clerk is instructed to push it immediately before ringing up the sale and then again after the sale is finalized. The dependent variable is Y, the time, in minutes, to complete the transaction; and a quantitative independent variable is X_1, the amount of the transaction, to the nearest dollar. The manager is also interested in determining whether the type of sales transaction—cash, credit card, or check—affects Y.

a. How many indicator variables would we need to define in order to account for all levels of the qualitative variable *type of sales transaction*?

b. Define appropriate indicator variables for the variable *type of sales transaction*.

c. Suppose the first four transactions in the sample were credit card, check, credit card, and cash. List the values of the indicator variables for these four entities.

d. Suppose the model $Y = \beta_0 + \beta_1 X_1 + \beta_2 X_2 + \beta_3 X_3 + \epsilon$ were fit to a set of $n = 16$ sales transactions, and the fitted model was

$$\hat{Y} = 4.278 + .013X_1 - 1.985X_2 - .483X_3$$

(Assume the variables X_2 and X_3 are the ones defined in part b.) Interpret β_2 and β_3.

e. If possible, estimate the difference in average time to complete a cash transaction and a credit card transaction, given the same sale amount.

Solution:

a. Since the qualitative variable has three levels or values, we need two indicator variables.

b. Let the indicator variables be defined as follows:

$$X_2 = \begin{cases} 1 & \text{if a cash transaction} \\ 0 & \text{otherwise} \end{cases}$$

$$X_3 = \begin{cases} 1 & \text{if a credit card transaction} \\ 0 & \text{otherwise} \end{cases}$$

c. The values of X_2 and X_3 are

Type of Transaction	X_2	X_3
Credit card	0	1
Check	0	0
Credit card	0	1
Cash	1	0

d. The average time to complete a cash transaction of $\$X_1$ is

$$E(Y \mid X_2 = 1 \text{ and } X_3 = 0) = \beta_0 + \beta_1 X_1 + \beta_2$$

while the average time to complete a check transaction of $\$X_1$ is

$$E(Y \mid X_2 = 0 \text{ and } X_3 = 0) = \beta_0 + \beta_1 X_1$$

The difference in these two averages is β_2; thus, β_2 represents the difference in average transaction time between a cash and a check transaction, given that both are for the same amount. The average time to complete a credit card transaction of $\$X_1$ is

$$E(Y \mid X_2 = 0 \text{ and } X_3 = 1) = \beta_0 + \beta_1 X_1 + \beta_3$$

and the average time to complete a check transaction of $\$X_1$ is

$$E(Y \mid X_2 = 0 \text{ and } X_3 = 0) = \beta_0 + \beta_1 X_1$$

Thus, β_3 is the difference in average transaction time between a credit card and a check purchase, given that both are for the same amount.

The average time to complete a cash transaction of $\$X_1$ is

$$E(Y \mid X_2 = 1 \text{ and } X_3 = 0) = \beta_0 + \beta_1 X_1 + \beta_2$$

and the average time to complete a credit card transaction of $\$X_1$ is

$$E(Y \mid X_2 = 0 \text{ and } X_3 = 1) = \beta_0 + \beta_1 X_1 + \beta_3$$

The difference $\beta_2 - \beta_3$ represents the difference in average transaction time between a cash and a credit card sale for the same amount. From the fitted model, we can estimate this difference to be $b_2 - b_3 = -1.985 - (-.483) = -1.502$. Roughly, the cash transaction is an average of 1½ minutes quicker than the credit card transaction, given the same dollar amount of a perfume sale.

□

Interaction Variables

Recall from our discussion in Chapter 14 that *interaction* between two independent variables X_1 and X_2 occurs when the effect of X_1 on Y is not constant for all values or levels of X_2. To demonstrate, examine Figure 17.14, where we have plotted Y = the amount of an order versus X_1 = the number of menu items selected per order for a fast food restaurant. The top regression line represents orders placed during lunch (11 A.M. to 2 P.M.) while the bottom line represents orders placed during breakfast (7 A.M. to 10 A.M.). Both lines show a direct relation between X_1 and Y, though the rate of change—that is, the slope of the line—is not constant for both lunch and breakfast orders. Hence, we say that X_1 interacts with a second variable, X_2, which is defined as:

$$X_2 = \begin{cases} 1 & \text{if the order is placed during lunch} \\ 0 & \text{if the order is placed during breakfast} \end{cases}$$

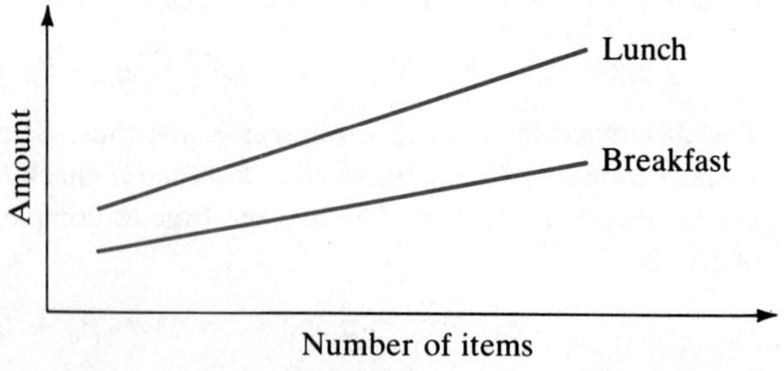

Figure 17.14 Interaction Between the Time of Day and Number of Items

To incorporate an interaction effect between two variables in a regression analysis, we introduce a *cross product* or *interaction variable* in the model. For the fast food restaurant, the interaction variable is X_1X_2 and the model would be $Y = \beta_0 + \beta_1X_1 + \beta_2X_2 + \beta_3X_1X_2 + \epsilon$. The interaction term $\beta_3X_1X_2$ contributes to the different slopes of the lines shown in Figure 17.14.

When an order is placed during lunch, $(X_2 = 1)$, the average amount will be

$$\begin{aligned} E(Y \mid X_2 = 1) &= \beta_0 + \beta_1X_1 + \beta_2 + \beta_3X_1 \\ &= (\beta_0 + \beta_2) + (\beta_1 + \beta_3)X_1 \end{aligned}$$

The average amount of a breakfast order $(X_2 = 0)$ is

$$E(Y \mid X_2 = 0) = \beta_0 + \beta_1X_1$$

Notice that the slope of the lunch line is $\beta_1 + \beta_3$, while the slope of the breakfast line is just β_1.

Since the variable X_1 is a quantitative variable, we interpret β_1 as a slope: β_1 is the change in average amount of a breakfast order for each additional menu item included in the order. If $\beta_3 = 0$, then the interaction term drops out of the model, and the slopes of the two lines in Figure 17.14 are the same. In this case, the interpretation of β_1 also extends to lunch time orders.

But when $\beta_3 \neq 0$, we cannot interpret β_3 separately. Instead, the quantity $\beta_1 + \beta_3$ is interpreted as the change in the average amount of a lunch order for each additional menu item included in the order.

Interaction variables are useful additions to a multiple regression model, especially when we suspect that the effects of X_1 on Y and X_2 on Y are not separate, or additive, effects. Without an interaction term in the model, we could not measure the different slopes of the lines in Figure 17.14, for instance, to assess the combined effects of X_1 and X_2 on Y. On the other hand, the interaction between two variables may be slight and have a negligible impact on Y. Therefore, we recommend justifying the inclusion of an interaction term (or terms) by testing the significance of the associated beta (or betas) using the testing procedure outlined on page 920 (or on page 933).

EXAMPLE 17.7

A bakery is experimenting with two types of cake mix and three baking temperatures. Specifically, cake texture (the measurement of which is not important here) is being investigated. From two batter mixes, 30 individual cakes are baked and rated. The dependent variable is Y, the texture rating of each cake. A quantitative independent variable is X_1, baking temperature in degrees Fahrenheit, and a qualitative independent variable is the type of cake mix—type A or type B.

a. Code the levels of the qualitative variable with an indicator variable X_2.

b. Write a model relating Y to X_1 and to X_2 with an interaction term.

c. Interpret β_1 and $\beta_1 + \beta_3$.

d. What does the quantity $\beta_2 + \beta_3 X_1$ represent?

e. Suppose the fitted model is

$$\hat{Y} = \underset{(5.354)}{-11.767} + \underset{(.0164)}{.084X_1} + \underset{(7.572)}{54.867X_2} - \underset{(.0233)}{.168X_1X_2}$$

Is the interaction term justified? Let $\alpha = .001$.

Solution:

a. The indicator variable X_2 is

$$X_2 = \begin{cases} 1 & \text{if type } A \text{ cake mix} \\ 0 & \text{if type } B \text{ cake mix} \end{cases}$$

b. The model is $Y = \beta_0 + \beta_1 X_1 + \beta_2 X_2 + \beta_3 X_1 X_2 + \epsilon$

c. For type A cakes ($X_2 = 1$), the average rating is

$$E(Y \mid X_2 = 1) = (\beta_0 + \beta_2) + (\beta_1 + \beta_3)X_1$$

For type B cakes ($X_2 = 0$), the average rating is

$$E(Y \mid X_2 = 0) = \beta_0 + \beta_1 X_1$$

Thus, β_1 is the change in the average rating of type B cakes for each increase of 1°F in the baking temperature. Similarly, $\beta_1 + \beta_3$ is the change in the average rating of type A cakes for each increase of 1°F in the baking temperature.

d. If we were to subtract $E(Y|X_2 = 0)$ from $E(Y|X_2 = 1)$ we would get $\beta_2 + \beta_3 X_1$. The quantity $\beta_2 + \beta_3 X_1$, therefore, represents the difference in average ratings between type A and type B cakes baked at a temperature of X_1°F.

e. We wish to test the significance of β_3:

H_0: $\beta_3 = 0$

H_a: $\beta_3 \neq 0$

TS: $t^* = \dfrac{b_3 - \beta_3}{S_{b_3}} = \dfrac{-.168 - 0}{.0233} = -7.21$ (rounded)

RR: For $\alpha = .001$ and $df = n - (k + 1) = 30 - 4 = 26$, the cutoff t-scores are ± 3.707.

C: Reject H_0.

The interaction term should remain in the model.

□

COMMENTS

1. Interaction is possible between almost all pairs of quantitative and/or qualitative (indicator) variables. Consider the following table of pairs of variables X_1 and X_2 of differing types:

X_1	X_2	Interaction Possible?
Quantitative	Quantitative	Yes
Quantitative	Indicator	Yes
Indicator	Indicator	Yes and no

If the two indicator variables represent levels of two *different* qualitative variables, interaction is possible. If the two indicator variables represent two levels of the *same* qualitative variable, interaction is not possible. Exercise 17.58 illustrates this point.

2. Adding all possible interaction variables and second-order variables to a model yields a *complete second-order model*. For instance, $Y = \beta_0 + \beta_1 X_1 + \beta_2 X_2 + \beta_3 X_1^2 + \beta_4 X_2^2 + \beta_5 X_1 X_2 + \epsilon$ is a complete second-order model for the quantitative variables X_1 and X_2.

CLASSROOM EXAMPLE 17.6

Analyzing Indicator and Interaction Variables in a Multiple Regression Model

At one of its plants, a manufacturer schedules a first and second shift to make liquid laundry detergent. The plant can operate up to five production lines per shift. However, the plant manager does not need to run at 100% capacity to meet demand. In an experiment to assess productivity, both shifts will operate with 2, 3, and then 4 lines of production for three-week periods. The dependent variable is Y, the average number of units of detergent produced per line per week. By the end of the study, the following values of Y were recorded:

	Number of Production Lines		
Shift	2	3	4
First	231	241	234
	233	242	230
	245	221	216
Second	255	230	216
	256	215	207
	236	223	212

a. There are two independent variables in this study: X_1, number of production lines, and X_2, shift. Classify each as quantitative or qualitative.

b. If necessary, set up an indicator variable for each qualitative variable you identified in part a.

c. Suppose the model $Y = \beta_0 + \beta_1 X_1 + \beta_2 X_2 + \beta_3 X_1 X_2 + \epsilon$ is proposed, where X_2 is an indicator variable. Interpret these quantities: β_1, $\beta_1 + \beta_3$, and $\beta_2 + \beta_3 X_1$.

d. Test the significance ($\alpha = .05$) of the interaction term if the fitted model is

$$\hat{Y} = \underset{(11.348)}{247.06} - \underset{(3.650)}{4.833X_1} + \underset{(16.048)}{36.722X_2} - \underset{(5.162)}{13.833X_1X_2}$$

SECTION 17.4 EXERCISES

17.40 Owning an automobile may be a poor investment due to the erosion of the vehicle's value over time. To investigate the resale value of cars, trucks, and utility vehicles, a study was conducted using a book of used car values to determine a vehicle's percentage of its original value retained after 1, 3, 5, and 7 years. Data for two classes of vehicles—luxury cars and utility vehicles—are listed in the following table. Let Y = percent of original value and X_1 = age, in years.

Luxury Cars: Percent of Original Value	Luxury Cars: Age (Years)	Utility Vehicles: Percent of Original Value	Utility Vehicles: Age (Years)
78%	1	92%	1
82	1	93	1
81	1	95	1
84	1	93	1
78	1	95	1
62	3	84	3
60	3	79	3
60	3	78	3
69	3	85	3
67	3	88	3
59	5	81	5
57	5	74	5
48	5	76	5
62	5	69	5
49	5	81	5
52	7	75	7
38	7	71	7
41	7	70	7
36	7	73	7
55	7	67	7

a. Plot the values of X_1 and Y for the luxury cars. Does a second order variable seem appropriate to describe the relation between Y and X_1?
b. Using the Regression processor, fit the model $Y = \beta_0 + \beta_1X_1 + \beta_2X_1^2 + \epsilon$ to the data for the luxury cars.
c. Test the significance of β_2, with $\alpha = .05$.

17.41 Refer to Exercise 17.40. Repeat parts a, b, and c using the data for the utility vehicles.

17.42 Refer to Exercise 17.40.

a. Create an indicator variable, X_2, for the qualitative variable *type of vehicle*.

b. Fit the model $Y = \beta_0 + \beta_1 X_1 + \beta_2 X_2 + \epsilon$ to the $n = 40$ data points.

c. Interpret β_2 and b_2. Is β_2 significantly different from 0 at the 10% level?

17.43 Refer to Exercises 17.40 and 17.42.

a. Fit the model $Y = \beta_0 + \beta_1 X_1 + \beta_2 X_2 + \beta_3 X_1 X_2 + \epsilon$ to the $n = 40$ data points, using the variable X_2 defined in part a of Exercise 17.42.

b. Are we justified in including the interaction term in the model? Let $\alpha = .01$.

c. Interpret $\beta_1 + \beta_3$, and $\beta_2 + \beta_3 X_1$.

17.44 Consider the following data set of size $n = 15$.

X	1	1	1	2	2	2	3	3	3	4	4	4	5	5	5
Y	4	5	2	5	4	6	8	7	7	12	11	13	18	16	18

a. Plot Y versus X. Should a second-order variable be included in a regression model relating Y to X?

b. Suppose a simple regression model and a second-order model were fit to the data, producing the following equations:

$$\text{Simple:}\quad \hat{Y} = \underset{(.9978)}{-1.233} + \underset{(.3008)}{3.433X}$$

$$R^2 = .9023;\ SSE = 35.30$$

$$\text{Second order:}\quad \hat{Y} = \underset{(1.2597)}{3.933} - \underset{(.9600)}{.995X} + \underset{(.1570)}{.738X^2}$$

$$R^2 = .9681;\ SSE = 12.419$$

By testing the significance of the second-order term, decide which model you would prefer. Use $\alpha = .01$.

17.45 The model $Y = \beta_0 + \beta_1 X_1 + \beta_2 X_2 + \beta_3 X_3 + \epsilon$ was fit to a set of $n = 15$ entities to yield the least-squares equation:

$$\hat{Y} = \underset{(30.438)}{8.220} - \underset{(4.0723)}{1.385X_1} + \underset{(2.2797)}{8.177X_2} + \underset{(.0046)}{.020X_3}$$

$$R^2 = .8738;\ F^* = 25.38;\ SSE = 2203.4$$

A second order model was proposed as well and generated the following equation:

$$\hat{Y} = \underset{(109.36)}{413.99} - \underset{(28.223)}{103.12X_1} - \underset{(3.976)}{17.051X_2} + \underset{(.0114)}{.049X_3} + \underset{(1.872)}{6.605X_1^2} + \underset{(.446)}{2.751X_2^2} - \underset{(.00000351)}{.00000995X_3^2}$$

$$R^2 = .9825;\ F^* = 74.95;\ SSE = 305.08$$

a. What test(s) would you conduct to determine whether the second-order variables are significant predictors of Y?

b. With $\alpha = .10$, execute the test(s) you specified in part a.

17.46 Write a second-order model for a multiple regression analysis having the following number of quantitative independent variables.

a. 1 **b.** 2 **c.** 3

17.47 Refer to Exercise 17.46. Write a *complete* second-order model for each part by including the appropriate interaction term(s) in the model.

17.48 Refer to Exercises 17.46 and 17.47. What is the minimum sample size needed to fit the complete second-order model in each part and still have at least five degrees of freedom available to estimate σ^2?

17.49 How many indicator variables are needed to represent the following qualitative variables in a multiple regression analysis? The levels of each variable are in parentheses.

a. investment objective of a mutual fund (maximum capital gains, growth, income, other).

b. category of itemized deduction on Schedule A of the federal income tax forms (medical, taxes, interest paid, casualty and theft losses, gifts to charity, and miscellaneous).

c. country or continent where a sports car is assembled (United States, Europe, Japan).

17.50 Refer to Exercise 17.49, part b. Suppose the model $Y = \beta_0 + \beta_1 X_1 + \beta_2 X_2 + \beta_3 X_3 + \beta_4 X_4 + \beta_5 X_5 + \beta_6 X_6 + \epsilon$ is proposed to relate Y = amount of largest deduction from Schedule A to X_1 = gross income and the qualitative independent variable "category of deduction" for a sample of itemized tax returns. The variables X_2 through X_6 in the model are indicator variables defined as

$$X_2 = \begin{cases} 1 & \text{if medical} \\ 0 & \text{otherwise} \end{cases}$$

$$\vdots$$

$$X_6 = \begin{cases} 1 & \text{if gifts to charity} \\ 0 & \text{otherwise} \end{cases}$$

Interpret these betas:

a. β_2 **b.** β_4

c. Which beta or combination of betas represent the difference in the average amount of the largest deduction between taxes and gifts to charity for taxpayers with the same gross incomes?

17.51 Refer to Exercise 17.49, part c. Suppose the model $Y = \beta_0 + \beta_1 X_1 + \beta_2 X_2 + \beta_3 X_3 + \beta_4 X_4 + \epsilon$ is proposed to relate Y = an index of customer satisfaction to X_3 = the price of the sports car, in thousands of dollars (\$25,000 means $X_3 = 25$), X_4 = engine size (in liters), and the qualitative independent variable *country or continent where the car is assembled*. Variables X_1 and X_2 are the following indicator variables:

$$X_1 = \begin{cases} 1 & \text{if the car is assembled in the United States} \\ 0 & \text{otherwise} \end{cases}$$

$$X_2 = \begin{cases} 1 & \text{if the car is assembled in Europe} \\ 0 & \text{otherwise} \end{cases}$$

Interpret these betas:

a. β_1 **b.** β_2 **c.** β_3

d. Which beta or combination of betas represent the difference in the average index between sports cars assembled in the United Sates and in Europe with the same engine size and price?

17.52 Consider the model in Exercise 17.51: $Y = \beta_0 + \beta_1 X_1 + \beta_2 X_2 + \beta_3 X_3 + \beta_4 X_4 + \epsilon$. Add the interaction terms $\beta_5 X_1 X_3$ and $\beta_6 X_2 X_3$ to the model.

a. Interpret $\beta_3 + \beta_5$.

b. What combination of betas represents the average index for Japanese sports cars with an engine size of 3.0 liters and a price of $20,000?

c. Answer the question posed in part d of Exercise 17.51.

17.53 Refer to Classroom Example 17.6. Suppose we wish to test whether there is a difference in the average productivity between the first shift and the second shift. What null hypothesis, if any, could you set up to conduct such a test under the following conditions:

a. the test for H_0: $\beta_3 = 0$ yields an accept H_0 conclusion.

b. the test for H_0: $\beta_3 = 0$ yields a reject H_0 conclusion.

17.54 An offbeat stock market theory holds that when a team from the original National Football League (NFL), which presently consists of all teams in the National Football Conference plus three teams in the American Football Conference, wins the Super Bowl, the stock market usually posts a gain for the year. When a team from the original American Football League (AFL), which consists of current American Football Conference teams, except Cleveland, Indianapolis (formerly Baltimore), and Pittsburgh, wins, the market tends to close down for the year. To investigate this theory, consider the following Super Bowl data from 1967 to 1987:

Year	Super Bowl Winner	Team Roots	Score	Change in NYSE Composite Index (%)
1967	Green Bay Packers	NFL	35–10	+23.1
1968	Green Bay Packers	NFL	33–14	+9.4
1969	New York Jets	AFL	16–7	−12.5
1970	Kansas City Chiefs	AFL	23–7	−2.5
1971	Baltimore Colts	NFL	16–13	+12.3
1972	Dallas Cowboys	NFL	24–3	+14.3
1973	Miami Dolphins	AFL	14–7	−19.6
1974	Miami Dolphins	AFL	24–7	−30.3
1975	Pittsburgh Steelers	NFL	16–6	+31.9
1976	Pittsburgh Steelers	NFL	21–17	+21.5
1977	Oakland Raiders	AFL	32–14	−9.3
1978	Dallas Cowboys	NFL	27–10	+2.1
1979	Pittsburgh Steelers	NFL	35–31	+15.5
1980	Pittsburgh Steelers	NFL	31–19	+25.7
1981	Oakland Raiders	AFL	27–10	−8.7
1982	San Francisco 49ers	NFL	26–21	+14.0
1983	Washington Redskins	NFL	27–17	+17.5
1984	Los Angeles Raiders	AFL	38–9	+1.3
1985	San Francisco 49ers	NFL	38–16	+26.2
1986	Chicago Bears	NFL	46–10	+14.0
1987	New York Giants	NFL	39–20	−0.3

Source: "Stocks and the big game," January 29, 1988. Copyright 1988, *USA Today*. Adapted with permission.

Let Y = the percentage change in the New York Stock Exchange (NYSE) composite index.

a. Define an indicator variable X_1 to categorize the conference or origin of each Super Bowl winner. Treat the NFC and NFL designations as the same category.

b. Fit the model $Y = \beta_0 + \beta_1 X_1 + \beta_2 X_2 + \epsilon$ to the data, where X_2 = the margin of victory.

c. Which term in the model in part b embodies the Super Bowl Theory? What null hypothesis would you set up to test the Super Bowl Theory?

d. Execute the test you prescribed in part c and comment on the Super Bowl Theory. Let $\alpha = .01$.

17.55 Consider the model in Exercise 17.54: $Y = \beta_0 + \beta_1 X_1 + \beta_2 X_2 + \epsilon$. Add an interaction variable to the model. Fit the model and test the significance of the interaction beta, with $\alpha = .10$.

17.56 Consider the model in Exercise 17.55: $Y = \beta_0 + \beta_1 X_1 + \beta_2 X_2 + \beta_3 X_1 X_2 + \epsilon$. Interpret the following:

a. $\beta_2 + \beta_3$

b. β_2

c. $\beta_1 + \beta_3 X_2$

17.57 A multinational corporation does business in three foreign countries: West Germany, Australia, and Japan. To model Y = quarterly sales (in millions of American dollars) in these countries, three variables are proposed:

1. Country in which the sales occur
2. Fiscal quarter of the year
3. Average value of the foreign currency versus the American dollar during the fiscal quarter.

Let X_1 and X_2 be indicator variables representing countries ($X_1 = 1$ if West Germany and $X_2 = 1$ if Australia). Let X_3, X_4, and X_5 be indicator variables representing the fiscal quarters ($X_3 = 1$ if the first quarter, $X_4 = 1$ if the second quarter, and $X_5 = 1$ if the third quarter). Assume X_6 is a quantitative independent variable representing the average value of the foreign currency per American dollar. Code the following sales information by assigning values to the dependent and all independent variables:

a. First quarter sales of \$10 million in Japan, when an American dollar averaged 127.06 yen.

b. Third quarter sales of \$1.3 million in West Germany, when an American dollar averaged 1.693 marks.

c. Fourth quarter sales of \$870,000 in Australia, when an American dollar averaged 1.348 Australian dollars.

d. Second quarter sales of \$8.3 million in Japan, when an American dollar averaged 131.30 yen.

e. Second quarter sales of \$2.7 million in West Germany, when an American dollar averaged 1.812 marks.

***17.58** Refer to Exercise 17.57. In general, if the values of an interaction variable $X_i X_j$ are identically equal to 0 when we multiply X_i times X_j, then the term $\beta_k X_i X_j$ cannot be included in the regression model, and we say that the interaction between the variables X_i and X_j is not possible. Note that this phenomenon will occur when two levels—that is, indicator variables—of the same qualitative variable are combined in an interaction term. Relative to the

*Optional

variables proposed in Exercise 17.57, indicate yes/no whether the following interactions are possible.

a. X_1X_6 **b.** X_1X_2
c. X_1X_3 **d.** X_2X_3
e. X_2X_6 **f.** X_3X_4
g. X_3X_6 **h.** X_4X_5

17.5 VARIABLE SELECTION PROCEDURES

As stated at the beginning of this chapter, one of the finer points of a regression analysis is to develop a concise set of predictor variables for Y. However, a complete model with second order variables, indicator variables and/or interaction variables may become intractable as the number of independent variables increases. To screen out nonsignificant predictors and keep the model manageable, we can use the t-test for an individual beta or the F-test for several betas. Alternatively, we could resort to a computer selection procedure that sequentially adds or removes one variable from the model depending on the variable's relative relation to Y.

There are many variable selection procedures available. The principle and the criterion for selecting variables are the key factors differentiating the procedures. Principle refers to whether we start with no variables in the model and progressively add one at each step, or start with all the variables in the model and successively remove one. Criterion is the basis by which a variable is included or excluded, as well as the specification at which point the procedure is halted. We will examine three popular selection procedures.

Forward Selection

As the name implies, forward selection is an add-in rather than a remove-from procedure. We start by fitting a simple regression model $Y = \beta_0 + \beta_1 X_i + \epsilon$ for each independent variable X_i. If there are 10 independent variables, for instance, then 10 models are fit. The t-test for β_1 is conducted for each model, and the test with the smallest p-value is sought. The associated independent variable is selected as the first variable in the model. For illustrative purposes, suppose it is X_5 from the pool of 10 independent variables.

Next, nine two-variable models of the form $Y = \beta_0 + \beta_1 X_5 + \beta_2 X_i + \epsilon$ are fit to the data. Again, t-tests are executed, but for β_2 only. The variable corresponding to the test with the smallest p-value is selected as the second variable in the model. Suppose it is variable X_9.

In selecting X_5 and X_9 in the first two steps by virtue of having the smallest p-values, we also would have required that the p-value for each be smaller than a predetermined α-risk. Though we often select $\alpha = .05$, we tend to be more generous in sequential procedures and allow α to be as large as .25, perhaps.

At the third step, 8 three-variable models, $Y = \beta_0 + \beta_1 X_5 + \beta_2 X_9 + \beta_3 X_i + \epsilon$, would be fit to the data. Following our p-value rule for inclusion in the model, we would look for a third variable to add to the model. The forward selection procedure would stop when none of the t-tests produce a p-value smaller than our predetermined α-risk. For instance, if at the third step no small p-values were found, the final model would involve only X_5 and X_9.

The main disadvantage to the forward selection procedure is that once a variable is selected and enters the model, it can never be removed. For instance, at the second stage we decided that, in the presence of X_5, X_9 was the most significant additional regressor. But it may be the case that X_5 and X_9 are not the most explanatory *pair* of variables; maybe X_9 and X_6 are best, even though X_5 was the first to enter the model. The forward selection procedure would be unable to consider the model $Y = \beta_0 + \beta_1 X_9 + \beta_2 X_6 + \epsilon$ at the second stage, once X_5 enters the model at the first stage.

Stepwise

A stepwise procedure, like the forward selection procedure, adds a variable to the model at each stage and also allows us to remove a previously entered variable. Thus, the scenario in which X_9 and X_6 are a better pair of predictors than X_5 and X_9 is possible under a stepwise procedure.

To explain the stepwise principle, let us again assume we have a pool of 10 independent variables. The first step is identical to that for forward selection: The most significant predictor is selected among the 10 simple regression models. Suppose it is X_5. In the second step, again we would fit nine two-variable models of the form $Y = \beta_0 + \beta_1 X_5 + \beta_2 X_i + \epsilon$. As before, we would discover that, given X_5 is already in the model, X_9 is the most significant predictor.

However, at this point, a stepwise procedure also conducts a t-test for each β_i to ensure that it should stay in the model. A second significance level is required for this test. Whereas we were lenient in allowing a variable to enter the model by choosing α as large as .25, we set a stricter threshold for removing a variable, with an α-risk of around 10%. It is therefore possible at the second stage to add a variable such as X_9, remove an already entered variable, perhaps X_5, and then add a different variable, perhaps X_6. In other words, at the end of the first step, we selected X_5 as the best single predictor; but by the end of the second step, we may prefer the pair X_6 and X_9.

A stepwise procedure improves on the principle of the forward selection procedure by allowing a variable to leave the model. If a variable is removed, it goes back into the pool of available variables and may reenter the model at a later stage. Stepwise regression requires two levels of significance, one for entering variables and one for removing variables. Other than the potential for fitting many more models than a forward selection procedure, a stepwise procedure has few drawbacks.

Backward Elimination

The backward elimination procedure starts with all variables in the model and sequentially removes the least significant variable remaining at each stage. When all remaining variables are significant, the process stops. A moderately large α-risk again is chosen to allow marginally important predictors to stay in the model while causing the removal of the truly nonsignificant ones. Often, α is in the neighborhood of .20 for this purpose.

At each step, the backward elimination procedure fits only one model, starting with the most complex one. Thus, the task of searching through a collection of three-variable, four-variable, and so on models is eliminated. For instance, with 10 independent variables, the model $Y = \beta_0 + \beta_1 X_1 + \cdots + \beta_{10} X_{10} + \epsilon$ is fit and 10 tests are conducted, one each for β_1 through β_{10}. We seek the largest *p*-value greater than .20, say. Suppose X_{10} had the largest *p*-value at .6317. Then X_{10} would be removed; and the model $Y = \beta_0 + \beta_1 X_1 + \cdots + \beta_9 X_9 + \epsilon$ would be fit, nine *t*-tests would be conducted, and we would drop the independent variable associated with the largest *p*-value exceeding .20. If none exceeded .20, the process stops and, in this case, the nine-variable model would be the final result.

Epilogue

Variable selection procedures like the ones discussed in this section are standard features on large, commercial software packages such as SAS, SPSS, and Minitab. Consult the user manuals to determine the criteria and default α-risks under which the procedures operate. It is possible to use the accompanying Regression processor to perform a forward selection or backward elimination routine by rerunning the program at each stage. However, to simulate a stepwise procedure may become tedious with a moderate pool of independent variables.

Finally, two caveats are worth mentioning. First, if we employed all three selection procedures on the same set of data and the same pool of independent variables, it is *likely* that we would end up with different sets of variables, since the selection criteria differ. Second, in light of the first comment, there is not a unique, *best* set of predictors. The best set according to one principle and criterion may not match the best set from another. Further, the "best set" is only as good as the original pool of potential variables. If an important variable is not included in the pool to start with, none of the selection procedures will subsequently uncover it for us.

17.6 SUMMARY

As we end our three-chapter investigation of regression, let us review the objectives that we have sought to achieve in a regression analysis:

1. Identify the dependent variable and potential independent variables.
2. Propose a statistical model to relate the variables.
3. Fit the model.
4. Test the model.
5. Analyze the residuals.
6. Use the model for prediction, if applicable.

Our work in this chapter has touched on the first, fourth, and fifth objectives.

In identifying potential variables, we introduced indicator variables as a means of coding the levels of a qualitative variable so that it may be included in a regression analysis. We also mentioned two ways of creating new variables from existing ones as the situation dictates: A second-order variable if Y and X_i have a curvilinear relation, and an interaction variable if the effect of X_i on Y is not the same for all levels or values of a second independent variable X_j.

In testing the model, we considered a test on an individual beta and a test of several betas. In both types of tests, the null hypothesis contained the assumption that the independent variables were either singly or jointly nonsignificant predictors of Y; thus, the associated beta was set equal to 0.

Finally, we addressed the assumptions about ϵ, the foundation upon which all the tests and interval estimates are based. We analyzed the residuals using the Lin-Mudholkar test to confirm the assumption that a normal population generated the observed fitting errors. We presented the Durbin-Watson test to check for serial correlation in the sequence of residuals, when the data are time-dependent.

As the dust settles, we hope you realize that model building in regression is more than just fitting a model to a set of data. We also seek to identify a subset of significant predictors from the pool of potential independent variables while simultaneously abiding by the assumptions surrounding the error term.

CHAPTER 17 EXERCISES

17.59 What is multicollinearity?

17.60 Based on a sample of size $n = 12$, a dependent variable Y is related to three independent variables by the following multiple regression equation:

$$\hat{Y} = \underset{(7.0141)}{14.486} + \underset{(.4747)}{.097X_1} - \underset{(.1861)}{.188X_2} + \underset{(2.5565)}{1.964X_3}$$

Test whether X_2 is related to Y, using $\alpha = .10$.

17.61 Refer to Exercise 17.60. Construct an 80% confidence interval for β_3. Based on this confidence interval, decide whether β_3 is significantly different from 0.

***17.62** (Background in matrix algebra required.) A multiple regression analysis involving $n = 13$ entities and two independent variables produced the following $(\mathbf{X'X})^{-1}$ matrix:

$$(\mathbf{X'X})^{-1} = \frac{1}{5181163}\begin{bmatrix} 6382775 & -56723 & -746842 \\ -56723 & 802 & 4805 \\ -746842 & 4805 & 112772 \end{bmatrix}$$

If $\hat{\sigma} = 20.471932$ and $\hat{Y} = 25.428 - .285X_1 + 1.948X_2$, do the following.

a. Test H_0: $\beta_1 = 0$ against a lower-tailed alternative, with $\alpha = .05$.

b. Construct a 98% confidence interval for β_2.

*Optional

17.63 Refer to Exercise 17.60. Suppose the correlation matrix is

$$\mathbf{C} = \begin{bmatrix} 1.0000 & .2543 & -.1313 & .2663 \\ .2543 & 1.0000 & .3629 & .7366 \\ -.1313 & .3629 & 1.0000 & .5415 \\ .2663 & .7366 & .5415 & 1.0000 \end{bmatrix}$$

Use Table 4.4 to flag pairs of independent variables that evidence strong positive or negative correlation. Which variable(s), if any, would you recommend deleting from the model because of the redundancy?

17.64 In Exercise 16.46, we modeled the dependent variable Y = the total bond return over a one-year period as a function of three independent variables:

X_1 = Bond term, in years
X_2 = Current yield of the bond
X_3 = Interest rate movement, in points (that is, percents)

Twenty-four data points produced the following fitted model:

$$\hat{Y} = \underset{(4.4010)}{-.729} + \underset{(.1059)}{.103X_1} + \underset{(.3543)}{1.006X_2} - \underset{(.5726)}{7.038X_3}$$

Construct a 98% confidence interval for β_2. Based on this interval, decide whether X_2 is a significant predictor of Y, in the presence of X_1 and X_3.

17.65 Refer to Exercise 17.64. Test whether interest rate movement is inversely related to the total bond return. Let $\alpha = .0005$. What is the p-value?

17.66 Refer to Exercise 17.64. Interpret $b_1 = .103$ in the fitted model.

***17.67** (Background in matrix algebra required.) A multiple regression analysis based on $n = 20$ entities generated the following $(\mathbf{X}'\mathbf{X})^{-1}$ matrix:

$$(\mathbf{X}'\mathbf{X})^{-1} = \begin{bmatrix} .67164 & -.16301 & -.01375 \\ -.16301 & .07448 & .00065 \\ -.01375 & .00065 & .00058 \end{bmatrix}$$

The fitted model was $\hat{Y} = 10.05 + 12.83X_1 - .06X_2$, with $\hat{\sigma} = 2.5742$.

a. Is β_1 significantly different from 0? Use $\alpha = .05$.
b. Find S_{b_2}.

17.68 Refer to Exercise 17.64. Suppose the correlation matrix is

$$\mathbf{C} = \begin{bmatrix} 1.0000 & .0333 & .2020 & -.9178 \\ .0333 & 1.0000 & -.1871 & .0000 \\ .2020 & -.1871 & 1.0000 & .0000 \\ -.9178 & .0000 & .0000 & 1.0000 \end{bmatrix}$$

Describe the separate relations of the independent variables to Y. Are there redundancies among the independent variables? Explain.

17.69 Determine the rejection region for the following tests:
a. H_0: $\beta_1 = \beta_3 = 0$; $n = 12$; $k = 3$; and $\alpha = .05$.
b. H_0: $\beta_1 = \beta_2 = \beta_4 = 0$; $n = 15$; $k = 5$; and $\alpha = .01$.

17.70 Refer to Exercise 17.69. Use the following values of SSE for the complete and reduced models to execute the tests in Exercise 17.69.

a. $SSE_c = 260.57$ and $SSE_r = 667.51$.
b. $SSE_c = 4.8957$ and $SSE_r = 6.1324$.

17.71 Investing in stocks and mutual funds for the long run used to be an optimum strategy for investors due to the investor's ally: the capital gains tax. Under pre-1986 tax laws that allowed capital gains exclusion, an investor's gain from the sale of a stock or fund was split into two parts, a taxable portion and a nontaxable portion. For example, in 1985 the split was 40% taxable and 60% nontaxable. The 1986 tax law changed the split into 100% taxable and 0% nontaxable. A study was conducted to determine, under the new tax law, a minimum number of years that an investor needed to hold the stock or fund before selling in order to overcome the higher tax rate. The variables included in the study were as follows:

Y = Breakeven number of years to hold the stock or fund
X_1 = Current income level
X_2 = Percent appreciation to date
X_3 = Percent of annual appreciation expected in the future
X_4 = Transaction costs

Based on a sample of 96 entities, the complete model yielded $SSE_c = 120.77$. It was thought that, in the presence of X_1 and X_3, the variables X_2 and X_4 offer limited additional explanation about Y. With those variables dropped from the model, the sum of squared fitting errors increased to 241.40. Using $\alpha = .01$, decide whether the percent appreciation to date and transaction costs should be removed from the model.

17.72 (Use of computer processor recommended.) An electric company is trying to determine the electrical energy needs for the residential customers in its service territory. By identifying certain customer and environmental characteristics, the utility hopes to build a multiple regression model for Y, monthly kilowatt hour usage. The independent variables under consideration are as follows:

X_1 = Square feet of living space
X_2 = Average monthly temperature (°F)
X_3 = Age of residence
X_4 = Proportion of available sunshine hours

Data for the variables X_2 and X_4 were obtained from the National Weather Service, with the values of X_4 determined by dividing the number of hours of actual sunshine by the total number of daylight hours. The utility used two years' worth of climatic data and monitored one residence selected at random for each month of the study. Following are the sample data:

Year	Month	Y	X_1	X_2	X_3	X_4
1	January	1805	950	35	31	.24
	February	1029	1600	41	27	.29
	March	1020	2100	48	30	.31
	April	515	1055	52	29	.39
	May	942	1250	60	18	.40
	June	764	925	71	12	.61
	July	3142	3150	86	8	.80
	August	1764	1075	86	4	.58
	September	2021	2775	83	15	.49
	October	879	1765	62	31	.25

(continues)

Year	Month	Y	X_1	X_2	X_3	X_4
	November	966	1900	51	17	.30
	December	818	1215	46	16	.28
2	January	1933	2050	35	11	.29
	February	1418	1630	33	12	.31
	March	2910	4100	45	2	.33
	April	1379	1815	50	18	.37
	May	1287	1710	58	10	.47
	June	1856	1925	69	19	.59
	July	1210	875	85	26	.76
	August	3217	2450	89	16	.70
	September	1677	2180	79	5	.62
	October	2400	3200	66	17	.45
	November	971	1425	54	32	.32
	December	1107	1320	42	4	.37

Use the Regression processor to fit the complete model to the data, and test the model for utility, using $\alpha = .01$. Assume the cutoff F-score for 4 and 19 df and $\alpha = .01$ is 4.50.

17.73 Refer to Exercise 17.72. Fit the reduced model $Y = \beta_0 + \beta_1 X_1 + \beta_2 X_4 + \epsilon$, and test the combined contribution of X_2 and X_3 to Y. Use $\alpha = .01$.

17.74 Refer to Exercise 17.72. In the complete model, is the age of the residence an important predictor of Y? Justify your answer with the results of a test or a confidence interval. Assume a 10% alpha-risk.

17.75 Refer to Exercise 17.72. The data from the National Weather Service (NWS) must be input into the electric company's data base file each month. Several transcription errors have occurred, and the manager of the utility's forecasting unit questions whether these data are needed. Determine the importance of the NWS variables in the complete model, using $\alpha = .01$.

17.76 Refer to Exercise 17.72. Decide if the data are time-dependent. If so, conduct the Durbin-Watson test for autocorrelation at a 5% level of significance.

17.77 Refer to Exercise 17.71. Suppose the correlation matrix for the capital gains tax study was as follows:

$$\mathbf{C} = \begin{bmatrix} 1.0000 & -.6099 & .1412 & -.5541 & -.3748 \\ -.6099 & 1.0000 & .0000 & .0000 & .0000 \\ .1412 & .0000 & 1.0000 & .0000 & .0000 \\ -.5541 & .0000 & .0000 & 1.0000 & .0000 \\ -.3748 & .0000 & .0000 & .0000 & 1.0000 \end{bmatrix}$$

a. Is multicollinearity a problem in this study?

b. Which variable(s), if any, would you recommend deleting from the study based on this correlation matrix?

17.78 What are the effects, if any, on the descriptive and inferential parts of a regression analysis if the Lin-Mudholkar test for normality yields a conclusion of reject H_0?

17.79 Construct a residual plot of $\hat{Y}_i$ versus e_i for the following $n = 18$ residuals. If $MSE = 221.6$, determine whether any of the residuals are outliers.

Entity	$\hat{Y}$	e_i	Entity	$\hat{Y}$	e_i
1	124	−4.0	10	166.9	2.1
2	132.7	−2.7	11	168.6	1.4
3	141.3	−6.3	12	171.3	3.7
4	151.6	−13.6	13	169.6	12.4
5	143.8	−1.8	14	169.2	20.8
6	150	−1.0	15	181.4	13.6
7	144.7	10.3	16	158	19.0
8	157.4	0.6	17	182.2	−29.2
9	154.8	5.2	18	169.5	−30.5

17.80 Refer to Exercise 17.79. Conduct the Durbin-Watson test for autocorrelation among the residuals, with a 5% risk of error. Assume there are two independent variables in the analysis.

17.81 Refer to Exercise 17.79. Conduct the Lin-Mudholkar test for normality, with a 1% risk of error. Round off the m_i values to three decimal places and the l_i values to four decimal places.

17.82 Determine the rejection region for the Durbin-Watson test for autocorrelation, based on the following information:

a. $\alpha = .05$; $n = 50$; and $k = 4$.
b. $\alpha = .01$; $n = 22$; and $k = 3$.

17.83 Refer to Exercise 17.82. Set up the rejection region for the Lin-Mudholkar test for normality for each part.

17.84 Refer to part a of Exercise 17.82. Complete the Durbin-Watson test, given the following information:

$$\Sigma e_i = 0$$
$$\Sigma e_i^2 = 7078338$$
$$\Sigma(e_i - e_{i-1})^2 = 8325543$$

Are the residuals serially correlated?

17.85 Refer to part b of Exercise 17.83. Complete the Lin-Mudholkar test, given the following information:

$$\Sigma m_i = .009 \qquad \Sigma l_i = 20.9615$$
$$\Sigma m_i^2 = .041193 \qquad \Sigma l_i^2 = 19.995932$$
$$\Sigma m_i l_i = .0043264$$

Are the residuals normally distributed?

17.86 Data on a grocery store's sales over a 24-hour period are listed in the following table:

Hour	Average Sale, Y	Number of Checkers, X_1	Number of Customers, X_2	Average Number of Items, X_3
12 A.M.– 1 A.M.	$ 4.31	1	10	6
1 A.M.– 2 A.M.	2.31	1	14	2
2 A.M.– 3 A.M.	.81	1	8	2
3 A.M.– 4 A.M.	1.84	1	1	2

(continues)

Hour	Average Sale, Y	Number of Checkers, X_1	Number of Customers, X_2	Average Number of Items, X_3
4 A.M.– 5 A.M.	2.14	1	3	4
5 A.M.– 6 A.M.	1.65	1	12	3
6 A.M.– 7 A.M.	2.48	1	20	7
7 A.M.– 8 A.M.	3.52	1	25	3
8 A.M.– 9 A.M.	1.34	2	33	4
9 A.M.–10 A.M.	9.37	3	42	8
10 A.M.–11 A.M.	21.14	4	54	24
11 A.M.–12 P.M.	15.78	5	110	14
12 P.M.– 1 P.M.	16.72	6	98	16
1 P.M.– 2 P.M.	15.13	6	131	14
2 P.M.– 3 P.M.	18.12	6	106	17
3 P.M.– 4 P.M.	13.65	6	154	12
4 P.M.– 5 P.M.	11.74	6	163	11
5 P.M.– 6 P.M.	13.75	7	205	13
6 P.M.– 7 P.M.	14.47	7	163	13
7 P.M.– 8 P.M.	12.56	6	107	13
8 P.M.– 9 P.M.	8.09	5	58	8
9 P.M.–10 P.M.	6.84	2	36	7
10 P.M.–11 P.M.	2.75	1	8	4
11 P.M.–12 A.M.	2.95	1	11	4

Fit the model $Y = \beta_0 + \beta_1 X_1 + \beta_2 X_2 + \beta_3 X_3 + \epsilon$, where Y = the average sale per hour. Test the assumption that the residuals are normally distributed. Use $\alpha = .05$. (*Hint:* Use as many decimal places as possible for the b_i's in order to generate accurate $\hat{Y}$'s. Round off the Y's to two decimal places, the m_i's to three decimal places, and the l_i's to four decimal places.)

17.87 Refer to Exercise 17.86. Test the assumption that the residuals are not autocorrelated. Use $\alpha = .05$.

17.88 Business travel by air almost always includes hassles, especially after the airplane lands at the destination airport. For instance, getting from the airport to the hotel may become an expedition, depending on the available ground transportation and the distance to be traveled. A study of transportation fees from a city's airport to its downtown area revealed the following data:

City	Miles	Cost		
		Limo	Taxi	Mass Transit
1	7	\$ 6.50	\$16.00	\$ .80
2	10	10.00	13.00	.60
3	7	5.00	9.00	1.05
4	7	7.25	10.50	.90
5	14	7.00	18.00	.90
6	8	5.00	14.00	4.00
7	16	5.00	32.50	3.25

(continues)

(continued)

City	Miles	Cost: Limo	Cost: Taxi	Cost: Mass Transit
8	8	\$ 6.00	\$12.00	\$2.00
9	5	5.00	8.50	.75
10	21	8.50	24.00	.90
11	4	5.00	6.70	1.05
12	15	6.30	27.50	.85
13	10	6.75	12.00	.60
14	5	4.50	9.50	.60
15	10	4.90	9.90	.70
16	5	3.50	10.00	.60
17	8	6.00	13.50	.60
18	15	8.00	26.00	6.50
19	10	8.00	14.50	1.00
20	13	5.00	23.25	1.00
21	10	5.00	12.00	1.00
22	19	7.00	30.00	2.90
23	9	5.00	16.50	.85
24	3	5.25	10.00	.60
25	13	5.90	18.00	1.00
26	8	5.00	10.75	.75
27	14	6.00	23.25	1.25
28	18	9.00	19.50	1.00
29	9	5.00	15.00	.95
30	7	6.75	11.00	.75

Define two indicator variables, X_2 for limos and X_3 for taxis, to represent the three modes of ground transportation.

17.89 Refer to Exercise 17.88. For the model $Y = \beta_0 + \beta_1 X_1 + \beta_2 X_2 + \beta_3 X_3 + \epsilon$, where Y = cost, X_1 = miles, and X_2 and X_3 are the indicator variables defined in the previous exercise, code the data for the first two cities. (*Hint:* Each city generates three values for Y, X_1, X_2, and X_3.)

17.90 Refer to Exercise 17.88. For the model in Exercise 17.89, interpret β_1 and β_2.

17.91 Refer to Exercise 17.88. Fit the model $Y = \beta_0 + \beta_1 X_1 + \beta_2 X_2 + \beta_3 X_3 + \epsilon$ to the data. Estimate the following quantities:

a. the difference in the average cost of traveling from the airport to the downtown area, using a taxi versus using mass transit.

b. the difference in the average cost of traveling from the airport to the downtown area, using a taxi versus using a limo.

17.92 Consider the model in Exercise 17.89: $Y = \beta_0 + \beta_1 X_1 + \beta_2 X_2 + \beta_3 X_3 + \epsilon$. Add the interaction terms $\beta_4 X_1 X_2$ and $\beta_5 X_1 X_3$ to the model. Interpret the following:

a. β_1 **b.** $\beta_1 + \beta_5$ **c.** $\beta_2 + \beta_4 X_1$

17.93 Consider the model in Exercise 17.92: $Y = \beta_0 + \beta_1 X_1 + \beta_2 X_2 + \beta_3 X_3 + \beta_4 X_1 X_2 + \beta_5 X_1 X_3 + \epsilon$. What null hypothesis would you set up to test that the rate of change in cost for each additional mile traveled is the same for limos and the mass transit mode of transportation?

17.94 Fit the model in Exercise 17.93 to the data in Exercise 17.88. Test whether the inclusion of the set of interaction terms is justified. Use $\alpha = .10$.

17.95 Write a complete second order model for a multiple regression analysis with the following:

a. one quantitative independent variable and one qualitative independent variable at three levels.

b. one quantitative independent variable and two qualitative independent variables, one at two levels and the other at three levels.

17.96 Refer to Exercise 17.95. What is the minimum sample size needed to fit the complete second order model in each part and still have at least 10 degrees of freedom available to estimate σ^2?

17.97 A group of wine connoisseurs participated in blind taste tests to rate the quality of wine on a scale of 0 to 100. High scores are desirable. Three independent variables were thought to be related to Y = the wine's average rating by the judges:

$$X_1 = \text{Price, in dollars}$$
$$X_2 = \text{Age of the wine, in years}$$
$$X_3 = \begin{cases} 1 & \text{if bottled in the United States} \\ 0 & \text{if bottled in Europe} \end{cases}$$

On the basis of 17 different wines, the following least-squares equation was obtained:

$$\hat{Y} = \underset{(6.061)}{84.333} + \underset{(.1296)}{.187X_1} - \underset{(1.118)}{.637X_2} + \underset{(2.853)}{9.292X_3}$$

$$R^2 = .4688;\ F^* = 3.824;\ SSE = 370.84$$

Before the taste tests were conducted, it was suspected that a high priced wine would be synonymous with a high rating. Is there sufficient evidence to conclude that price and rating are directly related? Let $\alpha = .10$.

17.98 Refer to Exercise 17.97. Interpret the following:
a. β_1 **b.** b_2 **c.** b_3

17.99 Refer to Exercise 17.97. The fitted second-order model was:

$$\hat{Y} = \underset{(18.468)}{88.025} + \underset{(.4207)}{.042X_1} - \underset{(7.5406)}{1.652X_2} + \underset{(3.6114)}{9.620X_3}$$
$$+ \underset{(.0078)}{.003X_1^2} + \underset{(.7153)}{.090X_2^2}$$

$$R^2 = .4758;\ F^* = 1.997;\ SSE = 365.95$$

Are the second-order variables needed? Use $\alpha = .10$.

17.100 Refer to Exercise 17.97. Based on the following fitted model, are the interaction terms justified if $\alpha = .05$?

$$\hat{Y} = \underset{(7.0632)}{84.934} + \underset{(.1458)}{.158X_1} - \underset{(1.294)}{.639X_2} + \underset{(15.316)}{10.226X_3}$$
$$+ \underset{(.5048)}{.475X_1X_3} - \underset{(3.4462)}{1.710X_2X_3}$$

$$R^2 = .5120;\ F^* = 2.308;\ SSE = 340.67$$

REFERENCES

Anscombe, F. J., and J. W. Tukey. 1963. "The Examination and Analysis of Residuals." *Technometrics, 5:* 141–160.

Daniel, W. W., and J. C. Terrell. 1986. *Business Statistics: Basic Concepts and Methodology,* 4th Edition. Houghton Mifflin Company, Boston.

Draper, N., and H. Smith. 1981. *Applied Regression Analysis,* 2nd Edition. John Wiley and Sons, New York.

Durbin, J., and G. S. Watson. 1951. "Testing for Serial Correlation in Least Squares Regression II." *Biometrika, 38:* 159–178.

Hildebrand, D. K., and L. Ott. 1987. *Statistical Thinking for Managers,* 2nd Edition. PWS-KENT, Boston.

Kleinbaum, D. G., and L. L. Kupper. 1978. *Applied Regression Analysis and Other Multivariate Methods*. PWS-KENT, Boston.

Lilliefors, H. W. 1967. "On the Kolmogorov-Smirnov Test for Normality with Mean and Variance Unknown." *Journal of the American Statistical Association, 62:* 399–402.

Lin, C. C., and G. S. Mudholkar. 1980. "A Simple Test for Normality Against Asymmetric Alternatives." *Biometrika, 67:* 455–461.

Mendenhall, W., J. E. Reinmuth, R. Beaver, and D. Duhan. *Statistics for Management and Economics,* 5th Edition. PWS-KENT, Boston.

Nelson, L. S. 1981. "A Simple Test for Normality," *Journal of Quality Technology, 13:* 76–77.

Neter, J., W. Wasserman, and M. K. Kutner. 1985. *Applied Linear Statistical Models,* 2nd Edition. Richard D. Irwin, Inc., Homewood, IL.

Neter, J., W. Wasserman, and G. A. Whitmore. 1988. *Applied Statistics,* 3rd Edition. Allyn and Bacon, Inc., Boston.

Pappas, J. L.; E. F. Brigham; and M. Hirschey. 1983. *Managerial Economics,* 4th Edition. The Dryden Press, Chicago.

Shapiro, S. S., and M. B. Wilk. 1965. "An Analysis of Variance Test for Normality." *Biometrika, 52:* 591–611.

Younger, M. S. 1979. *A Handbook for Linear Regression*. PWS-KENT, Boston.

Chapter Maxim *Times series forecasting is predicated on the premise that history repeats itself.*

CHAPTER 18
TIMES SERIES ANALYSIS

*Optional

Objectives

After studying this chapter and working the exercises, you should be able to

1. Define and graph a time series.
2. Describe the components of a time series.
3. Explain the assumptions that differentiate a multiplicative model from an additive model.
4. Isolate the trend component of a time series by generating a moving average or fitting a trend line.
5. Develop seasonal indices.
6. Derive and graph cyclical relatives.
7. Forecast Y_{t+1} using any one of the six simple forecasting techniques: naive, mean, moving average, trend projection, multiplicative model, and exponential smoothing.
8. Use the three error metrics—MAD, MAPE, and MSE—to measure the precision of a sequence of forecasts.
9. Distinguish fitting accuracy from forecasting accuracy.
10. Identify other qualitative and quantitative forecasting methods.

*11. Execute the Time Series processor to verify answers and/or solve problems.

All of the inferential procedures that we have discussed to this point apply to randomly sampled data from a population. However, data also occur merely as a result of "doing business," and are associated with specific times or dates. For example, a manager of a fast food restaurant may record the cash receipts on an hourly basis; a loan officer at a bank may count the number of mortgage loan applications received per week; and a sales representative may log the number of miles driven at the end of each day. Each variable produces values that are chronological. Rather than randomly select some of these values, we prefer to analyze the sequence of all values for trends or patterns.

As we mentioned in Chapter 2, sequentially recording numerical information on the basis of time generates time-related data, or time series data. Since time-related data are not obtained by the usual random sampling techniques described in Chapter 1, the analysis of the data is different. Yet, we still divide our work into the two types of business statistics procedures: description and inference.

*Applies to optional section.

A descriptive analysis of time-related data is referred to as **time series decomposition,** a process that entails identifying the historical long run direction of the data, the regularly recurring peaks and valleys, if any, and any periods of sustained high or low values of the variable. An inferential analysis of time-related data falls under the heading of **forecasting,** which involves predicting future values of the time series based on the past values of the variable.

Time series decomposition and forecasting methods will draw on familiar topics, such as averages and regression analyses, and will require some graphical skills. Also, we will measure the accuracy of our forecasts by computing quantities such as the mean square error and the average deviation. And, we will introduce some new techniques, terms, and formulas as we learn how to describe and predict time series data.

18.1 INTRODUCTION

Time-Related Data

The first step in analyzing time-related data is to understand the relationship of the variable to time: The variable realizes its values according to the passage of units of time. Therefore, time is the *entity* for the variable. Following are some examples:

Variable	Entity
T_t = Total dollar sales per hour in a grocery store	Hour
U_t = Daily closing price of the Dow Jones Industrial Average	Trading day
V_t = Weekly auction price of 6-month Treasury bills	Week
W_t = Monthly change in the Consumer Price Index	Month
X_t = Quarterly dividends of a share of common stock	Quarter
Y_t = Annualized yield on a one-year certificate of deposit	Year

In each case, the values of the variable occur after a unit of time has elapsed. Consequently, in time-related data studies, the variable often is subscripted with the letter t to emphasize its relationship to time.

Definition

A **time series** is a chronology of values for a variable in which the entity is a unit of time. The value of the variable at time t is denoted by y_t.

Our definition suggests several points about time series data:

1. They occur at regular intervals over time.
2. They do not result from a simple random sample.
3. They are ordered in sequence by time.
4. They cannot be rearranged in the sequence.

Time series data exist almost everywhere and provide us with a potentially valuable history lesson. We recommend beginning every analysis with a *line graph* of the data. Beyond the obvious zigs and zags that a graph of the data will produce, we are hoping to find a direction and/or pattern that the data follow. Clearly, our *goal* is to analyze the time series data. The following box lists two objectives that help us to achieve this goal.

Objectives of a Time Series Analysis

1. To describe the past behavior of the time series
2. To predict the future values of the time series

EXAMPLE 18.1

The controller of a large manufacturing company wishes to study the work-in-progress inventory, Y_t, on the quarterly balance sheets. For each quarter in the last four years the values of Y_t, in millions of dollars, are listed in the following table:

Year	Quarter	Y_t	Year	Quarter	Y_t
1987	1	15.2	1989	1	19.6
	2	18.4		2	22.1
	3	18.7		3	21.4
	4	22.2		4	24.4
1988	1	17.8	1990	1	19.6
	2	21.9		2	23.4
	3	21.9		3	24.1
	4	25.6		4	28.2

a. What is the entity for the variable Y_t?
b. Construct a line graph of these data.

Solution:

a. A quarter of a year is the entity for Y_t.
b. Figure 18.1 shows a line graph for the data.

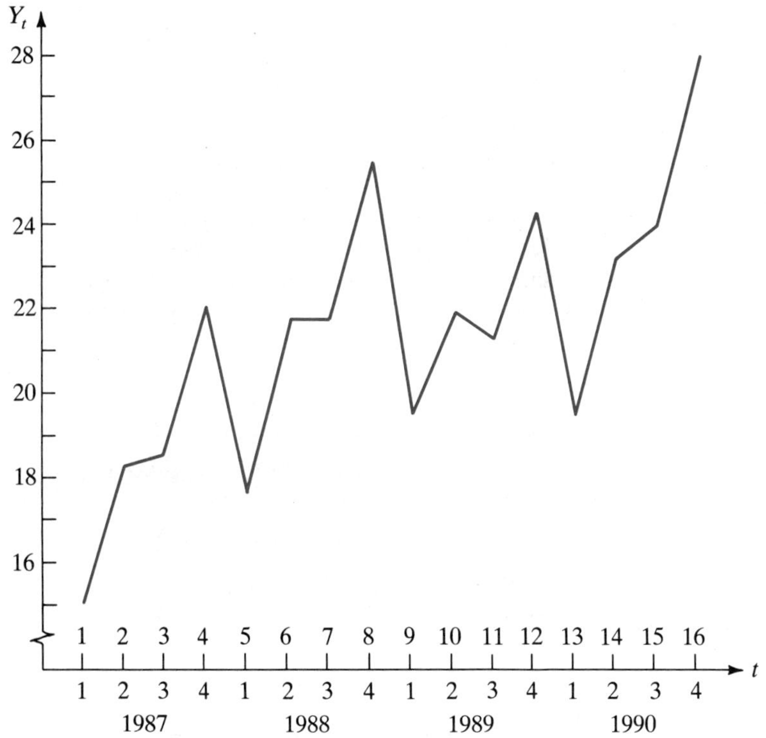

Figure 18.1 Quarterly Work-in-Progress Inventory (Millions of Dollars) for Example 18.1

Notice that the horizontal axis in Figure 18.1 represents time and is scaled in integers, starting at 1 for the first quarter of 1987 and ending at 16 for the fourth quarter of 1990. These numbers are artificial values for the units of time in lieu of the year-quarter designation. It is common to use a numbering system where the most distant time period is assigned the value 1 and succeeding time periods are matched to the appropriate integer value. Projecting forward, $t = 23$ would represent the third quarter of 1992, for example. The line graph is formed by connecting successive values of Y_t with line segments.

□

■ Time Series Components

The historical pattern of values for the variable Y_t is assumed to be a function of several factors or components. Table 18.1 lists these components and describes their relation to Y_t.

Table 18.1 Time Series Components

Component	Symbol	Definition
Trend	T_t	Long-term tendency of Y_t
Seasonal	S_t	Systematic fluctuations in Y_t of fixed duration due to seasonal influences
Cyclical	C_t	Intermittent fluctuations in Y_t of variable duration due to external economic influences
Irregular	I_t	Random variations in Y_t due to unpredictable influences

Trend The trend component describes the general long-run direction of Y_t. Often, the direction is characterized in a linear sense as either up, down, or stationary. Figure 18.2 shows these three characteristics. The bold line through the zigzags represents T_t.

Seasonal The seasonal component implies regular tendencies of Y_t to move up and down, according to the calendar or a clock. Although we naturally associate winter,

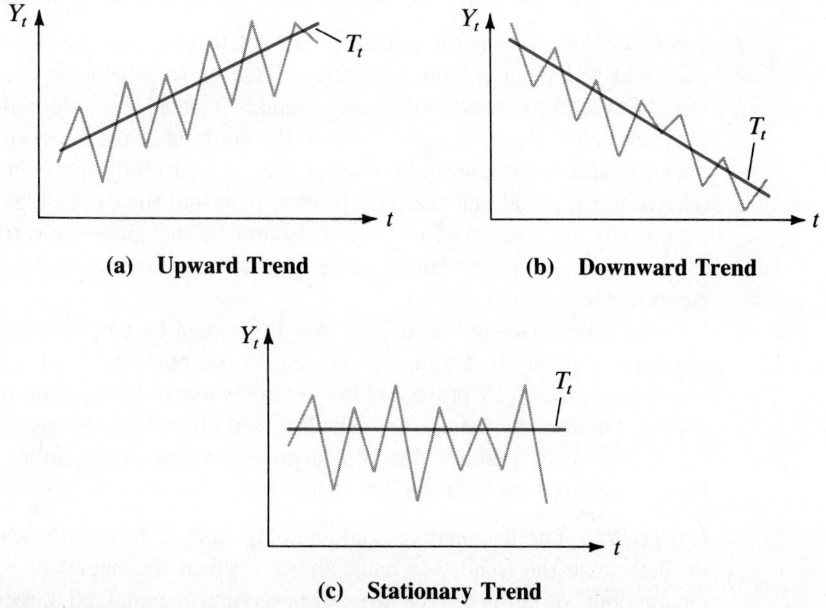

Figure 18.2 Trend Component

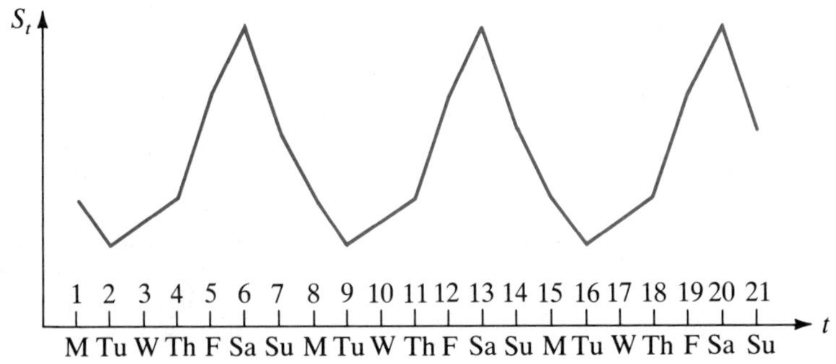

Figure 18.3 Seasonal Component

spring, summer, and fall with the notion of seasons, there are other meanings of *season*. For example, a restaurant tends to generate more revenue on weekend-days than on midweek-days, and this pattern repeats itself week after week. The days of the week for the restaurant represent its *seasons*. This means that we attribute relatively high or low revenues to the particular day of the week. This phenomenon is demonstrated in Figure 18.3 where the recurring highs and lows endure for approximately 3 and 4 days at a time, respectively. Note that $t = 1$ corresponds to a Monday. Also note that the vertical axis is labeled S_t. This figure is showing the seasonal component in isolation. In practice, it is mixed in with the other components.

Cyclical A cycle is the period of time it takes a time series to phase through a peak and valley. Unlike the regularity of the seasonal effects, cyclical effects occur irregularly and are unrelated to the calendar. Instead they are believed to be related to general business conditions. Further, the duration of the effects are typically longer than a year and for uneven periods of time. Consistently high and low values of Y_t, relative to the trend and seasonal factors, represent the cycle. Figure 18.4 shows the irregularity of cycles in a time series. Again note that since the vertical axis is labeled C_t, the cyclical component is being graphed in isolation of the other time series components.

The time series in Figure 18.4 was influenced by two relatively short periods of prosperity—intervals *A* and *C* indicated on the horizontal axis. These "up" cycles might be explained by prevailing low interest rates or by a period of declining budget deficits. On the other hand, the variable Y_t was affected adversely for a longer duration during period *B*, perhaps due to a depressed value of the dollar or to the Federal Reserve Board's monetary policy at the time.

Irregular The irregular or random component is essentially what's left over after we determine the trend, seasonal, and/or cyclical tendencies of Y_t and accounts for the random variation in the time series. Any unexplained forces or nonrecurring events, like a labor strike or a drought, usually are attributed to the irregular component.

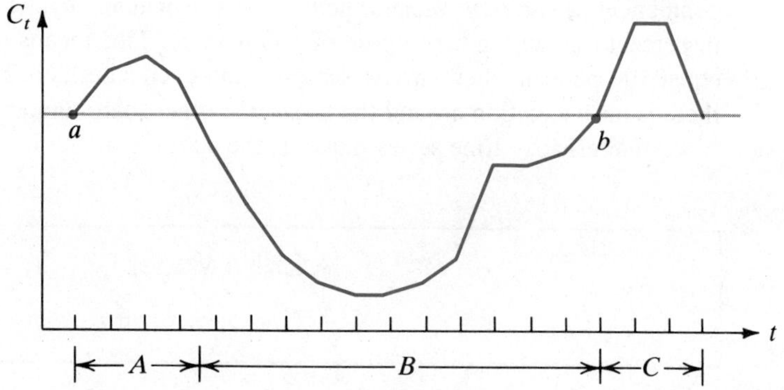

Figure 18.4 Cyclical Component

A time series may not contain all four components, although the irregular term is assumed to be a part of every time series. For instance, in Figure 18.4, the seasonal component is not visible and may or may not be present in the original time series. Yearly data typically exhibit a trend and a cyclical effect, but not a seasonal effect. Monthly and quarterly data are likely to show trend and seasonal influences, but usually not a cyclical component unless the data span many years.

COMMENTS

1. A time series with no trend is known as a *stationary* time series. Trend also may be characterized in a quadratic sense, though the linear characterization is perhaps more common.
2. Technically, a cycle is a complete revolution, such as from peak to peak or from valley to valley. For example, in Figure 18.4 the points labeled *a* and *b* represent the beginning and ending of the cycle.
3. Seasonal effects are viewed as repetitious cycles within a year's time. Cyclical vicissitudes span periods of more than one year, in general.

Time Series Models

As we previously mentioned, the variable Y_t is assumed to be a function of the trend, seasonal, cyclical, and irregular components. The most popular functional relationship for time series data is the multiplicative model in Equation 18–1.

Multiplicative Model

$$Y_t = T_t \cdot S_t \cdot C_t \cdot I_t \qquad (18\text{–}1)$$

The multiplicative model assumes the observed fluctuations in Y_t from the trend component are *percent changes;* hence, the seasonal and cyclical effects are expressed in percentages with a base figure of 100 percent. This means that whenever S_t and C_t equal 100 percent, there are no seasonal and cyclical effects at time t, and therefore, there is no fluctuation around the trend line other than random variation.

An alternative time series model is the additive model in Equation 18–2.

Additive Model

$$Y_t = T_t + S_t + C_t + I_t \qquad \textbf{(18–2)}$$

This model assumes that deviations in Y_t from the trend component are *absolute changes;* thus, the seasonal and cyclical effects are expressed in the original units of the data.

COMMENT The multiplicative model often is preferred, since a percentage change is a more stable means of describing business and economic data. For instance, if sales increase by a fairly constant 3 percent each first quarter of the year, then a multiplicative model with a first quarter seasonal effect of 103 percent accurately reflects the changing base from year to year. An additive model with a first quarter seasonal effect of, say, \$50,000 will be increasingly inaccurate over time. However, the additive model's advantage is that it is more flexible for inferential purposes.

In the next section we will present methods of describing the time series components for both types of models.

SECTION 18.1 EXERCISES

18.1 Indicate whether the following variables generate time-related data or time-free data.

a. X = The number of people in a sample of 38 who wear a watch on their left wrist.

b. X = The number of service calls received per week by a copier company, for a sample of 26 consecutive weeks.

c. X = The time between service calls to repair a copier at a specific firm, over a one-year period.

d. X = Monthly spread in rates between a 30-year fixed mortgage and a 15-year adjustable mortgage at a certain bank over a one-year period.

e. X = The number of new cars sold at a Honda dealership for a sample of 24 consecutive weeks.

18.2 Refer to Exercise 18.1. Identify an entity for each variable.

18.3 The development office at a small university monitors annual contributions to the university from alumni. Each year, a manager in the office determines the average (unrestricted) donation per donor. These figures, along with the percentage of all alumni who contributed, are shown in the following table.

Year	Average Contribution	Percent of Contributing Alumni
1970	$16.19	19.5
1971	17.64	19.7
1972	19.37	19.7
1973	20.53	18.6
1974	21.60	17.9
1975	26.32	17.9
1976	26.32	18.1
1977	29.55	18.4
1978	32.10	17.7
1979	36.59	18.0
1980	40.71	18.0
1981	46.42	18.4
1982	48.52	18.2
1983	53.54	18.9
1984	60.66	19.5
1985	66.06	20.1
1986	67.13	19.8
1987	58.30	16.6
1988	61.44	17.4
1989	64.75	18.4

a. Graph the average contribution data. Allow about 6 inches on the vertical axis for scaling. Which components, if any, does the time series appear to possess?

b. Graph the percentage of contributing alumni data. Allow about 4 inches on the vertical axis for scaling, beginning at the value 16. Which components, if any, does the time series appear to possess?

18.4 Identify two objectives in a time series analysis.

18.5 Differentiate a seasonal effect from a cyclical effect in terms of the following:

a. duration

b. regularity

c. assignable influences

18.6 Quarterly sales (in thousands of dollars) for the past five years for a drive-through car wash are listed as follows:

Year	Quarter	Sales	Year	Quarter	Sales
1	1	$19.8		3	$24.5
	2	26.5		4	28.5
	3	21.0	4	1	24.2
	4	25.0		2	25.2
2	1	21.7		3	25.7
	2	28.6		4	26.7
	3	23.3	5	1	24.7
	4	26.7		2	27.2
3	1	22.8		3	26.6
	2	30.3		4	27.9

Graph the data. Allow about 5 inches on the vertical axis for scaling, beginning at the value \$19.0. Are any components of the time series visible? If so, which one(s)?

18.7 Match the phrases "absolute change" and "relative change" to the two general time series model forms.

18.8 The prices received by farmers for various crops are monitored monthly and reported in the form of a price index. For the 1983–87 period, the commodity price index for cotton is listed in the following table:

Month	1983	1984	1985	1986	1987
January	481	529	441	456	440
February	487	549	418	480	392
March	525	592	474	490	401
April	510	567	481	500	425
May	537	614	485	494	507
June	528	574	509	494	559
July	567	556	511	519	577
August	545	568	478	395	538
September	530	554	472	404	548
October	533	544	484	398	544
November	566	524	477	453	549
December	559	474	454	462	542

Source: U.S. Department of Commerce, *Business Statistics, 1986–88*, supplement to the *Survey of Current Business*, Washington, D.C., 1988.

a. Graph the commodity price indices. Allow about 5 inches on the vertical axis for scaling, beginning at the value 400.
b. Describe the general trend, if any, in the time series.
c. Is a seasonal component apparent from the graph?

18.2 TIME SERIES DECOMPOSITION

One of our analysis objectives is to describe the past behavior of the time series. To do this, we assign behavior patterns of Y_t to the four components listed in Table 18.1. This is called *time series decomposition*. We isolate three factors—trend, seasonal, and cyclical—and quantify each factor's influence on Y_t. The residual variability that remains after this decomposition is assigned to the irregular factor.

Trend Isolation

There are two primary techniques for describing the general direction of the time series: a moving average and a regression-based trend line.

Moving Average When graphed, most time series resemble a series of jagged lines. These sharp one-step movements are part of an overall drift of the time series. *Averaging* is one means of smoothing the data so that we can see this long-run tendency. A moving average is a dynamic average in the sense that new observations of Y_t move in and replace old observations of Y_t that move out in order to maintain a constant divisor of n values. The average value necessarily is associated with or assigned to a specific time period. In this manner, the moving average generates an artificial sequence of values and becomes a time series itself.

Although moving averages are found by averaging n values of Y_t, there are two ways to do this. One way is to average the n most recent values; we call this (just) a **moving average**. Another approach is to average the $n/2$ leading values with the present and the $n/2$ most recent values; we call this a **centered moving average**. As we will see shortly in Example 18.2, a centered moving average cannot be determined for the first few time periods, since the $n/2$ most recent values are not available, nor for the last few time periods, because the $n/2$ leading values do not exist. Figure 18.5 shows the different types of moving averages for a hypothetical time series.

Traditionally, the centered moving average has been used in descriptive analyses, while the moving average is more popular for inferential purposes. We will discuss the centered moving average in this section and the moving average in Section 18.3.

The number of values in either type of moving average is called the *term* or *period* and usually is based on the entity defining the time series. For example, quarterly data suggest a 4-period (or 4-quarter) moving average; monthly data may require a 12-period (or 12-month) moving average.

Depending on whether the term involves an even or odd number of time periods, we compute the value of a centered moving average at time t, denoted by M_t, differently. For a four-period centered moving average, we find M_t as in Equation 18–3.

Four-Period Centered Moving Average

$$M_t = \frac{\frac{1}{2}y_{t-2} + y_{t-1} + y_t + y_{t+1} + \frac{1}{2}y_{t+2}}{4} \qquad (18\text{–}3)$$

Notice that M_t involves y_t, the two previous values, y_{t-1} and y_{t-2}, and the two succeeding values, y_{t+1} and y_{t+2}. Further, the extreme values receive a *weight* of only

½. Equation 18–3, therefore, is an example of a weighted average. For a three-period centered moving average, all values of Y_t are equally weighted, as shown in Equation 18–4.

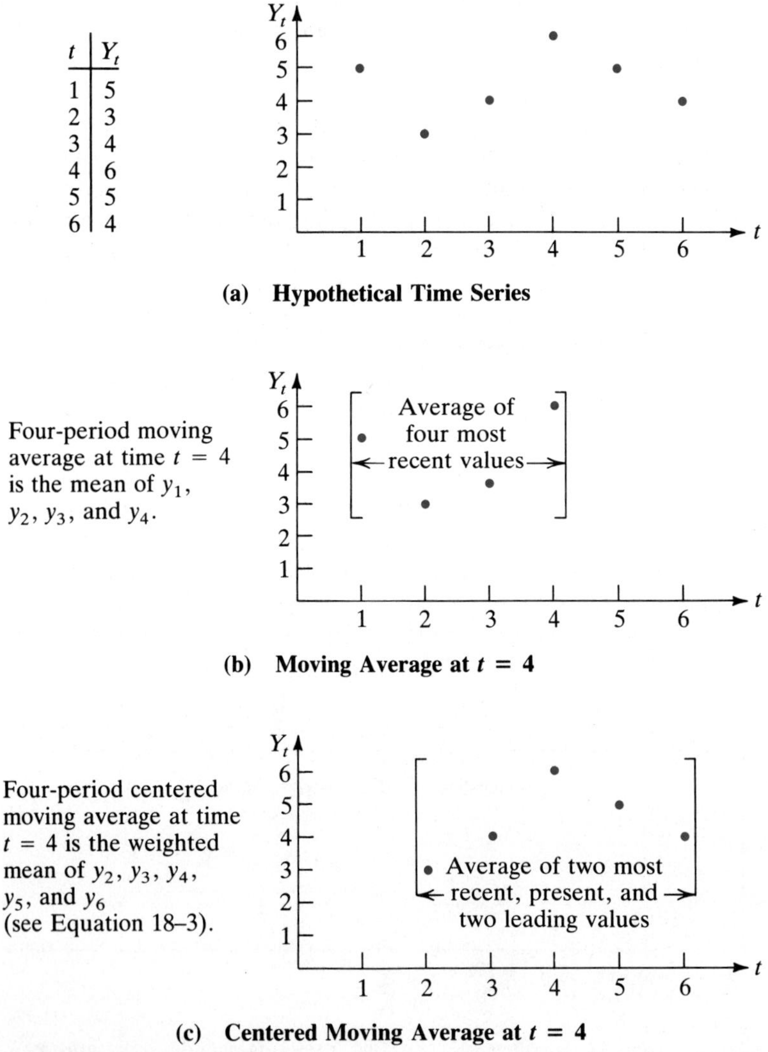

t | Y_t
1 | 5
2 | 3
3 | 4
4 | 6
5 | 5
6 | 4
(a) Hypothetical Time Series
Four-period moving average at time t = 4 is the mean of y1, y2, y3, and y4.
Average of four most recent values
(b) Moving Average at t = 4
Four-period centered moving average at time t = 4 is the weighted mean of y2, y3, y4, y5, and y6 (see Equation 18–3).
Average of two most recent, present, and two leading values
(c) Centered Moving Average at t = 4

Figure 18.5 Types of Moving Averages

Three-Period Centered Moving Average

$$M_t = \frac{y_{t-1} + y_t + y_{t+1}}{3} \tag{18–4}$$

COMMENTS

1. Equations 18–3 and 18–4 are specific to four- and three-period centered moving averages, respectively, but can be adapted easily to handle other periods. For example, a 12-period centered moving average is formed via

$$M_t = \frac{\frac{1}{2}y_{t-6} + y_{t-5} + \ldots + y_t + \ldots + y_{t+5} + \frac{1}{2}y_{t+6}}{12}$$

2. The computation of M_t for a centered moving average requires previous and future values of Y_t. Therefore, there will not be M_t values at the beginning of the time series and at the end of the time series. The number of missing M_t values depends on the term. For instance, in a time series that spans 20 time periods, M_1 and M_{20} will be missing in a three-period centered moving average, while M_1, M_2, M_{19}, and M_{20} will be missing in a four-period centered moving average.
3. A centered moving average M_t is viewed as an estimate of the trend and cyclical components, since the seasonal (and irregular) effects are assumed to average out. Symbolically, we write this as

$$M_t = T_t \cdot C_t$$

for the multiplicative model, and as

$$M_t = T_t + C_t$$

for the additive model. Centered moving averages are more likely to be associated with the multiplicative model, though, since they are used subsequently to isolate S_t in Equation 18–1.
4. The formula for a moving average, which will be denoted by M_t^* to distinguish it from the centered moving average, M_t, will be presented in the next section.

EXAMPLE 18.2

Construct a four-quarter centered moving average for the time series data in Example 18.1. Graph Y_t and M_t.

Solution:

The four-quarter centered moving average will start at M_3. The computation of M_3 from Equation 18–3 is

$$M_3 = \frac{(1/2)y_1 + y_2 + y_3 + y_4 + (1/2)y_5}{4}$$
$$= \frac{(1/2)15.2 + 18.4 + 18.7 + 22.2 + (1/2)17.8}{4}$$
$$= \frac{75.8}{4}$$
$$= 18.95$$

For M_4, we compute

$$M_4 = \frac{(1/2)18.4 + 18.7 + 22.2 + 17.8 + (1/2)21.9}{4}$$
$$= \frac{78.85}{4}$$
$$= 19.7125 \quad \text{or} \quad 19.71 \text{ (to two decimal places)}$$

The remaining values of M_t are found in a similar manner and are listed in the following table.

Year	Quarter	t	Y_t	M_t
1987	1	1	15.2	—
	2	2	18.4	—
	3	3	18.7	18.95
	4	4	22.2	19.71
1988	1	5	17.8	20.55
	2	6	21.9	21.38
	3	7	21.9	22.03
	4	8	25.6	22.28
1989	1	9	19.6	22.24
	2	10	22.1	22.03
	3	11	21.4	21.88
	4	12	24.4	22.04
1990	1	13	19.6	22.54
	2	14	23.4	23.35
	3	15	24.1	—
	4	16	28.2	—

Notice that the table has two missing M_t values—one at the beginning and one at the ending of the time series. The smoothing effect of M_t in reducing the sharp fluctuations in consecutive time periods is apparent in Figure 18.6, where both Y_t and M_t are plotted.

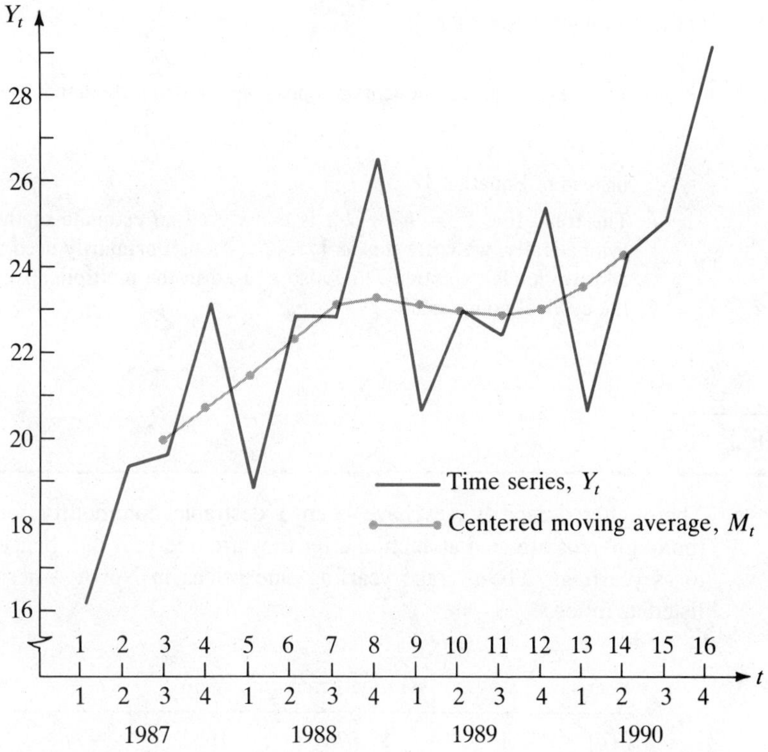

Figure 18.6 Quarterly Work-in-Progress Inventory (Millions of Dollars) for Example 18.2

□

Trend Line The movement of M_t through the peaks and valleys of Y_t suggests the second means of characterizing the long-run tendency of Y_t: a regression-based trend line. If we model the relationship between t and Y_t with a simple regression line as given in Equation 18–5, then the least-squares line forms an estimate of the trend in the time series.

Simple Regression Model

$$Y_t = \beta_0 + \beta_1 t + \epsilon \tag{18–5}$$

Since the independent variable t is a proxy for time, its values are just the integers 1, 2, and so on. The least-squares criterion produces the equation $\hat{Y}_t = b_0 + b_1 t$, where b_0 and b_1 are the estimated regression coefficients, t is the time variable, and $\hat{y}_t$ is the predicted value of Y_t.

COMMENTS

1. Quadratic trend components are possible by fitting the quadratic model

$$Y_t = \beta_0 + \beta_1 t + \beta_2 t^2 + \epsilon$$

instead of Equation 18–5.

2. The trend line $\hat{Y}_t = b_0 + b_1 t$ is considered an estimate of the trend component; symbolically, we write this as $\hat{Y}_t = T_t$. Though primarily used to estimate T_t in the additive model, the trend line also is used in the multiplicative model for isolating the cyclical component.

EXAMPLE 18.3

Thoroughbred race horses have been a desirable commodity for years. Typically, thoroughbreds are sold at auction after they are one-year old; hence, they are referred to as yearlings. The average yearling sale prices in North America since 1970 are listed as follows:

Year	t	Average Price	Year	t	Average Price
1970	1	$7,676	1978	9	19,846
1971	2	8,797	1979	10	24,768
1972	3	9,211	1980	11	29,683
1973	4	12,255	1981	12	35,146
1974	5	10,689	1982	13	32,991
1975	6	10,943	1983	14	41,258
1976	7	13,021	1984	15	41,396
1977	8	16,337	1985	16	41,311

Source: "They Race Horses, Don't They," Louisville *Courier-Journal*, November 17, 1986. Copyright 1986, Louisville *Courier-Journal*. Reprinted with permission.

Estimate the trend component by fitting a simple regression model to this time series.

Solution:

The regression analysis yields:

$$\hat{Y}_t = 69.225 + 2604.562t$$
$$r^2 = .9281$$
$$\hat{\sigma} = 3572.1142$$

A graph of Y_t and $\hat{Y}_t$ is shown in Figure 18.7.

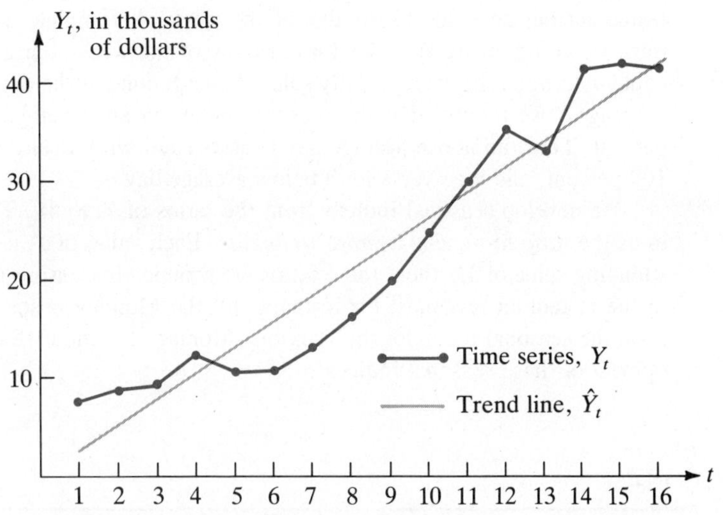

Figure 18.7 Average Yearling Sale Prices, 1970–1985

The linear model appears adequate, since about 93 percent of the variability in Y_t is explained by a linear function of the time variable.

□

Isolating the trend component with a regression line is appealing and relatively easy. Further, since we are *describing* the history of the time series, we need not worry about the distributional requirements of the error term in the regression model; they are of concern only when we are trying to make inferences.

The regression approach is particularly amenable to yearly data when there is no seasonal effect. The centered moving average approach is more suitable to seasonal time series. Focusing on the seasonal component is our next project.

Seasonal Indices

Recall that a seasonal influence causes alternating peaks and valleys and is attributed to the calendar or the clock. Assuming the pattern repeats on a regular basis (usually within a year), then our approach will be to isolate the seasonal effect for each season. The isolated effect is called an *index* for the multiplicative model and referred to as simply the *seasonal effect* for the additive model. The index or seasonal effect represents the change from a base level. Most often in business and economic time series, change is measured in a relative sense as a percentage of the base figure. Therefore, we will restrict our discussion to seasonal indices for the multiplicative model in the remainder of this section.

In Section 18.1, we discussed the regularity of high and low revenues in the restaurant business due to the day of the week. A complete cycle takes seven days; thus, we could create an index for each day of the week. For a seven-day period, we could determine the average daily sales. Though none of the days might be considered "average," we require that the seasonal index on an "average" day be 1.00 or 100 percent. Thus, an above average day is associated with a seasonal index greater than 100 percent, and vice-versa for a below average day.

We develop seasonal indices from the ratios of Y_t to M_t. The method is referred to as the *ratio-to-moving average technique*. Each value of Y_t is divided by the corresponding value of M_t; then, these ratios are grouped by seasons and averaged. Relative to the restaurant revenues, for instance, all the Monday ratios would be averaged to form the seasonal index for the "season" Monday. Example 18.4 illustrates the development of these seasonal indices.

EXAMPLE 18.4

Develop four seasonal indices—one for each quarter—for the work-in-progress inventory data of Example 18.1.

Solution:

The ratios of Y_t to M_t are listed in the last column of the following table:

Year	Quarter	t	Y_t	M_t	Y_t/M_t
1987	1	1	15.2	—	—
	2	2	18.4	—	—
	3	3	18.7	18.95	.9868
	4	4	22.2	19.71	1.1263
1988	1	5	17.8	20.55	.8662
	2	6	21.9	21.38	1.0243
	3	7	21.9	22.03	.9941
	4	8	25.6	22.28	1.1490
1989	1	9	19.6	22.24	.8813
	2	10	22.1	22.03	1.0032
	3	11	21.4	21.88	.9781
	4	12	24.4	22.04	1.1071
1990	1	13	19.6	22.54	.8696
	2	14	23.4	23.35	1.0021
	3	15	24.1	—	—
	4	16	28.2	—	—

To find the seasonal index for each quarter, we group the ratios for similar quarters, and then average the three values:

	Quarter			
Year	1	2	3	4
1987	—	—	.9868	1.1263
1988	.8862	1.0243	.9941	1.1490
1989	.8813	1.0032	.9781	1.1071
1990	.8696	1.0021	—	—
Average	.8724	1.0099	.9864	1.1275

The quarterly averages represent the "raw" seasonal indices. Recall that we introduced the concept of an index as representing a percentage change from a base line average of 100 percent. Presently, the raw indices average to 99.905 percent; thus, a slight technical adjustment is necessary for consistency. To achieve an average of 100 percent, we need the sum of the indices to be 4.0000; presently the sum is 3.9962. However, the ratio of 4.0000 to 3.9962 is proportional to the ratio of the adjusted seasonal index to the raw seasonal index:

$$\frac{4.0000}{3.9962} = \frac{\text{Adjusted seasonal index for season } i}{\text{Raw seasonal index for season } i}$$

Thus, to find the adjusted seasonal index we compute:

$$\text{Adjusted seasonal index for season } i = \frac{4}{3.9962} \times \text{Raw seasonal index for season } i$$

The adjusted seasonal indices are

	Quarter				
Index	1	2	3	4	Sum
Raw Seasonal	.8724	1.0099	.9864	1.1275	3.9962
Adjusted Seasonal	.8732	1.0109	.9873	1.1286	4.0000

We interpret each adjusted seasonal index relative to 100 percent, by subtracting 1.0000 from it and converting to percentages. For instance, the fourth quarter seasonal index of 1.1286 means that during the fourth quarter, the work-in-progress inventory is $1.1286 - 1.0000 = .1286$ or 12.86 percent higher, on average, than the value of the trend component at that time. Similarly, the third quarter adjusted seasonal index of .9873 means that Y_t is $.9873 - 1.0000 = -.0127$, or about 1.27% lower, on average, than the trend during the third quarter.

□

COMMENTS

1. The rationale for forming the ratio of Y_t to M_t is to eliminate the trend-cyclical effect. Recall that M_t estimates $T_t \cdot C_t$, and that Y_t is related to the four components by the multiplicative model of Equation 18–1. Thus,

$$\frac{Y_t}{M_t} = \frac{T_t \cdot C_t \cdot S_t \cdot I_t}{T_t \cdot C_t} = S_t \cdot I_t$$

When we average the ratios from the same season, we essentially remove the irregular component, leaving just S_t—the seasonal index.

2. When there are many ratios from the same season, we sometimes compute a modified mean by eliminating the smallest and largest values. For example, if we had ten years of quarterly data generating nine ratios for each quarter, we could delete the minimum and maximum ratio for each quarter and find the average of the remaining seven values. Alternatively, some authors recommend finding the median of the ratios rather than the mean or a modified mean.
3. You may read or hear the term "seasonally adjusted figures" in the media as in "seasonally adjusted unemployment rate" or "seasonally adjusted housing starts." Seasonally adjusted data are found by *deseasonalizing* the data. Sometimes denoted by D_t, deseasonalized data are computed via

$$D_t = \frac{Y_t}{S_t}$$

For example, in Example 18.4 the deseasonalized value for the first quarter of 1988 is

$$D_5 = \frac{Y_5}{S_5} = \frac{17.8}{.8732} = 20.4 \text{ (rounded)}$$

We report this figure as: The work-in-progress inventory for the first quarter of 1988 was 20.4 million dollars on a seasonally adjusted basis.

4. Adjusting the raw seasonal indices to average exactly 100.0 percent is usually necessary for most time series.
5. We adopt a different tack in interpreting the ratios Y_t/M_t (and hence the seasonal indices) when Y_t and M_t are negative numbers. For instance, if $Y_t = -.1$ and $M_t = -.5$, then the ratio is $(-.1)/(-.5) = .2$. To interpret this figure, we subtract it from 1.0000 rather than vice-versa, as we did in Example 18.4 where both Y_t and M_t were positive. Thus, the Y_t value of $-.1$ is $1.0000 - .2 = .8000$, or 80 percent *higher* than the value of the trend component ($M_t = -.5$) at time t. Exercise 18.72 demonstrates this reversal in interpreting the seasonal indices.

Cyclical Relatives

Cyclical fluctuations, like seasonal effects, are characterized by a sequence of above average values of Y_t followed by a string of below average values of Y_t (or vice-versa). Here, average is taken to be the trend line. A basic difference between cyclical fluctuations and seasonal effects is the time involved: A seasonal pattern lasts a year or less; a cyclical fluctuation takes longer.

To identify the cyclical component in the multiplicative model, we develop a **cyclical relative,** which represents the percentage that Y_t is above or below the trend line for each time period. Our method for finding cyclical relatives depends on the presence or absence of a seasonal factor. If the time series data are collected on a yearly basis, we assume there is no seasonal effect; but if the time series data are recorded quarterly, monthly, and so on, there may be a seasonal effect. In this latter case, we must *deseasonalize* Y_t by dividing by the appropriate seasonal index. At this point a deseasonalized time series and a yearly time series can be treated in an identical manner, since both are assumed to contain only the trend, cyclical, and irregular components.

Next, we *detrend* the time series. This involves three operations.

1. Fit a simple regression model to the yearly or deseasonalized time series data.
2. Use the fitted model to predict Y_t for each value of t.
3. Divide the time series value by the predicted value.

When we remove the trend effect from the time series, we are left with the cyclical and irregular components.

For yearly time series, the ratios found in Step 3 are treated as the cyclical relatives and denoted by C_t. That is, we assume that any short-term irregular effects already have been averaged out over the year it took to produce one value of Y_t; hence, the irregular component is negligible.

For a deseasonalized time series, the irregular effect still is present and, for purposes of computing the cyclical relatives, can be dampened by computing a three-period centered moving average. Example 18.5 demonstrates the development of cyclical relatives, C_t, for the seasonalized time series data of Example 18.1.

EXAMPLE 18.5

Using the quarterly work-in-progress inventory data from Example 18.1, isolate the cyclical component by determining C_t, the cyclical relative for each quarter. Graph the C_t values.

Solution:

The first step is to deseasonalize the data by forming Y_t/S_t. We will denote this ratio by D_t:

Year	Quarter	t	Y_t	S_t	$D_t = Y_t/S_t$
1987	1	1	15.2	.8732	17.41
	2	2	18.4	1.0109	18.20
	3	3	18.7	.9873	18.94
	4	4	22.2	1.1286	19.67
1988	1	5	17.8	.8732	20.38
	2	6	21.9	1.0109	21.66
	3	7	21.9	.9873	22.18
	4	8	25.6	1.1286	22.68

(continues)

(*continued*)

Year	Quarter	t	Y_t	S_t	$D_t = Y_t/S_t$
1989	1	9	19.6	.8732	22.45
	2	10	22.1	1.0109	21.86
	3	11	21.4	.9873	21.67
	4	12	24.4	1.1286	21.62
1990	1	13	19.6	.8732	22.45
	2	14	23.4	1.0109	23.15
	3	15	24.1	.9873	24.41
	4	16	28.2	1.1286	24.99

To detrend the time series, we first fit a simple regression model with D_t representing the dependent variable and t the independent variable. The fitted least-squares line is

$$\hat{D}_t = 18.066 + .402t$$

Substituting the values of 1 through 16 for t, we generate $\hat{D}_t$, the predicted trend values for the deseasonalized time series.

Year	Quarter	t	D_t	$\hat{D}_t$	$D_t/\hat{D}_t$	C_t
1987	1	1	17.41	18.47	.943	—
	2	2	18.20	18.87	.964	.963
	3	3	18.94	19.27	.983	.982
	4	4	19.67	19.67	1.000	.999
1988	1	5	20.38	20.08	1.015	1.024
	2	6	21.66	20.48	1.058	1.045
	3	7	22.18	20.88	1.062	1.062
	4	8	22.68	21.28	1.066	1.055
1989	1	9	22.45	21.68	1.036	1.031
	2	10	21.86	22.08	.990	.997
	3	11	21.67	22.49	.964	.966
	4	12	21.62	22.89	.945	.958
1990	1	13	22.45	23.29	.964	.962
	2	14	23.15	23.69	.977	.985
	3	15	24.41	24.09	1.013	1.003
	4	16	24.99	24.50	1.020	—

Next, we form the ratios $D_t/\hat{D}_t$; the values are listed in the penultimate column above. Finally, these ratios are combined into a three-period centered moving average to reduce any irregular effects, thus producing the C_t's. The graph of the C_t's is shown in Figure 18.8. As you can see, a definite cycle is visible from roughly the fourth quarter of 1987 ($t = 4$) to the third quarter of 1990 ($t = 15$).

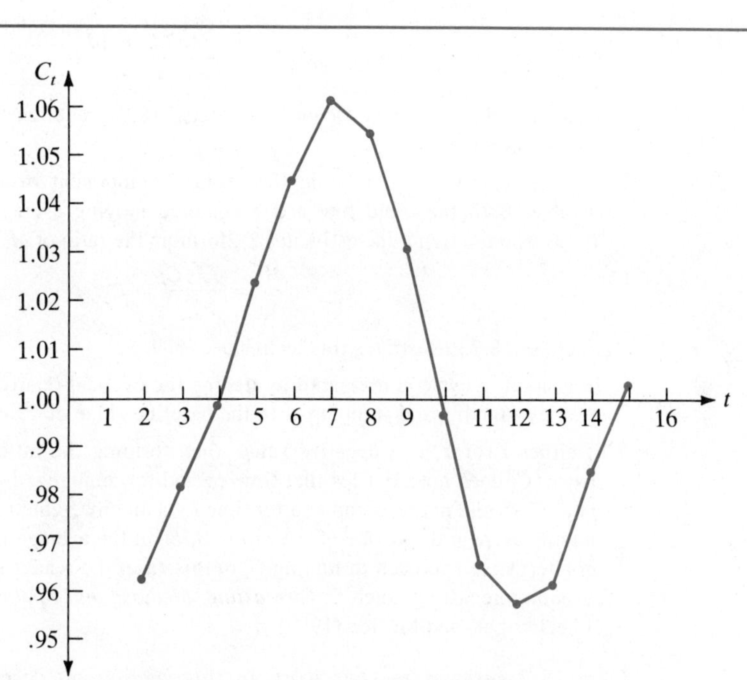

Figure 18.8 Cyclical Relatives for Work-in-Progress Inventory Time Series for Example 18.5

□

COMMENTS

1. In isolating the cyclical component in the multiplicative model, we assume there is no seasonal effect by working either with annualized data (Y_t) or with deseasonalized data (D_t). Hence, we view the time series model to be of the form $Y_t = T_t \cdot C_t \cdot I_t$ or $D_t = T_t \cdot C_t \cdot I_t$. When we detrend the time series by dividing by $\hat{Y}_t$, or by $\hat{D}_t$, we get

$$\frac{Y_t}{\hat{Y}_t} = \frac{D_t}{\hat{D}_t} = \frac{T_t \cdot C_t \cdot I_t}{T_t} = C_t \cdot I_t$$

For annualized data, we assume the I_t component is negligible; that is, I_t fluctuates so little from 100 percent that we ignore it. Hence, the ratio of Y_t to $\hat{Y}_t$ produces the cyclical relatives in this case. For deseasonalized data, though, the I_t factor is removed by computing a centered moving average to yield the C_t's.

2. Although we use more than one year as our guideline, the time frame for cyclical activity is defined as longer than two years in some texts.
3. Synonyms for "cyclical relatives" are *percents of trend* or *percentage relatives*.
4. A three-period centered moving average generally is sufficient to neutralize the irregular component for any type of seasonalized data. However, some texts may recommend computing a *weighted centered moving average*. A three-period weighted centered moving average assigns weights of 1, 2, and 1 to the $D_t/\hat{D}_t$ values (denoted by d_t for this explanation) in the following manner:

$$M_t = \frac{1d_{t-1} + 2d_t + 1d_{t+1}}{4}$$

Note that the divisor is 4, not 3. Exercise 18.77 requires this weighted centered moving average.

5. Another approach to isolating the cyclical component in a multiplicative model involves both the trend line and a centered moving average. Since M_t estimates $T_t \cdot C_t$ and the trend line estimates T_t, forming the ratio of M_t to $\hat{Y}_t$ should isolate C_t:

$$\frac{M_t}{\hat{Y}_t} = \frac{T_t \cdot C_t}{T_t} = C_t$$

Exercise 18.23 illustrates this technique.

6. In general, a cycle is measured by tracing the cyclical relatives through a peak and a valley, starting and stopping with the baseline value of $C_t = 1.000$.
7. If either Y_t or $\hat{Y}_t$ is a negative value, then forming the ratio $Y_t/\hat{Y}_t$ is meaningless; hence, C_t does not exist for that time period for annualized data. For seasonalized data, C_t should not be computed for time t via an unweighted or a weighted centered moving average if any one of the ratios $D_t/\hat{D}_t$ in the average is a negative value. For an alternative approach to finding C_t in this case, the reader is referred to advanced texts on forecasting, such as *Forecasting: Methods and Applications* by Makridakis, Wheelwright, and McGee (1983).

Table 18.2 summarizes our work in this section on describing the historical movement in a time series by decomposing its variability into a trend, seasonal, and/or cyclical component.

Table 18.2 Summary of Time Series Decomposition

Component	Name of Technique	Symbol	Applicable Model(s)	Procedure
Trend	Moving average	M_t	Multiplicative/ additive	Successively average n values of Y_t
	Trend line	$\hat{Y}_t$	Additive/ multiplicative	Fit a simple regression model to Y_t
Seasonal	Ratio-to-moving average seasonal index	S_t	Multiplicative	1. Form the ratio of Y_t to M_t 2. Average the ratios for the same season 3. Adjust the averages to 100 percent
Cyclical	Cyclical relatives	C_t	Multiplicative	1. Deseasonalize Y_t, if necessary 2. Detrend Y_t 3. Neutralize the irregular component, if necessary, with a centered moving average

Now that we have taken apart the time series to describe its past behavior, it seems natural to recombine the components to predict future values of Y_t. However, all four components will not or cannot be used. Clearly, the irregular component is beyond our control, so any attempt at predicting it would be futile. Similarly, the future cyclical activity is erratic and uncertain; simple forecasting techniques usually do not attempt to model it. To predict the next value, y_{t+1}, and future values in the time series, we will use the present and previous values of Y_t, the trend line, and the seasonal indices. These forecasting techniques are presented in the next section.

CLASSROOM EXAMPLE 18.1

Decomposing a Time Series

Quarterly sales, in millions of cases, of an instant soup mix are recorded in the following table for the period 1986–89. Our goal is to describe the four-year sales pattern by decomposing the time series into the trend, seasonal, and cyclical components.

Year	Quarter	t	Y_t	M_t	Y_t/M_t	S_t	D_t	$\hat{D}_t$	$D_t/\hat{D}_t$	C_t
1986	1	1	10.8							
	2	2	9.8							
	3	3	9.4							
	4	4	9.8							
1987	1	5	9.9							
	2	6	9.0							
	3	7	8.6							
	4	8	9.4							
1988	1	9	9.7							
	2	10	9.1							
	3	11	9.0							
	4	12	9.8							
1989	1	13	9.8							
	2	14	9.0							
	3	15	8.6							
	4	16	9.1							

a. Construct a four-quarter centered moving average, M_t, to depict the general trend in the time series. Use four decimal-place accuracy.

b. Plot Y_t and M_t on the same graph.

c. Find the seasonal index to four decimal-place accuracy for each quarter by doing the following:

(1) Determine the ratio of Y_t to M_t for each available quarter.

(2) Construct a table similar to the one that follows, and enter the ratios from the Y_t/M_t column.

(3) Find the average for each quarter.

Year	Quarter 1	2	3	4
1986				
1987				
1988				
1989				
Average				
Adjusted seasonal index				

(4) Adjust the quarterly averages to sum to 4.0000.

(5) Record the adjusted seasonal indices in the S_t column in a table similar to the one at the beginning of the example.

d. Develop the cyclical relatives for each quarter by doing the following:

(1) Deseasonalize the data. Enter the values in the D_t column, using two decimal-place accuracy.

(2) Fit a simple regression model to the deseasonalized data, using these intermediate calculations:

$$\Sigma t = 136 \qquad \Sigma D_t = 150.74$$
$$\Sigma t^2 = 1496 \qquad \Sigma t D_t = 1260.92$$
$$\bar{t} = 8.5 \qquad \bar{D}_t = 9.42125$$

Carry out the calculations of b_0 and b_1, to two decimal-place accuracy.

(3) Use the fitted regression equation to predict D_t for each quarter. Enter the predictions in the $\hat{D}_t$ column.

(4) Form the ratio $D_t/\hat{D}_t$.

(5) Find a three-period centered moving average. Enter the values in the C_t column.

e. Graph the cyclical relatives and indicate when the cycle begins and ends.

SECTION 18.2 EXERCISES

18.9 A major retailer tracks its sales of swimwear on a quarterly basis. For the past six years, following are the ratios of actual sales to a four-quarter centered moving average.

Year	Quarter 1	2	3	4
1985	—	—	1.4381	.7826
1986	.5964	1.1207	1.3992	.8114
1987	.6290	1.1165	1.3594	.8206

Year	Quarter 1	2	3	4
1988	.5909	1.1768	1.4535	.8413
1989	.5753	1.1576	1.5043	.7920
1990	.6033	1.1557	—	—

Develop seasonal indices for the retailer.

18.10 Refer to Exercise 18.9. Suppose the actual sales for the third quarter of 1990 were $85,000. Deseasonalize this value and report a seasonally adjusted figure.

18.11 Refer to Exercise 18.3.

a. Isolate the trend component for the average contribution data by fitting a simple regression model.

b. Use the fitted model to (point) predict the average contribution in 1989.

c. Relative to the actual 1989 average contribution, what is the percent error in the prediction from the regression model?

18.12 Refer to Exercise 18.6.

a. Develop a four-quarter centered moving average to estimate the trend component.

b. Plot M_t and Y_t on the same graph.

18.13 Refer to Exercise 18.6 and 18.12. Find the seasonal index for each quarter.

18.14 The monthly number of mobile homes manufactured and shipped, in thousands of units, is listed as follows:

Month	Year 1984	1985	1986	1987
January	20.0	18.6	18.9	16.0
February	22.2	19.7	18.7	16.3
March	25.6	24.0	20.5	19.6
April	25.9	26.2	22.9	21.1
May	29.0	28.0	22.5	20.2
June	27.8	25.1	21.4	21.5
July	24.6	24.3	20.3	21.0
August	30.0	27.8	21.6	22.0
September	24.4	24.6	21.5	21.7
October	27.7	27.7	23.1	22.2
November	21.9	21.0	17.2	17.0
December	16.6	16.9	15.8	14.3

Source: U.S. Department of Commerce, *Business Statistics, 1986–88*, supplement to the *Survey of Current Business*, Washington, D.C., 1988.

Construct a 12-month centered moving average for this time series.

18.15 Refer to Exercise 18.14.

a. Fit a trend line to the time series.

b. Find the trend line fitting errors for January and for August of each year.

c. From the fitting errors in part b, describe a general pattern in Y_t that the trend line fails to detect.

18.16 Using Equations 18–3 and 18–4 as guides, write the equation for M_t for the following:
a. a six-period centered moving average.
b. a five-period centered moving average.

18.17 Refer to Exercises 18.14 and 18.16. Find the seasonal index for each month. Use a six-month centered moving average to generate the M_t values.

18.18 Interpret the following seasonal indices.
a. $S_t = 105\%$
b. $S_t = 89\%$

18.19 Refer to Exercise 18.3.
a. Isolate the cyclical component for the percent of contributing alumni time series by finding cyclical relatives, C_t.
b. Graph the C_t's.

18.20 Suppose we have a time series of quarterly data and the trend line has a positive slope. If possible, describe the slope—positive, negative, or (essentially) zero—of a simple regression model fit through the sequence of values that result when we do the following:
a. detrend the time series.
b. deseasonalize the time series.
c. deseasonalize and then detrend the time series.

18.21 On the first Saturday in May, the granddaddy of horse races—the Kentucky Derby—is run at Churchill Downs in Louisville, Kentucky. The amount of money bet, in millions of dollars, on this race is given in the following table for the 60-year period from 1927 through 1986.

Year	Amount	Year	Amount	Year	Amount	Year	Amount
1927	$ .68	1942	$ .63	1957	$1.40	1972	$2.89
1928	.62	1943	.59	1958	1.64	1973	3.28
1929	.68	1944	.65	1959	1.50	1974	3.44
1930	.58	1945	.78	1960	1.49	1975	3.37
1931	.50	1946	1.20	1961	1.48	1976	3.45
1932	.28	1947	1.25	1962	1.55	1977	3.67
1933	.23	1948	.67	1963	1.82	1978	4.43
1934	.38	1949	1.03	1964	2.14	1979	4.01
1935	.41	1950	1.25	1965	2.23	1980	4.16
1936	.47	1951	1.29	1966	2.13	1981	4.57
1937	.59	1952	1.57	1967	1.93	1982	5.01
1938	.53	1953	1.53	1968	2.35	1983	5.55
1939	.58	1954	1.54	1969	2.63	1984	5.42
1940	.46	1955	1.68	1970	2.38	1985	5.77
1941	.65	1956	1.67	1971	2.65	1986	6.17

Source: "How the Betting Went," Louisville *Courier-Journal*, May 4, 1986. Copyright 1986, Louisville *Courier-Journal*. Reprinted with permission.

Develop the set of cyclical relatives for this time series.

18.22 Quarterly profits and dividends, as reported by manufacturing corporations, are collected and segregated into various industries by the Bureau of Economic Analysis in the Department of Commerce. For the stone, clay, and glass products industry the net profits after taxes, in millions of dollars, are as follows:

	Quarter			
Year	1	2	3	4
1984	161	613	614	482
1985	108	583	504	432
1986	187	664	655	614
1987	660	852	982	436

Source: U.S. Department of Commerce, *Business Statistics, 1986–88,* supplement to the *Survey of Current Business,* Washington, D.C., 1988.

Find the cyclical relative for each quarter. Use a three-period unweighted centered moving average to eliminate any irregular effects in $C_t \cdot I_t$. Identify the beginning and ending of any apparent cycles. In computing the D_t values, round to two decimal places.

***18.23** Refer to Exercise 18.22.

a. Estimate the trend in Y_t = net profits after taxes by fitting a simple regression model to the time series.

b. Isolate C_t by forming the ratio of M_t (from Exercise 18.22) to $\hat{Y}_t$.

c. Plot the cyclical relatives from Exercise 18.22 and the C_t's from part b on the same graph and compare.

18.3 TIME SERIES FORECASTING

Decomposing a time series to isolate its components provides a descriptive analysis of the past behavior of a variable Y_t. A natural follow-up to such an analysis is to predict the future movement in Y_t. This inferential activity is called *forecasting* and represents the second of our analysis objectives.

Forecasting is essential in business. Accurate sales forecasts can help managers to schedule production, make deliveries, maintain inventories, and facilitate cash flow, to name a few benefits. Forecasting can be as simple or as complex as the firm or manager wishes.

We will present six simple methods for predicting the next value in the time series, denoted by y_{t+1}. In general, the subscript of $t + 1$ simply means the upcoming quarter, month, year, and so forth in the sequence of time series values. The symbol $\hat{y}_{t+1}$ will signify the forecasted value of Y_{t+1}.

The basic principle guiding a simple forecasting technique is that the historical data patterns will repeat in the future. Based on this assumption, we generate a forecast by extrapolating the past and/or current values of Y_t. Many approaches are available to do this. What follows are six of the more popular techniques. After each has been explained, we will demonstrate them in forecasting Y_{t+1} in an example.

*Optional

COMMENTS

1. Forecasting the ensuing value in a monthly or quarterly time series is referred to as a *short-term forecast*. In Section 18.4, we will examine various time horizons in time series forecasting.
2. More sophisticated techniques than the ones we will present exist, but these advanced methods are beyond the scope of our book. The interested reader is referred to texts devoted to forecasting, such as *Forecasting: Methods and Applications* by Makridakis, Wheelwright, and McGee (1983).

Naive

A *naive forecast* for Y_{t+1} is simply the current value for the time series, y_t, as indicated in Equation 18–6.

Naive Forecast

$$\hat{y}_{t+1} = y_t \qquad (18\text{–}6)$$

The naive forecast is quick and easy to use and is intended for time series that are relatively stable. When a seasonal effect is present, however, the naive forecast is usually off the mark. (We will discuss the issue of forecasting accuracy in Section 18.4.) Also note that the forecast ignores all the historical observations in the time series except the most recent one.

COMMENT

A variation of the naive forecast in Equation 18–6 is $\hat{y}_{t+1} = ky_t$, where k is a constant growth (or contraction) factor. Choices for k include (one plus) the inflation rate or the average value of successive ratios, y_t/y_{t-1}, of the time series. Of course, the naive forecast is the special case when $k = 1$.

Mean

To utilize all past data, we could average successively the values into a *mean forecast*, as indicated in Equation 18–7.

Mean Forecast

$$\hat{y}_{t+1} = \frac{y_1 + y_2 + \ldots + y_t}{t} \qquad (18\text{–}7)$$

For stationary time series, a mean forecast may be adequate. But when a trend is present, the mean forecast consistently underpredicts Y_{t+1} for a positive trend and overpredicts for a negative trend.

An adaptation of the mean forecast for quarterly or monthly data is a *seasonalized mean forecast*. Rather than averaging all the Y_t values, we average only those from the same season. For quarterly data, we would average all the existing first quarter values at the time of the forecast, then all the existing second quarter values at the time of the forecast, and so on for the third and fourth quarters. Let $\overline{S}_1$ represent the average of the Y_t values from the first quarter, and define $\overline{S}_2$, $\overline{S}_3$, and $\overline{S}_4$ in a like fashion. A seasonalized mean forecast, where $\overline{S}_i$ is the seasonal average corresponding to the $(t + 1)$ time period, is given in Equation 18–8.

Seasonalized Mean Forecast

$$\hat{y}_{t+1} = \overline{S}_i \qquad \text{(18–8)}$$

COMMENTS

1. The seasonal indices that we developed in Section 18.2 are *not* the same as the $\overline{S}_i$'s.
2. A variation of Equation 18–8 is $\hat{y}_{t+1} = k \cdot \overline{S}_i$. As we mentioned in the naive forecast, the average of successive ratios of Y_t values from the same season often is used as the value for k. Exercise 18.38 illustrates this forecasting approach.

Moving Average

As Equations 18–3 and 18–4 indicate, a centered moving average uses leading and past values of Y_t in the computation of M_t. When our objective switches from description to prediction, we cannot use a centered moving average since the leading values are unavailable. Instead, we use a *moving average forecast,* which incorporates the most recent values in the computation. To distinguish a centered moving average, denoted by M_t, from a moving average, we will use M_t^* to denote the latter. Equation 18–9 gives a formula for finding a four-period moving average.

Four-Period Moving Average

$$M_t^* = \frac{y_{t-3} + y_{t-2} + y_{t-1} + y_t}{4} \qquad \text{(18–9)}$$

In a like manner, a three-period moving average would average the three most recent values, while the most recent 12 y_t's would be used in a 12-period moving average.

A moving average forecast for Y_{t+1} is simply the current value of the moving average as indicated in Equation 18–10.

Moving Average Forecast

$$\hat{y}_{t+1} = M_t^* \qquad (18\text{–}10)$$

In this forecasting technique, only the most recent values of Y_t are used to predict Y_{t+1}; all other historical values are ignored. For example, using a four-period average to forecast Y_{12} involves only the values y_8 through y_{11}; y_1 through y_7 are not utilized in $\hat{y}_{12}$. Generally, the accuracy of a moving average forecast is a function of the number of periods in the average. Forecasts based on a smaller number of periods may be less precise than those based on a larger number of periods.

COMMENTS

1. A one-period moving average is the same as the naive forecast.
2. Our differentiation between a centered moving average and a moving average may be blurred in applications and in courses where forecasting is the primary objective. A centered moving average is used only in description; a moving average is used in forecasting and also could be used in description. Hence, in some books only a moving average is presented.
3. A computationally easier form of Equation 18–9 for a four-period moving average is

$$M_t^* = M_{t-1}^* + \frac{y_t - y_{t-4}}{4}$$

This expression is useful for updating M_t^* provided $t > 4$.

Trend Projection

In Section 18.2, we related Y_t to t with a simple regression model (Equation 18–5) and called the fitted equation a *trend line*. Recall in our study of simple regression in Chapter 15 that we cautioned against predicting the dependent variable outside of the relevant range of values for the independent variable. Our reasoning was that the linear relationship between the variables may not continue beyond the sample domain. However, we said at the beginning of this section that time series forecasting is based on the belief that historical data patterns repeat in the future. Hence, if we *assume* that the linear trend in our time series data continues into time period $t + 1$, then we can use the fitted regression model to predict Y_{t+1}. This is called *trend projection*. Equation 18–11 represents the trend projection forecast for Y_{t+1}, where b_0 and b_1 are the estimated regression coefficients for the trend line.

Trend Projection Forecast

$$\hat{y}_{t+1} = b_0 + b_1(t + 1) \qquad (18\text{–}11)$$

For seasonalized data, a trend projection forecast may be considerably in error for each predicted time period, but produce a small overall average error. Information from all Y_t values is blended together to produce the values for b_0 and b_1, so technically all historical data is used to predict Y_{t+1}, though attaching a weight to a specific y_t is impossible.

COMMENTS

1. Unlike the three previous techniques, the trend projection forecast allows us to develop an interval estimate of Y_{t+1}, if desired. Assuming the error term (that is, the irregular component) is normally distributed, a prediction interval for Y_{t+1} can be constructed according to Equation 15–20. The time series variable t functions like the variable X in Chapter 15. Thus, $(t - \bar{t})^2$ replaces the term $(x - \bar{X})^2$ and SSt substitutes for SSX in Equation 15–20.
2. To adjust the trend projection for seasonal influences, we could fit a more sophisticated regression model. For instance, suppose we have quarterly data. The model $Y = \beta_0 + \beta_1 X_1 + \beta_2 X_2 + \beta_3 X_3 + \beta_4 t + \epsilon$ includes the variable t to account for the trend and three indicator variables—X_1, X_2, and X_3—to account for seasonal influences. Define X_1 to be 1 for the first quarter of the year; 0, otherwise. Define X_2 and X_3 similarly for the second and third quarters, respectively. Then the values of the independent variables for the first quarter of the second year, for example, would be $X_1 = 1, X_2 = 0, X_3 = 0$, and $t = 5$. Assuming additive effects of the time series components, this *multiple regression* model estimates $T_t + S_t$ and, once fit to the data, could be extrapolated to predict Y_{t+1}.
3. Another trouble spot with using the regression line to forecast Y_{t+1} is the assumptions surrounding the error term. In simple regression, we assumed that error random variables were independent for different values of X (see Chapter 15). But in time series, it is quite likely that for successive values of t, the errors are positively correlated. We introduced the Durbin-Watson test in Chapter 17 for detecting this problem of autocorrelation. If an interval forecast for Y_{t+1} is planned, we strongly recommend that you first conduct the Durbin-Watson test on the residuals. If autocorrelation is confirmed, do not proceed with Equation 15–20. Advanced forecasting methods are necessary in this case; consult an appropriate forecasting text.

Multiplicative Model

In the previous section, we decomposed a time series by isolating the trend, seasonal, and cyclical components. The trend factor charts the long-term direction of Y_t, while the seasonal effect, by definition, is a sequence of recurring peaks and valleys within a given interval of time. Cyclical fluctuations on the other hand, are of variable length and intensity and do not recur on a regular basis like seasonal effects. The cyclical relatives we developed reveal the historical lengths and intensities of each cycle. But, projecting these underlying economic conditions forward is more of a subjective assessment and less of a quantitative computation. Predicting C_{t+1} requires more advanced work than covered within our survey treatment. In a simple forecast, we will not attempt to take cyclical relatives into account. Likewise, the irregular component is unpredictable, so we shall assign $\hat{I}_{t+1} = 1.00$.

Assuming the time series components are related through Equation 18–1, a multiplicative model forecast for Y_{t+1} is given in Equation 18–12.

Multiplicative Model Forecast

$$\hat{y}_{t+1} = \hat{T}_{t+1} \cdot \hat{S}_{t+1} \quad (18\text{–}12)$$

The trend component for the $(t + 1)$ time period can be estimated with either a trend line projection or a moving average. The estimated seasonal component is merely the appropriate seasonal index for the time period $t + 1$. This forecasting method generally is more precise than the naive method and works fairly well for seasonal data. In addition, all the historical data are incorporated in the estimates of T_{t+1} and S_{t+1}.

COMMENTS

1. Should there be an estimate of C_{t+1} available, we would include it into the multiplicative model forecast by multiplying: $\hat{y}_{t+1} = \hat{T}_{t+1} \cdot \hat{S}_{t+1} \cdot \hat{C}_{t+1}$. Often, leading economic indicators serve as proxies for C_{t+1}.
2. In lieu of a trend projection based on Equation 18–11, some forecasters use the following alternative approach: Deseasonalize Y_t by computing $D_t = Y_t/S_t$; fit a simple regression model to the deseasonalized data to yield $\hat{D}_t = b_0 + b_1 t$; and project this trend line to time period $t + 1$. Then a (modified) multiplicative model forecast is

$$\hat{y}_{t+1} = \hat{D}_{t+1} \cdot \hat{S}_{t+1}$$

Exercise 18.37 demonstrates this forecasting technique.

3. For annual data with no seasonal indices, the multiplicative model forecast reduces to the trend projection forecast.

Exponential Smoothing

In the moving average forecast, we used only the n most recent values, giving each an equal weight. Thus, in a four-quarter moving average forecast for the next quarter, the contribution of last year's value of Y_t is the same as the contribution of the current quarter's value of Y_t. We may wish to place more emphasis on y_t than on y_{t-4}, since successive values of many economic variables may be highly correlated. Usually, the degree of correlation diminishes over time. For instance, this week's auction rate for three-month Treasury bills is likely to be close in value to last week's rate, but may be starkly different from the auction rate several weeks ago.

To enable a forecaster to be more sensitive to current data and less sensitive to distant data, we need a forecast with moving *weights*, not moving *averages*. Equation 18–13 is based on such a concept.

Exponentially Smoothed Forecast

$$\hat{y}_{t+1} = \hat{y}_t + \alpha(y_t - \hat{y}_t) \qquad \text{(18–13)}$$

An exponentially smoothed forecast (or "new" forecast), $\hat{y}_{t+1}$, consists of the most recent (or "old") forecast plus a correction factor. The correction factor is a constant, α, times the error in the old forecast: $y_t - \hat{y}_t$. The smoothing constant α is a number between 0 and 1 and must be chosen by the forecaster.

Although only the most recent value of the time series, y_t, appears explicitly in Equation 18–13, all previous values contribute in the forecast of Y_{t+1} as well. To see this, suppose the current time period is $t = 2$ and we wish to forecast Y_3. According to Equation 18–13, the forecast for the second time period is

$$\hat{y}_2 = \hat{y}_1 + \alpha(y_1 - \hat{y}_1)$$

There is no historical evidence with which to forecast Y_1, the first value of the time series, so we arbitrarily assign

$$\hat{y}_1 = y_1$$

in order to generate the forecast for Y_2.

Under this assumption, the forecast for Y_2 reduces to

$$\hat{y}_2 = y_1$$

For the third time period, we use Equation 18–13 again to predict

$$\hat{y}_3 = \hat{y}_2 + \alpha(y_2 - \hat{y}_2)$$

Substituting y_1 for $\hat{y}_2$ and then rearranging, we get

$$\begin{aligned}\hat{y}_3 &= y_1 + \alpha(y_2 - y_1)\\ &= \alpha y_2 + (1 - \alpha)y_1\end{aligned}$$

For three different choices of α, following are the relative contributions of y_1 and y_2 in predicting Y_3:

	Weights	
α	y_2	y_1
.1	.10	.90
.5	.50	.50
.9	.90	.10

Now assume we know y_3 and wish to predict Y_4. You can verify that the exponentially smoothed forecast for Y_4, as a function of α, y_1, y_2, and y_3, is

$$\begin{aligned}\hat{y}_4 &= \hat{y}_3 + \alpha(y_3 - \hat{y}_3)\\ &= \alpha y_3 + \alpha(1 - \alpha)y_2 + (1 - \alpha)^2 y_1\end{aligned}$$

Using the same three choices for α as before, following are the relative contributions of y_1, y_2, and y_3 in predicting Y_4:

	Weights		
α	y_3	y_2	y_1
.1	.10	.09	.81
.5	.50	.25	.25
.9	.90	.09	.01

Comparing this table with the previous one, we see that for a small value of α more weight is given to distant values. For a large value of α, more weight is placed on recent values. Though all Y_t values contribute to $\hat{y}_{t+1}$, the contribution of distant values diminishes with time, regardless of the value of α. For $\alpha = .1$, y_1 had a 90% weight in predicting Y_3, and then an 81% weight in predicting Y_4. When $\alpha = .9$, y_1's contribution decreased from 10 percent to 1 percent. Therefore, the size of α governs the rate of each y_t's diminishing influence in forecasting Y_{t+1}.

Two main factors contribute to the selection of α: the relative stability of the time series and the relative precision of the historical forecasts. For volatile time series, small values of α are preferred; for stable time series, larger values may be needed. Given a sequence of values for Y_t, we could use a computer simulation to derive the best choice for α that minimizes the historical errors of prediction. When we discuss the different ways of measuring forecasting error in Section 18.4, we will complete this explanation.

COMMENTS

1. An alternative form of Equation 18–13 that may be seen in other books is

$$\hat{y}_{t+1} = \alpha y_t + (1 - \alpha)\hat{y}_t$$

2. If $\alpha = 1$, the exponentially smoothed forecast coincides with the naive forecast.
3. Any value could be used for $\hat{y}_1$ to initialize the exponential smoothing process, but $\hat{y}_1 = y_1$ is the most convenient. Note that this is *not* actually a forecast, since the realized value, y_1, would not have existed at the time the forecasted value, $\hat{y}_1$, would have been made. Instead, $\hat{y}_1 = y_1$ is an initializing condition.
4. Exponential smoothing also could be used as a descriptive technique to estimate the trend component.
5. Our use of the Greek letter α to represent the smoothing constant is somewhat unfortunate. Recall that we previously used this letter to denote the probability of making a Type I error in a hypothesis testing situation. Generally, we prefer to avoid having one symbol mean two different things, depending on the problem context. But, in this case, α is more or less the standard statistical notation for both situations, so we have chosen to stick with tradition.
6. Exponentially smoothed forecasts are fairly accurate for stationary time series (those without a trend component), but are imprecise when a trend effect is present, regardless of the value of α. For example, in Exercise 18.63, part a, the data have a linear trend, but the exponentially smoothed forecasts consistently lag Y_t and the best choice for α to minimize the historical errors of prediction is essentially 1.00.

Now that we have introduced the simple forecasting techniques, let us demonstrate how they work.

EXAMPLE 18.6

In Example 18.1, we presented 16 quarterly values of the time series variable, Y_t, work-in-progress inventory for a large manufacturing company. The data are reproduced in the following table:

Year	Quarter	t	Y_t	Year	Quarter	t	Y_t
1987	1	1	15.2	1989	1	9	19.6
	2	2	18.4		2	10	22.1
	3	3	18.7		3	11	21.4
	4	4	22.2		4	12	24.4
1988	1	5	17.8	1990	1	13	19.6
	2	6	21.9		2	14	23.4
	3	7	21.9		3	15	24.1
	4	8	25.6		4	16	28.2

Predict Y_{17}, the inventory level for the first quarter of 1991, using the following:

a. a naive forecast.

b. a seasonalized mean forecast.

c. a four-quarter moving average forecast.

d. a trend projection forecast.

e. a multiplicative model forecast with a trend projection to estimate the T_{t+1} component.

f. an exponentially smoothed forecast, with $\alpha = .4$.

Solution:

a. The naive forecast by Equation 18–6 is

$$\hat{y}_{17} = y_{16} = 28.2$$

b. Since we have seasonal data, we will average the existing first quarter values of Y_t at the time of the forecast:

$$\bar{S}_1 = \frac{15.2 + 17.8 + 19.6 + 19.6}{4} = 18.05$$

The seasonalized mean forecast for the first quarter of 1991 is, according to Equation 18–8,

$$\hat{y}_{17} = \bar{S}_1 = 18.05$$

c. A four-quarter moving average for $t = 16$, according to Equation 18–9, is

$$M^*_{16} = \frac{y_{13} + y_{14} + y_{15} + y_{16}}{4}$$
$$= \frac{19.6 + 23.4 + 24.1 + 28.2}{4}$$
$$= \frac{95.3}{4} = 23.825$$

The moving average forecast, therefore, is

$$\hat{y}_{17} = M^*_{16} = 23.825$$

d. To develop a trend projection forecast, we must fit the simple regression model $Y_t = \beta_0 + \beta_1 t + \epsilon$ to the data. The fitted model is

$$\hat{Y}_t = 17.335 + .494t$$

For $t = 17$, the trend projection forecast is

$$\hat{y}_{17} = 17.335 + .494(17) = 25.733$$

e. The multiplicative model forecast is

$$\hat{y}_{17} = \hat{T}_{17} \cdot \hat{S}_{17}$$

From part d, we use the trend projection for $\hat{T}_{17}$. In Example 18.4, we developed the set of seasonal indices for these data. The seasonal index for the first quarter was $S_1 = .8732$, which we will use as $\hat{S}_{17}$. Thus, the multiplicative model forecast is

$$\hat{y}_{17} = (25.733)\,(.8732) = 22.47$$

f. An exponentially smoothed forecast with $\alpha = .4$ generates a sequence of forecasts like the following:

$$\hat{y}_2 = y_1$$
$$\hat{y}_3 = \hat{y}_2 + .4(y_2 - \hat{y}_2)$$
$$\hat{y}_4 = \hat{y}_3 + .4(y_3 - \hat{y}_3)$$

and so on. Before we can predict Y_{17}, we must evaluate the first 16 such forecasts. Following is a worksheet of the necessary computations, carried out to three decimal places:

t	Y_t	$\hat{Y}_t$	$(Y_t - \hat{Y}_t)$	$.4(Y_t - \hat{Y}_t)$
1	15.2	—	—	—
2	18.4	15.2	3.2	1.28
3	18.7	16.48	2.22	.888

t	Y_t	$\hat{Y}_t$	$(Y_t - \hat{Y}_t)$	$.4(Y_t - \hat{Y}_t)$
4	22.2	17.368	4.832	1.933
5	17.8	19.301	−1.501	−.600
6	21.9	18.701	3.199	1.280
7	21.9	19.981	1.919	.768
8	25.6	20.749	4.851	1.940
9	19.6	22.689	−3.089	−1.236
10	22.1	21.453	.647	.259
11	21.4	21.712	−.312	−.125
12	24.4	21.587	2.813	1.125
13	19.6	22.712	−3.112	−1.245
14	23.4	21.467	1.933	.773
15	24.1	22.240	1.860	.774
16	28.2	22.984	5.216	2.086
17		25.070		

Finally, for $t = 17$, the exponentially smoothed forecast is

$$\begin{aligned}\hat{y}_{17} &= \hat{y}_{16} + .4(y_{16} - \hat{y}_{16})\\ &= 22.984 + 2.086\\ &= 25.070\end{aligned}$$

To summarize Example 18.6, following are the six forecasts based on the time series data:

Technique	**Forecast for Y_{17}**
Naive	28.2
Seasonalized mean	18.05
Four-quarter moving average	23.825
Trend projection	25.733
Multiplicative model	22.47
Exponentially smoothed	25.070

□

Judging the accuracy of a time series forecast is our next order of business. We caution you not to select a forecasting technique solely on the precision of one point prediction. Instead, we should evaluate a forecasting method on the basis of average accuracy over a number of forecasts. Accuracy can be measured in several ways, as we will discover in the next section.

One final comment is appropriate before we leave this section, though. Given the variability in forecasted values for Y_{17} in Example 18.6, perhaps we would be wiser to consider synthesizing the several forecasts into a composite one rather than trying to select one "best" method. This idea of *combining forecasts* also will be discussed in the next section.

CLASSROOM EXAMPLE 18.2

Generating One-Period-Ahead Forecasts

In Classroom Example 18.1, we decomposed the time series of Y_t = quarterly sales, in millions of cases, of an instant soup mix. The data are reproduced as follows:

Year	Quarter	t	Y_t	Year	Quarter	t	Y_t
1986	1	1	10.8	1988	1	9	9.7
	2	2	9.8		2	10	9.1
	3	3	9.4		3	11	9.0
	4	4	9.8		4	12	9.8
1987	1	5	9.9	1989	1	13	9.8
	2	6	9.0		2	14	9.0
	3	7	8.6		3	15	8.6
	4	8	9.4		4	16	9.1

Forecast the sales for the first quarter of 1990, using the following techniques:

a. a naive forecast.

b. a seasonalized mean forecast.

c. a four-quarter moving average forecast.

d. a trend projection forecast, assuming $b_0 = 9.98$ and $b_1 = -.065$.

e. a multiplicative model forecast, with a trend projection to estimate the T_{t+1} component.

f. an exponentially smoothed forecast, with $\alpha = .2$, and assuming $\hat{y}_{15} = 9.4468$.

SECTION 18.3 EXERCISES

18.24 One of the barometers of the financial health of our country is the flow of money through the economy. The Federal Reserve Board has devised several measures of available cash, including an estimate of the billions of dollars in coins and currency in circulation at the end of each quarter. Following are these figures for the five-year period from 1983–87.

	Quarter			
Year	1	2	3	4
1983	155.7	162.0	162.8	171.9
1984	168.7	175.1	175.3	183.8
1985	179.2	185.9	187.3	197.5
1986	193.2	199.3	200.6	212.0
1987	207.8	215.2	216.8	230.2

Source: U.S. Department of Commerce, *Business Statistics, 1986–88*, supplement to the *Survey of Current Business*, Washington, D.C., 1988.

a. What is the naive forecast for the first quarter of 1988?
b. What is the seasonalized mean forecast for the first quarter of 1988?
c. What would have been the seasonalized mean forecast for the second quarter of 1985?
d. What would have been the mean forecast for the second quarter of 1985?

18.25 Refer to Exercise 18.24. Develop a four-quarter moving average forecast for the first quarter of 1988.

18.26 Refer to Exercise 18.24. Develop a trend projection forecast for the first quarter of 1988.

18.27 Refer to Exercise 18.24. Develop a multiplicative model forecast for the first quarter of 1988, using a trend projection to estimate the T_{t+1} component.

18.28 Refer to Exercise 18.24.
a. For which quarter(s) do you think an exponentially smoothed forecast using a large value for α would be most accurate? Least accurate? Why?
b. Develop an exponentially smoothed forecast with $\alpha = .8$, for the first quarter of 1988.
c. Using the sequence of forecasts developed in part b, re-examine your answers to part a.

18.29 Suppose we have four years of quarterly data and wish to forecast the time series for the next quarter—that is, Y_{17}. How much weight does y_{14} have in $\hat{y}_{17}$, if we use the following methods?
a. a naive forecast.
b. a seasonalized mean forecast.
c. a four-quarter moving average.
d. an exponentially smoothed forecast, with $\alpha = .6$.

18.30 What contribution does y_{25} make in an exponentially smoothed forecast of Y_{29}, with $\alpha = .8$?

18.31 Identify the weight of y_1 through y_8 in an exponentially smoothed forecast of Y_9, with $\alpha = .2$.

18.32 Suppose we have five years of monthly data and wish to forecast the time series for the next month—that is, Y_{61}. How much weight does y_{58} have in $\hat{y}_{61}$, if we use the following methods?
a. a naive forecast.
b. a seasonalized mean forecast.
c. a 12-month moving average.
d. an exponentially smoothed forecast, with $\alpha = .3$.

18.33 What contribution does y_{31} make in an exponentially smoothed forecast of Y_{35}, with $\alpha = .5$?

18.34 Refer to Exercises 18.14, 18.15, and 18.17. Predict the number of mobile homes manufactured and shipped in January 1988 using the following methods.
a. a naive forecast.
b. a seasonalized mean forecast.
c. a trend projection forecast.
d. a multiplicative model forecast with a 12-month moving average to estimate T_{t+1} (use the seasonal indices from Exercise 18.17).
e. a moving average forecast (use a six-month period in the moving average).

18.35 Refer to Exercises 18.3 and 18.11. Forecast the average contribution in 1990 with the following methods.
a. a naive forecast.
b. a trend projection.
c. a mean forecast.
d. a moving average (use a three-year period in the moving average).

18.36 Find $\hat{y}_{16}$ using an exponentially smoothed forecast with $\alpha = .8$. Assume $\hat{y}_{13} = 561.24$, $y_{13} = 187$, $y_{14} = 664$, and $y_{15} = 655$.

***18.37** Refer to Exercises 18.6 and 18.13.

a. Deseasonalize the original time series data by computing $D_t = Y_t/S_t$. Round the D_t values to three decimal places. (*Hint:* Study Example 18.5.)
b. Fit the simple regression model $D_t = \beta_0 + \beta_1 t + \epsilon$ to the deseasonalized data.
c. Develop a (modified) multiplicative model forecast for Y_{21} from $\hat{y}_{21} = \hat{D}_{21} \cdot \hat{S}_{21}$.

***18.38** Refer to Exercise 18.6. Form the ratio of successive Y_t values for the first quarter: y_5/y_1, y_9/y_5, y_{13}/y_9, and y_{17}/y_{13}. Find the average value of these ratios. Forecast Y_{21} with a variation of the seasonalized mean forecast, using the average value of the ratios as k in $\hat{y}_{21} = k \cdot \overline{S}_1$.

18.39 In a small subdivision a builder constructed three homes that included several energy efficient passive solar features. Over a five-year period, the average number of kilowatt hours (kwh) of electricity consumed per month by the three homes was recorded. Following are these data:

Month	1985	1986	1987	1988	1989	1990
January		2837	3011	1837	1924	2005
February		1982	1590	1750	1703	1863
March		1521	1149	1187	1079	1160
April		1069	995	1082	847	804
May	1087	1247	1091	1176	1339	
June	1697	1272	1011	1422	1240	
July	1445	1187	1459	1138	1359	
August	1354	983	1265	979	1263	
September	741	798	918	1057	844	
October	913	865	843	833	908	
November	1609	1820	1299	1378	1344	
December	2671	1527	2336	1899	2224	

Forecast the average kwh consumption for May 1990, using a naive forecast.

18.40 Refer to Exercise 18.39. Develop a 12-month moving average forecast for May 1990.

18.41 Refer to Exercise 18.39. Develop a seasonalized mean forecast for May 1990.

18.42 Refer to Exercise 18.39. What would you anticipate the overall trend pattern to be for this data set?

18.43 Refer to Exercise 18.39. Suppose May 1990 was dramatically colder than previous Mays, causing the residents to increase their electricity consumption.

a. Would a seasonalized mean forecast be too high or too low?
b. To which time series component would we attribute this unusual weather condition?
c. Which forecasting methods would be most sensitive to this unusual fluctuation when, one month later, we predict June's consumption? Least sensitive?

18.44 Refer to Exercises 18.39–18.41. Find the forecasting error, $y_{61} - \hat{y}_{61}$, for each technique, if the actual value was $y_{61} = 1050$.

*Optional

18.4 FORECASTING ANALYSIS

The list of criteria for selecting and analyzing a forecasting technique is probably as long as the number of available techniques. For example, forecasters should consider the time frame for which the forecast is intended, the past behavior of the time series, the cost of making a forecast, the historical accuracy of the forecasting technique, the desired accuracy of subsequent forecasts, the cost(s) of an inaccurate forecast, the complexity of the forecasting technique, and possibly the degree to which subjective inputs are incorporated into the forecasting process. So as not to become too overwhelmed, we will restrict our discussion to two factors in selecting and analyzing a forecast: the intended time frame of the forecast and the accuracy of the technique.

Time Horizon

How far into the future do we wish to predict? In the previous section, we presented forecasts for the next time period, $t + 1$. But, depending on the units of time, this could be the next month, quarter, or year. Generally, forecasts are classified according to three time frames: short-term, medium-range, or long-term, as indicated in Table 18.3.

Table 18.3 is a *general* characterization of forecasts and need not represent every situation. For instance, Y_t = hourly grocery store sales is "doubly seasonal" by the hour of the day and the days of the week. Daily sales usually peak between 11 A.M. and 2 P.M. Weekly sales are slow early in the week, usually cresting on Thursdays and remaining high on Fridays and Saturdays. A short-term forecast for a grocery store's sales is a matter of hours or days, not months into the future. Therefore, all short-term forecasts are not restricted to the 1 to 3 months time frame. The scope depends on the entity of time.

Nevertheless, the labels in Table 18.3 have evolved into relatively standard forecasting terms. Also in general, the seasonal component seems to impact the short-term forecast the most. In relating forecasting to a management hierarchy, we would expect to see short-term forecasts used most frequently by first line to mid-level managers. Typically low- to mid-level managers are concerned with the day-to-day operation of the firm and tend to focus on the short run details.

Table 18.3 Time Horizons in Forecasting

Name	Scope	Most Sensitive Component in Forecast
Short term	1 to 3 months	Seasonal
Medium range	3 months to 2 years	Cyclical
Long term	Over 2 years	Trend

Medium-range forecasts are less influenced by seasonal variation and more sensitive to cyclical effects. These forecasts tend to be the most difficult ones to make precisely. As we mentioned in connection with a multiplicative model forecast in Section 18.3, forecasting the cyclical component is tricky, and of course this uncertainly carries over to a medium-range forecast. Mid- and top-level managers utilize medium-range forecasts the most.

Long-term forecasts though are dominated by the trend component. Since we expect the seasonal and cyclical effects to have completed their cycles in a multi-year period, their effects on the overall direction of the time series are dampened. Being concerned with the "big picture," top-level managers are most interested in long-term forecasts.

As we might expect, accuracy wanes as the time frame waxes. Short-term forecasts are likely to be more precise than a long-term prediction, since many extraneous variables and economic conditions may alter the direction of the time series gradually but not instantly. The precision of the forecast is a second consideration in evaluating techniques.

COMMENT A fourth time frame—the immediate term—may be mentioned in some forecasting books. By our standards, an immediate term forecast would be a forecast for less than one month into the future.

Error Metrics

Forecasting error, denoted by e_t and represented in Equation 18–14, is the difference between the actual value of Y_t and the forecasted value.

Forecasting Error

$$e_t = y_t - \hat{y}_t \qquad \textbf{(18–14)}$$

The smaller the error, the more precise is the forecast. Though a forecasting technique may be very precise for a specific time period, we prefer to judge a technique on its overall precision in a sequence of forecasts, possibly extending the entire history of the time series. In repeated forecasts, we will see positive and negative errors, most of which will be offset when the errors are averaged. We have faced this predicament before and, as before, our approach will be to work with a squared error or the absolute value of an error. There are three common error metrics in this regard: squared error, absolute error, and percent error.

Mean Squared Error A familiar quantity from regression and analysis of variance is *mean squared error, MSE*. Each forecasting error is squared and then these squares are totaled and averaged. Equation 18–15 formalizes this concept. The denominator n represents the number of time periods for which e_t is computed.

Mean Squared Error

$$MSE = \frac{\Sigma e_t^2}{n} = \frac{\Sigma(y_t - \hat{y}_t)^2}{n} \qquad \text{(18–15)}$$

Mean Absolute Deviation Rather than square each deviation, we could find the absolute value of each forecasting error, and hence, the average error without regard to direction. We encountered such a measure in Chapter 3 when we studied the *average deviation,* also known as the *mean absolute deviation, MAD*.

Mean Absolute Deviation

$$MAD = \frac{\Sigma|e_t|}{n} = \frac{\Sigma|y_t - \hat{y}_t|}{n} \qquad \text{(18–16)}$$

Mean Absolute Percentage Error A third way of measuring precision is to determine the percentage that each forecast is in error relative to the actual value. Equation 18–17 shows how to compute a *percentage error,* denoted by p_t.

Percentage Error

$$p_t = \frac{y_t - \hat{y}_t}{y_t} \times 100\% \qquad \text{(18–17)}$$

Although we could find the average of the percentage errors, we will take the absolute value of each percentage error first and then find the average. The resulting measure, shown in Equation 18–18, is called the *mean absolute percentage error* and is denoted by *MAPE*.

Mean Absolute Percentage Error

$$MAPE = \frac{\Sigma\,|\,p_t\,|}{n} = \frac{\Sigma\left|\dfrac{y_t - \hat{y}_t}{y_t}\right|}{n} \times 100\% \qquad \text{(18–18)}$$

Of the three methods of measuring error, *MAPE* may be the most preferred method because it is easiest to interpret. The *MAPE* allows us to express error in percentage terms, facilitating comparisons. *MAPE,* then, is a measure of relative precision, not absolute precision as are *MSE* and *MAD*. Both *MSE* and *MAD* are sensitive to large errors and can be distorted more easily with one unusually high value of e_t.

Fitting Accuracy and Forecasting Accuracy

The three measures of accuracy can be applied to a time series in two ways: to historical values and to future values. When we compute the error metrics on the basis of a past sequence of data, we obtain a measure of *fitting accuracy*. Fitting accuracy is important in a model building sense. Chances are high that a forecasting technique with a small value of *MAD, MSE, or MAPE* based on historical data will continue to forecast well, since we are trying to extrapolate past behavior into the future. Therefore, we tend to use fitting accuracy to help select a forecasting technique for future predictions.

When we use a forecasting technique to predict subsequent, yet-to-be-realized, values of the time series, we also can apply the *MAD, MSE, or MAPE* criterion to the future sequence of data as it becomes known. Under these conditions, we obtain a measure of *forecasting accuracy*. In Example 18.7, we demonstrate the computation of each error metric on the historical values of the time series to compare the fitting accuracies of two techniques. Then, in Example 18.8, we extrapolate the trend line to predict the next four quarters of inventory levels and compare the percentage errors of these forecasts with the ensuing values to measure the forecasting accuracy of the trend line forecasts.

EXAMPLE 18.7

In Example 18.6, we used the work-in-progress inventory data—four years of quarterly data—to develop exponentially smoothed forecasts, with $\alpha = .4$. We also generated historical forecasts, using a trend projection with the fitted model $\hat{Y}_t = 17.335 + .494t$. Compare the fitting accuracies of the two forecasting techniques on the basis of the following:

a. *MSE* criterion
b. *MAD* criterion
c. *MAPE* criterion

Solution:

The estimated Y_t values for $t = 2$ through $t = 16$ are reproduced in the following table for the exponentially smoothed forecasts, along with the errors and percentage errors:

t	Y_t	$\hat{Y}_t$	e_t	p_t
1	15.2	—	—	—
2	18.4	15.2	3.2	17.39%
3	18.7	16.48	2.22	11.87
4	22.2	17.368	4.832	21.77
5	17.8	19.301	−1.501	−8.43
6	21.9	18.701	3.199	14.61
7	21.9	19.981	1.919	8.76
8	25.6	20.749	4.851	18.95
9	19.6	22.689	−3.089	−15.76
10	22.1	21.453	.647	2.93
11	21.4	21.712	−.312	−1.46
12	24.4	21.587	2.813	11.53
13	19.6	22.712	−3.112	−15.88
14	23.4	21.467	1.933	8.26
15	24.1	22.240	1.860	7.72
16	28.2	22.984	5.216	18.50

For the trend projection forecasts, following are the values of $\hat{Y}_t$, e_t, and p_t:

t	Y_t	$\hat{Y}_t$	e_t	p_t
1	15.2	17.829	−2.629	−17.30%
2	18.4	18.323	.077	.42
3	18.7	18.817	−.117	−.63
4	22.2	19.311	2.889	13.01
5	17.8	19.805	−2.005	−11.26
6	21.9	20.299	1.601	7.31
7	21.9	20.793	1.107	5.05
8	25.6	21.287	4.313	16.85
9	19.6	21.781	−2.181	−11.13
10	22.1	22.275	−.175	−.79
11	21.4	22.769	−1.369	−6.40
12	24.4	23.263	1.137	4.66
13	19.6	23.757	−4.157	−21.21
14	23.4	24.251	−.851	−3.64
15	24.1	24.745	−.645	−2.68
16	28.2	25.239	2.961	10.50

a. For the *MSE* criterion the sums of squared forecasting errors and *MSE* for both forecasting techniques are as follows:

Exponentially Smoothed	Trend Projection
$n = 15$	$n = 16$
$\Sigma e_t^2 = 140.27612$	$\Sigma e_t^2 = 76.830936$
$MSE = 9.3517$	$MSE = 4.8019$

Since the *MSE* is lower for the trend projection forecasts, we would prefer this technique over an exponentially smoothed forecast.

b. The $\Sigma|e_t|$ and value of *MAD* for both techniques are listed as follows:

Exponentially Smoothed	Trend Projection
$n = 15$	$n = 16$
$\Sigma\lvert e_t\rvert = 40.704$	$\Sigma\lvert e_t\rvert = 28.214$
$MAD = 2.7136$	$MAD = 1.7634$

Again, based on the *MAD* criterion, we prefer the trend projection forecasts.

c. In terms of *MAPE*, the trend projection forecasts are lower as well:

Exponentially Smoothed	Trend Projection
$n = 15$	$n = 16$
$\Sigma\lvert p_t\rvert = 183.82\%$	$\Sigma\lvert p_t\rvert = 132.84\%$
$MAPE = 12.25\%$	$MAPE = 8.30\%$

□

EXAMPLE 18.8

Refer to Example 18.7. Use a trend projection to forecast the next four quarters of inventory level. Assume the actual values are as follows:

Year	Quarter	Y_t
1991	1	20.1
	2	23.7
	3	23.9
	4	26.6

Compute the forecasting accuracy, using the *MAPE* criterion.

Solution:

The trend line is $\hat{Y}_t = 17.335 + .494t$. For the next four quarters, the values of the variable t are 17, 18, 19, and 20. The following table summarizes the forecasts and the errors:

Year	Quarter	Y_t	t	$\hat{Y}_t$	e_t	p_t
1991	1	20.1	17	25.733	−5.633	−28.02%
	2	23.7	18	26.227	−2.527	−10.66
	3	23.9	19	26.721	−2.821	−11.80
	4	26.6	20	27.215	−.615	−2.31

The mean absolute percentage error is

$$MAPE = \frac{\Sigma|p_t|}{4} = \frac{52.79}{4} = 13.1975 \quad \text{or } 13.20\%$$

This means that the trend projection forecasts were in error by an average of 13.2 percent for the four quarters of 1991.

□

COMMENTS

1. For some forecasting techniques—naive, mean, seasonalized mean, moving average, and exponential smoothing—it is not possible to compute e_t or p_t for all previous time periods, since $\hat{y}_t$ may not exist. Thus, for these techniques, the divisor n in Equations 18–15, 18–16, and 18–18 for *MSE, MAD,* and *MAPE,* respectively, is the number of time periods for which e_t or p_t exist.
2. In an exponentially smoothed forecast, when we change the value of α, we get a new value for *MSE, MAD,* and *MAPE*. For most time series, there will be a unique value of α that minimizes the fitting accuracy measure. The Time Series processor that we will introduce in Section 18.6 finds the optimal value of α to two decimal places for minimizing the *MAPE*. For instance, in Example 18.6 we can show that when α = .42, the minimum value of *MAPE* is 11.485 percent.
3. As the trend projections in Example 18.8 showed, we can forecast several periods ahead without seeing the actual values. This is not true for all forecasting techniques, though. For instance, an exponentially smoothed forecast for Y_{t+1} depends on the forecasting error at time t. Thus, an exponentially smoothed forecast can be made only one step ahead.
4. Technically, the correct way to measure the fitting accuracy of trend projection forecasts is to base successive y_t's on an updated trend line as each Y_t value occurs. In practice though, only one trend line is fit to the entire set of historical data and used to generate y_t's for each t.

In Example 18.7, we found that a trend projection forecast fared better than an exponentially smoothed forecast for the work-in-progress inventory data on each measure of overall fitting accuracy. Although in this example one technique bested another for each error metric, this phenomenon may not hold true in all cases.

Indeed, a good caveat to close this section is this: Do not expect one forecasting technique to fall out as the best for all forecasting applications. As the next section shows, qualitative forecasting techniques may offer some improvement in accuracy, and a combination of forecasts may be better yet.

CLASSROOM EXAMPLE 18.3

Comparing Fitting Accuracy

The average wholesale prices of large eggs, in cents per dozen, are recorded in the following table for the seven year period from 1980 through 1986.

Year	Average Wholesale Price (¢)
1980	62.8
1981	69.0
1982	66.8
1983	72.7
1984	78.6
1985	63.4
1986	68.1

Source: Commodity Research Bureau, New York, NY.

Compare the fitting accuracy of trend projection forecasts to those from a naive forecasting technique, for the 1981–86 time frame in terms of the *MAPE*. Assume the trend line is $\hat{Y}_t = 66.414 + .589t$. Round the $\hat{Y}_t$ values to two decimal places.

SECTION 18.4 EXERCISES

18.45 The civilian per capita consumption of chickens, in pounds, is indicated as follows:

Year	Consumption
1982	53.1
1983	53.8
1984	55.7
1985	58.0
1986	59.1
1987	62.9

Source: Commodity Research Bureau, New York, NY.

a. Find the *MAPE* for trend projection forecasts.
b. Find the *MAPE* for mean forecasts from 1983 through 1987.

18.46 Refer to Exercise 18.45. Repeat the instructions, using the *MSE* criterion instead.

18.47 Name two factors to be considered in selecting and analyzing a forecasting technique.

18.48 Why is a forecast for one year ahead not very sensitive to seasonal effects?

18.49 Suppose we wish to forecast the dollar exchange rate between the United States and Canada. Name the time horizon category if we plan to forecast the following:
a. next year's rate.
b. next quarter's rate.
c. next month's rate.
d. the rate three years from now.

18.50 Refer to Exercise 18.49. What time series component, if any, do you think will impact the forecast the most and the least when we forecast the following:

a. next year's rate.
b. next quarter's rate.
c. next month's rate.
d. the rate three years from now.

18.51 The federal debt, in billions of dollars, is shown in the following table for the years 1968–1987.

Year	Amount Outstanding	Year	Amount Outstanding
1968	369.8	1978	780.4
1969	367.1	1979	833.8
1970	382.6	1980	914.3
1971	409.5	1981	1003.9
1972	437.3	1982	1147.0
1973	468.4	1983	1381.9
1974	486.2	1984	1576.7
1975	544.1	1985	1827.5
1976	631.9	1986	2129.5
1977	709.1	1987	2354.3

Source: The Treasury Bulletin, December, 1987.

a. Find the *MSE* for trend projection forecasts.
b. Find the *MSE* for naive forecasts for the 19-year period 1969 through 1987.

18.52 Refer to Exercise 18.51. Compare the fitting accuracy of the mean forecast with a three-year moving average forecast starting with 1971 on the basis of the *MAPE* criterion. Round the $\hat{Y}_t$ values to two decimal places.

18.53 Refer to Exercise 18.51. Find the *MAD* for exponentially smoothed forecasts, with $\alpha = .9$.

18.54 Find the *MAPE* over the last two years for a multiplicative model forecast with a four-quarter moving average estimate of T_{t+1} based on the following data:

	Quarter			
Year	1	2	3	4
1	4.8	5.0	5.6	8.7
2	5.0	5.2	5.6	8.8
3	5.1	5.3	5.8	9.2

18.55 Refer to Exercise 18.54. Assume the actual values for each quarter of year 4 are 5.1, 5.4, 5.9, and 9.5 from quarter 1 to 4, respectively. Compare the forecasting accuracy of multiplicative model forecasts using a four-quarter moving average estimate of T_{t+1} with four-quarter moving average forecasts on the basis of the *MAPE* metric.

18.56 Refer to Exercises 18.24, 18.26, and 18.28.
a. Find the *MAPE* of trend projection forecasts.
b. Find the *MAPE* of exponentially smoothed forecasts using $\alpha = .8$.

18.57 Refer to the data in Exercise 18.24.

a. Find the *MSE* of naive forecasts.

b. Find the *MSE* of four-quarter moving average forecasts.

18.58 Refer to Exercise 18.3. Find the *MAD* for the sequence of mean forecasts for the percent of contributing alumni. Round the forecasts to two decimal places.

18.59 Consider the following quarterly time series data:

	Quarter			
Year	1	2	3	4
1	11	15	10	12
2	10	13	7	18
3	12	17	7	10

Trend line: $\hat{Y}_t = 12.197 - .056t$

Seasonal indices: $S_1 = .8859$ $S_2 = 1.2323$

$S_3 = .6924$ $S_4 = 1.1894$

Apply each of the simple forecasting techniques—naive, mean, seasonalized mean, moving average, trend projection, multiplicative model, and exponential smoothing—from Section 18.3 to these data to generate a sequence of historical forecasts. Use $\alpha = .1$ for the exponentially smoothed forecast. Use a trend projection to estimate T_{t+1} in the multiplicative model forecast. Round all forecasts to two decimal places. Compute the *MSE* for each to three decimal places, based on the last two years of data. Find the forecasting technique with the smallest and largest *MSE*.

18.60 Refer to Exercise 18.59. Use the trend line to forecast (to three decimal places) the time series for each quarter in year 4. Measure the forecasting accuracy, using the *MSE* criterion, if the actual values turn out to be: 11 (first quarter), 16, 9, and 14.

18.61 Explain the difference between fitting accuracy and forecasting accuracy.

18.62 Assume the following forecasts were made at the beginning of the year with the actual time series values recorded later:

Month	Actual	Forecasted
January	9.13	9.40
February	9.26	9.29
March	8.90	9.28
April	8.84	9.13
May	8.76	9.01
June	8.78	8.91

Determine the forecasting accuracy, with the following criteria:

a. *MAD* criterion **b.** *MSE* criterion **c.** *MAPE* criterion

***18.63** In Section 18.3, we mentioned that an exponentially smoothed forecast is most accurate for time series that are stationary (no trend component). When an upward trend is present in

*Optional

the time series, an exponentially smoothed forecast will always lag Y_t. That is, $\hat{y}_t$ generally will be smaller in each time period than y_t.

a. To see this lagging effect, generate the sequence of $\hat{y}_t$'s using an exponentially smoothed forecast with $\alpha = .9$ on the following data, which represent Y_t, the Soviet Union's yearly production of eggs, in billions of pounds. Round the forecasted values to three decimal places. Compare $\hat{y}_t$ with y_t. Verify that the *MAPE* is 2.42 percent.

Year	Y_t
1980	70.9
1981	70.9
1982	72.4
1983	75.1
1984	76.5
1985	77.3
1986	80.7
1987	83.1

Source: Commodity Research Bureau, New York, NY.

b. To see the effect of stationarity on an exponentially smoothed forecast, generate the sequence of $\hat{y}_t$'s on the following data, which represent Y_t = the United States yearly production of eggs, in billions of pounds. Round the forecasted values to three decimal places. Use $\alpha = .7$. Verify that the *MAPE* is .73 percent.

Year	Y_t
1980	69.7
1981	69.9
1982	69.6
1983	67.9
1984	68.5
1985	68.3
1986	68.6
1987	69.3

Source: Commodity Research Bureau, New York, NY.

18.5 OTHER FORECASTING TECHNIQUES

The six forecasting techniques we presented in Section 18.3 represent a starting point in forecasting time series. Though simple and easy to implement, they may be inadequate in some situations. Achieving a *MAPE* of, say, 10 percent may be sufficient for forecasting new car sales next year, but the same figure might be considered too large in forecasting the unemployment rate, for instance. More sophisticated methods may be sought. We will mention four quantitative approaches aimed at providing more precise forecasts, and then we will address a qualitative method of prediction. Lastly, we will broach the idea of combining forecasts to achieve greater precision.

Advanced Quantitative Forecasting Methods

Double Exponential Smoothing Exponentially smoothed forecasts are imprecise when a trend effect is present. Exercise 18.63 shows that the exponentially smoothed forecast tends to lag the trend, almost always producing a forecasted value that underestimates Y_t (positive trend assumed). To adjust, we can apply a *two-stage smoothing process*. The idea is to adjust the exponentially smooth forecast by a factor representing the trend component. There are two common ways of achieving this adjustment: Brown's linear exponentially smoothing forecast and Holt's two-parameter exponential smoothing forecast.

In Brown's method, we apply the smoothing process to the sequence of exponentially smoothed forecasts to yield a doubly smoothed forecast. Then the difference between the singly and doubly smoothed forecasts is added to the singly smoothed forecast as an estimate of the trend.

Holt's approach is to estimate the trend component separately based on the difference in successive exponentially smoothed forecasts. Then he combines the exponentially smoothed value of Y_{t+1} and the trend estimate to produce $\hat{y}_{t+1}$. Holt's method is similar to Brown's except that he uses a second smoothing constant, different from α, to estimate the trend.

Both methods work well for time series with a linear trend component, but both suffer when applied to a time series with a seasonal component as well. In such a case, Winter's seasonal exponential smoothing model, an adaptation of Holt's method, may be used. It involves three smoothing constants. For further information on these techniques, consult the references listed at the end of this chapter.

Autoregressive Models The lagging effect addressed by double and triple exponential smoothing models also can be treated with regression models. When a high degree of correlation exists between successive Y_t values, we can replace the independent variable t in Equation 18–5 with the previous value of the dependent variable Y_t to form the simple regression model in Equation 18–19.

First Order Autoregressive Model

$$Y_t = \beta_0 + \beta_1 Y_{t-1} + \epsilon \qquad \textbf{(18–19)}$$

Equation 18–19 is called a *first order autoregressive model*. The least-squares criterion is used to estimate β_0 and β_1.

COMMENT The term *autoregressive* implies that the dependent variable is regressed on itself with an appropriate time lag. Extending Equation 18–19 to a second order autoregressive model is intuitive: Add a term $\beta_2 Y_{t-2}$.

Moving Average Models Exponential smoothing also provides a basis for developing another type of model called a *moving average model*. Here, however, moving average does not refer to averaging the previous values of Y_t sequentially, but rather to finding a weighted average of the previous forecasting errors. For example, an exponentially smoothed forecast from Equations 18–13 and 18–14 is

$$\begin{aligned}\hat{y}_{t+1} &= \hat{y}_t + \alpha(y_t - \hat{y}_t)\\ &= \hat{y}_t + \alpha(e_t)\end{aligned}$$

Notice that one of the components in the forecast is $\alpha(e_t)$. If we iterate the process for $\hat{y}_t$, we get

$$\begin{aligned}\hat{y}_t &= \hat{y}_{t-1} + \alpha(y_{t-1} - \hat{y}_{t-1})\\ &= \hat{y}_{t-1} + \alpha(e_{t-1})\end{aligned}$$

Substituting this expression for $\hat{y}_t$ into the previous one for $\hat{y}_{t+1}$ yields

$$\hat{y}_{t+1} = \hat{y}_{t-1} + \alpha(e_{t-1}) + \alpha(e_t)$$

The last equation shows that a forecast could be viewed as a function of successive error terms. Expressed as a regression equation, a *first order moving average model* is presented in Equation 18–20.

First Order Moving Average Model

$$Y_t = \beta_0 + \beta_1 e_{t-1} + \epsilon \qquad \textbf{(18–20)}$$

COMMENT The β_0 and β_1 symbols in Equations 18–19 and 18–20 are *not* the same values; they simply denote the population regression coefficients of each model. In Equation 18–20, ϵ represents the (unknown) true forecasting error at time t, while e_{t-1} is the observed error from the previous time period. Extensions to higher order moving average models result when other lagged error terms, like e_{t-2}, and so on, are included.

Box-Jenkins Methodology George E. P. Box and Gwilym Jenkins contributed significantly to the time series literature. In the late 1960s, they developed a comprehensive procedure for denoting, recognizing, and using autoregressive and moving average models to forecast time series. First, they set up a classification system by combining autoregressive and moving average models with a third factor, differencing, that ensures the stationarity of a time series. The result was *ARIMA*—autoregressive integrated moving average—models. Their ARIMA system of model identification is now standard.

Second, they analyzed many different theoretical ARIMA models and discovered unique characteristics of certain categories of models. In a sense, they catalogued the blueprints by which specific ARIMA models behave.

Third, they proposed a three-step method for matching an empirical time series with the documented characteristics of their theoretical ARIMA models:

1. Identify a category of potential models for the time series.
2. Estimate the parameters of the model(s).
3. Perform diagnostic checks to ensure the error term is free of any patterns.

Once the proper model is verified, it can be used to predict future values.

Most of the advanced forecasting techniques we have mentioned here are adaptations or extensions of the simple forecasting techniques from Section 18.3. However, all of the methods to date are quantitative approaches to forecasting. We would be remiss if we did not mention the qualitative aspects of forecasting.

Qualitative Forecasting Approaches

Even with all of our analysis and modeling, forecasts still are in error. This is not to suggest that we should abandon our quantitative forecasting methods. However, it does raise the question as to whether subjective forecasts might perform as well.

Qualitative or judgmental forecasting techniques are based on personal assessments, not on mathematical formulas. In general, an expert within the firm looks at historical data, factors in related economic conditions, and injects her or his past experiences and "gut feelings" into a subjective forecast. An obvious drawback to this approach is that several experts, looking at the same data, may produce different forecasts.

Were it possible to harness the experts' opinions and synthesize them into a consensus, then a judgmental forecasting technique would seem more credible. One such strategy along these lines is called the *Delphi technique*.

In their article, "The Delphi Technique: An Application in Utility Planning" (*Proceedings of the Midwest Decision Sciences Institute*, 1988), Barbara Shiffler and Patrick S. Ryan summarize the Delphi principle:

> *Briefly, Delphi involves the formation of a panel of experts whose purpose is to provide a fairly unified opinion on one or more issues affecting some future event; for example, a utility's plans for future electric generation requirements. What makes Delphi unique is that panelists communicate only in writing.*
>
> *The Delphi technique is a desirable alternative in certain situations, primarily because of the anonymity of responses. The panel of experts answer questions, provide comments, and make recommendations* only on paper. *The responses are sent to a moderator, avoiding any face-to-face discussion with the other panel members. While perhaps appearing to be a cumbersome process, this method has several advantages. By guaranteeing anonymity, no particular individual can dominate the outcome—either by virtue of personality or title. The "group-think" phenomenon is mitigated. By controlling responses through a moderator, filibuster-type activities are avoided. Finally, since responses are usually redefined in terms of a numerical answer, the composite expert (subjective) opinion may be statistically analyzed to develop a quantifiable (objective) group decision.*
>
> *Of course, Delphi is not without its critics. Since each panel member completes the information request in the privacy of his or her own office, accountability is not*

as strong an influence as it would be in a typical committee meeting. Time constraints may result in "hurried" responses from some panel members. And, some members may yield their position simply to speed along consensus.

In this application, Shiffler and Ryan used a group of experts to help forecast the future demand for electricity based on the likelihood of three different growth scenarios.

The research literature comes down on both sides of the qualitative versus quantitative forecasting argument. Essam Mahmoud, writing in the *Journal of Forecasting*, (1984), conducted a thorough study on forecasting accuracy and concluded: "On the whole . . . quantitative methods outperform qualitative methods." On the other side, Moriarty and Adams (1984) found that a firm's judgmental forecasts for two types of durable goods had significantly lower *MAPE*'s than a sophisticated Box-Jenkins forecast they developed.

So the issue of a qualitative or a quantitative forecasting technique remains unsettled. Perhaps the best aspect of the issue is that a *combined forecast* may emerge as a compromise and be more precise than either a qualitative or a quantitative forecast alone.

Combined Forecasts

There is a fundamental law in finance: Don't put all your eggs in one basket. Basically, it can be shown that diversifying one's investments among different risky alternatives lowers the overall risk of the portfolio.

Interestingly, a similar proposition is emerging in time series. Researchers have discovered that a composite forecast that includes two (or more) individual forecasts may perform better, in terms of the three error metrics, than the separate forecasts. The simplest way of combining two forecasts is to take their average, or possibly a weighted average. Alternatively, we can combine forecasts by means of a multiple regression model. The dependent variable is Y_t; the forecasted values of Y_t from the different forecasting techniques serve as the independent variables. Example 18.9 demonstrates an improvement in accuracy by combining forecasts.

EXAMPLE 18.9

In Classroom Example 18.3, we compared the *MAPE*'s of a trend projection forecast and a naive forecast for the variable Y_t, average wholesale prices of large eggs over a six-year period. Suppose we wish to combine the trend projection forecast, denoted by $\hat{y}_{tp}$, with the naive forecast, denoted by $\hat{y}_n$, in the multiple regression model

$$Y_t = \beta_0 + \beta_1\hat{y}_{tp} + \beta_2\hat{y}_n + \epsilon$$

Find the fitting accuracy of the combined forecast, using the *MAPE*.

Solution:

The model was fit to the six-year period 1981–86. Following are the forecasted values of Y_t from the combined forecast:

Year	Y_t	$\hat{Y}_t$	$\lvert p_t \rvert$
1981	69.0	70.735	2.51%
1982	66.8	69.970	4.75
1983	72.7	70.047	3.65
1984	78.6	69.312	11.82
1985	63.4	68.577	8.17
1986	68.1	69.958	2.73
			$\Sigma\lvert p_t \rvert = 33.63\%$

The *MAPE* of the combined forecast is

$$MAPE = \frac{\Sigma\lvert p_t \rvert}{n} = \frac{33.63\%}{6} = 5.605 \quad \text{or } 5.61\% \text{ (rounded)}$$

In Classroom Example 18.3, we found the trend projection forecast to have a *MAPE* of 5.87 percent and the *MAPE* for the naive forecast to be 9.8 percent. Even with our small sample size of six, the combined forecast offers an improved value of *MAPE* and therefore is slightly more precise in terms of fitting accuracy than either of the separate forecasts.

□

All combined forecasts cannot reduce the relative or absolute error, but the notion of pooling information is appealing, especially when qualitative and quantitative inputs are combined. Clearly, the trend among forecasters is toward improving the precision of a forecast through composite forecasts, not on selecting one best forecasting technique.

*18.6 PROCESSOR: TIME SERIES

The processor titled Time Series performs most of the calculations described in this chapter. Two options appear on the submenu when we access the Time Series processor: Decomposition and Forecasting. Available from the Decomposition option are four choices: the trend line, a centered moving average, seasonal indices, and the cyclical relatives. The Forecasting option allows any of seven simple forecasting techniques to be used for forecasting Y_{t+1}. In addition, the fitting accuracy of each technique is given for the three error metrics *MAD*, *MSE*, and *MAPE*, as is the value of α that minimizes the *MAPE* metric for the exponential smoothing technique.

*Optional

Creating a Data Set

Before running the processor, we must create a data set of the Multivariate Raw Data type. As an illustration, we will use the data for Y_t = the work-in-progress inventory from Example 18.1. Following is a list of the data:

Year	Quarter	Y_t	Year	Quarter	Y_t
1987	1	15.2	1989	1	19.6
	2	18.4		2	22.1
	3	18.7		3	21.4
	4	22.2		4	24.4
1988	1	17.8	1990	1	19.6
	2	21.9		2	23.4
	3	21.9		3	24.1
	4	25.6		4	28.2

The number of existing values for Y_t—16 in this case—is the required input for n in creating the data set. We need to use only $v = 1$ variable—for Y_t—as input to the processor. The values of t are *not* needed.

Decomposition

After selecting the Decomposition option and viewing an Introduction screen of text, use the right arrow key to move across the menu to Vector. Here we encounter a Select Variable Info window, similar to what we saw in the Regression processor. The Time Series processor is set up to process the dependent variable Y_t by automatically assigning the integers from 1 to n as the values of the time variable t. Thus, we need to mark or confirm only the dependent variable. In general, the processor always marks the last variable listed in a Multivariate Raw Data type of file.

Proceed across the menu to Type and press the Enter key. Figure 18.9 shows the four choices of decomposition that can be selected. Although only one choice may be selected at a time, the choices are cumulative. This means, for instance, that if we

```
≡                          Decomposition                                    MENU
  Introduction  *ector [Type] Calculations  *esults  Exit              | F1=Help
                       Trend line
                       Centered moving average
                       Seasonal indices
                       Cyclical relatives
```

Figure 18.9 Decomposition Choices

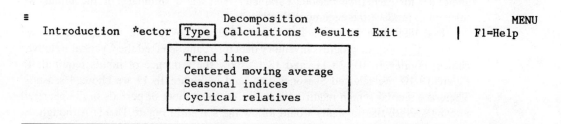

select Trend Line, only the trend line will be computed and displayed. But if we select Seasonal Indices, for example, we will generate the trend line, a centered moving average, *and* the seasonal indices.

As we access a choice, we may be required to provide input parameters in order to process the data correctly. Following is a list of inputs needed for each choice:

Choice	Inputs
Trend line	No inputs required
Centered moving average	Number of periods in the centered moving average
Seasonal indices	1. Number of seasons
	2. Number of periods in the centered moving average
Cyclical relatives	1. Type of data—annual or seasonal
	2. Number of seasons
	3. Number of periods in the centered moving average
	4. Type of moving average to neutralize the irregular effect—unweighted or weighted.

All inputs are entered by selecting the appropriate value or type from a menu. The Seasonal Indices choice requires two inputs: the number of seasons, which tells the processor how many seasonal indices to compute; and the number of periods in the centered moving average, which is used to generate M_t prior to forming the ratio-to-moving average Y_t/M_t. Normally the number of seasons and the number of periods are the same, but the processor is designed to handle different values here. (Exercise 18.17, for instance, requires such flexibility.)

The Cyclical Relatives choice may involve four inputs. First, we must indicate the type of data—annual or seasonal. If we have annual data, no other inputs are required. For seasonal data, as you recall, we must deseasonalize Y_t by dividing by the appropriate seasonal index. Thus, the number of seasons and the number of periods are required inputs for this computation. Finally, we should neutralize the irregular effect with a three-period centered moving average. The processor allows two types of centered moving averages here: unweighted and weighted (see Comment 4 immediately after Example 18.5).

In the Results section of the menu, there are two choices: Decomposition Values and Parameter Notes. Selecting the first yields a table of values for t, Y_t, $\hat{Y}_t$, M_t, S_t, and/or C_t for each time period. Parameter Notes is a summary of the inputs we selected to produce the decomposition values.

To demonstrate the inputs and outputs of Decomposition, let us suppose that we are using the work-in-progress inventory data for Y_t and selected the Cyclical Relatives choice. Figures 18.10, 18.11, and 18.12 show the sequence of inputs required. In Figure 18.10, we selected Seasonal data, and in Figure 18.11 we chose 4 seasons. There is a similar screen requiring a choice of the number of periods in the centered moving average that logically would follow the screen in Figure 18.11. Although we have not displayed it, we note that we selected 4 periods here as well. In Figure 18.12, we elected to use an unweighted centered moving average for the irregular effect.

```
≡                                  Decomposition                                MENU
   Introduction  *ector  [Type]  Calculations  *esults  Exit              |  F1=Help
                        Trend line
                        Centered moving average
                        Seasonal indices
                       [Cyclical relatives]
                          Annual
                         [Seasonal]
```

Figure 18.10 Type of Data Choices in Computing Cyclical Relatives

```
≡                                  Decomposition                                MENU
   Introduction  *ector  [Type]  Calculations  *esults  Exit              |  F1=Help
              3         Trend line
             [4]        Centered moving average
              5         Seasonal indices
              6        [Cyclical relatives]
              7

                        Select the number of seasons.
```

Figure 18.11 Number of Season Choices in Computing Cyclical Relatives

```
≡                                  Decomposition                                MENU
   Introduction  *ector  [Type]  Calculations  *esults  Exit              |  F1=Help
                        Trend line
                        Centered moving average
                        Seasonal indices
                       [Cyclical relatives]
                          Weighted
                         [Unweighted]
```

Figure 18.12 Types of Moving Averages to Neutralize the Irregular Effect in Computing Cyclical Relatives

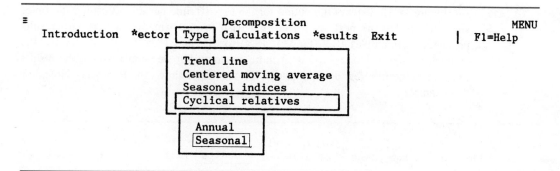

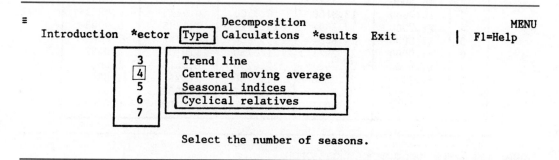

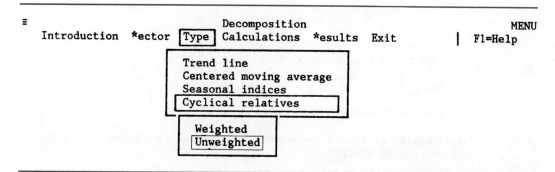

Figure 18.13 displays the output for the first 10 time periods. (Pushing the Enter key will display the next six.) You may wish to compare this screen with results presented in Examples 18.2 (for M_t), 18.4 (for S_t), and 18.5 (for C_t). Naturally, the computer generated values are a bit more accurate, but we should expect the results to agree to the first few decimal places. Figure 18.14 shows the Parameter Notes for this example.

≡ Decomposition READY
Introduction *ector Type Calculations | Results | Exit | F1=Help

t	Yt	$\hat{Y}$t	Mt	St	Ct
1	15.2000	17.8287	--	0.8732	--
2	18.4000	18.3224	--	1.0109	0.9632
3	18.7000	18.8160	18.9500	0.9873	0.9823
4	22.2000	19.3097	19.7125	1.1285	0.9993
5	17.8000	19.8034	20.5500	0.8732	1.0243
6	21.9000	20.2971	21.3750	1.0109	1.0452
7	21.9000	20.7907	22.0250	0.9873	1.0620
8	25.6000	21.2844	22.2750	1.1285	1.0544
9	19.6000	21.7781	22.2375	0.8732	1.0303
10	22.1000	22.2718	22.0250	1.0109	0.9963

Figure 18.13 Output for Decomposition Option

≡ Decomposition READY
Introduction *ector Type Calculations | Results | Exit | F1=Help

Notes:

The trend line is $\hat{Y}$t = 17.33500 + 0.49368 * t.

The number of periods in the centered moving average is 4.

The number of seasons is 4.

The irregular effect is dampened by a 3-period unweighted centered moving average.

Figure 18.14 Summary of Decomposition Inputs

Forecasting

Seven methods of forecasting Y_{t+1}—see Figure 18.15—are possible when we select the Forecasting option from the Time Series processor. Unlike the cumulative nature of the Decomposition choices, the Forecasting choices are not cumulative; we generate $\hat{Y}_t$ values only for the method selected. However, as before, certain inputs may be required. The following table summarizes the needed inputs:

Choice	Inputs
Naive	No inputs required
Mean	No inputs required
Seasonalized mean	The number of seasons
Moving average	The number of periods in the moving average
Trend projection	No inputs required
Multiplicative model	1. Type of estimate for T_{t+1}—trend projection or moving average (If moving average is selected, then the number of periods in the moving average also is required.) 2. The number of seasons
Exponential smoothing	The value for alpha

All inputs except the value for alpha are selected from menus. We must enter the intended value for α as a decimal, such as .50 or .32, as shown in Figure 18.16 (note that we entered $\alpha = .40$). The Multiplicative Model generates a forecast by multiplying the estimates of the trend and seasonal components, T_{t+1} and S_{t+1}, respectively. For the trend component estimate, we can use either a moving average or a trend projection.

```
≡                               Forecasting                                    MENU
  Introduction  *ector  [Type]  Calculations  *esults  Exit              | F1=Help
                        ┌──────────────────────────┐
                        │ Naive                    │
                        │ Mean                     │
                        │ Seasonalized mean        │
                        │ Moving average           │
                        │ Trend projection         │
                        │ Multiplicative model     │
                        │ Exponential smoothing    │
                        └──────────────────────────┘
```

Figure 18.15 Forecasting Choices

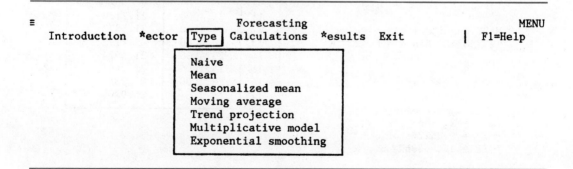

```
≡                              Forecasting                                   EDIT
  Introduction  *ector  [Type]  Calculations  *esults  Exit            | F1=Help

                        Naive
                        Mean
                        Seasonalized mean
                        Moving average
                        Trend projection
                        Multiplicative model
                        [Exponential smoothing]

                 ═══════════Value for alpha═══════════

                 Enter a value for α between 0 and 1: .40
```

Figure 18.16 Input Screen for Alpha in Exponential Smoothing

```
≡                              Forecasting                                  READY
  Introduction  *ector  Type  Calculations  [Results]  Exit            | F1=Help
```

The exponential smoothing forecast uses α = 0.40,

MAPE has its minimum value of 12.25074% when α = 0.42.

Error metrics

MAD = 2.71363
MSE = 9.35224
MAPE = 12.25396%

t	Yt	Exp. Smooth Forecast
11	21.4000	21.7120
12	24.4000	21.5872
13	19.6000	22.7123
14	23.4000	21.4674
15	24.1000	22.2404
16	28.2000	22.9843
17	--	25.0706

Figure 18.17 Output for Forecasting Option with Exponential Smoothing

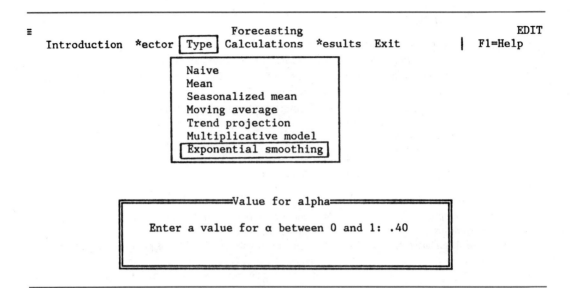

The output for each forecasting method consists of the following:

1. A table of historical forecasts—that is, $\hat{Y}_t$ values, where they exist, for $t = 1$ through $t = n$.
2. The one-period-ahead forecast, $\hat{y}_{t+1}$.
3. The error metrics *MAD*, *MSE*, and *MAPE* based on the fitting accuracies.
4. Miscellaneous notes about the forecast.

For the exponential smoothing method only, the processor also determines the value of alpha, to two decimal places, that minimizes the *MAPE* metric for the historical time series values. Figure 18.17 shows a partial output screen for this technique. (Touch the Escape key to see the exponential smoothing forecasts for the first 10 time periods.) You might wish to compare the output to the forecasts we presented in Example 18.6.

COMMENT Hand calculations may differ from those produced by the Time Series processor because the processor uses unrounded values in all its calculations. Consider the output from the Time Series processor to represent the higher degree of accuracy.

18.7 SUMMARY

When data are generated sequentially as the clock or calendar elapses, we call the chronology of values for the variable Y_t a time series. Analyzing a time series is done with two objectives in mind: describing the past data pattern and projecting future values of Y_t.

To describe the past, we decompose the time series into its components—trend, seasonal, cyclical, and irregular. A simple regression equation or a centered moving average reveals the general direction of Y_t. Seasonal indices capture the regular alternation of peaks and valleys, while the sequence of cyclical relatives exposes sustained favorable or unfavorable economic conditions affecting Y_t. Explaining the historical behavior of the time series establishes a basis for forecasting.

Time series forecasting is the process by which we extrapolate Y_t into the future. In Section 18.3, we introduced several simple forecasting techniques for doing so. Almost all of the techniques we presented yield a point prediction only. To make statistical inferences with a known confidence level or risk factor, we must model the irregular component of a time series with a probability distribution. Most forecasting techniques that do so are not simple ones and require a deeper treatment than we provided in Section 18.5. Yet intuitively, we can appreciate the power and sophistication of the Box-Jenkins approach, for example, which allows point and interval predictions.

We discussed three criteria for judging the accuracy of a forecast—*MAPE*, *MSE*, and *MAD*—with *MAPE* being preferred over the other two. However, forecasting is not purely quantitative. The literature shows that judgmental assessments play an important and sometimes accurate role in the forecasting process and should not be ignored.

In closing, we note that forecasters pursue the goal of improving forecast accuracy rather than contenting themselves with finding the right technique. Consequently, pooling information into a combined forecast is the emerging trend in time series forecasting.

18.8 TO BE CONTINUED . . .

. . . In Your College Courses

Time series data occur naturally in economics and business. As a result, the descriptive methods and inferential techniques we presented in this chapter are often-used tools in other business courses.

Production/Operations Management In most curricula, management majors are required to take a course (or two) in production and operations management. In Chapter 4 titled "Forecasting" of Heizer and Render's textbook, *Production and Operations Management* (1988), we read:

> *A time series is based on a sequence of evenly spaced (weekly, monthly, quarterly, and so on) data points. . . . Forecasting time series data implies that future values are predicted* only *from past values. . . . Analyzing time series means breaking down past data into components and then projecting them forward.*

They go on to discuss the four components we listed in Table 18.1 and the two time series models—multiplicative and additive—we formulated in Equations 18–1 and 18–2, respectively. They review briefly some of the forecasting procedures we discussed, such as the naive approach, moving averages, exponential smoothing, and trend projections.

Marketing Management A required course in most marketing disciplines is marketing management—the study of planning and executing the pricing, promotion, and distribution of goods and services. One part of the planning process is sales forecasting. Philip Kotler, in his textbook *Marketing Management* (1988), reviews both quantitative and qualitative forecasting techniques and has this to say about time series forecasting:

> *Many firms prepare their forecasts on the basis of past sales. . . . Time series analysis consists of decomposing the original sales series,* Y, *into the components,* T, C, S, *and* E. *Then these components are recombined to produce the sales forecast.*

(Kotler defines the irregular component as "erratic events" and symbolizes it with the letter E. We used the letter I.) The decomposition methods of Section 18.2 are clearly an integral part of sales forecasting in marketing.

Dalrymple and Parsons explore sales forecasting in more depth than Kotler with a discussion and illustration of seasonal adjustments in Chapter 3 of their *Marketing Management* (1986) text: "Perhaps the easiest way to improve the accuracy of sales forecasts is to make adjustments to eliminate seasonal effects." They demonstrate the development of seasonal indices exactly as we have and mention that seasonal adjustments tend to reduce forecasting errors.

Management Science Many management science courses include a discussion of time series analysis. "Because so many decisions depend on these estimates of the future, forecasting is undeniably one of the most important activities undertaken by most companies," writes Davis, McKeown, and Rakes in Chapter 4 in their text *Management Science: An Introduction* (1986). In this chapter the authors review moving averages, single and double exponential smoothing, and several regression-based techniques, as well as the *MAD* and *MSE* criteria for judging the accuracy of forecasts.

With these three courses as evidence, it is easy to see that the business curriculum is rich with applications of time series analysis. Though most widely used in sales forecasting, time series methods also might appear in budgeting applications in accounting or in strategic financial planning. Describing past data patterns and predicting future values of the variables are germane activities in most areas of business.

. . . In Business/The Media

Technical analyses of the economy and the financial markets, reported in newspapers, magazines, and on television, often use some form of time series methods.

For example, the biweekly magazine *Financial World* has a regular feature titled "FW Cockpit," which is a current summary of financial activities. One of the graphs is a 20-day moving average of the ratio of puts to calls on the Chicago Board Options Exchange. To some financial analysts, the put-call ratio is a harbinger of market direction.

Business Week magazine also has a regular report on business activity, primarily in the form of business indicators. At the bottom of the charts, we find this note: "Raw data in the production indicators are seasonally adjusted in computing the BW index (chart). . . ." The chart referred to in this note is reproduced in Figure 18.18. Notice that the production index is a four-week moving average.

Government agencies such as the Census Bureau, the Bureau of Economic Analysis, and the Federal Reserve Board are responsible for assimilating and disseminating huge amounts of time series data such as the monthly unemployment rate, the quarterly gross national product, and the yearly budget deficits. Entire sections of libraries are devoted exclusively to publications from these and other government agencies. The time series methodologies we introduced in this chapter are fundamental in the preparation of government documents and reports.

In-house forecasting units are common in medium to large companies. Utilities continually project electrical and natural gas demand five to ten years into the future, and then assess whether their current capacities are adequate to meet this demand. Even small businesses rely on forecasting—though it may be a judgmental or naive forecast—to plan for the future needs of the firm.

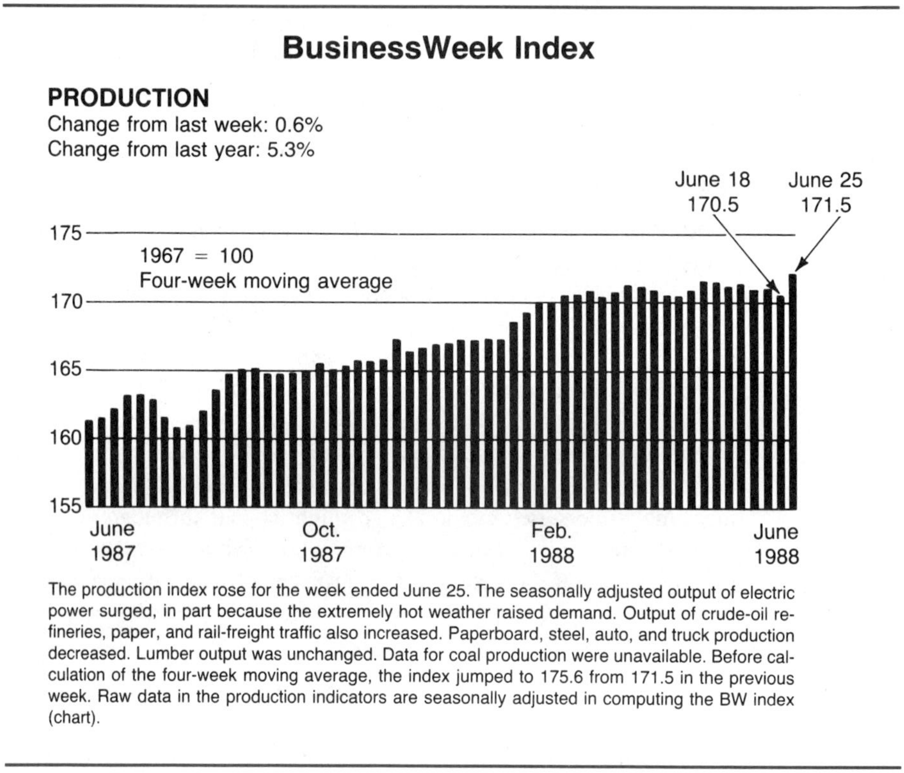

Figure 18.18 Example of Moving Average and Seasonally Adjusted Data

Reprinted from July 18, 1988, issue of *Business Week* by special permission, copyright © 1988 by McGraw-Hill, Inc.

As long as data are recorded by the calendar or the clock, time series analysis will be required. Since business-related data occur naturally in this fashion, be prepared to use the decomposition and forecasting tools from this chapter again.

CHAPTER 18 EXERCISES

18.64 What is the difference between seasonally adjusted data and deseasonalized data?

18.65 Indicate whether the following variables generate time-related data or time-free data.

a. Y = Number of registered conventioneers attending a national convention for each of the last 10 years.

b. Y = Monthly rainfall, in inches, for a city over a two-year period.

c. Y = Number of outpatient surgeries performed by each hospital in a certain city during the month of May.

d. Y = Number of passengers per flight for a major airline.

e. Y = Return for a mutual fund each year for a period of three years.

18.66 Refer to Exercise 18.65. Identify an entity for each variable.

18.67 Name and identify the four components of a time series.

18.68 Name the two types of descriptive time series models and explain how they are different.

18.69 The annualized returns for a mutual fund for the 15-year period from 1975 through 1989 are listed in the following table.

Year	Return	Year	Return
1975	11.8%	1983	18.1%
1976	21.3	1984	9.5
1977	7.0	1985	29.9
1978	13.6	1986	17.0
1979	29.4	1987	4.2
1980	37.7	1988	3.9
1981	6.5	1989	7.8
1982	24.2		

Graph these data. Identify, in words, the trend of this time series, if one exists.

18.70 Refer to Exercise 18.69. Forecast the time series for the years 1979 through 1990 with a four-year moving average forecast. Compute the *MSE* for the period 1979–1989.

18.71 What is plotted on the horizontal and vertical axes of a line graph?

18.72 Since the second quarter of 1983, the United States' net exports of goods and services as a percent of the Gross National Product (GNP) has been a negative number. The implication is that we are importing more than we are exporting. The figures on Y_t, net exports of goods and services as a percent of GNP, are as follows:

	Quarter			
Year	1	2	3	4
1983	0.7	−0.1	−0.6	−0.7
1984	−1.2	−1.7	−1.6	−1.8
1985	−1.4	−1.9	−2.0	−2.5
1986	−2.2	−2.4	−2.6	−2.7
1987	−2.7	−2.7	−2.7	−2.7

Source: U.S. Department of Commerce, Bureau of Economic Analysis, *Business Conditions Digest,* Washington, D.C., October 1988.

Develop seasonal indices for this time series. Interpret S_2 and S_3.

18.73 Refer to Exercise 18.72. Describe the trend in the time series by determining the trend line.

18.74 Refer to Exercises 18.72 and 18.73. Compare the fitting accuracy of a multiplicative model forecast with a four-quarter moving average estimate of the T_{t+1} component to a trend projection forecast, using the *MAPE* criterion.

18.75 One of the indicators of a local economy's robustness is the sales activity of homes. The numbers of homes sold by real estate agents in a two-county area are listed in the following table:

Year	Quarter 1	2	3	4
1981	904	1337	965	758
1982	859	991	1143	933
1983	1271	2259	1559	1132
1984	1441	1717	1643	1216
1985	1272	1615	1407	1411
1986	1306	1668	1475	1169
1987	1114	1323	1617	1585
1988	1563	2097	2157	1544

Source: Data from Bureau of Economic Research, University of Louisville, Louisville, KY.

Construct a four-quarter centered moving average to describe the trend in this time series.

18.76 Refer to Exercise 18.75. Find seasonal indices for each quarter.

18.77 Refer to Exercise 18.75 and 18.76. Construct the cyclical relatives for the time series and graph them. Identify any cycles from the graph. Use a three-period weighted centered moving average with weights of 1, 2, and 1 to eliminate any irregular effects.

18.78 Refer to Exercise 18.75. Forecast the first quarter of 1989 with an exponentially smoothed forecast. Use $\alpha = .2$ and assume $\hat{y}_{30} = 1458.07$.

18.79 Identify the weights of y_1 through y_4 in an exponentially smoothed forecast of Y_5, with $\alpha = .5$.

18.80 What contribution does y_{11} make in an exponentially smoothed forecast of Y_{13}, with $\alpha = .1$?

***18.81** Refer to Exercise 18.75. Use the Time Series processor to find the value of α that minimizes the *MAPE* for an exponentially smoothed forecast. What is the minimum *MAPE*?

***18.82** Refer to Classroom Example 18.3. Use the Time Series processor to find the value of α that minimizes the *MAPE* for an exponentially smoothed forecast. What is the minimum *MAPE*?

18.83 The change, in billions of dollars, in the Gross National Product (GNP) is indicated in the following table by quarters:

Year	Quarter 1	2	3	4
1985	73.8	53.4	68.0	60.9
1986	72.5	27.2	60.8	36.2
1987	87.2	92.4	83.8	94.8

Source: U.S. Department of Commerce, Bureau of Economic Analysis, *Business Conditions Digest*, Washington, D.C., October 1988.

*Optional

a. Determine the *MSE* for an exponentially smoothed forecast, with $\alpha = .4$.
b. Forecast Y_{13}.

18.84 Refer to Exercise 18.83. Compare the fitting accuracies for a naive forecast and an exponentially smoothed forecast, with $\alpha = .4$, on the basis of the *MAD* criterion.

***18.85** Refer to Exercise 18.83. Use the Time Series processor to find the value of α that minimizes the *MAPE* for an exponentially smoothed forecast. What is the minimum *MAPE*?

18.86 Suppose that the *MAPE* for an exponentially smoothed forecast is minimized when α is close to 1.00.
a. Which simple forecasting technique would this most closely resemble?
b. Which type of advanced forecasting model—autoregressive or moving average—seems appropriate in this situation?

18.87 From 1978 through 1987 the United States' exports of corn, in billions of dollars, are listed in the following table:

Year	Corn Exports
1978	5.3
1979	7.0
1980	8.5
1981	8.0
1982	5.6
1983	6.5
1984	7.0
1985	5.2
1986	2.7
1987	3.3

Source: U.S. Department of Commerce, *Business Statistics, 1986–88*, supplement to the *Survey of Current Business*, Washington, D.C., 1988.

Describe the trend component of the time series with a trend line. Find its *MAPE*.

18.88 Refer to Exercise 18.87. Find the fitting accuracies for a naive forecast and a mean forecast, using the *MAPE* metric.

***18.89** Refer to Exercises 18.87 and 18.88. Combine the naive and mean forecasts into the multiple regression model

$$Y_t = \beta_0 + \beta_1\hat{y}_n + \beta_2\hat{y}_m + \epsilon$$

where $\hat{y}_n$ is the naive forecast and $\hat{y}_m$ is the mean forecast. Fit the model to the data for the years 1979 through 1987. Find the *MAPE* for the combined forecast and compare with the individual *MAPE*s for the naive and mean forecasts in Exercise 18.88.

***18.90** The seasonally adjusted monthly unemployment rate for all workers, 16 years and older, from January 1983 to October 1988 is given in the following table. The figures represent percentages of the total work force.

Month	Year 1983	1984	1985	1986	1987	1988
January	10.4	8.0	7.4	6.7	6.7	5.8
February	10.4	7.8	7.2	7.2	6.6	5.7
March	10.3	7.8	7.2	7.1	6.5	5.6
April	10.2	7.7	7.3	7.1	6.3	5.4
May	10.1	7.4	7.2	7.2	6.3	5.6
June	10.1	7.2	7.3	7.1	6.1	5.3
July	9.4	7.5	7.4	7.0	6.0	5.4
August	9.5	7.5	7.1	6.9	6.0	5.6
September	9.2	7.3	7.1	7.0	5.9	5.4
October	8.8	7.4	7.1	6.9	6.0	5.3
November	8.5	7.2	7.0	6.9	5.9	
December	8.3	7.3	7.0	6.7	5.8	

Source: Data from Bureau of Economic Research, University of Louisville, Louisville, KY.

Our goal is to forecast the unemployment rate for November 1988.

a. Detrend the time series.

b. Use the Time Series processor to find the value of α that minimizes the *MAPE* for an exponentially smoothed forecast based on the detrended time series values.

c. Predict the November 1988 value of the detrended time series, using an exponentially smoothed forecast, with $\alpha = .4$.

d. Convert the forecast in part c back into the units of the raw data. (*Hint:* Multiply $\hat{y}_{t+1}$ by $\hat{D}_{t+1}$.) What is your resulting forecast for the November 1988 unemployment rate?

18.91 In terms of the multiplicative model of Equation 18–1, what is the symbolic representation of a deseasonalized time series value?

18.92 The farm production of beef, in millions of pounds, is indicated in the following table for the four-year period from 1983 through 1986.

Year	Quarter 1	2	3	4
1983	64	27	28	64
1984	61	26	26	61
1985	60	26	25	60
1986	55	24	24	55

Source: Commodity Research Bureau, New York, N.Y.

Compare the fitting accuracy of a multiplicative model forecast with a trend projection estimate of T_{t+1} for 1984 through 1986 with that of a four-quarter moving average forecast for the same three-year time frame on the basis of the *MAPE* criterion.

18.93 Refer to exercise 18.92. Compare the fitting accuracies of the two forecasts for 1984 through 1986, on the basis of the *MSE* criterion.

18.94 Suppose we have four years of quarterly data and wish to forecast Y_{17}. How much weight does y_{12} have in $\hat{y}_{17}$ if we use the following methods?

a. a naive forecast.

b. a seasonalized mean forecast.

c. a four-quarter moving average.

d. an exponentially smoothed forecast, with $\alpha = .7$.

18.95 Refer to Exercise 18.87. Forecast the United States' exports of corn for the three-year period 1988–90 by projecting the trend line. If the actual values of our corn exports were \$4.7, \$6.9, and \$7.3 billion, respectively, for the three-year period, determine the forecasting accuracy, using the *MSE* metric.

18.96 Natural gas consumption in residential homes in a Southern city follows a seasonal pattern, with low demand in the summer and high demand in the winter. Suppose we have the following data and three-year summaries on Y_t = the average residential natural gas consumption, in therms, for a small condominium community.

Recent Time Series Values	Three-Year Seasonal Means
$y_9 = 80.7$	$\overline{S}_1 = 118.9$
$y_{10} = 14.9$	$\overline{S}_2 = 19.5$
$y_{11} = 13.8$	$\overline{S}_3 = 15.3$
$y_{12} = 69.9$	$\overline{S}_4 = 60.2$

Exponentially smoothed ($\alpha = .9$) forecast, $\hat{y}_{12} = 43.4$.

Forecast the next four quarters of average natural gas consumption for the condominium community using these forecasting techniques: naive, seasonalized mean, four-quarter moving average, and exponentially smoothed ($\alpha = .9$). Assume the future values turn out to be $y_{13} = 84.0$, $y_{14} = 17.1$, $y_{15} = 7.2$, and $y_{16} = 45.9$. Measure the forecasting accuracy of each technique, using the *MAPE* criterion. Which forecasting technique had the smallest *MAPE*? Largest?

REFERENCES

Bails, D. G., and L. C. Peppers. 1982. *Business Fluctuations: Forecasting Techniques and Applications*. Prentice-Hall, Inc., Englewood Cliffs, NJ.

Brown, R. G. 1959. *Statistical Forecasting for Inventory Control*. McGraw-Hill, Inc, New York.

Dalrymple, D. J., and L. J. Parsons. 1986. *Marketing Management: Strategy and Cases*, 4th Edition. John Wiley & Sons, New York.

Daniel, W. W., and J. C. Terrell. 1983. *Business Statistics: Basic Concepts and Methodology*, 3rd Edition. Houghton Mifflin Company, Boston.

Davis, K. R., P. G. McKeown, and T. R. Rakes. 1986. *Management Science: An Introduction*. PWS-KENT Publishing Company, Boston.

Hanke, J. E., and A. G. Reitsch. 1989. *Business Forecasting*, 3rd Edition. Allyn and Bacon, Boston.

Heizer J., and B. Render. 1988. *Production and Operations Management*. Allyn and Bacon, Boston.

Holt, C. C. 1957. "Forecasting Seasonal and Trends by Exponentially Weighted Moving Averages." Office of Naval Research, Memorandum No. 52.

Kotler, P. 1988. *Marketing Management: Analysis, Planning, Implementation and Control,* 6th Edition. Prentice-Hall, Inc., Englewood Cliffs, NJ.

Mahmoud, E. 1984. "Accuracy in Forecasting: A Survey." *Journal of Forecasting, 3:* 139–159.

Makridakis, S., S. C. Wheelwright, and V. E. McGee. 1988. *Forecasting: Methods and Applications,* 2nd Edition. John Wiley & Sons, New York.

Moriarty, M., and A. J. Adams. 1984. "Management Judgment Forecasts, Composite Forecasting Models, and Conditional Efficiency." *Journal of Marketing Research, 21:* 239–250.

Neter, J., W. Wasserman, and G. A. Whitmore. 1988. *Applied Statistics,* 3rd Edition. Allyn and Bacon, Boston.

Shiffler, B., and P. S. Ryan. 1988. "The Delphi Technique: An Application in Utility Planning." *Proceedings of Midwest Decision Sciences Institute,* Louisville, KY.

Winters, P. R. 1960. "Forecasting Sales by Exponentially Weighted Moving Averages." *Management Science, 6:*324–342.

Chapter Maxim *For making an inference based on categorical or ranked data, use a nonparametric procedure.*

CHAPTER 19

CHI-SQUARE AND OTHER NONPARAMETRIC PROCEDURES

Objectives

After studying this chapter and working the exercises, you should be able to

1. Explain how nonparametric procedures differ from parametric tests.
2. Conduct tests for proportions in a multinomial population.
3. Determine whether variables in a cross-tabulation should be viewed as independent.
4. Compute the sample correlation for data expressed in the form of rankings.
5. Test the assumption that two sets of rankings are uncorrelated.
6. Use ranks to test the hypothesis that two populations have the same mean.
7. Describe two forms of nonrandomness in an ordered sequence.
8. Conduct a hypothesis test to establish whether a sequence of observations occurs randomly.
9. Use the runs test to check the regression model assumption of independent error terms.

We have now studied a variety of inference-making procedures. However, we have generally restricted ourselves to analyzing *quantitative* (as opposed to *qualitative*) data. Needless to say, not all situations that call for an inference involve quantitative data. A further restriction to this point is that for small samples ($n < 30$), we have generally assumed that our variable is normally distributed. For instance, use of the t-distribution with small samples presumes normality. This assumption is a fairly strong one and is not always warranted. (With large samples, the Central Limit Theorem applies for inferences about the mean, negating the need to assume that the individual observations are normally distributed.)

In this chapter, we will learn some inference procedures to deal with these gaps in our coverage. In particular, we will see how to test hypotheses about qualitative data—where the entities yield data either in the form of *categories*, or in the form of *rankings*. We also will learn a two-sample hypothesis test procedure that does not presume normality.

The statistical methods in this chapter are part of a set of procedures that are loosely referred to as nonparametric methods. To be called a **nonparametric method,** a procedure usually involves count information for categorical data; or rankings of observations; and/or a lack of specific assumptions, such as normality, about the population. In Section 11.3, you were introduced to one important nonparametric procedure, the signed ranks test. We presented this useful test alongside its parametric counterpart, the t-test. The signed ranks test does not assume that the individual values of X are normally distributed, and, as its name suggests, it employs rankings.

We already have used extensively another nonparametric procedure—the binomial distribution. Strictly speaking, a binomial inference is a nonparametric procedure since the sample observations yield count information for categorical data: Two labels or categories are possible for each binomial sample entity—success or failure.

We begin this chapter by considering tests for categorical data with more than two outcomes possible per observation. In so doing, we will use a sampling distribution encountered previously—the chi-square distribution.

19.1 CHI-SQUARE TESTS FOR CATEGORICAL DATA

Examples of Categorical Data

Categorical data can take a variety of forms; three examples are given as follows.

- A department store manager randomly samples purchases made in the young women's wear department and records the method of payment. She finds the following:

  Cash—30 purchases
  Check—52 purchases
  Credit card—68 purchases

  Since each entity (purchase) yields a word/phrase result—cash, check, or credit card—we have categorical data. The numbers given represent the sample frequencies of occurrence for each category. The manager could use these data to investigate a hypothesis, such as H_0: all categories (payment methods) are equally likely in the population of purchases.

- A new outlet of a discount department chain has been open about a year. A random sample of letters is taken from the store's files of written complaints received since its opening. Each complaint is labeled according to the writer's primary grievance:

  A—quality of item
  B—customer service/courtesy
  C—price of item
  D—miscellaneous

  For the discount chain as a whole, the complaint percentages are known to be $A = 40$ percent, $B = 25$ percent, $C = 20$ percent, $D = 15$ percent. Management may wish to obtain the category counts of the sample and then test the assumption that the new store complaint pattern is the same as that experienced chain-wide (H_0: the new store complaint percentages are $A = 40\%$, $B = 25\%$, $C = 20\%$, and $D = 15\%$).

- Surveys are a major source of categorical data since many questionnaires ask the respondent to choose from categorical alternatives. For instance, the American Medical Association might commission a study of 600 adults' attitudes towards permitting doctors to advertise. Suppose one statement in the survey is, "Physicians should be allowed to advertise their services and specialties as do other businesses," with these possible responses offered:

  ____strongly disagree
  ____disagree somewhat
  ____agree somewhat
  ____strongly agree

  Perhaps elsewhere in the survey instrument, the participant is asked to indicate her or his sex:

  ____female ____male

  If we wish to consider jointly the two variables (sex and opinion), we could arrange the counts in a cross-tab, or contingency table, as discussed in Section 4.4. The table might happen to have the frequencies shown as follows.

	Opinion			
Sex	**Strongly Disagree**	**Disagree Somewhat**	**Agree Somewhat**	**Strongly Agree**
Female	52	62	140	64
Male	54	60	128	40

  These are bivariate data, since we are considering two variables simultaneously. Both variables are categorical in nature: There are four opinion categories and, of course, two sex labels. If a researcher wished to make an inference as to whether the variables sex and opinion are independent (unrelated) in the population, he would test his hypothesis by making use of the sample frequencies given in the contingency table.

In this section we will learn two hypothesis testing procedures:

1. Tests about how the entities of a particular qualitative variable are distributed into three or more categories (in short, we will be generalizing the binomial distribution into the *multinomial* distribution).
2. Tests that can allow us to conclude whether two categorical variables in a cross-tab are related or unrelated to each other.

Tests for Multinomial Populations

In Chapter 6, we studied binomial sampling conditions. Multinomial conditions and assumptions are much the same, but with each of n sample entities now having three or more possible outcomes, or *cells*. We will use one of our earlier examples of categorical data to illustrate hypothesis testing for a multinomial population.

The manager of a department store wishes to see whether each of three payment methods (cash, check, or credit card) is equally likely to be used in the young women's wear department. A random sample of 150 purchases yields the following sample frequencies:

Cash—30 purchases
Check—52 purchases
Credit card—68 purchases

We wish to test, in words,

H_0: All categories (payment methods) are equally likely in the population of payments.

In symbols, we could express the null hypothesis as

$$H_0\text{: } \pi_{\text{Cash}} = \pi_{\text{Check}} = \pi_{\text{Credit card}} = 1/3$$

where π with a subscript represents the proportion of payments in the population having the indicated categorical outcome. For this example, we shall employ a Type I error risk of $\alpha = .05$. To help us make our inference, we need to ask ourselves how we would expect the counts to be divided among the payment methods if H_0 is true. These anticipated counts are denoted as theoretically expected frequencies. In this case, with a sample size of 150 payments and each payment method hypothesized to be equally likely, each of the three expected frequencies would be one-third of 150, or 50. Placing our sample evidence and our theoretically expected frequencies side-by-side, we have the following:

Payment Method	Observed or Sample Frequency, O	Expected Frequency If H_0 is True, E
Cash	30	50
Check	52	50
Credit card	68	50

Your intuition should tell you that if H_0 is true, the observed and expected frequencies should be similar across the cells. On the other hand, large discrepancies between observed and expected frequencies should suggest that H_0 is unlikely to be true. Our test statistic therefore will be a number that quantifies the overall extent of disagreement between observed and expected frequencies.

The test statistic is called *chi-square,* and it involves each cell's observed and expected frequencies.

Chi-Square Test Statistic

$$\chi^{2*} = \sum_{\text{all cells}} \frac{(\text{Observed frequency} - \text{Expected frequency})^2}{\text{Expected frequency}}$$

or abbreviated,

$$\chi^{2*} = \sum \frac{(O - E)^2}{E} \qquad \textbf{(19–1)}$$

The chi-square test statistic is sensitive to differences between the observed and expected frequencies. When observed and expected are close to each other—a likely result when H_0 is true—the test statistic will be relatively small. (Note that the case of perfect agreement of each O with its respective E will produce a test statistic equal to zero.) Large differences between O's and E's—a likely outcome when H_0 is false—will make the test statistic relatively large. It therefore should be apparent that we will reject our null hypothesis only when the chi-square test statistic is sufficiently large. In other words, the rejection region for the test will be placed in the right tail of the chi-square distribution.

The general form of the chi-square distribution is shown in Figure 19.1. We first introduced the chi-square distribution in Section 12.4, although for a different purpose at that time. You may wish to review the subsection "Use of the Chi-Square Tables" found in Chapter 12 to refamiliarize yourself with using the table given in Appendix G. (*A reminder:* For inferences involving categorical data, we will use only the upper-tail cutoff values on the chi-square distribution.)

The dividing line between accepting and rejecting H_0 depends on the Type I risk factor α and the degrees of freedom (df) associated with the sampling situation. For tests involving multinomial populations, the df will be the number of cells, or categories, minus one.

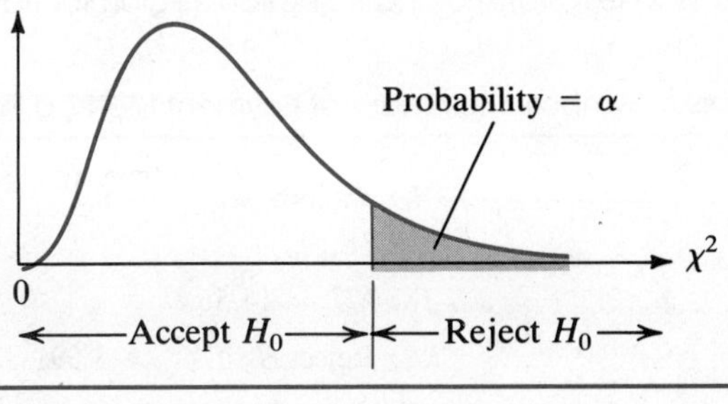

Figure 19.1 The Chi-Square Distribution for Tests Involving Categorical Data

Degrees of Freedom for Multinomial Tests

$$df = \text{Number of cells} - 1$$

For the payment method problem, there are three categories of payments, so the *df* is two. If we wish to test our hypothesis of equal proportions at the .05 level of significance, the chi-square table cutoff value is 5.991. The value of the test statistic is

$$\begin{aligned}\chi^{2*} &= \sum \frac{(O - E)^2}{E} \\ &= \frac{(30 - 50)^2}{50} + \frac{(52 - 50)^2}{50} + \frac{(68 - 50)^2}{50} \\ &= 8.00 + .08 + 6.48 = 14.56\end{aligned}$$

Since the test statistic exceeds the table cutoff value, we have strong evidence against the null hypothesis. We conclude that H_0 is false (see Figure 19.2). Table 19.1 summarizes this problem.

When we reject H_0, it is often of interest to see whether certain cells may have had a large impact on the test statistic. In the example above, for instance, the "cash" cell contributed 8.00 towards the sum of 14.56, while the "credit card" cell contrib-

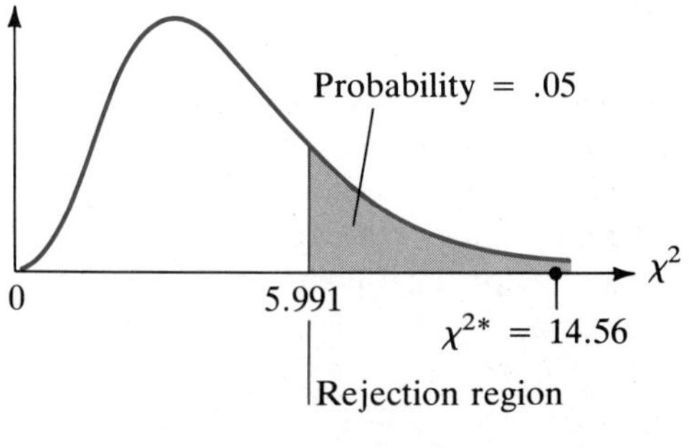

Figure 19.2 Test Statistic and Rejection Region for Payment Method Problem

Table 19.1 Symbolic Summary of Payment Method Problem

H_0: $\pi_{\text{Cash}} = \pi_{\text{Check}} = \pi_{\text{Credit card}} = 1/3$

H_a: not H_0

TS: $\chi^{2*} = 14.56$

RR: Reject H_0 if $\chi^{2*} > 5.991$

C: Reject H_0

uted 6.48. The cash cell's contribution is large since fewer purchases than expected were in this category; the credit card cell's impact is large since more purchases than expected appeared here. This suggests that $\pi_{\text{Cash}} < 1/3$ and that $\pi_{\text{Credit card}} > 1/3$. The "check" cell made a negligible contribution toward the sum of 14.56, because its observed and expected frequencies were relatively close to each other.

EXAMPLE 19.1

A new outlet of a discount department chain has been open about a year. A random sample of 90 letters is taken from the store's file of written complaints received since its opening. Each complaint is classified according to the writer's primary grievance, and category counts are made:

A—quality of service: 31 letters
B—customer service/courtesy: 25 letters
C—price of item: 17 letters
D—miscellaneous: 17 letters

For the discount chain as a whole, the complaint percentages are known to be $A = 40$ percent, $B = 25$ percent, $C = 20$ percent, and $D = 15$ percent. Test the assumption that the new store complaint pattern is the same as that of the chain as a whole, using $\alpha = .10$.

Solution:

In symbols, we can represent our hypotheses as

$$H_0\text{: } \pi_A = .40,\ \pi_B = .25,\ \pi_C = .20,\ \pi_D = .15$$

and

$$H_a\text{: not } H_0$$

We have a sample of 90 observed frequencies; the needed expected frequencies also must sum to 90. As with the payment method problem, we rely on the H_0 statement to point us toward the appropriate expected frequencies. Unlike that example, however, we do not wish to test for equally likely categories. Instead, we must distribute the 90 expected frequencies to the categories in accordance with the proportions presumed in H_0, namely .40, .25, .20, and .15. Our expected frequencies are, therefore, as follows:

$$\begin{aligned}
\text{Category } A\text{: } n\pi_A &= 90(.40) = 36 \text{ letters}\\
\text{Category } B\text{: } n\pi_B &= 90(.25) = 22.5 \text{ letters}\\
\text{Category } C\text{: } n\pi_C &= 90(.20) = 18 \text{ letters}\\
\text{Category } D\text{: } n\pi_D &= 90(.15) = 13.5 \text{ letters}
\end{aligned}$$

(Expected frequencies can be decimal numbers; observed frequencies must be integers.) Placing the expected and observed frequencies side-by-side, we have the following:

	Category			
Frequency	A	B	C	D
O (observed)	31	25	17	17
E (expected)	36	22.5	18	13.5

The test statistic then will be

$$\begin{aligned}\chi^{2*} &= \sum \frac{(O - E)^2}{E} \\ &= \frac{(31 - 36)^2}{36} + \frac{(25 - 22.5)^2}{22.5} + \frac{(17 - 18)^2}{18} + \frac{(17 - 13.5)^2}{13.5} \\ &= .694 + .278 + .056 + .907 = 1.935\end{aligned}$$

With four categories, the degrees of freedom will be three. Intersecting the $df = 3$ row with the column to place 10% risk in the right tail of the chi-square distribution, we find the cutoff value of 6.251. Since $1.935 < 6.251$, we accept H_0; we find the sample evidence to be consistent with chain-wide experience.

□

COMMENTS

1. Chi-square tests for categorical data have a minimum sample size requirement: Each expected frequency, E, should be at least five. In reality, when H_0 is true, our chi-square test statistic only approximately follows the chi-square distribution. This approximation works well for large sample sizes but can deteriorate when expected cell counts become small. One solution to having an expected frequency less than five is presented in Exercise 19.7.
2. The df for a multinomial test (number of cells minus one) represents the number of unrestricted cell frequencies. In Example 19.1 for instance, there were four cells: A, B, C, and D. If we know the frequency of any three of these, the fourth frequency is automatically known, since the four frequencies are constrained to add up to 90. In other words, the fourth frequency is not free to take on just any value—its value is a forced choice.
3. The chi-square test for a multinomial population also is called a *goodness-of-fit* or a *one dimensional chi-square test*. Since the multinomial null hypothesis has many ways to be false, we usually state the alternative hypothesis simply as not H_0.
4. An alternative form of the chi-square test statistic given in Equation 19–1 is

$$\chi^{2*} = \sum \frac{O^2}{E} - n$$

where n is the overall sample size.

5. Rounding expected frequencies to one decimal place is sufficient. However, we recommend rounding the values of $(O - E)^2/E$ or O^2/E to three decimal places.

CLASSROOM EXAMPLE 19.1

Testing for Goodness of Fit

In a self-study report to its accrediting institution, a business school stated that the following overall distribution of letter grades was given by its instructors over the past two years: 16 percent A's, 31 percent B's, 34 percent C's, 11 percent D's, and 8 percent F's. When the accrediting institution sent representatives to visit the business school and audit their programs, a random sample of letter grades was chosen from student records. The sample contained 37 A's, 67 B's, 61 C's, 24 D's, and 11 F's. Suppose it is of interest to test the hypothesis that the school's letter grade distribution is accurately stated in the self-study report.

a. Express the null hypothesis in symbols.

b. How many cells exist for this sampling situation?

c. The expected frequencies will sum to what number?

d. Supply the five expected frequencies.

e. Determine the numerical value of the test statistic.

f. For an inference with a 10% risk of Type I error, what cutoff value separates acceptance from rejection region?

g. Was your null hypothesis supported or contradicted?

Processor: Goodness of Fit Test We will use the setting of Example 19.1 to illustrate the Goodness of Fit processor. We begin by selecting Data from the main menu; then touch Create. For Example 19.1, we specify "X^2 Goodness of Fit." We then will be asked to provide the following: the number of categories, the category names, and the observed frequencies for each category. Note that the processor can accommodate up to ten categories.

When data entry is complete, move to Processors and select χ^2. A submenu with two choices will pop up; choose the first option. After reading the introductory screen, read in the Z-table (it is needed to compute the p-value), and move to Data. When we touch Enter, Figure 19.3 will appear, asking whether our null hypothesis is to assume that all categories are equally likely. If we choose the first option, we can proceed immediately to Calculations and Results. If we choose the second option, we will need to supply category probabilities for the null hypothesis statement.

Since Example 19.1 has a hypothesis with differing category probabilities, we choose the second option of Figure 19.3, and then furnish these probabilities. Figure 19.4 shows the screen as we are entering the fourth and final category probability. We

*Optional

```
≡                     Goodness Of Fit Tests                               MENU
   Introduction  *able [Data] *alculations  *esults  Exit           | F1=Help

                        All categories equally likely
                        Differing category probabilities
```

Figure 19.3 Null Hypothesis Types for a Goodness-of-Fit Test

```
≡                     Goodness Of Fit Tests                               EDIT
   Introduction  *able [Data] *alculations  *esults  Exit           | F1=Help

              π1  = 0.40                   π2  = 0.25
              π3  = 0.20
```

Figure 19.4 Entering Probabilities for the Null Hypothesis

```
≡                     Goodness Of Fit Tests                               READY
   Introduction  *able  Data  Calculations [Results] Exit           | F1=Help
```

Category Name	Observed Frequencies	Expected Frequencies
Quality	31	36
Courtesy	25	22.5
Price	17	18
Miscellaneous	17	13.5

The number of sample observations is n = 90

Figure 19.5 Table of Observed and Expected Frequencies

then advance to Calculations and Results. Two results screens are available. The Frequencies option of Results (shown in Figure 19.5) summarizes the observed and expected frequencies across the categories. The Test Summary option (not shown) provides a statement of H_0, the chi-square test statistic, the *df*, and the *p*-value.

Tests for Statistical Independence

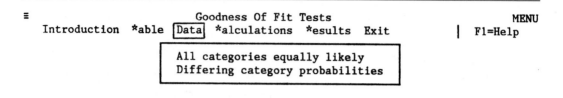

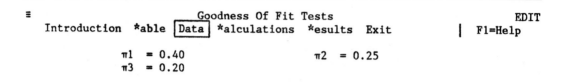

In this section, we will develop a chi-square test that can enable us to determine whether two variables in a cross-tabulation, or contingency table, are independent or dependent. Recall that a contingency table is a convenient way to present bivariate categorical data. Such a two-dimensional format displays one variable on the row, the other on the column.

As an example, consider the following 3 × 2 (3 rows, 2 columns) contingency table, where a random sample of 90 single-family residencies that have been sold in a medium-sized community have been cross-classified against the amount of time the home was on the market before it sold. Each of the 90 home sales, therefore, is classified into one of six cells.

	Days on the Market		
Selling Price	**60 or Less**	**61 or More**	**Total**
Under $75,000	18	12	30
$75,000 to $125,000	14	31	45
Over $125,000	4	11	15
Total	36	54	90

Often when we have a cross-tab, we wish to know whether the two variables (selling price and days on the market, in this case) are independent or dependent. What would it mean if selling price and days on the market were independent? As suggested in Chapter 5, independence implies that two variables are unrelated; independence would mean that how observations are categorized on one variable has no bearing on how they are categorized on the other variable. In this example, a finding of independence would mean that the number of days a home was on the market bears no relation to the home's selling price, and vice versa.

What would it mean if selling price and days on the market were dependent? Dependence means that observations' classifications on one variable to some degree are related to the classification on another variable. In this example, a finding of dependence would mean that number of days on the market before sale is somehow related to the selling price (perhaps lower priced homes sell quicker, for instance).

The hypotheses for a test of statistical independence are

$$H_0\text{: The two variables are independent}$$

and

$$H_a\text{: The two variables are dependent}$$

We will test the null hypothesis at the .05 level of significance in this example. The sample evidence in the 3 × 2 table constitutes the observed frequencies; to compute the chi-square test statistic, we need the expected frequencies for each of the six cells. As with the multinomial test, the expected frequencies are dictated by the null

Table 19.2 Observed and Expected Frequencies for Home Sale Example

Selling Price	Days on the Market: 60 or Less	Days on the Market: 61 or More	Total
Under $75,000	18 (12)	12 (18)	30
$75,000 to $125,000	14 (18)	31 (27)	45
Over $125,000	4 (16)	11 (9)	15
Total	36	54	90

hypothesis. Since H_0 declares independence, the expected frequencies must be those that (theoretically) would result under the condition of independence.

Following is an intuitive way to develop the expected frequencies. For our overall data set, 40 percent of homes (36 out of 90) sold in 60 days or less, while 60 percent (54 out of 90) took longer. Now if price is independent of time on market, then we should expect to see the same 40 percent and 60 percent reappear within each price category. In other words, the expected frequencies for the upper row should be $.40(30) = 12$, and $.60(30) = 18$. For the middle row $.40(45) = 18$, and $.60(45) = 27$; and for the bottom row $.40(15) = 6$, and $.60(15) = 9$. Table 19.2 shows our cross-tab with all six expected frequencies placed in parentheses next to their corresponding observed frequencies.

Here is another way to obtain expected frequencies. Letting i denote row number and j denote column number, the expected frequency for any given cell is given in the following box.

Expected Frequency for Cell$_{ij}$

$$E_{ij} = \frac{(\text{Row } i \text{ total})\ (\text{Column } j \text{ total})}{\text{Total sample size}} \tag{19–2}$$

Using this expression on the upper right cell (cell$_{12}$), for instance,

$$E_{12} = \frac{30(54)}{90} = 18 \quad \text{(as in Table 19.2)}$$

Two notes about our expected frequencies: For each row and for each column, the sum of the expected frequencies must agree with the sum of the observed frequencies; and we will require that all expected frequencies be at least five, as with the multinomial test.

Now that we have the expected frequencies, we can compute the chi-square test statistic. Since a 3 × 2 table has six cells, there will be six terms to add together to form the test statistic.

$$\begin{aligned}\chi^{2*} &= \sum \frac{(O - E)^2}{E} \\ &= \frac{(18 - 12)^2}{12} + \frac{(12 - 18)^2}{18} + \frac{(14 - 18)^2}{18} + \frac{(31 - 27)^2}{27} + \frac{(4 - 6)^2}{6} \\ &\quad + \frac{(11 - 9)^2}{9} \\ &= 3.0 + 2.0 + .889 + .593 + .667 + .444 \\ &= 7.593\end{aligned}$$

We need to compare the test statistic with the cutoff value to arrive at a conclusion. The df for a test of independence for a contingency table with r rows and c columns is $(r - 1)(c - 1)$.

Degrees of Freedom for a Test of Independence

$$df = (r - 1)(c - 1)$$

In this example, we have 3 rows and 2 columns in our table, so $df = (3 - 1)(2 - 1) = 2$. Using $\alpha = .05$ with $df = 2$, we find the cutoff value of 5.991. Since our test statistic exceeds the cutoff value, we reject H_0. We have found that the discrepancies between observed and expected frequencies are, as a group, too large to be consistent with having independent variables. Table 19.3 summarizes this example.

As with a multinomial chi-square test, when rejection occurs, it may be of interest to see which cells made a large contribution to the test statistic. For this example, it appears that the cells in the top row had the biggest impact (contributions of 3.0 and 2.0 toward 7.593). As a category, the inexpensive homes tended to sell more quickly than the other categories.

COMMENTS

1. As it is for a multinomial test, the df for a test of independence represents the number of unrestricted cell frequencies. For instance, in the home selling example where

Table 19.3 Symbolic Summary of Test of Independence

H_0: The variables (selling price and days on the market) are independent.
H_a: The variables are dependent.
TS: $\chi^{2*} = 7.593$
RR: Reject H_0 if $\chi^{2*} > 5.991$
C: Reject H_0

$df = 2$, only two expected frequencies are unrestricted—the other four are forced choices to meet the condition that each row's (and column's) expected frequency total must agree with the observed frequency total. For instance, suppose you have determined only two expected frequencies: $E_{22} = 27$ and $E_{23} = 9$. Given these, all other E's are available by subtraction. For example, E_{31} must be 6 (15 minus 9), so that the third row expected count equals the observed count of 15.

2. The chi-square test for statistical independence for categorical variables is roughly equivalent to testing the hypothesis of zero correlation for two quantitative variables. The chi-square test also is called a *two-dimensional chi-square test* or a *test for association*.
3. The requirement that all cell expected frequencies be at least five is generally regarded as conservative. Some statisticians say that when there are a large number of cells, the chi-square approximation works sufficiently well even if an expected frequency is as low as two.
4. As before, we recommend rounding expected cell frequencies to one decimal place, but carrying out (and rounding) computations of the chi-square test statistic to three decimal places.

EXAMPLE 19.2

A large hotel purchased 200 new color televisions several months ago: 80 of one brand and 60 of each of two other brands. Records were kept for each set as to how many service calls were required, resulting in the table that follows.

Number of Service Calls	TV Brand *S*	*R*	*T*	Total
None	10	16	14	40
One	27	47	26	100
Two or more	23	17	20	60
Total	60	80	60	200

Assume the TV sets are random samples of their brands. With 10% risk of Type I error, test for an association between TV brand and the number of service calls.

Solution:

For a contingency table test, the null hypothesis is always one that assumes the variables in question are statistically independent. We will have the following:

H_0: TV brand and number of service calls are independent.

and

H_a: TV brand and number of service calls are dependent.

The *df* for a 3 × 3 table will be (3 − 1)(3 − 1) = 4. For $\alpha = .10$, the chi-square cutoff value is 7.779. We need to allocate the 200 expected frequencies into the nine cells. Using Equation 19–2, following are the expected frequencies for four of the nine cells:

$$E_{11} = \frac{40(60)}{200} = 12 \qquad E_{12} = \frac{40(80)}{200} = 16$$

$$E_{21} = \frac{100(60)}{200} = 30 \qquad E_{22} = \frac{100(80)}{200} = 40$$

While we could use Equation 19–2 to determine all nine expected frequencies, it may be quicker to use it four times as indicated, and then obtain the remaining five expected frequencies by subtraction. (Since *df* = 4, we know that four is the minimum number of times we need to use the expected frequency formula.)

The following table shows all the cells' observed and expected frequencies, with the expected frequencies in parentheses.

Number of Service Calls	TV Brand *S*	*R*	*T*	Total
None	10 (12)	16 (16)	14 (12)	40
One	27 (30)	47 (40)	26 (30)	100
Two or more	23 (18)	17 (24)	20 (18)	60
Total	60	80	60	200

The test statistic will be the summation of nine terms:

$$\begin{aligned}\chi^{2*} &= \sum \frac{(O - E)^2}{E} \\ &= \frac{(10 - 12)^2}{12} + \cdots + \frac{(20 - 18)^2}{18} \\ &= .333 + \cdots + .222 \\ &= 6.378\end{aligned}$$

Since 6.378 is less than the cutoff value of 7.779, we do not reject H_0. We have not found convincing evidence of a dependent relation between TV brand and the number of service calls.

□

CLASSROOM EXAMPLE 19.2

Testing for Statistical Independence

A bank treasurer obtains a random sample of customer loan applications and cross-tabulates the lending decision that was made with the loan officer who acted on the application.

Loan Decision	Loan Officer		
	Hoffman	Cunningham	Burton
Approved	48 ( )	30 ( )	27 ( )
Denied	12 ( )	20 ( )	13 ( )

a. Construct a table similar to the preceding one, and determine the expected frequencies to test H_0: the variables *loan decision* and *loan officer* are statistically independent in the population. Place the values in parentheses in your table.

b. Compute the value of the test statistic.

c. What is the value of *df* for the test? What is the cutoff value from the chi-square table for $\alpha = .10$?

d. Is the null hypothesis contradicted or supported?

e. Does your conclusion mean that *loan officer* and *loan decision* are related to each other or unrelated to each other? Do you think the bank treasurer would like or dislike this finding? Why?

f. Which cell makes the largest contribution toward the value of the test statistic?

Processor: Test for Independence We will use the setting of Example 19.2 to illustrate the test for independence processor. We begin by selecting Data from the main menu; then touch Create. For Example 19.2 we move the highlight bar to Contingency Tables (χ^2). We then will be asked the number of rows and the number of columns for our crosstabulation (the maximum dimensions allowable are five rows and five columns). At this point we will enter our sample (observed) frequencies, row by row. Figure 19.6 shows the screen as we begin to enter the third row of data.

When all frequencies are entered, move to Processors and select χ^2; choose the second option—Test For Independence. After reading the introductory screen, read in the Z-table, touch Data to verify proper data type, and then perform Calculations. When we move to Results, two screens are available. The Frequencies display (not shown) recaps the observed and expected frequencies; the Test Summary screen is shown in Figure 19.7.

```
≡        Microcomputer Applications for INTRODUCTORY BUSINESS STATISTICS    EDIT
  Overview [Data] Files  Processors  Set Up  Exit                      | F1=Help

          Column 1                 Column 2                 Column 3
  1 10                    16                       14
  2 27                    47                       26
  3•23
```

Figure 19.6 Entering Observed Frequencies

```
≡                          Tests For Independence                          READY
  Introduction  *able  *ata  Calculations [Results] Exit                 | F1=Help

          Ho: The row and column variables are statistically
              independent in the population

          Ha: Not Ho

                           (O - E)²
          TS: X²* = Σ --------, the test statistic is 6.377778
                              E

              The degrees of freedom are 4

              The p-value for this test is 0.172656126313
```

Figure 19.7 Test for Independence Summary

SECTION 19.1 EXERCISES

19.1 For a chi-square test involving categorical data, find the table cutoff value for the following:

a. 4 *df*, and $\alpha = .01$.
b. 3 *df*, and $\alpha = .05$.
c. 7 *df*, and $\alpha = .10$.

19.2 What value of chi-square separates acceptance from rejection region for a multinomial test with the following:

a. five categories, and $\alpha = .025$.
b. four categories, and $\alpha = .01$.
c. six categories, and $\alpha = .05$.

19.3 For the following values of the chi-square test statistic, state whether H_0 should be rejected or not rejected.

a. $\chi^{2*} = 6.47$, $df = 2$, and $\alpha = .05$.
b. $\chi^{2*} = 10.12$, $df = 6$, and $\alpha = .10$.
c. $\chi^{2*} = 14.64$, $df = 5$, and $\alpha = .05$.

19.4 Use the chi-square goodness-of-fit test to see whether the discrepancies between observed and expected frequencies given in the following table are too large to be explained by random chance. Use the .05 level of significance.

	Category			
Frequency	*A*	*B*	*C*	*D*
E (expected)	26	30	25	15
O (observed)	24	44	20	8

19.5 A casino gambling die is rolled 90 times, with the following outcomes:

Number of Dots Showing	1	2	3	4	5	6
Count	14	22	18	14	9	13

Test the assumption that the die is fair, assuming 10% risk of a Type I error.

19.6 A mail order catalog company has introduced a new style of men's heavy winter overcoat in their fall/winter catalog. Based on past experience with other coats, the company is anticipating the following breakdown of orders by sizes: S = 12 percent, M = 45 percent, L = 35 percent, and XL = 8 percent. The first 300 coats ordered are S = 22, M = 125, L = 119, and XL = 34. The first 300 orders came in substantially sooner than expected, indicating a fast-moving item. It appears that the catalog company needs to obtain more of these coats from their supplier. Treating the first 300 orders as a random sample, does it seem reasonable to maintain the original proportions of coat size in any further orders from the supplier? Use α = .10.

***19.7** If a category has an expected frequency less than five, it can be combined with an adjacent one to form a redefined category. For instance, if the frequencies for five categories labeled H through L were as shown, category L would cause a problem since its expected frequency is only two.

Frequency	H	I	J	K	L
O (observed)	42	35	10	8	5
E (expected)	45	30	14	9	2

The categories could be redefined, however, in the following manner to bring all expected cell counts up to at least five:

Frequency	H	I	J	K or L
O (observed)	42	35	10	13
E (expected)	45	30	14	11

The process of combining cells reduces the degrees of freedom; in this case the original df = 4 becomes df = 3.

To obtain "random" bridge deals, major bridge tournaments often use "computer-dealt" hands instead of having players themselves shuffle the cards. (Naturally the computer doesn't deal the hands; instead it furnishes a listing of which positions—north, south, east, west—are to get which cards.) A study of 8024 computer-dealt bridge hands was done to see how closely the hands matched theoretically expected outcomes (reported in "Computer Hands: They Are Fair," by Sid Kilsheimer in the November 1981 issue of *The Contract Bridge BULLETIN*). The study focused on how many cards each player received in his longest suit. A bridge hand contains 13 cards. Since there are four suits, the player is guaranteed receiving

*Optional

at least four cards in some suit, and may of course have more than four. The following table resulted:

Player's Longest Suit	Sample Frequency	Category Probability
4 cards	2791	.35081
5 cards	3607	.44340
6 cards	1294	.16548
7 cards	294	.03527
8 cards	36	.00467
9 cards	2	.00037
	8024	1.0

a. Use the known probabilities of occurrence for each category to obtain the expected frequencies. Report expected frequencies to the nearest integer.
b. Combine any cells that have expected counts under five with an adjacent cell.
c. Compute the test statistic for H_0: the distribution of *longest suit counts* of computer-dealt hands is consistent with known probabilities.
d. State the table cutoff value and your conclusion, for $\alpha = .10$.

19.8 Six months ago, the three network television stations in a western city had the following market shares for the late evening news broadcasts:

CBS affiliate—45 percent
NBC affiliate—33 percent
ABC affiliate—22 percent

In the last six months, two of the three stations have had their late evening news "anchor" person leave to take a job elsewhere. A local ratings service now interviews a sample of 450 late evening news watchers, asking them to state their favorite channel. The sample counts are 208 for the CBS affiliate, 129 for NBC, and 113 for ABC. With 10% risk of incorrectly rejecting, test the hypothesis that there has been no change in the market shares for the late evening news broadcasts.

19.9 A metropolitan bus system samples rider counts on one of its express commuter routes for a week. Use the following data to establish whether ridership is evenly balanced by day of the week. Let $\alpha = .10$.

Day	Rider Count
Monday	57
Tuesday	63
Wednesday	64
Thursday	69
Friday	54

19.10 A retail clothing store classifies its customers' store charge accounts as current, 1–30 days late, 31–60 days late, and over 60 days late. Six months ago, the store's account status report showed the following conditions:

Status	Percent
Current	68
1–30 days late	17
31–60 days late	10
Over 60 days late	5

At the time of this report, the store instituted an increase of five percentage points in the interest rate assessed those accounts that became overdue. A random sample of 125 charge card accounts is now drawn to see whether the percentages given in the table still apply. Test the assumption of no change, using $\alpha = .10$. The sample shows 91 current, 19 1–30 days late, 11 31–60 days late, and 4 over 60 days late.

19.11 What is the table cutoff value for a test of independence when the following conditions exist (r = number of row categories, c = number of column categories)?

a. $r = 2$, $c = 5$, and $\alpha = .01$.
b. $r = 3$, $c = 3$, and $\alpha = .10$.
c. $r = 4$, $c = 2$, and $\alpha = .05$.

19.12 Find the expected frequencies that would be needed to conduct a test of independence between factors C and D.

	Factor C			
Factor D	C_1	C_2	C_3	Total
D_1				58
D_2				60
D_3				82
Total	100	80	20	

19.13 Use this sample to ascertain whether variables A and B are dependent or independent in the population. Tie your conclusions to a 1% risk of Type I error.

	Variable A		
Variable B	A_1	A_2	A_3
B_1	5	10	21
B_2	19	26	27

19.14 A sample of 489 adults completed a survey about their attitudes towards lawyers advertising. One question posed in the survey was, "I have a good idea of what most prices are for routine legal services." The responses to this question were cross-tabulated against the respondents' age (less than 48, or 48 and older).

Age	Response to Question: Agree or Strongly Agree	No Opinion	Disagree or Strongly Disagree
Younger	47	15	169
Older	54	41	163

Source: Adapted from Robert Hite and Edward Kiser, "Consumer's Attitudes Toward Lawyers with Regard to Advertising Professional Services," *Journal of the Academy of Marketing Science*, Spring, 1985.

Assuming the data are a random sample, test to see whether response to this question is independent of the age of the responder. Use the 10% level of significance.

19.15 A state cabinet official is contemplating running in her party's primary election for state governor. She sponsors a polling of registered party members to help assess her standing.

Education Level	Voter Rating: Unfavorable	Unknown or Neutral	Favorable
High School or Less	114	40	55
Some College or More	126	60	155

At $\alpha = .05$, do you find that voter rating and level of education are independent or dependent?

19.16 A random sample of young women was given two free bottles of shampoo in exchange for evaluating a new shampoo product. The shampoo formulation that the women evaluated was the same for all women except that three different colors were tested. One-third of the women received a tan color shampoo, one-third got a red color, and one-third got a green color. The women used the product for three weeks, then filled out a survey form. One of the questions on that form asked the shampoo users to rate the shampoo relative to other brands in its projected price range. Responses to this question were cross-tabulated against color as shown in the following table.

Rating	Color: Red	Green	Tan
Below Average	28	13	13
Average	20	28	36
Above Average	27	34	26

Are color and rating dependent or independent? Test at $\alpha = .05$.

19.17 A researcher analyzed a sample of 275 television commercials broadcast in the United States. The table shows classifications of adult male and female characters in the commercials with on-camera appearances of at least three seconds and/or at least one line of dialogue.

Character Age	Character Sex: Women	Men
Less than 35	68	26
35 thru 50	86	82
Over 50	15	24

Assume the commercials in the study are a random sample. With 1% risk of a Type I error, test the hypothesis that character age in all U.S. TV commercials is independent of character sex. (This exercise is based on an article by Mary C. Gilly, "Sex Roles in Advertising: A Comparison of Television Advertisements in Australia, Mexico, and the United States," in the *Journal of Marketing,* April 1988.)

19.18 A facsimile machine manufacturer purchases components from three different suppliers. Each component is tested for three types of defects before being placed into production. Over a two-month period, records are kept on type of defect found. Based on the data in the following table, does it appear that defect type and supplier are independent or dependent? Use the .10 level of significance.

Defect Type	Supplier: Allied Electric	Wilson Company	Circuits, Inc.
A	20	32	19
B	27	50	43
C	13	14	22

*19.19 When a contingency table has two rows and two columns, the test for independence is the same as the binomial test of H_0: $\pi_1 = \pi_2$. For instance, in Example 13.10, we tested the hypothesis that the loan delinquency rate for unmarried borrowers was the same as the rate for married borrowers, using the following data.

Marital Status	Sample Size, n	Number of Delinquent Loans, X
Unmarried	413	29
Married	537	47

We computed (to three decimal places) $Z^* = .975$ and accepted H_0 at $\alpha = .10$, since the absolute value of the test statistic did not exceed $Z = 1.65$. This same problem also can be formulated as a test for independence, using the contingency table shown.

Loan Status	Marital Status: Unmarried	Married	Total
Delinquent	29	47	76
Not Delinquent	384	490	874
Total	413	537	950

a. Conduct a test of independence for marital status and loan status, using $\alpha = .10$.
b. Within rounding error, verify that the binomial test statistic, when squared, is equal to the chi-square test statistic.
c. Within rounding error, verify that the binomial test cutoff value, when squared, is equal to the chi-square cutoff value. (The results of part b and c represent a relation between the binomial and chi-square distribution that applies in all such cases.)

19.20 The manager of three area shops specializing in VCR repairs sampled 50 customer invoices at random from each shop. One factor looked at was the proportion of the invoice that was for the repairperson's labor. The following table is developed.

	Labor, as a Percent of Invoice			
Service Center	Under 10%	10% up to 20%	20% up to 30%	Over 30%
1	7	18	14	11
2	8	21	16	5
3	12	24	9	5

At the $\alpha = .10$ risk level, can you conclude that, in the invoice population, the labor percent of invoice is related to the service center?

19.2 THE RANK CORRELATION COEFFICIENT

In Chapters 4 and 15, we discussed the sample correlation coefficient, r, as a measure of the strength of linear relation for quantitative bivariate variables, usually symbolized by X and Y. The correlation coefficient is a descriptive measure that can take on values from -1.0 to $+1.0$, as indicated in Figure 19.8. You may recall that a negative correlation results when the two variables are inversely related; a positive correlation follows from a direct relation. We also have learned how to use the sample correlation to make an inference about whether X and Y are correlated in the population. To do

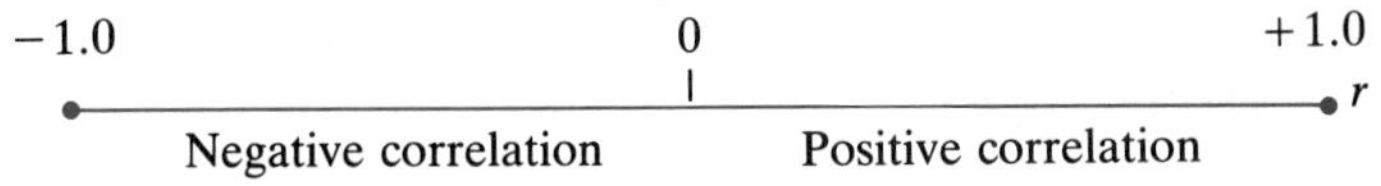

Figure 19.8 Range of Possible Values for the Correlation Coefficient

so, we test the null hypothesis that the population correlation is zero, which is to say that X and Y are not related.

As mentioned in Chapter 15, correlation analysis presumes *quantitative* data that follow the bivariate normal distribution. In this section, we will develop the rank correlation coefficient, also called Spearman's correlation. The major purpose of Spearman's correlation is to extend our ability to perform correlation analysis to a form of *qualitative* data: bivariate rankings.

As an example, consider two corporate vice presidents who have just completed the process of interviewing eight candidates for the position of personnel manager in the firm. Each vice president separately has contemplated the strengths and weaknesses of each candidate and has ranked the individuals from 1 = most promising to 8 = least promising. Table 19.4 shows the orderings. If we wish to assess the strength of the relation between the two sets of ranks, we will compute the sample rank correlation coefficient, r_s. Like its parametric counterpart, r_s is constrained to be between −1.0 and +1.0, inclusive. Its formula is given in Equation 19–3.

Spearman's Rank Correlation Coefficient

$$r_s = 1 - \frac{6\Sigma d^2}{n(n^2 - 1)} \qquad (19\text{–}3)$$

where $d = X - Y$ (or $Y - X$), and n = sample size. Table 19.5 illustrates the computation of the rank correlation coefficient for the set of rankings. The sample

Table 19.4 Rankings of Job Candidates by Two Different Interviewers

Candidate	Ranking of Vice President 1, X	Ranking of Vice President 2, Y
Feldhoff	2	4
Hancock	6	6
Johnson	5	7
Pringle	4	3
Reilly	3	1
Sayer	7	5
Stephan	1	2
Taylor	8	8

Table 19.5 Computation of Spearman's Rank Correlation

Candidate	X Ranks	Y Ranks	d	d^2
Feldhoff	2	4	−2	4
Hancock	6	6	0	0
Johnson	5	7	−2	4
Pringle	4	3	1	1
Reilly	3	1	2	4
Sayer	7	5	2	4
Stephan	1	2	−1	1
Taylor	8	8	0	0
				$\Sigma d^2 = 18$

$$r_s = 1 - \frac{6\Sigma d^2}{n(n^2 - 1)} = 1 - \frac{6(18)}{8(63)} = 1 - .214 = .786$$

result of $r_s = .786$ shows a positive correlation between the vice presidents' orderings. Spearman's correlation is also interpreted as a "coefficient of agreement" for preference ranks.

If we wish to make an inference as to whether ranked variables in a bivariate population are related, we will test the null hypothesis that the rank correlation in the population is zero. Specifically, letting ρ_s represent the population value of Spearman's correlation, we will test H_0: $\rho_s = 0$. H_a can be two-tailed or one-tailed.

To investigate the assumption that there is no correlation between job candidate rankings for the two vice presidents, we would view the eight rankings as a random sample, and test

$$H_0\colon\ \rho_s = 0 \text{ (the rankings are uncorrelated)}$$

versus

$$H_a\colon\ \rho_s > 0 \text{ (the rankings are positively correlated)}$$

Note that we recommend an upper-tail H_a in this example. This is because we anticipate that if a relation exists, it should be a positive one. We would expect the vice presidents to have similar rating criteria, and therefore tend to agree with each other.

Let us suppose we wish to test H_0 at the .05 risk level. To arrive at a conclusion, we need to compare our sample value ($r_s = .786$) against the appropriate value from the table of Spearman's correlation cutoffs given in Appendix L at the back of the book. Using the instructions for an upper-tail H_a at the bottom of this table, we find the $\alpha = .05$ risk value for $n = 8$ is .643. Since our sample value of $r_s = .786$ exceeds .643, we have strong evidence against H_0. We conclude that there is an overall positive relation between the two vice presidents' rankings of candidates. Accepting H_0 in this setting would imply that the raters had dissimilar or independent criteria.

EXAMPLE 19.3

In a random sample of college students, choices among the 10 top-rated prime-time television shows yielded the following preference rankings.

Show	Men's Ranking	Women's Ranking
A	8	3
B	9	6
C	2	5
D	7	9
E	1	4
F	5	1
G	6	8
H	3	2
I	10	7
J	4	10

At $\alpha = .10$, test the assumption of no correlation in rankings between men and women, using a two-tailed alternative hypothesis.

Solution:

The primary computation needed is Σd^2; note that in setting up a table to obtain this sum, it makes no difference which set of rankings is called X and which is called Y. In this case, we will form d by letting d = men's ranking minus women's ranking.

Show	Men	Women	d	d^2
A	8	3	5	25
B	9	6	3	9
C	2	5	−3	9
D	7	9	−2	4
E	1	4	−3	9
F	5	1	4	16
G	6	8	−2	4
H	3	2	1	1
I	10	7	3	9
J	4	10	−6	36
				$\Sigma d^2 = 122$

We then have

$$r_s = 1 - \frac{6\Sigma d^2}{n(n^2 - 1)}$$

$$= 1 - \frac{6(122)}{10(99)} = 1 - .739 = .261$$

For a two-tailed test with $n = 10$ and $\alpha = .10$, we find a Spearman's table cutoff value of .564. Since $r_s = .261$, we cannot reject the hypothesis of zero correlation between college women's and men's television preferences in the population.

In symbolic form, this example would appear as

$$H_0\colon \rho_s = 0$$

$$H_a\colon \rho_s \neq 0$$

$$TS\colon r_s = .261$$

$$RR\colon \text{Reject } H_0 \text{ if } |r_s| > .564$$

$$C\colon \text{Accept } H_0$$

□

COMMENTS

1. Remember that rankings, though numerical, are not quantitative data. Instead, they are coded qualitative data.
2. When we have quantitative bivariate data but cannot assume a bivariate normal distribution, we can convert the data to ranks and then determine r_s. This process is illustrated in Classroom Example 19.3.
3. It is possible to have one or more ties in a column of rankings. For example, a tie for the third position in a set of six rankings would result in the rankings being 1, 2, $3\frac{1}{2}$, $3\frac{1}{2}$, 5, and 6. The tied ranks get a midrank—an average of the ranks they would have had if they were adjacent but not tied. Whether a column of ranks has any ties or not, its sum always must be $n(n + 1)/2$.
4. If we enter rankings into the formula for the parametric correlation coefficient—see Equation 4–1—we obtain the same result that Equation 19–3 provides. The latter formula is dramatically easier to use. Equation 4–1 is preferred, however, when there are several ties in the rankings. Under these conditions, it is exact while Equation 19–3 is approximate.
5. Our table of cutoff values ends at $n = 30$. For larger sample sizes, there is a simple normal curve approximation. This is illustrated in Exercise 19.27.

CLASSROOM EXAMPLE 19.3

Correlating Price and Quality for Big-Screen TVs

Consumer Reports magazine (January 1986) rated different brands of 25-, 26-, and 27-inch TV sets. A panel of seasoned viewers judged each set on a variety of factors; these results were boiled down to an overall rating score. Following are some of the sets' ratings as well as the manufacturers' suggested retail price.

Brand	Overall Rating Score	Price
Philco	90	$ 750
Magnavox	88	899
RCA	86	800
Sony	84	1250
Panasonic	82	1099
Quasar	81	975
Zenith	77	960
Sears	68	780

Copyright 1986 by Consumers Union of United States, Inc., Mount Vernon, NY 10553. Excerpted by permission from *Consumers Reports*, January 1986.

a. Convert the overall rating scores to ranks, letting rank 1 be assigned to the highest score.

b. Convert the price data to ranks, letting rank 1 be assigned to the highest price.

c. Compute Spearman's rank correlation for these data:

$$r_s = 1 - \frac{6\Sigma d^2}{n(n^2 - 1)}$$

d. To test H_0: $\rho_s = 0$, what cutoff value would be used for $\alpha = .05$ and an upper-tailed alternative?

e. Do you accept or reject the null hypothesis?

f. Is your conclusion consistent with this statement: "Quality and price of large-screen TVs go hand in hand"?

SECTION 19.2 EXERCISES

19.21 Compute r_s for these data.

X Rankings	*Y* Rankings
4	7
3	6
5	4
7	2
1	5
6	3
2	1

19.22 State whether H_0 should be rejected or not rejected:

a. H_0: $\rho_s = 0$
H_a: $\rho_s \neq 0$ $\quad n = 25$, $\alpha = .05$, and $r_s = .474$

b. H_0: $\rho_s = 0$
H_a: $\rho_s > 0$ $\quad n = 17$, $\alpha = .05$, and $r_s = .364$

c. H_0: $\rho_s = 0$
H_a: $\rho_s < 0$ $\quad n = 18$, $\alpha = .025$, and $r_s = -.553$

19.23 a. In computing r_s, what value would Σd^2 have if the sample rankings had a perfect (positive) relation?

b. In computing r_s, what value would the term $6\Sigma d^2/n(n^2 - 1)$ have if the sample rankings had a perfect (negative) relation? (*Hint:* Use the rankings 1-2-3-4 for X and 4-3-2-1 for Y as a guide.)

c. In computing r_s, Σd is not needed, although verifying its value can serve as an arithmetic check. What must be the value of Σd if there are no ranking/arithmetic errors?

19.24 Test the null hypothesis that in the population the rank correlation coefficient is zero, against a lower-tailed alternative hypothesis. Let $\alpha = .01$.

Variable 1 Ranking	Variable 2 Ranking
7	2
5	4
1	6
4	3
3	8
2	9
6	7
8	1
9	5

19.25 A small panel of food and restaurant critics in a large city participated in a study where they rated a sample of dinner wines. The wines were served without the taster being aware of their identities or prices. Following are the panel's collective ratings of the wines (1 = best tasting), along with the wholesale prices of the wines. At the 5% level of significance, test the hypothesis of no correlation in the rankings versus an alternative which assumes that there is a positive relation between taste and price rankings.

Wine	Taste Rank	Price
A	7	\$8.95
B	4	9.89
C	11	6.95
D	9	8.49
E	3	10.75
F	1	10.60
G	6	7.95
H	2	9.60
I	10	7.45
J	8	6.65
K	5	9.79

19.26 Two women's fashion buyers attended a style show and ranked a sample of dress styles for sales potential in their company's mail order catalog. Test the assumption that the buyers' preferences are unrelated versus the proper one-tailed alternative, using $\alpha = .05$. In the preference orderings given below, the rank of 1 corresponds to the highest potential.

Dress Style	Buyer 1	Buyer 2
1	4	2
2	7	8
3	1	3
4	5	5
5	2	4
6	3	1
7	6	7
8	8	6

***19.27** For sample sizes above 30, a table of cutoff values for Spearman's correlation is not needed. Instead we can compute

$$Z^* = r_s\sqrt{n - 1}$$

and treat it like an ordinary large sample test statistic. For instance, suppose we have computed $r_s = .351$, based on 44 observations. To test a two-tailed null hypothesis of zero correlation in the population at $\alpha = .05$, the cutoff Z value is ± 1.96. Using the expression above, we would obtain

$$\begin{aligned} Z^* &= .351\sqrt{44 - 1} \\ &= .351(6.5574) = 2.30 \end{aligned}$$

Since $Z^* = 2.30 > 1.96$, H_0 would be rejected.

State whether H_0 should be rejected or not rejected.

a. H_0: $\rho_s = 0$ $\quad n = 71$, $\alpha = .01$, and $r_s = .239$
H_a: $\rho_s \neq 0$

b. H_0: $\rho_s = 0$ $\quad n = 55$, $\alpha = .01$, and $r_s = -.402$
H_a: $\rho_s < 0$

c. H_0: $\rho_s = 0$ $\quad n = 37$, $\alpha = .05$, and $r_s = .227$
H_a: $\rho_s > 0$

19.28 Different magazines are well known to appeal to readers with different education levels. For instance, readers of *Scientific American* have a substantially higher education level than do readers of *True Confessions*. Shuptrine and McVicker took random samples of advertisements from several magazines to see whether the ads they contained were matched, in terms of ease or difficulty of reading, to the readers' education level. The following rankings were obtained.

Magazine	Reader Education Rank	Advertisements' Readability Rank
Scientific American	1	1
Fortune Magazine	2	2

*Optional

(*continues*)

Magazine	Reader Education Rank	Advertisements' Readability Rank
The New Yorker	3	6
Sports Illustrated	4	9
Newsweek	5	8
People	6	7
National Enquirer	7	4
Grit	8	5
True Confessions	9	3

Source: F. Kelly Shuptrine and Daniel McVicker, "Readership Levels of Magazine Ads," *Journal of Advertising Research,* October 1981. Reprinted by permission.

The education ranks were obtained from *Target Group Index,* an annual industry publication that researches magazine readership demographics. The readability ranks were obtained by using the Gunning Fog Index, a widely used formula that takes into account factors such as average sentence length and the count of words with three or more syllables. In the rankings above, rank 1 was assigned to the most difficult level of advertising copy. At the 5% risk level, test the hypothesis that the rankings are uncorrelated, using the proper one-tailed alternative. Do you find a relation?

19.29 Two groups of professionals—researchers in market research agencies and business organization "in-house" researchers—were surveyed about ethical problems. Respondents were asked to identify major ethical problems of marketing researchers. For several major problems that were identified, abbreviated descriptions are given below. These categories are ranked by their frequency of selection by the two groups; note that one tie occurred. Treating the respondents in this survey as a random sample, test the hypothesis of no correlation between the two groups' rankings, using $\alpha = .05$. Employ a one-tailed alternative hypothesis that anticipates agreement between the two groups.

Ethical Problem Area	Agency Researchers	In-House Researchers
Research integrity	1	1
Treating outside clients fairly	3.5	2
Research confidentiality	3.5	3
Personnel issues	5	5
Treating respondents fairly	2	6
Treating others in the company fairly	6	7
Interviewer dishonesty	7	4

Source: Shelby D. Hunt, Lawrence B. Chonko, and James B. Wilcox, "Ethical Problems of Marketing Researchers," *Journal of Marketing Research, 21,* August, 1984. Reprinted by permission of the American Marketing Association.

19.30 Prior to the college basketball season, a panel of sportswriters predicted how the teams in a ten-team conference would fare in conference play. After the season, these predictions are compared with the actual results.

Team	Preseason Rank	Final Rank
A	7	7
B	6	8
C	5	3
D	1	2
E	10	9
F	4	1
G	9	10
H	3	5
I	8	6
J	2	4

a. Compute the rank correlation coefficient.
b. If you were to test the hypothesis of zero correlation between actual and predicted ranks, should the alternative be lower tail, upper tail, or two tail?
c. Carry out the test you recommended in part b, using a significance level of your own choosing.

19.3 THE TWO-SAMPLE RANK SUM TEST

Using Ranks to Test for Equal Means

In our chapter on small sample inference, we presented two options for testing H_0: μ = (some value). These choices were the (parametric) *t*-test and the (nonparametric) signed ranks test. The latter procedure has these advantages:

- Its computations are easier than those of the *t*-test, since they involve ranks instead of means and standard deviations.
- Its assumptions are less restrictive. A small sample *t*-test assumes a normal population, while the signed ranks method assumes only a symmetrical population.

In this section, we wish to extend the nonparametric one-sample procedure so that we can test H_0: $\mu_1 = \mu_2$ for independent samples. We already have studied the two-sample *t*-test for evaluating this statement (see Chapter 13). We now introduce its nonparametric equivalent—a two-sample procedure that uses ranks to evaluate H_0: $\mu_1 = \mu_2$. This test is known as the Mann-Whitney-Wilcoxon test, or more generically, simply the rank sum test.

To illustrate this procedure, we will consider a corporation's marketing manager who chooses independent random samples from each of two sales territories. The figures given are individual sales representative's sales volumes, in thousands of dollars, for the last two months.

Territory C	\$41.5	57.4	39.1	44.4	50.1
Territory D	\$47.7	61.2	62.1	70.6	

The marketing manager wants to know whether the average sales of all Territory C representatives are the same or different from the Territory D average.

If we chose to test H_0: $\mu_1 = \mu_2$ with a t-test, we would have to assume that both populations followed the normal distribution, and further, that both had the same variance. The rank sum test does not hinge on these assumptions, and it is easier to carry out.

The rank sum procedure begins by combining the observations from both samples into a single list arranged in order of magnitude, as shown in the first column of Table 19.6. These nine observations are then ranked from the smallest (rank = 1) to the largest (rank = 9), as in the center column of the table. The right column of the table, labeled sample membership, is needed to keep track of which sample (Territory C or D) each item in the combined listing came from.

This sample membership column can provide us with an intuitive advance look as to how this test operates. If the null hypothesis of equal means is true, then we would anticipate that the C's and D's should appear randomly mixed within the sample membership column. On the other hand, if we find the D's tending to cluster near the top of this column with most C's at the bottom, it would suggest that $\mu_C > \mu_D$. The opposite finding of many C's clustered at the top with most D's near the bottom would suggest that $\mu_C < \mu_D$. Our test statistic will summarize the placement of sample items in the combined listing.

To help us arrive at an inference, we specify the following terms:

n_1 = Number of items in the smaller sample. In this example, the smaller sample is size 4, from Territory D. Therefore, n_1 is equal to 4.

Table 19.6 Merging Two Samples for the Rank Sum Test

Combined Ordering	Rank	Sample Membership
\$39.1	1	C
41.5	2	C
44.4	3	C
47.7	4	D
50.1	5	C
57.4	6	C
61.2	7	D
62.1	8	D
70.6	9	D

n_2 = Number of items in the larger sample. With 5 items drawn from Territory C, $n_2 = 5$. If the two sample sizes are equal, it makes no difference which is labeled n_1 or n_2.

T_1 = Sum of the ranks for the n_1 items in the smaller sample. Here, $T_1 = 4 + 7 + 8 + 9 = 28$. T_1 will be our "test statistic." We will arrive at our accept/reject H_0 conclusion by comparing $T_1 = 28$ to values in a reference table. Note that T_1 would be a smaller number if the C's and D's were better mixed in the combined listing.

T_2 = Sum of the ranks for the n_2 items in the larger sample. Here $T_2 = 1 + 2 + 3 + 5 + 6 = 17$. T_2 serves no purpose other than as a math check on the ranking process. This check is that T_1 plus T_2 must equal $n(n + 1)/2$, where n is the combined sample size. In this example $T_1 + T_2 = 28 + 17 = 45$. This must equal $n(n + 1)/2 = 9(10)/2 = 45$.

We now compare our test statistic, $T_1 = 28$, to the appropriate numbers from the table of cutoff values. This table is found in Appendix M; an abbreviated version is shown in Table 19.7 in this section. Table 19.7 can be used for a two-tailed H_a with $\alpha = .10$ or for a one-tailed H_a when $\alpha = .05$. Let us suppose we wish to test H_0: $\mu_1 = \mu_2$ versus a two-tailed alternative, with 10% risk of Type I error. To use the table, we intersect the n_1 column with the n_2 row. For our example where $n_1 = 4$ and $n_2 = 5$, we locate the pair of values $T_L = 13$ and $T_U = 27$. These values indicate the lower and upper boundaries of the two-tailed rejection regions—see Figure 19.9. Since the cutoff values begin the rejection region, we would reject H_0 if $T_1 \leq 13$ or if $T_1 \geq 27$. Since $T_1 = 28 > T_U$, we do have sufficient cause to reject the null hypothesis at $\alpha = .10$. Our argument is that if H_0 were true, the probability is about .90 that T_1

Table 19.7 Cutoff Values for the Rank Sum Test: $\alpha = .10$ (Two Tail); $\alpha = .05$ (One Tail)

	n_1															
	3		4		5		6		7		8		9		10	
n_2	T_L	T_U	T_L	T_U	T_L	T_U	T_L	T_U	T_L	T_U	T_L	T_U	T_L	T_U	T_L	T_U
3	6	15	7	17	7	20	8	22	9	24	9	27	10	29	11	31
4	7	17	12	24	13	27	14	30	15	33	16	36	17	39	18	42
5	7	20	13	27	19	36	20	40	22	43	24	46	25	50	26	54
6	8	22	14	30	20	40	28	50	30	54	32	58	33	63	35	67
7	9	24	15	33	22	43	30	54	39	66	41	71	43	76	46	80
8	9	27	16	36	24	46	32	58	41	71	52	84	54	90	57	95
9	10	29	17	39	25	50	33	63	43	76	54	90	66	105	69	111
10	11	31	18	42	26	54	35	67	46	80	57	95	69	111	83	127

To test H_0: $\mu_1 = \mu_2$, let n_1 be the smaller sample size; let T_1 be the rank sum of the n_1 observations. *For Two-Tailed Tests,* reject H_0 if $T_1 \leq T_L$ or if $T_1 \geq T_U$. *For One-Tailed Tests,* require that μ_1 be associated with sample n_1. Then for H_a: $\mu_1 > \mu_2$, reject H_0 if $T_1 \geq T_U$. For H_a: $\mu_1 < \mu_2$, reject H_0 if $T_1 \leq T_L$.

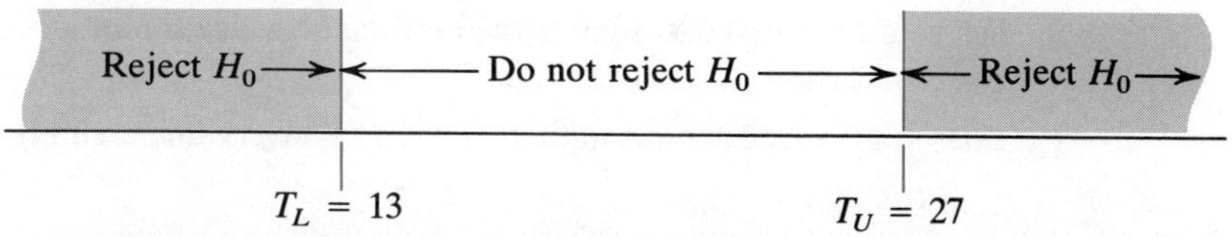

Figure 19.9 Rejection Regions for Rank Sum Test for $n_1 = 4$, $n_2 = 5$, $\alpha = .10$, Two-Tailed H_a

would be found between 13 and 27, noninclusive. The value of $T_1 = 28$ indicates that sample D items tend to be at the high end of the combined listing—an unlikely occurrence if the assumption of equal sales averages is really correct.

In symbolic form, this example would be represented as follows:

H_0: $\mu_C = \mu_D$
H_a: $\mu_C \neq \mu_D$
TS: $T_1 = 28$
RR: Reject H_0 if $T_1 \leq 13$ or $T_1 \geq 27$
C: Reject H_0

CLASSROOM EXAMPLE 19.4

Using Ranks to Compare Two Populations

There are two competing real estate brokers in a small city. Broker P claims in its advertising that the homes it sells are on the market less time than are homes listed through Broker S. To investigate this, you obtain a random sample of 15 recent real estate sales and then determine X, the length of time, in days, each home was on the market.

Broker P	62	28	14	113	71	29	53	54	34
Broker S	47	62	87	139	115	81			

a. The variable *time on the market* is believed to have a skewed right distribution; you decide to use the rank sum test instead of a t-test to evaluate H_0: $\mu_S = \mu_P$. If you wish H_a to embody Broker P's claim, should the alternative be that $\mu_S > \mu_P$ or that $\mu_S < \mu_P$?

b. Set up and complete a table with these headings: Combined X Listing, Rank, and Sample Membership. Use the midrank method to resolve any ranking ties.

c. Determine T_1, the rank sum for the smaller sample.

d. Determine T_2 and verify that $T_1 + T_2 = n(n + 1)/2$.

e. For your H_a statement, what is the table cutoff value for an $\alpha = .05$ test? What is your conclusion for the test?

f. Does your conclusion mean that Broker P's advertising claim is to be believed?

Handling Larger Samples

Our table of rank sum cutoff values cannot accommodate either sample size being larger than 10. There is, however, a normal curve approximation that eliminates the need for a rank sum table when either sample size exceeds 10. We still will determine T_1 as usual under these conditions, but we then will develop a Z-test statistic:

$$Z^* = \frac{T_1 - \text{Mean of } T_1}{\text{Standard deviation of } T_1}$$

When H_0 is true, the sampling distribution of T_1 is approximately normal. Its mean or expected value is $n_1(n + 1)/2$ and its standard deviation is $\sqrt{n_1 n_2(n + 1)/12}$. Therefore, a test statistic for the rank sum test for either sample size being greater than 10 will be as shown in Equation 19–4. Note that $n = n_1 + n_2$.

Large Sample Test Statistic for Rank Sum Test

$$Z^* = \frac{T_1 - \dfrac{n_1(n + 1)}{2}}{\sqrt{\dfrac{n_1 n_2(n + 1)}{12}}} \qquad \textbf{(19–4)}$$

Then Z^* is treated like an ordinary large sample test statistic, as illustrated in Example 19.4.

EXAMPLE 19.4

A Miami Airport limousine service firm recently placed 20 new vehicles into service; 8 were made by manufacturer F, 12 by manufacturer C. After a month of service, detailed records are kept for each vehicle for several days to measure fuel economy. The figures given are individual vehicles' miles per gallon. Treating these data as random samples, use the rank sum procedure to test the assumption of equal means. Use a two-tailed alternative hypothesis and a 10% risk factor.

Manufacturer *F*	9.14	11.40	8.76	10.12	10.16	9.01
	11.22	10.61				
Manufacturer *C*	10.17	11.61	9.86	10.14	11.60	10.17
	12.13	11.75	9.90	10.37	10.71	12.14

Solution:

A combined listing of fuel economy with ranks and sample membership is shown. The smaller sample is from manufacturer *F*:

Combined Listing	Rank	Sample Membership	Combined Listing	Rank	Sample Membership
8.76	1	*F*	10.37	11	*C*
9.01	2	*F*	10.61	12	*F*
9.14	3	*F*	10.71	13	*C*
9.86	4	*C*	11.22	14	*F*
9.90	5	*C*	11.40	15	*F*
10.12	6	*F*	11.60	16	*C*
10.14	7	*C*	11.61	17	*C*
10.16	8	*F*	11.75	18	*C*
10.17	9.5	*C*	12.13	19	*C*
10.17	9.5	*C*	12.41	20	*C*

n_1 is 8. T_1 is computed as

$$T_1 = 1 + 2 + 3 + 6 + 8 + 12 + 14 + 15 = 61$$

and the large sample test statistic is

$$Z^* = \frac{T_1 - \dfrac{n_1(n+1)}{2}}{\sqrt{\dfrac{n_1 n_2(n+1)}{12}}} = \frac{61 - \dfrac{8(21)}{2}}{\sqrt{\dfrac{8(12)(21)}{12}}} = \frac{61 - 84}{\sqrt{168}} = \frac{-23}{12.96} = -1.77$$

For $\alpha = .10$, we reject H_0 if $|Z^*| > 1.65$. In this case, we declare that our initial assumption of equal means is incorrect; a strong case has been made that the means differ (see Figure 19.10).

□

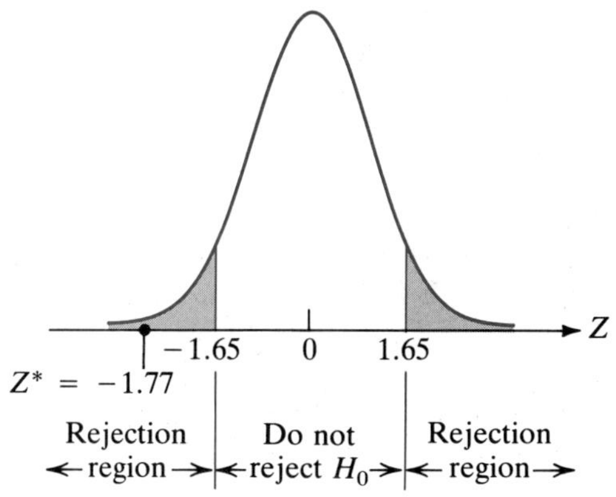

Figure 19.10 A Rank Sum Test Approximated by a Normal Distribution

COMMENTS

1. The rank sum test does not assume symmetry as does the signed ranks procedure. The rank sum null hypothesis is really more general than we have suggested—some textbooks will give the test statement as H_0: The two populations have identical probability distributions.
2. The rank sum test is a procedure for independent samples. Dependent, or matched, samples can be analyzed with a parametric t-test or the nonparametric signed ranks test (see Chapter 13).

SECTION 19.3 EXERCISES

19.31 State whether H_0 should be accepted or rejected.

a. H_0: $\mu_1 = \mu_2$ H_a: $\mu_1 \neq \mu_2$ $n_1 = 5$, $n_2 = 5$, $\alpha = .10$, and $T_1 = 22$

b. H_0: $\mu_1 = \mu_2$ H_a: $\mu_1 < \mu_2$ $n_1 = 7$, $n_2 = 8$, $\alpha = .05$, and $T_1 = 37.5$

c. H_0: $\mu_1 = \mu_2$ H_a: $\mu_1 > \mu_2$ $n_1 = 5$, $n_2 = 9$, $\alpha = .05$, and $T_1 = 60$

19.32 Test H_0: $\mu_1 = \mu_2$ against a two-tailed alternative hypothesis, using a 5% risk of Type I error.

Sample 1 Items	\$329	343	349	361	375	
Sample 2 Items	\$357	359	369	377	379	383

19.33 Test H_0: $\mu_1 = \mu_2$ against an upper-tailed alternative, using $\alpha = .05$.

Sample 1 Items	91	78	83	87	91
Sample 2 Items	84	77	73	80	82

19.34 State whether H_0 should be accepted or rejected.

a. H_0: $\mu_1 = \mu_2$ $\quad n_1 = 13$, $n_2 = 17$, $\alpha = .05$, and $T_1 = 147$
   H_a: $\mu_1 \neq \mu_2$

b. H_0: $\mu_1 = \mu_2$ $\quad n_1 = 9$, $n_2 = 10$, $\alpha = .025$, and $T_1 = 100$
   H_a: $\mu_1 > \mu_2$

c. H_0: $\mu_1 = \mu_2$ $\quad n_1 = 11$, $n_2 = 13$, $\alpha = .05$, and $T_1 = 170$
   H_a: $\mu_1 > \mu_2$

19.35 An aluminum fabrication plant operates two full shifts per day. For the past week, the amount of scrap material generated by each shift is weighed and recorded. Although the scrap can be melted down and recycled, it is desirable to generate as little scrap as possible. Following are the scrap amounts weighed per day, in pounds. Assume these data are a random sample. Test the assumption of equal shift means, using 10% risk of Type I error. Let your alternative be two-tailed.

7 A.M.–3 P.M.	946	1019	813	847	899
3 P.M.–11 P.M.	912	1056	1117	989	1011

19.36 The manufacturer of high percent cotton flannel shirts is experimenting with a new dye that holds promise of better resistance to fading in repeated washings. A sample of eight shirts is produced with the new dye. Along with a control sample of standard dye shirts, the new dye shirts are washed 25 times following label instructions. Each shirt is then given a score by an inspector to indicate its ability to retain its original colors. The inspector is unaware of which shirts are made with the new dye. The results are as follows:

Standard Dye	77	79	81	82	82	84	88	90
New Dye	83	83	86	87	88	89	91	91

Higher numbers indicate better fade resistance. The company is likely to adopt the new dye if a case can be made that it offers better fade resistance. Use the .05 level of significance to test the assumption of no mean difference in rating scores, against the appropriate alternative hypothesis.

19.37 Two experimental drugs for reducing blood cholesterol are given to random samples of candidate males over age 40. The subjects all have cholesterol counts from 280 to 300, and diet therapy to reduce their counts has been ineffective. The numbers given are cholesterol count reductions realized after three months of drug treatment.

Drug 1	Drug 2
31	35
18	19
26	28
29	11
49	37
41	14

At the .05 level of significance, test the assumption that the drugs are equally effective in reducing average cholesterol counts in subjects of this type.

19.38 A clerk in a mail-order clothing firm took a random sample of 36 recent orders. Twelve orders were paid for by check; 24 orders were paid via charge card. Suppose that you wish to test the assumption that average order amounts are the same for these two methods of payment. You plan to use the rank sum procedure with $\alpha = .10$.

a. Given your null hypothesis, what would be the expected value for the rank sum of check-paid orders?

b. What would be the cutoff value of the test statistic if the alternative hypothesis is to be H_a: $\mu_{\text{Check}} < \mu_{\text{Charge}}$?

c. What would be your test statistic and your conclusion if $T_1 = 172$?

19.39 A business school accrediting agency has developed an outcomes measure test for graduating business school students. The test asks last semester seniors a number of questions about topics in business and economics that they have studied during their course work. At one school, a random sample of 47 students took the test. Included in this group were 7 marketing majors and 11 accounting majors. With $\alpha = .10$, use the following data to test the hypothesis of no difference in average scores between the two academic majors.

Accounting Majors		Marketing Majors
85	74	81
71	77	87
68	83	92
79	70	76
93	61	84
88		83
		84

19.4 THE RUNS TEST FOR RANDOMNESS

Sequential Patterns: Present or Absent?

It is sometimes of interest to know whether there is a sequential pattern to a series of observations. In this final nonparametric section, we will present a test that can help us decide whether the order in which events occur is random. A finding of nonrandomness in a theoretically random sequence might suggest a cause or reason worthy of further investigation. The procedure we will learn is called the runs test for randomness. While there are parametric equivalents to the other nonparametric tests we have studied, there is no general counterpart to the runs test.

To introduce the terminology, we will consider a simple problem setting: tossing a coin 30 times. Using H to denote heads and T for tails, imagine that the 30 tosses give results in this order:

H H H H H T T T T T T T T T T T T T T H H H H H H H H H H H

While there is a reasonable balance of H's and T's, the pattern of observations might strike you as odd, since there is so much clustering of the H's and T's. Is the pattern consistent with the randomness or unpredictability we should expect to find in a sequence of 30 coin flips? We hope your intuition would cause you to say no because, in almost all cases, a given observation is the same as the preceding one. What about the following sequential pattern?

HTHTH THTHT HTHTH HTHTH THTHT HTHTH

Like the first listing, this one contains 16 H's and 14 T's. Is there anything unusual here? Again, we should be able to notice a pattern. Instead of clustering, there is a great deal of mixing of the H's and T's. This pattern is likewise not random. In most cases, a given observation is different from the preceding one.

These two simple examples illustrate two extremes of nonrandomness in an ordered sequence: clustering and mixing. We will approach the question of judging randomness by counting *runs*.

Definition

A **run** is a streak of like symbols in an ordered sequence that are followed and preceded by a different symbol (or no symbol).

The parenthetical part of the definition comes into play only at the beginning of a sequence (where there is no preceding symbol) and at the end (no following symbol). A run can be as short as one observation or, should all n observations in a sequence have the same symbol, as long as n. In the first coin tossing example above, there are three runs; the second example contains 29 runs. Length of a run is not of concern; what matters is the count of how many runs are present, relative to the sample size.

Our inference as to whether a sequence is random will be based on how many runs exist. We will use R to denote this count. While it is beyond our scope to develop the sampling distribution of R, we will appeal to your statistics knowledge to anticipate the following:

- For a random sequence, there is a mean or expected value for R. This mean value for R will depend on two factors: the total number of observations (30 in the coin toss example) and the symbol counts (16 and 14 in that example).
- When the sample count R is not significantly different from its mean or expected value, we will be willing to accept the null hypothesis of randomness.
- Unusually large or small values for R will imply a nonrandom sequence and rejection of the randomness hypothesis.

Figure 19.11 illustrates the spirit of our inference procedure. Although it is possible to use cutoff values of R to make the accept/reject decision as Figure 19.11 suggests, in practice it is simpler to state the cutoffs according to the standard normal distribution. We then can use the test statistic Z^* to arrive at our conclusion. This normal approximation works well when the overall sequence length is above 20 and

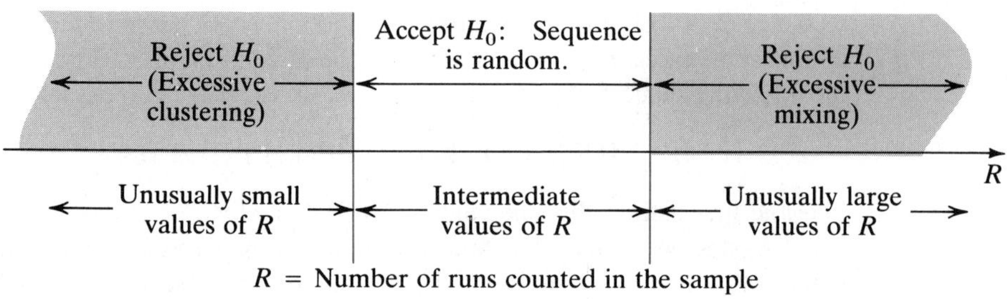

Figure 19.11 Relating *R*, the Number of Runs, to Various Conclusions

the symbol counts are both 10 or more. Our *Z*-test statistic will have the following form:

$$Z^* = \frac{R - \text{Mean of } R}{\text{Standard deviation of } R}$$

When the null hypothesis of randomness is true, the expected run count is

$$\mu_R = \text{Mean of } R = \frac{n + 2n_1n_2}{n}$$

where

n = total sample size
n_1 = number of symbols of one type
n_2 = number of symbols of a second type

The denominator in the test statistic is

$$\sigma_R = \text{Standard deviation of } R = \sqrt{\frac{(\mu_R - 1)(\mu_R - 2)}{n - 1}}$$

Equation 19–5 then gives us the test statistic for the runs test for randomness.

Test Statistic for the Runs Test

$$Z^* = \frac{R - \mu_R}{\sigma_R} \qquad (19\text{–}5)$$

EXAMPLE 19.5

A county police sergeant has compiled the following list for the past month for his district. Each N indicates no residential break-in reported for the day; each y indicates one or more break-ins did occur. At the .10 level of significance, is the sequence random?

N N N y y N y N N N y y N N N N y y y y N N N y y N N y N N y

Solution:

We are to test H_0: the sequence is random. Without prior reason to anticipate clustering or mixing of the break-ins, our alternative should be H_a: the sequence is not random. As such, H_a is two-tailed; it is sensitive to either form of nonrandomness.

A careful count shows 18 N's and 13 y's in the list of 31 observations. If H_0 is true, the expected number of runs is

$$\mu_R = \frac{n + 2n_1n_2}{n} = \frac{31 + 2(18)(13)}{31} = 16.097$$

and the standard deviation is

$$\sigma_R = \sqrt{\frac{(\mu_R - 1)(\mu_R - 2)}{n - 1}} = \sqrt{\frac{15.097(14.097)}{30}} = 2.663$$

The data set has $R = 14$ runs. Therefore, our test statistic is

$$Z^* = \frac{R - \mu_R}{\sigma_R} = \frac{14 - 16.097}{2.663} = -.79$$

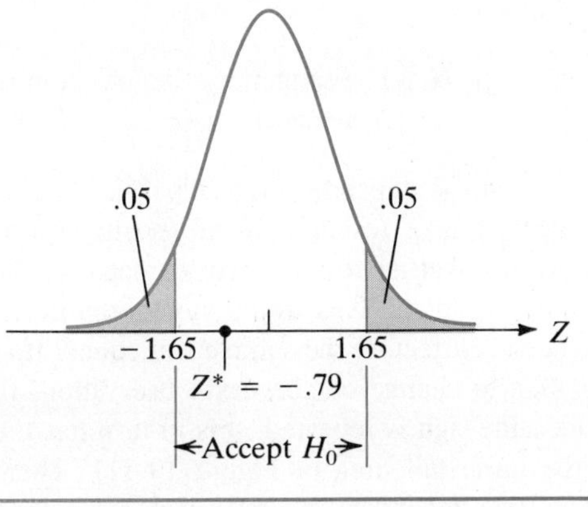

Figure 19.12 A Runs Test for Randomness Approximated by a Normal Distribution

At the .10 level of significance, we would retain H_0 since $Z^* = -.79$ is not in the rejection region (see Figure 19.12). The break-in occurrences are consistent with the assumption of randomness. If the sergeant was anticipating that an individual or group was operating according to some pattern, no evidence pointing in that direction was found.

□

COMMENTS

1. If we wish to express our runs test result in terms of R (as Figure 19.11 suggests), we can say that the unstandardized acceptance region for H_0 is $\mu_R \pm Z\sigma_R$. For Example 19.5, this would locate the acceptance region at 16.097 runs $\pm 1.65(2.663)$ runs, or from 11.70 runs to 20.49 runs. Since $R = 14$ is within this interval, the null hypothesis is accepted.
2. Tables for cutoff values for R do exist for sample sequences shorter than $n = 20$. (We do not recommend using the Z-test statistic under these conditions.) Since most applications of the runs test involve longer sequences, we have not introduced these tables.
3. Data in an ordered sequence may be numerical instead of categorical (symbols). Tests for randomness of a numerical sequence can be carried out by converting the observations into symbols. This is illustrated in Example 19.6.

The runs test can be used to check quickly a key assumption of the regression model—that the error terms, or residuals, are independent. Recall that a regression residual is defined as $e = (Y - \hat{Y})$; that is, a residual is the amount by which the actual value of Y differs from the fitted least-squares equation, $\hat{Y}$. Naturally some residuals will have positive signs while others have negative signs. If the regression model assumption of independence holds, the sequence of positives and negatives should be random. For most regression applications, we can verify the model assumption of independent error terms by testing for randomness with a two-tailed alternative:

H_0: The sequence of positive and negative residuals is random (the residuals are independent).

versus

H_a: The sequence is not random (the residuals are correlated).

However, when the dependent variable Y is time-related, there is a strong tendency for the succession of residuals to be (positively) autocorrelated. This means that a given residual, whether it is positive or negative, will tend to have the same sign as the preceding residual. This would violate the independence assumption. How would such a tendency affect R, the sample run count? If a plus or minus sign is more likely to recur than to change on the next observation, then there will tend to be streaks where the same sign is repeated; this in turn has the effect of reducing the run count R (see the lower-tail area of Figure 19.11). Therefore, for regression applications where Y is time-dependent, we will wish our procedure to have a one-tail alternative that is sensitive to a suppressed run count. We will evaluate as follows:

H_0: The residuals are independent.

versus

H_a: The residuals are positively correlated.

Since positive autocorrelation translates into fewer runs than expected under the randomness scenario, our alternative is a lower-tail test. Example 19.6 illustrates the runs test for regression residuals.

EXAMPLE 19.6

With 24 monthly observations for X = store advertising expenditures and Y = store revenue, a regression equation has been determined to relate X and Y. Before using the equation for predicting future values of Y, we wish to investigate the model assumption of independent error terms. A listing of the residuals, expressed in thousands of dollars, is given as follows:

Time Period	$Y - \hat{Y}$	Time Period	$Y - \hat{Y}$
1	\$2.3	13	−3.5
2	4.7	14	−3.1
3	0.6	15	−1.7
4	−1.1	16	−4.0
5	0.3	17	−4.3
6	5.4	18	−0.1
7	3.6	19	5.4
8	2.2	20	3.1
9	−2.8	21	2.0
10	−3.5	22	−0.9
11	−0.7	23	−1.9
12	1.3	24	−3.3

Using $\alpha = .05$, test the regression model assumption of independent error terms against an alternative that anticipates positive autocorrelation.

Solution:

By classifying each residual as positive or negative, we obtain this eight-run sequence:

+ + +   −   + + + +   − − −   +   − − − − − −   + + +   − − −

Since the dependent variable in the regression is time-related, our hypothesis statements will be as follows:

H_0: The residuals are independent.

H : The residuals are positively correlated.

With $n = 24$ and symbol counts of 11 and 13, the expected value for R is

$$\mu_R = \frac{n + 2n_1n_2}{n} = \frac{24 + 2(11)(13)}{24} = 12.917$$

with

$$\sigma_R = \sqrt{\frac{(\mu_R - 1)(\mu_R - 2)}{n - 1}} = \sqrt{\frac{11.917(10.917)}{23}} = 2.378$$

Since our sample run count is $R = 8$, the test statistic will be

$$Z^* = \frac{R - \mu_R}{\sigma_R} = \frac{8 - 12.917}{2.378} = -2.07$$

For $\alpha = .05$, lower-tail H_a, we will reject the null hypothesis if $Z^* < -1.65$. Since Z^* is -2.07, we have persuasive evidence that the residuals are positively correlated. □

COMMENTS

1. Some commercial software programs provide the runs tests results along with other regression information, such as the model coefficients, ANOVA table, and so forth.
2. The Durbin-Watson test presented in Chapter 17 is a parametric procedure that can detect correlated residuals. Its computations are considerably more involved than are those of the runs test, and it can yield inconclusive results.
3. If a residual should happen to equal zero, it is neither positive nor negative. We can ignore it when performing a runs test, although its omission would reduce the sample size.
4. We recommend rounding μ_R and σ_R to three decimal places.

CLASSROOM EXAMPLE 19.5

Is the Sequence of All-Star Game Winners Random?

Major league baseball began playing its annual All-Star game in 1933. Following is a list of winners (a = American League, N = National League) from 1933 through 1989. We wish to test the assumption that the sequence is random.

a a a N a   N a N a a   a N a a a   a N N N N   a N N a a   N a N N N
N a N N N   N N N N N   a N N N N   N N N N N   N N a N N   a N a a

(*Notes:* one tie game excluded; no game in 1945; two games played in 1959, 1960, 1961, 1962.)

a. Determine the overall sequence length n, as well as the values for each symbol's count. Let n_1 represent the American League.
b. What is the expected number of runs if the null hypothesis is true? Carry out to three decimal places.
c. To three decimal places, what is the standard deviation?

d. How many runs are in the sequence of All-Star game winners?

e. What is the value of the test statistic Z^*?

$$Z^* = \frac{R - \mu_R}{\sigma_R}$$

f. For $\alpha = .10$ and a two-tailed alternative hypothesis, do you determine that the sequence of All-Star game winners is random or not?

SECTION 19.4 EXERCISES

19.40 For the following sampling situations, state whether a two-tailed test for randomness in a sequence would result in acceptance or rejection of the null hypothesis:

a. $n = 40$, $n_1 = 12$, $n_2 = 28$, $R = 21$, and $\alpha = .10$.

b. $n = 200$, $n_1 = 18$, $n_2 = 182$, $R = 30$, and $\alpha = .05$.

c. $n = 44$, $n_1 = 22$, $n_2 = 22$, $R = 32$, and $\alpha = .01$.

19.41 At the .10 level of significance, test the following sequence for randomness.

c ddd c dd c d c d c d cc dd c d c d c ddd

19.42 The listing that follows indicates whether the share price of Dow Chemical common stock went up (+) or down (−) in value for 30 consecutive trading days on the New York Stock Exchange. At $\alpha = .10$, would you describe the sequence as random or nonrandom?

− − − − + − + − + + + − + − + + − + + + − − − + + − − + − +

19.43 A regression equation was computed using 30 observations of X and Y. An inspection of the residuals shows that 14 are negative and 16 are positive.

a. What is the expected number of runs in the residuals under the regression model assumption of independence?

b. What is the standard deviation?

c. Form a region of concentrated values for the number of runs.

d. Would a run count of 19 support the assumption of independence, at $\alpha = .05$?

19.44 A sociology instructor has made up a forty-item true-false examination. The answer key has the given appearance. Test the sequence for randomness, using $\alpha = .05$. If you find a significant departure from randomness, describe its nature.

FTFTT FTTFT FFTTT FTFTT TFTFF FTFTF TFTFF TFFTF

19.45 In Exercise 1.37, in the first chapter in the book, we offered these sets of 21 consecutive shots by three basketball players (X = hit, 0 = miss):

OXXXOXXXOXXOOOXOOXXOO
XOOXXXOOXOXOXOOOXOXXX
XOXOXOOOXXOXOXOOXXXOX

a. Are any of the three players not scoring in an unpredictable pattern? Choose a commonly used level of significance and test with an alternative sensitive to either form of nonrandomness in a sequence.

b. Some basketball players have the reputation of being "streak shooters." If such players really exist, how would this phenomenon influence the number of runs in a sequence of their shots?

19.46 A row of 32 corn plants is checked, and each plant is classified as pest-infested (p) or healthy (h). Can the sequence be viewed as a random one? Use $\alpha = .10$.

h h p p p h h h p p h h h h p h h h h p p p h h p p h h h h h h

19.5 SUMMARY

There are many nonparametric procedures; we have discussed only some of the most popular ones. Many nonparametric methods operate on qualitative data. For instance, the runs test and the chi-square tests for goodness of fit and for independence involve categorical, or nominal, data. As such, they have no direct parametric equivalent. Other methods such as the signed ranks test, the rank sum test, and Spearman's correlation involve ranked, or ordinal, data. Parametric counterparts do exist for the procedures that operate on ranks.

Compared to its parametric equivalent, a nonparametric procedure generally has weaker assumptions and less involved computations. Parametric *t*-tests, for instance, assume population normality and require parameter estimates, such as the mean and standard deviation. Nonparametric assumptions are less restrictive, and the number crunching is simpler. If the assumptions underlying a parametric method hold, however, the parametric test is more powerful and is therefore preferred.

CHAPTER 19 EXERCISES

19.47 State whether the null hypothesis would be accepted or rejected for the following values of the chi-square test statistic.

a. $\chi^{2*} = 12.12$, $\alpha = .05$, and $df = 5$.

b. $\chi^{2*} = 6.13$, $\alpha = .10$, multinomial, and five categories.

c. $\chi^{2*} = 11.94$, $\alpha = .01$, and a four-category variable cross-tabulated with a three-category variable.

19.48 A friend of yours says she has no use for random number tables since "I can generate digits in my head as random as those in any table." You are skeptical and ask her to "generate" 100 of her random digits quickly. As she announces digits, you tally up the counts, obtaining the following table.

Digit	Count	Digit	Count
0	15	5	16
1	9	6	6
2	12	7	17
3	4	8	10
4	6	9	5

When you express doubt about the distribution being consistent with randomness, she replies that variation in digit frequency should be expected due to chance variation in the sampling process. State and test a hypothesis to settle the issue, using $\alpha = .05$.

19.49 A safety engineer at a large production facility has categorized this year's OSHA-reportable worker injuries according to the day of the week that the injury occurred. Treating this year's experience as a random sample of plant experience, can you reject the hypothesis that accidents are evenly distributed across the week? Let $\alpha = .10$.

Day	Number of Injuries
Sunday	5
Monday	6
Tuesday	4
Wednesday	7
Thursday	8
Friday	7
Saturday	12

19.50 A warehouse containing several thousand pairs of women's shoes has suffered a fire. The owner of the stock of shoes is offering them at a substantial discount, stating that 25 percent of the pairs are ruined, 35 percent have minor smoke damage but can be sold, and that 40 percent are unharmed. Suppose a prospective buyer samples 128 pairs of shoes chosen at random, finding 41 pairs ruined, 51 damaged but saleable, and 36 unharmed. With 5% risk of a Type I error, does the sample evidence support or contradict the owner's claim about the condition of the shoes?

19.51 When a large employer decided not to renew its contract with its health insurance provider, the company employees were given a choice of three insurance options. Representatives from each of the three providers made presentations at group meetings to explain their offerings, and the employees then selected an insurer. The following table shows, for a random sample of 120 employees, a cross-tabulation of provider chosen with the employee's age.

Employee Age	Insurer Chosen		
	B	*H*	*P*
Under 30	16	10	5
30 up through 45	19	26	15
Over 45	5	12	12

a. At the .05 level, are the variables age and insurer chosen independent or dependent?
b. If you find a dependent relation, state the nature of that relation.

19.52 A survey of 370 shoppers in a southeastern city to investigate differences between single and multiple person household shoppers was done via personal interviews conducted in six supermarkets. ("Single and Multiple Person Shoppers: A Focus on Grocery Store Selection

Criteria and Grocery Shopping Attitudes and Behavior," by Hazel Ezell and Giselle Russell, Winter 1985 issue of *Journal of the Academy of Marketing Science*). One cross-tabulation of the results has the following setup:

Normally Shop in	Persons in Household	
	One	More Than One
Morning		
Afternoon		
Evening		

If we wished to test for independence between time of day that shopping occurs and the number of people in the household, determine the following.

a. What would be the degrees of freedom for the chi-square test statistic?

b. For $\alpha = .05$, what is the cutoff value from the chi-square table?

c. The actual chi-square value that resulted in the survey was 1.466. Does this mean the two variables are related or unrelated?

19.53 Refer to Exercise 19.52. The same study cross-tabulated persons in households against seven categories of shopper's education level. The resulting chi-square test statistic was 11.155. At the 5% level of significance, do you accept or reject the assumption of independence?

19.54 A statewide newspaper sampled adults in the state to ascertain their opinions on the governor's recent proposal to raise state income taxes. Contact was made by phone, using a random digit dialing method. The opinion was cross-classified with the individual's party membership.

Party Membership	Opinion		
	Oppose	Support	Undecided
None or Independent	37	16	5
Republican	30	22	8
Democrat	33	42	7

a. Test for the existence of a dependent relation between party membership and opinion on the tax proposal, using the .10 risk level.

b. If a dependent relation is found, state the nature of that relation.

19.55 A polling organization surveyed a random sample of adults over 25 years of age about their reading of periodicals. One question asked the respondent if they could read just one national weekly magazine, whether they would want it to be one that emphasized sports, general news, or entertainment. Responses to this question were cross-tabulated with level of education, as shown in the following table. At the .01 level, do you find evidence of a dependent relation?

Education Level	Magazine Type: Sports	General News	Entertainment
High School	22	24	44
Some College	35	34	45
College Graduate	33	62	61

19.56 Following is a listing of runner weights and the placement order in the finals of the 220-yard dash at a state high school track meet. Compute the rank correlation. Assign a rank of 1 to the lowest weight.

Runner	Finish Order	Weight
A	1	149
B	2	155
C	3	171
D	4	137
E	5	180
F	6	156
G	7	141
H	8	161

19.57 Two judges ranked eight gymnast contestants' routines on the uneven parallel bars as follows:

	Contestant							
Judge	*F*	*B*	*E*	*A*	*C*	*H*	*G*	*D*
1	1	2	3	4	5	6	7	8
2	2	1	3	4	7	6	5	8

a. Would you expect judges' preference orderings to agree, disagree, or not correlate in such a setting?

b. Compute the rank correlation.

c. Can the assumption of no correlation be rejected with 99% certainty?

19.58 The Rand McNally *Places Rated Almanac*, authored by Richard Boyer and David Savageau, offers a detailed look at 329 U.S. metropolitan areas. Each is evaluated on criteria such as climate, housing, crime, transportation, education, the arts, recreation, and economic outlook. The top 12 overall cities in the second edition of *Places Rated* are listed as follows:

Rankings Among the Top Twelve

Top Twelve Overall: Rank–City	The Arts	Housing Costs
1. Pittsburgh, PA	4	3
2. Boston, MA	1	10
3. Raleigh-Durham, NC	11	4
4. San Francisco, CA	3	11
5. Philadelphia, PA	2	5
6. Nassau-Suffolk, NY	8	9
7. St. Louis, MO	7	2
8. Louisville, KY	12	1
9. Norwalk, CT	10	12
10. Seattle, WA	6	8
11. Atlanta, GA	9	6
12. Dallas, TX	5	7

We randomly chose two of the criteria, housing and the arts, and obtained the cities' ranks relative to each other. (The arts rating depended on museums, public radio and TV offerings, symphony orchestras, opera companies, dance companies, universities offering degrees in the arts, and libraries. The housing ranking was a function of utility bills, property taxes, and mortgage payments.)

a. Determine the correlation of the arts and housing ranks given above.

b. Is your answer to part a significantly different from zero? Use a two-tailed test, with $\alpha = .10$.

19.59 Two versions of a sixty-second television commercial offering classic rock songs from the '60s were prepared by a recording company. In a test of their appeal, each version was aired in different markets of comparable size. Version *A* was broadcast in five markets, Version *B* in six. All commercials appeared between 5 and 6 P.M. on a weekday. Through use of different addresses/phone numbers for ordering, it was possible to know which commercial generated each order. Use a 10% risk factor to ascertain whether a strong case is made that either commercial is more effective in generating orders than the other. The numbers given are the order counts received from each test market.

Version *A*	31	40	28	21	47	
Version *B*	45	54	51	39	60	59

19.60 A labor economist wishes to compare small cities (population 8000 to 30,000) in the eastern and western parts of her state with respect to the percent of working mothers in the labor force. A random sample of 12 such cities from each half of the state is obtained. The figures given represent the percentage of women in each city's work force who have one child or more under 12 years of age.

Eastern Cities	39.8	40.3	43.5	46.7	46.8	47.2
	50.3	50.7	53.1	55.2	55.9	56.2
Western Cities	41.7	44.4	48.0	48.4	49.7	50.7
	52.5	55.1	55.4	56.8	57.1	59.7

Do you find or fail to find a significant difference in the group means? Tie your conclusion to a risk factor of your own choice.

19.61 An independent testing laboratory is asked to compare breaking strengths of two brands of basketball goal backboards. This is done by suspending a metal bar from the rim and adding weights in five-pound increments until the plastic shatters. Assume that the backboards were a random sample of each brand.

Brand	Weight Required to Shatter			
N	625	705	660	745
D	695	580	560	605

a. Test the hypothesis of no difference in means, using $\alpha = .10$, and a two-tailed alternative hypothesis.
b. If we wished to approach this inference problem with a *t*-test, what assumptions not required by the rank sum procedure would have to be made?

19.62 A regression model was developed using five years' worth of monthly data. A count of the residuals shows 28 positive and 32 negative residuals. Test the hypothesis that the residuals are independent against the appropriate one-tailed alternative, using $\alpha = .05$. What do you conclude if the residuals contain 22 runs?

19.63 A packaging process has been calibrated to dispense 2275 grams of sugar per sack. A quality control clerk records the following net weights for a sample of 24 consecutively filled sacks:

2271	2268	2288	2280	2290	2258
2266	2274	2280	2260	2277	2281
2290	2276	2259	2269	2261	2272
2284	2266	2277	2277	2270	2283

Classify each observation as an underfill (*U*) or as an overfill (*O*). Test the assumption of randomness in the sequence, using $\alpha = .10$.

19.64 Following is the sequence of Presidential election winners for this century. Ascertain whether the party winners should be viewed as a random or nonrandom sequence. Use $\alpha = .10$ and an alternative hypothesis which anticipates that a party, once in power, will tend to stay in power.

Year	Election Winner	Party
1900	McKinley	Republican
1904	T. Roosevelt	Republican
1908	Taft	Republican
1912	Wilson	Democrat
1916	Wilson	Democrat
1920	Harding	Republican
1924	Coolidge	Republican
1928	Hoover	Republican
1932	F. Roosevelt	Democrat
1936	F. Roosevelt	Democrat
1940	F. Roosevelt	Democrat
1944	F. Roosevelt	Democrat
1948	Truman	Democrat
1952	Eisenhower	Republican
1956	Eisenhower	Republican
1960	Kennedy	Democrat
1964	Johnson	Democrat
1968	Nixon	Republican
1972	Nixon	Republican
1976	Carter	Democrat
1980	Reagan	Republican
1984	Reagan	Republican
1988	Bush	Republican

REVIEW PROBLEMS CHAPTERS 17–19

R81 Are the heights of college basketball players exaggerated? Or, do college sports information directors provide accurate information to the media about how tall their players are? Let us define three possible outcomes when an unbiased measurement of a player's height is compared against his school's official listing:

Outcome 1: The unbiased measurement is within plus or minus $\frac{1}{4}$ inch of the official school listing.

Outcome 2: The measurement exceeds the listing by more than $\frac{1}{4}$ inch.

Outcome 3: The measurement is less than the listing by more than $\frac{1}{4}$ inch.

Now suppose that if schools make a reasonable attempt at being accurate, Outcome 1 will happen half the time while Outcomes 2 and 3 will each occur one-fourth of the time. For data, let us assume that the 52 collegians who tried out for the United States Pan American team in 1987 can be treated as a random sample of player measurements. All 52 were measured at the tryouts in Colorado Springs. Of these, 6 turned out to be the same height as listed by their schools. One player was more than $\frac{1}{4}$ inch taller. Forty-five were more than $\frac{1}{4}$ inch shorter, the most dramatic case being Iowa State's Jeff Grayer—listed at 6–6 but measured at 6–3$\frac{1}{2}$. Test the hypothesis that schools provide accurate listings, at the .01 level of significance. If you find evidence beyond a reasonable doubt against the hypothesis, state the form of bias that is found. (This problem is based on "Pan Am Tape: Players Shrunk," *The Sporting News,* June 29, 1987.)

R82 In "Businesswomen's Broader Latitude In Dress Codes Goes Just So Far" (*Wall Street Journal,* September 1, 1987), staff reporter Kathleen A. Hughes reopened the issue of appropriate attire for businesswomen in light of the return of the miniskirt in 1987. She cited a survey conducted by John T. Malloy, author of *Dress for Success,* of 298 women in corporate sales positions. Malloy asked the sales managers of these women to describe their style of dress and then tracked their career progress for three years, as follows:

	Dress			
Job Performance	Very Professional and Very Conservative	Appropriate	Sexy, Frilly, or Fashionable	Poorly Dressed
Top performers	12.5%	12.6%	11.5%	1.9%
Consistently above average	60.9	41.1	31.0	30.8
Average or below	21.9	44.2	54.0	50.0
Failing	4.7	2.1	3.4	17.3

Source: Reprinted by permission of *The Wall Street Journal,* © Dow Jones & Co., Inc., 1987. All Rights Reserved Worldwide.

Assume that the actual numbers of women in each dress category are 64, 95, 87, and 52, respectively, from left to right in the table. Analyze these data to determine whether dress and job performance are independent variables. Use $\alpha = .005$.

R83 Are the two different facets of a pro football team's kick return game related? Following are 1988 season data for the average kickoff return yards (X) and the average punt return yards (Y) for the 14 teams in the American Conference of the National Football League

Team	X = Kickoff Return Average	Y = Punt Return Average
Buffalo	18.7	5.8
Cincinnati	18.5	7.6
Cleveland	21.1	8.1
Denver	20.7	8.5
Houston	20.5	6.3
Indianapolis	19.9	9.8
Kansas City	16.5	6.7
L.A. Raiders	22.7	8.9
Miami	21.0	9.6
New England	21.9	10.5
N.Y. Jets	19.5	11.0
Pittsburgh	21.3	8.3
San Diego	25.2	9.0
Seattle	21.8	9.2

Source: Data from *The Sporting News,* January 9, 1989.

a. Convert X and Y to ranks and then compute Spearman's correlation coefficient. Assign a rank of 1 to the highest return average.

b. Treating the data as a random sample of NFL experience, test the assumption that the population correlation coefficient between kickoff and punt return rankings is zero. Use a two-tailed alternative, with a 10% risk of a Type I error.

c. Does your finding mean that there is or is not sufficient evidence to conclude that an association exists between NFL teams' punt and kickoff return averages?

d. What assumption about the population would be required to make an inference about r, the parametric correlation coefficient, instead of r_S as suggested here?

R84 Public television stations are viewer-supported, nonprofit entities. One such station is planning its annual solicitation campaign and would like to target those individuals most likely to contribute. The campaign coordinator obtained data on the following variables from a random sample of previous contributors:

$$Y = \text{Dollar amount of contribution}$$

$$X_1 = \text{Annual income, in thousands of dollars}$$

$$X_2 = \text{Years of education}$$

$$X_3 = \begin{cases} 1 & \text{if male } (M) \\ 0 & \text{if female } (F) \end{cases}$$

Y	X_1	X_2	X_3
\$1000	\$35	16	1
250	30	18	0
100	28	14	1
700	32	16	1
50	50	16	1
600	29	17	0
400	24	14	0
1500	40	16	1
200	40	14	1
800	32	16	1
1200	36	18	0
250	26	16	1
300	25	16	0
2500	41	18	1
50	20	13	1
500	29	18	0
2000	38	18	1
100	21	12	0
350	52	14	1
1100	37	18	0

Suppose the model $Y = \beta_0 + \beta_1 X_1 + \beta_2 X_2 + \beta_3 X_3 + \epsilon$ is proposed.

a. Interpret β_1, β_2, and β_3.

b. Fit the model to the data and write out the equation of the fitted model.

c. Is there a significant difference in the average contribution between males and females? Justify your answer with a test, having a 5% α-risk.

R85 Refer to R84. Suppose the interaction terms $\beta_4 X_1 X_3$ and $\beta_5 X_2 X_3$ are added to the model. What null hypothesis would you set up to test the situations in parts a and b?

a. That the change in the average amount of contribution for each additional thousand dollars of annual income is the same for males and females, assuming the contributors have the same years of education?

b. That the change in the average amount of contribution for each additional year of education is the same for males and females, assuming the contributors have the same annual income?

c. Fit the complete model with the interaction terms and then conduct a test to determine if the interaction terms are significantly different from 0. Use $\alpha = .10$.

R86 A produce wholesaler randomly chooses 6 apples from each of two suppliers and obtains their weights, in grams:

Supplier *K*	360	441	416	404	351	379
Supplier *V*	390	399	438	411	470	414

With $\alpha = .10$, test the assumption of no mean difference in the populations from which these samples were obtained. Use a nonparametric procedure, with a two-tailed alternative.

R87 An industrial safety engineer examined company records in the year before and after a major assembly line was partially robotized. One purpose of the study was to test to see if the pattern of worker injuries along the assembly line was affected by the change. The following table shows the count of four commonly occurring worker injuries cross-tabulated against whether the injuries happened before or after the automation.

Injury Type (ANSI Code and Description)	Automation Status	
	Before	**After**
160—Contusion, Crush, Bruise	153	143
170—Cut, Laceration, Puncture	189	135
210—Fracture	15	9
260—Inflammation: Joint, Tendon, Muscle	45	98

Source: Data from "Safety Performance Analysis of a Dishwasher Assembly Plant Before and After Automation," by John R. Primovic, Master's Degree Thesis, August 1987, University of Louisville.

Assume that the data can be viewed as a random sample of company-wide experience before and after automating assembly lines.

a. Would the null hypothesis be that injury type is related to automation status, or that injury type is unrelated to automation status?

b. For 10% risk of incorrect rejection of the null hypothesis, what is the table cutoff value?
c. Determine the value of the test statistic and state the conclusion for the test.

R88 The changes in business inventories as a percent of the Gross National Product are indicated as follows for the years 1984–1987:

	Quarter			
Year	1	2	3	4
1984	2.6	1.8	1.7	1.2
1985	0.5	0.5	0.0	0.2
1986	1.1	0.5	0.0	0.0
1987	0.9	0.7	0.3	1.5

Source: U.S. Department of Commerce, Bureau of Economic Analysis, *Business Conditions Digest,* Washington, D.C., October 1988.

a. Construct a four-quarter centered moving average to estimate the trend component.
b. Develop a seasonal index for each quarter.
c. Isolate the cyclical component by finding cyclical relatives. Use a three-period weighted centered moving average, with weights of 1, 2, and 1 to eliminate the irregular component.

R89 After obtaining the least-squares equation for a data set with 72 observations, you wish to test your regression residuals for autocorrelation. Since the data are time related, you anticipate that if autocorrelation is found in the residuals, it will be positive autocorrelation. Inspection of the residuals reveals 34 negative residuals, 38 positive residuals, with a total of 33 runs. Test the assumption that the residuals are uncorrelated versus the proper one-tailed alternative, using $\alpha = .05$. Does it appear that the regression model assumption of uncorrelated residuals is justified?

R90 Members of a stock investment club evaluated a group of nine high technology corporations with respect to the investment objectives of potential for long-term growth and safety/steady income. Rankings are as follows, with the most desirable opportunities receiving the lowest number rankings.

Company	Long-Term Growth	Safety/ Steady Income
A	4	6
B	9	7
C	1	3
D	8	5
E	2	8
F	3	4
G	6	1
H	5	9
I	7	2

a. Compute the nonparametric measure of strength of relation between the rankings.
b. Is your result significantly different from zero? Use $\alpha = .10$, with a two-tailed alternative hypothesis.

R91 A lithotriper is a machine that creates shock waves to smash kidney stones. No surgery is required, the treatment is less painful than traditional surgery, and patients recuperate in a couple of days. After installing such a machine, a hospital treated the following numbers of patients in the first 12 months:

Month	1	2	3	4	5	6	7	8	9	10	11	12
Number of Patients	150	135	133	107	91	72	57	64	81	88	97	104

a. Fit the model $Y_t = \beta_0 + \beta_1 t + \beta_2 t^2 + \epsilon$, where t represents months and Y_t is the number of patients per month.
b. For month 2 through month 12, compare the fitting accuracies of the forecasts generated by the regression model from part a with those from an exponentially smoothed forecast ($\alpha = .3$) using the *MSE* metric.
c. Suppose the actual number of patients treated with the lithotriper in the next four months is 110, 108, 121, and 135. Compare the forecasting accuracies of the regression model and exponentially smoothed forecasts ($\alpha = .3$) using the *MAPE* metric.

R92 Assume the following residuals resulted from a multiple regression analysis with five independent variables:

i	e_i	i	e_i
1	18.3	11	38.3
2	−50.0	12	−46.7
3	−6.1	13	−46.0
4	−51.1	14	71.9
5	−81.7	15	22.1
6	−1.2	16	−34.7
7	66.5	17	25.2
8	62.8	18	−6.9
9	−9.4	19	27.2
10	1.6	20	−0.1

a. Test the assumption that the residuals are not autocorrelated, using the Durbin-Watson test. Let $\alpha = .01$.
b. Test the assumption that the residuals are normally distributed, using the Lin-Mudholkar test. Let $\alpha = .05$.

R93 An advertising researcher analyzed a sample of Australian television commercials and a sample of commercials aired in the United States. The table shows, for the five most common product categories, a count of each country's commercials.

Product Category	Country	
	U.S.	Australia
Food, Snacks, Soda	85	17
Restaurant and Retail Outlets	24	26
Household Cleaners	30	11
Personal and Beauty Care	30	6
Drugs and Medicine	36	4

Source: Data from Mary C. Gilly, "Sex Roles in Advertising: A Comparison of Television Advertisements in Australia, Mexico, and the United States," *Journal of Marketing*, April, 1988.

Assume the commercials in the study are a random sample. With 5% risk of a Type I error, are commercial product category and country of commercial broadcast independent or dependent?

R94 An industrial psychologist interviews random samples of a company's sales force to assess their degree of job satisfaction. Some salespeople interviewed work in the company's technical products division while the others do not. The psychologist uses a point system to arrive at a total score measure of job satisfaction. In the data that follow, higher scores indicate greater degree of satisfaction.

Technical Products Salespeople	76	93	64	84	91	89	67	81
Other Salespeople	78	57	58	72	95			
	61	67	52	88	68	69		

At $\alpha = .10$, test a hypothesis of no difference in population means, with a two-tailed alternative hypothesis. Since the score populations cannot be presumed to follow normal distributions with equal variance, use a procedure that does not require these assumptions. Do you find strong evidence that the populations from which these samples came have different means?

R95 In the mid-1980s the most popular color of automobile was red, representing 20% of the new cars sold. Other popular choices during the mid-80s are given in the accompanying table. Approximately five years later, a follow-up study was conducted to determine if consumer preferences had changed. Conduct the appropriate test based on the following random sample of colors of 386 new cars sold. Let $\alpha = .01$.

Color	Mid-1980s Percentages	No. of New Cars Sold
Red	20.0	68
White	12.6	52
Dark blue	12.4	34
Light blue	9.0	49
Silver	7.8	40
Other	38.2	143

R96 Suppose the color percentages of peanut M&M's are brown 30 percent, yellow 20 percent, orange 10 percent, red 20 percent and green 20 percent (only plain M&M's have a tan color). These are overall percentages; however, how closely individual bags match these figures will depend on how well the colors are mixed in the packaging process. Suppose you buy a one-pound bag of peanut M&M's to see how well the bag color distribution conforms with the given percentages.

a. Which null hypothesis could you test?

1. The bag contents came from a population that is not well mixed.
2. The bag contents came from a population that is well mixed.

b. Carry out your hypothesis test, at $\alpha = .01$. Your sample evidence is the color counts found in your bag: 60 brown, 35 yellow, 15 orange, 29 red, and 51 green.

R97 Are stress and salary directly related? To find out, a random sampling from a list of occupations in *The Jobs Rated Almanac* was taken and produced the following data:

Occupation	Average Salary in Thousands of Dollars	Stress Ranking
1. State police officer	$35.0	2
2. Bank officer	31.2	13
3. Chemist	41.0	16
4. Economist	43.1	10
5. Computer programmer	29.5	17
6. Optician	29.7	14
7. Air traffic controller	38.6	3
8. Astronaut	38.3	1
9. Veterinarian	50.1	9
10. Lawyer	43.5	5
11. Electrician	29.5	11
12. Rabbi	54.5	8
13. Major league baseball player	424.9	7
14. Postal inspector	45.9	12
15. Fashion model	32.7	6
16. Statistician	28.8	15
17. Member of Congress	89.5	4

Source: "Congress moving up?" *USA Today*, January 16, 1989. Copyright 1989, *USA Today*. Adapted with permission.

Rank the salaries from high (rank = 1) to low (rank = 17). Test the hypothesis that salary and stress rankings are unrelated. Let $\alpha = .10$.

R98 Sara H. Dinius and Robert B. Rogow in their article, "Applications of the Delphi Method in Identifying Characteristics Big Eight Firms Seek in Entry-Level Accountants" (*Journal of Accounting Education, 6,* 1988), used the Delphi Method to elicit the most important perceived characteristics of entry-level accounting applicants from a panel of personnel partners or managers from seven of the Big Eight accounting firms. They reported that the opinion of the panel members seemed to converge by the third round. One of the characteristics of interest was an applicant's overall GPA. Each panelist was asked to rate his or her perceived degree of importance of this factor on a 1 = low to 7 = high scale. Suppose the data are as follows:

	Importance						
Round	1	2	3	4	5	6	7
1	2	0	2	8	12	19	13
2	0	2	1	6	8	28	10
3	0	0	0	1	14	33	6

The numbers in each row represent the number of panelists who selected that degree of importance. Test the null hypothesis that panelists' opinions are not affected by the feedback—that is, opinions are independent—from round to round. Use $\alpha = .01$.

R99 A major appliance manufacturer used Census Bureau X-11, a sophisticated forecasting model, to predict monthly sales of its refrigerators. In a test of forecasting accuracy, subjective predictions developed by senior sales/marketing managers were compared with the X-11 predictions.

Month	Subjective Forecast	X-11 Forecast	Actual Unit Sales
January	3540	3488	3627
February	2550	2562	2492
March	3490	3396	3340
April	3690	3597	3761
May	4140	4204	4014
June	4250	4414	4166

Each forecast was made one month in advance. For instance, the June forecasts shown were developed the first day or two in June after the May sales had become known.

a. Determine the mean absolute percentage error (*MAPE*) for each forecasting method over the six-month sample period.

b. Use the rank sum procedure to test the hypothesis of equal *MAPE*'s for the two methods. Use the 10% level of significance and a two-tailed alternative.

R100 The Rose Bowl football game dates back to 1902. Since 1947, it has been a matchup of teams from the Big 10 and the Pacific 10 conferences. Following is a listing of winners from 1947 through 1989 (b = Big 10, p = Pac 10).

b b b b b   b p b b b   b b b p p   b p b b p   b p b p p
p p b p p   p p p p b   p p p p p   p b b

At $\alpha = .10$, does the sequence of winners appear random?

APPENDICES

APPENDIX A BINOMIAL TABLES

Accurate to four places, these tables give binomial probabilities, $P(X = x)$, for various combinations of n and π. For $\pi \leq .50$, use the column headings at top of page together with row headings at left margin. For $\pi \geq .50$, use the column indicators at bottom of page in conjunction with the row headings at the right margin.

Examples: (1) Find $P(X = 0)$ when $n = 4$ and $\pi = .15$: Using the left row and top column headings, you should locate .5220. (2) Find $P(X = 1)$ when $n = 3$ and $\pi = .60$: Using the bottom indicators for columns and right row headings, you should find .2880.

		π											
n	x	.05	.10	.15	.20	.25	.30	.35	.40	.45	.50		
2	0	.9025	.8100	.7225	.6400	.5625	.4900	.4225	.3600	.3025	.2500	2	
	1	.0950	.1800	.2550	.3200	.3750	.4200	.4550	.4800	.4950	.5000	1	
	2	.0025	.0100	.0225	.0400	.0625	.0900	.1225	.1600	.2025	.2500	0	2
3	0	.8574	.7290	.6141	.5120	.4219	.3430	.2746	.2160	.1664	.1250	3	
	1	.1354	.2430	.3251	.3840	.4219	.4410	.4436	.4320	.4084	.3750	2	
	2	.0071	.0270	.0574	.0960	.1406	.1890	.2389	.2880	.3341	.3750	1	
	3	.0001	.0010	.0034	.0080	.0156	.0270	.0429	.0640	.0911	.1250	0	3
4	0	.8145	.6561	.5220	.4096	.3164	.2401	.1785	.1296	.0915	.0625	4	
	1	.1715	.2916	.3685	.4096	.4219	.4116	.3845	.3456	.2995	.2500	3	
	2	.0135	.0486	.0975	.1536	.2109	.2646	.3105	.3456	.3675	.3750	2	
	3	.0005	.0036	.0115	.0256	.0469	.0756	.1115	.1536	.2005	.2500	1	
	4	.0000	.0001	.0005	.0016	.0039	.0081	.0150	.0256	.0410	.0625	0	4
5	0	.7738	.5905	.4437	.3277	.2373	.1681	.1160	.0778	.0503	.0312	5	
	1	.2036	.3280	.3915	.4096	.3955	.3602	.3124	.2592	.2059	.1562	4	
	2	.0214	.0729	.1382	.2048	.2637	.3087	.3364	.3456	.3369	.3125	3	
	3	.0011	.0081	.0244	.0512	.0879	.1323	.1811	.2304	.2757	.3125	2	
	4	.0000	.0004	.0022	.0064	.0146	.0284	.0488	.0768	.1128	.1562	1	
	5	.0000	.0000	.0001	.0003	.0010	.0024	.0053	.0102	.0185	.0312	0	5
6	0	.7351	.5314	.3771	.2621	.1780	.1176	.0754	.0467	.0277	.0156	6	
	1	.2321	.3543	.3993	.3932	.3560	.3025	.2437	.1866	.1359	.0938	5	
	2	.0305	.0984	.1762	.2458	.2966	.3241	.3280	.3110	.2780	.2344	4	
	3	.0021	.0146	.0415	.0819	.1318	.1852	.2355	.2765	.3032	.3125	3	
	4	.0001	.0012	.0055	.0154	.0330	.0595	.0951	.1382	.1861	.2344	2	
	5	.0000	.0001	.0004	.0015	.0044	.0102	.0205	.0369	.0609	.0938	1	
	6	.0000	.0000	.0000	.0001	.0002	.0007	.0018	.0041	.0083	.0156	0	6
7	0	.6983	.4783	.3206	.2097	.1335	.0824	.0490	.0280	.0152	.0078	7	
	1	.2573	.3720	.3960	.3670	.3115	.2471	.1848	.1306	.0872	.0547	6	
	2	.0406	.1240	.2097	.2753	.3115	.3177	.2985	.2613	.2140	.1641	5	
	3	.0036	.0230	.0617	.1147	.1730	.2269	.2679	.2903	.2918	.2734	4	
	4	.0002	.0026	.0109	.0287	.0577	.0972	.1442	.1935	.2388	.2734	3	
	5	.0000	.0002	.0012	.0043	.0115	.0250	.0466	.0774	.1172	.1641	2	
	6	.0000	.0000	.0001	.0004	.0013	.0036	.0084	.0172	.0320	.0547	1	
	7	.0000	.0000	.0000	.0000	.0001	.0002	.0006	.0016	.0037	.0078	0	7
		.95	.90	.85	.80	.75	.70	.65	.60	.55	.50	x	n
		π											

APPENDIX A Continued

		π											
n	**x**	**.05**	**.10**	**.15**	**.20**	**.25**	**.30**	**.35**	**.40**	**.45**	**.50**		
8	**0**	.6634	.4305	.2725	.1678	.1001	.0576	.0319	.0168	.0084	.0039	**8**	
	1	.2793	.3826	.3847	.3355	.2670	.1977	.1373	.0896	.0548	.0312	**7**	
	2	.0515	.1488	.2376	.2936	.3115	.2965	.2587	.2090	.1569	.1094	**6**	
	3	.0054	.0331	.0839	.1468	.2076	.2541	.2786	.2787	.2568	.2188	**5**	
	4	.0004	.0046	.0185	.0459	.0865	.1361	.1875	.2322	.2627	.2734	**4**	
	5	.0000	.0004	.0026	.0092	.0231	.0467	.0808	.1239	.1719	.2188	**3**	
	6	.0000	.0000	.0002	.0011	.0038	.0100	.0217	.0413	.0703	.1094	**2**	
	7	.0000	.0000	.0000	.0001	.0004	.0012	.0033	.0079	.0164	.0312	**1**	
	8	.0000	.0000	.0000	.0000	.0000	.0001	.0002	.0007	.0017	.0039	**0**	**8**
9	**0**	.6302	.3874	.2316	.1342	.0751	.0404	.0207	.0101	.0046	.0020	**9**	
	1	.2985	.3874	.3679	.3020	.2253	.1556	.1004	.0605	.0339	.0176	**8**	
	2	.0629	.1722	.2597	.3020	.3003	.2668	.2162	.1612	.1110	.0703	**7**	
	3	.0077	.0446	.1069	.1762	.2336	.2668	.2716	.2508	.2119	.1641	**6**	
	4	.0006	.0074	.0283	.0661	.1168	.1715	.2194	.2508	.2600	.2461	**5**	
	5	.0000	.0008	.0050	.0165	.0389	.0735	.1181	.1672	.2128	.2461	**4**	
	6	.0000	.0001	.0006	.0028	.0087	.0210	.0424	.0743	.1160	.1641	**3**	
	7	.0000	.0000	.0000	.0003	.0012	.0039	.0098	.0212	.0407	.0703	**2**	
	8	.0000	.0000	.0000	.0000	.0001	.0004	.0013	.0035	.0083	.0176	**1**	
	9	.0000	.0000	.0000	.0000	.0000	.0000	.0001	.0003	.0008	.0020	**0**	**9**
10	**0**	.5987	.3487	.1969	.1074	.0563	.0282	.0135	.0060	.0025	.0010	**10**	
	1	.3151	.3874	.3474	.2684	.1877	.1211	.0725	.0403	.0207	.0098	**9**	
	2	.0746	.1937	.2759	.3020	.2816	.2335	.1757	.1209	.0763	.0439	**8**	
	3	.0105	.0574	.1298	.2013	.2503	.2668	.2522	.2150	.1665	.1172	**7**	
	4	.0010	.0112	.0401	.0881	.1460	.2001	.2377	.2508	.2384	.2051	**6**	
	5	.0001	.0015	.0085	.0264	.0584	.1029	.1536	.2007	.2340	.2461	**5**	
	6	.0000	.0001	.0012	.0055	.0162	.0368	.0689	.1115	.1596	.2051	**4**	
	7	.0000	.0000	.0001	.0008	.0031	.0090	.0212	.0425	.0746	.1172	**3**	
	8	.0000	.0000	.0000	.0001	.0004	.0014	.0043	.0106	.0229	.0439	**2**	
	9	.0000	.0000	.0000	.0000	.0000	.0001	.0005	.0016	.0042	.0098	**1**	
	10	.0000	.0000	.0000	.0000	.0000	.0000	.0000	.0001	.0003	.0010	**0**	**10**
11	**0**	.5688	.3138	.1673	.0859	.0422	.0198	.0088	.0036	.0014	.0005	**11**	
	1	.3293	.3835	.3248	.2362	.1549	.0932	.0518	.0266	.0125	.0054	**10**	
	2	.0867	.2131	.2866	.2953	.2581	.1998	.1395	.0887	.0513	.0269	**9**	
	3	.0137	.0710	.1517	.2215	.2581	.2568	.2254	.1774	.1259	.0806	**8**	
	4	.0014	.0158	.0536	.1107	.1721	.2201	.2428	.2365	.2060	.1611	**7**	
	5	.0001	.0025	.0132	.0388	.0803	.1321	.1830	.2207	.2360	.2256	**6**	
	6	.0000	.0003	.0023	.0097	.0268	.0566	.0985	.1471	.1931	.2256	**5**	
	7	.0000	.0000	.0003	.0017	.0064	.0173	.0379	.0701	.1128	.1611	**4**	
	8	.0000	.0000	.0000	.0002	.0011	.0037	.0102	.0234	.0462	.0806	**3**	
	9	.0000	.0000	.0000	.0000	.0001	.0005	.0018	.0052	.0126	.0269	**2**	
	10	.0000	.0000	.0000	.0000	.0000	.0000	.0002	.0007	.0021	.0054	**1**	
	11	.0000	.0000	.0000	.0000	.0000	.0000	.0000	.0000	.0002	.0005	**0**	**11**
		.95	**.90**	**.85**	**.80**	**.75**	**.70**	**.65**	**.60**	**.55**	**.50**	**x**	**n**
		π											

(continues)

APPENDIX A Continued

		π											
n	x	.05	.10	.15	.20	.25	.30	.35	.40	.45	.50		
12	0	.5404	.2824	.1422	.0687	.0317	.0138	.0057	.0022	.0008	.0002	12	
	1	.3413	.3766	.3012	.2062	.1267	.0712	.0368	.0174	.0075	.0029	11	
	2	.0988	.2301	.2924	.2835	.2323	.1678	.1088	.0639	.0339	.0161	10	
	3	.0173	.0852	.1720	.2362	.2581	.2397	.1954	.1419	.0923	.0537	9	
	4	.0021	.0213	.0683	.1329	.1936	.2311	.2367	.2128	.1700	.1208	8	
	5	.0002	.0038	.0193	.0532	.1032	.1585	.2039	.2270	.2225	.1934	7	
	6	.0000	.0005	.0040	.0155	.0401	.0792	.1281	.1766	.2124	.2256	6	
	7	.0000	.0000	.0006	.0033	.0115	.0291	.0591	.1009	.1489	.1934	5	
	8	.0000	.0000	.0001	.0005	.0024	.0078	.0199	.0420	.0762	.1208	4	
	9	.0000	.0000	.0000	.0001	.0004	.0015	.0048	.0125	.0277	.0537	3	
	10	.0000	.0000	.0000	.0000	.0000	.0002	.0008	.0025	.0068	.0161	2	
	11	.0000	.0000	.0000	.0000	.0000	.0000	.0001	.0003	.0010	.0029	1	
	12	.0000	.0000	.0000	.0000	.0000	.0000	.0000	.0000	.0001	.0002	0	12
13	0	.5133	.2542	.1209	.0550	.0238	.0097	.0037	.0013	.0004	.0001	13	
	1	.3512	.3672	.2774	.1787	.1029	.0540	.0259	.0113	.0045	.0016	12	
	2	.1109	.2448	.2937	.2680	.2059	.1388	.0836	.0453	.0220	.0095	11	
	3	.0214	.0997	.1900	.2457	.2517	.2181	.1651	.1107	.0660	.0349	10	
	4	.0028	.0277	.0838	.1535	.2097	.2337	.2222	.1845	.1350	.0873	9	
	5	.0003	.0055	.0266	.0691	.1258	.1803	.2154	.2214	.1989	.1571	8	
	6	.0000	.0008	.0063	.0230	.0559	.1030	.1546	.1968	.2169	.2095	7	
	7	.0000	.0001	.0011	.0058	.0186	.0442	.0833	.1312	.1775	.2095	6	
	8	.0000	.0000	.0001	.0011	.0047	.0142	.0336	.0656	.1089	.1571	5	
	9	.0000	.0000	.0000	.0001	.0009	.0034	.0101	.0243	.0495	.0873	4	
	10	.0000	.0000	.0000	.0000	.0001	.0006	.0022	.0065	.0162	.0349	3	
	11	.0000	.0000	.0000	.0000	.0000	.0001	.0003	.0012	.0036	.0095	2	
	12	.0000	.0000	.0000	.0000	.0000	.0000	.0000	.0001	.0005	.0016	1	
	13	.0000	.0000	.0000	.0000	.0000	.0000	.0000	.0000	.0000	.0001	0	13
14	0	.4877	.2288	.1028	.0440	.0178	.0068	.0024	.0008	.0002	.0001	14	
	1	.3593	.3559	.2539	.1539	.0832	.0407	.0181	.0073	.0027	.0009	13	
	2	.1229	.2570	.2912	.2501	.1802	.1134	.0634	.0317	.0141	.0056	12	
	3	.0259	.1142	.2056	.2501	.2402	.1943	.1366	.0845	.0462	.0222	11	
	4	.0037	.0349	.0998	.1720	.2202	.2290	.2022	.1549	.1040	.0611	10	
	5	.0004	.0078	.0352	.0860	.1468	.1963	.2178	.2066	.1701	.1222	9	
	6	.0000	.0013	.0093	.0322	.0734	.1262	.1759	.2066	.2088	.1833	8	
	7	.0000	.0002	.0019	.0092	.0280	.0618	.1082	.1574	.1952	.2095	7	
	8	.0000	.0000	.0003	.0020	.0082	.0232	.0510	.0918	.1398	.1833	6	
	9	.0000	.0000	.0000	.0003	.0018	.0066	.0183	.0408	.0762	.1222	5	
	10	.0000	.0000	.0000	.0000	.0003	.0014	.0049	.0136	.0312	.0611	4	
	11	.0000	.0000	.0000	.0000	.0000	.0002	.0010	.0033	.0093	.0222	3	
	12	.0000	.0000	.0000	.0000	.0000	.0000	.0001	.0005	.0019	.0056	2	
	13	.0000	.0000	.0000	.0000	.0000	.0000	.0000	.0001	.0002	.0009	1	
	14	.0000	.0000	.0000	.0000	.0000	.0000	.0000	.0000	.0000	.0001	0	14
		.95	.90	.85	.80	.75	.70	.65	.60	.55	.50	x	n
		π											

APPENDIX A Continued

		π											
n	x	.05	.10	.15	.20	.25	.30	.35	.40	.45	.50		
15	0	.4633	.2059	.0874	.0352	.0134	.0047	.0016	.0005	.0001	.0000	15	
	1	.3658	.3432	.2312	.1319	.0668	.0305	.0126	.0047	.0016	.0005	14	
	2	.1348	.2669	.2856	.2309	.1559	.0916	.0476	.0219	.0090	.0032	13	
	3	.0307	.1285	.2184	.2501	.2252	.1700	.1110	.0634	.0318	.0139	12	
	4	.0049	.0428	.1156	.1876	.2252	.2186	.1792	.1268	.0780	.0417	11	
	5	.0006	.0105	.0449	.1032	.1651	.2061	.2123	.1859	.1404	.0916	10	
	6	.0000	.0019	.0132	.0430	.0917	.1472	.1906	.2066	.1914	.1527	9	
	7	.0000	.0003	.0030	.0138	.0393	.0811	.1319	.1771	.2013	.1964	8	
	8	.0000	.0000	.0005	.0035	.0131	.0348	.0710	.1181	.1647	.1964	7	
	9	.0000	.0000	.0001	.0007	.0034	.0116	.0298	.0612	.1048	.1527	6	
	10	.0000	.0000	.0000	.0001	.0007	.0030	.0096	.0245	.0515	.0916	5	
	11	.0000	.0000	.0000	.0000	.0001	.0006	.0024	.0074	.0191	.0417	4	
	12	.0000	.0000	.0000	.0000	.0000	.0001	.0004	.0016	.0052	.0139	3	
	13	.0000	.0000	.0000	.0000	.0000	.0000	.0001	.0003	.0010	.0032	2	
	14	.0000	.0000	.0000	.0000	.0000	.0000	.0000	.0000	.0001	.0005	1	
	15	.0000	.0000	.0000	.0000	.0000	.0000	.0000	.0000	.0000	.0000	0	15
16	0	.4401	.1853	.0743	.0281	.0100	.0033	.0010	.0003	.0001	.0000	16	
	1	.3706	.3294	.2097	.1126	.0535	.0228	.0087	.0030	.0009	.0002	15	
	2	.1463	.2745	.2775	.2111	.1336	.0732	.0353	.0150	.0056	.0018	14	
	3	.0359	.1423	.2285	.2463	.2079	.1465	.0888	.0468	.0215	.0085	13	
	4	.0061	.0514	.1311	.2001	.2252	.2040	.1553	.1014	.0572	.0278	12	
	5	.0008	.0137	.0555	.1201	.1802	.2099	.2008	.1623	.1123	.0667	11	
	6	.0001	.0028	.0180	.0550	.1101	.1649	.1982	.1983	.1684	.1222	10	
	7	.0000	.0004	.0045	.0197	.0524	.1010	.1524	.1889	.1969	.1746	9	
	8	.0000	.0001	.0009	.0055	.0197	.0487	.0923	.1417	.1812	.1964	8	
	9	.0000	.0000	.0001	.0012	.0058	.0185	.0442	.0840	.1318	.1746	7	
	10	.0000	.0000	.0000	.0002	.0014	.0056	.0167	.0392	.0755	.1222	6	
	11	.0000	.0000	.0000	.0000	.0002	.0013	.0049	.0142	.0337	.0667	5	
	12	.0000	.0000	.0000	.0000	.0000	.0002	.0011	.0040	.0115	.0278	4	
	13	.0000	.0000	.0000	.0000	.0000	.0000	.0002	.0008	.0029	.0085	3	
	14	.0000	.0000	.0000	.0000	.0000	.0000	.0000	.0001	.0005	.0018	2	
	15	.0000	.0000	.0000	.0000	.0000	.0000	.0000	.0000	.0001	.0002	1	
	16	.0000	.0000	.0000	.0000	.0000	.0000	.0000	.0000	.0000	.0000	0	16
		.95	.90	.85	.80	.75	.70	.65	.60	.55	.50	x	n
		π											

(continues)

APPENDIX A Continued

		π											
n	*x*	.05	.10	.15	.20	.25	.30	.35	.40	.45	.50		
18	0	.3972	.1501	.0536	.0180	.0056	.0016	.0004	.0001	.0000	.0000	18	
	1	.3763	.3002	.1704	.0811	.0338	.0126	.0042	.0012	.0003	.0001	17	
	2	.1683	.2835	.2556	.1723	.0958	.0458	.0190	.0069	.0022	.0006	16	
	3	.0473	.1680	.2406	.2297	.1704	.1046	.0547	.0246	.0095	.0031	15	
	4	.0093	.0700	.1592	.2153	.2130	.1681	.1104	.0614	.0291	.0117	14	
	5	.0014	.0218	.0787	.1507	.1988	.2017	.1664	.1146	.0666	.0327	13	
	6	.0002	.0052	.0301	.0816	.1436	.1873	.1941	.1655	.1181	.0708	12	
	7	.0000	.0010	.0091	.0350	.0820	.1376	.1792	.1892	.1657	.1214	11	
	8	.0000	.0002	.0022	.0120	.0376	.0811	.1327	.1734	.1864	.1669	10	
	9	.0000	.0000	.0004	.0033	.0139	.0386	.0794	.1284	.1694	.1855	9	
	10	.0000	.0000	.0001	.0008	.0042	.0149	.0385	.0771	.1248	.1669	8	
	11	.0000	.0000	.0000	.0001	.0010	.0046	.0151	.0374	.0742	.1214	7	
	12	.0000	.0000	.0000	.0000	.0002	.0012	.0047	.0145	.0354	.0708	6	
	13	.0000	.0000	.0000	.0000	.0000	.0002	.0012	.0045	.0134	.0327	5	
	14	.0000	.0000	.0000	.0000	.0000	.0000	.0002	.0011	.0039	.0117	4	
	15	.0000	.0000	.0000	.0000	.0000	.0000	.0000	.0002	.0009	.0031	3	
	16	.0000	.0000	.0000	.0000	.0000	.0000	.0000	.0000	.0001	.0006	2	
	17	.0000	.0000	.0000	.0000	.0000	.0000	.0000	.0000	.0000	.0001	1	
	18	.0000	.0000	.0000	.0000	.0000	.0000	.0000	.0000	.0000	.0000	0	18
20	0	.3585	.1216	.0388	.0115	.0032	.0008	.0002	.0000	.0000	.0000	20	
	1	.3774	.2702	.1368	.0576	.0211	.0068	.0020	.0005	.0001	.0000	19	
	2	.1887	.2852	.2293	.1369	.0669	.0278	.0100	.0031	.0008	.0002	18	
	3	.0596	.1901	.2428	.2054	.1339	.0716	.0323	.0123	.0040	.0011	17	
	4	.0133	.0898	.1821	.2182	.1897	.1304	.0738	.0350	.0139	.0046	16	
	5	.0022	.0319	.1028	.1746	.2023	.1789	.1272	.0746	.0365	.0148	15	
	6	.0003	.0089	.0454	.1091	.1686	.1916	.1712	.1244	.0746	.0370	14	
	7	.0000	.0020	.0160	.0545	.1124	.1643	.1844	.1659	.1221	.0739	13	
	8	.0000	.0004	.0046	.0222	.0609	.1144	.1614	.1797	.1623	.1201	12	
	9	.0000	.0001	.0011	.0074	.0271	.0654	.1158	.1597	.1771	.1602	11	
	10	.0000	.0000	.0002	.0020	.0099	.0308	.0686	.1171	.1593	.1762	10	
	11	.0000	.0000	.0000	.0005	.0030	.0120	.0336	.0710	.1185	.1602	9	
	12	.0000	.0000	.0000	.0001	.0008	.0039	.0136	.0355	.0727	.1201	8	
	13	.0000	.0000	.0000	.0000	.0002	.0010	.0045	.0146	.0366	.0739	7	
	14	.0000	.0000	.0000	.0000	.0000	.0002	.0012	.0049	.0150	.0370	6	
	15	.0000	.0000	.0000	.0000	.0000	.0000	.0003	.0013	.0049	.0148	5	
	16	.0000	.0000	.0000	.0000	.0000	.0000	.0000	.0003	.0013	.0046	4	
	17	.0000	.0000	.0000	.0000	.0000	.0000	.0000	.0000	.0002	.0011	3	
	18	.0000	.0000	.0000	.0000	.0000	.0000	.0000	.0000	.0000	.0002	2	
	19	.0000	.0000	.0000	.0000	.0000	.0000	.0000	.0000	.0000	.0000	1	
	20	.0000	.0000	.0000	.0000	.0000	.0000	.0000	.0000	.0000	.0000	0	20
		.95	.90	.85	.80	.75	.70	.65	.60	.55	.50	*x*	*n*
		π											

APPENDIX A Continued

		π											
n	*x*	.05	.10	.15	.20	.25	.30	.35	.40	.45	.50		
25	0	.2774	.0718	.0172	.0038	.0008	.0001	.0000	.0000	.0000	.0000	25	
	1	.3650	.1994	.0759	.0236	.0063	.0014	.0003	.0000	.0000	.0000	24	
	2	.2305	.2659	.1607	.0708	.0251	.0074	.0018	.0004	.0001	.0000	23	
	3	.0930	.2265	.2174	.1358	.0641	.0243	.0076	.0019	.0004	.0001	22	
	4	.0269	.1384	.2110	.1867	.1175	.0572	.0224	.0071	.0018	.0004	21	
	5	.0060	.0646	.1564	.1960	.1645	.1030	.0506	.0199	.0063	.0016	20	
	6	.0010	.0239	.0920	.1633	.1828	.1472	.0908	.0442	.0172	.0053	19	
	7	.0001	.0072	.0441	.1108	.1654	.1712	.1327	.0800	.0381	.0143	18	
	8	.0000	.0018	.0175	.0623	.1241	.1651	.1607	.1200	.0701	.0322	17	
	9	.0000	.0004	.0058	.0294	.0781	.1336	.1635	.1511	.1084	.0609	16	
	10	.0000	.0000	.0016	.0118	.0417	.0916	.1409	.1612	.1419	.0974	15	
	11	.0000	.0000	.0004	.0040	.0189	.0536	.1034	.1465	.1583	.1328	14	
	12	.0000	.0000	.0000	.0012	.0074	.0268	.0650	.1140	.1511	.1550	13	
	13	.0000	.0000	.0000	.0003	.0025	.0115	.0350	.0760	.1236	.1550	12	
	14	.0000	.0000	.0000	.0000	.0007	.0042	.0161	.0434	.0867	.1328	11	
	15	.0000	.0000	.0000	.0000	.0002	.0013	.0064	.0212	.0520	.0974	10	
	16	.0000	.0000	.0000	.0000	.0000	.0004	.0021	.0088	.0266	.0609	9	
	17	.0000	.0000	.0000	.0000	.0000	.0001	.0006	.0031	.0115	.0322	8	
	18	.0000	.0000	.0000	.0000	.0000	.0000	.0001	.0009	.0042	.0143	7	
	19	.0000	.0000	.0000	.0000	.0000	.0000	.0000	.0002	.0013	.0053	6	
	20	.0000	.0000	.0000	.0000	.0000	.0000	.0000	.0000	.0001	.0016	5	
	21	.0000	.0000	.0000	.0000	.0000	.0000	.0000	.0000	.0000	.0004	4	
	22	.0000	.0000	.0000	.0000	.0000	.0000	.0000	.0000	.0000	.0001	3	
	23	.0000	.0000	.0000	.0000	.0000	.0000	.0000	.0000	.0000	.0000	2	
	24	.0000	.0000	.0000	.0000	.0000	.0000	.0000	.0000	.0000	.0000	1	
	25	.0000	.0000	.0000	.0000	.0000	.0000	.0000	.0000	.0000	.0000	0	25
		.95	.90	.85	.80	.75	.70	.65	.60	.55	.50	*x*	*n*
		π											

APPENDIX B POISSON TABLES

These tables give Poisson probabilities, $P(X = x)$, for different values of the Poisson distribution mean λ.

Examples: (1) Find $P(X = 3)$ when the mean number of occurrences per interval is .50: The result is .0126. (2) Find $P(X \leq 1)$ when $\lambda = 3.5$: The result is .0302 + .1057 = .1359.

	λ									
x	.1	.2	.3	.4	.5	.6	.7	.8	.9	1.0
0	.9048	.8187	.7408	.6703	.6065	.5488	.4966	.4493	.4066	.3679
1	.0905	.1637	.2222	.2681	.3033	.3293	.3476	.3595	.3659	.3679
2	.0045	.0164	.0333	.0536	.0758	.0988	.1217	.1438	.1647	.1839
3	.0002	.0011	.0033	.0072	.0126	.0198	.0284	.0383	.0494	.0613
4	.0000	.0001	.0003	.0007	.0016	.0030	.0050	.0077	.0111	.0153
5	.0000	.0000	.0000	.0001	.0002	.0004	.0007	.0012	.0020	.0031
6	.0000	.0000	.0000	.0000	.0000	.0000	.0001	.0002	.0003	.0005
7	.0000	.0000	.0000	.0000	.0000	.0000	.0000	.0000	.0000	.0001

	λ									
x	1.1	1.2	1.3	1.4	1.5	1.6	1.7	1.8	1.9	2.0
0	.3329	.3012	.2725	.2466	.2231	.2019	.1827	.1653	.1496	.1353
1	.3662	.3614	.3543	.3452	.3347	.3230	.3106	.2975	.2842	.2707
2	.2014	.2169	.2303	.2417	.2510	.2584	.2640	.2678	.2700	.2707
3	.0738	.0867	.0998	.1128	.1255	.1378	.1496	.1607	.1710	.1804
4	.0203	.0260	.0324	.0395	.0471	.0551	.0636	.0723	.0812	.0902
5	.0045	.0062	.0084	.0111	.0141	.0176	.0216	.0260	.0309	.0361
6	.0008	.0012	.0018	.0026	.0035	.0047	.0061	.0078	.0098	.0120
7	.0001	.0002	.0003	.0005	.0008	.0011	.0015	.0020	.0027	.0034
8	.0000	.0000	.0001	.0001	.0001	.0002	.0003	.0005	.0006	.0009
9	.0000	.0000	.0000	.0000	.0000	.0000	.0001	.0001	.0001	.0002

	λ									
x	2.1	2.2	2.3	2.4	2.5	2.6	2.7	2.8	2.9	3.0
0	.1225	.1108	.1003	.0907	.0821	.0743	.0672	.0608	.0550	.0498
1	.2572	.2438	.2306	.2177	.2052	.1931	.1815	.1703	.1596	.1494
2	.2700	.2681	.2652	.2613	.2565	.2510	.2450	.2384	.2314	.2240
3	.1890	.1966	.2033	.2090	.2138	.2176	.2205	.2225	.2237	.2240
4	.0992	.1082	.1169	.1254	.1336	.1414	.1488	.1557	.1622	.1680
5	.0417	.0476	.0538	.0602	.0668	.0735	.0804	.0872	.0940	.1008
6	.0146	.0174	.0206	.0241	.0278	.0319	.0362	.0407	.0455	.0504
7	.0044	.0055	.0068	.0083	.0099	.0118	.0139	.0163	.0188	.0216
8	.0011	.0015	.0019	.0025	.0031	.0038	.0047	.0057	.0068	.0081
9	.0003	.0004	.0005	.0007	.0009	.0011	.0014	.0018	.0022	.0027
10	.0001	.0001	.0001	.0002	.0002	.0003	.0004	.0005	.0006	.0008
11	.0000	.0000	.0000	.0000	.0000	.0001	.0001	.0001	.0002	.0002
12	.0000	.0000	.0000	.0000	.0000	.0000	.0000	.0000	.0000	.0001

APPENDIX B Continued

	λ									
x	3.1	3.2	3.3	3.4	3.5	3.6	3.7	3.8	3.9	4.0
0	.0450	.0408	.0369	.0334	.0302	.0273	.0247	.0224	.0202	.0183
1	.1397	.1304	.1217	.1135	.1057	.0984	.0915	.0850	.0789	.0733
2	.2165	.2087	.2008	.1929	.1850	.1771	.1692	.1615	.1539	.1465
3	.2237	.2226	.2209	.2186	.2158	.2125	.2087	.2046	.2001	.1954
4	.1734	.1781	.1823	.1858	.1888	.1912	.1931	.1944	.1951	.1954
5	.1075	.1140	.1203	.1264	.1322	.1377	.1429	.1477	.1522	.1563
6	.0555	.0608	.0662	.0716	.0771	.0826	.0881	.0936	.0989	.1042
7	.0246	.0278	.0312	.0348	.0385	.0425	.0466	.0508	.0551	.0595
8	.0095	.0111	.0129	.0148	.0169	.0191	.0215	.0241	.0269	.0298
9	.0033	.0040	.0047	.0056	.0066	.0076	.0089	.0102	.0116	.0132
10	.0010	.0013	.0016	.0019	.0023	.0028	.0033	.0039	.0045	.0053
11	.0003	.0004	.0005	.0006	.0007	.0009	.0011	.0013	.0016	.0019
12	.0001	.0001	.0001	.0002	.0002	.0003	.0003	.0004	.0005	.0006
13	.0000	.0000	.0000	.0000	.0001	.0001	.0001	.0001	.0002	.0002
14	.0000	.0000	.0000	.0000	.0000	.0000	.0000	.0000	.0000	.0001

	λ									
x	4.1	4.2	4.3	4.4	4.5	4.6	4.7	4.8	4.9	5.0
0	.0166	.0150	.0136	.0123	.0111	.0101	.0091	.0082	.0074	.0067
1	.0679	.0630	.0583	.0540	.0500	.0462	.0427	.0395	.0365	.0337
2	.1393	.1323	.1254	.1188	.1125	.1063	.1005	.0948	.0894	.0842
3	.1904	.1852	.1798	.1743	.1687	.1631	.1574	.1517	.1460	.1404
4	.1951	.1944	.1933	.1917	.1898	.1875	.1849	.1820	.1789	.1755
5	.1600	.1633	.1662	.1687	.1708	.1725	.1738	.1747	.1753	.1755
6	.1093	.1143	.1191	.1237	.1281	.1323	.1362	.1398	.1432	.1462
7	.0640	.0686	.0732	.0778	.0824	.0869	.0914	.0959	.1002	.1044
8	.0328	.0360	.0393	.0428	.0463	.0500	.0537	.0575	.0614	.0653
9	.0150	.0168	.0188	.0209	.0232	.0255	.0280	.0307	.0334	.0363
10	.0061	.0071	.0081	.0092	.0104	.0118	.0132	.0147	.0164	.0181
11	.0023	.0027	.0032	.0037	.0043	.0049	.0056	.0064	.0073	.0082
12	.0008	.0009	.0011	.0013	.0016	.0019	.0022	.0026	.0030	.0034
13	.0002	.0003	.0004	.0005	.0006	.0007	.0008	.0009	.0011	.0013
14	.0001	.0001	.0001	.0001	.0002	.0002	.0003	.0003	.0004	.0005
15	.0000	.0000	.0000	.0000	.0001	.0001	.0001	.0001	.0001	.0002

(continues)

APPENDIX B Continued

x	λ 5.5	6.0	6.5	7.0	7.5	8.0	9.0	10.0	12.0	15.0
0	.0041	.0025	.0015	.0009	.0006	.0003	.0001	.0000	.0000	.0000
1	.0225	.0149	.0098	.0064	.0041	.0027	.0011	.0005	.0001	.0000
2	.0618	.0446	.0318	.0223	.0156	.0107	.0050	.0023	.0004	.0000
3	.1133	.0892	.0688	.0521	.0389	.0286	.0150	.0076	.0018	.0002
4	.1558	.1339	.1118	.0912	.0729	.0573	.0337	.0189	.0053	.0006
5	.1714	.1606	.1454	.1277	.1094	.0916	.0607	.0378	.0127	.0019
6	.1571	.1606	.1575	.1490	.1367	.1221	.0911	.0631	.0255	.0048
7	.1234	.1377	.1462	.1490	.1465	.1396	.1171	.0901	.0437	.0104
8	.0849	.1033	.1188	.1304	.1373	.1396	.1318	.1126	.0655	.0194
9	.0519	.0688	.0858	.1014	.1144	.1241	.1318	.1251	.0874	.0324
10	.0285	.0413	.0558	.0710	.0858	.0993	.1186	.1251	.1048	.0486
11	.0143	.0225	.0330	.0452	.0585	.0722	.0970	.1137	.1144	.0663
12	.0065	.0113	.0179	.0263	.0366	.0481	.0728	.0948	.1144	.0829
13	.0028	.0052	.0089	.0142	.0211	.0296	.0504	.0729	.1056	.0956
14	.0011	.0022	.0041	.0071	.0113	.0169	.0324	.0521	.0905	.1024
15	.0004	.0009	.0018	.0033	.0057	.0090	.0194	.0347	.0724	.1024
16	.0001	.0003	.0007	.0014	.0026	.0045	.0109	.0217	.0543	.0960
17	.0000	.0001	.0003	.0006	.0012	.0021	.0058	.0128	.0383	.0847
18	.0000	.0000	.0001	.0002	.0005	.0009	.0029	.0071	.0256	.0706
19	.0000	.0000	.0000	.0001	.0002	.0004	.0014	.0037	.0161	.0557
20	.0000	.0000	.0000	.0000	.0001	.0002	.0006	.0019	.0097	.0418
21	.0000	.0000	.0000	.0000	.0000	.0001	.0003	.0009	.0055	.0299
22	.0000	.0000	.0000	.0000	.0000	.0000	.0001	.0004	.0030	.0204
23	.0000	.0000	.0000	.0000	.0000	.0000	.0000	.0002	.0016	.0133
24	.0000	.0000	.0000	.0000	.0000	.0000	.0000	.0001	.0008	.0083
25	.0000	.0000	.0000	.0000	.0000	.0000	.0000	.0000	.0004	.0050
26	.0000	.0000	.0000	.0000	.0000	.0000	.0000	.0000	.0002	.0029
27	.0000	.0000	.0000	.0000	.0000	.0000	.0000	.0000	.0001	.0016
28	.0000	.0000	.0000	.0000	.0000	.0000	.0000	.0000	.0000	.0009
29	.0000	.0000	.0000	.0000	.0000	.0000	.0000	.0000	.0000	.0004
30	.0000	.0000	.0000	.0000	.0000	.0000	.0000	.0000	.0000	.0002
31	.0000	.0000	.0000	.0000	.0000	.0000	.0000	.0000	.0000	.0001
32	.0000	.0000	.0000	.0000	.0000	.0000	.0000	.0000	.0000	.0001

APPENDIX C THE Z-TABLE

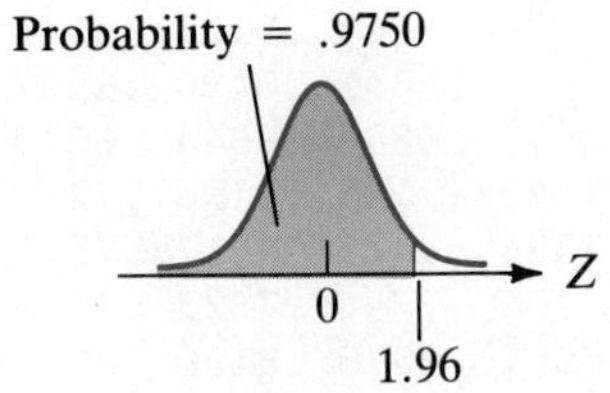

Entries in body of table are cumulative probabilities from $-\infty$ to Z (see shaded area of figure).

Example: $P(-\infty < Z \leq 1.96) = .9750$.

Z	.00	.01	.02	.03	.04	.05	.06	.07	.08	.09
−3.4	.0003	.0003	.0003	.0003	.0003	.0003	.0003	.0003	.0003	.0002
−3.3	.0005	.0005	.0005	.0004	.0004	.0004	.0004	.0004	.0004	.0003
−3.2	.0007	.0007	.0006	.0006	.0006	.0006	.0006	.0005	.0005	.0005
−3.1	.0010	.0009	.0009	.0009	.0008	.0008	.0008	.0008	.0007	.0007
−3.0	.0013	.0013	.0013	.0012	.0012	.0011	.0011	.0011	.0010	.0010
−2.9	.0019	.0018	.0018	.0017	.0016	.0016	.0015	.0015	.0014	.0014
−2.8	.0026	.0025	.0024	.0023	.0023	.0022	.0021	.0021	.0020	.0019
−2.7	.0035	.0034	.0033	.0032	.0031	.0030	.0029	.0028	.0027	.0026
−2.6	.0047	.0045	.0044	.0043	.0041	.0040	.0039	.0038	.0037	.0036
−2.5	.0062	.0060	.0059	.0057	.0055	.0054	.0052	.0051	.0049	.0048
−2.4	.0082	.0080	.0078	.0075	.0073	.0071	.0069	.0068	.0066	.0064
−2.3	.0107	.0104	.0102	.0099	.0096	.0094	.0091	.0089	.0087	.0084
−2.2	.0139	.0136	.0132	.0129	.0125	.0122	.0119	.0116	.0113	.0110
−2.1	.0179	.0174	.0170	.0166	.0162	.0158	.0154	.0150	.0146	.0143
−2.0	.0228	.0222	.0217	.0212	.0207	.0202	.0197	.0192	.0188	.0183
−1.9	.0287	.0281	.0274	.0268	.0262	.0256	.0250	.0244	.0239	.0233
−1.8	.0359	.0351	.0344	.0336	.0329	.0322	.0314	.0307	.0301	.0294
−1.7	.0446	.0436	.0427	.0418	.0409	.0401	.0392	.0384	.0375	.0367
−1.6	.0548	.0537	.0526	.0516	.0505	.0495	.0485	.0475	.0465	.0455
−1.5	.0668	.0655	.0643	.0630	.0618	.0606	.0594	.0582	.0571	.0559
−1.4	.0808	.0793	.0778	.0764	.0749	.0735	.0721	.0708	.0694	.0681
−1.3	.0968	.0951	.0934	.0918	.0901	.0885	.0869	.0853	.0838	.0823
−1.2	.1151	.1131	.1112	.1093	.1075	.1056	.1038	.1020	.1003	.0985
−1.1	.1357	.1335	.1314	.1292	.1271	.1251	.1230	.1210	.1190	.1170
−1.0	.1587	.1562	.1539	.1515	.1492	.1469	.1446	.1423	.1401	.1379
− .9	.1841	.1814	.1788	.1762	.1736	.1711	.1685	.1660	.1635	.1611
− .8	.2119	.2090	.2061	.2033	.2005	.1977	.1949	.1922	.1894	.1867
− .7	.2420	.2389	.2358	.2327	.2297	.2266	.2236	.2206	.2177	.2148
− .6	.2743	.2709	.2676	.2643	.2611	.2578	.2546	.2514	.2483	.2451
− .5	.3085	.3050	.3015	.2981	.2946	.2912	.2877	.2843	.2810	.2776
− .4	.3446	.3409	.3372	.3336	.3300	.3264	.3228	.3192	.3156	.3121
− .3	.3821	.3783	.3745	.3707	.3669	.3632	.3594	.3557	.3520	.3483
− .2	.4207	.4168	.4129	.4090	.4052	.4013	.3974	.3936	.3897	.3859
− .1	.4602	.4562	.4522	.4483	.4443	.4404	.4364	.4325	.4286	.4247
− .0	.5000	.4960	.4920	.4880	.4840	.4801	.4761	.4721	.4681	.4641

(continues)

Z	.00	.01	.02	.03	.04	.05	.06	.07	.08	.09
.0	.5000	.5040	.5080	.5120	.5160	.5199	.5239	.5279	.5319	.5359
.1	.5398	.5438	.5478	.5517	.5557	.5596	.5636	.5675	.5714	.5753
.2	.5793	.5832	.5871	.5910	.5948	.5987	.6026	.6064	.6103	.6141
.3	.6179	.6217	.6255	.6293	.6331	.6368	.6406	.6443	.6480	.6517
.4	.6554	.6591	.6628	.6664	.6700	.6736	.6772	.6808	.6844	.6879
.5	.6915	.6950	.6985	.7019	.7054	.7088	.7123	.7157	.7190	.7224
.6	.7257	.7291	.7324	.7357	.7389	.7422	.7454	.7486	.7517	.7549
.7	.7580	.7611	.7642	.7673	.7703	.7734	.7764	.7794	.7823	.7852
.8	.7881	.7910	.7939	.7967	.7995	.8023	.8051	.8078	.8106	.8133
.9	.8159	.8186	.8212	.8238	.8264	.8289	.8315	.8340	.8365	.8389
1.0	.8413	.8438	.8461	.8485	.8508	.8531	.8554	.8577	.8599	.8621
1.1	.8643	.8665	.8686	.8708	.8729	.8749	.8770	.8790	.8810	.8830
1.2	.8849	.8869	.8888	.8907	.8925	.8944	.8962	.8980	.8997	.9015
1.3	.9032	.9049	.9066	.9082	.9099	.9115	.9131	.9147	.9162	.9177
1.4	.9192	.9207	.9222	.9236	.9251	.9265	.9279	.9292	.9306	.9319
1.5	.9332	.9345	.9357	.9370	.9382	.9394	.9406	.9418	.9429	.9441
1.6	.9452	.9463	.9474	.9484	.9495	.9505	.9515	.9525	.9535	.9545
1.7	.9554	.9564	.9573	.9582	.9591	.9599	.9608	.9616	.9625	.9633
1.8	.9641	.9649	.9656	.9664	.9671	.9678	.9686	.9693	.9699	.9706
1.9	.9713	.9719	.9726	.9732	.9738	.9744	.9750	.9756	.9761	.9767
2.0	.9772	.9778	.9783	.9788	.9793	.9798	.9803	.9808	.9812	.9817
2.1	.9821	.9826	.9830	.9834	.9838	.9842	.9846	.9850	.9854	.9857
2.2	.9861	.9864	.9868	.9871	.9875	.9878	.9881	.9884	.9887	.9890
2.3	.9893	.9896	.9898	.9901	.9904	.9906	.9909	.9911	.9913	.9916
2.4	.9918	.9920	.9922	.9925	.9927	.9929	.9931	.9932	.9934	.9936
2.5	.9938	.9940	.9941	.9943	.9945	.9946	.9948	.9949	.9951	.9952
2.6	.9953	.9955	.9956	.9957	.9959	.9960	.9961	.9962	.9963	.9964
2.7	.9965	.9966	.9967	.9968	.9969	.9970	.9971	.9972	.9973	.9974
2.8	.9974	.9975	.9976	.9977	.9977	.9978	.9979	.9979	.9980	.9981
2.9	.9981	.9982	.9982	.9983	.9984	.9984	.9985	.9985	.9986	.9986
3.0	.9987	.9987	.9987	.9988	.9988	.9989	.9989	.9989	.9990	.9990
3.1	.9990	.9991	.9991	.9991	.9992	.9992	.9992	.9992	.9993	.9993
3.2	.9993	.9993	.9994	.9994	.9994	.9994	.9994	.9995	.9995	.9995
3.3	.9995	.9995	.9995	.9996	.9996	.9996	.9996	.9996	.9996	.9997
3.4	.9997	.9997	.9997	.9997	.9997	.9997	.9997	.9997	.9997	.9998

Commonly Used Z-Table Values

Confidence Intervals	**Confidence**	**80%**	**90%**	**95%**	**98%**	**99%**	**99.9%**	
	Z-value	±1.28	±1.65	±1.96	±2.33	±2.58	±3.29	
Hypothesis Testing	**Type I Error Risk = α**	**.20**	**.10**	**.05**	**.02**	**.01**	**.001**
	Two-tailed test Z-value	±1.28	±1.65	±1.96	±2.33	±2.58	±3.29
	One-tailed test Z-value*	.84	1.28	1.65	2.05	2.33	3.09

*Minus sign needs to be provided for lower-tailed tests.

APPENDIX D THE EXPONENTIAL DISTRIBUTION TABLE

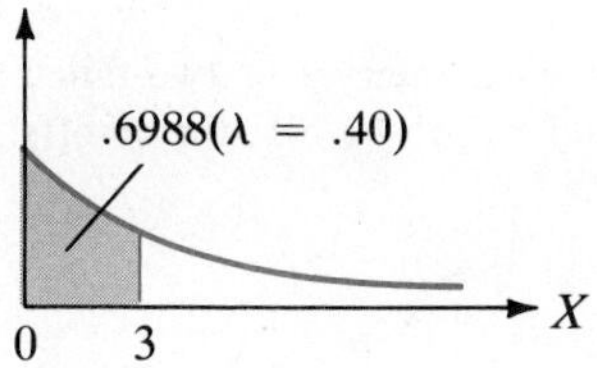

Entries in body of table are cumulative probabilities from 0 to X (see shaded portion of figure) for selected combinations of λ and X.

Example: $P(X \leq 3)$ when $\lambda = .40$ is .6988. (Table values are equivalent to $1 - e^{-y}$, where $y = \lambda x$.)

	X										
λ	**.20**	**.50**	**1.0**	**1.5**	**2.0**	**3.0**	**4.0**	**5.0**	**6.0**	**8.0**	**10.0**
.05	.0100	.0247	.0488	.0723	.0952	.1393	.1813	.2212	.2592	.3297	.3935
.10	.0198	.0488	.0952	.1393	.1813	.2592	.3297	.3935	.4512	.5507	.6321
.15	.0296	.0723	.1393	.2015	.2592	.3624	.4512	.5276	.5934	.6988	.7769
.20	.0392	.0952	.1813	.2592	.3297	.4512	.5507	.6321	.6988	.7981	.8647
.25	.0488	.1175	.2212	.3127	.3935	.5276	.6321	.7135	.7769	.8647	.9179
.30	.0582	.1393	.2592	.3624	.4512	.5934	.6988	.7769	.8347	.9093	.9502
.40	.0769	.1813	.3297	.4512	.5507	.6988	.7981	.8647	.9093	.9592	.9817
.50	.0952	.2212	.3935	.5276	.6321	.7769	.8647	.9179	.9502	.9817	.9933
.60	.1131	.2592	.4512	.5934	.6988	.8347	.9093	.9502	.9727	.9918	.9975
.70	.1306	.2953	.5034	.6501	.7534	.8775	.9392	.9698	.9850	.9963	.9991
.80	.1479	.3297	.5507	.6988	.7981	.9093	.9592	.9817	.9918	.9983	.9997
.90	.1647	.3624	.5934	.7408	.8347	.9328	.9727	.9889	.9955	.9993	.9999
1.0	.1813	.3935	.6321	.7769	.8647	.9502	.9817	.9933	.9975	.9997	1
1.5	.2592	.5276	.7769	.8946	.9502	.9889	.9975	.9994	.9999	1	1
2.0	.3297	.6321	.8647	.9502	.9817	.9975	.9997	1	1	1	1
2.5	.3935	.7135	.9179	.9765	.9933	.9994	1	1	1	1	1
3.0	.4512	.7769	.9502	.9889	.9975	.9999	1	1	1	1	1
4.0	.5507	.8647	.9817	.9975	.9997	1	1	1	1	1	1
5.0	.6321	.9179	.9933	.9994	1	1	1	1	1	1	1
6.0	.6988	.9502	.9975	.9999	1	1	1	1	1	1	1
8.0	.7981	.9817	.9997	1	1	1	1	1	1	1	1
10	.8647	.9933	1	1	1	1	1	1	1	1	1

Source: Computed by R. E. Shiffler and A. J. Adams.

APPENDIX E *t*-TABLE CUTOFF VALUES

For confidence intervals and two-tailed hypothesis tests, find the risk column of interest, α, at the top of the table; for one-tailed tests, find the risk column at the bottom. The cutoff value is then found at the intersection of the risk column and the *df* row.

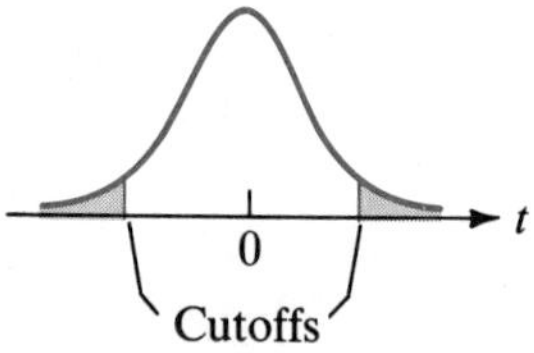

Example: For a confidence interval or two-tailed test where $df = 10$ and α is to be .05, the cutoffs are $t = \pm 2.228$.

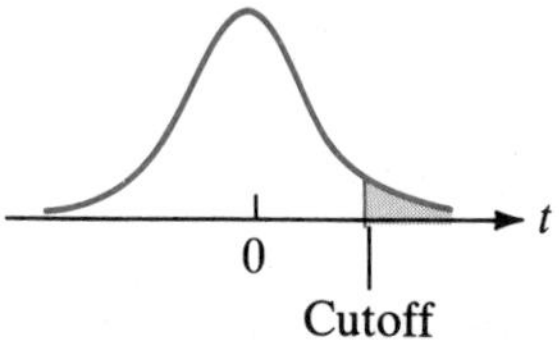

Example: For an upper-tailed H_a with $\alpha = .10$ and $df = 9$, the cutoff is $t = 1.383$. Since all values are given as positive, a minus sign needs to be provided if H_a is lower tailed.

Degrees of Freedom, *df*	Two-Tailed α (Probability in Both Tails Combined)					
	.20	.10	.05	.02	.01	.001
1	3.078	6.314	12.706	31.821	63.657	636.619
2	1.886	2.920	4.303	6.965	9.925	31.598
3	1.638	2.353	3.182	4.541	5.841	12.941
4	1.533	2.132	2.776	3.747	4.604	8.610
5	1.476	2.015	2.571	3.365	4.032	6.859
6	1.440	1.943	2.447	3.143	3.707	5.959
7	1.415	1.895	2.365	2.998	3.499	5.405
8	1.397	1.860	2.306	2.896	3.355	5.041
9	1.383	1.833	2.262	2.821	3.250	4.781
10	1.372	1.812	2.228	2.764	3.169	4.587
11	1.363	1.796	2.201	2.718	3.106	4.437
12	1.356	1.782	2.179	2.681	3.055	4.318
13	1.350	1.771	2.160	2.650	3.012	4.221
14	1.345	1.761	2.145	2.624	2.977	4.140
15	1.341	1.753	2.131	2.602	2.947	4.073
16	1.337	1.746	2.120	2.583	2.921	4.015
17	1.333	1.740	2.110	2.567	2.898	3.965
18	1.330	1.734	2.101	2.552	2.878	3.922
19	1.328	1.729	2.093	2.539	2.861	3.883
20	1.325	1.725	2.086	2.528	2.845	3.850
21	1.323	1.721	2.080	2.518	2.831	3.819
22	1.321	1.717	2.074	2.508	2.819	3.792
23	1.319	1.714	2.069	2.500	2.807	3.767
24	1.318	1.711	2.064	2.492	2.797	3.745
25	1.316	1.708	2.060	2.485	2.787	3.725
26	1.315	1.706	2.056	2.479	2.779	3.707
27	1.314	1.703	2.052	2.473	2.771	3.690
28	1.313	1.701	2.048	2.467	2.763	3.674
29	1.311	1.699	2.045	2.462	2.756	3.659
≥30	1.282	1.645	1.960	2.326	2.576	3.291
	.10	.05	.025	.01	.005	.0005
	One-Tailed α (Probability in One Tail Only)					

Source: Computed by R. E. Shiffler and A. J. Adams.

APPENDIX F CUTOFF VALUES FOR SIGNED RANKS TEST

After stating H_0 and H_a, rank the observations and then verify that T_+ and T_- sum to $n(n + 1)/2$. For two-tailed tests, find the risk column of interest, α, at the top of the table; for one-tailed tests, use the α at the bottom. For all cases, reject H_0 if the sum of interest (T_+ or T_-) is equal to or less than the table value.

Two-Tailed Tests: Reject H_0 if the smaller of T_+ or T_- is equal to or less than the table value.

Example: for $n = 10$, two-tailed, with $\alpha = .05$, if the smaller total were 11, we would accept H_0 since $11 > 8$.

Upper-Tailed H_a: Reject H_0 if T_- is equal to or less than the table value.

Example: For $n = 7$, upper tailed, with $\alpha = .05$, if T_- were 3, we would reject H_0 since $3 \leq 3$.

Lower-Tailed H_a: Reject H_0 if T_+ is equal to or less than the table value.

Sample Size, n	Two-Tailed α .20	.10	.05	.02	.01
4	0	*	*	*	*
5	2	0	*	*	*
6	3	2	0	*	*
7	5	3	2	0	*
8	8	5	3	1	0
9	10	8	5	3	1
10	14	10	8	5	3
11	17	13	10	7	5
12	21	17	13	9	7
13	26	21	17	12	9
14	31	25	21	15	12
15	36	30	25	19	15
16	42	35	29	23	19
17	48	41	34	27	23
18	55	47	40	32	27
19	62	53	46	37	32
20	69	60	52	43	37
21	77	67	58	49	42
22	86	75	65	55	48
23	94	83	73	62	54
24	104	91	81	69	61
25	113	100	89	76	68
26	124	110	98	84	75
27	134	119	107	92	83
28	145	130	116	101	91
29	157	140	126	110	100
30	169	151	137	120	109
One-Tailed α	.10	.05	.025	.01	.005

*Sample size is not large enough to reject H_0 at indicated α.

Source: Adapted with permission from D. K. Hildebrand and L. Ott, *Statistical Thinking for Managers*, Boston: PWS-KENT Publishers, © 1987.

APPENDIX G CHI-SQUARE DISTRIBUTION CUTOFF VALUES

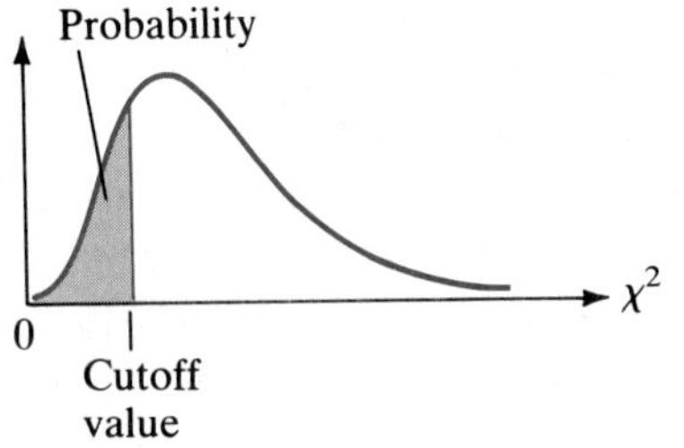

Note: Entries on this page are cutoff values to place a specified probability amount in the left tail. For example, to have probability = .05 in the left tail when *df* = 9, the table value is χ^2 = 3.325.

Degrees of Freedom, *df*	Probability in Left Tail				
	.005	.01	.025	.05	.10
1	0.0000393	0.000157	0.000982	0.00393	0.0158
2	0.01003	0.02010	0.05063	0.1026	0.211
3	0.0717	0.1148	0.2158	0.352	0.584
4	0.207	0.297	0.484	0.711	1.064
5	0.412	0.554	0.831	1.145	1.610
6	0.676	0.872	1.237	1.635	2.204
7	0.989	1.239	1.690	2.167	2.833
8	1.344	1.646	2.180	2.733	3.490
9	1.735	2.088	2.700	3.325	4.168
10	2.156	2.558	3.247	3.940	4.865
11	2.603	3.053	3.816	4.575	5.578
12	3.074	3.571	4.404	5.226	6.304
13	3.565	4.107	5.009	5.892	7.042
14	4.075	4.660	5.629	6.571	7.790
15	4.601	5.229	6.262	7.261	8.547
16	5.142	5.812	6.908	7.962	9.312
17	5.697	6.408	7.564	8.672	10.085
18	6.265	7.015	8.231	9.390	10.865
19	6.844	7.633	8.907	10.117	11.651
20	7.434	8.260	9.591	10.851	12.443
21	8.034	8.897	10.283	11.591	13.240
22	8.643	9.542	10.982	12.338	14.041
23	9.260	10.196	11.689	13.091	14.848
24	9.886	10.856	12.401	13.848	15.659
25	10.520	11.524	13.120	14.611	16.473
26	11.160	12.198	13.844	15.379	17.292
27	11.808	12.879	14.573	16.151	18.114
28	12.461	13.565	15.308	16.928	18.939
29	13.121	14.256	16.047	17.708	19.768
30	13.787	14.953	16.791	18.493	20.599
50	27.991	29.707	32.357	34.764	37.689
60	35.535	37.485	40.482	43.188	46.459
80	51.172	53.540	57.153	60.391	64.278
100	67.328	70.065	74.222	77.929	82.358

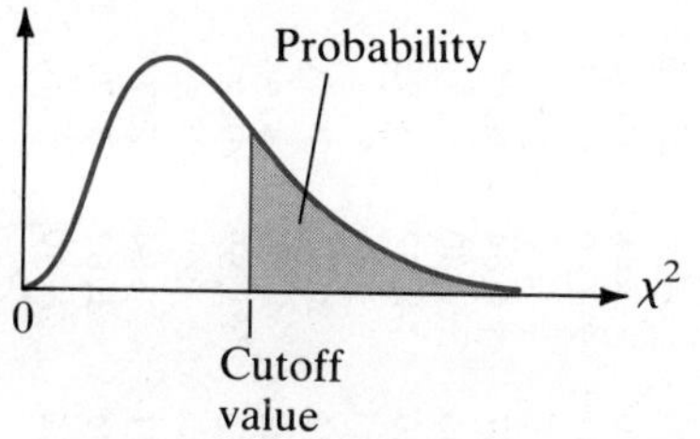

Note: Entries on this page are cutoff values to place a specified probability amount in the right tail. For example, to have probability = .10 in the right tail when df = 4, the table value is χ^2 = 7.779.

Degrees of Freedom, df	Probability in Right Tail				
	.10	.05	.025	.01	.005
1	2.706	3.841	5.024	6.635	7.879
2	4.605	5.991	7.378	9.210	10.597
3	6.251	7.815	9.348	11.345	12.838
4	7.779	9.488	11.143	13.277	14.860
5	9.236	11.070	12.833	15.086	16.750
6	10.645	12.592	14.449	16.812	18.548
7	12.017	14.067	16.013	18.475	20.278
8	13.362	15.507	17.535	20.090	21.955
9	14.684	16.919	19.023	21.666	23.589
10	15.987	18.307	20.483	23.209	25.188
11	17.275	19.675	21.920	24.725	26.757
12	18.549	21.026	23.337	26.217	28.300
13	19.812	22.362	24.736	27.688	29.819
14	21.064	23.685	26.119	29.141	31.319
15	22.307	24.996	27.488	30.578	32.801
16	23.542	26.296	28.845	32.000	34.267
17	24.769	27.587	30.191	33.409	35.718
18	25.989	28.869	31.526	34.805	37.156
19	27.204	30.144	32.852	36.191	38.582
20	28.412	31.410	34.170	37.566	39.997
21	29.615	32.671	35.479	38.932	41.401
22	30.813	33.924	36.781	40.289	42.796
23	32.007	35.172	38.076	41.638	44.181
24	33.196	36.415	39.364	42.980	45.558
25	34.382	37.652	40.647	44.314	46.928
26	35.563	38.885	41.923	45.642	48.290
27	36.741	40.113	43.194	46.963	49.645
28	37.916	41.337	44.461	48.278	50.993
29	39.087	42.557	45.722	49.588	52.336
30	40.256	43.773	46.979	50.892	53.672
50	63.167	67.505	71.420	76.154	79.490
60	74.397	79.082	83.298	88.379	91.952
80	96.578	101.879	106.629	112.329	116.321
100	118.498	124.342	129.561	135.807	140.169

For two-tailed hypothesis tests and confidence intervals for σ^2 (Chapter 12), obtain cutoff points from both tails. An upper-tailed inference requires only a right-tail value; vice-versa if H_a is lower tailed. For goodness-of-fit tests and tests for statistical independence (Chapter 19), use right-tail values only.

Source: Computed by R. E. Shiffler and A. J. Adams.

APPENDIX H F-DISTRIBUTION CUTOFF VALUES

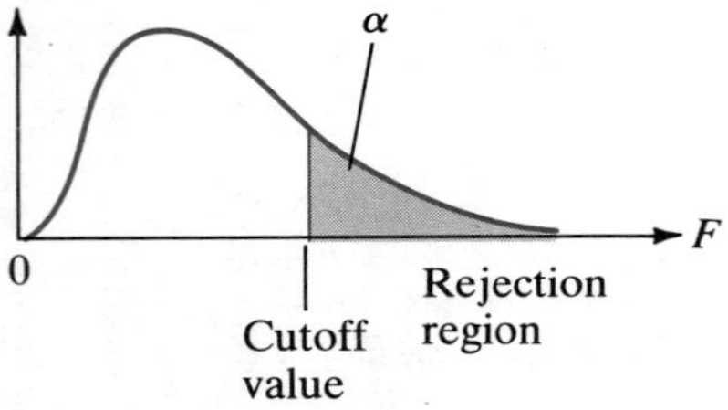

Note: Entries in this table are cutoff values to place probability α in the rejection region. For example, to have α = .05 when numerator df = 2 and denominator df = 7, the table value is F = 4.74.

Denominator *df*	α	Numerator *df* 1	2	3	4	5	6	7	8	10	12	15	20	40	120
1	.10	39.86	49.50	53.59	55.83	57.24	58.20	58.91	59.44	60.19	60.71	61.22	61.74	62.53	63.06
	.05	161.4	199.5	215.7	224.6	230.2	234.0	236.8	238.9	241.9	243.9	245.9	248.0	251.1	253.3
	.01	4052	5000	5403	5625	5764	5859	5928	5982	6056	6106	6157	6209	6287	6339
2	.10	8.53	9.00	9.16	9.24	9.29	9.33	9.35	9.37	9.39	9.41	9.42	9.44	9.47	9.48
	.05	18.51	19.00	19.16	19.25	19.30	19.33	19.35	19.37	19.40	19.41	19.43	19.45	19.47	19.49
	.01	98.50	99.00	99.17	99.25	99.30	99.33	99.36	99.37	99.40	99.42	99.43	99.45	99.47	99.49
3	.10	5.54	5.46	5.39	5.34	5.31	5.28	5.27	5.25	5.23	5.22	5.20	5.18	5.16	5.14
	.05	10.13	9.55	9.28	9.12	9.01	8.94	8.89	8.85	8.79	8.74	8.70	8.66	8.59	8.55
	.01	34.12	30.82	29.46	28.71	28.24	27.91	27.67	27.49	27.23	27.05	26.87	26.69	26.41	26.22
4	.10	4.54	4.32	4.19	4.11	4.05	4.01	3.98	3.95	3.92	3.90	3.87	3.84	3.80	3.78
	.05	7.71	6.94	6.59	6.39	6.26	6.16	6.09	6.04	5.96	5.91	5.86	5.80	5.72	5.66
	.01	21.20	18.00	16.69	15.98	15.52	15.21	14.98	14.80	14.55	14.37	14.20	14.02	13.75	13.56
5	.10	4.06	3.78	3.62	3.52	3.45	3.40	3.37	3.34	3.30	3.27	3.24	3.21	3.16	3.12
	.05	6.61	5.79	5.41	5.19	5.05	4.95	4.88	4.82	4.74	4.68	4.62	4.56	4.46	4.40
	.01	16.26	13.27	12.06	11.39	10.97	10.67	10.46	10.29	10.05	9.89	9.72	9.55	9.29	9.11
6	.10	3.78	3.46	3.29	3.18	3.11	3.05	3.01	2.98	2.94	2.90	2.87	2.84	2.78	2.74
	.05	5.99	5.14	4.76	4.53	4.39	4.28	4.21	4.15	4.06	4.00	3.94	3.87	3.77	3.70
	.01	13.75	10.92	9.78	9.15	8.75	8.47	8.26	8.10	7.87	7.72	7.56	7.40	7.14	6.97
7	.10	3.59	3.26	3.07	2.96	2.88	2.83	2.78	2.75	2.70	2.67	2.63	2.59	2.54	2.49
	.05	5.59	4.74	4.35	4.12	3.97	3.87	3.79	3.73	3.64	3.57	3.51	3.44	3.34	3.27
	.01	12.25	9.55	8.45	7.85	7.46	7.19	6.99	6.84	6.62	6.47	6.31	6.16	5.91	5.74

8	.10	3.46	3.11	2.92	2.81	2.73	2.67	2.62	2.59	2.54	2.50	2.46	2.42	2.36	2.32
	.05	5.32	4.46	4.07	3.84	3.69	3.58	3.50	3.44	3.35	3.28	3.22	3.15	3.04	2.97
	.01	11.26	8.65	7.59	7.01	6.63	6.37	6.18	6.03	5.81	5.67	5.52	5.36	5.12	4.95
9	.10	3.36	3.01	2.81	2.69	2.61	2.55	2.51	2.47	2.42	2.38	2.34	2.30	2.23	2.18
	.05	5.12	4.26	3.86	3.63	3.48	3.37	3.29	3.23	3.14	3.07	3.01	2.94	2.83	2.75
	.01	10.56	8.02	6.99	6.42	6.06	5.80	5.61	5.47	5.26	5.11	4.96	4.81	4.57	4.40
10	.10	3.29	2.92	2.73	2.61	2.52	2.46	2.41	2.38	2.32	2.28	2.24	2.20	2.13	2.08
	.05	4.96	4.10	3.71	3.48	3.33	3.22	3.14	3.07	2.98	2.91	2.85	2.77	2.66	2.58
	.01	10.04	7.56	6.55	5.99	5.64	5.39	5.20	5.06	4.85	4.71	4.56	4.41	4.17	4.00
11	.10	3.23	2.86	2.66	2.54	2.45	2.39	2.34	2.30	2.25	2.21	2.17	2.12	2.05	2.00
	.05	4.84	3.98	3.59	3.36	3.20	3.09	3.01	2.95	2.85	2.79	2.72	2.65	2.53	2.45
	.01	9.65	7.21	6.22	5.67	5.32	5.07	4.89	4.74	4.54	4.40	4.25	4.10	3.86	3.69
12	.10	3.18	2.81	2.61	2.48	2.39	2.33	2.28	2.24	2.19	2.15	2.10	2.06	1.99	1.93
	.05	4.75	3.89	3.49	3.26	3.11	3.00	2.91	2.85	2.75	2.69	2.62	2.54	2.43	2.34
	.01	9.33	6.93	5.95	5.41	5.06	4.82	4.64	4.50	4.30	4.16	4.01	3.86	3.62	3.45
13	.10	3.14	2.76	2.56	2.43	2.35	2.28	2.23	2.20	2.14	2.10	2.05	2.01	1.93	1.88
	.05	4.67	3.81	3.41	3.18	3.03	2.92	2.83	2.77	2.67	2.60	2.53	2.46	2.34	2.25
	.01	9.07	6.70	5.74	5.21	4.86	4.62	4.44	4.30	4.10	3.96	3.82	3.66	3.43	3.25
14	.10	3.10	2.73	2.52	2.39	2.31	2.24	2.19	2.15	2.10	2.05	2.01	1.96	1.89	1.83
	.05	4.60	3.74	3.34	3.11	2.96	2.85	2.76	2.70	2.60	2.53	2.46	2.39	2.27	2.18
	.01	8.86	6.51	5.56	5.04	4.69	4.46	4.28	4.14	3.94	3.80	3.66	3.51	3.27	3.09
15	.10	3.07	2.70	2.49	2.36	2.27	2.21	2.16	2.12	2.06	2.02	1.97	1.92	1.85	1.79
	.05	4.54	3.68	3.29	3.06	2.90	2.79	2.71	2.64	2.54	2.48	2.40	2.33	2.20	2.11
	.01	8.68	6.36	5.42	4.89	4.56	4.32	4.14	4.00	3.80	3.67	3.52	3.37	3.13	2.96
16	.10	3.05	2.67	2.46	2.33	2.24	2.18	2.13	2.09	2.03	1.99	1.94	1.89	1.81	1.75
	.05	4.49	3.63	3.24	3.01	2.85	2.74	2.66	2.59	2.49	2.42	2.35	2.28	2.15	2.06
	.01	8.53	6.23	5.29	4.77	4.44	4.20	4.03	3.89	3.69	3.55	3.41	3.26	3.02	2.84
18	.10	3.01	2.62	2.42	2.29	2.20	2.13	2.08	2.04	1.98	1.93	1.89	1.84	1.75	1.69
	.05	4.41	3.55	3.16	2.93	2.77	2.66	2.58	2.51	2.41	2.34	2.27	2.19	2.06	1.97
	.01	8.29	6.01	5.09	4.58	4.25	4.01	3.84	3.71	3.51	3.37	3.23	3.08	2.84	2.66
20	.10	2.97	2.59	2.38	2.25	2.16	2.09	2.04	2.00	1.94	1.89	1.84	1.79	1.71	1.64
	.05	4.35	3.49	3.10	2.87	2.71	2.60	2.51	2.45	2.35	2.28	2.20	2.12	1.99	1.90
	.01	8.10	5.85	4.94	4.43	4.10	3.87	3.70	3.56	3.37	3.23	3.09	2.94	2.69	2.52

(continues)

APPENDIX H Continued

Denominator df	α	Numerator df 1	2	3	4	5	6	7	8	10	12	15	20	40	120
22	.10	2.95	2.56	2.35	2.22	2.13	2.06	2.01	1.97	1.90	1.86	1.81	1.76	1.67	1.60
	.05	4.30	3.44	3.05	2.82	2.66	2.55	2.46	2.40	2.30	2.23	2.15	2.07	1.94	1.84
	.01	7.95	5.72	4.82	4.31	3.99	3.76	3.59	3.45	3.26	3.12	2.98	2.83	2.58	2.40
24	.10	2.93	2.54	2.33	2.19	2.10	2.04	1.98	1.94	1.88	1.83	1.78	1.73	1.64	1.57
	.05	4.26	3.40	3.01	2.78	2.62	2.51	2.42	2.36	2.25	2.18	2.11	2.03	1.89	1.79
	.01	7.82	5.61	4.72	4.22	3.90	3.67	3.50	3.36	3.17	3.03	2.89	2.74	2.49	2.31
26	.10	2.91	2.52	2.31	2.17	2.08	2.01	1.96	1.92	1.86	1.81	1.76	1.71	1.61	1.54
	.05	4.23	3.37	2.98	2.74	2.59	2.47	2.39	2.32	2.22	2.15	2.07	1.99	1.85	1.75
	.01	7.72	5.53	4.64	4.14	3.82	3.59	3.42	3.29	3.09	2.96	2.81	2.66	2.42	2.23
28	.10	2.89	2.50	2.29	2.16	2.06	2.00	1.94	1.90	1.84	1.79	1.74	1.69	1.59	1.52
	.05	4.20	3.34	2.95	2.71	2.56	2.45	2.36	2.29	2.19	2.12	2.04	1.96	1.82	1.71
	.01	7.64	5.45	4.57	4.07	3.75	3.53	3.36	3.23	3.03	2.90	2.75	2.60	2.35	2.17
30	.10	2.88	2.49	2.28	2.14	2.05	1.98	1.93	1.88	1.82	1.77	1.72	1.67	1.57	1.50
	.05	4.17	3.32	2.92	2.69	2.53	2.42	2.33	2.27	2.16	2.09	2.01	1.93	1.79	1.68
	.01	7.56	5.39	4.51	4.02	3.70	3.47	3.30	3.17	2.98	2.84	2.70	2.55	2.30	2.11
40	.10	2.84	2.44	2.23	2.09	2.00	1.93	1.87	1.83	1.76	1.71	1.66	1.61	1.51	1.42
	.05	4.08	3.23	2.84	2.61	2.45	2.34	2.25	2.18	2.08	2.00	1.92	1.84	1.69	1.58
	.01	7.31	5.18	4.31	3.83	3.51	3.29	3.12	2.99	2.80	2.66	2.52	2.37	2.11	1.92
60	.10	2.79	2.39	2.18	2.04	1.95	1.87	1.82	1.77	1.71	1.66	1.60	1.54	1.44	1.35
	.05	4.00	3.15	2.76	2.53	2.37	2.25	2.17	2.10	1.99	1.92	1.84	1.75	1.59	1.47
	.01	7.08	4.98	4.13	3.65	3.34	3.12	2.95	2.82	2.63	2.50	2.35	2.20	1.94	1.73
120	.10	2.75	2.35	2.13	1.99	1.90	1.82	1.77	1.72	1.65	1.60	1.55	1.48	1.37	1.26
	.05	3.92	3.07	2.68	2.45	2.29	2.17	2.09	2.02	1.91	1.83	1.75	1.66	1.50	1.35
	.01	6.85	4.79	3.95	3.48	3.17	2.96	2.79	2.66	2.47	2.34	2.19	2.03	1.76	1.53
∞	.10	2.71	2.30	2.08	1.94	1.85	1.77	1.72	1.67	1.60	1.55	1.49	1.42	1.30	1.17
	.05	3.84	3.00	2.60	2.37	2.21	2.10	2.01	1.94	1.83	1.75	1.67	1.57	1.39	1.22
	.01	6.63	4.61	3.78	3.32	3.02	2.80	2.64	2.51	2.32	2.18	2.04	1.88	1.59	1.32

Source: Computed by R. E. Shiffler and A. J. Adams.

APPENDIX I VALUES FOR Q FOR THE TUKEY *T* METHOD

$$T \text{ range} = Q\sqrt{\frac{MSE}{n_j}}$$

Note: Entries below are values of Q to place in the equation at left to determine the T range in ANOVA follow-up tests.

Example: For a follow-up analysis with $\alpha = .05$ when the F-ratio df are 2 and 9, the table value is $Q = 3.95$.

Denominator		Numerator *df*											
df	**α**	**1**	**2**	**3**	**4**	**5**	**6**	**7**	**8**	**9**	**10**	**12**	**18**
	.10	8.93	13.4	16.4	18.5	20.2	21.5	22.6	23.6	24.5	25.2	26.5	29.3
1	.05	18.0	27.0	32.8	37.1	40.4	43.1	45.4	47.4	49.1	50.6	53.2	58.8
	.01	90.0	135	164	186	202	216	227	237	246	253	266	294
	.10	4.13	5.73	6.77	7.54	8.14	8.63	9.05	9.41	9.72	10.0	10.5	11.5
2	.05	6.08	8.33	9.80	10.9	11.7	12.4	13.0	13.5	14.0	14.4	15.1	16.6
	.01	14.0	19.0	22.3	24.7	26.6	28.2	29.5	30.7	31.7	32.6	34.1	37.5
	.10	3.33	4.47	5.20	5.74	6.16	6.51	6.81	7.06	7.29	7.49	7.83	8.58
3	.05	4.50	5.91	6.82	7.50	8.04	8.48	8.85	9.18	9.46	9.72	10.2	11.1
	.01	8.26	10.6	12.2	13.3	14.2	15.0	15.6	16.2	16.7	17.1	17.9	19.5
	.10	3.01	3.98	4.59	5.03	5.39	5.68	5.93	6.14	6.33	6.49	6.78	7.41
4	.05	3.93	5.04	5.76	6.29	6.71	7.05	7.35	7.60	7.83	8.03	8.37	9.13
	.01	6.51	8.12	9.17	9.96	10.6	11.1	11.5	11.9	12.3	12.6	13.1	14.2
	.10	2.85	3.72	4.26	4.66	4.98	5.24	5.46	5.65	5.82	5.97	6.22	6.79
5	.05	3.64	4.60	5.22	5.67	6.03	6.33	6.58	6.80	6.99	7.17	7.47	8.12
	.01	5.70	6.97	7.80	8.42	8.91	9.32	9.67	9.97	10.2	10.5	10.9	11.8
	.10	2.75	3.56	4.07	4.44	4.73	4.97	5.17	5.34	5.50	5.64	5.87	6.40
6	.05	3.46	4.34	4.90	5.30	5.63	5.90	6.12	6.32	6.49	6.65	6.92	7.51
	.01	5.24	6.33	7.03	7.56	7.97	8.32	8.61	8.87	9.10	9.30	9.65	10.4
	.10	2.68	3.45	3.93	4.28	4.55	4.78	4.97	5.14	5.28	5.41	5.64	6.13
7	.05	3.34	4.16	4.68	5.06	5.36	5.61	5.82	6.00	6.16	6.30	6.55	7.10
	.01	4.95	5.92	6.54	7.01	7.37	7.68	7.94	8.17	8.37	8.55	8.86	9.55
	.10	2.63	3.37	3.83	4.17	4.43	4.65	4.83	4.99	5.13	5.25	5.46	5.93
8	.05	3.26	4.04	4.53	4.89	5.17	5.40	5.60	5.77	5.92	6.05	6.29	6.80
	.01	4.74	5.63	6.20	6.63	6.96	7.24	7.47	7.68	7.87	8.03	8.31	8.94
	.10	2.59	3.32	3.76	4.08	4.34	4.54	4.72	4.87	5.01	5.13	5.33	5.79
9	.05	3.20	3.95	4.41	4.76	5.02	5.24	5.43	5.59	5.74	5.87	6.09	6.58
	.01	4.60	5.43	5.96	6.35	6.66	6.91	7.13	7.32	7.49	7.65	7.91	8.49
	.10	2.56	3.27	3.70	4.02	4.26	4.47	4.64	4.78	4.91	5.03	5.23	5.67
10	.05	3.15	3.88	4.33	4.65	4.91	5.12	5.30	5.46	5.60	5.72	5.93	6.40
	.01	4.48	5.27	5.77	6.14	6.43	6.67	6.87	7.05	7.21	7.36	7.60	8.15
	.10	2.54	3.23	3.66	3.96	4.20	4.40	4.57	4.71	4.84	4.95	5.15	5.57
11	.05	3.11	3.82	4.26	4.57	4.82	5.03	5.20	5.35	5.49	5.61	5.81	6.27
	.01	4.39	5.14	5.62	5.97	6.25	6.48	6.67	6.84	6.99	7.13	7.36	7.88

(*continues*)

APPENDIX I Continued

Denominator df	α	Numerator df 1	2	3	4	5	6	7	8	9	10	12	18
12	.10	2.52	3.20	3.62	3.92	4.16	4.35	4.51	4.65	4.78	4.89	5.08	5.49
	.05	3.08	3.77	4.20	4.51	4.75	4.95	5.12	5.27	5.39	5.51	5.71	6.15
	.01	4.32	5.04	5.50	5.84	6.10	6.32	6.51	6.67	6.81	6.94	7.17	7.66
13	.10	2.50	3.18	3.59	3.88	4.12	4.30	4.46	4.60	4.72	4.83	5.02	5.43
	.05	3.06	3.73	4.15	4.45	4.69	4.88	5.05	5.19	5.32	5.43	5.63	6.05
	.01	4.26	4.96	5.40	5.73	5.98	6.19	6.37	6.53	6.67	6.79	7.01	7.48
14	.10	2.49	3.16	3.56	3.85	4.08	4.27	4.42	4.56	4.68	4.79	4.97	5.37
	.05	3.03	3.70	4.11	4.41	4.64	4.83	4.99	5.13	5.25	5.36	5.55	5.97
	.01	4.21	4.89	5.32	5.63	5.88	6.08	6.26	6.41	6.54	6.66	6.87	7.33
15	.10	2.48	3.14	3.54	3.83	4.05	4.23	4.39	4.52	4.64	4.75	4.93	5.32
	.05	3.01	3.67	4.08	4.37	4.59	4.78	4.94	5.08	5.20	5.31	5.49	5.90
	.01	4.17	4.83	5.25	5.56	5.80	5.99	6.16	6.31	6.44	6.55	6.76	7.20
16	.10	2.47	3.12	3.52	3.80	4.03	4.21	4.36	4.49	4.61	4.71	4.89	5.28
	.05	3.00	3.65	4.05	4.33	4.56	4.74	4.90	5.03	5.15	5.26	5.44	5.84
	.01	4.13	4.78	5.19	5.49	5.72	5.92	6.08	6.22	6.35	6.46	6.66	7.09
18	.10	2.45	3.10	3.49	3.77	3.98	4.16	4.31	4.44	4.55	4.65	4.83	5.21
	.05	2.97	3.61	4.00	4.28	4.49	4.67	4.82	4.96	5.07	5.17	5.35	5.74
	.01	4.07	4.70	5.09	5.38	5.60	5.79	5.94	6.08	6.20	6.31	6.50	6.91
20	.10	2.44	3.08	3.46	3.74	3.95	4.12	4.27	4.40	4.51	4.61	4.78	5.16
	.05	2.95	3.58	3.96	4.23	4.45	4.62	4.77	4.90	5.01	5.11	5.28	5.66
	.01	4.02	4.64	5.02	5.29	5.51	5.69	5.84	5.97	6.09	6.19	6.37	6.76
24	.10	2.42	3.05	3.42	3.69	3.90	4.07	4.21	4.34	4.44	4.54	4.71	5.07
	.05	2.92	3.53	3.90	4.17	4.37	4.54	4.68	4.81	4.92	5.01	5.18	5.55
	.01	3.96	4.54	4.91	5.17	5.37	5.54	5.69	5.81	5.92	6.02	6.19	6.56
30	.10	2.40	3.02	3.39	3.65	3.85	4.02	4.16	4.28	4.38	4.47	4.64	4.99
	.05	2.89	3.49	3.85	4.10	4.30	4.46	4.60	4.72	4.82	4.92	5.08	5.43
	.01	3.89	4.45	4.80	5.05	5.24	5.40	5.54	5.65	5.76	5.85	6.01	6.36
60	.10	2.36	2.96	3.31	3.56	3.75	3.91	4.04	4.16	4.25	4.34	4.49	4.82
	.05	2.83	3.40	3.74	3.98	4.16	4.31	4.44	4.55	4.65	4.73	4.88	5.20
	.01	3.76	4.28	4.60	4.82	4.99	5.13	5.25	5.36	5.45	5.53	5.67	5.98
120	.10	2.34	2.93	3.28	3.52	3.71	3.86	3.99	4.10	4.19	4.28	4.42	4.74
	.05	2.80	3.36	3.68	3.92	4.10	4.24	4.36	4.47	4.56	4.64	4.78	5.09
	.01	3.70	4.20	4.50	4.71	4.87	5.01	5.12	5.21	5.30	5.38	5.51	5.79
∞	.10	2.33	2.90	3.24	3.48	3.66	3.81	3.93	4.04	4.13	4.21	4.35	4.65
	.05	2.77	3.31	3.63	3.86	4.03	4.17	4.29	4.39	4.47	4.55	4.68	4.97
	.01	3.64	4.12	4.40	4.60	4.76	4.88	4.99	5.08	5.16	5.23	5.35	5.61

Source: Computed by R. E. Shiffler and A. J. Adams.

APPENDIX J LIN-MUDHOLKAR TEST FOR NORMALITY: CUTOFF VALUES OF *r*

This table contains the cutoff values of the test statistic r for testing the null hypothesis that the population from which a set of sample data was randomly selected is normal, against a two-tailed alternative hypothesis (see Chapter 17). The critical values are found by intersecting the appropriate sample size row with the α-risk column. All cutoffs are $\pm$ values.

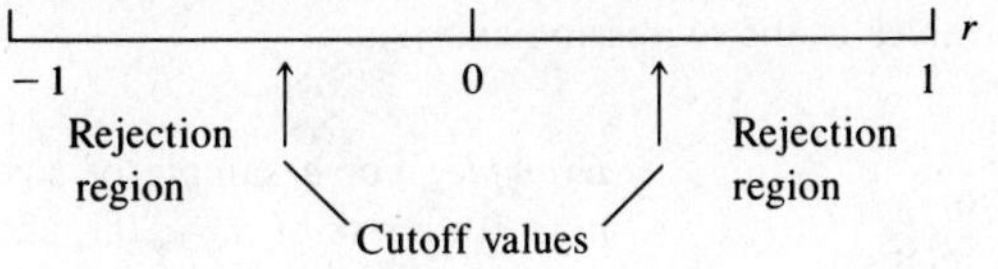

Example: For a sample size of $n = 20$ and a risk of $\alpha = .05$, the cutoff values are $r = \pm .615$.

Sample Size, n	Two-Tailed α (Total Probability in Rejection Region) .10	.05	.01
10	.705	.767	.844
11	.682	.746	.826
12	.661	.726	.809
13	.642	.708	.794
14	.625	.692	.780
15	.610	.677	.767
16	.596	.663	.755
17	.582	.650	.744
18	.570	.638	.733
19	.558	.626	.723
20	.548	.615	.713
21	.538	.605	.704
22	.528	.595	.695
23	.519	.586	.686
24	.510	.578	.678
25	.502	.569	.670
26	.494	.561	.663
27	.487	.553	.655
28	.480	.546	.648
29	.473	.539	.642
30	.467	.532	.635
35	.438	.502	.605
40	.414	.476	.579
45	.394	.455	.556
50	.376	.435	.535
60	.347	.403	.501
70	.324	.378	.472
80	.305	.356	.448
90	.289	.338	.427
100	.275	.322	.408

Source: "A Simple Test for Normality," Lloyd S. Nelson, *Journal of Quality Technology, 13* (No. 1, 1981):76–77. Reprinted with permission.

APPENDIX K DURBIN-WATSON TEST FOR AUTOCORRELATION: CUTOFF VALUES OF *d*

This table contains the cutoff values of the test statistic d (see Equation 17–4) for testing the null hypothesis that the residuals are not autocorrelated against the alternative hypothesis that the residuals are positively autocorrelated. The cutoff values are found by intersecting the appropriate sample size and α-risk row with the d_L and d_U columns. The symbol k represents the number of independent variables in the regression analysis.

Positive Autocorrelation	Inconclusive	No Autocorrelation
	d_L	d_U

Example: For a sample of size $n = 20$, a risk of $\alpha = .05$, and a multiple regression analysis with $k = 3$ independent variables, the cutoff values are $d_L = 1.00$ and $d_U = 1.68$.

		$k = 1$		$k = 2$		$k = 3$		$k = 4$		$k = 5$	
n	α	d_L	d_U	d_L	d_U	d_L	d_U	d_L	d_U	d_L	d_U
15	.05	1.08	1.36	0.95	1.54	0.82	1.75	0.69	1.97	0.56	2.21
	.01	0.81	1.07	0.70	1.25	0.59	1.46	0.49	1.70	0.39	1.96
16	.05	1.10	1.37	0.98	1.54	0.86	1.73	0.74	1.93	0.62	2.15
	.01	0.84	1.09	0.74	1.25	0.63	1.44	0.53	1.66	0.44	1.90
17	.05	1.13	1.38	1.02	1.54	0.90	1.71	0.78	1.90	0.67	2.10
	.01	0.87	1.10	0.77	1.25	0.67	1.43	0.57	1.63	0.48	1.85
18	.05	1.16	1.39	1.05	1.53	0.93	1.69	0.82	1.87	0.71	2.06
	.01	0.90	1.12	0.80	1.26	0.71	1.42	0.61	1.60	0.52	1.80
19	.05	1.18	1.40	1.08	1.53	0.97	1.68	0.86	1.85	0.75	2.02
	.01	0.93	1.13	0.83	1.26	0.74	1.41	0.65	1.58	0.56	1.77
20	.05	1.20	1.41	1.10	1.54	1.00	1.68	0.90	1.83	0.79	1.99
	.01	0.95	1.15	0.86	1.27	0.77	1.41	0.68	1.57	0.60	1.74
21	.05	1.22	1.42	1.13	1.54	1.03	1.67	0.93	1.81	0.83	1.96
	.01	0.97	1.16	0.89	1.27	0.80	1.41	0.72	1.55	0.63	1.71
22	.05	1.24	1.43	1.15	1.54	1.05	1.66	0.96	1.80	0.86	1.94
	.01	1.00	1.17	0.91	1.28	0.83	1.40	0.75	1.54	0.66	1.69
23	.05	1.26	1.44	1.17	1.54	1.08	1.66	0.99	1.79	0.90	1.92
	.01	1.02	1.19	0.94	1.29	0.86	1.40	0.77	1.53	0.70	1.67
24	.05	1.27	1.45	1.19	1.55	1.10	1.66	1.01	1.78	0.93	1.90
	.01	1.04	1.20	0.96	1.30	0.88	1.41	0.80	1.53	0.72	1.66

APPENDIX K Continued

		k = 1		k = 2		k = 3		k = 4		k = 5	
n	α	d_L	d_U	d_L	d_U	d_L	d_U	d_L	d_U	d_L	d_U
25	.05	1.29	1.45	1.21	1.55	1.12	1.66	1.04	1.77	0.95	1.89
	.01	1.05	1.21	0.98	1.30	0.90	1.41	0.83	1.52	0.75	1.65
26	.05	1.30	1.46	1.22	1.55	1.14	1.65	1.06	1.76	0.98	1.88
	.01	1.07	1.22	1.00	1.31	0.93	1.41	0.85	1.52	0.78	1.64
27	.05	1.32	1.47	1.24	1.56	1.16	1.65	1.08	1.76	1.01	1.86
	.01	1.09	1.23	1.02	1.32	0.95	1.41	0.88	1.51	0.81	1.63
28	.05	1.33	1.48	1.26	1.56	1.18	1.65	1.10	1.75	1.03	1.85
	.01	1.10	1.24	1.04	1.32	0.97	1.41	0.90	1.51	0.83	1.62
29	.05	1.34	1.48	1.27	1.56	1.20	1.65	1.12	1.74	1.05	1.84
	.01	1.12	1.25	1.05	1.33	0.99	1.42	0.92	1.51	0.85	1.61
30	.05	1.35	1.49	1.28	1.57	1.21	1.65	1.14	1.74	1.07	1.83
	.01	1.13	1.26	1.07	1.34	1.01	1.42	0.94	1.51	0.88	1.61
35	.05	1.40	1.52	1.34	1.58	1.28	1.65	1.22	1.73	1.16	1.80
	.01	1.19	1.31	1.14	1.37	1.08	1.44	1.03	1.51	0.97	1.59
40	.05	1.44	1.54	1.39	1.60	1.34	1.66	1.29	1.72	1.23	1.79
	.01	1.25	1.34	1.20	1.40	1.15	1.46	1.10	1.52	1.05	1.58
45	.05	1.48	1.57	1.43	1.62	1.38	1.67	1.34	1.72	1.29	1.78
	.01	1.29	1.38	1.24	1.42	1.20	1.48	1.16	1.53	1.11	1.58
50	.05	1.50	1.59	1.46	1.63	1.42	1.67	1.38	1.72	1.34	1.77
	.01	1.32	1.40	1.28	1.45	1.24	1.49	1.20	1.54	1.16	1.59
60	.05	1.55	1.62	1.51	1.65	1.48	1.69	1.44	1.73	1.41	1.77
	.01	1.38	1.45	1.35	1.48	1.32	1.52	1.28	1.56	1.25	1.60
70	.05	1.58	1.64	1.55	1.67	1.52	1.70	1.49	1.74	1.46	1.77
	.01	1.43	1.49	1.40	1.52	1.37	1.55	1.34	1.58	1.31	1.61
80	.05	1.61	1.66	1.59	1.69	1.56	1.72	1.53	1.74	1.51	1.77
	.01	1.47	1.52	1.44	1.54	1.42	1.57	1.39	1.60	1.36	1.62
90	.05	1.63	1.68	1.61	1.70	1.59	1.73	1.57	1.75	1.54	1.78
	.01	1.50	1.54	1.47	1.56	1.45	1.59	1.43	1.61	1.41	1.64
100	.05	1.65	1.69	1.63	1.72	1.61	1.74	1.59	1.76	1.57	1.78
	.01	1.52	1.56	1.50	1.58	1.48	1.60	1.46	1.63	1.44	1.65

Source: Reprinted by permission of the Biometrika Trustees from J. Durbin and G. S. Watson, "Testing for Serial Correlation in Least Squares Regression II," *Biometrika,* Vol. 38 (1951), pp. 173 and 175.

APPENDIX L CUTOFF VALUES FOR SPEARMAN'S CORRELATION

To test H_0: $\rho_s = 0$, compare the sample value of r_s with the cutoff value for the proper sample size, H_a statement, and α. For a two-tailed H_a, use the α column and instructions at left below; for a one-tailed H_a, see right below.

Two-Tailed Tests: Reject H_0 if the absolute value of r_s exceeds the cutoff value.

Example: If H_a is to be $\rho_s \neq 0$ with $n = 10$ and $\alpha = .10$, the cutoff value is .564. If the sample data yield $r_s = .710$, reject H_0.

Upper-Tailed Test: Reject H_0 if $r_s >$ cutoff value.

Example: if H_a is that $\rho_s > 0$ with $n = 12$ and $\alpha = .05$, the cutoff value is .497. If $r_s = .411$, do not reject H_0.

Lower-Tailed Test: Reject H_0 if $r_s <$ cutoff value.

Example: if H_a is that $\rho_s < 0$, suppose r_s is computed to be $-.504$. If $n = 10$ and $\alpha = .01$, do not reject H_0 since $-.504$ is not more negative than $-.745$.

Sample Size, n	Two-Tailed α .10	.05	.02	.01
5	.900	*	*	*
6	.829	.886	.943	*
7	.714	.786	.893	*
8	.643	.738	.833	.881
9	.600	.683	.783	.833
10	.564	.648	.745	.794
11	.523	.623	.736	.818
12	.497	.591	.703	.780
13	.475	.566	.673	.745
14	.457	.545	.646	.716
15	.441	.525	.623	.689
16	.425	.507	.601	.666
17	.412	.490	.582	.645
18	.399	.476	.564	.625
19	.388	.462	.549	.608
20	.377	.450	.534	.591
21	.368	.438	.521	.576
22	.359	.428	.508	.562
23	.351	.418	.496	.549
24	.343	.409	.485	.537
25	.336	.400	.475	.526
26	.329	.392	.465	.515
27	.323	.385	.456	.505
28	.317	.377	.448	.496
29	.311	.370	.440	.487
30	.305	.364	.432	.478
One-Tailed α	.05	.025	.01	.005

*Sample size is not large enough to reject H_0 at this α.

Source: Computed by R. E. Shiffler and A. J. Adams.

APPENDIX M CUTOFF VALUES FOR THE RANK SUM TEST

To test H_0: $\mu_1 = \mu_2$, let n_1 be the smaller sample size; let T_1 be the rank sum of the n_1 observations. **Two-tailed tests:** reject H_0 if $T_1 \leq T_L$ or if $T_1 \geq T_U$. **One-tailed tests:** μ_1 must be associated with sample n_1. Then for H_a: $\mu_1 > \mu_2$, reject H_0 if $T_1 \geq T_U$. For H_a: $\mu_1 < \mu_2$, reject H_0 if $T_1 \leq T_L$.

$\alpha = .10$ (Two Tail) and $\alpha = .05$ (One Tail)

	n_1															
	3		4		5		6		7		8		9		10	
n_2	T_L	T_U	T_L	T_U	T_L	T_U	T_L	T_U	T_L	T_U	T_L	T_U	T_L	T_U	T_L	T_U
3	6	15	7	17	7	20	8	22	9	24	9	27	10	29	11	31
4	7	17	12	24	13	27	14	30	15	33	16	36	17	39	18	42
5	7	20	13	27	19	36	20	40	22	43	24	46	25	50	26	54
6	8	22	14	30	20	40	28	50	30	54	32	58	33	63	35	67
7	9	24	15	33	22	43	30	54	39	66	41	71	43	76	46	80
8	9	27	16	36	24	46	32	58	41	71	52	84	54	90	57	95
9	10	29	17	39	25	50	33	63	43	76	54	90	66	105	69	111
10	11	31	18	42	26	54	35	67	46	80	57	95	69	111	83	127

$\alpha = .05$ (Two Tail) and $\alpha = .025$ (One Tail)

	n_1															
	3		4		5		6		7		8		9		10	
n_2	T_L	T_U	T_L	T_U	T_L	T_U	T_L	T_U	T_L	T_U	T_L	T_U	T_L	T_U	T_L	T_U
3	5	16	6	18	6	21	7	23	7	26	8	28	8	31	9	33
4	6	18	11	25	12	28	12	32	13	35	14	38	15	41	16	44
5	6	21	12	28	18	37	19	41	20	45	21	49	22	53	24	56
6	7	23	12	32	19	41	26	52	28	56	29	61	31	65	32	70
7	7	26	13	35	20	45	28	56	37	68	39	73	41	78	43	83
8	8	28	14	38	21	49	29	61	39	73	49	87	51	93	54	98
9	8	31	15	41	22	53	31	65	41	78	51	93	63	108	66	114
10	9	33	16	44	24	56	32	70	43	83	54	98	66	114	79	131

APPENDIX N USING THE IBS SOFTWARE

N.1 Introduction

The IBS Software (short for Microcomputer Applications for Introductory Business Statistics) that accompanies this text is intended to help you understand business statistics better. For almost every chapter, we have designed what we call a *processor*. A processor accepts data, parameter values or summary statistics as input, processes the information, and displays the results of its computations as output.

We expect that you will find two main uses for the software. First, it may serve as a check on calculations that you have performed by hand or with a calculator. Second, as you develop expertise with the software, you may use it to solve problems directly. As with any software package, you first must learn how it works before you can expect it to perform as it is intended.

In the sleeve at the back of the book you should find two disks, one labeled Program Disk 1 and the second labeled Program Disk 2. Let us call these two disks included with the book the *master disks*.

System Requirements The following is a list of the requirements you will need to be able to use the accompanying IBS software:

1. An IBM personal computer, Personal System/2 or anything compatible. (Sorry, the IBS software will not run on Apple computers or an IBM PCjr.)
2. The computer must have two 5¼-inch disk drives or one 5¼-inch disk drive and a hard drive. The IBS software will run on a computer with one disk drive only, but this will require some switching of disks if you choose to save or retrieve data in a file. (The software also is available on a 3½-inch size diskette. If you prefer this size, ask your bookstore to order the 3½-inch disks from PWS-KENT Publishing Company.)
3. The computer must have at least 512K of accessible memory. (If you have less than 512K of memory, the IBS software will not be read from the Program Disk to your computer, and a message such as "Program too big to fit in memory" will appear.)
4. A version of DOS numbered 2.00 or higher. (DOS usually is purchased with the computer and often resides on a disk called the System Disk, the Boot Disk, or the DOS Disk. We cannot provide DOS with our software.) If you do not know what version of DOS you have, type 'ver' after the DOS prompt A> and then press the Enter or Return key. (Do *not* type the apostrophes surrounding 'ver'.)

Although it would be most convenient to have a printer attached to your computer, this is not necessary. We have provided a method to enable you to save results on a Data Disk (to be explained in the next subsection), and then print the results from the Data Disk when you are able to work on a computer with a printer attached. A color monitor is nice to work on since the IBS software produces colorful screens, but it is not necessary for our software.

Making a Backup Copy of the Program Disks We highly recommend that the first thing you do with the IBS software is to make a backup copy. Once made, the backup copies should be used and the master disks set aside in a safe place.

For floppy disk drive users: To make a backup copy, you will need three blank disks and your System Disk that has DOS on it. First, we must format the blank disks. Put your System Disk in drive A and turn on the computer. Either update the date and time or press the Enter or Return key twice. When the DOS prompt A> appears, insert a blank disk in drive B and type 'format B:' after the DOS prompt. (Leave a space between the word "format" and the letter "B", but do not leave a space between "B" and the colon ":".) Press the Enter or Return key and wait 30 to 45 seconds until you see a message such as "Format complete" and a question such as "Format another (Y/N)?". Remove the freshly formatted disk from drive B, insert another blank disk in drive B, and repeat the process until you have a total of three formatted disks. *Note:* If you are using a machine with a high density disk drive such as an IBM AT or a 286 compatible machine, then to format a double density disk in the high density drive, type 'format A:/4', assuming drive A is the high density drive.

Second, label the three formatted disks as follows: (1) Program Disk 1 Backup, (2) Program Disk 2 Backup, and (3) Data Disk. We eventually will need the third disk—the Data Disk—so we recommend creating it now.

Third, we must copy the contents of the master disks onto the backup disks. Remove the System Disk from drive A and insert the master disk called Program Disk 1 (the one from the sleeve in the back of the book). Insert the disk labeled Program Disk 1 Backup in drive B. At the DOS prompt A>, type 'copy A:*.* B:'. Leave a space between the word "copy" and the letter "A" and between the second asterisk and the letter "B". Press the Enter or Return key and wait until the DOS prompt A> reappears.

Fourth, remove both disks, insert the master disk labeled Program Disk 2 in drive A, and insert the disk labeled Program Disk 2 Backup in drive B. Again type 'copy A:*.* B:', press the Enter or Return key, and wait for the DOS prompt to reappear.

At this point you should have five disks: the two master disks, the two backup disks, and the Data Disk. Store the master disks.

For hard disk drive users: To copy the IBS software onto the hard disk, do the following. Make drive C the current drive and create a subdirectory, such as IBS, by typing 'md IBS' after the DOS prompt C:\>. Now change to that subdirectory by typing 'cd IBS.' Insert the master disks (one at a time) into drive A and copy the files to drive C. This will install the IBS software; the same subdirectory can serve as your data disk.

N.2 Getting Started

Loading the IBS Software We are now ready to begin using the IBS software. To load the software, follow these steps:

1. If necessary, remove the System (DOS) Disk from drive A and insert the disk labeled Program Disk 1 Backup. (*Hard drive users:* Make the IBS directory the current directory on your hard drive.)
2. Type 'start' (capital or lower case letters, it doesn't matter) after the DOS prompt A> and press the Enter or Return key.
3. At this point, the IBS software gives you the option of updating the date and time, as shown in Figure N.1. Either enter the date and time or press the Enter or Return key twice to begin loading the IBS software from the Program Disk 1 Backup into the computer's memory. *Note:* The second line in Figure N.1 shows "A>echo off". The echo off message appears automatically after we type 'start', as we did in the first line. The user does *not* need to type 'echo off'. (*Hard drive users:* You should see Figure N.2 appear briefly and then Figure N.4. Skip steps 4, 5, and 6.)
4. Wait until the instruction shown in Figure N.3 appears on the screen. It may take up to 45 seconds to read in the program from Program Disk 1 Backup, so be patient. The copyright screen in Figure N.2 will appear briefly before the one shown in Figure N.3.

```
A>start

A>echo off
Current date is Wed  8-16-1989
Enter new date:
Current time is  3:35:41.00
Enter new time:

Starting...Microcomputer Applications for INTRODUCTORY BUSINESS STATISTICS
Please wait ...
```

Figure N.1 Loading the IBS Software

```
Software written by John K. Hedges
                    Riverdale Software Systems

     in consultation with
                    Dr. Ronald Shiffler
                and Dr. Arthur Adams.

(C)Copyright 1990 by PWS-KENT Publishing.  All Rights Reserved.
No part of this publication may be reproduced or transmitted in
any form or by any means, electronic or mechanical, including
photocopy, recording, or any information storage or retrieval
system, without permission in writing from the publisher.

PWS-KENT Publishing Company, Inc.
Statler Office Building
20 Park Plaza                                             VBeta
Boston, MA  02116                                   Aug 18 1989
```

Figure N.2 Copyright Screen

Do *not* touch the computer until Figure N.3 appears. (*Hard drive users:* The message in Figure N.3 should *not* appear. If it does, then quit and check that both Program Disks were copied onto drive C and that the IBS directory C:\IBS\ is the current directory.)

5. When instructed by the message on the screen, remove the disk labeled Program Disk 1 Backup from drive A, insert the disk labeled Program Disk 2 Backup in drive A, and press the Enter or Return key. When Figure N.4 appears, the IBS software is ready to use. Do not remove Program Disk 2 Backup during your session.
6. Insert the Data Disk in drive B at this time, in case we might need it during our impending use of the software.

```
≡        Microcomputer Applications for INTRODUCTORY BUSINESS STATISTICS  READY
  Overview  Data  Files  Processors  Set Up  Exit                      | F1=Help




             ═══════════ Overlay file needed ═══════════
             Insert Program disk #2 into drive A & press a key.

                       Press Ctrl-Break to quit.
```

Figure N.3 Instruction to Switch Program Disks

```
≡        Microcomputer Applications for INTRODUCTORY BUSINESS STATISTICS  READY
 [Overview]  Data  Files  Processors  Set Up  Exit                     | F1=Help

     This program accompanies the
   INTRODUCTORY BUSINESS STATISTICS
   text by Shiffler and Adams.  It is
   to aid in learning about the appli-
   cation of statistics to business
   problems.

   Press the Enter key to continue or
  F8 to proceed directly to the Menu.
```

Figure N.4 First Overview Screen

General Information About Screen Layout Figure N.5 shows the first Overview screen for the IBS software. Refer to this figure as we discuss the major features of the screen.

1. Program Name. On the top line of the screen, we display the program name. For instance, in Figure N.5 the program name is "Microcomputer Applications for INTRODUCTORY BUSINESS STATISTICS." When we access a processor, the name of the selected processor will replace the program name.

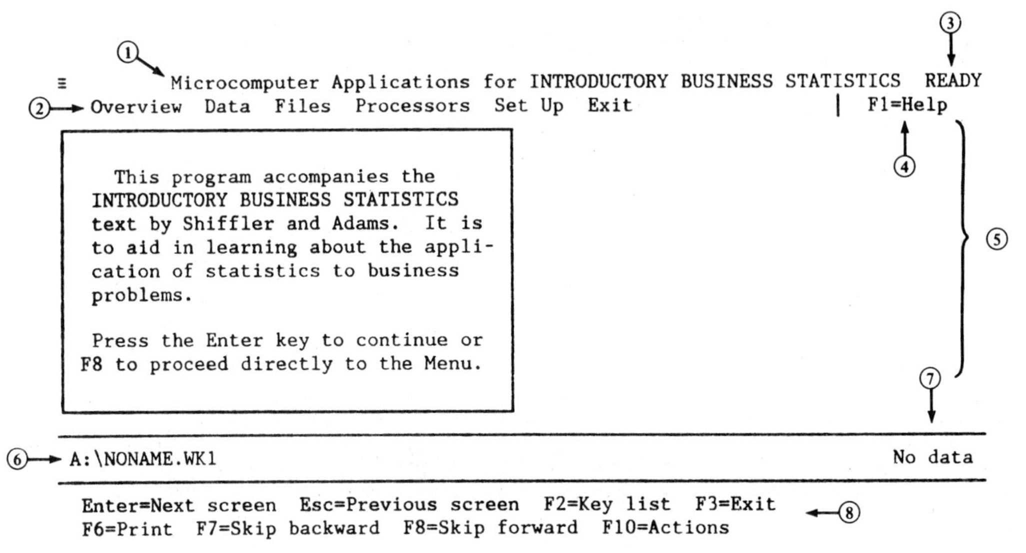

Figure N.5 General Screen Layout

2. Main Menu. The program main menu stretches horizontally across the screen on the second line. In Figure N.5 we see six options on the menu beginning with Overview and ending with Exit. ("Set Up" is the name of one option, not two.) To access one of the options, we must move the highlight box across the menu using the left or right arrow keys, and then press the Enter or Return key.

3. Status. In the upper right corner of the screen is the status of the program; it shows 'READY' in Figure N.5. Other messages such as 'MENU' or 'WAIT' also may appear depending on actions taken by the user. The software will accept a keyboard input, such as pressing the Enter or Return key, when the status is 'READY'. However, if you press a key when the 'WAIT' message is visible, the intended action of the key will be delayed or ignored altogether.

4. Help. Below the Status message is the notation 'Fl = Help'. This means that by pressing the Fl function key a "help screen" will pop up. Although we cannot anticipate all the problems that might arise while running the software, we have tried to provide additional comments in the help windows to further explain what happens next or what is required of the user.

5. Work Area. Between the top two lines and the bottom five lines, which always remain on the screen, is the work area. When we access an option from the menu, windows will appear in the work area. In the windows, we will present information such as the boxed text in Figure N.5, pull-down menus of choices within the selected option, formulas and/or computations, or results in the form of tables, figures, or numbers. The work area is where the main action takes place, and where most of your attention no doubt will be focused.

6. Drive Designation and File Name. The data set drive (drive A), subdirectory path (it is the root directory in Figure N.5, but if there had been a subdirectory, it would have followed the backslash symbol), and file name (NONAME.WK1) appear here. The drive is where the program will go to save and retrieve data files. On a two-drive system we will

```
≡           Microcomputer Applications for INTRODUCTORY BUSINESS STATISTICS   MENU
     Overview [Data] Files  Processors  Set Up  Exit                    | F1=Help
              Create
                Multivariate raw data
                Organized ungrouped frequency
                Organized grouped frequency
                Probability distribution
                X² goodness of fit
                Contingency tables (X²)
                One way ANOVA variables
                Randomized block variables
                Two way ANOVA variables
```

Figure N.6 Types of Data Sets That Can Be Created

want to use drive B for this purpose. Section 3.6 explains how to change the drive designation from A to B. The file named NONAME.WK1 contains no data. It was a default name we chose to indicate that no file has been accessed. When we save or retrieve an actual data set, the new data set's file name will replace NONAME.WK1.

7. Data File Type. Presently the message 'No data' appears in Figure N.5 at item 7. This is because we have not created a data set or retrieved a data file yet. When we do, the type of data will be described here. (The next section illustrates the different data types.)

8. Legend of Active Keys. At the bottom of the screen on the last two lines is a legend of keys and the activity that each will perform when pressed. For instance, pressing the Enter key in Figure N.5 will remove the present screen of text and display the next screen of text. The function keys are denoted by F2, F3, and so on.

N.3 Options on the Main Menu

The main menu for the IBS software is listed horizontally on the second line of the screen (see Figure N.5). What follows is a brief description of each option on this menu.

Overview The Overview option presents two screens of introductory text. No inputs from the user are required. Simply read the text (or press the F8 function key to skip forward) and go on.

Data The Data option is used to create, view, and/or edit data sets. Most processors are set up to accept data sets as input, so it is a good idea to create the data set before accessing the desired processor.

Figure N.6 shows the nine choices available in creating a data set. Each data type is explained in more detail in the text. For example, to learn how to create a Multivariate Raw Data set, refer to Section 2.5 (for a univariate data set) or to Section 4.4 (for a bivariate data set).

Files The Files option enables us to save a data set to and retrieve a data file from the Data Disk in drive B. Figure N.7 shows the choices with the Files option. Notice the description of the type of data set—Multivariate Raw Data—in the lower right corner of Figure N.7. This signifies that we have created a Multivariate Raw Data set and that it is the resident data set in memory.

```
≡        Microcomputer Applications for INTRODUCTORY BUSINESS STATISTICS   MENU
  Overview  Data [Files] Processors  Set Up  Exit                 | F1=Help
                 Retrieve
                 Save
                 List
                 Rename
                 Erase
                 Directory
                 OS
```

Figure N.7 Types of Activities Involving Data Sets That Reside in the Computer

```
≡        Microcomputer Applications for INTRODUCTORY BUSINESS STATISTICS   MENU
  Overview  Data [Files] Processors  Set Up  Exit                 | F1=Help
                 Retrieve
                 *ave
                 List
                 Rename
                 Erase
                 Directory
                 OS
```

Figure N.8 Types of Activities Involving Data Sets That Reside in a File on a Data Disk, But Not in the Computer

Without a data set in memory, the notation 'No data' would appear as in Figure N.8. Also notice in Figure N.8 that the "S" in the Save choice is replaced with an asterisk. This is our way of denoting that that option is, at the moment, inappropriate or unavailable. The reason that 'Save' is available in Figure N.7 is that we created a data set, which then could be saved. At the time we printed Figure N.8, there were 'No data' on hand that could be saved.

The List, Rename, Erase, and Directory commands are DOS commands and allow us to perform the named activity directly to the files stored on the Data Disk without having to go to the operating system. The last choice—OS—permits us to leave the IBS software and go to DOS and the prompt A>. (*Note:* The DOS program "Command.com" is required to use the OS menu choice.)

Saving a data set is explained in Section 2.5, while file retrieval is discussed in Section 3.6.

Processors Figure N.9 displays the list of processors. Within most processors, there are further choices to be made. For instance, the Simple Random Sample processor has two

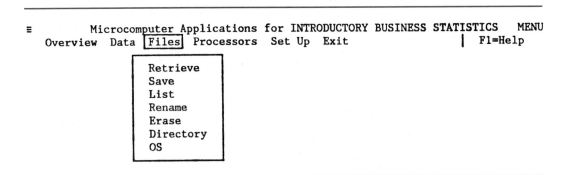

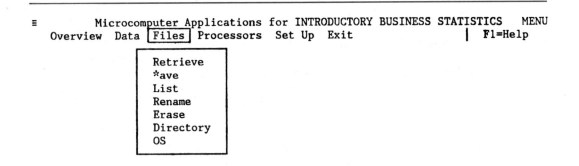

```
≡        Microcomputer Applications for INTRODUCTORY BUSINESS STATISTICS    MENU
  Overview  Data  Files  [Processors]  Set Up  Exit                   | F1=Help

                         Simple random sample
                         Frequency distributions
                         Descriptive statistics
                         Least squares line
                         Discrete probability distributions
                         Normal probability distribution
                         One sample inference
                         Two sample inference
                         Analysis of variance (ANOVA)
                         Regression
                         Time series
                         X² (chi-square)
```

Figure N.9 Menu of Processors

```
≡        Microcomputer Applications for INTRODUCTORY BUSINESS STATISTICS    MENU
  Overview  Data  Files  Processors  [Set Up]  Exit                   | F1=Help

                                   IBS data information
                                   Hardware information
                                   Confirm setting
                                   Program information
```

Figure N.10 Choices Within the Set Up Option

choices, one for Sample Entity Numbers and a second for Sample Data Points. Each processor is explained in a starred section in the chapter to which it is tied.

Set Up This option provides information about the hardware, available memory, and the status of the "Confirm Setting" option, as indicated in Figure N.10. Only one of the choices in this option—Confirm Setting—is an executable choice. The others merely display information. Confirm Setting is an option to have the program ask you to confirm intended inputs and actions. For instance, after typing an input to a processor, the program may respond with the query "Is this correct? (y/n)" when the Confirm Setting is switched "On." If the Confirm Setting is switched to "Off", the program automatically proceeds when you press the Enter or Return Key. Novice users may wish to switch the Confirm Setting to "On"; experienced users no doubt will prefer to switch the Confirm Setting to "Off". The default setting is "Off".

Exit The Exit option returns us to the operating system and the DOS prompt A>. If you have not saved your data set at this point, it will be lost if you choose to exit the IBS software.

N.4 Printing

When using the IBS software, there will be occasions when you wish to have a "hard copy" of certain screens. If you are at a computer that is connected to a printer, you can press the Print Screen (PrtSc) key. (On some keyboards, press the PrtSc key with the Shift key depressed.) This will print whatever appears on the screen.

There are other printing features to the IBS software, however, including the ability to store a screen on a Data Disk for printing at a later time. This capability would be useful if you were doing work on a computer that was not connected to a printer. To access these additional features, press the F6 key—also known as the PRINT key for the IBS software—when you are at a point where you wish to print the screen. Use of the F6 key will bring the pop-up menu shown in Figure N.11 onto the screen. Below is a brief explanation of the six menu choices on the Print menu.

The first choice, 'Set Project Name or Title,' will enable you to add text of your choosing (up to 25 characters are available) to the bottom of the current screen. For instance, you might title a problem, enter your name, specify the date, and so forth.

The 'Print Screen' choice has a two-item submenu (see Figure N.12). If our computer is connected to a printer and we choose the 'To Printer' option, we will print the current screen without the Print Menu. If we are working on a computer without an attached printer, then choosing the 'To Disk File' option enables us to save the contents of the screen on a Data Disk. We will be asked to specify a file name and to state the drive in which our Data Disk resides (in a two-drive system, presumably this would be drive B). The print file then can be retrieved and printed from the Data Disk whenever we have access to a printer, using the 'Print File From Disk' option.

The 'Print Current Data Set' choice will print the data set currently in the computer's memory.

The 'Set Printer Options' choice lets the user change default values on such variables as printer port, form length, form width, and form type. The printer port default is line printer number 1 (LPT1); if you were attached to multiple printers you could route the output to printer 2 or 3. The form length default is 66 lines per page; the form width default is 80 characters per line. The form type default is continuous; to use separately fed single pages, change the default to 'Hand Fed.'

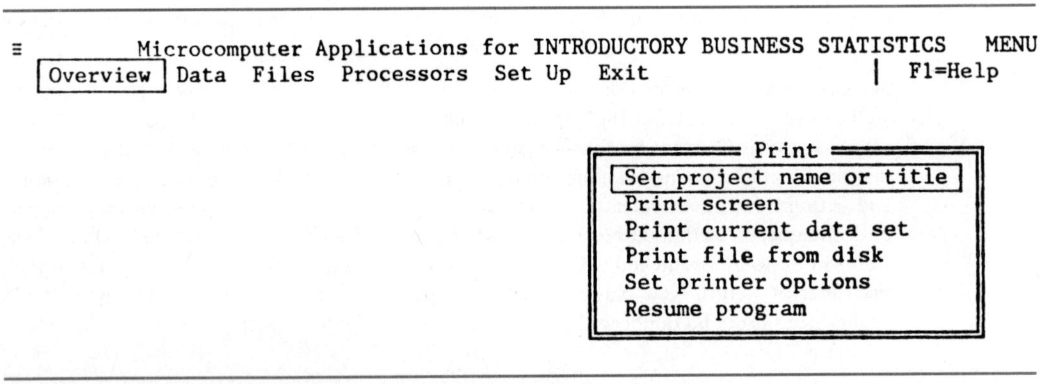

Figure N.11 The Print Menu

```
≡        Microcomputer Applications for INTRODUCTORY BUSINESS STATISTICS   MENU
 Overview  Data  Files  Processors  Set Up  Exit                 | F1=Help

                                        ═══════ Print ═══════
                                         Set project name or title
                                         Print screen
                                                        a set
                                          To printer    isk
                                          To disk file  ns
```

Figure N.12 Print Screen Options

Finally, the 'Resume Program' option is used to clear the Print Menu off the screen and return to the program.

N.5 Advanced Features

Mouse Support The IBS software will accept a mouse pointing device as input. It simulates the basic keyboard controls on all menus. Move the mouse slowly to effect left–right and up–down movement. The left button selects and the right button cancels. Some mice have a middle button, which toggles the HELP screens on or off.

Screen Rendition The IBS software will configure itself to any common text screen size. The DOS mode command may support 25, 43, or 50 rows of text, depending on hardware and DOS version. The Paradise VGA Card extended text modes also are supported for 132 columns × 25 rows and 132 columns × 43 rows. Set these modes prior to loading the IBS software. Refer to the appropriate reference manual for your equipment to set the desired screen rendition.

Data Format Information for Use with Other Software The IBS software will accept files created in a spreadsheet program, such as Lotus, or in a data base program, such as dBASE, provided the data are arranged properly. In addition, an IBS data file can be saved with a WKS (worksheet) extension or with a DBF (data base) extension for export to spreadsheet or data base programs.

WKS Files. To interface with spreadsheet programs using the Lotus file format, certain information about the data set must be stored in specific cell locations, depending on the type of data set. There are nine data set types, shown in Figure N.6. Table N.1 lists the required cell information.

In addition, certain parameters are needed in cells B1, C1, and, for data types 7 and 9, the cells in row 2 of the spreadsheet. Table N.2 shows these inputs according to data type.

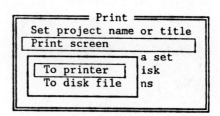

Table N.1 Cell Information for WKS Files

Cell	Contents	Type of Input	Choices
A1	Data Type	Number	Numbers 1 through 9
A3	Data description or name type	Label	Up to 30 characters
A4, B4, C4 . . .	Column names	Label	Up to 20 characters
A5, B5, C5 . . .	Data	Number	Any number
A6, B6, C6 . . .	Data	Number	Any number
. . .	Data	Number	Any number

As an example, suppose we created a Multivariate raw data set of size $n = 5$. The $v = 2$ variable names are X, engine size, and Y, miles per gallon, and the data are as follows:

X	263	242	148	130	180
Y	23	24	31	30	27

Table N.2 Inputs for WKS Files

	Cells				
Data Type	B1	C1	A2	B2	C2 . . .
1—Multivariate raw data	n = sample size	v = number of variables	Blank	Blank	Blank
2—Organized ungrouped frequency	Number of different values of X	2	Blank	Blank	Blank
3—Organized grouped frequency	Number of different midpoints	2	Blank	Blank	Blank
4—Probability distribution	Number of different values of X	2	Blank	Blank	Blank
5—χ^2 goodness of fit	1	Number of categories	Blank	Blank	Blank
6—Contingency tables	Number of rows	Number of columns	Blank	Blank	Blank
7—One way ANOVA variables	Largest sample size among treatments	Number of treatments	Sample size for treatment 1	Sample size for treatment 2	Sample size for treatment 3
8—Randomized block variables	Number of blocks	Number of treatments	Blank	Blank	Blank
9—Two way ANOVA variables	Number of rows	Number of columns	Number of values per cell	Blank	Blank

If we save this data set with a WKS extension, here is what it would look like in terms of a Lotus spreadsheet:

	A	B	C
1	1	5	2
2			
3	Multivariate raw data		
4	Engine Size	Miles Per Gallon	
5	263	23	
6	242	24	
7	148	31	
8	130	30	
9	180	27	

In general, row 2 of the spreadsheet is blank, unless we are dealing with type 7 (One Way ANOVA Variables) or type 9 (Two Way ANOVA Variables) data sets. For type 7 data sets, row 2 of the spreadsheet is filled with the sample sizes per treatment. For type 9 data sets, insert in cell A2 the cell count in the analysis of variance. Leave the rest of the row 2 blank.

No other information may be in the spreadsheet, since the cell data are processed sequentially. A cell should be left blank for missing data in type 7 data sets. Any variation or unexpected data will result in an interpretation error message of "retrieve failure."

DBF Files. The dBASE II and III data formats are supported in a limited way. The import DBF file must conform to the simple outline below. Please note that any non-numeric fields are inconsistent with the IBS software and will cause an import "read failure."

The first record and first field must contain the data type, numbered 1 through 9. If data type 7 is created, then the treatment entity count for each set of fields (that is, a treatment) must appear in the second record. If data type 9 is being imported, then the cell count must appear in the first field of the second record. All other records contain data in numeric fields only.

Record Number	Field 1 . . . Field *n*
1	Data type
2	Data or actual counts for type 7 or cell count for type 9
3	Data . . .
.	. . .
.	. . .
.	. . .
m	Data

The field names serve as the column names when importing data, as the data type needs them.

TXT Files. The fixed length data format used in the text format allows the user to create, read, and print the data using standard ASCII text methods. The data is formatted right justified as follows:

Line	Contents
1	5 digit data type, 5 digit row count, 5 digit column count (optional 5 digit entity counts for data type 7 for each treatment; cell count for data type 9)
2	Up to 30 characters (including right side trailing blanks)
3	Up to 20 characters for column labels (one per column)
4	Up to 20 (right justified) numbers
.	
.	
.	
$n + 1$	A carriage return

The five digit "row count" and "column count" referred to in line 1 should contain the same information as that required by the WKS file format for cells B1 and C1, respectively. Refer to the previous chart for the needed inputs for each cell. Data type 9 is a special case where the "row count" is the number of entities per cell in the ANOVA and the "column count" is the number of rows multiplied by the number of columns.

As an example, suppose we wish to save the Multivariate raw data set of size $n = 5$ involving the variables X = Engine Size and Y = Miles Per Gallon with a TXT extension. (Or, suppose we wish to create the data set using an ASCII text editor.) The TXT file would look like the following. Note that "b" represents a blank space.

Line	
1	bbbb1bbbb5bbbb2
2	bMultivariatebRawbData
3	bbbbbbbbbEnginebSizebbbbMilesbPerbGallon
4	bbbbbbbbbbbbbbbbb263bbbbbbbbbbbbbbbbbb23
5	bbbbbbbbbbbbbbbbb242bbbbbbbbbbbbbbbbbb24
6	bbbbbbbbbbbbbbbbb148bbbbbbbbbbbbbbbbbb31
7	bbbbbbbbbbbbbbbbb130bbbbbbbbbbbbbbbbbb30
8	bbbbbbbbbbbbbbbbb180bbbbbbbbbbbbbbbbbb27
9	(carriage return)

N.6 Developmental Notes

Author John K. Hedges of Riverdale Software Systems developed the software in consultation with the text authors. He worked primarily on an IBM AT compatible computer with 2.5 Mb of memory and a 40 Mb hard disk drive running PC-DOS 4.00. He used the C language, including proposed ANSI standard additions.

Software Tools Below is a list of vendors of software tools and libraries used in the development of the IBS software:

1. *Turbo C*, Version 2.0 from Borland International
   Created using *Turbo C*, Copyright © Borland 1987, 1988.
2. *Plink86-Plus*, Version 2.22 from Phoenix Technologies Ltd
   Copyright 1983, 1984, 1985.

3. *POLYTRON Version Control System,* Version 2.0c from POLYTRON Corporation
   Copyright 1985, 1986, 1988.
4. *POLYMAKE,* Version 3.0 from POLYTRON Corporation
   Copyright 1984, 1987, 1988.
5. A customized version of *Turbo C Tools,* Version 5.00 from Blaise Computing Inc.
   Copyright 1987.
6. *The WKS Library,* Version 2.00 from Tenon Software Services, Inc.
   Copyright 1987, 1988.

N.7 Trademark Acknowledgments

1. IBM, IBM AT, IBM PC, IBM PCjr, and IBM PS/2 are registered trademarks of International Business Machines Corporation.
2. PLINK86-PLUS is a trademark of Phoenix Software Associates Ltd.
3. PC-DOS is a trademark of International Business Machines Corporation.
4. POLYTRON and POLYWINDOWS are registered trademarks of POLYTRON Corporation.
5. Turbo C TOOLS is a trademark of Blaise Computing Inc.
6. Turbo C is a registered trademark of Borland International, Inc.
7. Lotus 1-2-3 is a trademark of Lotus Development Corporation.
8. dBASE is a registered trademark of Ashton-Tate.
9. Paradise is a registered trademark of Paradise Systems, Inc.

ANSWERS TO ODD-NUMBERED EXERCISES

Chapter 1

1.1 Qualitative and quantitative

1.3 **a.** Qualitative **b.** Quantitative **c.** Quantitative **d.** Qualitative **e.** Qualitative **f.** Qualitative

1.5 **a.** 0, 1, 2, . . . **b.** Highest degree earned **c.** Employee **d.** Business **e.** 5%, 5½%, . . . ; bank

1.7 **a.** Quantitative **b.** Quantitative **c.** Qualitative **d.** Qualitative **e.** Qualitative **f.** Qualitative

1.9 **a.** All of the houses on Galway Lane **b.** All of the months in the previous year **c.** All of the students at this university **d.** All employees of the insurance company **e.** All of the subsidiary accounts for the merchandise inventory account **f.** All mutual funds

1.11 Plan, execute, report, evaluate, act

1.13 Descriptive statistics and inferential statistics

1.15 Description

1.17 **a.** Inference **b.** Description **c.** Description **d.** Inference

1.19 Judgment

1.21

	Benefit	Drawback
a.	Representative	May be more costly
b.	May be more efficient	May involve more time
c.	Representative	Selection of entities may be subject to personal biases
d.	Convenient	May include (or exclude) "similar" entities

1.23 **a.** 11, 08, 02, 66, 05, 33, 88, 34, 39, 64, 62, 68 **b.** 47, 13, 32, 42, 37 **c.** 208, 440, 273, 487, 183, 599, 510, 473, 257, 388

1.25 Simple random sample

1.27 Census; promotes quality of the product

1.29 Primary: advantage—data are current; disadvantage—cost
Secondary: advantage—saves time in collecting data; disadvantage–data may not fit the problem

1.31 **a.** Quantitative **b.** Quantitative **c.** Quantitative **d.** Quantitative **e.** Quantitative **f.** Qualitative

1.33 **a.** Inference **b.** Inference **c.** Inference **d.** Inference **e.** Description

1.35 An *entity* is the "thing" that possesses or generates data; a *variable* is a quality or quantity that, when applied to an entity, yields data points.

1.37 No; no obvious patterns appear

1.39 Static—descriptive; dynamic—inferential

1.41 A sampling error is the difference between a sample-based computation and the corresponding population-based computation; a nonsampling error is usually a tainted data point.

1.43 **a.** 040, 308, 189, 267, 444, 248, 561, 552, 129, 301, 491 **b.** 57, 65, 53, 40, 83, 25, 81, 50, 61 **c.** 847, 285, 314, 910, 871

1.45 Yes, merely add 50 to the lowest invoice number and 86 thereafter; the first 3 invoice numbers would be 14,572, 14,658, and 14,744.

1.47 **a.** Either a cluster or a stratified sample would be most appropriate since the population entities are naturally segregated into groups (months). **b.** January, 10; February, 10; March, 5; April, 7; May, 6; June, 2; July, 2; August, 5; September, 3; October, 3; November, 3; December, 6 **c.** January: 253, 276, 128, 395, 224, 038, 405, 242, 049, 292; February: 027, 141, 366, 173, 149, 312, 341, 283, 122, 170; March: 041, 073, 141, 119, 151; April: 055, 231, 237, 174, 236, 207, 265; May: 064, 212, 158, 252, 094, 231; June: 04, 35; July: 008, 004; August: 101, 142, 176, 031, 061; September: 051, 022, 043; October: 040,

Note: Some exercises, when done with the IBS software, may have slightly different answers due to no rounding in the intermediate calculations.

023, 016; November: 086, 062, 011; December: 122, 170, 153, 121, 077, 071.

1.49 The different results are due to a random sampling (*Newsday* poll) as opposed to a nonrandom sampling (*Ann Landers'* poll).

1.51 Nonsampling

Chapter 2

2.1 a. Discrete b. Discrete c. Continuous

2.3 Each piece of data must fall into one and only one category.

2.5 a. Continuous b. Discrete c. Continuous

2.7 The principle of exclusion is violated.

2.9 The principle of inclusion is violated when, for instance, a student's final average is 89.5.

2.11 Decide upon a standard degree of accuracy for the data, develop a system for dealing with missing data, check the data for unusual responses, code the data, and create a data file.

2.13 a. $n = 54$ b. $rf = .259$ c. Ungrouped frequency distribution

d.

X	0	1	2	3	4	5
cf	10	27	41	50	53	54

e. 28th through the 41st

2.15 a. $c = 6$ b. $w = 3.8$ c. 61.1 and 64.9

2.17 Yes, w is the difference between any two consecutive midpoints.

2.19 $w = 8$; the endpoints of the first class are 51 and 59.

2.21 a. $cf = 47$

b.

Class No.	1	2	3	4	5	6
rf	.028	.099	.239	.296	.268	.070

c. $n = 71$ d. $M = -1.3$

2.23

Classes	f
17–24	11
24–31	10
31–38	4
38–45	3
45–52	2
	$n = 30$

2.25 a.

Classes	f
778–1182	13
1182–1586	17
1586–1990	9
1990–2394	7
2394–2798	2
2798–3202	2
	$n = 50$

b. \$1384 c. \$1182–\$1586

2.27

Class No.	rf	crf
1	.028	.028
2	.099	.127
3	.239	.366
4	.296	.662
5	.268	.930
6	.070	1.000

2.29 Histogram:

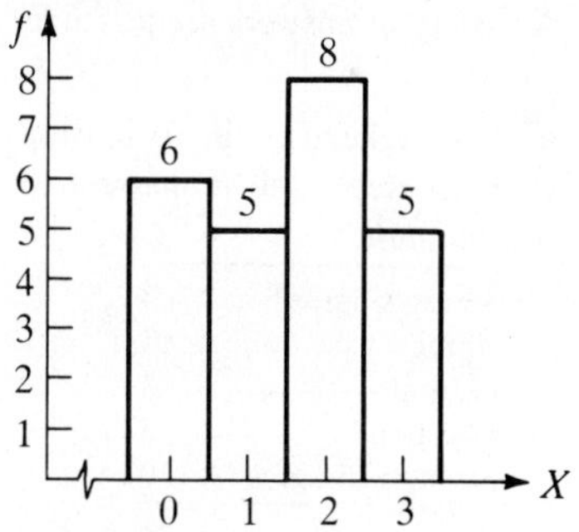

2.31 Histogram:

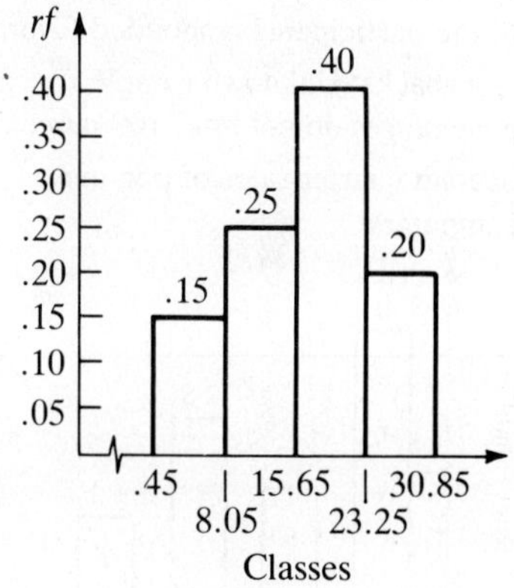

2.33 a. Continuous

b.

Classes	f
0.0– 9.0	14
9.0–18.0	11
18.0–27.0	12
27.0–36.0	8
36.0–45.0	3
45.0–54.0	2

c. Debt as a percentage of capital for a sample of 50 companies:

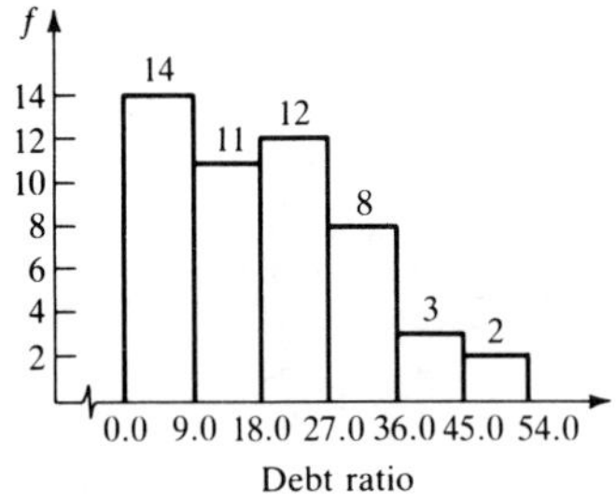

d. Yes; asymmetrical, positive

2.35 A variety of answers are possible.

2.37 14–18

2.39 **a.** Time-related **b.** Time-free **c.** Time-free **d.** Time-related

2.41 **a.**

Response	f
Strongly agree	3
Agree	14
Neutral	8
Disagree	3
Strongly disagree	0
	$n = 28$

b. Yes, roughly 61 percent (17 out of 28) of the participants responded favorably.

2.43 Drawbacks: Unknown sample size and percentages do not total 100 percent.

2.45 Second quarter sales of personal computers:

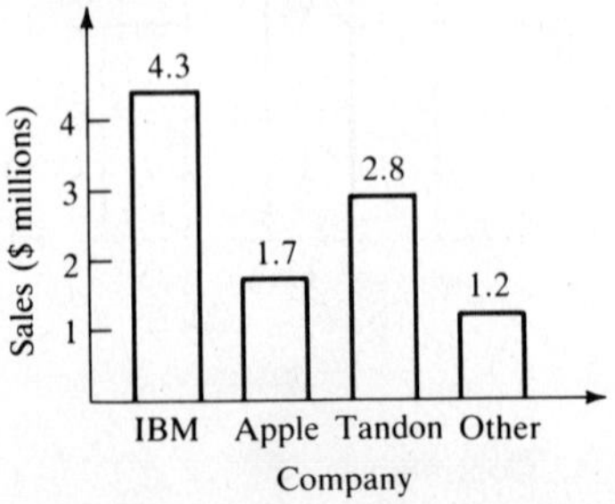

2.47 Restaurant customers' preferences among evening entrees:

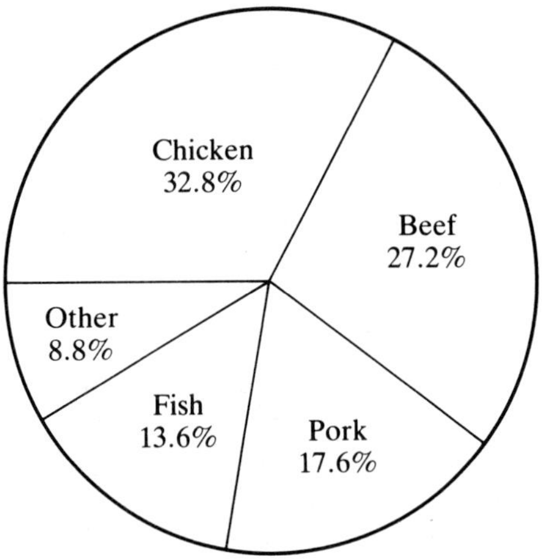

2.49 **a.**

Category	% of Tax Dollars	Tax Rate, per $100 Valuation
State	34.09	.331
County	17.57	.171
School	44.15	.429
Fire and sewer	4.19	.041
	100	.972

b. $.046 **c.** No **d.** $.977 per $100 of assessed valuation

e. Proposed distribution of tax dollars:

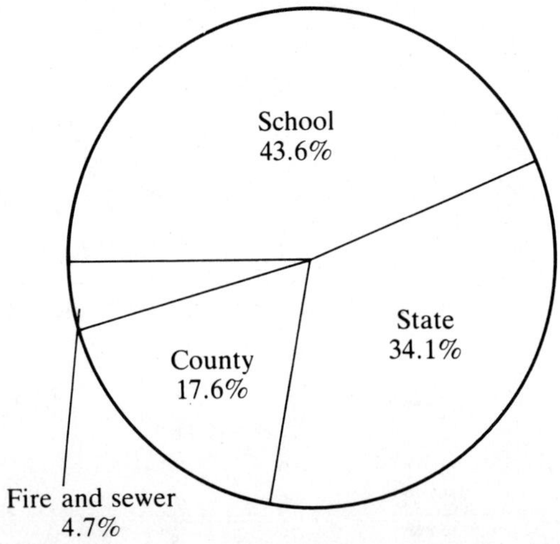

51.92
56.10
108.02

2.51 Sales by regions:

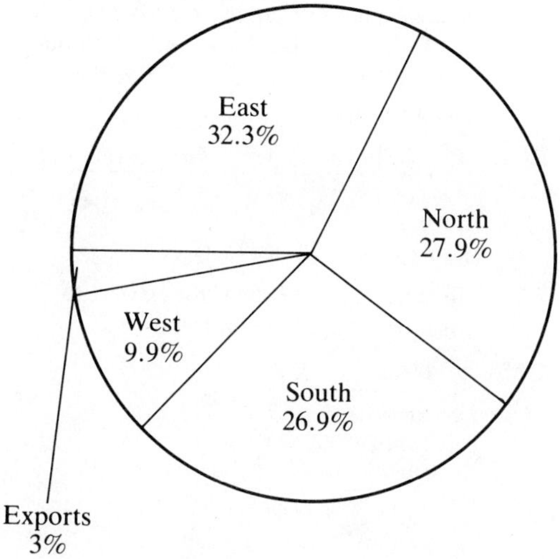

2.53 A variety of answers are possible.

2.55 **a.** \$360 million **b.** Pie chart or bar graph **c.** Time-free **d.** 1985 soft-drink market shares:

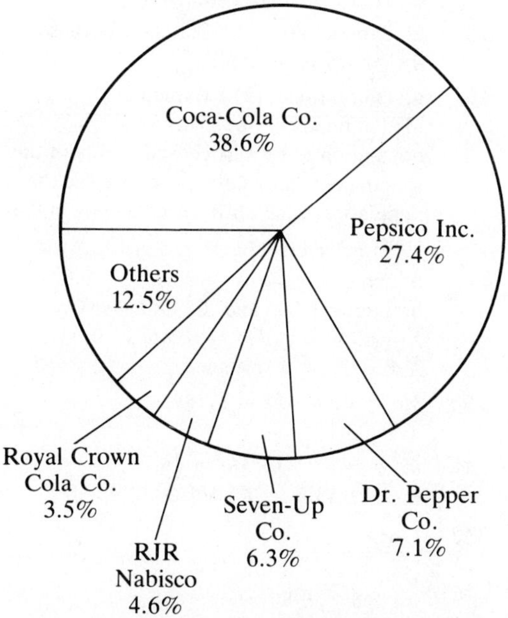

2.57 **a.** $c = 7$ **b.** $w = 4.18$
c. and **d.**

Classes	f	cf
51.92–56.10	5	5
56.10–60.28	10	15
60.28–64.46	7	22
64.46–68.64	12	34
68.64–72.82	25	59
72.82–77.00	21	80
77.00–81.18	16	96
	$n = 96$	

e. 15

2.59 **a.** 5 **b.** 6 **c.** 7

2.61 **a.** Those who gave no response are not included. **b.** Not an oversight; open-ended class intended for extremely large responses **c.** approximately 361

2.63 **a.** Categories are adequate. **b.** United States—pie chart on left; Japan—pie chart on right

2.65 **a.** $w = 4.3$ **b.** $w = 3.4$
c. $w = 2.9$

2.67 **a.** and **b.**

Classes	f	rf
4– 41	21	.429
41– 78	21	.429
78–115	4	.082
115–152	2	.041
152–189	0	.000
189–226	1	.020
	$n = 49$	1.000

c. 78

d.

Classes	f
4– 25	9
25– 46	19
46– 67	10
67– 88	5
88–109	3
109–130	2
130 and up	1
	$n = 49$

The concentration of values between 4 and 78 is refined to give us more detail about the distribution.

2.69 **a.** .555, 1.195, and 1.835
b. $w = .640$

2.71 **a.**

Classes	f
20.8–23.1	1
23.1–25.4	6
25.4–27.7	19

2.71 **a.**

Classes	f
27.7–30.0	16
30.0–32.3	16
32.3–34.6	5
	$n = 63$

b. No, the percentage of grids in the sample outside the desired range is much higher than 5 percent.

2.73 Daily revenue of shoe boutique:

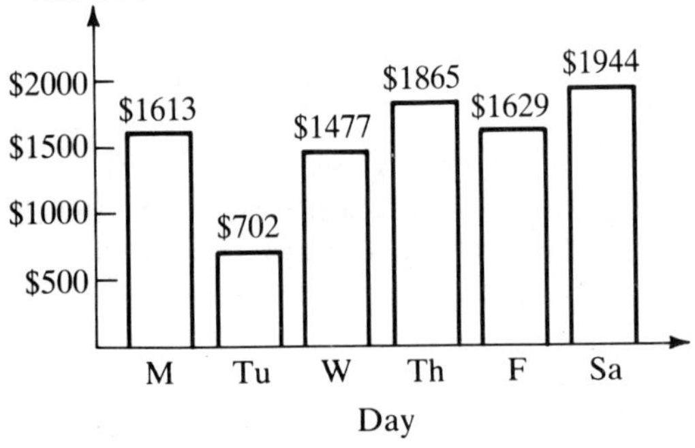

2.75 U.S. military active duty personnel as of September 30, 1986:

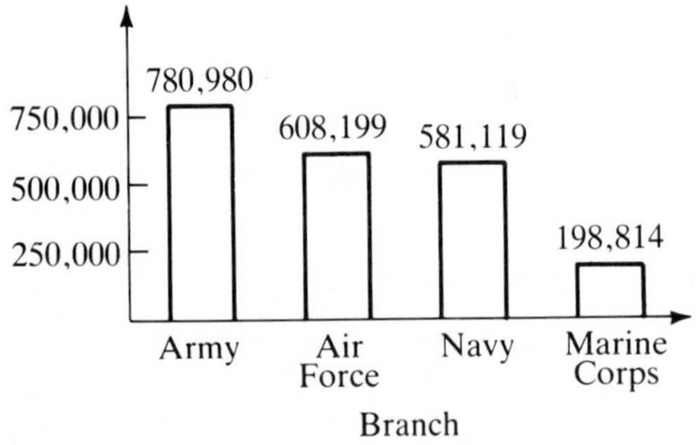

2.77 **a.** $c = 6$ **b.** $w = 1.10$
c. and **d.**

Classes	M
13.18–14.28	13.73
14.28–15.38	14.83
15.38–16.48	15.93
16.48–17.58	17.03
17.58–18.68	18.13
18.68–19.78	19.23

2.79 **a.** Method of transportation to winter or summer vacation site **b.** Qualitative **c.** Yes, if we assume that only one response per person was recorded **d.** Train, boat, bicycle **e.** Break out the Auto/Truck/RV category into the 3 specified modes.

2.81 **a.**

X	0	1	2	3	4
cf	3	14	16	21	23

b. 60.9% **c.** 1.9 or less

2.83 To summarize the data; for clarity; to visualize shapes and/or trends

2.85 **a.**

Meal	f
Breakfast	27
Lunch	25
Dinner	8
	$n = 60$

b. 45%

Chapter 3

3.1 **a.** Statistic **b.** Statistic **c.** Statistic **d.** Parameter

3.3 **a.** Sample; no **b.** Statistic, because the survey is a sample

3.5 **a.** (1) Statistic, (2) Parameter, (3) Parameter **b.** Define the population to be all licensed agents of the specific company only. **c.** Define the population to be all licensed agents in the United States.

3.7 **a.** No mode; $Md = -2$, $\bar{X} = -2.6$ **b.** bimodal: .03 and .05; $Md = .05$, $\bar{X} = .94$ **c.** $Mo = 0$, $Md = 0$, $\bar{X} = 1/7$, or .1 (rounded)

3.9 $Mo = 40.4$; $Md = 37.89$; $\bar{X} = 35.10$

3.11 **a.** $w = 4$ **b.** $Mo = 10$ **c.** $\bar{X} = 13.6$ **d.** $Md = 13.4$

3.13

Classes	f
.54–.61	5
.61–.68	8
.68–.75	8
.75–.82	18
.82–.89	10
.89–.96	2
	$n = 51$

$\overline{X} = .751$, $Md = .768$, $Mo = .785$; values are not identical because accuracy is lost by grouping.

3.15 $\overline{X} = 11.03$; $Md = 11.27$

3.17 **a.** yes, $n = 134$ **b.** $\overline{X} = 2.8$ **c.** $Md = 3$ **d.** Yes, $X = 2$ is the mode.

3.19 **a.** $X =$ the number of videos rented per household last month **b.** household **c.** $Md = 3.5$ **d.** $4,634

3.21 $\overline{X} = \$994.37$; $Mg = \$1,067.40$

3.23 $GM = 1.24\%$

3.25 $GM = 5.06\%$

3.27 **a.** 10%; −9.09% **b.** $\overline{X} = .455\%$ **c.** $GM = 0\%$ **d.** geometric mean

3.29

	Rg	AD	S^2	S
a.	.456	.172	.0466	.2158
b.	4.29	1.32	2.786	1.669
c.	12	3.6	19.9	4.5
d.	6	1.9	5.8	2.4
e.	1	.2	.1	.3
f.	4	2	8	2.8

3.31 **a.** $\overline{X} = 5$ **b.** $S = 3.0$ **c.** 4 **d.** 1.33

3.33 **a.** $\overline{X} = 9$ **b.** $S = 2.4$ **c.** $AD = 1.8$

3.35 **a.** $\overline{X} = 50$; $Md = 52$ **b.** $\Sigma|X - Md| = 28$; $\Sigma|X - \overline{X}| = 30$ **c.** $\Sigma(X - \overline{X})^2 = 272$; $\Sigma(X - Md)^2 = 296$

3.37 $Rg = 1$; $AD = .2$; $S^2 = .1$; $S = .3$

3.39 **a.** $Rg = 12.6$ **b.** $\Sigma|X - Md| = 28$; $\Sigma|X - \overline{X} = 30$ **c.** $\Sigma(X - \overline{X})^2 = 272$; $\Sigma(X - Md)^2 = 296$

3.41 $S^2 = 16$ for each formula

3.43 $S = 2.5$; one large P–E ratio inflates the value of S to 5.5.

3.45 $AD = 1.7$; $S = 2.1$

3.47 **a.** $\overline{X} = 33$ **b.** $S = 35.7$ **c.** $CV = 1.0818$, or 108.18% **d.** $g_1 = .873$

3.49 $g_1 = .534$

3.51 $g_1 = .878$; skewed

3.53 $g_1 = .430$

3.55 **a.** A **b.** CV's are essentially equal.

3.57 Investment 2

3.59 $g_1 = 1.544$; skewed to the right

3.61 $IR = 49$

3.63 $g_2 = -.489$

3.65 $\overline{X} = 3$; $Md = 2$; $Mo = 2$; positive

3.67 **a.**

Classes	f
2.89–2.92	6
2.92–2.95	15
2.95–2.98	17
2.98–3.01	14
3.01–3.04	6
3.04–3.07	2
	$n = 60$

b. Weight of 3 oz. bag of dog food treats:

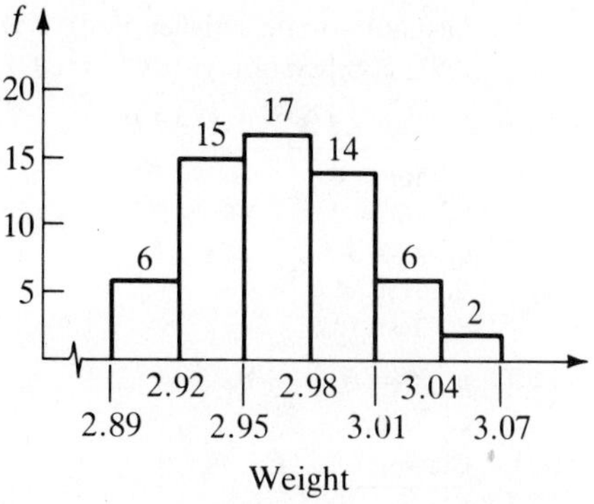

c. Skewed to the right **d.** $\overline{X} = 2.968$; $Md = 2.966$; $Mo = 2.965$; yes **e.** $S = .038$ **f.** $\overline{X} \pm S = 2.930$ to 3.006; 44; 73.3%. $\overline{X} \pm 2S = 2.892$ to 3.044; 58; 96.7%

3.69 **a.** $\overline{X} = 27$; $Md = 23.5$; no mode **b.** $Rg = 33$; $AD = 8.6$; $S = 10.3$ **c.** Yes **d.** No, because there is no mode **e.** Positive; $g_1 = .541$

3.71 **a.** 68% **b.** 81.5% **c.** 2.35%

3.73 Sixteenth percentile $= \overline{X} - S$; 84th percentile $= \overline{X} + S$

3.75 **a.** $\overline{X} = 47.97$; $S = 35.78$ **b.** −23.59 to 119.53 **c.** Percentage change in building permits:

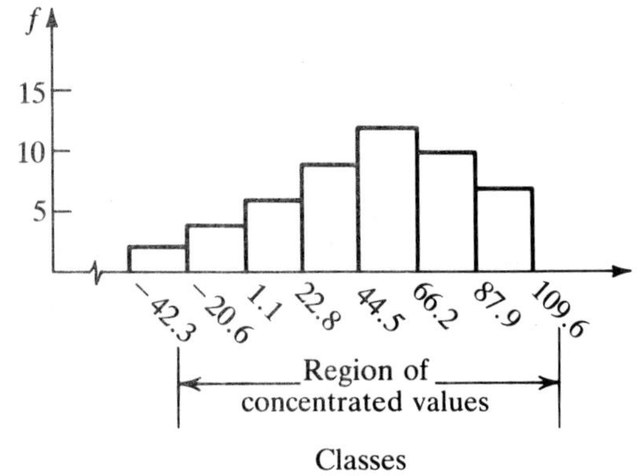

d. 48 values at a minimum, or 96%

3.77 a. $\overline{X} = 3; S = 3.5; Z_{max} = .86$; yes
b. $\overline{X} = 9; S = 3.162277\ldots$;
$Z_{min} = -2.8460499\ldots$

3.79 $\overline{X} \cong 90; S \cong 14$

3.81 Inequality to be satisfied is $.011 \leq S \leq .091$; computation yields $S = .038$.

3.83

	Test Ad *A*	Test Ad *B*
a.	$Md = 4$	$Md = 4$
b.	$Mo = 3$	$Mo = 5$
c.	$\overline{X} = 3.5$	$\overline{X} = 3.5$
d.	Test Ad *B*	
e.	Test Ad *B*	

3.85 a. $\overline{X} = 3.6$ b. $S^2 = 1.9$
c. $S = 1.4$

3.87

Classes	*f*
104–137	2
137–170	4
170–203	9
203–236	6
236–269	4
	$n = 25$

$\overline{X} = 194.4; Md = 193.8$

3.89 a. 83.85% b. 95% c. 84%

3.91 $X = -2.9$

3.93 a. $AD = 2.1$ b. $S^2 = 6.3$
c. $S = 2.5$

3.95 a. $\overline{X} = 3.4$

b.

X	*f*
0	14
1	5
2	8
3	2
4	2
5	9
6	7
7	4
8	4
9	1
	$n = 56$

c. $\overline{X} = 3.4$; yes; no accuracy is lost in an ungrouped frequency distribution.
d. $Md = 3$ e. $Mo = 0$ f. No, according to the Dispersion Inequality, the standard deviation must be less than or equal to 60 percent of the range.

3.97 a. $\overline{X} = 16.92\%$ b. $S = 14.67\%$
c. About one-half of a standard deviation below the mean; $Z = -.47$
d. $X = 48.5\%$ e. Yes, fairly representative, although a bit more stable

3.99 $\mu = -7.8$

3.101 a. $\overline{X} = 26.8; S = 8.2$ b. 100%; exceeds the guideline; not surprising, since data set is small and not very mound-shaped c. $CV = .306$ or 30.6% d. No, according to the Dispersion Inequality, the standard deviation must be less than or equal to 60 percent of the range.

3.103 a. $\overline{X} = 2.38; Md = 2.3$ b. $\overline{X} = 3.08$; $Md = 3.0$ c. Mean: $3.08 - 2.38 = .7$; median: $3.0 - 2.3 = .7$
d. Mean $= \overline{X} + d$; median $= Md + d$, where $\overline{X}$ and Md are the mean and median of the original data set e. Zero

3.105 $\overline{X} = 42.92$

3.107 $GM = 1.138\%$

3.109 $\overline{X} = 6.41; S^2 = 15.42; S = 3.93$

3.111 $\overline{X}_5 = 11.24; S_5 = 1.52$

3.113 Standard deviation = 2.4

3.115 a. $w = \$15$ b. Midpoints are 17.5, 32.5, 47.5, 62.5, 77.5, and 92.5.
c. $Md = \$37.29$; no; median depends on the position occupied, not on the value of the largest observation d. $\overline{X} = \$45.12$;

yes; mean depends on the values of all observations

3.117 Mean = 3.4, indicating the mean response tended toward "disagree."

3.119 The reported value of S does not satisfy the inequality given in Exercise 3.80.

3.121 $\overline{X}$ = 4.5 minutes; S = 1.75 minutes

3.123 Original data: $\overline{X}$ = 1.66; Md = 1.8; new data: $\overline{X}$ = 3.56; Md = 1.9. The mean is very sensitive to extreme values; the median is not.

3.125 g_1 = .575; skewed to the right

3.127 a. $\overline{X}$ = 75.5 b. Md = 75.3 c. S^2 = 86.5

3.129 S = 4.8

3.131 Histogram:

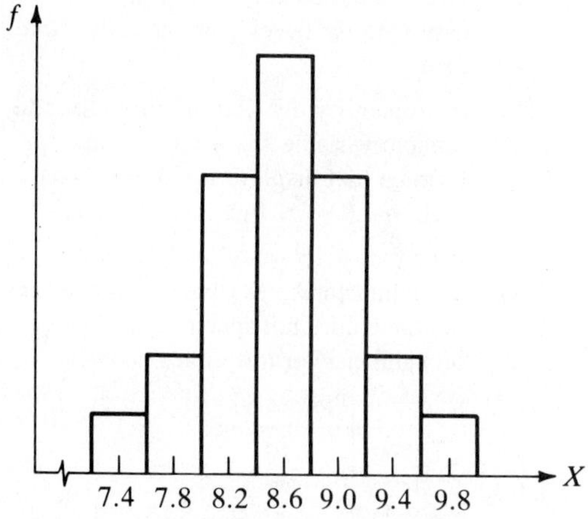

3.133 a. at most .111 . . . b. at least .555 . . . c. at most .16

Chapter 4

4.1 a. No b. Yes

4.3 a. Inverse b. Direct c. No relation d. Direct

4.5 a. L is lowest; M is highest. b. Y c. Direct

4.7 Direct

4.9 a. Games won

4.11 a. .757 b. −.771 c. .791

4.13 Conclude X and Y are uncorrelated

4.15 a. .324 b. No

4.17 b. .849 c. .890

4.19 .243

4.21 a. Yes b. .660 c. .660 < .707; we infer S and T are not correlated.

4.23 (x,y) is an actual data point somewhere on the scatterplot; $(x,\hat{y})$ is a predicted data point on the regression line.

4.25 $\hat{Y} = -1.64 + 1.5X$; yes

4.27 a. $\hat{Y} = -.59 + .87X$ b. $\hat{Y} = 2.9$ c. $Y - \hat{Y} = 1.6$ d. Each additional dependent is associated with an $87 increase in family average annual dental bill.

4.29 a. Yes b. $r = .894$ c. $\hat{Y} = 166.15 + 20.93X$ e. $\hat{Y}$ = 480 lbs.

4.31 Slope = .836; intercept = .359

4.33 a. $\hat{Y} = -22.53 + 10.867X$ b. 31.8

4.35 a. 1500

b.

.062	.110	.180	.352
.271	.223	.153	.647
.333	.333	.333	.999

c. Nonsmokers are more likely in the no accident group and less likely in the high accident group; vice versa for smokers.

4.37 a. Direct b. Direct c. Inverse

4.39 a. Undefined since a term in denominator of Equation 4–1 is zero b. −1 or +1 depending on slope of connecting line

4.41 a. Little or no apparent relation b. $r = -.026$ c. no

4.43 a. Direct b. $\hat{Y} = -3.02 + .91X$ c. 24 d. $r = .812$

4.45 $r = .909$

4.47 Random points

4.49 a. 90% b. 14%

4.51 Letting P be the independent variable, $\hat{Y} = .547 + .003X$. Each 1% increase in positions held by females improves the female-to-male pay ratio by an average of .003. $r = .685$, exceeding the cutoff value, thereby providing evidence of a relation in the population.

Chapter 5

5.1 A probability is a number between 0 and 1 that quantifies uncertainty.

5.3 **a.** (1) No; no (2) Yes; yes **b.** (1) Yes (2) No **c.** (1) No (2) Yes, if hourly wage is less than $9.38 per hour and overtime is possible; no, if hourly wage is more than $9.38 per hour and 40-hour work weeks are the norm

5.5 .467

5.7 **a.** Unconditional **b.** Conditional **c.** Joint

5.9 **a.** .65 **b.** .8 **c.** .651

5.11 **a.** .21 **b.** .25 **c.** .05

5.13 **a.** 114/191 = .597 **b.** 6/191 = .031 **c.** 5/77 = .065 **d.** 119/191 = .623

5.15 **a.** Discrete **b.** Discrete **c.** Continuous **d.** Continuous

5.17 **a.** X = number of consumers preferring round-shaped tablets, or Y = preference (1 = round, 0 = oblong) of each consumer **b.** X = rating per participant, or Y = number of participants rating the seminar a "1" **c.** X = rate of complaints per 100,000 passengers filed during the month of May per airline carrier

5.19 **a.** Continuous **b.** Discrete **c.** Continuous

5.21

x	$P(X = x)$
0	6/15
1	8/15
2	1/15

5.23 **a.** X = inflation rate for the upcoming year; discrete **b.** x **c.** $P(X = x)$ **d.** .15 **e.** 0

5.25 **a.** .10 **b.** .82 **c.** Additive

5.27 **a.** 0 **b.** 15/86 = .174 **c.** Histogram:

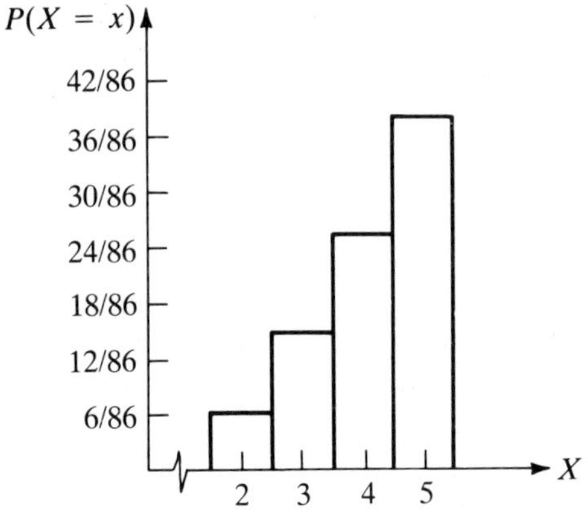

5.29 **a.** .312 **b.** .938 **c.** .313 **d.** Additive

5.31 Derive theoretically from a set of assumptions; develop empirically from data

5.33 A probability distribution for a discrete random variable X is a table, graph, or formula that displays x and $P(X = x)$ such that (i) $0 \le P(X = x) \le 1$ and (ii) $\Sigma P(X = x) = 1$.

5.35 **a.** Values for X are listed or implied, but the table does not specify the probabilities for the values.

b.

x	$P(X = x)$
1700	.10
1800	.30
1900	.50
2000	.10

c. Probability distribution for T = T-bill forecasts (one of many possible answers):

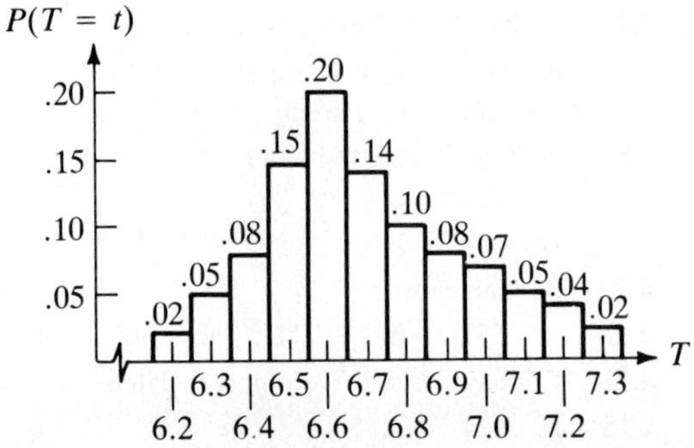

d. Many answers are possible; here is an example:

CPI	Probability
1.5	.02
1.6	.04
1.7	.05
1.8	.12
1.9	.17
2.0	.20
2.1	.17
2.2	.12
2.3	.05
2.4	.04
2.5	.02

e. A possible probability distribution for Y is

y	$P(Y = y)$
220	.05
230	.15
240	.30
250	.45
260	.05

5.37 **a.** Yes **b.** No **c.** No

5.39 $\mu = 11$; $\sigma = 5.7$

5.41 **a.** .4 **b.** 1 **c.** .8

5.43 Guess = \$50; Actual = \$52.50

5.45 $\mu = 2/11 = .2$

5.47 Because a probability distribution models a population not a sample

5.49 **a.** $\mu = 2.6$ **b.** $\mu = 2.6$

5.51 **a.** $\mu = \$25.75$ **b.** $\sigma = \$37.92$ **c.** .30

5.53 .05

5.55 **a.** Continuous **b.** Discrete **c.** Discrete **d.** Continuous **e.** Discrete **f.** Continuous

5.57 **a.** 0 **b.** .125 **c.** .984 **d.** 0 **e.** 1 **f.** 0

5.59 P: $CV = 78.64\%$; Q: $CV = 62.92\%$; recommend Q

5.61 K is least risky; J is most risky.

5.63 Expected return = 11.07% or 11% (rounded)

5.65 To quantify the uncertainty of sampling from a population

5.67 Table, graph, formula

5.69 **a.** .12 **b.** Probability distribution:

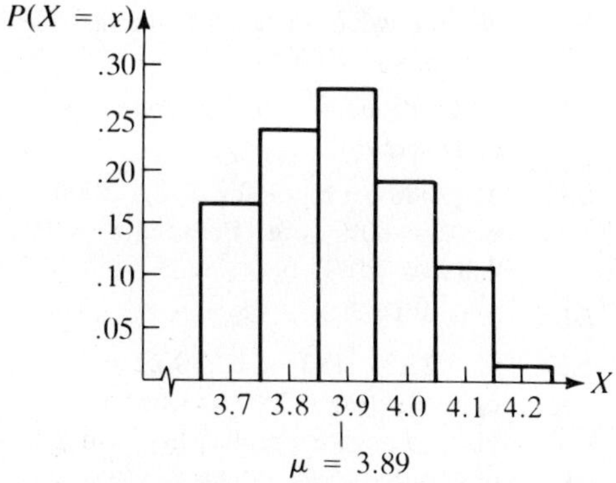

c. No, skewed right **d.** $\mu = 3.89$

5.71 $\mu = 2.7$; $\sigma = 5.4$

5.73 **a.** $4/14 = .286$

b.

x	$P(X = x)$
1	2/14
2	4/14
3	8/14

c. Probability distribution:

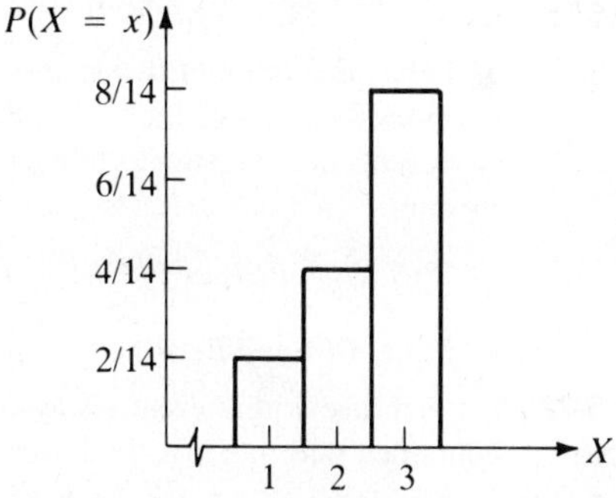

5.75 **a.** \$13,500 **b.** \$6,500 **c.** Anything greater than .39

5.77 **a.** Continuous **b.** Probability distribution:

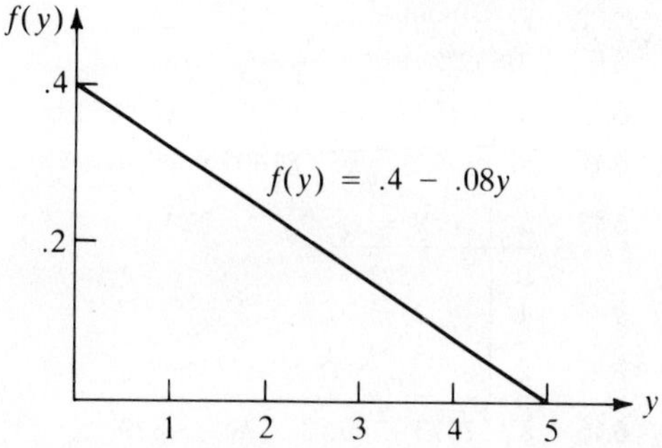

c. .84 d. Yes, $P(Y \leq 4) = .96$

5.79 No, the values of the random variable X are not specified.

5.81 a. Continuous b. Discrete c. Discrete

5.83 a. .2000 b. .3125 c. .5000 d. $X = 60\%$ e. Between $X = 10\%$ and $X = 80\%$

5.85 $\mu = 4.13$

5.87 a. 4/22 = .182 b. 10/22 = .456 c. Evaluating $P(X = x)$ when $x = 0$ yields a negative probability, which is impossible.

5.89 Probability is a number between 0 and 1; a probability distribution is a table, graph, or formula containing several probabilities, plus values of a random variable.

5.91 Classical

5.93 Mean for return; standard deviation for risk

5.95 a. 6/11 = .545 b. 7/11 = .636

5.97 a. False, the union of two events A and B is denoted by P(A or B). b. False, if two events are statistically independent, then

$$P(B|A) = P(B) \text{ or } P(A|B) = P(A).$$

c. False, $P(A \text{ and } B) = 0$.

5.99 No; because a joint event always is contained within a marginal event

5.101 $\mu = 5.4\%$, $\sigma^2 = 7.3$; $\sigma = 2.7\%$

Chapter 6

6.3 a. Not binomial b. Binomial c. Binomial

6.5 .0632

6.7 a. .2153 b. .8496 c. .0512

6.11 a. .3432 b. .8160 c. .0127

6.13

x	$P(X = x)$
0	.0593
1	.2783
2	.4354
3	.2270

6.15 a. .2373 b. 1 c. .0879

6.17 .7560

6.19 $\mu = 3.6$
$\sigma = 1.7$
Skewed right

6.21 a. 1.25 or 1.3 (rounded) b. .97 or 1.0 (rounded) c. .3955

6.23 a. 325.9 b. 17.5 c. 291 through 360 d. Yes

6.25 a. Poisson b. Binomial c. Poisson

6.27 a. .5767 b. .8152 c. $x = 2$, $x = 3$ equally likely

6.29 .9, .9 (or .9487), .4066

6.31 a. .6 (or .63) b. .3297 c. .0072 d. .0383

6.33 a. Binomial b. 6 c. .0025 d. .0838

6.35

π	Poisson Standard Deviation	Binomial Standard Deviation	% Error
.01	.70711	.70356	.504
.05	1.58114	1.54110	2.598
.10	2.23607	2.12132	5.409
.20	3.16228	2.82843	11.803
.50	5.0	3.53553	41.421

6.37 5, .0055

6.39 a. .5404 b. .0988 c. .4596

6.41 $P(X = 10) = .0000$; nearly impossible to happen if π is really .90

6.43 a. .5000 (.4999 using table) b. .3036 c. .2122

6.45 a. .0754 b. .2060

6.47 a. Poisson b. Neither c. Binomial

6.49 a. 1.3 b. .2165 c. .2019 d. .0408

6.51 a. .4912 b. .4211 c. .0035

6.55 .0055

6.57 6, 8

6.59 85.91%

6.61 .0345

6.63 a. .60 b. 48 through 72 c. A, 71 is inside RCV

6.65 a. .0146 b. .0994 c. .73

6.67 a. .83 and .67 b. .1502 c. .8545 d. .3749

6.69 a. Binomial b. $n \leq 20$, $\pi \leq .05$ c. .60 d. 0, 1, 2 e. In control

6.71 a. 1 b. .0010

6.73 a. .0155 b. .0003 c. .3522

6.75 a. .3998 b. .9464 c. 2.7, 1.5

Chapter 7

7.1 a. Yes b. Yes c. Yes d. Yes e. No f. Yes g. No h. Yes

7.3 Half (.50)

7.5 a. $Z = .50$ b. $Z = 0$ c. $Z = -.75$

7.7 a. $X = 29$ b. $X = 47.6$ c. $X = 63.2$

7.9 444

7.11 a. .1492 b. .9944 c. .4404 d. .0000

7.13 a. $Z = -.84$ b. $Z = .52$ c. $Z = -.25$ and $+.25$

7.15 a. .0228 b. .3108 c. .1574

7.17 a. .0475 b. .8190 c. 13.35% d. About 224 days

7.19 a. .1335 b. .0132 c. .5788

7.21 a. .2709 b. .0143 c. .0019

7.23 a. .0006 b. .1362 c. .0264

7.25 Point probabilities for a normal variable are zero.

7.27 a. .0192 b. 1.0 c. .7764

7.29 a. .25 b. .84 c. $-.92$

7.31 From $Z = -.67$ to $+.67$, or 1.34 standard deviations

7.33 a. .0336 b. .0436

7.35 .5793

7.37 .1038

7.39 .1210

7.41 .2483 with no continuity correction

7.43 a. 3.33 for both b. .2592 c. .1653

7.45 a. .0821 b. .0067

7.49 a. .6849 b. .0397 c. .0216

7.51 a. .19 b. $-.86$ c. 1.39

7.53 2700

7.55 a. $-.92$ b. 1.96 c. $\pm.39$

7.57 a. .62% b. 1.983 liters c. 10.56% d. 2.037 liters

7.59 a. .0013 b. .0244 c. .3783

7.61 169 lbs.

7.63 .1977

7.65 .6528 with no continuity correction; .6680 with continuity correction

7.67 a. .0401 b. 19.84 days c. .9593

7.69 .1175

7.71 a. 2.0 minutes b. 1.15 minutes c. .25 d. .625

7.73 Mean is \$32.19; standard deviation = \$7.18

7.75 a.

Bonus Category	% of Managers in Category
0	77
2,000	9
5,000	10
10,000	4

b. About \$4,700

7.77 a. -25 and 25 b. .20 c. Zero d. 14.43

7.79 a. 270, 14.88 b. About 9% c. $X = 305$ (no continuity correction)

C7.1 a. .4484 b. .0985 c. .0174 d. \$342.52 e. .8159 f. \$497.02 g. .2247 h. \$968.76 i. \$85 and \$1,045, to the nearest dollar j. .8212

Chapter 8

8.1 Mean = 840; standard error = 15.556; normal shape

8.3 Approximately normal with standard error about \$1,905 (or \$1,905.13)

8.5 a. 1802 (or 1802.4983) b. About 160

8.7 a. 1.44 seconds
b. About 117.12 − 122.88 seconds
c. Standard error = 3.95 seconds; RCV = 112.1 to 127.9 seconds

8.9 a. 6.25 grams b. 12.5 grams

8.11 a. μ_{Md} b. σ_{Rg} c. μ_{S^2} d. σ_S

8.13 a. Normal b. Approximately normal c. Approximately normal d. Non-normal e. Approximately normal

8.15 a. 132.76 b. 153.18 c. 128.91

8.17 a. $49,308 and $52,892 b. $54,362

8.19 a. 28,000 ± 1,025 miles b. From 27657 to 28343 miles c. *RCV*: 28000 ± 592 miles; from 27802 to 28198 miles

8.21 a. .3520, using $Z = -.38$ b. .0122 c. From 92 to 108 d. 64 e. About 93

8.23 a. Approximately normal b. No, we do not know shape of the population. c. About .8340

8.25 a. 3.857, 2.403, 1.655, 1.086
b. 3.891, 2.461, 1.740, 1.212
c. .88%, 2.41%, 5.14%, 11.60%

8.27 a. Yes b. 8.212 c. .9652
d. No, finding one value in 48 above the *RCV* is not unusual.

8.29 a. 1, σ b. 100%, 0, 0

8.31 Mean = 5.60%;
standard error = .548%;
normal sampling distribution

8.33 a. Approximately normal, 7.59
b. Approximately normal, 4.80
c. Approximately normal, 2.52

8.35 a. About 14 points (or 14.142)
b. About 28 points (or 27.719)

8.37 a. $1,520 to $1,760 b. .0038 c. .7888 d. .9232

8.39 a. .5098 b. .9164
c. $P(\bar{X} \geq 100) = .0003$;
a mean this far from 90 (more than 3 standard deviations) is unlikely;
there may be reason to doubt that the mean really is 90.

8.41 .9818, .7620

8.43 a. 21 ± 2.9 titles b. 21 ± 1.5 titles, using *FPC*

8.47 a. .78 b. −1.30 c. .11
d. 2.67

8.49 To 2 decimals places, ± 1.28, ± 1.96, and ± 2.58 (2.575 rounded)

Chapter 9

9.1 Estimation and hypothesis testing

9.3 a. Correct b. Correct c. Not correct d. Not correct

9.5 Objective = to test a hypothesis about the standard deviation of the terms of 36-month loans; population = file of all expired 36-month loans; random sample = the one hundred expired loans that were selected

9.7 Any interval estimate based on a sample represents only a portion of the population. It cannot be a perfect representation of the population, and thus we cannot have 100% confidence in it.

9.9 .27 ± .11, or .16 to .38

9.11 a. No b. No c. Constant

9.13 Null hypothesis, alternative hypothesis, test statistic, rejection region, conclusion

9.15 a. Null b. Null c. Alternative
d. Alternative e. Null

9.17 Null hypothesis: The moisture content is within specifications. Alternative hypothesis: The moisture content is too low or too high.

9.19 Null hypothesis: The volume of traffic is not heavy. Alternative hypothesis: The volume of traffic is heavy.

9.21 a. Type I error: The manager fails to offer the job to a potentially good employee. Type II error: The manager offers the job to a potentially bad employee. b. Type II error

9.23 An objective, an accessible population, and a random sample

9.25 (i) Some of the population data are inaccessible; (ii) the sample data set is biased; and (iii) an objective is missing or miscommunicated.

9.27 Let π = the error rate in the company's small invoices. Test H_0: $\pi \leq .01$ (or H_0: $\pi = .01$) versus H_a: $\pi > .01$.

9.29 a. Type I error: Buying when we should have been selling. Type II error: Selling when we should have been buying.
b. Type I error

9.31 a. H_a is FTC b. Brown & Williamson's claim (H_0)

9.33 H_0: $\sigma = 6.8$ (for the 36-month loan); H_a:$\sigma \neq 6.8$

9.35 a. Type II error b. Correct decision c. Correct decision d. Type I error

Chapter 10

10.1 a. ± 1.15 b. ± 1.44 c. ± 2.81

10.3 a. 85.1 ± 3.2 b. Extremely unlikely c. $\overline{X}$

10.5 8.95 ± 1.20

10.7 Degree of confidence, narrows; standard deviation, narrows; sample size, widens

10.9 7.4 to 10.2 books

10.11 b. \$66.34 to \$73.48

10.13 77.2 ± 8.9 ppm

10.15 Wider; the 85% confidence interval

10.17 236

10.19 200

10.21 171 (using $\hat{\sigma} = .333$)

10.23 \$117, \$281.25

10.25 \$79.36

10.27 $Z^* = -2.41 < -1.65$; reject H_0.

10.29 $Z^* = 2.63 > 1.65$; reject H_0.

10.31 H_0 could not be rejected.

10.33 a. Two tailed test of H_0: $\mu = \$1{,}800$ b. $Z^* = -.38$; accept H_0 for typically used levels of significance. c. support.

10.35 $Z^* = -2.21 < -1.65$; reject H_0, the difference is statistically significant.

10.37 a. ± 2.58 b. -1.28 c. 2.05

10.39 $Z^* = 1.14 < 1.28$; do not reject H_0.

10.41 $Z^* = -4.69 < -2.58$; reject H_0.

10.43 $Z^* = -5.34$; reject for any commonly used α.

10.45 \$935 ± \$8

10.47 $Z^* = 3.56 > 2.33$; reject H_0.

10.49 a. H_0: $\mu = 5.0$ percent; upper tail H_a c. $Z^* = .72 < 1.65$; do not reject H_0.

10.51 About .0002

10.53 a. Accept b. Reject c. Accept d. Reject

10.55 $\overline{X}$ should be μ.

10.57 USDA concluding the population mean exceeds 5% fat when in fact the mean is 5% or less fat content

10.59 a. Probability of correctly rejecting H_0 (power) b. Probability of correctly accepting H_0

10.61 Type II error probability = power = .50

10.63

Value of μ	P (Reject H_0)
65	.0150
64 or 66	.1112
63 or 67	.5000
62 or 68	.8888
61 or 69	.9925
60 or 70	1.0 (approximately)

10.65 491.72 ± 23.03

10.69 6.6 days ± 1.0 day

10.71 11.6 columns ± .5 column

10.73 a. ± 1.04 b. ± 1.55 or ± 1.56 c. ± 2.24

10.75 \$1,191 ± \$136

10.77 26 ± 2 minutes

10.79 63 students

10.81 508 boxes

10.83 a. 41 employees b. Larger

10.85 a. -1.41 b. ± 1.75 c. 2.81

10.87 $Z^* = 3.78 > 2.33$; the discrepancy is statistically significant.

10.89 $Z^* = -2.64 < -1.96$; reject H_0.

10.91 $29.60 \pm .36$ mpg

10.93 .0039

10.95 a. Accept b. Reject c. Accept

10.97 a. Type II b. Type I c. Type II

10.99 Concluding that the population mean is 20 grains when in fact it is not

10.101 In no way

10.103 $Z^* = 3.64 > 1.65$; conclude that the average price per item in the catalog is higher than desired.

10.105 Estimate S as $S = 4.30$; assume all observations are within $\pm 3\ S$ from $\overline{X}$: 32.7 and 58.5 mg.

10.107 **a.** $\overline{X} = 94.05, S = 2.85$ **b.** $\pm .88$ percent **c.** $Z^* = 3.44 > 1.65$; reject H_0. **d.** .0003

Chapter 11

11.1 σ is unknown, n less than 30, normal distribution of X's assumed

11.3 **a.** ± 1.761 **b.** -1.833 **c.** 2.365

11.5 **a.** -3.499 **b.** ± 2.365 **c.** ± 3.499

11.7 **a.** T **b.** F **c.** T **d.** T **e.** F

11.9 $12.3 \pm .7$ ppm

11.11 **b.** $49.75 \pm .97$ lbs.

11.13 **a.** A 1 oz. serving **c.** $t^* = .91 < 3.106$; accept H_0.

11.15 **a.** H_0: $\mu_{\text{new}} = 78$; upper tail H_a **b.** $t = 1.717$ **c.** $t^* = 2.64$; reject H_0; yes

11.17 **a.** 13 **b.** 41 **c.** 5

11.19 Compute $T_+ = 16\frac{1}{2} < 30$; reject H_0.

11.21 $T_+ = 158.5$; reject H_0 at .10 level, but not at .05 or less.

11.23 $T_- = 55 < 60$; reject H_0.

11.25 Smaller sum = 49; table value = 30; accept H_0.

11.27 **a.** Cutoff (absolute) = 1.109; accept H_0. **b.** Cutoff (absolute) = .942; reject H_0. **c.** Cutoff (absolute) = .880; accept H_0.

11.29 $g_1 = .0447$; cutoff = .942; accept H_0.

11.31 H_0 is that the process is in control. Type I error is concluding that the process is out of control when in reality it is in control.

11.33 $UCL = 118.735$; $LCL = 117.265$ degrees Fahrenheit

11.35 **a.** $LCL = 311.95$; $UCL = 328.05$ calories; all points are in control. **b.** Largest range is 16; in control

11.37 **a.** ± 2.110 **b.** -2.718 **c.** ± 2.539 **d.** 1.96

11.39 $\$79.40 \pm \$1.50$

11.41 **b.** 19.77 **c.** ± 1.708 **d.** 19.77 ± 1.24 micrograms

11.43 $t^* = -.94 > -1.812$; do not reject H_0.

11.45 **a.** Concluding that the average consumption has not gone down when in fact is has **b.** $.025 < p\text{-value} < .05$

11.47 **a.** That the individual X's form a normal population **b.** $.01 < p\text{-value} < .025$

11.49 Quicker/easier calculations, weaker assumptions, higher power

11.51 $T_- = 2$; table value = 3; reject H_0.

11.53 $T_+ = 28$; reject H_0 at .10 or .05 level; do not reject for $\alpha = .025$ or less.

11.55 $LCL = 1.4985$ cm; $UCL = 1.5015$; yes

11.57 About 2.98 ml

11.59 **a.** $g_1 = -.35$ **b.** Cutoffs are ± 1.239; accept H_0—symmetry

11.61 **a.** Use a t-test. $t^* = 2.99$; table value = ± 2.048; reject H_0. **b.** Use the signed ranks test. $T_- = 95$; table value = 126; reject H_0. **c.** We can use the signed ranks test if we can justify the assumption of symmetry; $g_1 = .271$; table value = $\pm .818$ (using $n = 29$ cutoffs); accept H_0 symmetry.

11.63 **a.** Conceivably a t-test or the signed ranks test **b.** $t^* = 2.08$; table value = 2.201; do not reject H_0. For signed ranks, $T_- = 16$; table value = 13; do not reject H_0. **c.** t-test assumes population normality; signed ranks assumes population symmetry.

Chapter 12

12.1 **a.** Not **b.** Binomial **c.** Not

12.3 .80, .761

12.5 **a.** Yes **b.** No **c.** Yes **d.** No

12.7 **a.** Yes **b.** .0177 **c.** $.06 \pm .0354$ **d.** .0455

12.9 .2451

12.11 $.1211 \pm .0116$

12.13 $.69 \pm .033$

12.15 Smaller one

12.17 $.4125 \pm .0623$

12.19 **a.** $.50 \pm .075$ **b.** 51 through 69 **c.** 65 is in the *RCV* of a random guesser.

12.21 a. 1537 b. 984 c. 984 d. 150

12.23 2,897 households

12.25 373 records

12.27 a. π should be a decimal number.
b. The equal sign does not belong in H_a.
c. The equal sign belongs in H_0, not H_a.
d. p should be π.

12.29 $Z^* = 2.38$; significant at .05 and .01

12.31 .0125

12.33 a. The intent is to raise the proportion.
b. H_0: $\pi = .39$; upper tail H_a
c. $Z^* = 2.07$; reject H_0; the desired effect is realized.

12.35 .4532

12.37 c is most risky; d is safest.

12.39 $Z^* = -1.15 > -1.65$; do not reject H_0.

12.41 See text listing

12.43 a. Generally OK, although confidence is not stated. b. $Z = 1.73$, roughly 91.6% confidence c. .67 ± .027

12.45 a. 95% is reported; Equation 12–7 suggests 96.8%. b. .45 ± .027

12.47 .957

12.49 a. 1.635 and 12.592 b. 4.575 and 19.675 c. 7.261 and 24.996

12.53 $120.31 \le \sigma^2 \le 173.10$, $10.97 \le \sigma \le 13.16$

12.55 a. $.133 \le \sigma^2 \le .530$
b. $.365 \le \sigma \le .728$
c. Normal population of X's d. The sampling distribution is not symmetrical.

12.57 $.42\% \le \sigma \le .80\%$

12.59 $\chi^{2*} = 22.040$; table value = 30.144; do not reject H_0.

12.61 p-value < .01

12.63 p-value < .005

12.65 The probability of accepting the assumption that $\sigma = .08$ when that assumption is in fact correct.

12.67 a. .68 ± .066 b. .0170

12.69 a. Mean = .13; standard error = .018
b. .7330 c. .0475

12.71 Shorter game

12.73 .369 ± .046

12.75 a. ±.025 ($2\frac{1}{2}\%$) b. No, the 1 percent change observed is smaller than the potential sampling error.

12.77 .795 ± .023

12.79 $4,812

12.81 182

12.83 $Z^* = 1.02$; table value = 1.28; do not reject H_0.

12.85 a. .33 b. About 3,680

12.87 a. $Z^* = 9.83$, table value = 1.65; reject H_0.

12.89 $Z^* = 1.34$; table value = ±1.65; not unusual

12.91 $Z^* = -2.39$; reject H_0 for $\alpha = .01$ or larger.

12.93 b. $3.37 \le \sigma \le 6.35$

12.95 H_0: $\sigma = \$5$, or H_0: $\sigma^2 = 25$

12.97 $\chi^{2*} = 6.05$; table value = 10.851; reject H_0.

12.99 b. $\chi^{2*} = 50.071$; table value = 49.588; reject H_0. c. two d. $Z^* = .28$; accept H_0.

12.101 $UCL = .156$ or 15.6% defective; the sample is 13.75% defective; not convincing evidence

12.103 Case A: $P(X \le 3) = .0105$ from binomial tables
Case B: $P(X \le 60) = .0146$
Case C: $P(X \le 666) = .0094$;
C is strongest case.

Chapter 13

13.1 $3.50 \le (\mu_1 - \mu_2) \le 10.50$

13.3 $0 \le (\mu_B - \mu_E) \le .72$

13.5 Accepted

13.7 .0192

13.9 $Z^* = 1.87$, table value = 1.28; reject H_0.

13.11 p-value = .0096

13.13 Normal populations with equal dispersion

13.15 2.71 ± .30 grams

13.17 b. $t^* = 2.98$, table value = 1.86; reject H_0.

13.19 1.10 ± 1.87

13.21 a. 4.5 b. $t^* = 3.09$; accept H_0 only if Type I error risk factor is .02 or less.

13.23 Smaller sum = 1; reject H_0 at .10 level; accept H_0 at .05 or .01 level.

13.25 $t^* = 2.95$; table value = 1.383; reject H_0.

13.27 a. .05 ± .076

13.29 a. .08 b. $Z^* = -1.05$; cutoffs are ±1.65; no c. .2938

13.31 Women's smoking rate is between 4% and 12% higher than men's.

13.33 $\bar{p} = .07$; $Z^* = 2.04$; table value = ±1.96; reject H_0.

13.35 \$1.44 ± \$1.23

13.37 No

13.39 .2709

13.41 a. $t^* = 2.43$; table value = 2.048; yes

13.43 \$13 ± \$8, to nearest \$

13.45 $Z^* = 1.55$; no

13.47 b. 15 ± 3 packages c. Sales at the same stores on the same day of the week

13.49 .22 ± .065

13.51 $\bar{p} = .134$; $Z^* = .93$; no

13.53 29.9 ± 1.5 rev/sec

13.55 a. Choice 1, due to outbreak pattern b. Ineffective c. Fear of contracting polio from the vaccine

Chapter 14

14.1 F^* becomes large.

14.3 a. .10 b. .99 c. .05

14.5 a. $.05 < p\text{-value} < .10$ b. $p\text{-value} > .10$ c. $p\text{-value} < .01$

14.7 a.

Source	*SS*	*df*	*MS*	*F*
Treatments	3.3524	2	1.6762	2.23
Error	6.0260	8	.75325	
Total	9.3784	10		

b. no

14.9 $F^* = 1.17$; table value = 2.61; yes

14.11 $F^* = 1.60$; table value = about 2.64; accept H_0.

14.13

Source	*SS*	*df*	*MS*	*F*
Treatments	173.33	2	86.67	11.30
Error	92.00	12	7.67	
Total	265.33	14		

14.15 a.

Source	*SS*	*df*	*MS*	*F*
Treatments	26	4	6.50	1.32
Error	74	15	4.93	
Total	100	19		

b. Table value = 2.36; no

14.17 $F^* = 1.20$; table value = 2.35; yes

14.19 a. $SST = 11.298095$; $SSTR = 7.798095$
b. H_0: $\mu_1 = \mu_2 = \mu_3 = \mu_4 = \mu_5$
H_a: not H_0
TS: $F^* = 8.91$
RR: $F^* > 4.77$
C: reject H_0

14.21 a. 3.61 b. 8.32

14.23 a. About 3.07; .059 b. Significant difference between D and K, and between D and L; nonsignificant difference between K and L

14.25

Source	*SS*	*df*	*MS*	*F*
Treatments	55.0	4	13.75	1.08
Blocks	211.2	2	105.60	
Error	102.2	8	12.775	
Total	368.4	14		

Table value = 2.81; accept H_0.

14.27 a.

Source	*SS*	*df*	*MS*	*F*
Treatments	26	2	13	1.95
Blocks	510	3	170	
Error	40	6	6.67	
Total	576	11		

b. Table value = 5.14; accept H_0.

14.29 $F^* = 1.43$; table value = 3.11; accept H_0.

14.31 a. Stores with largest sales could end up in the same treatment, etc. Differences in store volume could swamp differences due to display location. b. Blocks balance display location across volume types.

c.

Source	*SS*	*df*	*MS*	*F*
Treatments	726.0	3	242.0	27.66
Blocks	1459.5	2	729.75	
Error	52.5	6	8.75	
Total	2238.0	11		

d. Yes

14.33 Interaction

14.35 **a.** $F^* = 3.60$; table value $= 2.54$; reject H_0. **b.** $F^* = 1.60$; table value $= 2.93$; accept H_0. **c.** $F^* = 1.20$; table value $= 2.54$; accept H_0.

14.37 **a.** 24, 2, 3, 4 **b.** 20.46 **c.** .19845 **d.** .000267, .028975, .011858 **e.** .15735

f.

Source	*SS*	*df*	*MS*	*F*
Columns	.028975	2	.0144875	1.657
Rows	.000267	1	.000267	.031
Interaction	.011858	2	.005929	.678
Error	.15735	18	.0087417	
Total	.19845	23		

g. No **h.** No

14.39 **a.** No **b.** 9 **c.** Yes; accept H_0 **d.** No

14.41 **a.** Row means are 15.80 and 15.53; column means are 15.2, 16.6, and 15.2; overall mean is 15.667; and treatment (cell) means are 17.4, 13.0, 16.8, 16.4, 13.2, and 17.2.

b.

Source	*SS*	*df*	*MS*	*F*
Temperatures	13.07	2	6.53	5.09
Mixes	.53	1	.53	.42
Interaction	88.27	2	44.13	34.40
Error	30.80	24	1.28	
Total	132.67	29		

c. For temperatures, $F^* = 5.09$; table value $= 3.40$; reject H_0. For mixes, $F^* = .42$; table value $= 4.26$; accept H_0. **d.** 325 degrees **e.** Yes **f.** S-300 and T-350; no

14.43 **a.**

Source	*SS*	*df*	*MS*	*F*
Temperature	232.56	1	232.56	281.42
Brands	4.34	1	4.34	5.25
Interaction	.34	1	.34	.41
Error	26.44	32	.83	
Total	263.70	35		

b. No **c.** Both F^*'s exceed the table value of 2.88; yes **d.** Additive

14.45 $F^* = 1.53$; table value $= 2.59$; accept H_0.

14.47 $F^* = 1.29$; table value $= 6.39$; accept H_0.

14.49 $F^* = 2.57$; table value $= 3.87$; accept H_0.

14.51 **a.** Yes **b.** Yes **c.** No

14.53

Source	*SS*	*df*	*MS*	*F*
Treatments	88.12	2	44.06	2.83
Error	388.65	25	15.55	
Total	476.77	27		

$F^* = 2.83$; table value = about 3.38; accept H_0.

14.55 **a.** $H_0: \mu_1 = \mu_2 = \mu_3$ **b.** 2 and 276 **c.** $F = 4.79$ or $F = 4.61$ **d.** Yes

14.57 **a.** Table value is 2.29, which is less than $F^* = 4.11$; reject H_0. **b.** About 3.40 **c.** T range $= 5.11$; no

14.59 Only catalysts 3 an 4

14.61 **a.**

Source	*SS*	*df*	*MS*	*F*
Treatments	250	2	125	4.00
Blocks	3090	4	772.5	
Error	250	8	31.25	
Total	3590	14		

b. Yes

14.63 **a.** 5 **b.** For columns, $F^* = 3.45$; table value = about 3.32; reject H_0. For rows, $F^* = 3.81$; reject H_0. **c.** $F^* = 1.60$ for interaction; table value is not exceeded; additive

14.65 **a.** .375 **b.** 26

14.67 **a.** $F^* = 1.77$ is less than 3.92 (using 1 and 120 *df*); no **b.** $F^* = .52$; accept

H_0. **c.** $F^* = 8.54$, well in excess of the table value; yes, men smoking Frontiersman rate it higher than men smoking April; opposite for women.

14.69 **a.** $.01 < p\text{-value} < .05$
b. $p\text{-value} > .10$
c. $.01 < p\text{-value} < .05$

14.71 **a.** $F^* = 1.29$; table value $= 2.75$ (using 1 and 120 *df*); no **b.** $F^* = 4.12$; yes
c. $F^* = 2.92$; yes

14.73 Not useful

Chapter 15

15.1 **a.** Graph of the sample data and the least-squares line:

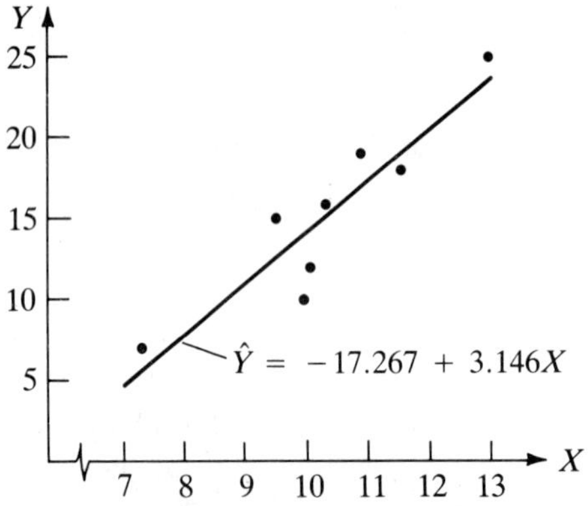

The fitting errors are 1.9756, .8632, 1.3012, 2.0654, −1.2266, 1.3690, −3.8784 and −2.5076, respectively, from $X = 10.9$ through $X = 10.1$.
b. −.004775; yes; assume average value of error term is zero. Thus, average value of fitting errors should be zero, or close to zero due to rounding errors.
c. $SSE = 35.315861$
d. $\hat{\sigma} = 2.42610$ **e.** Direct

15.3 **a.** Independent = age of vehicle; dependent = percentage of retained value
b. Independent = temperature; dependent = attendance
c. Independent = GNP, dependent = sales revenue

15.5 $\hat{\sigma} = 3.82971$

15.7 $\hat{\sigma} = 2.76812$

15.9 **a.**

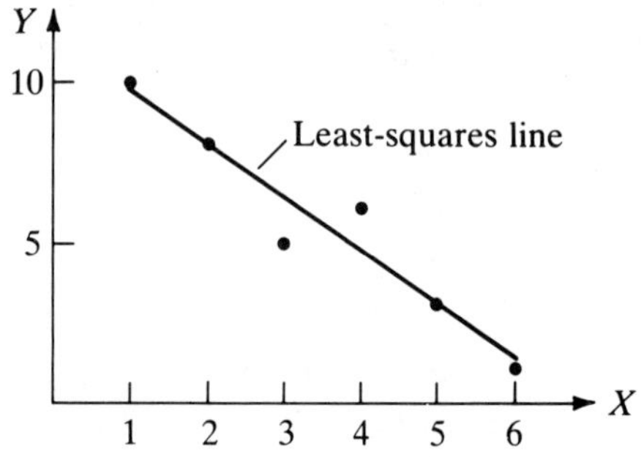

b. $Y = \beta_0 + \beta_1 X + \epsilon$ **c.** $\hat{Y} = 11.4 - 1.686X$ **d.** The line is plotted on the graph in part a. **e.** $\hat{\sigma} = .97101$

15.11

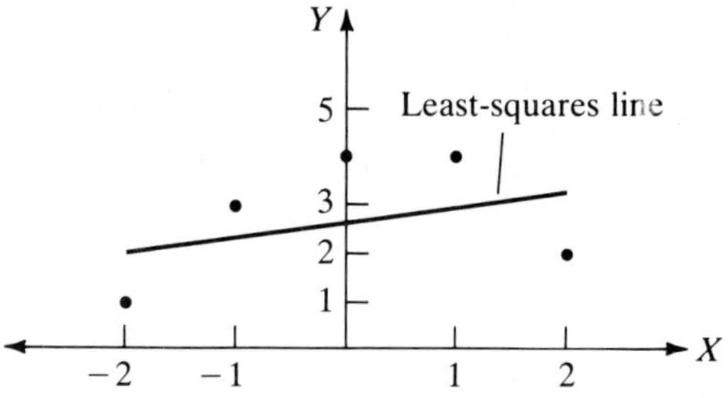

$\hat{Y} = 2.8 + .3X$; X and Y appear to have a curvilinear (parabolic) relationship instead of a linear one.

15.13 $\hat{Y} = 2.8 + .37X$

15.15 **a.** Assign the year 1980 to $X = 1$, etc.
b. $Y = \beta_0 + \beta_1 X + \epsilon$

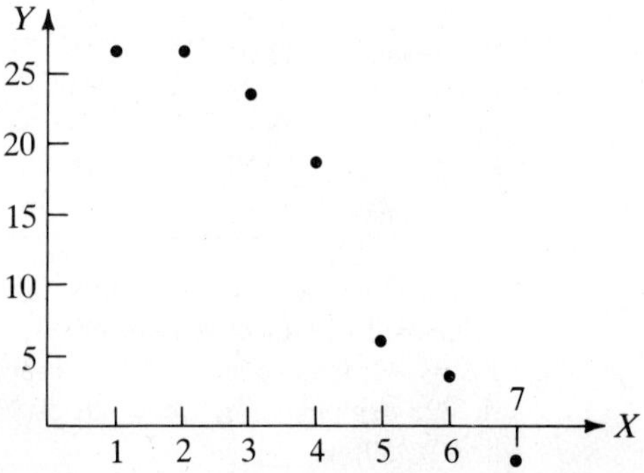

c. Negative, because Y decreases as X increases **d.** $\hat{Y} = 36.314 - 5.411X$
e. -1.037

15.17 $\dfrac{\partial SSE}{\partial b_0} = -2\Sigma Y + 2\Sigma(b_0 + b_1X)$

$\dfrac{\partial SSE}{\partial b_1} = -2\Sigma XY + 2\Sigma(b_0 + b_1X)X$

15.19 **a, b,** and **c.** Verification required
d. The two intercepts will be $b_1\bar{X}$ units apart.

15.21 $.146 \pm .746$; no, since 0 is in the interval we accept H_0: $\beta_1 = 0$.

15.23

Source	SS	df	MS	F
Regression	9.1	1	9.1	25.59
Error	3.2	9	.3556	
Total	12.3	10		

$F^* = 25.59$ – reject H_0: $\beta_1 = 0$; $r^2 = .7398$; about 74% of the variability in the Y-values is explained by the regression equation.

15.25 $t^* = -2.69$; reject H_0: $\beta_1 = 0$

15.27 **a.** $b_1 = 2.527$; $S_{b_1} = .63922$
b. 2.527 ± 1.598 **c.** Between \$929 and \$4,125 **d.** $r^2 = .4046$; about 40% of the variability in annual salary is attributable to the number of years of schooling beyond high school.

15.29 **a.** $\hat{Y} = .36 + .027X$

b.

Source	SS	df	MS	F
Regression	.133461	1	.133461	30.82
Error	.034639	8	.00433	
Total	.1681	9		

c. $F^* = 30.82$ – reject H_0: $\beta_1 = 0$
d. Between 9¢ and 18¢

15.31 Verification required

15.33 **a.** $\hat{Y} = 195.865 - 2.209X$
b. -2.209 ± 1.879 **c.** Yes
d. Yes; supply of available rooms
e. 44.82%

15.35 **a.** $r^2 = 79.42\%$ **b.** $r^2 = 40.39\%$
c. $r^2 = 4.19\%$

15.37 Verification required

15.39 **a.** Model is not useful since $F^* < 10.04$.
b. Model is useful since $F^* > 7.56$.
c. Model is not useful since $F^* < 34.12$.

15.41 **a.** 99.393 ± 98.251; we are 95% confident that Y will fall between 1.142 and 197.644 when $X = 100$.
b. 99.393 ± 35.672; we are 95% confident that the average value of Y will fall between 63.721 and 135.065 when $X = 100$. **c.** 63.7% smaller

15.43 **a.** $X = 1$: 4.6 ± 2.789;
$X = 2$: 3.2 ± 1.826;
$X = 2.5$: 2.5 ± 1.667;
$X = 3$: 1.8 ± 1.826;
$X = 4$: $.4 \pm 2.789$ **b.** Lower limits and upper limits form a curve or band
c. No; thinnest part of the band occurs at $X = 2.5$ (the mean), while thickest part of the band occurs at $X = 1$ and $X = 4$.

15.45 **a.** 22.368 ± 10.888
b. 29.913 ± 1.402

15.47 **a.** Objective 1—plot the data:

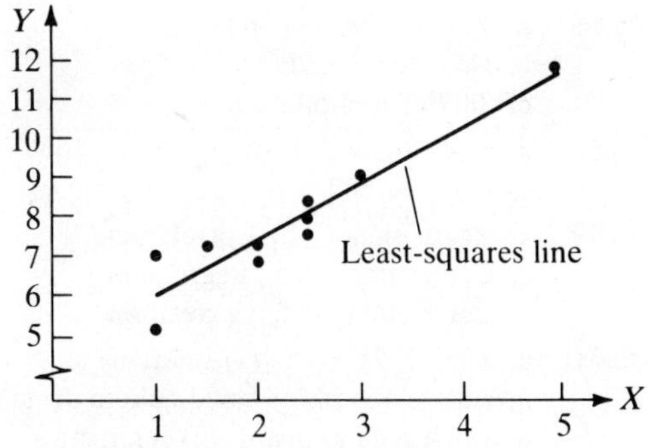

Objective 2: $Y = \beta_0 + \beta_1X + \epsilon$
Objective 3: $\hat{Y} = 4.591 + 1.378X$
Objective 4: $t^* = 7.96$;
reject H_0: $\beta_1 = 0$. (Model is useful.)
b. 8.036 ± 1.831; we are 98% confident that the show will have a rating between 6.205 and 9.867. **c.** Positive—larger than anticipated rating; negative—smaller than anticipated rating **d.** $10.103 \pm .808$

15.49 **a.** $\hat{Y} = -27.649 + 1.112X$

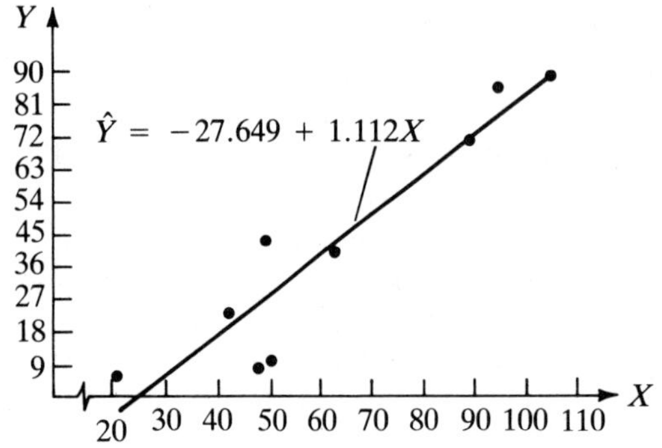

b. $t^* = 6.96$ – reject H_0: $\beta_1 = 0$
c. Prediction interval, because airline is interested in staffing needs at one airport; 0 to 52.127; we are 95% confident that between 0 and 52 employees will be needed. **d.** −3.185; point prediction is unreliable; prediction interval would include positive values for Y.

15.51 **a.** $t^* = -.05$ – accept H_0: $\beta_1 = 0$
b. 10.4 ± 12.488

15.53 **a.** $\hat{Y} = .494 + .321X$;
$t^* = 5.41$; reject H_0: $\beta_1 = 0$
b. 1.136 ± .206 **c.** No

15.55 **a.** $\hat{Y} = 43.899 - .360X$
b. No, $t^* = -.61$
c. 38.290 ± 4.532

15.57 **a.** $r = .8429$ **b.** $r = .2046$
c. $r = -.7887$

15.59 **a.** Regression **b.** Correlation
c. Correlation **d.** Regression
e. Correlation **f.** Correlation

15.61 **a.** $r = .7521$ **b.** (1) Inferring an inverse relationship should help to decide in which fund to invest. (2) Inferring a direct relationship provides no help in deciding in which fund to invest.

15.63 $t^* = -8.88$; reject H_0: $\rho = 0$;
p-value $< .0005$

15.65 **a.** ±.632 **b.** $t^* = \pm 2.3066308$
c. .05 **d.** 5% and the ≠ alternative

15.67 **a.** $r = .6363$ **b.** $t^* = 4.36$; reject H_0: $\rho = 0$; yes, directly related

15.69 **a.** $r = -.8660$ **b.** $r = -.8660$
c. $r = -.8660$ **d.** $r = -.8660$
e. Yes

15.71 **a.** $b_1 = .9$ **b.** $S_X = 1.29099$;
$S_Y = 1.25831$ **c.** $r = .92338$; values agree to first 5 decimal places.

15.73 Test for model utility is probably not significant; thus, the negative value for b_1 is plausible as sampling error from $\beta_1 = 0$.

15.75 $\hat{Y} = 2.667 + 1.714X$; \$14.665 million

15.77 X and Y have a bivariate normal probability distribution.

15.79 $\hat{Y} = 9.89 + 1.25X$

15.81 $b_0 = \overline{Y}$; horizontal line; X and Y are not linearly related.

15.83 **a.** No; p-value exceeds alpha risk of 5%.
b. $\overline{Y} = 2$ **c.** 2 ± 2.376

15.85 **a.** Decrease **b.** Decrease
c. Increase **d.** Decrease
e. Decrease

15.87 **a.** Scatter plot

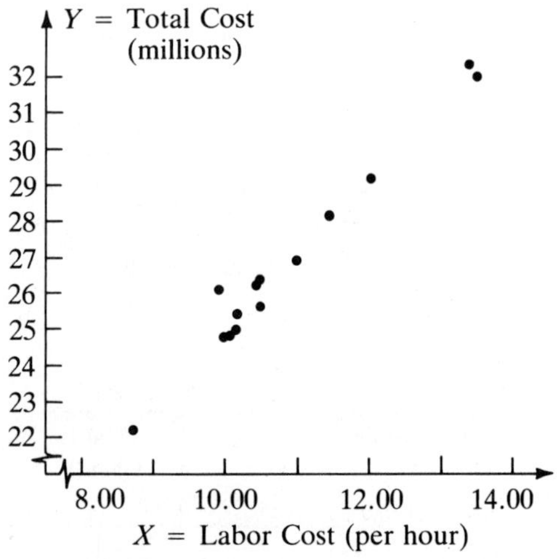

b. $\hat{Y} = 4.280 + 2.075X$ **c.** Yes;
$t^* = 22.11$; reject H_0: $\beta_1 = 0$
d. 22.955 ± .666

15.89 $r = .2151$

15.91 **a.** $\hat{Y} = -.049 + .017X$; $t^* = 20.62$; reject H_0: $\beta_1 = 0$ **b.** −.219 ± .048
c. $r = .9503$; yes, $t^* = 14.64$; reject H_0: $\rho = 0$

15.93 a.

Source	SS	df	MS	F
Regression	1415013.6	1	1415013.6	12.10
Error	935422.5	8	116927.81	
Total	2350436.1	9		

b. $F^* = 12.10$; reject H_0: $\beta_1 = 0$
c. $r^2 = .6020$; approximately 60% of the variability in the Y-values is explained by the regression equation.

15.95 Y is an observed value of the dependent variable; $\bar{Y}$ is the average of the Y-values; $\hat{Y}$ is a predicted value of the dependent variable.

15.97 $t^* = 2.17$; reject H_0: $\beta_1 = 0$

15.99 $r = .65395$; $t^* = 2.59$; reject H_0: $\rho = 0$

15.101 $r = -.927$; yes, $t^* = -12.85$; reject H_0: $\rho = 0$

15.103 a. $\hat{Y} = 943.763 - 217.379X$
b. $8660.52

15.105 $\hat{Y} = X$

Chapter 16

16.1 $Y = \beta_0 + \beta_1 X_1 + \beta_2 X_2 + \ldots + \beta_k X_k + \epsilon$; Y is the dependent variable; $X_1, X_2, \ldots, X_k$ are the independent variables; $\beta_0, \beta_1, \beta_2, \ldots, \beta_k$ are the regression coefficients; and ϵ is the error term.

16.3 a.

$$\mathbf{Y} = \begin{bmatrix} 7 \\ 5 \\ 4 \\ 1 \\ 3 \\ 1 \\ -1 \\ -2 \\ -1 \end{bmatrix} \quad \boldsymbol{\beta} = \begin{bmatrix} \beta_0 \\ \beta_1 \\ \beta_2 \end{bmatrix} \quad \boldsymbol{\epsilon} = \begin{bmatrix} \epsilon_1 \\ \epsilon_2 \\ \epsilon_3 \\ \epsilon_4 \\ \epsilon_5 \\ \epsilon_6 \\ \epsilon_7 \\ \epsilon_8 \\ \epsilon_9 \end{bmatrix}$$

b.

$$\mathbf{X} = \begin{bmatrix} 1 & -4 & 9 \\ 1 & -3 & 5 \\ 1 & -2 & 6 \\ 1 & -1 & 2 \\ 1 & 0 & 2 \\ 1 & 1 & 2 \\ 1 & 2 & 1 \\ 1 & 3 & 4 \\ 1 & 4 & 2 \end{bmatrix}$$

c. $\mathbf{Y}$, 9×1; $\boldsymbol{\beta}$, 3×1; $\boldsymbol{\epsilon}$, 9×1; $\mathbf{X}$, 9×3

16.5 a. Y = the daily semiprivate room rate
b. Yes, maintain a common degree of accuracy for all values of the same variable.

16.7 X_1 = size of the lot; X_2 = square feet of living space; X_3 = number of bedrooms; X_4 = age of the home (other answers are possible)

16.9 a. 28×1 b. 5×1 c. 28×5

16.11

$$\mathbf{Y} = \begin{bmatrix} 10 \\ 7 \\ 15 \\ 22 \\ 18 \\ 12 \\ 24 \\ 19 \\ 13 \\ 7 \end{bmatrix} \quad \boldsymbol{\beta} = \begin{bmatrix} \beta_0 \\ \beta_1 \\ \beta_2 \end{bmatrix} \quad \boldsymbol{\epsilon} = \begin{bmatrix} \epsilon_1 \\ \epsilon_2 \\ \epsilon_3 \\ \epsilon_4 \\ \epsilon_5 \\ \epsilon_6 \\ \epsilon_7 \\ \epsilon_8 \\ \epsilon_9 \\ \epsilon_{10} \end{bmatrix}$$

$$\mathbf{X} = \begin{bmatrix} 1 & 0.0 & 50 \\ 1 & 6.0 & 60 \\ 1 & 2.0 & 40 \\ 1 & 0.0 & 30 \\ 1 & 8.5 & 33 \\ 1 & 8.5 & 50 \\ 1 & 4.0 & 25 \\ 1 & 0.0 & 25 \\ 1 & 0.0 & 50 \\ 1 & 4.5 & 45 \end{bmatrix}$$

$\mathbf{Y}$, 10×1; $\boldsymbol{\beta}$, 3×1; $\boldsymbol{\epsilon}$, 10×1; $\mathbf{X}$, 10×3

16.13 Because highly correlated independent variables are undesirable

16.15 **a.** True **b.** True **c.** False **d.** True **e.** True

16.17 **a.** $\hat{\sigma} = .53116$ **b.** $\hat{\sigma} = 2.15079$ **c.** $\hat{\sigma} = 2.85936$

16.19 **a.** $\hat{Y} = 12.661 - .012X_1 - .209X_2 + 2.997X_3$ **b.** $\hat{\sigma} = 6.44817$ **c.** Fitting error = 2.65 **d.** Computer activity **e.** b_0 is *not* interpretable as a predicted Y-value because 0 is not included in the relevant range for each independent variable.

16.21 **a.** $\hat{Y} = 36.994 - 1.280X_1 + 1.366X_2 + .081X_3 - 5.199X_4$ **b.** $\hat{\sigma} = 1.65164$

16.23 **a.** 24×1 **b.** 24×4 **c.** 4×24 **d.** 4×1 **e.** 4×4 **f.** 4×1

16.25

a. $\mathbf{X'Y} = \begin{bmatrix} 298.9 \\ 9270.5 \\ 1674 \end{bmatrix}$

b. Verification activity **c.** $\hat{Y} = 25.428 + .285X_1 + 1.948X_2$

16.27 **a.** The multiple regression model fits the set of sample data perfectly. **b.** The multiple regression model does not fit the set of sample data.

16.29

Source	*SS*	*df*	*MS*	*F*
Regression	36.062	2	18.031	9.07
Error	13.923	7	1.989	
Total	49.985	9		

16.31 Reject H_0: $\beta_1 = \beta_2 = 0$ since $F^* = 553.14 > 5.72$. (The model is useful.)

16.33 Reject H_0: $\beta_1 = \beta_2 = 0$ since $F^* = 14.73 > 4.74$. (The model is useful.)

16.35 H_0: $\beta_1 = \beta_2 = 0$; H_a: not H_0; *TS*: $F^* = 9.06$; *RR*: Reject H_0 if $F^* > 9.55$; *C*: Accept H_0.

16.37 Accept H_0: $\beta_1 = \beta_2 = 0$ since $F^* = 3.51 < 4.74$. (The model is not useful.)

16.39 **a.** Chest score for vehicle 17 is 46; left thigh for vehicle 44 is 449; right thigh for vehicle 11 is 960. **b.** Computer activity **c.** $\hat{Y} = 952.26 + 12.391X_1 - .0514X_2 - .252X_3 - .103X_4$ **d.** Reject H_0: $\beta_1 = \beta_2 = \beta_3 = \beta_4 = 0$ since $F^* = 3.26 > 2.09$ (roughly). (The model is useful.)

16.41 The prediction interval pertains to predicting weekly unit sales at a specific grocery store; the confidence interval pertains to estimating the average weekly unit sales at all grocery stores in the salesman's territory.

16.43 21.97 ± 4.32, or 17.65 through 26.29; no, because 25 is in the prediction interval.

16.45 Decrease, because $X_1 = 8\%$ is closer to $\overline{X}_1$ than is $X_1 = 7\%$

16.47 **a.** 1.745 ± 2.311, or $-.566$ through 4.056 **b.** $1.745 \pm .981$, or .764 through 2.726

16.49 **a.** 20×6 **b.** 6×1 **c.** 1×6 **d.** 1×1

16.51 **a.** 49.14 ± 4.88, or 44.26 through 54.02 **b.** 49.14 ± 2.34, or 46.80 through 51.48

16.53 (i) All variables are random variables; and (ii) all variables behave according to a multivariate normal distribution.

16.55 **a.** $R^2 = .9845$

b. $\mathbf{C} = \begin{bmatrix} 1.000 & -.69425 & .87639 \\ -.69425 & 1.000 & -.28307 \\ .87639 & -.28307 & 1.000 \end{bmatrix}$

c. X_2

16.57 **a.** (i) Accept H_0: $\beta_1 = \beta_2 = 0$ since $F^* = 12.12 < 19$. (The model is not useful.) (ii) Reject H_0: $\beta_1 = \beta_2 = \beta_3 = 0$ since $F^* = 18.02 > 2.60$ (roughly). (The model is useful.) **b.** (i) $R^2 =$

.9238 (ii) $R^2 = .0479$ c. We should not judge the adequacy of a fitted multiple regression model solely on the basis of its R^2 value.

16.59 $\mathbf{C} = \begin{bmatrix} 1.000 & .32516 & -.07871 \\ .32516 & 1.000 & -.18822 \\ -.07871 & -.18822 & 1.000 \end{bmatrix}$

16.61 $\mathbf{C} =$

$$\begin{bmatrix} 1.000 & .03334 & .20203 & -.91784 \\ .03334 & 1.000 & -.18707 & 0 \\ .20203 & -.18707 & 1.000 & 0 \\ -.91784 & 0 & 0 & 1.000 \end{bmatrix}$$

X_3 is most strongly correlated with Y

16.63 X_2; X_1

16.65 X_1 = weight; X_2 = number of cylinders; X_3 = price; X_4 = engine size (other answers are possible)

16.67 X_1 = temperature; X_2 = amount of rain forecast; X_3 = day of the week (other answers are possible)

16.69 (i) ϵ is a random variable; (ii) ϵ has a normal distribution with a mean of 0 and a variance of σ^2; (iii) for each different pair of settings for the variables $X_1, X_2, \ldots, X_k$, the error random variables are independent.

16.71 a.

$$\mathbf{X} = \begin{bmatrix} 1 & 1 & 0 \\ 1 & 16 & 3 \\ 1 & 32 & 6 \\ 1 & 3 & 0 \\ 1 & 2 & 0 \\ 1 & 6 & 2 \\ 1 & 4 & 0 \\ 1 & 10 & 0 \\ 1 & 15 & 5 \\ 1 & 16 & 1 \\ 1 & 10 & 1 \\ 1 & 3 & 2 \end{bmatrix} \quad \mathbf{Y} = \begin{bmatrix} 3 \\ 3 \\ 0 \\ 4 \\ 4 \\ 3 \\ 18 \\ 15 \\ 1 \\ 29 \\ 11 \\ 3 \end{bmatrix}$$

b. $\mathbf{X}' =$

$$\begin{bmatrix} 1 & 1 & 1 & 1 & 1 & 1 & 1 & 1 & 1 & 1 & 1 & 1 \\ 1 & 16 & 32 & 3 & 2 & 6 & 4 & 10 & 15 & 16 & 10 & 3 \\ 0 & 3 & 6 & 0 & 0 & 2 & 0 & 0 & 5 & 1 & 1 & 2 \end{bmatrix}$$

c. $\mathbf{X}'\mathbf{Y} = \begin{bmatrix} 94 \\ 909 \\ 66 \end{bmatrix}$

$$\mathbf{X}'\mathbf{X} = \begin{bmatrix} 12 & 118 & 20 \\ 118 & 2036 & 359 \\ 20 & 359 & 80 \end{bmatrix}$$

d. $(\mathbf{X}'\mathbf{X})^{-1} = (1/174148) \times$

$$\begin{bmatrix} 33999 & -2260 & 1642 \\ -2260 & 560 & -1948 \\ 1642 & -1948 & 10508 \end{bmatrix}$$

e. $\mathbf{b} = (1/174148) \begin{bmatrix} 1249938 \\ 168032 \\ -922856 \end{bmatrix}$

f. $\hat{Y} = 7.177 + .965X_1 - 5.299X_2$

16.73 The value of X_2 is used to determine the value of X_1; Y has a functional relationship with X_1 and X_2, namely, $Y = X_1X_2$.

16.75 a. $\hat{Y} = -.923 + .0224X_1 + 3.106X_2 + .00852X_3$ b. $\hat{\sigma} = 1.56754$ c. Reject H_0: $\beta_1 = \beta_2 = \beta_3 = 0$ since $F^* = 2120.31 > 2.80$. (The model is useful.) d. 45.03 ± 3.14, or 41.89 through 48.17 e. Yes

f. $\mathbf{C} =$

$$\begin{bmatrix} 1.000 & .95696 & .97927 & .85126 \\ .95696 & 1.000 & .93253 & .76149 \\ .97927 & .93253 & 1.000 & .76528 \\ .85126 & .76149 & .76528 & 1.000 \end{bmatrix}$$

No, all pairs of independent variables are highly correlated, which suggests redundancy.

16.77 .1446

16.79 a. $n \times (k + 1)$ b. $(k + 1) \times (k + 1)$ c. $(k + 1) \times 1$ d. 1×1

16.81 a. 12×1 b. 12×3 c. 3×12 d. 3×1 e. 3×3 f. 3×1 g. 3×1 h. 1×1

16.83 10.702 ± 3.004, or 7.698 through 13.706

16.85 $\mathbf{C} = \begin{bmatrix} 1.000 & -.00717 & .21431 \\ -.00717 & 1.000 & 0 \\ .21431 & 0 & 1.000 \end{bmatrix}$

16.87 a.

$$\mathbf{X} = \begin{bmatrix} 1 & -2 \\ 1 & -1 \\ 1 & 0 \\ 1 & 1 \\ 1 & 2 \end{bmatrix} \quad \mathbf{Y} = \begin{bmatrix} 5 \\ 7 \\ 8 \\ 10 \\ 12 \end{bmatrix}$$

b. $\mathbf{X'X} = \begin{bmatrix} 5 & 0 \\ 0 & 10 \end{bmatrix}$

$(\mathbf{X'X})^{-1} = (1/50)\begin{bmatrix} 10 & 0 \\ 0 & 5 \end{bmatrix}$

$\mathbf{X'Y} = \begin{bmatrix} 42 \\ 17 \end{bmatrix}$ $\quad \mathbf{b} = (1/50)\begin{bmatrix} 420 \\ 85 \end{bmatrix}$

$\hat{Y} = 8.4 + 1.7X$

c. Same equation

16.89 **a.** Do not round the data; declare all observations to be accurate to two decimal places. In baking, fractions of teaspoons and cups can make a big difference in taste and texture; rounding is not advisable. **b.** Computer activity **c.** $\hat{Y} = 7.115 + .090X_1 - .015X_2 + .542X_3 + 1.986X_4 + .015X_5 + .400X_6 - 1.049X_7$

d.

Source	*SS*	*df*	*MS*	*F*
Regression	35.57737	7	5.082	4.60
Error	8.83700	8	1.105	
Total	44.41437	15		

e. Reject H_0: $\beta_1 = \beta_2 = \beta_3 = \beta_4 = \beta_5 = \beta_6 = \beta_7 = 0$ since $F^* = 4.60 > 2.62$. (The model is useful.) **f.** 8.47 ± 3.80, or 4.67 through 12.27 **g.** X_4 **h.** X_3 **i.** No, this strategy would be advisible if the independent variables were uncorrelated. However, the large pairwise correlations among the independent variables precludes such a strategy.

16.91 **a.** $\hat{Y} = -1.6 + 3.1X_1 + 1.9X_2$ **b.** $\hat{Y} = 34.8$ **c.** 7.84 **d.** Less than; for SSE to equal 1000, given the squared error contribution of 7.84 for one data point, the squared error for the majority of the other 24 data points must be more than 7.84.

16.93 **a.** $\hat{Y} = .8511 + .0004154X_1 + .001502X_2$ **b.** 1.277 ± .111, or 1.166 through 1.388 **c.** Narrower; the value $X_1 = 260$ is closer to $\overline{X}_1$ than is the value $X_1 = 200$

16.95 **a.** $\hat{\sigma} = .69300$ **b.** $\hat{\sigma} = 28.67166$ **c.** $\hat{\sigma} = .5$

16.97 **a.** 7.14 ± 1.92, or 5.22 through 9.06 **b.** 6.93 ± .69, or 6.24 through 7.62

16.99 X_1 = square feet of selling area; X_2 = number of competitors within a one-mile radius; X_3 = monthly advertising expenses

16.101

Source	*SS*	*df*	*MS*	*F*
Regression	195	4	48.75	7.70
Error	151.92	24	6.33	
Total	346.92	28		

$\alpha = .01$

16.103 Reject H_0: $\beta_1 = \beta_2 = \beta_3 = \beta_4 = 0$ since $F^* = 38.43 > 3.01$. (The model is useful.)

16.105 **a.** Recall product and recall claim **b.** Recall brand and recognition claim **c.** All but recall product and recognition claim

Chapter 17

17.1 Accept H_0: $\beta_1 = 0$ since $t^* = 1.25 < 1.325$. X_1 is not directly related to Y, given the presence of X_2, X_3, and X_4.

17.3 $b_3 = -59.609$ means the average money earned will decrease by an estimated \$59,609 for each additional stroke per round (on average), given that the number of holes played, the percentage of greens reached in regulation, and the average number of putts per round remain constant.

17.5 **a.** Accept H_0: $\beta_2 = 0$ since $t^* = -.29$ is between ±2.132. X_2 is unrelated to Y, given the presence of X_1. **b.** $S_{b_1} = .2172$ **c.** −1.107 ± .603, or −1.710 through −.504; X_1 should be retained in the model.

17.7 −.902 ± .974, or −1.876 through .072. No, X_1 could be deleted from the model.

17.9 X_1 and X_2 are inversely related to Y. However, X_1's relation is strong, while X_2's relation is weak. They are not contributing redundant information because they are uncorrelated ($r_{12} = 0$).

17.11 **a.** Model 1: $\hat{Y} = 31.864 + 4.564X_1 + 10.405X_2$; Model 2: $\hat{Y} = 38.850 +$

$8.800X_1$; Model 3: $\hat{Y} = 37.088 + 15.432X_2$ **b.** Model 1: $b_1 = 4.564$; $S_{b_1} = 3.3906$; Model 2: $b_1 = 8.800$; $S_{b_1} = 2.6563$ **c.** Model 1: $b_2 = 10.405$; $S_{b_2} = 5.5768$; Model 2: $b_2 = 15.432$; $S_{b_2} = 4.2174$ **d.** Model 1: $\hat{Y} = 66.366$; Model 2: $\hat{Y} = 65.250$; Model 3: $\hat{Y} = 67.952$

17.13 **a.** Reject H_0 if $F^* > 2.40$. **b.** Reject H_0 if $F^* > 3.68$. **c.** Reject H_0 if $F^* > 5.665$ (interpolated).

17.15 Reject H_0: $\beta_4 = \beta_7 = 0$ since $F^* = 8.47 > 4.46$.

17.17 **a.** Accept H_0: $\beta_2 = \beta_3 = 0$ since $F^* = .37 < 3.98$. **b.** Reject H_0: $\beta_1 = \beta_3 = 0$ since $F^* = 4.09 > 5.785$ (interpolated). **c.** Reject H_0: $\beta_2 = \beta_3 = \beta_4 = 0$ since $F^* = 2.68 > 2.22$ (interpolated).

17.19 Accept H_0: $\beta_3 = \beta_4 = \beta_6 = \beta_9 = 0$ since $F^* = 1.14 < 3.06$.

17.21 No, accept H_0: $\beta_1 = \beta_3 = \beta_4 = \beta_5 = \beta_8 = 0$ since $F^* = 2.06 < 4.56$.

17.23 $Y = \beta_0 + \beta_1 X_1 + \beta_2 X_3 + \epsilon$; remove X_2 (because it is correlated with X_1, and $r_{Y2} < r_{Y1}$) and X_4 (because it is uncorrelated with Y).

17.25 **a.** Accept H_0: $\beta_3 = 0$ since $F^* = 1.35 < 4.325$ (interpolated). **b.** Accept H_0: $\beta_3 = 0$ since $t^* = 1.16 < 2.080$. **c.** $(1.16)^2 \cong 1.35$.

17.27 20

17.29 Residual plot:

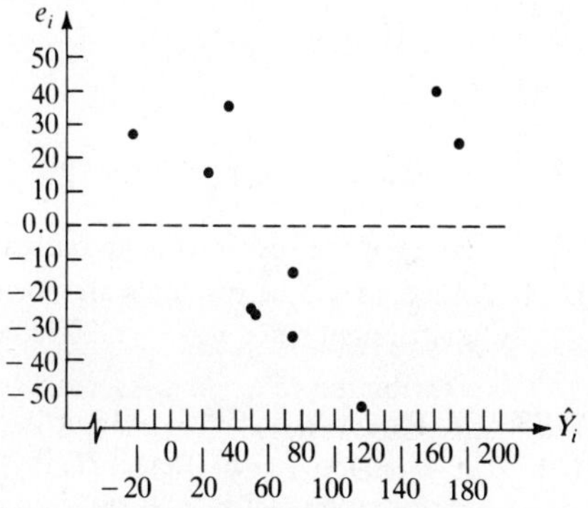

No troublesome patterns are apparent.

17.31 Accept H_0: The residuals are normally distributed, since $r = .388$ is between $\pm .663$.

17.33 **a.** Reject H_0 if $d < 1.10$; accept H_0 if $d > 1.66$; the test is inconclusive if $1.10 \le d \le 1.66$. **b.** Reject H_0 if $d < 1.29$; accept H_0 if $d > 1.72$; the test is inconclusive if $1.29 \le d \le 1.72$. **c.** Reject H_0 if $d < .77$; accept H_0 if $d > 1.25$; the test is inconclusive if $.77 \le d \le 1.25$.

17.35 None

17.37 No, $d = 1.52$ is between 1.38 and 1.67; the test is inconclusive.

17.39 Accept H_0: The residuals are not autocorrelated since $d = 1.43 > 1.25$ (d_U).

17.41 **a.** Yes, graph showing percent of original value of utility vehicles as a function of age:

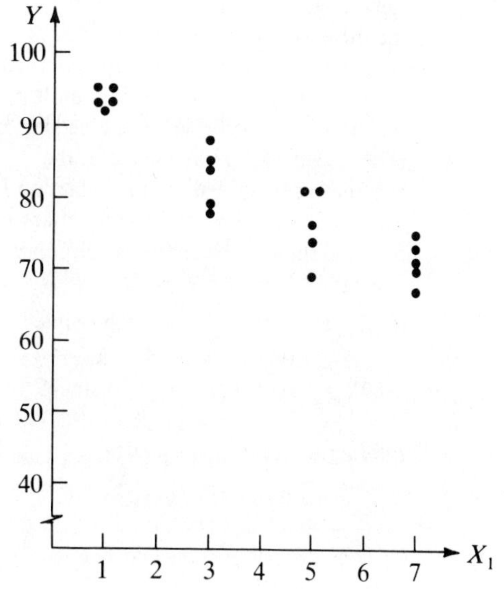

b. $\hat{Y} = 99.697 - 6.590X_1 + .363X_1^2$ **c.** Accept H_0: $\beta_2 = 0$ since $t^* = 1.81 < 2.110$. The inclusion of the variable X_1^2 is not justified.

17.43 **a.** $\hat{Y} = 95.710 - 3.690X_1 - 11.370X_2 - 2.170X_1X_2$ **b.** Yes, reject H_0: $\beta_3 = 0$ since $t^* = -3.10 < -2.576$.

c. $\beta_1 + \beta_3$ is the change in the average percentage of original value retained by luxury cars for each additional year of age; $\beta_2 + \beta_3 X_1$ is the difference in the average percentage of original value retained between luxury cars and utility vehicles that are both X_1 years old.

17.45 **a.** H_0: $\beta_4 = \beta_5 = \beta_6 = 0$ **b.** Reject H_0: $\beta_4 = \beta_5 = \beta_6 = 0$ since $F^* = 16.59 > 2.92$.

17.47 **a.** $Y = \beta_0 + \beta_1 X_1 + \beta_2 X_1^2 + \epsilon$ **b.** $Y = \beta_0 + \beta_1 X_1 + \beta_2 X_2 + \beta_3 X_1^2 + \beta_4 X_2^2 + \beta_5 X_1 X_2 + \epsilon$ **c.** $Y = \beta_0 + \beta_1 X_1 + \beta_2 X_2 + \beta_3 X_3 + \beta_4 X_1^2 + \beta_5 X_2^2 + \beta_6 X_3^2 + \beta_7 X_1 X_2 + \beta_8 X_1 X_3 + \beta_9 X_2 X_3 + \epsilon$

17.49 **a.** 3 **b.** 5 **c.** 2

17.51 **a.** β_1 is the difference in the average index of customer satisfaction between sports cars built in the United States and sports cars built in Japan, assuming the same engine size and price. **b.** β_2 is the difference in the average index of customer satisfaction between sports cars built in Europe and sports cars built in Japan, assuming the same engine size and price. **c.** β_3 is the change in the average index of customer satisfaction for each increase of one thousand dollars in price, assuming the engine size remains constant. **d.** $\beta_1 - \beta_2$

17.53 **a.** H_0: $\beta_2 = 0$ **b.** None possible

17.55 $\hat{Y} = -24.093 + 41.979X_1 + .770X_2 - .882X_1X_2$; accept H_0: $\beta_3 = 0$ since $t^* = -1.44$ is between ± 1.740. The interaction term is not justified.

17.57

	Y	X_1	X_2	X_3	X_4	X_5	X_6
a.	10	0	0	1	0	0	127.060
b.	1.3	1	0	0	0	1	1.693
c.	.87	0	1	0	0	0	1.348
d.	8.3	0	0	0	1	0	131.300
e.	2.7	1	0	0	1	0	1.812

17.59 A condition of high correlation between pairs of independent variables

17.61 1.964 ± 3.571, or -1.607 through 5.535; β_3 is not significantly different from 0 since the confidence interval includes zero.

17.63 X_1 and X_3 evidence strong positive correlation; delete X_1.

17.65 Reject H_0: $\beta_3 = 0$, since $t^* = -12.29 < -3.850$; p-value $< .0005$.

17.67 **a.** Yes; reject H_0: $\beta_1 = 0$ since $t^* = 18.26 > 2.110$. **b.** $S_{b_2} = .06199$

17.69 **a.** Reject H_0 if $F^* > 4.46$. **b.** Reject H_0 if $F^* > 6.99$.

17.71 Reject H_0: $\beta_2 = \beta_4 = 0$, since $F^* = 45.45 > 4.88$ (interpolated). X_2 and X_4 should not be removed from the model.

17.73 Accept H_0: $\beta_2 = \beta_3 = 0$, since $F^* = .38 < 5.93$ (interpolated). X_2 and X_3 could be removed from the model.

17.75 Accept H_0: $\beta_2 = \beta_4 = 0$, since $F^* = 4.15 < 5.78$. The NWS variables could be removed from the model.

17.77 **a.** No **b.** None

17.79 Residual plot:

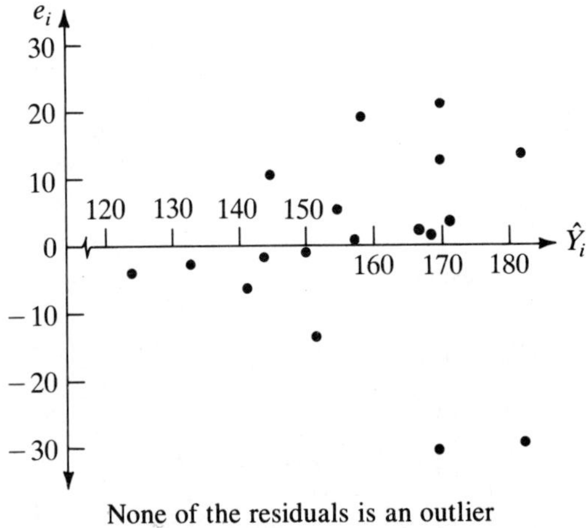

None of the residuals is an outlier

17.81 Accept H_0: The residuals are normally distributed since $r = -.526$ is between $\pm .733$.

17.83 **a.** Reject H_0 if $r < -.435$ or if $r > +.435$. **b.** Reject H_0 if $r < -.695$ or if $r > +.695$.

17.85 Yes, accept H_0: The residuals are normally distributed, since $r = -.135$ is between $\pm.695$.

17.87 Accept H_0: The residuals are not autocorrelated, since $d = 2.77 > 1.66$ (d_U).

17.89

Y	X_1	X_2	X_3
6.50	7	1	0
16.00	7	0	1
.80	7	0	0
10.00	10	1	0
13.00	10	0	1
.60	10	0	0

17.91 a. $b_3 = \$14.56$ b. $b_3 - b_2 = \$9.78$

17.93 H_0: $\beta_4 = 0$

17.95 a. $Y = \beta_0 + \beta_1X_1 + \beta_2X_2 + \beta_3X_3 + \beta_4X_1^2 + \beta_5X_1X_2 + \beta_6X_1X_3 + \epsilon$
b. $Y = \beta_0 + \beta_1X_1 + \beta_2X_2 + \beta_3X_3 + \beta_4X_4 + \beta_5X_1^2 + \beta_6X_1X_2 + \beta_7X_1X_3 + \beta_8X_1X_4 + \beta_9X_2X_3 + \beta_{10}X_2X_4 + \epsilon$

17.97 Yes, reject H_0: $\beta_1 = 0$ since $t^* = 1.44 > 1.35$.

17.99 No, accept H_0: $\beta_4 = \beta_5 = 0$ since $F^* = .07 < 2.86$.

Chapter 18

18.1 a. Time-free b. Time-related
c. Time-free d. Time-related
e. Time-related

18.3 a. Annual contributions, 1970–1989:

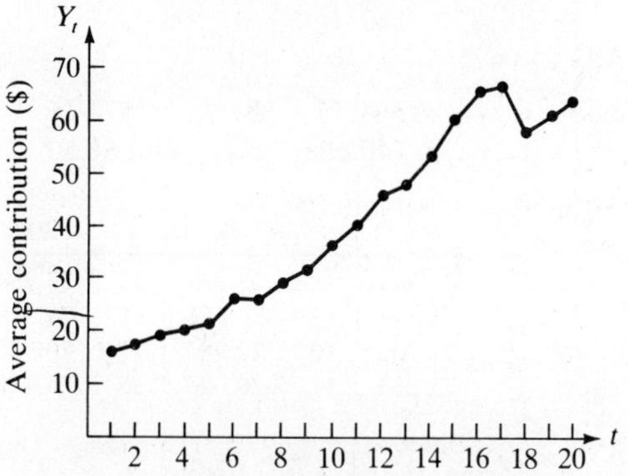

Trend and irregular b. Graph of percentage of contributing alumni, 1970–1989:

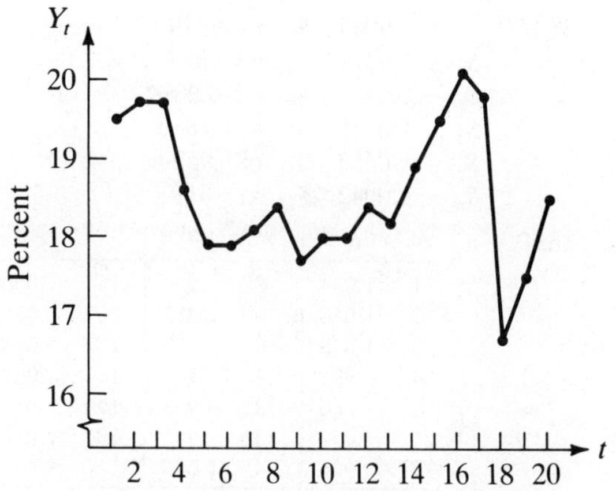

Trend, irregular, and possibly cyclical

18.5

	Seasonal Effect	Cyclical Effect
a. Duration	Within a year's time	Spans periods of more than a year
b. Regularity	Regular	Irregular
c. Assignable influences	Calendar or clock	External business, economic conditions

18.7 Absolute change; additive model, relative change; multiplicative model

18.9 $S_1 = .6012$; $S_2 = 1.1498$; $S_3 = 1.4363$; $S_4 = .8126$

18.11 a. $\hat{Y}_t = 9.206 + 2.998t$
b. $\hat{Y}_{20} = \$69.17$ c. 6.83%

18.13 $S_1 = .9173$; $S_2 = 1.0828$; $S_3 = .9416$; $S_4 = 1.0583$

18.15 a. $\hat{Y}_t = 25.722 - .151t$

b.

	Trend Line Fitting Error	
Year	January	August
1984	−5.571	5.486
1985	−5.159	5.098
1986	−3.047	.710
1987	−4.135	2.922

c. The fitting errors are all negative for the month of January and all positive for the month of August. There is a seasonal effect in Y_t that the trend line fails to capture.

18.17 $S_1 = .9045$ $S_7 = .9419$
$S_2 = .9214$ $S_8 = 1.0687$
$S_3 = 1.0372$ $S_9 = 1.0286$
$S_4 = 1.0730$ $S_{10} = 1.1835$
$S_5 = 1.0655$ $S_{11} = .9524$
$S_6 = 1.0042$ $S_{12} = .8191$

18.19 **a.**

t	C_t	t	C_t	t	C_t
1	1.034	8	.988	15	1.060
2	1.046	9	.952	16	1.095
3	1.048	10	.970	17	1.081
4	.991	11	.972	18	.908
5	.956	12	.995	19	.953
6	.957	13	.986	20	1.010
7	.970	14	1.026		

b. Graph of cyclical relatives for the percent of contributing alumni:

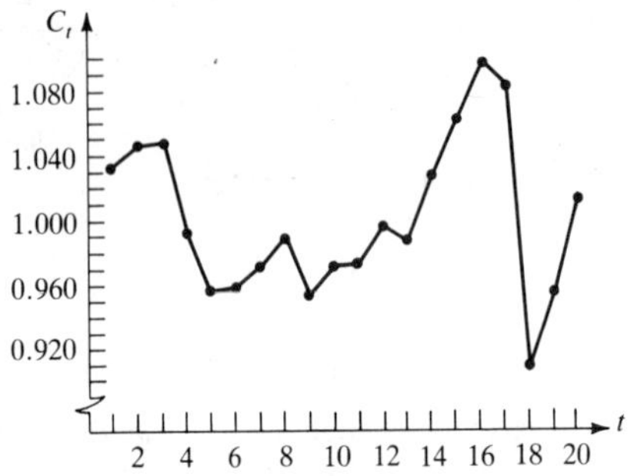

18.21 **a.**

t	C_t	t	C_t	t	C_t
1	—	21	1.041	41	.675
2	—	22	.522	42	.798
3	—	23	.753	43	.869
4	—	24	.862	44	.765
5	—	25	.841	45	.830
6	—	26	.972	46	.882
	5.897	27	.901	47	.976
8	3.115	28	.864	48	.999
9	2.000	29	.901	49	.956
10	1.632	30	.857	50	.956
11	1.590	31	.689	51	.994
12	1.167	32	.776	52	1.174
13	1.080	33	.683	53	1.040
14	.742	34	.654	54	1.056
15	.925	35	.626	55	1.136
16	.802	36	.634	56	1.220
17	.679	37	.720	57	1.325
18	.683	38	.819	58	1.269
19	.754	39	.827	59	1.325
20	1.073	40	.767	60	1.390

18.23 **a.** $\hat{Y}_t = 318.2 + 25.410t$

b.

t	C_t	t	C_t
1	—	9	.851
2	—	10	.886
3	1.168	11	.986
4	1.073	12	1.078
5	1.972	13	1.135
6	1.877	14	1.120
7	1.840	15	—
8	1.837	16	—

c. Cyclical relatives for Y_t = net profit after taxes from Exercise 18.22.

t	C_t	t	C_t
1	—	9	.802
2	1.207	10	.820
3	1.285	11	.928
4	.996	12	1.277
5	.882	13	1.303
6	.767	14	1.353
7	.887	15	.901
8	.782	16	—

18.25 $\hat{y}_{21} = M^*_{20} = 217.5$

18.27 $\hat{y}_{21} = 221.72$

18.29 **a.** None **b.** None **c.** .25 **d.** .096

18.31

y_t	Weight	y_t	Weight
y_1	.2097152	y_5	.1024
y_2	.0524288	y_6	.128
y_3	.065536	y_7	.16
y_4	.08192	y_8	.20

18.33 .0625

18.35 **a.** $\hat{y}_{21} = \$64.75$ **b.** $\hat{y}_{21} = \$72.16$
c. $\hat{y}_{21} = \$40.69$ **d.** $\hat{y}_{21} = \$61.50$

18.37 **a.**

t	D_t	t	D_t	t	D_t
1	21.585	8	25.229	15	27.294
2	24.474	9	24.856	16	25.229
3	22.302	10	27.983	17	26.927
4	23.623	11	26.020	18	25.120

t	D_t	t	D_t	t	D_t
5	23.656	12	26.930	19	28.250
6	26.413	13	26.382	20	26.363
7	24.745	14	23.273		

b. $\hat{D}_t = 23.239 + .199t$
c. $\hat{y}_{21} = 25.15$

18.39 $\hat{y}_{61} = 804$

18.41 $\hat{y}_{61} = 1188$

18.43 a. Too low b. Irregular c. Most sensitive: naive, exponential smoothing (with large alpha); least sensitive: mean, seasonalized mean, trend projection, multiplicative model using a trend projection

18.45 a. .95% b. 5.93%

18.47 Intended time frame and accuracy

18.49 a. Medium range b. Short term c. Short term d. Long term

18.51 a. 48644.14 b. 18984.843

18.53 $MAD = 114.82966$

18.55 Multiplicative model $MAPE = 1.91\%$; moving average $MAPE = 20.70\%$

18.57 a. 45.292 b. 87.924

18.59

Technique	*MSE*	
Naive	42.5	(largest)
Mean	16.574813	
Seasonal mean	11.0625	
Moving average	19.328125	
Trend projection	14.91595	
Multiplicative model	5.6396875	(smallest)
Exponential smoothing	16.574125	

18.61 Fitting accuracy describes the performance of a sequence of forecasts for past time periods, while forecasting accuracy describes the performance of a sequence of forecasts for future time periods.

18.63 a. Verification activity b. Verification activity

18.65 a. Time-related b. Time-related c. Time-free d. Time-free e. Time-related

18.67 The four components are: (1) trend, the long term tendency of Y_t; (2) seasonal, the systematic fluctuations in Y_t of fixed duration due to "seasonal" influences; (3) cyclical, the intermittent fluctuations in Y_t of variable duration due to external economic influences; and (4) irregular, the random variations in Y_t due to unpredictable influences.

18.69 Graph of annualized returns for a mutual fund for the 15-year period from 1975 to 1989:

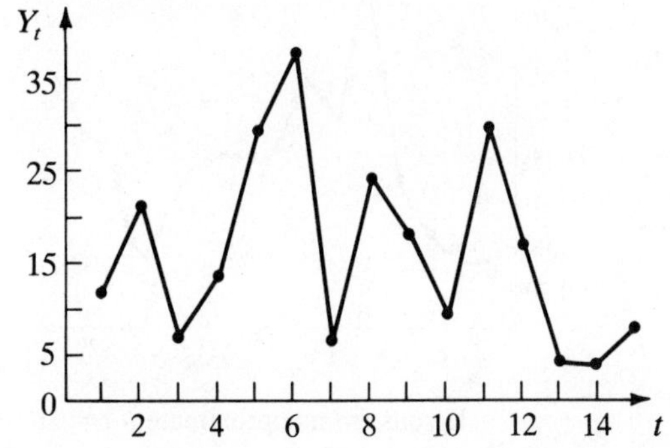

No trend is visible.

18.71 Horizontal axis − t(time); vertical axis − Y_t

18.73 $\hat{Y}_t = -.181 - .152t$

18.75

t	M_t	t	M_t	t	M_t
1	—	12	1530.000	23	1380.500
2	—	13	1472.750	24	1313.375
3	985.375	14	1493.750	25	1288.000
4	936.500	15	1483.125	26	1357.750
5	915.500	16	1449.250	27	1465.875
6	959.625	17	1407.000	28	1618.750
7	1033.000	18	1401.875	29	1783.000
8	1243.000	19	1430.500	30	1845.375
9	1453.500	20	1441.375	31	—
10	1530.375	21	1456.500	32	—
11	1576.500	22	1434.750		

18.77

t	C_t	t	C_t	t	C_t
1	—	12	1.132	23	.906
2	.965	13	1.112	24	.848
3	.911	14	1.154	25	.781
4	.828	15	1.093	26	.805
5	.793	16	1.055	27	.918
6	.835	17	.998	28	1.023
7	.858	18	.963	29	1.064
8	.992	19	1.002	30	1.088
9	1.206	20	.997	31	1.089
10	1.293	21	1.005	32	—
11	1.252	22	.937		

Graph of cyclical relatives for quarterly sales activity of homes:

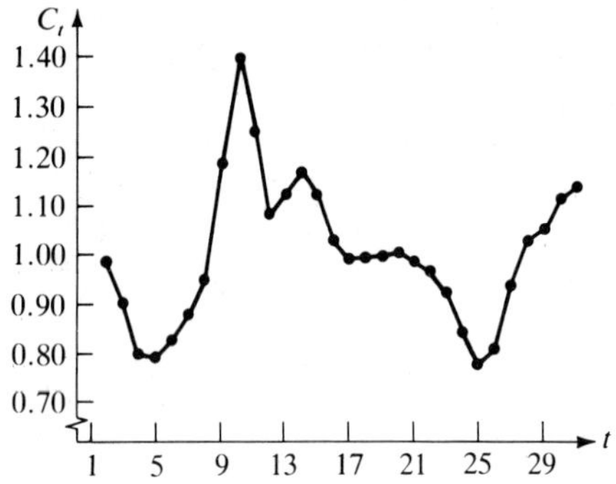

Cycle runs from approximately $t = 2$ through $t = 19$.

18.79

y_t	Weight
y_1	.125
y_2	.125
y_3	.25
y_4	.5

18.81 $MAPE = 15.86444\%$ when $\alpha = .17$

18.83 **a.** $MSE = 479.85308$ **b.** $\hat{y}_{13} = 85.04$

18.85 $MAPE = 34.75029\%$ when $\alpha = .47$

18.87 $\hat{Y}_t = 8.133 - .404t$; $MAPE = 22.95\%$

18.89 Combined $MAPE = 21.07\%$; naive $MAPE = 28.602\%$; mean $MAPE = 40.85\%$

18.91 $D_t = T_t \cdot C_t \cdot I_t$

18.93 Multiplicative model $MSE = .53432$; moving average $MSE = 280.49479$

18.95 $MSE = 23.45$

Chapter 19

19.1 **a.** 13.277 **b.** 7.815 **c.** 12.017

19.3 **a.** Reject H_0. **b.** Accept H_0. **c.** Reject H_0.

19.5 Table value is 9.236; $\chi^{2*} = 6.667$; accept H_0.

19.7 **a.**

Longest	Expected	Observed
4	2815	2791
5	3558	3607
6	1328	1294
7	283	294
8	37	36
9	3	2

b. Define a category "8 or 9" to combine the last 2 categories. **c.** $\chi^{2*} = 2.277$ **d.** Cutoff value is 7.779; accept H_0.

19.9 $\chi^{2*} = 2.30$; table value is 7.779; accept H_0.

19.11 **a.** 13.277 **b.** 7.779 **c.** 7.815

19.13 $\chi^{2*} = 4.531$; table value is 9.210; accept H_0.

19.15 $\chi^{2*} = 21.794$; table value is 5.991; dependent.

19.17 $\chi^{2*} = 16.640$; table value is 9.210; reject H_0.

19.19 **a.** $\chi^{2*} = .9499$; table value is 2.706; accept H_0. **b.** $.975^2$ is very close to .9499. **c.** 1.65^2 is close to 2.706.

19.21 $r_s = -.250$

19.23 **a.** 0 **b.** 2 **c.** 0

19.25 $r_s = .836$; table value is .523; reject H_0.

19.27 **a.** Cutoff is ± 2.58; $Z^* = 2.00$; do not reject. **b.** Cutoff is -2.33; $Z^* = -2.95$; reject. **c.** Cutoff is 1.65; $Z^* = 1.36$; do not reject.

19.29 $r_s = .491$; do not reject H_0.

19.31 **a.** Accept H_0. **b.** Reject H_0. **c.** Reject H_0.

19.33 $T_1 = 36$; reject H_0, since cutoff is 36.

19.35 $T_1 = 19$ (or 36); cutoffs are 19 and 36; reject H_0.

19.37 $T_1 = 46$ (or 32); cutoffs are 26 and 52; accept H_0.

19.39 $Z^* = 1.45$; cutoffs are ± 1.65; accept H_0.

19.41 $Z^* = 2.13$; table values are ± 1.65; reject H_0.

19.43 **a.** 15.933 **b.** 2.679
c. 15.933 ± 5.357 runs
d. 19 is within RCV; support

19.45 **a.** For $\alpha = .10$ or less, accept the randomness hypothesis in all cases
b. Would lower the number of runs.

19.47 **a.** Reject H_0. **b.** Accept H_0.
c. Accept H_0.

19.49 $\chi^{2*} = 5.714$; table value is 10.645; accept H_0.

19.51 **a.** $\chi^{2*} = 9.777$; table value is 9.488; conclude that the variables are dependent.
b. Plan B more favored by young and less by old; opposite for plan P

19.53 Table value is 12.592; accept H_0.

19.55 $\chi^{2*} = 7.465$; table value is 13.277; no.

19.57 **a.** Agree **b.** $r_s = .881$ **c.** Cutoff is .833; reject H_0 at $\alpha = .01$.

19.59 **a.** $T_1 = 18$; cutoffs are 20 and 40; reject H_0.

19.61 **a.** Cutoffs are 12 and 24;
$T_1 = 12$ (or 24); reject H_0.
b. Normal populations with equal variances

19.63 $Z^* = -.42$; cutoffs are ± 1.65; accept H_0.

Review Problems

R1 **a.** $\hat{Y} = 14.56 + .039X$; yes, since $r = .733 > .707$ **b.** 38,400 votes, to the nearest hundred **c.** X and Y are linearly related.

R3

X	f	Relative Frequency	Cumulative Frequency
0	5	.104	5
1	9	.188	14
2	12	.250	26
3	5	.104	31
4	9	.188	40
5	3	.063	43
6	2	.042	45
7	2	.042	47
8	1	.021	48

c. 2

R5 **a.** Yes **b.** $r = .862$ **c.** Yes

R7 **a.** 5 **b.** \$491 **c.** \$8637.50

R9 **a.** 2 calls **b.** $\bar{X} = 8$ and $\bar{Y} = 2$; least squares line passes through $(\bar{X},\bar{Y})$.

R11 **a.** Ungrouped **b.** 225 through 257
c. 1st **d.** 3rd **e.** 4th

R13

X_2	f
8	3
9	3
10	6
11	2
12	2
13	2
14	4
15	2
16	5
17	1

R15 **a.**

Style of Music	f
Easy listening	4
Classical	4
Country	7
Rock	9
Top 40	6

b.

	Radio Band	
Style of Music	AM	FM
Easy listening	2	2
Classical	0	4
Country	2	5
Rock	1	8
Top 40	5	1

R17 **a.**

Class Number	M
5	.225
4	.175
3	.125
2	.075
1	.025

b. .060 (rounded); we assume the midpoint .225 represents a typical value in class number 5. **c.** .045 (rounded); no assumptions are necessary.

R19 **a.** Bank credit card **b.** Estimate of the missing value is 16.3%.

c.

Annual Fee ($)	f
0	2
12	1
15	5
20	11
24	1

mode = $20

d.

Interest rates (%)	f
12.4–14.6	4
14.6–16.8	4
16.8–19.0	9
19.0–21.2	3

median = 17.3 (rounded)

R21 **a.** Discrete **b.** 1, 1, 1.7 (rounded) **c.** About 1.2 **d.** .0640 **e.** 3

R23 **a.** 15 **b.** .0082 (.0119 with c.c.)

R25 **a.** Binomial **b.** May win more often at home vs. on the road, etc. **c.** (1) .0000035; (2) .0005585; (3) .0000134

R27 **a.** 1.1 call/week **b.** Poisson **c.** .0995 **d.** 1 **e.** .1108

R29 **a.** .9312 **b.** .07% (.0007) **c.** 224.386 cm

R31 **a.** Binomial with $n = 720$, $\pi = .0011$ **b.** Yes, Poisson approximation sampling conditions are satisfied. **c.** .792 **d.** Accuracy rate is unchanged because $X = 2$ is in the region of concentrated values.

R33 **a.**

	Purchased Service Contract	
Brand	Yes	No
Sanyo	8	44
Philco	8	34
Sony	12	44

b. 42/150 = .28 (rounded) **c.** 122/150 = .813 **d.** 8/42 = .190 (rounded)

R35 **a.** Yes, her average rate of return was 7.3% (rounded). **b.** 3.0% (rounded)

R37 **a.** (4) **b.** .1587

R39 $6,503.80

R41 **a.** 43 **b.** 31.50 ± 2.24 yrs; wider **c.** $\sigma \neq S$

R43 **a.** 1 **b.** H_0: $\pi = 1/3$; H_a: $\pi > 1/3$ **c.** Type I: Subjects are only guessing but we conclude that they are not guessing.

R45 $\$1.92 \leq \sigma \leq \$3.99$

R47 1,666 ± 117 sq. ft.

R49 **a.** H_0: $\pi = .03$; H_a: $\pi > .03$ **b.** $Z^* = 1.33$; cutoff is $Z = 1.28$; buy the rights.

R51 **a.** Most efficient—stratified or cluster sampling; least efficient—simple random sample **b.** $n = 120$ **c.** .317 ± .051 **d.** Since all possible values for π in the 80% confidence interval exceed the 25% requirement, the Center should qualify for the grant.

R53 Yes, reject H_0: $\mu = 850$, since $Z^* = 2.40 > 2.33$. (Hint: Use FPC.)

R55 **a.** False **b.** True **c.** False

R57 No, p-value of .2420 indicates a 24.2% risk of erroneously rejecting H_0: $\mu = 37$, based on the observed data.

R59 **a.** .37 **b.** 35 through 55 **c.** $n = 55$; $k = 46$ **d.** Between .0000003 and .0000034 **e.** Uphold the election result

R61 **a.** We would need candidates for bypass surgery to be assigned to treatment and control groups. Any such study would take years to evaluate. **b.** The null hypothesis would be that bypass surgery is no more effective than current practice (see Chapter 13 Maxim).

R63 **a.** $\hat{Y} = 1.0 + 2.5X$ **b.** 2.50 ± .23 **c.** .909

R65 **a.** 22 **b.** $F^* = 449.2$ **c.** Table value of 8.10 is exceeded; yes. **d.** 95.7%

R67 H_0: average sales are equal for each of the three promotion methods. $F^* = 4.77$; table value is 4.26; H_0 is rejected.

R69 a. $Z^* = 3.29$; reject H_0 for the .01 level or larger. b. $Z^* = .93$; accept H_0 for any commonly used level of significance.

R71 a. Yes, accept H_0: $\mu_1 = \mu_2 = \ldots = \mu_{10}$, since $F^* = 2.35 < 2.70$.
b. $\hat{Y} = -5.643 + 6.250X$
c. No, accept H_0: $\rho = 0$ since $t^* = .80$ is between -2.365 and $+2.365$

R73 a. Reject H_0: $\beta_1 = \beta_2 = 0$ since $F^* = 38.08 > 5.49$ (roughly) and declare the fitted model useful.
b. $r_{Y1} = -.8302$; $r_{12} = .8967$
c. $5,056

R75 a. Yes, reject H_0: $\mu_A = \mu_B = \mu_C$, since $F^* = 25.87 > 3.15$. b. Yes, reject H_0: $\mu_1 = \mu_2 = \ldots = \mu_{10}$ since $F^* = 99.59 > 2.045$ (roughly).

R77 a. Reject H_0: $\mu_{NL} = \mu_L$, since $t^* = 1.90 > 1.313$. b. $.5 \pm 1.76$
c. No, because zero is in the confidence interval.

R79 No, accept H_0: $\mu_S = \mu_L$, since $Z^* = -.90$ is not less than -1.28.

R81 Reject H_0: $\pi_1 = .25$; $\pi_2 = .50$; $\pi_3 = .25$, since $\chi^{2*} = 105.231 > 9.210$.

R83 a. $r_s = .407$ b. Accept H_0: $\rho_s = 0$, since $r_s = .407$ is between $\pm .457$.
c. There is insufficient evidence to conclude that an association exists between NFL teams' punt and kickoff return averages. d. Assume X and Y have a bivariate normal distribution.

R85 a. H_0: $\beta_4 = 0$ b. H_0: $\beta_5 = 0$
c. $\hat{Y} = -717.007 + 93.735X_1 - 87.533X_2 - 5158.743X_3 - 93.391X_1X_3 + 514.589X_2X_3$; reject H_0: $\beta_4 = \beta_5 = 0$, since $F^* = 5.85 > 2.73$.

R87 a. H_0: Injury type is unrelated to automation status b. 6.251
c. Reject H_0, since $\chi^{2*} = 30.608 > 6.251$.

R89 Accept H_0: The residuals are independent (uncorrelated), since $Z^* = -.93$ is not less than -1.645; yes.

R91 a. $\hat{Y} = 188.795 - 30.133t + 1.944t^2$
b. Regression: $MSE = 75.94$; exponential smoothing: $MSE = 812.22$
c. Regression: $MAPE = 36.63\%$; exponential smoothing: $MAPE = 15.62\%$

R93 Reject H_0: Country and product category are independent, since $\chi^{2*} = 30.206 > 9.488$.

R95 Accept H_0: $\pi_1 = .200$; $\pi_2 = .126$; $\pi_3 = .124$; $\pi_4 = .090$; $\pi_5 = .078$; $\pi_6 = .382$, since $\chi^{2*} = 14.654 < 15.086$.

R97 Reject H_0: $\rho_s = 0$, since $r_s = .420$ is greater than .412.

R99 a. Subjective forecast: $MAPE = 2.71\%$; X-11 forecast: $MAPE = 3.89\%$
b. Accept H_0: $MAPE_{sub} = MAPE_{X-11}$ since $T_1 = 31$ is between $T_L = 28$ and $T_U = 50$.

INDEX

Index of Exercises

t-TABLE CUTOFF VALUES

For confidence intervals and two-tailed hypothesis tests, find the risk column of interest, α, at the top of the table; for one-tailed tests, find the risk column at the bottom. The cutoff value is then found at the intersection of the risk column and the *df* row.

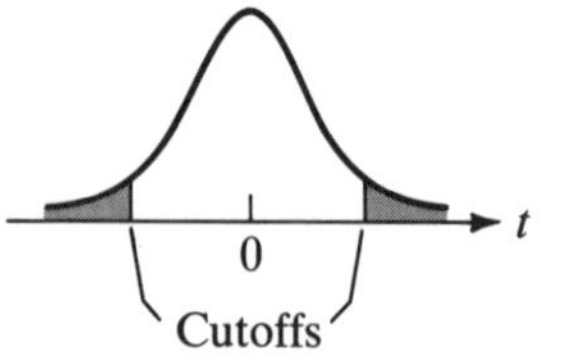

Example: For a confidence interval or two-tailed test where $df = 10$ and α is to be .05, the cutoffs are $t = \pm 2.228$.

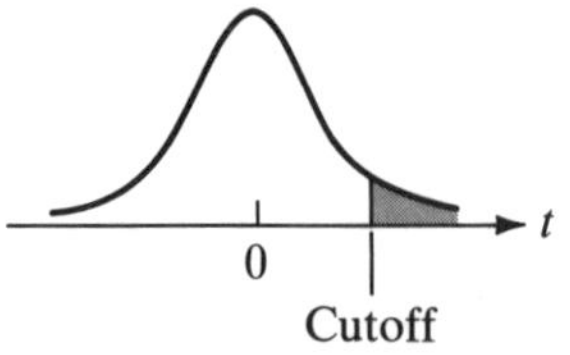

Example: For an upper-tailed H_a with $\alpha = .10$ and $df = 9$, the cutoff is $t = 1.383$. Since all values are given as positive, a minus sign needs to be provided if H_a is lower tailed.

Degrees of Freedom, *df*	Two-Tailed α (Probability in Both Tails Combined)					
	.20	.10	.05	.02	.01	.001
1	3.078	6.314	12.706	31.821	63.657	636.619
2	1.886	2.920	4.303	6.965	9.925	31.598
3	1.638	2.353	3.182	4.541	5.841	12.941
4	1.533	2.132	2.776	3.747	4.604	8.610
5	1.476	2.015	2.571	3.365	4.032	6.859
6	1.440	1.943	2.447	3.143	3.707	5.959
7	1.415	1.895	2.365	2.998	3.499	5.405
8	1.397	1.860	2.306	2.896	3.355	5.041
9	1.383	1.833	2.262	2.821	3.250	4.781
10	1.372	1.812	2.228	2.764	3.169	4.587
11	1.363	1.796	2.201	2.718	3.106	4.437
12	1.356	1.782	2.179	2.681	3.055	4.318
13	1.350	1.771	2.160	2.650	3.012	4.221
14	1.345	1.761	2.145	2.624	2.977	4.140
15	1.341	1.753	2.131	2.602	2.947	4.073
16	1.337	1.746	2.120	2.583	2.921	4.015
17	1.333	1.740	2.110	2.567	2.898	3.965
18	1.330	1.734	2.101	2.552	2.878	3.922
19	1.328	1.729	2.093	2.539	2.861	3.883
20	1.325	1.725	2.086	2.528	2.845	3.850
21	1.323	1.721	2.080	2.518	2.831	3.819
22	1.321	1.717	2.074	2.508	2.819	3.792
23	1.319	1.714	2.069	2.500	2.807	3.767
24	1.318	1.711	2.064	2.492	2.797	3.745
25	1.316	1.708	2.060	2.485	2.787	3.725
26	1.315	1.706	2.056	2.479	2.779	3.707
27	1.314	1.703	2.052	2.473	2.771	3.690
28	1.313	1.701	2.048	2.467	2.763	3.674
29	1.311	1.699	2.045	2.462	2.756	3.659
≥30	1.282	1.645	1.960	2.326	2.576	3.291
	.10	.05	.025	.01	.005	.0005
	One-Tailed α (Probability in One Tail Only)					

Source: Computed by R. E. Shiffler and A. J. Adams.